General data and fundamental constants

Quantity	Symbol	Value
Speed of light†	c	$2.997\,924\,58 \times 10^8 \text{ m s}^{-1}$
Elementary charge	e	$1.602\,177 \times 10^{-19} \text{ C}$
Faraday constant	$F = eN_A$	$9.6485 \times 10^4 \text{ C mol}^{-1}$
Boltzmann constant	k	$1.380\,66 \times 10^{-23} \text{ J K}^{-1}$
Gas constant	$R = kN_A$	$8.314\,51 \text{ J K}^{-1} \text{ mol}^{-1}$
		$8.205\,78 \times 10^{-2} \text{ dm}^3 \text{ atm K}^{-1} \text{ mol}^{-1}$
		$62.364 \text{ L Torr K}^{-1} \text{ mol}^{-1}$
Planck constant	h	$6.626\,08 \times 10^{-34} \text{ J s}$
	$\hbar = h/2\pi$	$1.054\,57 \times 10^{-34} \text{ J s}$
Avogadro constant	N_A	$6.022\,14 \times 10^{23} \text{ mol}^{-1}$
Atomic mass unit	u	$1.660\,54 \times 10^{-27} \text{ kg}$
Mass of		
electron	m_e	$9.109\,39 \times 10^{-31} \text{ kg}$
proton	m_p	$1.672\,62 \times 10^{-27} \text{ kg}$
neutron	m_n	$1.674\,93 \times 10^{-27} \text{ kg}$
Vacuum permeability†	μ_0	$4\pi \times 10^{-7} \text{ J s}^2 \text{ C}^{-2} \text{ m}^{-1}$
		$4\pi \times 10^{-7} \text{ T}^2 \text{ J}^{-1} \text{ m}^3$
Vacuum permittivity	$\varepsilon_0 = 1/c^2\mu_0$	$8.854\,19 \times 10^{-12} \text{ J}^{-1} \text{ C}^2 \text{ m}^{-1}$
	$4\pi\varepsilon_0$	$1.112\,65 \times 10^{-10} \text{ J}^{-1} \text{ C}^2 \text{ m}^{-1}$
Bohr magneton	$\mu_B = e\hbar/2m_e$	$9.274\,02 \times 10^{-24} \text{ J T}^{-1}$
Nuclear magneton	$\mu_N = e\hbar/2m_p$	$5.050\,79 \times 10^{-27} \text{ J T}^{-1}$
Electron g value	g_e	$2.002\,32$
Bohr radius	$a_0 = 4\pi\varepsilon_0\hbar^2/m_e e^2$	$5.291\,77 \times 10^{-11} \text{ m}$
Rydberg constant	$R_\infty = m_e e^4/8h^3 c\varepsilon_0^2$	$1.097\,37 \times 10^5 \text{ cm}^{-1}$
Fine structure constant	$\alpha = \mu_0 e^2 c/2h$	$7.297\,35 \times 10^{-3}$
Gravitational constant	G	$6.672\,59 \times 10^{-11} \text{ N m}^2 \text{ kg}^{-2}$
Standard acceleration of free fall†	g	$9.806\,65 \text{ m s}^{-2}$

D1581243

† Exact (defined) values

Prefixes

f	p	n	μ	m	c	d	k	M	G
femto	pico	nano	micro	milli	centi	deci	kilo	mega	giga
10^{-15}	10^{-12}	10^{-9}	10^{-6}	10^{-3}	10^{-2}	10^{-1}	10^3	10^6	10^9

Physical Chemistry

Physical Chemistry

Fourth Edition

P. W. ATKINS

*University Lecturer and Fellow of
Lincoln College, Oxford*

Oxford   Melbourne   Tokyo
OXFORD UNIVERSITY PRESS

Oxford University Press, Walton Street, Oxford OX2 6DP

Oxford New York Toronto
Delhi Bombay Calcutta Madras Karachi
Kuala Lumpur Singapore Hong Kong Tokyo
Melbourne Auckland Madrid

and associated companies in
Berlin Ibadan

Oxford is a trade mark of Oxford University Press

© *P. W. Atkins 1978, 1982, 1986, 1990*

First edition 1978
Second edition 1982
Third edition 1986
Fourth edition 1990
Reprinted (with corrections), 1990, 1991, 1992 (twice)

British Library Cataloguing in Publication Data

Atkins, P. W. (Peter William), 1940–
Physical chemistry.—4th ed
1. Physical chemistry
I. Title
541.3
ISBN 0–19–855283–1
ISBN 0–19–855284–X (paperback)

Printed in Great Britain by Richard Clay Ltd,
Bungay, Suffolk

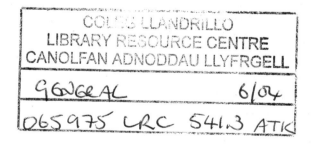

Preface

My indebtedness to my far-flung colleagues continues with this edition, for they have stimulated me to make a number of changes and have guided my hand as they have done in previous editions. Some faces among them are new, others have contributed previously. I thank them in what I hope is not a too ungracious bulk in the acknowledgements that follow this preface.

With this edition I have a threefold aim. Foremost is my determination to keep the text in line with changing attitudes towards and progress in the subject. However, I am very conscious that I must not merely collect techniques. I regard it as essential to show our students that physical chemistry is a vigorous and evolving subject but not to overburden them with details. A textbook provides the foundation for instruction: it should not strive either to be a review article or to take too popularist a line. Later I shall identify the principal new topics in this edition.

Second, I am conscious that students have found some aspects of the earlier editions difficult, and in this edition I have made a considerable effort to simplify the presentation. However, I have tried to do this without damaging the integrity of the development or reducing the text's rigour. To achieve this difficult balancing act I have eliminated a number of side-tracking details, simplified arguments, adopted a more open typographical style for equations, and dispatched some of the more intrusive background material to end-of-chapter appendices which, to seem more inviting, I have called *Further information*. Thus, almost all the material of earlier editions is present, but it is possible to circumvent the arcane more easily.

In the same spirit of trying to help, I have increased by nearly half the number of worked *Examples* in the text, and have made them more instructive. These *Examples* are an integral part of the development of the text, as I hope readers will come to appreciate, for they not only illustrate how the immediately preceding material may be used in a numerical calculation but show how to build on it too. The same spirit of simplification will be found at the ends of the chapters, for the generally welcomed 'Introductory problems' that were provided by Professor Charles Morrow for the third edition have been taken as the nucleus of another innovation. Now all the simpler problems—about twenty a chapter, twice as many as before—have been grouped together in a section called *Exercises,* and the more difficult problems, which often require more work or incorporation of material from other chapters, are now grouped together as *Problems*. New exercises and problems have been added, others combined, improved, refined, simplified, or rejected.

Third, I have restructured parts of the text to make it easier to use and to give it a more logical progression. Thus, the thermochemistry chapter of earlier editions has been absorbed into Chapter 2, so now the reader can proceed to the relatively elementary applications of thermodynamics without first having to wade through a chapter on formal manipulations of thermodynamic state functions. The latter are there, but follow this elementary material. Likewise, I have also restructured the electrochemistry chapters, and have combined old Chapters 11 and 12 into one (Chapter 10).

In the earlier editions it was necessary to negotiate the rapids of Debye–Hückel theory before coping with the relatively elementary material on electrochemical cells. Now all that is changed, but the Debye–Hückel theory is still present. The new chapter conveys the qualitative basis of the limiting law, and the theory itself is present in a *Further information* section. I have also totally recast the material leading to the Nernst equation to make it more accessible, and the material on Brønsted acids and bases is now treated as an aspect of equilibrium thermodynamics (Chapter 9) rather than electrochemistry.

Similar large-scale modifications will be found elsewhere in the text. In each case, I have tried to bring the reader to the heart of the topic directly, so that there is an immediate sense of achievement. A typical example of the type of change I have made is the simplified account of statistical thermodynamics, where now I drive hard towards calculating internal energy and entropy, and leave until later more complicated aspects of the subject.

The principal new material that I have included in the revision is a virtually entirely new chapter on magnetic resonance, which concentrates on Fourier transform techniques, and a greatly extended discussion of lasers, specifically their operation and their applications in chemistry. I have included a lot of other smaller topics throughout the text, which would be too intrusive to list here in detail.

And what have I dropped? Cuts have resulted in a reduction in overall text length. The fact that the text now has slightly more pages than in the third edition is, I think, largely a result of taking more room to display the equations more clearly. Gone is Chapter 0, for I no longer feel it serves a useful purpose; some of its introductory material has been absorbed into other chapters. Gone is Mössbauer spectroscopy, for we have found that it was little needed. Gone too is lot of discursive discussion throughout the text, with slimmed-down arguments and a more sinewy presentation of the essentials. I have aimed at a simpler style, with less fat, but avoiding terseness.

All in all, I have reconsidered every word. I hope it is a hallmark of the text that when it comes to revisions I do not simply tinker but address the compilation of a new edition as a challenge in depth. And there is only one way that that depth can be achieved: by responding to my unseen and often unknown friends around the world, with all their different languages, attitudes, and experiences. To them, as always, I am deeply grateful.

Oxford, 1989   P.W.A.

Acknowledgements

I wish to thank the following people who have contributed so much and in numerous ways to the preparation of this new edition:

Dr D. Bassett, Imperial College, London
Dr A. E. Beezer, Royal Holloway and Bedford New College
Professor R. Buck, University of North Carolina, Chapel Hill
Dr V. Cheng, Baptist College, Hong Kong
Dr W. Clegg, University of Newcastle upon Tyne
Professor A. K. Covington, University of Newcastle upon Tyne
Professor H. B. Dunford, University of Alberta, Edmonton
Dr W. R. Flavell, Imperial College, London
Dr D. O. Hayward, Imperial College, London
Dr P. Jones, University of Newcastle upon Tyne
Professor N. Kestner, Louisiana State University, Baton Rouge
Professor W. Leenstra, University of Vermont, Burlington
Dr B. Levitt, Imperial College, London
Dr D. Nicholson, Imperial College, London
Professor L. G. Pederson, University of North Carolina, Chapel Hill
Professor K. Sandros, Chalmers University of Technology, Gothenburg
Professor R. L. Scott, University of California, Los Angeles
Dr J. G. Smith, University of Newcastle upon Tyne
Dr D. Tildesley, University of Southampton
Professor C. Trapp, University of Louisville, Kentucky
Dr R. G. Wooley, Trent Polytechnic, Nottingham

I am very grateful to Michael Clugston, who has read the entire text and has made numerous wise suggestions.

Colleagues who contributed significantly to the earlier editions included: W. J. Albery (London, now Oxford), G. Allen (London, now Unilever), H. C. Andersen (Stanford), M. D. Archer (Cambridge), L. D. Barron (Glasgow), B. W. Bassett (London), S. M. Blinder (Ann Arbor), L. Brewer (Berkeley), R. D. Brown (Monash), A. D. Buckingham (Cambridge), J. B. Cruickshank (Bristol), A. L. Denio (Wisconsin), T. M. Dunn (Michigan), R. A. Dwek (Oxford), R. G. Edgell (London), L. Epstein (Pittsburg), D. H. Everett (Bristol), A. H. Francis (Michigan), M. H. Freemantle (Jordan, now IUPAC), P. J. Gans (New York), C. J. Gilmore (Glasgow), L. P. Gold (Pennsylvania State), A. Hamnett (Oxford, now Newcastle), F. Herring (Vancouver), B. J. Howard (Oxford), B. Huddle (Roanoke), R. J. Hunter (Sydney), D. Husain (Cambridge), D. A. King (Liverpool, now Cambridge), B. P. Levitt (London), G. Lowe (Oxford), R. M. Lynden-Bell (Cambridge), A. J. Macdermott (Oxford), M. L. McGlashan (London), I. C. McNeill (Glasgow), I. M. Mills (Reading), J. Murto (Helsinki), D. Nicholson (London), J. Ogilvie (Bahrain), A. D. Pethybridge (Reading), M. J. Pilling (Oxford, now Leeds), D. W. Pratt (Pittsburg), C. Pritchie (Tulane), C. K. Prout (Oxford), H. Reiss (UCLA), H. S. Rossotti (Oxford), J. S. Rowlinson (Oxford), M. Schwartz (North Texas State), G. Shalhoub (La Salle), J. Simons (Utah), M. Spiro (London), R. C. Stern (Lawrence Livermore), B. Stevens (Tampa), J. M. Thomas (Cambridge, now The Royal Institution), D. J. Waddington (York), W. A. Wakeham (London), S. M. Walker (Liverpool), R. Westrom (Michigan State), D. H. Whiffen (Newcastle), D. J. Williams (London), J. S. Winn (Berkeley, now Dartmouth), M. Wolfsberg (Irvine), R. W. Zuehlke (Bridgeport).

Contents

Contents

Contents

Units and Notation

SI units are used throughout, but the units atmosphere (1 atm = 101.325 kPa) and Torr (1 atm = 760 Torr) are retained where appropriate as convenient units of pressure in addition to bar (1 bar = 10^5 Pa); standard pressure is taken as 1 bar. Concentrations are expressed in mol dm^{-3}, abbreviated to M (1 M = 1 mol dm^{-3}). We use the symbols (and the exact values)

$$p^{\ominus} = 1 \text{ bar} \quad m^{\ominus} = 1 \text{ mol kg}^{-1} \quad \mathcal{T} = 298.15 \text{ K} \quad 1 \text{ L} = 1 \text{ dm}^3$$

Some conversions are listed below. Others are given inside the front cover.

SI units

Force (newton, N): $1 \text{ N} = 1 \text{ J m}^{-1} = 1 \text{ kg m s}^{-2} = (10^5 \text{ dyne})$
Pressure (pascal, Pa): $1 \text{ Pa} = 1 \text{ N m}^{-2} = 1 \text{ J m}^{-3}$
Energy (joule, J): $1 \text{ J} = 1 \text{ kg m}^2 \text{ s}^{-2} \; (= 10^7 \text{ erg})$
Power (watt, W) $1 \text{ W} = 1 \text{ J s}^{-1}$
Current (ampere, A) $1 \text{ A} = 1 \text{ C s}^{-1}$
Potential (volt, V) $1 \text{ V} = 1 \text{ J C}^{-1}$
Magnetic flux density (tesla, T): $1 \text{ T} = 1 \text{ V s m}^{-2} = 1 \text{ J C}^{-1} \text{ s m}^{-2}$

Non-SI units

Length (angstrom, Å): $1 \text{ Å} = 10^{-10} \text{ m} = 10^{-8} \text{ cm} = 100 \text{ pm}$
Energy (calorie, cal; electronvolt, eV): 1 cal† = 4.184 J

$1 \text{ eV} = 1.602\,19 \times 10^{-19} \text{ J} \triangleq 96.485 \text{ kJ mol}^{-1} \triangleq 8065.5 \text{ cm}^{-1}$

$1 \text{ cm}^{-1} \triangleq 1.986 \times 10^{-23} \text{ J} \triangleq 11.96 \text{ J mol}^{-1} \triangleq 0.1240 \text{ meV}$

Dipole moment (debye, D): $1 \text{ D} = 3.335\,64 \times 10^{-30} \text{ C m}$
Pressure (atmosphere, atm; torr, Torr; millimetres of mercury, mmHg):

$1 \text{ atm} = 101.325 \text{ kPa} = 760 \text{ Torr}†$

$1 \text{ mmHg} = 133.322\,4 \text{ Pa} \qquad 1 \text{ Torr} = 133.322\,2 \text{ Pa}$

Numerical calculations

When showing intermediate arithmetical steps using a calculator, the figures beyond those justified by the data are shown in smaller type, as in $2.6/2.3 = 1.13_{04}$.

† Exact (defined) values.

Part 1

Equilibrium

Part 1 of the text develops the concepts that are needed for the discussion of equilibria in chemistry. Equilibria include physical change, such as fusion and vaporization, and chemical change, including electrochemistry. The discussion is in terms of thermodynamics, and particularly the roles of enthalpy and entropy. We shall see that a unified view of equilibrium and the direction of spontaneous change is obtained in terms of the chemical potential of substances. The chapters deal with the properties of matter in bulk; Part 2 will show how these properties stem from the behaviour of individual atoms.

The properties of gases

Check-list of key ideas

1. The concept of *state* and *equation of state* (Section 1.1).

2. The definition of *pressure*, its measurement, and its units (Section 1.1).

3. Introduction of the concept of temperature in terms of *thermal equilibrium* and the summary of experience called the *Zeroth Law of thermodynamics* (Section 1.1).

4. The definition of the *thermodynamic temperature scale* (Section 1.1).

5. The introduction of the *perfect gas law* (Section 1.2, eqn 1) as a *limiting law*.

6. The special cases of the perfect gas law, specifically *Boyle's law* (eqn 3), *Gay-Lussac's law* (eqn 4), and *Avogadro's principle*.

7. The extension of the description of gases to include mixtures of gases in terms of *Dalton's law* for perfect gases (eqn 7) and *partial pressures* in general (eqn 8).

8. The properties of *real gases* expressed in terms of the *isotherms* and the *compression factor* (eqn 9) and summarized by the *virial equation of state* (eqn 10).

9. The *van der Waals equation* as an approximate equation of state of real gases (Section 1.4) and some of the conclusions that can be drawn from it, particularly the values of the *critical constants* of gases.

10. The unification of the description of real gases using the *principle of corresponding states* (Section 1.5), and an illustration of the principle using the van der Waals equation.

We begin by examining the properties of the simplest state of matter, a **gas**, a substance that fills any container it occupies. We shall see how to describe the physical properties of gases in terms of the laws they obey, and then use these laws as a foundation for the development of thermodynamics in the following chapters. Initially we shall consider only pure gases, but later we shall see that the same ideas and equations apply to mixtures of gases too.

The worked examples early in the chapter serve a dual role. As well as illustrating a point made in the text (their normal purpose throughout the text), they introduce some of the conventions we shall adopt concerning notation and significant figures.

The perfect gas

There is no need to know the molecular origin of the bulk properties we meet in thermodynamics. However, it is often helpful in practice to have a molecular model in mind, for then it is easier to interpret the equations and to develop an insight into the origins of properties. We shall, in fact, find it helpful to picture a gas as a collection of molecules in continuous random, chaotic motion, with speeds that increase as the temperature is raised. A gas differs from a liquid (in which the molecules are also moving chaotically) in that its molecules are widely separated from each other, except during collisions, and move largely independently of each other. A part of the purpose of studying physical chemistry is to learn how to translate a model into a quantitative expression for an observable, measurable property, and in Part 3 we shall see that this model is quantitatively consistent with the properties we are about to describe.

1.1 The states of gases

Apart from the volume V occupied by the gas and its amount of substance (the number of moles, n), the fundamental properties for the study of gases are its **pressure** p and its **temperature** T. The first part of this section will be taken up with explaining these last two properties. The concept of the mole is reviewed in the *Further information* section at the end of the chapter.

Once we know the values of V, n, p, and T of a sample of any pure substance we know its **state**, and can be confident that whenever that substance is in that state it will have exactly the same properties (for example, the same density, heat capacity, colour, and so on). However, almost the first experiments done in the subject that has become physical chemistry showed that not all the four variables V, n, p, and T are independent, and that any one of them can be expressed in terms of the other three. In other words, there exists an **equation of state** that relates the volume, amount, pressure, and temperature of any pure substance, and there are only *three* independent variables needed to specify its state. Much of the rest of this chapter will describe some of the equations of state of gases that have been obtained experimentally or proposed theoretically.

Pressure

Pressure is force per unit area, and the greater the force acting on a given surface, the greater the pressure. The SI unit of pressure, the **pascal** (Pa), is defined as 1 newton per square metre:

$$1\,\text{Pa} = 1\,\text{N m}^{-2}$$

However, several other units are still widely used. These include the **bar** and the **atmosphere** (atm):

$$1\,\text{bar} = 100\,\text{kPa exactly}$$

$$1\,\text{atm} = 101.325\,\text{kPa exactly}$$

Example 1.1: *Calculating pressure*

Suppose Isaac Newton weighed 65 kg. Calculate the pressure he exerted on the ground when wearing (a) boots, with soles of total area 250 cm² in contact with the ground, (b) ice skates, of total area 2.0 cm².

Answer. The downward force an object (Isaac) of mass m exerts on the surface of the earth is $F = mg$ where g is the acceleration of free fall, $9.81\,\text{m s}^{-2}$. In Isaac's case,

$$F = 65\,\text{kg} \times 9.81\,\text{m s}^{-2} = 638\,\text{N}$$

since $1\,\text{N} = 1\,\text{kg m s}^{-2}$. The force is the same whatever his footwear. The pressure he exerts in each case is $p = F/A$, where A is the area over which the force acts. Hence:

$$\text{(a)}\quad p = \frac{638\,\text{N}}{2.50 \times 10^{-2}\,\text{m}^2} = 2.6 \times 10^4\,\text{Pa, 26 kPa}$$

$$\text{(b)}\quad p = \frac{638\,\text{N}}{2.0 \times 10^{-4}\,\text{m}^2} = 3.2 \times 10^6\,\text{Pa, 3.2 MPa}$$

Comment. A pressure of 26 kPa corresponds to 0.26 atm and a pressure of 3.2 MPa corresponds to 31 atm. The small 8 in 638 is a 'non-significant figure' that we carry through the intermediate stages of a calculation.

Exercise. Calculate the pressure exerted by a mass of 1.0 kg pressing through the point of a pin of area $1.0 \times 10^{-2}\,\text{mm}^2$ on the surface of the earth.

[981 MPa, 9.7×10^3 atm]

A pressure of 1 bar is the standard pressure for reporting thermodynamic data, and we shall denote it $p^{\ominus}$:

$$p^{\ominus} = 1\,\text{bar exactly}$$

The **torr** (the name of the unit; its symbol, for the pleasure of pedants, is Torr) is defined so that

$$1\,\text{atm} = 760\,\text{Torr exactly}$$

The torr is almost exactly equal to the **millimetre of mercury** (mmHg), which is defined so that 1 mmHg is the pressure exerted by a 1 mm high column of mercury. For most purposes, mmHg and Torr can be used interchangeably.

The pressure of the atmosphere is measured with a **barometer**. The original version of a barometer (which was invented by Torricelli) was an inverted tube of mercury, the height of the mercury column being proportional to the external pressure.

Example 1.2: *Calculating the pressure exerted by a column of liquid*

Calculate the pressure at the base of a column of liquid of density ρ and height h at the surface of the earth.

Answer. Suppose the column has radius r, then its cross-sectional area is πr^2 and its volume is $\pi r^2 h$. Since the mass of this volume of liquid is $m = \rho \pi r^2 h$, the force it exerts at its base is

$$F = mg = \rho \pi r^2 h g$$

The pressure is this force divided by the area on which it acts, which is the cross-section of the column:

$$p = \frac{F}{A} = \frac{\rho \pi r^2 h g}{\pi r^2}$$

That is, for any liquid of density ρ, the pressure at the base of a column of height h is

$$p = \rho g h$$

Comment. Note that the pressure is independent of the radius of the column.

Exercise. Calculate the pressure at the base of a column of length l held at an angle θ to the vertical.

$$[\rho g l \cos \theta]$$

The pressure of a sample of gas in a container is measured with a **manometer** (Fig. 1.1). In its simplest form, a manometer is a U-tube filled with some liquid of low volatility (such as silicone oil). The pressure of the gas is proportional to the difference in heights of the liquid in the two arms (plus the external pressure if one tube is open to the atmosphere). More sophisticated techniques are used at lower pressures. Methods that avoid the complication of having to account for the vapour from the manometer fluid are also available. These include monitoring the deflection of a diaphragm, either mechanically or electrically, or monitoring the change in some pressure-sensitive electrical property.

Temperature

Everyone knows intuitively what 'temperature' means, but our task here is to place this intuition on a firm foundation. A full appreciation of this elusive property will come much later when we have done more thermodynamics and, later still, have dealt with molecular energies.

The concept of temperature springs from the observation that energy can flow from one substance to another when they are in contact (as when a red-hot metal is plunged into water). The 'temperature' is the property that tells us the *direction* of the flow of energy. Thus, if energy flows from A to B, we say that A has a higher temperature than B. If no energy flows from A to B when they are in contact, we say that they have the same temperature and have reached a state of **thermal equilibrium**.

Suppose a substance A (which we can think of as a block of iron) is in thermal equilibrium with a substance B (a block of copper), and that B is also in thermal equilibrium with another substance C (a flask of water). Then it has been found experimentally that A and C are also in thermal equilibrium when they are put in contact. This is summarized by a statement somewhat grandiosely called the **Zeroth Law of thermodynamics**:

The Zeroth Law of thermodynamics: If A is in thermal equilibrium with B, and B is in thermal equilibrium with C, then C is also in thermal equilibrium with A.

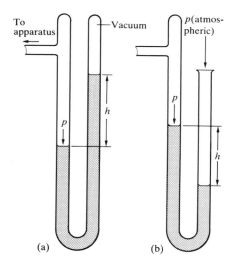

Fig. 1.1 Two versions of a manometer used to measure the pressure of a sample of gas. (a) The height difference h of the two columns in the sealed-tube manometer is directly proportional to the pressure of the sample, and $p = \rho g h$, where ρ is the density of the liquid. (b) The difference in heights of the columns in the open-tube manometer is proportional to the *difference* in pressure between the sample and the atmosphere. In this case, the pressure of the sample is lower than that of the atmosphere.

The Zeroth Law is the fundamental principle that allows us to build **thermometers**, which are devices that display a change in temperature as a change in some physical property (such as the length of a column of mercury). Thus, suppose that B is a glass capillary containing mercury then, when A is in contact with B, the mercury column in the latter has a certain length. According to the Zeroth Law, if the mercury column in B has the same length when it is placed in contact with another substance C, we can predict that no energy will flow between A and C when they are in contact whatever the identities of the substances. Moreover, we can use the length of the mercury column as a measure of the temperatures of A and C.

The relation of the numerical value of the temperature to the property selected for monitoring it is arbitrary. In the early days of thermometry, temperatures were related to the length of a column of liquid, and the difference in lengths shown when the thermometer was first in contact with melting ice and then with boiling water was divided into 100 steps called *degrees*, the lowest point being labelled 0. That led to the **Celsius scale**. However, since different liquids expand to different extents, and do not always expand uniformly over a given range, thermometers constructed from different materials showed different numerical values of the temperature between their fixed points. Thus, while A, B, C, ... may all have been ascribed the temperature 28.7°C when a mercury-in-glass thermometer was used, they may all have been ascribed 28.8°C when an alcohol-in-glass thermometer was used. The variation of quoted temperatures is even greater when electrical measurements, such as the resistance of a wire, are included.

The pressure of a gas at constant volume, however, can be used to construct a temperature scale that is almost independent of the identity of the gas. Furthermore, this near uniformity becomes exact as the density of the gas is reduced to zero. This uniformity lets us establish the **thermodynamic temperature scale**, which we define shortly.

Temperatures on the thermodynamic (or Kelvin) scale are denoted T and are normally reported in **kelvin**, K. When we want to signify temperatures on the Celsius scale we use the symbol θ. The two scales are related by

$$T/\text{K} = \theta/°\text{C} + 273.15 \text{ exactly}$$

That is, 0°C corresponds to 273 K.

Example 1.3: *Converting temperatures*

Express 25°C as a temperature in kelvins.

Answer. This is a numerically trivial example, but it is included to give an opportunity to demonstrate the significance of the manner in which the expression above has been written. We have used (and use throughout the text) the procedure called 'quantity calculus' in which a physical quantity (such as θ) is the product of a numerical value (25) and a unit (°C). That is, 25°C stands for $25 \times °\text{C}$. Therefore, $\theta/°\text{C}$ stands for the physical quantity $(25 \times °\text{C})$ divided by the unit (°C):

$$\theta/°\text{C} = 25$$

The conversion then reads

$$T/\text{K} = 25 + 273.15 = 298$$

Multiplication of both sides by the unit K then gives

$$T = 298 \text{ K}$$

Comment. Quantity calculus is a very powerful way of keeping track of units, and we shall use it in all numerical work.

Exercise. Use the result obtained in Example 1.2, that $p = \rho g h$, to calculate (using quantity calculus) the pressure at the base of a 1.0 m high column of water of density 1.0 g cm^{-3}.

[9.8 kPa]

1.2 The gas laws

A large number of measurements on gases (which we shall summarize below) have shown that at low pressures the pressure, volume, temperature, and amount of gas are related by the expression

$$pV = nRT \qquad (1)°$$

where the **gas constant** R is the same for every gas. This equation is an example of an equation of state of the kind mentioned in Section 1.1 since it is a relation between the variables of a sample of a substance, and is called the **perfect gas equation of state**.

Equation 1 is obeyed quite well for most gases at room temperature and pressure (close to 25°C and 1 atm). All gases obey it increasingly closely as the pressure is decreased. That is, eqn 1 is a **limiting law** in the sense that all gases obey it in the limit of zero pressure. A gas that obeys eqn 1 exactly is called a **perfect gas** or an **ideal gas**. A **real gas** is an actual gas, such as hydrogen, oxygen, or air, which does not obey eqn 1 exactly except in the limit of zero pressure. When an equation applies only to a perfect gas we attach a superscript ° to the equation number.

The value of the gas constant can be obtained by evaluating pV/nT for a gas in the limit of zero pressure (to guarantee that it is behaving perfectly). However, a more accurate value can be obtained by measuring the speed of sound in a low-pressure gas and extrapolating its value to zero pressure:

$$R = 8.314 \text{ J K}^{-1} \text{ mol}^{-1}$$

Some values in other units are given in Table 1.1.

Table 1.1. The gas constant in various units

$R = 8.314 \text{ J K}^{-1} \text{ mol}^{-1}$
   $8.206 \times 10^{-2} \text{ L atm K}^{-1} \text{ mol}^{-1}$
   $62.36 \text{ L Torr K}^{-1} \text{ mol}^{-1}$
   $1.987 \text{ cal K}^{-1} \text{ mol}^{-1}$

The response to pressure

By setting n and T constant in eqn 1 we obtain **Boyle's law**:

$$pV = \text{constant (at constant } n, T) \qquad (2)°$$

This law was first proposed in 1662 by Robert Boyle who, acting on the suggestion of his assistant John Townley, verified that at constant temperature the volume of a fixed amount of gas is inversely proportional to its pressure (Fig. 1.2). Each of the curves in the illustration corresponds to a single temperature and hence is called an **isotherm** (from the Greek, meaning 'equal heat'). According to Boyle's law, the isotherms of gases are hyperbolas (curves that satisfy $xy = $ constant); however, real gases have hyperbolic isotherms only in the limit of $p \to 0$.

Boyle's law is used to predict the pressure of a gas when its volume is changed (or vice versa). If the initial values of pressure and volume are p_1 and V_1, then because the product pV is a constant, the final values p_2 and V_2

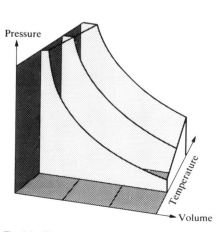

Fig. 1.2 The pressure–volume dependence of a fixed amount of perfect gas at three different temperatures. Each curve is a hyperbola ($p \propto 1/V$) and is called an *isotherm*.

must satisfy

$$p_2 V_2 = p_1 V_1 \quad \text{(at constant } n, T) \tag{3}°$$

The molecular explanation of Boyle's law can be traced to the fact that if the volume of a sample is reduced by half, there are twice as many molecules per unit volume. Twice as many molecules strike the walls in a given period of time, and so the average force they exert is doubled. Hence, when the volume is halved the pressure of the gas is doubled, and $p \times V$ is a constant. Moreover, at very low pressures the molecules are so far apart that they exert negligible forces on each other on average. This accounts for the observation that the law is 'universal' in the sense of applying to any gas without reference to its chemical composition.

The response to temperature

Another special case of the perfect gas law is obtained by holding constant the pressure p and amount n of gas. This gives **Gay-Lussac's law**:

$$V \propto T \quad \text{(at constant } n, p) \tag{4a}°$$

Similarly, by holding the amount and volume constant instead:

$$p \propto T \quad \text{(at constant } n, V) \tag{4b}°$$

The linear variation of pressure with temperature is illustrated in Fig. 1.3. These two equations can be used to predict the volume of a perfect gas as a fixed amount is heated (or cooled) at constant pressure or at constant volume, for they imply that

$$\frac{V_1}{V_2} = \frac{T_1}{T_2} \quad \text{(at constant } n, p) \tag{5a}°$$

$$\frac{p_1}{p_2} = \frac{T_1}{T_2} \quad \text{(at constant } n, V) \tag{5b}°$$

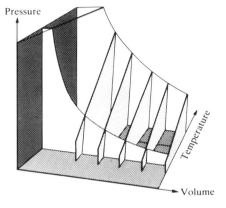

Fig. 1.3 The pressure–temperature dependence of a fixed amount of a perfect gas at different molar volumes. The straight lines ($p \propto T$) are called *isochores* (from the Greek, 'equal space').

Example 1.4: *Using the gas laws*

In an industrial process, nitrogen is heated to 500 K in a vessel of constant volume. If it enters the vessel at a pressure of 100 atm and a temperature of 300 K, what pressure does it exert at the working temperature?

Answer. In the absence of more detailed instructions, assume that the gas is perfect. Since the volume is constant, rearrange the equation above to

$$p_2 = \frac{T_2}{T_1} \times p_1$$

$$= \frac{500\ \text{K}}{300\ \text{K}} \times 100\ \text{atm} = 167\ \text{atm}$$

Comment. Experiment shows that the pressure is actually 183 atm under these conditions, so the assumption that the gas is perfect leads to a 10 per cent error.

Exercise. What temperature would result in the same sample exerting a pressure of 300 atm?   [900 K]

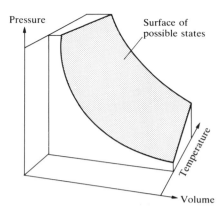

Fig. 1.4 A region of the *p, V, T* surface of a fixed amount of perfect gas. The points forming the surface represent the only states of the gas that can exist.

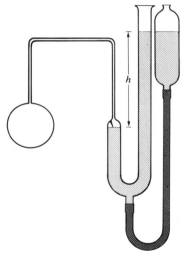

Fig. 1.5 A constant-volume gas thermometer. The bulb on the left is put in contact with the sample and the pressure of the enclosed gas is measured by noting the height *h* of the mercury column.

The molecular explanation of Gay-Lussac's law lies in the fact that raising the temperature of a gas increases the average speed of its molecules. The molecules collide with the walls more frequently, and do so with greater impact. Therefore they exert on them a greater average force, and hence exert a greater pressure.

We can combine Fig. 1.3 with Fig. 1.2 to show the variation of pressure with volume and temperature in a single diagram, Fig. 1.4. The graph is an example of a ***p, V, T* surface**, a diagrammatic portrayal of the perfect gas equation of state. The only states in which a perfect gas can exist are those on the surface. Real gases may exist in different states from those of a perfect gas, and are described by *p, V, T* surfaces with different shapes, but which coincide with the perfect-gas surface at low pressures. We shall see an example later.

The thermodynamic temperature scale

Temperature can be measured with a **constant-volume gas thermometer** (Fig. 1.5) by comparing the pressures of the gas it contains when it is in thermal contact first with the sample of interest and then with a standard. The standard is taken as water at its **triple point**, the unique condition of temperature and pressure when ice, water, and water vapour coexist at equilibrium. The temperature of the triple point is defined as $T_3 = 273.16$ K exactly.[1]

If the pressure measured when the gas thermometer is in contact with the sample is p, and the pressure when it is at the temperature of the triple point of water T_3 is p_3, then the temperature of the sample is $T \approx (p/p_3)T_3$. Since this is exact only when the gas is behaving perfectly, readings are taken with decreasing amounts of gas in the thermometer, and the results extrapolated to zero pressure. The thermodynamic temperature is then given by

$$T = \lim_{p \to 0} T(p) \quad \text{with} \quad T(p) = \frac{p}{p_3} \times T_3 \quad \text{and} \quad T_3 = 273.16 \text{ K}$$

Ordinary, imperfect, but easier to use thermometers may then be calibrated against this measurement.

Avogadro's principle

According to eqn 1, at constant pressure and temperature the volume of a perfect gas is proportional to the amount of gas present:

$$V \propto n \quad \text{(at constant } p, T)$$

This expression is the essential content of the principle stated by Amedeo Avogadro, that equal volumes of gases at the same temperature and pressure contain the same number of molecules. His principle implies that the **molar volume** V_m of a gas, the volume it occupies per mole of molecules

$$V_m = \frac{V}{n}$$

[1] The perfect-gas temperature scale is the fundamental scale of temperature, and other scales are defined in terms of it. Thus, the zero of the Celsius scale is now *defined* as lying at exactly 0.01 K below the temperature of the triple point of water, so 0°C lies at exactly 273.15 K. Water is found to freeze at 273.1500 ± 0.0003 K under 1 atm pressure.

should be the same for all gases so long as the temperature and pressure are the same. Indeed, from eqn 1, for any gas behaving perfectly,

$$V_m = \frac{RT}{p} \qquad (6)°$$

Two sets of conditions are currently used as 'standard' values for reporting data. One is **standard temperature and pressure** (STP), which corresponds to 0°C and 1 atm. The other is **standard ambient temperature and pressure** (SATP), which corresponds to 25°C (more precisely, to 298.15 K) and 1 bar (that is, $p^{\ominus}$). When each set of values is substituted into eqn 6 we find that (where $1 \, L = 1 \, dm^3$)

$$\text{STP:} \quad V_m = 22.414 \, L \, mol^{-1}$$

$$\text{SATP:} \quad V_m = 24.790 \, L \, mol^{-1}$$

We denote the molar volume at SATP by $V_m^{\ominus}$.

Dalton's law

So far, we have dealt with a single, pure gas. Now we turn to a mixture of gases, such as the atmosphere, and will see that very similar equations apply to them. This should not be surprising: since the same limiting law $pV = nRT$ applies to every pure gas, we can expect it to apply to a mixture of gases too. In this chapter we limit ourselves to gases that do not react when they mix.

The kind of question we need to answer when dealing with gaseous mixtures is the contribution that each component gas makes to the total pressure of the sample. In the nineteenth century John Dalton[2] made observations that provide the answer and summarized them in a law:

Dalton's law: The pressure exerted by a mixture of perfect gases is the sum of the pressures exerted by the individual gases occupying the same volume alone.

Therefore, if a certain amount of H_2 exerts 0.25 atm when present alone in a container, and an amount of N_2 exerts 0.80 atm when present alone in the same container at the same temperature, the total pressure when both are present is 1.05 atm.

More generally, if an amount n_A of a perfect gas A occupies a container of volume V at a temperature T, its pressure is $p_A = n_A(RT/V)$. If instead an amount n_B of another perfect gas B occupies the container, its pressure is $p_B = n_B(RT/V)$. When both are present simultaneously, Dalton's law tells us that the total pressure is

$$p = p_A + p_B$$

If the mixture consists of several gases A, B, C, ... present in the amounts

---

[2] John Dalton (1766–1844) is the Dalton of the atomic hypothesis, and also the Dalton of *daltonism*, or colour-blindness (from which he suffered and which he described). He was described as 'an indifferent experimenter, and singularly wanting in the language and power of illustration'. He paid another price: 'Into society he rarely went, and amusement he had none, with the exception of a game of bowls on Thursday afternoons'.

$n_A, n_B, n_C, \ldots$, the total pressure is

$$p = p_A + p_B + p_C + \ldots = \sum_J p_J \quad \text{with} \quad p_J = \frac{n_J RT}{V} \qquad (7)°$$

Throughout the text we use the symbol $\sum$ to denote a sum.

Example 1.5: *Using Dalton's law*

A container of volume $10.0\,L$ holds $1.00\,mol\,N_2$ and $3.00\,mol\,H_2$ at $298\,K$. What is the total pressure?

Answer. We assume the gases to be perfect and calculate the two individual pressures from the perfect gas law. The total pressure they exert is the sum of the individual pressures. Since under the stated conditions

$$\frac{RT}{V} = \frac{8.206 \times 10^{-2}\,L\,atm\,K^{-1}\,mol^{-1} \times 298\,K}{10.0\,L} = 2.445\,atm\,mol^{-1}$$

$$p(N_2) = 1.00\,mol \times 2.445\,atm\,mol^{-1} = 2.445\,atm$$

$$p(H_2) = 3.00\,mol \times 2.445\,atm\,mol^{-1} = 7.335\,atm$$

The total pressure is therefore $9.78\,atm$.

Exercise. $1.00\,mol\,N_2$ and $2.00\,mol\,O_2$ were added to the same container (with the nitrogen and hydrogen still inside). Calculate the total pressure at $298\,K$.

[17.1 atm]

Mole fractions and partial pressures

We can take a step closer to the discussion of mixtures of real gases by introducing the **mole fraction** x_J of each component J. The mole fraction of J in a mixture is the number of moles of J molecules present (n_J) expressed as a fraction of the total number of moles of molecules (n) in the sample:

$$x_J = \frac{n_J}{n} \quad \text{with} \quad n = n_A + n_B + \ldots \qquad (8a)$$

When no J molecules are present, $x_J = 0$; when only J molecules are present, $x_J = 1$. A mixture of $1.0\,mol\,N_2$ and $3.0\,mol\,H_2$, and therefore of $4.0\,mol$ molecules in all, consists of mole fractions 0.25 of N_2 and 0.75 of H_2. It follows from the definition of x_J that, whatever the composition of the mixture,

$$x_A + x_B + \ldots = \sum_J x_J = 1 \qquad (8b)$$

Next, we define the **partial pressure** p_J of a gas in a mixture (*any* gas, not just a perfect gas), as

$$p_J = x_J p \qquad (8c)$$

where p is the total pressure of the mixture. It follows from eqn 8b that the sum of the partial pressures is equal to the total pressure (for real and perfect gases):

$$\sum_J p_J = \sum_J x_J p = \left(\sum_J x_J\right)p = p$$

Figure 1.6 shows how the partial pressures of a binary (two-component) mixture contribute to the total pressure as the mole fraction of one component increases from 0 to 1.

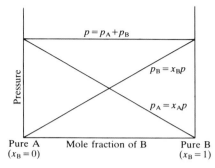

Fig. 1.6 The partial pressures p_A and p_B of a binary mixture of (real or perfect) gases of total pressure p as the composition changes from pure A to pure B. The sum of the partial pressures is equal to the total pressure.

Example 1.6: *Calculating partial pressures*

The mass percentage composition of dry air at sea level is approximately N_2 75.5; O_2 23.2; Ar 1.3. What is the partial pressure of each component when the total pressure is 1.000 atm?

Answer. To use eqn 8 we need the mole fractions of the components. The number of moles of molecules of molar mass M in a sample of mass m is m/M. Therefore, the numbers of moles of each type of molecule present in 100.0 g of air are

$$n(N_2) = \frac{100.0\text{ g} \times 0.755}{28.02\text{ g mol}^{-1}} = 2.694\text{ mol}$$

$$n(O_2) = \frac{100.0\text{ g} \times 0.232}{32.00\text{ g mol}^{-1}} = 0.725\text{ mol}$$

$$n(Ar) = \frac{100.0\text{ g} \times 0.013}{39.95\text{ g mol}^{-1}} = 0.032\text{ mol}$$

Since overall $n = 3.451$ mol, the mole fractions and partial pressures are as follows:

	N_2	O_2	Ar
Mole fraction	0.781	0.210	0.0093
Partial pressure/atm	0.781	0.210	0.0093

Comment. We have not had to assume that the gases are perfect: partial pressures are *defined* as $p_J = x_J p$ for any gas.

Exercise. When carbon dioxide is taken into account the mass percentages are 75.52, 23.15, and 1.28 for N_2, O_2, and Ar and 0.046 for CO_2. What are the partial pressures when the total pressure is 0.900 atm?
$$[0.703, 0.189, 0.0084, 0.0003\text{ atm}]$$

Partial pressures are *defined* as being proportional to the mole fractions, and hence necessarily add together to give the total pressure. However, in the case of a mixture of perfect gases, a partial pressure is also the pressure that a gas would exert if it were alone in the container. To show this, we set $p = nRT/V$ and $x_J = n_J/n$ in eqn 8c and get

$$p_J = \frac{n_J}{n} \times \frac{nRT}{V} = \frac{n_J RT}{V}$$

in accord with eqn 7 and Dalton's law.

Real gases

Real gases are imperfect; that is, they do not obey the perfect gas law exactly. The deviations from the law are particularly important at high

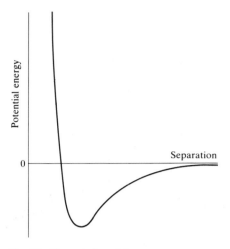

Fig. 1.7 The variation of the potential energy of two molecules on their separation. High positive potential energy (at very small separations) indicates that the interactions between them are strongly repulsive at these distances. At intermediate separations, where the potential energy is negative, the attractive interactions dominate. At large separations (on the right) the potential energy is zero and there is no interaction between the molecules.

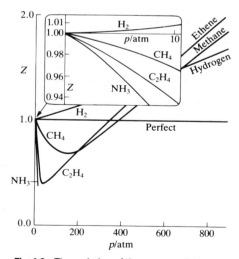

Fig. 1.8 The variation of the compression factor $Z = pV_m/RT$ with pressure for several gases at 0°C. A perfect gas has $Z = 1$ at all pressures. Notice that although the curves approach 1 as $p \rightarrow 0$, they do so with different slopes. At the *Boyle temperature*, which varies from gas to gas, the slope of the curve is zero at $p = 0$ (like that of a perfect gas).

pressures and low temperatures, especially when the gas is on the point of condensing to liquid.

1.3 Molecular interactions

Real gases show deviations from the perfect gas law because molecules interact with each other: repulsive forces between molecules assist expansion and attractive forces assist compression.

Repulsive forces between neutral molecules are significant only when the molecules are almost in contact: they are short-range interactions, even on a scale measured in molecular diameters (Fig. 1.7). Since they are short-range interactions, repulsions can be expected to be important only when the molecules are close together on average. This is the case at high pressure, when a large number of molecules occupy a small volume. On the other hand, attractive intermolecular forces have a relatively long range and are effective over several molecular diameters. They are important when the molecules are fairly close together but not necessarily touching (at the intermediate separations in Fig. 1.7). Attractive forces are ineffective when the molecules are far apart (well to the right in Fig. 1.7).

It follows that at low pressure, when the molecules occupy a large volume, the molecules are so far apart for most of the time that the intermolecular forces play no significant role, and the gas behaves perfectly. At moderate pressure, when the molecules are on average only a few molecular diameters apart, the attractive forces dominate the repulsive forces. In this case, the gas can be expected to be more readily compressible than a perfect gas. At high pressure, when the molecules are on average close together, the repulsive forces dominate and the gas can be expected to be less readily compressible.

The compression factor

That real gases reflect this distance-dependence of the forces can be demonstrated by plotting the **compression factor** Z against pressure, where

$$Z = \frac{pV_m}{RT} \tag{9}$$

Since for a perfect gas $Z = 1$ under all conditions, deviation of Z from 1 is a measure of imperfection.

Some results are plotted in Fig. 1.8. At very low pressures, all the gases shown have $Z \approx 1$ and are behaving nearly perfectly. At high pressures, all the gases have $Z > 1$, signifying that they are more difficult to compress than a perfect gas (the product pV_m is greater than RT). Repulsive forces are now dominant. At intermediate pressures, some of the gases have $Z < 1$, indicating that the attractive forces are dominant and favour compression.

Virial coefficients

Figure 1.9 shows some experimental isotherms for carbon dioxide. At large molar volumes and high temperatures the real and perfect isotherms do not differ greatly. The small differences suggest that the perfect gas law is correct at low pressures and is in fact the first term in an expression of the form

$$pV_m = RT(1 + B'p + C'p^2 + \dots) \tag{10a}$$

In many applications, a more convenient expansion is

$$pV_m = RT\left(1 + \frac{B}{V_m} + \frac{C}{V_m^2} + \ldots\right) \qquad (10b)$$

These expressions are two versions of the **virial equation of state** (the name comes from the Latin word for force). $B, C, \ldots$, which depend on the temperature, are the second, third, $\ldots$ **virial coefficients** (Table 1.2). The third virial coefficient C is usually less important than the second B in the sense that at typical molar volumes $C/V_m^2 \ll B/V_m$. The virial equation is the first example of a common procedure in physical chemistry, when a simple law (in this case $pV = nRT$) is treated as the first term in a series in powers of a variable (in this case p or V_m).

The virial equation can be used to demonstrate the important point that although the equation of state of a real gas may coincide with the perfect gas law as $p \to 0$, all its *properties* do not necessarily coincide with those of a perfect gas. Consider, for example, the value of dZ/dp, the slope of the graph of compression factor against pressure. For a perfect gas $dZ/dp = 0$, but for a real gas

$$\frac{dZ}{dp} = B' + 2pC' + \ldots \to B' \quad \text{as} \quad p \to 0$$

However, B' is not necessarily zero. Therefore, although for a real gas $Z \to 1$ as $p \to 0$ (and, more generally, the equation of state of a real gas coincides with the perfect gas law as $p \to 0$), the *slope* of Z against p does not approach 0 (the perfect gas value). Since other properties (as we shall see) also depend on derivatives, the properties of real gases do not always coincide with the perfect gas values at low pressures. By a similar argument based on eqn 10b, the variation of Z with molar volume at large molar volumes (the equivalent of low pressures) is

$$\frac{dZ}{d(1/V_m)} \to B \quad \text{as} \quad V_m \to \infty \quad (\text{or } p \to 0)$$

Since the virial coefficients depend on the temperature, there may be a temperature at which Z does approach 1 with zero slope at low pressure or high molar volume. At this temperature, which is called the **Boyle temperature** T_B, the properties of the real gas do coincide with those of a perfect gas as $p \to 0$ or $V_m \to \infty$. According to the relation above, Z has zero slope as $V_m \to \infty$ if $B = 0$, so we can conclude that at the Boyle temperature $B = 0$. It then follows from eqn 10b that $pV_m \approx RT$ over a more extended range of pressures than at other temperatures because the first term after 1 in the virial equation is zero and C/V_m^2 and higher terms are negligibly small. For helium $T_B = 22.64$ K; for air $T_B = 346.8$ K; some more values are given in Table 1.3.

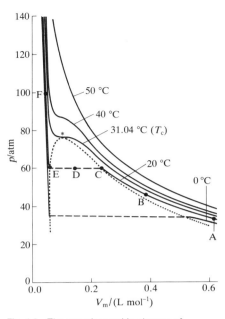

Fig. 1.9 The experimental isotherms of carbon dioxide at several temperatures. The *critical isotherm*, the isotherm at the critical temperature, is at 31.04°C. The critical point is marked with a star.

Table 1.2. Second virial coefficients, $B/(\text{cm}^3 \text{mol}^{-1})$†

	273 K	600 K
Ar	−21.7	11.9
CO_2	−149.7	−12.4
N_2	−10.5	21.7
Xe	−153.7	−19.6

† Most tables in the text are abbreviated versions of the tables collected together in the *Data section* at the end of the book.

Table 1.3. Critical constants of gases

	p_c/atm	$V_c/(\text{cm}^3 \text{mol}^{-1})$	T_c/K	Z_c	T_B/K
Ar	48.0	75.3	150.7	0.292	411.5
CO_2	72.9	94.0	304.2	0.274	714.8
He	2.26	57.8	5.2	0.305	22.6
N_2	33.5	90.1	126.3	0.292	327.2

Condensation

Now consider what happens when the volume of a sample of gas initially in the state marked A in Fig. 1.9 is decreased at constant temperature (by pushing in a piston). Near A, the pressure of the gas rises in approximate agreement with Boyle's law. Serious deviations from that law begin to appear when the volume has been reduced to B.

At C (which corresponds to about 60 atm in the case of carbon dioxide), all similarity to perfect behaviour is lost, for suddenly the piston slides in without any further rise in pressure: this is represented by the horizontal line CDE. Examination of the contents of the vessel shows that just to the left of C a liquid appears, and there are two phases separated by a sharply defined surface. As the volume is decreased from C through D to E, the amount of liquid increases. There is no additional resistance to the piston at this stage because the gas can respond by condensing. The pressure corresponding to the line CDE, when both liquid and vapour are present in equilibrium, is called the **vapour pressure** of the liquid at the temperature of the experiment.

At E, the sample is entirely liquid and the piston rests on its surface. Any further reduction of volume requires the exertion of considerable pressure. This is reflected by the sharply rising line to the left of E. Even a small reduction of volume from E to F requires a great increase in pressure.

Critical constants

The isotherm at the temperature T_c (304.19 K, 31.04°C for CO_2) plays a special role in the theory of the states of matter. An isotherm at a fraction of a kelvin below T_c behaves as we have already described: at some pressure a liquid condenses from the gas and is distinguishable from it by the presence of a visible surface. If, however, the compression takes place at T_c itself, a surface separating two phases does not appear and the volumes at each end of the horizontal part of the isotherm have merged to a single point, the **critical point** of the gas. The temperature, pressure, and molar volume at the critical point are called the **critical temperature** T_c, **critical pressure** p_c, and **critical molar volume** V_c of the substance. Collectively, T_c, p_c, and V_c are the **critical constants** (Table 1.3).

At and above T_c the sample has a single phase that occupies the entire volume of the container. Such a phase is, by definition, a gas. Hence, the liquid phase of a substance does not form above the critical temperature. The critical temperature of oxygen, for instance, signifies that it is impossible to produce liquid oxygen by compression alone if its temperature is greater than 154.8 K: to liquefy it—to obtain a fluid phase that does not occupy the entire volume—the temperature must first be lowered to below 154.8 K, and then the gas compressed isothermally.

1.4 The van der Waals equation

Conclusions can be drawn from the virial equations of state only by inserting specific values of the coefficients. It is often useful to have a broader, if less precise, view of all gases. Therefore, we introduce the approximate equation of state suggested by Johannes van der Waals in 1873. It is an excellent example of an equation that can be obtained by thinking scientifically about a mathematically complicated but physically simple problem. Van der Waals himself proposed it on the basis of experimental evidence available to him in conjunction with rigorous thermodynamic arguments.

Table 1.4. Van der Waals coefficients at 298 K

	$a/(\text{atm L}^2\,\text{mol}^{-2})$	$b/(10^{-2}\,\text{L mol}^{-1})$
Ar	1.345	3.22
CO_2	3.592	4.267
He	0.034	2.37
N_2	1.390	3.913

Constructing the equation

The repulsive interactions between molecules are taken into account by supposing that they cause the molecules to behave as small but impenetrable spheres. The non-zero volume of the molecules implies that instead of moving in a volume V they are restricted to a smaller volume $V - nb$, where nb is approximately the total volume taken up by the molecules themselves. This argument suggests that the perfect gas law $p = nRT/V$ should be replaced by

$$p = \frac{nRT}{V - nb}$$

The pressure depends on both the frequency of collisions with the walls and the force of each collision. Both the frequency of the collisions and their force are reduced by the attractive forces, which act with a strength roughly proportional to the molar concentration n/V of molecules in the sample. Therefore, the pressure is reduced in proportion to the *square* of this concentration. If the reduction of pressure is written as $-a(n/V)^2$, where a is a constant characteristic of each gas, the combined effect of the repulsive and attractive forces is the **van der Waals equation**:

$$p = \frac{nRT}{V - nb} - a\left(\frac{n}{V}\right)^2 \tag{11a}$$

This equation is often written in terms of the molar volume $V_m = V/n$ as

$$p = \frac{RT}{V_m - b} - \frac{a}{V_m^2} \tag{11b}$$

The term a/V_m^2 is called the **internal pressure** of the gas. It is sometimes convenient to reorganize the equation into a form resembling $pV = nRT$:

$$\left(p + \frac{an^2}{V^2}\right)(V - nb) = nRT \tag{11c}$$

Example 1.7: *Using the van der Waals equation*

Estimate the molar volume of CO_2 at 500 K and 100 atm by treating it as a van der Waals gas.

Answer. We rearrange eqn 11b into an equation for V_m:

$$V_m^3 - \left(b + \frac{RT}{p}\right)V_m^2 + \left(\frac{a}{p}\right)V_m - \frac{ab}{p} = 0$$

According to Table 1.4, $a = 3.592\,\text{L}^2\,\text{atm mol}^{-2}$ and $b = 4.267 \times 10^{-2}\,\text{L mol}^{-1}$. Therefore, since

$$\frac{RT}{p} = \frac{8.206 \times 10^{-2}\,\text{L atm K}^{-1}\,\text{mol}^{-1} \times 500\,\text{K}}{100\,\text{atm}} = 0.4103\,\text{L mol}^{-1}$$

the coefficients in this equation are

$$b + RT/p = 0.4530 \text{ L mol}^{-1}$$

$$a/p = 3.592 \times 10^{-2} \text{ (L mol}^{-1})^2$$

$$ab/p = 1.533 \times 10^{-3} \text{ (L mol}^{-1})^3$$

Then, on writing $x = V_m/(\text{L mol}^{-1})$, we must solve

$$x^3 - 0.4530\, x^2 + (3.592 \times 10^{-2})x - (1.533 \times 10^{-3}) = 0$$

The most elementary way of solving this cubic equation is by starting from the perfect gas value $x = 0.410$ and then looking for the value of x that actually solves the equation. (A better way is to use a computer or programmable calculator.) We find $x = 0.366$, which implies that $V_m = 0.366 \text{ L mol}^{-1}$.

Comment. Cubic equations can be solved analytically: the formula is given in Section 3.8.2 of *Handbook of mathematical functions*, M. Abramowitz and I. Stegun, Dover (1965), a rich source of this kind of information.

Exercise. Calculate the molar volume of Ar at 100°C and 100 atm on the assumption that it is a van der Waals gas. $\qquad$ [0.298 L mol^{-1}]

Table 1.5. Equations of state

	Equation	Critical constants		
		p_c	V_c	T_c
Perfect gas	$p = \dfrac{RT}{V_m}$			
Van der Waals	$p = \dfrac{RT}{V_m - b} - \dfrac{a}{V_m^2}$	$\dfrac{a}{27b^2}$	$3b$	$\dfrac{8a}{27bR}$
	$p_r = \dfrac{8T_r}{3V_r - 1} - \dfrac{3}{V_r^2}$			
Berthelot	$p = \dfrac{RT}{V_m - b} - \dfrac{a}{TV_m^2}$	$\dfrac{1}{12}\left(\dfrac{2aR}{3b^3}\right)^{1/2}$	$3b$	$\dfrac{2}{3}\left(\dfrac{2a}{3bR}\right)^{1/2}$
	$p_r = \dfrac{8T_r}{3V_r - 1} - \dfrac{3}{T_r V_r^2}$			
Dieterici	$p = \dfrac{RT e^{-a/RTV_m}}{V_m - b}$	$\dfrac{a}{4e^2 b^2}$	$2b$	$\dfrac{a}{4Rb}$
	$p_r = \dfrac{e^2 T_r e^{-2/T_r V_r}}{2V_r - 1}$			
Beattie–Bridgman	$p = \dfrac{(1 - \gamma)RT(V_m + \beta) - \alpha}{V_m^2}$			
	with $\alpha = a_0\left(1 + \dfrac{a}{V_m}\right)$			
	$\beta = b_0\left(1 - \dfrac{b}{V_m}\right)$			
	$\gamma = \dfrac{c_0}{V_m T^3}$			
Virial (Kammerlingh Onnes)	$p = \dfrac{RT}{V_m}\left\{1 + \dfrac{B(T)}{V_m} + \dfrac{C(T)}{V_m^2} + \dots\right\}$			

We have built the van der Waals equation using vague arguments about the volumes of molecules and the effects of forces. It can be derived in other ways, but the present method has the advantage that it shows how to derive the form of an equation out of general ideas. The derivation also has the advantage of keeping imprecise the significance of the coefficients a and b: they are much better regarded as empirical parameters than as precisely defined molecular properties.

The reliability of the equation

We now examine to what extent the van der Waals equation predicts the behaviour of real gases. It is too optimistic to expect a single, simple expression to account for the p, V, T behaviour of all substances, and accurate work on gases must resort to the virial equation, use tabulated values of the coefficients at various temperatures, and analyse the systems numerically. The advantage of the van der Waals equation is that it is analytical and allows us to draw some general conclusions about real gases. When the equation fails we must use one of the other equations of state that have been proposed (some are listed in Table 1.5), invent a new one, or go back to the virial equation.

That having been said, we can begin to judge the reliability of the equation by comparing the isotherms it predicts with the experimental isotherms in Fig. 1.9. Some calculated isotherms are shown in Fig. 1.10, and apart from the oscillations below the critical temperature they do resemble experimental isotherms quite well. The oscillations, the **van der Waals loops**, are unrealistic because they suggest that under some conditions an increase in pressure results in an increase in volume. Therefore they are replaced by horizontal lines drawn so the loops define equal areas above and below the lines: this is called the **Maxwell construction**. The van der Waals coefficients are found by fitting the calculated curves to the experimental, and the values for some gases are listed in Table 1.4.

The features of the equation

The principal features of the van der Waals equation are as follows:

(1) *Perfect gas isotherms are obtained at high temperatures and large molar volumes.* When the temperature is high, RT may be so large that the first term in eqn 11b greatly exceeds the second. Furthermore, if the molar volume is large (in the sense $V_m \gg b$), we can replace the denominator $V_m - b$ by V_m. Hence, the equation reduces to $p = RT/V_m$, the perfect gas equation.

(2) *Liquids and gases coexist when cohesive and dispersing effects are in balance.* The van der Waals loops occur when both terms in eqn 11b have similar magnitudes. The first term arises from the kinetic energy of the molecules and their repulsive interactions; the second represents the effect of the attractive interactions.

(3) *The critical constants are related to the van der Waals coefficients.* For $T < T_c$, the calculated isotherms oscillate and each one passes through a minimum followed by a maximum. These extrema converge as $T \to T_c$ and coincide at $T = T_c$, and at the critical point the curve has a flat inflexion. From the properties of curves, we know that an inflexion of this type occurs when both the first and second derivatives are zero. Hence, we can find the critical constants by calculating these derivatives and setting them equal to

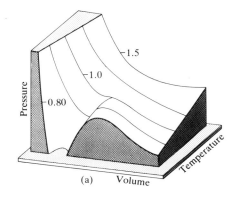

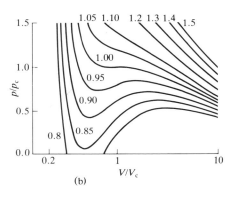

Fig. 1.10 The van der Waals isotherms at several values of T/T_c. (a) The shape of the surface (compare it with the perfect gas surface in Fig. 1.4). (b) A selection of individual isotherms. The van der Waals 'loops' are normally replaced by horizontal straight lines. The critical isotherm is the one at $T/T_c = 1$.

zero:

$$\left.\begin{aligned} \frac{dp}{dV_m} &= \frac{-RT}{(V_m - b)^2} + \frac{2a}{V_m^3} = 0 \\ \frac{d^2p}{dV_m^2} &= \frac{2RT}{(V_m - b)^3} - \frac{6a}{V_m^4} = 0 \end{aligned}\right\} \quad \text{at} \quad p_c, V_c, T_c$$

Solving these two equations gives

$$V_c = 3b \qquad p_c = \frac{a}{27b^2} \qquad T_c = \frac{8a}{27Rb} \tag{12a}$$

These relations can be tested by noting that the **critical compression factor** Z_c is predicted to be equal to

$$Z_c = \frac{p_c V_c}{R T_c} = \frac{3}{8} \tag{12b}$$

for all gases. We see from Table 1.3 that although $Z_c < \frac{3}{8}$ (or 0.375), it is approximately constant (at 0.3) and the discrepancy is reasonably small.

(4) *The Boyle temperature is related to the critical temperature.* The van der Waals equation can be expanded into a virial equation. The first step is to express eqn 11b as

$$p = \frac{RT}{V_m} \left\{ \frac{1}{1 - \dfrac{b}{V_m}} - \frac{a}{RTV_m} \right\}$$

So long as $b/V_m < 1$, the first term inside the brackets can be expanded using $(1-x)^{-1} = 1 + x + x^2 + \dots$, which gives

$$p = \frac{RT}{V_m} \left\{ 1 + \left(b - \frac{a}{RT} \right) \frac{1}{V_m} + \dots \right\}$$

We can now identify the second virial coefficient as

$$B = b - \frac{a}{RT} \tag{13}$$

Since at the Boyle temperature $B = 0$,

$$T_B = \frac{a}{bR} = \frac{27 T_c}{8} \tag{14}$$

This relation can be tested using the information in Table 1.3.

1.5 The principle of corresponding states

An important technique for comparing the properties of objects is to choose a related fundamental property of the same kind and to set up a relative scale on that basis. We have seen that the critical constants are characteristic properties of gases, and so it may be that a scale can be set up using them as yardsticks. We therefore introduce the **reduced variables** by dividing the actual variable by the corresponding critical constant:

Reduced pressure: $\qquad p_r = \dfrac{p}{p_c}$

Reduced volume: $\qquad V_r = \dfrac{V_m}{V_c}$

Reduced temperature: $\qquad T_r = \dfrac{T}{T_c}$

Van der Waals, who first tried this, hoped that gases confined to the same reduced volume at the same reduced temperature would exert the same reduced pressure. The hope was largely fulfilled. Figure 1.11 shows the dependence of the compression factor Z on the reduced pressure for a variety of gases at various reduced temperatures. The success of the procedure is strikingly clear: compare this graph with Fig. 1.8, where similar data are plotted without using reduced variables.

The observation that real gases at the same volume and temperature exert the same reduced pressure is called the **principle of corresponding states**. It is only an approximation, and works best for gases composed of spherical molecules; it fails, sometimes badly, when the molecules are nonspherical or polar.

Example 1.8: *Using the principle of corresponding states*

A sample of argon of molar volume 17.2 L mol^{-1} is maintained at 10.0 atm and 280 K. At what molar volume, pressure, and temperature would a sample of nitrogen be in a corresponding state?

Answer. First, we use the data in Table 1.3 to express the properties of the Ar sample as reduced variables:

$$V_r = \frac{17.2 \text{ dm}^3 \text{ mol}^{-1}}{75.3 \text{ cm}^3 \text{ mol}^{-1}} = 228.4$$

$$p_r = \frac{10.0 \text{ atm}}{48.0 \text{ atm}} = 0.2083$$

$$T_r = \frac{280 \text{ K}}{150.7 \text{ K}} = 1.858$$

Then we use the critical constants for N_2 to find the variables for which it has these reduced values:

$$V_m = 228.4 \times 90.1 \text{ cm}^3 \text{ mol}^{-1} = 20.6 \text{ L mol}^{-1}$$

$$p = 0.2083 \times 33.5 \text{ atm} = 6.98 \text{ atm}$$

$$T = 1.858 \times 126.3 \text{ K} = 235 \text{ K}$$

Exercise. A sample of helium of molar volume 10.5 L mol^{-1} is maintained at 0.10 atm and 10.0 K. At what molar volume, pressure, and temperature would a sample of CO_2 be in a corresponding state?

[17.1 L mol^{-1}, 3.2 atm, 585 K]

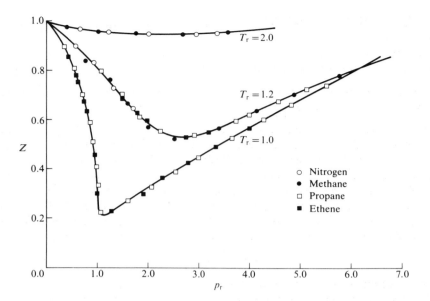

Fig. 1.11 The compression factors of four gases plotted using reduced variables. The use of reduced variables organizes the data on to single curves.

The van der Waals equation sheds some light on the principle. First, we express the equation in terms of the reduced variables, which gives

$$p_r p_c = \frac{RT_r T_c}{V_r V_c - b} - \frac{a}{V_r^2 V_c^2}$$

Then we express the critical constants in terms of a and b

$$\frac{ap_r}{27b^2} = \frac{8aT_r}{27b(3bV_r - b)} - \frac{a}{9b^2 V_r^2}$$

which reorganizes into

$$p_r = \frac{8T_r}{3V_r - 1} - \frac{3}{V_r^2} \tag{15}$$

This equation has the same form as the original, but the coefficients a and b, which differ from gas to gas, have disappeared. It follows that if the isotherms are plotted in terms of the reduced variables (as we did in fact in Fig. 1.10 without drawing attention to the fact), then the same curves are obtained whatever the gas. This is precisely the content of the principle of corresponding states, and so the van der Waals equation is compatible with it.

Looking for too much significance in this apparent triumph is mistaken, because other equations of state also accommodate the principle (Table 1.5). In fact, all we need are two parameters playing the roles of a and b, for then the equation can always be manipulated into reduced form. The observation that real gases obey the principle approximately amounts to saying that the effects of the attractive and repulsive interactions can each be approximated in terms of a single parameter. The importance of the principle is then not so much its theoretical interpretation but the way that it enables the properties of a range of gases to be coordinated on to a single diagram (e.g. Fig. 1.11 instead of Fig. 1.8).

Further information: the mole

When we wish to report the number of atoms, molecules, ions, or formula units in a sample we use the unit **mole** (mol):

> 1 mol of a substance contains as many atoms (or molecules, etc.) as there are atoms in exactly 12 g of ^{12}C.

This number is approximately 6.022×10^{23}. Therefore 1 mol H_2 molecules consists of 6.022×10^{23} H_2 molecules and twice that number of H atoms. Similarly 1 mol NaCl consists of 6.022×10^{23} NaCl formula units and therefore 6.022×10^{23} Na^+ ions and 6.022×10^{23} Cl^- ions. In general it is essential to specify the 'entities' when reporting the number of moles. It is ambiguous to say '1 mol oxygen': we must specify that the sample consists of 1 mol O atoms or 1 mol O_2 molecules. We shall sometimes find it convenient to use the millimole (mmol) when dealing with small amounts of substance:

$$1 \text{ mmol} = 10^{-3} \text{ mol}$$

Like all units, the mole is the unit for reporting a physical quantity. In its case, the formal name for the physical quantity is **amount of substance** n. Therefore, formally we should say that the amount of substance of H_2O molecules in a sample is, for example, 1.5 mol H_2O. However, the simpler term **amount** is also officially sanctioned, and we can speak of a sample as containing an amount of H_2O molecules. In practice, though, chemists have shown a reluctance to depart from the colloquial term 'number of moles', and most still use expressions such as 'the number of moles in the sample is n'.

If a sample contains N specified entities (atoms, molecules, and so on), the amount of substance (number of moles) of the entities that it contains is

$$n = \frac{N}{N_A}$$

where N_A is **Avogadro's constant**:

$$N_A = 6.022 \times 10^{23} \text{ mol}^{-1}$$

(Note that N_A has units, and so is a constant, not a number.) Conversely, if we know the amount of substance in a sample, the number of entities present is

$$N = n \times N_A$$

A **molar quantity** is the quantity per mole. For example, the molar volume V_m is the volume per mole and the molar mass M is the mass per mole:

$$V_m = \frac{V}{n} \qquad M = \frac{m}{n}$$

where V is the volume of a sample, m its mass, and n the amount of substance it contains. The molar masses of atoms and molecules are still widely called atomic or molecular weights, but we shall avoid these obsolescent terms. Molar mass is normally reported in g mol^{-1}, as in 18.02 g mol^{-1} for H_2O, but kg mol^{-1} is useful for macromolecules. Thus a polymer might have a molar mass of 50 kg mol^{-1} ($50\,000 \text{ g mol}^{-1}$).

The quantity widely called **molar concentration** c (and still often 'molarity') is the amount of substance per unit volume:

$$c = \frac{n}{V}$$

For a pure substance, the molar concentration is the inverse of the molar volume. The official name of c is either simply 'concentration' (if there is no risk of ambiguity) or 'amount concentration'. Although the former is widely encountered, the latter has not yet been widely accepted. Molar concentration is generally reported in $mol\,dm^{-3}$ (or $mol\,L^{-1}$); these units are so widely used that they are commonly abbreviated to M (read 'molar'):

$$1\,\text{M} = 1\,mol\,dm^{-3}$$

Further reading

B. H. Flowers and E. Mendoza, *Properties of matter.* Wiley, London (1970).

D. Tabor, *Gases, liquids, and solids.* Cambridge University Press (1979).

A. J. Walton, *Three phases of matter.* Oxford University Press (1983).

G. W. Thomson, and D. R. Douslin, Determination of pressure and volume. In *Techniques of chemistry,* Vol. V (eds. A. Weissberger and B. W. Rossiter). Wiley, New York (1971).

J. A. Hall, *The measurement of temperature.* Barnes and Noble, New York (1966).

J. B. Ott, J. R. Coates, and H. T. Hall, Comparisons of equations of state. *J. chem. Educ.* **48,** 515 (1971).

J. H. Dymond and E. B. Smith, *The virial coefficients of pure gases and mixtures.* Oxford University Press (1980).

International critical tables, Vol. 3 (*p, V, T* data). McGraw-Hill (1928).

Exercises

1.1 A sample of air occupies 1.0 L at 25°C and 1.00 atm. What pressure is needed to compress it to $100\,cm^3$ at this temperature?

1.2 (a) Could 131 g of Xe in a vessel of volume 1.0 L exert a pressure of 20 atm at 25°C if it behaved as a perfect gas? If not, what pressure would it exert? (b) What pressure would it exert if it behaved as a van der Waals gas?

1.3 A perfect gas undergoes isothermal compression, which reduces its volume by 2.20 L. The final pressure and volume of the gas are 3.78×10^3 Torr and 4.65 L respectively. Calculate the original pressure of the gas in (a) Torr, (b) atm.

1.4 To what temperature must a 1.0 L sample of a perfect gas be cooled from room temperature in order to reduce its volume to $100\,cm^3$?

1.5 A car tyre (i.e. an automobile tire) was inflated to a pressure of $24\,lb\,in^{-2}$ ($1\,atm = 14.7\,lb\,in^{-2}$) on a winter's day when the temperature was −5°C. What pressure will be found, assuming no leaks have occurred and that the volume is constant, on a subsequent summer's day when the temperature is 35°C? What complications should be taken into account in practice?

1.6 A perfect gas at 340 K is heated at constant pressure until its volume has increased by 14 per cent. What is the final temperature of the gas?

1.7 At sea level, where the pressure was 755 Torr, the gas in a balloon occupied $2.0\,m^3$. To what volume will the balloon expand when it has risen to an altitude where the pressure is (a) 100 Torr, (b) 10 Torr? Assume constant temperature, and that the material of the balloon is infinitely extensible.

1.8 A sample of 255 mg of neon occupies 3.00 L at 122 K. Use the perfect gas law to calculate the pressure of the gas.

1.9 A gas mixture consists of 320 mg of methane, 175 mg of argon, and 225 mg of neon. The partial pressure of Ne at 300 K is 66.5 Torr. Calculate (a) the volume and (b) the total pressure of the mixture.

1.10 The density of a gaseous compound was found to be $1.23\,g\,L^{-1}$ at 330 K and 150 Torr. What is the molar mass of the compound?

1.11 In an experiment to measure the molar mass of a gas, $250\,cm^3$ of the gas was confined in a glass vessel. The pressure was 152 Torr at 298 K and, after correcting for buoyancy effects, the mass of the gas was 33.5 mg. What is the molar mass of the gas?

1.12 Calculate the pressure exerted by 1.0 mol C_2H_6 behaving as (a) a perfect gas, (b) a van der Waals gas when it is confined under the following conditions: (i) at 273.15 K in 22.414 L, (ii) at 1000 K in $100\,cm^3$. Use the data in Table 1.4 despite the fact that they refer to 25°C.

1.13 Estimate the critical constants of a gas with van der Waals parameters $a = 0.751 \text{ atm L}^2 \text{ mol}^{-2}$ and $b = 0.0226 \text{ L mol}^{-1}$.

1.14 A gas at 250 K and 15 atm has a molar volume 12 per cent smaller than that calculated from the perfect gas law. Calculate (a) the compression factor under these conditions and (b) the molar volume of the gas. Which are dominating in the sample, the attractive or the repulsive forces?

1.15 At 300 K and 20 atm, the compression factor of a gas is 0.86. Calculate (a) the volume occupied by 8.2 mmol of the gas under these conditions and (b) an approximate value of the second virial coefficient B at 300 K.

1.16 A vessel of volume 22.4 L contains 2.0 mol H_2 and 1.0 mol N_2 at 273.15 K. Calculate (a) the mole fractions of each component, (b) their partial pressures, and (c) their total pressure.

1.17 The critical constants of methane are $p_c = 45.6 \text{ atm}$, $V_c = 98.7 \text{ cm}^3 \text{ mol}^{-1}$, and $T_c = 190.6 \text{ K}$. Calculate the van der Waals parameters of the gas and estimate the radius of the molecules.

1.18 Use the van der Waals parameters of Cl_2 to calculate approximate values of (a) the Boyle temperature of Cl_2 and (b) the radius of a Cl_2 molecule regarded as a sphere.

1.19 Suggest the pressure and temperature that 1 mol of (a) NH_3, (b) Xe, (c) He will be in at states corresponding to a mol H_2 at 1.0 atm and 25°C.

Problems

Numerical problems

1.1 A diving bell has an air space of 3 m³ when on the deck of a boat. What is the volume of the air space when the bell has been lowered to a depth of 50 m? Take the mean density of sea water to be 1.025 g cm^{-3} and assume that the temperature is the same as on the surface.

1.2 What pressure difference must be generated across the length of a 15 cm vertical drinking straw in order to drink a water-like liquid of density 1.0 g cm^{-3}?

1.3 A meteorological balloon had a radius of 1.0 m when released at sea level at 20°C and expanded to a radius of 3.0 m when it had risen to its maximum altitude where the temperature was −20°C. What is the pressure inside the balloon at that altitude?

1.4 Deduce the relation between the pressure and density ρ of a perfect gas of molar mass M. Confirm graphically, using the following data on dimethyl ether at 25°C, that perfect behaviour is reached at low pressures and find the molar mass of the gas.

p/Torr	91.74	188.98	277.3	452.8	639.3	760.0
$\rho/(\text{g L}^{-1})$	0.225	0.456	0.664	1.062	1.468	1.734

1.5 Investigate some of the technicalities of ballooning using the perfect gas law. Suppose your balloon has radius of 3.0 m and that it is spherical. (a) How much H_2 (in moles) is needed to inflate it to a pressure of 1.0 atm in an ambient temperature of 25°C at sea level? (b) What mass can the balloon lift at sea level, where the density of air is 1.22 kg m^{-3}? (c) What would be the payload if He were used instead of H_2?

1.6 The molar mass of a newly synthesized fluorocarbon was measured in a 'gas microbalance'. This consists of a glass bulb forming one end of a beam, the whole surrounded by a closed container. The beam is pivoted, and the balance point is attained by raising the pressure of gas in the container, so increasing the buoyancy of the enclosed bulb. In one experiment, the balance point was reached when the fluorocarbon pressure was 327.10 Torr; for the same setting of the pivot, a balance was reached when CHF_3 (molar mass $70.014 \text{ g mol}^{-1}$) was introduced at 423.22 Torr. A repeat of the experiment with a different setting of the pivot required a pressure of 293.22 Torr of the fluorocarbon and 427.22 Torr of the CHF_3. What is the molar mass of the fluorocarbon? Suggest a molecular formula.

1.7 A constant-volume perfect gas thermometer indicates a pressure of 50.2 Torr at the triple point temperature of water (273.16 K). (a) What change of pressure indicates a change of 1 K at this temperature? (b) What pressure indicates a temperature of 100.00°C? (c) What change of pressure indicates a change of 1 K at the latter temperature?

1.8 A vessel of volume 22.4 L contains 2.0 mol H_2 and 1.0 mol N_2 at 273.15 K initially. All the H_2 reacted with sufficient N_2 to form NH_3. Calculate the partial pressures and the total pressure of the final mixture.

1.9 Calculate the molar volume of Cl_2 at 350 K and 2.30 atm using (a) the perfect gas law and (b) the van der Waals equation. Use the answer to (a) to calculate a first approximation to the correction term for attraction and then use successive approximations to obtain a numerical answer for part (b).

1.10 The critical volume and critical pressure of a certain gas are $160 \text{ cm}^3 \text{ mol}^{-1}$ and 40 atm respectively. Estimate the critical temperature by assuming that the gas obeys the Berthelot equation of state. Estimate the radii of the gas molecules on the assumption that they are spheres.

1.11 Estimate the coefficients a and b in the Dieterici equation of state from the critical constants of xenon. Calculate the pressure exerted by 1.0 mol Xe when it is confined to 1.0 L at 25°C.

Theoretical problems

1.12 Express the van der Waals equation of state as a virial expansion in powers of $1/V_m$ and obtain expressions for B and C in terms of the parameters a and b. The expansion you will need is

$$\frac{1}{1-x} = 1 + x + x^2 + \ldots$$

Measurements on argon gave $B = -21.7 \, \text{cm}^3 \, \text{mol}^{-1}$ and $C = 1200 \, \text{cm}^6 \, \text{mol}^{-2}$ for the virial coefficients at 273 K. What are the values of a and b in the corresponding van der Waals equation of state?

1.13 Express the Dieterici equation of state as a virial expansion in powers of $1/V_m$ and obtain expressions for B and C in terms of the parameters a and b. Use the data in Problem 1.12 to deduce values of a and b for Ar at 273 K.

1.14 A scientist with a simple view of life proposed the following equation of state:

$$p = \frac{RT}{V_m} - \frac{B}{V_m^2} + \frac{C}{V_m^3}$$

Show that the equation leads to critical behaviour. Find the critical constants of the gas in terms of B and C and an expression for the critical compression factor.

1.15 Equations 10a and 10b are expansions in p and $1/V_m$. Find the relation between B, C and B', C'.

1.16 The second virial coefficient B' can be obtained from measurements of the density ρ of a gas at a series of pressures. Show that the graph of p/ρ against p should be a straight line with slope proportional to B'. Use the data on dimethyl ether in Problem 1.4 to find the values of B' and B at 25°C.

1.17 The 'barometric formula'

$$p = p_0 e^{-Mgh/RT}$$

relates the pressure of a gas of molar mass M at some altitude h to its pressure p_0 at sea level. Derive this relation by showing that the change in pressure dp for an infinitesimal change in altitude dh where the density is ρ is $dp = -\rho g \, dh$. Remember that ρ depends on the pressure. Evaluate the pressure difference between the top and bottom of (a) a laboratory vessel of height 15 cm, and (b) the World Trade Center, 1350 ft. Ignore temperature variations.

The First Law: the concepts

2

Check-list of key ideas

1. The introduction of the concepts of a *system* and its *surroundings* and of *work*, *heat*, and *internal energy* (Sections 2.1 and 2.2).

2. The classification of processes as *exothermic* and *endothermic* (Section 2.1).

3. The statement of the *First Law of thermodynamics* in a simple but practical form, and a second formulation of the law that shows how the concepts of heat and energy are based on and are measured in terms of work (Section 2.2).

4. The introduction of the internal energy as a *state property* (Section 2.2).

5. The derivation of the expression for *expansion work* (eqn 4) and its application to expansion against constant external pressure (eqn 5) and *reversible expansion* (eqn 6).

6. An introduction to the principles of *calorimetry* (Section 2.5).

7. The introduction of the state property *enthalpy* (eqn 9) and its relation to the energy transferred as heat at constant pressure (eqn 10).

8. The definition of the *standard state* of a substance and of the *standard enthalpy* change for a physical transformation and a chemical reaction (Section 2.6).

9. An introduction to a number of different types of enthalpy change, including the *enthalpy of phase transition*, of *solution*, of *ionization*, and of *dissolution* (Section 2.6).

10. The statement of *Hess's law* as a special case of the enthalpy being a state property and the manipulation of *thermochemical equations* (Section 2.6).

11. The concept of a *thermodynamic cycle* and the particular case of a *Born–Haber cycle* for the discussion of *lattice enthalpy* (Section 2.7).

12. An explanation of how *standard reaction enthalpies* are expressed in terms of *standard enthalpies of formation*, including a method of expressing chemical equations that will be used throughout the text (eqns 12 and 13).

13. The definitions of *heat capacity at constant volume* and *heat capacity at constant pressure* (Section 2.8).

14. The variation of the reaction enthalpy with temperature and *Kirchhoff's law* (eqn 17).

The release of energy can be used to heat the surroundings when a fuel burns in a furnace, to produce mechanical work when a fuel burns in an engine, and electrical work when a chemical reaction pumps electrons through a circuit. In chemistry, we encounter reactions that can be harnessed to provide heat and work, reactions whose liberated energy might be squandered but which give products we require, and reactions that constitute the processes of life. **Thermodynamics**, the study of the transformations of energy, enables us to discuss all these matters quantitatively.

The basic concepts

A **system** is the part of the world in which we have a special interest. It may be a reaction vessel, an engine, an electrochemical cell, and so on. Around the system is its **surroundings**, where we make our observations. The two parts may be in contact and are separated by a boundary, and specifying the system and its surroundings amounts to a careful specification of the boundary between them. When matter can be transferred through the boundary between the system and its surroundings the system is **open**; otherwise it is **closed**. An **isolated system** is a closed system that has neither mechanical nor thermal contact with its surroundings.

2.1 Work, heat, and energy

Work, heat, and energy are the basic concepts of thermodynamics, and of these the most fundamental is work. As we shall see, all measurements of heat and changes in energy amount to measurements of work.

Work is done during a process if that process could be used to bring about a change in the height of a weight somewhere in the surroundings. Examples include a gas that pushes out a piston and raises a weight, stretching a strip of rubber, and winding a spring. Driving an electric current through a resistance is also an example of work because the same current could be driven through a motor and used to raise a weight. When we charge a battery with a current from an external source we do work on it.

We shall say that work is done *by* the system if a weight has been raised in the surroundings, and that it has been done *on* the sysem if a weight has been lowered. When we need to measure the amount of work we use its definition as *force × distance*, which we develop below.

Energy is the capacity of a system to do work. When we do work on an otherwise isolated system (for instance, by compressing a gas or winding a spring), we increase its capacity to do work, and hence increase its energy. When the system does work (when the piston moves out or the spring unwinds), its energy is reduced because it can do less work than before.

Experiments have shown that the energy of a system (its ability to do work) may be changed by means other than work itself. When the energy of a system changes as a result of a temperature difference between it and its surroundings we say that energy has been transferred as **heat**. When a beaker of water (the system) stands on a hot plate, the capacity of the system to do work increases, so its energy has increased; since the increase has occurred as a result of a temperature difference, that energy has been transferred to the system as heat.[1]

Not all substances permit the transfer of energy even though there is a temperature difference between the system and its surroundings. Walls that do permit energy transfer as heat (such as steel and glass) are called **diathermic** ('dia' is the Greek word for 'through'). Walls that do not permit energy transfer as heat are called **adiabatic**. A Dewar flask is a good approximation to an adiabatic container.

A process that releases energy as heat is called **exothermic**. All combustion reactions are exothermic. Processes that absorb energy as heat are called **endothermic**. An example is the vaporization of water. An endothermic process in an adiabatic container results in a lowering of temperature of the system; an exothermic process results in a rise of temperature. An endothermic process taking place in a diathermic container under isothermal conditions results in energy flowing into the system as heat. An exothermic process in a diathermic container results in a release of energy as heat into the surroundings.

In molecular terms, the process of heating is the transfer of energy that makes use of the differences in **thermal motion**—the chaotic, random motion of the molecules—between the system and the surroundings. The greater thermal motion of the molecules in the hotter surroundings may stimulate the slower molecules in the cooler system to move more energetically, and the energy of the system is increased. When a system heats its surroundings, its molecules stimulate the thermal motion of the molecules in the surroundings at the expense of its own energy.

In molecular terms, work is the transfer of energy that makes use of *organized motion*. When a weight is raised or lowered, its atoms move in an organized way. The atoms in a spring move in an orderly way; the electrons in an electric current move in an orderly direction. When a system does work it drives some of the atoms or electrons in its surroundings in an organized way. Likewise, when we do work on a system, we transfer energy to it in an organized way, as the atoms in a weight are lowered or a current of electrons is passed.

The distinction between work and heat must be made in the surroundings. When we do work on a system, the energy leaves the surroundings in an orderly way, but it does not always arrive in the system that way. Thus, the work done on an electric heater may end up as thermal motion. That a falling weight may stimulate thermal motion in the system is irrelevant: work is identified as energy transfer making use of the organized motion of atoms *in the surroundings*, and heat is energy transfer making use of thermal motion

---

[1] If the heater is immersed in the water and is regarded as a part of the system, then the transfer of energy is a result of doing *work* on the system. In this case, there is no temperature difference between the system and its surroundings, and the external source has driven the current through the system, a form of work. Electric heaters are devices for converting work into an increase in temperature.

in the surroundings. In the compression of a gas, for instance, work is done as the particles of the compressing weight descend, but the effect of the incoming piston is to accelerate the gas molecules to higher average speeds. Since collisions between molecules quickly randomize their directions, the organized motion of the weight is in effect stimulating thermal motion in the gas. We observe the falling weight, the organized descent of its atoms, and report that work is being done even though it is stimulating thermal motion.

2.2 The First Law

From now on, we shall call the total energy of a system its **internal energy** U. It is impossible to know the *absolute* value of the internal energy of any system, but in thermodynamics we deal only with *changes* in the internal energy. We shall denote by ΔU the change in internal energy when a system changes from an initial state i with internal energy U_i to a final state f with internal energy U_f:

$$\Delta U = U_f - U_i$$

The internal energy is a **state property** in the sense that its value depends only on the current state of the system, and is independent of how that state has been prepared. A 1 L sample of hydrogen at 500 K and 10 atm, for instance, has the same internal energy however the hydrogen has been prepared. Other examples of state properties are temperature, pressure, and density. State properties are also called **state functions**.

The internal energy is also an **extensive property**, a property that depends on the size (extent) of the system. Thus, the internal energy of 2 kg of water is twice that of 1 kg of water at the same temperature and pressure. Other examples of extensive properties include mass and volume. In contrast, the temperature of a system is an **intensive property**, for it is independent of the size of the system, and 2 kg of water has the same temperature as 1 kg of water taken from the same source. Other examples of intensive properties are the density, the pressure, and all molar quantities, such as the molar volume and the molar internal energy, the internal energy per mole.

Internal energy, heat, and work are all measured in the same unit, the **joule** (J), which is defined as

$$1 \, J = 1 \, kg \, m^2 \, s^{-2}$$

For example, a mass of 2 kg travelling at $1 \, m \, s^{-1}$ has a kinetic energy of

$$\tfrac{1}{2}mv^2 = \tfrac{1}{2} \times 2 \, kg \times (1 \, m \, s^{-1})^2 = 1 \, kg \, m^2 \, s^{-2} = 1 \, J$$

Each beat of the human heart uses about 1 J of energy. We shall normally express changes in internal energy in kilojoules, where

$$1 \, kJ = 10^3 \, J$$

and changes in molar internal energy in $kJ \, mol^{-1}$. Typical values in chemistry are often about $100 \, kJ \, mol^{-1}$.

The conservation of energy

Experiments have established two further characteristics of the internal energy. The first is that the internal energy of an *isolated* system is constant. This observation is often summarized by the remark that 'energy is conserved'. The evidence for the conservation of energy is the impossibility of constructing perpetual motion machines that run without fuel: it is impossible to create (or destroy) energy.

The second characteristic of internal energy (which takes us beyond the simple conservation of energy law of physics) is that whereas *we* may know how energy has been transferred (because we can see whether a weight was raised or lowered in the surroundings, or whether ice has melted), the system is blind to the mode employed. Heat and work are *equivalent* ways of changing a system's energy: energy is energy, however it is gained or lost. A thermodynamic system is like a bank: it accepts deposits in either currency, but stores its reserves as internal energy.

These two characteristics of the internal energy are summarized in a statement called the **First Law of thermodynamics**:

The internal energy of a system is constant unless it is changed by doing work or by heating.

If we write w for the work done on a system, q for the energy transferred as heat to it, and ΔU for the resulting change in internal energy, the mathematical form of the First Law is

$$\Delta U = q + w \qquad (1)$$

This equation states that the change in internal energy of a closed system is equal to the energy that passes through its boundary as heat and work. If none passes through (when the system is isolated), then $\Delta U = 0$. The sign convention implied by eqn 1 is explained in Box 2.1.

Example 2.1: *Calculating a change in internal energy*

A certain electric motor produced 15 kJ of energy each second as mechanical work and lost 2 kJ as heat to the surroundings. What is the change in the internal energy of the motor and its power supply each second?

Answer. Since energy is lost from the system as work, w is negative, and $w = -15$ kJ. Energy is also lost as heat, so $q = -2$ kJ. The total change in internal energy is therefore

$$\Delta U = -2\,\text{kJ} - 15\,\text{kJ} = -17\,\text{kJ}$$

Comment. If we had decided to call the motor alone the system, then its internal energy would not have changed: the motor lost 17 kJ to the surroundings but the power supply did 17 kJ of work on the motor. The net loser is the power supply; the motor is a device that converts one form of energy into another.

Exercise. When a spring was wound, 100 J of work was done on it, but 15 J escaped to the surroundings as heat. What is the change in internal energy of the spring?   [+85 J]

In eqn 1 we have an expression of the First Law that is adequate for most purposes in thermodynamics. However, there are several unsatisfactory features about it, such as how we measure 'heat'. The following section gives a more sophisticated version of the law, and shows that all aspects of eqn 1 can be put on a firmer foundation. As the remainder of the text does not depend on this material, it is possible to omit it and go immediately to the next section ('Work and heat').

Box 2.1 The sign convention

We use the convention that dw denotes work done *on* a system and dq denotes heat supplied *to* a system. When dw or dq is positive, it signifies that energy has been supplied to the system as work or as heat, and has contributed to the internal energy ($dU = dw + dq$). When dw or dq is negative, it signifies that energy has left the system as work or as heat, and has contributed to a reduction of the internal energy. The convention applies to measurable changes as well as to infinitesimal changes. Thus $w = +10\,kJ$ signifies that 10 kJ of energy has been supplied to the system by doing work on it, and $q = -10\,kJ$ signifies that 10 kJ of energy has left the system as heat

The formal statement of the First Law

We shall start by pretending that we do not know what we mean by 'energy'. All we know, we pretend, is what is meant by work, because we can observe a weight being raised or lowered in the surroundings. We also know how to measure work by noting how high the weight is raised. Throughout this section, *work* will be the fundamental, measurable quantity, and we shall define energy, heat, and the First Law in terms of it alone.

In an adiabatic system, the same rise in temperature is brought about by the same amount of any kind of work we do on the system. Thus, if we do 1 kJ of mechanical work (stirring it with rotating paddles), or 1 kJ of electrical work (by passing an electric current through a heater), and so on, the same rise in temperature is produced. The statement of the **First Law of thermodynamics** that we shall use in this section is a summary of this experience:

> The work needed to change an adiabatic system from one specified state to another is the same however the work is done.

This looks completely different from the form we gave before, but we shall now see how it implies that law.

Suppose we do work w_{ad} on an adiabatic system to change it from an initial state i to a final state f. To be definite, we might be changing the temperature of 1 kg of water in an adiabatic container from 20°C to 30°C at constant pressure by doing work on it. The work may be of any kind (mechanical or electrical) and may take the system through different intermediate states (different temperatures and pressures, for instance). This would seem to require us to label w_{ad} with the path and to write $w_{ad(mechanical)}$ or $w_{ad(electrical)}$. However, the First Law tells us that w_{ad} is the same for all paths and depends only on the initial and final states. This independence of path is analogous to climbing a mountain: the height we must climb between any two points is independent of the path we take. In mountain climbing we can attach a number—called the altitude A—to each point on the mountain and express the height h of the climb as a difference in altitudes:

$$h = A_f - A_i = \Delta A$$

That is, the fact that h is independent of the path taken implies the existence of the state property A. The First Law has exactly the same implication. The fact that w_{ad} is independent of the path implies that to each state of the system we can attach a number—we call it the **internal energy**

U—and express the work as a difference in internal energies:

$$w_{ad} = U_f - U_i = \Delta U$$

This equation also shows that we can measure the change in the internal energy of a system by measuring the work needed to bring about the change in an adiabatic system.

The mechanical definition of heat

The First Law (in the second form) does not mention heat. However, we shall now show that not only does it imply the existence of this concept but that it also provides a fundamental definition of it in terms of measurable quantities.

Suppose we strip away the insulation around the system and make it diathermic. The system is now in thermal contact with its surroundings as we drive it from the same initial state to the same final state (such as changing the temperature of 1 kg of water from 20°C to 30°C at constant pressure). The change in internal energy is the same as before, because U is a state property, but we might find that the work we must do is not the same as before. Thus, whereas we might have needed to do 42 kJ of work when the system was in an adiabatic container, to achieve the same change we might now have to do 50 kJ of work. The difference between the work done in the two cases is *defined* as the heat absorbed in the process:

$$q = w_{ad} - w$$

In the present case, we would conclude that

$$q = 42 \text{ kJ} - 50 \text{ kJ} = -8 \text{ kJ}$$

and report that 8 kJ of energy had left the system as heat. We see that we now have a purely *mechanical* definition of heat in terms of work, which we know how to measure.

Finally, we can express the last relation in a more familiar way. Since we already know that ΔU is (by definition) equal to w_{ad}, the expression for the energy transferred to the system as heat is

$$q = \Delta U - w$$

However, this is equivalent to eqn 1, the mathematical form of the First Law that we saw earlier.

Work and heat

We can open the way to powerful methods of calculation by switching attention to *infinitesimal* changes of state (such as infinitesimal changes in temperature) and infinitesimal changes in the internal energy dU. Then, if the work done on a system is the infinitesimal amount dw and the energy supplied to it as heat is dq, in place of $\Delta U = q + w$ we have:

$$dU = dq + dw \qquad (2)$$

To use eqn 2 we must be able to relate dq and dw to events taking place in the surroundings. We begin by discussing **mechanical work**, which includes the work of compressing gases and the work done by a gas as it expands and drives back the atmosphere. Many chemical reactions result in the generation or consumption of gases, and the thermodynamic characteristics of the reaction, such as the heat it generates, depend on the work done while it takes place.

2.3 Mechanical work

The calculation of work starts from the definition used in physics. This states that the work done to move an object a distance dz against an opposing force F is

$$dw = -F\,dz$$

The negative sign tells us that when the system moves an object against an opposing force, the internal energy of the system will decrease. The total work done when the system moves the object from z_i to z_f is the sum of all the infinitesimal contributions along the path:

$$w = -\int_{z_i}^{z_f} F\,dz$$

If the force is independent of position and has the value F everywhere on the path, the integral evaluates to

$$w = -(z_f - z_i)F$$

In the case of a gravitational field acting on a mass m on the surface of the earth, the force opposing lifting is equal to the weight mg, where g is the acceleration of free fall ($9.81\ \mathrm{m\ s^{-2}}$). The work done by a system that lifts the mass through a height $h = z_f - z_i$ is therefore

$$w = -mgh \qquad (3)$$

Example 2.2: *Calculating the work for a constant force*

Calculate the work done when a system raises a column of water of radius 5.0 mm through 10 cm.

Answer. Since the centre of mass of the column lies halfway along its length l, the centre of mass is raised to a height $h = \tfrac{1}{2}l$. If the mass of the column is m,

$$w = -mg \times \tfrac{1}{2}l = -\tfrac{1}{2}mgl$$

The mass of the column is $\rho \times \pi r^2 l$, where ρ is the density of water. Therefore,

$$w = -\tfrac{1}{2}\pi\rho r^2 g l^2$$

$$= -3.9\ \mathrm{g\ m^2\ s^{-2}} = -3.9\ \mathrm{mJ}$$

Comment. The negative sign shows that the system undergoes a reduction in internal energy.

Exercise. Calculate the work done when you, the system, lift this book (of mass approximately 1.5 kg) through a height of 10 cm. [−1.5 J]

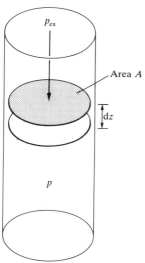

Fig. 2.1 When a piston of area A moves out through a distance dz, it sweeps out a volume $dV = A\,dz$. The external pressure p_{ex} is equivalent to a weight pressing on the piston, and the force opposing expansion is $F = p_{ex}A$.

2.4 Work of compression and expansion

In thermodynamics we are often concerned with the work done on or by a system as it expands. This can be calculated by considering the arrangement shown in Fig. 2.1, in which one wall of a system is a massless, frictionless, rigid, perfectly fitting piston of area A. If the external pressure is p_{ex}, the force on the outer face of the piston is $p_{ex}A$; this force is equivalent to a constant weight of magnitude $p_{ex}A$ pressing down on the system.

We shall suppose that the motion of the piston is **quasistatic**, or slow compared with any processes that spread energy and matter through the surroundings, so that no regions of turbulence are generated. Another way of expressing this is to say that the surroundings must remain in **internal equilibrium** in the sense that there will be no flow of energy or matter from one region to another in the surroundings when the piston halts. Quasistatic motion ensures that the surroundings are in internal equilibrium.

When the system expands quasistatically through a distance dz it raises the weight $p_{ex}A$ through a distance dz, so the work done is $dw = -p_{ex}A \times dz$. But $A\, dz$ is the volume swept out in the course of the expansion, which we write dV. Therefore, the work done when the system expands through dV against a constant pressure p_{ex} is

$$dw = -p_{ex}\, dV \qquad (4)$$

If instead the system is compressed, a weight of magnitude $p_{ex}A$ is lowered in the surroundings, so work of magnitude $p_{ex}A \times |dz|$ is now done on the system. It is important to note that it is still the *external* pressure that determines the magnitude of the work even though it is the system that is opposing the insertion of the piston. This is because the act of compression corresponds to the lowering of a weight of magnitude $p_{ex}A$ in the surroundings. In either case, therefore, the work done on the system is given by eqn 4, but in a compression dV is negative (a reduction of volume) and so dw is positive. Work is done on the system by compression and, so long as no other energy changes take place, its internal energy rises.

Other types of work (e.g. electrical work) have analogous expressions, with each one the product of an intensive factor (the pressure, for instance) and an extensive factor (the change in volume). Some are collected in Table 2.1. For the present we continue with the work associated with changing the volume, the **expansion work**, and see what we can extract from eqn 4.

Table 2.1. Varieties of work†

Type of work	dw	Comments	Units§
Expansion	$-p_{ex}\, dV$	p_{ex} is the external pressure dV is the change in volume	Pa m³
Surface expansion	$\gamma\, d\sigma$	γ is the surface tension $d\sigma$ is the change in area	N m⁻¹ m²
Extension	$f\, dl$	f is the tension dl is the change of length	N m
Electrical	$\phi\, dq$	ϕ is the electric potential dq is the change in charge	V C

† In general, the work done on a system can be expressed in the form $dw = -F\, dz$, where F is a 'generalized force' and dz is a 'generalized displacement'.
§ For work in J. Note that 1 N m = 1 J and 1 V C = 1 J.

Free expansion

Free expansion occurs when $p_{ex} = 0$ and there is no opposing force. No weight is raised, so no work is done, and $dw = 0$ for each stage of the

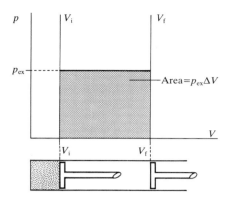

Fig. 2.2 The work done by a gas when it expands against a constant external pressure is equal to the shaded area in this 'indicator diagram'.

expansion. Hence, overall:

$$w = 0$$

Expansion against constant pressure

Since the external pressure p_{ex} is constant throughout the expansion (e.g., the piston is pressed on by the atmosphere of the outside world), the work done as the system passes quasistatically through each successive infinitesimal displacement dV is $dw = -p_{ex} dV$. The total work done in the expansion from V_i to V_f is the sum of all these equal contributions:

$$w = -\int_{V_i}^{V_f} p_{ex} \, dV = -p_{ex} \int_{V_i}^{V_f} dV = -p_{ex}(V_f - V_i)$$

Therefore, writing the change in volume as $\Delta V = V_f - V_i$,

$$w = -p_{ex} \Delta V \tag{5}$$

This result is illustrated graphically in Fig. 2.2: the magnitude of w is equal to the area beneath the horizontal line at $p = p_{ex}$ lying between the initial and final volumes. This type of graph is called an **indicator diagram** (James Watt first used one to indicate aspects of the operation of his steam engine).

Reversible expansion

A **reversible change** in thermodynamics is a change that can be reversed by an infinitesimal modification of a variable. The key word 'infinitesimal' sharpens the everyday meaning of the word 'reversible' as something that can change direction. We say that a system is in **equilibrium** with its surroundings if an infinitesimal change in the conditions in opposite directions results in opposite changes in its state. One example of reversibility that we have encountered already is the thermal equilibrium of two bodies with the same temperature. The transfer of energy between the two is reversible because if the temperature of either body is lowered by an infinitesimal amount, energy flows into it, but if it is raised by an infinitesimal amount, energy flows out of it.

Suppose a gas is confined by a piston and that the external pressure p_{ex} is set equal to the pressure of the confined gas p. Such a system is in **mechanical equilibrium** with its surroundings, since an infinitesimal change in the external pressure in either direction causes changes in volume in opposite directions. If the external pressure is made infinitesimally less than the internal pressure, the gas expands slightly. If the external pressure is increased infinitesimally, the gas is compressed slightly. In either case the change is reversible in the thermodynamic sense. If, on the other hand, the external pressure differs measurably from the internal pressure, then changing p_{ex} infinitesimally will not decrease it below the pressure of the gas, and so will not change the direction of the process. The system is not in mechanical equilibrium with its surroundings and the expansion is thermodynamically **irreversible**.

To achieve reversible expansion we must match p_{ex} to p at each stage:

$$dw = -p_{ex} dV = -p \, dV \tag{6}_r$$

(Equations valid only for reversible processes are labelled $_r$.) Although the pressure inside the system appears in this expression for the work, it does so only because p_{ex} has been set equal to p to ensure reversibility. The total work of reversible expansion is therefore

$$w = -\int_{V_i}^{V_f} p\, dV$$

We can evaluate the integral once we know how the pressure of the confined gas depends on its volume. This is the link with the material covered in Chapter 1, for if we know the equation of state of the gas, we can express p in terms of V.

Isothermal reversible expansion

We shall illustrate the use of an equation of state to evaluate the work by considering the isothermal reversible expansion of a perfect gas. The expansion is made isothermal by keeping the system in thermal contact with its surroundings (such as a constant-temperature bath). Since the equation of state is $pV = nRT$, we know that at each stage $p = nRT/V$, with V the current volume. The temperature T is constant in an isothermal expansion, and so may be taken outside the integral. It follows that the work of reversible isothermal expansion of a perfect gas from V_i to V_f at a temperature T is

$$w = -nRT \int_{V_i}^{V_f} \frac{dV}{V} = -nRT \ln \frac{V_f}{V_i} \qquad (7)_r^\circ$$

When the final volume is greater than the initial volume, as in an expansion, the logarithm in eqn 7 is positive and hence $w < 0$. In this case, the system has done work on the outside world and the internal energy of the system has decreased as a result of the work it has done. The equations also show that more work is done for a given change of volume when the temperature is increased. This is because the greater pressure of the confined gas needs a higher opposing pressure in order to ensure reversibility.

The result of the calculation can be expressed graphically, for the work is equal to the area under the isotherm $p = nRT/V$ (Fig. 2.3). Superimposed on the diagram is the rectangular area obtained in the case of irreversible expansion against a constant external pressure fixed at the same final value as is reached in the reversible expansion. We obtain more work when the expansion is reversible (the area is greater) because matching the external pressure to the internal pressure at each stage ensures that none of the system's pushing power is wasted. We cannot obtain more work than the reversible amount because increasing the external pressure even infinitesimally results in contraction. We may infer from this discussion that, because some pushing power is wasted when $p > p_{ex}$, the maximum work available from a system operating between specified initial and final states and passing along a specified path is obtained when it is operating reversibly.

We have introduced the connection between reversibility and maximum work for the special case of a perfect gas undergoing expansion. Later (Section 4.8) we shall see that it applies to all substances and to all kinds of work.

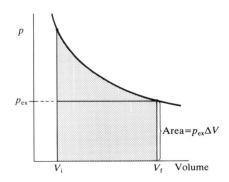

Fig. 2.3 The work done during the reversible isothermal expansion of a perfect gas is equal to the area under the isotherm. The work done during the irreversible expansion against the same final pressure is equal to the rectangular area.

Example 2.3: *Calculating the work of gas production*

Calculate the work done when 50 g of iron reacts with hydrochloric acid in (a) a closed vessel of fixed volume, (b) an open beaker at 25°C.

Answer. In (a) the volume cannot change, so no work is done and $w = 0$. In (b) the gas drives back the atmosphere and therefore $w = -p_{ex} \Delta V$. We can neglect the initial volume because the final volume (after the production of gas) is so much larger and $\Delta V = V_f - V_i \approx V_f = nRT/p_{ex}$ where n is the amount of H_2 produced. Therefore,

$$ w = -p_{ex} \times \frac{nRT}{p_{ex}} = -nRT $$

Since the reaction is

$$ Fe(s) + 2HCl(aq) \rightarrow FeCl_2(aq) + H_2(g) $$

1 mol $H_2(g)$ is generated when 1 mol $Fe(s)$ is consumed, and n can therefore be taken as the amount of Fe atoms that react. Since the molar mass of Fe is 55.85 g mol^{-1},

$$ w = \frac{-50\,g}{55.85\,g\,mol^{-1}} \times 8.314\,J\,K^{-1}\,mol^{-1} \times 298.15\,K = -2.2\,kJ $$

The system, the reaction mixture, does 2.2 kJ of work driving back the atmosphere.

Comment. Note that the external pressure does not affect the final result: the lower the pressure, the larger the volume occupied by the gas, so the effects cancel.

Exercise. Calculate the expansion work done when 50 g of water is electrolyzed under constant pressure at 25°C.   [-10 kJ]

2.5 Heat and enthalpy

In general, the change in internal energy of a system is

$$ dU = dq + dw_e + dw_{exp} $$

where dw_e is work in addition to the expansion work, dw_{exp}. For instance, dw_e might be the electrical work of driving a current through a circuit. A system kept at constant volume can do no expansion work, and so $dw_{exp} = 0$. If the system is also incapable of doing any other kind of work (if it is not, for instance, an electrochemical cell connected to an electric motor), $dw_e = 0$ too. Under these circumstances:

$$ dU = dq \text{ at constant volume, no additional work} \tag{8a} $$

We express this by writing $dU = dq_V$. Hence, in measuring the energy supplied to a constant-volume system as heat ($q > 0$) or obtained from it ($q < 0$) when it undergoes a change of state, we are in fact measuring the change in its internal energy:

$$ \Delta U = \int_i^f dq \text{ for a change at constant volume, no additional work} \tag{8b} $$

Calorimetry

The most important device for measuring ΔU is the adiabatic bomb calorimeter (Fig. 2.4). The change of state—which may be a chemical reaction—is initiated inside a constant-volume container, the bomb. The bomb is immersed in a stirred water bath, and the whole device is the calorimeter. The calorimeter is also immersed in an outer water bath. The temperature of the water in the calorimeter and of the outer bath are both monitored and adjusted to the same value: this ensures that there is no net loss of heat from the calorimeter to the surroundings (the bath) and hence that the calorimeter is adiabatic.

The change in temperature ΔT of the calorimeter that results from the reaction is proportional to the energy the reaction releases or absorbs as heat. Therefore, by measuring ΔT we can determine q_V and hence find ΔU. The conversion of ΔT to q_V involves knowing the **heat capacity** C of the calorimeter. This is the coefficient of proportionality between the energy supplied as heat and the rise in temperature it causes:

$$q = C \times \Delta T$$

To measure C, we pass an electric current through a heater in the calorimeter and determine the electrical *work* that we do on it. All the energy we supply as work passes through the boundary between the heater and the calorimeter as heat, and causes a rise in temperature. Since we are indirectly measuring the heat supplied, and can measure the rise in temperature of the calorimeter it causes, we can deduce the value of C and use that value to interpret the rise in temperature produced by the reaction. For instance, if we pass a current of 1.0 A from a 12 V supply for 3000 s, the work we do on the calorimeter is

$$w = I \times V \times t = 36\,000 \text{ A V s} = 36 \text{ kJ}$$

(because 1 A V s = 1 J). If the temperature rises by 5.5 K, the heat capacity of the calorimeter is

$$C = \frac{q}{\Delta T} = \frac{36 \text{ kJ}}{5.5 \text{ K}} = 6.5 \text{ kJ K}^{-1}$$

An alternative approach is to compare the temperature rise of the test sample and a standard measured under the same conditions. We can then calculate the heat produced by the sample using

$$\frac{q_{\text{sample}}}{q_{\text{standard}}} = \frac{\Delta T_{\text{sample}}}{\Delta T_{\text{standard}}}$$

However, at some stage the value of q_{standard} must have been determined by a measurement of the heat capacity of a calorimeter. In thermodynamics we never measure heat directly but infer its value from measurements of work or changes in temperature.

Enthalpy

When a system is free to change its volume against a constant external pressure, the change in internal energy is no longer equal to the energy supplied as heat. In effect, some of that energy supplied as heat is converted into the work required to drive back the surroundings, so $dU < dq$. We shall show, however, that at constant pressure the heat supplied is equal to the

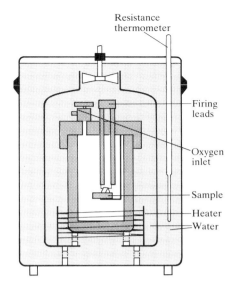

Fig. 2.4 A constant-volume bomb calorimeter. The 'bomb' is the central vessel, which is massive enough to withstand high pressures. The calorimeter (for which the heat capacity must be known) is the entire assembly shown here. In order to ensure adiabaticity, the calorimeter may be immersed in a water bath with a temperature continuously readjusted to that of the calorimeter at each stage of the combustion.

change in another thermodynamic property of the system, the **enthalpy** H. This is defined as

$$H = U + pV \qquad (9)$$

p is the pressure of the system, and pV is a part of the definition of H for any system, and does not imply a restriction to perfect gases. Like the internal energy, the enthalpy depends only on the current state of the system and is therefore a state function. As for any state function, the change in enthalpy between any pair of initial and final states is independent of the path between them.

We shall now show that dH is equal to the energy supplied as heat at constant pressure to a system that can do no additional work. For a general change in the state of the system, U changes by dU and the product pV changes by $d(pV)$, so H changes by

$$dH = dU + d(pV) = dU + p\,dV + V\,dp$$

If we substitute

$$dU = dq + dw_e - p_{ex}\,dV$$

into this expression, where $-p_{ex}\,dV$ is the expansion work and dw_e is any other kind of work, we get

$$dH = dq + dw_e - p_{ex}dV + p\,dV + V\,dp$$

If the system is in mechanical equilibrium with its surroundings at a pressure p (so $p_{ex} = p$), this simplifies to

$$dH = dq + dw_e + V\,dp$$

Now we impose the conditions that there is no additional work ($dw_e = 0$) and that the heating occurs at constant pressure ($dp = 0$). Then

$$dH = dq \text{ at constant pressure, no additional work} \qquad (10a)$$

This result, which is often written $dH = dq_p$, is the result we set out to confirm. Therefore, when a system is heated at constant pressure, and only expansion work can occur, the change in H is equal to the energy supplied as heat:

$$\Delta H = \int_i^f dq \text{ for a change at constant pressure, no additional work} \qquad (10b)$$

For example, if we supply 36 kJ of energy through an electric heater immersed in an open beaker of water, the enthalpy of the water increases by 36 kJ and we write $\Delta H = +36$ kJ.

The measurement of an enthalpy change

The enthalpy change accompanying a physical or chemical change can be measured with a calorimeter by monitoring the temperature change that accompanies a process occurring at constant pressure. One way of doing this for a combustion reaction is to use an **adiabatic flame calorimeter** (Fig. 2.5) and to measure ΔT when a given amount of substance burns in a supply of oxygen and using the heat capacity as a conversion factor.

Another route to ΔH is to measure the internal energy change using a bomb calorimeter, and then to convert the value of ΔU to ΔH. Since solids

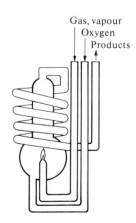

Gas, vapour
Oxygen
Products

Fig. 2.5 A constant-pressure flame calorimeter consists of this element immersed in a stirred water bath. Combustion occurs as a known amount of reactant is passed through to fuel the flame, and the rise of temperature is monitored.

and liquids have small molar volumes, and pV is so small for them, ΔH and ΔU are almost identical for reactions in which no gases are involved.

Example 2.4: *Relating ΔH and ΔU (1)*

The internal energy change in the conversion of 1.0 mol of the calcite form of $CaCO_3$ to the aragonite form is $+0.21$ kJ. Calculate the enthalpy change when the pressure is 1.0 bar given that the densities of the solids are 2.71 g cm^{-3} and 2.93 g cm^{-3} respectively.

Answer. The change in enthalpy in the transformation is

$$\Delta H = H(\text{aragonite}) - H(\text{calcite})$$

$$= \{U(a) + pV(a)\} - \{U(c) + pV(c)\}$$

$$= \Delta U + p\{V(a) - V(c)\} = \Delta U + p\,\Delta V$$

The volume of 1.0 mol $CaCO_3$ (100 g) as aragonite is 34 cm^3, and that of 1.0 mol $CaCO_3$ as calcite is 37 cm^3. Therefore,

$$p\,\Delta V = 1.0 \times 10^5\,\text{Pa} \times (34 - 37) \times 10^{-6}\,\text{m}^3$$

$$= -0.3\,\text{J}$$

Hence, ΔH and ΔU differ insignificantly (by 0.1 per cent).

Comment. We can almost always ignore the difference between the enthalpy and internal energy of condensed phases, except at very high pressures, when pV is no longer negligible.

Exercise. Calculate the difference between ΔH and ΔU when 1.0 mol of grey tin (density 5.75 g cm^{-3}) changes to white tin (density 7.31 g cm^{-3}) under 10.0 bar pressure. At 298 K, $\Delta H = +2.1$ kJ.   [$\Delta H - \Delta U = -4.4$ J]

If a reaction produces gases, we ignore the pV terms for everything else, and assume that each gas is perfect. Then the enthalpy of each gas is

$$H = U + pV = U + nRT$$

This implies that the change of enthalpy in the reaction is

$$\Delta H = \Delta U + \Delta n_g RT \qquad (11)$$

where Δn_g is the change in the amount of gas in the reaction. For example, in the reaction

$$2H_2(g) + O_2(g) \rightarrow 2H_2O(l) \quad \Delta n_g = -3\,\text{mol}$$

because 3 mol of gas molecules are replaced by 2 mol of liquid.

Example 2.5: *Relating ΔH and ΔU (2)*

The enthalpy change accompanying the formation of 1.00 mol $NH_3(g)$ from its elements at 298 K is -46.1 kJ. Estimate the change in internal energy.

Answer. The chemical equation is

$$\tfrac{3}{2}H_2(g) + \tfrac{1}{2}N_2(g) \rightarrow NH_3(g).$$

The change in the amount of gas-phase molecules is

$$\Delta n_g = 1.00\,\text{mol} - 0.50\,\text{mol} - 1.50\,\text{mol} = -1.00\,\text{mol}$$

Since $RT = 2.48$ kJ mol^{-1} at 298 K,

$$\Delta U = \Delta H - (-1.00\,\text{mol}) \times RT = -43.6\,\text{kJ}$$

Comment. The difference is now 5 per cent. One of the most serious sources of error in this procedure is the assumption that the gases are perfect.

Exercise. Calculate ΔU for the combustion of 1 mol of propene given that $\Delta H = -2058$ kJ. [-2052 kJ]

Example 2.6: *Calculating a change in enthalpy*

Water is brought to the boil under a pressure of 1.0 atm. When an electric current of 0.50 A from a 12 V supply is passed for 300 s through a resistance in thermal contact with it, it is found that 0.798 g of water is vaporized. Calculate the molar internal energy and enthalpy changes at the boiling point (373.15 K).

Answer. Since the vaporization occurs at constant pressure, the enthalpy change is equal to the work done on the heater (which enters the water as heat):

$$\Delta H = 0.50 \text{ A} \times 12 \text{ V} \times 300 \text{ s} = +1.8 \text{ kJ}$$

Since 0.798 g of water is 0.0443 mol H_2O, the molar enthalpy of vaporization is

$$\Delta H_m = \frac{1.8 \text{ kJ}}{0.0443 \text{ mol}} = +41 \text{ kJ mol}^{-1}$$

Since the volume of the vapour is so much greater than that of the liquid, and the pressure is constant, we use

$$\Delta(pV) = pV(g) - pV(l) \approx pV(g)$$

in

$$\Delta U = \Delta H - \Delta(pV)$$

If the vapour is taken to be a perfect gas, $pV(g)$ may be replaced by nRT, so $\Delta U = \Delta H - nRT$ and, dividing through by n to obtain molar quantities,

$$\Delta U_m = \Delta H_m - RT = +38 \text{ kJ mol}^{-1}$$

Comment. The $+$ sign is added to positive quantities to emphasize that they represent an increase in internal energy or enthalpy. Notice that the internal energy change is smaller than the enthalpy change because energy has been used to drive back the surrounding atmosphere.

Exercise. The molar enthalpy of vaporization of benzene at its boiling point (353.2 K) is 30.8 kJ mol^{-1}. What is the molar internal energy change? For how long would the same 12 V source need to supply a 0.50 A current in order to vaporize a 10 g sample? [$+27.9$ kJ mol^{-1}, 594 s]

Enthalpy and internal energy changes may also be measured by non-calorimetric methods: these are described in Chapters 8 and 10.

Thermochemistry

The study of the heat produced or required by chemical reactions is called **thermochemistry**. Thermochemistry is a branch of thermodynamics because a reaction vessel and its contents form a system. Thus we can measure (indirectly, by measuring work or a temperature rise) the energy a reaction produces as heat and identify q, depending on the conditions, with a change in internal energy or in enthalpy. Conversely, if we know the ΔU or ΔH for a reaction, we can predict the amount of energy it can produce as heat.

2.6 Standard enthalpy changes

Changes in enthalpy when a system undergoes a physical or chemical change are normally reported for the process taking place under a set of standard conditions. In most of our discussions we shall consider the **standard enthalpy change** $\Delta H^{\ominus}$, the change in enthalpy for a process in which the initial and final substances are in their **standard states**:

> The standard state of a substance at a specified temperature is its pure form at 1 bar pressure.[2]

Thus the standard state of liquid ethanol at 298 K is the pure liquid at 298 K and 1 bar, and the standard state of hydrogen gas at 500 K is the pure gas at 500 K and 1 bar.

As an example of a standard enthalpy change, the **standard enthalpy of vaporization** $\Delta H_{vap}^{\ominus}$ is the enthalpy change per mole when a pure liquid at 1 bar pressure vaporizes to a gas at 1 bar pressure, as in

$$H_2O(l) \rightarrow H_2O(g) \quad \Delta H_{vap}^{\ominus}(373\ \text{K}) = +40.66\ \text{kJ mol}^{-1}$$

Similarly, the **standard reaction enthalpy** $\Delta H^{\ominus}$ (a quantity that we shall use so frequently that we shall not give it a subscript) is the change in enthalpy when the reactants in their standard states change to products in their standard states, as in

$$CH_4(g) + 2O_2(g) \rightarrow CO_2(g) + 2H_2O(l) \quad \Delta H^{\ominus}(298\ \text{K}) = -890\ \text{kJ mol}^{-1}$$

This standard value refers to the reaction in which 1 mol of CH_4 molecules in the form of pure methane gas at 1 bar reacts completely with 2 mol of O_2 molecules in the form of pure oxygen gas to produce 1 mol of CO_2 molecules as pure carbon dioxide at 1 bar and 2 mol of H_2O molecules as pure liquid water at 1 bar, all the substances being at 298 K. In general, the standard reaction enthalpy refers to the overall process (unmixed reactants) $\rightarrow$ (unmixed products). However, except in the case of ionic reactions in solution, the enthalpy changes accompanying mixing and unmixing are insignificant in comparison with the contribution from the reaction itself. The recommended temperature for reporting thermodynamic data is 298.15 K (corresponding to 25°C): since this temperature is frequently mentioned, it is given the special symbol $\mathcal{T}$, standing for 'conventional temperature'.

Enthalpies of physical change

The standard enthalpy change that accompanies a change of physical state is called the **standard enthalpy of transition** and denoted $\Delta H_{trs}^{\ominus}$ (Table 2.2). The enthalpy of vaporization is one example. Two others are the **standard**

---

[2] The choice of 1 bar as the standard pressure (in place of 1 atm) for reporting thermodynamic data is recent. At the level of accuracy of most data, especially those relating to liquids and solids, the differences in values are usually negligible except in work of the highest precision. A meticulous analysis of the consequences of the change has been given by R. D. Freeman in *Bull. chem. Thermodynamics* **25,** 523 (1982). Modern tables of data (e.g. the tables published by the National Bureau of Standards) now widely, but not universally, adopt the new standard.

Table 2.2 Standard enthalpies of fusion and vaporization at the transition temperature, $\Delta H^{\ominus}/(\text{kJ mol}^{-1})$

	T_f/K	Fusion	T_b/K	Vaporization
Ar	83.81	1.188	87.29	6.51
C_6H_6	278.61	10.59	353.2	30.8
H_2O	273.15	6.008	373.15	40.656
				44.016 at $\maltese$
He	3.5	0.021	4.22	0.084

Fusion: $X(s) \rightarrow X(l)$; Vaporization: $X(l) \rightarrow X(g)$.

enthalpy of fusion $\Delta H^{\ominus}_{\text{fus}}$, as in

$$H_2O(s) \rightarrow H_2O(l) \quad \Delta H^{\ominus}_{\text{fus}}(273\text{ K}) = +6.01\text{ kJ mol}^{-1}$$

in which ice at 1 bar pressure melts to liquid water at 1 bar pressure, and the **standard enthalpy of sublimation** $\Delta H^{\ominus}_{\text{sub}}$, the standard enthalpy for a process in which a solid vaporizes, as in

$$C(s, \text{graphite}) \rightarrow C(g) \quad \Delta H^{\ominus}_{\text{sub}}(\maltese) = +716.68\text{ kJ mol}^{-1}$$

We have remarked that a change in enthalpy is independent of the path between the two states. This feature is of great importance in thermochemistry, for it implies that the same value of $\Delta H^{\ominus}$ is obtained however the change is brought about (so long as the initial and final states are the same). For example, we can picture sublimation of a substance A as occurring directly:

$$A(s) \rightarrow A(g) \quad \Delta H^{\ominus}_{\text{sub}}(T)$$

However, the same overall result is obtained if the solid is thought of as melting at the temperature T and then vaporizing at that temperature:

$$A(s) \rightarrow A(l) \quad \Delta H^{\ominus}_{\text{fus}}(T)$$
$$A(l) \rightarrow A(g) \quad \Delta H^{\ominus}_{\text{vap}}(T)$$
$$\text{Overall:} \quad A(s) \rightarrow A(g) \quad \Delta H^{\ominus}_{\text{fus}}(T) + \Delta H^{\ominus}_{\text{vap}}(T)$$

Since the overall result is the same, the overall enthalpy change is the same in each case, and we can conclude that[3]

$$\Delta H^{\ominus}_{\text{sub}}(T) = \Delta H^{\ominus}_{\text{fus}}(T) + \Delta H^{\ominus}_{\text{vap}}(T)$$

An immediate conclusion is that, because the enthalpy of fusion is always positive, the enthalpy of sublimation of a substance is always greater than its enthalpy of vaporization (at a given temperature).

A consequence of H being a state function is that the standard enthalpy of a forward process and its reverse must be equal and opposite (Fig. 2.6):

$$\Delta H^{\ominus}(\text{reverse}) = -\Delta H^{\ominus}(\text{forward})$$

For instance, the standard enthalpy of condensation of a substance at a temperature T must be the negative of its standard enthalpy of vaporization

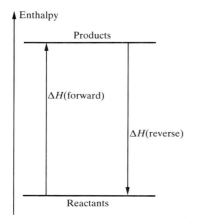

Fig. 2.6 The enthalpy change of a reverse process (downwards in the illustration) is the negative of the enthalpy change of the forward process (upwards in this case). This must be the case, because enthalpy is a state function and a forward process followed by its reverse must leave H unchanged.

[3] This relation is exactly true only at the triple point, for only then are the pressures as well as the temperatures identical for the phase transitions.

at that temperature. Thus, since the vaporization of water is endothermic by 44 kJ mol^{-1} at 298 K, its condensation at that temperature is exothermic by 44 kJ mol^{-1}.

The **standard enthalpy of solution** $\Delta H_{sol}^{\ominus}$ of a substance is the standard enthalpy change when it dissolves in a specified amount of solvent. The *limiting* enthalpy of solution is the standard enthalpy change when the substance dissolves in an infinite amount of solvent and the interactions between the ions (or solute molecules if it is a non-electrolyte) are therefore negligible. For HCl:

$$HCl(g) \rightarrow HCl(aq) \quad \Delta H_{sol}^{\ominus} = -75.14 \text{ kJ mol}^{-1}$$

and so 75 kJ of energy is released as heat when 1.0 mol of HCl(g) dissolves to produce an infinitely dilute solution. Some values for other substances are given in Table 2.3.

Table 2.3. Limiting enthalpies of solution, $\Delta H_{sol}^{\ominus}(\mp)/(\text{kJ mol}^{-1})$

NaF(s)	+1.90	NaCl(s)	+3.89
KF(s)	−17.74	KCl(s)	+17.22
NH$_4$NO$_3$(s)	+25.69	H$_2$SO$_4$(l)	−95.28

Solution: X(g, l, or s) → X(aq).

Enthalpies of ionization

Two very important enthalpy changes are those accompanying the formation of cations and anions of gas phase atoms and molecules. The **enthalpy of ionization** $\Delta H_i^{\ominus}$ is the standard enthalpy change for the removal of an electron:

$$E(g) \rightarrow E^+(g) + e^-(g) \quad \Delta H_i^{\ominus}$$

Since 1 mol of gaseous reactants give 2 mol of gaseous products, the internal energy and enthalpy of ionization differ by about RT:[4]

$$\Delta H_i^{\ominus} = \Delta U_i^{\ominus} + RT$$

The **ionization energy** E_i is the change in internal energy for the same process at $T = 0$ (Table 2.4). Since the ionization energy at ordinary temperatures is quite similar to that at $T = 0$, it is normally good enough to use the relation

$$\Delta H_i^{\ominus} = E_i + RT$$

Since RT is only about 2.5 kJ mol^{-1} at room temperature and ionization energies are typically more than 100 times larger, it is normally safe to ignore the differences between $\Delta H_i^{\ominus}$, $\Delta U_i^{\ominus}$, and E_i.

A cation may itself be ionized, in which case the change in internal energy (at $T = 0$) is called the **second ionization energy** $E_{i,2}$. The second ionization energy (and the corresponding enthalpy change) is always larger than the first. Ionization energies and, from them, enthalpies of ionization, are widely obtained from spectroscopic measurements, as we shall explain in Chapter 12.

The standard enthalpy change accompanying electron attachment to an atom, ion, or molecule in the gas phase is the **electron gain enthalpy** $\Delta H_{ea}^{\ominus}$:

$$E(g) + e^-(g) \rightarrow E^-(g) \quad \Delta H_{ea}^{\ominus}$$

The *negative* of the corresponding internal energy change at $T = 0$ is widely called the **electron affinity** E_{ea} of E (Table 2.5). With this change of sign, a positive electron affinity corresponds to exothermic electron gain. For

Table 2.4. First and second ionization energies, $E_i/(\text{kJ mol}^{-1})$

H	1312	
He	2372	5250
Mg	738	1451
Na	496	4562

Ionization:
E(g) → E$^+$(g) + e$^-$(g) $E_i = \Delta H^{\ominus}(0)$.

Table 2.5. Electron affinities, $E_{ea}/(\text{kJ mol}^{-1})$

Cl	349		
F	322		
H	73		
O	141	O$^-$	−844

Electron gain:
E(g) + e$^-$(g) → E$^-$(g) $E_{ea} = -\Delta H_{ea}^{\ominus}(0)$.

[4] It may appear questionable to treat a gas composed of cations and electrons at 1 bar pressure as even remotely perfect. However, in Section 5.3 we shall see that the standard states of real gases are defined to be the equivalent perfect gas.

example, the attachment of one electron to an O atom is exothermic (Table 2.5), so O has a positive electron affinity. However, the attachment of a second electron is strongly endothermic, and so O^- has a negative electron affinity. As in the case of cation formation, the internal energy change and the enthalpy change differ by RT:

$$\Delta H_{ea}^{\ominus} = -E_{ea} - RT$$

Since electron gain is the reverse of electron loss, the electron gain enthalpy of a neutral atom is the negative of the ionization enthalpy of its anion:

$$E(g) + e^-(g) \rightarrow E^-(g) \quad \Delta H_{ea}^{\ominus}(E)$$

$$E^-(g) \rightarrow E(g) + e^-(g) \quad \Delta H_i^{\ominus}(E^-) = -\Delta H_{ea}^{\ominus}(E)$$

That is, at $T = 0$,

$$E_i(E^-) = E_{ea}(E)$$

and the electron affinity of E is equal to the ionization energy of E^-. Hence, measuring the ionization energy of a negative ion can be used to determine the electron affinity of the neutral species.

Enthalpies of bond formation and dissociation

The **bond dissociation enthalpy** $\Delta H^{\ominus}(A{-}B)$ is the standard reaction enthalpy for the process in which the A—B bond is broken:

$$A{-}B(g) \rightarrow A(g) + B(g) \quad \Delta H^{\ominus}(A{-}B)$$

A and B may be atoms or groups of atoms, as in

$$CH_3OH(g) \rightarrow CH_3(g) + OH(g) \quad \Delta H^{\ominus}(CH_3{-}OH) = +380 \text{ kJ mol}^{-1}$$

Some experimental values are listed in Table 2.6.

The dissociation enthalpy of a given bond depends on the structure of the rest of the molecule: for H_2O, for instance, $\Delta H(HO{-}H) = +499 \text{ kJ mol}^{-1}$ and $\Delta H(O{-}H) = +428 \text{ kJ mol}^{-1}$. Removing the first and second H atoms results in different enthalpy changes because the electronic structure of the molecule adjusts after the first atom is removed. The **mean bond enthalpy** $B(A{-}B)$ is the value of the bond dissociation enthalpy of the A—B bond averaged over a series of related compounds (Table 2.7). For instance, the O—H bond enthalpy is calculated using the data for H_2O and similar compounds, such as $\Delta H(CH_3O{-}H) = +437 \text{ kJ mol}^{-1}$ in methanol. Mean bond enthalpies are useful because they let us make estimates of enthalpy changes in reactions where data might not be available (see Example 2.7 below).

The standard enthalpy change that accompanies the separation of all the atoms in a substance (which may be an element or a compound) is called the **enthalpy of atomization** $\Delta H_a^{\ominus}$. For example, the enthalpy of atomization of H_2O is the sum of the HO—H and H—O bond dissociation enthalpies, which from Table 2.6 is $+920 \text{ kJ mol}^{-1}$. Values for other substances are given in Table 2.8. For an elemental solid that evaporates to a monatomic gas, the enthalpy of atomization is the same as its enthalpy of sublimation:

$$Na(s) \rightarrow Na(g) \quad \Delta H_a^{\ominus} = \Delta H_{sub}^{\ominus} = +107 \text{ kJ mol}^{-1}$$

The two enthalpy changes are not the same if the vapour consists of molecules, as in the sublimation of solid N_2.

Table 2.6. Bond dissociation enthalpies, $\Delta H^{\ominus}(A{-}B)/(\text{kJ mol}^{-1})$ at 298 K

H—CH$_3$	435
H—Cl	431
H—H	436
H—OH	492
H—O	428
H$_3$C—CH$_3$	368

Dissociation: $AB(g) \rightarrow A(g) + B(g)$.

Table 2.7. Mean bond enthalpies, $B(A{-}B)/(\text{kJ mol}^{-1})$

	H	C	N	O
H	436			
C	412	348		
		612		
		838		
N	388	305	163	
		613	409	
			944	
O	463	360	157	146
				497

Subsequent values are for single, double, and triple bonds respectively.

Table 2.8. Standard enthalpies of atomization, $\Delta H_a^{\ominus}(\mp)/(\text{kJ mol}^{-1})$

C(s)	716.7
Cu(s)	338.3
K(s)	89.2
Na(s)	107.3

Atomization: $X(s) \rightarrow X(g)$.

Example 2.7: *Using mean bond enthalpy data*

Use mean bond enthalpy and enthalpy of atomization data to estimate the standard enthalpy of the reaction

$$C(s, gr) + 2H_2(g) + \tfrac{1}{2}O_2(g) \rightarrow CH_3OH(l)$$

where gr denotes the graphite form of carbon.

Answer. We first atomize the reactant molecules and then form the product, the reverse of its atomization. Note that the gaseous product must be condensed to form the stated product. Use data from Tables 2.7 and 2.8. The enthalpy change for

$$C(s, gr) + 2H_2(g) + \tfrac{1}{2}O_2(g) \rightarrow C(g) + 4H(g) + O(g)$$

is

$$\Delta H^{\ominus} = \Delta H_a^{\ominus}(C, s) + 2\Delta H^{\ominus}(H-H) + \tfrac{1}{2}\Delta H^{\ominus}(O{=}O) = +1837 \text{ kJ mol}^{-1}$$

The enthalpy change when these atoms form $CH_3OH(g)$ in the reaction

$$C(g) + 4H(g) + O(g) \rightarrow CH_3OH(g)$$

is

$$\Delta H^{\ominus} = -\{3B(C-H) + B(C-O) + B(O-H)\} = -2059 \text{ kJ mol}^{-1}$$

The enthalpy of the overall reaction is therefore the sum of the two enthalpy changes, -222 kJ mol^{-1}. Since the enthalpy of vaporization of $CH_3OH(l)$ is $+37.99$ kJ mol^{-1}, the enthalpy of condensation of $CH_3OH(g)$ is -37.99 kJ mol^{-1}. The overall enthalpy of the reaction is therefore -260 kJ mol^{-1}.

Comment. The measured value is -239.0 kJ mol^{-1}. The discrepancy stems from the use of mean bond enthalpies in the calculation of the enthalpy of construction of the CH_3OH molecule.

Exercise. Estimate the standard enthalpy change of the reaction in which liquid ethanol is formed from its elements.   [-287 kJ mol^{-1}]

Enthalpies of chemical change

The **standard enthalpy of combustion** $\Delta H_c^{\ominus}$ is the standard reaction enthalpy for the oxidation of an organic substance to CO_2 and H_2O in the case of compounds of C, H, and O, and to N_2 if N is also present. Some values are listed in Table 2.9.

$$C_6H_{12}O_6(s) + 6O_2(g) \rightarrow 6CO_2(g) + 6H_2O(l) \quad \Delta H_c^{\ominus} = -2808 \text{ kJ mol}^{-1}$$

Here and henceforth, numerical values are for $\mathbf{\mathcal{F}}$ unless otherwise stated.

The **standard enthalpy of hydrogenation** is the standard reaction enthalpy for the hydrogenation of an unsaturated organic compound. Two especially important cases are the hydrogenations of ethene and benzene:

$$CH_2{=}CH_2(g) + H_2(g) \rightarrow CH_3CH_3(g) \quad \Delta H^{\ominus} = -137 \text{ kJ mol}^{-1}$$

$+ 3H_2(g) \rightarrow$ $\quad \Delta H^{\ominus} = -205 \text{ kJ mol}^{-1}$

The interest in these two values lies in the observation that the second is not three times the first, as might be expected on the basis that benzene

Table 2.9. Standard enthalpies of formation and combustion of organic compounds, $\Delta H^{\ominus}(\mathbf{\mathcal{F}})/(\text{kJ mol}^{-1})$

	Formation	Combustion
Benzene, $C_6H_6(l)$	+49.0	−3268
Ethane, $C_2H_6(g)$	−84.7	−1560
Glucose, $C_6H_{12}O_6(s)$	−1274	−2808
Methane, $CH_4(g)$	−74.8	−890
Methanol, $CH_3OH(l)$	−238.7	−726

Formation: $C(s, gr)$ + elements in reference states → compound.

contains three double bonds. The value for benzene is less than $3 \times (-137\,\mathrm{kJ})$ by 206 kJ, which therefore represents a thermochemical stabilization of benzene (so benzene lies closer in energy than expected to the fully hydrogenated, stable form). This stabilization is explained in Part 2.

Hess's law

We can combine the standard enthalpies of individual reactions to obtain the enthalpy of another reaction. This application of the First Law is widely called **Hess's law**:

> The overall reaction enthalpy is the sum of the reaction enthalpies of the individual reactions into which a reaction may be divided.

The individual steps need not be realizable in practice—they may be hypothetical reactions, the only requirement being that they should balance. The thermodynamic basis of the law is the path-independence of the value of $\Delta H^{\ominus}$ and the implication that we may take the specified reactants, pass through any (possibly hypothetical) set of reactions to the specified products, and overall obtain the same change of enthalpy.

Example 2.8: *Using Hess's law*

The standard reaction enthalpy for the hydrogenation of propene,

$$CH_2{=}CHCH_3(g) + H_2(g) \rightarrow CH_3CH_2CH_3(g)$$

is $-124\,\mathrm{kJ\,mol^{-1}}$. The standard reaction enthalpy for the combustion of propane,

$$CH_3CH_2CH_3(g) + 5O_2(g) \rightarrow 3CO_2(g) + 4H_2O(l)$$

is $-2220\,\mathrm{kJ\,mol^{-1}}$. Calculate the standard reaction enthalpy for the combusion of propene.

Answer. Add and subtract the reactions given, together with any others needed, so as to reproduce the reaction required. Then add and subtract the reaction enthalpies in the same way. Additional data are in Tables 2.9 and 2.10. The combustion reaction we require is

$$C_3H_6(g) + \tfrac{9}{2}O_2(g) \rightarrow 3CO_2(g) + 3H_2O(l).$$

This can be recreated from the following sum:

	$\Delta H^{\ominus}/(\mathrm{kJ\,mol^{-1}})$
$C_3H_6(g) + H_2(g) \rightarrow C_3H_8(g)$	-124
$C_3H_8(g) + 5O_2(g) \rightarrow 3CO_2(g) + 4H_2O(l)$	-2220
$H_2O(l) \rightarrow H_2(g) + \tfrac{1}{2}O_2(g)$	$+286$
$C_3H_6(g) + \tfrac{9}{2}O_2(g) \rightarrow 3CO_2(g) + 3H_2O(l)$	-2058

Comment. These chemical equations with their associated enthalpies are called 'thermochemical equations'. The skill to develop is the ability to assemble a given equation from others.

Exercise. Calculate the enthalpy of hydrogenation of benzene from its enthalpy of combustion and the enthalpy of combustion of cyclohexane.

$$[-224\,\mathrm{kJ\,mol^{-1}}]$$

2.7 Enthalpies of formation

Thermochemical data are often reported in terms of the enthalpy of a compound relative to its elements under standard conditions. Thus the **standard enthalpy of formation** $\Delta H_f^{\ominus}$ of a substance is the standard reaction enthalpy for its formation from its elements in their **reference states**. The reference state of an element is its most stable state at the specified temperature and 1 bar pressure. Thus, at 298 K the reference state of nitrogen is a gas of N_2 molecules, that of mercury is the liquid, that of carbon is graphite, and that of tin is the white (metallic) form. The standard enthalpy of formation of liquid benzene at 298 K, for example, is the standard enthalpy of the reaction

$$6C(s, \text{gr}) + 3H_2(g) \rightarrow C_6H_6(l) \quad \Delta H_f^{\ominus}(C_6H_6, l) = +49.0 \text{ kJ mol}^{-1}$$

There is one exception to this general prescription of reference states: that of phosphorus is taken to be white phosphorus despite this allotrope not being the most stable form but simply the more reproducible form of the element. The standard enthalpies of formation of elements in their reference states are zero at all temperatures since they are the enthalpies of such 'null' reactions as

$$N_2(g) \rightarrow N_2(g) \quad \Delta H_f^{\ominus}(N_2, g) = 0$$

Some enthalpies of formation are listed in Tables 2.9 and 2.10.

The Born–Haber cycle

The enthalpy of formation of a solid may be analysed into several contributions. In the case of NaCl, for instance, the overall reaction $Na(s) + \frac{1}{2}Cl_2(g) \rightarrow NaCl(s)$ can be regarded as the outcome of five steps:

1. $Na(s) \rightarrow Na(g)$ $\Delta H_{sub}^{\ominus}(Na)$
2. $Na(g) \rightarrow Na^+(g) + e^-(g)$ $\Delta H_i^{\ominus}(Na)$
3. $\frac{1}{2}Cl_2(g) \rightarrow Cl(g)$ $\frac{1}{2}\Delta H^{\ominus}(Cl\text{-}Cl)$
4. $Cl(g) + e^-(g) \rightarrow Cl^-(g)$ $\Delta H_{ea}^{\ominus}(Cl)$
5. $Na^+(g) + Cl^-(g) \rightarrow NaCl(s)$

The enthalpy change of Step 5 is the negative of the **lattice enthalpy** ΔH_L. In general, the lattice enthalpy is the standard enthalpy change accompanying the formation of a gas of ions from the crystalline solid:

$$MX(s) \rightarrow M^+(g) + X^-(g) \quad \Delta H_L^{\ominus}$$

All lattice enthalpies are positive (Table 2.11). The **lattice energy** is the lattice enthalpy at $T = 0$.

The sequence of steps is depicted in Fig. 2.7. When

$$Na(s) + \frac{1}{2}Cl_2(g) \rightarrow NaCl(s) \quad \Delta H_f^{\ominus}(NaCl, s)$$

is included, we obtain a **cycle**, a closed path, known as a **Born–Haber cycle**. The importance of a cycle is that the sum of enthalpy changes round a cycle is zero (because the initial and final states are the same). It follows that if all but one of the enthalpy changes in a cycle are known, the unknown may be found.

Table 2.10. Standard enthalpies of formation of inorganic compounds, $\Delta H_f^{\ominus}(\mp)/(\text{kJ mol}^{-1})$

$H_2O(l)$	−285.8	$H_2O_2(l)$	−187.8
$NH_3(g)$	−46.1	$N_2H_4(l)$	+50.6
$NO_2(g)$	+33.2	$N_2O_4(g)$	+9.2
$NaCl(s)$	−411.2	$KCl(s)$	−436.8

Formation: Elements in reference states → compound.

Table 2.11. Lattice enthalpies at 298 K, $\Delta H_L^{\ominus}/(\text{kJ mol}^{-1})$

NaF	926
NaCl	787
KCl	717
MgO	3850
MgS	3406

Lattice enthalpy:
$$XY(s) \rightarrow X^+(g) + Y^-(g) \quad \Delta H_L^{\ominus}$$

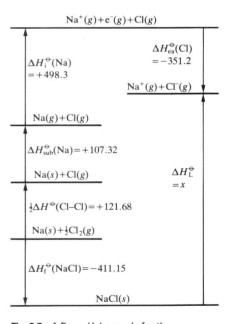

Fig. 2.7 A Born–Haber cycle for the determination of the lattice enthalpy. The sum of the enthalpy changes round the cycle is zero. In other words, the distance up on the left must be equal to the distance up on the right. From this equality, x must be determined.

Example 2.9: *Using a Born–Haber cycle*

Calculate the lattice enthalpy of NaCl(s) at 298 K using the Born–Haber cycle in Fig. 2.7.

Answer. We must find x in the illustration from the condition that the distance up on the left must equal the distance up on the right. The following data may be assembled from the tables:

$$\tfrac{1}{2}\Delta H^{\ominus}(\text{Cl–Cl}) = +121.68 \text{ kJ mol}^{-1} \quad \Delta H^{\ominus}_{\text{sub}}(\text{Na}, s) = +107.32 \text{ kJ mol}^{-1}$$

$$\Delta H^{\ominus}_{\text{i}} = +498.3 \text{ kJ mol}^{-1} \quad \Delta H^{\ominus}_{\text{ea}} = -351.2 \text{ kJ mol}^{-1}$$

$$\Delta H^{\ominus}_{\text{f}}(\text{NaCl}, s) = -411.15 \text{ kJ mol}^{-1}$$

The distance up on the left, from NaCl(s) to the top of the cycle, is

$$(411.15 + 121.68 + 107.32 + 498.3) \text{ kJ mol}^{-1} = 1138.4 \text{ kJ mol}^{-1}$$

The distance up on the right is $(x + 351.2) \text{ kJ mol}^{-1}$. The two distances are equal; therefore $x = 787.2$. Since the lattice enthalpy is positive (it corresponds to the up-going arrow in the cycle), $\Delta H^{\ominus}_{\text{L}} = +787.2 \text{ kJ mol}^{-1}$

Exercise. Calculate the lattice enthalpy of $CaBr_2$. [2148 kJ mol^{-1}]

The reaction enthalpy in terms of enthalpies of formation

Conceptually we can regard a reaction as proceeding by 'unforming' the reactants into their elements, and then forming those elements into the products. The value of $\Delta H^{\ominus}$ for the overall reaction is the sum of the 'unforming' and forming enthalpies. Since 'unforming' is the reverse of forming, the enthalpy of an unforming step is the negative of the enthalpy of formation. Hence, in the enthalpies of formation of substances, we have enough information to calculate the enthalpy of any reaction. For example, the reaction enthalpy of

$$2HN_3(l) + 2NO(g) \rightarrow H_2O_2(l) + 4N_2(g)$$

is the sum of the following contributions:

$$2HN_3(l) \rightarrow 3N_2(g) + H_2(g) \quad -2 \times \Delta H^{\ominus}_{\text{f}}(\text{HN}_3, l)$$

$$2NO(g) \rightarrow N_2(g) + O_2(g) \quad -2 \times \Delta H^{\ominus}_{\text{f}}(\text{NO}, g)$$

$$H_2(g) + O_2(g) \rightarrow H_2O_2(l) \quad \Delta H^{\ominus}_{\text{f}}(\text{H}_2\text{O}_2, l)$$

and from the data in Table 2.10 is $-896.3 \text{ kJ mol}^{-1}$.

The calculation we have just done can be expressed more generally using a notation we shall employ frequently in the following chapters. First, we write the chemical equation

$$aA + bB \rightarrow cC + dD$$

in the form

$$0 = cC + dD - aA - bB$$

or, in general,

$$0 = \sum_{\text{J}} \nu_{\text{J}} S_{\text{J}} \tag{12}$$

where S_{J} is a particular substance and ν_{J} its stoichiometric coefficient in the reaction. Note that v_J is *positive* for *p*roducts and negative for reactants. The standard reaction enthalpy is the sum of the standard enthalpies of formation of the products less the sum of the standard enthalpies of formation of the reactants. That is,

$$\Delta H^{\ominus} = \{c\Delta H_f^{\ominus}(C) + d\Delta H_f^{\ominus}(D)\} - \{a\Delta H_f^{\ominus}(A) + b\Delta H_f^{\ominus}(B)\}$$

More succinctly and generally:

$$\Delta H^{\ominus} = \sum_J v_J \Delta H_f^{\ominus}(J) \tag{13}$$

where $\Delta H_f^{\ominus}(J)$ is the standard enthalpy of formation of substance S_J.

Example 2.10: *Calculating a standard reaction enthalpy*
Express the standard reaction enthalpy of

$$2HN_3(l) + 2NO(g) \rightarrow H_2O_2(l) + 4N_2(g)$$

in terms of the standard enthalpies of formation of the components.

Answer. We express the reaction in the form of eqn 12, identify the stoichiometric coefficients, and then use eqn 13 to express $\Delta H^{\ominus}$ in terms of the $\Delta H_f^{\ominus}$. The reaction is

$$0 = H_2O_2(l) + 4N_2(g) - 2HN_3(l) - 2NO(g)$$

Therefore, $v(H_2O_2) = +1$, $v(N_2) = +4$, $v(HN_3) = -2$, $v(NO) = -2$. The standard reaction enthalpy is therefore

$$\Delta H^{\ominus} = \Delta H_f^{\ominus}(H_2O_2) + 4\Delta H_f^{\ominus}(N_2) - 2\Delta H_f^{\ominus}(HN_3) - 2\Delta H_f^{\ominus}(NO)$$

This is the same as before when we note that $\Delta H_f^{\ominus}(N_2, g) = 0$.

Exercise. Express the standard reaction enthalpy of

$$2C_3H_6(g) + 9O_2(g) \rightarrow 6CO_2(g) + 6H_2O(l)$$

in terms of enthalpies of formation.

$$[6\Delta H_f^{\ominus}(CO_2) + 6\Delta H_f^{\ominus}(H_2O) - 2\Delta H_f^{\ominus}(C_3H_6) - 9\Delta H_f^{\ominus}(O_2)]$$

The enthalpies of formation of substances in solution

The enthalpy of formation of a substance in solution is the standard enthalpy change that accompanies reactions such as

$$\tfrac{1}{2}H_2(g) + \tfrac{1}{2}Cl_2(g) \rightarrow HCl(aq) \quad \Delta H_f^{\ominus}(HCl, aq) = -167 \text{ kJ mol}^{-1}$$

$$Na(s) + \tfrac{1}{2}Cl_2(g) \rightarrow NaCl(aq) \quad \Delta H_f^{\ominus}(NaCl, aq) = -407 \text{ kJ mol}^{-1}$$

We can analyse the enthalpy of formation in solution using a cycle like the one we used for the formation of the solid. The only difference is that in place of the lattice formation step (Step 5 in Fig. 2.7) we use

$$5'. \quad Na^+(g) + Cl^-(g) \rightarrow Na^+(aq) + Cl^-(aq) = NaCl(aq)$$

In this case, the standard enthalpy change is the **enthalpy of hydration** $\Delta H_{hyd}^{\ominus}$ of the gaseous ions (and the enthalpy of solvation, $\Delta H_{solv}^{\ominus}$, in general). Since all the other enthalpy changes in the cycle are known, this quantity may be determined: the distance up on the left of Fig. 2.8

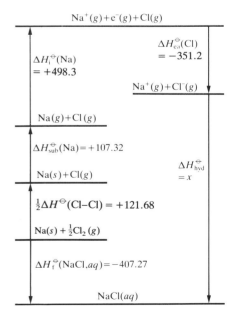

Fig. 2.8 A similar thermodynamic cycle for the determination of the enthalpy of hydration of Na^+ and Cl^- ions. The distance up on the left is equal to the distance up on the right.

Table 2.12. Standard molar enthalpies of hydration at infinite dilution, $\Delta H_{hyd}^{\ominus}(\mbox{\textcent})/(kJ\,mol^{-1})$

	Li$^+$	Na$^+$	K$^+$
F$^-$	−1026	−911	−828
Cl$^-$	−884	−783	−685
Br$^-$	−856	−742	−658

Table 2.13. Limiting enthalpies of formation of ions in aqueous solution, $\Delta H_f^{\ominus}(\mbox{\textcent})/(kJ\,mol^{-1})$

Cations		Anions	
H$^+$	0	OH$^-$	−230.0
Na$^+$	−240.1	Cl$^-$	−167.2
Cu^{2+}	+64.8	SO$_4^{2-}$	−909.3
Al^{3+}	−531	PO$_4^{3-}$	−1277.4

Formation: Elements in reference states → ion in water, and $\Delta H_f^{\ominus}(H^+, aq) \equiv 0$.

Table 2.14. Standard ion hydration enthalpies, $\Delta H_{hyd}^{\ominus}(\mbox{\textcent})/(kJ\,mol^{-1})$

Li$^+$	−520	F$^-$	−506
Na$^+$	−405	Cl$^-$	−364
K$^+$	−321	Br$^-$	−337

Hydration: $X^+(g) \to X^+(aq)$ based on $H^+(g) \to H^+(aq)$ $\Delta H^{\ominus} = -1090\,kJ\,mol^{-1}$.

$(1134.6\,kJ\,mol^{-1})$ is equal to the distance up on the right $(351.2\,kJ\,mol^{-1} + x)$. Hence $x = 783.4\,kJ\,mol^{-1}$. The solvation process is the step down on the right, and so the enthalpy of hydration of NaCl is $-x$, or $-783.4\,kJ\,mol^{-1}$. Other values may be obtained in this way, and some are listed in Table 2.12. The values there refer to the formation of the solution at zero concentration (infinite dilution), since we are not taking into account the interactions between the ions that are important at normal concentrations.[5]

The enthalpies of formation of individual ions in solution

We can think of the enthalpy of formation of a fully ionized compound in solution as the sum of the enthalpies of formation of solutions of the constituent ions. Thus, we can think of the enthalpy of formation of HCl(aq) as the sum of the enthalpies of formation of H$^+$(aq) and Cl$^-$(aq).

The problem is to decide how the overall value is to be apportioned between the two kinds of ion. This is resolved by taking the arbitrary decision to set the standard enthalpy of formation of H$^+$(aq) equal to zero:

$$\tfrac{1}{2}H_2(g) \to H^+(aq) \quad \Delta H_f^{\ominus}(H^+, aq) = 0 \text{ at all temperatures}$$

It follows that the standard enthalpy of formation of Cl$^-$(aq) is $-167\,kJ\,mol^{-1}$. Combining that value with the observed enthalpy of formation of NaCl(aq) then leads to a value for Na$^+$(aq), and so on. This procedure leads to the values in Table 2.13.

Just as we analysed the enthalpy of hydration of NaCl in terms of a thermodynamic cycle, we can do the same for the individual hydration enthalpies of the ions. However, we cannot adopt an arbitrary convention about the hydration enthalpy of the proton because it would conflict with the choice already made about its enthalpy of formation: we have selected zero for the step $\tfrac{1}{2}H_2(g) \to H^+(aq)$, and we cannot simultaneously select zero for $H^+(g) \to H^+(aq)$. We are therefore forced to estimate the actual value of the enthalpy change in the step by calculating the energy of interaction of a proton and the surrounding water molecules. There is some agreement (from mass spectrometry and the energy of proton attachment to small clusters of water molecules) that

$$H^+(g) \to H^+(aq) \quad \Delta H^{\ominus} \approx -1090\,kJ\,mol^{-1}.$$

With this value accepted, the value for Cl$^-$(g)$\to$Cl$^-$(aq) can be obtained from data on HCl, and then the value for Cl$^-$ combined with the value of ΔH_{hyd} for NaCl as obtained above, to arrive at the value of about $-400\,kJ\,mol^{-1}$ for Na$^+$(g)$\to$Na$^+$(aq). The individual ion hydration enthalpies in Table 2.14 have been obtained in this way.

The data in Table 2.14 conform with common sense (and with calculation, as we see in Chapter 9). Thus, small, highly charged ions have the most negative (exothermic) hydration enthalpies because they attract the solvent so strongly. The hydration enthalpy of the electron, the enthalpy of the process e$^-$(g)$\to$e$^-$(aq), is only about $-170\,kJ\,mol^{-1}$. This value suggests that when the electron is trapped in water (as when water is irradiated with high energy radiation), the electron occupies a volume similar to that of about two H$_2$O molecules.

---

[5] We explain what is meant by the standard state of a solution in Chapter 10.

The variation of enthalpy with temperature

The enthalpy of a substance increases as it is heated. Hence a reaction enthalpy changes with temperature because the enthalpy of each substance in a reaction varies in a characteristic way.

2.8   The heat capacities of substances

We can discuss the dependence of the enthalpy on the temperature, and through that the dependence of the reaction enthalpy, in terms of heat capacities (Section 2.5). This discussion will also introduce us to a concept and a notation—partial differentiation—that we shall use extensively in the following chapters.

The heat capacity at constant volume

The heat capacity of a substance depends on the conditions. First, suppose the system is constrained to have constant volume and is unable to do any kind of work. The heat required to bring about a change in temperature dT is $dq_V = C_V \, dT$, where C_V is the **heat capacity at constant volume**. However, since $dU = dq_V$, we can write

$$dU = C_V \, dT \text{ at constant volume}$$

and identify C_V with dU/dT with the volume constant. When, as in this definition, one or more variables are held constant during the change of another, the derivatives are called 'partial derivatives' with respect to the changing variable. The d is replaced by ∂ and the variables held constant are added as a subscript. In this case it follows that

$$C_V = \left(\frac{\partial U}{\partial T}\right)_V \tag{14}$$

This is the *definition* of C_V.

Example 2.11: *Evaluating partial derivatives*

A function of several variables we have already encountered is the van der Waals equation of state, Table 1.5:

$$p = \frac{nRT}{V - nb} - a\left(\frac{n}{V}\right)^2$$

Find the partial derivatives of p with respect to V and T.

Answer. To find $(\partial p/\partial T)_V$, differentiate p with respect to T, regarding n and V as constants (like a, b, and R):

$$\left(\frac{\partial p}{\partial T}\right)_V = \frac{nR}{V - nb}$$

To find $(\partial p/\partial V)_T$, differentiate with respect to V, holding n and T constant:

$$\left(\frac{\partial p}{\partial V}\right)_T = \frac{-nRT}{(V - nb)^2} + \frac{2an^2}{V^3}$$

Comment. The two partial derivatives are themselves functions of V and T, and so we could go on to find the four second derivatives. Later we shall find it useful to note that the order of differentiation is unimportant, and that $(\partial^2 p/\partial V \, \partial T) = (\partial^2 p/\partial T \, \partial V)$, as may be checked in this instance.

Exercise. Calculate $(\partial p/\partial V)_T$ and $(\partial p/\partial T)_V$ for the Dieterici equation of state in Table 1.5.

$$[\{nRT/(V-nb)\}\{(an/RTV^2) + 1/(V-nb)\}e^{-an/RTV};$$
$$\{nR/(V-nb)\}\{1 + an/RTV\}e^{-an/RTV}]$$

The heat capacity at constant pressure

Now suppose that the system is subjected to constant pressure and is allowed to expand or contract as it is heated. The heat required to bring about the same change in temperature is $dq_p = C_p\, dT$, where C_p is the **heat capacity at constant pressure**. In this case the system may have changed its volume; therefore some of the energy supplied as heat may have been returned to the surroundings as work and not used exclusively to raise the system's temperature. Hence in general C_V and C_p are different. Since $dq_p = dH$, it follows that

$$C_p = \left(\frac{\partial H}{\partial T}\right)_p \qquad (15)$$

This is the definition of C_p.

Since the expansion work is done at constant external pressure, for a given supply of heat, the temperature rise is less at constant pressure than at constant volume. That is, $C_p > C_V$ for a given sample. The difference is greater for gases than for liquids or solids because the latter change their volumes only slightly when heated, and therefore generally do little work. In Chapter 3 we derive the following simple relation between the two heat capacities of a perfect gas:

$$C_p - C_V = nR \qquad (16)°$$

Molar heat capacities are of the order of R itself (see Table 2.15), and so for gases the difference between the two heat capacities is important.

Table 2.15. Molar heat capacities at 25°C and 1 atm, $C/(\text{J K}^{-1}\,\text{mol}^{-1})$

	C_V	C_p
Ar	12.48	20.79
CO_2	28.46	37.11
He	12.48	20.79
N_2	20.74	29.12

2.9 The temperature-dependence of reaction enthalpies

The reaction enthalpies of many important reactions have been measured over a range of temperatures, and for serious work these accurate data must be used. However, in the absence of this information, reaction enthalpies at different temperatures may be estimated from heat capacities and the reaction enthalpy at a reference temperature.

When the temperature is changed through an infinitesimal amount dT, the enthalpy of a substance changes by $C_p\, dT$. Therefore, for a change of temperature from T_1 to T_2, the enthalpy of a substance changes from $H(T_1)$ to

$$H(T_2) = H(T_1) + \int_{T_1}^{T_2} C_p\, dT$$

Since this equation applies to each substance in the reaction, the standard reaction enthalpy changes from $\Delta H^{\ominus}(T_1)$ to

$$\Delta H^{\ominus}(T_2) = \Delta H^{\ominus}(T_1) + \int_{T_1}^{T_2} \Delta C_p\, dT \qquad (17)$$

where

$$\Delta C_p = \{cC_p(\text{C}) + dC_p(\text{D})\} - \{aC_p(\text{A}) + bC_p(\text{B})\}$$

in which $C_p(\text{J})$ is the molar heat capacity of substance S_J. More succinctly,

$$\Delta C_p = \sum_\text{J} \nu_\text{J} C_p(\text{J})$$

Equation 17 is known as **Kirchhoff's law**. It is normally good enough to assume that ΔC_p is independent of the temperature, at least over reasonably limited ranges. This is illustrated in the following example. However, in some cases the temperature dependence of heat capacities is taken into account by writing

$$C_{p,\text{J}}(T) = a_\text{J} + b_\text{J} T + \frac{c_\text{J}}{T^2}$$

A selection of values of the temperature-independent coefficients a, b, and c are listed in Table 2.16.

Table 2.16. Temperature variation of molar heat capacities, $C_p/(\text{J K}^{-1}\,\text{mol}^{-1}) = a + bT + c/T^2$

	a	$b/(10^{-3}\,\text{K}^{-1})$	$c/(10^5\,\text{K}^2)$
C(s, gr)	16.86	4.77	−8.54
$CO_2(g)$	44.22	8.79	−8.62
$H_2O(l)$	75.29	0	0
$N_2(g)$	28.58	3.77	−0.50

Example 2.12: *Using Kirchhoff's law*

The standard enthalpy of formation of gaseous water at 25°C is − 241.82 kJ mol^{-1}. Estimate its value at 100°C given the following values of the molar heat capacities at constant pressure: $H_2O(g)$: 33.58 J K^{-1} mol^{-1}; $H_2(g)$: 28.84 J K^{-1} mol^{-1}; $O_2(g)$: 29.37 J K^{-1} mol^{-1}. Assume that the heat capacities are independent of temperature.

Answer. If ΔC_p is independent of temperature in the range T_1 to T_2, the integral in eqn 17 evaluates to $\Delta C_p \times (T_2 - T_1)$, or more simply $\Delta C_p \Delta T$. Therefore,

$$\Delta H^\ominus(T_2) = \Delta H^\ominus(T_1) + \Delta C_p \Delta T$$

The reaction is

$$H_2(g) + \tfrac{1}{2}O_2(g) \rightarrow H_2O(g)$$

so

$$\Delta C_p = C_p(H_2O, g) - C_p(H_2, g) - \tfrac{1}{2}C_p(O_2, g)$$

$$= -9.94\ \text{J K}^{-1}\,\text{mol}^{-1}$$

Then, since $\Delta T = +75$ K,

$$\Delta H_f^\ominus(373\ \text{K}) = \Delta H^\ominus(\text{?}) + (75\ \text{K}) \times (-9.94\ \text{J K}^{-1}\,\text{mol}^{-1})$$

$$= -242.6\ \text{kJ mol}^{-1}$$

Exercise. Estimate the standard enthalpy of formation of $NH_3(g)$ at 400 K from the data in Table 2.10.　　　　　　　　　　　　$[-48.4\ \text{kJ mol}^{-1}]$

Further reading

General background

E. B. Smith, *Basic chemical thermodynamics*. Oxford University Press (1982).

J. B. Fenn, *Engines, energy, and entropy*. W. H. Freeman & Co., New York (1982).

S. W. Angrist, Perpetual motion machines. *Scientific American*, **218**(1), 114 (1968).

P. A. Rock, *Chemical thermodynamics*. University Science Books and Oxford University Press (1983).

M. L. McGlashan, *Chemical thermodynamics*. Academic Press, London (1979).

L. K. Nash, Bibliography of thermodynamics. *J. chem. Educ.* **64,** 42 (1965).

J. M. Sturtevant, Calorimetry. In *Techniques of chemistry* (ed. A. Weissberger and B. W. Rossiter) Volume V, p. 347, Wiley-Interscience, New York (1971).

H. A. Skinner (ed.), *Experimental thermochemistry*. Wiley-Interscience, New York (1962).

J. D. McCullough and D. W. Scott, *Experimental thermodynamics*. Butterworths, London (1968).

Applications

W. E. Dasent, *Inorganic energetics*. Cambridge University Press (1982).

D. A. Johnson, *Some thermodynamic aspects of inorganic chemistry* (2nd edn). Cambridge University Press (1982).

A. Lehninger, *Bioenergetics* (2nd edn). Benjamin-Cummings, Menlo Park (1971).

Data

NBS tables of thermodynamic properties. Supplement to Vol. 2, *J. Phys and Chem. Ref. Data* (1982).

J. B. Pedley, R. D. Naylor, and S. P. Kirby, *Thermochemical data of organic compounds*. Chapman and Hall, London (1986).

J. A. Dean (ed.), *Handbook of organic chemistry*. McGraw-Hill, New York (1987).

C. E. Moore, *Ionization potentials and ionization limits derived from the analysis of optical spectra*. United States Department of Commerce, Washington (1970).

R. C. Weast (ed.), *Handbook of chemistry and physics,* Volume 70. CRC Press, Boca Raton (1989).

R. D. Freeman, Conversion of standard (1 atm) thermodynamic data to the new standard-state pressure, 1 bar (10^5 Pa). *Bull. chem. Thermodynamics,* **25,** 523 (1982).

J. Emsley, *The elements*. Oxford University Press (1989).

Exercises

Assume all gases are perfect unless stated otherwise. To two significant figures, 1.0 atm is the same as 1.0 bar. Unless otherwise stated, thermochemical data are for 298 K.

2.1 Calculate the work done to raise a mass of 1.0 kg through 10 m on the surface of (a) the earth ($g = 9.81$ m s^{-2}) and (b) the moon ($g = 1.60$ m s^{-2}).

2.2 Calculate the work needed for a 65 kg person to climb through 4.0 m on the surface of the earth.

2.3 A chemical reaction takes place in a container of cross-sectional area 100 cm^2; the container has a loosely fitted piston at one end. As a result of the reaction, the piston is pushed out through 10 cm against an external pressure of 1.0 atm. Calculate the work done by the system.

2.4 A sample of 4.50 g of methane occupies 12.7 L at 310 K. (a) Calculate the work done when the gas expands isothermally against a constant external pressure of 200 Torr until its volume has increased by 3.3 L. (b) Calculate the work that would be done if the same expansion occurred reversibly.

2.5 In the isothermal reversible compression of 52.0 mmol of a perfect gas at 260 K, the volume of the gas is reduced to one-third its initial value. Calculate w for this process.

2.6 A certain liquid sample occupies 0.450 L at 0°C and 1 bar, and shrinks by 0.67 per cent when subjected to compression under constant external pressure of 95 bar. Calculate w.

2.7 A strip of magnesium of mass 15 g is dropped into a beaker of dilute hydrochloric acid. Calculate the work done by the system as a result of the reaction. The atmospheric pressure is 1.0 atm and the temperature 25°C.

2.8 Calculate the amount of heat required to melt 750 kg of sodium metal at 370.95 K.

2.9 When 229 J of energy is supplied as heat to 3.0 mol of Ar(g) at constant pressure, the temperature of the sample increases by 2.55 K. Calculate the molar heat capacities at constant volume and constant pressure of the gas.

2.10 The heat capacity of air at room temperature and pressure is approximately 21 J K^{-1} mol^{-1}. How much heat is required to raise the temperature of a 5.0 m × 5.0 m × 3.0 m room by 10°C? If losses are neglected, how long will it take a 1.0 kW heater to achieve that increase given that $1\,W = 1\,J\,s^{-1}$?

2.11 25 g of a liquid is cooled from 290 K to 275 K at constant pressure by the extraction of 1.2 kJ of energy as heat. Calculate q and ΔH and estimate the heat capacity of the sample.

2.12 When 3.0 mol of O_2 is heated at a constant pressure of 3.25 atm, its temperature increases from 260 K to 285 K. Given that the molar heat capacity of O_2 at constant pressure is 29.4 J K^{-1} mol^{-1}, calculate q, ΔH, and ΔU.

2.13 A certain liquid has $\Delta H^{\ominus}_{vap} = +26.0$ kJ mol^{-1}. Calculate q, w, ΔH, and ΔU when 0.50 mol is vaporized at 250 K and 750 Torr.

2.14 The standard enthalpy of formation of ethylbenzene is -12.5 kJ mol^{-1}. Calculate its standard enthalpy of combustion.

2.15 Calculate the standard enthalpy of hydrogenation of 1-hexene to hexane given that its standard enthalpy of combustion of 1-hexene is -4003 kJ mol^{-1}.

2.16 Calculate the standard internal energy of formation of liquid methyl acetate from its standard enthalpy of formation, which is -442 kJ mol^{-1}.

2.17 The standard enthalpy of combustion of naphthalene is -5157 kJ mol^{-1}. Calculate its standard enthalpy of formation.

2.18 The temperature of a bomb calorimeter rose by 1.617 K when a current of 3.20 A was passed for 27.0 s from a 12.0 V source. Calculate the heat capacity of the calorimeter.

2.19 When 120 mg of naphthalene, $C_{10}H_8(s)$, was burned in a bomb calorimeter the temperature rose by 3.05 K. Calculate the heat capacity of the calorimeter. By how much will the temperature rise when 100 mg of phenol, $C_6H_5OH(s)$, is burned in the calorimeter under the same conditions?

2.20 When 0.3212 g of glucose was burned in a bomb calorimeter of heat capacity 641 J K^{-1} the temperature rose by 7.793 K. Calculate (a) the standard molar enthalpy of combustion, (b) the standard internal energy of combustion, and (c) the standard enthalpy of formation of glucose.

2.21 Calculate the standard enthalpy of solution of AgCl(s) in water from the enthalpies of formation of the solid and the aqueous ions.

2.22 The standard enthalpy of decomposition of the yellow complex NH_3–SO_2 into NH_3 and SO_2 is $+40$ kJ mol^{-1}. Calculate the standard enthalpy of formation of NH_3–SO_2.

2.23 Given that the enthalpy of combustion of graphite is -393.51 kJ mol^{-1} and that of diamond is -395.41 kJ mol^{-1}, calculate the enthalpy of the graphite $\rightarrow$ diamond transition.

2.24 The mass of a typical sugar cube is 1.5 g. Calculate the energy released as heat when a cube is burned in air. To what height could you climb on the energy a cube provides assuming 25 per cent of the energy is available for work?

2.25 The standard enthalpy of combustion of propane gas is -2220 kJ mol^{-1} and the standard enthalpy of vaporization of the liquid is $+15$ kJ mol^{-1}. Calculate (a) the standard enthalpy and (b) the standard internal energy of combustion of the liquid.

2.26 Classify as endothermic or exothermic the following reactions:

(a) $CH_4(g) + 2O_2(g) \rightarrow CO_2(g) + 2H_2O(l)$
    $\Delta H^{\ominus} = -890$ kJ mol^{-1}

(b) $2C(s) + H_2(g) \rightarrow C_2H_2(g)$ $\Delta H^{\ominus} = +227$ kJ mol^{-1}

(c) $NaCl(s) \rightarrow NaCl(aq)$ $\Delta H^{\ominus} = +3.9$ kJ mol^{-1}

2.27 Express the reactions in Exercise 2.26 in the form $0 = \sum_J \nu_J S_J$ and identify the stoichiometric coefficients.

2.28 Use standard enthalpies of formation to calculate the standard enthalpies of the following reactions:

(a) $2NO_2(g) \rightarrow N_2O_4(g)$

(b) $NH_3(g) + HCl(g) \rightarrow NH_4Cl(s)$

(c) Cyclopropane(g) $\rightarrow$ propene(g)

(d) $HCl(aq) + NaOH(aq) \rightarrow NaCl(aq) + H_2O(l)$

2.29 Calculate the standard enthalpy of formation of N_2O_5 from the following data:

$2NO(g) + O_2(g) \rightarrow 2NO_2(g)$ $\Delta H^{\ominus} = -114.1$ kJ mol^{-1}

$4NO_2(g) + O_2(g) \rightarrow 2N_2O_5(g)$ $\Delta H^{\ominus} = -110.2$ kJ mol^{-1}

$N_2(g) + O_2(g) \rightarrow 2NO(g)$ $\Delta H^{\ominus} = +180.5$ kJ mol^{-1}

2.30 Calculate the standard enthalpies of formation of $KClO_3(s)$ from the enthalpy of formation of KCl, $NaHCO_3(s)$ from the enthalpies of formation of CO_2 and NaOH, and NOCl(g) from the enthalpy of formation of NO given in Table 2.10, together with the following information:

$2KClO_3(s) \rightarrow 2KCl(s) + 3O_2(g)$ $\Delta H^{\ominus} = -89.4$ kJ mol^{-1}

$NaOH(s) + CO_2(g) \rightarrow NaHCO_3(s)$ $\Delta H^{\ominus} = -127.5$ kJ mol^{-1}

$2NOCl(g) \rightarrow 2NO(g) + Cl_2(g)$ $\Delta H^{\ominus} = +75.5$ kJ mol^{-1}

2.31 Use the information in Table 2.10 to predict the standard reaction enthalpy of $2NO_2(g) \rightarrow N_2O_4(g)$ at 100°C from its value at 25°C.

2.32 Set up a thermodynamic cycle for determining the enthalpy of hydration of Mg^{2+} ions using the following data: enthalpy of sublimation of Mg, $+167.2$ kJ mol^{-1}; first and second ionization energies of Mg, 7.646 eV and 15.035 eV; dissociation enthalpy of Cl_2, $+241.6$ kJ mol^{-1}; electron affinity of Cl, $+3.78$ eV; enthalpy of solution of $MgCl_2$, -150.5 kJ mol^{-1}; enthalpy of hydration of Cl$^-$, -383.7 kJ mol^{-1}.

Problems

Assume all gases are perfect unless stated otherwise. To two significant figures, 1.0 atm is the same as 1.0 bar. Unless otherwise stated, thermochemical data are for 298 K.

Numerical problems

2.1 A 5.0 g block of solid carbon dioxide is allowed to evaporate in a vessel of volume 100 cm^3 maintained at 20°C. Calculate the work done when the system expands (a) isothermally against a pressure of 1.0 atm, and (b) isothermally and reversibly to the same volume as in (a).

2.2 1.0 mol $CaCO_3$ was heated to 800°C when it decomposed. The heating was carried out in a container fitted with a piston which was initially resting on the solid. Calculate the work done during complete decomposition at 1.0 atm. What work would be done if instead of having a piston the container was open to the atmosphere?

2.3 A sample consisting of 2.0 mol CO_2 occupies a fixed volume of 15.0 L at 300 K. When it is supplied with 2.35 kJ of energy as heat its temperature increases to 341 K. Assume that CO_2 is described by the van der Waals equation of state, and calculate w, ΔU, and ΔH.

2.4 A sample of 70 mmol Kr(g) expands reversibly and isothermally at 373 K from 5.25 cm^3 to 6.29 cm^3, and the internal energy of the sample is known to increase by 83.5 J. Use the virial equation of state up to the second coefficient $B = -28.7$ cm^3 mol^{-1} to calculate w, q, and ΔH for this change of state.

2.5 A piston exerting a pressure of 1.0 atm rests on the surface of water at 100°C. The pressure is reduced infinitesimally, and as a result 10 g of water evaporates and absorbs 22.2 kJ of heat. Calculate w, ΔU, ΔH, and ΔH_m.

2.6 A new fluorocarbon of molar mass 102 g mol^{-1} was placed in an electrically heated vessel. When the pressure was 650 Torr, the liquid boiled at 78°C. After the boiling point had been reached, it was found that a current of 0.232 A from a 12.0 V supply passed for 650 s vaporized 1.871 g of the sample. Calculate the molar enthalpy and molar internal energy of vaporization.

2.7 An object is cooled by the evaporation of liquid methane at its normal boiling point (112 K). What volume of gas at 1.00 atm pressure must be formed from the liquid in order to remove 32.5 kJ of energy as heat from the object?

2.8 The molar heat capacity of ethane is represented in the temperature range 298 K to 400 K by the empirical expression

$$C_p/(\text{J K}^{-1}\,\text{mol}^{-1}) = 14.73 + 0.1272(T/\text{K})$$

The corresponding expressions for C(s) and H$_2$(g) are given in Table 2.16. Calculate the standard enthalpy of formation of ethane at 350 K from its value at 298 K.

2.9 Several characteristics of a hydrocarbon must be considered when it is being considered as a fuel. Among them are the specific enthalpy, the heat evolved per unit mass, for the advantage of a high molar enthalpy of combustion may be eliminated if a large mass of fuel is to be transported. Use the data in Tables 2.9 and 2.10 to calculate the following information on butane, pentane, and octane: (a) the heat output per mole, (b) the heat output per gram.

2.10 Geophysical conditions are sometimes so extreme that quantities neglected in normal laboratory experiments take on an overriding importance. For example, consider the formation of diamond under geophysically typical conditions. The density of graphite is 2.27 g cm^{-3} and that of diamond is 3.52 g cm^{-3} at a certain temperature and 500 kbar. By how much does ΔU differ from ΔH for the graphite → diamond transition?

2.11 A sample of the sugar D-ribose ($C_5H_{10}O_5$) of mass 0.727 g was weighed into a calorimeter and then ignited in the presence of excess oxygen. The temperature rose by 0.910 K. In a separate experiment in the same calorimeter, the combustion of 0.825 g of benzoic acid, for which the internal energy of combustion is -3251 kJ mol^{-1}, gave a temperature rise of 1.940 K. Calculate the internal energy of combustion of D-ribose and its enthalpy of formation.

2.12 The standard enthalpy of formation of the sandwich compound bis(benzene)chromium, in which benzene is the bread and chromium is the meat, was measured in a calorimeter. It was found that for the reaction

$$\text{Cr}(C_6H_6)_2(s) \rightarrow \text{Cr}(s) + 2C_6H_6(g)$$

$\Delta U(583\text{ K}) = +8.0$ kJ mol^{-1}. Find the corresponding reaction enthalpy and estimate the standard enthalpy of formation of the compound at 583 K. The molar heat capacity of benzene is 140 J K^{-1} mol^{-1} in its liquid range and 28 J K^{-1} mol^{-1} as a gas.

2.13 The standard enthalpy of combustion of sucrose is -5645 kJ mol^{-1}. What is the advantage (in kJ mol^{-1} of energy released as heat) of complete aerobic oxidation compared with anaerobic hydrolysis of sucrose to lactic acid?

2.14 In an experiment to measure the enthalpy of solution of KF in glacial acetic acid (J. Emsley, *J. chem. Soc.*, 2702 (1971)), a known mass of the anhydrous salt was added to a known mass of acid in a Dewar flask fitted with a heating coil, stirrer, and thermometer. The experiment was then repeated using the salt KF·CH$_3$COOH. The following is a reconstruction of an experiment (m denotes molality; mass of solvent = 1.000 kg):

(1) KF; heat capacity of calorimeter 4.168 kJ K^{-1}

m/(mol KF/kg CH$_3$COOH)	0.194	0.590	0.821	1.208
ΔT/K	1.592	4.501	5.909	8.115

(2) KF·CH$_3$COOH; heat capacity of calorimeter 4.203 kJ K^{-1}

m/(mol KF/kg CH$_3$COOH)	0.280	0.504	0.910	1.190
ΔT/K	-0.227	-0.432	-0.866	-1.189

Calculate the enthalpies of solvation of the two compounds at these molalities and at infinite dilution. Find the best straight line of the form $\Delta H = a + bm$. Account for the difference in enthalpies of the two compounds.

2.15 The hydrogen bond between F$^-$ and CH$_3$COOH is very strong (J. Emsley, *J. chem. Soc.*, 2702 (1971)) and its strength may be analysed by setting up a Born–Haber cycle with the following data: Lattice enthalpy of KF·CH$_3$COOH 734 kJ mol^{-1}; enthalpy of vaporization of CH$_3$COOH 20.8 kJ mol^{-1}; enthalpy of solution of KF +35.2 kJ mol^{-1}; enthalpy of solution of KF·CH$_3$COOH -3.1 kJ mol^{-1}. Find the energy of the hydrogen bond between F$^-$ and CH$_3$COOH in the gas phase.

Theoretical problems

2.16 In a machine of a particular design, the opposing force acting on a mass m varies as $F \sin(\pi x/a)$. Calculate the work needed to move the mass (a) from $x = 0$ to $x = a$ and (b) from $x = 0$ to $x = 2a$.

2.17 Calculate the work done during the isothermal reversible expansion of a gas that satisfies the virial equation of state

$$\frac{pV_m}{RT} = 1 + \frac{B}{V_m} + \frac{C}{V_m^2} + \ldots$$

Evaluate (a) the work for 1.0 mol Ar at 273 K (for data, see Table 1.2) and (b) the same amount of a perfect gas. Let the expansion be from 500 cm^3 to 1000 cm^3 in each case.

2.18 Calculate the work done during the isothermal reversible expansion of a van der Waals gas. Account physically for the way in which the coefficients a and b appear in the final expression. Plot on the same graph the indicator diagrams for the isothermal reversible expansion of (a) a perfect gas, (b) a van der Waals gas in which $a = 0$ and $b = 5.11 \times 10^{-2}$ mol^{-1}, and (c) $a = 4.2$ L^2 atm mol^{-2} and $b = 0$. (Use other values too if you have a computer available.) The values selected exaggerate the imperfections but give rise to significant effects on the indicator diagrams. Take $V_i = 1.0$ L, $n = 1.0$ mol, and $T = 298$ K.

2.19 Express the work of isothermal reversible expansion of a van der Waals gas in reduced variables and find a definition of 'reduced work' that makes the overall expression independent of the identity of the gas. Calculate the work of isothermal reversible expansion along the critical isotherm from V_c to $x \times V_c$.

2.20 Use (a) the perfect gas equation and (b) the Dieterici equation of state (Table 1.5) to evaluate $(\partial p/\partial T)_V$ and $(\partial p/\partial V)_T$. Go on to confirm that $\partial^2 p/\partial V\, \partial T = \partial^2 p/\partial T\, \partial V$.

2.21 The molar heat capacity of a substance can often be expressed as

$$C_p = a + bT + \frac{c}{T^2}$$

Deduce an expression for the enthalpy of the reaction $0 = \sum_J \nu_J S_J$ at T' in terms of its enthalpy at T and the three coefficients a, b, and c. Estimate the error involved in ignoring the temperature variation of C_p for the formation of H$_2$O(l) at 100°C.

3

The First Law: the machinery

Check-list of key ideas

1. Introduction of the concepts of *exact and inexact differentials* and their significance in thermodynamics (Section 3.1).

2. The derivation of an expression for the *variation of internal energy with temperature* when the pressure is held constant (eqn 5).

3. The *Joule experiment* to measure the variation of the internal energy of a gas with volume (Section 3.1).

4. The derivation of an expression for the *variation of enthalpy with temperature* when the volume is held constant (eqn 11).

5. The *Joule-Thomson experiment* to measure the change in temperature of a gas when it expands adiabatically (Section 3.2).

6. The relation between the heat capacities at constant volume and constant pressure for any substance (eqn 13) and for the special case of a perfect gas.

7. The derivation of an expression for the *change in temperature* when a perfect gas expands adiabatically either irreversibly (eqn 16) or reversibly (eqn 17).

8. The derivation of perfect gas *adiabats* showing how the pressure and volume of a perfect gas are related when it undergoes reversible adiabatic expansion (eqn 19).

9. A summary of the different types of change that a gas may undergo and the work done during each of them (Section 3.5).

This chapter develops the concepts introduced in Chapter 2. We shall see how the concept of 'state property', which has already been introduced in connection with the internal energy and enthalpy, can be developed into a very powerful calculational device. We shall also develop ways of manipulating thermodynamic expressions and show how they can relate apparently unrelated experimental quantities. In particular, we shall see that one of the powerful aspects of thermodynamics is that a property can be measured indirectly by measuring others and then combining their values.

State functions and differentials

Properties that are independent of how the sample was prepared are called **state functions**. The pressure, internal energy, and heat capacity are examples, for they depend on the current state of the system and not its previous history. Properties that relate to the preparation of the state are called **path functions**. Examples include the work that is done in preparing a state or the energy transferred as heat. We do not speak of a system in a particular state as possessing work or heat. In each case, the energy transferred as work or heat relates to the path being taken, not the current state itself.

3.1   State functions

We can begin to see the importance of the distinction between state and path functions by considering its implications for the First Law.

Consider a system undergoing the changes depicted in Fig. 3.1. The initial state of the system is p_i, V_i, T_i (together with whatever variables are needed to specify its composition), and in this state the internal energy is U_i. Work is done on the system to change it adiabatically to a state p_f, V_f, T_f. In this state it has an internal energy U_f and the work done on the system as it traversed the path from i to f is w. Notice our use of language: U is a property of the state; w is a property of the path. Now consider another process in which the initial and final states are the same but in which the compression is not adiabatic. The internal energy of both the initial and final states are the same as before (because U is a state function). However, in the second path an energy q' enters the system as heat and the work w' is not the same as w. The work and the heat are path functions.

Exact and inexact differentials

If a system is taken along a path (e.g. by compressing it isothermally), U changes from U_i to U_f, and the overall change is the sum of all the infinitesimal changes along the path:

$$\Delta U = \int_i^f dU = U_f - U_i$$

The value of ΔU depends on the initial and final states but is independent of the path between them. We express this path independence of the integral by saying that dU is an **exact differential**.

If a system is heated, the total energy transferred as heat is the sum of all

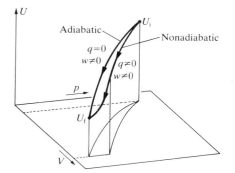

Fig. 3.1 As the pressure and volume of a system are changed (as indicated by the paths in the p, V plane), the internal energy changes (vertical axis). An adiabatic and a non-adiabatic path are shown: they correspond to different values of q and w but to the same value of ΔU. (The paths on the plane can be regarded as indicator diagrams for the process, the work in each case being equal to the area beneath them and the V-axis, as explained in Chapter 2.)

individual contributions at each point of the path:

$$q = \int_{i,\,path}^{f} dq$$

Notice the difference between this and the preceding equation. First, we do not write Δq, because q is not a state function and the energy supplied as heat cannot be expressed as $q_f - q_i$. Secondly, we must specify the path of integration because q depends on the path selected (an adiabatic path has $q = 0$, a nonadiabatic path between the same two states would have $q \neq 0$). We express this path dependence by saying that dq is an **inexact differential** ('inexact' because to integrate it we must supply further information, the specification of the path). Often dq is written $đq$ to emphasize that it is inexact in this sense.

We know that the work done depends on the path selected; therefore we know immediately that dw is an inexact differential. It is often written $đw$.

Example 3.1: *Calculating work, heat, and internal energy*

Consider a perfect gas inside a cylinder fitted with a piston. Let the initial state be T, V_i and the final state be T, V_f. The change of state can be brought about in many ways, of which the two simplest are the following. *Path 1*, in which there is free, irreversible expansion against zero external pressure. *Path 2*, in which there is reversible, isothermal expansion. Calculate w, q, and ΔU for each process. We need to know that the internal energy of a perfect gas is independent of its volume (a result we prove later).

Answer. Since the internal energy of a perfect gas is independent of volume for an isothermal process, $\Delta U = 0$ for both paths. Since $\Delta U = q + w$, in each case $q = -w$. The work of irreversible expansion is zero (since $p_{ex} = 0$, Section 2.4); so in *Path 1*, $w = 0$ and $q = 0$. For *Path 2*, the work is given by eqn 7 of Section 2.4, so $w = -nRT \ln (V_f/V_i)$ and $q = nRT \ln (V_f/V_i)$.

Exercise. Calculate the values of q, w, and ΔU for an irreversible isothermal expansion against a constant non-zero external pressure.

$$[q = +p_{ex}\Delta V, \ w = -p_{ex}\Delta V, \ \Delta U = 0]$$

Changes in internal energy

We shall now begin to unfold the consequences of dU being an exact differential by noting that for a closed system of constant composition (the only type we consider in this chapter), U is a function of volume and temperature.[1] When V changes to $V + dV$ at constant temperature, U changes to

$$U' = U + \left(\frac{\partial U}{\partial V}\right)_T dV$$

The coefficient $(\partial U/\partial V)_T$, the slope of U with respect to V at constant temperature, is the partial derivative of U with respect to V. If instead, T

---

[1] U can be regarded as a function of V, T, and p; but since there is an equation of state it is possible to express p in terms of V and T, and so p is not an independent variable. We could choose p, T or p, V as independent variables, but V, T fit our purpose.

changes to $T + dT$ at constant volume, the internal energy changes to

$$U' = U + \left(\frac{\partial U}{\partial T}\right)_V dT$$

Now suppose that *both* V and T change infinitesimally. The new internal energy, neglecting second-order infinitesimals, is

$$U' = U + \left(\frac{\partial U}{\partial V}\right)_T dV + \left(\frac{\partial U}{\partial T}\right)_V dT$$

The internal energy U' differs from U by the infinitesimal amount dU. Therefore, from the last equation we obtain the very important result that

$$dU = \left(\frac{\partial U}{\partial V}\right)_T dV + \left(\frac{\partial U}{\partial T}\right)_V dT \qquad (1)$$

The interpretation of this equation is that in a closed system of constant composition, any infinitesimal change in the internal energy is proportional to the infinitesimal changes of volume and temperature, the coefficients of proportionality being the partial derivatives.

Most of the partial derivatives we meet have an easily identifiable physical meaning, and thermodynamics gets shapeless and difficult only when that meaning is not kept in sight. In the present case, we have already met $(\partial U/\partial T)_V$ in Chapter 2: it is the heat capacity at constant volume C_V:

$$C_V = \left(\frac{\partial U}{\partial T}\right)_V$$

Hence

$$dU = \left(\frac{\partial U}{\partial V}\right)_T dV + C_V dT$$

The coefficient $(\partial U/\partial V)_T$ plays a major role in thermodynamics since it is a measure of the dependence of the internal energy of a substance on the volume it occupies. We shall denote it π_T:

$$\pi_T = \left(\frac{\partial U}{\partial V}\right)_T \qquad (2)$$

Then

$$dU = \pi_T dV + C_V dT \qquad (3)$$

If the internal energy increases ($dU > 0$) as the sample is compressed isothermally ($dV < 0$), $\pi_T < 0$. If there are no interactions between the molecules, the internal energy should be independent of their separation and hence independent of the volume the sample occupies. This suggests that for a perfect gas

$$\pi_T = \left(\frac{\partial U}{\partial V}\right)_T = 0$$

In fact, it is sometimes convenient to regard the statement $\pi_T = 0$ as the *definition* of a perfect gas, for later we shall see that it implies the equation of state $pV = nRT$. Later we shall also see that for a van der Waals gas,

$$\pi_T = \frac{a}{V_m^2}$$

This shows that π_T can be identified as the internal pressure of the gas (Section 1.4).

Example 3.2: *Estimating a change in internal energy*

From the van der Waals equation for ammonia, $\pi_T = 840 \text{ J m}^{-3} \text{ mol}^{-1}$ at 300 K and $C_V = 27.32 \text{ J K}^{-1} \text{ mol}^{-1}$. What is the change of molar internal energy of ammonia when it is heated through 2.0 K and compressed through 100 cm³?

Answer. Infinitesimal changes in volume and temperature result in an infinitesimal change of internal energy as given by eqn 3. Since the changes in the present problem are small, we may write

$$\Delta U \approx \pi_T \Delta V + C_V \Delta T$$
$$\approx -0.084 \text{ J mol}^{-1} + 55 \text{ J mol}^{-1} = 55 \text{ J mol}^{-1}$$

Comment. Note that the change is dominated by the effect of temperature. If ammonia behaved as a perfect gas, π_T would be zero and the volume change would have no effect on U.

Exercise. Confirm that π_T has the same dimensions as pressure, and express the value for ammonia in atmospheres. [8.3×10^{-3} atm]

The Joule experiment

James Joule thought that he could measure π_T by observing the change in temperature of a gas when it is allowed to expand into a vacuum. He used two metal vessels immersed in a water bath (Fig. 3.2). One was filled with air at 22 atm, the other was evacuated. He then tried to measure the change in temperature of the water of the bath when a stopcock was opened and the air expanded into a vacuum. He observed no change.

The thermodynamic implications of the experiment are as follows. No work was done in the expansion into a vacuum, and so $w = 0$. No heat entered or left the system (the gas) because the temperature of the bath did not change, so $q = 0$. Consequently, within the accuracy of the experiment, $\Delta U = 0$. It follows that U does not change much when a gas expands isothermally. However, Joule's experiment was crude, for the heat capacity of the water bath and the metal vessels was so large that the temperature change that gases do in fact cause was too small to measure. His experiment was on a par with Boyle's: he extracted an essential limiting property of a gas, a property of a perfect gas, without detecting the small deviations characteristic of real gases.

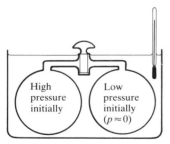

Fig. 3.2 A schematic diagram of the apparatus used by Joule in an attempt to measure the change in internal energy when a gas expands isothermally. The heat absorbed by the gas is proportional to the change in temperature of the bath.

Changes in internal energy at constant pressure

Partial derivatives have many useful properties (Box 3.1) which we shall draw on frequently. Skilful use of them can often turn some unfamiliar quantity into something we can recognize and interpret. As an example, suppose we want to find out how the internal energy depends on the temperature when the *pressure* of the system is kept constant. Then we can

---

Box 3.1   Relations between partial derivatives

If f is a function of x and y, then when x and y change by dx and dy, f changes by

$$df = \left(\frac{\partial f}{\partial x}\right)_y dx + \left(\frac{\partial f}{\partial y}\right)_x dy$$

Partial derivatives may be taken in any order:

$$\frac{\partial^2 f}{\partial x \, \partial y} = \frac{\partial^2 f}{\partial y \, \partial x}$$

In the following, z is a variable on which x and y depend (for example, x, y, and z might correspond to p, V, and T).

Relation No. 1. When x is changed at constant z:

$$\left(\frac{\partial f}{\partial x}\right)_z = \left(\frac{\partial f}{\partial x}\right)_y + \left(\frac{\partial f}{\partial y}\right)_x \left(\frac{\partial y}{\partial x}\right)_z$$

Relation No. 2 (the Inverter).

$$\left(\frac{\partial x}{\partial y}\right)_z = \frac{1}{\left(\dfrac{\partial y}{\partial x}\right)_z}$$

Relation No. 3 (the Permuter).

$$\left(\frac{\partial x}{\partial y}\right)_z = -\left(\frac{\partial x}{\partial z}\right)_y \left(\frac{\partial z}{\partial y}\right)_x$$

By combining this and Relation No. 2 we obtain *Euler's chain relation*:

$$\left(\frac{\partial x}{\partial y}\right)_z \left(\frac{\partial y}{\partial z}\right)_x \left(\frac{\partial z}{\partial x}\right)_y = -1$$

Relation No. 4. This relation establishes whether or not df is an exact differential.

$$df = g \, dx + h \, dy \text{ is exact if } \left(\frac{\partial g}{\partial y}\right)_x = \left(\frac{\partial h}{\partial x}\right)_y$$

If df is exact, its integral between specified limits is independent of the path.

---

use the relations in Box 3.1 to extract an expression for $(\partial U/\partial T)_p$ from eqn 3.

Relation No. 1 takes the equation, divides through by dT (to give dU/dT on the left), and then imposes the condition of constant pressure. It yields

$$\left(\frac{\partial U}{\partial T}\right)_p = \pi_T \left(\frac{\partial V}{\partial T}\right)_p + C_V$$

It is usually sensible in thermodynamics to inspect the output of a manipulation like this to see if it contains any recognizable physical quantity. The differential coefficient on the right in this expression is the change of volume with increase of temperature (at constant pressure). This is a readily accessible property which is normally tabulated as the **expansion coefficient** α, defined as

$$\alpha = \frac{1}{V}\left(\frac{\partial V}{\partial T}\right)_p \tag{4}$$

That is, α is the rate of change of the volume with temperature per unit volume.

Example 3.3: *Calculating the expansion coefficient of a gas*

Calculate the volume change that occurs when $50 \, \text{cm}^3$ of neon, treated as a perfect gas, is heated through $5.0 \, \text{K}$ at $298 \, \text{K}$.

Answer. If neon behaves as a perfect gas, we can calculate α by substituting $pV = nRT$ into eqn 4, which gives

$$\alpha = \frac{1}{V}\left(\frac{\partial}{\partial T}\frac{nRT}{p}\right)_p = \frac{nR}{pV} = \frac{1}{T}$$

The increase in volume when the temperature is raised by the small amount ΔT is therefore

$$\Delta V \approx \left(\frac{\partial V}{\partial T}\right)_p \times \Delta T = \alpha V \Delta T = \frac{V \Delta T}{T}$$

Substitution of the data gives

$$\Delta V \approx \frac{50\ cm^3 \times 5.0\ K}{298\ K} = 0.84\ cm^3$$

Exercise. For copper, $\alpha = 5.01 \times 10^{-5}\ K^{-1}$; calculate the change in volume that occurs when a 50 cm³ block is heated through 5.0 K. [12 mm³]

Introduction of the general definition of α into the equation for $(\partial U/\partial T)_p$ gives

$$\left(\frac{\partial U}{\partial T}\right)_p = \alpha \pi_T V + C_V \qquad (5a)$$

This equation is entirely general (so long as the system is closed and its composition constant). It expresses the dependence of the internal energy on the temperature at constant pressure in terms of C_V, which can be measured in one experiment, α, which can be measured in another (Table 3.1), and the ubiquitous quantity $\pi_T = (\partial U/\partial V)_T$. For a perfect gas, $\pi_T = 0$, so

$$\left(\frac{\partial U}{\partial T}\right)_p = C_V \qquad (5b)°$$

That is, the constant-volume heat capacity of a perfect gas is equal to the slope of internal energy with temperature at constant pressure as well as, by definition, to its slope at constant volume.

Table 3.1. Expansion coefficients α and isothermal compressibilities κ_T

Substance	$\alpha/(10^{-4}\ K^{-1})$	$\kappa_T/(10^{-6}\ atm^{-1})$
Benzene	12.4	92.1
Diamond	0.030	0.187
Lead	0.861	2.21
Water	2.1	49.6

3.2 The temperature dependence of the enthalpy

We can carry out a similar set of operations on the enthalpy H

$$H = U + pV$$

U, p, and V are all state functions; therefore H is also a state function and hence dH is an exact differential. If we regard H as a function of p and T, by the same argument as for U (but with p in place of V) we find that for a closed system of constant composition,

$$dH = \left(\frac{\partial H}{\partial p}\right)_T dp + \left(\frac{\partial H}{\partial T}\right)_p dT \qquad (6a)$$

The second coefficient is the constant-pressure heat capacity

$$C_p = \left(\frac{\partial H}{\partial T}\right)_p$$

Therefore,

$$dH = \left(\frac{\partial H}{\partial p}\right)_T dp + C_p\, dT \qquad (6b)$$

The variation of the enthalpy at constant volume

So far, we know how U varies with temperature at constant pressure and constant volume. We also know how H varies with temperature at constant pressure (this is the constant-pressure heat capacity). The only missing variation is that of H with temperature at constant volume, $(\partial H/\partial T)_V$. We can obtain this coefficient from the last equation by repeating the development that we have just carried through for U. Indeed, it is often helpful to keep in mind when manipulating U and H that V and p play analogous roles in the two functions.

First, we divide eqn 6b through by dT and impose constant volume:

$$\left(\frac{\partial H}{\partial T}\right)_V = \left(\frac{\partial H}{\partial p}\right)_T \left(\frac{\partial p}{\partial T}\right)_V + C_p$$

The third differential coefficient looks like something we ought to recognize, and is perhaps related to $(\partial V/\partial T)_p$, the expansion coefficient. Relation No. 3 shuffles p, V, and T around inside partial differentials, and acting on $(\partial p/\partial T)_V$ it produces

$$\left(\frac{\partial p}{\partial T}\right)_V = \frac{-1}{\left(\frac{\partial T}{\partial V}\right)_p \left(\frac{\partial V}{\partial p}\right)_T}$$

Unfortunately, $(\partial T/\partial V)_p$ occurs instead of $(\partial V/\partial T)_p$, but relation No. 2 inverts partial differentials, and leads to

$$\left(\frac{\partial p}{\partial T}\right)_V = -\frac{\left(\frac{\partial V}{\partial T}\right)_p}{\left(\frac{\partial V}{\partial p}\right)_T} = -\frac{\alpha V}{\left(\frac{\partial V}{\partial p}\right)_T}$$

The coefficient $(\partial V/\partial p)_T$ is a measure of the change of volume under the influence of pressure at constant temperature. It is normally reported in terms of the **isothermal compressibility** κ_T, defined as

$$\kappa_T = -\frac{1}{V}\left(\frac{\partial V}{\partial p}\right)_T \tag{7}$$

(The negative sign ensures that κ_T is positive, because an increase of pressure, positive dp, brings about a reduction of volume, negative dV). Some values of κ_T are listed in Table 3.1; for a perfect gas,

$$\kappa_T = \frac{1}{p}$$

and the higher the pressure of the gas, the lower its compressibility.

Example 3.4: *Using the isothermal compressibility*
The isothermal compressibility of water at 20°C and 1 atm is $4.96 \times 10^{-5}\,\text{atm}^{-1}$. What change of volume occurs when a $50\,\text{cm}^3$ sample is subjected to an additional 1000 atm pressure?

Answer. For an infinitesimal change of pressure, the volume changes by $dV = (\partial V/\partial p)_T\, dp = -\kappa_T V\, dp$. Therefore, for the specified change, integrate both sides and obtain

$$\Delta V = -\int \kappa_T V\, dp$$

If V and κ_T are effectively constant over the pressure range,

$$\Delta V = -\kappa_T V \int dp = -\kappa_T V\, \Delta p = -2.5\ \text{cm}^3$$

Comment. The assumption of constant V and κ_T is probably marginally acceptable. Note that very high pressures are needed to bring about significant changes of volume.

Exercise. A $50\ \text{cm}^3$ sample of copper is subjected to an additional pressure of $100\ \text{atm}$ and a temperature increase of $5.0\ \text{K}$. Estimate the total change in volume.   [$8.8\ \text{mm}^3$]

If we now collect all the fragments of the preceding calculation, we find that

$$\left(\frac{\partial H}{\partial T}\right)_V = \frac{\alpha}{\kappa_T}\left(\frac{\partial H}{\partial p}\right)_T + C_p \tag{8}$$

The coefficient $(\partial H/\partial p)_T$ is the analogue of $(\partial U/\partial V)_T$. We might suspect that it can be measured in a similar way. The next few paragraphs explore this point.

First, we change $(\partial H/\partial p)_T$ into something recognizable. Using Relation No. 3 we find

$$\left(\frac{\partial H}{\partial p}\right)_T = \frac{-1}{\left(\frac{\partial p}{\partial T}\right)_H \left(\frac{\partial T}{\partial H}\right)_p}.$$

A double use of the inverter, Relation No. 2, then gives

$$\left(\frac{\partial H}{\partial p}\right)_T = -\left(\frac{\partial T}{\partial p}\right)_H \left(\frac{\partial H}{\partial T}\right)_p = -\left(\frac{\partial T}{\partial p}\right)_H C_p$$

We seem to be on the right track because one term generated is the heat capacity C_p, and we have not generated something more complicated than we started with (an occupational hazard in thermodynamics: when it happens be prepared to start again). The other coefficient in the equation is called the **Joule–Thomson coefficient** μ:

$$\mu = \left(\frac{\partial T}{\partial p}\right)_H \tag{9}$$

We see how to measure it shortly. Hence

$$\left(\frac{\partial H}{\partial p}\right)_T = -\mu C_p \tag{10}$$

It follows that

$$\left(\frac{\partial H}{\partial T}\right)_V = \left(1 - \frac{\alpha\mu}{\kappa_T}\right)C_p \qquad (11)$$

This is the final equation for $(\partial H/\partial T)_V$. It applies to any substance. Since all the quantities that appear in it can be measured in suitable experiments, we now know how H varies with T.

The Joule–Thomson effect

We need to see how to measure $(\partial T/\partial p)_H$, the change in temperature accompanying a change of pressure at constant enthalpy. The cunning required to impose that constraint was supplied by Joule and William Thomson (later Lord Kelvin), who not only solved this problem but found a way of making the experiment very sensitive. They recognized that the fault with Joule's original experiment was that the heat capacity of the apparatus was too great, which resulted in the temperature changes being unmeasurably small.

Joule and Thomson had the good idea of using the gas (with its low heat capacity) as its own heat bath and to establish a steady state flow so that the effect was easier to measure (Fig. 3.3). They let the gas expand through a throttle from one constant pressure to another and monitored the difference in temperature. The whole apparatus was insulated so that the process was adiabatic. They observed a lower temperature on the low pressure side, the difference being proportional to the pressure difference they maintained. This cooling by adiabatic expansion is now called the **Joule–Thomson effect**.

The thermodynamic analysis of the experiment depends on the system being adiabatic, so $q = 0$. In order to calculate the work done as the gas passes through the throttle, consider the passage of a fixed amount from the high pressure side, where the pressure is p_i, the temperature T_i, and the gas occupies a volume V_i. It emerges on the low pressure side, where the same amount of gas has a pressure p_f and a temperature T_f and occupies a volume V_f. The gas on the left (Fig. 3.4) is compressed isothermally by the upstream

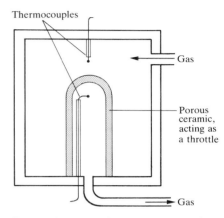

Fig. 3.3 A diagram of the apparatus used for measuring the Joule–Thomson effect. The gas expands through the throttle, and the whole apparatus is thermally insulated. As explained in the text, this arrangement corresponds to an isenthalpic expansion (expansion at constant enthalpy). Whether the expansion results in a heating or a cooling of the gas depends on the conditions.

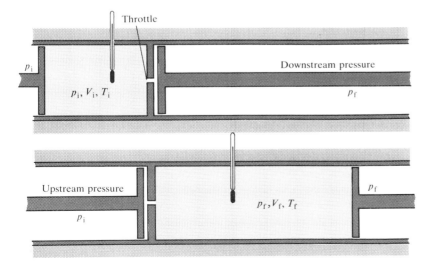

Fig. 3.4 A diagram representing the thermodynamic basis of Joule–Thomson expansion. The pistons represent the upstream and downstream gases, which maintain constant pressures either side of the throttle. The transition from the upper diagram to the lower, which represent the passage of a given amount of gas through the throttle, occurs without change of enthalpy.

Fig. 3.5 A schematic diagram of the apparatus used for measuring the isothermal Joule–Thomson coefficient. The electrical heating required to offset the cooling arising from expansion is interpreted as ΔH and used to calculate $(\partial H/\partial p)_T$, which is then converted to μ as explained in the text.

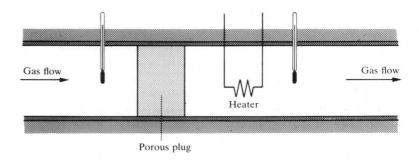

gas acting as a piston. The relevant pressure is p_i and the volume changes from V_i to 0; therefore, the work done on the gas is $-p_i(0 - V_i)$, or $p_i V_i$. The gas expands isothermally on the right of the throttle (but possibly at a different constant temperature) against the pressure p_f provided by the downstream gas acting as a piston to be driven out. The volume changes from 0 to V_f, so the work done on the gas in this stage is $-p_f(V_f - 0)$, or $-p_f V_f$. The total work done on the gas is the sum of these two quantities, or $p_i V_i - p_f V_f$. It follows that the change in internal energy of the gas as it moves from one side of the throttle to the other is

$$U_f - U_i = w = p_i V_i - p_f V_f$$

Reorganizing this expression gives

$$U_f + p_f V_f = U_i + p_i V_i, \quad \text{or} \quad H_f = H_i$$

Table 3.2. Inversion temperatures, normal freezing and boiling points, and Joule–Thomson coefficients at 1 atm and 298 K

	T_i/K	T_f/K	T_b/K	$\mu/(\text{K atm}^{-1})$
Ar	723	83.8	87.3	
CO_2	1500		194.7	1.11
He	40		4.2	−0.060
N_2	621	63.3	77.4	0.25

Therefore, the expansion occurs without change of enthalpy: it is **isenthalpic**. The quantity observed is the temperature change per unit change of pressure, $\Delta T/\Delta p$. Adding the constraint of constant enthalpy and taking the limit of small p implies that the thermodynamic quantity measured is $(\partial T/\partial p)_H$.

The modern method of measuring μ is indirect, and involves measuring the **isothermal Joule–Thomson coefficient**, the quantity $(\partial H/\partial p)_T$. The two coefficients are related by eqn 10. The procedure is to pump the gas continuously at a steady pressure through a heat exchanger (which brings it to the required temperature), and then through a throttle inside a thermally insulated container. The steep pressure drop is measured, and the cooling effect is exactly offset by an electric heater placed immediately after the throttle (Fig. 3.5). The energy provided by the heater is monitored. Since the heat can be identified with the value of ΔH for the gas, and the pressure change Δp is known, $(\partial H/\partial p)_T$ can be obtained from the limiting value of $\Delta H/\Delta p$ as $p \to 0$, and then converted to μ. Some values obtained in this way are listed in Table 3.2.

Real gases have non-zero Joule–Thomson coefficients in general (even in the limit of zero pressure) and the sign of the coefficient may be either positive or negative. A positive sign implies that dT is negative when dp is negative, in which case the gas cools on expansion. The sign and magnitude of μ depend on the gas and the conditions. Gases that show a heating effect ($\mu < 0$) show a cooling effect ($\mu > 0$) when the temperature is lowered to below their **inversion temperature**, T_I (Table 3.2).

The technological importance of the Joule–Thomson effect lies in its application to the cooling and liquefaction of gases. The **Linde refrigerator** works on the principle that below its inversion temperature a gas cools on expansion. It is then essential to know the conditions under which μ is positive: this is illustrated for three gases in Fig. 3.6. The principle is illustrated in Fig. 3.7: after recirculation of the cooling gas, and for a big enough pressure drop across the throttle, the temperature falls below the condensation temperature, and the liquid forms. Note the importance of working beneath the inversion temperature, and therefore the need to cool some gases by other means initially: using helium at room temperature would turn the refrigerator into an expensive oven.

For a perfect gas, $\mu = 0$ and its temperature is unchanged on Joule–Thomson expansion. This points clearly to the involvement of intermolecular forces in determining the size of the effect. However, the Joule–Thomson coefficient of a real gas does not necessarily approach zero as the pressure is reduced: the coefficient is an example of one of the properties mentioned in Section 1.3 that depend on derivatives and not on p, V, and T themselves.

3.3 The relation between C_V and C_p

The constant pressure heat capacity C_p differs from the constant volume heat capacity C_V by the work needed to change the volume of the system when the pressure is held constant. This work arises in two ways. One is the work of driving back the atmosphere, the other is the work of stretching the bonds in the material, including the weak intermolecular interactions. In the case of a perfect gas, the second makes no contribution but the first contributes whatever the identity of the gas. We shall now derive a general relation between the two heat capacities, and show that it reduces to the perfect gas result in the absence of intermolecular forces.

The relation for a perfect gas

First, we carry through the calculation for a perfect gas. In this special case, we can use eqn 5b to express both heat capacities in terms of derivatives at constant pressure:

$$C_p - C_V = \left(\frac{\partial H}{\partial T}\right)_p - \left(\frac{\partial U}{\partial T}\right)_p$$

Then we introduce

$$H = U + pV = U + nRT$$

into the first term, which results in

$$C_p - C_V = \left(\frac{\partial U}{\partial T}\right)_p + nR - \left(\frac{\partial U}{\partial T}\right)_p$$
$$= nR$$

This is the result we quoted in Section 2.8.

The general case

When doing a problem in thermodynamics a useful rule is to go back to first principles. In the present problem we do this twice, first by expressing C_p and C_V in terms of their definitions:

$$C_p - C_V = \left(\frac{\partial H}{\partial T}\right)_p - \left(\frac{\partial U}{\partial T}\right)_V$$

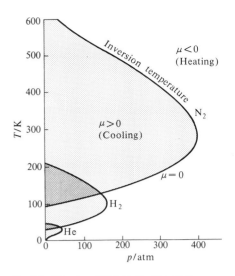

Fig. 3.6 The sign of the Joule–Thomson coefficient μ depends on the conditions. Inside the boundary, the shaded areas, it is positive and outside it is negative. The temperature corresponding to the boundary at a given pressure is the 'inversion temperature' of the gas at that pressure. For a given pressure, the temperature must be below a certain value if cooling is required, but if it becomes too low, the boundary is crossed again and heating occurs.

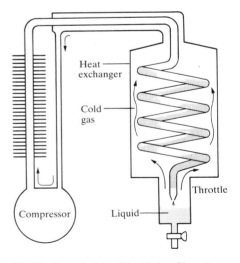

Fig. 3.7 The principle of the Linde refrigerator is shown in this diagram. The gas is recirculated, and so long as it is beneath its inversion temperature it cools on expansion through the throttle. The cooled gas cools the high-pressure gas, which cools still further as it expands. Eventually liquefied gas drips from the throttle.

and then by inserting the definition $H = U + pV$:

$$C_p - C_V = \left(\frac{\partial U}{\partial T}\right)_p + \left(\frac{\partial pV}{\partial T}\right)_p - \left(\frac{\partial U}{\partial T}\right)_V$$

We have already calculated the difference of the first and third quantities (eqn 5):

$$\left(\frac{\partial U}{\partial T}\right)_p - \left(\frac{\partial U}{\partial T}\right)_V = \alpha \pi_T V$$

αV gives the change of volume when the temperature is raised, and $(\partial U/\partial V)_T$ converts this change in volume into a change in internal energy. We can simplify the remaining term by noting that, since p is constant,

$$\left(\frac{\partial pV}{\partial T}\right)_p = p\left(\frac{\partial V}{\partial T}\right)_p = \alpha pV$$

The middle term of this expression identifies it as the contribution to the work of pushing back the atmosphere: $(\partial V/\partial T)_p$ gives the change in volume caused by a change of temperature, and multiplication by p converts this volume change into work.

Collecting the two contributions gives

$$C_p - C_V = \alpha(p + \pi_T)V \tag{12}$$

This is an entirely general relation, applicable to all materials. We can quickly check that it reduces to the perfect gas expression by using $\alpha = 1/T$ and $\pi_T = 0$.

At this point we can go further using the result we prove in Chapter 5 that

$$\pi_T = T\left(\frac{\partial p}{\partial T}\right)_V - p$$

When this is inserted in eqn 12 we obtain

$$C_p - C_V = \alpha TV\left(\frac{\partial p}{\partial T}\right)_V$$

The same coefficient was encountered in the work leading to eqn 7; therefore

$$C_p - C_V = \frac{\alpha^2 TV}{\kappa_T} \tag{13}$$

This is a **thermodynamic expression**, which means that it applies to any substance (i.e. that it is 'universally true'). Setting $\alpha = 1/T$ and $\kappa_T = 1/p$ results in the perfect gas expression. For the relation between molar heat capacities, the V on the right hand side of eqn 13 is replaced by the molar volume.

Since thermal expansivities α of liquids and solids are small, it is tempting to deduce from eqn 13 that for them $C_p \approx C_V$. But this is not always so, because the compressibility κ_T might also be small and so α^2/κ_T might be large. For example, substituting the data for water at 25°C gives $C_p = 75.3\,\text{J K}^{-1}\,\text{mol}^{-1}$ compared with $C_V = 74.8\,\text{J K}^{-1}\,\text{mol}^{-1}$. In some cases, in fact, the two heat capacities differ by as much as 30 per cent.

Example 3.5: *Evaluating the difference between C_p and C_V*

Estimate the difference between C_p and C_V for carbon tetrachloride at 25°C, when $C_p = 132 \, \text{J K}^{-1} \, \text{mol}^{-1}$. At this temperature, its density ρ is $1.59 \, \text{g cm}^{-3}$, its expansion coefficient is $1.24 \times 10^{-3} \, \text{K}^{-1}$, and its isothermal compressibility is $9.05 \times 10^{-5} \, \text{atm}^{-1}$.

Answer. Since, for a pure liquid of molar mass M, the molar volume and the density are related by

$$V_m = \frac{M}{\rho}$$

eqn 13 becomes

$$C_p - C_V = \frac{\alpha^2 T M}{\rho \kappa_T}$$

The compressibility converts to

$$\kappa_T = \frac{9.05 \times 10^{-5} \, \text{atm}^{-1}}{1.013 \times 10^5 \, \text{Pa atm}^{-1}} = 8.93 \times 10^{-10} \, \text{Pa}^{-1}$$

Substitution of the data gives

$$C_p - C_V = \frac{(1.24 \times 10^{-3} \, \text{K}^{-1})^2 \times 298 \, \text{K} \times (153.82 \, \text{g mol}^{-1})}{1.59 \, \text{g cm}^{-3} \times 8.93 \times 10^{-10} \, \text{Pa}^{-1}}$$

$$= 4.96 \times 10^7 \, \text{Pa cm}^3 \, \text{K}^{-1} \, \text{mol}^{-1}$$

$$= 4.96 \times 10 \, \text{Pa m}^3 \, \text{K}^{-1} \, \text{mol}^{-1} = 49.6 \, \text{J K}^{-1} \, \text{mol}^{-1}$$

because $1 \, \text{Pa m}^3 = 1 \, \text{N m} = 1 \, \text{J}$.

Comment. The difference in heat capacities is 38 per cent of C_p itself.

Exercise. Repeat the calculation for benzene, for which $\rho = 0.88 \, \text{g cm}^{-3}$, $\alpha = 1.24 \times 10^{-3} \, \text{K}^{-1}$, and $\kappa_T = 9.21 \times 10^{-5} \, \text{atm}^{-1}$.　　　　　$[45 \, \text{J K}^{-1} \, \text{mol}^{-1}]$

Work of adiabatic expansion

In Section 2.4 we calculated the work of isothermal expansion of a perfect gas. We are now equipped to deal with another type of work, the work of adiabatic expansion. The approach we use illustrates a useful rule in thermodynamics, that when calculating a property it may be possible to find a state function related to it, and to calculate the change in that function by the most convenient path. In the present case, because the change is adiabatic, $dq = 0$ at each stage of the expansion. Consequently, $dU = dw$. Therefore, instead of calculating the work done during the expansion, we can calculate the change in internal energy between the same initial and final states:

$$w = \int_i^f dU$$

The equation just given applies to any closed, adiabatic system. The remainder of this section specializes to the case of a perfect gas. This allows us to relate dU to the change in volume in a simple way, since we can use eqn 3 with $\pi_T = 0$:

$$w = \int_i^f C_V \, dT \qquad\qquad (14)°$$

For many gases, C_V is almost independent of temperature and so the integration is very simple:

$$w = C_V \int_{T_i}^{T_f} \mathrm{d}T = C_V(T_f - T_i) = C_V \, \Delta T \qquad (15)°$$

Hence, the work done during an adiabatic expansion is proportional to the temperature difference between the initial and final states. This equation applies to any adiabatic expansion or contraction of a perfect gas, reversible or irreversible, so long as the expansion is quasistatic and the surroundings are in internal equilibrium.

3.4 Special cases

A general conclusion from eqn 15 is that if $w < 0$ (so the system has done work), then $\Delta T < 0$ regardless of whether the change was reversible or irreversible. This should not be surprising: since no heat can enter when an adiabatic system does work, the internal energy must fall. However, because the internal energy of a perfect gas is unaffected by changes of volume alone, a reduction in internal energy must mean that the temperature has fallen too.

Irreversible adiabatic expansion

If the perfect gas expands against zero external pressure, it does no work. Therefore, $w = 0$, and consequently $\Delta T = 0$. This is a peculiar case: the expansion is simultaneously adiabatic and isothermal.

If the expansion occurs against fixed external pressure then the work done is $w = -p_{ex} \Delta V$. This is a general result, and applies equally to isothermal and adiabatic changes. Since this work must be the same as the work calculated from eqn 15 for a perfect gas, we can find the change of temperature that accompanies this irreversible adiabatic expansion:

$$\Delta T = \frac{-p_{ex} \Delta V}{C_V} \qquad (16)°$$

Notice that on expansion $\Delta V > 0$ so $\Delta T < 0$ and the temperature falls. The signs are looking after themselves, but it is important to learn their language. Also notice that if the external pressure is zero, the temperature does not change when the expansion occurs: this accords with the first case we considered. (It is always sensible to check an equation by seeing whether it reduces to a simpler, known result.)

Example 3.6: *Calculating the final temperature of a gas*

A sample of 2.0 mol Ar in a cylinder of 5.0 cm² cross-sectional area at a pressure of 5.0 atm is allowed to expand adiabatically against an external pressure of 1.0 atm. During the expansion it pushes a piston through 1.0 m. If the initial temperature is 300.0 K, what is the final temperature of the gas?

Answer. The expansion is (a) adiabatic, (b) irreversible; therefore use eqn 16 with C_V from Table 2.15, which gives $\Delta T = -2.0$ K. The temperature falls to 298.0 K.

Comment. The temperature falls because energy has been extracted as work and none has returned as heat.

Exercise. Calculate the final temperature and the change in internal energy when 2.0 mol NH_3 is used instead, the other conditions being unchanged.

[299.1 K, −51 J]

Reversible adiabatic expansion

If the expansion is reversible at every stage, the external pressure is matched to the internal throughout the process. Consider some stage during which the pressure (inside and out) is p, then when the volume changes by dV the work done is $dw = -p\,dV$. Since $dq = 0$, it follows that $dU = -p\,dV$. However, by the argument above, for a perfect gas $dU = C_V\,dT$. These two terms must be equal (dU is an exact differential, independent of path), and so at each stage

$$C_V\,dT = -p\,dV$$

Since at each stage the perfect gas obeys $pV = nRT$, this equality becomes

$$\frac{C_V\,dT}{T} = \frac{-nR\,dV}{V}$$

Since C_V may be taken to be independent of temperature (this is true for monatomic perfect gases, and approximately true for others), both sides may be integrated between the initial and final states:

$$C_V \int_{T_i}^{T_f} \frac{dT}{T} = -nR \int_{V_i}^{V_f} \frac{dV}{V}$$

or

$$C_V \ln \frac{T_f}{T_i} = -nR \ln \frac{V_f}{V_i}$$

On writing $c = C_V/nR$ and doing a little rearranging, we get

$$\ln \left(\frac{T_f}{T_i}\right)^c = \ln \frac{V_i}{V_f}$$

which implies that

$$V_f T_f^c = V_i T_i^c \tag{17a}_r^\circ$$

We can now predict the temperature of a perfect gas that has expanded (or contracted) adiabatically and reversibly from a volume V_i and temperature T_i to a volume V_f:

$$T_f = \left(\frac{V_i}{V_f}\right)^{1/c} T_i \tag{17b}_r^\circ$$

The work done as the volume changes is obtained by substituting this result into $w = C_V\,\Delta T$:

$$w = C_V(T_f - T_i) = C_V T_i \left\{\left(\frac{V_i}{V_f}\right)^{1/c} - 1\right\} \tag{18}_r^\circ$$

Example 3.7: *Calculating the work of adiabatic expansion*

A sample of argon at 1.0 atm pressure and 25°C expands reversibly and adiabatically from 0.50 L to 1.00 L. Calculate its final temperature, the work done during the expansion, and the change in internal energy.

Answer. The final temperature is calculated from eqn 17a and the work from eqn 18. Take C_V from Table 2.15, which gives $c = 1.501$, and calculate n from $pV = nRT$, which gives $n = 0.0204$ mol. Then, from eqn 17b, $T_f = 188$ K and from eqn 18, $w = -28$ J. Since $q = 0$ for an adiabatic change, $\Delta U = -28$ J.

Exercise. Calculate the final temperature, the work done, and the change in internal energy when ammonia is used in a reversible adiabatic expansion from 0.50 L to 2.00 L, the other initial conditions being the same.

[196 K, −57 J, −57 J]

Perfect gas adiabats

It is now easy to find the relation between p and V when a perfect gas undergoes reversible abiabatic change. From the perfect gas equation

$$\frac{p_i V_i}{p_f V_f} = \frac{T_i}{T_f}$$

and from eqn 17 we have

$$\frac{T_i}{T_f} = \left(\frac{V_f}{V_i}\right)^{1/c}$$

Combining the two gives

$$p_i V_i^\gamma = p_f V_f^\gamma \qquad (19)_r^\circ$$

where

$$\gamma = \frac{nR}{C_V} + 1 = \frac{nR + C_V}{C_V}$$

For a perfect gas $C_V + nR = C_p$, so γ is the ratio of heat capacities:

$$\gamma = \frac{C_p}{C_V} \qquad (20)$$

This expression is henceforth taken to be the definition of γ for all substances. Equation 19 is often expressed in the form

$$pV^\gamma = \text{constant} \qquad (21)_r^\circ$$

For all gases $\gamma > 1$. For a perfect monatomic gas $\gamma = \frac{5}{3}$, a result we establish in Part 2. This means that an **adiabat**, the curve showing the dependence of the pressure on the volume (Fig. 3.8) falls more steeply $(p \propto 1/V^\gamma)$ than the corresponding isotherm $(p \propto 1/V)$. The physical reason is that in an isothermal expansion, energy flows into the system as heat and maintains the temperature, and so the pressure does not fall as much as in an adiabatic expansion.

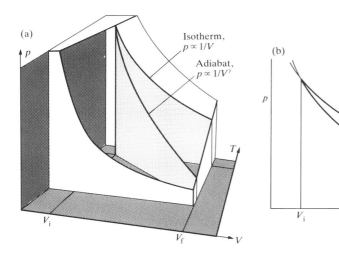

(a)

Isotherm,
$p \propto 1/V$

Adiabat,
$p \propto 1/V^{\gamma}$

(b)

Isotherm

Adiabat

Fig. 3.8 The isotherms and adiabats corresponding to the expansion of a perfect gas are paths on the p, V, T surface, as shown in (a). The projections of these paths on to the p, V plane, (b), show that the pressure decreases more rapidly along an adiabat than an isotherm, and therefore that the work corresponding to the process (the area under the curves) is less for adiabatic reversible expansion than for isothermal reversible expansion between the same initial and final volumes.

Example 3.8: *Calculating the pressure change in an adiabatic expansion*

A sample of argon at 1.00 atm expands reversibly and adiabatically to twice its initial volume. Calculate its final pressure.

Answer. We use eqn 19 rearranged into

$$p_f = p_i \times \left(\frac{V_i}{V_f}\right)^{\gamma}$$

and $\gamma = \frac{5}{3}$. This gives

$$p_f = 1.00 \text{ atm} \times \left(\frac{1}{2}\right)^{5/3} = 0.31 \text{ atm}$$

Comment. If the initial temperature is 298 K, the final temperature will be 188 K (from eqn 17).

Exercise. Calculate the final pressure when neon at 1.0 atm is compressed reversibly and adiabatically to 75 per cent of its initial volume.   [1.6 atm]

3.5 Comments on types of change

The results for the different types of change are collected in Table 3.3. It will be helpful to remember the following points.

Table 3.3. Work done on expansion†

Type of work	w	q	ΔU	ΔT
Expansion against $p=0$				
Isothermal	0	$0°$	$0°$	0
Adiabatic	0	0	0	0
Expansion against constant pressure				
Isothermal	$-p_{ex}\Delta V$	$p_{ex}\Delta V°$	$0°$	0
Adiabatic	$-p_{ex}\Delta V$	0	$-p_{ex}\Delta V$	$\dfrac{-p_{ex}\Delta V°}{C_V}$
Reversible expansion or compression				
Isothermal	$-nRT \ln \dfrac{V_f}{V_i}°$	$nRT \ln \dfrac{V_f}{V_i}°$	$0°$	0
Adiabatic	$C_V\Delta T°$	0	$C_V\Delta T°$	$T_i\left\{\left(\dfrac{V_i}{V_f}\right)^{1/c} - 1\right\}°$

† The entries marked ° are for a perfect gas; the rest apply to any substance. $c = C_V/nR$.

The work done on a system during *any* infinitesimal change is $dw = -p_{ex} dV$, where p_{ex} is the pressure acting on the system. In a *reversible change*, the external pressure is always matched to the internal: $p_{ex} = p$, and p depends on the temperature and volume of the gas (and the amount present). In an *irreversible expansion*, $p_{ex} < p$, and the work done by the system is less than in the case of reversible expansion.

In an *isothermal expansion*, the temperature of the gas is maintained by the thermal bath representing the surroundings. For a perfect gas, the internal energy is independent of the volume during an isothermal change, which implies that as much energy enters the system as heat as leaves the system as work: $dq = -dw$. In an *adiabatic expansion*, no heat enters the system, but work is done. Therefore $dq = 0$, and the internal energy change is equal to the work done: $dU = dw$.

In the reversible expansion of a perfect gas through a given volume from the same initial state, the isothermal process produces more work than the adiabatic process. This is because the internal energy is continuously replenished by the flow of energy that maintains the initial temperature.

Further reading

S. M. Blinder, Mathematical methods in elementary thermodynamics. *J. chem. Educ.*, **43**, 85 (1966).

H. Margenau and G. M. Murphy, *The mathematics of physics and chemistry*. Van Nostrand, New York (1956).

P. A. Rock, *Chemical thermodynamics*. University Science Books and Oxford University Press (1983).

K. E. Bett, J. S. Rowlinson, and G. Saville, *Thermodynamics for chemical engineers*. Athlone Press, London (1975).

M. L. McGlashan, *Chemical thermodynamics*. Academic Press, London (1979).

Exercises

Assume that all gases are perfect and that all data refer to 298 K unless stated otherwise.

3.1 Show that the following functions have exact differentials: (a) $x^2 y + 3y^2$, (b) $x \cos (xy)$, (c) $t(t + e^s) + 25s$.

3.2 Express $(\partial C_V / \partial V)_T$ as a second derivative of U, and find its relation to $(\partial U / \partial V)_T$. From this relation, show that $(\partial C_V / \partial V)_T = 0$ for a perfect gas.

3.3 By direct differentiation of $H = U + pV$, obtain a relation between $(\partial H / \partial U)_p$ and $(\partial U / \partial V)_p$. Confirm the result by expressing $(\partial H / \partial U)_p$ as the ratio of two derivatives with respect to volume and then using the definition of enthalpy.

3.4 Write an expression for dV given that V is a function of p and T. Deduce an expression for $d \ln V$ in terms of the expansion coefficient and the isothermal compressibility.

3.5 The internal energy of a perfect monatomic gas relative to its value at $T = 0$ is $\frac{3}{2} nRT$. Calculate $(\partial U / \partial V)_T$ and $(\partial H / \partial V)_T$ for the gas.

3.6 The volume of 1.00 g of a certain liquid varies with temperature as

$$V = V'\{0.75 + 3.9 \times 10^{-4}(T/K) + 1.48 \times 10^{-6}(T/K)^2\}$$

where V' is its volume at 300 K. Given that its density at 300 K is 0.875 g cm^{-3}, calculate its expansion coefficient at 320 K.

3.7 The isothermal compressibility of copper at 293 K is 7.35×10^{-7} atm^{-1}. Calculate the pressure that must be applied in order to increase its density by 0.08 per cent.

3.8 Given that $\mu = 0.25$ K atm^{-1} for N_2, calculate the value of its isothermal Joule–Thomson coefficient. Calculate the energy that must be supplied as heat to maintain constant temperature when 15.0 mol N_2 flow through a throttle in an isothermal Joule–Thomson experiment and the pressure drop is 75 atm.

3.9 A sample of 4.0 mol O_2 is originally confined in 20 L at 270 K and then undergoes adiabatic expansion against a constant pressure of 600 Torr until the volume has tripled. Calculate q, w, ΔT, ΔU, and ΔH.

3.10 A sample of 3.0 mol of gas at 200 K and 2.00 atm is compressed reversibly and adiabatically until the temperature reaches 250 K. Given that its molar constant-volume heat capacity is 27.5 J K^{-1} mol^{-1}, calculate q, w, ΔU, ΔH, and the final pressure and volume.

3.11 A sample of 1.0 mol of perfect gas with $C_p = 20.8$ J K^{-1} is initially at 3.25 atm and 310 K. It undergoes reversible adiabatic expansion until its pressure reaches 2.50 atm. Calculate the final volume and temperature and the work done.

3.12 Estimate the changes in volume that occur when 1.0 cm^3 blocks of (a) mercury and (b) diamond are heated through 5 K at room temperature.

3.13 In order to design a particular kind of refrigerator we need to know the temperature drop brought about by adiabatic expansion of the refrigerant gas. For one type of freon, $\mu = 1.2$ K atm^{-1}. What pressure difference is needed to produce a temperature drop of 5.0 K?

3.14 The temperature decreases measured when a freon expanded from a pressure p to 1.0 atm at 0°C were as follows:

p/atm	32	24	18	11	8	5
ΔT/K	−22	−18	−15	−10	−7.4	−4.6

Determine μ for the gas at this temperature.

3.15 Consider a system consisting of 2.0 mol of CO_2 gas (assumed perfect) at 25°C confined to a cylinder of cross section 10 cm^2 at 10 atm. The gas is allowed to expand adiabatically and reversibly. Calculate w, q, ΔU, ΔH, and ΔT when the piston has moved 20 cm.

3.16 A sample consisting of 65.0 g of Xe is confined in a container at 2.00 atm and 298 K and then allowed to expand adiabatically (a) reversibly to 1.00 atm, (b) against a constant pressure of 1.00 atm. Calculate the final temperature in each case.

Problems

Assume that all gases are perfect and that all data refer to 298 K unless stated otherwise.

Numerical problems

3.1 The isothermal compressibility of lead is 2.3×10^{-6} atm^{-1}. Express this value in Pa^{-1}. A cube of lead of side 10 cm at 25°C was to be inserted in the keel of an underwater exploration TV camera, and its designers needed to know the stresses in the equipment. Calculate the change in volume of the cube at a depth of 1000 m (disregarding the effects of temperature). Take the mean density of sea water as 1.03 g cm^{-3}. Given that the expansion coefficient of lead is 8.61×10^{-5} K^{-1} and that the temperature where the camera operates is −5°C, calculate the volume of the block taking the temperature into account too.

3.2 Evaluate the difference in molar heat capacities at constant pressure and constant volume for (a) copper, (b) ethanol at 25°C. Estimate the difference in the energy required as heat to increase the temperature of a 500 g sample of each to 75°C under conditions of constant pressure and constant volume.

3.3 Calculate the change in (a) the molar internal energy and (b) the molar enthalpy of water when its temperature is raised by 10 K. Account for the difference between the two quantities.

3.4 The constant-volume heat capacity of a gas can be measured by observing the decrease in temperature when it expands adiabatically and reversibly. If the decrease in pressure is also measured we can use it to infer the value of γ and hence, by combining the two values, deduce the constant-pressure heat capacity. When a fluorocarbon gas was allowed to expand reversibly and adiabatically to twice its volume, the temperature fell from 298.15 K to 248.44 K and its pressure fell from 1522.2 Torr to 613.85 Torr. Evaluate C_p.

3.5 Estimate γ for xenon at 100°C and 1.00 atm on the assumption that it is a van der Waals gas.

Theoretical problems

3.6 Consider a system in which the volume and the pressure of a perfect gas both depend on the time t. Show that the rate of change of the pressure is related to that of the volume and temperature by

$$\frac{\mathrm{d}\ln p}{\mathrm{d}t} = \frac{\mathrm{d}\ln T}{\mathrm{d}t} - \frac{\mathrm{d}\ln V}{\mathrm{d}t}$$

Now suppose that the system cools exponentially towards $T = 0$ with a time constant τ_T and, at the same time, a piston is driven into the container in such a way that the volume is reduced exponentially towards $V = 0$ with a time constant τ_V. How rapidly does the pressure change? What is the time dependence of the pressure when τ_T and τ_V are equal?

3.7 The pressure of a given amount of a van der Waals gas depends on T and V. Find an expression for dp in terms of dT and dV.

3.8 Rearrange the van der Waals equation of state to give an expression for T as a function of p and V (with n constant). Calculate $(\partial T/\partial p)_V$ and confirm that $(\partial T/\partial p)_V = 1/(\partial p/\partial T)_V$. Go on to confirm Euler's chain relation.

3.9 Calculate the isothermal compressibility and the expansion coefficient of a van der Waals gas. Show, using Euler's chain relation, that

$$\kappa_T R = \alpha(V_m - b)$$

Go on to show that, in terms of reduced variables,

$$8\kappa_r = \alpha_r(3V_r - 1)$$

3.10 Given that

$$\mu C_p = T\left(\frac{\partial V}{\partial T}\right)_p - V$$

derive an expression for μ in terms of the van der Waals coefficients a and b, and express it in terms of reduced variables. Evaluate μ at 25°C and 1.0 atm, when its molar volume is 24.6 L mol^{-1}. Use the expression obtained to derive a formula for the inversion temperature of a van der Waals gas in terms of reduced variables and evaluate it for the xenon sample.

3.11 One thermodynamic equation of state is given following eqn 12. Derive its partner

$$\left(\frac{\partial H}{\partial p}\right)_T = -T\left(\frac{\partial V}{\partial T}\right)_p + V$$

from it and the general relations between partial differentials.

3.12 Show that the Joule–Thomson coefficient μ of a gas that is described by the virial equation of state $pV_m/RT =$

$1 + B/V_m$ is given by

$$\mu = \frac{T^2}{C_p}\left(\frac{\partial}{\partial T}\frac{B}{T}\right)_p$$

if $4pB/RT \ll 1$. Estimate the value of μ for Ar at 25°C.

3.13 Show that for a van der Waals gas,

$$C_p - C_V = \lambda R \quad \text{with} \quad \frac{1}{\lambda} = 1 - \frac{(3V_r - 1)^2}{4V_r^3 T_r}$$

and evaluate the difference for Xe at 25°C and 10.0 atm.

3.14 Show that the value of ΔH for the adiabatic expansion of a perfect gas may be calculated by integration of $dH = V\,dp$, and evaluate the integral for the reversible adiabatic expansion of a perfect gas.

3.15 The speed of sound in a gas of molar mass M is related to the ratio of heat capacities γ by

$$c = \left(\frac{\gamma RT}{M}\right)^{1/2}$$

Show that $c = (\gamma p/\rho)^{1/2}$, where ρ is the density of the gas. Calculate the speed of sound in argon at 25°C.

The Second Law: the concepts

4

Check-list of key ideas

1. The identification of the property of an isolated system that determines whether a change is *spontaneous* or not.

2. The statement of the *Second Law of thermodynamics* (Section 4.2).

3. The statistical definition of *entropy* in terms of *Boltzmann's formula* (eqn 2) and its relation to the disorder in a system.

4. The *thermodynamic definition* of entropy in terms of the energy transferred as heat to the surroundings (eqn 3) and the adaptation of this definition to the calculation of the entropy change of the system (eqn 4).

5. The derivation of the *Clausius inequality* for the change of entropy (eqn 5).

6. The confirmation that the entropy of an isolated system does increase when a gas expands irreversibly and an object cools spontaneously (Section 4.3).

7. The calculation of the entropy change of a substance when it undergoes a phase transition (eqn 6).

8. The calculation of the entropy change when a perfect gas expands isothermally (eqn 7).

9. The entropy of a substance at one temperature in terms of its entropy at another temperature (eqn 8).

10. The measurement of the entropy of a substance calorimetrically (Section 4.4).

11. The *Nernst heat theorem* for the entropy change accompanying a transformation at $T = 0$ (Section 4.5).

12. The *Third Law of thermodynamics, Third Law entropies* of substances, and the definition of the *standard reaction entropy* (eqn 9).

13. The *efficiency* of heat engines as a function only of the temperature (eqn 11) and its illustration for the special case of the *Carnot cycle* (Section 4.6).

14. The use of the thermodynamic efficiency to establish the *thermodynamic temperature scale* (Section 4.6).

15. The *coefficient of performance* of refrigerators (eqn 15) and the power needed to sustain low temperatures (eqn 16).

16. The principle of *adiabatic demagnetization* for reaching low temperatures and the operation of *heat pumps* (Section 4.7).

17. The *criteria for spontaneous change* in terms of the entropy, the internal energy, and the enthalpy (eqns 17 and 18).

18. The definition of the *Helmholtz function* and the *Gibbs function* (eqn 19) and the criteria for spontaneous change.

19. The relation of the Helmholtz function to the *maximum work* done in a process (eqn 21).

20. The Gibbs function as the *free energy* and its relation to the *maximum nonexpansion work* (eqn 22).

21. The definition of the *standard reaction Gibbs function* (eqn 23) and its expression in terms of the *standard Gibbs function of formation* (eqn 24) of substances.

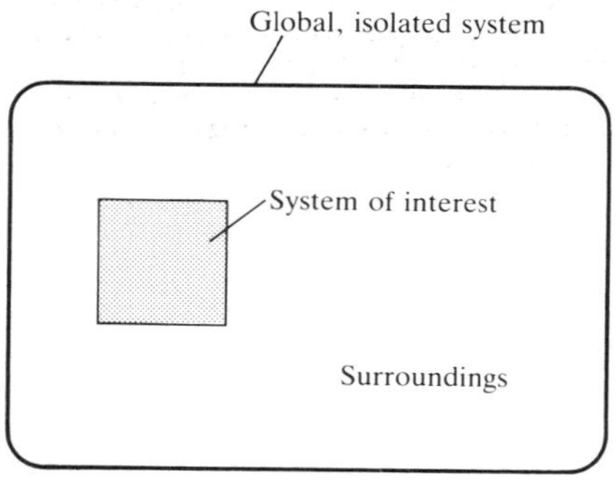

Fig. 4.1 The global isolated system that we consider in this chapter, and its division into the system of interest and that system's surroundings.

Some things happen naturally, some things don't. A gas expands to fill the available volume, a hot body cools to the temperature of its surroundings, and a chemical reaction runs in one direction rather than another. Some aspect of the world determines the **spontaneous** direction of change, the direction of change that does not require work to bring it about. We can confine a gas to a smaller volume, we can cool an object with a refrigerator, and we can force some reactions to go in reverse (as in the electrolysis of water). However, none of these happens spontaneously; they must be brought about by doing work.

We shall begin by identifying the characteristic of a system that determines the direction of spontaneous change. To be definite in what follows, we shall consider an isolated system inside which the events are taking place. We shall divide this total, isolated system into two parts that are distinguished by the boundary between them. One part of the global, isolated system is the system of interest, such as a reaction mixture (Fig. 4.1). The other is the immediate surroundings of that system, which we shall usually take to be an infinitely large thermal reservoir. Although we seem to be dealing with a lot of different systems and surroundings, it will always be clear whether we are referring to the global, isolated system or the system of interest inside it.

The direction of spontaneous change

What determines the direction of spontaneous change? It is not the total energy of the isolated system. The First Law of thermodynamics states that energy is conserved in any process, and we cannot disregard that law now and say that everything tends towards the state of lowest energy: the total energy of an isolated system is constant.

Is it perhaps the energy of the system of interest that tends towards a minimum? Two arguments show that this cannot be so. First, a perfect gas expands spontaneously into a vacuum, yet its internal energy remains constant. Secondly, if the energy of a system does happen to decrease during a spontaneous change, the energy of its surroundings must increase by the same amount (by the First Law). The increase in energy of the surroundings is just as spontaneous a process as the decrease in energy of the system.

When a change occurs, the total energy of the isolated system remains constant but it is parcelled out in different ways. Can it be, therefore, that the direction of change is related to the *distribution* of energy? We shall see that this is the key: spontaneous changes are always accompanied by a reduction in the 'quality' of energy in the sense that it is degraded into a more widely dispersed, chaotic form.

4.1 The dispersal of energy

The role of the distribution of energy can be illustrated by thinking about a ball (the system of interest) bouncing on a floor (the surroundings within the global, isolated system). We all know that after each bounce the ball does not rise as high. This is because there are frictional losses in the materials of the ball and floor. The direction of spontaneous change is towards a state in which the ball is at rest with all its energy degraded into the thermal motion of the atoms of the virtually infinite floor (Fig. 4.2).

A ball resting quietly on a warm floor has never been observed to start bouncing. For bouncing to begin, something rather special would need to happen. In the first place, some of the thermal motion of the floor would have to accumulate in a single, small object, the ball. This requires a spontaneous localization of energy from the myriad of vibrations of the atoms of the floor into the much smaller number that constitute the ball. Furthermore, whereas the thermal motion is disorderly, in order for the ball to move upwards, its atoms must all move in the same direction. This localization of random motion as *orderly* motion is so unlikely that we can dismiss it as virtually impossible.

We have found the signpost of spontaneous change: we look for the direction of change that leads to the greater chaotic dispersal of the total energy of the isolated system. Dispersal accounts for the direction of change of the bouncing ball, because its energy is dissipated into the thermal motion of the floor. The reverse process is not spontaneous because it is extremely unlikely that the chaotic distribution of energy will become coordinated by chance into local, uniform motion. A gas does not spontaneously contract, because in order to do so the chaotic motion of its molecules would have to take them all into the same region of the container; the opposite change is a natural consequence of increasing chaos. A cool object does not spontaneously become warmer than its surroundings: it is extremely improbable that the jostling of randomly vibrating atoms in the surroundings will lead to the accumulation of excess thermal motion in the object. The opposite change, the spreading of the object's energy into the surroundings as thermal motion is a natural consequence of chaos.

4.2 Entropy

The First Law of thermodynamics led to the introduction of the internal energy, U. This is a state function that lets us assess whether a change is

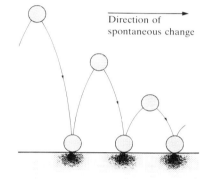

Fig. 4.2 The direction of spontaneous change for a ball bouncing on a floor. On each bounce some of its energy is degraded into the thermal motion of the atoms of the floor, and that energy disperses. The reverse has never been observed.

permissible: only those changes may occur for which the internal energy of an isolated system remains constant. The law that gives us the signpost of spontaneous change, the Second Law of thermodynamics, is expressed in terms of another state function, the 'entropy', S. We shall see that the entropy (which we shall define shortly) lets us assess whether one state is accessible from another by a *spontaneous* change. The First Law uses the internal energy to identify *permissible* changes (those that conserve energy); the Second Law uses the entropy to identify the *spontaneous* changes among these permissible changes:

Second Law: The entropy of an isolated system increases in the course of a spontaneous change:

$$\Delta S_{tot} > 0 \tag{1}$$

where S_{tot} is the total entropy of all parts of the isolated system.

Irreversible processes (like cooling to the temperature of the surroundings and the free expansion of gases) are spontaneous processes and hence are accompanied by an increase in entropy. We can express this by saying that irreversible processes generate entropy. On the other hand, reversible processes are finely balanced changes, with the system in equilibrium with its surroundings at every stage. Each infinitesimal step along a path is reversible, and occurs without dispersing energy chaotically and hence without increasing the entropy: reversible processes do not generate entropy. At most, reversible processes transfer entropy from one part of an isolated system to another.

The statistical definition of entropy

One definition of entropy will be presented in detail in Chapter 19, but it is helpful to be aware of it from the outset since it helps us visualize the thermodynamic definition on which we concentrate in this chapter. This **statistical definition** of entropy supposes that we can actually calculate the degree of disorder in a system. Specifically, the entropy is calculated using a formula proposed by Ludwig Boltzmann in 1896:

$$S = k \ln W \tag{2}$$

where k is Boltzmann's constant

$$k = 1.381 \times 10^{-23} \, \text{J K}^{-1}$$

This constant is related to the gas constant by $R = N_A k$. The quantity W is the number of different ways in which the energy of the system can be achieved by rearranging the atoms or molecules among their available states. The units of entropy are the same as those of k. It follows that the units of molar entropy, the entropy per mole, are $\text{J K}^{-1} \text{mol}^{-1}$ (the same as the units of R and of molar heat capacity).

As an illustration of how eqn 2 may be used, a solid of N HCl molecules at $T = 0$ has its lowest possible energy when all the molecules are perfectly ordered (Fig. 4.3a). Then $W = 1$ because there is only one way of achieving a perfectly ordered sample and $S = 0$ (because $\ln 1 = 0$). This perfectly ordered system has zero entropy.

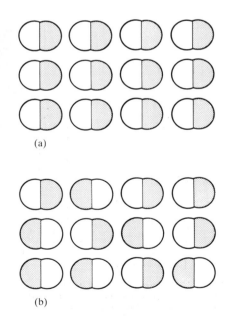

(a)

(b)

Fig. 4.3 (a) According to the Boltzmann formula, the entropy of a perfectly ordered solid is zero since $W = 1$. (b) The entropy of a disorderly solid in which each diatomic molecule can adopt either of two orientations at $T = 0$ is greater than 0 (specifically $Nk \ln 2$).

Example 4.1: *Calculating the entropy using the Boltzmann formula*

Calculate the entropy of $1.00\,\text{mol}\,CO(s)$ at $T = 0$ on the assumption that each CO molecule can adopt either of two orientations (Fig. 4.3b) without affecting the energy.

Answer. The sample consists of N CO molecules with $N = 6.022 \times 10^{23}$. Since each molecule can lie in two orientations in the crystal without affecting the energy, the total number of ways of achieving the same energy for N molecules is

$$W = 2 \times 2 \times \ldots = 2^N$$

The entropy of this sample is

$$\begin{aligned}
S &= k \ln 2^N = Nk \ln 2 \\
&= 6.022 \times 10^{23} \times 1.381 \times 10^{-23}\,\text{J K}^{-1} \times \ln 2 \\
&= 5.76\,\text{J K}^{-1}
\end{aligned}$$

Comment. CO has a much smaller electric dipole moment than HCl, and the energy of a sample is virtually the same if the molecules are head to tail or head to head. This disorderly arrangement has a higher entropy.

Exercise. The $FClO_3$ molecule may take up four orientations in the solid at $T = 0$ without affecting the energy. Calculate the entropy of $1.00\,\text{mol}\,FClO_3$.

$$[11.5\,\text{J K}^{-1}]$$

We take this statistical approach to the entropy further in Chapter 19. The point it has introduced, and which will be useful to bear in mind in what follows, is that as the disorder of a system increases, so does its entropy.

The thermodynamic definition of entropy

The thermodynamic approach to entropy, the main subject of this chapter, concentrates on the *change* in entropy dS during a process, not its absolute value S. The definition of dS is based on the view that a change in the extent of dispersal of energy can be related to the energy transferred as heat when a process occurs. In Chapter 19 we shall see that the statistical and thermodynamic definitions coincide: that is an exciting moment in physical chemistry, for it establishes a link between bulk properties (the concern of thermodynamics) and the underworld of atoms.

We begin by defining the **entropy change in the surroundings**, dS'. In this chapter, the prime will always denote the surroundings of the actual system within the global, isolated system. We represent the surroundings by a large thermal reservoir (a water bath in practice) that remains at the temperature T. Suppose a falling weight is coupled to the reservoir (for example, by driving a generator connected to a heater, as in Fig. 4.4), and that when the weight falls a quantity of heat dq' is transferred to the reservoir. The greater the amount of heat transferred to the reservoir, the greater the thermal motion that is stimulated in it, and hence the greater the dispersal of energy that occurs. This suggests that we write

$$dS' \propto dq'$$

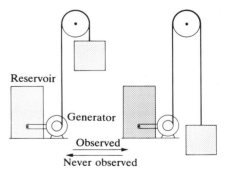

Fig. 4.4 The fundamental spontaneous process is represented by a falling weight and a thermal reservoir. The potential energy of the weight is transferred to the thermal motion of the reservoir, never vice versa spontaneously. Note that a reservoir is a sink of infinite extent, and its temperature remains constant however much energy is transferred as heat.

The 'quality' of the energy is degraded more completely if it is transferred to a cold reservoir than to a hot one. In the latter case we could go on to extract work from an engine driven by the flow of dq' from the cold reservoir to a colder one, but this is not possible if the energy is transferred directly to the colder reservoir. It follows that less entropy is generated when a given quantity of energy is transferred as heat to hot surroundings than to cold surroundings. The simplest way of taking this dependence on temperature into account is to write

$$dS' = \frac{dq'}{T'} \qquad (3a)$$

where T' is the temperature at which the transfer of heat takes place. For a finite change with the reservoir at a constant temperature

$$\Delta S' = \frac{q'}{T'} \qquad (3b)$$

Large changes in entropy occur when a lot of thermal motion is generated at low temperature.

Equation 3 makes it very simple to calculate the changes in entropy of the surroundings that accompany any process. For instance, in the case of any adiabatic change,

$$\Delta S' = 0 \quad \text{when} \quad q' = 0$$

This is true however the change takes place, reversibly or irreversibly, so long as no local hot spots are formed in the surroundings. That is, it is true so long as the surroundings remain in internal equilibrium. If hot spots do form, the localized energy may subsequently disperse spontaneously and hence generate entropy. When a chemical reaction takes place in the system with enthalpy change ΔH, the heat that enters the surroundings at constant pressure is $q' = -\Delta H$, so the entropy change of the surroundings is

$$\Delta S' = -\frac{\Delta H}{T'}$$

We shall see later that this relation plays a vital role in determining the direction of spontaneous chemical changes.

Example 4.2: *Calculating the entropy change in the surroundings*

Calculate the entropy change in the surroundings when 1.00 mol $H_2O(l)$ is formed from its elements under standard conditions at 298 K.

Answer. The standard enthalpy of the reaction $H_2(g) + \frac{1}{2}O_2(g) \rightarrow H_2O(l)$ is the standard enthalpy of formation of $H_2O(l)$; use the data in Table 2.9. Since $\Delta H_f^\ominus = -285.8 \text{ kJ mol}^{-1}$, for 1.00 mol $H_2O(l)$, $\Delta H = -286 \text{ kJ}$. Therefore,

$$\Delta S' = -\frac{\Delta H}{T'} = -\frac{(-286 \times 10^3 \text{ J})}{298.15 \text{ K}}$$

$$= +959 \text{ J K}^{-1}$$

Comment. The entropy of the surroundings increases in the course of an exothermic reaction. For an endothermic reaction, their entropy decreases.

Exercise. Calculate the entropy change in the surroundings when 1.00 mol $N_2O_4(g)$ is formed from 2.00 mol $NO_2(g)$ under standard conditions at ⊖. $[-192\ \text{J K}^{-1}]$

The entropy change in the system

Now we adapt the definition of the entropy change in the surroundings to the calculation of the entropy change of the system itself. The strategy we adopt is to use the surroundings to restore the system to its initial state *reversibly* with no further generation of entropy. Then we inspect the surroundings to see how much entropy has been transferred in the process. We shall set the temperature of the surroundings equal to that of the system, so that they are in thermal equilibrium, and henceforth write $T' = T$.

Let the original change in the entropy of the system when the process of interest occurs be dS (this is the change we want to measure). The process need not be reversible, but we suppose that we can find a path that joins the same two states and which is reversible. For example, the change may be the isothermal, irreversible expansion of a gas through a volume dV, in which case the reversible path will be the reversible isothermal expansion of the gas through the same volume. The same change of entropy is obtained in each case because S is a state function, but the energy absorbed as heat, dq in general and dq_{rev} for the reversible path, might be different. Now suppose that we restore the system to its initial state reversibly. Its entropy changes by $-dS$ (since the entropy is a state function, and its value must return to what it was originally if the state is restored). The energy we must supply as heat is also the negative of the change in the forward step, and hence is equal to $-dq_{rev}$. Since this energy comes from the surroundings, they undergo a change of energy $dq' = dq_{rev}$ and so their entropy changes by $dS' = dq_{rev}/T$. However, the total entropy change of the global, isolated system during the restoration is zero (because it is carried out reversibly). Therefore:

$$-dS + \frac{dq_{rev}}{T} = 0$$

or

$$dS = \frac{dq_{rev}}{T} \tag{4a}$$

That is, we can determine the entropy change when a system changes between two specified states by finding the heat necessary to take it along a *reversible* path between the same two states. For a measurable change, the entropy change is the sum (integral) of the infinitesimal changes:

$$\Delta S = \int_i^f \frac{dq_{rev}}{T} \tag{4b}$$

Example 4.3: *Calculating the entropy change during the isothermal expansion of a perfect gas*

Calculate the entropy change when a perfect gas expands isothermally.

Answer. Since the temperature is constant

$$\Delta S = \int \frac{dq_{rev}}{T} = \frac{1}{T} \int dq_{rev} = \frac{q_{rev}}{T}$$

The calculations in Chapter 2 led to the result that for the isothermal reversible expansion of a gas from V_i to V_f,

$$q_{rev} = nRT \ln \frac{V_f}{V_i}$$

Therefore, the entropy change accompanying this change of state is

$$\Delta S = nR \ln \frac{V_f}{V_i}$$

Comment. As an illustration of this formula, when 1.00 mol of any perfect gas doubles its volume at any temperature

$$\Delta S = 1.00 \text{ mol} \times 8.314 \text{ J K}^{-1} \text{mol}^{-1} \times \ln 2$$
$$= +5.76 \text{ J K}^{-1}$$

Exercise. Calculate the change in entropy when the pressure of a perfect gas is changed isothermally from p_i to p_f.   [$\Delta S = nR \ln (p_i/p_f)$]

4.3 The entropy of irreversible change

Consider a system in thermal and mechanical contact with its surroundings. We shall arrange for the system and the surroundings to be in thermal equilibrium (at the same temperature). However, they are not necessarily in mechanical equilibrium (a gas might have a greater pressure than its surroundings). Any change of state is accompanied by a change in entropy dS of the system and dS' of the surroundings. The overall change in entropy is greater than zero in general, because the process might be irreversible:

$$dS + dS' \geq 0, \quad \text{or} \quad dS \geq -dS'$$

(The equality applies if the process is reversible.) Since $dS' = -dq/T$, where dq is the heat supplied to the system, it follows that for any change

$$dS \geq \frac{dq}{T} \tag{5}$$

This is the **Clausius inequality**. We shall now use this result to show that the entropy does indeed increase for two of the cases mentioned in the introduction, the free expansion of a gas and the cooling of a hot substance.

Spontaneous expansion

First, suppose a system undergoes an irreversible adiabatic change. Then $dq = 0$, and by the Clausius inequality,

$$dS > 0$$

That is, for this type of spontaneous change the entropy of the *system* has increased. (We consider the entropy of the surroundings shortly.) Now consider the case of isothermal irreversible expansion. We saw in Chapter 3 that when a perfect gas expands isothermally, its internal energy remains constant: $(\partial U/\partial V)_T = 0$. Therefore, according to the First Law,

$$dU = dq + dw = 0, \quad \text{so} \quad dq = -dw$$

If the gas expands freely into a vacuum, it does no work; so $dw = 0$, which implies that $dq = 0$ too. According to the Clausius inequality, therefore

$$dS > 0$$

Now we consider the surroundings. In both cases $dq = 0$ and no heat is transferred into the surroundings. Since eqn 3 gives us the change in entropy of the surroundings however the change comes about (so long as the surroundings remain in internal equilibrium), for both types of change

$$dS' = 0$$

The overall entropy change is the sum of the changes in the system and its surroundings. Since $dS > 0$ and $dS' = 0$ for both, we conclude that for irreversible adiabatic expansion and for free isothermal expansion of a perfect gas

$$dS_{tot} > 0$$

Hence, the processes are spontaneous.

Spontaneous cooling

Consider a transfer of energy as heat[1] $|q|$ from one large reservoir—the source—at a high temperature T_h to another—the sink—at a lower temperature T_c. When $|q|$ leaves the hot source, the entropy of the source changes by $-|q|/T_h$ (a decrease). When $|q|$ enters the cold reservoir, its entropy changes by $+|q|/T_c$ (an increase). The overall change in entropy is therefore

$$\Delta S_{tot} = \frac{|q|}{T_c} - \frac{|q|}{T_h} = |q| \left(\frac{1}{T_c} - \frac{1}{T_h} \right)$$

which is positive (because $T_h \geq T_c$). Hence, cooling (the transfer of heat from hot to cold) is spontaneous, as we know from experience. When the temperatures of the two reservoirs are equal, $\Delta S_{tot} = 0$: the reservoirs are at thermal equilibrium.

4.4 Entropy changes accompanying specific processes

We shall now see how to calculate the entropy change accompanying some simple processes.

The entropy of phase transition at the transition temperature

Since a change in the degree of molecular order occurs when a substance freezes or boils, we should expect a change of entropy. At the transition temperature T_t (such as the freezing point), the system is in equilibrium (for instance, at the freezing point solid and liquid are at equilibrium) and the

---

[1] $|q|$ is the absolute magnitude of q; so $|q| = 10\,\text{kJ}$ if $q = +10\,\text{kJ}$ or if $q = -10\,\text{kJ}$.

Table 4.1. Entropies (and temperatures) of phase transitions at 1 atm pressure, $\Delta S_{trs}/(\text{J K}^{-1}\,\text{mol}^{-1})$

	Fusion (at T_f)	Vaporization (at T_b)
Ar	14.2 (at 83.8 K)	74.5 (at 87.3 K)
C_6H_6	38.0 (at 279 K)	87.2 (at 353 K)
H_2O	22.0 (at 273.15 K)	109.0 (at 373.15 K)
He	6.0 (at 3.5 K)	19.9 (at 4.22 K)

temperatures of the system and the surroundings are the same. Energy is therefore transferred reversibly and isothermally between the system and its surroundings as heat. Since at constant pressure $q_{rev} = \Delta H_{trs}$, the change in entropy of the system is

$$\Delta S = \frac{\Delta H_{trs}}{T_t} \tag{6}$$

If the phase transition is exothermic ($\Delta H_{trs} < 0$, as in freezing), the entropy change is negative. This is consistent with the system becoming more orderly. If the transition is endothermic ($\Delta H_{trs} > 0$, as in melting), the entropy change is positive, which is consistent with the system becoming more disorderly. Melting and vaporizing are endothermic processes, and so both are accompanied by an increase in the system's entropy. This is consistent with liquids being more disorderly than solids, and gases more disorderly than liquids. Some experimental entropies of transition are listed in Table 4.1.

In Table 4.2 we list in more detail the molar enthalpies of vaporization of several liquids. We see that a wide range of liquids give approximately the same molar entropy of vaporization (about $85\,\text{J K}^{-1}\,\text{mol}^{-1}$): this is **Trouton's rule.** The explanation of the rule is that a comparable amount of disorder is generated when a liquid evaporates and becomes a gas. Some liquids, however, deviate sharply from the rule. This is often because the molecules in the liquid are arranged in a partially orderly manner, and so a greater change of disorder occurs when it evaporates. An example is water, where the large entropy change reflects the presence of structure arising from hydrogen bonding in the liquid. Hydrogen bonds tend to organize the molecules in the liquid so that they are less random than, for example, the molecules in liquid hydrogen sulphide. Methane has an unusually low entropy of vaporization. A part of the reason for this is that the entropy of the gas itself is unusually low. As we shall see in Chapter 19, a part of the reason for its low entropy is the low moment of inertia of the molecule, which makes it difficult to excite into rotation.

Table 4.2. The molar entropies of vaporization of liquids

	$\Delta H_{vap}/(\text{kJ mol}^{-1})$	$\theta_b/°C$	$\Delta S_{vap}/(\text{J K}^{-1}\,\text{mol}^{-1})$
Benzene	+30.8	80.1	+87.2
Carbon tetrachloride	+30.00	76.1	+85.9
Cyclohexane	+30.1	80.7	+85.1
Hydrogen sulphide	+18.7	−60.4	+87.9
Methane	+8.18	−161.5	+73.2
Water	+40.7	100.0	+109.1

Example 4.4: *Using Trouton's rule*

Predict the molar enthalpy of vaporization of bromine given that it boils at 59.2°C.

Answer. Since there is no hydrogen bonding in bromine, we can use Trouton's rule in the form

$$\Delta H_{vap} = T_b \times 85 \text{ J K}^{-1}\text{ mol}^{-1}$$
$$= 28 \text{ kJ mol}^{-1}$$

Comment. The experimental value is 29.45 kJ mol^{-1}.

Exercise. Predict the enthalpy of vaporization of ethane from its boiling point, −88.6°C.   [16 kJ mol^{-1}]

The expansion of a perfect gas

We established in Example 4.3 that the change in entropy of a perfect gas that expands isothermally from V_i to V_f is

$$\Delta S = nR \ln \frac{V_f}{V_i} \qquad (7)°$$

This expression applies whether the change of state occurs reversibly or irreversibly (because S is a state function and independent of the path). If the change is reversible, the entropy change in the surroundings (which are in thermal and mechanical equilibrium with the system) must be such as to give $\Delta S_{tot} = 0$. Therefore, in this case, the change in entropy of the surroundings is

$$\Delta S' = -nR \ln \frac{V_f}{V_i}$$

However, if the expansion occurs freely and irreversibly, if no work is done ($w = 0$), and if the temperature remains constant (implying $\Delta U = 0$), there is no energy transferred between the system and its surroundings as heat ($q = 0$). Consequently, the entropy of the surroundings does not change. The entropy of the system changes by the same amount as before (eqn 7, because entropy is a state function), and so the overall change in entropy of the global isolated system is

$$\Delta S_{tot} = nR \ln \frac{V_f}{V_i}$$

This is an explicit expression for calculating the extent to which the entropy of the global isolated system has increased in the course of an irreversible change.

The entropy change when a system is heated

We can use eqn 4b to calculate the entropy of a system at a temperature T_f from its value at T_i:

$$S(T_f) = S(T_i) + \int_{T_i}^{T_f} \frac{dq_{rev}}{T}$$

We shall be particularly interested in the entropy change when the system is subjected to constant pressure (such as from the atmosphere) during the heating. Then from the definition of heat capacity (Section 2.8),

$$dq_{rev} = C_p \, dT$$

so long as the system is doing no non-expansion work. Consequently, at constant pressure:

$$S(T_f) = S(T_i) + \int_{T_i}^{T_f} \frac{C_p \, dT}{T} \tag{8a}$$

Similarly, at constant volume

$$S(T_f) = S(T_i) + \int_{T_i}^{T_f} \frac{C_V \, dT}{T} \tag{8b}$$

These expressions let us find the entropy of a substance at any temperature provided the heat capacity has been measured in the range of interest. For gases it is often the case that C_p is independent of temperature over moderate ranges, and so then, at constant pressure:

$$S(T_f) = S(T_i) + C_p \int_{T_i}^{T_f} \frac{dT}{T} = S(T_i) + C_p \ln \frac{T_f}{T_i}$$

Example 4.5: *Calculating the entropy change*

Calculate the entropy change when argon at 25°C and 1.00 atm pressure in a container of volume 500 cm³ is allowed to expand to 1000 cm³ and is simultaneously heated to 100°C.

Answer. Since S is a state function, choose the most convenient path from the initial state (n, 500 cm³, 298 K) to the final state (n, 1000 cm³, 373 K). One path is (a) isothermal expansion from 500 cm³ to 1000 cm³ at 298 K followed by (b) constant-volume heating from 298 K to 373 K at the final volume, 1000 cm³. For (a) use eqn 7; for (b) use eqn 8b. Calculate n from $pV = nRT$, which gives 0.0204 mol and take the molar C_V from Table 2.15.

Stage (a): $\Delta S = nR \ln 2 = +0.118 \text{ J K}^{-1}$

Stage (b): $\Delta S = C_V \ln \frac{373 \text{ K}}{298 \text{ K}} = +0.057 \text{ J K}^{-1}$

The overall entropy change is the sum of these two changes: $\Delta S = +0.175 \text{ J K}^{-1}$.

Exercise. Calculate the entropy change when the same initial sample is compressed to 50.0 cm³ and cooled to −25°C. $[-0.44 \text{ J K}^{-1}]$

The measurement of entropies

The entropy of a system at a temperature T can be related to its entropy at $T = 0$ by measuring its heat capacity C_p at different temperatures and evaluating the integral in eqn 8. At each phase transition, the entropy of transition ($\Delta H_{trs}/T_t$) must be added. For example, if a substance melts at T_f

and boils at T_b, its entropy above its boiling temperature is given by

$$S(T) = S(0) + \int_0^{T_f} \frac{C_p(\text{solid}) \, dT}{T} + \frac{\Delta H_{\text{fus}}}{T_f} + \int_{T_f}^{T_b} \frac{C_p(\text{liquid}) \, dT}{T}$$
$$+ \frac{\Delta H_{\text{vap}}}{T_b} + \int_{T_b}^{T} \frac{C_p(\text{gas}) \, dT}{T}$$

All the properties required, except $S(0)$, can be measured calorimetrically, and the integrals can be evaluated either graphically or, as is now more usual, by numerical integration on a computer. This is illustrated in Fig. 4.5: the area under the curve of C_p/T against T is the integral required. Since $dT/T = d \ln T$, an alternative procedure is to plot C_p against $\ln T$. Examples of this procedure are given in the Problems.

One problem with the measurement of entropy is the difficulty of measuring heat capacities near $T = 0$. There are good theoretical grounds for assuming that the heat capacity is proportional to T^3 when T is low, and this is the basis of the **Debye extrapolation** (which is justified in Part 2). In this method, C_p is measured down to as low a temperature as possible, and a curve of the form aT^3 is fitted to the data. That determines the value of a, and the expression $C_p = aT^3$ is assumed valid down to $T = 0$.

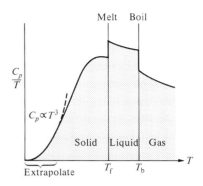

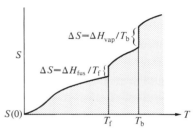

Fig. 4.5 The calculation of entropy from heat capacity data. The upper diagram shows the variation of C_p/T with the temperature for a sample. The lower curve shows the entropy, which is equal to the area beneath the upper curve up to the corresponding temperature, plus the entropy of each phase transition passed.

Example 4.6: *Calculating the entropy at low temperatures*

The molar heat capacity of gold at 10 K is 0.43 J K^{-1} mol^{-1}. What is its molar entropy at that temperature?

Answer. We can assume that at low temperatures the heat capacity is equal to aT^3, in which case

$$S(T) - S(0) = \int_0^T \frac{C_p \, dT}{T} = a \int_0^T T^2 \, dT = \tfrac{1}{3}aT^3$$

However, since aT^3 is the heat capacity at the temperature T,

$$S(T) = S(0) + \tfrac{1}{3}C_p$$
$$= S(0) + 0.14 \text{ J K}^{-1} \text{mol}^{-1}$$

Comment. Since the heat capacity is very small at low temperatures, only small errors arise from the Debye extrapolation.

Exercise. For metals there is also a contribution to the heat capacity from the electrons which is linearly proportional to T when the temperature is low. Find its contribution to the entropy.
$$[S = S(0) + C_p]$$

As an example, the standard entropy of nitrogen gas at $\mathcal{F}$ has been calculated from the following data:

	$S_m/(\text{J K}^{-1} \text{mol}^{-1})$
Debye extrapolation (0 to 10 K)	1.92
Integration, eqn 8	25.25
Phase transition (at 35.61 K)	6.43
Integration, eqn 8	23.38
Phase transition (melting, 63.14 K)	11.42
Integration, eqn 8	11.41
Phase transition (vaporization, 77.32 K)	72.13

93

Perfect gas expression, eqn 7 from 77.32 K to 298.15 K	39.20
Correction for gas imperfection[2]	0.92
Total:	192.06

Hence,

$$S_m^{\ominus}(\mathbf{F}) = S_m(0) + 192.1 \text{ J K}^{-1} \text{ mol}^{-1}$$

We deal with the value of $S(0)$ below.

4.5 The Third Law of thermodynamics

The question remaining is the value of $S(0)$. At $T = 0$ all energy of thermal motion has been quenched. In the case of a perfect crystal at $T = 0$, all the particles are in a regular, uniform array. The absence of disorder and thermal motion suggests that such materials also have zero entropy. This is consistent with the Boltzmann formula, for if $W = 1$ (only one way of arranging the molecules), $S = 0$.

The Nernst heat theorem

The thermodynamic observation that turns out to be consistent with this conclusion is known as the **Nernst heat theorem**:

> The entropy change of a transformation approaches zero as the temperature approaches zero:
> $$\Delta S \rightarrow 0 \quad \text{as} \quad T \rightarrow 0$$

For example, the entropy of the transition between orthorhombic sulphur, $S(\alpha)$, and monoclinic sulphur, $S(\beta)$, can be measured by determining its enthalpy (-402 J mol^{-1}) at the transition temperature (369 K):

$$\Delta S_m = S_m(\alpha) - S_m(\beta)$$

$$= \frac{-402 \text{ J mol}^{-1}}{369 \text{ K}} = -1.09 \text{ J K}^{-1} \text{ mol}^{-1}$$

The two individual entropies can also be determined by measuring the heat capacities from $T = 0$ up to $T = 369$ K (the rate of transformation is so slow at low temperatures that the relative thermodynamic instability of the monoclinic form is unimportant; the forms are internally thermodynamically stable). It is found that

$$S_m(\alpha) = S_m(\alpha, 0) + 37 \text{ J K}^{-1} \text{ mol}^{-1}$$
$$S_m(\beta) = S_m(\beta, 0) + 38 \text{ J K}^{-1} \text{ mol}^{-1}$$

implying that at the transition temperature

$$\Delta S = S_m(\alpha, 0) - S_m(\beta, 0) - 1 \text{ J K}^{-1} \text{ mol}^{-1}$$

On comparing this value with the one above, we conclude that

$$S_m(\alpha, 0) - S_m(\beta, 0) = 0$$

in accord with the theorem.

It follows from the Nernst theorem, that if we arbitrarily ascribe the value zero to the entropies of elements in their perfect crystalline form at $T = 0$,

---

[2] Gas imperfections are discussed in Section 5.3.

then all perfect crystalline compounds also have zero entropy at $T = 0$ (because the change in entropy that accompanies the formation of the compounds, like all transformations, is zero). Hence, *all* perfect crystals may be taken to have zero entropy at $T = 0$. This conclusion is summarized by the **Third Law of thermodynamics**:

Third Law: If the entropy of every element in its stable state at $T = 0$ is taken as zero, every substance has a positive entropy which at $T = 0$ may become zero, and does become zero for all perfect crystalline substances, including compounds.

Note that a non-crystalline perfect state, such as the superfluid state of He, is included by the opening phrase. Note too, that the Third Law does not state that entropies *are* zero at $T = 0$: it merely implies that all perfect materials have the same entropy then. As far as thermodynamics is concerned, choosing this common value as zero is a matter of convenience.

Third Law entropies

The choice $S(0) = 0$ for perfect crystals will be made from now on. Entropies reported on that basis are called **Third Law entropies** (and often just 'entropies'). When the substance is in its standard state at the temperature T, the standard (Third Law) entropy is denoted $S^{\ominus}(T)$. A list of values at 298 K is given in Table 4.3.

The **standard reaction entropy** $\Delta S^{\ominus}$ is defined, like the standard reaction enthalpy, as the difference between the entropies of the pure, separated products and the pure, separated reactants, all substances being in their standard states at the specified temperature. Thus, in general:

$$\Delta S^{\ominus} = \sum_J v_J S_J^{\ominus} \tag{9}$$

Table 4.3. Standard Third Law entropies at 298 K

	$S_m^{\ominus}/(\text{J K}^{-1}\,\text{mol}^{-1})$
Solids:	
C(s, graphite)	5.7
C(s, diamond)	2.4
$C_{12}H_{22}O_{11}(s)$	360.2
$I_2(s)$	116.1
Liquids:	
$C_6H_6(l)$	173.3
$H_2O(l)$	69.9
Hg(l)	76.0
Gases:	
$C(CH_3)_4(g)$	306
$CH_4(g)$	186.3
$CO_2(g)$	213.7
$H_2(g)$	130.7
He(g)	126.2
$IF_7(g)$	346.5
$NH_3(g)$	192.5

Example 4.7: *Calculating a standard reaction entropy*

Calculate the standard reaction entropy of $H_2(g) + \frac{1}{2}O_2(g) \rightarrow H_2O(l)$ at $\mathbf{f}$.

Answer. Identify the stoichiometric coefficients in the reaction written in the form $0 = H_2O(l) - H_2(g) - \frac{1}{2}O_2(g)$. Then use eqn 9 with standard entropies taken from Table 4.3. The coefficients are $v(H_2O) = 1$, $v(H_2) = -1$, and $v(O_2) = -\frac{1}{2}$. Therefore:

$$\Delta S^{\ominus} = v(H_2O)S_m^{\ominus}(H_2O, l) + v(H_2)S_m^{\ominus}(H_2, g) + v(O_2)S_m^{\ominus}(O_2, g)$$
$$= (69.9 - 130.7 - \tfrac{1}{2} \times 205.1)\,\text{J K}^{-1}\,\text{mol}^{-1}$$
$$= -163.4\,\text{J K}^{-1}\,\text{mol}^{-1}$$

Comment. The large decrease in entropy is largely due to the formation of a liquid from two gases.

Exercise. Calculate the standard reaction entropy for the combustion of $CH_4(g)$ at $\mathbf{f}$. $[-243\,\text{J K}^{-1}\,\text{mol}^{-1}]$

The efficiencies of thermal processes

Thermodynamics grew out of the study of ways of improving the efficiencies of **heat engines**, which are devices for converting heat into work (such as steam engines and jet engines). While these early applications of thermo-dynamics are apparently far removed from chemistry, we shall see that analogous considerations are relevant to discussions of chemical processes, such as the use of chemical reactions in electrochemical cells to produce electric currents and the biosynthesis of proteins. They are also relevant to considerations of how it is possible to achieve low temperatures.

4.6   The efficiency of heat engines

We begin with the calculation of the maximum work that can be obtained from a heat engine operating between a hot reservoir, the hot source, at a temperature T_h and a cold reservoir, the cold sink, at a temperature T_c.

Maximum work

If the engine is to produce work, it must do so spontaneously (an engine is worthless if it has to be driven!). Hence, the flow of energy as heat from the hot source to the cold sink must be accompanied by an overall increase in entropy.

We saw earlier (in Section 4.3) that the change in entropy when an energy $|q|$ leaves the source as heat and enters the sink is

$$\Delta S = \frac{|q|}{T_c} - \frac{|q|}{T_h}$$

Since the temperature of a cold reservoir is lower than that of a hot one, an overall increase in entropy may also be obtained even if *less than* $|q|$ is transferred to the cold sink. Since the removal of $|q_h|$ from the hot source changes its entropy by $-|q_h|/T_h$, and a transfer of $|q_c|$ to the cold sink increases its entropy by $|q_c|/T_c$, the overall entropy change in general is

$$\Delta S = \frac{|q_c|}{T_c} - \frac{|q_h|}{T_h}$$

This is not negative if

$$|q_c| \geq |q_h| \times \frac{T_c}{T_h}$$

Therefore, we are free to use some kind of device, a heat engine, to draw off the difference $|q_h| - |q_c|$ as work, yet still have a spontaneous process (Fig. 4.6). It then follows that the maximum work the engine can do is

$$|w_{max}| = |q_h| - |q_{c,min}| = |q_h|\left(1 - \frac{T_c}{T_h}\right)$$

The **Carnot efficiency** ϵ is defined as

$$\epsilon = \frac{\text{Maximum work generated}}{\text{Energy supplied as heat}} = \frac{|w_{max}|}{|q_h|} \qquad (10)$$

Therefore,

$$\epsilon = 1 - \frac{T_c}{T_h} \qquad (11)_r$$

This result is independent of the working substance.

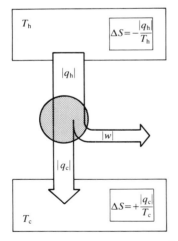

Fig. 4.6   The thermodynamic basis of engines. When a quantity of energy $|q_h|$ is withdrawn from a hot reservoir the entropy decreases by $|q_h|/T_h$, and when $|q_c|$ is added to a cold reservoir its entropy is increased by $|q_c|/T_c$. So long as $T_c < T_h$, the overall entropy change may still be positive even if less energy is returned to the cold reservoir than was removed from the hot. The difference $|q_h| - |q_c|$ may be withdrawn as work, the overall process being spontaneous.

The Carnot cycle

We can check that the Carnot efficiency is obtained for the special case of an engine using a perfect gas as the working fluid. The **Carnot cycle** is illustrated in Fig. 4.7; each stage is reversible. The stage from A to B is an isothermal expansion; that from B to C is an adiabatic expansion; C to D is an isothermal compression; D to A is an adiabatic compression. The work done and the heat transferred in each stage can be taken from Table 3.3, and the resulting expressions are given in the illustration. The work done on going round the cycle is the sum of the work done in each stage:

$$|w| = nRT_h \ln \frac{V_B}{V_A} + nRT_c \ln \frac{V_D}{V_C}$$

(The adiabatic stages cancel.) The energy supplied as heat by the hot source is

$$|q_h| = nRT_h \ln \frac{V_B}{V_A}$$

We can simplify the ratio $|w|/|q_h|$ by using the relations between temperature and volume for reversible adiabatic processes (i.e. $V_B/V_C = (T_c/T_h)^c$, eqn 17 of Section 3.4), which leads to

$$\epsilon = 1 - \frac{T_c}{T_h}$$

which is exactly the Carnot efficiency.

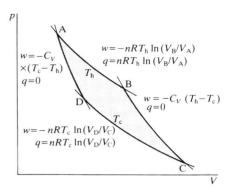

Fig. 4.7 The Carnot cycle for a perfect gas. All stages are reversible. A to B is an isothermal expansion at a temperature T_h, B to C an adiabatic expansion which lowers the temperature to T_c. C to D is an isothermal compression at T_c. D to A completes the cycle with an adiabatic compression that raises the temperature to T_h again. The work done at each stage and the heat absorbed are shown.

Example 4.8: *Using the Carnot cycle*

Use the Carnot cycle to demonstrate explicitly that the integral of dq_{rev}/T around a cycle is zero.

Answer. The integral in this case is the sum of the heat transferred in the two isothermal stages (the adiabatic stages contribute zero). Use the information on reversible adiabatic changes derived in Section 3.4 to relate volumes and temperatures. From Fig. 4.7:

$$\int \frac{dq}{T} = nR \ln \frac{V_B}{V_A} + nR \ln \frac{V_D}{V_C}$$

$$= nR \ln \frac{V_B V_D}{V_A V_C}$$

From eqn 17 of Section 3.4, the ratio of volumes is

$$\frac{V_B}{V_C} \times \frac{V_D}{V_A} = \left(\frac{T_c}{T_h}\right)^c \times \left(\frac{T_h}{T_c}\right)^c = 1$$

and so the integral is zero.

Exercise. Evaluate the Carnot efficiency explicitly using the information in Fig. 4.7 and Section 3.4.   [eqn 11]

Selecting the most efficient conditions

Equation 11 is of major economic significance, for it shows that an engine cannot convert heat into work with 100 per cent efficiency (except at $T_c = 0$).

The limit implied by the Carnot efficiency applies to all engines, whatever their construction, and whatever their working substance (its derivation made no mention of any specific substance). It is an upper limit to their conversion efficiency, and technological deficiencies (e.g. friction) reduce the actual efficiencies of real engines. Its positive feature is that it indicates how high conversion efficiencies may be achieved:

$$\epsilon \to 1 \quad \text{as} \quad T_c \to 0 \quad \text{or} \quad T_h \to \infty$$

Lowering the temperature of the cold sink is not a practicable option, and so engines and turbines are designed to run at high temperatures. A typical generating station using superheated steam at about 550°C and a cold sink at about 100°C has a thermodynamic efficiency of only 55 per cent. The remaining 45 per cent of heat taken from the hot source is discarded into the surroundings to ensure that enough entropy is generated to make the overall process spontaneous. An internal combustion engine operates between about 3200 K (the high temperature being brought about by the combustion of fuel) and 1400 K (the temperature in the exhaust manifold). Its thermodynamic efficiency is therefore only 56 per cent, but other losses reduce this ideal value to around 25 per cent in practice.

The thermodynamic temperature scale

Heat engines are of major theoretical significance, since they allow us to define a fundamental temperature scale, one that is independent of the thermometric substance.

Suppose we have an engine that is working reversibly with a hot source at a temperature T_h and a cold sink at a temperature T, then we know that

$$\frac{|q_c|}{|q_h|} \geq \frac{T}{T_h}$$

This expression enabled Kelvin to define the **thermodynamic temperature scale** in terms of the ratio of the heat withdrawn from the hot source and supplied to the cold sink, both of which can be measured. The zero of the thermodynamic temperature scale is defined as the value of T at which the Carnot efficiency becomes equal to 1:

$$\epsilon = 1 \quad \text{at} \quad T = 0$$

The size of the units is defined by choosing a single fixed point the triple point of water and defining it as $T_3 = 273.16$ K exactly. Then, if the heat engine has a hot source at this temperature, the temperature of the cold sink—the object with the temperature we want to measure—can be found by measuring q_h and q_c and using

$$T = \frac{q_c}{q_h} \times T_3 \tag{12}$$

This result is independent of the working substance. Moreover, we saw in Chapter 2 that the heat transferred can, in principle, be measured mechanically (in terms of the raising of a weight). Therefore, we can, in principle at least, use the distance moved by a weight to measure temperature.

4.7 The energetics of refrigeration

Work must be done in order to transfer heat from a system to its warmer surroundings and to go against the natural tendency of change, the spontaneous flow from hot to cold. Cooling against a temperature gradient is achieved using a **refrigerator**—a heat engine operating in reverse.

The work to achieve low temperatures

The idea behind refrigeration, and the calculation of the minimum amount of work that must be done, are illustrated in Fig. 4.8. When an amount of energy $|q_c|$ is removed from a cool body, the entropy is lowered. When the same energy is released into a hot reservoir as heat, the entropy increases, and the overall change is

$$\Delta S = -\frac{|q_c|}{T_c} + \frac{|q_c|}{T_h} = -|q_c|\left(\frac{1}{T_c} - \frac{1}{T_h}\right) < 0$$

This is not a spontaneous process because not enough entropy is generated in the hot sink to overcome the loss from the cold source (Fig. 4.8a). In order to generate more entropy, the energy released as heat must exceed that absorbed. This can be achieved by adding to the stream of energy by doing work (Fig. 4.8b). Our task is to find the minimum quantity of work that has to be added in order for the process to take place. The outcome is expressed as the **coefficient of performance** c:

$$c = \frac{|q_c|}{|w|} \tag{13}$$

where $|q_c|$ is the energy to be removed from the cold object and $|w|$ the work required to bring that removal about. The less the work required, the more efficient the operation, and the greater the coefficient of performance.

If $|q_c|$ is withdrawn from the cold object and work $|w|$ is done, the energy that must be dissipated into the hot reservoir (such as the surrounding room) as heat is $|q_h| = |q_c| + |w|$. Therefore,

$$\frac{1}{c} = \frac{|q_h| - |q_c|}{|q_c|} = \frac{|q_h|}{|q_c|} - 1$$

The refrigerator is at its most efficient when it is working reversibly, because then $|w|$ is a minimum. The best coefficient of performance c° is therefore given by

$$\frac{1}{c^\circ} = \left(\frac{|q_h|}{|q_c|}\right)_{\text{rev}} - 1 \tag{14}$$

The next step is to express the ratio of qs in terms of the temperatures of the hot and cold parts of the system. To do this we make use of the fact that the entropy of the cold object changes by $-(|q_c|/T_c)_{\text{rev}}$, that of the hot reservoir changes by $+(|q_h|/T_h)_{\text{rev}}$, but (because the process is occurring reversibly) there is no net entropy production. In other words, just enough work is being done to ensure that overall the process occurs without

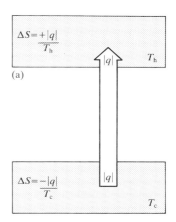

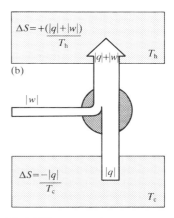

Fig. 4.8 (a) When energy leaves a reservoir as heat and enters a hotter one, there is an overall decrease in entropy, and so the process is not spontaneous. (b) If energy is added to the stream as work and is dispersed into the hot sink as heat, then just enough extra entropy may be generated to cancel the reduction in entropy of the cold source, and the overall process is spontaneous.

reduction of entropy. Therefore

$$\Delta S = \left(\frac{|q_h|}{T_h}\right)_{rev} - \left(\frac{|q_c|}{T_c}\right)_{rev} = 0$$

which solves to

$$\left(\frac{|q_h|}{|q_c|}\right)_{rev} = \frac{T_h}{T_c}$$

It follows that the coefficient of performance of a perfect refrigerator working reversibly between the temperatures T_c and T_h is

$$c^\circ = \frac{T_c}{T_h - T_c} \qquad (15)_r$$

The ideal value makes no reference to the type of refrigerator or the working substance: it depends only on the temperatures of the hot and cold reservoirs. Practical refrigerators (which do not work reversibly) have coefficients of performance lower than c°.

The work to sustain low temperatures

A related problem is the work required to sustain a low temperature once it has been reached. No thermal insulation is perfect, and so there is always a flow of heat from the warm surroundings back into the sample. According to Newton's law of cooling, the rate of this re-entry of energy as heat is proportional to the temperature difference $T_h - T_c$. In order to sustain the low temperature, heat must be removed from the cold object at the same rate as it leaks in. If the rate at which heat leaks in is

$$\frac{d\,|q_c|}{dt} = A(T_h - T_c)$$

where A is a constant that depends on the size of the sample and the thermal conductivity of its insulation, the minimum power P, the minimum rate at which work must be done to sustain the temperature difference, is

$$P = \frac{d\,|w|}{dt} = \frac{1}{c^\circ} \times \frac{d\,|q_c|}{dt} = \frac{A(T_h - T_c)}{c^\circ}$$

$$= \frac{A(T_h - T_c)^2}{T_c} \qquad (16)$$

We see that the power increases as the *square* of the temperature difference we are trying to sustain. For this reason, air-conditioners are much more expensive to run on hot days than on days that are merely warm. Notice too that the power depends inversely on the temperature of the cold object: very high powers must be dissipated when the temperature is very low.

Example 4.9: *Assessing the power needed to sustain low temperatures*

Assess the relative powers that must be expended to maintain the same object at (a) 0°C with surroundings at 20°C and (b) 1.0 mK with surroundings at 1.0 K.

Answer. We calculate the ratio of powers for each case using eqn 16. For (a), $P = A \times 1.5\,K$; for (b), $P = A \times 1.0 \times 10^3\,K$. Therefore, the ratio of powers required is $P(b)/P(a) = 7 \times 10^2$.

Comment. The result means that if 150 W is needed in (a), 100 kW is needed in (b). In practice the experimenter would use a much smaller sample in (b), and take great care with its insulation.

Exercise. An air conditioner required 50 W to maintain a room at 20°C when the surroundings were at 22°C; calculate the power needed when the surroundings were at 32°C. [1.8 kW]

Adiabatic demagnetization

The world record low temperature stands at about $20\,nK$ $(2 \times 10^{-8}\,K)$. Gases may be cooled by Joule–Thomson expansion below their inversion temperatures, and temperatures as low as about 4 K (the boiling point of helium) may be reached without great difficulty. Temperatures lower than 4 K can be reached by evaporating liquid helium by pumping rapidly through large diameter pipes: as the helium evaporates it withdraws energy from the object being cooled. Temperatures as low as about 1 K can be reached, but below this temperature the volatility of helium is too low for the process to be effective, and the superfluid phase begins to interfere with the cooling by creeping around the apparatus.

The method used to reach very low temperatures is **adiabatic demagnetization**, cooling making use of the magnetic properties of a substance. In Part 2 we shall see that magnetic properties arise because electrons behave as tiny magnets. In normal circumstances in a paramagnetic material (one composed of molecules in which not all the electrons are paired), these magnets are orientated at random. However, in a magnetic field more of them are aligned along the field than against it (Fig. 4.9). In thermodynamic terms, the application of a magnetic field lowers the entropy of the paramagnetic system on account of the ordering it induces. Therefore, a sample has different entropy/temperature curves, depending

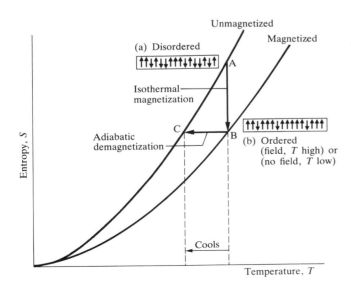

Fig. 4.9 The technique of adiabatic demagnetization is used to attain very low temperatures. The upper curve shows the variation of the entropy of a paramagnetic system in the absence of an applied field. The lower curve shows the variation in entropy when a field is applied and has made the electron magnets more orderly. The isothermal magnetization step is from A to B; the adiabatic demagnetization step (at constant entropy) is from B to C.

101

on the magnetic field it experiences. At a given temperature, the entropy is lower when the field is on than when it is off.

A sample of paramagnetic material, such as a d- or f-metal complex, is cooled to about 1 K in the way already described. (Gadolinium sulphate is often used because each gadolinium ion carries several electrons, but is separated from its neighbours by a sheath of solvating water molecules.) Then the sample is magnetized by the application of a strong magnetic field. This magnetization occurs while the sample is surrounded by helium gas, which provides a thermal contact with a cold reservoir. The magnetization is therefore isothermal, and heat leaves the sample as the electron magnets adopt lower energy by aligning with the applied field. This stage is represented by the line AB in Fig. 4.9.

Thermal contact between the sample and surroundings is now broken by pumping away the gas, and then the magnetic field is reduced slowly to zero. Since there is no flow of heat in this last reversible, adiabatic step, the entropy of the sample remains constant. Hence, the state of the sample changes from B to C in Fig. 4.9. When the field has reached zero, the sample has a lower entropy than it had initially, but is otherwise the same. Therefore, it now also has a lower temperature: adiabatic demagnetization has cooled the sample.

Even lower temperatures can be reached if instead of electronic magnetic moments the magnetic moments of nuclei are used. This process of **adiabatic nuclear demagnetization** works on the same principle as the electronic method, and has been used to establish the world record (in copper).

From the Third Law of thermodynamics, we know that entropy curves of substances coincide as $T \to 0$. This implies that adiabatic demagnetization— and, in fact, any process—cannot be used to cool an object to absolute zero in a finite number of steps. For suppose the law were false, then the entropy might behave as in Fig. 4.10a. The last step indicated could cool the system to $T = 0$. However, the Third Law asserts that the curves actually coincide at $T = 0$ as shown in Fig. 4.10b, and so no finite sequence of steps can cool the system to $T = 0$. By this method (and any method), absolute zero is unattainable. This is summarized by an alternative version of the Third Law of thermodynamics:

Third Law: It is impossible to reach $T = 0$ in a finite number of steps.

No one has ever succeeded in cooling any system to $T = 0$, and the Third Law generalizes this experience into a principle.

Heat pumps

Refrigeration is closely related to the operation of the **heat pump**, where work is done to pump heat from a cold reservoir (such as a river or the surrounding land) into a hot sink (such as a house). A heat pump is a refrigerator operating in reverse. Since work is supplied to the energy stream and delivered to the hot sink (the house), the total heat delivered when the pump is working reversibly is

$$|q_h| = |q_c| + |w| = (1 + c^\circ)\,|w|$$

If indoors the temperature is $T_h = 300$ K and outside $T_c = 290$ K, then $c^\circ = 29$. This means that for each 1 kJ of energy used to drive the pump

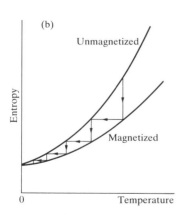

(a) Unmagnetized / Magnetized

Entropy / Temperature

(b) Unmagnetized / Magnetized

Entropy / Temperature

Fig. 4.10 The connection between the Nernst theorem and the unattainability of $T = 0$ is illustrated in this pair of diagrams. In (a) we assume that the theorem is false, and that the entropies do not coincide as $T \to 0$: we see that $T = 0$ can be reached in a finite number of steps. In (b) we see that the steps down in temperature become progressively smaller if, as the Nernst theorem asserts, the entropies do coincide as $T \to 0$. A finite number of steps does not now take the system to $T = 0$.

($w = 1\,\text{kJ}$), 30 kJ of heat will be supplied to the house. A 1 kW pump would therefore be a 30 kW heater if it worked reversibly. Commercial heat pumps have $c \approx 5$, which still amounts to a good yield in terms of the energy used to drive the equipment.

Concentrating on the system

Entropy is the basic concept for discussing the direction of natural change, but in order to use it we have to investigate changes in both the system and its surroundings. We have seen that it is always very simple to calculate the entropy change in the latter, and we shall now see that it is possible to devise a simple method for taking that contribution into account automatically. This focuses our attention on the system and simplifies discussions.

4.8 The Helmholtz and Gibbs functions

Consider a system in thermal equilibrium with its surroundings at a temperature T. The Clausius inequality, eqn 5, then reads

$$\mathrm{d}S - \frac{\mathrm{d}q}{T} \geq 0$$

This inequality can be developed in two ways.

First, consider heat transfer at constant volume. Then, so long as there is no non-expansion work, $\mathrm{d}q_V = \mathrm{d}U$, and

$$\mathrm{d}S - \frac{\mathrm{d}U}{T} \geq 0$$

The importance of the inequality in this form is that it expresses the criterion for spontaneous change solely in terms of the properties of the system. The inequality rearranges to

$$T\,\mathrm{d}S \geq \mathrm{d}U \quad \text{(constant } V\text{, no non-expansion work)} \qquad (17\mathrm{a})$$

At either constant internal energy ($\mathrm{d}U = 0$) or constant entropy ($\mathrm{d}S = 0$) this becomes

$$\mathrm{d}S_{U,V} \geq 0 \qquad \mathrm{d}U_{S,V} \leq 0 \qquad (17\mathrm{b})$$

These are the criteria for natural changes in terms of properties relating to the system. The first states that in a system at constant volume (such as an isolated system), the entropy increases in a spontaneous change. That is essentially the content of the Second Law. The second inequality is less obvious, for it says that if the entropy and volume of a system are constant, then the internal energy must decrease in a spontaneous change. We should interpret this as implying that if the entropy of the system is unchanged, then there must be an increase in entropy of the surroundings, which can be achieved only if the energy of the system decreases as energy flows out as heat.

When heat is transferred at constant pressure, and there is no work other than expansion work, we can write $\mathrm{d}q_p = \mathrm{d}H$ and obtain

$$T\,\mathrm{d}S \geq \mathrm{d}H \quad \text{(constant } p\text{, no non-expansion work)}. \qquad (18\mathrm{a})$$

At either constant enthalpy or constant entropy this becomes

$$dS_{H,p} \geq 0 \qquad dH_{S,p} \leq 0 \qquad \text{(18b)}$$

The interpretations of these inequalities are similar to those of eqn 17. The entropy of the system must increase if its enthalpy remains constant (for there can then be no change in entropy of the surroundings). Alternatively, the enthalpy must decrease if the entropy of the system is constant, for then it is essential to have an increase in entropy of the surroundings.

Since eqns 17 and 18 have the forms $dU - T\,dS \leq 0$ and $dH - T\,dS \leq 0$ respectively, they can be expressed more simply by introducing two more thermodynamic functions, the **Helmholtz function** A and the **Gibbs function** G:

$$\text{Helmholtz function: } A = U - TS \qquad \text{(19a)}$$

$$\text{Gibbs function: } G = H - TS \qquad \text{(19b)}$$

All the symbols refer to the system. When the state of the system changes at constant temperature,

$$dA = dU - T\,dS; \qquad dG = dH - T\,dS$$

When we introduce eqn 17 into the first of these relations and eqn 18 into the second, we obtain the criteria of spontaneous change as

$$dA_{T,V} \leq 0; \qquad dG_{T,p} \leq 0 \qquad \text{(20)}$$

(It must be understood that no non-expansion work is being done in either case.) These inequalities are the most important conclusions from thermodynamics for chemistry. They are developed in subsequent sections and chapters.

Some remarks on the Helmholtz function

A change in a system at constant temperature and volume is spontaneous if $dA_{T,V} \leq 0$. That is, a change under these conditions is spontaneous if it corresponds to a decrease in the Helmholtz function. Such systems move spontaneously towards states of lower A if a path is available, and the criterion of equilibrium is $dA_{T,V} = 0$.

The expression $dA = dU - T\,dS$ and $dA < 0$ is sometimes interpreted as follows. A negative value of dA is favoured by a negative value of dU and a positive value of $T\,dS$, which suggests that the tendency of a system to move to lower A is due to its tendency to move towards states of lower internal energy and higher entropy. This interpretation is false (even though it is a good rule of thumb for remembering the expression for dA) because the tendency to lower A is solely a tendency towards states of greater overall entropy. Systems change spontaneously if in doing so the total entropy increases, not because they tend to lower internal energy. The form of dA gives the impression that systems favour lower energy, but that is misleading. dS is the entropy change of the system, and $-dU/T$ is the entropy change of the surroundings: their total tends to a maximum.

Maximum work

It turns out that A carries a greater significance than being simply a signpost. If we know the value of dA for a change (and of ΔA for a measurable change), then we can also state the maximum amount of work the system can do. This is why A is sometimes called the **maximum work function**, or the **work function** (*Arbeit* is the German word for work; hence A). The proof is in two stages.

First we prove that a system does maximum work when it is working reversibly. (This was demonstrated in Chapter 2 for the expansion of a perfect gas; now we prove its universal validity.) We combine the Clausius inequality d$S \geq$ dq/T in the form T d$S \geq$ dq with the First Law d$U =$ d$q +$ dw, and obtain

$$\mathrm{d}U \leq T\,\mathrm{d}S + \mathrm{d}w$$

(because we are replacing dq by something that in general is larger). This rearranges to

$$-\mathrm{d}w \leq -\mathrm{d}U + T\,\mathrm{d}S$$

We are interested in the maximum work that we can obtain from the system, and so we are interested in cases when dw is negative and hence $-$dw is positive. It will simplify keeping track of signs therefore if we deal with the absolute value of the work, $|\mathrm{d}w|$, and write the last inequality as

$$|\mathrm{d}w| \leq |-\mathrm{d}U + T\,\mathrm{d}S|$$

Then, since T, S, and U are all state functions, it follows that the maximum value of $|\mathrm{d}w|$ is given by

$$|\mathrm{d}w_{\mathrm{max}}| = |-\mathrm{d}U + T\,\mathrm{d}S|$$

and that this work is done only when the path is traversed reversibly (because then the equality applies). Since at constant temperature d$A =$ d$U - T$ dS, we conclude that $|\mathrm{d}w_{\mathrm{max}}| = |\mathrm{d}A|$. For a measurable change when the Helmholtz function changes by ΔA, it follows that

$$|w_{\mathrm{max}}| = |\Delta A| \qquad (21)_{\mathrm{r}}$$

That is, the maximum work for a specified change of state is given by the change in A.

We can write eqn 21 in the form

$$|w_{\mathrm{max}}| = |\Delta U - T\,\Delta S|$$

We see that in some cases, depending on the sign of $T\,\Delta S$, not all the change in internal energy may be available for doing work. If the change occurs with a decrease in entropy (of this system), so $T\,\Delta S < 0$, the maximum work is less than $|\Delta U|$. This is because, in order for the change to be spontaneous, some of the energy must escape as heat in order to generate enough entropy in the surroundings to overcome the reduction in the system (Fig. 4.11a). In this case Nature is demanding a tax on the internal energy as it is converted into work. This is the origin of the name (Helmholtz) **free energy** for A, since ΔA is that part of the change in internal energy that we are free to use to do work.

If the change occurs with an increase in entropy of the system (so that

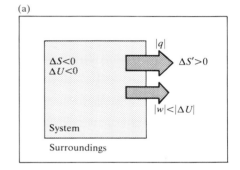

(a)

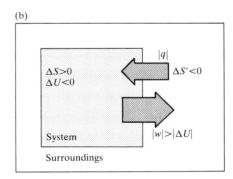

(b)

Fig. 4.11 In a system not isolated from its surroundings, the work done may be different from the change in internal energy. Moreover, the process is spontaneous if overall the entropy of the global, isolated system increases. In (a) the entropy of the system decreases, so that of the surroundings must increase in order for the process to be spontaneous, which means that energy must pass from the system to the surroundings as heat. Therefore, less work than ΔU can be obtained. In (b) the entropy of the system increases, hence we can afford to lose some entropy of the surroundings; that is, some of their energy may be lost as heat to the system. This energy can be returned to them as work. Hence the work done can exceed ΔU.

$T \Delta S > 0$), then the maximum work is greater than $|\Delta U|$. The explanation of this apparent paradox is that the system is not isolated, and energy may flow in as heat or work is done. Since the entropy of the system increases, we can afford a reduction in entropy of the surroundings yet still have, overall, a spontaneous process. Therefore, some heat (no more than the value of $T \Delta S$) may leave the surroundings and contribute to the work the change is generating (Fig. 4.11b). Nature is now providing a tax refund.

Example 4.10: *Calculating the maximum available work*

When glucose is oxidized to carbon dioxide and water at 25°C according to the equation

$$C_6H_{12}O_6(s) + 6O_2(g) \rightarrow 6CO_2(g) + 6H_2O(l)$$

calorimetric measurements give $\Delta U^{\ominus} = -2808 \text{ kJ mol}^{-1}$ and $\Delta S^{\ominus} = +182.4 \text{ J K}^{-1} \text{ mol}^{-1}$ at 𝔗. How much of this energy change can be extracted as (a) heat, (b) work?

Answer. Use $\Delta U = q_V$, $\Delta H = q_p$, and $\Delta H = \Delta U$ (since the number of gas molecules is unchanged by the reaction). For the available work, use eqn 21. (a) Under conditions of constant volume or constant pressure, the quantity of energy available as heat is 2808 kJ mol^{-1}. (b) Since $T = 298$ K, the value of ΔA is

$$\Delta A = \Delta U - T \Delta S = -2862 \text{ kJ mol}^{-1}$$

Therefore, the combustion of glucose can be used to produce up to 2862 kJ mol^{-1} of work.

Comment. The maximum work available is greater than the change in internal energy on account of the positive entropy of reaction (which is partly due to the generation of a large number of small molecules from one big one). The system can therefore draw in energy from the surroundings (so reducing their entropy) and make it available for doing work.

Exercise. Repeat the calculation for the combustion of methane under the same conditions using enthalpy data from Table 2.9 and $\Delta S^{\ominus} = -140.3 \text{ J K}^{-1}$.
$$[q_V = -887 \text{ kJ}, q_p = -890 \text{ kJ}, |w_{max}| = 845 \text{ kJ}]$$

Some remarks on the Gibbs function

The Gibbs function, which is also known as the **Gibbs free energy**, is more common in chemistry than the Helmholtz function because, at least in laboratory chemistry, we are usually more interested in changes occurring at constant pressure, than at constant volume. The criterion $(dG)_{T,p} \leq 0$ carries over into chemistry as the remark that, at constant temperature and pressure, chemical reactions are spontaneous in the direction of decreasing Gibbs function. Therefore, if we want to know whether a reaction will run in some direction, the pressure and temperature being constant, we assess the change of Gibbs function for the reaction. If G decreases as the reaction proceeds, the reaction has a tendency to convert the reactants into products. If G increases, the reverse reaction is spontaneous.

The spontaneity of endothermic reactions

An illustration of the role of G is provided by the existence of spontaneous endothermic reactions. Since in such reactions H increases, the system rises spontaneously to states of higher enthalpy, and $dH > 0$. That the reaction is spontaneous implies that $\mathrm{d}G < 0$ despite $\mathrm{d}H > 0$. This must mean that the entropy of the system increases so much that $T\,\mathrm{d}S$ is strongly positive and outweighs $\mathrm{d}H$ in $\mathrm{d}G = \mathrm{d}H - T\,\mathrm{d}S$. Endothermic reactions are therefore driven by the increase in entropy of the system, and this entropy change overcomes the reduction in entropy brought about in the surroundings by the inflow of heat into the system ($\mathrm{d}S' = -\mathrm{d}H/T$).

Example 4.11: *Calculating the reaction Gibbs function*

Calculate the change in Gibbs function when $N_2O_4(g)$ forms $NO_2(g)$ under standard conditions at $\mathcal{F}$. The change of entropy is $+175.8\,\mathrm{J\,K^{-1}\,mol^{-1}}$ for the reaction.

Answer. The reaction is $N_2O_4(g) \rightarrow 2NO_2(g)$. Use the enthalpies of formation (Table 2.10) to calculate the standard reaction enthalpy. Then combine the values to give $\Delta G = \Delta H - T\,\Delta S$, which follows from eqn 19b at constant temperature. The standard reaction enthalpy is $\Delta H^{\ominus} = +57.2\,\mathrm{kJ\,mol^{-1}}$. Therefore, at $\mathcal{F}$,

$$\Delta G^{\ominus} = \Delta H^{\ominus} - T\,\Delta S^{\ominus}$$
$$= +57.2\,\mathrm{kJ\,mol^{-1}} - 298.15\,\mathrm{K} \times 175.8\,\mathrm{J\,K^{-1}\,mol^{-1}}$$
$$= +4.8\,\mathrm{kJ\,mol^{-1}}$$

Comment. The reaction as specified is not spontaneous at $\mathcal{F}$, but we see that as the temperature increases, the negative contribution of $T\,\Delta S$ becomes more important and eventually may dominate the endothermic enthalpy change. We discuss this question in detail in Chapter 9 when we shall see that partial dissociation occurs even at $\mathcal{F}$.

Exercise. Is the oxidation of iron to $Fe_2O_3(s)$ spontaneous given that the formation of $Fe_2O_3(s)$ is accompanied by an entropy change of $-272\,\mathrm{J\,K^{-1}\,mol^{-1}}$? $[\Delta G^{\ominus} = -741\,\mathrm{kJ\,mol^{-1}};\ \text{yes}]$

Maximum non-expansion work

The analogue of the maximum work interpretation of ΔA, and the origin of the name (Gibbs) *free* energy, can be found for ΔG. In a general change,

$$\mathrm{d}H = \mathrm{d}q + \mathrm{d}w + \mathrm{d}(pV)$$

When the change is reversible $\mathrm{d}q = T\,\mathrm{d}S$ and

$$\mathrm{d}G = T\,\mathrm{d}S + \mathrm{d}w + \mathrm{d}(pV) - T\,\mathrm{d}S = \mathrm{d}w + \mathrm{d}(pV)$$

The work consists of expansion work, which for a reversible change is given by $-p\,\mathrm{d}V$, and possibly some other kind of work (e.g. the electrical work of pushing electrons through a circuit or of raising a column of liquid); this non-expansion work we denote $\mathrm{d}w_e$. Therefore, with $\mathrm{d}(pV) = p\,\mathrm{d}V + V\,\mathrm{d}p$,

$$\mathrm{d}G = -p\,\mathrm{d}V + \mathrm{d}w_e + p\,\mathrm{d}V + V\,\mathrm{d}p = \mathrm{d}w_e + V\,\mathrm{d}p$$

If the change occurs at constant pressure (as well as constant temperature), the last term disappears, and $\mathrm{d}G = \mathrm{d}w_e$. Therefore, in terms of the magnitudes of the changes,

$$|\mathrm{d}w_e| = |\mathrm{d}G| \qquad \text{(temperature and pressure constant)}$$

However, since the process is reversible, $|\mathrm{d}w_e|$ must have its maximum value, and so for a measurable change we can conclude that

$$|w_{e,\max}| = |\Delta G| \qquad (T, p \text{ constant}) \tag{22}_r$$

That is, the maximum non-expansion work we can obtain from a process at constant pressure and temperature is given by the value of ΔG for the process. Equation 22 is particularly useful for assessing the electrical work that may be produced by fuel cells and electrochemical cells, and we shall see many applications of it.

Example 4.12: *Calculating the maximum non-expansion work of a reaction*

How much energy is available for sustaining muscular and nervous activity from the combustion of glucose molecules under standard conditions at 37°C (blood temperature)? The standard entropy change is $+182.4\,\text{J K}^{-1}\,\text{mol}^{-1}$ for the reaction as stated.

Answer. The non-expansion work available is obtained from the change in Gibbs function accompanying the combustion reaction. Ignore the temperature dependence of the reaction enthalpy and obtain $\Delta H^{\ominus}$ from Table 2.9. Then use $\Delta G = \Delta H - T\,\Delta S$.

$$\Delta G^{\ominus} = -2808\,\text{kJ mol}^{-1} - 310\,\text{K} \times 182.4\,\text{J K}^{-1}\,\text{mol}^{-1}$$

$$= -2864\,\text{kJ mol}^{-1}$$

Therefore, $|w_{e,\max}| = 2864\,\text{kJ mol}^{-1}$, and the combustion of glucose molecules can do up to $2864\,\text{kJ mol}^{-1}$ of non-expansion work.

Comment. A 70 kg person would need to do 2.1 kJ of work to climb vertically through 3 m; therefore, at least 0.13 g of glucose is needed to complete the task (and in practice significantly more).

Exercise. How much non-expansion work can be obtained from the combustion of $CH_4(g)$ under standard conditions at $\mathbf{F}$? Use $\Delta S^{\ominus} = -140\,\text{J K}^{-1}\,\text{mol}^{-1}$.

$$[849\,\text{kJ mol}^{-1}]$$

4.9 Standard molar Gibbs functions

Standard reaction entropies and reaction enthalpies can be combined to obtain the **standard reaction Gibbs function** $\Delta G^{\ominus}$:

$$\Delta G^{\ominus} = \Delta H^{\ominus} - T\,\Delta S^{\ominus} \tag{23}$$

As in the case of reaction enthalpies, it is convenient to define the **standard Gibbs function of formation** $\Delta G_f^{\ominus}$:

The standard Gibbs function of formation is the standard reaction Gibbs function for the formation of a compound from its elements in their reference states.

Standard Gibbs functions of formation of the elements are zero, since their formation is a 'null' reaction, as in

$$Cl_2(g) \rightarrow Cl_2(g) \qquad \Delta G_f^{\ominus} = 0$$

A selection of values for substances is given in Table 4.4. From the values

Table 4.4. Standard Gibbs functions of formation at 298 K

	$\Delta G_f^{\ominus}/(\text{kJ mol}^{-1})$
C(s, diamond)	+2.9
$C_6H_6(l)$	+124.3
$CH_4(g)$	−50.7
$CO_2(g)$	−394.4
$H_2O(l)$	−237.1
$NH_3(g)$	−16.5
NaCl(s)	−384.1

there, it is a simple matter to obtain the standard reaction Gibbs function by taking the appropriate combination:

$$\Delta G^{\ominus} = \sum_{J} v_J \, \Delta G_f^{\ominus}(J) \qquad (24)$$

Example 4.13: *Calculating a standard reaction Gibbs function*

Calculate the standard reaction Gibbs function for $CO(g) + \frac{1}{2}O_2(g) \rightarrow CO_2(g)$ at $\mathcal{T}$.

Answer. Write the reaction in the form $0 = CO_2(g) - CO(g) - \frac{1}{2}O_2(g)$, identify the stoichiometric coefficients, and use eqn 24 with data from Table 4.4. The stoichiometric coefficients are $v(CO_2) = 1$, $v(CO) = -1$, and $v(O_2) = -\frac{1}{2}$. Therefore:

$$\Delta G^{\ominus}(\mathcal{T}) = v(CO_2)\Delta G_f^{\ominus}(CO_2, g) + v(CO)\Delta G_f^{\ominus}(CO, g)$$
$$+ v(O_2)\Delta G_f^{\ominus}(O_2, g)$$
$$= \{-394.4 - (-137.2) - \tfrac{1}{2}(0)\} \text{ kJ mol}^{-1}$$
$$= -257.2 \text{ kJ mol}^{-1}$$

Comment. $\Delta G^{\ominus}$ is the change of Gibbs function in a reaction in which the pure, separated gases are converted *completely* into the pure product. In due course we shall have to deal with changes of Gibbs function for systems in which there are mixtures of reactants and products.

Exercise. Calculate the standard reaction Gibbs function for the combustion of $CH_4(g)$ at $\mathcal{T}$. $[-818 \text{ kJ mol}^{-1}]$

Calorimetry (for ΔH directly, and for S via heat capacities) is only one of the ways of determining the values of Gibbs functions. They may also be obtained from equilibrium constants (Chapter 8) and electrochemical measurements (Chapter 10), and they may be calculated using data from spectroscopic observations (Chapter 20). The information in Table 4.4, however, together with the machinery we shall now construct, is all we need in order to draw far-reaching conclusions about reactions and other processes of interest in chemistry.

Further reading

P. W. Atkins, *The second law*. Scientific American Books, New York (1984).

P. A. Rock, *Chemical thermodynamics*. University Science Books and Oxford University Press (1983).

J. B. Fenn, *Engines, energy, and entropy*. W. H. Freeman & Co., New York (1982).

M. W. Zemansky and R. H. Dittman, *Heat and thermodynamics*. McGraw-Hill, New York (1981).

L. K. Nash, Bibliography of thermodynamics. *J. chem. Educ.*, **42**, 71 (1965).

Exercises

Assume that all gases are perfect and that data refer to 298 K unless otherwise stated.

4.1 Calculate the change in entropy when 25 kJ of energy is transferred reversibly and isothermally as heat to a large block of iron at (a) 0°C, (b) 100°C.

4.2 Calculate the molar entropy of a constant volume sample of $Ne(g)$ at 500 K given that it is 146.22 J K^{-1} mol^{-1} at 298 K.

4 | The Second Law: the concepts

4.3 A 1.75 kg sample of aluminium is cooled at constant pressure from 300 K to 265 K. Calculate the amount of energy that must be removed as heat and the change in entropy of the sample.

4.4 A 25 g sample of methane gas at 250 K and 18.5 atm expands isothermally until its pressure is 2.5 atm. Calculate the change in entropy of the gas.

4.5 A sample of perfect gas that initially occupies 15.0 L at 250 K and 1.00 atm is compressed isothermally. To what volume must be the gas be decreased in order to reduce its entropy by $5.0\,\mathrm{J\,K^{-1}}$?

4.6 Calculate the change in entropy when 50 g of water at 80°C is poured into 100 g of water at 10°C in an insulated vessel given that $C_p = 75.5\,\mathrm{J\,K^{-1}\,mol^{-1}}$.

4.7 The enthalpy of vaporization of chloroform ($CHCl_3$) is $29.4\,\mathrm{kJ\,mol^{-1}}$ at its normal boiling point of 334.88 K. Calculate the entropy of vaporization of chloroform at this temperature. What is the entropy change in the surroundings?

4.8 Calculate the standard reaction entropy at 298 K of

(a) $2CH_3CHO(g) + O_2(g) \rightarrow 2CH_3COOH(l)$
(b) $2AgCl(s) + Br_2(l) \rightarrow 2AgBr(s) + Cl_2(g)$
(c) $Hg(l) + Cl_2(g) \rightarrow HgCl_2(s)$
(d) $Zn(s) + Cu^{2+}(aq) \rightarrow Zn^{2+}(aq) + Cu(s)$
(e) $C_{12}H_{22}O_{11}(s) + 12O_2(g) \rightarrow 12CO_2(g) + 11H_2O(l)$

4.9 Combine the reaction entropies calculated in Exercise 4.8 with the reaction enthalpies and calculate the standard reaction Gibbs functions of the reactions at 298 K.

4.10 Use standard Gibbs functions of formation to calculate the standard reaction Gibbs function at 298 K of the reactions in Exercise 4.8.

4.11 The standard enthalpy of combustion of solid phenol (C_6H_5OH) is $-3054\,\mathrm{kJ\,mol^{-1}}$ at 298 K and its standard molar entropy is $144.0\,\mathrm{J\,K^{-1}\,mol^{-1}}$. Calculate the standard Gibbs function of formation of phenol at 298 K.

4.12 Calculate the entropy change in the system and the surroundings, and the total change, when a 14 g sample of $N_2(g)$ at 298 K and 1.00 bar doubles its volume in (a) an isothermal reversible expansion, (b) an isothermal irreversible expansion against $p_{ex} = 0$, and (c) an adiabatic reversible expansion.

4.13 Calculate the change in entropy when a perfect gas is compressed to half its volume and simultaneously heated to twice its initial temperature.

4.14 Calculate the maximum non-expansion work per mole that may be obtained from a fuel cell in which the chemical reaction is the combustion of methane at 298 K.

4.15 Calculate the Carnot efficiency of a primitive steam engine operating on steam at 100°C and discharging at 60°C. Repeat the calculation for a modern steam turbine that operates with steam at 300°C and discharges at 80°C.

4.16 The enthalpy of the graphite $\rightarrow$ diamond phase transition, which under 100 kbar occurs at 2000 K, is $+1.9\,\mathrm{kJ\,mol^{-1}}$. Calculate the entropy change of the transition.

4.17 How much work must be done in order to cool the air in an otherwise empty room of dimensions 5.0 m × 5.0 m × 3.0 m from 30°C to 22°C when the ambient temperature is (a) 20°C, (b) 30°C? (Assume that $c°$ is constant over the range and that $C_p = 29\,\mathrm{J\,K^{-1}\,mol^{-1}}$ for the air of mean density 1.2 mg cm^{-3} and molar mass 29 g mol^{-1}.)

4.18 A refrigerator operating reversibly extracts 45 kJ of energy as heat from a cold source and delivers 67 kJ to a hot sink at 300 K. Calculate the temperature of the cold source.

4.19 A refrigerator operates reversibly between 80 K and 200 K. Calculate the work required to remove 2.10 kJ of energy from the cold source to the hot sink as heat.

4.20 Calculate the coefficient of performance of a perfect refrigerator operating in a room at 20°C when the interior is at (a) 0°C, (b) −10°C.

4.21 Calculate the minimum work needed to freeze 250 g of water originally at 0°C standing in a room at 20°C. What would be the minimum time required in a refrigerator operating ideally at 100 W?

Problems

Assume that all gases are perfect and that data refer to 298 K unless otherwise stated.

Numerical problems

4.1 Calculate the difference in molar entropy (a) between liquid water and ice at −5°C, (b) between liquid water and its vapour at 95°C and 1.00 atm. The differences in heat capacities on melting and on vaporization are $37.3\,\mathrm{J\,K^{-1}\,mol^{-1}}$ and $-41.9\,\mathrm{J\,K^{-1}\,mol^{-1}}$ respectively. Distinguish between the entropy of the sample, the surroundings, and the total system, and discuss the spontaneity of the transitions at the two temperatures.

4.2 The heat capacity of chloroform (trichloromethane) in the range 240 K to 330 K is given by

$$C_p/(\mathrm{J\,K^{-1}\,mol^{-1}}) = 91.47 + 7.5 \times 10^{-2}\,(T/\mathrm{K})$$

In a particular experiment, 1.00 mol $CHCl_3$ is heated from 273 K to 300 K. Calculate the change in molar entropy of the sample.

4.3 The standard molar entropy of $NH_3(g)$ is $192.4\,\mathrm{J\,K^{-1}\,mol^{-1}}$ at 298 K, and its heat capacity is given by

$$C_p = a + bT + \frac{c}{T^2}$$

with the coefficients given in Table 2.16. Calculate the standard molar entropy at (a) 100°C and (b) 500°C.

4.4 A block of copper of mass 500 g and initially at 293 K is in thermal contact with an electric heater of resistance 1.00 kΩ and negligible mass. A current of 1.00 A is passed for 15.0 s. Calculate the change in entropy of the copper, taking $C_p = 24.4 \, J \, K^{-1} \, mol^{-1}$. The experiment is then repeated with the copper immersed in a stream of water that maintains its temperature at 293 K. Calculate the change in entropy of the copper and the water in this case.

4.5 Calculate the standard Helmholtz function of formation of $CH_3OH(l)$ at 298 K from the standard Gibbs function of formation and the assumption that H_2 and O_2 are perfect gases.

4.6 Calculate the change in entropy when 200 g of (a) water at 0°C, (b) ice at 0°C is added to 200 g of water at 90°C in an insulated container.

4.7 Calculate (a) the maximum work and (b) the maximum non-expansion work that can be obtained from the freezing of supercooled water at −5°C and 1.0 atm. The densities of water and ice are $0.999 \, g \, cm^{-3}$ and $0.917 \, g \, cm^{-3}$ respectively at −5°C.

4.8 The heat capacity of lead varies with temperature as follows:

T/K	10	15	20	25	30	50
$C_p/(J \, K^{-1} \, mol^{-1})$	2.8	7.0	10.8	14.1	16.5	21.4

T/K	70	100	150	200	250	298
$C_p/(J \, K^{-1} \, mol^{-1})$	23.3	24.5	25.3	25.8	26.2	26.6

Calculate the standard Third Law entropy of lead at (a) 0°C and (b) 25°C.

4.9 Suppose that an internal combustion engine runs on octane, for which the enthalpy of combustion is $-5512 \, kJ \, mol^{-1}$ and take the mass of 1 gallon of fuel as 3 kg. What is the maximum height, neglecting all forms of friction, to which a 1000 kg car can be driven on 1 gallon of fuel given that the engine cylinder temperature is 2000°C and the exit temperature is 800°C?

4.10 The standard reaction Gibbs function of

$$K_4[Fe(CN)_6] \cdot 3H_2O(s) \rightarrow 4K^+(aq)$$

$$+ [Fe(CN)_6]^{4-}(aq) + 3H_2O(l)$$

is $+26.120 \, kJ \, mol^{-1}$ (I. R. Malcolm, L. A. K. Staveley, and R. D. Worswick, *J. chem. Soc. Faraday Trans.* I, 1532 (1973)). The enthalpy of solution of the trihydrate is $+55.000 \, kJ \, mol^{-1}$. Calculate the standard molar entropy of solution of the hexacyanoferrate(II) ion in water given that the standard molar entropy of the solid trihydrate is $599.7 \, J \, K^{-1} \, mol^{-1}$ and that of the K^+ ion in water is $102.5 \, J \, K^{-1} \, mol^{-1}$.

4.11 The heat capacity of anhydrous potassium hexacyanoferrate(II) varies with temperature as follows:

T/K	10	20	30	40	50	60	70	80
$C_p/(J \, K^{-1} \, mol^{-1})$	2.09	14.43	36.44	62.55	87.03	111.0	131.4	149.4

T/K	90	100	110	150	160	170	180
$C_p/(J \, K^{-1} \, mol^{-1})$	165.3	179.6	192.8	237.6	247.3	256.5	265.1

T/K	190	200
$C_p/(J \, K^{-1} \, mol^{-1})$	273.0	280.3

Calculate the molar enthalpy relative to its value at $T = 0$ and the Third Law entropy at each of these temperatures.

4.12 The compound 1,3,5-trichloro-2,4,6-trifluorobenzene is an intermediate in the conversion of hexachlorobenzene to hexafluorobenzene, and its thermodynamic properties have been examined by measuring its heat capacity over a wide temperature range (R. L. Andon and J. F. Martin, *J. chem. Soc. Faraday Trans.* I, 871 (1973)). Some of the data are as follows:

T/K	14.14	16.33	20.03	31.15	44.08	64.81
$C_p/(J \, K^{-1} \, mol^{-1})$	9.492	12.70	18.18	32.54	46.86	66.36

T/K	100.90	140.86	183.59	225.10	262.99	298.06
$C_p/(J \, K^{-1} \, mol^{-1})$	95.05	121.3	144.4	163.7	180.2	196.4

Calculate the molar enthalpy relative to its value at $T = 0$ and the Third Law entropy of the compound at these temperatures.

4.13 Calculate the minimum work needed to reduce the temperature of a 1.0 g block of copper from 1.10 K to 0.10 K, the surroundings being at 1.20 K. Proceed by supposing that the heat capacity remains constant at $39 \, \mu J \, K^{-1} \, mol^{-1}$ and that the coefficient of performance can be evaluated at the mean temperature of the block. Then go on to do a more realistic calculation in which we assume that

$$C_p = AT^3 + BT$$

with $A = 48.2 \, \mu J \, K^{-4} \, mol^{-1}$ and $B = 688 \, \mu J \, K^{-2} \, mol^{-1}$, and taking into account the variation of the coefficient of performance with temperature.

Theoretical problems

4.14 Show that the integral of dq/T round a Carnot cycle, denoted $\oint dq/T$, is zero. Then show that if the isothermal reversible expansion stage is replaced by an isothermal irreversible expansion that $\oint dq/T < 0$.

4.15 Represent the Carnot cycle on a temperature/entropy diagram and show that the area enclosed by the cycle is equal to the work done.

4.16 Find an expression for the change in entropy when two blocks of the same substance and of equal mass, one at the temperature T_h and the other at T_c, are brought into thermal contact and allowed to reach equilibrium. Evaluate the change for two 500 g blocks of copper with $C_p = 24.4 \, J \, K^{-1} \, mol^{-1}$, taking $T_h = 500 \, K$ and $T_c = 250 \, K$.

4.17 The definitions of the enthalpy, Gibbs function, and

Helmholtz function have all been of the form

$$g = f + yz$$

Show that the addition of the product yz is a general way of converting a function of x and y to a function of x and z in the sense that if $df = a\,dx - z\,dy$ then

$$dg = a\,dx + y\,dz$$

4.18 Calculate the minimum work needed to cool an object of finite size from T_i to T_f using a refrigerator in a room at temperature T_h. (Hint: write $|dw| = |dq|/c°$, and express dq in terms of dT using the heat capacity, which is assumed to be constant. Then integrate the expression.) Find the work needed to freeze 250 g of water initially at 20°C, the temperature of the room. What work would be required if the water was initially at 25°C (and the room still at 20°C)?

The Second Law: the machinery

5

Check-list of key ideas

1. The formulation of the *fundamental equation* (eqn 1) expressing a change in internal energy in terms of changes in the volume and the entropy.

2. The derivation of the *Maxwell relations* (Table 5.1) between differential coefficients.

3. The derivation of a *thermodynamic equation of state* (eqn 4) relating the internal pressure of a sample to its pressure and temperature.

4. The variation of the Gibbs function with the temperature and the pressure (eqn 6) and the derivation of the *Gibbs–Helmholtz equation* (eqn 8).

5. The pressure dependence of the Gibbs function of a solid and liquid (eqn 10) and of a perfect gas (eqn 11).

6. The introduction of the *chemical potential* (eqn 13) and the variation with pressure of the chemical potential of a perfect gas (eqn 14).

7. The definition of the *fugacity* and the *standard state* of a real gas (Section 5.3) and the calculation of the *fugacity coefficient* from compressibility data (eqn 17).

8. The chemical potential of a substance in a mixture (Section 5.4) and the derivation of the *fundamental equation of chemical thermodynamics* (eqn 20).

9. The relation of the *chemical potential* to the composition dependence of *thermodynamic functions* (Section 5.5).

In this chapter we begin to derive some of the rich consequences of combining the First and Second Laws of thermodynamics. In particular, we explore the Gibbs function—the Gibbs free energy—and show how it varies with temperature, pressure, and composition. We also meet the 'chemical potential', the quantity on which almost all the most important applications of thermodynamics to chemistry are based.

Combining the First and Second Laws

We have seen that the First Law of thermodynamics may be written

$$dU = dq + dw$$

For a reversible change in a closed system (one in which there is no change of composition), and in the absence of any non-expansion work,

$$dw = -p \, dV \qquad dq = T \, dS$$

Therefore,

$$dU = T \, dS - p \, dV \qquad (1)$$

However, since dU is an exact differential, its value is independent of path, and so the same value of dU is obtained whether the change is brought about irreversibly or reversibly. Therefore, eqn 1 applies to any change—reversible or irreversible—of a closed system that does no non-expansion work. We shall call this combination of the First and Second Laws the **fundamental equation**.

The fact that the fundamental equation applies to both reversible and irreversible changes may appear puzzling at first sight. The reason is that only in the case of a reversible change may $T \, dS$ be identified with dq and $-p \, dV$ with dw. When the change is irreversible, $T \, dS > dq$ (the Clausius inequality) and $p \, dV > dw$. The *sum* of dw and dq remains equal to the *sum* of $T \, dS$ and $-p \, dV$, provided the composition is constant.

5.1 Properties of the internal energy

Equation 1 shows that the internal energy of a closed system changes in a simple way when S and V are changed ($dU \propto dS$ and $dU \propto dV$). This suggests that U may be regarded as a function of S and V and that we should write it $U(S, V)$. We could regard U as a function of other variables, such as S and p or T and V, because they are all interrelated; but the simplicity of the fundamental equation suggests that $U(S, V)$ is the best choice when the system is closed and no non-expansion work is done.

The *mathematical* consequence of U being a function of S and V is that a change dU can be expressed in terms of the changes dS and dV by

$$dU = \left(\frac{\partial U}{\partial S}\right)_V dS + \left(\frac{\partial U}{\partial V}\right)_S dV \qquad (2)$$

When this is compared to the *thermodynamic* relation, eqn 1, it follows that for systems of constant composition,

$$\left(\frac{\partial U}{\partial S}\right)_V = T \qquad -\left(\frac{\partial U}{\partial V}\right)_S = p \qquad (3)$$

The first of these two equations is a purely thermodynamic definition of temperature as the ratio of the changes in the internal energy and entropy of a constant volume closed system. We are beginning to generate relations between the properties of a system and to discover the power of thermodynamics for establishing unexpected relations.

The Maxwell relations

Since the fundamental equation is an expression for an exact differential, the coefficients of dS and dV must pass the test of Relation 4 in Box 3.1 (the test for exact differentials). With $a = T$ and $b = -p$ it follows that

$$\left(\frac{\partial T}{\partial V}\right)_S = -\left(\frac{\partial p}{\partial S}\right)_V$$

We have generated a relation between quantities which, at first sight, would not seem to be related.

The equation just derived is an example of a **Maxwell relation**. However, apart from being unexpected, it does not look particularly interesting. Nevertheless, it does suggest that there may be other similar relations that are more useful. Indeed, the fact that H, G, and A are state functions can be used to derive three more Maxwell relations. The argument to obtain them runs in the same way in each case: since H, G, and A are state functions, the expressions for dH, dG, and dA satisfy Relation 4 in Box 3.1. All four relations are listed in Table 5.1. In the next section we derive one of them, but as no new principles are involved we shall not derive them all.

Table 5.1. The Maxwell relations

$$\left(\frac{\partial T}{\partial V}\right)_S = -\left(\frac{\partial p}{\partial S}\right)_V$$

$$\left(\frac{\partial T}{\partial p}\right)_S = \left(\frac{\partial V}{\partial S}\right)_p$$

$$\left(\frac{\partial p}{\partial T}\right)_V = \left(\frac{\partial S}{\partial V}\right)_T$$

$$\left(\frac{\partial V}{\partial T}\right)_p = -\left(\frac{\partial S}{\partial p}\right)_T$$

The variation of internal energy with volume

The coefficient

$$\pi_T = \left(\frac{\partial U}{\partial V}\right)_T$$

played a central role in the manipulation of the First Law, and in Section 3.3 we used the relation

$$\pi_T = T\left(\frac{\partial p}{\partial T}\right)_V - p \tag{4}$$

which was to be proved in this chapter. We are now ready to derive it from the relations we have just established.

We can obtain the coefficient π_T from eqn 2 using Relation 1 of Box 3.1:

$$\left(\frac{\partial U}{\partial V}\right)_T = \left(\frac{\partial U}{\partial S}\right)_V\left(\frac{\partial S}{\partial V}\right)_T + \left(\frac{\partial U}{\partial V}\right)_S = T\left(\frac{\partial S}{\partial V}\right)_T - p$$

This is already beginning to look like the expression we want. One of the Maxwell relations does the job of turning $(\partial S/\partial V)_T$ into something else:

$$\left(\frac{\partial S}{\partial V}\right)_T = \left(\frac{\partial p}{\partial T}\right)_V$$

and substituting this relation completes the proof that

$$\pi_T = T\left(\frac{\partial p}{\partial T}\right)_V - p$$

This is called a **thermodynamic equation of state** because it expresses a quantity in terms of the two variables T and p and applies to any material and any phase.

Example 5.1: *Deriving a thermodynamic relation*

Show thermodynamically that $\pi_T = 0$ for a perfect gas, and compute its value for a van der Waals gas.

Answer. Proving a result 'thermodynamically' means basing it entirely on general thermodynamic relations and equations of state, without drawing on molecular arguments (such as the existence of intermolecular forces). For a perfect gas, $p = nRT/V$. Therefore, from eqn 4 with $(\partial p / \partial T)_V = nR/V$,

$$\pi_T = T\left(\frac{nR}{V}\right) - p = 0$$

For a van der Waals gas,

$$p = \frac{nRT}{V - nb} - \frac{n^2 a}{V^2}$$

Therefore, since

$$\left(\frac{\partial p}{\partial T}\right)_V = \frac{nR}{V - nb}$$

we can write

$$\pi_T = \frac{nRT}{V - nb} - \left(\frac{nRT}{V - nb}\right) + \frac{n^2 a}{V^2} = \frac{a}{V_m^2}$$

Comment. This calculation shows that the internal energy of a van der Waals gas *increases* when it expands isothermally (because $\pi_T > 0$), and that the increase is related to the parameter that models the attractive interactions between the particles: a larger molar volume implies a weaker mean attraction.

Exercise. Calculate π_T for a gas that obeys the virial equation of state.

$$[RT^2(\partial B / \partial T)_V / V_m^2 + \dots]$$

5.2 Properties of the Gibbs function

The same arguments as we applied to the fundamental equation for U may be applied to the Gibbs function $G = H - TS$. When the system changes its state, G may change because H, T, and S change. For infinitesimal changes in each property,

$$dG = dH - T\,dS - S\,dT$$

Since $H = U + pV$, we have

$$dH = dU + p\,dV + V\,dp$$

For a closed system doing no non-expansion work, dU can be replaced by the fundamental equation $dU = T\,dS - p\,dV$. The result of these steps is

$$dG = (T\,dS - p\,dV) + p\,dV + V\,dp - T\,dS - S\,dT$$

That is, for a closed system doing no non-expansion work (and therefore at equilibrium internally),

$$dG = V\,dp - S\,dT \tag{5}$$

This expression suggests that G may be regarded as a function of p and T, and that we should write it $G(p, T)$. This confirms that it is an important quantity in chemistry because the pressure and temperature are usually the variables under our control. Hence, G carries around the combined consequences of the First and Second Laws in a way that makes it particularly suitable for chemical applications.

The same argument that led to eqn 3, when applied to the exact differential dG, now gives

$$\left(\frac{\partial G}{\partial T}\right)_p = -S \qquad \left(\frac{\partial G}{\partial p}\right)_T = V \qquad (6)$$

Example 5.2: *Deriving thermodynamic relations*

Derive the two relations in eqn 6.

Exercise. If we regard G as a function of p and T, a change in its value can be expressed as

$$dG = \left(\frac{\partial G}{\partial p}\right)_T dp + \left(\frac{\partial G}{\partial T}\right)_p dT$$

However, dG is also given by eqn 5:

$$dG = V\,dp - S\,dT$$

Since it is an exact differential, the coefficients of dT and dp must be equal in the two cases, so

$$\left(\frac{\partial G}{\partial p}\right)_T = V \qquad \left(\frac{\partial G}{\partial T}\right)_p = -S$$

as we were asked to prove.

Exercise. Find the corresponding relations based on A treated as a function of T and V.

$$[(\partial A/\partial V)_T = -p, \ (\partial A/\partial T)_V = -S]$$

The relations in eqn 6 show how the Gibbs function varies with temperature and pressure. Since S is positive, it follows that G decreases when the temperature is raised at constant pressure and composition. Moreover, the relation shows that G decreases most sharply when the entropy of the system is large. Therefore, the Gibbs function of the gaseous phase of a substance, which has a high molar entropy, is more sensitive to temperature than its liquid and solid phases. Since V is positive, G always increases when the pressure of the system is increased at constant temperature (and composition). Since the molar volumes of gases are large, G is more sensitive to changes of pressure for the gaseous phase of a substance than for its liquid and solid phases.

Example 5.3: *Calculating the effect of pressure on the Gibbs function*

Calculate the change in the molar Gibbs function of (a) liquid water treated as an incompressible fluid and (b) water vapour treated as a perfect gas when the pressure is increased isothermally from 1.0 bar to 2.0 bar at 298 K.

Answer. Since the increase in pressure is isothermal, eqn 5 becomes (for molar quantities)

$$\Delta G_m = \int V_m \, dp$$

For the incompressible liquid, V_m is constant at $18.0 \text{ cm}^3 \text{ mol}^{-1}$, so

$$\Delta G_m = V_m \Delta p = 18.0 \times 10^{-6} \text{ m}^3 \text{ mol}^{-1} \times 1.0 \times 10^5 \text{ Pa}$$

$$= +1.8 \text{ J mol}^{-1}$$

(because $1 \text{ Pa m}^3 = 1 \text{ N m} = 1 \text{ J}$). For the perfect gas, the molar volume changes with pressure, and we write $V_m = RT/p$:

$$\Delta G_m = RT \int_{p_i}^{p_f} \frac{dp}{p} = RT \ln \frac{p_f}{p_i}$$

$$= 2.48 \text{ kJ mol}^{-1} \times \ln 2.0$$

$$= +1.7 \text{ kJ mol}^{-1}$$

Comment. Note that G increases in both cases, and the increase for a gas is 1000 times greater than for the liquid.

Exercise. Calculate the change in G_m for ice at $-10°C$, when it has density 0.917 g cm^{-3}, when the pressure is increased from 1.0 bar to 2.0 bar.

$$[+2.0 \text{ J mol}^{-1}]$$

When we examine eqn 5 with the test for an exact differential using relation 4 of Box 3.1, and set $a = V$ and $b = -S$, we find that

$$\left(\frac{\partial V}{\partial T} \right)_p = -\left(\frac{\partial S}{\partial p} \right)_T \qquad (7)$$

This is another of the Maxwell relations in Table 5.1.

The temperature dependence of the Gibbs function

The temperature dependence of G given by eqn 6 can be expressed in terms of the enthalpy because we can replace S by

$$S = \frac{H - G}{T}$$

which follows from the definition of G; then

$$\left(\frac{\partial G}{\partial T} \right)_p = \frac{G - H}{T}$$

We can turn this expression into a simpler and more useful form. First, we write it as

$$\left(\frac{\partial G}{\partial T} \right)_p - \frac{G}{T} = \frac{-H}{T}$$

The term on the left is simplified by noting that

$$\left(\frac{\partial G}{\partial T} \right)_p - \frac{G}{T} = T \left(\frac{\partial}{\partial T} \frac{G}{T} \right)_p$$

This is proved using the rules for differentiating a product:

$$\left(\frac{\partial}{\partial T}\frac{G}{T}\right)_p = \frac{1}{T}\left(\frac{\partial G}{\partial T}\right)_p + G\left(\frac{\partial}{\partial T}\frac{1}{T}\right)_p$$

$$= \frac{1}{T}\left(\frac{\partial G}{\partial T}\right)_p - \frac{G}{T^2}$$

$$= \frac{1}{T}\left\{\left(\frac{\partial G}{\partial T}\right)_p - \frac{G}{T}\right\}$$

In this way we obtain the **Gibbs–Helmholtz equation**:

$$\left(\frac{\partial}{\partial T}\frac{G}{T}\right)_p = \frac{-H}{T^2} \tag{8a}$$

(G, H is a helpful way of remembering what this equation relates.) It shows that if the enthalpy of the system is known, then the temperature dependence of G/T is also known.

Example 5.4: *Manipulating the Gibbs–Helmholtz equation*

Show that

$$\left(\frac{\partial(G/T)}{\partial(1/T)}\right)_p = H \tag{8b}$$

Answer. This is an exercise in manipulating partial differentials. We note that the left hand side of eqn 8a can be written

$$\left(\frac{\partial(G/T)}{\partial T}\right)_p = \left(\frac{\partial(G/T)}{\partial(1/T)}\right)_p\left(\frac{\partial(1/T)}{\partial T}\right)_p$$

$$= \left(\frac{\partial(G/T)}{\partial(1/T)}\right)_p \times \left(-\frac{1}{T^2}\right)$$

because

$$\frac{\mathrm{d}(1/T)}{\mathrm{d}T} = \frac{\mathrm{d}(T^{-1})}{\mathrm{d}T} = -\frac{1}{T^2}$$

Therefore, substitution of this result into eqn 8a and multiplication of both sides by T^2 gives eqn 8b.

Comment. The result shows that, if H is independent of temperature over a limited range, then a plot of G/T against $1/T$ should be a straight line of slope H. We shall see the usefulness of this result in Chapter 9.

Exercise. Find the equation for the temperature dependence of A that corresponds to eqn 8b. $[(\partial(A/T)/\partial(1/T))_V = U]$

The Gibbs–Helmholtz equation is most useful when it is applied to changes, including changes of physical state and chemical reactions at constant pressure. Then, with $\Delta G = G_f - G_i$ for the change of Gibbs function between the final and initial states, since the equation applies to both G_f and G_i, we can write

$$\left(\frac{\partial}{\partial T}\frac{\Delta G}{T}\right)_p = \frac{-\Delta H}{T^2} \tag{8c}$$

where ΔG and ΔH are for changes at constant pressure.

It may seem that $\Delta G/T$ is a clumsy quantity to deal with, but in Chapter 8 we shall see that it is exactly what is needed when discussing chemical equilibria. That the expression has an important role to play can be appreciated by noting that $-\Delta G/T$ is equal to ΔS_{tot}, the change in the total entropy of an isolated system. Therefore, the Gibbs–Helmholtz equation is a disguised form of the expression for the manner in which this total entropy changes as the temperature is changed:

$$\left(\frac{\partial \Delta S_{tot}}{\partial T}\right)_p = \frac{\Delta H}{T^2}$$

It can therefore be expected to play a major role in discussions of the effect of temperature on equilibria.

The pressure dependence of the Gibbs function

We can integrate eqn 5 to find the Gibbs function at one pressure in terms of its value at another:

$$G_f = G_i + \int_{p_i}^{p_f} V \, dp \tag{9}$$

In the case of a liquid or solid, the volume changes only slightly as the pressure changes, and so V may be treated as a constant and taken outside the integral. Then, for molar quantities,

$$G_{m,f} = G_{m,i} + (p_f - p_i)V_m$$

$$= G_{m,i} + V_m \Delta p \tag{10}$$

Under normal laboratory conditions $V_m \Delta p$ is very small, as we saw in Example 5.3, and may be neglected, and we may usually suppose that the Gibbs functions of solids and liquids are independent of pressure. However, if we are interested in geophysical problems, then since pressures in the earth's interior are huge, their effect on the Gibbs function cannot be ignored and we have to use eqn 10. If the pressures are so great that there are substantial volume changes, we must use the complete expression, eqn 9.

Example 5.5: *Calculating the effect of pressure and temperature on ΔG*

The pressure deep inside the earth is probably greater than 3×10^3 kbar, and the temperature there is around $4 \times 10^3\,°C$. Estimate the change in ΔG on going from crust to core for a process in which $\Delta V = +1.0\,cm^3\,mol^{-1}$ and $\Delta S = +2.1\,J\,K^{-1}\,mol^{-1}$.

Answer. Since $(\partial \Delta G/\partial p)_T = \Delta V$ and $(\partial \Delta G/\partial T)_p = -\Delta S$, we can use

$$\delta \Delta G \approx \Delta V \, \delta p \qquad \delta \Delta G \approx -\Delta S \, \delta T$$

and hence estimate the total change as

$$\Delta G(\text{core}) - \Delta G(\text{crust}) = [p(\text{core}) - p(\text{crust})]\Delta V - [T(\text{core}) - T(\text{crust})]\Delta S$$

$$= 3 \times 10^2 \text{ kJ mol}^{-1} - 8 \text{ kJ mol}^{-1}$$

$$= 3 \times 10^2 \text{ kJ mol}^{-1}$$

Comment. The effect of pressure dominates, and results in a very large modification of ΔG for the process. This is the thermodynamic reason why materials change their forms at great depths in the earth's interior.

Exercise. Calculate the difference in molar Gibbs function between the top and bottom of a column of mercury in a barometer. The density of mercury is 13.6 g cm^{-3}. $[1.5 \text{ J mol}^{-1}]$

The molar volumes of gases are large, and so the correction term may be large even if the pressure difference is small. Furthermore, since the volume depends strongly on the pressure, we cannot treat it as a constant in the integral in eqn 9. For a perfect gas we substitute $V = nRT/p$ into the integral, and find

$$G(p_f) = G(p_i) + nRT \int_{p_i}^{p_f} \frac{dp}{p}$$

$$= G(p_i) + nRT \ln \frac{p_f}{p_i} \qquad (11)°$$

This expression shows that when the pressure is increased tenfold at room temperature, the Gibbs function increases by about 6 kJ mol^{-1}.

The chemical potential of a perfect gas

A perfect gas is in its standard state when its pressure is $p^{\ominus} = 1$ bar; its Gibbs function is then $G^{\ominus}$. According to eqn 11, at any other pressure p its Gibbs function G is

$$G = G^{\ominus} + nRT \ln \frac{p}{p^{\ominus}} \qquad (12a)°$$

The molar Gibbs function, $G_m = G/n$, is therefore

$$G_m = G_m^{\ominus} + RT \ln \frac{p}{p^{\ominus}} \qquad (12b)°$$

We define the **chemical potential** μ of a pure substance as

$$\mu = \left(\frac{\partial G}{\partial n}\right)_{p,T} \qquad (13)$$

That is, the chemical potential shows how the Gibbs function of a system changes when the substance is added to it: if $\mu > 0$, the Gibbs function

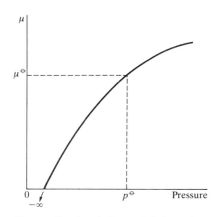

Fig. 5.1 The chemical potential of a perfect gas is proportional to ln p and the standard state is reached when $p = p^{\ominus}$. Note that as $p \to 0$, μ becomes negatively infinite.

increases as n is increased. For a pure substance, the Gibbs function is simply $G = n \times G_m$, so

$$\mu = \left(\frac{\partial nG_m}{\partial n}\right)_{p,T} = G_m$$

and the chemical potential is the same as the molar Gibbs function. Thus, for a perfect gas it follows from eqn 12b that

$$\mu = \mu^{\ominus} + RT \ln \frac{p}{p^{\ominus}} \qquad (14)°$$

This variation of the chemical potential with the pressure of a perfect gas is illustrated in Fig. 5.1.

Although it may seem that all we have done is to rename the molar Gibbs function, in due course (when μ has been defined for mixtures) we shall see that the chemical potential plays a central role in thermodynamics. We can suspect that μ will have an important role to play in discussions of chemical equilibria, because it carries the information about how G changes as the amount of substance changes, and hence how G changes as the composition of a system changes.

Example 5.6: *Calculating the change in chemical potential*

Calculate the change in chemical potential when water vaporizes at 1 bar and 25°C.

Answer. For a pure substance, the change in chemical potential is the same as the change in molar Gibbs function. Therefore, since both the liquid and the vapour are in their standard states, we can use the data in Table 2.10 for the transformation $H_2O(l, \mathcal{F}) \to H_2O(g, \mathcal{F})$ and obtain

$$\Delta G_m = \Delta G_f^{\ominus}(H_2O, g) - \Delta G_f^{\ominus}(H_2O, l)$$
$$= (-228.57 \text{ kJ mol}^{-1}) - (-237.13 \text{ kJ mol}^{-1})$$
$$= +8.56 \text{ kJ mol}^{-1}$$

That is, the chemical potential of water vapour at 1 bar and 25°C is 8.56 kJ mol^{-1} greater than that of liquid water under the same conditions.

Comment. The greater chemical potential of the vapour is consistent with the chemical intuition of water vapour as being more 'potent' than the liquid. We shall make this idea precise later.

Exercise. Calculate the change in chemical potential of benzene when it vaporizes at its boiling point. [0]

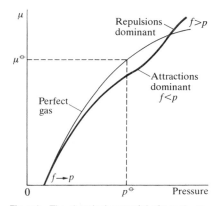

Fig. 5.2 The chemical potential of a real gas. As $p \to 0$, μ coincides with the value for a perfect gas. When attractive forces are dominant (at intermediate pressures), the chemical potential is less than that of a perfect gas and the molecules have a lower 'escaping tendency'. At high pressures, when repulsive forces are dominant, the chemical potential of a real gas is greater than that of a perfect gas. Then the 'escaping tendency' is increased.

5.3 Real gases: the fugacity

Equation 14 applies only to perfect gases. However, the same type of formula can be used to express the pressure dependence of the chemical potentials of real gases, which might resemble that shown in Fig. 5.2, and through them the Gibbs functions of systems composed of real gases. Thus we replace the true pressure p by an effective pressure, called the **fugacity**

f, and write

$$\mu = \mu^{\ominus} + RT \ln \frac{f}{p^{\ominus}} \tag{15}$$

The name 'fugacity' comes from the Latin for 'fleetness' in the sense of 'escaping tendency'; f has the same dimensions as pressure. In later chapters we derive thermodynamically exact expressions in terms of chemical potentials, and therefore in terms of fugacities. However, the results will be of practical use only if we can relate fugacities to pressures. This is the task we deal with in the remainder of this section.

Standard states of real gases

A perfect gas is in its standard state when its pressure is $p^{\ominus}$ (that is, 1 bar): the pressure arises solely from the kinetic energy of the molecules and there are no intermolecular forces to take into account. We aim to recapture this 'kinetic energy only' definition for a real gas by picturing it as a hypothetical state in which all the intermolecular forces have been extinguished:

The standard state of a real gas is a hypothetical state in which the gas is at a pressure $p^{\ominus}$ and behaving perfectly.

The advantage of this definition is that the standard state of a real gas has the simple properties of a perfect gas: if we had defined the standard state as the one for which $f = 1$ bar, the standard states of different gases would have had relatively complex properties. The choice of a hypothetical standard state literally standardizes the interactions between the particles by setting them to zero.[1] Then differences of standard chemical potential of different gases arise solely from the internal structure and properties of the molecules, not from the way they interact with each other.

The relation between fugacity and pressure

We shall write the fugacity as

$$f = \gamma p \tag{16}$$

where γ is the dimensionless **fugacity coefficient**; in general, γ depends on the identity of the gas, the pressure, and the temperature. Then

$$\mu = \mu^{\ominus} + RT \ln \frac{p}{p^{\ominus}} + RT \ln \gamma$$

As $\mu^{\ominus}$ relates to a hypothetical 'kinetic energy only' gas, and the term $\ln p/p^{\ominus}$ is the same as for a perfect gas, $RT \ln \gamma$ must express the entire effect of all the intermolecular forces. Since all gases become perfect as the pressure approaches zero (so $f \to p$ as $p \to 0$), we know that

$$\gamma \to 1 \quad \text{as} \quad p \to 0$$

We shall now derive an expression for γ. Equation 9 is true for all gases,

---

[1] The alternative choice, of selecting as the standard state the gas at zero pressure, at which it certainly behaves perfectly, runs into difficulty because (by eqn 14), $\mu \to -\infty$ as $p \to 0$.

whether real or perfect. Expressing it in terms of molar quantities and then using eqn 15 gives

$$\int_{p'}^{p} V_m \, dp = \mu - \mu' = RT \ln \frac{f}{f'}$$

In this expression, f is the fugacity when the pressure is p and f' the fugacity when the pressure is p'. If the gas were perfect we could write

$$\int_{p'}^{p} V_m^\circ \, dp = \mu^\circ - \mu^{\circ\prime} = RT \ln \frac{p}{p'}$$

where the superscript $^\circ$ denotes quantities relating to a perfect gas. The difference of the two equations is

$$\int_{p'}^{p} (V_m - V_m^\circ) \, dp = RT \left(\ln \frac{f}{f'} - \ln \frac{p}{p'} \right)$$

or

$$\ln \frac{fp'}{pf'} = \frac{1}{RT} \int_{p'}^{p} (V_m - V_m^\circ) \, dp$$

We can improve the look of this untidy expression. As $p' \to 0$, the gas behaves perfectly and f' becomes equal to the pressure p'. Therefore, $p'/f' \to 1$ as $p' \to 0$. If we take this limit (which means setting $p'/f' = 1$ on the left and $p' = 0$ on the right), the last equation becomes

$$\ln \frac{f}{p} = \frac{1}{RT} \int_{0}^{p} (V_m - V_m^\circ) \, dp$$

Then, with $\gamma = f/p$,

$$\ln \gamma = \frac{1}{RT} \int_{0}^{p} (V_m - V_m^\circ) \, dp$$

For a perfect gas, $V_m^\circ = RT/p$. For a real gas, $V_m = RTZ/p$, where Z is the compression factor (Section 1.3). Therefore

$$\ln \gamma = \int_{0}^{p} \left(\frac{Z-1}{p} \right) dp \qquad (17)$$

This is an explicit expression for the fugacity coefficient at any pressure p and therefore, through eqn 16, for the fugacity of the gas at that pressure.

To evaluate γ, we need experimental data on the compression factor from very low pressures up to the pressure of interest. Some information of this kind is available in numerical tables, in which case the integral may be evaluated numerically. Sometimes an algebraic expression is available for Z (for instance, from one of the equations of state, Table 1.5) and it may be possible to evaluate the integral analytically. Thus, if we know the virial coefficients for the gas we can obtain the fugacity using

$$\ln \gamma = B'p + \tfrac{1}{2}C'p^2 + \dots$$

This expression was obtained by explicit evaluation of eqn 17.

Example 5.7: *Calculating a fugacity*

Suppose that the attractive interactions between gas particles can be neglected, and find an expression for the fugacity of a van der Waals gas in terms of the pressure. Estimate its value for ammonia at 10.00 atm and 298.15 K.

Answer. If we neglect a in the van der Waals equation,

$$p = \frac{RT}{V_m - b}$$

and hence

$$Z = 1 + \frac{pb}{RT}$$

The integral required is therefore

$$\int_0^p \left(\frac{Z-1}{p}\right) dp = \int_0^p \frac{b\,dp}{RT} = \frac{bp}{RT}$$

Consequently, from eqns 16 and 17, the fugacity at the pressure p is

$$f = p\,e^{pb/RT}$$

From Table 1.4, $b = 3.707 \times 10^{-2}\,\text{L mol}^{-1}$, and so $pb/RT = 0.015$, giving

$$f = 10.00\,\text{atm} \times e^{0.015} = 10.15\,\text{atm}$$

Comment. The effect of the repulsive term (as represented by the b term in the van der Waals equation) is to increase the fugacity above the pressure, and so the effective pressure of the gas—its escaping tendency—is greater than if it were perfect.

Exercise. Find an expression for the fugacity coefficient when the attractive interaction is dominant in a van der Waals gas, and the pressure is low enough to make the approximation $4ap/(RT)^2 \ll 1$. Evaluate the fugacity for ammonia, as above.　　　　$[\ln \gamma = -ap/(RT)^2,\ 9.33\,\text{atm}]$

It is clear from Fig. 1.6 that for most gases $Z < 1$ up to moderate pressures, but that $Z > 1$ at higher pressures. If $Z < 1$ throughout the range of integration, the integrand in eqn 17 is negative and $\gamma < 1$. This implies that $f < p$ (the molecules tend to stick together) and that the chemical potential of the gas is less than that of a perfect gas (Fig. 5.2). At higher pressures the integration range where $Z > 1$ may dominate the range where it is less than unity. The integral is then positive, $\gamma > 1$, and $f > p$ (the repulsive interactions are dominant and tend to drive the particles apart). Now the chemical potential of the gas is greater than that of the perfect gas at the same pressure (Fig. 5.2).

Figure 5.3, which has been calculated using the full van der Waals equation of state, shows how the fugacity depends on the pressure in terms of the reduced variables. Since the critical constants are available in Table 1.2, the graphs can be used for quick estimates of the fugacities of a wide range of gases. Table 5.2 gives some explicit values for nitrogen.

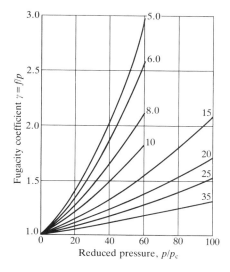

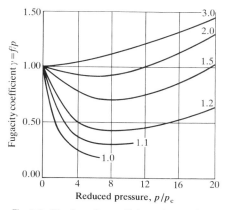

Fig. 5.3 The fugacity coefficient of a van der Waals gas plotted using the reduced variables of the gas. The curves are labelled with the reduced temperature $T_r = T/T_c$.

Table 5.2. The fugacity of nitrogen at 273 K

p/atm	f/atm
1	0.99955
10	9.9560
100	97.03
1000	1839

Example 5.8: *Estimating the fugacity of a gas*

Estimate the fugacity of nitrogen at 500 atm and 0°C.

Answer. Since the critical pressure and temperature of nitrogen are 33.5 atm and 126.2 K, the reduced pressure and temperature of the sample are

$$p_r = \frac{500 \text{ atm}}{33.5 \text{ atm}} = 14.9$$

$$T_r = \frac{273 \text{ K}}{126.2 \text{ K}} = 2.16$$

These values correspond to approximately $\gamma = 1.15$ in Fig. 5.3b, so the fugacity of nitrogen is approximately

$$f = 1.15 \times 500 \text{ atm} = 575 \text{ atm}$$

under the stated conditions.

Comment. Since $\gamma > 1$, the repulsive contributions are dominant in N_2 at 500 atm.

Exercise. Estimate the fugacity of carbon dioxide at 90°C and 580 atm.

[230 atm]

Open systems and changes of composition

In an open system (one in which the composition may vary) the Gibbs function depends on the composition as well as the pressure and the temperature. Thus it may change when p, T, and the composition change, and for a binary system

$$dG = \left(\frac{\partial G}{\partial p}\right)_{T,n} dp + \left(\frac{\partial G}{\partial T}\right)_{p,n} dT + \left(\frac{\partial G}{\partial n_1}\right)_{p,T,n_2} dn_1 + \left(\frac{\partial G}{\partial n_2}\right)_{p,T,n_1} dn_2 \quad (18)$$

This fearsome expression simply says that G may change because any of the four variables that define the state may change. Clearly, our first job is to simplify the notation.

5.4 The chemical potential

As a first step, consider how G changes when the composition is constant but the pressure and temperature change infinitesimally:

$$dG = \left(\frac{\partial G}{\partial p}\right)_{T,n} dp + \left(\frac{\partial G}{\partial T}\right)_{p,n} dT$$

We already know that $dG = V\,dp - S\,dT$ under the same conditions. Therefore, since dG is an exact differential, we may identify the coefficients:

$$\left(\frac{\partial G}{\partial p}\right)_{T,n} = V \qquad \left(\frac{\partial G}{\partial T}\right)_{p,n} = -S$$

These relations are the same as those in eqn 6 but are dressed more elaborately because the constant composition is stated explicitly.

The remaining differential coefficients in eqn 18 should be recognized as slightly more elaborate versions of the chemical potential of a single substance. Specifically, the coefficients with respect to the composition are by definition the chemical potentials of substances in a mixture:

$$\left(\frac{\partial G}{\partial n_1}\right)_{p,T,n_2} = \mu_1 \qquad \left(\frac{\partial G}{\partial n_2}\right)_{p,T,n_1} = \mu_2$$

That is, the chemical potential μ_1 expresses how G changes as substance 1 is added to the system (the pressure, temperature, and amount of substance 2 being constant); μ_2 does the same for the addition of substance 2. The chemical potentials depend on the composition of the mixture. For instance, adding 10^{-3} mol CH_3OH (almost an infinitesimal amount of methanol) to 1 L of a 20 per cent methanol/water mixture leads to a change in the overall G which is different from the change brought about by the addition of the same amount of CH_3OH to an 80 per cent mixture.

In general, we define the chemical potential of a substance J as

$$\mu_J = \left(\frac{\partial G}{\partial n_J}\right)_{p,T,n'} \qquad (19)$$

where the subscript n' signifies that the amounts of all the other components (those other than J) are constant. The introduction of these conclusions into eqn 18 leads to

$$dG = V\,dp - S\,dT + \mu_1\,dn_1 + \mu_2\,dn_2 + \ldots$$
$$= V\,dp - S\,dT + \sum_J \mu_J\,dn_J \qquad (20)$$

This is the **fundamental equation of chemical thermodynamics**. Its implications and consequences are explored and developed in the next five chapters.

At constant pressure and temperature eqn 20 simplifies to

$$dG = \mu_1\,dn_1 + \mu_2\,dn_2 + \ldots = \sum_J \mu_J\,dn_J$$

We saw in Section 4.8 that under the same conditions $dG = dw_{e,max}$. Therefore,

$$dw_{e,max} = \sum_J \mu_J\,dn_J \quad \text{(at constant } p, T) \qquad (21)$$

That is, non-expansion work can arise from the changing composition of a system that is not at internal equilibrium. For instance, in an electrochemical cell, the chemical reaction is arranged to take place in two distinct sites (at the two electrodes). The cell is not at internal equilibrium, and the electrical work it performs can be traced to its changing composition as products are formed from reactants.

5.5 The wider significance of μ

The chemical potential does more than show how G varies with composition. Since

$$G = U + pV - TS$$

a general infinitesimal change in U for a system of variable composition can be written

$$dU = -p\,dV - V\,dp + S\,dT + T\,dS + dG$$

$$= -p\,dV - V\,dp + S\,dT + T\,dS + (V\,dp - S\,dT + \mu_1\,dn_1 + \mu_2\,dn_2 + \ldots)$$

$$= -p\,dV + T\,dS + (\mu_1\,dn_1 + \mu_2\,dn_2 + \ldots)$$

This is the generalization of the fundamental equation (eqn 1) to systems in which the composition may change. It follows that at constant volume and entropy,

$$dU = \mu_1\,dn_1 + \mu_2\,dn_2 + \ldots = \sum_J \mu_J\,dn_J$$

and hence that

$$\mu_J = \left(\frac{\partial U}{\partial n_J}\right)_{V,S,n'} \tag{22a}$$

Therefore, not only does the chemical potential show how G changes when the composition changes, it also shows how the internal energy changes (but under a different set of conditions). In the same way it is easy to deduce that

$$\mu_J = \left(\frac{\partial H}{\partial n_J}\right)_{p,S,n'} \tag{22b}$$

$$\mu_J = \left(\frac{\partial A}{\partial n_J}\right)_{V,T,n'} \tag{22c}$$

Thus we see that the μ_J show how all the extensive thermodynamic properties U, H, A, and G depend on the composition. This is why the chemical potential is so central to chemistry.

Further reading

P. A. Rock, *Chemical thermodynamics*. University Science Books and Oxford University Press (1983).

M. L. McGlashan, *Chemical thermodynamics*. Academic Press, London (1979).

M. W. Zemansky and R. H. Dittman, *Heat and thermodynamics*. McGraw-Hill, New York (1981).

G. N. Lewis and M. Randall, *Thermodynamics*. (Revised by K. S. Pitzer and L. Brewer); McGraw-Hill, New York (1961).

B. D. Wood, *Applications of thermodynamics*. Addison-Wesley, New York (1982).

Exercises

Assume all gases are perfect and that the temperature is 298 K unless stated otherwise.

5.1 3.0 mmol of $N_2(g)$ occupy 36 cm³ at 300 K and expand to 60 cm³ isothermally. Calculate ΔG for the process.

5.2 The change in the Gibbs function of a certain constant pressure process was found to fit the expression

$$\Delta G/\text{J} = -85.40 + 36.5(T/\text{K})$$

Calculate the value of ΔS for the process.

5.3 When the pressure on a 35 g sample of a liquid was

increased isothermally from 1 atm to 3000 atm, the Gibbs function increased by 12 kJ. Calculate the density of the liquid.

5.4 When 2.00 mol of a gas at 330 K and 3.50 atm is subjected to isothermal compression, its entropy decreases by 25.0 J K^{-1}. Calculate the final pressure of the gas and ΔG for the compression.

5.5 Calculate the change in chemical potential of a perfect gas that is compressed isothermally from 1.8 atm to 29.5 atm at 40°C.

5.6 The fugacity coefficient of a certain gas at 200 K and 50 bar is $\gamma = 0.72$. Calculate the difference of its chemical potential from that of a perfect gas in the same state.

5.7 At 373 K, the second virial coefficient B of xenon is -81.7 cm^3 mol^{-1}. Calculate the value of B' and hence estimate the fugacity coefficient of xenon at 50 atm and 373 K.

5.8 Estimate the change in the Gibbs function of 1.0 L of benzene when the pressure acting on it is increased from 1.0 atm to 100 atm.

5.9 Calculate the change in the molar Gibbs function of hydrogen gas when it is compressed isothermally from 1.0 atm to 100.0 atm at 298 K.

Problems

Numerical problems

5.1 Calculate $\Delta G^{\ominus}(375 \text{ K})$ for the reaction

$$2\text{CO}(g) + \text{O}_2(g) \rightarrow 2\text{CO}_2(g)$$

from the value of $\Delta G^{\ominus}(\text{𝔽})$, $\Delta H^{\ominus}(\text{𝔽})$, and the Gibbs–Helmholtz equation.

5.2 Estimate the standard reaction Gibbs function of

$$\text{N}_2(g) + 3\text{H}_2(g) \rightarrow 2\text{NH}_3(g)$$

at (a) 500 K, (b) 1000 K from the value at 298 K.

5.3 At 298 K the standard enthalpy of combustion of sucrose is -5645 kJ mol^{-1} and the standard Gibbs function of the reaction is -5798 kJ mol^{-1}. Estimate the additional non-expansion work that may be obtained by raising the temperature to blood temperature, 37°C.

5.4 At 200 K, the compression factor of oxygen varies with pressure as shown below. Evaluate the fugacity of oxygen at this temperature and 100 atm.

p/atm	1.0000	4.00000	7.00000	10.0000	40.00	70.00	100.0
Z	0.99701	0.98796	0.97880	0.96956	0.8734	0.7764	0.6871

Theoretical problems

5.5 Show that, for a perfect gas, $(\partial U/\partial S)_V = T$ and $(\partial U/\partial V)_S = -p$.

5.6 Two of the four Maxwell relations were derived in the text, but two were not. Complete their derivation by showing that $(\partial S/\partial V)_T = (\partial p/\partial T)_V$ and $(\partial T/\partial p)_S = (\partial V/\partial S)_p$.

5.7 Use the Maxwell relations and Euler's chain relation to express $(\partial p/\partial S)_V$ and $(\partial V/\partial S)_p$ in terms of the heat capacities, the expansion coefficient, and the isothermal compressibility.

5.8 Use the Maxwell relations to show that the entropy of a perfect gas depends on the volume as $S \propto R \ln V$.

5.9 Derive the thermodynamic equation of state

$$\left(\frac{\partial H}{\partial p}\right)_T = V - T\left(\frac{\partial V}{\partial T}\right)_p$$

Derive an expression for $(\partial H/\partial p)_T$ for (a) a perfect gas and (b) a van der Waals gas. In the latter case, estimate its value for 1.0 mol of Ar(g) at 298 K and 10 atm assuming $b/V_m \ll 1$. By how much does the enthalpy of the argon change when the pressure is increased isothermally to 11 atm?

5.10 Show that if $B(T)$ is the second virial coefficient of a gas, and $\Delta B = B(T'') - B(T')$, $\Delta T = T'' - T'$, and T is the mean of T'' and T', then

$$\pi_T \approx \frac{RT^2 \Delta B}{V_m^2 \Delta T}$$

Estimate π_T for argon given that $B(250 \text{ K}) = -28.0$ cm^3 mol^{-1} and $B(300 \text{ K}) = -15.6$ cm^3 mol^{-1} at 275 K and (a) 1.0 atm, (b) 10.0 atm.

5.11 (a) Prove that the heat capacities C_V and C_p of a perfect gas are independent of both volume and pressure. May they depend on the temperature? (b) Deduce an expression for the dependence of C_V on volume of a gas that is described by $pV_m/RT = 1 + B/V_m$.

5.12 The 'Joule coefficient' μ_J is defined as $\mu_J = (\partial T/\partial V)_U$. Show that

$$\mu_J C_V = p - \frac{\alpha T}{\kappa_T}$$

5.13 Evaluate π_T for a Dieterici gas (Table 1.5). Justify the form of the expression obtained.

5.14 Instead of assuming that the volume of a condensed phase is constant when pressure is applied, assume only that the compressibility is constant. Show that when the pressure is changed isothermally by Δp, G changes to

$$G' = G + V_m \Delta p(1 - \tfrac{1}{2}\kappa_T \Delta p)$$

Assess the error in assuming that a solid is incompressible by applying this expression to the compression of copper when $\Delta p = 500$ atm. (For copper at 25°C, $\kappa_T = 0.8 \times 10^{-6}$ atm^{-1} and $\rho = 8.93$ g cm^{-3}.)

5.15 Derive an expression for the reaction Gibbs function $\Delta G^{\ominus\prime}$ at a temperature T' in terms of its value $\Delta G^{\ominus}$ at T using the Gibbs–Helmholtz equation and (a) assuming that ΔH does not vary with temperature, (b) assuming instead that ΔC_p does not vary with temperature and using Kirchhoff's law.

5.16 The adiabatic compressibility κ_S is defined like κ_T but at constant entropy. Show that for a perfect gas $p\gamma\kappa_S = 1$ (where γ is the ratio of heat capacities).

5.17 Show that, if S is regarded as a function of T and V, then

$$T \, dS = C_V \, dT + T\left(\frac{\partial p}{\partial T}\right)_V dV$$

Calculate the energy that must be transferred as heat to a van der Waals gas that expands reversibly and isothermally from V_i to V_f.

5.18 Suppose that S is regarded as a function of p and T. Show that

$$T \, dS = C_p \, dT - \alpha TV \, dp$$

Hence show that the energy transferred as heat when the pressure on an incompressible liquid or solid is increased by Δp is equal to $-\alpha TV \, \Delta p$. Evaluate q when the pressure acting on 100 cm^3 of mercury at 0°C is increased by 1.0 kbar. ($\alpha = 1.82 \times 10^{-4}$ K^{-1}.)

5.19 The volume of a newly synthesized polymer was found to depend exponentially on the pressure as $V = V_0 e^{-p/p^*}$ where p is the excess pressure and p^* is a constant. Deduce an expression for the Gibbs function of the polymer as a function of excess pressure. What is the natural direction of change of the compressed material when the pressure is relaxed?

5.20 Find an expression for the fugacity coefficient of a gas that obeys the equation of state

$$\frac{pV_m}{RT} = 1 + \frac{B}{V_m} + \frac{C}{V_m^2}$$

Use the resulting expression to estimate the fugacity of Ar at 1.00 atm and 0°C using $B = -21.13$ cm^3 mol^{-1} and $C = 1054$ cm^6 mol^{-2}.

5.21 Derive an expression for the fugacity coefficient of a gas that obeys the equation of state

$$\frac{pV_m}{RT} = 1 + \frac{BT}{V_m}$$

and plot γ against $4pB/R$.

Changes of state: physical transformations of pure substances

6

Check-list of key ideas

1. The use of a *phase diagram* to depict the temperatures and pressures at which different phases are stable (Section 6.1) and the significance of the *phase boundaries*.

2. The significance of the *melting point*, the *boiling point*, the *critical temperature*, and the *triple point* (Section 6.1).

3. The interpretation of representative *phase diagrams* of four substances (Section 6.2).

4. The expression of phase equilibria in terms of the chemical potential of the phases and the response of the phase equilibrium to the temperature (eqn 1) and the pressure (eqn 2).

5. The calculation of the vapour pressure when an additional pressure is applied to a liquid (eqn 3).

6. The calculation of the phase boundaries, using the *Clapeyron equation* (eqn 4) for the solid/liquid phase boundary and the approximate *Clausius–Clapeyron equation* (eqn 8) for the liquid/vapour and solid/vapour phase boundaries.

7. The classification of phase transitions as *first order* or *second order* and the meaning of the term λ *transition* (Section 6.5).

8. The use of the *surface tension* to describe the thermodynamic properties of liquid surfaces (Section 6.6).

9. The derivation of the *Laplace equation* (eqn 12) for the pressure difference across a curved surface and of the *Kelvin equation* (eqn 13) for the vapour pressure of droplets.

10. The role of *nucleation* in droplet formation (Section 6.7).

11. The description of *capillary action* (Section 6.8) and the rise of liquids in narrow tubes (eqn 15), including the effect of *contact angle*.

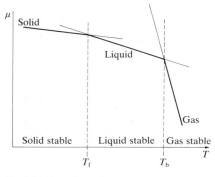

Fig. 6.1 The schematic temperature dependence of the chemical potential of the solid, liquid, and gas phases of water. The phase with the lowest μ at a specified temperature is the most stable one at that temperature. The transition temperatures, the melting and boiling temperatures, are the temperatures at which the chemical potentials of two phases are equal.

Boiling, freezing, and the conversion of graphite to diamond are all examples of changes of phase without change of composition. In this chapter we describe such processes thermodynamically using as our guiding principle the tendency of systems at constant temperature and pressure to lower their Gibbs function. Since we are dealing with pure substances, the molar Gibbs function of the system is the same as the chemical potential μ, and so the tendency to change is in the direction of decreasing chemical potential.

It must never be forgotten that these alternative ways of expressing the direction all stem from the tendency of an isolated system that contains the substance and its surroundings to change in the direction of increasing entropy. Although a *substance* may take on greater order and its entropy decrease, such as when it freezes, there is an increase in entropy of its surroundings (as result of the heat released into them) and the entropy of the global, isolated system is greater overall.

A phase transition occurs at a characteristic temperature for a given pressure. Thus, at 1 bar pressure, ice is the stable phase of water below 0°C but above 0°C the liquid is more stable. This indicates that below 0°C the chemical potential of ice is lower than that of the liquid, $\mu(s) < \mu(l)$ (Fig. 6.1), and that above 0°C $\mu(l) < \mu(s)$. The transition temperature is the temperature at which the two chemical potentials coincide and $\mu(s) = \mu(l)$. However, we must always distinguish between the thermodynamics of phase transitions and their rates, and a phase transition that is predicted from thermodynamics may occur too slowly to be significant in practice. For instance, at normal temperatures and pressures the chemical potential of graphite is lower than that of diamond, and so there is a thermodynamic tendency for diamonds to convert into graphite. But for this to take place, the C atoms must change their locations, and this is an unmeasurably slow process except at high temperatures. The rate of attainment of equilibrium is a kinetic problem, and is outside the range of thermodynamics. In gases and liquids the mobilities of the molecules allow phase transitions to occur rapidly, but in solids thermodynamic instability may be frozen in.

Phase diagrams

A **phase diagram** of a substance shows the regions of pressure and temperature at which its various phases are thermodynamically stable (Fig. 6.2). The boundaries between regions, the **phase boundaries**, show the values of p and T at which two phases coexist in equilibrium.

6.1 Phase boundaries

The pressure of a vapour in equilibrium with its condensed phase at a specified temperature is called the **vapour pressure** of the substance at that temperature. Hence, the phase boundaries between the liquid and the vapour and between the solid and the vapour show how the vapour pressures of the two condensed phases vary with temperature.

Critical points and boiling points

We must distinguish the behaviour of a liquid in an open vessel from one in a sealed vessel. In an open vessel, the vapour pressure rises and, when it

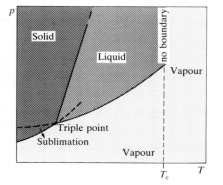

Fig. 6.2 The general regions of pressure and temperature where solid, liquid, or gas is stable (i.e. has lowest chemical potential) are shown on this phase diagram. For example, the solid phase is the most stable phase at low temperatures and high pressures. In the following sections we locate the precise boundaries between the regions.

matches the external pressure, vaporization can occur throughout the bulk of the liquid and the vapour can expand into the surroundings. That is, a liquid boils when its vapour pressure is equal to the external pressure. The temperature at which the vapour pressure of the liquid is equal to the ambient pressure is called the **boiling temperature**. When the external pressure is 1 atm, the boiling temperature is called the **normal boiling point**, T_b. With the replacement of 1 atm by 1 bar as standard pressure, there is some advantage in modifying the definition so that the transition temperature refers to that pressure; the term **standard boiling point** is then used. Since 1 bar is slightly less than 1 atm (1.00 bar = 0.987 atm), the standard boiling point is slightly lower than the normal boiling point. The normal boiling point of water is 100.0°C, its standard boiling point is 99.6°C.

In a sealed vessel, although the vapour pressure rises as the sample is heated, the density of the vapour increases because it is confined to a fixed volume (Fig. 6.3). There comes a stage when the density of the vapour is equal to that of the remaining liquid and the surface between the two phases disappears. The temperature at which the surface disappears is the **critical temperature** T_c which we first encountered in Section 1.3. The corresponding vapour pressure is the critical pressure p_c. At and above this temperature, a single uniform phase fills the container and an interface no longer exists.

Melting points and triple points

The temperature at which, under a specified pressure, liquid and solid coexist in equilibrium is called the **melting temperature**. Since a substance melts at exactly the same temperature as it freezes (except in some bizarre systems), the melting temperature is the same as the **freezing temperature**. The melting temperature when the pressure is 1 atm is called the **normal melting point** T_f, and that when the pressure is 1 bar is called the **standard melting point**; the normal and standard melting points are negligibly different for most purposes.

There is a set of conditions under which three different phases (typically solid, liquid, and vapour) all simultaneously coexist in equilibrium. It is represented by the **triple point**, where the three phase boundaries coincide (Fig. 6.2). The location of the triple point of a pure substance is outside our control: it occurs at a single definite pressure and temperature characteristic of the substance. For water it lies at 273.16 K and 6.11 mbar (4.58 Torr), and the three phases of water coexist in equilibrium at no other combination of pressure and temperature. This invariance of the triple point is the basis of its use in the definition of the thermodynamic temperature scale (Sections 1.2 and 4.6).

As we see from Fig. 6.2, the triple point marks the lowest pressure at which a liquid phase of a substance can exist. If the slope of the solid/liquid phase boundary is as shown in the diagram, the triple point also marks the lowest temperature at which the liquid can exist; the critical temperature is the upper limit.

6.2 Phase diagrams of single substances

We shall now show how these general features appear in the phase diagrams of pure substances. In Chapter 8 we consider the more elaborate phase

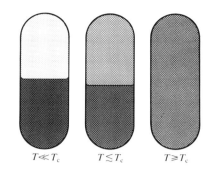

$T \ll T_c$ $T \lesssim T_c$ $T \gtrsim T_c$

Fig. 6.3 When a liquid is heated in a sealed container, the density of the vapour phase increases and that of the liquid decreases slightly. There comes a stage at which the two densities are equal and the interface between the fluids disappears. This occurs at the critical temperature. The container needs to be strong: the critical temperature of water is 374°C and the vapour pressure is then 218 atm.

diagrams that summarize the phase equilibria of mixtures and for which composition is an additional variable.

Water

Figure 6.4 is the phase diagram for water. The liquid–vapour line summarizes how the vapour pressure of liquid water varies with temperature. Conversely, it also summarizes how the boiling temperature varies with pressure. The solid/liquid line shows how the melting temperature depends on the pressure, and indicates that enormous pressures are needed to bring about significant changes. Notice that the line has a negative slope, which means that the melting temperature falls as the pressure is raised. The reason can be traced to the decrease in volume that occurs on melting, and hence it being more favourable for the solid to transform into the liquid as the pressure is raised. The decrease in volume is a result of the very open structure of the ice crystal lattice: the water molecules are held apart (as well as together) by the hydrogen bonds between them, but the structure partially collapses on melting and the liquid is denser than the solid.

A practical consequence of the decrease in melting temperature with pressure may be the motion of glaciers: where ice is pressed against the sharp edges of stones and rocks it melts, and the glacier inches forwards. However, for many substances surface melting occurs below the normal melting point, and the explanation of glacier motion (and ice skating) may

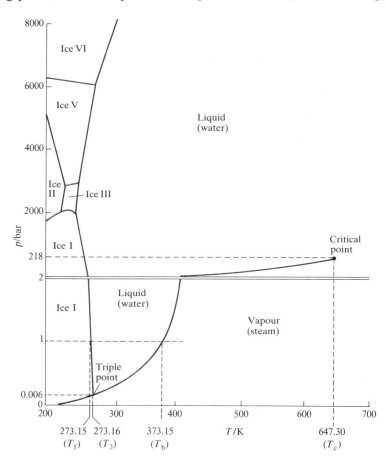

Fig. 6.4 The experimental phase diagram for water showing the different solid phases. Note the change of vertical scale at 2 bar.

be more subtle. The reduction in chemical potential of the water below that of the ice may stem from differences in the surface tensions of the ice/rock, ice/water, and rock/water interfaces.

At high pressures, different solid phases of water come into stability as the bonds between molecules are modified. Some of these phases (which are called ice-II, III, V, VI, and VII: ice-IV was an illusion, like the once-fashionable alternative liquid called 'polywater') melt at high temperatures. Ice-VII, for instance, melts at 100°C, but exists only above 25 kbar.

Carbon dioxide

The phase diagram for carbon dioxide is shown in Fig. 6.5. The features to notice include the 'orthodox' slope of the solid–liquid boundary, which indicates that the melting temperature of solid carbon dioxide rises as the pressure is increased. Notice also that as the triple point lies above 1 atm, the liquid cannot exist at normal atmospheric pressures whatever the temperature, and the solid sublimes when left in the open (hence the name 'dry ice'). To obtain the liquid, it is necessary to exert a pressure of at least 5.11 bar. Cylinders of carbon dioxide generally contain the liquid or compressed gas; at room temperature that implies a vapour pressure of 67 bar if both gas and liquid are present in equilibrium. When the gas squirts through the throttle it cools by the Joule–Thomson effect, so when it emerges into a region where the pressure is only 1 atm, it condenses into a finely divided snowlike solid.

Carbon

Figure 6.6 is the phase diagram for carbon. It is ill-defined because the various phases come into stability at extremes of temperature and pressure, and gathering the data is very difficult. For instance, at atmospheric pressure, carbon gas is the stable phase only at temperatures well over 4000 K. To obtain liquid carbon it is necessary to work at about 4500 K and 1 kbar (about 1000 atm), or at 2000 K and 1000 kbar. Making diamonds is a minor problem in comparison, because the diamond phase comes into

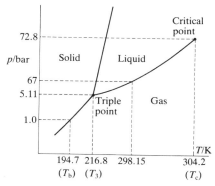

Fig. 6.5 The experimental phase diagram for carbon dioxide. Note that, as the triple point lies at pressures well above atmospheric, liquid carbon dioxide does not exist under normal conditions (a pressure of at least 5.11 bar must be applied).

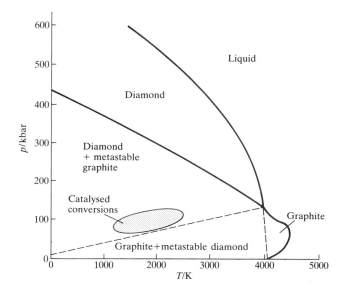

Fig. 6.6 The phase diagram for carbon. Note that the pressure axis is in kilobars (1 kbar is about 1000 atm). There are very large uncertainties about the precise form of this phase diagram because the data are so difficult to obtain.

stability at 10^4 bar and 1000 K. Small diamonds are synthesized and are widely used in industry, but the phase diagram does not reveal the full problem. The rate of conversion is an important factor, and pure graphite changes into diamond at a useful rate only when the temperature is around 4000 K and the pressure exceeds 200 kbar; but then the apparatus tends to disappear first. Therefore catalysts are added in commercial synthesis, and then the conversion proceeds at 100 kbar and 2000 K, which are attainable conditions. The contamination by the metal catalysts such as molten Ni (which also acts as a solvent for the carbon) enables commercial and natural diamonds to be distinguished. This high pressure procedure has been partially replaced by one that makes use of decomposition of methane and the preferential decomposition of the graphite phase.[1]

Helium

The phase diagram of He is shown in Fig. 6.7. Helium behaves unusually at low temperatures. For instance, the solid and gas phases are never in equilibrium however low the temperature. This is because the He atoms are so light that they vibrate with a large amplitude motion even at very low temperatures, and the solid simply shakes itself apart. Solid He can be obtained, but only by holding the atoms together by applying pressure. A second peculiarity is that pure ^{4}He has a liquid–liquid phase transition at its **λ-line** (the reason for this name is explained in Section 6.5). The liquid phase marked He-I behaves like a normal liquid. The other phase, He-II, is a **superfluid**, so called because it flows without viscosity.

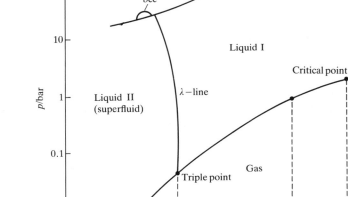

Fig. 6.7 The phase diagram for helium (^{4}He). The λ-line marks the conditions under which the two liquid phases are in equilibrium; He-II is the superfluid phase. Note that a pressure of over 20 bar must be exerted before solid helium can be obtained.

The isotope ^{3}He has a phase diagram quite similar to that of ^{4}He, but until recently it was thought that it lacked the λ-line and had only a normal liquid phase. Recent observations, however, have shown that there is a

---

[1] The mechanism appears to be obscure. It appears that both diamond and graphite phases are formed, but that the radicals in the gas, which include H atoms and CH radicals, react more rapidly with the graphite phase and remove it, leaving the diamond phase. This is an example of a non-equilibrium, kinetically controlled reaction.

superfluid phase, and evidence is still being accumulated. The isotope is also unusual in that the entropy of the liquid is lower than that of the solid, and melting is an exothermic process.

Phase stability and phase transitions

We shall now see how thermodynamic considerations can account for the features—including the anomalies—of the phase diagrams we have just described.

We shall base our discussion on the following consequence of the Second Law:

> At equilibrium, the chemical potential of a substance is the same throughout a sample, regardless of how many phases are present.

Thus, when the liquid and solid phases of a substance are in equilibrium, the chemical potential of the substance is the same throughout the liquid and throughout the solid, and is the same in the solid as in the liquid.

The proof is as follows. Consider a system in which the chemical potential of a substance is μ_1 at one point and μ_2 at another. When an amount dn of the substance is transferred from one point to the other, the Gibbs function changes by $-\mu_1 \, dn$ in the first step and by $+\mu_2 \, dn$ in the second. The overall change is therefore $dG = (\mu_2 - \mu_1) \, dn$. If the chemical potential at point 1 is higher than at point 2, the transfer is accompanied by a decrease in G, and so has a spontaneous tendency to occur. Only if $\mu_1 = \mu_2$ is there no change in G, and only then is the system at equilibrium. The argument applies when the transfer of matter occurs between different phases as well as between different locations within a single phase.

6.3 The dependence of stability on the conditions

We denote the chemical potentials of the solid, liquid, and gas phases of a substance $\mu(s)$, $\mu(l)$, and $\mu(g)$ respectively. At a given pressure, a phase is thermodynamically stable over the range of temperatures at which it has a lower chemical potential than any other phase. At low temperatures the solid phase has the lowest chemical potential and is the most stable. However, the chemical potentials of the three phases change with temperature in different ways, and it is possible that as the temperature is raised the chemical potential of another phase falls below that of the solid. When that happens, a phase transition occurs if it is kinetically feasible to do so.

The temperature dependence of phase stability

In Section 5.2 we saw that the temperature dependence of the chemical potential is

$$\left(\frac{\partial \mu}{\partial T}\right)_p = -S_m \tag{1}$$

(because $\mu = G_m$ for a pure substance). This relation shows that as the temperature is raised, the chemical potential of a pure substance decreases (because $S_m > 0$ always). It also implies that the slope is steeper for gases than for liquids, because $S_m(g) > S_m(l)$, and steeper for a liquid than the corresponding solid, because $S_m(l) > S_m(s)$ almost always. The steep negative slope of $\mu(l)$ results in its falling below $\mu(s)$ when the temperature is high enough (Fig. 6.1), and then the liquid becomes the stable phase: the solid melts. The chemical potential of the gas phase plunges downwards as the temperature is raised (because the entropy of the vapour is so high), and there comes a temperature at which it lies lowest. Then the gas is the stable phase and the liquid vaporizes.

The response of melting to applied pressure

When pressure is applied to a sample, most substances melt at a higher temperature. It is as though the pressure is preventing the formation of the less dense liquid phase. Exceptions to this behaviour include water, for which the liquid is denser than the solid, and the application of pressure to water encourages the formation of the liquid phase. That is, water freezes at a lower temperature when it is under pressure.

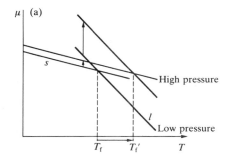

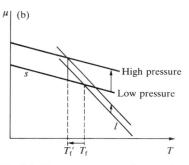

Fig. 6.8 The pressure dependence of the chemical potential of a substance depends on the molar volume of the phase. The curves show the effect of increasing pressure on the chemical potentials of the solid and liquid phases, and the corresponding effects on the melting temperatures. (a) In this case the molar volume of the solid is less than that of the liquid and $\mu(s)$ increases less than $\mu(l)$. As a result, the melting point rises. (b) Here the molar volume is greater for the solid than the liquid (as for water), $\mu(s)$ increases more strongly than $\mu(l)$, and the melting temperature is lowered.

Example 6.1: *Assessing the effect of pressure on the chemical potential*

Calculate the effect on the chemical potentials of increasing the pressure from 1.00 bar to 2.00 bar on ice and water at 0°C (their densities are $0.917\,\mathrm{g\,cm^{-3}}$ and $0.999\,\mathrm{g\,cm^{-3}}$ respectively).

Answer. From eqn 10 of Chapter 5 we know that the change in chemical potential of an incompressible substance is

$$\Delta\mu = V_m\,\Delta p$$

The molar volumes of ice and water (of molar mass $18.02\,\mathrm{g\,mol^{-1}}$) are $19.7\,\mathrm{cm^3\,mol^{-1}}$ and $18.0\,\mathrm{cm^3\,mol^{-1}}$; therefore

$$\Delta\mu(\text{ice}) = 1.97 \times 10^{-5}\,\mathrm{m^3\,mol^{-1}} \times 1.00 \times 10^5\,\mathrm{Pa}$$

$$= +1.97\,\mathrm{J\,mol^{-1}}$$

$$\Delta\mu(\text{water}) = 1.80 \times 10^{-5}\,\mathrm{m^3\,mol^{-1}} \times 1.00 \times 10^5\,\mathrm{Pa}$$

$$= +1.80\,\mathrm{J\,mol^{-1}}$$

Comment. The chemical potential of ice rises more than that of water; so if they are initially in equilibrium, there will be a tendency for the ice to melt.

Exercise. Calculate the effect of an increase in pressure of 1.00 bar on a liquid and a solid of molar mass $44.0\,\mathrm{g\,mol^{-1}}$ that are in equilibrium with densities $2.35\,\mathrm{g\,cm^{-3}}$ and $2.50\,\mathrm{g\,cm^{-3}}$ respectively.
[$\Delta\mu(l) = +1.87\,\mathrm{J\,mol^{-1}}$, $\Delta\mu(s) = +1.76\,\mathrm{J\,mol^{-1}}$; solid forms]

We can rationalize the response of melting temperatures to pressure in terms of the different pressure dependence of the chemical potentials of the solid and liquid phases, and the displacement of the temperature at which their graphs cross (Fig. 6.8). The chemical potential depends on pressure as

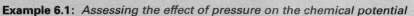

$$\left(\frac{\partial\mu}{\partial p}\right)_T = V_m \qquad (2)$$

An increase in pressure raises the chemical potential of any pure substance (because $V_m > 0$). In most cases, $V_m(l) > V_m(s)$ and the equation predicts that an increase in pressure increases the chemical potential of the liquid more than that of the solid. As shown in Fig. 6.8a, the effect is to raise the melting temperature slightly. For water, $V_m(l) < V_m(s)$, and an increase in pressure increases the chemical potential of the *solid* more than that of the liquid (Fig. 6.8b). In this case, the melting temperature is lowered slightly.

The effect of applied pressure on vapour pressure

We have considered the effect of applying pressure *simultaneously* to the solid and liquid phases of a substance. Now we consider the effect of applying pressure solely to the condensed phase and leaving the partial pressure of the gas phase unchanged (except for any effects that stem from the compression of the condensed phase). This can be achieved by compressing the condensed phase mechanically or subjecting it to the applied pressure of an inert gas. When pressure is applied to the condensed phase, its vapour pressure rises: in effect, molecules are squeezed out of the condensed phase and escape as a gas. A complication (which we ignore here) is that if the condensed phase is a liquid, the pressurizing gas might dissolve and change its properties.

We can calculate the vapour pressure of a pressurized liquid by using the fact that at equilibrium $\mu(l) = \mu(g)$. Therefore, for any change that preserves equilibrium, the change in $\mu(l)$ must be equal to the change in $\mu(g)$, and we can write $d\mu(g) = d\mu(l)$. When the pressure P on the liquid is increased by dP, the chemical potential of the liquid changes by

$$d\mu(l) = V_m \, dP$$

The chemical potential of the vapour changes by

$$d\mu(g) = V_m \, dp$$

where dp is the change in its vapour pressure p. If we treat the vapour as a perfect gas,

$$d\mu(g) = \frac{RT \, dp}{p}$$

and by equating the two changes in chemical potential we obtain

$$\frac{RT \, dp}{p} = V_m \, dP$$

This expression can be integrated once we know the limits of integration. When there is no additional pressure acting on the liquid, P (the pressure experienced by the liquid) is equal to the normal vapour pressure p^*; so when $P = p^*$, $p = p^*$ too. When there is an additional pressure ΔP on the liquid, so $P = p^* + \Delta P$, the vapour pressure is p (the value we want to find). The integrations required are therefore as follows:

$$RT \int_{p^*}^{p} \frac{dp}{p} = \int_{p^*}^{p^* + \Delta P} V_m \, dP$$

We now assume that the molar volume of the liquid is the same throughout the small range of pressures involved. Then

$$RT \ln \frac{p}{p^*} = V_m \Delta P \qquad (3a)$$

which rearranges to

$$p = p^* e^{V_m \Delta P / RT} \qquad (3b)$$

This equation shows how the vapour pressure increases when the pressure acting on the condensed phase is increased.

Example 6.2: *Estimating the effect of pressure on the vapour pressure*

Derive an expression from eqn 3b that is valid for small changes in p and calculate the percentage increase of the vapour pressure of water for an increase in pressure of 10 bar at 25°C.

Answer. If $V_m \Delta P / RT \ll 1$, we can expand the exponent in eqn 3b (using $e^x = 1 + x + \dots$) to $1 + V_m \Delta P / RT$ and so obtain

$$\frac{p - p^*}{p^*} \approx \frac{V_m \Delta P}{RT}$$

For water (which has density 0.997 g cm^{-3} at 25°C and therefore molar volume $18.07 \text{ cm}^3 \text{ mol}^{-1}$),

$$\frac{V_m \Delta P}{RT} = \frac{1.807 \times 10^{-5} \text{ m}^3 \text{ mol}^{-1} \times 1.0 \times 10^6 \text{ Pa}}{8.314 \text{ J K}^{-1} \text{ mol}^{-1} \times 298 \text{ K}} = 7.3 \times 10^{-3}$$

Since $V_m \Delta P / RT \ll 1$, the approximate formula can be used, and

$$\frac{p - p^*}{p^*} \times 100 \text{ per cent} = 0.73 \text{ per cent}$$

Exercise. Calculate the percentage effect of an increase in pressure of 100 bar on the vapour pressure of benzene at 25°C, for which $\rho = 0.879 \text{ g cm}^{-3}$.

[43 per cent]

6.4 The location of phase boundaries

We can find the precise locations of the phase boundaries—the pressures and temperatures at which two phases can coexist—by making use of the fact that since the two phases are then in equilibrium, their chemical potentials are equal. Therefore, where the phases α and β are in equilibrium,

$$\mu_\alpha(p, T) = \mu_\beta(p, T)$$

By solving this equation for p in terms of T we shall get an equation for the phase boundary.

The slopes of the phase boundaries

It is simplest to discuss the phase boundaries in terms of their slopes, and so we begin by finding an equation for dp/dT.

Let p and T be changed infinitesimally, but in such a way that the two phases α and β remain in equilibrium. The chemical potentials of the phases were equal, and they remain equal; therefore the changes in them must be equal and we can write $d\mu_\alpha = d\mu_\beta$. Since, from eqn 5 in Section 5.2, we know that

$$d\mu = -S_m\,dT + V_m\,dp$$

for each phase, it follows that

$$-S_{\alpha,m}\,dT + V_{\alpha,m}\,dp = -S_{\beta,m}\,dT + V_{\beta,m}\,dp$$

where $S_{\alpha,m}$ and $S_{\beta,m}$ are the molar entropies of the phases and $V_{\alpha,m}$ and $V_{\beta,m}$ their molar volumes. Hence

$$(V_{\beta,m} - V_{\alpha,m})\,dp = (S_{\beta,m} - S_{\alpha,m})\,dT$$

which rearranges into the **Clapeyron equation**:

$$\frac{dp}{dT} = \frac{\Delta S_m}{\Delta V_m} \tag{4}$$

$\Delta S_m = S_{\beta,m} - S_{\alpha,m}$ and $\Delta V_m = V_{\beta,m} - V_{\alpha,m}$ are the changes in molar entropy and molar volume when the transition occurs. This important result for the slope of the phase boundary at any point is exact; it applied to any phase transition of a pure substance.

The solid–liquid boundary

Melting (fusion) is accompanied by a molar enthalpy change ΔH_{fus} and occurs at a temperature T. The molar entropy of melting at T is therefore $\Delta H_{fus}/T$, and the Clapeyron equation becomes

$$\frac{dp}{dT} = \frac{\Delta H_{fus}}{T\,\Delta V_{fus}} \tag{5}$$

where ΔV_{fus} is the molar volume change on melting. The enthalpy of melting is positive (about the only exception is ^3He, which melts exothermically), and the volume change is usually positive but always small. This means that the slope dp/dT is steep and usually positive. The curve itself can be obtained by integrating dp/dT assuming that ΔH_{fus} and ΔV_{fus} barely change with temperature and pressure. If the melting temperature is T^* when the pressure is p^*, and is T when the pressure is p, the integration required is

$$\int_{p^*}^{p} dp = \frac{\Delta H_{fus}}{\Delta V_{fus}} \int_{T^*}^{T} \frac{dT}{T}$$

Therefore, the approximate equation of the solid/liquid boundary is

$$p = p^* + \frac{\Delta H_{fus}}{\Delta V_{fus}} \ln \frac{T}{T^*} \tag{6a}$$

This equation was originally obtained by yet another Thomson—James, the brother of William, Lord Kelvin. When T is close to T^*, we can approximate the logarithm using

$$\ln \frac{T}{T^*} = \ln \left(1 + \frac{T - T^*}{T^*}\right) \approx \frac{T - T^*}{T^*}$$

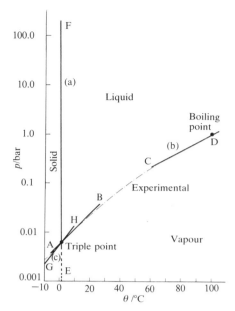

Fig. 6.9 The phase boundaries for water as calculated in Examples 6.3, 6.5, and 6.6. Note that the vertical axis is logarithmic, and so it has the effect of squashing down the upper parts of the diagram (compare with the lower part of Fig. 6.4, which is on a linear scale).

because $\ln(1 + x) \approx x$ when x is small; then

$$p = p^* + \frac{\Delta H_{fus}(T - T^*)}{T^* \, \Delta V_{fus}} \tag{6b}$$

This is the equation of a steep straight line (Fig. 6.9a).

Example 6.3: *Constructing a solid/liquid phase boundary*

Construct the ice/liquid phase boundary for water at temperatures between −1°C and 0°C. What is the melting temperature of ice under a pressure of 1.5 kbar?

Answer. For the first part we use eqn 6a; for the fixed point we take the normal melting point at $p^* = 1.0$ atm (≈ 1.0 bar) and $T^* = 273.15$ K. From Table 2.2 we have $\Delta H_{fus} = +6.008$ kJ mol^{-1} and from Example 6.1, $\Delta V_m = -1.7$ cm^3 mol^{-1}. Hence,

$$p/\text{bar} = 1.0 - 3.53 \times 10^4 \ln \frac{T}{273.15 \text{ K}}$$

This formula gives the following values:

$\theta/°C$	−1.0	−0.8	−0.6	−0.4	−0.2	0.0
p/bar	130	105	79	53	27	1.0

These points are plotted in Fig. 6.9 as the line EF. For the second part, we rearrange the formula into

$$T = 273.15 \text{ K} \times \exp\left(\frac{1.0 - p/\text{bar}}{3.53 \times 10^4}\right)$$

Then, with $p = 1.5$ kbar, $T = 262$ K, or −11°C.

Comment. Note the decrease in melting temperature with increasing pressure: water is denser than ice, and so ice responds to pressure by tending to melt.

Exercise. Estimate the value of the difference between the standard melting point of ice and its normal melting point. [About +0.1 mK]

The liquid–vapour boundary

The molar entropy of vaporization at a temperature T is equal to $\Delta H_{vap}/T$; the Clapeyron equation for the liquid–vapour boundary is therefore

$$\frac{dp}{dT} = \frac{\Delta H_{vap}}{T \, \Delta V_{vap}} \tag{7}$$

The enthalpy of vaporization is positive; ΔV_{vap} is large and positive. Therefore dp/dT is always positive, but it is much smaller than for the solid–liquid boundary.

Example 6.4: *Estimating the effect of pressure on the boiling point*

Use Troutons's law to estimate the effect of increasing pressure on the boiling point of a liquid.

Answer. According to Trouton's law (Section 4.4),

$$\Delta H_{vap} \approx 85 \text{ J K}^{-1} \text{mol}^{-1} \times T_b$$

We also need the value of ΔV_m. Since the molar volume of a gas is so much greater than the molar volume of a liquid, we can write

$$\Delta V_{vap} = V_m(g) - V_m(l) \approx V_m(g)$$

and take for $V_m(g)$ the molar volume of a perfect gas, which is about 30 L mol^{-1} at 1 atm and near but above room temperature. Therefore,

$$\frac{dp}{dT} \approx \frac{85 \text{ J K}^{-1} \text{mol}^{-1}}{30 \times 10^{-3} \text{ m}^3 \text{ mol}^{-1}} = 2.8 \times 10^3 \text{ Pa K}^{-1}$$

$$\approx 0.03 \text{ atm K}^{-1}$$

This corresponds to a dT/dp slope of about 30 K/atm; hence a change of pressure of $+0.1$ atm can be expected to change a boiling temperature by about $+3$ K.

Exercise. Estimate dT/dp for water at its normal boiling point using the information in Table 2.2 and $V_m(g) \approx RT/p$.
[28 K atm^{-1}]

As in the example, because the molar volume of a gas is so much greater than the molar volume of a liquid, we can write $\Delta V_{vap} \approx V_m(g)$. Moreover, if the gas behaves perfectly, $V_m(g) = RT/p$. These two approximations turn the exact Clapeyron equation into the **Clausius–Clapeyron equation**:

$$\frac{d \ln p}{dT} = \frac{\Delta H_{vap}}{RT^2} \qquad (8a)°$$

(We have used $dx/x = d \ln x$.) If we also assume that the enthalpy of vaporization is independent of temperature this equation integrates to

$$p = p^* e^{-C} \quad \text{with} \quad C = \frac{\Delta H_{vap}}{R} \left(\frac{1}{T} - \frac{1}{T^*} \right) \qquad (8b)°$$

where p^* is the vapour pressure when the temperature is T^* and p its value when the temperature is T. This is the curve plotted as the liquid–vapour boundary in Fig. 6.9b. The line does not extend beyond the critical temperature T_c because above this temperature the liquid does not exist (Section 1.3).

Example 6.5: *Constructing a vapour pressure curve*

Construct the vapour pressure curve for water between $-5°C$ and $100°C$.

Answer. Use eqn 8b, taking as the fixed point (p^*, T^*) the triple point (6.11 mbar, 273.16 K) for temperatures near the lower end of the range and as the normal boiling point (1.01 bar, 373.15 K) for temperatures near the upper end of the range. Use $\Delta H_{vap}(273 \text{ K}) = +45.05 \text{ kJ mol}^{-1}$ and $\Delta H_{vap}(373 \text{ K}) = +40.66 \text{ kJ mol}^{-1}$ for the respective ranges. Evaluation of eqn 8b for several temperatures gives the following values:

$\theta/°C$	-5	0	5	10	20	30	...	70	80	90	100
$p/$atm	0.004	0.006	0.009	0.012	0.024	0.044		0.32	0.48	0.70	1.0

The curve is plotted as AB and CD in Fig. 6.9, and compared there with the experimental vapour pressure.

Comment. The negative curvature of the curve as plotted comes from the use of a logarithmic scale. Note that we can use eqn 8 in conjunction with experimental vapour pressure data to determine enthalpies of vaporization.

Exercise. Calculate the standard boiling point of water from its normal boiling point. [99.6°C]

The solid–vapour boundary

The only difference between this case and the last is the replacement of the enthalpy of vaporization by the enthalpy of sublimation, ΔH_{sub}. The approximations that led to the Clausius–Clapeyron equation give the following expressions for the temperature dependence of the sublimation vapour pressure:

$$\frac{d \ln p}{dT} = \frac{\Delta H_{sub}}{RT^2} \qquad (9a)°$$

$$p = p^* e^{-C} \quad \text{with} \quad C = \frac{\Delta H_{sub}}{R}\left(\frac{1}{T} - \frac{1}{T^*}\right) \qquad (9b)°$$

Since the enthalpy of sublimation is greater than the enthalpy of vaporization, the equation predicts a steeper slope for the sublimation curve than for the vaporization curve near where they meet (Fig. 6.9c).

Example 6.6: *Constructing a solid/vapour phase boundary*

Construct the ice/vapour phase boundary over the range −10°C to +5°C, using the information that at 273 K, $\Delta H_{vap} = +45.05 \text{ kJ mol}^{-1}$ and $\Delta H_{fus} = +6.01 \text{ kJ mol}^{-1}$.

Answer. In order to use eqn 9b, we need to know the enthalpy of sublimation. From the First Law, and the fact that H is a state function, we can write

$$\Delta H_{sub} = \Delta H_{fus} + \Delta H_{vap}$$

The sublimation vapour pressures at several temperatures are as follows:

$\theta/°C$	−10	−5	0	5
p/bar	0.003	0.004	0.006	0.009

These points are plotted as GH in Fig. 6.9 and compared with the experimental curve.

Comment. The small discrepancy between the experimental and calculated curves is a result of assuming that the enthalpy of sublimation is a constant.

Density of solid: 1.53 g cm⁻³
Density of liquid: 0.78 g cm⁻³
ΔH_{sub}: 25.2 kJ mol⁻¹
ΔH_{fus}: 8.3 kJ mol⁻¹
Triple point: −57°C, 5.2 bar
Critical constants: 31°C, 74 bar

Exercise. Construct the phase diagram for carbon dioxide using the data given in the margin. [Fig. 6.5]

6.5 The classification of phase transitions

We can use thermodynamic properties of substances, and in particular the behaviour of the chemical potential, to classify phase transitions into different types.

First-order transitions

Many familiar phase transitions, like melting and vaporization, are accompanied by changes of enthalpy and volume. At the transition from one phase α to another β,

$$\left(\frac{\partial \mu_\beta}{\partial p}\right) - \left(\frac{\partial \mu_\alpha}{\partial p}\right) = V_{\beta,m} - V_{\alpha,m} = \Delta V$$

$$\left(\frac{\partial \mu_\beta}{\partial T}\right) - \left(\frac{\partial \mu_\alpha}{\partial T}\right) = -S_{\beta,m} + S_{\alpha,m} = -\Delta S = \frac{-\Delta H}{T}$$

Since ΔV and ΔS are non-zero for melting and vaporization, it follows that the slopes of the chemical potential plotted against either pressure or temperature are different either side of the transition (Fig. 6.10a). In other words, the first derivatives of the chemical potentials are discontinuous at the transition, and the transitions are classified as **first-order phase transitions**.

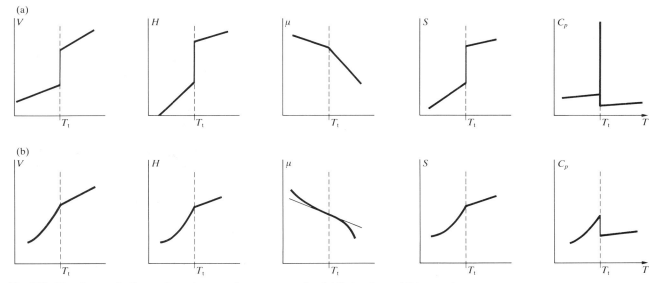

Fig. 6.10 The changes in thermodynamic properties accompanying (a) first-order and (b) second-order phase transitions.

The heat capacity C_p of a substance is the slope of the enthalpy with respect to temperature. At a first-order phase transition, H changes by a finite amount for an infinitesimal change of temperature (Fig. 6.10a). At the transition the slope of H and therefore the heat capacity are infinite. The physical reason is that heating drives the transition rather than raising the temperature. For example, boiling water stays at the same temperature even though heat is being supplied. It follows that a first-order phase transition is also characterized by an infinite heat capacity at the transition temperature (Fig. 6.10a).

Second-order and λ transitions

A **second-order phase transition** is one in which the first derivative of μ is continuous but its second derivative is discontinuous. A continuous slope of μ (a graph with the same slope on either side of the transition) implies that

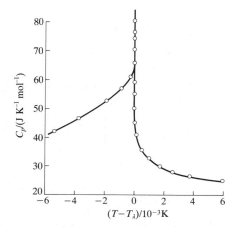

Fig. 6.11 The λ-curve for helium, where the heat capacity rises to infinity. The shape of this curve is the origin of the name 'λ-transition'.

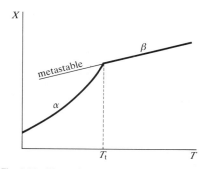

Fig. 6.12 The variation of thermodynamic properties close to the temperature of a second-order transition. The loss of the α phase is complete when T_t is reached from below, but the β phase may persist as a metastable phase when T_t is passed from above.

Table 6.1. Surface tensions of liquids at 293 K

	$\gamma/(\text{N m}^{-1})$†
Benzene	2.888×10^{-2}
Mercury	47.2×10^{-2}
Methanol	2.26×10^{-2}
Water	7.275×10^{-2}

† An alternative form of this unit is J m^{-2}.

the volume and entropy (and hence the enthalpy) do not change at the transition (Fig. 6.10b). Second-order transitions include order–disorder transitions in alloys, the onset of ferromagnetism, and the fluid–superfluid transition of helium. Typically the heat capacity begins to grow well before the transition: compare Figs. 6.10 and 6.11. The shape of the heat capacity curve resembles the Greek letter λ, and so this type of phase change is also called a **λ-transition**.

All systems showing second-order transitions have thermodynamic properties that vary as shown in Fig. 6.12, but the slope of the α phase may be infinite at the transition. In each case the β phase may be extrapolated to $T < T_t$, where it is metastable. However, the α phase cannot be extrapolated to the metastable region above the transition temperature. Physically, a second-order transition is one that occurs gradually as T_t is approached from below, and is complete at T_t.

The liquid surface

In this section we consider the interfacial region, the region of a sample where one phase ends and another begins. We concentrate on the liquid–vapour interface, which is interesting because it is so mobile. Interfaces between solids and other phases are dealt with in Chapter 29.

6.6 Surface tension

Liquids tend to adopt shapes that minimize their surface area, for then the maximum number of molecules are in the bulk and surrounded by the greatest number of neighbours. Droplets of liquids therefore tend to be spherical, because a sphere is the geometrical object with the smallest surface/volume ratio. However, there may be other forces present that compete against the tendency to form this ideal shape, and in particular gravity may flatten spheres into puddles or oceans.

We can express surface effects in the language of Helmholtz and Gibbs functions. The link between these functions and the surface area is the work needed to change the area by a given amount and the fact that dA and dG are equal (under different conditions) to the work done in changing the energy of a system. The work we need to do to change the surface area of a sample by an infinitesimal amount dσ is proportional to dσ, and we write

$$dw = \gamma \, d\sigma \qquad (10)$$

The coefficient γ is called the **surface tension**; its dimensions are energy/area (J m^{-2}). However, as in Table 6.1, values of γ are usually reported in N m^{-1} (because $1 \text{ J} = 1 \text{ N m}$). At constant volume and temperature we can identify the work of surface formation with the change in the Helmholtz function, and write

$$dA = \gamma \, d\sigma \qquad (11)$$

Since the Helmholtz function decreases (d$A < 0$) if the surface area decreases (d$\sigma < 0$), surfaces have a natural tendency to contract. This is a more formal way of expressing what we have already described.

Example 6.7: *Using the surface tension*

Calculate the work needed to raise a wire of length l and to stretch the surface of a liquid through a height h in the arrangement shown in Fig. 6.13. Disregard gravitational potential energy.

Answer. When the wire of length l is raised through a height h it increases the area of the liquid by twice the area of the rectangle (because there is a surface on each side). The total increase is therefore $2lh$. The work done is therefore $2\gamma lh$.

Comment. The work can be expressed as a force × distance by writing it as $2\gamma l \times h$, and identifying $2\gamma l$ as the opposing force on the wire of length l. This is why γ is called a tension and why its units are often chosen to be N m^{-1} (so γl is a force in newtons).

Exercise. Calculate the work of creating a cavity of radius r in a liquid of surface tension γ. $[4\pi r^2\gamma]$

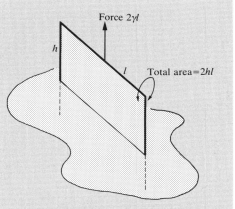

Fig. 6.13 The model used for calculating the work of forming a liquid film when a wire of length l is raised and pulls the surface with it through a height h.

6.7 The vapour pressure above a curved surface

We shall now see that there are two consequences of the surface tension that are relevant to the properties of pure phases. One is that the vapour pressure of a liquid depends on the curvature of its surface. The other is the capillary rise (or fall) of liquids in narrow tubes.

Bubbles, cavities, and droplets

By a **bubble** we mean a region in which vapour (and possibly air too) is trapped by a thin film; by a **cavity** we mean a vapour-filled hole in a liquid. What are widely called 'bubbles' in liquids are therefore strictly cavities: true bubbles have two surfaces (one on each side of the film); cavities have only one. The treatments of both are similar, but a factor of 2 is required for bubbles in order to take into account the doubled surface area. By a **droplet** we mean a small volume of liquid at equilibrium surrounded by its vapour (and possibly also air).

The cavities in a liquid are at equilibrium when the tendency for their surface area to decrease is balanced by the rise of internal pressure which would then result. If the pressure inside a cavity is p_{in} and its radius is r, the outward force is pressure × area $= 4\pi r^2 p_{in}$. The force inwards arises from the sum of the external pressure p_{out} and the surface tension. The change in surface area when the radius of a sphere changes from r to $r + dr$ is

$$d\sigma = 4\pi(r + dr)^2 - 4\pi r^2 = 8\pi r\, dr$$

The work done when we stretch the surface by this amount is therefore

$$dw = 8\pi\gamma r\, dr$$

As force × distance is work, the force opposing stretching through a distance dr when the radius is r is

$$F = 8\pi\gamma r$$

When the outward and inward forces are balanced,

$$4\pi r^2 p_{in} = 4\pi r^2 p_{out} + 8\pi\gamma r$$

which rearranges into the **Laplace equation**:

$$p_{in} = p_{out} + \frac{2\gamma}{r} \qquad (12)$$

The Laplace equation shows that, because $2\gamma/r > 0$, the pressure inside a curved surface (on the concave side of the interface) is always greater than the pressure outside. It also shows that the difference decreases to zero as the radius of curvature becomes infinite (when the surface is flat). Small cavities have very small radii of curvature, and so the pressure difference across their surface is quite large. For instance, a 0.10 mm radius 'bubble' (actually, a cavity) in champagne implies a pressure difference of 1.5 kPa, which is enough to sustain a 15 cm column of water.

We saw in Section 6.3 that the vapour pressure of a liquid depends on the pressure applied to the liquid (eqn 3). Since curving a surface gives rise to a pressure differential of $\Delta P = 2\gamma/r$, we expect the vapour pressure above a curved surface to be different from that above a flat surface. By substituting this value of ΔP into eqn 3b, we obtain the **Kelvin equation** for the vapour pressure of a liquid when it is dispersed as droplets of radius r:

$$p = p^* e^{2\gamma V_m/RTr} \qquad (13a)$$

The analogous expression for the vapour pressure inside a cavity can be written at once. The pressure of the liquid outside the cavity is less than the pressure inside, and so the only change is in the sign of the exponent in the last expression:

$$p = p^* e^{-2\gamma V_m/RTr} \qquad (13b)$$

Nucleation

In the case of droplets of water of radius 10^{-3} mm and 10^{-6} mm the ratios p/p^* at 25°C are about 1.001 and 3.0 respectively. The second figure, although quite large, is unreliable because at that radius the droplet is less than about 10 molecules in diameter and the basis of the calculation is suspect. The first figure shows that the effect is usually small; nevertheless it may have important consequences. Consider, for example, the formation of a cloud. Warm, moist air rises into the cooler regions higher in the atmosphere. At some altitude the temperature is so low that the vapour becomes thermodynamically unstable with respect to the liquid and we expect it to condense into a cloud of liquid droplets. The initial step can be imagined as a swarm of water molecules sticking together into a microscopic droplet. Since the initial droplet is so small it has an enhanced vapour pressure. Therefore, instead of growing it evaporates. This effect stabilizes the vapour because an initial tendency to condense is overcome by a heightened tendency to evaporate. The vapour phase is then said to be **supersaturated** because it is thermodynamically unstable yet is prevented from condensing by kinetic effects.

Clouds do form, and so there must be a mechanism. Two processes are responsible. The first is that a sufficiently large number of molecules might stick together into a droplet so big that the enhanced evaporative effect is unimportant. The chance of one of these **spontaneous nucleation** centres forming is low, and in rain formation it is not a dominant mechanism. The more important process depends on the presence of minute dust particles or other kinds of foreign matter. These **nucleate** the condensation by providing surfaces to which the water molecules can attach. The **cloud chamber** used to track the paths of elementary particles works on the same principle. In a very clean environment a supersaturated mixture of water vapour and air does not condense, but when an ionizing, swiftly moving elementary particle passes through, the ions formed in its path nucleate the condensation and the trajectory is mapped as a streak of condensed water.

Liquids may be **superheated** above their boiling points and **supercooled** below their freezing points. In each case the thermodynamically stable phase is not achieved on account of the kinetic stabilization that occurs in the absence of nucleation centres. For example, superheating occurs because the vapour pressure inside a cavity is artificially low, and so any cavity that does form tends to collapse. This is encountered when an unstirred beaker of water is heated, for its temperature may be raised above its boiling point. Violent bumping often ensues as spontaneous nucleation leads to bubbles big enough to survive. In order to ensure smooth boiling at the true boiling temperature, nucleation centres, such as small pieces of sharp-edged glass or bubbles (cavities) of air, should be introduced. The **bubble chamber**, the more modern way of tracking elementary particles, works on a similar principle, but depends on the nucleation of the evaporation of superheated liquid hydrogen by ionizing radiation.

6.8 Capillary action

The tendency of liquids to rise up capillary tubes is a consequence of surface tension. Consider what happens when a glass capillary tube is first immersed in water or any liquid that has a tendency to adhere to the walls. The energy is lowest when a thin film covers as much of the glass as possible. As this film creeps up the inside wall it has the effect of curving the surface of the liquid inside the tube (Fig. 6.14a). The pressure just beneath the curving meniscus is less than the atmospheric pressure by approximately $2\gamma/r$, where r is the radius of the tube and we assume a hemispherical surface. The pressure immediately under the flat surface outside the tube is P, the atmospheric pressure; but inside the tube under the curved surface it is only $P - 2\gamma/r$. The excess external pressure presses the liquid up the tube until hydrostatic equilibrium (equal pressures at equal depths, another consequence of the equality of chemical potential) has been reached (Fig. 6.14b).

Capillary rise

As we established in Example 1.2, the pressure exerted by a column of liquid of density ρ and height h is

$$p = \rho g h \tag{14}$$

This hydrostatic pressure matches the pressure difference $2\gamma/r$ at equilibrium. Therefore, the height of the column at equilibrium is obtained by equating $2\gamma/r$ and $\rho g h$, which gives

$$h = \frac{2\gamma}{\rho g r} \tag{15}$$

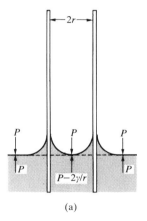

(a)

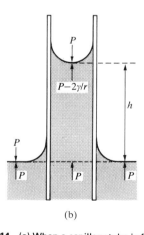

(b)

Fig. 6.14 (a) When a capillary tube is first stood in a liquid, the latter climbs up the walls, so curving the surface. The pressure just under the meniscus is then less than that due to the atmosphere by an amount $2\gamma/r$ (by the Laplace equation). (b) The equality of pressures at equal heights throughout the liquid is achieved if the liquid rises in the tube. The condition of equilibrium is when the hydrostatic pressure at the foot of the column ($\rho g h$) cancels the pressure difference arising from the curved surface.

This simple expression provides an accurate way of measuring the surface tension of liquids.

Example 6.8: *Measuring the surface tension*

In an experiment to measure the surface tension of water over a range of temperatures, a capillary tube of internal diameter 0.40 mm was supported vertically in the sample. The density of the liquid was measured in separate experiments. The following results were obtained:

$\theta/°C$	10	15	20	25	30
h/cm	7.56	7.46	7.43	7.36	7.29
$\rho/(g\,cm^{-3})$	0.9997	0.9991	0.9982	0.9971	0.9957

Find the temperature variation of the surface tension.

Answer. Use eqn 15 in the form $\gamma = \frac{1}{2}\rho hgr$, with $g = 9.81\ m\,s^{-2}$ and $r = 0.20$ mm:

$\theta/°C$	10	15	20	25	30
$\gamma/(mJ\,m^{-2})$	74	73	73	72	71

Comment. Surface tension generally decreases with increasing temperature.

Exercise. Suppose that, over a limited range, the surface tension and the density decrease linearly with temperature. Is there a temperature at which the height of a capillary column is a maximum (or a minimum)?
[For $\gamma = \gamma_0(1 - \alpha T)$ and $\rho = \rho_0(1 - \beta T)$, $dh/dT \propto \beta - \alpha$]

When the adhesive forces between the liquid and the material of the capillary wall are weaker than the cohesive forces within the liquid (as for mercury in glass), the liquid in the tube retracts from the walls. This curves the surface with the concave, high pressure side downwards. In order to equalize the pressure at the same depth throughout the liquid, the surface must fall to compensate for the heightened pressure arising from its curvature. This results in a capillary depression.

The contact angle

In many cases there is a non-zero angle between the edge of the meniscus and the wall. If this **contact angle** is θ_c, eqn 15 is modified by multiplying the right hand side by $\cos \theta_c$.

The origin of the contact angle can be traced to the balance of forces at the line of contact between the liquid and the solid (Fig. 6.15). If the solid/gas, solid/liquid, and liquid/gas surface tensions (essentially the energy needed to create unit area of each of the interfaces) are denoted γ_{sg}, γ_{sl}, and γ_{lg} respectively, the forces are in balance if

$$\gamma_{sg} = \gamma_{sl} + \gamma_{lg} \cos \theta_c$$

This solves to

$$\cos \theta_c = \frac{\gamma_{sg} - \gamma_{sl}}{\gamma_{lg}}$$

If $0 < \theta_c < 90°$ (which occurs when $\gamma_{sl} = 0$), the liquid 'wets' (spreads over) the surface fully. In this case, no work is needed to create the solid/liquid interface. If $\theta_c \approx \pi$ (which occurs when $\gamma_{sg} = 0$, and no work is needed to prepare the gas/solid interface), a drop of liquid on the surface of

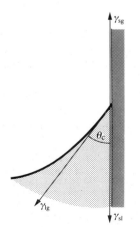

Fig. 6.15 The balance of forces that result in a contact angle θ_c.

a solid remains separated from it by a thin film of vapour. For mercury in contact with glass, $\gamma_c = 140°$, which requires $\gamma_{sg} < \gamma_{sl}$. This occurs if less energy is needed to create a solid/gas interface than a solid/liquid interface, and is attributed to the strong cohesive forces within the liquid.

Further reading

Experimental techniques

E. L. Skau and J. C. Arthur, Determination of melting and freezing temperatures. In *Techniques of chemistry* (ed. A. Weissberger and B. W. Rossiter) V, 105, Wiley-Interscience, New York (1971).

J. R. Anderson, Determination of boiling and condensation temperatures. In *Techniques of chemistry* (ed. A. Weissberger and B. W. Rossiter) V, 199, Wiley-Interscience, New York (1971).

A. E. Alexander and J. B. Hayter, Determination of surface and interfacial tension. In *Techniques of chemistry* (ed. A. Weissberger and B. W. Rossiter) V, 501, Wiley-Interscience, New York (1971).

Properties

T. E. Jordan, *Vapor pressure of organic compounds.* Interscience, New York (1954).

J. S. Rowlinson and B. Widom, *Molecular theory of capillarity.* Clarendon Press, Oxford (1982).

H. E. Stanley, *Introduction to phase transitions and critical phenomena.* Clarendon Press, Oxford (1971).

Data

J. Timmermans (ed.), *Physico-chemical constants of pure organic compounds.* Elsevier, Amsterdam (1956).

T. E. Jordan, *Vapor pressure of organic compounds.* Wiley-Interscience, New York (1954).

R. C. Weast (ed.), *Handbook of chemistry and physics.* CRC Press, Boca Raton (1990).

Exercises

6.1 The vapour pressure of dichloromethane at 24.1°C is 400 Torr and its enthalpy of vaporization is 28.7 kJ mol^{-1}. Estimate the temperature at which its vapour pressure is 500 Torr.

6.2 The molar volume of a certain solid is 161.0 cm^3 mol^{-1} at 1.00 atm and 350.75 K, its melting temperature. The molar volume of the liquid at this temperature and pressure is 163.3 cm^3 mol^{-1}. At 100 atm the melting temperature changes to 351.26 K. Calculate the molar enthalpy and entropy of fusion of the solid.

6.3 The vapour pressure of a liquid in the temperature range 200 K to 260 K were found to fit the expression

$$\ln\,(p/\text{Torr}) = 16.255 - \frac{2501.8}{T/\text{K}}$$

Calculate the enthalpy of vaporization of the liquid.

6.4 Calculate the work done when the surface of liquid water at 25°C is increased from 150 cm^2 to 2500 cm^3.

6.5 Calculate the surface tension of a liquid of density 0.871 g cm^{-3} given that it rises 1.20 cm in a capillary tube of diameter 0.80 mm.

6.6 At 20°C the vapour pressure of CCl$_4$ is 0.90 Torr greater in the form of droplets than as the bulk liquid, for which it is 87.05 Torr. Given that the density of the liquid is 1.60 g cm^{-3}, calculate the radii of the droplets.

6.7 When benzene freezes at 5.5°C its density changes from 0.879 g cm^{-3} to 0.891 g cm^{-3}. Its enthalpy of fusion is 10.59 kJ mol^{-1}. Estimate the freezing point of benzene at 1000 atm.

6.8 In July in Los Angeles, the incident sunlight at ground level has a power density of 1.2 kW m^{-2} at noon. A swimming pool of area 50 m^2 is directly exposed to the sun. What is the maximum rate of loss of water?

6.9 An open vessel containing (a) water, (b) benzene, (c) mercury stands in a laboratory measuring 5 m × 5 m × 3 m at 25°C. What mass of each substance will be found in the air if there is no ventilation? (The vapour pressures are (a) 24 Torr, (b) 98 Torr, (c) 1.7 mTorr.)

6.10 On a cold, dry morning after a frost, the temperature was $-5°C$ and the partial pressure of water in the atmosphere fell to 2 Torr. Will the frost sublime? What partial pressure of water would ensure that the frost remained?

6.11 Refer to Fig. 6.9 and describe the changes that would be observed when water vapour at 1.0 atm and 400 K is cooled at constant pressure to 260 K. Sketch the appearance of a plot of temperature against time.

6.12 Refer to Fig. 6.9 and describe the changes that would

be observed when cooling takes place at the pressure of the triple point.

6.13 Use the phase diagram in Fig. 6.5 to state what would be observed when a sample of carbon dioxide, initially at 1.0 atm and 298 K is subjected to the following cycle: (a) isobaric (constant pressure) heating to 320 K, (b) isothermal compression to 100 atm, (c) isobaric cooling to 210 K, (d) isothermal decompression to 1.0 atm, isobaric heating to 298 K.

6.14 By how much is the vapour pressure of benzene changed when it is dispersed in the form of droplets of radius (a) $10\,\mu m$, (b) $0.10\,\mu m$ at 25°C?

Problems

Numerical problems

6.1 The enthalpy of vaporization of a certain liquid is found to be $14.4\,kJ\,mol^{-1}$ at 180 K, its normal boiling point. The molar volumes of the liquid and the vapour at the boiling point are $115\,cm^3\,mol^{-1}$ and $14.5\,dm^3\,mol^{-1}$ respectively. Estimate dp/dT from the Clapeyron equation and estimate the percentage error in its value if the Clausius–Clapeyron equation is used instead.

6.2 Calculate the difference in slope of the chemical potential against temperature on either side of (a) the normal freezing point of water and (b) the normal boiling point of water. By how much does the chemical potential of water supercooled to $-5.0°C$ exceed that of ice at that temperature?

6.3 Calculate the difference in slope of the chemical potential against pressure on either side of (a) the normal freezing point of water and (b) the normal boiling point of water. The densities of ice and water at 0°C are $0.917\,g\,cm^{-3}$ and $1.000\,g\,cm^{-3}$, and those of water and water vapour at 100°C are $0.958\,g\,cm^{-3}$ and $0.598\,g\,L^{-1}$. By how much does the chemical potential of water vapour exceed that of liquid water at 1.2 atm and 100°C?

6.4 The enthalpy of fusion of mercury is $2.292\,kJ\,mol^{-1}$, and its normal freezing point is 234.3 K with a change in molar volume of $+0.517\,cm^3\,mol^{-1}$. At what temperature will the bottom of a 10 m high column of mercury (of density $13.6\,g\,cm^{-3}$) be expected to freeze?

6.5 50.0 L of dry air was slowly bubbled through a thermally insulated beaker containing 250 g of water initially at 25°C. Calculate the final temperature. (The vapour pressure of water is approximately constant at 23.8 Torr throughout, and its heat capacity is $75.5\,J\,K^{-1}\,mol^{-1}$. Assume that the air is not heated or cooled and that water vapour is a perfect gas.)

6.6 The vapour pressure p of nitric acid varies with temperature as follows:

$\theta/°C$	0	20	40	50	70	80	90	100
$p/$Torr	14.4	47.9	133	208	467	670	937	1282

What is (a) the normal boiling point and (b) the enthalpy of vaporization of nitric acid?

6.7 The vapour pressure of the ketone carvone ($M = 150.2\,g\,mol^{-1}$), a component of spearmint, is as follows:

$\theta/°C$	57.4	100.4	133.0	157.3	203.5	227.5
$p/$Torr	1.00	10.0	40.0	100	400	760

What is the normal boiling point and the enthalpy of vaporization of carvone?

6.8 Construct the phase diagram for benzene near its triple point at 36 Torr and 5.50°C using the following data: $\Delta H_{fus} = 10.6\,kJ\,mol^{-1}$, $\Delta H_{vap} = 30.8\,kJ\,mol^{-1}$, $\rho(s) = 0.891\,g\,cm^{-3}$, $\rho(l) = 0.879\,g\,cm^{-1}$.

6.9 The assumption that the number of molecules on the surface of a droplet is much less than the total number might fail when the sample is dispersed as very small droplets. Estimate the ratio of the number of water molecules ($r \approx 120\,pm$) on the surface of a spherical droplet to the number in the droplet when the radius of the droplets is (a) $10^{-5}\,mm$, (b) $10^{-2}\,mm$, (c) 1 mm.

6.10 A sample of benzene of mass 100 g is dispersed as droplets of radius $1.0\,\mu m$. The surface tension of benzene is $2.8 \times 10^{-2}\,N\,m^{-1}$ and its density is $0.88\,g\,cm^{-3}$. Calculate the change in the Helmholtz function. What is the minimum work needed to bring about the dispersal?

6.11 At 30°C, the surface tension of ethanol in contact with its vapour is $2.189 \times 10^{-2}\,N\,m^{-1}$ and its density is $0.780\,g\,cm^{-3}$. How far up a tube of internal diameter 0.20 mm will it rise? What pressure is needed to push the mensiscus back level with the surrounding liquid?

6.12 A glass tube of internal diameter 1.00 cm surrounds a glass rod of diameter 0.98 cm. How high will water rise in the space between them at 25°C?

Theoretical problems

6.13 Show that for a transition between two solid phases of the same density, that ΔG is independent of the pressure.

6.14 The change in enthalpy is given by $dH = C_p\, dT + V\, dp$. The Clapeyron equation relates dp and dT at equilibrium, and so in combination the two equations can be used to find how the enthalpy changes along a phase boundary as the temperature changes and the two phases remain in equilibrium. Show that

$$d\left(\frac{\Delta H}{T}\right) = \Delta C_p\, d\ln T$$

6.15 In the 'gas saturation method' for the measurement of vapour pressure, a volume V of gas (as measured at a temperature T and a pressure P) is bubbled slowly through the liquid that is maintained at the temperature T and a mass loss m is measured. Show that the vapour pressure p of the liquid is related to its molar mass M by

$$p = \frac{AmP}{1 + Am} \qquad A = \frac{RT}{MPV}$$

The vapour pressure of geraniol ($M = 154.2\ \text{g mol}^{-1}$), which is a component of oil of roses, was measured at 110°C. It was found that when 5.00 L of N_2 at 760 Torr was passed slowly through the heated liquid, the loss of mass was 0.32 g. Calculate the vapour pressure of geraniol.

6.16 Combine the barometric formula (stated in Problem 1.17) for the dependence of the pressure on altitude with the Clausius–Clapeyron equation and predict how the boiling temperature of a liquid depends on the altitude and the ambient temperature. Take the mean ambient temperature temperature as 20°C and predict the boiling temperature of water at 3000 m.

6.17 Calculate the capillary rise for a liquid with a contact angle θ_c by assuming that the meniscus is a part of a sphere of radius greater than the radius of the capillary. Then derive the same formula by balancing the forces acting on the liquid and supposing that only the vertical component of the surface tension draws the liquid up the tube.

6.18 We saw how to derive Maxwell relations in Chapter 5. Now regard μ as a function of the surface area σ as well as of T and p and deduce the relation $(\partial V/\partial\sigma)_{p,T} = (\partial\gamma/\partial p)_{\sigma,T}$ using the fact that $d\mu$ is an exact differential. Evaluate $\partial V/\partial\sigma$ for a spherical droplet of radius r and hence show that the Maxwell relation leads at once to the Laplace equation.

6.19 Show that the velocity of the formation of a hole in the skin of a bubble is of the order of $(2\gamma/\rho\delta)^{1/2}$, where γ is the surface tension of the fluid, ρ its density, and δ is the thickness of the film. Estimate the velocity in soapy water for which $\gamma = 2.6 \times 10^{-2}\ \text{N m}^{-1}$.

7

Changes of state: physical transformations of simple mixtures

Check-list of key ideas

1. The definition and measurement of *partial molar properties* (Section 7.1).

2. The total volume of a mixture as the sum of partial molar volumes (eqn 3) and the total Gibbs function as a sum of chemical potentials (eqn 5).

3. The *Gibbs–Duhem equation* relating the changes in the chemical potentials of all the substances in a mixture (eqn 6).

4. The *Gibbs function of mixing* (eqn 7) of two perfect gases, their *entropy of mixing* (eqn 8), and their *enthalpy of mixing*.

5. The *chemical potential of a liquid* in terms of the partial pressure of its vapour (eqn 9) and, through *Raoult's law* (eqn 10), in terms of its mole fraction (eqn 11).

6. The definition of an *ideal solution* (Section 7.3) and of an ideal dilute solution through *Henry's law* (eqn 12).

7. The *thermodynamic mixing functions* of an ideal solution and the definition and significance of *excess functions* of real solutions (Section 7.4).

8. The description of *colligative properties* in terms of the chemical potential of the solvent (Section 7.5), and the calculation of the *elevation of boiling point* (eqn 15), the *depression of freezing point* (eqn 16), and the *osmotic pressure* in terms of the *van't Hoff equation* (eqn 19).

9. The estimation of the *solubility* of a substance that forms an ideal solution in terms of its enthalpy of fusion (eqn 17).

10. The description of mixtures of volatile liquids in terms of *vapour pressure diagrams* (Section 7.6) and their interpretation in terms of the *lever rule* (eqn 23).

11. The construction of *temperature–composition diagrams* and their use in the description of *distillation and fractional distillation* (Section 7.7).

12. The formation of *azeotropes* (Section 7.7) and the distillation of *immiscible liquids* (Section 7.8).

13. The description of *real solutions* in terms of the *activities* of the solvent and solute (eqn 25a).

14. The definition of the *solvent activity* and the *activity coefficient* in terms of Raoult's law (eqn 25b) and the definition of the *solute activity* in terms of Henry's law (eqn 26b).

15. The determination of activities by measurement of vapour pressure (Examples 7.11 and 7.12).

We now leave pure materials and the limited but important changes they can undergo, and examine mixtures. We shall consider only unreactive mixtures here, and leave the consequences of reactions until Chapter 9. However, both discussions will be based on the chemical potential, and so we shall see how this simple concept unifies much of chemistry. At this stage we deal mainly with **binary mixtures**, or mixtures consisting of two components. This means that we shall often be able to simplify equations by making use of the relation $x_A + x_B = 1$ for the mole fractions.

Another restriction of this chapter is that we shall consider mainly **non-electrolyte solutions**, where the solute is not present as ions. We shall delay until Chapter 10 the special problems of **electrolyte solutions**, in which the solute is ionized and where the ions generally interact strongly with each other.

The thermodynamic description of mixtures

In this and the following chapters we need a set of concepts that enable us to apply thermodynamics to mixtures of variable composition. We have already seen that the partial pressure, the contribution to the total pressure of one component in a gaseous mixture, is used to discuss the properties of mixtures of gases. For a more general description of the thermodynamics of mixtures we need to introduce other 'partial' properties.

7.1 Partial molar quantities

The easiest partial molar property to visualize is the 'partial molar volume', the contribution to the volume that a component in a sample makes to the total volume.

Partial molar volume

Imagine a huge volume of pure water. When a further 1 mol of H_2O is added, the volume increases by 18 cm^3 and we can report that 18 cm^3 mol^{-1} is the molar volume of pure water. However, when we add 1 mol of H_2O to a huge volume of pure ethanol, the volume increases by only 14 cm^3. The reason for the different increase in volume is that the volume occupied by a given number of water molecules depends on the molecules that surround them. Now there is so much ethanol present that each H_2O molecule is surrounded by pure ethanol, and the packing of the molecules results in them occupying only 14 cm^3. The quantity 14 cm^3 mol^{-1} is the **partial molar**

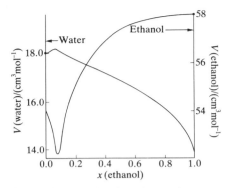

Fig. 7.1 The partial molar volumes of water and ethanol at 25°C. Note the different scales (water on the left, ethanol on the right).

volume of water in pure ethanol, the volume of the mixture that can be ascribed to one of the components.

The partial molar volumes of the components of a mixture vary with composition because the environment of each type of molecule changes as the composition changes from pure A to pure B. It is this changing molecular environment, and the consequential modification in the forces acting between molecules, that results in the variation of the thermodynamic properties of a mixture as its composition is changed. The partial molar volumes of water and ethanol across the full composition range at 25°C are shown in Fig. 7.1.

The partial molar volume V_J of a substance J at some general composition is defined formally as follows:

$$V_J = \left(\frac{\partial V}{\partial n_J} \right)_{p,T,n'} \tag{1}$$

where n_J is the amount (the number of moles) of J and the subscript n' signifies that the amounts of all other substances present are held constant. The partial molar volume is the slope of the graph of the total volume as the amount of J is changed, the pressure, temperature, and amount of the other components being constant. Its value depends on the composition, as we saw for water and ethanol. The definition implies that when the composition of the mixture is changed by the addition of dn_A of A and dn_B of B, the total volume of the mixture changes by

$$dV = \left(\frac{\partial V}{\partial n_A} \right)_{p,T,n_B} dn_A + \left(\frac{\partial V}{\partial n_B} \right)_{p,T,n_A} dn_B$$

$$= V_A \, dn_A + V_B \, dn_B \tag{2}$$

Once we know the partial molar volumes of the two components of a mixture at the composition (and temperature) of interest we can state the total volume V of the mixture using

$$V = n_A V_A + n_B V_B \tag{3}$$

The reasoning behind this simple result is as follows. Consider a very large sample of the mixture of the specified composition. Then, when an amount n_A of A is added, the composition remains virtually unchanged, V_A is constant, and the volume of the sample changes by $n_A V_A$. When n_B of B is added, the volume changes by $n_B V_B$ for the same reason. The total change of volume is therefore $n_A V_A + n_B V_B$. The sample now occupies a larger volume, but the proportions of the components are still the same. Now scoop out of this enlarged volume a sample containing an amount n_A of A and n_B of B. Its volume is $n_A V_A + n_B V_B$. Because V is a state function, the same sample could have been prepared simply by mixing the appropriate amounts of A and B. This justifies eqn 3.

Partial molar volumes (and partial molar quantities in general) can be measured in several ways. One method is to measure the dependence of the volume on the composition and to determine the slope dV/dn at the composition of interest. A better technique is the 'method of intercepts', which is outlined in the *Further information* section.

Example 7.1: *Using partial molar volumes*

A corrupt barman attempts to prepare $100.0 \, \text{cm}^3$ of a drink by mixing $30.0 \, \text{cm}^3$ of ethanol with $70.0 \, \text{cm}^3$ of water at 25°C. What volumes should have been mixed in order to arrive at a mixture of the same strength but of the required volume?

Answer. The barman seems to have understood neither the concept nor the importance of partial molar quantities. To find the total volume of such a mixture, we should use eqn 3 with partial molar volumes taken from Fig. 7.1 for the specified composition. The mole fraction composition of the mixture is determined from the densities (listed in the Data Section) and molar masses of the components. The amounts that correspond to the stated volumes are $3.87 \, \text{mol}$ of H_2O and $0.514 \, \text{mol}$ of C_2H_5OH. Therefore, $x(H_2O) = 0.883$ and $x(C_2H_5OH) = 0.117$. From Fig. 7.1, the partial molar volumes at this composition are $V(H_2O) = 18.0 \, \text{cm}^3 \, \text{mol}^{-1}$ and $V(C_2H_5OH) = 53.6 \, \text{cm}^3 \, \text{mol}^{-1}$, and the total volume of the mixture is

$$V = 3.87 \, \text{mol} \times 18.0 \, \text{cm}^3 \, \text{mol}^{-1} + 0.514 \, \text{mol} \times 53.6 \, \text{cm}^3 \, \text{mol}^{-1}$$
$$= 97.3 \, \text{cm}^3$$

A mixture of the same relative composition but of total volume $100.0 \, \text{cm}^3$ will have the same mole fractions of the components, and therefore the same partial molar volumes, but a different overall amount n. It follows that we must scale up the volumes mixed in the ratio $100.0/97.3 = 1.028$, and therefore mix $72.0 \, \text{cm}^3$ of water and $30.8 \, \text{cm}^3$ of ethanol.

Comment. It would probably be unwise to attempt to explain this to the barman.

Exercise. At 25°C, the density of a 50 per cent by mass ethanol/water solution is $0.914 \, \text{g cm}^{-3}$. Given that the partial molar volume of water in the solution is $17.4 \, \text{cm}^3 \, \text{mol}^{-1}$, what is the partial molar volume of the ethanol?

$$[56.4 \, \text{cm}^3 \, \text{mol}^{-1}]$$

Partial molar Gibbs functions

The concept of a partial molar quantity can be extended to any of the extensive state functions. One already encountered, but under a different name, is the partial molar Gibbs function, the chemical potential:

$$\mu_J = \left(\frac{\partial G}{\partial n_J} \right)_{p, T, n'} \tag{4}$$

By the same argument that led to eqn 3, the total Gibbs function of a mixture is

$$G = n_A \mu_A + n_B \mu_B \tag{5}$$

where μ_A and μ_B are the chemical potentials at the composition of the mixture.

The Gibbs–Duhem equation

Since the total Gibbs function of a mixture is given by eqn 5 and the chemical potentials depend on the composition, when the compositions are changed infinitesimally we might expect G to change by

$$dG = \mu_A \, dn_A + \mu_B \, dn_B + n_A \, d\mu_A + n_B \, d\mu_B$$

However, we saw in Section 5.2 that at constant pressure and temperature the Gibbs function changes by

$$dG = \mu_A \, dn_A + \mu_B \, dn_B$$

Since G is a state function, these two equations must be equal to each other, which implies that at constant temperature and pressure

$$n_A \, d\mu_A + n_B \, d\mu_B = 0$$

This equation is a special case of the **Gibbs–Duhem equation**:

$$\sum_J n_J \, d\mu_J = 0 \qquad (6)$$

The significance of this result is that the chemical potentials of a mixture cannot change independently: in a binary mixture, if one increases the other must decrease (see the following example). The same line of reasoning applies to all partial molar quantities.

Example 7.2: *Using the Gibbs–Duhem equation*

Show that the analogue of the Gibbs–Duhem equation for molar volume is that

$$\sum_J n_J \, dV_J = 0$$

and hence when the partial molar volumes of the components in a binary mixture are plotted, that where one increases the other must decrease.

Answer. From eqn 3 (written for several components) we can write

$$dV = d \sum_J n_J V_J = \sum_J n_J \, dV_J + \sum_J V_J \, dn_J$$

However, we know from eqn 2 that

$$dV = \sum_J V_J \, dn_J$$

Since V is a state function, the two expressions are equal, which implies that

$$\sum_J n_J \, dV_J = 0$$

For a binary mixture this becomes

$$n_A \, dV_A + n_B \, dV_B = 0$$

which rearranges to

$$dV_A = -\frac{n_B}{n_A} dV_B$$

Therefore, where one partial molar volume increases ($dV_B > 0$) the other must decrease ($dV_A < 0$).

Comment. This can be seen in Fig. 7.1, where increases in the partial molar volume of water are mirrored by decreases in the partial molar volume of ethanol, and vice versa.

Exercise. Show that if the chemical potential of a component in a binary mixture increases, the other must decrease. $\qquad [n_A \, d\mu_A = -n_B \, d\mu_B]$

Negative partial molar volumes

One final word of warning: molar volumes and molar entropies are always positive, but the corresponding partial molar quantities need not be. For example, the limiting partial molar volume of $MgSO_4$ (its partial molar volume in the limit of zero concentration) is $-1.4\,cm^3\,mol^{-1}$, which means that the addition of 1 mol of $MgSO_4$ to a large volume of water results in a *decrease* in volume of $1.4\,cm^3$. The contraction occurs because the salt breaks up the open structure of water as the ions become hydrated, and it collapses slightly.

7.2 The thermodynamics of mixing

The dependence of the Gibbs function of a mixture on its composition is given by eqn 5, and we know that systems tend towards lower Gibbs function. This is the link we need in order to apply thermodynamics to the discusssion of spontaneous changes of composition. One simple example of a spontaneous mixing process is that of two gases introduced into the same container. The mixing is spontaneous, and so must correspond to a decrease in G. We shall now see how to express this idea quantitatively.

The Gibbs function of mixing

Let the amounts of two perfect gases in the two containers be n_A and n_B; both are at a temperature T and a pressure p. At this stage, the chemical potentials of the two gases have their 'pure' values and the Gibbs function of the total system is

$$G_i = n_A\mu_A + n_B\mu_B$$

$$= n_A\left(\mu_A^{\ominus} + RT\ln\frac{p}{p^{\ominus}}\right) + n_B\left(\mu_B^{\ominus} + RT\ln\frac{p}{p^{\ominus}}\right)$$

After mixing, each gas exerts a partial pressure p_J, with $p_A + p_B = p$. The total Gibbs function changes to

$$G_f = n_A\left(\mu_A^{\ominus} + RT\ln\frac{p_A}{p^{\ominus}}\right) + n_B\left(\mu_B^{\ominus} + RT\ln\frac{p_B}{p^{\ominus}}\right)$$

The difference $G_f - G_i$, the **Gibbs function of mixing** ΔG_{mix}, is therefore

$$\Delta G_{mix} = n_A RT\ln\frac{p_A}{p} + n_B RT\ln\frac{p_B}{p}$$

We may replace n_J by $x_J n$ and use Dalton's law (Section 1.2) to write $p_J/p = x_J$ for each component, which gives

$$\Delta G_{mix} = nRT(x_A\ln x_A + x_B\ln x_B) \qquad (7)°$$

Since mole fractions are never greater than 1, the logarithms in this equation are negative, and $\Delta G_{mix} < 0$. This confirms that perfect gases mix spontaneously in all proportions. However, the equation extends common sense by allowing us to discuss the process quantitatively. We see, for instance, that ΔG_{mix} is directly proportional to the temperature but is independent of the pressure.

Example 7.3: *Calculating a Gibbs function of mixing*

A container is divided into two compartments. One contains 3.0 mol of H_2 at 1.0 atm and 25°C, the other contains 1.0 mol of N_2 at 3.0 atm and 25°C. Calculate the Gibbs function of mixing when the partition is removed.

Answer. The initial Gibbs function is

$$G_i = 3.0 \text{ mol} \times \{\mu^{\ominus}(H_2) + RT \ln 1.0\} + 1.0 \text{ mol} \times \{\mu^{\ominus}(N_2) + RT \ln 3.0\}$$

The initial volumes occupied are

$$V(H_2) = \frac{3.0 \text{ mol} \times RT}{1.0 \text{ atm}} \qquad V(N_2) = \frac{1.0 \text{ mol} \times RT}{3.0 \text{ atm}}$$

and so the total volume of the container is

$$V = RT\left(3.0 + \frac{1.0}{3.0}\right) \times \frac{1 \text{ mol}}{1 \text{ atm}} = \frac{10.0 \text{ mol} \times RT}{3.0 \text{ atm}}$$

The final pressure (with $n = 4.0$ mol) is therefore

$$p = \frac{nRT}{V} = 1.2 \text{ atm}$$

The final partial pressures, using $x(H_2) = 0.75$ and $x(N_2) = 0.25$ are therefore $p(H_2) = 0.90$ atm and $p(N_2) = 0.30$ atm and the final Gibbs function is

$$G_f = 3.0 \text{ mol} \times \{\mu^{\ominus}(H_2) + RT \ln 0.90\} + 1.0 \text{ mol} \times \{\mu^{\ominus}(N_2) + RT \ln 0.30\}$$

The Gibbs function of mixing is the difference

$$G_f - G_i = -2.62 \text{ mol} \times RT$$

which evaluates to -6.5 kJ.

Comment. The value of ΔG_{mix} is the sum of two contributions: the mixing itself, and the changes in pressure of the two gases. When 3.0 mol of H_2 mixes with 1.0 mol of N_2 at the *same* pressure, the change of Gibbs function is given by eqn 7 as -5.6 kJ independent of the initial common pressure.

Exercise. 2.0 mol of H_2 at 2.0 atm and 25°C and 4.0 mol of N_2 at 3.0 atm and 25°C were used instead. Calculate ΔG_{mix}. What would be the value of ΔG_{mix} had the pressures been identical initially?   [-9.8 kJ, -9.5 kJ]

Other thermodynamic mixing functions

The quantitative expression for ΔG_{mix} lets us compute the **entropy of mixing** ΔS_{mix}. Since $(\partial G / \partial T)_{p,n} = -S$, it follows immediately from eqn 7 that, for a mixture of perfect gases,

$$\Delta S_{mix} = -\left(\frac{\partial \Delta G_{mix}}{\partial T}\right)_{p,n_A,n_B}$$

$$= -nR(x_A \ln x_A + x_B \ln x_B) \tag{8}°$$

Since $\ln x < 0$, $\Delta S_{mix} > 0$. This is what we expect when one gas disperses into the other and the system becomes more chaotic. The entropy of mixing in the last example is readily found to be $+22$ J K^{-1}.

Example 7.4: *Calculating the entropy of mixing*

Calculate the composition of a mixture for which the entropy of mixing is a maximum.

Answer. We need to differentiate eqn 8 with respect to x_A and look for the value of x_A at which the derivative is zero. Since $x_B = 1 - x_A$, we need to differentiate

$$\Delta S_{mix} = -nR\{x_A \ln x_A + (1 - x_A) \ln (1 - x_A)\}$$

This gives (using $d \ln x/dx = 1/x$),

$$\frac{d\Delta S_{mix}}{dx_A} = -nR\{\ln x_A + 1 - \ln (1 - x_A) - 1\}$$

$$= -nR \ln \frac{x_A}{1 - x_A}$$

which is zero when $x_A = \frac{1}{2}$. Hence, the maximum entropy of mixing occurs for the preparation of a mixture that contains equal mole fractions of the two components.

Exercise. Calculate the maximum value of the entropy of mixing.   [$nR \ln 2$]

The **enthalpy of mixing** ΔH_{mix} of two perfect gases may be found from $\Delta G = \Delta H - T \Delta S$ (because $\Delta T = 0$ for the isothermal process). From eqns 7 and 8, we find

$$\Delta H_{mix} = 0 \qquad \text{(constant } p, T\text{)}$$

The enthalpy of mixing is zero, as we should expect for a system in which there are no interactions between particles. It follows that the whole of the driving force for mixing comes from the increase in entropy of the system (since $\Delta H_{mix} = 0$, the entropy of the surroundings is unchanged).

The change in volume that accompanies mixing, ΔV_{mix}, can be found using $(\partial G/\partial p)_{T,n} = V$. Since for perfect gases ΔG_{mix} is independent of pressure,

$$\Delta V_{mix} = 0 \quad \text{(constant } p, T\text{)}$$

This is also to be expected for a system without interactions. Since there is neither volume nor enthalpy change on mixing, and at constant pressure $\Delta H = \Delta U + p \Delta V$, it follows that the internal energy of mixing ΔU_{mix} is also zero:

$$\Delta U_{mix} = 0 \quad \text{(constant } p, T\text{)}$$

7.3 The chemical potentials of liquids

In order to discuss the equilibrium properties of liquid mixtures we need to know how the chemical potential of a liquid depends on its composition. To calculate its value, we use the fact that the chemical potential of a substance present as a dilute vapour, which from eqn 14 of Chapter 5 we know to be $\mu = \mu^{\ominus} + RT \ln (p/p^{\ominus})$, must be equal to its chemical potential in the liquid at equilibrium.

Ideal solutions

We denote quantities relating to pure substances by the superscript *, so the chemical potential of pure liquid A is $\mu_A^*(l)$. Since the vapour pressure of

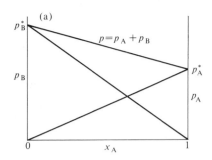

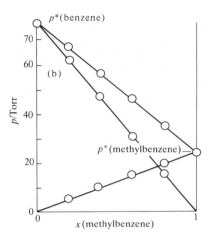

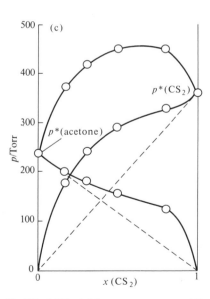

Fig. 7.2 (a) The total vapour pressure and the two partial vapour pressures of an ideal binary mixture are proportional to the mole fractions of the components. (b) Two similar liquids, in this case benzene and toluene (methylbenzene), behave almost ideally, but (c) strong deviations from ideality are shown by dissimilar liquids (in this case carbon disulphide and acetone).

the pure liquid is p_A^*, the chemical potential of A in the vapour is $\mu_A^{\ominus} + RT \ln (p_A^*/p^{\ominus})$. These two chemical potentials are equal at equilibrium, so we can write

$$\mu_A^*(l) = \mu_A^{\ominus} + RT \ln \frac{p_A^*}{p^{\ominus}}$$

If another substance is also present in the liquid, the chemical potential of A in the liquid is $\mu_A(l)$ and its vapor pressure is p_A. In this case

$$\mu_A(l) = \mu_A^{\ominus} + RT \ln \frac{p_A}{p^{\ominus}}$$

Next, we combine the two equations to eliminate the standard chemical potential of the gas, and obtain

$$\mu_A(l) = \mu_A^*(l) + RT \ln \frac{p_A}{p_A^*} \qquad (9)°$$

The final step draws on experimental information about the relation between the ratio of vapour pressures and the composition of the liquid. In a series of experiments on mixtures of closely related liquids (such as benzene and toluene), the French chemist François Raoult found that the ratio p_A/p_A^* is proportional to the mole fraction of A in the liquid. **Raoult's law**, as this relation is now called, is normally written

$$p_A = x_A p_A^* \qquad (10)°$$

This is illustrated in Fig. 7.2a. Some mixtures obey Raoult's law very well, especially when the components are chemically similar (Fig. 7.2b). Mixtures that obey the law throughout the composition range from pure A to pure B are called **ideal solutions**. When we write equations relating to ideal solutions we shall label them with the superscript °, as in eqn 10.

For an ideal solution, it follows from eqns 9 and 10 that

$$\mu_A(l) = \mu_A^*(l) + RT \ln x_A \qquad (11)°$$

This important equation can be turned round and used as the *definition* of an ideal solution (so that it implies Raoult's law rather than stemming from it). It is in fact a better definition than eqn 10 because it does not assume that the gas is perfect.

Some solutions depart significantly from Raoult's law (Fig. 7.2c). Nevertheless, even in these cases the law is obeyed increasingly closely for the component in excess (the solvent) as it approaches purity (Fig. 7.2c). This means that the law is a good approximation for the solvent so long as the solution is dilute.

Ideal dilute solutions

In ideal solutions the solute, as well as the solvent, obeys Raoult's law. However, in real solutions at low concentrations, although the vapour pressure of the solute is proportional to its mole fraction, the slope is not equal to the vapour pressure of the pure substance (Fig. 7.3). This linear but different dependence was discovered by the English chemist William Henry, and is now called **Henry's law**. We express it quantitatively by writing

$$p_B = x_B K_B \qquad (12)°$$

x_B is the mole fraction of the solute and K_B is a constant (with the dimensions of pressure) chosen so that the plot of the vapour pressure of B against its mole fraction is tangent to the experimental curve at $x_B = 0$ (Fig. 7.3). Mixtures obeying Henry's law are ideal in a different sense from those obeying Raoult's law and are called **ideal dilute solutions**. We shall also label equations with ° when they have been derived from Henry's law. Over the range of compositions where the solvent obeys Raoult's law, the solute obeys Henry's law.

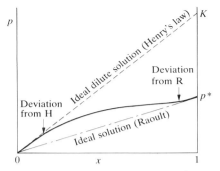

Fig. 7.3 When a component (the solvent) is nearly pure, it behaves according to Raoult's law and has a vapour pressure that is proportional to mole fraction with a slope p^*. When it is the minor component (the solute) its vapour pressure is still proportional to the mole fraction, but the constant of proportionality is now K. This is Henry's law.

Example 7.5: *Verifying Raoult's and Henry's laws*

The vapour pressures of each component in a mixture of propanone (acetone, A) and chloroform (trichloromethane, C) were measured at 35°C with the following results:

x_C	0	0.20	0.40	0.60	0.80	1
p_C/Torr	0	35	82	142	219	293
p_A/Torr	347	270	185	102	37	0

Confirm that the mixture conforms to Raoult's law for the component in large excess and to Henry's law for the minor component. Find the Henry's law constants.

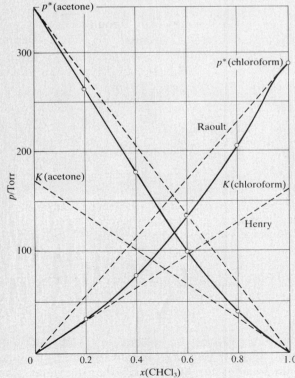

Fig. 7.4 The experimental partial vapour pressures of a mixture of chloroform and acetone based on the data in Example 7.5. The values of K are obtained by extrapolating the dilute solution vapour pressures as explained in the Example.

Answer. Plot the partial vapour pressures against mole fraction. Raoult's law is tested by comparing the data with the straight line $p = xp^*$ for each component in the region in which it is in excess (and acting as the solvent). Henry's law is tested by finding a straight line $p = xK$ that is tangent to each partial vapour pressure at low x where the component can be treated as the solute. The data are plotted in Fig. 7.4 together with the Raoult's and Henry's law lines. Henry's law requires $K_A = 175$ Torr and $K_C = 165$ Torr.

Comment. Notice how the data deviate from both Raoult's and Henry's laws even for quite small departures from $x = 1$ and $x = 0$ respectively. We deal with these deviations in Section 7.5.

Exercise. The vapour pressure of methyl chloride at various mole fractions in a mixture at 25°C was found to be as follows:

x	0.005	0.009	0.019	0.024
p/Torr	205	363	756	946

Estimate Henry's law constant. $\qquad [4 \times 10^4 \text{ Torr}]$

Table 7.1. Henry's law constants for gases in water at 298 K

	K/Torr
CO_2	1.25×10^6
H_2	5.34×10^7
N_2	6.51×10^7
O_2	3.30×10^7

Some Henry's law data are listed in Table 7.1. They are often used in calculations relating to gas solubilities (where the solute is a gas), as the following example illustrates.

Example 7.6: *Using Henry's law*

Estimate the solubility of oxygen in water at 25°C and a partial pressure of 190 Torr.

Answer. The mole fraction of solute is given by Henry's law as $x = p/K$ where p is its partial pressure. Calculate the amount of O_2 dissolved in 1.00 kg water. Since the amount dissolved is small, approximate the mole fraction by writing

$$x(O_2) = \frac{n(O_2)}{n(O_2) + n(H_2O)} \approx \frac{n(O_2)}{n(H_2O)}$$

Hence,

$$n(O_2) = x(O_2)n(H_2O) = \frac{pn(H_2O)}{K}$$

$$= \frac{190 \text{ Torr} \times 55.5 \text{ mol}}{3.30 \times 10^7 \text{ Torr}}$$

$$= 3.2 \times 10^{-4} \text{ mol}$$

The molality is therefore 3.2×10^{-4} mol kg^{-1}, corresponding to a concentration of approximately 3.2×10^{-4} M.

Comment. A knowledge of Henry's law constants for gases in fats and lipids is important for the discussion of respiration, especially when the partial pressure of oxygen is abnormal, as in diving and mountaineering.

Exercise. Calculate the concentration of nitrogen in water exposed to air at 25°C; partial pressures were calculated in Example 1.6. $[5.1 \times 10^{-4}$ M]

7.4 Liquid mixtures

Ideal mixing functions

The Gibbs function of mixing of two liquids to form an ideal solution is calculated in much the same way as for two gases. When the liquids are separate the total Gibbs function is

$$G_i = n_A \mu_A^*(l) + n_B \mu_B^*(l)$$

When they are mixed, the individual chemical potentials are given by eqn 11 and the total Gibbs function is

$$G_f = n_A\{\mu_A^*(l) + RT \ln x_A\} + n_B\{\mu_B^*(l) + RT \ln x_B\}$$

Consequently, the Gibbs function of mixing is

$$\Delta G_{mix} = nRT(x_A \ln x_A + x_B \ln x_B) \qquad (13)°$$

where $n = n_A + n_B$.

Equation 13 is the same as that for two perfect gases, and all the conclusions drawn there are valid here: the driving force for mixing is the increasing entropy of the system as the particles mingle, the enthalpy and internal energy of mixing are zero, and there is no change of volume. It should be noted, however, that solution ideality means something different from gas perfection. In the latter there are no interactions between particles. In ideal solutions there are interactions, but the average A–B interactions in the mixture are the same as the average A–A and B–B interactions in the pure liquids. The dependences of the Gibbs function, entropy, and enthalpy of mixing on the composition of the mixture are shown in Fig. 7.5.

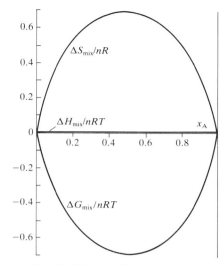

Fig. 7.5 The Gibbs function, enthalpy, and entropy of mixing of an ideal binary mixture, calculated using eqn 13. The entropy is expressed as a multiple of nR, the Gibbs function and enthalpy as a multiple of nRT.

Excess functions

Real solutions are composed of particles for which A–A, A–B, and B–B interactions are all different. Not only may there be an enthalpy change when liquids mix, but there may also be an additional contribution to the entropy change arising from the way that the particles of one type might cluster together instead of mingling freely with the others. If the enthalpy change is large and positive (and mixing is endothermic) or if the entropy change is adverse (because of a reorganization of the particles that results in an orderly mixture), the Gibbs function might be positive for mixing. In that case, separation is spontaneous and the liquids may be immiscible. Alternatively, the liquids might be **partially miscible**, which means that they are miscible only over a certain range of compositions.

The thermodynamic properties of real solutions may be expressed in terms of the **excess functions** (G^E, S^E, etc.), the difference between the observed thermodynamic function of mixing and the function for an ideal solution. In the case of the excess entropy, for example,

$$S^E = \Delta S_{mix} + nR(x_A \ln x_A + x_B \ln x_B) \qquad (14)$$

Deviations of the excess functions from zero then indicate the extent to which the solutions are non-ideal. In this connection a useful model system is the **regular solution**, in which $H^E \neq 0$ but $S^E = 0$. A regular solution can be thought of as one in which the two kinds of particles are distributed

Fig. 7.6 Experimental excess functions at 25°C. (a) H^E for benzene/cyclohexane; this shows that the mixing is endothermic (because $\Delta H_{mix} = 0$ for an ideal solution). (b) V^E for tetrachloroethene/cyclopentane; this shows that there is a contraction at low tetrachloroethene mole fractions, but an expansion at high mole fractions (because $\Delta V_{mix} = 0$ for a perfect mixture).

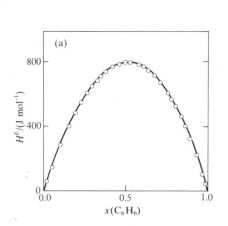

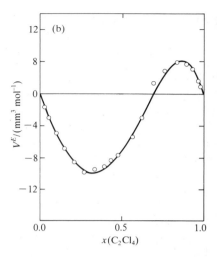

randomly but have different interactions with each other. Two examples of the composition dependence of excess functions are shown in Fig. 7.6.

7.5 Colligative properties

In this section we see how to calculate the effect of a solute on the boiling and freezing points of mixtures, and how a solute gives rise to an osmotic pressure. In dilute solutions, all the properties we consider depend only on the number of solute particles present, not their identity, and for this reason they are called **colligative properties** (denoting 'depending on the collection').

The common features of colligative properties

We make two assumptions. The first is that the solute is not volatile, and so it does not contribute to the vapour. The second is that the solute does not dissolve in the solid solvent. The latter assumption is quite drastic, although it is true of many mixtures; it can be avoided at the expense of more algebra, but that introduces no new principles. Both assumptions are removed in the general but qualitative discussion of mixtures in Chapter 8.

Colligative properties stem from the reduction of the chemical potential of the liquid solvent as a result of the presence of solute. The reduction is from $\mu_A^*(l)$ for the pure solvent to $\mu_A^*(l) + RT \ln x_A$ when a solute is present (since $x_A < 1$, $\ln x_A$ is negative). The chemical potentials of the vapour and solid are unchanged by the presence of the non-volatile, insoluble solute. As can be seen from Fig. 7.7, this implies that the liquid–vapour equilibrium occurs at a higher temperature (the boiling point is raised) and the solid–liquid equilibrium occurs at a lower temperature (the freezing point is lowered).

The physical basis of the lowering of the chemical potential is not the energy of interaction of the solute and solvent particles, because the lowering occurs even in ideal solutions (which have zero enthalpy of mixing). If it is not an enthalpy effect it must be an entropy effect. In the absence of a solute, the pure liquid solvent has an entropy that reflects its disorder. Its vapour pressure reflects the tendency of the isolated system

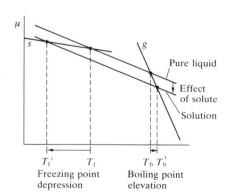

Fig. 7.7 The chemical potential of a solvent in the presence of a solute. The lowering of the liquid's chemical potential has a greater effect on the freezing point than on the boiling point because of the angles at which the lines intersect (which are determined by entropies, recall Fig. 6.1).

containing the solution to change towards greater entropy, which can be achieved if the liquid vaporizes to form a more chaotic gas. When a solute is present, there is an additional contribution to the entropy of the liquid, even in an ideal solution, and the tendency of the solvent to escape is not so compelling. When there is already an additional randomness in the liquid, the global system reaches its maximum entropy when less liquid evaporates than when the solute, and the randomness it causes, is absent. The effect of the solute appears as a lowered vapour pressure, and hence a higher boiling point. Similarly, the enhanced randomness of the solution opposes the tendency to freeze and a lower temperature must be reached before equilibrium between solid and solution is achieved. Hence, the freezing point is lowered.

The strategy for the quantitative discussion of the elevation of boiling point and the depression of freezing point is to look for the temperature at which one phase (the pure vapour or the pure solid) has the same chemical potential as the solvent in the solution. This is the new equilibrium temperature.

The elevation of boiling point

The heterogeneous equilibrium of interest when considering boiling is between the solvent vapour and the solvent in solution (Fig. 7.8). We denote the solvent by A and the solute by B. The equilibrium is established at a temperature for which

$$\mu_A^*(g) = \mu_A^*(l) + RT \ln x_A$$

This equation rearranges into

$$\ln (1 - x_B) = \frac{\mu_A^*(g) - \mu_A^*(l)}{RT} = \frac{\Delta G_{vap}}{RT}$$

where ΔG_{vap} is the Gibbs function of vaporization of the pure solvent and x_B is the mole fraction of the solute; we have used $x_A + x_B = 1$. We now write

$$\Delta G_{vap} = \Delta H_{vap} - T \Delta S_{vap}$$

and ignore the small temperature dependence of ΔH and ΔS. Then at a general mole fraction x_B,

$$\ln (1 - x_B) = \frac{\Delta G_{vap}}{RT} = \frac{\Delta H_{vap}}{RT} - \frac{\Delta S_{vap}}{R}$$

When $x_B = 0$, the boiling point is that of the pure liquid, T^*, and

$$\ln 1 = \frac{\Delta G_{vap}}{RT^*} = \frac{\Delta H_{vap}}{RT^*} - \frac{\Delta S_{vap}}{R}$$

Since $\ln 1 = 0$, the difference of the two equations is

$$\ln (1 - x_B) = \frac{\Delta H_{vap}}{R} \left(\frac{1}{T} - \frac{1}{T^*} \right)$$

We now suppose that the amount of solute present is so small that $x_B \ll 1$.

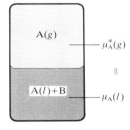

Fig. 7.8 The heterogeneous equilibrium involved in the calculation of the elevation of boiling point is between A in the pure vapour and A in the mixture. A being the solvent and B an involatile solute.

This allows us to write $\ln(1 - x_B) \approx -x_B$ and hence to obtain

$$x_B = \frac{\Delta H_{vap}}{R}\left(\frac{1}{T^*} - \frac{1}{T}\right)$$

Since $T \approx T^*$, we can also write

$$\frac{1}{T^*} - \frac{1}{T} = \frac{T - T^*}{TT^*} \approx \frac{\Delta T}{T^{*2}} \qquad \Delta T = T - T^*$$

which gives

$$\Delta T = \left(\frac{RT^{*2}}{\Delta H_{vap}}\right)x_B \qquad\qquad (15a)°$$

Equation 15 for the elevation of boiling point ΔT makes no reference to the identity of the solute, only to its mole fraction. This identifies the elevation of boiling point as a colligative property. The value of ΔT does depend on the properties of the solvent, and the biggest changes occur for solvents with high boiling points but low enthalpies of vaporization. We show in the following example that, for dilute solutions, the elevation of boiling point may be written

$$\Delta T = K_b m_B \qquad\qquad (15b)$$

where K_b is the **ebullioscopic constant** of the solvent and m_B is the molality of the solution (the number of moles of solute per kilogram of solvent).

Example 7.7: *Evaluating the elevation of boiling point*

Show that, when the solution is dilute, the elevation of boiling point is given by eqn 15b. Evaluate K_b for benzene as solvent.

Answer. Since the mole fraction of B is small, $n_A \gg n_B$ and

$$x_B = \frac{n_B}{n_A + n_B} \approx \frac{n_B}{n_A}$$

The amount (number of moles) of solvent molecules in 1 kg of solvent of molar mass M is

$$n_A = \frac{1\,\text{kg}}{M}$$

Therefore,

$$x_B = \frac{n_B}{n_A} = \frac{n_B M}{1\,\text{kg}} = m_B \times M$$

Hence, from eqn 15a,

$$\Delta T = \left(\frac{RT^{*2}M}{\Delta H_{vap}}\right)m_B$$

and we can identify the constant as

$$K_b = \frac{RT^{*2}M}{\Delta H_{vap}}$$

For benzene, $T^* = 353.2\,\text{K}$, $M = 78.11\,\text{g mol}^{-1}$, and $\Delta H_{vap} = 30.8\,\text{kJ mol}^{-1}$ (Table 2.2), and substitution of these values gives $K_b = 2.63\,\text{K}/(\text{mol kg}^{-1})$.

Comment. The experimental value for benzene is $2.53\,\text{K}/(\text{mol kg}^{-1})$.

Exercise. Evaluate the ebullioscopic constant for water. $\qquad$ [$0.51\,\text{K}/(\text{mol kg}^{-1})$]

The ebullioscopic constants of solvents are best regarded as empirical constants to be determined experimentally. Once measured they could be used to determine the molar masses of solutes. The technique is called **ebullioscopy**, but it is now rarely used.

The depression of freezing point

The heterogeneous equilibrium now of interest is between pure solid solvent and the solution with solute present at a mole fraction x_B (Fig. 7.9). At the freezing point, the chemical potentials of A in the two phases are equal:

$$\mu_A^*(s) = \mu_A^*(l) + RT \ln x_A$$

The only difference between this calculation and the last is the appearance of the solid's chemical potential in place of the vapour's. Therefore we can write the result directly from eqn 15:

$$\Delta T = \left(\frac{RT^{*2}}{\Delta H_{fus}}\right) x_B \tag{16a}°$$

where ΔT is the freezing point depression, $T^* - T$, and ΔH_{fus} is the enthalpy of fusion of the solvent. Large depressions are observed in solvents with low enthalpies of fusion and high melting points. When the solution is dilute, the mole fraction is proportional to the molality, and we can write (by an analogous argument to that in Example 7.7)

$$\Delta T = K_f m_B \tag{16b}°$$

K_f is the **cryoscopic constant** (Table 7.2) which, like the ebullioscopic constant, is best regarded as an empirical parameter. Once the cryoscopic constant of a solvent is known, the depression of freezing point may be used to measure the molar mass of a solute in the method known as **cryoscopy**.

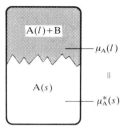

Fig. 7.9 The heterogeneous equilibrium involved in the calculation of the lowering of freezing point is between A in the pure solid and A in the mixture, A being the solvent and B a solute that is insoluble in solid A.

Table 7.2. Cryoscopic and ebullioscopic constants

	$K_f/(K/mol\ kg^{-1})$	$K_b/(K/mol\ kg^{-1})$
Benzene	5.12	2.53
Camphor	40	
Phenol	7.27	3.04
Water	1.86	0.51

Example 7.8: *Using cryoscopy*

Very precise measurements in the freezing point of a $1.00\ mmol\ kg^{-1}$ $CH_3COOH(aq)$ solution gave a freezing point depression of 2.1 mK. What can be inferred?

Answer. The expected depression can be calculated from the information in Table 7.2: from $K_f = 1.86\ K/(mol\ kg^{-1})$ and $m_B = 1.00 \times 10^{-3}\ mol\ kg^{-1}$ we predict $\Delta T = 1.86\ mK$. The discrepancy from the observed value indicates that a different number of particles is present in solution than the molality itself indicates. The additional particles arise from the ionization

$$CH_3COOH(aq) \rightleftharpoons CH_3CO_2^-(aq) + H^+(aq)$$

For each CH_3COOH molecule that ionizes, one $CH_3CO_2^-$ and one H^+ are produced, and so the actual molality of the solution is $(1 - \alpha)m_B + \alpha m_B + \alpha m_B = (1 + \alpha)m_B$ if a fraction $1 - \alpha$ remain non-ionized. Therefore, to account for a depression of 2.1 mK instead of 1.86 mK, we require $1 + \alpha = 2.1/1.86 = 1.1$, suggesting that 10 per cent of the acid molecules are ionized.

Exercise. 120 mg of benzoic acid (C_6H_5COOH) was dissolved in 100 g of water, and a depression of freezing point of 20 mK was observed. What proportion of the molecules are ionized? [9 per cent]

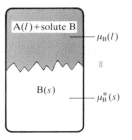

Fig. 7.10 The heterogeneous equilibrium involved in the calculation of the solubility is between pure solid B and B in the mixture.

Solubility

Although it is not strictly a colligative property, the solubility of a solute may be estimated by the same techniques as we have been using. If a solid solute is left in contact with a solvent, it dissolves until the solution is saturated. Saturation implies equilibrium, and at equilibrium the chemical potential of the pure solid solute, $\mu_B^*(s)$, and the chemical potential of B in solution, μ_B, are equal (Fig. 7.10). Since the latter is

$$\mu_B = \mu_B^*(l) + RT \ln x_B$$

we can write

$$\mu_B^*(s) = \mu_B^*(l) + RT \ln x_B$$

This is the same as the starting equation of the last section, except that the quantities refer to the solute B, not the solvent A.

The starting point is the same but the aim is different. In the present case we want to find the mole fraction of B in solution at equilibrium when the temperature is T. Therefore, we rearrange the last equation to

$$\ln x_B = \frac{\mu_B^*(s) - \mu_B^*(l)}{RT} = -\frac{\Delta G_{fus}}{RT}$$

$$= \frac{-\Delta H_{fus}}{RT} + \frac{\Delta S_{fus}}{R}$$

At the melting point of the solute, T^*, we know that $\Delta G_{fus} = 0$, and so $\Delta G_{fus}/RT^* = 0$ too, and it may be added to the right hand side. Taking the enthalpy and entropy of melting as constant over the temperature range of interest, the entropy terms cancel, and in much the same way as before we obtain

$$\ln x_B = \frac{-\Delta H_{fus}}{R}\left(\frac{1}{T} - \frac{1}{T^*}\right) \tag{17}°$$

This shows that the solubility of B decreases exponentially as the temperature is lowered from its melting point. Moreover, solutes with high melting points and large enthalpies of melting have low solubilities at normal temperatures.

Example 7.9: *Estimating the effect of temperature on solubility*

Find an expression for the change in solubility (expressed as a ratio of mole fractions) when the temperature is changed by a small amount ΔT at the temperature T, and evaluate the change for a 1.0 K increase at 298 K for a substance with $\Delta H_{fus} = 5 \text{ kJ mol}^{-1}$.

Answer. If x_B is the mole fraction in the saturated solution at a temperature T, and x'_B its mole fraction at a temperature T', subtraction of two equations of the form of eqn 17 gives

$$\ln \frac{x'_B}{x_B} = \frac{-\Delta H_{fus}}{R}\left(\frac{1}{T'} - \frac{1}{T}\right)$$

Since $T \approx T'$ and $\Delta T = T' - T$, this simplifies to

$$\ln \frac{x'_B}{x_B} \approx \frac{\Delta H_{fus}\,\Delta T}{RT^2}$$

Since $\Delta H_{fus} = 5\,\text{kJ mol}^{-1}$,

$$\ln \frac{x'_B}{x_B} \approx \frac{5\times 10^3\,\text{J mol}^{-1}\times 1\,\text{K}}{8.314\,\text{J K}^{-1}\,\text{mol}^{-1}\times(298\,\text{K})^2} = 0.0068$$

Hence, $x'_B/x_B \approx 1.01$, a change of 1 per cent.

Exercise. What is the solubility of a compound at its melting point?

[Miscible in all proportions]

Osmosis

The phenomenon of **osmosis** (from the Greek for 'push') is the passage of a pure solvent into a solution separated from it by a **semipermeable membrane**, a membrane permeable to the solvent but not to the solute (Fig. 7.11). The **osmotic pressure** Π is the pressure that must be applied to the solution to stop the flow. One of the most important examples of osmosis is transport of fluids through cell membranes, but it is also the basis of **osmometry**, the determination of molar mass by measurement of osmotic pressures, especially of macromolecules.

In the simple arrangement shown in Fig. 7.11b, the opposing pressure arises from the head of solution that the osmosis itself produces, and equilibrium is reached when the hydrostatic pressure of the column of solution matches the osmotic pressure. The complication of this arrangement is that the entry of solvent into the solution results in its dilution, and so it is more difficult to treat than the arrangement in Fig. 7.11a where there is no flow and the concentrations remain unchanged.

The thermodynamic treatment of osmosis depends on noting that, at equilibrium, the chemical potential of the solvent must be the same on each side of the membrane. On the pure solvent side the chemical potential of the solvent, which is at a pressure p, is $\mu_A^*(p)$. On the solution side, the chemical potential is lowered by the presence of the solute at a mole fraction x_A but is raised on account of the greater pressure $p + \Pi$ that the solution experiences. At equilibrium the two are equal, and we can write

$$\mu_A^*(p) = \mu_A(x_A, p + \Pi)$$

The presence of solute is taken into account in the normal way:

$$\mu_A(x_A, p + \Pi) = \mu_A^*(p + \Pi) + RT \ln x_A$$

We saw in Section 5.2 how to take the effect of pressure into account:

$$\mu_A^*(p + \Pi) = \mu_A^*(p) + \int_p^{p+\Pi} V_m\, dp$$

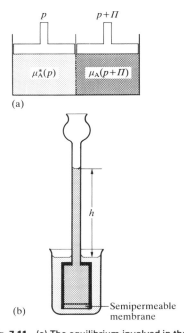

Fig. 7.11 (a) The equilibrium involved in the calculation of osmotic pressure Π is between pure solvent A at a pressure p on one side of the semipermeable membrane and A as a component of the mixture on the other side of the membrane, where the pressure is $p + \Pi$. (b) In a simple version, A is at equilibrium on each side of the membrane when enough has passed into the solution to cause a hydrostatic pressure difference.

where V_m is the molar volume of the pure solvent. When the last three equations are combined we get

$$-RT \ln x_A = \int_p^{p+\Pi} V_m \, dp \qquad (18)°$$

For dilute solutions, $\ln x_A$ may be replaced by $\ln(1 - x_B) \approx -x_B$. We may also assume that the pressure range in the integration is so small that the molar volume of the solvent is a constant. That being so, V_m may be taken outside the integral, giving

$$RTx_B = \Pi V_m$$

When the solution is dilute, $x_B \approx n_B/n_A$. Moreover, since $n_A V_m = V$, the total volume of the solvent, the equation simplifies to the **van't Hoff equation**:

$$\Pi V = n_B RT \qquad (19a)°$$

Since $n_B/V = [B]$, the molar concentration of the solute, a simpler form of this equation is

$$\Pi = [B]RT \qquad (19b)°$$

Equation 19 applies only to such dilute solutions that we can be confident are behaving ideally. One of the most common applications of osmometry is to the measurement of molar masses of macromolecules (proteins and synthetic polymers). As these huge molecules dissolve to produce solutions that are far from ideal, it is assumed that the van't Hoff equation is only the first term of a virial-like expansion:

$$\Pi = [B]RT\{1 + B[B] + \ldots\} \qquad (19c)$$

The additional terms take the non-ideality into account. The osmotic pressure is measured at a series of concentrations, and a plot of $\Pi/[B]$ against $[B]$ is used to find the molar mass of B.

Example 7.10: *Using osmometry*

The osmotic pressures of solutions of polyvinyl chloride (PVC) in cyclohexanone at 298 K are given below. The pressures are expressed in terms of the heights of solution (of density $\rho = 0.980 \text{ g cm}^{-3}$) in balance with the osmotic pressure. Find the molar mass of the polymer.

$c/(\text{g dm}^{-3})$	1.00	2.00	4.00	7.00	9.00
h/cm	0.28	0.71	2.01	5.10	8.00

Answer. We use eqn 19c with $[B] = c/M$ where c is the mass concentration and M is the molar mass of the polymer; the osmotic pressure is related to the hydrostatic pressure by $\Pi = \rho g h$ (Example 1.2) with $g = 9.81 \text{ m s}^{-2}$. Since

$$\frac{h}{c} = \frac{RT}{\rho g M}\left(1 + \frac{Bc}{M} + \ldots\right)$$

$$= \frac{RT}{\rho g M} + \left(\frac{RTB}{\rho g M^2}\right)c + \ldots$$

we should plot h/c against c, and expect a straight line with intercept $RT/\rho g M$ at $c = 0$. The data give the following:

$c/(\text{g dm}^{-3})$	1.00	2.00	4.00	7.00	9.00
$(h/c)/(\text{cm/g dm}^{-3})$	0.28	0.36	0.503	0.729	0.889

The points are plotted in Fig. 7.12. The intercept is at 0.21. Therefore,

$$M = \frac{RT}{\rho g} \times \frac{1}{0.21 \text{ cm g}^{-1} \text{ dm}^3} = 1.2 \times 10^5 \text{ g mol}^{-1}$$

Comment. Osmometry is a very important technique for the measurement of molar masses of macromolecules, partly because the other colligative techniques give such small effects. We take up the discussion again in Chapter 23.

Exercise. Estimate the depression of freezing point of the most concentrated of these solutions, taking K_f as about $10 \text{ K}/(\text{mol kg}^{-1})$. [0.8 mK]

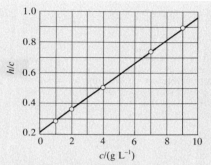

Fig. 7.12 The plot involved in the determination of molar mass by osmometry (Example 7.10). The molar mass is calculated from the intercept at $c = 0$; in Chapter 23 we shall see that additional information comes from the slope.

Mixtures of volatile liquids

In this section we consider the relation of the boiling point of a binary liquid mixture to its composition. In the process we see how to break away from ideal solutions and describe the properties of real mixtures.

7.6 Vapour pressure diagrams

In an ideal solution of two liquids, the vapour pressures of the components are related to the composition by Raoult's law:

$$p_A = x_A p_A^* \qquad p_B = x_B p_B^* \qquad (20a)°$$

where p_A^* is the vapour pressure of pure A and p_B^* that of pure B. The total vapour pressure p of the mixture is therefore

$$p = p_A + p_B = x_A p_A^* + x_B p_B^*$$
$$= p_B^* + (p_A^* - p_B^*)x_A \qquad (20b)°$$

This shows that the total vapour pressure (at some fixed temperature) changes linearly with the composition from p_B^* to p_A^* (Fig. 7.13). All points above the line (where the pressure acting on the system exceeds the vapour pressure) correspond to the liquid being the stable phase. All points below the line correspond to the vapour being stable.

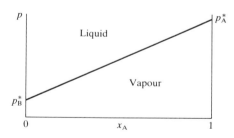

Fig. 7.13 The dependence of the total vapour pressure of a binary mixture on the mole fraction of A in the liquid when Raoult's law is obeyed.

The composition of the vapour

The compositions of liquid and vapour in equilibrium are not necessarily the same. Common sense suggests that the vapour should be richer in the more volatile component. This can be confirmed as follows. The partial pressures of the components are given by eqn 20a. It follows from Dalton's law that the mole fractions in the gas, y_A and y_B, are

$$y_A = \frac{p_A}{p} \qquad y_B = \frac{p_B}{p} \qquad (21a)$$

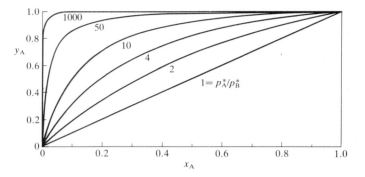

Fig. 7.14 The mole fraction of A in the vapour of a binary ideal solution expressed in terms of its mole fraction in the liquid, calculated using eqn 21 for various values of p_A^*/p_B^* with A more volatile than B. In all cases the vapour is richer than the liquid in A.

The partial pressures and the total pressure may be expressed in terms of the mole fractions in the liquid using eqn 20, which gives

$$y_A = \frac{x_A p_A^*}{p_B^* + (p_A^* - p_B^*)x_A} \qquad y_B = 1 - y_A \qquad (21b)°$$

We now have to show that, if A is the more volatile component, its mole fraction in the vapour, y_A, is greater than its mole fraction in the liquid, x_A. Figure 7.14 shows the dependence of y_A on x_A for various values of $p_A^*/p_B^* > 1$: we see that in all cases $y_A > x_A$, as we expect. Note that if B is non-volatile, so that $p_B^* = 0$ at the temperature of interest, then it makes no contribution to the vapour ($y_B = 0$).

Equation 21 shows how the total vapour pressure depends on the composition of the liquid. Since we can relate the composition of the liquid to the composition of the vapour through eqn 20, we can now also relate the vapour pressure to the composition of the vapour itself using Dalton's law. This leads to

$$p = \frac{p_A^* p_B^*}{p_A^* + (p_B^* - p_A^*)y_A} \qquad (22)°$$

This curve is plotted in Fig. 7.15.

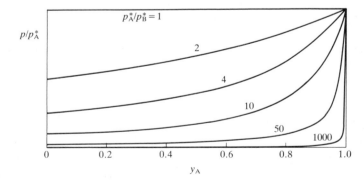

Fig. 7.15 The dependence of the vapour pressure of the same system as in Fig. 7.14, but expressed in terms of the mole fraction of A in the vapour using eqn 22.

The interpretation of the diagrams

We can use either diagram to discuss the phase equilibria of the mixture. However, if we are interested in distillation, both the vapour and the liquid compositions are of equal interest. It is then sensible to combine both diagrams into one. This is done in Fig. 7.16, where the composition axis is labelled z_A, the overall mole fraction of A in the system. On and above the upper line, only liquid is present and then $z_A = x_A$; on and below the lower line only vapour is present, and then $z_A = y_A$.

A point in the unshaded region of the diagram indicates not only qualitatively that both liquid and vapour are present, but represents quantitatively the relative amounts of each. Suppose that the overall composition is $z_A = a$ and the pressure is p_3. Since the point (a, p_3) lies in the unshaded sector, we know at once that both phases are present in equilibrium.

The lever rule

To find the relative amounts, we measure the distances l and l' along the horizontal **tie line**, and then use the **lever rule** (Fig. 7.17):

$$n'l' = nl \qquad (23)$$

where n is the amount of liquid and n' the amount of vapour. In the present case, since $l' = \frac{2}{3}l$, the amount of liquid is about $\frac{2}{3}$ the amount of vapour.

To prove the lever rule we write $n'' = n + n'$ and the overall amount of A as $n''z_A$. The overall amount of A is also the sum of its amounts in the two phases:

$$n''z_A = nx_A + n'y_A$$

Since also

$$n''z_A = nz_A + n'z_A$$

by equating these two expressions it follows that

$$n'(z_A - y_A) = n(x_A - z_A)$$

or

$$n'l' = nl$$

as was to be proved.

In order to see in more detail how the rule is used, consider what happens when a mixture of composition a_1 in Fig. 7.16 is subjected to decreasing pressure. The sample remains entirely liquid until the pressure is reduced to p_2, when liquid and vapour can coexist. At p_2 the liquid has composition a_2 and the vapour has composition a_2'. Since l'/l is almost infinite for this tie line, there is only a trace of vapour present. When the pressure is reduced to p_3, the composition of the liquid is a_3 and that of the vapour is a_3', but overall the composition is still a. The relative amounts of liquid and vapour are given by the value of l'/l measured along the tie line at p_3. This means that the amount of liquid is about $\frac{2}{3}$ the amount of vapour. Furthermore, the vapour (of composition a_3') is richer in A, the more volatile component, than the liquid (which has composition a_3). When the pressure has fallen to p_4, the sample is almost completely gaseous: the vapour's composition is a_4' and the trace of liquid that remains (trace, because $l'/l \approx 0$) has the composition a_4 and so is rich in B. Only an infinitesimal reduction in pressure is now needed to eliminate the last trace of liquid, and from then on the sample is entirely gaseous and of composition a.

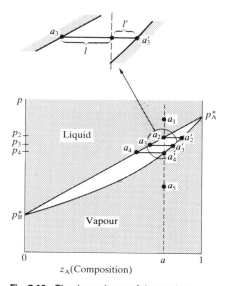

Fig. 7.16 The dependence of the total vapour pressure of an ideal solution on the mole fraction of A in the entire system. A point in the white sector corresponds to both liquid and vapour being present with proportions given by the lever rule. The enlargement shows the case discussed in the text, where the amount of liquid is about $\frac{2}{3}$ the amount of vapour (i.e. $l'/l = \frac{2}{3}$).

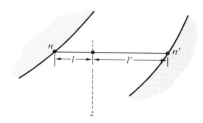

Fig. 7.17 The lever rule. The distances l and l' are used to find the proportions of the amounts of vapour and liquid present at equilibrium. The lever rule is so called because a similar rule relates the masses at two ends of a lever to their distances from a pivot ($ml = m'l'$ for balance).

7.7 Temperature–composition diagrams

Reducing the pressure at constant temperature is one way of doing distillation, but it is more common to distil at constant pressure by raising the temperature. To discuss distillation in this way we need a **temperature–composition diagram**, a diagram in which the boundaries show the composition of the phases that are in equilibrium at various temperatures (and a given pressure, typically 1 atm). An example is shown in Fig. 7.18. Note that the liquid phase region now lies in the lower part of the diagram.

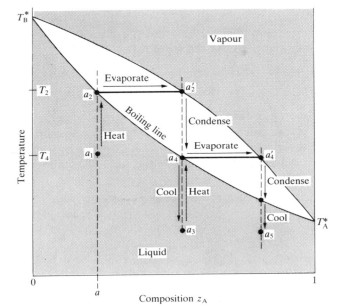

Fig. 7.18 The temperature–composition diagram corresponding to an ideal mixture with A more volatile than B. Successive boilings and condensations of a liquid originally of composition a_1 lead to a condensate that is pure A. This is the process of fractional distillation.

The distillation of mixtures

Using a temperature–composition diagram involves the same kind of interpretation as we have already seen in connection with the pressure–composition diagram. Consider what happens when a liquid of composition a is heated. Initially its state is a_1. It boils when the temperature reaches T_2. Then the liquid has composition a_2 and the vapour (which is present only as a trace) has composition a_2'. The vapour is richer in the more volatile component A, as common sense leads us to expect. From the location of a_2' we can state the vapour's composition at the boiling point, and from the location of the tie line joining a_2 and a_2' we can read off the boiling temperature of the original liquid mixture.

In a simple **distillation**, the vapour is withdrawn and condensed. If the vapour in this example is drawn off and completely condensed, then the first drop gives a liquid of composition a_3, which is richer in the more volatile component than the original liquid. In **fractional distillation**, the boiling and condensation cycle is repeated successively. We can follow the changes that occur by seeing what happens when the condensate of composition a_3 is reheated. The phase diagram shows that this mixture boils at T_4 and yields a vapour of composition a_4' which is even richer in the more volatile component. That vapour is drawn off, and the first drop condenses to a liquid of composition a_5. The cycle can then be repeated until in due course almost pure A is obtained.

Azeotropes

Whereas many liquids have temperature–composition phase diagrams resembling the ideal version in Fig. 7.18, in a number of important cases there are marked deviations. A maximum in the phase boundary (Fig. 7.19) may occur when the interactions between the components reduce the vapour pressure of the mixture below the ideal value: in effect, the A–B interactions stabilize the liquid. In such cases the excess Gibbs function G^E (Section 7.4) is negative (more favourable to mixing than ideal). Examples of this behaviour include chloroform/acetone and nitric acid/water mixtures. Phase boundaries showing a minimum (Fig. 7.20) indicate that the mixture is destabilized relative to the ideal solution, the A–B interactions then being unfavourable. For such mixtures G^E is positive (less favourable to mixing than ideal), and there may be contributions from both enthalpy and entropy effects. Examples include dioxane/water and ethanol/water.

Deviations from ideality are not always so strong as to lead to a maximum or minimum in the phase boundaries, but when they do there are important consequences for distillation. Consider a liquid of composition a on the left of the maximum in Fig. 7.19. The vapour (at a_2') of the boiling mixture (at a_2) is richer in B. If that vapour is removed (and condensed elsewhere), the remaining liquid will have composition a_3 and its vapour will be a_3'. If that vapour is removed, the composition of the boiling liquid shifts to a_4 and its vapour to a_4'. Hence, as evaporation proceeds, the composition of the remaining liquid shifts towards A as B is drawn off. The boiling point of the liquid rises, and the vapour becomes richer in A. When so much B has been evaporated that the liquid has reached the composition b, the vapour has the same composition as the liquid. Evaporation then occurs without change of composition. The mixture is said to form an **azeotrope** (which comes from the Greek words for 'boiling without changing').

When the azeotropic composition has been reached, distillation cannot separate the two liquids because the condensate retains the composition of the liquid. One example of azeotrope formation is hydrochloric acid/water, which is azeotropic at 80 per cent water (by mass) and boils unchanged at 108.6°C.

The system shown in Fig. 7.20 is also azeotropic, but shows it in a different way. Suppose we start with a mixture of composition a_1 and follow the changes in the vapour that rises through a **fractionating column**, essentially a vertical glass tube packed with glass rings to give a large surface area. The mixture boils at a_2 to give a vapour of composition a_2'. This vapour condenses in the column to a liquid of the same composition (now marked a_3). That liquid reaches equilibrium with its vapour at a_3', which condenses higher up the tube to give a liquid of the same composition, which we now call a_4. The fractionation therefore shifts the vapour towards the azeotropic composition, but not beyond, and the azeotropic vapour emerges from the top of the column. An example is ethanol/water, which boils unchanged when the water content is 4 per cent and the temperature is 78°C.

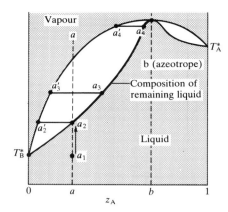

Fig. 7.19 A high-boiling azeotrope. When the mixture at a_1 is distilled, the composition of the remaining liquid changes towards b but no further.

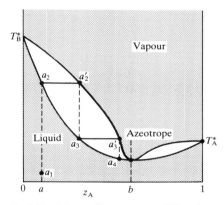

Fig. 7.20 A low-boiling azeotrope. When the mixture at a_1 is fractionally distilled, the vapour in equilibrium in the fractionating column moves towards b.

7.8 Immiscible liquids

Finally we consider the distillation of two immiscible liquids, such as oil and water. As they are immiscible we can regard their 'mixture' as unscrambled with each component in a separate vessel (Fig. 7.21). If the vapour pressures of the two pure components are p_A and p_B, the total vapour pressure is $p = p_A + p_B$ and the mixture boils when $p = 1$ atm. The presence of the second component means that the 'mixture' boils at a lower temperature than either would alone because boiling begins when the total pressure reaches 1 atm, not when either vapour pressure reaches 1 atm. This is the basis of steam distillation, which enables some heat-sensitive organic compounds to be distilled at a lower temperature than their normal boiling point. The only snag is that the composition of the condensate is in proportion to the vapour pressures of the components, and so oils of low volatility distil in low abundance.

Fig. 7.21 The distillation of two immiscible liquids can be regarded as the joint distillation of the separated components, and boiling occurs when the sum of the partial pressures equals 1 atm.

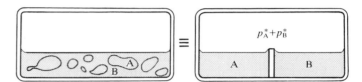

Real solutions

So far, our work on real solutions has been largely qualitative. Now we see how to make it quantitative so that calculations that we did for ideal solutions earlier in the chapter can be carried through for real solutions too.

7.9 The solvent activity

The general form of the chemical potential of a (real or ideal) solute or solvent is

$$\mu_A(l) = \mu_A^*(l) + RT \ln \frac{p_A}{p_A^*} \tag{24}$$

where p_A^* is the vapour pressure of pure A and p_A its vapour pressure when in a mixture. In the case of an ideal solution, both solvent and solute obey Raoult's law at all concentrations and we write

$$\mu_A(l) = \mu_A^*(l) + RT \ln x_A$$

The standard state of the solvent or the solute is the pure liquid (at 1 bar pressure), and is obtained when $x_A = 1$. When the solution does not obey Raoult's law, the form of the last equation can be preserved by writing

$$\mu_A(l) = \mu_A^*(l) + RT \ln a_A \tag{25a}$$

a_A is the **activity** of A, a kind of 'effective' mole fraction just as the fugacity (Section 5.3) is an effective pressure.

Since eqn 24 is true for both real and ideal solutions (the only approximation being the use of pressures rather than fugacities), we can conclude that

$$a_A = \frac{p_A}{p_A^*} \tag{25b}$$

The activity of a component in a mixture can therefore be determined experimentally by measuring its vapour pressure.

Example 7.11: *Measuring the activity of a solvent*

The vapour pressure of an $0.5\,\text{M}\,KNO_3(aq)$ solution at 100°C is 749.7 Torr. What is the activity of the water at this temperature?

Answer. We simply use eqn 25b. Since the vapour pressure of pure water at 100°C is 760.0 Torr, we obtain

$$a(H_2O) = \frac{749.7\,\text{Torr}}{760.0\,\text{Torr}} = 0.9864$$

Exercise. The vapour pressure of a sample of sea water is 19.02 kPa at 18°C. Calculate the activity of the water in the solution given that the vapour pressure of pure water at 18°C is 19.38 kPa.   [0.9814]

Since all solvents obey Raoult's law (that $p_A/p_A^* = x_A$) increasingly closely as the concentration of solute approaches zero, the activity of the solvent approaches the mole fraction as $x_A \to 1$:

$$a_A \to x_A \quad \text{as} \quad x_A \to 1$$

As in the case of real gases, a convenient way of expressing this convergence is to introduce the **activity coefficient** γ by the definition

$$a_A = \gamma_A x_A \qquad \gamma_A \to 1 \quad \text{as} \quad x_A \to 1 \tag{25c}$$

The chemical potential of the solvent is then

$$\mu_A = \mu_A^* + RT \ln x_A + RT \ln \gamma_A$$

an equation closely resembling that for the chemical potential of a real gas (eqn 16 in Section 5.3). The standard state of the solvent, the pure liquid solvent at 1 bar pressure, is established when $x_A = 1$.

7.10   The solute activity

The problem with defining activity coefficients and standard states for solutes is that they approach ideal dilute (Henry's law) behaviour at *low* concentrations, not as $x_B \to 1$ (corresponding to pure solute). We shall show how to set up the definitions for a solute that obeys Henry's law exactly, and then show how to allow for deviations.

Ideal dilute solutes

A solute B that satisfies Henry's law has a vapour pressure given by $p_B = K_B x_B$, where K_B is an empirical constant. In this case, the chemical potential of B is

$$\mu_B = \mu_B^* + RT \ln \frac{p_B}{p_B^*}$$

$$= \mu_B^* + RT \ln \frac{K_B}{p_B^*} + RT \ln x_B$$

Both K_B and p_B^* are constant characteristics of the solute, and so the second term may be combined with the first to give a new standard chemical potential which we denote $\mu^\dagger$:

$$\mu_B = \mu_B^\dagger + RT \ln x_B \quad \text{with} \quad \mu_B^\dagger = \mu_B^* + RT \ln \frac{K_B}{p_B^*}$$

Real solutes

We now permit deviations from ideal dilute, Henry's law behaviour. For the solvent activity we introduced a_A in place of x_A into Raoult's law, and obtained eqn 25a for the chemical potential. For the *solute*, we introduce a_B in place of x_B in Henry's law, and obtain

$$\mu_B = \mu_B^\dagger + RT \ln a_B \tag{26a}$$

The standard state remains unchanged in this last stage, and all the deviations from ideality are captured in the activity a_B. The value of the activity at any concentration can be obtained in the same way as for the solvent, but in place of eqn 25b we use

$$a_B = \frac{p_B}{K_B} \tag{26b}$$

As in the case of the solvent, it is sensible to introduce an activity coefficient through

$$a_B = \gamma_B x_B \tag{26c}$$

Now all the deviations from ideality are captured in the activity coefficient γ_B. Since the solute obeys Henry's law as its concentration goes to *zero*, it follows that

$$a_B \to x_B \quad \text{and} \quad \gamma_B \to 1 \quad \text{as} \quad x_B \to 0$$

Deviations of the solute from ideality disappear at zero concentration, but for the solvent the deviations disappear as purity is approached.

Example 7.12: *Measuring activity*

Use the information in Example 7.5 to calculate the activity and activity coefficient of chloroform in acetone at 25°C treating it as both a solvent and a solute.

Answer. For the activity of chloroform as a solvent (the Raoult's law activity), form $a = p/p^*$ and $\gamma = a/x$; for its activity as a solute (the Henry's law activity), form $a = p/K$ and $\gamma = a/x$. Since $p^* = 293$ Torr and $K = 165$ Torr, we can construct the following table:

x	0	0.20	0.40	0.60	0.80	1	
a	0	0.12	0.28	0.48	0.75	1.00	Raoult
γ	—	0.60	0.70	0.80	0.94	1.00	
a	0	0.21	0.50	0.86	1.33	1.78	Henry
γ	1	1.05	1.25	1.43	1.66	1.78	

Comment. Notice how $\gamma \to 1$ as $x \to 1$ in the Raoult's law case, but $\gamma \to 1$ as $x \to 0$ in the Henry's law case.

Exercise. Calculate the activities and activity coefficients for acetone according to the two conventions.

Activities in terms of molalities

The compositions of mixtures are often expressed as molalities m in place of mole fractions. It therefore proves convenient to introduce yet another definition of activity, but one that follows naturally from what we have done so far. First, we note that in dilute solutions the amount of solute is much less than the amount of solvent ($n_B \ll n_A$) and so to a good approximation $x_B \approx n_B/n_A$. Since n_B is proportional to the molality m_B, we can write

$$x_B = \frac{km_B}{m^\ominus} \qquad m^\ominus \equiv 1 \text{ mol kg}^{-1}$$

k is a constant and $m^\ominus$ has been introduced so that the right hand side is dimensionless (it cancels the units in m itself). For an ideal dilute solution we can therefore write

$$\mu_B = \mu_B^\dagger + RT \ln k + RT \ln \frac{m_B}{m^\ominus}$$

We now define a new standard chemical potential by combining $\ln k$ with $\mu^\dagger$:

$$\mu_B^\ominus = \mu_B^\dagger + RT \ln k$$

which allows us to write

$$\mu_B = \mu_B^\ominus + RT \ln \frac{m_B}{m^\ominus}$$

According to this definition, the chemical potential of the solute has its standard value $\mu_B^\ominus$ when the molality of B is equal to $m^\ominus$ (i.e., at 1 mol kg^{-1}).

Now, as before, we incorporate deviations from ideality by introducing a dimensionless activity a_B, a dimensionless activity coefficient γ_B, and writing

$$a_B = \frac{\gamma_B m_B}{m^\ominus} \qquad \gamma_B \to 1 \quad \text{as} \quad m_B \to 0 \qquad (27a)$$

The standard state remains unchanged in this last stage and, as before, all the deviations from ideality are captured in the activity coefficient γ_B. We then arrive at the following succinct expression for the chemical potential of a real solute at any molality:

$$\mu = \mu^{\ominus} + RT \ln a \tag{27b}$$

It is important to be aware of the different definitions of standard states and activities, and they are summarized in Box 7.1. We shall put them to work in the next few chapters, when we shall see that using them is much easier than defining them.

Box 7.1   Standard states

For the *solvent* (the major component), use Raoult's law and take the standard state as the pure solvent. Then

$$\mu = \mu^* + RT \ln a$$

with $a = p/p^*$ and $a = \gamma x$. Note that $\gamma \to 1$ as $x \to 1$ (pure solvent).

For the *solute* (the minor component), use Henry's law, with the standard state as a hypothetical state of the pure solute. Then

$$\mu = \mu^{\dagger} + RT \ln a$$

with $a = p/K$ and $a = \gamma x$. Note that $\gamma \to 1$ as $x \to 0$ (zero concentration). Alternatively, use

$$\mu = \mu^{\ominus} + RT \ln a$$

with $a = \gamma m/m^{\ominus}$, $m^{\ominus} = 1\ \text{mol kg}^{-1}$. In this case $\gamma \to 1$ as $m \to 0$ (zero concentration). The standard state is now a hypothetical state of the solution at molality $m^{\ominus}$.

Further information: the method of intercepts

Consider some extensive thermodynamic property X (such as the volume or the Gibbs function of a system). Let the binary system contain amounts n_A of A and n_B of B, the total amount being $n = n_A + n_B$ and the mole fractions $x_A = n_A/n$ and $x_B = n_B/n$. Define the mean molar property as $X_m = X/n$. The partial molar properties (which we seek to determine) are $X_A = (\partial X/\partial n_A)_{n_B}$ and $X_B = (\partial X/\partial n_B)_{n_A}$. Constant temperature and pressure are understood throughout.

The partial molar quantity X_A can be written

$$X_A = \left(\frac{\partial X}{\partial n_A}\right)_{n_B} = \left(\frac{\partial n X_m}{\partial n_A}\right) = X_m + n\left(\frac{\partial X_m}{\partial n_A}\right)_{n_B}$$

Now, since $x_B = n_B/n$, it follows that

$$\left(\frac{\partial X_m}{\partial n_A}\right)_{n_B} = \left(\frac{\partial x_B}{\partial n_A}\right)_{n_B} \frac{dX_m}{dx_B} = \frac{-n_B}{n^2}\frac{dX_m}{dx_B}$$

and so

$$X_A = X_m - \frac{n n_B}{n^2}\frac{dX_m}{dx_B}$$

or

$$X_m = X_A + \frac{dX_m}{dx_B} \times x_B$$

The last equation is the key result. It is the equation of a straight line of slope (dX_m/dx_B) and intercept X_A at $x_B = 0$. Therefore, we should plot X_m

against x_B, and at the value of x_B of interest draw a tangent to the experimental curve (Fig. 7.22). This is the line of the equation, and so we can take the partial molar quantity from the intercept at $x_B = 0$. The same argument, with A and B interchanged, shows that the intercept at $x_B = 1$ gives the value of X_B at the selected composition.

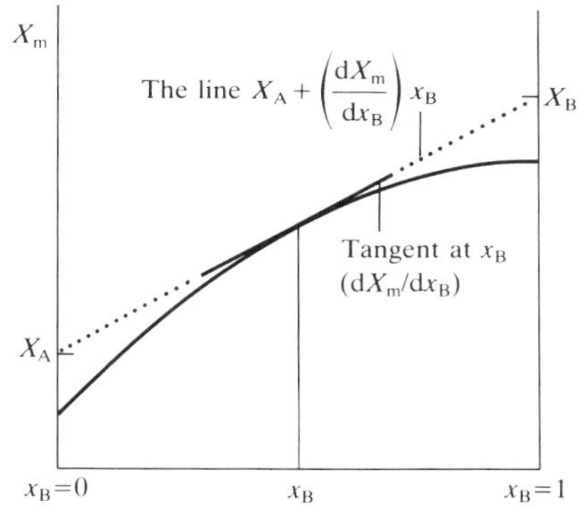

Further reading

Experimental techniques

J. R. Overton, *Determination of osmotic pressure*. In *Techniques of chemistry* (ed. A. Weissberger and B. W. Rossiter) V, 309, Wiley-Interscience, New York (1971).

W. J. Mader and L. T. Brady, *Determination of solubility*. In *Techniques of chemistry* (ed. A. Weissberger and B. W. Rossiter) V, 257, Wiley-Interscience, New York (1971).

Properties

J. S. Rowlinson and F. L. Swinton, *Liquids and liquid mixtures*. Butterworths, London (1982).

J. H. Hildebrand and R. L. Scott, *Solubilities of non-electrolytes*. Reinhold, New York (1950).

J. H. Hildebrand, J. M. Prausnitz and R. L. Scott, *Regular and related solutions*. Van Nostrand Reinhold, New York (1970).

J. N. Murrell and E. A. Boucher, *Properties of liquids and solutions*. Wiley-Interscience, New York (1982).

M. R. J. Dack (ed.), *Solutions and solubilities*. In *Techniques of chemistry* (ed. A. Weissberger) VIII, Wiley-Interscience, New York (1975).

H. F. Franzen, The freezing point depression law in physical chemistry. In *J. chem. Educ.* **65,** 1077 (1988). This article deals with the cases when the solute can dissolve in the solid solvent.

Surfaces

A. W. Adamson, *Physical chemistry of surfaces*. Wiley-Interscience, New York (1970).

M. J. Rosen, *Surfactants and interfacial phenomena*. Wiley-Interscience, New York (1978).

J. R. Bickerman, *Physical surfaces*. Academic Press, New York (1970).

Data

International critical tables (Vol. 4). McGraw-Hill, New York (1927).

Landolt–Börnstein tables, Springer (many volumes).

J. Timmermans, *Physico-chemical constants of binary systems in concentrated solutions* (Vols. 1–4). Interscience, New York (1959).

Fig. 7.22 The extrapolation needed to find the partial molar quantities X_A and X_B at the composition x_B.

Exercises

7.1 The partial molar volumes of acetone and chloroform in a mixture in which the mole fraction of $CHCl_3$ is 0.4693 are 74.166 $cm^3 \, mol^{-1}$ and 80.235 $cm^3 \, mol^{-1}$ respectively. What is the volume of a solution of mass 1.000 kg?

7.2 At 300 K, the vapour pressure of dilute solutions of HCl in liquid $GeCl_4$ are as follows:

$x(HCl)$	0.005	0.012	0.019
p/kPa	32.0	76.9	121.8

Show that the solution obeys Henry's law in this range of mole fractions and calculate Henry's law constant at 300 K.

7.3 Predict the vapour pressure of HCl above a solution in liquid $GeCl_4$ of molality 0.10 mol kg^{-1}.

7.4 At 90°C, the vapour pressure of toluene is 400 Torr and that of o-xylene is 150 Torr. What is the composition of the liquid mixture that boils at 90°C when the pressure is 0.50 atm? What is the composition of the vapour produced?

7.5 Calculate the cryoscopic and ebullioscopic constants of carbon tetrachloride.

7.6 The vapour pressure of a 500 g sample of benzene is 400 Torr at 60.6°C, but it fell to 386 Torr when 19.0 g of an involatile organic compound was dissolved in it. Calculate the molar mass of the compound.

7.7 Benzene freezes at 5.50°C and its enthalpy of fusion is 10.59 kJ mol^{-1}. Consider an ideal solution that is saturated with benzene and has mole fraction 0.905 in benzene. What is the temperature of the solution?

7.8 The addition of 100 g of a compound to 750 g of CCl_4 lowered the freezing point of the solvent by 10.5 K. Calculate the molar mass of the compound.

7.9 The osmotic pressure of an aqueous solution at 300 K is 120 kPa. Calculate the freezing point of the solution.

7.10 The vapour pressure of pure liquid A at 300 K is 575 Torr and that of pure liquid B is 390 Torr. These two compounds form ideal liquid and gaseous mixtures. Consider the equilibrium composition of a mixture in which the mole fraction of A in the vapour is 0.350. Calculate the total pressure of the vapour and the composition of the liquid mixture.

7.11 Consider a container of volume 5.0 L that is divided into two compartments of equal size. In the left compartment there is N_2 gas at 1.0 atm and 25°C; in the right compartment there is H_2 at the same temperature and pressure. Calculate the entropy and Gibbs function of mixing when the partition is removed. Assume that the gases are perfect.

7.12 Air is a mixture with a composition given in Example 1.6. Calculate the entropy of mixing when it is prepared from the pure gases.

7.13 Calculate the Gibbs function, entropy, and enthalpy of mixing when 500 g of hexane is mixed with 500 g of heptane at 298 K.

7.14 What proportions of hexane and heptane should be mixed (a) by mole fraction, (b) by mass in order to achieve the greatest entropy of mixing?

7.15 Use Henry's law and the data in Table 7.1 to calculate the solubility of CO_2 in water at 25°C when its partial pressure is (a) 0.10 atm, (b) 1.00 atm.

7.16 The mole fractions of N_2 and O_2 in air at sea level are approximately 0.78 and 0.21. Calculate the molalities of the solution formed in an open flask of water at 25°C.

7.17 A water carbonating plant is available for use in the home and operates by providing carbon dioxide at 5.0 atm. Estimate the molar concentration of the soda water it produces.

7.18 Calculate the freezing point of a 250 cm^3 glass of water sweetened with 7.5 g of sucrose.

7.19 The enthalpy of fusion of anthracene is 28.8 kJ mol^{-1} and its melting point is 217°C. Calculate its ideal solubility in benzene at 25°C.

7.20 Predict the ideal solubility of lead in bismuth at 280°C given that its melting point is 327°C and its enthalpy of fusion is 5.2 kJ mol^{-1}.

7.21 The osmotic pressure of solutions of polystyrene in toluene were measured at 25°C and the pressure was expressed in terms of the height of the solvent of density 1.004 g cm^{-3}:

$c/(g\ L^{-1})$	2.042	6.613	9.521	12.602
h/cm	0.592	1.910	2.750	3.600

Calculate the molar mass of the polymer.

7.22 The molar mass of an enzyme was determined by dissolving it in water, measuring the osmotic pressure at 20°C, and extrapolating the data to zero concentration. The following data were obtained:

$c/(mg\ cm^{-3})$	3.221	4.618	5.112	6.722
h/cm	5.746	8.238	9.119	11.990

Calculate the molar mass of the enzyme.

7.23 The following temperature/composition data were obtained for a mixture of octane (O) and toluene (T) at 760 Torr, where x is the mole fraction in the liquid and y the mole fraction in the vapour in equilibrium.

θ/°C	110.9	112.0	114.0	115.8	117.3	119.0	120.0	123.0
x_T	0.908	0.795	0.615	0.527	0.408	0.300	0.203	0.097
y_T	0.923	0.836	0.698	0.624	0.527	0.410	0.297	0.164

The boiling points are 110.6°C for T and 125.6°C for O. Plot the temperature/composition diagram for the mixture. What is the composition of the vapour in equilibrium with the liquid of composition (a) $x_T = 0.250$ and (b) $x_O = 0.250$?

Problems

Numerical problems

7.1 The following table gives the mole fraction of methylbenzene (A) in liquid and gaseous mixtures with butanone at equilibrium at 303.15 K and the total pressure p. Take the vapour to be perfect and calculate the partial pressures of the two components. Plot them against their respective mole fractions in the liquid mixture and find the Henry's law constants for the two components.

x_A	0	0.0898	0.2476	0.3577	0.5194
y_A	0	0.0410	0.1154	0.1762	0.2772
p/kPa	36.066	34.121	30.900	28.626	25.239

x_A	0.6036	0.7188	0.8019	0.9105	1
y_A	0.3393	0.4450	0.5435	0.7284	1
p/kPa	23.402	20.698	18.592	15.496	12.295

7.2 The volume of an aqueous solution of NaCl at 25°C was measured at a series of molalities m and it was found that the volume fitted the expression

$$V/cm^3 = 1003 + 16.62m + 1.77m^{3/2} + 0.12m^2$$

where m should be interpreted as $m/(mol\ kg^{-1})$ and V is the volume of a solution formed from 1.000 kg of water. Calculate the partial molar volume of the components at $m = 0.100\ mol\ kg^{-1}$.

7.3 At 18°C the total volume of a solution formed from $MgSO_4$ and 1.000 kg of water fits the expression

$$V/cm^3 = 1001.21 + 34.69(m - 0.070)^2$$

Calculate the partial molar volumes of the salt and the solvent when $m = 0.050\ mol\ kg^{-1}$.

7.4 Use the method of intercepts to plot the partial molar volume of HNO_3 in aqueous solution at 20°C using the following data, where w is the mass percentage of HNO_3.

w	2.162	10.98	20.80	30.00	39.20	51.68
$\rho/(g\ cm^{-3})$	1.01	1.06	1.12	1.18	1.24	1.32

w	62.64	71.57	82.33	93.40	99.60
$\rho/(g\ cm^{-3})$	1.38	1.42	1.46	1.49	1.51

7.5 The densities of aqueous solutions of copper(II) sulphate at 20°C were measured as set out below. Determine and plot the partial molar volume of $CuSO_4$ in the range of the measurements.

Mass $CuSO_4/(100\ g\ solution)$	5	10	15	20
$\rho/(g\ cm^{-3})$	1.051	1.107	1.167	1.230

7.6 What proportions of ethanol and water should be mixed in order to produce 100 cm³ of a mixture containing 50 per cent by mass of ethanol? What change in volume is brought about by adding 1.00 cm³ of ethanol to the mixture? (Use data from Fig. 7.1.)

7.7 When chloroform is added to acetone at 25°C, the volume of the mixture varies with composition as follows:

x	0	0.194	0.385	0.559	0.788	0.889	1.000
$V_m/(cm^3\ mol^{-1})$	73.99	75.29	76.50	77.55	79.08	79.82	80.67

x is the mole fraction of the chloroform. Determine the partial molar volumes of the two components at these compositions and plot the result.

7.8 Consider a container of volume 5.0 L that is divided into two compartments of equal size. In the left compartment there is N_2 gas at 3.0 atm and 25°C; in the right compartment there is H_2 at 1.0 atm and 25°C. Calculate the entropy and Gibbs function of mixing when the partition is removed. Assume that the gases are perfect.

7.9 A fluorocarbon is believed to be either $CF_3(CF_2)_3CF_3$ or $CF_3(CF_2)_4CF_3$. To what precision must the freezing point depression be measured when 1.00 g of the fluorocarbon is dissolved in 100.0 g of camphor if the two possibilities are to be distinguished?

7.10 Potassium fluoride is very soluble in glacial acetic acid and the solutions have a number of unusual properties. In an attempt to understand them, freezing point depression data were obtained by taking a solution of known molality and then diluting it several times (J. Emsley, *J. chem. Soc.* A, 2702 (1971).) The following data were obtained:

$m/(mol\ kg^{-1})$	0.015	0.037	0.077	0.295	0.602
$\Delta T/K$	0.115	0.295	0.470	1.381	2.67

Calculate the apparent molar mass of the solute and suggest an interpretation. Use $\Delta H_{fus} = 11.4\ kJ\ mol^{-1}$ and $T_f^* = 290\ K$.

7.11 In a study of the properties of an aqueous solution of $Th(NO_3)_4$ (by A. Apelblat, D. Azoulay, and A. Sahar; *J. chem. Soc. Faraday Trans.* I, 1618 (1973)), a freezing point depression of 0.0703 K was observed for a 9.6 mmol kg⁻¹ aqueous solution. What is the apparent number of ions per formula unit?

7.12 The table below lists the vapour pressures of mixtures of iodoethane (I) and ethyl acetate (A) at 50°C. Find the activity coefficients of both components on (a) the Raoult's law basis, (b) the Henry's law basis with I as solute.

x_I	0	0.0579	0.1095	0.1918	0.2353	0.3718
$p_I/Torr$	0	28.0	52.7	87.7	105.4	155.4
$p_A/Torr$	280.4	266.1	252.3	231.4	220.8	187.9

x_I	0.5478	0.6349	0.8253	0.9093	1.0000
$p_I/Torr$	213.3	239.1	296.9	322.5	353.4
$p_A/Torr$	144.2	122.9	66.6	38.2	0

7.13 Plot the vapour pressure data for a mixture of benzene (B) and acetic acid (A) given below and plot the vapour pressure/composition curve for the mixture at 50°C. Then confirm that Raoult's and Henry's laws are obeyed in the appropriate regions. Deduce the activities and activity coefficients of the components on the Raoult's law basis and then, taking B as the solute, its activity and activity coefficient on a Henry's law basis. Finally, evaluate the excess Gibbs function of the mixture over the composition range spanned by the data.

x_A	0.0160	0.0439	0.0835	0.1138	0.1714	0.2973
$p_A/Torr$	3.63	7.25	11.51	14.2	18.4	24.8
$p_B/Torr$	262.9	257.2	249.6	244.8	231.8	211.2

x_A	0.3696	0.5834	0.6604	0.8437	0.9931
$p_A/Torr$	28.7	36.3	40.2	50.7	54.7
$p_B/Torr$	195.6	153.2	135.1	75.3	3.5

7.14 The excess Gibbs function of solutions of methyl-cyclohexane (MCH) and tetrahydrofuran (THF) at 303.15 K were found to fit the expression

$$G^E = RTx(1-x)\{0.4857 - 0.1077(2x-1) + 0.0191(2x-1)^2\}$$

where x is the mole fraction of the methylcyclohexane. Calculate the Gibbs function of mixing when a mixture of 1 mol of MCH and 3 mol of THF is prepared.

Theoretical problems

7.15 The excess Gibbs function of a certain binary mixture is equal to $gRTx_A(1-x_A)$ where g is a constant. Find an expression for the chemical potential of A in the mixture and sketch its dependence on the composition.

7.16 Use the Gibbs–Duhem equation to derive the 'Gibbs–Duhem–Margules equation'

$$\left(\frac{\partial \ln f_A}{\partial \ln x_A}\right)_{p,T} = \left(\frac{\partial \ln f_B}{\partial \ln x_B}\right)_{p,T}$$

where f is the fugacity. Use the relation to show that when the fugacities are replaced by pressures, then if Raoult's law applies to one component in a mixture it must also apply to the other.

7.17 Use the Gibbs–Duhem equation to show that the partial molar volume (or any partial molar property) of a component B can be obtained if the partial molar volume (or other property) of A is known for all compositions up to the one of interest. Do this by proving that

$$V_B = V_B^* + \int_{V_A^*}^{V_A} \frac{x_A}{1-x_A} dV_A$$

Use the data in Problem 7.7 to evaluate the integral graphically to find the partial molar volume of chloroform at $x = 0.500$.

7.18 Use the Gibbs–Helmholtz equation to find an expression for $d \ln x_A$ in terms of dT. Integrate $d \ln x_A$ from $x_A = 0$ to the value of interest, and integrate the right hand side from the transition temperature for the pure liquid A to the value in the solution. Show that, if the enthalpy of transition is constant, eqns 15 and 16 are obtained.

7.19 The 'osmotic coefficient' ϕ is defined as

$$\phi = -\frac{x_A}{x_B} \ln a_A$$

By writing $r = x_B/x_A$ and using the Gibbs–Duhem equation, show that we can calculate the activity of B from the activities of A over a composition range using the formula

$$\ln \frac{a_B}{r} = \phi - \phi(0) + \int_0^r \frac{\phi-1}{r} dr$$

7.20 Show that the osmotic pressure of a real solution is given by

$$\Pi V_m = -RT \ln a_A$$

Go on to show that, so long as the concentration of the solution is low, this takes the form

$$\Pi V = \phi[B]RT$$

and hence that the osmotic coefficient ϕ (which is defined in Problem 7.19) may be determined from osmometry.

Changes of state: the phase rule

8

Check-list of key ideas

1. The definition of the terms *phase*, *component*, and *variance* of a system (Sections 8.1 and 8.2).

2. The statement of the *phase rule* (eqn 1) and its application to the *phase diagrams* of one-component systems.

3. The use of *thermal analysis* to determine phase boundaries (Section 8.2).

4. The derivation of the phase rule from the equality of chemical potentials of a substance in every phase (Section 8.2).

5. The description of *partially miscible liquids* in terms of phase diagrams (Section 8.3) and the use of the lever rule in their interpretation.

6. The significance of the *upper critical temperature* and the *lower critical temperature* of a binary liquid mixture (Section 8.3).

7. The description of the *distillation* of partially miscible liquids in terms of their phase diagrams (Section 8.4).

8. The phase diagrams of binary solid systems and the formation of *eutectics* (Section 8.5).

9. The phase diagrams of mixtures that form compounds and the process of *incongruent melting* (Section 8.5).

10. The use of *fractional crystallization* and *zone refining* to purify solids and of *zone levelling* to dope them selectively.

11. The use of *triangular phase diagrams* for the description of ternary systems (Section 8.7).

In this chapter we describe a systematic way of discussing the changes mixtures undergo when they are heated and cooled and when their compositions are changed. In particular, we see how to construct and interpret phase diagrams. These diagrams let us judge whether two or three substances are mutually miscible, whether an equilibrium can exist over a range of conditions, or whether the system must be brought to a definite pressure, temperature, and composition before equilibrium is established.

Phases, components, and degrees of freedom

The word 'phase' comes from the Greek word for appearance. A **phase** is a state of matter that is uniform throughout, not only in chemical composition but also in physical state. (The words are Gibbs's.) Thus we speak of the solid, liquid, and gas phases of a substance, and of its various solid phases (as for black phosphorus and white phosphorus). By a **component** we mean a species present in the system, as for the solute and solvent in a binary solution.

8.1 Definitions

The number of phases

The number of phases in a system is denoted P. A gas, or a gaseous mixture, is a single phase, a crystal is a single phase, and two totally miscible liquids form a single phase. Ice is a single phase ($P = 1$) even though it might be chipped into small fragments. A slurry of ice and water is a two-phase ($P = 2$) system even though it is difficult to map the boundaries between the phases.

An alloy of two metals is a two-phase system ($P = 2$) if the metals are immiscible, but a single phase system ($P = 1$) if they are miscible. This example shows that it is not always easy to decide whether a system consists of one phase or of two. A **solution** of solid A in solid B—a homogeneous mixture of the two components—is uniform on a molecular scale. In a solution, atoms of A are surrounded by atoms of A and B, and any sample cut from the solid, however small, is representative of the composition of the whole.

A **dispersion** is uniform on a macroscopic but not a microscopic scale, for it consists of grains or droplets of one component in a matrix of the other. A small sample could come entirely from one of the minute grains of pure A and would not be representative of the whole (Fig. 8.1). Such dispersions are important because in many advanced materials (including steels), heat treatment cycles are used to achieve the precipitation of a fine dispersion of particles of one phase (such as a carbide phase) within a matrix formed by a saturated solid solution phase. It is the ability to control this microstructure resulting from phase equilibrium that makes it possible to tailor the mechanical properties of the materials to a particular application.

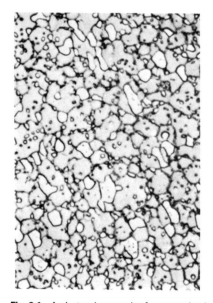

Fig. 8.1 A photomicrograph of a superplastic NiCrFe alloy. Its properties are discussed in the article by H. W. Hayden, R. C. Gibson, and J. H. Brophy in *Scientific American,* March 1969.

The number of components

The number of components in a system C is the minimum number of *independent* species necessary to define the composition of *all* the phases present in the system. The definition is easy to apply when the species present in a system do not react, for then we simply count their number. For instance, pure water is a one-component system ($C = 1$) and a mixture of ethanol and water is a two-component system ($C = 2$).

If species do react, and are at equilibrium, we must take into account the significance of the phrase 'all the phases' in the definition. Thus, for ammonium chloride in equilibrium with its vapour

$$NH_4Cl(s) \rightleftharpoons NH_3(g) + HCl(g)$$

both phases have the formal composition 'NH_4Cl' and the system has one component. If extra $HCl(g)$ is added, the system has two components because now the relative amounts of HCl and NH_3 are arbitrary. In contrast, calcium carbonate in equilibrium with its vapour

$$CaCO_3(s) \rightleftharpoons CaO(s) + CO_2(g)$$

is a two-component system because '$CaCO_3$' does not describe the composition of the vapour. (Since the three species are related by the reaction stoichiometry, the concentration of calcium oxide is not an independent variable.) In this case $C = 2$ whether we start from pure calcium carbonate, or equal amounts of calcium oxide and carbon dioxide, or arbitrary amounts of all three.

Example 8.1: *Counting components*

How many components are present in the following systems: (a) sucrose in water; (b) sodium chloride in water; (c) aqueous phosphoric acid?

Answer. Identify the number S of different kinds of species (e.g. ions) present in each single phase system. Identify the number R of relations between the species (e.g. reactions at equilibrium, charge neutrality). Then the number of components is the number of species less the number of relations: $C = S - R$. (a) The species present are water molecules and sucrose molecules, so $S = 2$. There are no relations between them, so $R = 0$. Therefore, $C = 2$. (b) The species present are H_2O molecules, Na^+ ions, and Cl^- ions, so $S = 3$. Since the solution is electrically neutral, the numbers of Na^+ and Cl^- ions are the same. Therefore, there is one relation, and $R = 1$. Consequently, $C = 3 - 1 = 2$. (c) The species present in aqueous phosphoric acid are H_2O, H_3PO_4, $H_2PO_4^-$, HPO_4^{2-}, PO_4^{3-}, H^+, so $S = 6$. However, the acid ionizations are at equilibrium:

$$H_3PO_4(aq) \rightleftharpoons H_2PO_4^-(aq) + H^+(aq)$$

$$H_2PO_4^-(aq) \rightleftharpoons HPO_4^{2-}(aq) + H^+(aq)$$

$$HPO_4^{2-}(aq) \rightleftharpoons PO_4^{3-}(aq) + H^+(aq)$$

and there are three relations between the species. There is also overall charge neutrality, and so the total number of cation charges must equal the total number of anion charges, whatever their kind. The total number of relations is therefore $R = 4$, and so the number of components is $C = 6 - 4 = 2$.

Comment. Even if we were to include the vapour phase in each case, since that is 'water', and one of the components already specified is water, all three systems remain two-component systems.

Exercise. Give the number of components in the following systems: (a) water, allowing for its ionization, (b) aqueous acetic acid, (c) magnesium carbonate in equilibrium with its vapour.   [(a) 1, (b) 2, (c) 2]

8.2 The phase rule

In a single-component system ($C = 1$), the pressure and temperature may be changed independently if only one phase is present ($P = 1$). If we define the **variance** F of the system as the number of intensive variables that can be changed independently without disturbing the number of phases in equilibrium, then $F = 2$. That is, the system is **bivariant** and has two degrees of freedom. In one of the most elegant calculations in the whole of chemical thermodynamics, J. W. Gibbs[1] deduced the **phase rule**, which is a general relation between the variance F, the number of components C, and the number of phases at equilibrium P for a system of any composition:

$$F = C - P + 2 \qquad (1)$$

We shall see that this rule summarizes what we already know about one-component systems, explain how it is derived, and then apply it to more complex cases.

One-component systems

For a one-component system, such as pure water,

$$F = 3 - P$$

When only one phase is present, $F = 2$ and both p and T can be varied independently. In other words, a single phase is represented by an *area* on a phase diagram. When two phases are in equilibrium $F = 1$, which implies that pressure is not freely variable if we have set the temperature. That is, the equilibrium of two phases is represented by a *line* in the phase diagram. Instead of selecting the temperature, we can select the pressure, but having done so the two phases come into equilibrium at a single definite temperature. Therefore, freezing (or any other phase transition) occurs at a definite temperature at a given pressure.

When three phases are in equilibrium $F = 0$. This special invariant condition can therefore be established only at a definite temperature and pressure. The equilibrium of three phases is therefore represented by a *point*, the triple point, on the phase diagram. Four phases cannot be in equilibrium in a one-component system because F cannot be negative.

These features are illustrated by the phase diagram for water shown in Fig. 8.2 and the events that take place as a sample at a is cooled at constant pressure. The sample remains entirely gaseous until the temperature reaches b, when liquid appears. Two phases are now in equilibrium and $F = 1$. Since we have decided to specify the pressure, which uses up the single degree of freedom, the temperature at which this equilibrium occurs is not under our control. Lowering the temperature takes the system to c in the one-phase, liquid region. The temperature can now be varied around the point c at will, and only when ice appear at d does the variance become 1 again.

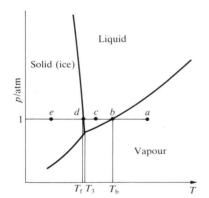

Fig. 8.2 The phase diagram for water, a simplified version of Fig. 6.4. T_3 marks the temperature of the triple point, T_b the normal boiling point, and T_f the normal freezing point.

[1] Josiah Willard Gibbs spent most of his working life at Yale, and may justly be regarded as the originator of chemical thermodynamics. He reflected for years before publishing his conclusions, and then did so in precisely expressed papers in an obscure journal (*The Transactions of the Connecticut Academy of Arts and Sciences*). He needed interpreters before the power of his work was recognized and before it could be applied to industrial processes. He is regarded by many as one of the first great American scientists.

Experimental procedures

Detecting a phase change is not always as simple as seeing a kettle boil, and so special techniques have been developed. One is **thermal analysis**, which takes advantage of the effect of the enthalpy change during a first-order transition (Section 6.5). In this method a sample is allowed to cool, and its temperature is monitored. At a first-order transition, heat is evolved and the cooling stops until the transition is complete. The cooling curve along the isobar *cde* in Fig. 8.2 therefore has the shape shown in Fig. 8.3. The transition temperature is obvious, and is used to mark point *d* on the phase diagram. This technique is useful for solid–solid transitions, where simple visual inspection of the sample may be inadequate.

Modern work on phase transitions often deals with systems at very high pressures, and more sophisticated detection procedures must be adopted. Some of the highest pressures currently attainable are produced in the **diamond-anvil cell** illustrated in Fig. 8.4. The sample is placed in a minute cavity between two gem-quality diamonds, and then pressure is exerted simply by turning the screw. The advance in design this represents is quite remarkable, for with a turn of the screw pressures of up to about 1 Mbar can be reached which a few years ago could not be reached with equipment weighing tons.

The pressure is monitored spectroscopically by observing the shift of spectral lines in small pieces of ruby added to the sample, and the properties of the sample itself are observed optically through the diamond anvils. One application of the technique is to study the transition of covalent solids to metallic solids. Iodine, for instance, becomes metallic at around 200 kbar while remaining as I_2, but makes a transition to a monatomic metallic solid at around 210 kbar. Studies such as these are relevant to the structure of material deep inside the earth (at the centre of the earth the pressure is around 5 Mbar) and in the interiors of the giant planets, where even hydrogen is believed to be metallic.

The general case

The behaviour of a one-component system can be interpreted by drawing on the result obtained in Section 6.2, that the chemical potential μ of a substance is the same at equilibrium in every phase in which it appears. When two phases are present at equilibrium the equality $\mu_\alpha(p, T) = \mu_\beta(p, T)$ is an equation for p in terms of T, and so the temperature is set by the choice of the pressure. When three phases are present at equilibrium there are three equations between the three chemical potentials, but one of them is redundant (because $A = B$ with $B = C$ implies $A = C$). The two

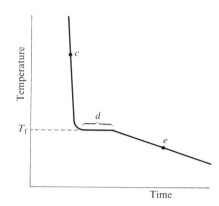

Fig. 8.3. The cooling curve for the isobar *cde* in Fig. 8.2. The halt marked *d* corresponds to the pause in the fall of temperature while the first-order exothermic transition (freezing) occurs. This enables T_f to be located even if the transition cannot be observed visually.

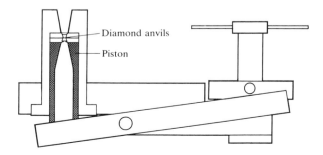

Fig. 8.4 Ultra-high pressures (up to about 1 Mbar) can be achieved using a diamond anvil. The sample, together with a ruby for pressure measurement and a drop of liquid for pressure transmission, are placed between two gem-quality diamonds. The principle of its action is like a nutcracker: the pressure is exerted by turning the screw by hand.

remaining equations $\mu_\alpha(p, T) = \mu_\beta(p, T)$ and $\mu_\beta(p, T) = \mu_\gamma(p, T)$ have a solution only for fixed values of p and T (just like any pair of two simultaneous equations in two unknowns), and so $F = 0$: there are no degrees of freedom. We see that the variance reflects the number of equations that must be satisfied, and the more equations there are, the smaller F is.

We shall now see that similar arguments can be used to derive the phase rule for a system in which there are C components and P phases. (An alternative derivation is carried through in the next example.) We begin by counting the total number of *intensive* variables. The pressure p and temperature T count as 2. We can specify the composition of a phase by giving the mole fractions of only $C - 1$ components. We need specify $C - 1$ and not all C mole fractions because $x_1 + x_2 + \ldots + x_C = 1$, and all mole fractions are known if all except one are specified. Since there are P phases, the total number of composition variables is $P(C - 1)$. At this stage, the total number of intensive variables is $P(C - 1) + 2$.

At equilibrium, the chemical potential of a component J must be the same in every phase:

$$\mu_{J,\alpha} = \mu_{J,\beta} = \ldots \text{ for } P \text{ phases}$$

That is, there are $P - 1$ equations to be satisfied for each component J. As there are C components, the total number of equations is $C(P - 1)$. Each equation reduces our freedom to vary one of the $P(C - 1) + 2$ intensive variables. It follows that the total number of degrees of freedom is

$$F = P(C - 1) + 2 - C(P - 1) = C - P + 2$$

which is the phase rule, eqn 1.

Example 8.2: *Deriving the phase rule*

Derive the phase rule from the general definition of dG and the equality of the Gibbs function in all P phases.

Answer. The change in the Gibbs function of any phase in general is given by

$$dG = V\,dp - S\,dT + \sum_J \mu_J\,dn_J$$

Since there are C components J, there are $C + 2$ variables in this expression (p, T, and the n_J). There is one such equation for each of the P phases present at equilibrium. For the phases to remain in equilibrium when any variation occurs, all P values of dG must be the same. Since there are P equations and $C + 2$ variables, the number of degrees of freedom is

$$F = C + 2 - P$$

which is the phase rule.

Two-component systems

When two components are present in a system, $C = 2$ and

$$F = 4 - P$$

For simplicity, we shall keep the pressure constant (at 1 atm, for instance), which uses up one of the degrees of freedom, and write $F' = 3 - P$ for the remaining variance. One of these remaining degrees of freedom is the temperature, the other is the composition (as expressed by the mole fraction of one component). Hence we should be able to depict the phase equilibria of the system on a temperature–composition diagram of the kind introduced in Section 7.7. A vertical line in the diagram represents a system with the same composition at different temperatures and is called an **isopleth** (from the Greek words for 'equal abundance').

8.3 Liquid–liquid phase diagrams

We begin by considering binary systems consisting of pairs of **partially miscible liquids**, which are liquids that do not mix in all proportions at all temperatures. An example is hexane and nitrobenzene.

Phase separation

Suppose we add a little of a liquid B to a sample of another liquid A at some temperature T'. It dissolves completely, and the binary system remains a single phase. As more B is added, a stage comes when no more dissolves. The sample now consists of two phases in equilibrium with each other, the most abundant one consisting of A saturated with B, the minor one a trace of B saturated with A. In the temperature-composition diagram drawn in Fig. 8.5, the composition of the former is represented by the point a' and that of the latter by the point a''. The horizontal line joining a' and a'' is a **tie line**, a line joining two phases that are in equilibrium with each other. As in the phase diagrams encountered in Chapter 7, the relative abundances of the two phases joined by a tie line are given by the lever rule (eqn 23 in Section 7.6), which in this case may be written

$$\frac{l''}{l'} = \frac{\text{amount of phase of composition } a'}{\text{amount of phase of composition } a''} \tag{2}$$

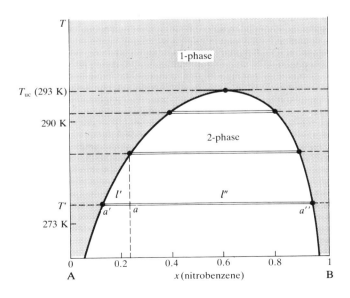

Fig. 8.5 The temperature–composition diagram for hexane and nitrobenzene at 1 atm. The upper critical temperature (T_{uc}) is the temperature above which no phase separation occurs. For this system it lies at 293 K.

When more B is added, A dissolves in it slightly. The compositions of the two phases in equilibrium remain a' and a'', but the amount of the second phase increases at the expense of the first. A stage is reached when so much B is present that it can dissolve all the A, and the system reverts to a single phase. The addition of more B now simply dilutes the solution, and from then on it remains a single phase.

The temperature affects the compositions of the two phases at equilibrium. In the case of hexane and nitrobenzene, raising the temperature increases their miscibility. The two-phase system is therefore less extensive, because each phase in equilibrium is richer in both components: the A-rich phase is richer in B and the B-rich phase is richer in A. The entire phase diagram can be constructed by repeating the observations at different temperatures and drawing the envelope of the two-phase region.

Example 8.3: *Interpreting a liquid–liquid phase diagram*

A mixture of 50 g (0.59 mol) of hexane and 50 g (0.41 mol) of nitrobenzene was prepared at 290 K. What are the compositions of the phases, and in what proportions do they occur? To what temperature must the sample be heated in order to obtain a single phase?

Answer. We denote hexane by A and nitrobenzene by B. Refer to Fig. 8.5 and use the lever rule. The point $x_B = 0.41$, $T = 290$ K occurs in the two-phase region of the phase diagram in Fig. 8.5. The horizontal tie-line cuts the phase boundary at $x_B = 0.37$ and $x_B = 0.83$, and so those are the compositions of the two phases. The ratio of amounts of each phase is equal to the ratio of the distances l' and l'' for the tie-line at 290 K:

$$\frac{l''}{l'} = \frac{0.83 - 0.41}{0.41 - 0.37} = \frac{0.42}{0.04} = 11$$

Heating the sample to 293 K takes it into the single phase region.

Comment. Since the phase diagram has been constructed experimentally, these conclusions are 'exact'. They would be modified if the system were subjected to a different pressure.

Exercise. Repeat the problem for 50 g hexane and 100 g nitrobenzene at 273 K.   [$x_B = 0.09$ and 0.95 in ratio 1:1.3; 293 K]

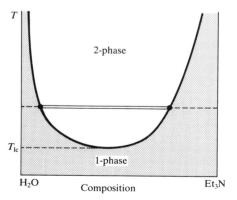

Fig. 8.6 The temperature–composition diagram for water and triethylamine. This shows a lower critical temperature (the temperature below which phase separation does not occur) at $T_{lc} = 292$ K.

Critical temperatures

The **upper critical temperature** T_{uc} is the upper limit of temperatures at which phase separation occurs. Above the upper critical temperature the two components are fully miscible. This temperature exists because the greater thermal motion leads to greater miscibility of the two components.

Some systems show a **lower critical temperature** T_{lc} below which they mix in all proportions and above which they form two phases. An example is water and triethylamine (Fig. 8.6). In this case, at low temperatures the two components are more miscible because they form a weak complex; at higher temperatures the complexes break up and the two components are less miscible.

Some systems have both upper and lower critical temperatures. This is because after the weak complexes have been disrupted, leading to partial miscibility, the thermal motion at higher temperatures homogenizes the

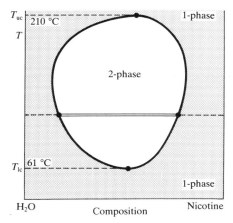

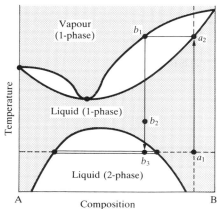

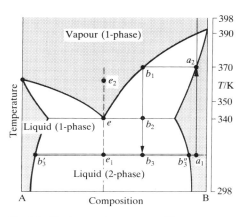

Fig. 8.7 The temperature–composition diagram for water and nicotine, which has both upper and lower critical temperatures. Note the high temperatures for the liquid: the diagram corresponds to a sample under pressure.

Fig. 8.8 The temperature–composition diagram for a binary system in which the upper critical temperature is less than the boiling point at all compositions. The mixture forms a low-boiling azeotrope.

Fig. 8.9 The temperature–composition diagram for a binary system in which boiling occurs before the two liquids are fully miscible.

mixture again, just as in the case of ordinary partially miscible liquids. The most famous example is nicotine and water, which are partially miscible between 61°C and 210°C (Fig. 8.7).

8.4 The distillation of partially miscible liquids

We now consider what happens when the conditions are such that a vapour may be present too. We shall consider a pair of liquids that are partially miscible and form a low-boiling azeotrope (Section 7.7). This combination of properties is quite common because both properties reflect the tendency of the two kinds of molecule to avoid each other. There are two possibilities, one when the liquids become fully miscible before they boil, the other when boiling occurs before mixing is complete.

Miscibility before boiling

Figure 8.8 shows the phase diagram for two components that become fully miscible before they boil. Distillation of a mixture of composition a_1 leads to a vapour of composition b_1 which condenses to the completely miscible single-phase solution at b_2. Phase separation occurs only when this distillate is cooled towards b_3. This applies only to the first drop of distillate. If distillation continues, the composition of the remaining liquid changes towards pure B. In the end, when the whole sample has evaporated and condensed, the composition is back to a_1.

Boiling before mixing completely

Figure 8.9 shows the second possibility, in which there is no upper critical temperature. The distillate obtained from a liquid initially of composition a_1 has composition b_3 and is a two-phase mixture. One phase has composition b_3' and the other has composition b_3''.

The behaviour of a system of composition e is interesting. A system at e_1 forms two phases, which persist (but with changing proportions) up to the boiling point at e. The vapour of this mixture has the same composition as the liquid (the liquid is an azeotrope). Similarly, condensing a vapour of composition e_2 gives a liquid of the same composition. The mixture vaporizes and condenses like a single substance.

Example 8.4: *Interpreting a phase diagram*

State the changes that occur when a mixture of composition $x_B = 0.95$ (a_1) in Fig. 8.9 is boiled and the vapour condensed.

Answer. The area occupied by the point gives the number of phases; the compositions of the phases are given by the points at the intersections of the horizontal tie-line with the phase boundaries; the relative abundances are given by the lever rule (eqn 2). The initial point is in the one-phase region. When heated it boils at 370 K (a_2) giving a vapour of composition $x_B = 0.66$ (b_1). The liquid gets richer in B, and the last drop (of pure B) evaporates at 392 K. The boiling range of the liquid is therefore 370 to 392 K. If the initial vapour is drawn off, it has a composition $x_B = 0.66$. This would be maintained if the sample were very large, but for a finite sample it shifts to higher values and ultimately to $x_B = 0.95$. Cooling the distillate corresponds to moving down the $x_B = 0.66$ isopleth. At 350 K, for instance, the liquid phase has composition $x_B = 0.87$, the vapour $x_B = 0.49$, in relative proportions 1 : 1.3. At 340 K the sample is entirely liquid, and consists of three phases, the vapour, and two liquids, one of composition $x_B = 0.14$, the other of composition $x_B = 0.84$ in the ratio 0.85 : 1. Further cooling moves the system into the two-phase region, and at 298 K the compositions are 0.05 and 0.93 in the ratio 0.46 : 1. As further distillate boils over, the overall composition of the distillate becomes richer in B. When the last drop has been condensed the phase composition is the same as at the beginning.

Exercise. Repeat the discussion, beginning at the point $x_B = 0.04$, $T = 298$ K.

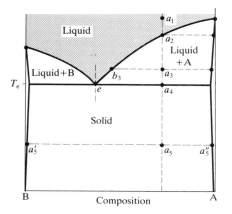

Fig. 8.10 The temperature–composition phase diagram for two almost immiscible solids and their completely miscible liquids. Note the similarity to Fig. 8.9. The isopleth through e corresponds to the eutectic composition, the mixture with lowest melting point.

8.5 Liquid–solid phase diagrams

Solid and liquid phases may both be present in a system at temperatures below the boiling point. An example is a pair of metals that are almost completely immiscible right up to their melting points (such as antimony and bismuth). The phase diagram is shown in Fig. 8.10; note how closely it resembles Fig. 8.9, but instead of liquid and vapour phases we have solid and liquid phases.

Consider the two-component liquid of composition a_1. When it is cooled to a_2 it enters the two-phase region labelled 'Liquid + A'. Almost pure solid A begins to come out of solution and the remaining liquid becomes richer in B. On cooling to a_3, more of the solid forms, and the relative amounts of the solid and liquid (which are in equilibrium) are given by the lever rule: at this stage there are roughly equal amounts of each. The liquid phase is richer in B than before (its composition is given by b_3) because A has been deposited. At a_4 there is less liquid than at a_3, and its composition is given by e. This liquid now freezes to give a two-phase system of almost pure A and almost pure B. At a_5, for example, the compositions of the two phases are a_5' and a_5''.

Eutectics

The isopleth at e in Fig. 8.10 corresponds to the **eutectic** composition, the word coming from the Greek words for 'easily melted'. A liquid with the eutectic composition freezes at a single temperature, without previously depositing solid A or B. A solid with the eutectic composition melts, without change of composition, at the lowest temperature of any mixture. Solutions of composition to the right of e deposit A as they cool, and solutions to the left deposit B: only the eutectic mixture (apart from pure A or pure B) solidifies at a *single* definite temperature ($F' = 0$ when $C = 2$ and $P = 3$) without gradually unloading one or other of the components from the liquid.

One technologically important eutectic is solder, which consists of 67 per cent Sn and 33 per cent Pb by mass and melts at 183°C. The euctectic formed by 23 per cent NaCl and 77 per cent H_2O melts at −21°C. When salt is added to ice under isothermal conditions (e.g. when spread on an icy road) the mixture melts if the temperature is above −21°C (and the eutectic composition has been achieved). When salt is added to ice under adiabatic conditions (e.g. to ice in a vacuum flask) the ice melts, but in doing so it absorbs heat from the rest of the mixture. The temperature of the system falls, and if enough salt is added, cooling continues down to the eutectic temperature. Eutectic formation occurs in the great majority of binary alloy systems, and is of great importance for the microstructure of solid materials, for although a eutectic solid is a two-phase system, it crystallizes out in a nearly homogeneous mixture of microcrystals. The two microcrystalline phases can be distinguished by microscopy and structural techniques such as X-ray diffraction.

Thermal analysis is a very useful practical way of detecting eutectics. We can see how it is used by considering the rate of cooling down the isopleth at a_1 in Fig. 8.10. The liquid cools steadily (Fig. 8.11) until it reaches a_2, when A begins to be deposited. Cooling is now slower because the solidification of A is exothermic and retards the cooling. When the remaining liquid reaches the eutectic composition, the temperature remains constant ($F' = 0$) until the whole sample has solidified: this is the **eutectic halt**. If the liquid has the eutectic composition e initially, then the liquid cools steadily down to the freezing temperature of the eutectic, when there is a long eutectic halt as the entire sample solidifies (like the freezing of a pure liquid).

Monitoring the cooling curves at different overall compositions gives a clear indication of the structure of the phase diagram. The solid–liquid boundary is given by the points at which the rate of cooling changes. The longest eutectic halt gives the location of the eutectic composition and its melting temperature.

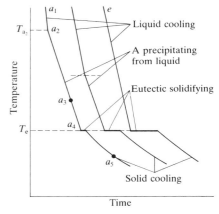

Fig. 8.11 The cooling curves for the system shown in Fig. 8.10. For isopleth a the rate of cooling slows at a_2 because solid A comes out of solution. There is a complete halt at T_e while the eutectic solidifies. This halt is longest for the eutectic isopleth, e. The cooling curves are used to construct the phase diagram.

Reacting systems

Many binary mixtures react to produce compounds, and technologically important examples of this behaviour include the III/V semiconductors, such as gallium arsenide, which forms the compound GaAs. Although three substances may be present, $C = 2$ because the relative amounts of the three species are related by the stoichiometry of the reaction $A + B \rightarrow C$. We shall illustrate some of the principles involved with a system that forms a compound C which also forms eutectic mixtures with the components A and B.

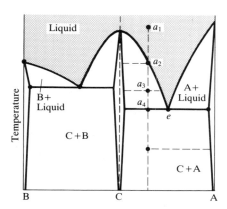

Fig. 8.12 The phase diagram for a system in which A and B react to form a compound C. This resembles two versions of Fig. 8.10 in each half of the diagram. C signifies a true compound, not just an equimolar mixture.

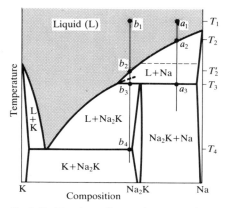

Fig. 8.13 The phase diagram for an actual system like that shown in Fig. 8.12, but with two differences. One is that the compound is Na_2K, corresponding to A_2B and not AB as in that illustration. The second is that the compound exists only as the solid, not as the liquid. The transformation of the compound at its melting point is an example of incongruent melting.

A system prepared by mixing an excess of A with B consists of C and unreacted A. This is a binary C, A system, which we suppose forms a eutectic, the principal change from the eutectic phase diagram in Fig. 8.10 being that the whole of the diagram is squeezed into the range of compositions lying between pure A and equal amounts of A and B ($x_A = 0.5$, Fig. 8.12). On the left of the phase diagram there is another binary eutectic system in which B is in excess. The interpretation of the information in the complete diagram is obtained in the same way as for Fig. 8.10. The solid deposited on cooling along the isopleth at a_1 in Fig. 8.12 is the compound C slightly contaminated with A, and the two-phase solid that exists when the temperature is below a_4 consists of C and A (each one slightly contaminated by the other).

Incongruent melting

In some cases the compound C is not stable as a liquid. An example is the alloy Na_2K which survives only as a solid (Fig. 8.13).

Consider what happens as a liquid at a_1 is cooled. At a_2 some solid Na (slightly contaminated with K) is deposited, and the remaining liquid is richer in K. Just below a_3 the sample is entirely solid, and consists of solid Na and solid Na_2K (each slightly contaminated by the other). Now consider the isopleth at b_1. At b_2 solid Na deposits; however, at b_3 a reaction occurs, forming Na_2K: this compound is formed by the K atoms diffusing into the solid Na. At this stage the liquid Na/K mixture is in equilibrium with a little solid Na_2K, but there is still no liquid compound. As cooling continues, the amount of solid compound increases until at b_4 the liquid reaches its eutectic composition. It then solidifies to give a two-phase solid consisting of solid K and solid Na_2K.

If the solid is reheated, the sequence of events is reversed. No liquid Na_2K forms at any stage because it is too unstable to exist as a liquid. This behaviour is an example of **incongruent melting**, in which a compound melts into its components and does not itself form a liquid phase.

8.6 Ultrapurity and controlled impurity

Advances in technology have called for materials of extreme purity. For example, semiconductor devices consist of almost perfectly pure Si or Ge doped to a precisely controlled extent. For these materials to operate successfully the impurity level must be kept down to less than 1 in 10^9 (which corresponds to about one grain of salt in 5 tons of sugar).

Consider a liquid of composition a in Fig. 8.14: it is mainly B with some A impurity. On cooling to a_1, a solid of composition b_1 appears. Removing that solid gives a slightly purer material than the original, but not much of it (by the lever rule). That solid could be used as the starting substance for a second stage of this **fractional crystallization** process. In each stage, the composition is shifted towards pure B, in the manner of fractional distillation; but the procedure is slow and wasteful.

We should recognize, however, that Fig. 8.14 applies when the freezing is so slow that the composition of the solid is uniform and has its equilibrium composition. In a real system this is not the case, because A does not have time to disperse throughout the whole solid sample. The technique of **zone refining** makes use of the non-equilibrium properties of the system. It relies on the impurities being more soluble in the molten sample than in the solid, and sweeps them up by passing a molten zone repeatedly from one end to the other along a sample.

Consider a liquid (this represents the molten zone) on the isopleth at a, and let it cool without the entire sample coming to overall equilibrium. If the temperature falls to a_2 a solid of composition b_2 is deposited and the remaining liquid (the zone where the heater has moved on) is at a_2'. Cooling that liquid down an isopleth passing through a_2' deposits solid of composition b_3 and leaves liquid at a_3'. The process continues until the last drop of liquid to solidify is heavily contaminated with A. There is plenty of everyday evidence that impure liquids freeze in this way. For example, an ice cube is clear near the surface but misty in the core. This is because the water used to make ice normally contains dissolved air; freezing proceeds from the outside, and air is accumulated in the retreating liquid phase. It cannot escape from the interior of the cube, and so when that freezes it occludes the air in a mist of tiny bubbles.

In zone refining the sample is in the form of a narrow cylinder. This is heated in a thin disk-like zone which is swept from one end of the sample to the other. The advancing liquid zone accumulates the impurities as it passes. In practice a train of hot and cold zones are swept repeatedly from one end to the other. The zone at the end of the sample is the impurity dump: when the heater has gone by, it cools to a dirty solid which can be discarded.

A modification of zone refining is **zone levelling**. It is used to introduce controlled amounts of impurity (for example, of In into Ge). A sample rich in the required dopant is put at the head of the main sample, and made molten. The zone is then dragged repeatedly in alternate directions through the sample, where it deposits a uniform distribution of the impurity.

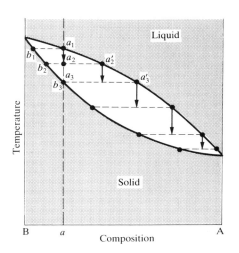

Fig. 8.14 A binary temperature–composition diagram can be used to discuss zone refining as explained in the text.

Three-component systems

For a three-component system, $F = 5 - P$, and hence the variance may reach 4. Holding the temperature and pressure constant leaves two degrees of freedom (the mole fractions of two of the components). One of the best ways of showing how phase equilibria vary with the composition of the system is to use a triangular phase diagram. This section explains how these are constructed and interpreted and gives two simple examples.

8.7 Triangular phase diagrams

The mole fractions of the three components of a ternary system ($C = 3$) satisfy

$$x_A + x_B + x_C = 1$$

A phase diagram drawn as an equilateral triangle ensures that this property is satisfied automatically because the sum of the distances to a point inside an equilateral triangle measured parallel to the edges is equal to the length of the side of the triangle (Fig. 8.15), which may be taken to have unit length.

Figure 8.15 shows how this works in practice. The edge AB corresponds to $x_C = 0$, and likewise for the other two edges. Hence, each of the three edges corresponds to one of the three binary systems (A, B), (B, C), and (C, A). An interior point corresponds to a system in which all three

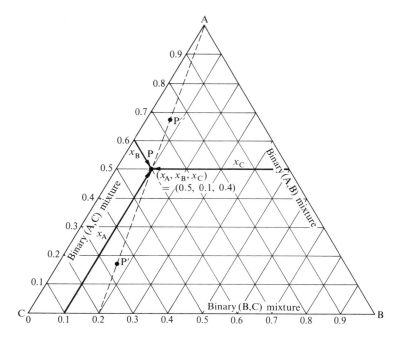

Fig. 8.15 The triangular coordinates used for the discussion of three-component systems. The edges correspond to binary systems. All points along the broken line correspond to mole fractions of C and B in the same ratio.

substances are present. The point P, for instance, represents $x_A = 0.50$, $x_B = 0.10$, and $x_C = 0.40$.

Any point of a straight line joining an apex to a point on the opposite edge (the broken line in Fig. 8.15) represents a composition that is progressively richer in A the closer the point is to the A apex but which has the same proportions of B and C. Therefore, if we wish to represent the changing composition of a system as A is added, we draw a line from the A apex to the point on BC representing the initial binary system. Any ternary system formed by adding A then lies at some point on this line.

Example 8.5: *Marking points on a ternary phase diagram*

Mark the following points on a triangular composition diagram:

(a) $x_A = 0.20$, $x_B = 0.80$, $x_C = 0$
(b) $x_A = 0.42$, $x_B = 0.26$, $x_C = 0.32$
(c) $x_A = 0.80$, $x_B = 0.10$, $x_C = 0.10$
(d) $x_A = 0.10$, $x_B = 0.20$, $x_C = 0.70$
(e) $x_A = 0.20$, $x_B = 0.40$, $x_C = 0.40$
(f) $x_A = 0.30$, $x_B = 0.60$, $x_C = 0.10$

Answer. x_A is measured along either edge leading to apex A; likewise for x_B and B; x_C takes care of itself (but it is sensible to check). The points are plotted in Fig. 8.16.

Comment. Note that the points (d), (e), and (f) have $x_A/x_B = 0.50$, and fall on a straight line, as stated in the text.

Exercise. Plot the following points:

(a') $x_A = 0.25$, $x_B = 0.25$, $x_C = 0.50$
(b') $x_A = 0.50$, $x_B = 0.25$, $x_C = 0.25$

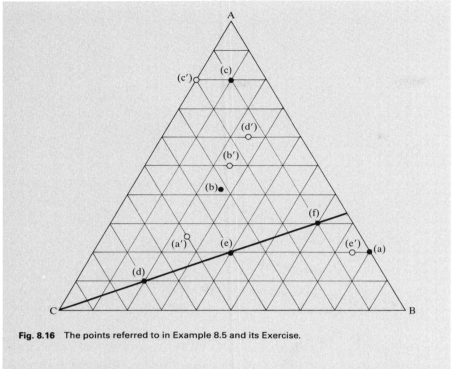

Fig. 8.16 The points referred to in Example 8.5 and its Exercise.

(c') $x_A = 0.80$, $x_B = 0$, $x_C = 0.20$
(d') $x_A = 0.60$, $x_B = 0.25$, $x_C = 0.15$
(e') $x_A = 0.20$, $x_B = 0.75$, $x_C = 0.05$

8.8 Partially miscible liquids

Water and acetic acid are fully miscible, as are chloroform and acetic acid. Water and chloroform are only partially miscible. What happens when all three are present together?

The phase diagram for this ternary system at room temperature and pressure is shown in Fig. 8.17. It shows that the two fully miscible pairs form single-phase regions and that the water/chloroform system (along the base of the triangle) has a two-phase region. The base of the triangle corresponds to one of the horizontal lines in a two-component phase diagram, such as Fig. 8.5.

A single-phase system is formed when enough acetic acid is added to the binary water/chloroform mixture. This is shown by following the line a_1, a_4 in Fig. 8.17. Starting at a_1 we have a two-phase system, and the relative amounts of two phases can be read off in the usual way (using the lever rule). Adding acetic acid takes the system along the line joining a_1 to the acetic acid apex. At a_2 the solution still has two phases, but there is more water in the chloroform phase (a_2') and more chloroform in the water (a_2'') because the acid helps both to dissolve. The phase diagram shows that there is more acetic acid in the water-rich phase than in the other (a_2'' is closer than a_2' to the acetic acid apex). At a_3 two phases are present, but the chloroform-rich layer is present only as a trace. Further addition of acid takes the system towards a_4, and only a single phase is present.

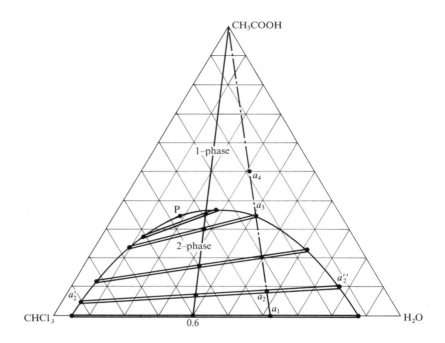

Fig. 8.17 The phase diagram, at fixed temperature and pressure, of the three-component system acetic acid, chloroform, and water. Only some of the tie lines have been drawn in the two-phase region. All points along the line *a* correspond to chloroform and water present in the same ratio.

Example 8.6: *Interpreting a ternary phase diagram* (1)

A mixture is prepared consisting of chloroform ($x_C = 0.60$) and water ($x_W = 0.40$). Describe the changes that occur when acetic acid is added to the mixture.

Answer. We base the answer on Fig. 8.17. The relative proportions of chloroform and water remain constant, and so the addition of acetic acid (A) corresponds to motion along the line from the point $x_C = 0.60$ on the base line opposite the A apex to the apex itself. The tie-lines give the compositions of the phases at their intersections with the boundaries; the lever rule gives their proportions. We shall denote compositions in the order (x_C, x_W, x_A). The initial composition is $(0.60, 0.40, 0)$. This point lies in the two-phase region, the phase compositions being $(0.95, 0.05, 0)$ and $(0.12, 0.88, 0)$ with relative proportions 1.3:1. Addition of acetic acid takes the system along the straight line to A. When sufficient has been added to raise its mole fraction to 0.18 the overall composition is $(0.49, 0.33, 0.18)$ and the system consists of two phases of compositions $(0.82, 0.06, 0.12)$ and $(0.17, 0.60, 0.23)$ in almost equal abundance. When enough acid has been added to raise its mole fraction to 0.37 the system consists of a trace of a phase of composition $(0.64, 0.11, 0.25)$ and a dominating phase of composition $(0.35, 0.28, 0.37)$. Further addition of acid takes the system into the single-phase region, where it remains right up to the point corresponding to pure acid.

Exercise. Repeat the question, starting with a mixture of composition $(0.70, 0, 0.30)$ to which water is added.

The point marked P in Fig. 8.17 is called the **plait point**. It is yet another example of a critical point. At the plait point, the compositions of the two phases in equilibrium become identical.

8.9 The role of added salts

The presence of one solute may affect the solubility of another. The **salting-out effect** is the reduction of the solubility of a gas (or other non-ionic substance) in water when a salt is added. A **salting-in effect** may also occur, in which the ternary system is more concentrated (in the sense of having less water) than in the binary system. A salt may also affect the solubility of another electrolyte, as we can see by examining the ternary system consisting of ammonium chloride, ammonium sulphate, and water (Fig. 8.18).

The point b indicates the solubility of the chloride in water: a mixture of composition b_1 consists of the undissolved chloride and a saturated solution of composition b. The point c similarly indicates the solubility of the sulphate. A system of composition a_1 is unsaturated, and forms a single phase. As water is evaporated, the composition moves along the line a_1 to a_4. At a_2 it enters the two-phase region, and some solid chloride crystallizes (all the tie-lines ending on that boundary also end at the pure chloride apex). The liquid becomes richer in sulphate, and its composition moves towards d. When enough water has been removed to bring the overall composition to a_3 the liquid composition is d. At this point (which is joined to the chloride apex and to the sulphate apex), the system consists of saturated solution in equilibrium with the two solids. Note that this point, which corresponds to the joint solubility of the two solids, corresponds to a smaller mole fraction of water than in either of the binary systems b and c. This means that the two salts form a more concentrated solution overall than either does alone.

If more water is removed after the system has arrived at d, all that happens is that the amount of solution decreases, but its composition

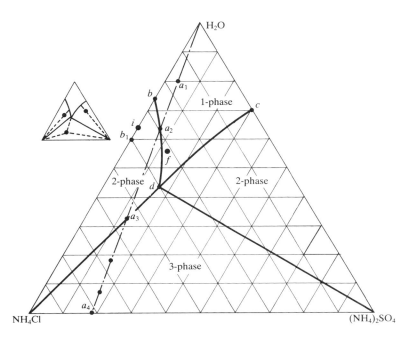

Fig. 8.18 The phase diagram, at constant temperature and pressure, for the ternary system $NH_4Cl/(NH_4)_2SO_4/H_2O$. The points i and f are the ones mentioned in Example 8.7. All tie-lines in the two-phase regions terminate at an apex, and all tie-lines in the three-phase region terminate at the three corners of the three-phase triangular area, as shown in the insert.

remains constant (at d, the saturated solution). Both solids are precipitated, and the system has three phases: each point in the three-phase region is tied to d and the two solid apexes. When the composition arrives at a_4 we have a binary system, consisting of a mixture of the two solids, and all the water has gone.

Example 8.7: *Interpreting a ternary phase diagram* (2)

A solution of 50 g of ammonium chloride in 30 g of water is prepared at room temperature. 45 g of ammonium sulphate is then added. Describe the initial and final states.

Answer. We use Fig. 8.18 after converting compositions to mole fractions. Molar masses are as follows: H_2O, 18.02; NH_4Cl, 53.49; $(NH_4)_2SO_4$, 132.1 g mol^{-1}. We shall write compositions in the order (x_W, x_C, x_S) for water (W), chloride (C), and sulphate (S) and use the lever rule for proportions of each phase. The initial amounts are $n_W = 1.66$ mol and $n_C = 0.93$ mol, and so the initial composition is $(0.64, 0.36, 0)$. In the final state the amounts of W and C are the same but $n_S = 0.34$ mol. Therefore the final composition is $(0.57, 0.31, 0.12)$. From Fig. 8.18 we see that $(0.64, 0.36, 0)$ corresponds to a two-phase system consisting of solid C with a saturated solution of composition $(0.64, 0.36, 0)$ in relative proportions $0.16:1$. After addition of S there is only one phase.

Exercise. Describe the initial and final states of a solution of compositions $(0.80, 0, 0.20)$ and $(0.40, 0.50, 0.10)$.

Although phase diagrams might look complicated, they convey simple, experimentally established information. In order to interpret them it is helpful to think operationally. That is, definite processes should be imagined, and the diagram should be considered bearing in mind how it was constructed originally: phases come and go, systems boil and freeze, and relative amounts of different phases change. It is also wise to concentrate on the lines rather than the areas. The points at the ends of the tie-lines passing give the compositions of the phases in equilibrium with each other. The distances of the points from the isopleth give (through the lever rule) their relative abundances.

Further reading

R. C. Parker and D. S. Kristol, Freezing points, triple points, and phase equilibria. In *J. chem. Educ.* **51**, 658 (1974).

A. Jayaraman, The diamond-anvil high-pressure cell. In *Scientific American* **250** (4), 42 (1984).

F. E. Wetmore and D. J. LeRoy, *Principles of phase equilibria.* (1951).

R. Ginell, *Association theory: phases of matter and their transformations.* Elsevier, Amsterdam (1979).

A. Reisman, *Phase equilibria.* Academic Press, New York (1970).

Exercises

8.1 State the number of components in the following systems. (a) NaH_2PO_4 in water at equilibrium with water vapour but disregarding the fact that the salt is ionized. (b) The same, but taking into account the ionization of the salt. (c) $AlCl_3$ in water, noting that hydrolysis and precipitation of $Al(OH)_3$ occur.

8.2 Blue $CuSO_4 \cdot 5H_2O$ crystals release their water of hydration when heated. How many phases and components are present in an otherwise empty heated container?

8.3 Ammonium chloride decomposes when it is heated. How many components and phases are present when the salt is heated in an otherwise empty container? Now suppose that additional ammonia is also present. How many components and phases are present?

8.4 A saturated solution of Na_2SO_4, with excess of the solid, is present at equilibrium with its vapour in a closed vessel. How many phases and components are present. What is the variance of the system? Identify the independent variables.

8.5 Now suppose that the solution referred to in Exercise 8.4 is not saturated. How many phases and components are present. What is the variance of the system? Identify the independent variables.

8.6 Sketch the phase diagram of the system NH_3/N_2H_4 given that the two substances do not form a compound with each other, that NH_3 freezes at $-78°C$ and N_2H_4 freezes at $+2°C$, and that a eutectic is formed when the mole fraction of N_2H_4 is 0.07 and that the eutectic melts at $-80°C$.

8.7 Figure 8.8 shows the phase diagram for two partially miscible liquids, which can be taken to be that for water and 2-methyl-1-propanol (B). Describe what will be observed when a mixture of composition b_3 is heated, at each stage giving the number, composition, and relative amounts of the phases present.

8.8 Figure 8.19 is the phase diagram for silver/tin. Label the regions, and describe what will be observed when liquids of compositions a and b are cooled to 200 K.

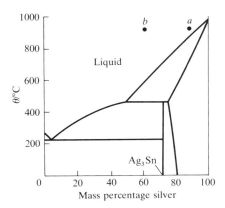

Fig. 8.19 The phase diagram for a binary silver/tin system.

8.9 Indicate on the phase diagram in Fig. 8.19 the feature that denotes incongruent melting. What is the composition of the eutectic mixture and at what temperature does it melt?

8.10 Sketch the cooling curves for the isopleths a and b in Fig. 8.19.

8.11 Use the phase diagram in Fig. 8.19 to state (a) the solubility of Ag in Sn at 800°C and (b) the solubility of Ag_3Sn in Ag at 460°C, (c) the solubility of Ag_3Sn in Ag at 300°C.

8.12 Uranium tetrafluoride and zirconium tetrafluoride melt at 1035°C and 912°C respectively. They form a continuous series of solid solutions with a minimum melting temperature of 765°C and composition $x(ZrF_4) = 0.77$. At 900°C, the liquid solution of composition $x(ZrF_4) = 0.28$ is in equilibrium with a solid solution of composition $x(ZrF_4) = 0.14$. At 850°C the two compositions are 0.87 and 0.90 respectively. Sketch the phase diagram for this system and state what is observed when a liquid of composition $x(ZrF_4) = 0.40$ is cooled slowly from 900°C to 500°C.

8.13 Methane (melting point 91 K) and tetrafluoromethane (melting point 89 K) do not form solid solutions with each other and as liquids they are only partially miscible. The upper critical temperature of the liquid mixture is 94 K at $x(CF_4) = 0.43$ and the eutectic temperature is 84 K at $x(CF_4) = 0.88$. At 86 K, the phase in equilibrium with the tetrafluoromethane-rich solution changes from solid methane to a methane-rich liquid. At that temperature, the two liquid solutions that are in mutual equilibrium have the compositions $x(CF_4) = 0.10$ and $x(CF_4) = 0.80$. Sketch the phase diagram.

8.14 Describe the phase changes that take place when a liquid mixture of 4.0 mol of B_2H_6 (melting point 131 K) and 1.0 mol of CH_3OCH_3 (melting point 135 K) is cooled from 140 K to 90 K. These substances form a compound $(CH_3)_2OB_2H_6$ that melts congruently at 133 K. The system exhibits one eutectic at $x(B_2H_6) = 0.25$ and 123 K and another at $x(B_2H_6) = 0.90$ and 104 K.

8.15 Refer to the information in Exercise 8.14 and sketch the cooling curves for liquid mixtures in which $x(B_2H_6)$ is (a) 0.10, (b) 0.30, (c) 0.50, (d) 0.80, and (e) 0.95.

8.16 Hexane and perfluorohexane show partial miscibility below 22.70°C. The critical concentration at the upper critical temperature is $x = 0.355$, where x is the mole fraction of C_6F_{14}. At 22°C the two solutions in equilibrium have $x = 0.24$ and $x = 0.48$ respectively, and at 21.5°C the mole fractions are 0.22 and 0.51. Sketch the phase diagram. Describe the phase changes that occur when perfluorohexane is added to a fixed amount of hexane at (a) 23°C, (b) 22°C.

8.17 Mark the following features on triangular coordinates: (a) the point (0.2, 0.2, 0.6), (b) the point (0, 0.2, 0.8), (c) the point at which all three mole fractions are the same.

8.18 Mark the following points on a ternary phase diagram for the system $NaCl/Na_2SO_4 \cdot 10H_2O/H_2O$: (a) 25 per cent by mass NaCl, 25 per cent $Na_2SO_4 \cdot 10H_2O$, and the rest H_2O, (b) the line denoting the same relative composition of the salts but with changing amounts of water.

8.19 Refer to the ternary phase diagram in Fig. 8.17. How many phases are present, and what are their compositions and relative abundances, in a mixture that contains 2.3 g of water,

9.2 g of chloroform, and 3.1 g of acetic acid? Describe what happens when (a) water, (b) acetic acid is added to the mixture.

8.20 Figure 8.18 shows the phase diagram for the ternary system $NH_4Cl/(NH_4)_2SO_4/H_2O$ at 25°C. Identify the number of phases present for mixtures of compositions (a) (0.2, 0.4, 0.4), (b) (0.4, 0.4, 0.2), (c) (0.2, 0.1, 0.7), (d) (0.40, 0.16, 0.44). The numbers are mole fractions of the three components in the order (NH_4Cl, $(NH_4)_2SO_4$, H_2O).

8.21 Referring to Fig. 8.18, deduce the solubility of (a) NH_4Cl, (b) $(NH_4)_2SO_4$ in water at 25°C.

8.22 Describe what happens when (a) $(NH_4)_2SO_4$ is added to a saturated solution of NH_4Cl in water in the presence of excess NH_4Cl, (b) water is added to a mixture of 25 g of NH_4Cl and 75 g of $(NH_4)_2SO_4$.

8.23 At a certain temperature, the solubility of I_2 in liquid CO_2 is $x(I_2) = 0.03$. At the same temperature its solubility in nitrobenzene is 0.04. Liquid carbon dioxide and nitrobenzene are miscible in all proportions, and the solubility of I_2 in the mixture varies linearly with the proportion of nitrobenzene. Sketch a phase diagram for the ternary system.

Problems

Numerical problems

8.1 The compound p-azoxyanisole forms a liquid crystal. 5.0 g of the solid was placed in a tube, which was then evacuated and sealed. Use the phase rule to prove that the solid will melt at a definite temperature and that the liquid crystal phase will make a transition to a normal liquid phase at a definite temperature.

8.2 Magnesium oxide and nickel oxide withstand high temperatures. However, they do melt when the temperature is high enough and the behaviour of mixtures of the two is of considerable interest to the ceramics industry. Draw the temperature/composition diagram for the system using the data below, where x is the mole fraction of MgO in the solid and y its mole fraction in the liquid.

θ/°C	1960	2200	2400	2600	2800
x	0	0.35	0.60	0.83	1.00
y	0	0.18	0.38	0.65	1.00

State (a) the melting point of a mixture with $x = 0.30$, (b) the composition and proportion of the phases present when a solid of composition $x = 0.30$ is heated to 2200°C, (c) the temperature at which a liquid of composition $y = 0.70$ will begin to solidify.

8.3 The Bi/Cd phase diagram is of interest in metallurgy, and its general form can be estimated using the equation for the depression of freezing point. Construct the diagram using the following data: $T_f(Bi) = 544.5$ K, $T_f(Cd) = 594$ K, $\Delta H_{fus}(Bi) = 10.88$ kJ mol^{-1}, $\Delta H_{fus}(Cd) = 6.07$ kJ mol^{-1}. The metals are mutually insoluble as solids. Use a phase diagram to state what would be observed when a liquid of composition $x(Bi) = 0.70$ is cooled slowly from 550 K. What are the relative abundances of the liquid and solid at (a) 460 K and (b) 350 K. Sketch the cooling curve for the mixture.

8.4 Solutions of m-toluidine (3-methylphenylamine) were made up in glycerol and then warmed from room temperature. The mixture became turbid at θ_1 and then cleared at θ_2. Plot the phase diagram using the data below, and find the upper and lower critical temperatures.

%	18	20	40	60	80	85
θ_1/°C	48	18	8	10	19	25
θ_2/°C	53	90	120	118	83	53

% denotes the mass percentage composition of m-toluidine. State what happens as m-toluidine is added dropwise to glycerol at 60°C. State the number of phases present at each composition and their relative amounts.

8.5 At 25°C, aqueous solutions containing Li_2SiF_6 and $(NH_4)_2SiF_6$ can be in equilibrium with the two virtually pure solids or with solid $Li(NH_4)SiF_6$. It is also found that solid Li_2SiF_6 can be in equilibrium with solutions ranging from mass percentage 29.5% Li_2SiF_6 and 70.5% H_2O to 29.3% Li_2SiF_6 and 67.6% H_2O. Also $(NH_4)_2SiF_6$ can be in equilibrium with solutions ranging from 18.7% $(NH_4)_2SiF_6$ and 81.3% H_2O to 16.5% $(NH_4)_2SiF_6$ and 70.4% H_2O. Sketch the phase diagram for this ternary system and include some tie-lines. Describe the changes that occur when a solution that is initially 3.0% Li_2SiF_6 and 83.0% H_2O is slowly dehydrated.

8.6 Sketch the phase diagram for the Mg/Cu system using the following information: $T_f(Mg) = 648$°C, $T_f(Cu) = 1085$°C; two intermetallic compounds are formed with $T_f(MgCu_2) = 800$°C and $T_f(Mg_2Cu) = 580$°C; eutectics of mass percentage Mg composition and melting points 10% (690°C), 33% (560°C), and 65% (380°C). A sample of Mg/Cu alloy containing 25% Mg by mass was prepared in a crucible heated to 800°C in an inert atmosphere. Describe what will be observed if the melt is cooled slowly to room temperature. Specify the composition and relative abundances of the phases and sketch the cooling curve.

8.7 Methanol (M), diethyl ether (E), and water (W) form a partially miscible ternary system. The phase diagram at 20°C was determined by adding methanol to various binary ether/water mixtures of mole fraction x in ether and noting the mole fractions of methanol y at which complete miscibility occurred. Plot the phase diagram using the following data.

x	0.10	0.20	0.30	0.40	0.50	0.60	0.70	0.80	0.90
y	0.20	0.27	0.30	0.28	0.26	0.22	0.17	0.12	0.07

How many phases will be present in a mixture that consists of 5.0 g of methanol, 30.0 g of ether, and 50.0 g of water? What mass of water would have to be added or removed to change the number of phases?

8.8 Iron(II) chloride (melting point 677°C) and potassium chloride (melting point 776°C) form the compounds $KFeCl_3$ and K_2FeCl_4 at elevated temperatures. $KFeCl_3$ melts congruently at 399°C and K_2FeCl_4 melts incongruently at 380°C. Eutectics are formed with compositions $x = 0.38$ (melting point 351°C) and $x = 0.54$ (melting point 393°C), where x is the mole fraction of $FeCl_2$. The KCl solubility curve intersects the K_2FeCl_4 curve at $x = 0.34$. Sketch the phase diagram. State the phases that are in equilibrium when a mixture of composition $x = 0.36$ is cooled from 400°C to 300°C.

8.9 The binary system nitroethane/decahydronaphthalene (DEC) shows partial miscibility, with the two-phase region lying between $x = 0.08$ and $x = 0.84$, where x is the mole fraction of nitroethane. The binary system liquid carbon dioxide/DEC is also partially miscible, with its two-phase region lying between $y = 0.36$ and $y = 0.80$, where y is the

mole fraction of DEC. Nitroethane and liquid carbon dioxide are miscible in all proportions. The addition of liquid carbon dioxide to mixtures of nitroethane and DEC increases the range of miscibility, and the plait point is reached when z, the mole fraction of CO_2, is 0.18 and $x = 0.53$. The addition of nitroethane to mixtures of carbon dioxide and DEC also results in another plait point at $x = 0.08$ and $y = 0.52$. (a) Sketch the phase diagram for the ternary system. (b) For some binary mixtures of nitroethane and liquid carbon dioxide, the addition of arbitrary amounts of DEC will not cause phase separation. Find the range of concentrations for such binary mixtures.

Theoretical problems

8.10 Show that two phases are in thermal equilibrium only if their temperatures are the same and that they are in mechanical equilibrium only if their pressures are equal.

8.11 Prove that a straight line from the apex A of a ternary phase diagram to the opposite edge BC represents mixtures of constant ratio of B and C however much A is present.

9 Changes of state: chemical reactions

Check-list of key ideas

1. The definition of the *extent of reaction* and of the *reaction Gibbs function* (eqn 1).

2. The meaning of the terms *exergonic* and *endergonic* (Section 9.1).

3. The reaction Gibbs function in terms of the *reaction quotient* and the standard reaction Gibbs function (eqn 4), and the *equilibrium constant* in terms of the standard reaction Gibbs function (eqn 5).

4. The relation between the *thermodynamic equilibrium constant* and the *practical equilibrium constant* (eqn 7).

5. The variation of the equilibrium constant with pressure (eqn 8), *Le Chatelier's principle* (Section 9.3), and the variation of the equilibrium composition with pressure (eqn 10).

6. The *van't Hoff equation* for the variation of the equilibrium constant with temperature (eqn 11) and its use for the determination of reaction enthalpy (Example 9.4).

7. The use of the *Giauque functions* (eqn 13) for discussing equilibria at different temperatures (eqn 14).

8. Thermodynamic considerations in the extraction of metals from their ores and the interpretation of an *Ellingham diagram* (Section 9.5).

9. The discussion of acids and bases in terms of *Brønsted equilibria* (Section 9.6) and the definition of *acidity constants* (eqn 16).

10. The distinction between *weak acids* and *strong acids* and the role of the *autoprotolysis equilibrium* of water (Section 9.6).

11. The use of equilibrium constants to calculate the pH in the course of an acid/base titration and the derivation of the *Henderson−Hasselbalch equation* (eqn 24).

12. The equilibria responsible for *buffer action* and *indicator detection* of the equivalence point (Section 9.6).

13. The definition of the *biological standard state* (Section 9.7) and its relation to the conventional standard state (eqn 27).

14. Thermodynamic aspects of metabolism and respiration involving the reactions of ATP (Section 9.7).

Chemical reactions move towards a dynamic equilibrium in which both reactants and products are present but have no further tendency to undergo net change. Sometimes the concentration of products is so much greater than the concentration of unchanged reactants in the equilibrium mixture that for all practical purposes the reaction is 'complete'. However, in many important cases the equilibrium mixture has significant concentrations of both reactants and products. In this chapter we see how to use thermodynamics to predict the equilibrium composition under any reaction conditions.

In industry it is obviously worse than useless to build a sophisticated plant if the overall reaction has a tendency to run in the wrong direction. If a plant is to be run economically we must know how to maximize yields. Does that mean raising or lowering the temperature or the pressure? Thermodynamics gives a very simple recipe for deciding what to do. We might also be interested in the way that food is used in the complicated series of biochemical reactions involved in warming the body, powering muscular contraction, and energizing the nervous system. Some reactions (such as the oxidation of carbohydrates) are spontaneous and may be coupled to others to drive them in non-spontaneous but necessary directions (as in the biosynthesis of proteins). With thermodynamics we can sort out the reactions that need to be driven, and calculate the spare driving force of reactions that occur spontaneously. Another field in which equilibria are of great importance is that of the reactions of acids and bases, and we deal with them towards the end of the chapter.

However, as always in thermodynamics, we must bear in mind the possibility that kinetic factors might upset our predictions. Thermodynamics can tell us the direction of spontaneous change and the composition at equilibrium. It cannot tell us whether a kinetically viable pathway exists for that change to occur or for the equilibrium to be reached.

Spontaneous chemical reactions

We have seen that the direction of spontaneous change at constant temperature and pressure is towards lower values of the Gibbs function G. The idea is entirely general, and in this chapter we apply it to the discussion of reactions. In the case of the reaction $A + B \rightarrow C + D$, for instance, we can decide whether there is a spontaneous tendency for A and B to form C and D by calculating the Gibbs function of the mixture at various compositions (Fig. 9.1) and identifying the location of the minimum of the graph, which occurs at the composition corresponding to equilibrium. We can then predict whether A and B will react to form C and D virtually completely (as in a), partially (as in b), or virtually not at all (as in c).

9.1 The Gibbs function minimum

We begin with the simplest possible chemical equilibrium: $A \rightleftharpoons B$. Even though this looks trivial, there are many examples of it, such as the isomerization of pentane to 2-methylbutane and the conversion (racemization) of L-alanine to D-alanine.

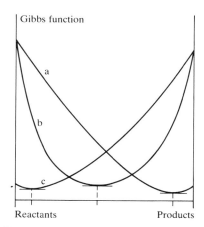

Fig. 9.1 The direction of spontaneous chemical change is towards the minimum Gibbs function. This illustration shows three possibilities: (a) equilibrium lies close to pure products (the reaction 'goes to completion'), (b) equilibrium corresponds to reactants and products present in similar proportions, and (c) equilibrium lies close to pure reactants (the reaction 'does not go').

The extent of reaction

Suppose an infinitesimal amount $d\xi$ of A turns into B, then we can write

$$\text{Change in amount of A present } dn_A = -d\xi$$

$$\text{Change in amount of B present } dn_B = +d\xi$$

ξ is a measure of the progress of the reaction, and is called the **extent of reaction**. We shall choose it so that (for the reaction $A \rightarrow B$), pure A corresponds to $\xi = 0$ and $\xi = 1 \, mol$ signifies that $1 \, mol$ of A has been destroyed and $1 \, mol$ of B has been formed.

We can express the change that $d\xi$ of reaction causes in G of the entire system in terms of the chemical potentials (the partial molar Gibbs functions) of the substances in the mixture. At constant temperature and pressure

$$dG = \mu_A \, dn_A + \mu_B \, dn_B = -\mu_A \, d\xi + \mu_B \, d\xi$$

This equation can be reorganized into

$$\left(\frac{\partial G}{\partial \xi}\right)_{p,T} = \mu_B - \mu_A$$

which is an expression for the slope of the Gibbs function as the extent of reaction changes.

Since the chemical potentials depend on the composition, the slope of G changes as the reaction proceeds. Moreover, since the reaction proceeds in the direction of decreasing G,

$$\text{If } \mu_A > \mu_B, \text{ the reaction proceeds } A \rightarrow B$$

$$\text{If } \mu_A < \mu_B, \text{ the reaction proceeds } A \leftarrow B$$

This is illustrated in Fig. 9.2. When $\mu_A = \mu_B$, the slope of G is zero. This occurs at the minimum of the curve in Fig. 9.2, and corresponds to the position of chemical equilibrium:

$$\text{If } \mu_A = \mu_B, \text{ the reaction is at equilibrium}$$

Fig. 9.2 As the reaction advances (represented by motion from left to right along the horizontal axis) the slope of the Gibbs function changes. Equilibrium corresponds to zero slope, at the foot of the valley.

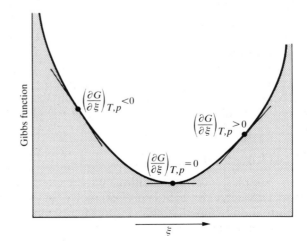

The reaction Gibbs function

The **reaction Gibbs function** ΔG_r at a specified composition of the reaction mixture is defined as the slope of the Gibbs function:

$$\Delta G_r = \left(\frac{\partial G}{\partial \xi}\right)_{p,T} \tag{1}$$

Although Δ normally signifies a *difference* in values, in this context it signifies a *differential coefficient*, the slope of G with respect to ξ. The subscript r on ΔG will remind us that ΔG_r is a slope and not a difference.

We have seen that for A $\rightarrow$ B the differential coefficient is equal to the difference $\mu_B - \mu_A$ *at a specified composition*. Therefore, in this case,

$$\Delta G_r = \mu_B - \mu_A$$

It follows that ΔG_r can be interpreted as the change in G when 1 mol of reactants forms 1 mol of products at a *fixed composition* of the reaction mixture. (The composition is effectively constant if the change takes place in such a large system that loss of 1 mol of A and formation of 1 mol of B has a negligible effect.) It follows that we can write the condition for equilibrium as

$$\Delta G_r = 0$$

This equation states that the slope of the Gibbs function is zero when products are formed from reactants at the equilibrium composition of the reaction mixture.

We must be careful to distinguish the reaction Gibbs function, the slope of G at a specified composition, from the standard reaction Gibbs function $\Delta G^{\ominus}$:

$$\Delta G^{\ominus} = \mu_B^{\ominus} - \mu_A^{\ominus}$$

This is the change per mole of reaction when 1 mol of A in its standard state is converted to 1 mol of B in its standard state (Section 4.9). For gases, liquids, and solids (but not solutions), the standard states correspond to the pure reactants and the pure products. In this case Δ has its normal meaning. As we saw in Section 4.9, the difference in standard molar Gibbs functions of the products and reactants is equal to the difference in their standard Gibbs functions of formation, and in practice we calculate $\Delta G^{\ominus}$ from

$$\Delta G^{\ominus} = \Delta G_f^{\ominus}(B) - \Delta G_f^{\ominus}(A)$$

Exergonic and endergonic reactions

The reaction Gibbs function can be used to rewrite the conditions for a reaction's being spontaneous at some specified composition.

> If $\Delta G_r < 0$, the reaction has a tendency to proceed A $\rightarrow$ B
> If $\Delta G_r > 0$, the reaction has a tendency to proceed A $\leftarrow$ B

Reactions for which $\Delta G_r < 0$ are called **exergonic**, which signifies that, since they are spontaneous, they can be used to drive other processes (such as other reactions, or used to do non-expansion work). Reactions for which $\Delta G_r > 0$ are called **endergonic**. They are spontaneous in the reverse (exergonic) direction. Reactions at equilibrium are spontaneous in neither direction: they are neither exergonic nor endergonic.

9.2 The composition of reactions at equilibrium

We have seen that the condition for equilibrium can be expressed in terms of the chemical potentials of the reactants and products. We saw in Chapter 5 how to express chemical potentials in terms of concentrations and partial pressures. In this section we show how to combine these relations and predict the composition of the reaction mixture at equilibrium.

Perfect gas equilibria

We start with the simple A $\rightleftharpoons$ B equilibrium and suppose that A and B are perfect gases. Expressing the chemical potentials in terms of their partial pressures gives

$$\Delta G_r = \mu_B - \mu_A$$

$$= \left(\mu_B^\ominus + RT \ln \frac{p_B}{p^\ominus}\right) - \left(\mu_A^\ominus + RT \ln \frac{p_A}{p^\ominus}\right)$$

$$= \Delta G^\ominus + RT \ln \frac{p_B}{p_A}$$

The ratio of partial pressures in the logarithm is an example of a **reaction quotient** Q_p, and so

$$\Delta G_r = \Delta G^\ominus + RT \ln Q_p \qquad (2)^\circ$$

At equilibrium $\Delta G_r = 0$. The value of the reaction quotient at equilibrium is called the **equilibrium constant** K_p of the reaction

$$K_p = (Q_p)_{\text{equilibrium}} = \left(\frac{p_B}{p_A}\right)_{\text{eq}}$$

Then, with $\Delta G_r = 0$ when $Q_p = K_p$, we conclude that

$$RT \ln K_p = -\Delta G^\ominus \qquad (3)^\circ$$

This is a special case of one of the most important equations in chemical thermodynamics: it is the link between tables of thermodynamic data, such as those in Table 2.10, and the chemically important equilibrium constant. We see that when $\Delta G^\ominus > 0$, $K_p < 1$; and so at equilibrium the partial pressure of A exceeds that of B, which means that the reactant A is favoured in the equilibrium. When $\Delta G^\ominus < 0$, $K_p > 1$, and so at equilibrium the partial pressure of B exceeds that of A. Now the product B is favoured in the equilibrium.

The general case of a reaction

The derivation of the relation between G and the equilibrium constant of a general reaction runs in a similar way. We shall develop the special and general cases in parallel and consider the reaction

$$2A + 3B \rightarrow C + 2D$$

This is a special case of the general reaction[1]

$$0 = \sum_J v_J S_J$$

in which $v_A = -2$, $v_B = -3$, $v_C = +1$, $v_D = +2$.

When the reaction advances by $d\xi$, the amounts of reactants and products change as follows:

$$dn_A = -2d\xi \qquad dn_B = -3d\xi \qquad dn_C = +d\xi \qquad dn_D = +2d\xi$$

$$dn_J = v_J \, d\xi \quad \text{in general}$$

The change in the Gibbs function at constant temperature and pressure is

$$dG = \mu_A \, dn_A + \mu_B \, dn_B + \mu_C \, dn_C + \mu_D \, dn_D$$

$$= (-2\mu_A - 3\mu_B + \mu_C + 2\mu_D) \, d\xi$$

$$= \left(\sum_J v_J \mu_J \right) d\xi \quad \text{in general}$$

Therefore, the reaction Gibbs function, the slope of G as ξ changes, is

$$\Delta G_r = \left(\frac{\partial G}{\partial \xi} \right)_{p,T} = -2\mu_A - 3\mu_B + \mu_C + 2\mu_D$$

$$= \sum_J v_J \mu_J \quad \text{in general}$$

We can express the chemical potentials in terms of activities using

$$\mu_J = \mu_J^{\ominus} + RT \ln a_J$$

where we interpret a_J as a fugacity if J is a gas (specifically, $a_J = f_J/p^{\ominus}$), and it is easy to deduce that

$$\Delta G_r = \Delta G^{\ominus} + RT \ln Q \tag{4}$$

where the standard reaction Gibbs function is

$$\Delta G^{\ominus} = -2\mu_A^{\ominus} - 3\mu_B^{\ominus} + \mu_C^{\ominus} + 2\mu_D^{\ominus}$$

$$= \sum_J v_J \mu_J^{\ominus} \quad \text{in general}$$

and

$$Q = \frac{a_C a_D^2}{a_A^2 a_B^3} = \prod_J a_J^{v_J} \quad \text{in general}$$

In this expression, $\prod$ is the sign for multiplication: since the reactants have negative stoichiometric coefficients, they automatically appear as the denominator when the product is written out explicitly.[2]

---

[1] This way of writing reactions was introduced in Section 2.7. Recall that the stoichiometric coefficients of products are positive and those of reactants are negative.

[2] The value of the reaction quotient depends on the choice of standard pressure (e.g. $p^{\ominus} = 1$ bar, as here, or $p^{\ominus} = 1$ atm, as previously). This is why it is sometimes written as $Q^{\ominus}$, the superscript being there to signify that a standard pressure (and in general, a standard state for systems other than gases) has been selected. Similarly, K is written $K^{\ominus}$ because its value also depends on the choice of standard in general. However, adding superscripts to equations already heavily burdened with them makes them look even more complicated, and so we shall not adopt this recommended practice even though it is wise. The reason underlying the recommendation, that the numerical values of Q and K depend on the choice of standard, must be borne in mind.

Example 9.1: *Writing a general reaction quotient*

Use the general formula for Q to write the reaction quotient for

$$4NH_3(g) + 5O_2(g) \rightarrow 4NO(g) + 6H_2O(g)$$

Answer. We write the equation in the form

$$0 = 4NO(g) + 6H_2O(g) - 4NH_3(g) - 5O_2(g)$$

which lets us identify the stoichiometric coefficients as

$$\nu(NO) = +4, \qquad \nu(H_2O) = +6, \qquad \nu(NH_3) = -4, \qquad \nu(O_2) = -5$$

and hence to write

$$Q = \prod_J a_J^{\nu_J} = a(NO)^4 a(H_2O)^6 a(NH_3)^{-4} a(O_2)^{-5}$$

$$= \frac{a(NO)^4 a(H_2O)^6}{a(NH_3)^4 a(O_2)^5}$$

Since all the substances are gases, we replace the activities by fugacities, and obtain

$$Q = \frac{\left(\dfrac{f(NO)}{p^{\ominus}}\right)^4 \left(\dfrac{f(H_2O)}{p^{\ominus}}\right)^6}{\left(\dfrac{f(NH_3)}{p^{\ominus}}\right)^4 \left(\dfrac{f(O_2)}{p^{\ominus}}\right)^5} = \frac{f(NO)^4 f(H_2O)^6}{f(NH_3)^4 f(O_2)^5 p^{\ominus}}$$

Comment. At low pressures, it may be acceptable to replace the fugacities by the partial pressures of the gases.

Exercise. Repeat the question for the reaction

$$2H_2S(g) + 3O_2(g) \rightarrow 2SO_2(g) + 2H_2O(g)$$

$$\left[Q = \frac{f(SO_2)^2 f(H_2O)^2 p^{\ominus}}{f(H_2S)^2 f(O_2)^3}\right]$$

At equilibrium the slope of G is zero: $\Delta G_r = 0$. The activities then have their equilibrium values, and we can write

$$K = (Q)_{equilibrium} = \left(\frac{a_C a_D^2}{a_A^2 a_B^3}\right)_{eq}$$

$$= \left(\prod_J a_J^{\nu_J}\right)_{eq} \quad \text{in general}$$

From now on, we shall not write the 'equilibrium' subscript explicitly, and will rely on the context to make it clear that for K we use equilibrium values and for Q we use specified values. Setting $\Delta G_r = 0$ and replacing Q by K in eqn 4 leads at once to

$$RT \ln K = -\Delta G^{\ominus} \tag{5a}$$

with

$$K = \prod_J a_J^{\nu_J} \tag{5b}$$

This K is called the **thermodynamic equilibrium constant**. We can evaluate the standard reaction Gibbs function in the usual way from tables of standard Gibbs functions of formation at the temperature of the reaction:

$$\Delta G^{\ominus} = \sum_{J} \nu_{J} \Delta G_{f}^{\ominus}(J)$$

In the case of a gas-phase reaction, each activity can be replaced by a fugacity (more precisely, by $f/p^{\ominus}$), and we write

$$K = \prod_{J} \left(\frac{f_{J}}{p^{\ominus}}\right)^{\nu_{J}} \qquad (6a)$$

If we can treat the gases as perfect, we replace the fugacities by the partial pressures, giving

$$K_{p} = \prod_{J} \left(\frac{p_{J}}{p^{\ominus}}\right)^{\nu_{J}} \qquad (6b)^{\circ}$$

Example 9.2: *Calculating an equilibrium constant*

Calculate the equilibrium constant of the reaction

$$N_{2}(g) + 3H_{2}(g) \rightarrow 2NH_{3}(g)$$

at 298 K.

Answer. Identify the stoichiometric coefficients, then use eqns 5 and 6 with data from Table 2.10. The reaction is

$$0 = 2NH_{3}(g) - N_{2}(g) - 3H_{2}(g)$$

so $\nu(NH_{3}) = +2$, $\nu(N_{2}) = -1$, and $\nu(H_{2}) = -3$. Therefore:

$$K_{p} = \left(\frac{p(NH_{3})}{p^{\ominus}}\right)^{2} \left(\frac{p(N_{2})}{p^{\ominus}}\right)^{-1} \left(\frac{p(H_{2})}{p^{\ominus}}\right)^{-3}$$

$$= \frac{p(NH_{3})^{2} p^{\ominus 2}}{p(N_{2}) p(H_{2})^{3}}$$

$$\Delta G^{\ominus} = 2\Delta G_{f}^{\ominus}(NH_{3}) - \Delta G_{f}^{\ominus}(N_{2}) - 3\Delta G_{f}^{\ominus}(H_{2})$$

$$= 2\Delta G_{f}^{\ominus}(NH_{3}) = -2 \times 16.5 \text{ kJ mol}^{-1}$$

Therefore, since $RT = 2.48 \text{ kJ mol}^{-1}$,

$$\ln K_{p} = \frac{33.0 \text{ kJ mol}^{-1}}{2.48 \text{ kJ mol}^{-1}} = 13.3$$

Hence, $K_{p} = 6.0 \times 10^{5}$.

Comment. Note that K_{p} and Q_{p} are dimensionless. This will be the case with all the equilibrium constants in the following pages.

Exercise. Evaluate the equilibrium constant for $N_{2}O_{4}(g) \rightarrow 2NO_{2}(g)$ at 298 K.

[0.15]

The relation between the thermodynamic and the practical equilibrium constants

The only remaining problem is to express the thermodynamic equilibrium constant in terms of the mole fractions x_{J} or molalities m_{J} of the substances.

To do so, we need to know the activity coefficients, and then to use $a_J = \gamma_J x_J$ or $a_J = \gamma_J m_J/m^{\ominus}$ (recalling that the activity coefficients depend on the choice). For example, in the latter case, for an equilibrium of the form $A + B \rightleftharpoons C + D$, where all four species are solutes, we write

$$K = \frac{a_C a_D}{a_A a_B} = \frac{\gamma_C \gamma_D}{\gamma_A \gamma_B} \times \frac{m_C m_D}{m_A m_B} = K_{\gamma} K_m \tag{7}$$

The activity coefficients must be evaluated at the equilibrium composition of the mixture, which may involve a complicated calculation, because the latter is known only if the equilibrium composition is already known. In elementary applications (and to begin the iterative calculation of the concentrations in a real example) the assumption is often made that the activity coefficients are all so close to unity that $K_{\gamma} = 1$. Then we obtain the result widely used in elementary chemistry that $K \approx K_m$, and equilibria are discussed in terms of molalities (or concentrations) themselves. In Chapter 10 we shall see a way of making better estimates of activity coefficients for equilibria involving ions.

The response of equilibria to the conditions

The equilibrium constant of a reaction is unaffected by the presence of a catalyst or an enzyme (a biological catalyst), which merely increases the rate at which equilibrium is attained. However, in industry reactions rarely reach equilibrium, partly on account of the rates at which reactants mix, and under these non-equilibrium conditions catalysts can have some odd effects and may change the composition of the reaction mixture. For example, in the commercially important Fischer–Tropsch synthesis of hydrocarbon fuels from carbon monoxide and hydrogen, the choice of catalyst influences the distribution of chain lengths in the product.

9.3 The response of equilibria to pressure

The equilibrium constant depends on $\Delta G^{\ominus}$, which is defined at a single, standard pressure. It is a constant, independent of the pressure at which the equilibrium is established. The same is therefore true of K, and so K is independent of pressure.[3] Formally we may express this independence as

$$\left(\frac{\partial K}{\partial p}\right)_T = 0 \tag{8}$$

Le Chatelier's principle

The conclusion that K is independent of pressure does not mean that the equilibrium *composition* is independent of the pressure. In the case of the

---

[3] This is not quite true for reactions in solution, for which

$$\left(\frac{\partial \ln K}{\partial p}\right)_T = \frac{-\Delta V^{\ominus}}{RT}$$

where $\Delta V^{\ominus}$ is the standard volume of reaction, the change in volume between the standard states of the reactants and products, which is usually small. For our purposes we shall suppose that $\Delta V^{\ominus} = 0$.

perfect gas equilibrium $A \rightleftharpoons 2B$, for example, increasing the pressure decreases the number of B molecules and increases the number of A molecules, but it does so in such a way that the equilibrium constant remains unchanged. This behaviour is a special case of **Le Chatelier's principle**:

> A system at equilibrium, when subjected to a disturbance, responds in a way that tends to minimize the effect of the disturbance.

The principle implies that when pressure is applied to a system at equilibrium, the reaction will adjust so as to minimize the increase. This it can do by reducing the number of particles in the gas phase, which implies a shift $A \leftarrow 2B$.

The variation of composition with pressure

The response to the pressure can be expressed quantitatively. The calculation also provides a justification of Le Chatelier's principle for this type of disturbance.

First, consider the special case of the $A \rightleftharpoons 2B$ equilibrium. Suppose that there is an amount n of A present initially (and no B). At equilibrium the amount of A is $(1 - \alpha)n$ and the amount of B is $2\alpha n$. It follows that the mole fractions present at equilibrium are

$$x_A = \frac{(1-\alpha)n}{(1-\alpha)n + 2\alpha n} = \frac{1-\alpha}{1+\alpha} \qquad x_B = \frac{2\alpha}{1+\alpha}$$

The equilibrium constant is

$$K_p = \frac{(p_B/p^\ominus)^2}{(p_A/p^\ominus)} = \frac{(x_B p/p^\ominus)^2}{(x_A p/p^\ominus)}$$

$$= \frac{x_B^2}{x_A} \times \frac{p}{p^\ominus}$$

$$= \frac{4\alpha^2}{1-\alpha^2} \times \frac{p}{p^\ominus}$$

which rearranges to

$$\alpha = \left(\frac{K_p}{K_p + 4p/p^\ominus} \right)^{\frac{1}{2}} \tag{9}^\circ$$

This formula shows that even though K_p is independent of pressure, the amounts of A and B *do* depend on pressure (Fig. 9.3). It also shows that as p is increased, α decreases, in accord with Le Chatelier's principle.

We can express the effect of pressure in a general way by considering the general perfect gas equilibrium

$$0 = \sum_J \nu_J S_J \qquad K_p = \prod_J \left(\frac{p_J}{p^\ominus} \right)^{\nu_J}$$

If we replace each partial pressure by the mole fraction and total pressure p

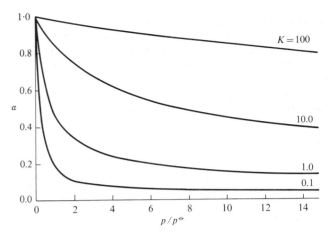

Fig. 9.3 The pressure dependence of the equilibrium composition of an $A(g) \rightleftharpoons 2B(g)$ reaction calculated using eqn 9 for different values of the equilibrium constant. $\alpha = 0$ corresponds to pure A, $\alpha = 1$ corresponds to pure B.

using $p_J = x_J p$, we find

$$K_p = \prod_J \left(\frac{x_J p}{p^\ominus}\right)^{\nu_J} = x_A^{\nu_A} x_B^{\nu_B} \cdots \left(\frac{p}{p^\ominus}\right)^{\nu_A} \left(\frac{p}{p^\ominus}\right)^{\nu_B} \cdots$$

$$= \prod_J x_J^{\nu_J} \times \left(\frac{p}{p^\ominus}\right)^{\nu_A + \nu_B + \cdots}$$

$$= K_x \times \left(\frac{p}{p^\ominus}\right)^{\nu} \qquad \nu = \sum_J \nu_J \qquad (10)^\circ$$

Since the reactants have negative stoichiometric coefficients, ν is the *difference* between the numbers of gas particles in the products and the reactants.

Now, since K_p is independent of pressure, it follows that K_x must be proportional to $p^{-\nu}$. This means that if ν is positive (an increase of gas phase particles) then K_x decreases with increasing pressure, implying a shift towards reactants. If ν is negative, then the opposite is true. Only if $\nu = 0$ (the same numbers of gas phase particles on each side of the equation) is the equilibrium composition independent of pressure (and $K_x = K_p$).

Example 9.3: *Predicting the effect of pressure on an equilibrium*

Predict the effect of a ten-fold increase of pressure on the composition of the ammonia synthesis at equilibrium.

Answer. We can use Le Chatelier's principle to predict the qualitative effect, and eqn 10 to predict the quantitative effect on K_x. Since in the forward direction of the reaction

$$N_2(g) + 3H_2(g) \rightleftharpoons 2NH_3(g)$$

the number of gas molecules decreases (from 4 to 2), Le Chatelier's principle predicts that an increase of pressure will favour the product. Since $\nu = 2 - 1 - 3 = -2$,

$$K_p = K_x \times \left(\frac{p}{p^\ominus}\right)^{-2} \quad \text{implying} \quad K_x = \left(\frac{p}{p^\ominus}\right)^2 \times K_p$$

Since K_p is independent of pressure, K_x will increase 100-fold when the pressure is increased ten-fold.

Comment. The Haber synthesis of ammonia is run at high pressure in order to make use of this result. In precise work it is necessary to base the argument on the pressure independence of the thermodynamic equilibrium constant, and therefore to take note of the pressure dependence of the fugacity coefficients when discussing K_x.

Exercise. Predict the effect of a ten-fold pressure increase on the equilibrium composition of the reaction $3N_2(g) + H_2(g) \rightarrow 2N_3H(g)$.

[100-fold increase in K_x]

9.4 The response of equilibria to temperature

Le Chatelier's principle predicts that an equilibrium will tend to shift in the endothermic direction if the temperature is raised, for then energy is absorbed as heat. Likewise, an equilibrium can be expected to shift in the exothermic direction if the temperature is lowered, for then the reduction in temperature is opposed. These conclusions can be summarized as follows:

Exothermic reactions: increased temperature favours the reactants.
Endothermic reactions: increased temperature favours the products.

We shall now justify these remarks and see how to express the changes quantitatively.

The van't Hoff equation
Since

$$\ln K = \frac{-\Delta G^{\ominus}}{RT}$$

the variation of $\ln K$ with temperature is

$$\frac{d \ln K}{dT} = -\frac{1}{R} \frac{d}{dT} \left(\frac{\Delta G^{\ominus}}{T} \right)$$

The differentials are complete because K and $\Delta G^{\ominus}$ depend only on temperature, not on pressure. To develop this equation we use the Gibbs–Helmholtz equation (Section 5.2)

$$\frac{d}{dT} \left(\frac{\Delta G^{\ominus}}{T} \right) = \frac{-\Delta H^{\ominus}}{T^2}$$

where $\Delta H^{\ominus}$ is the standard reaction enthalpy at the temperature T. Combining the two equations gives the **van't Hoff equation**:

$$\frac{d \ln K}{dT} = \frac{\Delta H^{\ominus}}{RT^2} \qquad (11a)$$

This is exact. Another form of the equation is obtained by noting that as

$$\frac{d}{dT} \left(\frac{1}{T} \right) = \frac{-1}{T^2} \quad \text{or} \quad \frac{dT}{T^2} = -d\left(\frac{1}{T} \right)$$

we can write

$$\frac{d \ln K}{d(1/T)} = \frac{-\Delta H^{\ominus}}{R} \qquad (11b)$$

219

Equation 11a shows that d ln K/d$T < 0$ (and therefore dK/d$T < 0$) for a reaction that is exothermic under standard conditions ($\Delta H^{\ominus} < 0$). A negative slope means that ln K, and therefore K itself, decreases as the temperature rises. Therefore, as asserted above, in the case of an exothermic reaction the equilibrium shifts away from products. The opposite occurs in the case of endothermic reactions.

Some insight into this behaviour can be found in the expression $\Delta G = \Delta H - T\,\Delta S$ written in the form $-\Delta G/T = -\Delta H/T + \Delta S$, which emphasizes that we are dealing with the entropy of the global, isolated system that contains the reaction system. When the reaction is exothermic, $-\Delta H/T$ corresponds to a positive change of entropy of the surroundings of the reaction system, and is a driving force for the formation of products. When the temperature is raised, $-\Delta H/T$ decreases and the increasing entropy of the surroundings is a less potent driving force; as a result, the equilibrium lies less to the right. When the reaction is endothermic, the principal driving force is the increasing entropy of the reaction system. The importance of the unfavourable change of entropy of the surroundings is reduced if the temperature is raised (because then $\Delta H/T$ is smaller), and so the reaction is able to shift towards products.

Example 9.4: *Measuring a reaction enthalpy*

The data below show how the equilibrium constant K_p depends on the temperature for the reaction

$$Ag_2CO_3(s) \rightleftharpoons Ag_2O(s) + CO_2(g)$$

Calculate the reaction enthalpy of the decomposition.

T/K	350	400	450	500
K_p	3.98×10^{-4}	1.41×10^{-2}	1.86×10^{-1}	1.48

Answer. If we plot $-\ln K_p$ against $1/T$, it follows from eqn 11b that the slope will be $\Delta H^{\ominus}/R$. We therefore draw up the following table:

T/K	350	400	450	500
$(10^3\,\text{K})/T$	2.86	2.50	2.22	2.00
$-\ln K_p$	7.83	4.26	1.68	−0.39

These points are plotted in Fig. 9.4. The slope of the graph is $+9.6 \times 10^3$, so

$$\Delta H^{\ominus} = +9.6 \times 10^3\ \text{K} \times R = +80\ \text{kJ mol}^{-1}$$

Comment. This is a non-calorimetric method of determining $\Delta H^{\ominus}$. A drawback is that the reaction enthalpy is temperature dependent, and so the plot is not expected to be perfectly linear. However, the temperature dependence is weak in many cases, and so the plot is reasonably straight. In practice, the method is not very accurate, but it is often the only method available.

Exercise. The equilibrium constant of the reaction $2SO_2(g) + O_2(g) \rightleftharpoons 2SO_3(g)$ is 4.0×10^{24} at 300 K, 2.5×10^{10} at 500 K, and 3.0×10^4 at 700 K. Estimate the reaction enthalpy at 500 K. [-200 kJ mol^{-1}]

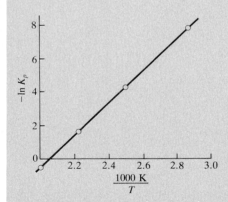

Fig. 9.4 When $-\ln K$ is plotted against $1/T$, a straight line is expected with slope equal to $\Delta H^{\ominus}/R$. This is a non-calorimetric method for the measurement of reaction enthalpies. The data are those for Example 9.4.

The value of K at different temperatures

If we wish to find the value of the equilibrium constant at a temperature T_2 in terms of its value K_1 at another temperature T_1, we would integrate eqn 11b:

$$\int_{\ln K_1}^{\ln K_2} d \ln K = \frac{-1}{R} \int_{1/T_1}^{1/T_2} \Delta H^{\ominus} \, d\left(\frac{1}{T}\right)$$

The integral on the left evaluates easily to $\ln K_2 - \ln K_1$. If we suppose that $\Delta H^{\ominus}$ varies only slightly with temperature over the temperature range of interest, we may take it outside the integral. Then

$$\ln K_2 - \ln K_1 = \frac{-\Delta H^{\ominus}}{R} \left(\frac{1}{T_2} - \frac{1}{T_1}\right) \qquad (12)$$

Example 9.5: *Calculating the equilibrium constant at another temperature*

The equilibrium constant for the synthesis of ammonia at 298 K was calculated in Example 9.2. Estimate its value at 500 K.

Answer. The standard reaction enthalpy is $2\Delta H_f^{\ominus}(NH_3) = -92.2 \, \text{kJ mol}^{-1}$ (Table 2.10), and assumed to be constant over the temperature range in question. Since at 298 K, $K_p = 6.0 \times 10^5$, by eqn 12,

$$\ln K_2 = \ln (6.0 \times 10^5) - \frac{(-92.2 \, \text{kJ mol}^{-1})}{8.314 \, \text{J K}^{-1} \, \text{mol}^{-1}} \times \left(\frac{1}{500 \, \text{K}} - \frac{1}{298 \, \text{K}}\right)$$

$$= -1.73$$

so $K_2 = 0.18$.

Comment. Note the considerable decrease in the value of the equilibrium constant for this exothermic reaction. This is in accord with Le Chatelier's principle, but now expressed quantitatively.

Exercise. The equilibrium constant for $N_2O_4(g) \rightleftharpoons 2NO_2(g)$ was calculated in the Exercise of Example 9.2. Estimate its value at 100°C. [15]

Giauque functions

A more accurate procedure for predicting the effect of temperature on equilibria is to tabulate $\Delta G^{\ominus}(T)$ in such a way that accurate interpolations may be made between the listed values. Then K may be calculated directly from eqn 5a at any temperature in the range listed without having to rely on dubious approximations. It is therefore common to tabulate a quantity related to $\Delta G^{\ominus}$ but which varies more slowly with temperature (so it can be interpolated more accurately). The quantity is sometimes called the **Giauque function** Φ. There are actually two such functions:

$$\Phi_0 = \frac{G_m^{\ominus}(T) - H_m^{\ominus}(0)}{T}$$

$$= \frac{H_m^{\ominus}(T) - H_m^{\ominus}(0)}{T} - S_m^{\ominus}(T) \qquad (13a)$$

Table 9.1. Giauque functions, $-\Phi_0/(\text{J K}^{-1}\text{mol}^{-1})$

	500 K	1000 K	$\{H_m^{\ominus}(\mathbb{F}) - H_m^{\ominus}(0)\}/(\text{kJ mol}^{-1})$
$H_2(g)$	116.9	137.0	8.468
$H_2O(g)$	172.8	196.7	9.908
$N_2(g)$	177.5	197.9	8.669
$NH_3(g)$	176.9	203.5	9.92

and

$$\Phi_T = \Phi_0 - \frac{H_m^{\ominus}(\mathbb{F}) - H_m^{\ominus}(0)}{T} \tag{13b}$$

The difference in enthalpies needed for the latter relation is normally listed in tables (such as in Table 9.1). By combining eqns 13a and 13b it follows that

$$\frac{G_m^{\ominus}(T)}{T} = \Phi_T + \frac{H_m^{\ominus}(\mathbb{F})}{T} \tag{13c}$$

The values of Φ may be compiled either by calculation (using the techniques of statistical thermodynamics described in Part 2) or by measurement based on determining heat capacities down to very low temperatures, as explained in Section 4.4.

The form of the Giauque functions makes them quite easy to use to calculate the equilibrium constant of a reaction at any temperature. As we have seen, the quantity controlling the magnitude of K is $\Delta G^{\ominus}/T$. From eqn 13c this is

$$\frac{\Delta G^{\ominus}}{T} = \Delta\Phi_T + \frac{\Delta H^{\ominus}(\mathbb{F})}{T} \tag{14a}$$

so

$$\ln K = \frac{-\Delta\Phi_T}{R} - \frac{\Delta H^{\ominus}(\mathbb{F})}{RT} \tag{14b}$$

Example 9.6: *Using the Giauque functions*

Calculate the equilibrium constant for the ammonia synthesis at 500 K.

Answer. We use the reaction as written in the preceding examples, and take Φ_0 from Table 9.1 and $\Delta H^{\ominus}(\mathbb{F}) = -92.2 \text{ kJ mol}^{-1}$ from Table 2.10. From Table 9.1,

$$\Delta\Phi_0 = \{2 \times (-176.9) - 3 \times (-116.9) - (-177.5)\} \text{ J K}^{-1}\text{mol}^{-1}$$

$$= +174.4 \text{ J K}^{-1}\text{mol}^{-1}$$

Then, from eqn 13c and Table 9.1

$$\Delta\Phi_T = 174.4 \text{ J K}^{-1}\text{mol}^{-1} - \frac{(2 \times 9.92 - 3 \times 8.468 - 8.669) \text{ kJ mol}^{-1}}{500 \text{ K}}$$

$$= 202.9 \text{ J K}^{-1}\text{mol}^{-1}$$

Consequently, from eqn 14b,

$$\ln K = \frac{-202.9\,\text{J K}^{-1}\,\text{mol}^{-1}}{R} + \frac{92.2\,\text{kJ mol}^{-1}}{R \times 500\,\text{K}}$$

$$= -2.23, \quad \text{so } K = 0.11$$

Exercise. Evaluate the equilibrium constant at 1000 K. $[9.3 \times 10^{-6}]$

Applications to selected systems

In this section we look at some of the conclusions that can be drawn from the existence of equilibrium constants and from the equation $\Delta G^{\ominus} = -RT \ln K$. Since thermodynamic data are often listed for $\mathcal{F}$, equilibrium constants are often calculated at that temperature. Since $R\mathcal{F} = 2.48\,\text{kJ mol}^{-1}$, if we write $\Delta G^{\ominus} = g\,\text{kJ mol}^{-1}$, then $\ln K = -g/2.48$. Furthermore, since $\lg K = (\ln K)/2.303$:

$$\text{At } 25°\text{C:} \quad K = 10^{-g/5.71} \tag{15}$$

Equation 15 gives a quick way of calculating K near room temperature. Note that when $\Delta G^{\ominus}$ (and therefore g) is negative, $K > 1$ and products dominate reactants.

9.5 The extraction of metals from their oxides

Metals can be obtained from their oxides by reduction with carbon if either of the equilibria

$$\text{MO}(s) + \text{C}(s) \rightleftharpoons \text{M}(s) + \text{CO}(g)$$

$$\text{MO}(s) + \tfrac{1}{2}\text{C}(s) \rightleftharpoons \text{M}(s) + \tfrac{1}{2}\text{CO}_2(g)$$

lies to the right. These equilibria can be discussed in terms of the thermodynamic functions for the reactions

$$\begin{aligned}
&\text{(i)} \quad \text{M}(s) + \tfrac{1}{2}\text{O}_2(g) \;\rightarrow\; \text{MO}(s) \\
&\text{(ii)} \quad \tfrac{1}{2}\text{C}(s) + \tfrac{1}{2}\text{O}_2(g) \;\rightarrow\; \tfrac{1}{2}\text{CO}_2(g) \\
&\text{(iii)} \quad \text{C}(s) + \tfrac{1}{2}\text{O}_2(g) \;\rightarrow\; \text{CO}(g) \\
&\text{(iv)} \quad \text{CO}(g) + \tfrac{1}{2}\text{O}_2(g) \rightarrow \text{CO}_2(g)
\end{aligned}$$

The temperature dependences of the standard Gibbs functions of these reactions depend on the reaction entropy through $(\partial \Delta G^{\ominus}/\partial T)_p = -\Delta S^{\ominus}$. Since in reaction (iii) there is a net increase in the amount of gas, the standard reaction entropy is large and positive; therefore, $\Delta G^{\ominus}$ decreases sharply with increasing temperature. In reaction (iv), there is a similar net decrease in the amount of gas, and so $\Delta G^{\ominus}$ increases sharply with increasing temperature. In reaction (ii), the amount of gas is constant, and so the entropy change is small and $\Delta G^{\ominus}$ changes only slightly with temperature. These remarks are summarized in Fig. 9.5, which is called an **Ellingham diagram** (note that $\Delta G^{\ominus}$ decreases upwards!).

The standard Gibbs function of reaction (i) indicates the metal's affinity

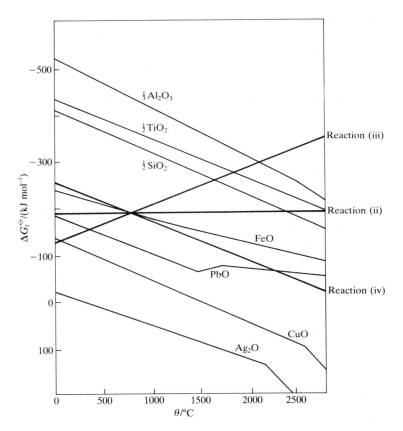

Fig. 9.5 An Ellingham diagram for the discussion of metal ore reduction. Note that $\Delta G^{\ominus}$ is most negative at the top of the diagram.

for oxygen. At room temperature the contribution of the reaction entropy to $\Delta G^{\ominus}$ is dominated by the reaction enthalpy, and so the order of increasing $\Delta G^{\ominus}$ is the same as the order of increasing $\Delta H^{\ominus}$. This gives the order of values on the left of the diagram (Al_2O_3 is most exothermic, Ag_2O is least). The standard reaction entropy is similar for all metals because in each case gaseous oxygen is eliminated and a compact, solid oxide is formed. This implies that the temperature dependence of the standard Gibbs function of oxidation should be similar for all metals, as is shown by the similar slopes of the lines in the diagram. The kinks at high temperatures correspond to the evaporation of the metals; less pronounced kinks occur at the melting temperatures of the metals and the oxides.

Reduction of the oxide depends on the competition of the carbon for the oxygen bound to the metal. The standard Gibbs functions for the reductions can be expressed in terms of the standard Gibbs functions for the reactions above:

$$MO(s) + C(s) \rightarrow M(s) + CO(g) \qquad \Delta G^{\ominus} = \Delta G^{\ominus}(iii) - \Delta G^{\ominus}(i)$$

$$MO(s) + \tfrac{1}{2}C(s) \rightarrow M(s) + \tfrac{1}{2}CO_2(g) \qquad \Delta G^{\ominus} = \Delta G^{\ominus}(ii) - \Delta G^{\ominus}(i)$$

$$MO(s) + CO(g) \rightarrow M(s) + CO_2(g) \qquad \Delta G^{\ominus} = \Delta G^{\ominus}(iv) - \Delta G^{\ominus}(i).$$

The equilibrium lies to the right if $\Delta G^{\ominus} < 0$. This is the case when the line for reaction (i) lies below (is more positive than) the line for one of the carbon reactions (ii) to (iv).

We can predict the success of a reduction at any temperature simply by looking at the diagram: a metal oxide is reduced by any carbon reaction lying above it because the overall reaction then has $\Delta G^{\ominus} < 0$. For example, CuO can be reduced to Cu at any temperature above room temperature. Even in the absence of carbon, Ag_2O decomposes when heated above 200°C because then the standard Gibbs function for reaction (i) becomes positive (and the reverse reaction is then spontaneous). On the other hand, Al_2O_3 is not reduced by carbon until the temperature has been raised to above 2000°C. It is a stable material up to that temperature.

9.6 Acids and bases

One of the most important examples of chemical equilibrium is the one that exists when acids and bases are present in solution. According to the **Brønsted–Lowry classification**:

> An **acid** HA is a proton donor, $HA \rightarrow H^+ + A^-$
> A **base** B is a proton acceptor, $B + H^+ \rightarrow BH^+$

HCl is an acid, because it can donate a proton (that is, an H^+ ion) to another molecule. NH_3 is a base, because it can accept a proton from another molecule and become NH_4^+. These definitions make no mention of the solvent (and apply even if no solvent is present): however, by far the most important medium is aqueous solution, and we confine our attention to that.

Acid–base equilibria in water

An acid HA takes part in the following equilibrium in water:

$$HA(aq) + H_2O(l) \rightleftharpoons A^-(aq) + H_3O^+(aq) \qquad K = \frac{a(H_3O^+)a(A^-)}{a(HA)a(H_2O)}$$

The H_3O^+ ion, the hydrated proton, is called a **hydronium ion**. An example of an acid ionization equilibrium is the one established when HF is present in water:

$$HF(aq) + H_2O(l) \rightleftharpoons F^-(aq) + H_3O^+(aq)$$

The proton acceptor A^- (in this case the F^- ion) remaining after a Brønsted acid has donated a proton is called the **conjugate base** of the acid HA.

For a base B in water, the characteristic equilibrium is

$$B(aq) + H_2O(l) \rightleftharpoons BH^+(aq) + OH^-(aq) \qquad K = \frac{a(BH^+)a(OH^-)}{a(B)a(H_2O)}$$

The proton donor formed when the base accepts a proton and becomes BH^+ is called the **conjugate acid** of the base B. An example of a base in water is NH_3:

$$NH_3(aq) + H_2O(aq) \rightleftharpoons NH_4^+(aq) + OH^-(aq)$$

The conjugate acid of NH_3 is the ammonium ion, NH_4^+. Since BH^+ is a Brønsted acid in its own right, it also participates in a proton transfer equilibrium

$$BH^+(aq) + H_2O(l) \rightleftharpoons B(aq) + H_3O^+(aq) \qquad K = \frac{a(H_3O^+)a(B)}{a(BH^+)a(H_2O)}$$

225

An example is the equilibrium when NH_4^+ ions are present in water:

$$NH_4^+(aq) + H_2O(l) \rightleftharpoons NH_3(aq) + H_3O^+(aq)$$

The importance of this relation is that we can treat bases in the same way as we treat acids, so long as we focus on the properties of the conjugate acid, BH^+ (e.g. on NH_4^+ when dealing with aqueous ammonia).

The general form of the equilibrium for proton transfer equilibria is

$$Acid(aq) + H_2O(l) \rightleftharpoons Base(aq) + H_3O^+(aq) \qquad K = \frac{a(H_3O^+)a(Base)}{a(Acid)a(H_2O)}$$

where 'Acid' may be a substance that we conventionally regard as an acid (such as HF) or the conjugate acid of a base (such as NH_4^+). In the Brønsted–Lowry theory, there is no fundamental distinction between acids and conjugate acids or between bases and conjugate bases: an acid is an acid and a base is a base.

Acidity constants

The equilibrium of an acid in water can be expressed in terms of an equilibrium constant, the **acidity constant** K_a, in which the activity of the solvent water is assumed to be unity (because the solutions are dilute):

$$K_a = \frac{a(H_3O^+)a(Base)}{a(Acid)} \tag{16}$$

For example, for aqueous acetic acid

$$CH_3COOH(aq) + H_2O(l) \rightleftharpoons CH_3CO_2^-(aq) + H_3O^+(aq)$$

$$K_a = \frac{a(H_3O^+)a(CH_3CO_2^-)}{a(CH_3COOH)}$$

For aqueous ammonia

$$NH_4^+(aq) + H_2O(l) \rightleftharpoons NH_3(aq) + H_3O^+(aq)$$

$$K_a = \frac{a(H_3O^+)a(NH_3)}{a(NH_4^+)}$$

Values of acidity constants fall over a wide range (e.g., $K_a = 5.6 \times 10^{-10}$ for NH_4^+ and 0.16 for HIO_3 at 298 K). It is therefore convenient to list them as their logarithms, and we introduce

$$pK_a = -\lg K_a. \tag{17}$$

Since the logarithm is to the base 10, p denotes the negative power of 10. In this notation, $pK_a = 9.25$ for NH_4^+ and 0.8 for HIO_3. It is important to remember that the higher the pK_a of an acid, the smaller its K_a and hence the weaker its proton donating power to water. A list of values is given in Table 9.2.

We shall see that the use of pK_a in place of K_a simplifies the appearance of a number of equations. This simplification stems from the fact that the K_a of the acid ionization equilibrium is related to the Gibbs function of the

Table 9.2. Acidity constants in water at 298 K

	pK_{a1}	pK_{a2}	pK_{a3}
Acetic acid, CH_3COOH	4.75		
Ammonium ion, NH_4^+	9.25		
Carbonic acid, H_2CO_3	6.37	10.25	
Phosphoric acid, H_3PO_4	2.12	7.21	12.67

proton donation reaction by

$$\Delta G^{\ominus} = -RT \ln K_a = 2.303 \, RT \times pK_a$$

Hence, manipulations of pK_a values are in fact manipulations of $\Delta G^{\ominus}$ values in disguise.

Autoionization and pH

Water is **amphiprotic**: it can act as both an acid and a base:

$$\underset{\text{Acid}}{HF(aq)} + \underset{\text{Base}}{H_2O(l)} \rightarrow H_3O^+(aq) + F^-(aq)$$

$$\underset{\text{Acid}}{H_2O(l)} + \underset{\text{Base}}{NH_3(aq)} \rightarrow NH_4^+(aq) + OH^-(aq)$$

A special case of this amphiprotic character occurs with water itself, for one H_2O molecule may act as an acid by donating a proton to another H_2O molecule acting as a base. This is an example of an **autoprotolysis equilibrium**, a proton transfer equilibrium involving a single substance

$$\underset{\text{Acid}}{H_2O(l)} + \underset{\text{Base}}{H_2O(l)} \rightleftharpoons H_3O^+(aq) + OH^+(aq)$$

$$K_w = a(H_3O^+)a(OH^-) \qquad pK_w = -\log K_w \qquad (18)$$

K_w is the **autoprotolysis constant** of water (as before, the ion concentrations are so small that the water activity is almost exactly 1 and does not appear in the denominator). At 25°C, $K_w = 1.008 \times 10^{-14}$ (p$K_w = 14.00$), showing that only a few of the water molecules are ionized. The activities of ions in pure water can therefore be calculated by noting that the concentrations of H_3O^+ and OH^- are equal in pure water, so it is reasonable to suppose that their activities are also equal[4] and therefore to write

$$a(H_3O^+) = K_w^{1/2} = 1.004 \times 10^{-7} \text{ at } 298 \text{ K}$$

Since the activity is so low, so long as no other ions are present in the solution, the concentrations of H_3O^+ and OH^- ions in pure water are each about 1.0×10^{-7} M.

The H_3O^+ ion activity plays a central role in many processes, and its magnitude can vary over a wide range. For instance, in 1 M $HCl(aq)$, $a(H_3O^+) = 0.81$, in pure water it is about 10^{-7}, and in 1 M $NaOH(aq)$ it is about 10^{-14}. The wide span of values is compressed if we use the **pH scale**, where

$$pH = -\lg a(H_3O^+) \qquad (19a)$$

---

[4] The activities of individual ions and the sharing of the departures from ideality between them will be clarified when we consider mean ionic activity coefficients in Section 10.2.

The higher the pH of a solution, the *lower* the hydrogen ion activity. Note that there is nothing mysterious about a negative pH, for it merely corresponds to an activity greater than unity. For example, in 2.00 M HCl(aq), in which the hydrogen ion activity is 2.02, pH $= -0.31$. We shall see that it is also convenient to use pOH, which is defined as

$$pOH = -\lg a(OH^-) \tag{19b}$$

By taking logarithms of the expression for K_w and changing signs throughout, we find

$$pK_w = pH + pOH \tag{20}$$

Hence, if the pH of a solution increases, the pOH must decrease to keep their sum equal to pK_w (which is 14.00 at 25°C). This relation is very useful for calculating the pH of a solution if the concentration of base is given. For example, in an 0.001 M NaOH(aq) solution, in which pOH is approximately 3.0, the pH is

$$pH = pK_w - pOH = 11.0$$

Since in pure water pH $=$ pOH (because the activities of the two ions are equal), at 298 K

$$pH = \tfrac{1}{2}pK_w = 7.00$$

Hence, pH $= 7.00$ corresponds to neutrality at 298 K. (At blood temperature, 37°C, when $pK_w = 13.68$, neutrality corresponds to pH $= 6.84$.) In an acidic aqueous solution, $a(H_3O^+)$ is greater than in pure water, and so then pH < 7; in basic solutions, pH > 7.

The autoprotolysis constant of water also provides a link between the equilibrium constant for proton transfer to a base:

$$B(aq) + H_2O(l) \rightleftharpoons BH^+(aq) + OH^-(aq) \qquad K_b = \frac{a(BH^+)a(OH^-)}{a(B)}$$

and the acidity constant of the conjugate acid so formed:

$$BH^+(aq) + H_2O(l) \rightleftharpoons H_3O^+(aq) + B(aq) \qquad K_a = \frac{a(H_3O^+)a(B)}{a(BH^+)}$$

When we multiply out the two expressions we get

$$K_a K_b = a(H_3O^+)a(OH^-) = K_w$$

and hence by taking logarithms

$$pK_a + pK_b = pK_w \tag{21}$$

It follows that the weaker the base (the larger the pK_b), the stronger its conjugate acid (the smaller its pK_a).

pH calculations

In very dilute aqueous solutions, activities are almost equal to molalities (in the sense $a \approx m/m^\ominus$, where $m^\ominus = 1$ mol kg^{-1}), and molalities are almost exactly the same as molar concentrations (in the sense $m/m^\ominus \approx [X]/M$).

Hence, when it simplifies the discussion, we shall make the approximations of replacing the activities in acidity constants by concentrations and writing

$$K_a \approx \frac{[H_3O^+][\text{Base}]}{[\text{Acid}]}$$

However, this is legitimate only if *all* the ions are present at low concentration, not merely the ions of interest, because (as we shall see quantitatively in Section 10.2) *all* ions contribute to the departures from ideality. Concentrations must be *very* low for this to be permissible, since even for an 0.001 M solution of a 1:1 electrolyte in water at 25°C, activity coefficients are about 0.96, and their neglect introduces an error of approaching 10 per cent into the values of equilibrium constants. When concentrations are not low enough for concentrations to be used, activity coefficients can be found from tables (or estimated from the equations we give in Section 10.2).

Example 9.7: *Calculating the pH of an acid*

Calculate the pH of an 0.20 M HCN(*aq*) solution.

Answer. The equilibrium to consider is

$$HCN(aq) + H_2O(l) \rightleftharpoons CN^-(aq) + H_3O^+(aq) \qquad K_a = \frac{a(H_3O^+)a(CN^-)}{a(HCN)}$$

Each CN^- ion produced is accompanied by the formation of an H_3O^+ ion, so their concentrations are equal, and it is reasonable to write $a(H_3O^+) \approx a(CN^-)$. In a dilute solution, the activity of the HCN molecules is approximately equal to their concentration, so we write $a(HCN) \approx [HCN]/M$. Since K_a is so small (from Table 9.2, $K_a = 4.9 \times 10^{-10}$), we can suppose that very few of the HCN molecules are ionized, and therefore that $[HCN] \approx 0.20$ M. Hence

$$a(H_3O^+) \approx (K_a \times [HCN]/M)^{1/2}$$

and therefore, taking logarithms,

$$pH \approx \tfrac{1}{2}pK_a - \tfrac{1}{2}\lg([HCN]/M)$$

$$\approx \tfrac{1}{2} \times 9.31 - \tfrac{1}{2}\lg 0.20 = 5.0$$

Comment. Since we have made several approximations, the numerical conclusions are not reliable to more than one decimal place, and even that may be over optimistic.

Exercise. Calculate the pH of 0.10 M $NH_3(aq)$. [11.1]

Strong and weak acids

A **strong acid** is a strong proton donor, and its acidity constant is effectively infinite in water (examples are the ionization of HCl in water and the first ionization of H_2SO_4):

$$HCl(aq) + H_2O(l) \rightarrow H_3O^+(aq) + Cl^-(aq) \text{ virtually completely}$$

$$H_2SO_4(aq) + H_2O(l) \rightarrow H_3O^+(aq) + HSO_4^-(aq) \text{ virtually completely}$$

A **strong base** is a strong proton acceptor, and is virtually fully protonated in water. An example is the O^{2-} ion, which does not exist as such in water because it is fully protonated:

$$O^{2-}(aq) + H_2O(l) \rightarrow 2OH^-(aq) \text{ virtually completely}$$

Saying that a base is strong is equivalent to saying that its conjugate acid is a very weak proton donor; hence its acidity constant is very small indeed. Thus, OH^- is an extremely weak proton donor (to H_2O).

A **weak acid** is a Brønsted acid that is incompletely ionized in solution and in particular has $K_a < 1$. An example is acetic acid in water

$$CH_3COOH(aq) + H_2O(l) \rightleftharpoons H_3O^+(aq) + CH_3CO_2^-(aq) \qquad pK_a = 4.75$$

Another example is the HSO_4^- ion in water

$$HSO_4^-(aq) + H_2O(l) \rightleftharpoons H_3O^+(aq) + SO_4^{2-}(aq) \qquad pK_a = 1.92$$

Thus H_2SO_4 is a strong acid in water but its conjugate base HSO_4^- is a weak acid. Since the pK_a of HSO_4^- is smaller than that of CH_3COOH, HSO_4^- is a stronger weak acid than CH_3COOH in water (in the sense that is more fully ionized).

A **weak base** is a Brønsted base that is only partly protonated. An example is NH_3:

$$NH_3(aq) + H_2O(l) \rightleftharpoons NH_4^+(aq) + OH^-(aq)$$

or in terms of proton donation by the conjugate acid NH_4^+:

$$NH_4^+(aq) + H_2O(l) \rightleftharpoons NH_3(aq) + H_3O^+(aq)$$

The fact that both NH_3 and NH_4^+ coexist in the solution implies that NH_4^+ (and, in general, the conjugate acid of any weak base) is a weak acid. For NH_3 in water, for example, $pK_a = 9.25$.

Example 9.8: *Calculating the pH of a solution*

Calculate the pH of 0.15 M $NH_4Cl(aq)$.

Answer. The cation NH_4^+ is a weak acid, so we can expect its presence to lower the pH of the solution. The Cl^- ion is far too weak a base to have any effect. The equilibrium to consider is therefore

$$NH_4^+(aq) + H_2O(l) \rightleftharpoons NH_3(aq) + H_3O^+(aq) \qquad K_a = \frac{[NH_3][H_3O^+]}{[NH_4^+]}$$

Since an NH_3 molecule is produced for each H_3O^+ ion formed, we can write $[NH_3] \approx [H_3O^+]$. The NH_4^+ is such a weak acid that $[NH_4^+]$ is insignificantly different from the nominal concentration of the solution. Therefore, with $K_a = 5.6 \times 10^{-10}$,

$$[H_3O^+] \approx (K_a \times 0.15)^{1/2} = 9.2 \times 10^{-6} \text{ M}$$

Hence,

$$pH = -\lg(9.2 \times 10^{-6}) = 5.04$$

Comment. We have ignored any hydronium ions that might come from the water itself; as we shall see below, this contribution is usually negligible except in pure water.

Exercise. Calculate the pH of 0.15 M $NaCH_3CO_2$. [9.0]

The distinction between weak and strong acids and bases is an illustration of the different types of behaviour shown in Fig. 9.1: strong acids and bases are those for which the Gibbs function minimum lies close to the (ionized) products; weak acids and bases are those for which the minimum lies close to the (non-ionized) reactants.

Polyprotic acids

With **polyprotic acids**, which are substances with more than one acid proton, it is necessary to distinguish the successive acidity constants. For a substance with two acidic protons (such as H_2SO_4), the equilibria are

$$H_2A(aq) + H_2O(l) \rightleftharpoons HA^-(aq) + H_3O^+(aq) \qquad K_{a_1} = \frac{a(H_3O^+)a(HA^-)}{a(H_2A)}$$

in which HA^- is the conjugate base, and

$$HA^-(aq) + H_2O(l) \rightleftharpoons A^{2-}(aq) + H_3O^+(aq) \qquad K_{a_2} = \frac{a(H_3O^+)a(A^{2-})}{a(HA^-)}$$

in which HA^- now acts as the acid and A^{2-} is its conjugate base. Generally $K_{a_2} < K_{a_1}$, typically by three orders of magnitude, because the second proton is more difficult to remove, partly on account of the negative charge on HA^-.

Example 9.9: *Calculating the concentration of carbonate ion in carbonic acid*

Calculate the molar concentration of CO_3^{2-} ions in carbonic acid.

Answer. The CO_3^{2-} ion is produced in the equilibrium

$$HCO_3^-(aq) + H_2O(l) \rightleftharpoons CO_3^{2-}(aq) + H_3O^+(aq) \qquad K_{a_2} = \frac{[CO_3^{2-}][H_3O^+]}{[HCO_3^-]}$$

Hence,

$$[CO_3^{2-}] = \frac{[HCO_3^-]K_{a_2}}{[H_3O^+]}$$

We assume that, because K_{a_2} is small, the second ionization hardly affects the concentration of HCO_3^- ions produced in the first ionization, and that the bulk of the H_3O^+ concentration stems from this ionization, which implies that $[H_3O^+] \approx [HCO_3^-]$. Hence

$$[CO_3^{2-}] \approx K_{a_2}$$

Since from Table 9.2 we know that $pK_{a_2} = 10.25$, it follows that $[CO_3^{2-}] \approx 5.6 \times 10^{-11}$ M.

Exercise. Calculate the concentration of S^{2-} ions in $H_2S(aq)$. [7.1×10^{-15} M]

Acid–base titrations

One application in which acidity constants play an important role is in acid–base titrations, for they can be used to decide the value of the pH that signals the **equivalence point**, the stage when a stoichiometrically equivalent amount of acid has been added to a base.

In the case of a strong acid/strong base titration, the ions present at equivalence (the cations from the strong base, such as Na^+ from NaOH, and the anions from the strong acid, such as Cl^- from HCl) barely affect the pH. The solution consists of these ions, water, and H_3O^+ and OH^- ions from the autoprotolysis of water. Since the autoprotolysis guarantees that pH = pOH, it follows that pH = 7 at the equivalence point.

In a titration of a weak acid (such as CH_3COOH) and strong base (NaOH), at equivalence the solution contains $CH_3CO_2^-$ ions and Na^+ ions together with any ions stemming from autoprotolysis. The presence of the Brønsted base $CH_3CO_2^-$ means that we can expect pH > 7. In a titration of a weak base (such as NH_3) and a strong acid (HCl), the solution contains NH_4^+ ions and Cl^- ions at equivalence. Since Cl^- is only a very weak Brønsted base and NH_4^+ is a weak Brønsted acid (Example 9.8), the solution is acidic and its pH will be less than 7.

The next few paragraphs develop this argument and show how to predict the pH at any stage of an acid–base titration. We shall do this in two stages. In the first, we show how to plot the complete pH curve during a titration. This procedure is well-fitted to a computer, but it gives unwieldy equations. Therefore, we shall also show how to make a series of approximations that are useful in practice.

The complete pH curve

We shall suppose that we are titrating a volume V_A of a weak acid of nominal concentration A_0 with a strong base MOH of concentration B. At some stage in the calculation we have added a volume V_B of base to the analyte (the solution in the flask), so bringing its volume to $V = V_A + V_B$. Since the solution is electrically neutral at all stages, we know that at all times

$$[M^+] + [H_3O^+] = [A^-] + [OH^-]$$

We also know that the total number of A groups (as HA and A^-) is constant and equal to $A_0 \times V_A$. However, their concentration is changing because the volume of the solution is changing. At any stage, therefore,

$$[HA] + [A^-] = A \quad \text{with} \quad A = \frac{A_0 V_A}{V_A + V_B}$$

Similarly, the concentration of M^+ cations is changing, since the amount added at any stage is $B \times V_B$ and the volume the cations occupy is $V_A + V_B$. Therefore, at any stage

$$[M^+] = S \quad \text{with} \quad S = \frac{B V_B}{V_A + V_B}$$

(S denotes the current concentration of the salt that is the product of the reaction.) At all stages of the titration we know that [HA] and $[A^-]$ are related by

$$K_a = \frac{[H_3O^+][A^-]}{[HA]}$$

and that the water autoprotolysis relates $[H_3O^+]$ and $[OH^-]$.

Now we bring all these features together. The conservation of A groups can be expressed in terms of $[A^-]$ alone by making use of the acidity constant. This gives

$$[A^-] = \frac{AK_a}{K_a + [H_3O^+]}$$

Next, we express the electrical neutrality condition in terms of $[H_3O^+]$ alone, since we now have expressions for $[M^+]$ in terms of the volume of added base, for $[A^-]$ in terms of $[H_3O^+]$, and for $[OH^-]$ in terms of K_w. The condition becomes

$$[H_3O^+] + \frac{BV_B}{V_A + V_B} = \frac{A_0V_AK_a}{(V_A + V_B)(K_a + [H_3O^+])} + \frac{K_w}{[H_3O^+]}$$

This is an equation for $[H_3O^+]$ in terms of the volume of base added. Unfortunately, it is a cubic equation in $[H_3O^+]$, which is very awkward to solve. However, it can be solved very readily for V_B as a function of $[H_3O^+]$, which enables us to find the volume of base needed to achieve any pH:

$$\frac{V_B}{V_A} = \frac{(K_a + [H_3O^+])(K_w - [H_3O^+]^2) + K_wA_0[H_3O^+]}{(K_a + [H_3O^+])(B[H_3O^+] + [H_3O^+]^2 - K_w)} \qquad (22)$$

Some of the curves obtained using this formula are shown in Fig. 9.6.

Although eqn 22 is very general, and gives the relation between pH and the volume of base added at all stages of a titration, it is far from transparent and, without a computer, not very easy to use. We shall therefore seek to identify the main features of the curve by making a series of approximations.

The pH in the course of a weak acid—strong base titration

The approximations we shall make are based on the fact that the acid is weak, and therefore that HA is more abundant than any A^- ions in the solution. Furthermore, when HA is present, it provides hydronium ions that greatly outnumber any that stem from the autoprotolysis of water. Finally, when excess base is present, the OH^- ions it provides dominate any that come from the water autoprotolysis.

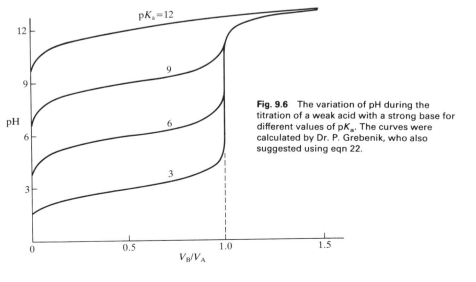

Fig. 9.6 The variation of pH during the titration of a weak acid with a strong base for different values of pK_a. The curves were calculated by Dr. P. Grebenik, who also suggested using eqn 22.

Example 9.10: *Estimating the pH at the start of a titration*

Estimate the pH of a solution of a weak acid of concentration A_0. What is the pH of 0.010 M $HOCl(aq)$?

Answer. We can proceed as in Example 9.7, but we shall make the approximation of working in concentrations. The additional approximations to identify are that the acid is mostly present as HA molecules (because it is weak) but any H_3O^+ ions that it does form swamp any stemming from the autoprotolysis of water. The first of these approximations lets us write $[HA] \approx A_0$. We can also write $[A^-] \approx [H_3O^+]$ because HA is the only important source of both ions. Hence,

$$K_a = \frac{[H_3O^+][A^-]}{[HA]} \approx \frac{[H_3O^+]^2}{A_0}$$

which rearranges into

$$pH = \tfrac{1}{2}pK_a - \tfrac{1}{2}\lg A_0 \tag{23}$$

From Table 9.2, for HOCl $pK_a = 7.43$, so

$$pH = \tfrac{1}{2} \times 7.43 - \tfrac{1}{2}\lg 0.010 = 4.7$$

Exercise. Find an expression for the pH of a weak base of concentration B_0 and calculate the pH of 0.010 M $NH_3(aq)$.

$$[pH = \tfrac{1}{2}pK_w + \tfrac{1}{2}pK_a + \tfrac{1}{2}\lg B_0, \ pH = 10.6]$$

After the addition of some base (but before the equivalence point is reached), the concentration of A^- ions stems almost entirely from the salt that is present, for the weak acid present provides only a few A^- ions. Therefore $[A^-] \approx S$. The number of HA molecules that remain is the original number $A_0 V_A$ less the HA that have been converted to salt by the addition of base, so the concentration of HA is $A' = A - S$. This ignores the small additional loss of HA as a result of its ionization in solution. Hence

$$K_a \approx \frac{[H_3O^+]S}{A'}$$

which rearranges into the **Henderson–Hasselbalch equation**

$$pH = pK_a - \lg \frac{A'}{S} \tag{24}$$

When the concentrations of acid and salt are equal,

$$pH = pK_a \quad \text{when} \quad S = A'$$

Hence the pK_a of the acid can be measured directly from the pH of the mixture. In practice this is done by recording the pH during a titration and then examining the record for the pH half way to the equivalence point.

At the equivalence point the H_3O^+ ions in the solution stem from the Brønsted equilibrium

$$A^-(aq) + H_2O(l) \rightleftharpoons HA(aq) + OH^-(aq) \qquad K_b \approx \frac{[HA][OH^-]}{[A^-]}$$

and the influence of the OH^- ions it produces on the water autoprotolysis equilibrium.

Example 9.11: *Calculating the pH at the equivalence point*

Calculate the pH at the equivalence of the titration of a weak acid with a strong base. Calculate its value for the titration of 25.00 mL of 0.100 M HOCl(aq) with 0.100 M NaOH(aq).

Answer. We must identify the appropriate approximations when the solution is that of the salt alone and only M^+ and A^- ions are nominally present. Since only a small amount of HA is formed in this way, the concentration of A^- ions is almost exactly that of the salt and we can write $[A^-] \approx S$. The OH^- ions that arise from the Brønsted equilibrium written above greatly outnumber those produced by the water autoprotolysis, so $[HA] \approx [OH^-]$. Therefore,

$$K_b = \frac{[OH^-]^2}{S} \quad \text{or} \quad [OH^-] = (SK_b)^{1/2}$$

Therefore,

$$pOH = \tfrac{1}{2}pK_b - \tfrac{1}{2}\lg S$$

Since $pH = pK_w - pOH$ and $pK_b = pK_w - pK_a$, it follows that

$$pH = \tfrac{1}{2}pK_a + \tfrac{1}{2}pK_w + \tfrac{1}{2}\lg S \tag{25a}$$

At the equivalence point of the titration described, the concentration of NaOCl is 0.050 M (because the volume of the solution has increased from 25.00 mL to 50.00 mL), so

$$pH = \tfrac{1}{2} \times 7.43 + \tfrac{1}{2} \times 14.00 + \tfrac{1}{2}\lg 0.050 = 10.1$$

Exercise. Show that the pH at the equivalence point of a titration of a strong acid with a weak base is given by

$$pH = \tfrac{1}{2}pK_a - \tfrac{1}{2}\lg S \tag{25b}$$

where pK_a is that of the conjugate acid of the weak base. Calculate the pH for the equivalence point of a titration of 25.00 mL of 0.200 M NH$_3$(aq) with 0.300 M HCl(aq).   [5.09]

When so much strong base has been added that the titration has been carried well past the equivalence point, the pH is determined by the excess base present. Since

$$[H_3O^+] = \frac{K_w}{[OH^-]}$$

if we write the concentration of excess base as B' we find

$$pH = pK_w + \lg B' \tag{26}$$

In this expression, as in all the preceding ones, the concentrations must take into account the change of volume that occurs as the base is added to the acid sample.

The general form of the pH curve throughout a titration is illustrated in Fig. 9.7. The pH rises slowly from the value given by the 'weak acid alone' formula (eqn 23) following the values given by the Henderson–Hasselbalch equation (eqn 24) until the equivalence point is approached. It then changes

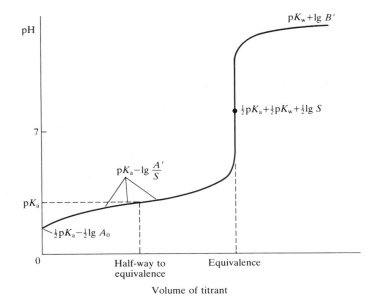

Fig. 9.7 A summary of the regions of the pH curve of the titration of a weak acid with a strong base and the equations used in different regions.

rapidly to and through the value given by the 'salt alone' formula (eqn 25a). It then climbs less rapidly towards the value given by the 'base in excess' formula (eqn 26). The equivalence point can be detected easily by observing where the pH changes rapidly through the value given by the 'salt alone formula' (eqn 25a).

Buffers and indicators

The slow variation of the pH in the vicinity of $S = A'$, when the salt and acid concentrations are equal, is the basis of **buffer action**, the ability to oppose changes in pH when small amounts of acids and bases are added to the solution. The mathematical basis of buffer action is the logarithmic dependence given by the Henderson–Hasselbalch equation (eqn 24), which is quite flat near $pH = pK_a$. The physical basis of buffer action is that the existence of an abundant supply of A^- ions (because a salt is present) can remove any H_3O^+ ions brought by additional acid, and the existence of an abundant supply of HA molecules can supply H_3O^+ ions to react with any base that is added. This stabilization of a dynamic equilibrium against an outside perturbation can be regarded as another example of Le Chatelier's principle.

Example 9.12: *Calculating the pH of a buffer solution*

Calculate the pH of a buffer formed from $0.200\,\mathrm{M}$ $KH_2PO_4(aq)$ and $0.100\,\mathrm{M}$ $K_2HPO_4(aq)$.

Answer. Our aim is to calculate the pH of the solution of an acid and its salt using the Henderson–Hasselbalch equation. To do so, we must identify the acid HA and its conjugate base A^-. In this case the acid is the anion $H_2PO_4^-$ and its conjugate base is the anion HPO_4^{2-}:

$$H_2PO_4^-(aq) + H_2O(l) \rightleftharpoons H_3O^+(aq) + HPO_4^{2-}(aq)$$

The acidity constant we require is therefore pK_{a_2} for H_3PO_4, which from Table 9.2 is 7.21. Then, with $A' = 0.200\,\mathrm{M}$ and $S = 0.100\,\mathrm{M}$, eqn 24 gives the pH of the solution as

$$pH = 7.21 - \lg\frac{0.200}{0.100} = 6.91$$

Hence, the solution should buffer close to $pH = 7$.

Exercise. Calculate the pH of a buffer solution which is $0.100\,\text{M}\,NH_3(aq)$ and $0.200\,\text{M}\,NH_4Cl(aq)$.   [8.95; more realistically: 9]

The rapid change of pH near the equivalence point is the basis of indicator detection. An **acid–base indicator** is normally some large, water-soluble, organic molecule which can exist as acid (HIn) or conjugate base (In$^-$) forms that differ in colour. The two forms are in equilibrium in solution:

$$HIn(aq) + H_2O(l) \rightleftharpoons In^-(aq) + H_3O^+(aq) \qquad K_{In} = \frac{a(In^-)a(H_3O^+)}{a(H_2O)a(HIn)}$$

At the equivalence point, the pH changes sharply through several pH units, so $a(H_3O^+)$ changes through several orders of magnitude. The indicator equilibrium changes so as to accommodate the change of pH, with HIn the dominant species on the acid side of the equivalence point and In$^-$ dominant on the basic side. The accompanying colour change signals the equivalence point of the titration. The colour in fact changes over a range of pH, and the pH half way through its colour change is the **end point** of the titration. With a well-chosen indicator, the end point coincides with the equivalence point of the titration.

Care must be taken to use an indicator that changes colour at the pH appropriate to the type of titration. Thus, in a weak acid–strong base titration, the equivalence point lies at the pH given by eqn 25a, and so an indicator that changes at that pH must be selected. Similarly, in a strong acid–weak base titration, an indicator changing near the pH given by eqn 25b should be used.

9.7 Biological activity: the thermodynamics of ATP

An important biochemical is adenosine triphosphate, ATP (**1**). Its function is to store the energy made available when food is metabolized and then to supply it on demand to a wide variety of processes, including muscular contraction, reproduction, and vision. The essence of ATP's action is its ability to lose its terminal phosphate group by hydrolysis and to form adenosine diphosphate (ADP):

$$ATP(aq) + H_2O(l) \rightarrow ADP(aq) + P_i^-(aq) + H^+(aq)$$

(P_i^- denotes an inorganic phosphate group, such as $H_2PO_4^-$). This reaction is exergonic and can drive an endergonic reaction if suitable enzymes are available.

Biological standard states

The conventional standard state of hydrogen ions (unit activity, $pH = 0$) is not appropriate to normal biological conditions. Therefore, in biochemistry it is common to adopt the **biological standard state**, in which $pH = 7$ (an activity of 10^{-7}, neutral solution). We shall adopt this convention in this

1

section, and label the corresponding standard thermodynamic functions as $G^{\oplus}$, $H^{\oplus}$, and $S^{\oplus}$ (some texts use $X^{\circ\prime}$).

Example 9.13: *Using biological standard states*

Calculate the value of $\Delta G^{\oplus}$ for the reaction

$$\text{NADH}(aq) + \text{H}^+(aq) \rightarrow \text{NAD}^+(aq) + \text{H}_2(g)$$

at 37°C given that $\Delta G^{\ominus} = -21.8\,\text{kJ mol}^{-1}$. NADH is the reduced form of nicotinamide adenine dinucleotide and NAD$^+$ is its oxidized form; the molecules play an important role in the later stages of the respiratory process.

Answer. The Gibbs function for a reaction of the form

$$\text{A} + n\text{H}^+(aq) \rightarrow \text{B} + \text{C}$$

is

$$\Delta G_r = \mu(\text{B}) + \mu(\text{C}) - \mu(\text{A}) - n\mu(\text{H}^+)$$

If all the species other than H$^+$ are in their standard states, this becomes

$$\Delta G_r = \mu^{\ominus}(\text{B}) + \mu^{\ominus}(\text{C}) - \mu^{\ominus}(\text{A}) - n\mu(\text{H}^+).$$

Since

$$\mu(\text{H}^+) = \mu^{\ominus}(\text{H}^+) + RT \ln a(\text{H}^+)$$

$$= \mu^{\ominus}(\text{H}^+) - 2.303\,RT \times \text{pH}$$

this becomes

$$\Delta G_r = \mu^{\ominus}(\text{B}) + \mu^{\ominus}(\text{C}) - \mu^{\ominus}(\text{A}) - n\mu^{\ominus}(\text{H}^+) + 2.30\,nRT \times \text{pH}$$

$$= \Delta G^{\ominus} + 2.30\,nRT \times \text{pH}$$

At pH = 7.00 the reaction Gibbs function has its biological standard value, and so

$$\Delta G^{\oplus} = \Delta G^{\ominus} + 16.1 \times nRT \tag{27}$$

For the reaction as written, it follows that

$$\Delta G^{\oplus} = -21.8\,\text{kJ mol}^{-1} + 16.1 \times 8.314\,\text{JK}^{-1}\,\text{mol}^{-1} \times 310\,\text{K}$$

$$= +19.7\,\text{kJ mol}^{-1}$$

Comment. The biological standard value is the same as the usual standard value when hydrogen ions (or OH$^-$ ions) are not involved in the reaction (since then $n = 0$).

Exercise. Calculate $\Delta G^{\oplus}$ for the hydrolysis of ATP (the reaction written in the text above) given that at 37°C (blood temperature) $\Delta G^{\ominus} = -72\,\text{kJ mol}^{-1}$.

$$[-30\,\text{kJ mol}^{-1}]$$

The standard values for the ATP hydrolysis at 37°C (310 K, blood temperature) are $\Delta G^{\oplus} = -30\,\text{kJ mol}^{-1}$, $\Delta H^{\oplus} = -20\,\text{kJ mol}^{-1}$, and $\Delta S^{\oplus} = +34\,\text{J K}^{-1}\,\text{mol}^{-1}$. The hydrolysis is therefore exergonic ($\Delta G^{\oplus} < 0$) under these conditions, and 30 kJ mol^{-1} is available for driving other reactions. Moreover, because the reaction entropy is large, the reaction Gibbs function is sensitive to temperature. On account of its exergonicity the ADP-phosphate bond has been called a **high-energy phosphate bond**. The name is intended to signify a high tendency to undergo reaction, and should not be confused with 'strong' bond. In fact, even in the biological sense it is not of very 'high energy'. The action of ATP depends on it being intermediate in activity. Thus it acts as a phosphate donor to a number of acceptors (e.g. glucose), but is recharged by more powerful phosphate donors in the respiration cycle.

Anaerobic and aerobic metabolism

The efficiency of some biological processes can be gauged in terms of the value of $\Delta G^{\ominus}$ given above. The energy source of anaerobic cells is glycolysis, and at blood temperature $\Delta G^{\ominus} = -218 \text{ kJ mol}^{-1}$. The standard reaction enthalpy is -120 kJ mol^{-1}, the exergonicity exceeding the exothermicity on account of the large increase of entropy accompanying the fragmentation of the glucose molecule. The glycolysis is coupled to a reaction in which two ADP molecules are converted into two ATP molecules:

$$\text{Glucose} + 2\text{P}_i^- + 2\text{ADP} \rightarrow 2\text{Lactate}^- + 2\text{ATP} + 2\text{H}_2\text{O}$$

The standard reaction Gibbs function is $(-218 \text{ kJ mol}^{-1}) - 2(-30 \text{ kJ mol}^{-1}) = -158 \text{ kJ mol}^{-1}$. The reaction is exergonic, and therefore spontaneous: the metabolism of the food has been used to 'recharge' the ATP.

Metabolism by aerobic respiration is much more efficient. The standard Gibbs function of combustion of glucose is $-2880 \text{ kJ mol}^{-1}$, and so terminating its oxidation at lactic acid is a poor use of resources. In aerobic respiration the oxidation is carried out to completion, and an extremely complex set of reactions preserves as much of the energy released as possible. In the overall reaction, 38 ATP molecules are generated for each glucose molecule consumed. Each mole of ATP extracts 30 kJ from the 2880 kJ supplied by 1 mol of $C_6H_{12}O_6$ (180 g of glucose), and so 1140 kJ has been stored for later use.

Each ATP molecule can be used to drive an endergonic reaction for which $\Delta G^{\ominus}$ does not exceed 30 kJ mol^{-1}. For example, the biosynthesis of sucrose from glucose and fructose can be driven (if a suitable enzyme system is available) because the reaction is endergonic to the extent $\Delta G^{\ominus} = +23 \text{ kJ mol}^{-1}$. The biosynthesis of proteins is strongly endergonic, not only on account of the enthalpy change but also on account of the large decrease in entropy that occurs when many amino acids are assembled into a precisely determined sequence. For instance, the formation of a peptide link is endergonic, with $\Delta G^{\ominus} = +17 \text{ kJ mol}^{-1}$, but the biosynthesis occurs indirectly and is equivalent to the consumption of three ATP molecules for each link. In a moderately small protein like myoglobin, with about 150 peptide links, the construction alone requires 450 ATP molecules, and therefore about 12 mol of glucose molecules for 1 mol of protein molecules.

Further reading

General background

I. M. Klotz and R. M. Rosenberg, *Chemical thermodynamics* (2nd edn). Benjamin, Menlo Park (1972).

P. A. Rock, *Chemical thermodynamics*. University Science Books and Oxford University Press (1983).

G. N. Lewis and M. Randall, *Thermodynamics*. Revised by K. S. Pitzer and L. Brewer, McGraw-Hill, New York (1961).

K. G. Denbigh, *The principles of chemical equilibrium* (4th edn). Cambridge University Press (1981).

P. A. H. Wyatt, *A thermodynamic bypass*: *GOTO log K*. Royal Society of Chemistry, London (1982). This short book explores the simplification that stems from expressing all $\Delta G^{\ominus}$ values as pK.

Applications

D. A. Johnson, *Some thermodynamic aspects of inorganic chemistry* (2nd edn). Cambridge University Press (1982).

W. E. Dasent, *Inorganic energetics*. Cambridge University Press (1982).

I. Klotz, *Energy changes in biochemical reactions*. Academic Press, New York (1967).

A. Lehninger, *Bioenergetics* (2nd edn). Benjamin-Cummings, Menlo Park (1971).

F. M. Harold, *The vital force*: *a study of bioenergetics*. W. H. Freeeman & Co., New York (1986).

P. W. Atkins, *General chemistry*. Scientific American Books, New York (1989). This text describes acid–base equilibria in greater detail.

Exercises

9.1 The equilibrium constant for the isomerization of *cis*-2-butene to *trans*-2-butene is $K = 2.07$ at 400 K. Calculate the standard reaction Gibbs function.

9.2 The standard reaction Gibbs function of the isomerization of *cis*-2-pentene to *trans*-2-pentene at 400 K is $-3.67 \text{ kJ mol}^{-1}$. Calculate the equilibrium constant of the isomerization.

9.3 The standard reaction enthalpy of $Zn(s) + H_2O(g) \rightarrow ZnO(s) + H_2(g)$ is approximately constant at $+224 \text{ kJ mol}^{-1}$ from 920 K up to 1280 K. The standard reaction Gibbs function is $+33 \text{ kJ mol}^{-1}$ at 1280 K. Assuming that both quantities remain constant, estimate the temperature at which the equilibrium constant becomes greater than 1.

9.4 The equilibrium constant of the reaction

$$2C_3H_6(g) \rightleftharpoons C_2H_4(g) + C_4H_8(g)$$

is found to fit the expression

$$\ln K = -1.04 - \frac{1088 \text{ K}}{T} + \frac{1.51 \times 10^5 \text{ K}^2}{T^2}$$

between 300 K to 600 K. Calculate the standard reaction enthalpy and standard reaction entropy at 400 K.

9.5 The standard reaction Gibbs function of the isomerization of borneol to isoborneol in the gas phase at 503 K is $+9.4 \text{ kJ mol}^{-1}$. Calculate the reaction Gibbs function in a mixture consisting of 0.15 mol of borneol and 0.30 mol of isoborneol when the total pressure is 600 Torr.

9.6 The equilibrium pressure of H_2 over solid uranium and uranium hydride at 500 K is 1.04 Torr. Calculate the standard Gibbs function of formation of $UH_3(s)$ at 500 K.

9.7 Calculate the percentage change in the equilibrium constant K_x of the following reactions when the total pressure is increased from 1.0 bar to 2.0 bar at the same temperature:

(a) $H_2CO(g) \rightleftharpoons CO(g) + H_2(g)$

(b) $CH_3OH(g) + NOCl(g) \rightleftharpoons HCl(g) + CH_3NO_2(g)$

9.8 The equilibrium constant for the gas-phase isomerization of borneol ($C_{10}H_{17}OH$) to isoborneol at 503 K is 0.106. A mixture consisting of 7.50 g of borneol and 14.0 g of isoborneol in a 5.0 L container is heated to 503 K and allowed to come to equilibrium. Calculate the mole fractions of the two substances at equilibrium.

9.9 Use the data in Table 2.10 to decide which of the following reactions have $K > 1$ at 298 K.

(a) $HCl(g) + NH_3(g) \rightarrow NH_4Cl(s)$

(b) $2Al_2O_3(s) + 3Si(s) \rightarrow 3SiO_2(s) + 4Al(s)$

(c) $Fe(s) + H_2S(g) \rightarrow FeS(s) + H_2(g)$

(d) $FeS_2(s) + 2H_2(g) \rightarrow Fe(s) + 2H_2S(g)$

(e) $2H_2O_2(l) + H_2S(g) \rightarrow H_2SO_4(l) + 2H_2(g)$

9.10 Which of the equilibria in Exercise 9.9 are favoured (that is, K increases) by a rise in temperature at constant pressure?

9.11 What is the standard enthalpy of a reaction for which the equilibrium constant is (a) doubled, (b) halved when the temperature is increased by 10 K at 298 K?

9.12 Suppose you make an error of 10 per cent in the determination of an equilibrium constant at 298 K, what error does that imply for $\Delta G^{\ominus}$? Conversely, suppose that $\Delta G^{\ominus}$ is in error by 10 per cent: what is the percentage error in K?

9.13 The standard Gibbs function of formation of $NH_3(g)$ is $-16.5 \text{ kJ mol}^{-1}$ at 298 K. What is the reaction Gibbs function when the partial pressure of N_2, H_2, and NH_3 (treated as perfect gases) are 3.0 bar, 1.0 bar, and 4.0 bar respectively? What is the spontaneous direction of the reaction in this case?

9.14 The dissociation vapour pressure of NH_4Cl at 427°C is 608 kPa but at 459°C it has risen to 1115 kPa. Calculate (a) the equilibrium constant, (b) the standard reaction Gibbs function, (c) the standard enthalpy, (d) the standard entropy of dissociation, all at 427°C. Assume that the vapour behaves as a perfect gas and that $\Delta H^{\ominus}$ and $\Delta S^{\ominus}$ are independent of temperature in the range given.

9.15 At 20°C, $pK_w = 14.17$, at 25°C it is 14.00, and at 30°C it is 13.84. Calculate the standard enthalpy of the autoprotolysis reaction at 25°C.

9.16 Estimate the temperature at which (a) $CaCO_3$ decomposes and (b) $CuSO_4 \cdot 5H_2O$ undergoes dehydration.

9.17 Use the Giauque functions to evaluate $\Delta G^{\ominus}$ and K for the reactions (a) $N_2(g) + 3H_2(g) \rightarrow 2NH_3(g)$ at 1000 K and (b) $CO(g) + H_2O(g) \rightarrow H_2(g) + CO_2(g)$ at 500 K and 2000 K.

9.18 At the half-way point in the titration of a weak acid with a strong base the pH was measured as 5.40. What is the acidity constant and the pK_a of the acid? What is the pH of the solution that is 0.015 M in the acid?

9.19 Calculate the pH of (a) $0.10 \text{ M } NH_4Cl(aq)$, (b) $0.10 \text{ M } NaCH_3CO_2$, (c) $0.100 \text{ M } CH_3COOH(aq)$.

9.20 Calculate the pH at the equivalence point of the titration of 25.00 mL of 0.100 M lactic acid with 0.150 M $NaOH(aq)$.

9.21 Sketch the pH curve of a solution containing $0.10 \text{ M } NaCH_3CO_2(aq)$ and a variable amount of acetic acid.

9.22 From the information in Tables 9.2 and 9.3, select suitable buffers for (a) pH = 2.2 and (b) pH = 7.0.

Problems

Numerical problems

9.1 The equilibrium pressure of H_2 over $U(s)$ and $UH_3(s)$ between 450 K and 715 K fits the expression

$$\ln(p/\text{Pa}) = 69.32 - \frac{14.64 \times 10^3 \text{ K}}{T} - 5.65 \ln(T/\text{K})$$

Find an expression for the standard enthalpy of formation of $UH_3(s)$ and from it calculate ΔC_p.

9.2 The standard reaction enthalpy of the decomposition of $CaCl_2 \cdot NH_3(s)$ into $CaCl_2(s)$ and $NH_3(g)$ is nearly constant at $+78 \text{ kJ mol}^{-1}$ between 350 K and 470 K. The equilibrium pressure of the NH_3 in the presence of $CaCl_2 \cdot NH_3$ is 12.8 Torr at 400 K. Find an expression for the temperature dependence of $\Delta G^{\ominus}$ in the same range.

9.3 Calculate the equilibrium constant of the reaction

$$CO(g) + H_2(g) \rightleftharpoons H_2CO(g)$$

given that for the production of liquid formaldehyde $\Delta G^{\ominus} = +28.95 \text{ kJ mol}^{-1}$ at 298 K and that the vapour pressure of formaldehyde is 1500 Torr.

9.4 Acetic acid was evaporated in a container of volume 21.45 cm^3 at 437 K and at an external pressure of 764.3 Torr, which was then sealed. The mass of acid present in the sealed container was 0.0519 g. The experiment was repeated with the same container but at 471 K, and it was found that 0.0380 g of acetic acid was present. Calculate the equilbrium constant for the dimerization of the acid in the vapour and the enthalpy of vaporization.

9.5 Hydrogen and carbon monoxide have been investigated for use in fuel cells, and so their solubilities in molten salts are of interest. Their solubilities in a molten $NaNO_3/KNO_3$ mixture was examined (E. Desimoni and P. G. Zambonin, *J. chem. Soc. Faraday Trans.* I, 2014 (1973)) with the following results:

$$\lg s(H_2) = -5.39 - \frac{768 \text{ K}}{T}$$

$$\lg s(CO) = -5.98 - \frac{980 \text{ K}}{T}$$

where s is the solubility in $\text{mol cm}^{-3} \text{bar}^{-1}$. Calculate the standard molar enthalpies of solution of the two gases at 570 K.

9.6 A sealed container was filled with $0.300 \text{ mol } H_2(g)$, $0.400 \text{ mol } I_2(g)$, and $0.200 \text{ mol } HI(g)$ and total pressure 1.00 bar. Calculate the amounts of the components in the mixture at equilibrium given that $K_p = 870$ for $H_2(g) + I_2(g) \rightleftharpoons 2HI(g)$.

9.7 In the gas phase reaction $2A + B \rightarrow 3C + 2D$ it was found that when 1.00 mol of A, 2.00 mol of B, and 1.00 mol of D were mixed and came to equilibrium at 1.00 bar, the resulting mixture contained 0.90 mol of C. Calculate the equilibrium constant.

9.8 Triethylamine (TEA) and 2,4-dinitrophenol (DNP) form a complex in chlorobenzene, and the equilibrium constant for its formation has been measured over a range of temperatures (K. J. Ivin, J. J. McGarvey, E. L. Simmons, and R. Small, *J. chem. Soc. Faraday Trans.* I, 1016 (1973)):

$\theta/°C$	17.5	25.2	30.0	35.5	39.5	45.0
K	29670 ± 1230	14450 ± 560	9270 ± 70	5870 ± 120	3580 ± 30	2670 ± 70

Calculate the standard enthalpy and entropy of formation of the complex from TEA and DNP at 20°C.

9.9 The dissociation of I_2 can be monitored by measuring the total pressure, and three sets of results are as follows:

T/K	973	1073	1173
$100p/\text{atm}$	6.244	7.500	9.181
$10^4 n_I$	2.4709	2.4555	2.4366

where n_I is the nominal amount of I_2 molecules in the mixture, which occupied 342.68 cm^3. Calculate the equilibrium constants of the dissociation and the standard enthalpy of dissociation at the mean temperature.

9.10 Boron trifluoride acts as a catalyst for the equilibrium between acetaldehyde (CH_3CHO) and paraldehyde, a trimer of acetaldehyde. The partial pressures of the components in the trimerization reaction are too low for the accurate determination of the equilibrium constant by direct measurement, but this can be overcome by ensuring that liquid forms of the two substances are always present (W. K. Busfield, R. M. Lee, and D. Merigold, *J. chem. Soc. Faraday Trans.* I, 936 (1973)). Assume that the gases are perfect, and show that the equilibrium constant for the trimerization can be written

$$K = \frac{p_P(p_A - p)(p_A - p_P)^2}{p_A^3(p - p_P)^3}$$

where p_A is the vapour pressure of acetaldehyde, p_P that of paraldehyde, and p is the total pressure. Use the enthalpy of vaporization of acetaldehyde and paraldehyde, which are respectively 25.6 kJ mol^{-1} and 41.5 kJ mol^{-1} and the following data to calculate the enthalpy and entropy of trimerization of acetaldehyde in the gas phase.

$\theta/°C$	20.0	22.0	26.0	28.0	30.0	32.0	34.0	36.0	38.0	40.0
p/kPa	23.9	27.3	36.5	42.6	49.9	56.9	65.1	74.3	85.0	96.2

You also need to know that the vapour pressures of the two components are given by

$$\ln(p/\text{kPa}) = a - \frac{\Delta H_{vap}}{RT}$$

with $a = 15.1$ for acetaldehyde and $b = 17.2$ for paraldehyde. Given that the boiling points of acetaldehyde and paraldehyde are 294 K and 398 K respectively, calculate the enthalpy and entropy of the trimerization in the liquid phase.

Theoretical problems

9.11 Show that if K_p increases with temperature, then K_γ must decrease.

9.12 Express the equilibrium constant of a gas phase reaction $A + 3B \rightarrow 2C$ in terms of the equilibrium value of the extent of reaction given that initially A and B were present in stoichiometric proportions. Find an expression for ξ as a function of the total pressure p of the reaction mixture and sketch a graph of the expression obtained.

9.13 When light passes through a cell of length l containing an absorbing gas at a pressure p, the absorption is proportional to pl. Consider the equilibrium $N_2O_4 \rightleftharpoons 2NO_2$, with NO_2 the absorbing species. Show that when two cells of lengths l_1 and l_2 are used, and the pressures needed to obtain equal absorptions are p_1 and p_2 respectively, then the equilibrium constant is given by

$$K = \frac{(\rho^2 p_1 - p_2)^2}{\rho(\rho - 1)(p_2 - \rho p_1)p^\ominus}$$

with $\rho = l_1/l_2$. The following data were obtained (R. J. Nordstrum and W. H. Chan, *J. phys. Chem.* **80**, 847 (1976)):

Absorbance	p_1/Torr	p_2/Torr
0.05	1.00	5.47
0.10	2.10	12.00
0.15	3.15	18.65

with $l_1 = 395 \text{ mm}$ and $l_2 = 75 \text{ mm}$. Find the equilibrium constant of the reaction.

9.14 Find an expression for the standard reaction Gibbs function at a temperature T' in terms of its value at another temperature T and the coefficients a, b, and c in the expression for the heat capacity listed in Table 2.16. Evaluate the standard Gibbs function of formation of $H_2O(l)$ at 372 K from its value at 298 K.

Equilibrium electrochemistry

10

Check-list of key ideas

1. The use and definition of the standard thermodynamic functions of formation of ions in solution (Section 10.1).

2. The contributions to the *standard Gibbs function of formation* of ions in solution and the use of the Born equation (eqn 1) to judge trends.

3. The *activities* of ions in solution and the significance of the *mean activity coefficient* (Section 10.2 and eqn 4).

4. The physical basis of the *Debye–Hückel limiting law* (eqn 5) for calculating mean activity coefficients and the definition and significance of *ionic strength* (eqn 6).

5. The use of *half-reactions* to discuss *redox reactions* (Section 10.3) and oxidation and reduction half-reactions at electrodes (Section 10.3).

6. The half-reactions at different types of electrode, including the *gas electrode,* the *insoluble-salt electrode,* and the *redox electrode* (Section 10.3).

7. The classification of *electrochemical cells* and the role of the *salt bridge* (Section 10.4).

8. The *cell reaction* and the *cell potential,* and the link between the *zero-current cell potential* and the *reaction Gibbs function* (eqn 10).

9. The derivation of the *Nernst equation* (eqn 11) for the dependence of the zero-current cell potential on the composition of the cell.

10. The definition of the *standard cell potential* and its relation to the equilibrium constant of the cell reaction (eqn 12).

11. The definition of *standard reduction potential* and its dependence on the composition (Section 10.5).

12. The measurement of standard reduction potentials and activity coefficients (Section 10.5).

13. The formulation and significance of the *electrochemical series* (Section 10.6).

14. The calculation of *solubility products* and of *solubilities* from standard reduction potentials (Section 10.7).

15. The use of electrochemical measurements to determine pH and pK_a (Section 10.8).

16. The principles of *potentiometric titrations* (Section 10.9) and the variation of potential as an oxidizing agent is added to a solution (eqn 16).

17. The electrochemical measurement of the standard Gibbs function, enthalpy, and entropy of reaction from the *temperature dependence* of the zero-current cell potential (Section 10.10).

The thermodynamic properties of electrolyte solutions are discussed in terms of chemical potentials and activities in much the same way as solutions of non-electrolytes. However, ions interact strongly with each other through their electric charges, so deviations from ideality are important even at very low concentrations.

Many reactions of ions involve the transfer of electrons. They can be studied (and utilized) by allowing them to take place in an electrochemical cell, for then the reaction produces an electric current in an external circuit. Measurements like the ones we describe in this chapter lead to a collection of data that are very useful for discussing the characteristics of electrolyte solutions and of equilibria in solution, as we shall see.

In common with the preceding chapters, we concentrate here on the thermodynamics of electrochemical processes. Their kinetic aspects are described in Chapter 30.

The thermodynamic properties of ions in solution

Many of the concepts described in previous chapters carry over without change into the discussion of electrolyte solutions. However, departures from ideality are important except in extremely dilute solutions (less concentrated than about 10^{-3} M), and we must know how to take them into account.

10.1 Thermodynamic functions of formation

We can express the standard enthalpy and Gibbs function of a reaction involving ions in solution using the standard enthalpies and Gibbs functions of formation listed in Table 2.10. We use these functions in exactly the same way as for neutral compounds, as the following example illustrates.

Example 10.1: *Using Gibbs functions of formation*

Calculate the solubility of silver chloride in water at 25°C.

Answer. The 'reaction' is

$$AgCl(s) \rightarrow Ag^+(aq) + Cl^-(aq)$$

and in the saturated solution (at equilibrium)

$$K = a(Ag^+)a(Cl^-)$$

Since one Ag^+ ion is formed for each Cl^- ion, their concentrations are equal, and we can write

$$a(Ag^+) = K^{1/2}$$

Since AgCl is only very sparingly soluble, activities may be replaced by molalities, giving

$$m(Ag^+) \approx K^{1/2} \times m^{\ominus}$$

and the molality of the Ag^+ ions in the saturated solution is equal to the solubility S of the solid:

$$S \approx K^{1/2} \times m^{\ominus}$$

We can calculate the equilibrium constant from

$$RT \ln K = -\Delta G^{\ominus}$$

and obtain $\Delta G^{\ominus}$ from

$$\begin{aligned}
\Delta G^{\ominus} &= \Delta G_f^{\ominus}(Ag^+, aq) + \Delta G_f^{\ominus}(Cl^-, aq) - \Delta G_f^{\ominus}(AgCl, s) \\
&= +77.11 + (-131.23) - (-109.79) \text{ kJ mol}^{-1} \\
&= +55.67 \text{ kJ mol}^{-1}
\end{aligned}$$

Therefore, since $RT = 2.4790 \text{ kJ mol}^{-1}$,

$$\ln K = \frac{-55.67 \text{ kJ mol}^{-1}}{2.4790 \text{ kJ mol}^{-1}} = -22.46$$

so $K = 1.77 \times 10^{-10}$ and $S = 1.33 \times 10^{-5} \text{ mol kg}^{-1}$.

Exercise. Use the data in Table 2.10 to calculate the solubility of Hg_2Cl_2 in water at 25°C. $[3.07 \times 10^{-7} \text{ mol kg}^{-1}]$

The values of $\Delta H_f^{\ominus}$ and $\Delta G_f^{\ominus}$ refer to the formation of solutions of ions from the reference states of the parent elements (Section 4.9). However, a special problem with ions arises from the fact that we cannot form solutions of cations without their accompanying anions. Thus, although the standard enthalpy of an overall reaction such as

$$Ag(s) + \tfrac{1}{2}Cl_2(g) \rightarrow Ag^+(aq) + Cl^-(aq)$$
$$\Delta H^{\ominus} = \Delta H_f^{\ominus}(Ag^+, aq) + \Delta H_f^{\ominus}(Cl^-, aq) = -61.58 \text{ kJ mol}^{-1}$$

is meaningful, the enthalpies of the individual formation reactions

$$Ag(s) - e^- \rightarrow Ag^+(aq) \quad \text{and} \quad \tfrac{1}{2}Cl_2(g) + e^- \rightarrow Cl^-(aq)$$

are not.

The enthalpies of formation of ions

The problem is solved by *defining* one ion, conventionally the hydrogen ion, to have zero enthalpy of formation:

$$\Delta H_f^{\ominus}(H^+, aq) = 0 \text{ at all temperatures}$$

Then in the reaction

$$\tfrac{1}{2}H_2(g) + \tfrac{1}{2}Cl_2(g) \rightarrow H^+(aq) + Cl^-(aq) \qquad \Delta H^{\ominus} = -167.16 \text{ kJ mol}^{-1}$$

we can write

$$\Delta H^{\ominus} = \Delta H_f^{\ominus}(H^+, aq) + \Delta H_f^{\ominus}(Cl^-, aq) = \Delta H_f^{\ominus}(Cl^-, aq)$$

Table 10.1. Standard thermodynamic functions of ions in solution at 25°C

Ion	$\Delta H_f^{\ominus}/(\text{kJ mol}^{-1})$	$S^{\ominus}/(\text{J K}^{-1}\,\text{mol}^{-1})$	$\Delta G_f^{\ominus}/(\text{kJ mol}^{-1})$
Cl^-	−167.2	+56.5	−131.2
Cu^{2+}	+64.8	−99.6	+65.5
H^+	0	0	0
K^+	−252.4	+102.5	−283.3
Na^+	−240.1	+59.0	−261.9
PO_4^{3-}	−1277	−221.8	−1019

and hence identify $\Delta H_f^{\ominus}(Cl^-)$ as $-167.16\,\text{kJ mol}^{-1}$ since the chloride ion is (in terms of the convention) responsible for the entire reaction enthalpy. Then, with $\Delta H_f^{\ominus}(Cl^-, aq)$ established, we can find the value of $\Delta H_f^{\ominus}(Ag^+, aq)$ and any other ion (Tables 10.1 and 2.10).

Example 10.2: *Calculating a standard enthalpy of formation of an ion*

Calculate the enthalpy of formation of $Ag^+(aq)$ from the information given above.

Answer. The reaction and its standard enthalpy are

$$Ag(s) + \tfrac{1}{2}Cl_2(g) \rightarrow Ag^+(aq) + Cl^-(aq)$$
$$\Delta H^{\ominus} = \Delta H_f^{\ominus}(Ag^+, aq) + \Delta H_f^{\ominus}(Cl^-, aq) = -61.58\,\text{kJ mol}^{-1}$$

Therefore,

$$\Delta H_f^{\ominus}(Ag^+, aq) = -61.58\,\text{kJ mol}^{-1} - \Delta H_f^{\ominus}(Cl^-, aq)$$
$$= -61.58 - (-167.16)\,\text{kJ mol}^{-1}$$
$$= +105.58\,\text{kJ mol}^{-1}$$

Exercise. The standard enthalpy of formation of $AgNO_3(aq)$ is -99.4 kJ mol^{-1}. Calculate the standard enthalpy of formation of the nitrate ion in water. $[-205.0\,\text{kJ mol}^{-1}]$

The Gibbs functions of formation of ions

We can use the same procedure to define the standard Gibbs functions of formation and the standard entropies of ions in solution. The standard Gibbs function of formation of the H^+ ion in water is defined as zero:

$$\Delta G_f^{\ominus}(H^+, aq) = 0 \text{ at all temperatures}$$

Then in the reaction

$$\tfrac{1}{2}H_2(g) + \tfrac{1}{2}Cl_2(g) \rightarrow H^+(aq) + Cl^-(aq) \qquad \Delta G^{\ominus} = -131.23\,\text{kJ mol}^{-1}$$

we can write

$$\Delta G^{\ominus} = \Delta G_f^{\ominus}(H^+, aq) + \Delta G_f^{\ominus}(Cl^-, aq) = \Delta G_f^{\ominus}(Cl^-, aq)$$

and hence identify $\Delta G_f^{\ominus}(Cl^-)$ as $-131.23\,\text{kJ mol}^{-1}$ just as we did for its enthalpy of formation. Then, with $\Delta G_f^{\ominus}(Cl^-, aq)$ established, we can find the value of $\Delta G_f^{\ominus}(Ag^+, aq)$ from

$$Ag(s) + \tfrac{1}{2}Cl_2(g) \rightarrow Ag^+(aq) + Cl^-(aq) \qquad \Delta G^{\ominus} = -54.12\,\text{kJ mol}^{-1}$$

which leads to $\Delta G_f^{\ominus}(\text{Ag}^+, aq) = +77.11 \text{ kJ mol}^{-1}$, the value used in Example 10.1. All the values in Tables 10.1 and 2.10 are calculated in the same way. Later in the chapter we see how we measure the reaction Gibbs functions on which they are based.

Contributions to the Gibbs function of formation

We can identify the factors responsible for the magnitude of a Gibbs function of formation of an ion in solution by analysing it in terms of a thermodynamic cycle. As an illustration, we consider the reasons for the difference between the standard Gibbs functions of formation of Cl^- and Br^- in water, which are -131 kJ mol^{-1} and -104 kJ mol^{-1} respectively. We do so by treating their formation as the outcome of the following sequence (with values taken from Table 2.10)[1].

		$\Delta G^{\ominus}/(\text{kJ mol}^{-1})$	
		$X = \text{Cl}$	$X = \text{Br}$
Dissociation of H_2:	$\frac{1}{2}H_2(g) \rightarrow H(g)$	$+203$	$+203$
Ionization of H:	$H(g) \rightarrow H^+(g) + e^-$	$+1318$	$+1318$
Hydration of H^+:	$H^+(g) \rightarrow H^+(aq)$	x	x
Dissociation of X_2:	$\frac{1}{2}X_2(g) \rightarrow X(g)$	$+106$	$+82$
Electron gain by X:	$X(g) + e^- \rightarrow X^-(g)$	-355	-331
Hydration of X^-:	$X^-(g) \rightarrow X^-(aq)$	y	y'
Overall:	$\frac{1}{2}H_2(g) + \frac{1}{2}X_2(g) \rightarrow H^+(aq) + X^-(aq)$	$\Delta G_f^{\ominus}(\text{Cl}^-)$	$\Delta G_f^{\ominus}(\text{Br}^-)$

According to the convention we have adopted, the overall Gibbs function is equal to $\Delta G_f^{\ominus}(X^-, aq)$. Its value is the sum of the Gibbs functions of the individual steps:

$$\Delta G_f^{\ominus}(\text{Cl}^-, aq) = x + y + 1272 \text{ kJ mol}^{-1}$$
$$\Delta G_f^{\ominus}(\text{Br}^-, aq) = x + y' + 1272 \text{ kJ mol}^{-1}$$

(It is coincidental that the number 1272 appears for both ions, since the difference in electron affinities is cancelled by the difference in dissociation enthalpies.) An important point to note is that the value of $\Delta G_f^{\ominus}(X^-, aq)$ is not determined by the properties of X alone but includes contributions from the dissociation, ionization, and hydration of hydrogen.

The difference between the two values is due (in this example) to the difference in the hydration of the ions:

$$\Delta G_f^{\ominus}(\text{Cl}^-, aq) - \Delta G_f^{\ominus}(\text{Br}^-, aq) = y - y'$$

The two unknown quantities y and y', are the Gibbs functions of hydration $\Delta G_H^{\ominus}$ of the ions, and in general their **Gibbs function of solvation** $\Delta G_S^{\ominus}$. The latter is the standard Gibbs function for

$$M^+(g) \rightarrow M^+(\text{solution}) \qquad \Delta G_S^{\ominus}$$

Gibbs functions of solvation may be estimated from an equation derived by Max Born, who identified $\Delta G_S^{\ominus}$ with the electrical work of transferring an ion from a vacuum into the solvent treated as a continuous dielectric of

---

[1] The standard Gibbs functions of formation of the gas phase ions are unknown. We have therefore used their enthalpies of formation, and have assumed that the entropy of ionization of H is largely cancelled by the entropy of electron gain of Cl. Partly for this reason, we have reduced the numbers of significant figures in the calculation from those available in Table 2.10.

10.1 | Equilibrium electrochemistry

Table 10.2. Relative permittivities (dielectric constants) at 25°C

	ε_r
Ammonia	16.9
	22.4 (at −33°C)
Benzene	2.274
Ethanol	24.30
Water	78.54

relative permittivity ε_r (Table 10.2). The resulting **Born equation** is

$$\Delta G_S^{\ominus} = \frac{-z_i^2 e^2 N_A}{8\pi\varepsilon_0 r_i}\left(1 - \frac{1}{\varepsilon_r}\right) \tag{1}$$

where z_i is the charge number of the ion and r_i its radius (N_A is Avogadro's constant). Note that $\Delta G_S^{\ominus} < 0$, and that it is strongly negative for small, highly charged ions in media of high relative permittivity.

Example 10.3: *Accounting for the difference in Gibbs functions of formation*

Account for the difference in the values of $\Delta G_f^{\ominus}$ for Cl^- and Br^- in water at 25°C given their radii as 181 pm and 196 pm respectively.

Answer. We know that Cl^- has the more negative value (by $27\,kJ\,mol^{-1}$), which suggests that the Gibbs function of hydration is greater than for Br^-, which is consistent with the smaller radius of Cl^-. For water at 25°C, when $\varepsilon_r = 78.54$, the Born equation is

$$\Delta G_H^{\ominus} = \frac{-z_i^2}{r_i/pm} \times 6.86 \times 10^4 \,kJ\,mol^{-1}$$

Hence, since $z_i^2 = 1$ in both cases, we find $\Delta G_H^{\ominus}(Cl^-) = -379\,kJ\,mol^{-1}$ and $\Delta G_H^{\ominus}(Br^-) = -350\,kJ\,mol^{-1}$. Therefore

$$\Delta G_f^{\ominus}(Cl^-, aq) - \Delta G_f^{\ominus}(Br^-, aq) = -29\,kJ\,mol^{-1}$$

in good agreement with the experimental difference ($-27\,kJ\,mol^{-1}$).

Comment. The Gibbs function of formation of an ion is a subtle balance of many contributions, as the thermodynamic cycle shows, and it is possible to predict trends only by taking into account all of them.

Exercise. Estimate the value of $\Delta G_f^{\ominus}(I^-, aq)$ from the value for Cl^- given the radius of I^- as 220 ppm and other data in Table 2.10.

[$-67\,kJ\,mol^{-1}$ ($-52\,kJ\,mol^{-1}$ actual)]

The entropies of ions in solution

The entropy of an ion in solution is reported on a scale in which the entropy of the H^+ ions in water is taken as zero:

$$S^{\ominus}(H^+, aq) = 0 \text{ at all temperatures}$$

Some standard values based on this choice are listed in Table 10.1 and Table 2.10. The values are in fact *partial* molar entropies, in the sense described in Section 8.1, because only then can the entropy of a solute depend on the properties of the solvent.

According to this convention, a positive entropy means that an ion has a higher entropy than H^+ in water and a negative entropy means that the ion has a lower entropy than H^+ in water. For instance, the entropy of $Cl^-(aq)$ is $+57\,J\,K^{-1}\,mol^{-1}$ and that of $Mg^{2+}(aq)$ is $-138\,J\,K^{-1}\,mol^{-1}$. Ionic entropies vary as expected on the basis that they are related to the degree to which the ions order the water molecules in their vicinity in the solution. Small, highly charged ions induce local structure in the surrounding water, and the entropy of the solution is decreased more than in the case of large, singly-charged ions. The absolute, Third Law partial molar entropy of the proton in water can be estimated by proposing a model of the structure it induces, and there is some agreement on the value $-21\,\text{J K}^{-1}\,\text{mol}^{-1}$. The negative value indicates that the proton induces order in the solvent.

10.2 Ion activities

For non-electrolytes (Chapter 7) we can make the approximation that solute activities can be replaced by their molalities (in the sense $a \approx m/m^{\ominus}$, with $m^{\ominus} = 1\,\text{mol kg}^{-1}$). However, in ionic solutions, the interactions between ions are so strong that we can make this approximation only in very dilute solutions (less than $10^{-3}\,\text{M}$) and in precise work we must use the activities themselves. For example, we have seen that the thermodynamic equilibrium constant K (the one calculated from $\Delta G^{\ominus}$) is related to the equilibrium constant in terms of molalities K_m (or the corresponding constants in terms of concentrations or mole fractions) by

$$K = K_\gamma K_m$$

(this is eqn 7 of Section 9.2). Therefore, to interpret K correctly, we must know how the activity coefficients depend on the molality of the solution.

The definition of activity

We saw in Section 7.10 that the chemical potential of a solute in a real solution is related to its activity a by

$$\mu = \mu^{\ominus} + RT \ln a$$

where the standard state is a hypothetical solution of molality $m^{\ominus}$ in which the ions are behaving ideally. The activity is related to the molality by

$$a = \frac{\gamma m}{m^{\ominus}}$$

where the activity coefficient γ depends on the composition, molality, and temperature of the solution. (The $m^{\ominus}$ simply cancels the units of the molality m.) As the solution approaches ideality (in the sense of obeying Henry's law) at low molalities, the activity coefficient tends towards 1:

$$\gamma \to 1 \quad \text{and} \quad a \to m/m^{\ominus} \quad \text{as} \quad m \to 0$$

Since all the deviations from ideality are carried in the activity coefficient, the chemical potential can be written

$$\mu = \mu^{\ominus} + RT \ln \frac{m}{m^{\ominus}} + RT \ln \gamma$$

$$= \mu^{\circ} + RT \ln \gamma \tag{2}$$

where μ° is the chemical potential of the ideal dilute solution of the same molality.

Mean activity coefficients

If the chemical potential of a univalent cation M^+ is denoted μ_+ and that of a univalent anion X^- is denoted μ_-, the total Gibbs function of the ions in the electrically neutral solution is the sum of these partial molar quantities.

The Gibbs function of an ideal solution is

$$G^\circ = \mu_+^\circ + \mu_-^\circ$$

However, for a real solution of M^+ and X^- of the same molality,

$$
\begin{aligned}
G &= \mu_+ + \mu_- \\
&= \mu_+^\circ + \mu_-^\circ + RT \ln \gamma_+ + RT \ln \gamma_- \\
&= G^\circ + RT \ln \gamma_+ \gamma_-
\end{aligned}
$$

All the deviations from ideality are contained in the last term.

There is no way of disentangling the product $\gamma_+ \gamma_-$ experimentally and of assigning one part of the non-ideality to the cations and another part to the anions. The best we can do experimentally is to assign responsibility for the non-ideality equally to both kinds of ion. Therefore, for a 1,1-electrolyte, we write

$$\gamma_\pm = (\gamma_+ \gamma_-)^{1/2} \tag{3a}$$

and express the individual chemical potentials of the ions as

$$\mu_+ = \mu_+^\circ + RT \ln \gamma_\pm \qquad \mu_- = \mu_-^\circ + RT \ln \gamma_\pm \tag{3b}$$

The sum of these two chemical potentials is the same as before, but now the non-ideality is shared equally.

This approach can be generalized to the case of a salt M_pX_q that dissolves to give a solution of p cations and q anions. The total Gibbs function of the ions is the sum of the partial molar Gibbs functions:

$$
\begin{aligned}
G &= p\mu_+ + q\mu_- \\
&= G^\circ + pRT \ln \gamma_+ + qRT \ln \gamma_-
\end{aligned}
$$

If we introduce the **mean activity coefficient**

$$\gamma_\pm = (\gamma_+^p \gamma_-^q)^{1/s} \qquad s = p + q \tag{4a}$$

and write the chemical potential of each ion as

$$\mu_i = \mu_i^\circ + RT \ln \gamma_\pm \tag{4b}$$

we get the same expression as above for G when we write

$$G = p\mu_+ + q\mu_- \tag{4c}$$

However, now both types of ion share equal responsibility for the non-ideality.

The Debye–Hückel limiting law

The long range and strength of the coulombic interaction between ions means that it is likely to be primarily responsible for the departures from ideality in ionic solutions and to dominate all the other contributions to non-ideality. This domination is the basis of the **Debye–Hückel theory** of ionic solutions, which was devised by Peter Debye and Erich Hückel in

1923. We give here a qualitative acccount of the theory and its principal conclusions. The calculations are outlined in the *Further information* section at the end of the chapter.

Since oppositely charged ions attract each other, cations and anions are not uniformly distributed in solutions: anions are more likely to be found near cations, and vice versa (Fig. 10.1). Overall the solution is electrically neutral, but near any given ion there is an excess of **counter ions**, the ions of opposite charge. Averaged over time, more counter ions than like ions pass by any given ion, and they come and go in all directions. This time-averaged, spherical haze around a given ion has a net charge equal in magnitude but opposite in sign to that on the central ion, and is called its **ionic atmosphere**. The energy, and therefore the chemical potential, of any given central ion is lowered as a result of its coulombic interaction with its ionic atmosphere. This lowering of energy appears as the difference between the Gibbs function G and the ideal value $G°$ of the solution, and hence can be identified with $RT \ln \gamma_{\pm}$.

The model leads to the result (*Further information*) that at very low concentrations the activity coefficient can be calculated from the **Debye–Hückel limiting law**

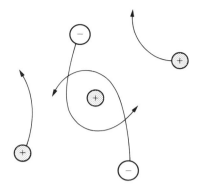

Fig. 10.1 The picture underlying the Debye–Hückel theory is of a tendency for anions to be found around cations, and of cations to be found anions. The average spherical non-zero charge distribution around an ion is called the ionic atmosphere.

$$\lg \gamma_{\pm} = -|z_+ z_-| A I^{1/2} \qquad (5)$$

where $A = 0.509/(\text{mol kg}^{-1})^{1/2}$ for an aqueous solution at 25°C (in general, A depends on the relative permittivity and the temperature) and I is the **ionic strength** of the solution

$$I = \tfrac{1}{2} \sum_i z_i^2 m_i \qquad (6a)$$

z_i is the charge number of an ion i and m_i is its molality. The ionic strength occurs widely wherever ionic solutions are discussed, as we shall see. The sum extends over all the ions present in the solution. In this case of two types of ion at molalities m_+ and m_-

$$I = \tfrac{1}{2}(m_+ z_+^2 + m_- z_-^2) \qquad (6b)$$

I emphasizes the charges of the ions because the charge numbers occur as their squares. Table 10.3 summarizes the relation of ionic strength and molality in an easily usable form.

Table 10.3. Ionic strength and molality, $I = k \times m$

	X^-	X^{2-}	X^{3-}	X^{4-}
M^+	1	3	6	10
M^{2+}	3	4	15	12
M^{3+}	6	15	9	42
M^{4+}	10	12	42	16

For example, the ionic strength of an M_2X_3 solution of molality m, which is understood to give M^{3+} and X^{2-} ions in solution, is 15m.

Example 10.4: *Estimating the mean ionic activity coefficient*

Estimate the mean activity coefficient of 0.00500 mol kg^{-1} KCl(aq) at 25°C.

Answer. For a completely dissociated (1, 1)-electrolyte, $|z_+| = 1$ and $|z_-| = 1$, the ionic strength is

$$I = \tfrac{1}{2}(m_+ + m_-) = m$$

where m is the molality of the solution (and $m_+ = m_- = m$); hence, in this case, from eqn 5,

$$\lg \gamma_{\pm} = \frac{-0.509}{(\text{mol kg}^{-1})^{1/2}} \times (0.00500 \text{ mol kg}^{-1})^{1/2} = -0.0360$$

Table 10.4. Mean activity coefficients in water at 298 K

$m/(\text{mol kg}^{-1})$	KCl	CaCl$_2$
0.001	0.966	0.888
0.01	0.902	0.732
0.1	0.770	0.524
1.0	0.607	0.725

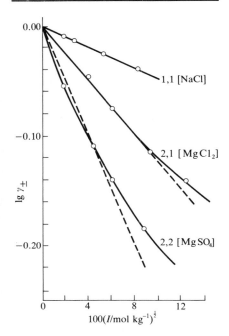

Fig. 10.2 An experimental test of the Debye–Hückel limiting law. Although there are marked deviations for moderate ionic strengths, the limiting slopes as $I \rightarrow 0$ are in good agreement with the theory, and so it can be used for extrapolating data to low molalities.

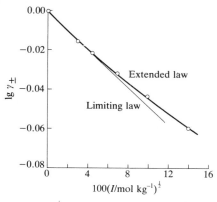

Fig. 10.3 The extended Debye–Hückel law gives agreement with experiment over a wider range of molalities (as shown here for a 1,1-electrolyte), but it fails at higher molalities.

Hence, $\gamma_\pm = 0.920$.

Comment. The experimental value is 0.927 (Table 10.4, Data section).

Exercise. Calculate the ionic strength and the mean activity coefficient of $0.00100 \text{ mol kg}^{-1}$ CaCl$_2(aq)$ at 25°C. $\qquad [0.00300 \text{ mol kg}^{-1}, 0.880]$

The name 'limiting law' is applied to eqn 5 for the same reason as for gases. Ionic solutions of moderate molalities may have activity coefficients that differ from the values given by this expression, yet all solutions are expected to conform in the limit of arbitrarily low molalities. Some experimental values of activity coefficients for salts of various valence types are listed in Table 10.4. Figure 10.2 shows some of these values plotted against $I^{1/2}$ and compares them with the theoretical straight lines calculated from eqn 5. The agreement at low molalities (less than 0.01 to 0.001 mol kg^{-1}, depending on charge type) is impressive, and convincing evidence in support of the model. Nevertheless, the departures from the theoretical curves above these molalities are large, and show that the approximations are valid only at very low concentrations.

When the ionic strength of the solution is too high for the limiting law to be valid, it is found that the activity coefficient may be estimated from the **extended Debye–Hückel law**:

$$\lg \gamma_\pm = -\frac{A\,|z_+ z_-|\,I^{1/2}}{1 + BI^{1/2}} \qquad (7)$$

where B is another constant. Although B can be interpreted as a measure of the closest approach of the ions, it is best regarded as an adjustable parameter. A curve drawn in this way is shown in Fig. 10.3. It is clear that eqn 7 accounts for some activity coefficients over a moderate range of dilute solutions (up to about 0.1 mol kg^{-1}); nevertheless it remains very poor near 1 mol kg^{-1} and no current theory is reliable at this molality.

Electrochemical cells

Now we turn to the investigation of reactions in solution in terms of electrical measurements. The basic piece of apparatus is an **electrochemical cell**. This consists of two **electrodes**—metallic conductors—dipping into an **electrolyte**—an ionic conductor (which may be a solution, a liquid, or a solid). An electrode and its electrolyte comprise an **electrode compartment**. The two electrodes may share the same compartment. If the electrolytes are different, the two compartments may be joined by a **salt bridge**, which is an electrolyte solution that completes the electrical circuit and enables the cell to function.

An electrochemical cell that produces electricity as a result of the spontaneous reaction occuring inside it is called a **galvanic cell**. An electrochemical cell in which a non-spontaneous reaction is driven by an external source of current is called an **electrolytic cell**. We consider only the thermodynamics of galvanic cells in this section. The kinetic problems, which govern the current output of galvanic cells and play an important role in the functioning of electrolytic cells, are considered in Chapter 30.

10.3 Half-reactions and electrodes

A **redox reaction** is a reaction in which there is a transfer of electrons from one substance to another. The **reducing agent** (or 'reductant') is the electron donor and the **oxidizing agent** (or 'oxidant') is the electron acceptor. The electron transfer may be accompanied by other events, such as atom or ion transfer, but the net effect is the change in oxidation number of an element. Examples include combustion

$$2Mg(s) + O_2(g) \rightarrow 2MgO(s)$$

in which Mg is the reducing agent and O_2 the oxidizing agent, reaction with hydrogen

$$CuO(s) + H_2(g) \rightarrow Cu(s) + H_2O(g)$$

in which H_2 is the reducing agent and CuO the oxidizing agent, and the displacement of copper from solution

$$Cu^{2+}(aq) + Zn(s) \rightarrow Cu(s) + Zn^{2+}(aq)$$

in which Cu^{2+} ions are the oxidizing agents and Zn metal is the reducing agent.

Half-reactions

Any redox reaction may be expressed as the sum of two **half-reactions**, which are conceptual reactions showing the loss and gain of electrons. For example, we can express the reduction of Cu^{2+} ions by Zn as the sum of the following two half-reactions:

Reduction of Cu^{2+}: $Cu^{2+}(aq) + 2e^- \rightarrow Cu(s)$
Oxidation of Zn: $Zn(s) \rightarrow Zn^{2+}(aq) + 2e^-$
Overall (sum): $Cu^{2+}(aq) + Zn(s) \rightarrow Cu(s) + Zn^{2+}(aq)$

It is common practice, however, to write all half-reactions as reductions, and then the overall reaction is t. *difference* of the two:

Reduction of Cu^{2+}: $Cu^{2+}(aq) + 2e^- \rightarrow Cu(s)$
Reduction of Zn^{2+}: $Zn^{2+}(aq) + 2e^- \rightarrow Zn(s)$
Overall (difference): $Cu^{2+}(aq) + Zn(s) \rightarrow Cu(s) + Zn^{2+}(aq)$

The reduced and oxidized substances in a half-reaction form a redox **couple**, denoted Ox/Red. Thus, the redox couples mentioned so far are Cu^{2+}/Cu and Zn^{2+}/Zn. In general we shall write a couple as Ox/Red and the corresponding reduction half-reaction as

$$Ox + \nu e^- \rightarrow Red \qquad (8)$$

(This is in some respects the electronic analogue of the relationship between a Brønsted base and its conjugate acid, Section 9.6, in which a proton is transferred.)

We shall often find it useful to express the composition of an electrode compartment in terms of the reaction quotient Q for the half-reaction. This is defined like the reaction quotient for the overall reaction, but ignoring the electrons. Thus, for the two half-reactions given above, we would write

$$Cu^{2+}(aq) + 2e^- \rightarrow Cu(s) \qquad Q = \frac{1}{a(Cu^{2+})}$$

$$Zn^{2+}(aq) + 2e^- \rightarrow Zn(s) \qquad Q = \frac{1}{a(Zn^{2+})}$$

because the pure metal (the standard state of the element) has unit activity.

The overall reaction need not be a redox reaction for it to be expressed in terms of half-reactions. For instance, the expansion of a gas

$$H_2(g, p_i) \rightarrow H_2(g, p_f)$$

is not a redox reaction but it can be expressed as the difference of two reductions:

$$2H^+(aq) + 2e^- \rightarrow H_2(g, p_f)$$
$$2H^+(aq) + 2e^- \rightarrow H_2(g, p_i)$$

The two couples in this case are both H^+/H_2.

Example 10.5: *Expressing a reaction in terms of half-reactions*

Express the dissolution of $AgCl(s)$ as the difference of two reduction half-reactions.

Answer. The overall process is

$$AgCl(s) \rightarrow Ag^+(aq) + Cl^-(aq)$$

If we select as one half-reaction the reduction of AgCl (more precisely, the reduction of the Ag(I) in AgCl to the metal),

$$AgCl(s) + e^- \rightarrow Ag(s) + Cl^-(aq)$$

the second half-reaction, when subtracted from this one, must give the overall reaction. It is therefore the reduction of $Ag^+(aq)$ to Ag metal:

$$Ag^+(aq) + e^- \rightarrow Ag(s)$$

Comment. The dissolution of AgCl is another example of an overall reaction that is not a redox reaction (there is no net change of oxidation number) but which can be expressed as the difference of two reduction half-reactions.

Exercise. Express the formation of H_2O from H_2 and O_2 in acid solution as the difference of two reduction half-reactions.
$$[4H^+(aq) + 4e^- \rightarrow 2H_2(g), \ O_2(g) + 4H^+(aq) + 4e^- \rightarrow 2H_2O(l)]$$

Reactions at electrodes

In an electrochemical cell, one half-reaction takes place in one electrode compartment and the other takes place in the other compartment. In this way, the reduction and oxidation processes responsible for the overall spontaneous reaction are separated in space. As the reaction proceeds, the electrons released in the half-reaction

$$Red_1 \rightarrow Ox_1 + ve^-$$

in one compartment travel through the external circuit, and enter the cell through the other electrode. There they are used to reduce the oxidized member of the couple in that compartment:

$$Ox_2 + ve^- \rightarrow Red_2$$

The electrode where oxidation occurs is called the **anode**; the electrode where reduction occurs is called the **cathode**:

> Anode reaction (oxidation): $Red_1 \rightarrow Ox_1 + ve^-$
>
> Cathode reaction (reduction): $Ox_2 + ve^- \rightarrow Red_2$

In a galvanic cell, the cathode has a higher potential than the anode. This is because the species undergoing reduction, Ox_2, withdraws electrons from its electrode (the cathode, Fig. 10.4), so leaving a relative positive charge on it (corresponding to a high potential). At the anode, oxidation results in the transfer of electrons to the electrode, so giving it a relative negative charge (corresponding to a low potential). In an electrolytic cell, the anode is also the location of oxidation (by definition), but now electrons must be withdrawn from the species in that compartment since it does not occur spontaneously, and at the cathode there must be a supply of electrons to drive the reduction. Therefore, in an electrolytic cell the potential of the anode must be made relatively positive to that of the cathode.

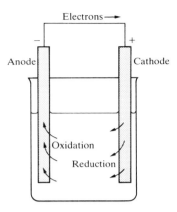

Fig. 10.4 When a spontaneous reaction takes place in a galvanic cell, electrons are deposited in one electrode (the site of oxidation, the anode) and collected from another (the site of reduction, the cathode), and so there is a net flow of current which can be used to do work.

Varieties of electrode

In a **gas electrode** (Fig. 10.5), a gas is in equilibrium with a solution of its ions in the presence of an inert metal. The inert metal (which is often Pt) acts as a source or sink of electrons, but takes no other part in the reaction other than acting as a catalyst for it. One example is the **hydrogen electrode**, in which hydrogen is bubbled through a solution of hydrogen ions and the redox couple is H^+/H_2. This electrode is denoted

$$Pt\,|H_2(g)|\,H^+(aq)$$

where the vertical lines denote interfaces between phases. Note that the electrode description runs in the order $Red\,|\,Ox$, which is opposite to the order in which the couple is denoted.

The hydrogen electrode may be either a cathode or an anode, depending on the other electrode in the cell and the spontaneous direction of the overall reaction. The reaction at the electrode when it is acting as a cathode is

$$2H^+(aq) + 2e^- \rightarrow H_2(g) \qquad Q = \frac{f(H_2)/p^\ominus}{a(H^+)^2}$$

where f is the fugacity (Section 5.3). In elementary work, the fugacity is replaced by the pressure p, and then $p/p^\ominus$ is the numerical value of the pressure in bars.

> **Example 10.6:** *Writing the half-reaction for a gas electrode*
>
> Write the half-reaction quotient for an oxygen reduction in dilute acidic solution.
>
> **Answer.** The reduction of O_2 in acidic solution produces H_2O according to the half-reaction
>
> $$O_2(g) + 4H^+(aq) + 4e^- \rightarrow 2H_2O(l)$$

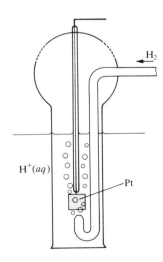

Fig. 10.5 In a gas electrode the gas is bubbled over the inert (but catalytic) metal surface, and the equilibrium is between it and its ions (e.g. between H_2 and H^+ or between Cl_2 and Cl^-).

The reaction quotient for the half-reaction is therefore

$$Q = \frac{a(H_2O)^2 p^{\ominus}}{a(H^+)^4 f(O_2)} \approx \frac{p^{\ominus}}{a(H^+)^4 p(O_2)}$$

where the approximations used in the second step are that the activity of water is 1 (since the solution is dilute) and the oxygen behaves like a perfect gas.

Exercise. Write the half-reaction and the reaction quotient for a chlorine gas electrode. $\quad [Cl_2(g) + 2e^- \rightarrow 2Cl^-(aq), \ Q = a(Cl^-)^2 p^{\ominus}/f(Cl_2)]$

An **insoluble-salt electrode** consists of a metal M covered by a porous layer of insoluble salt MX, the whole immersed in a solution containing X^- ions. The electrode is denoted $M\,|\,MX|\,X^-$, a common example being $Ag\,|\,AgCl|\,Cl^-$. The reduction half-reaction for the electrode is typically

$$MX(s) + e^- \rightarrow M(s) + X^-(aq) \qquad Q = a(X^-)$$

or something similar. This is the case for the silver/silver chloride electrode:

$$AgCl(s) + e^- \rightarrow Ag(s) + Cl^-(aq) \qquad Q = a(Cl^-)$$

Example 10.7: *Writing the half-reaction for an insoluble-salt electrode*

Write the half-reaction and the reaction quotient for the lead/lead sulphate electrode of the lead–acid battery.

Answer. The electrode is $Pb\,|PbSO_4(s)|\,SO_4^{2-}(aq)$, in which Pb(II) is reduced to metallic lead. The reduction half-reaction is

$$PbSO_4(s) + 2e^- \rightarrow Pb(s) + SO_4^{2-}(aq)$$

and the reaction quotient, since the two pure solids have unit activity, is $Q = a(SO_4^{2-})$.

Exercise. Write the half-reaction and the reaction quotient for the calomel electrode $Hg(l)\,|Hg_2Cl_2(s)|\,Cl^-(aq)$.
$$[Hg_2Cl_2(s) + 2e^- \rightarrow 2Hg(l) + 2Cl^-(aq), \ Q = a(Cl^-)^2]$$

All electrodes depend on oxidation and reduction, but the term **oxidation–reduction electrode**, or **redox electrode**, is normally reserved for the case in which a species exists in solution in two oxidation states. The equilibrium is

$$Ox + ve^- = Red \qquad Q = \frac{a(Red)}{a(Ox)}$$

The redox electrode is denoted $M\,|\,Red, Ox$, where M is an inert metal making electrical contact with the solution. An example is

$$Pt\,|\,Fe^{2+}(aq), Fe^{3+}(aq)$$

$$Fe^{3+}(aq) + e^- \rightarrow Fe^{2+}(aq) \qquad Q = \frac{a(Fe^{2+})}{a(Fe^{3+})}$$

10.4 Varieties of cell

The simplest type of cell has a single electrolyte common to both electrodes (Fig. 10.6). In some cases it is necessary to immerse the electrodes in

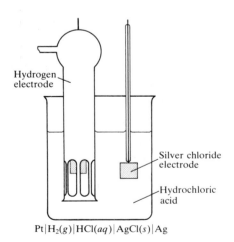

Hydrogen electrode

Silver chloride electrode

Hydrochloric acid

$Pt|H_2(g)|HCl(aq)|AgCl(s)|Ag$

Fig. 10.6 A simple electrochemical cell without a liquid junction. Hydrogen is bubbled over the platinum electrode, which shares a common electrolyte (hydrochloric acid) with the other electrode (a silver/silver chloride electrode).

different electrolytes, as in the **Daniell cell** (Fig. 10.7) in which the redox couple at one electrode is Cu^{2+}/Cu and at the other is Zn^{2+}/Zn. In an **electrolyte concentration cell** (Fig. 10.8), the electrode compartments are identical except for the concentrations of the electrolytes. In an **electrode concentration cell** the electrodes themselves have different concentrations, either because they are gas electrodes operating at different pressures or because they are amalgams (solutions in mercury) with different concentrations.

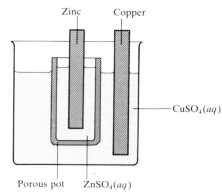

Fig. 10.7 One version of the Daniell cell.

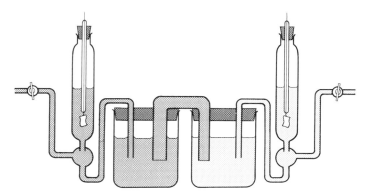

Fig. 10.8 In an electrolyte concentration cell, the two electrolytes (e.g., hydrochloric acid) are at different concentrations, but the electrodes are otherwise the same. The bridge—a concentrated salt solution—completes the electrical circuit by allowing ions to migrate between the two solutions.

Liquid junction potentials

In a cell with two different electrolyte solutions in contact, as in the Daniell cell, there is an additional source of potential difference, the **liquid junction potential** E_{lj}, across the interface of the two electrolytes. Another example of a junction potential is that between different concentrations of hydrochloric acid. At the junction, the mobile H^+ ions diffuse into the more dilute solution. The bulkier Cl^- ions follow, but initially do so more slowly, which results in a potential difference at the junction. However, the potential settles down to a value such that, after that brief initial period, the ions diffuse at the same rates. Electrolyte concentration cells always have a liquid junction; electrode concentration cells do not.

The contribution of the liquid junction to the potential can be reduced (to about 1 to 2 mV) by joining the electrolyte compartments through a salt bridge consisting of a saturated electrolyte solution (usually KCl) in agar jelly (Fig. 10.9). The reason for the success of the salt bridge is that the liquid junction potentials at either end are largely independent of the concentrations of the two dilute solutions, and so nearly cancel.

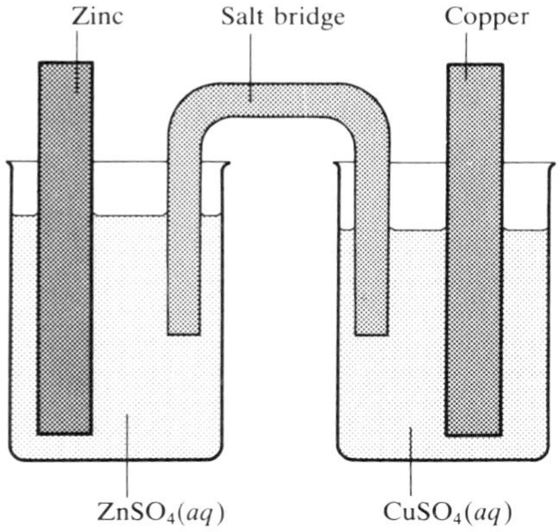

Fig. 10.9 The salt bridge, essentially an inverted U-tube full of concentrated salt solution in a jelly, has two opposing liquid junction potentials which almost cancel.

Notation

In the notation for cells, phase boundaries are denoted by vertical bars. For example, the cell in Fig. 10.6 is denoted

$$Pt\,|H_2(g)|\,HCl(aq)\,|\,AgCl(s)|\,Ag$$

A liquid junction is denoted by $\vdots$, so the cell in Fig. 10.7, is written

$$Zn(s)\,|\,ZnSO_4(aq)\,\vdots\,CuSO_4(aq)\,|\,Cu(s)$$

A double vertical line $\parallel$ denotes an interface for which it is assumed that the junction potential has been eliminated. Thus the cell in Fig. 10.9 is denoted

$$Zn(s) \mid ZnSO_4(aq) \parallel CuSO_4(aq) \mid Cu(s)$$

An electrolyte concentration cell in which the liquid junction potential is assumed to be eliminated is denoted

$$Pt \mid H_2(g) \mid HCl(aq, m_L) \parallel HCl(aq, m_R) \mid H_2(g) \mid Pt$$

The cell reaction

The current produced by a galvanic cell arises from the spontaneous reaction taking place inside it. The **cell reaction** is the reaction in the cell written *on the assumption* that the right-hand electrode is the cathode and hence that the spontaneous reaction is one in which reduction is taking place in the right-hand compartment. Later we see how to predict if the right-hand electrode is in fact the cathode; if it is, the cell reaction is spontaneous as written. If the left-hand electrode turns out to be the cathode, the reverse of the cell reaction is spontaneous and we should rewrite the cell accordingly if we want to preserve the normal convention for chemical equations, that A $\rightarrow$ B is spontaneous.

In order to write the cell reaction corresponding to the cell diagram, we first write the right-hand half-reaction as a reduction (since we have assumed that to be spontaneous). Then we subtract from it the left-hand reduction half-reaction (for, by implication, that electrode is the site of oxidation). Thus, in the cell

$$Zn(s) \mid ZnSO_4(aq) \parallel CuSO_4(aq) \mid Cu(s)$$

the two electrodes and their reduction half-reactions are

$$\begin{aligned} \text{Right:} \quad & Cu^{2+}(aq) + 2e^- \rightarrow Cu(s) \\ \text{Left:} \quad & Zn^{2+}(aq) + 2e^- \rightarrow Zn(s) \end{aligned}$$

Hence, the overall cell reaction is the difference:

$$\text{Overall (R}-\text{L):} \quad Cu^{2+}(aq) + Zn(s) \rightarrow Cu(s) + Zn^{2+}(aq)$$

Had we written the cell in the opposite order,

$$Cu(s) \mid CuSO_4(aq) \parallel ZnSO_4(aq) \mid Zn(s)$$

we would have written

$$\begin{aligned} \text{Right:} \quad & Zn^{2+}(aq) + 2e^- \rightarrow Zn(s) \\ \text{Left:} \quad & Cu^{2+}(aq) + 2e^- \rightarrow Cu(s) \\ \text{Overall (R}-\text{L):} \quad & Cu(s) + Zn^{2+}(aq) \rightarrow Cu^{2+}(aq) + Zn(s) \end{aligned}$$

This overall reaction is the reverse of the one above. If we know (at this stage by experiment) that zinc displaces copper and not vice versa, the first arrangement gives the spontaneous reaction correctly and is the cell diagram we should adopt.

The cell potential

A cell that has not reached chemical equilibrium can do electrical work as the reaction inside it drives electrons through an external circuit. The work that a given transfer of electrons can accomplish depends on the potential difference between the two electrodes. This potential difference is called the **cell potential** and is measured in volts, V. When the cell potential is large, a given number of electrons travelling between the electrodes can do a large amount of electrical work; when the cell potential is small, the same number of electrons can do only a small amount of work. A cell in which the reaction is at equilibrium can do no work, and then the cell potential is zero.

According to the discussion in Section 4.8, we know that the *maximum* electrical work that a system (the cell) can do is given by the value of ΔG, and in particular that for a spontaneous process (in which both ΔG and w are negative)

$$w_{e,max} = \Delta G \text{ at constant temperature and pressure} \qquad (9)$$

Therefore, to make thermodynamic measurements on the cell by measuring the work it can do, we must ensure that it is operating reversibly. Only then is it producing maximum work and only then can eqn 9 be used to relate that work to ΔG. Moreover, we saw in Section 9.1 that the reaction Gibbs function ΔG_r is actually a differential coefficient evaluated at a specified composition of the reaction mixture. Therefore, to measure ΔG_r we must ensure that the cell is operating reversibly at a specific, constant composition. Both these conditions are achieved by measuring the cell potential when it is balanced by an opposing source of potential. The resulting potential difference is called the **zero-current cell potential** E (formerly, the 'electromotive force', or emf, of the cell).

Sources of irreversibility in an electrochemical cell include the slowness with which the reactions take place at the electrodes, and therefore the slowness of the response of these reactions to changes of potential. This suggests one way of testing for reversibility using a supply to balance the cell's potential, and an electronic microammeter. If the electrode reactions are slow, then little current flows when the cell is slightly off balance, and so an indication of irreversibility is that very little current is produced near the balance point. Reversible cells, where the electrode reactions are rapid and responsive to changes of potential, often give appreciable currents when just off-balance.

The relation between E and ΔG

We can derive the relation between the reaction Gibbs function and the zero-current cell potential by considering the change in G when the cell reaction advances by an infinitesimal amount $d\xi$ at some composition. By the same argument as in Section 9.2, G changes by

$$dG = \sum_J \nu_J \mu_J \, d\xi = \Delta G_r \, d\xi$$

where the reaction Gibbs function at the specified composition is

$$\Delta G_r = \left(\frac{\partial G}{\partial \xi} \right)_{p,T} = \sum_J \nu_J \mu_J$$

The maximum work that the reaction can do as it advances by $d\xi$ at constant temperature and pressure is therefore

$$dw_e = \Delta G_r \, d\xi$$

This work is infinitesimal, and the composition of the system is virtually constant when it occurs.

Suppose that when the reaction advances by an amount $d\xi$, an amount $v\,d\xi$ of electrons must travel from the anode to the cathode. The total charge transported between the electrodes when this occurs is $-veN_A\,d\xi$ (because $v\,d\xi$ is the amount of electrons transferred and the charge of an electron is $-e$). The product of e and N_A is called the **Faraday constant** F:

$$F = 96.485 \text{ kC mol}^{-1}$$

F is the magnitude of the charge per mole of electrons. Hence, the total charge transported is $-vF\,d\xi$.

The work done when an infinitesimal charge $-vF\,d\xi$ travels from the anode to the cathode is equal to the product of the charge and the potential difference E (Table 2.1):

$$dw_e = -vF\,d\xi \times E$$

When we equate this work to the expression for dG, the advancement $d\xi$ cancels, and we find

$$-vFE = \Delta G_r \qquad\qquad (10a)$$

Therefore, knowing the reaction Gibbs function at a specified composition, we can state the zero-current cell potential at that composition. Note that a negative reaction Gibbs function, corresponding to a spontaneous cell reaction, corresponds to a positive zero-current cell potential.

Example 10.8: *Estimating a cell potential*

Estimate the cell potential that can be expected for a typical cell reaction.

Answer. We have calculated a number of reaction Gibbs functions in previous chapters and sections of this chapter, and have seen that they are of the order of -100 kJ mol^{-1}. Therefore, taking v as 1, we can expect a typical cell potential to be of the order of

$$E \approx -\frac{(-100 \times 10^3 \text{ J mol}^{-1})}{96 \times 10^3 \text{ C mol}^{-1}} = 1 \text{ V}$$

since $1 \text{ J} = 1 \text{ C V}$.

Exercise. Estimate the zero-current cell potential for a cell in which the reaction is $H^+(aq, 0.020 \text{ mol kg}^{-1}) \rightarrow H^+(aq, 0.010 \text{ mol kg}^{-1})$ at 25°C.
$$[E = (RT/F) \ln 2.0 = 18 \text{ mV}]$$

The Nernst equation

We can go on to relate the zero-current cell potential to the activities of the participants in the cell reaction. We know from eqn 4 in Section 9.2 that the reaction Gibbs function is related to the composition by

$$\Delta G_r = \Delta G^\ominus + RT \ln Q \qquad Q = \prod_J a_J^{v_J}$$

Therefore,

$$E = -\frac{\Delta G^\ominus}{vF} - \frac{RT}{vF} \ln Q$$

The first term on the right of this equation is called the **standard cell potential** and denoted $E^{\ominus}$:

$$-\nu F E^{\ominus} = \Delta G^{\ominus} \qquad (10b)$$

It is the standard reaction Gibbs function of the reaction expressed as a potential (in volts). It follows that

$$E = E^{\ominus} - \frac{RT}{\nu F} \ln Q \qquad (11)$$

This is the **Nernst equation** for the zero-current cell potential at any cell composition. Since $RT/F = 25.7$ mV at 25°C, a practical form of eqn 11 is

$$E = E^{\ominus} - \frac{25.7 \text{ mV}}{\nu} \ln Q$$

Hence, for a reaction in which $\nu = 1$, if Q is increased by a factor of 10, the cell potential increases by 59.2 mV.

Example 10.9: *Using the Nernst equation to calculate a cell potential*

Calculate the zero-current cell potential of a Daniell cell at 25°C in which the molality of copper(II) sulphate is 1.0 mmol kg^{-1} and that of zinc sulphate is 3.0 mmol kg^{-1}.

Answer. The cell reaction (the reduction of Cu^{2+} ions by zinc) is

$$Cu^{2+}(aq) + Zn(s) \rightarrow Cu(s) + Zn^{2+}(aq) \qquad Q = \frac{a(Zn^{2+})}{a(Cu^{2+})} \qquad \nu = 2$$

We need the mean activity coefficients in each of the solutions. Since they are both 2,2-electrolytes, the ionic strength is $4m$ (Table 10.3), and by the Debye–Hückel limiting law (eqn 5),

$$\lg \gamma_{\pm} = \frac{-0.509}{(\text{mol kg}^{-1})^{1/2}} \times 4 \times (4m)^{1/2}$$

which evaluates to 0.74 for the copper(II) sulphate and 0.60 for the zinc sulphate. Hence,

$$Q = \frac{0.60 \times 0.0030 \text{ mol kg}^{-1}}{0.74 \times 0.0010 \text{ mol kg}^{-1}} = 2.43$$

The standard reaction Gibbs function is calculated from the standard Gibbs functions of formation:

$$\Delta G^{\ominus} = \Delta G_f^{\ominus}(Zn^{2+}, aq) - \Delta G_f^{\ominus}(Cu^{2+}, aq)$$
$$= -147.1 - (+65.6) \text{ kJ mol}^{-1} = -212.7 \text{ kJ mol}^{-1}$$

Hence we find

$$E^{\ominus} = \frac{-(-212.7 \text{ kJ mol}^{-1})}{2 \times 96.485 \text{ kC mol}^{-1}} = +1.102 \text{ V}$$

Therefore, from the Nernst equation,

$$E = 1.102 \text{ V} - (25.7 \times 10^{-3} \text{ V}) \ln 2.43 = 1.08 \text{ V}$$

Exercise. Calculate the zero-current potential of the cell $Pt \mid H_2(g, 2.0 \text{ bar}) \mid HCl(aq, 0.10 \text{ mol kg}^{-1}) \mid Hg_2Cl_2(s) \mid Hg(l)$ at 25°C. Take activity coefficients from Table 10.4. [0.400 V]

Cells at equilibrium

A special case of the Nernst equation has great importance in electrochemistry. Suppose the reaction has reached equilibrium; then $Q = K$, where K is the equilibrium constant of the cell reaction. However, a chemical reaction at equilibrium cannot do work, and hence generates zero potential difference between the electrodes of a galvanic cell. Therefore, setting $E = 0$ and $Q = K$ in the Nernst equation gives

$$\ln K = \frac{vFE^{\ominus}}{RT} \tag{12}$$

This very important equation lets us predict equilibrium constants from cell potentials. For example, since the standard potential of the zinc/copper cell is

$$\text{Zn}(s) \mid \text{ZnSO}_4(aq) \parallel \text{CuSO}_4(aq) \mid \text{Cu}(s) \qquad E^{\ominus} = +1.10 \text{ V}$$

the equilibrium constant for the cell reaction (for which $v = 2$) is

$$\text{Cu}^{2+}(aq) + \text{Zn}(s) \rightarrow \text{Cu}(s) + \text{Zn}^{2+}(aq) \qquad K = \frac{a(\text{Zn}^{2+})}{a(\text{Cu}^{2+})} = 1.5 \times 10^{37}$$

Hence, the displacement of copper by zinc goes virtually to completion.

Concentration cells

We can use the same procedure to derive an expression for the potential of an electrolyte concentration cell. Consider the cell

$$\text{M} \mid \text{M}^+(aq, \text{L}) \parallel \text{M}^+(aq, \text{R}) \mid \text{M}$$

where the solutions L and R have different molalities. The cell reaction is

$$\text{M}^+(aq, \text{R}) \rightarrow \text{M}^+(aq, \text{L}) \qquad Q = \frac{a_\text{L}}{a_\text{R}} \qquad v = 1$$

The standard cell potential is zero, because the cell cannot drive a current through a circuit when the two electrode compartments are identical. Therefore, the cell potential is

$$E = \frac{-RT}{F} \ln \frac{a_\text{L}}{a_\text{R}} \approx \frac{-RT}{F} \ln \frac{m_\text{L}}{m_\text{R}}$$

If R is the more concentrated solution, $E > 0$. Physically, this arises because positive ions tend to be reduced, so withdrawing electrons from the electrode, and this process is dominant in the right-hand electrode compartment.

One important example of a membrane system that resembles this description is the biological cell wall, which has different permeabilities for K^+ and Na^+ ions. The concentration of K^+ inside the cell is about 20 to 30 times that on the outside, and is maintained at the level by a specific pumping operation fuelled by ATP and governed by enzymes. If the system is approximately at equilibrium, the potential difference between the two sides is predicted to be

$$E \approx 25.7 \text{ mV} \ln \tfrac{1}{20} = -77 \text{ mV}$$

This accords quite well with the measured value.

This potential difference plays a particularly interesting role in the transmission of nerve impulses. K^+ and Na^+ pumps occur throughout the nervous system, and when the nerve is inactive there is a high K^+ concentration inside the cells and a high Na^+ concentration outside. The potential difference across the cell wall is about $-70\,mV$. When the cell wall is subjected to a pulse of about $20\,mV$, the structure of the membrane adjusts and it becomes permeable to Na^+. This causes a decrease in membrane potential as the Na^+ ions flood into the interior of the cell. The change in potential difference triggers the adjacent part of the cell wall, and the pulse of collapsing potential passes along the nerve. Behind the pulse the sodium and potassium pumps restore the concentration difference ready for the next pulse.

10.5 Reduction potentials

A galvanic cell consists of the conjunction of two electrodes, and each one can be considered as making a characteristic contribution to the cell potential. Although it is not possible to measure the contribution of a single electrode, we can define one as having a particular value and then assign values to others on that basis. The specially selected electrode is the **standard hydrogen electrode** (SHE), which is assigned the value zero:

$$Pt \mid H_2(g) \mid H^+(aq) \quad E^\ominus = 0 \text{ at all temperatures}$$

The **standard reduction potential** $E^\ominus$ of another couple is then assigned by constructing a cell in which it is the right-hand electrode and the standard hydrogen electrode is the left-hand electrode. For example, the standard reduction potential of the Ag^+/Ag couple is the standard potential of the following cell:

$$Pt \mid H_2(g) \mid H^+(aq) \parallel Ag^+(aq) \mid Ag(s)$$
$$E^\ominus(Ag^+/Ag) = E^\ominus = +0.80 \text{ V}$$

and the standard potential of the $AgCl/Ag, Cl^-$ couple is the standard potential of the cell at some specified temperature (typically 298 K):

$$Pt \mid H_2(g) \mid H^+(aq) \parallel Cl^-(aq) \mid AgCl(s) \mid Ag(s)$$
$$E^\ominus(AgCl/Ag, Cl^-) = E^\ominus = +0.22 \text{ V}$$

The standard potential of a cell in terms of reduction potentials

We can calculate the standard potential of a cell formed from any two electrodes by taking the difference of their standard reduction potentials. This recipe follows from the fact that a cell such as

$$Ag(s) \mid Ag^+(aq) \parallel Cl^-(aq) \mid AgCl(s) \mid Ag(s)$$

is equivalent to two cells joined back-to-back:

$$Ag(s) \mid Ag^+(aq) \parallel H^+(aq) \mid H_2(g) \mid Pt \text{---}$$
$$\text{---} Pt \mid H_2(g) \mid H^+(aq) \parallel Cl^-(aq) \mid AgCl(s) \mid Ag(s)$$

The overall potential of this composite cell, and therefore of the cell of interest, is

$$E^\ominus = E^\ominus(AgCl/Ag, Cl^-) - E^\ominus(Ag^+/Ag) = -0.58 \text{ V}$$

Table 10.5. Standard reduction potentials at 25°C

Couple	$E^{\ominus}/V$
$Na^+(aq) + e^- \rightarrow Na(s)$	-2.71
$Zn^{2+}(aq) + 2e^- \rightarrow Zn(s)$	-0.76
$2H^+(aq) + 2e^- \rightarrow H_2(g)$	0
$AgCl(s) + e^- \rightarrow Ag^+(aq) + Cl^-(aq)$	$+0.22$
$Cu^{2+}(aq) + 2e^- \rightarrow Cu(s)$	$+0.34$
$Ce^{4+}(aq) + e^- \rightarrow Ce^{3+}(aq)$	$+1.61$

The standard reduction potentials in Table 10.5 can all be used in the same way, and the standard cell potential is the difference right–left of the corresponding standard reduction potentials. Since $\Delta G^{\ominus} = -\nu F E^{\ominus}$, it then follows that if the result gives $E^{\ominus} > 0$, then the corresponding cell reaction is spontaneous in the direction written (in the sense that $K > 1$).

Example 10.10: *Identifying the spontaneous direction of a reaction*

One of the reactions important in corrosion in an acidic environment is

$$Fe(s) + 2H^+(aq) + \tfrac{1}{2}O_2(g) \rightarrow Fe^{2+}(aq) + H_2O(l)$$

Does the equilibrium constant favour the formation of $Fe^{2+}(aq)$?

Answer. The two reduction half-reactions are

(a) $Fe^{2+}(aq) + 2e^- \rightarrow Fe(s)$ $E^{\ominus} = -0.44$ V
(b) $2H^+(aq) + \tfrac{1}{2}O_2(g) + 2e^- \rightarrow H_2O(l)$ $E^{\ominus} = +1.23$ V

The difference (b) − (a) is

$$Fe(s) + 2H^+(aq) + \tfrac{1}{2}O_2(g) \rightarrow Fe^{2+}(aq) + H_2O(l) \quad E^{\ominus} = +1.67 \text{ V}$$

Therefore, since $E^{\ominus} > 0$, $\Delta G^{\ominus} < 0$ and the reaction has $K > 1$, favouring products.

Exercise. Can zinc displace copper from solution when the ions are at unit activity?   [Yes, $Zn(s) + Cu^{2+}(aq) \rightarrow Zn^{2+}(aq) + Cu(s)$]

Example 10.11: *Calculating an equilibrium constant*

Calculate the equilibrium constant for the disproportionation $2Cu^+(aq) \rightarrow Cu(s) + Cu^{2+}(aq)$ at 298 K.

Answer. The equilibrium constant of the reaction is

$$K = \frac{a(Cu^{2+})}{a(Cu^+)^2}$$

with the activities evaluated at equilibrium. The cell is

$$Pt \mid Cu^+(aq), Cu^{2+}(aq) \parallel Cu^+(aq) \mid Cu(s)$$

and so the standard reduction potentials we require are

R: $Cu^+(aq) + e^- \rightarrow Cu(aq)$ $E^{\ominus} = +0.52$ V
L: $Cu^{2+}(aq) + e^- \rightarrow Cu^+(aq)$ $E^{\ominus} = +0.16$ V

The standard cell potential is therefore

$$E^{\ominus} = +0.52 \text{ V} - 0.16 \text{ V} = +0.36 \text{ V}$$

Since $v = 1$,

$$\ln K = \frac{+0.36 \text{ V}}{0.02569 \text{ V}} = 14.0$$

Hence, $K = 1.2 \times 10^6$.

Comment. The equilibrium lies strongly towards the right of the reaction as written, and so Cu^+ disproportionates almost totally in solution.

Exercise. Calculate the equilibrium constant of the reaction $Sn^{2+}(aq) + Pb(s) \rightarrow Sn(s) + Pb^{2+}(aq)$ at 298 K.
[2.0]

The composition dependence of individual reduction potentials

We know that the overall cell potential depends on composition in accord with the Nernst equation, eqn 11. Similar equations may be written for the individual reduction potentials by considering the left-hand electrode to be a standard hydrogen electrode. For example, the reaction in the cell

$$Pt \mid H_2(g) \mid H^+(aq) \parallel Ag^+(aq) \mid Ag(s)$$

is

$$H_2(g) + 2Ag^+(aq) \rightarrow 2H^+(aq) + 2Ag(s) \qquad Q = \frac{a(H^+)^2 p^{\ominus}}{f(H_2)a(Ag^+)^2}$$

and the Nernst equation is

$$E = E^{\ominus} - \frac{RT}{2F} \ln Q$$

$$= E^{\ominus}(Ag^+/Ag) - \frac{RT}{2F} \ln Q$$

When the hydrogen electrode has its standard composition, $a(H^+) = 1$ and $f(H_2) = p^{\ominus}$, but the Ag^+/Ag electrode has an arbitrary composition, the cell potential can be regarded as arising solely from the Ag^+/Ag electrode. We can write

$$E(Ag^+/Ag) = E^{\ominus}(Ag^+/Ag) - \frac{RT}{2F} \ln \frac{1}{a(Ag^+)^2}$$

$$= E^{\ominus}(Ag^+/Ag) - \frac{RT}{F} \ln \frac{1}{a(Ag^+)}$$

The same expression is obtained by writing the half-reaction at the right-hand electrode as the reduction

$$Ag^+(aq) + e^- \rightarrow Ag(s) \qquad Q = \frac{1}{a(Ag^+)}$$

and using the Nernst equation directly:

$$E(Ag^+/Ag) = E^{\ominus}(Ag^+/Ag) - \frac{RT}{F} \ln Q$$

Similarly, for the cell

$$\text{Pt} \mid \text{H}_2(g) \mid \text{H}^+(aq) \parallel \text{Cl}^-(aq) \mid \text{AgCl}(s) \mid \text{Ag}(s)$$

in which the reaction is

$$\text{H}_2(g) + 2\text{AgCl}(s) \rightarrow 2\text{HCl}(aq) + 2\text{Ag}(s) \quad Q = \frac{a(\text{H}^+)^2 a(\text{Cl}^-)^2}{f(\text{H}_2)/p^\ominus}$$

when the hydrogen electrode has its standard composition the cell potential arises solely from the AgCl/Ag, Cl^- couple and we can write

$$E(\text{AgCl}/\text{Ag}, \text{Cl}^-) = E^\ominus(\text{AgCl}/\text{Ag}, \text{Cl}^-) - \frac{RT}{2F} \ln a(\text{Cl}^-)^2$$

$$= E^\ominus(\text{AgCl}/\text{Ag}, \text{Cl}^-) - \frac{RT}{F} \ln a(\text{Cl}^-)$$

The same expression is obtained by writing the reaction at the right-hand electrode as

$$\text{AgCl}(s) + \text{e}^- \rightarrow \text{Ag}(s) + \text{Cl}^-(aq) \quad Q = a(\text{Cl}^-)$$

and using the Nernst equation directly.

Example 10.12: *Calculating the potential of an electrode*

Calculate the change that takes place in the potential of a silver/silver chloride electrode when an excess of $0.010 \text{ mol kg}^{-1} \text{ KCl}(aq)$ is added at 298 K. The activity of Cl^- ions is 1.3×10^{-5} in a saturated silver chloride solution at this temperature.

Answer. The potential difference is given by eqn 10 and depends on the Cl^- ion activity. When silver chloride is present alone, the saturated solution has a very low Cl^- activity. When potassium chloride is added the $\text{Cl}^-(aq)$ molality rises to $0.010 \text{ mol kg}^{-1}$ and we use the data in Table 10.4 to convert this to an activity. The mean activity coefficient is 0.906 for $0.010 \text{ mol kg}^{-1} \text{ KCl}(aq)$, and so $a = 0.00906$. The change in electrode potential is therefore

$$\Delta E = \frac{-RT}{F} \ln \frac{\gamma_{\pm f} m_f}{\gamma_{\pm i} m_i} = -0.17 \text{ V}$$

Exercise. Calculate ΔE when an $0.05 \text{ mol kg}^{-1} \text{ KCl}(aq)$ solution is added to a calomel electrode compartment at 298 K (for which, initially, $a = 8.7 \times 10^{-7}$).

[−0.28 V]

The hydrogen electrode and pH

The potential of a hydrogen electrode is

$$E(\text{H}^+/\text{H}_2) = E^\ominus(\text{H}^+/\text{H}_2) - \frac{RT}{2F} \ln Q$$

$$= \frac{RT}{F} \ln \frac{a(\text{H}^+)p^{\ominus 1/2}}{f(\text{H}_2)^{1/2}}$$

This expression makes sense physically. Increasing the activity of the hydrogen ions (decreasing the pH) increases the tendency of the positive ions to discharge at the electrode, and so we should expect its potential to become more positive. Since the ion activity occurs in the numerator of the logarithmic term, the equation does predict that E increases as a is increased.

The potential of the hydrogen electrode is directly proportional to the pH of the solution. Thus, if $f = p^{\ominus}$ (so the hydrogen fugacity is 1 bar), we can use

$$\ln a(H^+) = \ln 10 \times \log a(H^+) = -2.303 \, pH$$

to obtain

$$E(H^+/H_2) = -\frac{2.303RT}{F} pH$$

At 25°C, when $RT/F = 25.69 \, \text{mV}$, this becomes

$$E(H^+/H_2) = -59.16 \, \text{mV} \times pH$$

Hence, each unit decrease in pH increases the electrode potential by 59 mV.

Example 10.13: *Calculating the pressure dependence of an electrode potential*

Chlorine gas is bubbled over a platinum electrode dipping into aqueous sodium chloride at 298 K. Calculate the change in electrode potential when the chlorine pressure is increased from 1.0 atm to 2.0 atm.

Answer. We assume that the gas behaves perfectly at the pressures specified, and so replace fugacity by pressure. The electrode half-reaction is

$$Cl_2(g) + 2e^- \rightarrow 2Cl^-(aq) \quad Q = \frac{a(Cl^-)^2}{f(Cl_2)/p^{\ominus}}$$

Therefore

$$E(Cl_2/Cl^-) = E^{\ominus}(Cl_2/Cl^-) - \frac{RT}{2F} \ln Q$$

$$= E^{\ominus}(Cl_2/Cl^-) - \frac{RT}{F} \ln \frac{a}{(f/p^{\ominus})^{1/2}}$$

where a is the Cl^- activity. Therefore, the change in potential at 298 K (when $RT/F = 25.69 \, \text{mV}$) is

$$E(2 \, \text{atm}) - E(1 \, \text{atm}) = \frac{RT}{F} \ln 2^{1/2} = +8.9 \, \text{mV}$$

Comment. The potential increases when the gas pressure is increased because the equilibrium shifts away from $Cl_2(g)$ towards $Cl^-(aq)$, and the resulting withdrawal of electrons from the electrode makes it more positive.

Exercise. Find an expression for the potential of an electrode at which the half-reaction is $G_2(g) + 4e^- \rightarrow 2G^{2-}(aq)$. $[E = E^{\ominus} - (RT/2F) \ln a/(f/p^{\ominus})^{1/2}]$

Finally, for a redox electrode, eqn 10 gives

$$E(Fe^{3+}/Fe^{2+}) = E^{\ominus}(Fe^{3+}/Fe^{2+}) - \frac{RT}{F} \ln \frac{a(Fe^{2+})}{a(Fe^{3+})}$$

An increase in Fe^{3+} tends to increase the electrode potential and an increase in Fe^{2+} tends to lower it. However, the wider importance of this equation is that instead of treating the potential difference as arising *from* the equilibrium we can imagine *controlling* the equilibrium by modifying the electrode potential. If we can ensure that the potential of the cell

$$\text{SHE} \| \text{Fe}^{3+}(aq), \text{Fe}^{2+}(aq) \,|\, \text{Pt}$$

is $E(\text{Fe}^{3+}/\text{Fe}^{2+})$, or the appropriate value if the electrode on the left is something other than a SHE, then the concentrations of reduced and oxidized species will adjust so that $a(\text{Fe}^{2+})/a(\text{Fe}^{3+})$ has a value that satisfies the equation for E. This gives us electrical control over the composition of solutions.

The measurement of standard reduction potentials

The procedure for measuring a standard reduction potential can be illustrated by considering a specific case, the silver chloride electrode. The measurement is made on the **Harned cell**:

$$\text{Pt} \,|\, \text{H}_2(g) \,|\, \text{HCl}(aq) \,|\, \text{AgCl}(s) \,|\, \text{Ag}(s)$$
$$\tfrac{1}{2}\text{H}_2(g) + \text{AgCl}(s) \rightarrow \text{HCl}(aq) + \text{Ag}(s)$$

for which

$$E = E^{\ominus}(\text{AgCl/Ag, Cl}^-) - \frac{RT}{F} \ln \frac{a(\text{H}^+)a(\text{Cl}^-)}{(f/p^{\ominus})^{1/2}}$$

We shall set $f = p^{\ominus}$ from now on. The activities can be expressed in terms of the molality m and the mean activity coefficient $\gamma_\pm$ through eqn 4:

$$E = E^{\ominus}(\text{AgCl/Ag, Cl}^-) - \frac{RT}{F} \ln \left(\frac{m}{m^{\ominus}}\right)^2 - \frac{RT}{F} \ln \gamma_\pm^2$$

which rearranges to

$$E + \frac{2RT}{F} \ln \frac{m}{m^{\ominus}} = E^{\ominus}(\text{AgCl/Ag, Cl}^-) - \frac{2RT}{F} \ln \gamma_\pm \qquad (13a)$$

From the Debye–Hückel limiting law for a 1,1-electrolyte,

$$\lg \gamma_\pm = -Am^{1/2} \qquad A = 0.509/(\text{mol kg}^{-1})^{1/2}$$

Therefore,

$$E + \frac{2RT}{F} \ln \frac{m}{m^{\ominus}} = E^{\ominus}(\text{AgCl/Ag, Cl}^-) + \frac{2.34RT}{F} \times \left(\frac{m}{m^{\ominus}}\right)^{1/2} \qquad (13b)$$

(In precise work, the $m^{1/2}$ term is brought to the left, and a higher order correction term from the extended Debye–Hückel law is used on the right.) The expression on the left is evaluated at a range of cell molalities, plotted against $m^{1/2}$, and extrapolated to $m = 0$. The intercept is the value of $E^{\ominus}(\text{AgCl/Ag, Cl}^-)$, as shown in Fig. 10.10.

The measurement of activity coefficients

Once the standard potential of an electrode in a cell is known, the activities of the ions with respect to which it is reversible can be determined simply by measuring the cell potential with the ions at the concentration of interest. For example, the mean activity coefficient of the ions in hydrochloric acid of

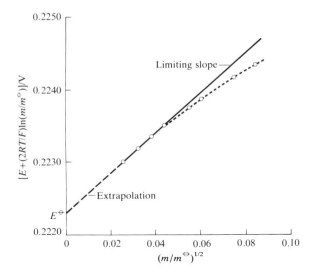

Fig. 10.10 The plot and the $m^{1/2}$ extrapolation used for the experimental measurement of a standard cell potential. The intercept at $m^{1/2} = 0$ is $E^{\ominus}$.

molality m is obtained from eqn 13a in the form

$$\ln \gamma_{\pm} = \frac{F}{2RT} \times \{E^{\ominus}(\text{AgCl/Ag, Cl}^-) - E\} - \ln \frac{m}{m^{\ominus}} \qquad (13c)$$

once E has been measured.

Applications of reduction potentials

The measurement of zero-current cell potential is a convenient source of data on the Gibbs functions, enthalpies, and entropies of reactions. In practice the standard values of these quantities are the ones normally determined.

10.6 The electrochemical series

We have seen that for two redox couples Ox_1/Red_1 and Ox_2/Red_2, and the cell

$$\text{Red}_1, \text{Ox}_1 \parallel \text{Red}_2, \text{Ox}_2 \quad E^{\ominus} = E_2^{\ominus} - E_1^{\ominus}$$

that the cell reaction

$$\text{Red}_1 + \text{Ox}_2 \rightarrow \text{Ox}_1 + \text{Red}_2$$

is spontaneous as written if $E^{\ominus} > 0$ and therefore if $E_2^{\ominus} > E_1^{\ominus}$. Since in the cell reaction Red_1 is reducing Ox_2, we can conclude that

> Red_1 has a thermodynamic tendency to reduce Ox_2 if $E_1^{\ominus}$ is lower than $E_2^{\ominus}$.

More briefly: low reduces high. For example,

$$E^{\ominus}(\text{Zn}^{2+}/\text{Zn}) = -0.76 \text{ V} < E^{\ominus}(\text{Cu}^{2+}/\text{Cu}) = +0.34 \text{ V}$$

and Zn has a thermodynamic tendency to reduce Cu^{2+}. Hence the reaction

$$\text{Zn}(s) + \text{CuSO}_4(aq) \rightarrow \text{ZnSO}_4(aq) + \text{Cu}(s)$$

can be expected to have $K > 1$ (in fact, as we have seen, $K = 1.5 \times 10^{37}$ at 298 K).

Table 10.5 shows a part of the **electrochemical series** of the elements, their redox couples arranged in the order of their reducing power. The reduced member of a couple lower in the table (with lower standard reduction potential) reduces the oxidized member of couples higher up. This is a qualitative conclusion. The quantitative value of K is then obtained by doing the calculations we have introduced previously.

Example 10.14: *Using the electrochemical series*

Can zinc displace magnesium from aqueous solutions at 298 K?

Answer. Displacement corresponds to a reduction of $Mg^{2+}(aq)$ by $Zn(s)$. This may occur if $E^{\ominus}(Zn^{2+}/Zn) < E^{\ominus}(Mg^{2+}/Mg)$ and hence the couple Zn^{2+}/Zn is lower in the table than Mg^{2+}/Mg. From Table 10.5 we see that the Zn^{2+}/Zn couple lies above Mg^{2+}/Mg. Therefore, zinc cannot displace magnesium.

Comment. There may be kinetic factors which result in immeasurably slow rates of reaction.

Exercise. Can lead displace (a) iron, (b) copper from solution at 298 K?

[(a) No, (b) Yes]

10.7 Solubility products

We can discuss the solubility S (the molality of the saturated solution) of a sparingly soluble salt MX in terms of the equilibrium

$$MX(s) \rightleftharpoons M^+(aq) + X^-(aq) \qquad K_s = a(M^+)a(X^-) \qquad (14a)$$

where the activities are those at equilibrium (that is, in the saturated solution) and we have used $a = 1$ for a pure solid. The equilibrium constant K_s is called the **solubility product** of the salt. When the solubility is so low that $\gamma_\pm \approx 1$ even in the saturated solution we can write $a = m/m^{\ominus}$; moreover, since both molalities are equal to S in the saturated solution,

$$K_s = \left(\frac{S}{m^{\ominus}}\right)^2, \quad \text{so } S = K_s^{1/2} m^{\ominus} \qquad (14b)$$

It follows that we can predict S from the standard potential of a cell with a reaction corresponding to the solubility equilibrium.

Example 10.15: *Evaluating a solubility product*

Evaluate the solubility of $AgCl(s)$ from cell potential data at 298 K.

Answer. We need to find an electrode combination that reproduces the solubility equilibrium, and then identify the solubility product with the equilibrium constant of the cell reaction. The equilibrium is

$$AgCl(s) \rightleftharpoons Ag^+(aq) + Cl^-(aq) \quad K_s = a(Ag^+)a(Cl^-)$$

and we saw in Example 10.5 that it can be expressed as the difference of the following two half-reactions:

$$AgCl(s) + e^- \rightarrow Ag(s) + Cl^-(aq) \quad E^{\ominus} = +0.22 \text{ V}$$
$$Ag^+(aq) + e^- \rightarrow Ag(s) \quad E^{\ominus} = +0.80 \text{ V}$$

The cell potential is therefore -0.58 V. Since $v = 1$,

$$\ln K_s = \frac{FE}{RT} = -22._6$$

Therefore, $K_s = 1.5 \times 10^{-10}$ and $S = 1.2 \times 10^{-5}$ mol kg^{-1}.

Exercise. Calculate the solubility product of mercury(I) chloride at 298 K.

[1.3×10^{-18}]

10.8 The measurement of pH and pK

The measurement of pH is simple in principle, for it is based on the measurement of the potential of a hydrogen electrode immersed in the solution. The left-hand electrode of the cell is typically a saturated calomel reference electrode with potential $E(\text{cal})$; the right-hand electrode is the hydrogen electrode with potential

$$E(\text{H}^+/\text{H}_2) = -59.16 \text{ mV} \times \text{pH}$$

The potential of the cell is therefore

$$E = -59.16 \text{ mV} \times \text{pH} - E(\text{cal})$$

(at 25°C) and therefore[2]

$$\text{pH} = \frac{E + E(\text{cal})}{-59.16 \text{ mV}} \tag{15}$$

In practice, indirect methods are much more convenient, and the hydrogen electrode is replaced by the **glass electrode**. This electrode (Fig. 10.11) is sensitive to hydrogen ion activity, and has a potential proportional to pH. It is filled with a phosphate buffer containing Cl$^-$ ions, and conveniently has $E = 0$ when the external medium is at pH $= 7$. The glass electrode is much more convenient to handle than the gas electrode itself, and can be calibrated using solutions of known pH.

The pK_a of an acid may now be measured electrically as explained in Section 9.7, by noting the pH of a solution containing equal amounts of the acid and its conjugate base, for then pH $= pK_a$.

10.9 Potentiometric titrations

In a **redox titration**, the reduced form of an ion (e.g., Fe^{2+}) is oxidized by the addition of an oxidant (e.g., Ce^{4+}). In a **potentiometric titration**, the equivalence point of a redox titration is detected by monitoring the

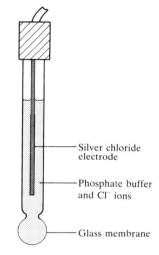

Fig. 10.11 The glass electrode. It is usually in conjunction with a calomel electrode that makes contact with the test solution through a salt bridge.

Silver chloride electrode

Phosphate buffer and Cl$^-$ ions

Glass membrane

[2] The practical definition of the pH of a solution X is

$$\text{pH(X)} = \text{pH(S)} + \frac{FE}{2.3026RT}$$

where E is the potential of the cell

$$\text{Pt} \mid \text{H}_2 \mid \text{X}(aq) \parallel 3.5 \text{ M KCl}(aq) \parallel \text{S}(aq) \mid \text{H}_2 \mid \text{Pt}$$

and S is a solution of standard pH. The currently recommended primary standards include a saturated aqueous solution of potassium hydrogen tartrate, which has pH $= 3.557$ at 25°C and 0.0100 mol kg^{-1} disodium tetraborate, which has pH $= 9.180$ at that temperature.

potential of the cell formed by a platinum electrode in the mixture and another electrode in electrical contact with the mixture through a salt bridge. As we shall see, there is a sharp change of cell potential at the equivalence point, when exactly enough oxidant has been added to oxidize all the reduced ion.

We illustrate the technique with the redox reaction

$$\text{Fe}^{2+}(aq) + \text{Ce}^{4+}(aq) \rightarrow \text{Fe}^{3+}(aq) + \text{Ce}^{3+}(aq)$$

for which the two half-reactions are

$$\text{Ce}^{4+}(aq) + e^- \rightarrow \text{Ce}^{3+}(aq) \qquad E^\ominus(\text{Ce}^{4+}/\text{Ce}^{3+}) = +1.61 \text{ V}$$
$$\text{Fe}^{3+}(aq) + e^- \rightarrow \text{Fe}^{2+}(aq) \qquad E^\ominus(\text{Fe}^{3+}/\text{Fe}^{2+}) = +0.77 \text{ V}$$

The equilibrium constant for the reaction (which has $v = 1$) is therefore 1.52×10^{14}, and so Ce^{4+} is a good choice because it ensures that the equilibrium lies strongly towards products (Fe^{3+}). During the course of the titration the presence of the iron couple gives rise to an electrode potential $E(\text{Fe}^{3+}/\text{Fe}^{2+})$ and the cerium couple gives rise to $E(\text{Ce}^{4+}/\text{Ce}^{3+})$. However, these couples share the same electrode, and therefore the potentials must be equal (a single electrode cannot be at two different potentials simultaneously). Consequently, at any stage of the titration the electrode potential may be written in either of the following forms:

$$E = E^\ominus(\text{Ce}^{4+}/\text{Ce}^{3+}) - \frac{RT}{F} \ln \frac{a(\text{Ce}^{3+})}{a(\text{Ce}^{4+})}$$

$$E = E^\ominus(\text{Fe}^{3+}/\text{Fe}^{2+}) - \frac{RT}{F} \ln \frac{a(\text{Fe}^{2+})}{a(\text{Fe}^{3+})}$$

and whichever is more convenient may be used. We shall simplify the following discussion by supposing that activities may be replaced by molalities, and we shall also ignore the dilution of the sample as the cerium solution is added: these are minor technical complications which do not affect the main conclusions.

Suppose that initially the amount of Fe(II) present in the sample is f. As Ce(IV) is added this is reduced to $(1 - x)f$; simultaneously, the amount of Fe(III) rises to xf, where x depends on the amount of Ce(IV) added (if K were infinite, xf would be equal to the amount of oxidant added). It follows that at an intermediate stage of the titration, the electrode's potential is

$$E = E^\ominus(\text{Fe}^{3+}/\text{Fe}^{2+}) + \frac{RTg}{F} \qquad g = \ln \frac{x}{1-x} \qquad (16a)$$

The function g occurs widely in electrochemistry, and is plotted in Fig. 10.12: it has a large negative value when $x \ll 1$, is zero when $x = \frac{1}{2}$, and rises to a large positive value as x approaches 1. An initial rapid rise, a slow variation around $x = \frac{1}{2}$, and a further rapid rise, is characteristic of many electrochemical measurements, and it can usually be traced back to the properties of g. We saw another example in the discussion of acid-base titrations in Section 9.7.

In the context of a potentiometric titration, the shape of g indicates that the electrode potential is initially strongly negative, rises rapidly towards $E^\ominus(\text{Fe}^{3+}/\text{Fe}^{2+})$ as Ce(IV) is added, and is exactly equal to that value when

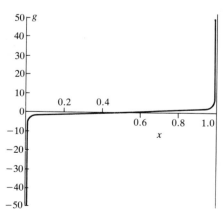

Fig. 10.12 The function g in eqn 16. This quick–slow–quick variation occurs widely in electrochemistry.

enough has been added to equalize the molalities (actually the activities) of Fe(II) and Fe(III). As more Ce(IV) is added the electrode potential rises slowly at first, but then the rapid second rise of g takes over. The line in Fig. 10.12 barely leaves the zero axis (on the scale used) until x has reached about 0.98, and so the equivalence point of the titration is signalled by the abrupt rise in potential away from $E^{\ominus}(Fe^{3+}/Fe^{2+})$.

When enough Ce(IV) has been added to eliminate all but a trace of the Fe(II), it is more convenient to discuss the potential in terms of the Ce(IV) molality. If an amount c of Ce(IV) has been added, the amount present in solution after the equivalence point is about $c - f$, because virtually all the Fe(II) has been oxidized (since the equilibrium constant is so large). The amount of Ce(III) present is about f however much Ce(IV) is added after the equivalence point. Therefore the potential of the electrode is

$$E = E^{\ominus}(Ce^{4+}/Ce^{3+}) + \frac{RTh}{F} \qquad h = \ln\frac{c - f}{f} \qquad (16b)$$

E rises to $E^{\ominus}(Ce^{4+}/Ce^{3+})$ at $c = 2f$ (about twice as much cerium as was needed to reach the equivalence point), and then climbs only very slowly as more cerium is added because when $c \gg f$, $h \approx \ln c/f$, which increases only very slowly as c increases.

The complete curve is sketched in Fig. 10.13: it is clear that the equivalence point can be detected by noting where there is a rapid change of potential from the vicinity of $E^{\ominus}(Fe^{3+}/Fe^{2+})$ to the vicinity of $E^{\ominus}(Ce^{4+}/Ce^{3+})$ or, in general, for other redox systems, from the standard reduction potential of the couple in the analyte to the standard reduction potential of the couple in the titrant.

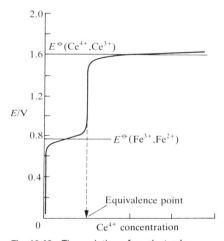

Fig. 10.13 The variation of an electrode potential during the titration of Fe(II) with Ce(IV). The equivalence point is recognized by the sudden change from the standard potential of the Fe^{3+}/Fe^{2+} couple to that of the Ce^{4+}/Ce^{3+} couple.

10.10 Thermodynamic functions from cell potential measurements

The standard cell potential is related to the standard reaction Gibbs function through eqn 10:

$$\Delta G^{\ominus} = -\nu F E^{\ominus}$$

Therefore, simply by measuring $E^{\ominus}$ we can obtain the standard reaction Gibbs function. This value can then be used to calculate the Gibbs function of formation of ions using the convention explained in Section 10.1. For example, the cell reaction of

$$H_2 \mid H^+(aq) \parallel Ag^+(aq) \mid Ag \qquad E^{\ominus} = +0.7996 \text{ V}$$

is

$$Ag^+(aq) + \tfrac{1}{2}H_2(g) \rightarrow H^+(aq) + Ag(s) \qquad \Delta G^{\ominus} = -\Delta G_f^{\ominus}(Ag^+, aq)$$

Therefore, with $\nu = 1$, we find

$$\Delta G_f^{\ominus}(Ag^+, aq) = F E^{\ominus} = +77.1 \text{ kJ mol}^{-1}$$

as in Table 10.1.

Example 10.16: *Evaluating a reduction potential from two others*

Given that the standard reduction potentials of the Cu^{2+}/Cu and Cu^+/Cu couples are $E^{\ominus}(Cu^{2+}/Cu) = +0.340$ V and $E^{\ominus}(Cu^+/Cu) = +0.522$ V, evaluate $E^{\ominus}(Cu^{2+}/Cu^+)$.

Answer. The reaction Gibbs functions may be added (because G is a state function). Therefore, convert the $E^{\ominus}$ values to $\Delta G^{\ominus}$ values using eqn 10, add them appropriately, and convert the overall $\Delta G^{\ominus}$ to the required $E^{\ominus}$ using eqn 10 again. The electrode reactions are as follows:

(a) $Cu^{2+}(aq) + 2e^- \rightarrow Cu(s)$ $E^{\ominus} = +0.340$ V

$$\Delta G^{\ominus} = -2F \times 0.340 \text{ V} = -0.68 \text{ V} \times F$$

(b) $Cu^+(aq) + e^- \rightarrow Cu(s)$ $E^{\ominus} = +0.522$ V

$$\Delta G^{\ominus} = -0.522 \text{ V} \times F$$

The required reaction is

(c) $Cu^{2+}(aq) + e^- \rightarrow Cu^+(aq)$ $E^{\ominus}$

Since (c) = (a) − (b), the standard Gibbs function of reaction (c) is

$$\Delta G^{\ominus} = \Delta G^{\ominus}(a) - \Delta G^{\ominus}(b) = -0.16 \text{ V} \times F$$

Therefore, since now $v = 1$,

$$E^{\ominus} = -\frac{\Delta G^{\ominus}}{F} = +0.16 \text{ V}$$

Comment. Note that we cannot combine the $E^{\ominus}$ values directly, as they are not extensive properties, and we must always work via $\Delta G^{\ominus}$.

Exercise. Calculate the standard potential of the Fe^{3+}/Fe couple from the values for the Fe^{3+}/Fe^{2+} and Fe^{2+}/Fe couples.   [−0.016 V]

The temperature coefficient of the cell potential gives the entropy of the cell reaction. This follows from the thermodynamic relation $(\partial G/\partial T)_p = -S$ and eqn 10b, which combine to give

$$\left(\frac{\partial E^{\ominus}}{\partial T}\right)_p = \frac{\Delta S^{\ominus}}{vF} \tag{17}$$

Hence we have an electrochemical technique for obtaining standard reaction entropies and through them the entropies of ions in solution.

Finally, we can combine the results obtained so far and use them to obtain the standard reaction enthalpy:

$$\Delta H^{\ominus} = \Delta G^{\ominus} + T \Delta S^{\ominus} = -vF\left\{E^{\ominus} - T\left(\frac{\partial E^{\ominus}}{\partial T}\right)_p\right\}. \tag{18}$$

This is a noncalorimetric method of measuring $\Delta H^{\ominus}$ and, through the convention $\Delta H_f^{\ominus}(H^+, aq) = 0$, the standard enthalpies of formation of ions in solution. Thus, electrical measurements can be used to calculate all the thermodynamic properties with which this chapter began.

Example 10.17: *Using the temperature coefficient of the cell potential*

The standard cell potential of

$$Pt \mid H_2(g) \mid HCl(aq) \mid Hg_2Cl_2(s) \mid Hg(l)$$

was found to be $+0.2699 \, V$ at $293 \, K$ and $+0.2669 \, V$ at $303 \, K$. Evaluate the standard reaction Gibbs function, enthalpy, and entropy at $298 \, K$.

Answer. The cell reaction is

$$Hg_2Cl_2(s) + H_2(g) \rightarrow 2Hg(l) + 2HCl(aq) \qquad \nu = 2$$

We find $\Delta G^{\ominus}$ from eqn 10b at $298 \, K$ by linear interpolation between the two temperatures (i.e. in this case take the mean $E^{\ominus}$ because $298 \, K$ lies midway between $293 \, K$ and $303 \, K$). Since the mean standard cell potential is $+0.2684 \, V$,

$$\Delta G^{\ominus} = -2FE^{\ominus} = -51.8 \, kJ \, mol^{-1}$$

The temperature coefficient of $E^{\ominus}$ is

$$\left(\frac{\partial E^{\ominus}}{\partial T} \right)_p \approx \frac{0.2669 \, V - 0.2699 \, V}{10 \, K} = -3.0 \times 10^{-4} \, V \, K^{-1}$$

Therefore, from eqn 17,

$$\Delta S^{\ominus} = 2 \times F \times (-3.0 \times 10^{-4} \, V \, K^{-1}) = -58 \, J \, K^{-1} \, mol^{-1}$$

We now form

$$\Delta H^{\ominus} = \Delta G^{\ominus} + T \, \Delta S^{\ominus} = -69 \, kJ \, mol^{-1}$$

Comment. One difficulty with this procedure lies in the accurate measurement of small temperature coefficients of cell potential. Nevertheless, it is another example of the striking ability of thermodynamics to relate the apparently unrelated, in this case to relate electrical measurements to thermal properties.

Exercise. Predict the standard potential of the Harned cell at $303 \, K$ from tables of thermodynamic data.   [0.2191 V]

Further information: the Debye–Hückel theory

We imagine a solution in which all the ions have their actual positions, but in which their interactions have been turned off. The difference in chemical potential between the ideal and real solutions is then identified with w_e, the electrical work of charging the system. Therefore, for a salt M_pX_q, from eqn 4

$$\ln \gamma_{\pm} = \frac{w_e}{sRT} \qquad s = p + q \qquad \text{(A1)}$$

It follows that we must (a) find the final distribution of the ions and (b) the work of charging them in that distribution.

The distribution of ions

The Coulomb potential at a distance r from an isolated ion of charge z_ie in a medium of permittivity ε is

$$\phi_i = \frac{Z_i}{r} \quad \text{with} \quad Z_i = \frac{z_ie}{4\pi\varepsilon}$$

The effect of the ionic atmosphere is to cause the potential to decay with distance more sharply, and we write

$$\phi_i = \frac{Z_i}{r} e^{-r/r_D} \qquad (A2)$$

where r_D is called the **Debye length**. When r_D is large the shielded potential is virtually the same as the unshielded potential. When r_D is small, the shielded potential is much smaller than the unshielded potential, even for short distances.

To calculate r_D we need to know how the **charge density** ρ of the ionic atmosphere, the charge per unit volume, varies with distance from the ion. Charge density and electrostatic potential are related by **Poisson's equation**, which for a spherically symmetrical charge distribution is

$$\frac{1}{r^2}\frac{d}{dr}\left(r^2 \frac{d\phi_i}{dr}\right) = \frac{-\rho_i}{\varepsilon}$$

Substituting eqn A2 results in

$$r_D^2 = \frac{-\varepsilon \phi_i}{\rho_i} \qquad (A3)$$

To solve this equation we need to relate ρ_i and ϕ_i.

The energy of an ion of charge $z_j e$ at a distance where it experiences the potential ϕ_i of the central ion i relative to its energy when it is far away in the bulk solution is

$$E = z_j e \phi_i$$

Therefore, according to the Boltzmann distribution (Section 19.1) the ratio of the concentration c_j of ions at a distance r and the concentration in the bulk, c_j° is:

$$\frac{c_j}{c_j^\circ} = e^{-E/kT}$$

The charge density ρ_i at a distance r from the ion i is the concentration of each type of ion multiplied by their charges:

$$\rho_i = c_+ z_+ F + c_- z_- F$$
$$= c_+^\circ F z_+ e^{-z_+ e\phi_i/kT} + c_-^\circ F z_- e^{-z_- e\phi_i/kT}$$

Since the average electrostatic interaction energy is small compared with kT we may write the last equation as

$$\rho_i = (c_+^\circ z_+ + c_-^\circ z_-)F - (c_+^\circ z_+^2 + c_-^\circ z_-^2)\left(\frac{F^2\phi_i}{RT}\right) + \ldots$$

The first term in the expansion is zero because it is the charge density in a uniform solution, and the solution is electrically neutral. The unwritten terms are assumed to be too small to be significant. Expressed in terms of the ionic strength, the one remaining term is

$$\rho_i = -\frac{2\rho F^2 I \phi_i}{RT} \qquad (A4)$$

We can now solve eqn A3 for r_D:

$$r_D = \left(\frac{\varepsilon RT}{2\rho F^2 I}\right)^{1/2} \qquad (A5)$$

The work of charging

To calculate the activity coefficient we need to find the electrical work of charging the central ion when it is surrounded by its atmosphere. To do so, we need to know the potential at the ion due to its atmosphere, ϕ_{atmos}. This is the difference between the total potential, as given by eqn A4, and the potential due to the central ion itself:

$$\phi_{atmos} = \phi_i - \phi_{central\ ion} = Z_i \left(\frac{e^{-r/r_D}}{r} - \frac{1}{r} \right)$$

The potential at the ion (at $r = 0$) is obtained by taking the limit of this expression as $r \to 0$, and is

$$\phi_{atmos}(0) = -\frac{Z_i}{r_D}$$

If the charge of the central ion were q and not $z_i e$, the potential due to its atmosphere would be

$$\phi_{atmos}(0) = -\frac{q}{4\pi\varepsilon} \times \frac{1}{r_D}$$

The work of adding a charge dq to a region where the electrical potential is $\phi_{atmos}(0)$ is

$$dw_e = \phi_{atmos}(0)\,dq$$

Therefore, the total work of fully charging (per mole of ions) is

$$w_e = N_A \int_0^{z_i} \phi_{atmos}(0)\,dq = \frac{-N_A}{4\pi\varepsilon r_D} \int_0^{z_i} q\,dq$$

$$= \frac{-z_i^2 e^2 N_A}{8\pi\varepsilon r_D} = \frac{-z_i^2 F^2}{8\pi\varepsilon N_A r_D} \tag{A6}$$

It follows from eqn A1 that the mean activity coefficient of the ions is

$$\ln \gamma_\pm = \frac{p w_{e,+} + q w_{e,-}}{sRT} = \frac{-F^2(p z_+^2 + q z_-^2)}{8\pi\varepsilon s N_A R T r_D}$$

However, for neutrality $p z_+ + q z_- = 0$, and so

$$\ln \gamma_\pm = \frac{-|z_+ z_-| F^2}{8\pi\varepsilon r_D N_A R T}$$

When we replace r_D using eqn A5, and convert the logarithms to base 10, this gives

$$\lg \gamma_\pm = -A\,|z_+ z_-|\,I^{1/2} \tag{A7}$$

where

$$A = \frac{F^3}{4\pi N_A} \left(\frac{\rho}{2\varepsilon^3 R^3 T^3} \right)^{1/2} \ln 10 \tag{A8}$$

This is the limiting law. For water at 298 K, when $\rho = 0.997\ \mathrm{g\ cm}^{-3}$ and $\varepsilon_r = 78.54$, $A = 0.509/(\mathrm{mol\ kg}^{-1})^{1/2}$.

Further reading

N. J. Selley, *Experimental approach to electrochemistry*. Edward Arnold, London (1977).

R. A. Robinson and R. H. Stokes, *Electrolyte solutions* (2nd edn). Academic Press, New York (1959).

J. O'M. Bockris and A. K. N. Reddy, *Modern electrochemistry*. Plenum, New York (1970).

J. Koryta, *Ions, electrodes, and membranes*. Wiley-Interscience, New York (1982).

D. R. Crow, *Principles and applications of electrochemistry* (3rd edn). Chapman and Hall, London (1988).

D. C. Harris, *Quantitative chemical analysis*. W. H. Freeman & Co., New York (1987).

J. O'M. Bockris, B. E. Conway *et al.* (ed.), *Comprehensive treatise on electrochemistry* (Vols 1–7). Plenum, New York (1980).

A. J. Bard, R. Parsons and R. Jordan, *Standard potentials in aqueous solution*. Marcel Dekker, New York (1985).

A. K. Covington, R. G. Bates and D. A. Durst, Definition of pH scales, standard reference values, and related terminology. In *Pure and Appl. Chem.* **57**, 531 (1985).

Exercises

10.1 Calculate the solubility of mercury(II) chloride at 25°C from standard Gibbs functions of formation.

10.2 Estimate the standard Gibbs function of formation of $F^-(aq)$ from the value for $Cl^-(aq)$, taking the radius of F^- as 131 pm.

10.3 Confirm that the ionic strengths of KCl, $MgCl_2$, $FeCl_3$, $Al_2(SO_4)_3$, and $CuSO_4$ solutions are related to the molalities m of the solution by $I(KCl) = m$, $I(MgCl_2) = 3m$, $I(FeCl_3) = 6m$, $I(Al_2(SO_4)_3) = 15m$, and $I(CuSO_4) = 4m$.

10.4 Calculate the ionic strength of a solution that is $0.10\ mol\ kg^{-1}$ in $KCl(aq)$ and $0.20\ mol\ kg^{-1}$ $CuSO_4(aq)$.

10.5 Calculate the ionic strength of a solution that is $0.040\ mol\ kg^{-1}$ in $K_3[Fe(CN)_6](aq)$, $0.030\ mol\ kg^{-1}$ in $KCl(aq)$, and $0.050\ mol\ kg^{-1}$ in $NaBr(aq)$.

10.6 Calculate the masses of (a) $Ca(NO_3)_2$ or of (b) NaCl to add to a $0.150\ mol\ kg^{-1}$ solution of $KNO_3(aq)$ containing 500 g of solvent to raise its ionic strength to $0.250\ mol\ kg^{-1}$.

10.7 What molality of $CuSO_4$ has the same ionic strength as $1.00\ mol\ kg^{-1}\ KCl(aq)$?

10.8 Express the mean activity coefficient of an ion in a solution of $CaCl_2$ in terms of the activity coefficients of the individual ions.

10.9 The mean activity coefficient in an $0.500\ mol\ kg^{-1}$ $LaCl_3(aq)$ solution is 0.303 at 25°C. What is the percentage error in the value predicted by the Debye–Hückel limiting law?

10.10 The mean activity coefficients of HBr in three dilute aqueous solutions at 25°C are 0.930 (at $5.0\ mmol\ kg^{-1}$), 0.907 (at $10.0\ mmol\ kg^{-1}$), and 0.879 (at $20.0\ mmol\ kg^{-1}$). Estimate the value of B in the extended Debye–Hückel law.

10.11 For CaF_2, $K_s = 3.9 \times 10^{-11}$ at 25°C and the standard Gibbs function of formation of $CaF_2(s)$ is $-1167\ kJ\ mol^{-1}$. Calculate the standard Gibbs function of formation of $CaF_2(aq)$.

10.12 Consider a hydrogen electrode in aqueous HBr solution at 25°C operating at 1.15 atm. Calculate the change in the electrode potential when the molality of the acid is changed from $5.0\ mmol\ kg^{-1}$ to $20.0\ mmol\ kg^{-1}$. Activity coefficients are given in Exercise 10.10.

10.13 Devise a cell in which the cell reaction is $Mn(s) + Cl_2(g) \rightarrow MnCl_2(aq)$. Give the half-reactions for the electrodes and from the standard cell potential of 2.54 V deduce the standard potential of the Mn^{2+}/Mn couple.

10.14 Write the cell reactions and electrode half-reactions for the following cells:
(a) $Zn \mid ZnSO_4(aq) \parallel AgNO_3(aq) \mid Ag$
(b) $Cd \mid CdCl_2(aq) \parallel HNO_3(aq) \mid H_2(g) \mid Pt$
(c) $Pt \mid K_3[Fe(CN)_6](aq), K_4[Fe(CN)_6](aq) \parallel CrCl_3(aq) \mid Cr$
(d) $Pt \mid Cl_2(g) \mid HCl(aq) \parallel K_2CrO_4(aq) \mid Ag_2CrO_4(s) \mid Ag$
(e) $Pt \mid Fe^{3+}(aq), Fe^{2+}(aq) \parallel Sn^{4+}(aq), Sn^{2+}(aq) \mid Pt$
(f) $Cu \mid Cu^{2+}(aq) \parallel Mn^{2+}(aq), H^+(aq) \mid MnO_2(s) \mid Pt$

10.15 Devise cells in which the following are the reactions:
(a) $Zn(s) + CuSO_4(aq) \rightarrow ZnSO_4(aq) + Cu(s)$
(b) $2AgCl(s) + H_2(g) \rightarrow 2HCl(aq) + 2Ag(s)$
(c) $2H_2(g) + O_2(g) \rightarrow 2H_2O(l)$
(d) $2Na(s) + 2H_2O(l) \rightarrow 2NaOH(aq) + H_2(g)$
(e) $H_2(g) + I_2(s) \rightarrow 2HI(aq)$

10.16 Use standard reduction potentials to calculate the standard potentials of the cells in Exercises 10.14 and 10.15.

10.17 (a) Calculate the standard cell potential of $Hg(l) \mid HgCl_2(aq) \parallel TlNO_3(aq) \mid Tl$ at 25°C. (b) Calculate the cell potential when the activity of the Hg^{2+} ion is 0.150 and that of the Tl^+ ion is 0.93.

10.18 Calculate the standard Gibbs functions at 25°C of the following reactions from the reduction potential data in Table 10.5:
(a) $2Na(s) + 2H_2O(l) \rightarrow 2NaOH(aq) + H_2(g)$
(b) $2K(s) + 2H_2O(l) \rightarrow 2KOH(aq) + H_2(g)$
(c) $K_2S_2O_8(aq) + 2KI(aq) \rightarrow I_2(s) + 2K_2SO_4(aq)$
(d) $Pb(s) + Zn(NO_3)_2(aq) \rightarrow Pb(NO_3)_2(aq) + Zn(s)$

10.19 The standard Gibbs function of the reaction

$$K_2CrO_4(aq) + 2Ag(s) + 2FeCl_3(aq)$$
$$\rightarrow Ag_2CrO_4(s) + 2FeCl_2(aq) + 2KCl(aq)$$

is $-62.5 \text{ kJ mol}^{-1}$ at 298 K. (a) Calculate the standard potential of the corresponding galvanic cell and (b) the standard reduction potential of the $Ag_2CrO_4/Ag, CrO_4^{2-}$ couple.

10.20 Use the Debye–Hückel limiting law and the Nernst equation to estimate the potential of the cell

$$Ag \mid AgBr(s) \mid KBr(aq, 0.050 \text{ mol kg}^{-1}) \parallel$$
$$Cd(NO_3)_2(aq, 0.010 \text{ mol kg}^{-1}) \mid Cd$$

at 25°C.

10.21 Use the information in Tables in 10.1 and 10.5 to calculate the standard potential of the cell $Ag \mid AgNO_3(aq) \parallel Fe(NO_3)_2(aq) \mid Fe$ and the standard Gibbs function and enthalpy of the cell reaction at 25°C. Estimate the value of $\Delta G^{\ominus}$ at 35°C.

10.22 The solubility product of $Cu_3(PO_4)_2$ is 1.3×10^{-37}. Calculate (a) the solubility of $Cu_3(PO_4)_2$, (b) the potential of the cell

$$Pt \mid H_2(g) \mid HCl(aq, pH = 0) \parallel Cu_3(PO_4)_2(aq, \text{satd.}) \mid Cu$$

at 25°C.

10.23. Calculate the equilibrium constants of the following reactions at 25°C from reduction potential data:
(a) $Sn(s) + Sn^{4+}(aq) \rightleftharpoons 2Sn^{2+}(aq)$
(b) $Sn(s) + 2AgCl(s) \rightleftharpoons SnCl_2(aq) + 2Ag(s)$
(c) $2Ag(s) + Cu(NO_3)_2(aq) \rightleftharpoons Cu(s) + 2AgNO_3(aq)$
(d) $Sn(s) + CuSO_4(aq) \rightleftharpoons Cu(s) + SnSO_4(aq)$
(e) $Cu^{2+}(aq) + Cu(s) \rightleftharpoons 2Cu^+(aq)$

10.24. The solubilities of $AgCl$ and $BaSO_4$ in water are $1.34 \times 10^{-5} \text{ mol kg}^{-1}$ and $9.51 \times 10^{-4} \text{ mol kg}^{-1}$ respectively at 25°C. Calculate their solubility products. Is there any significant effect when activity coefficients are ignored?

10.25. Derive an expression for the potential of an electrode for which the half-reaction is the reduction of $Cr_2O_7^{2-}$ ions to Cr^{3+} ions in acidic solution.

10.26. The zero-current potential of the cell $Pt \mid H_2(g) \mid HCl(aq) \mid AgCl(s) \mid Ag$ was 0.322 V at 25°C. What is the pH of the electrolyte solution?

10.27. The solubility of $AgBr$ is $2.6 \ \mu\text{mol kg}^{-1}$ at 25°C. What is the zero-current potential of the cell $Ag \mid AgBr(aq) \mid AgBr(s) \mid Ag$ at that temperature?

10.28. The standard potential of the cell $Ag \mid AgI(s) \mid AgI(aq) \mid Ag$ is 0.9509 V at 25°C. Calculate (a) the solubility of AgI and (b) its solubility product.

Problems

Numerical problems

10.1 Devise a cell in which the overall reaction is

$$Pb(s) + Hg_2SO_4(s) \rightarrow PbSO_4(s) + 2Hg(l)$$

What is its potential when the electrolyte is saturated with both salts at 25°C?

10.2 A fuel cell develops an electric potential from the chemical reaction between reagents supplied from an outside source. What is the zero-current potential of a cell fuelled by (a) hydrogen and oxygen, (b) the combustion of butane at 1.0 atm and 298 K?

10.3 Hydrogen behaves imperfectly at high pressure, and its equation of state may be summarized by the virial equation

$$\frac{pV_m}{RT} = 1 + 5.37 \times 10^{-4} \times (p/\text{atm}) + 3.5 \times 10^{-8} \times (p/\text{atm})^2$$

The zero-current potential of the cell

$$Pt \mid H_2(g) \mid HCl(aq, 0.10 \text{ mol kg}^{-1}) \mid Hg_2Cl_2(s) \mid Hg(l)$$

has been measured up to 1000 atm (W. R. Hainsworth, H. J. Rowley, and D. A. McInnes, *J. Am. chem. Soc.* **46,** 1437 (1924)), and good agreement between experiment and a theoretical expression was achieved up to 600 atm on the assumption that the only significant volume change was that associated with the hydrogen. Above 600 atm the compressibility of the other components became significant. What is the cell potential at 500 atm? (Use activity coefficients from Table 10.4.)

10.4 The fugacity of a gas can be determined electrochemically from the pressure dependence of the reduction potential of a gas electrode. The zero-current potential of the cell

$$Pt \mid H_2(g, p^{\ominus}) \mid HCl(aq, 0.010 \text{ mol kg}^{-1}) \mid Cl_2(g, p) \mid Pt$$

was as follows at 298 K:

p/bar	1.000	50.00	100.0
E/V	1.5962	1.6419	1.6451

10 | Equilibrium electrochemistry

Calculate the fugacities of chlorine at the three pressures. (Use activity coefficients from Table 10.4.)

10.5 The zero-current potential of the cell

$$Pt \,|\, H_2(g, p^{\ominus}) \,|\, HCl(aq, m) \,|\, Hg_2Cl_2(s) \,|\, Hg(l)$$

has been measure with high precision (G. J. Hills and D. J. G. Ives, *J. chem. Soc.* 311 (1951)), with the following results at 25°C:

$m/(\text{mmol kg}^{-1})$	1.6077	3.0769	5.0403	7.6938	10.9474
E/V	0.60080	0.56825	0.54366	0.52267	0.50532

Find the standard potential of the cell and the mean activity coefficient of HCl at these molalities. (Make a least-squares fit of the data to the best straight line.)

10.6 Careful measurements of the potential of the cell

$$Pt \,|\, H_2(g, p^{\ominus}) \,|\, NaOH(aq, 0.0100 \text{ mol kg}^{-1}),$$
$$NaCl(aq, 0.01125 \text{ mol kg}^{-1}) \,|\, AgCl(s) \,|\, Ag$$

have been reported (C. P. Bezboruah, M. F. G. F. C. Camoes, A. K. Covington, and J. V. Dobson, *J. chem. Soc. Faraday Trans.* I, **69**, 949 (1973)). Among the data is the following information:

$\theta/°C$	20.0	25.0	30.0
E/V	1.04774	1.04864	1.04942

Calculate pK_w at these temperatures and the standard enthalpy and entropy of the autoprotolysis of water at 25.0°C.

10.7 Measurements of the potentials of cells of the type

$$Ag \,|\, AgX(s) \,|\, MX(m_1) \,|\, M_xHg \,|\, MX(m_2) \,|\, AgX(s) \,|\, Ag$$

where M_xHg denotes an amalgam and the electrolyte is an alkali metal halide dissolved in ethylene glycol have been reported (U. Sen, *J. chem. Soc. Faraday Trans.* I, **69**, 2006 (1973)) and a selection of values for LiCl are given below. Estimate the activity coefficient at the concentration marked * and then use this value to calculate activity coefficients from the measured cell potential at the other concentrations. Base your answer on the following version of the extended Debye–Hückel law:

$$\lg \gamma_{\pm} = \frac{-Am^{1/2}}{1 + Bm^{1/2}} + km$$

with $A = 1.461/(\text{mol kg}^{-1})^{1/2}$, $B = 1.70/(\text{mol kg}^{-1})$, and $k = 0.20/(\text{mol kg}^{-1})$. For $m_2 = 0.09141 \text{ mol kg}^{-1}$,

$m_1/(\text{mol kg}^{-1})$	0.0555	0.09141	0.1652	0.2171	1.040	1.350
E/V	−0.0220	0.0000	0.0263	0.0379	0.1156	0.1336

10.8 Suppose the extended Debye–Hückel law for a 1,1-electrolyte is written in the simplified form

$$\lg \gamma_{\pm} = -0.509(m/m^{\ominus})^{1/2} + k(m/m^{\ominus})$$

where k is a constant. Show that a plot of y against $m/m^{\ominus}$,

where

$$y = E + 0.1183 \lg (m/m^{\ominus}) - 0.0602(m/m^{\ominus})^{1/2}$$

should give a straight line with intercept $E^{\ominus}$ and slope $-0.1183k$. Apply the technique to the following data (at 25°C) on the cell

$$Pt \,|\, H_2(g, p^{\ominus}) \,|\, HCl(aq, m) \,|\, AgCl(s) \,|\, Ag$$

$m/(\text{mmol kg}^{-1})$	123.8	25.63	9.138	5.619	3.215
E/mV	341.99	418.24	468.60	492.57	520.53

(a) Find the standard cell potential and the standard reduction potential of the $AgCl/Ag, Cl^-$ couple. (b) The zero-current cell potential was measured as 352.4 mV when $m = 100.0 \text{ mmol kg}^{-1}$. What is the pH and the mean ionic activity coefficient?

10.9 The mean activity coefficients for aqueous solutions of NaCl at 25°C are given below. Confirm that they support the Debye–Hückel limiting law and that an improved fit is obtained with the extended law.

$m/(\text{mmol kg}^{-1})$	1.0	2.0	5.0	10.0	20.0
$\gamma_{\pm}$	0.9649	0.9519	0.9275	0.9024	0.8712

10.10 The standard reduction potential of the $AgCl/Ag, Cl^-$ couple has been measured very carefully over a range of temperature (R. G. Bates and V. E. Bowers, *J. Res. Nat. Bur. Stand.* **53**, 283 (1954)) and the results were found to fit the expression

$$E^{\ominus}/V = 0.23659 - 4.8564 \times 10^{-4}(\theta/°C)$$
$$- 3.4205 \times 10^{-6}(\theta/°C)^2 + 5.869 \times 10^{-9}(\theta/°C)^3$$

Calculate the standard Gibbs function and enthalpy of formation of $Cl^-(aq)$ and its entropy at 298 K.

10.11 The osmotic coefficient ϕ (Problem 7.19) is related (a) to the depression of freezing point of a 1,1-electrolyte by $\phi = \Delta T/2mK_f$, and (b) to the mean activity coefficient $\gamma_{\pm}$ by

$$-\ln \gamma_{\pm} = 1 - \phi + \int_0^m \frac{1-\phi}{m} dm$$

Deduce the mean activity coefficient of $0.050 \text{ mol kg}^{-1}$ $KCl(aq)$ from the following information:

$m/(\text{mol kg}^{-1})$	0.010	0.020	0.030	0.040	0.050
$\Delta T/K$	0.0355	0.0697	0.1031	0.137	0.172

Use the Debye–Hückel limiting law for the integration as $m \to 0$.

10.12 Use the data below to confirm that the Debye–Hückel limiting law correctly predicts the limiting values of the mean activity coefficient of acetic acid by demonstrating that pK_a', where $K_a = K_a'K_{\gamma}$, plotted against $(\alpha m)^{1/2}$, where m is the molality of the acid and α its degree of ionization, should be a straight line.

$m/(\text{mmol kg}^{-1})$	0.0280	0.1114	0.2184	1.0283	2.414	5.9115
α	0.5393	0.3277	0.2477	0.1238	0.0829	0.0540

10.13 The $Sb\,|\,Sb_2O_3(s)\,|\,OH^-(aq)$ electrode is reversible with respect to OH^- ions. Derive an expression for its potential in terms of (a) the pOH and (b) the pH of the solution. By how much does the potential change when the molality of $NaOH(aq)$ in the electrode compartment is increased from 0.010 mol kg^{-1} to 0.050 mol kg^{-1} at 25°C? Use the Debye–Hückel limiting law to estimate any activity coefficients required.

10.14 Superheavy elements are now of considerable interest. Shortly before it was (falsely) believed that the first had been discovered, an attempt was made to predict the chemical properties of ununpentium (element 115, O. L. Keller, C. W. Nestor, and B. Fricke, *J. phys. Chem.* **78**, 1945 (1974)). In one part of the paper the standard enthalpy and entropy of the reaction

$$Uup^+(aq) + \tfrac{1}{2}H_2(g) \rightarrow Uup(s) + H^+(aq)$$

were estimated from the following data: $\Delta H_{sub}(Uup) = 1.5\,eV$, $E_i(Uup) = 5.2\,eV$, $\Delta H_H^\ominus(Uup^+, aq) = -3.22\,eV$, $S^\ominus(Uup^+, aq) = 1.34\,meV\,K^{-1}$, $S^\ominus(Uup, s) = 0.69\,meV\,K^{-1}$. Estimate the expected standard reduction potential of the Uup^+/Uup couple.

Theoretical problems

10.15 Show that the solubility S of a sparingly soluble 1:1 salt is related to its solubility product by

$$S = K_s^{1/2}e^{1.172S^{1/2}}$$

10.16 Suppose that a sparingly soluble salt MX has solubility product K_s and solubility S. Show that in an ideal solution that is of concentration C of a freely soluble salt NX the solubility of MX is changed to

$$S' = \tfrac{1}{2}(C^2 + 4K_s)^{1/2} - \tfrac{1}{2}C$$

and that $S' \approx K_s/C$ when K_s is small (in a sense to be specified).

10.17 Show that if the ionic strength of a solution of the sparingly soluble salt MX and the freely soluble salt NX is dominated by the concentration of the latter, and if it is valid to use the Debye–Hückel limiting law, then the solubility S' in the mixed solution is given by

$$S' \approx \frac{K_s e^{4.606AC^{1/2}}}{C}$$

when K_s is small (in a sense to be specified).

10.18 Show that the freezing-point depression of a real solution in which the solvent of molar mass M has activity a_A obeys

$$\frac{d\ln a_A}{d(\Delta T)} = \frac{-M}{K_f}$$

and use the Gibbs–Duhem equation to show that

$$\frac{d\ln a_B}{d(\Delta T)} = \frac{1}{m_B K_f}$$

where a_B is the solute activity and m_B is its molality. Use the Debye–Hückel limiting law to show that the osmotic coefficient ϕ (Problem 7.19) is given by

$$\phi = 1 - \tfrac{1}{3}A'm^{1/2}$$

with $A' = 2.303A$.

Part 2

Structure

In Part 1 we examined the properties of bulk matter from the viewpoint of thermodynamics. In Part 2 we turn to the study of individual atoms and molecules from the viewpoint of quantum mechanics. We shall see the viewpoints merge in Chapter 19.

Quantum theory: introduction and principles

Check-list of key ideas

1. The central concepts of *classical physics*, particularly that of a *trajectory* calculated using *Newton's laws of motion* (Section 11.1).

2. The classical description of *rotation* and *harmonic vibration*, and the excitation of these modes of motion to any energy (Section 11.1).

3. The classical *equipartition theorem* for calculating the mean energy of any mode of motion (Section 11.1).

4. The characteristics of *black-body radiation*, the failure of the classical *Rayleigh–Jeans law* (eqn 10) to account for it, and the success of the *Planck distribution* (eqn 12) based on the *quantization* of energy.

5. The classical interpretation of the *Dulong and Petit Law* for heat capacity (eqn 13), its failure at low temperatures, and the success of the *Einstein formula* (eqn 14) based on the quantization of material oscillators.

6. The interpretation of the *photoelectric effect* in terms of photons of radiation (eqn 15).

7. The interpretation of the *Compton effect* in terms of the linear momentum of photons (eqn 16).

8. The *de Broglie relation* (eqn 19) between the momentum of a particle and its wavelength and its confirmation by the diffraction of electrons.

9. The *wavefunction* of a system and the *Schrödinger equation* for calculating the wavefunction of a system (eqn 20).

10. The interpretation of the curvature of the wavefunction in terms of the *kinetic energy* of the particle (eqn 23).

11. The *Born interpretation* of the wavefunction as a *probability amplitude* for the location of a particle (Section 11.4).

12. The *constraints* on physically acceptable wavefunctions, and their implication that the energy of a system is quantized (Section 11.4).

13. The role of *operators* in quantum mechanics and the significance of *eigenfunctions* and *eigenvalues* (Section 11.5).

14. The construction and interpretation of *superpositions* (eqn 32) and the significance of an *expectation value* (eqn 33) as a quantum mechanical average value.

15. The *uncertainty principle* (eqn 34), its significance, and its interpretation in terms of superpositions.

16. The concept of *complementary observables* in quantum mechanics (Section 11.7).

To understand the behaviour of individual atoms and molecules, and of the electrons and nuclei inside them, we need to know how particles move in response to the forces acting on them. It was once thought that the motion of atomic and subatomic particles could be expressed using the laws of **classical mechanics** introduced in the 17th century by Isaac Newton, for these laws were very successful at explaining the motion of everyday objects and planets. However, towards the end of the 19th century, experimental evidence accumulated showing that classical mechanics failed when it was applied to very small particles, and it took until about 1926 to discover the appropriate concepts and equations for describing them. We describe the concepts of this new mechanics, which is called **quantum mechanics,** in this chapter, and apply them throughout the remainder of the text.

Classical mechanics

In this section we introduce some of the central concepts of classical mechanics and illustrate how they are used in simple cases. Then we shall see how observations led to the realization that the concepts were inapplicable to atoms and molecules and that they had to be replaced.

11.1 The equations of classical physics

We can see how classical mechanics describes objects by considering the implications of two equations. One equation is constructed on the basis that the total energy is constant. The other is constructed in terms of the response of particles to the forces acting on them.

The trajectory in terms of the energy

The total energy of a particle is the sum of its **kinetic energy** E_K, the energy arising from its motion, and its **potential energy** V, the energy arising from its position:

$$E = E_K + V \tag{1a}$$

Since $E_K = \frac{1}{2}mv^2$, where v is its speed at the time of interest and m is its mass,

$$E = \frac{1}{2}mv^2 + V \tag{1b}$$

In terms of the **linear momentum,** $p = mv$, we can write

$$E = \frac{p^2}{2m} + V \qquad (1c)$$

We can use these equations in a number of ways. For example, since $v = \mathrm{d}x/\mathrm{d}t$, we can rearrange eqn 1b into a differential equation for x as a function of t:

$$\frac{\mathrm{d}x}{\mathrm{d}t} = \left\{ \frac{2(E-V)}{m} \right\}^{1/2} \qquad (2)$$

The solution of this equation for a given total energy gives the position of the particle as a function of time. Then, by substituting that value of x into the appropriate formula for the potential energy V, we can calculate the speed (and hence the momentum) of the particle at that instant. Since a statement of both $x(t)$ and $p(t)$ is called the **trajectory** of the particle, eqn 2 allows us to predict the trajectory of the particle exactly.

The simplest example of the procedure is that of a particle in a uniform, constant potential, with V independent of x and t. We can set $V = 0$ for simplicity, and then solve

$$\frac{\mathrm{d}x}{\mathrm{d}t} = \left(\frac{2E}{m} \right)^{1/2}$$

The solution is

$$x(t) = x(0) + \left(\frac{2E}{m} \right)^{1/2} t$$

Since the constant energy E is related to the initial momentum $p(0)$ by $E = p(0)^2/2m$, the trajectory is

$$x(t) = x(0) + \frac{p(0)t}{m} \qquad p(t) = m\frac{\mathrm{d}x}{\mathrm{d}t} = p(0) \qquad (3)$$

Hence, if we know the initial position and momentum, we can predict all later positions and momenta.

Newton's second law

The second basic equation of classical mechanics is **Newton's second law of motion,** that the rate of change of momentum is equal to the force F acting on the particle:

$$\frac{\mathrm{d}p}{\mathrm{d}t} = F \qquad (4a)$$

Since $p = m\,\mathrm{d}x/\mathrm{d}t$, it is sometimes more convenient to write this equation as

$$m\frac{\mathrm{d}^2x}{\mathrm{d}t^2} = F \qquad (4b)$$

and to note that d^2x/dt^2 is the **acceleration** of the particle. It follows that if we know the force acting everywhere and at all times, then solving eqn 4 will also give the trajectory. This calculation is equivalent to the one based on E, but is more suitable in some applications.

For example, consider the case of a particle that is subjected to a constant force F for a time τ, and is then allowed to travel freely. Equation 4a becomes

$$\frac{dp}{dt} = F, \text{ a constant, for times between } t = 0 \text{ and } t = \tau$$

$$\frac{dp}{dt} = 0 \text{ for times later than } t = \tau$$

The first equation has the solution

$$p(t) = p(0) + Ft \quad (0 < t < \tau)$$

and at the end of the period τ the momentum is

$$p(\tau) = p(0) + F\tau$$

The second equation has the solution $p = $ constant; so, at all times later than τ, the particle has the momentum $p(\tau)$. For simplicity, we shall suppose that the particle is initially at rest, and set $p(0) = 0$. Since the kinetic energy is $p^2/2m$, in this example its value is

$$E = \frac{F^2\tau^2}{2m}$$

at all times after the force ceases to act. We see that the total energy of the accelerated particle is increased to E by the force. Moreover, since the applied force F and the time τ for which it acts may take any value, the energy may be increased to any value.

Rotational motion

The same type of calculation may be applied to free rotational motion. This is an important type of motion that we shall encounter throughout the following chapters, particularly when discussing the motion of molecules in the gas phase. The **angular momentum** J of a particle is

$$J = I\omega \tag{5}$$

where ω is its **angular velocity,** its rate of change of angular position, and I is the **moment of inertia.**[1] For a point particle of mass m moving in a circle of radius r, $I = mr^2$. To accelerate a rotation it is necessary to apply a **torque** T, a twisting force, and Newton's equation is

$$\frac{dJ}{dt} = T$$

If a constant torque is applied for a time τ, the rotational energy of an

[1] The analogous roles of m and I, of v and ω, and of p and J in the translational and rotational cases respectively should be remembered, because they provide a ready way of constructing and recalling equations.

initially stationary body is increased to

$$E = \frac{T^2 \tau^2}{2I}$$

The implication of this equation is that an appropriate torque and period for which it is applied can excite the rotation to an arbitrary energy.

The harmonic oscillator

A third fundamental type of motion is oscillatory motion, such as the vibration of atoms in a bond. A **harmonic oscillator** (the only type we consider at this stage) consists of a particle that experiences a restoring force proportional to its displacement:

$$F = -kx \tag{6}$$

The constant of proportionality k is called the **force constant,** and the stronger the spring the greater the force constant. The negative sign in F signifies that the direction of the force is opposite to that of the displacement: when x is positive (displacement to the right), the force is negative (pushing towards the left), and vice versa.

Example 11.1: *Estimating a force constant*

Estimate the force constant of a spring that extends by 1.0 cm when a 100-g mass is attached to it (on the surface of the earth).

Answer. The gravitational force on the mass is

$$F_G = 100 \text{ g} \times 9.81 \text{ m s}^{-2} = 0.981 \text{ kg m s}^{-2}$$
$$= 0.981 \text{ N}$$

(since $1 \text{ N} = 1 \text{ kg m s}^{-2}$). At equilibrium this force is balanced by the force arising from the extension of the spring, so equating F_G and kx when $x = 1.0$ cm gives

$$k = \frac{0.981 \text{ N}}{1.0 \text{ cm}} = 98 \text{ N m}^{-1}$$

Comment. A typical force constant of a chemical bond is about 500 N m^{-1}.

Exercise. Calculate the force needed to stretch an HCl molecule by 5.0 pm given that $k = 516 \text{ N m}^{-1}$.   [2.6 nN]

Equation 4b becomes

$$m \frac{d^2 x}{dt^2} = -kx$$

and a solution is

$$x = A \sin \omega t \qquad p = m\omega A \cos \omega t \qquad \omega = \left(\frac{k}{m}\right)^{1/2} \tag{7}$$

We see that the position of the particle varies **harmonically** (as $\sin \omega t$) with a frequency $v = \omega/2\pi$ and that it is stationary when the displacement x has its maximum value A, which is called the **amplitude** of the motion.

Example 11.2: *Using Newton's law*

Devise the relation between the total energy of a harmonic oscillator and the amplitude of its motion.

Answer. The total energy is the sum of the kinetic and potential energies. The potential energy is related to the force by $F = -dV/dx$, and since in this case $F = -kx$, it follows that

$$V = \tfrac{1}{2}kx^2$$

if $V = 0$ at $x = 0$. The total energy is therefore

$$E = \frac{p^2}{2m} + V = \frac{p^2}{2m} + \tfrac{1}{2}kx^2$$

$$= \tfrac{1}{2}kA^2$$

by substitution of the expressions for the momentum and displacement and use of $\sin^2 \omega \tau + \cos^2 \omega \tau = 1$.

Exercise. At what displacement is the acceleration of the oscillator greatest?

$$[x = \pm A]$$

We saw in Example 11.2 that the total energy of a harmonic oscillator at any instant is

$$E = \tfrac{1}{2}kA^2$$

This equation implies that the energy of the oscillating particle can be raised to any value by a suitably controlled impulse that knocks it to an amplitude A. It is important to note that the frequency of the motion depends only on the structure of the oscillator (as represented by k and m) and is independent of the energy; the amplitude governs the energy, through $E = \tfrac{1}{2}kA^2$, and is independent of the frequency.

The equipartition theorem

One very useful conclusion from classical mechanics that we shall use several times in the chapter concerns the average energy of a collection of particles in thermal equilibrium at a temperature T. The **equipartition theorem** states:

At a temperature T, the average value of each quadratic term in the expression for the total energy is $\tfrac{1}{2}kT$

By a 'quadratic term' we mean one of the form x^2, p^2, and so on. The constant k is **Boltzmann's constant,**

$$k = 1.381 \times 10^{-23} \, \text{J K}^{-1}$$

k will figure extensively later in the text, and we shall see that the gas constant R is the product of k and Avogadro's constant, $R = kN_A$.

It follows from the equipartition theorem that the average energy of a collection of particles that can move freely in one dimension is $\tfrac{1}{2}kT$, because in this case the energy has the form $p^2/2m$, with p the linear momentum along x. For motion in three dimensions (as for a gas in a container), the total energy is the sum of three such quadratic terms:

$$E = \frac{1}{2m}(p_x^2 + p_y^2 + p_z^2)$$

Therefore, the average energy of a particle is $\frac{3}{2}kT$.

Example 11.3: *Using the equipartition theorem*

What is the average energy of a collection of harmonic oscillators in thermal equilibrium at a temperature T?

Answer. The expression for the total energy is given in Example 11.2:

$$E = \frac{p^2}{2m} + \frac{1}{2}kx^2$$

It has two quadratic terms: one is proportional to p^2 and another to x^2; therefore, according to the equipartition theorem, the average energy of the oscillators is $2 \times \frac{1}{2}kT$, or kT.

Exercise. What is the average energy of a collection of objects that are free to rotate around a single axis? $[\frac{1}{2}kT]$

11.2 The failures of classical physics

The lessons of the examples we have described are that classical physics (1) predicts a precise trajectory and (2) allows the translational, rotational, and vibrational modes of motion to be excited to any energy simply by controlling the forces, torques, and impulses we apply. These conclusions agree with everyday experience. Everyday experience, however, does not extend to individual atoms, and careful experiments of the type we describe below have shown that the laws of classical mechanics fail when we deal with the transfers of very small quantities of energy. Classical mechanics is in fact only an approximate description of the motion of particles, and fails when small masses, small moments of inertia, and small transfers of energy are involved.

Black-body radiation

A hot object emits electromagnetic radiation. At high temperatures, an appreciable proportion of the radiation is in the visible region of the spectrum, and a higher proportion of short-wavelength blue light is generated as the temperature is raised. This behaviour is seen when a heated iron bar glowing red hot becomes white hot when heated further. The precise dependence is illustrated in Fig. 11.1, which shows how the energy output depends on the wavelength at several temperatures. The curves are those of an ideal emitter called a **black body,** which is an object capable of emitting and absorbing all frequencies of radiation uniformly. A good approximation to a black body is a pin-hole in a container, because the radiation leaking out of the hole has been absorbed and reemitted inside so many times that it has come to thermal equilibrium with the walls.

Figure 11.1 shows two main features. The first is that the peak shifts to smaller wavelengths as the temperature is raised, and the short-wavelength tail spreads through the whole of the visible region. As a result, the

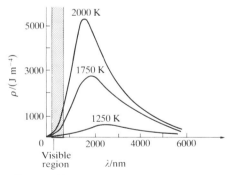

Fig. 11.1 The energy per unit volume per unit wavelength range in a black-body cavity at several temperatures. Note how the energy density increases in the visible region as the temperature is raised, and how the peak shifts to shorter wavelengths. The total energy density (the area under the curve) increases as the temperature is increased (as T^4).

291

perceived colour shifts towards the blue, as already mentioned. An analysis of the data led Wilhelm Wien (in 1893) to a conclusion now known as **Wien's Displacement Law** and which summarizes this behaviour:

$$T\lambda_{max} = \tfrac{1}{5}c_2 \qquad c_2 = 1.44 \text{ cm K} \tag{8}$$

The constant c_2 is called the **second radiation constant.** Using its value, we can predict that $\lambda_{max} \approx 2900$ nm at 1000 K.

The second feature of the radiation had been noticed in 1879 by Josef Stefan, who considered the **total energy density** $\mathcal{U}$, the total energy per unit volume in the electromagnetic field. He concluded that the total energy density is proportional to the fourth power of the temperature, a result now known as **Stefan's Law:**

$$\mathcal{U} = aT^4 \tag{9a}$$

An alternative form of this law is in terms of the **excitance** M, the power emitted per unit area. Since the energy density of the radiation and the excitance of a black body are proportional to each other, M is also proportional to T^4 and we can write

$$M = \sigma T^4 \qquad \sigma = 5.67 \times 10^{-8} \text{ W m}^{-2} \text{ K}^{-4} \tag{9b}$$

σ is called the **Stefan–Boltzmann constant.** Stefan's law implies that 1 cm² of the surface of a black body at 1000 K radiates about 5.7 W when all wavelengths are taken into account.

Lord Rayleigh studied black-body radiation from a classical viewpoint, and thought of the electromagnetic field as a collection of harmonic oscillators. One oscillator was needed for each frequency of light, and Rayleigh regarded the presence of light of a certain frequency ν (and therefore of wavelength $\lambda = c/\nu$) as the excitation of the electromagnetic oscillator of that frequency. Rayleigh used the classical equipartition theorem for the average energy of the oscillators and, with minor help from James Jeans, arrived at the **Rayleigh–Jeans law:**

$$d\mathcal{U} = \rho \, d\lambda \qquad \rho = \frac{8\pi kT}{\lambda^4} \tag{10}$$

The factor ρ is the energy per unit volume per unit wavelength.

Unfortunately (for Rayleigh and Jeans), although the Rayleigh–Jeans formula is quite successful at large wavelengths and low frequencies, it fails badly at high frequencies. Thus, as λ decreases, ρ increases without going through a maximum (Fig. 11.2). The equation therefore predicts that oscillators of extremely small wavelength (high frequency, corresponding to ultraviolet light, X-rays, and even γ-rays) are strongly excited even at room temperature. This absurd result, the implication that a large amount of energy accumulates in the high-frequency region of the electromagnetic spectrum, is the **ultraviolet catastrophe.** Hence, according to classical physics, objects should glow in the dark: there should in fact be no darkness.

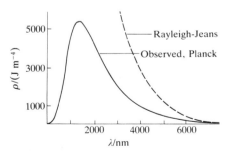

Fig. 11.2 Theoretical attempts to account for black-body radiation. The Rayleigh–Jeans law (eqn 10) leads to an infinite energy density at short wavelengths called the 'ultraviolet catastrophe'. The Planck distribution (eqn 12) is in good agreement with experiment.

The Planck distribution

Max Planck studied black-body radiation from the viewpoint of thermo-dynamics, in which he was an expert. In 1900 he found that he could account for the experimental observations by proposing that the energy of each electromagnetic oscillator is limited to discrete values and cannot be varied arbitrarily. The limitation of energy to discrete values is called the **quantization** of energy (from the Latin word *quantum*, amount). In particular, he proposed that the permitted energies of an oscillator of frequency v are integral multiplies of hv, where h is a fundamental constant now known as **Planck's constant**:

$$E = nhv \qquad h = 6.626 \times 10^{-34} \text{ J Hz}^{-1} \qquad (11)$$

with $n = 0, 1, 2, \ldots$.

Since an oscillator of frequency v can possess only the energies 0, hv, $2hv, \ldots$, Planck's work led to the view that a ray of light of that frequency can be thought of as consisting of $0, 1, 2, \ldots$ particles, each particle having an energy hv. These particles are called **photons**. If the electromagnetic field of frequency v has an energy E in some region, the number of photons of that frequency in the region is E/hv.

Example 11.4: *Calculating the number of photons*

Calculate the number of photons emitted by a 100-W yellow lamp in 1.0 s. Take the wavelength of yellow light as 560 nm and assume 100 per cent efficiency.

Answer. A 100-W lamp emits 100 J of energy in 1.0 s (if all the energy it consumes is converted into radiation). Since 560-nm photons have frequency 5.4×10^{14} Hz, each one has an energy

$$E = hv = 6.626 \times 10^{-34} \text{ J Hz}^{-1} \times 5.4 \times 10^{14} \text{ Hz}$$

$$= 3.7 \times 10^{-19} \text{ J}$$

The number of photons required to carry away 100 J of energy is therefore

$$N = \frac{100 \text{ J}}{3.7 \times 10^{-19} \text{ J}} = 2.7 \times 10^{20}$$

Exercise. How many 1000-nm photons does a 1-mW infrared rangefinder emit in 0.1 s?

[5×10^{14}]

It is quite easy to see why Planck's hypothesis is likely to be successful. The atoms in the walls of the black body are in thermal motion, and this motion excites the oscillators of the electromagnetic field. According to classical mechanics, all the oscillators of the field share equally in the energy supplied by the walls, and so even the highest frequencies are excited. According to quantum mechanics, however, the oscillators are excited only if they can acquire an energy of at least hv. This is too large for the walls to supply in the case of the high-frequency oscillators, and so the latter remain unexcited. The effect of quantization is to eliminate the contribution from the high-frequency oscillators, for they cannot be excited with the energy available.

Detailed calculation (of the kind we do in Chapter 19) shows that the energy density in the range λ to $\lambda + d\lambda$ is given by the **Planck distribution**:

$$d\mathscr{U} = \rho \, d\lambda \qquad \rho = \frac{8\pi hc}{\lambda^5}\left\{\frac{1}{e^{hc/\lambda kT} - 1}\right\} \qquad (12)$$

This expression fits the experimental curve very well at all wavelengths (Fig. 11.2) and h, which is an adjustable parameter in the theory, may be obtained by varying its value until a best fit is obtained.

The Planck distribution resembles the Rayleigh–Jeans law (eqn 10) apart from the all-important exponential factor. For short wavelengths, $hc/\lambda kT$ is large and $e^{hc/\lambda kT} \to \infty$; therefore

$$\rho \to 0 \quad \text{as} \quad \lambda \to 0 \quad \text{and} \quad \nu \to \infty$$

Hence the energy density approaches zero at high frequencies, in agreement with observation. For long wavelengths, $hc/\lambda kT \ll 1$, and

$$e^{hc/\lambda kT} - 1 = 1 + \frac{hc}{\lambda kT} + \ldots - 1 \approx \frac{hc}{\lambda kT}$$

In this case, the Planck formula reduces to the Rayleigh–Jeans law:

$$\rho \to \frac{8\pi kT}{\lambda^4} \quad \text{as} \quad \lambda \to \infty \quad \text{and} \quad \nu \to 0$$

The classical expression is also obtained if h is erroneously set equal to zero (take the limit of the Planck distribution as $h \to 0$ in the numerator and the denominator):

$$\rho \to \frac{8\pi kT}{\lambda^4} \quad \text{as} \quad h \to 0$$

Example 11.5: *Using the Planck distribution*

A spherical cavity of volume 1.0 cm^3 (e.g. the interior of a small lamp bulb) is heated to 1500 K. Calculate the energy inside the cavity due to radiation in the wavelength range 550 to 575 nm.

Answer. We use eqn 12 to calculate ρ in the middle of the wavelength range (562.5 nm):

$$\frac{8\pi hc}{\lambda^5} = 8.866 \times 10^7 \text{ J m}^{-4}$$

$$e^{hc/\lambda kT} = 2.545 \times 10^7$$

Therefore,

$$\rho = 8.866 \times 10^7 \text{ J m}^{-4} \times \frac{1}{2.545 \times 10^7 - 1} = 3.483 \text{ J m}^{-4}$$

Then we multiply by $\Delta\lambda = 25$ nm for the energy density:

$$\mathscr{U} = 3.483 \text{ J m}^{-4} \times 25 \times 10^{-9} \text{ m} = 8.7 \times 10^{-8} \text{ J m}^{-3}$$

Finally, we multiply by the volume of the cavity to obtain the total energy of the radiation in the range 550 to 575 nm:

$$\mathscr{U} = 8.7 \times 10^{-8} \text{ J m}^{-3} \times 1.0 \times 10^{-6} \text{ m}^3 = 8.7 \times 10^{-14} \text{ J}$$

Comment. Since each photon carries an energy hc/λ, it follows that there are about 3×10^5 yellow photons in the cavity. For more precise calculations, and when the wavelength range is wide, the energy density must be evaluated by integration of eqn 12 over the specified range (e.g. numerically).

Exercise. Calculate the energy of infrared radiation in the wavelength range 1000 to 1025 nm, and the number of infrared photons, in the same cavity.

$$[9.0 \times 10^{-12} \text{ J}, 5 \times 10^7]$$

The Planck distribution accounts for the Stefan and Wien laws. The former is obtained by integrating the energy density from all wavelengths from $\lambda = 0$ to $\lambda = \infty$, which gives

$$\mathcal{U} = \int \rho \, d\lambda = aT^4$$

with

$$a = \frac{4\sigma}{c} \quad \text{and} \quad \sigma = \frac{\pi^2 k^4}{60c^2 h^3}$$

Substitution of the values of the fundamental constants then gives

$$\sigma = 5.670 \times 10^{-8} \text{ W m}^{-2} \text{ K}^{-4}$$

in accord with the experimental value. The Wien law is obtained by looking for the wavelength at which $d\mathcal{U}/d\lambda = 0$; at high temperatures this occurs at

$$\lambda_{\max} T = \frac{hc}{5k}$$

which implies that the second radiation constant c_2 is

$$c_2 = \frac{hc}{k}$$

The value of c_2 works out as 1.439 cm K, in good agreement with experiment.

Heat capacities

Accounting for black-body radiation involves examining how energy is taken up by the electromagnetic field. Accounting for the heat capacities of solids involves examining how energy is taken up by the vibrations of atoms (the only way they can store energy if they cannot rotate or travel through the sample). We can suspect that a study of heat capacities might also show evidence of the quantization of energy.

If classical physics were valid, the mean energy of vibration of each atom oscillating in one dimension in a solid would be kT (Example 11.3). Since the N atoms in a block are free to vibrate in three dimensions, the total vibrational energy of a block is expected to be $3NkT$. The contribution of the vibrational energy to the molar internal energy is therefore

$$U_{\text{m}} = 3N_{\text{A}} kT = 3RT$$

Since the heat capacity at constant volume (Section 2.8) is $C_V = (\partial U_{\text{m}}/\partial T)_V$, classical physics predicts that

$$C_V = 3R \tag{13}$$

independent of the temperature. This result is known as **Dulong and Petit's Law**, for it had been proposed by them on the basis of some somewhat slender experimental evidence. In the early days of physical chemistry the law was used to estimate the molar masses of elements from measurements of heat capacities.

Einstein turned his attention to heat capacities when technological advances made it possible to test Dulong and Petit's Law at low temperatures. Significant deviations were observed: all metals were found to have molar heat capacities lower than $3R$ at low temperatures, and the values appeared to approach zero as $T \to 0$. To account for these observations, Einstein assumed that each atom oscillated about its equilibrium position with a single frequency v, and invoked Planck's hypothesis to assert that the energy of any oscillation is nhv, where n is an integer. First he calculated the molar vibrational energy of the metal (by a method we describe in Section 20.4) and obtained

$$U_m = \frac{3N_A hv}{e^{hv/kT} - 1}$$

(Notice the similarity of this to the Planck distribution.) Then he found the heat capacity by differentiating with respect to T, and obtained a result now known as the **Einstein formula**:

$$C_V = 3R\left(\frac{hv}{kT}\right)^2 \left\{\frac{e^{hv/2kT}}{e^{hv/kT} - 1}\right\}^2 \qquad (14)$$

At high temperatures ($kT \gg hv$), the exponentials in the Einstein formula can be expanded as $1 + hv/kT + \ldots$ and higher terms ignored. The result is

$$C_V = 3R\left(\frac{hv}{kT}\right)^2 \left\{\frac{1 + hv/2kT + \ldots}{1 + hv/kT \ldots - 1}\right\}^2 = 3R$$

in agreement with the classical result. At low temperatures,

$$e^{hv/kT} \to \infty \quad \text{implying} \quad C_V \to 0$$

and so Einstein's formula accounts for the decrease of heat capacity at low temperatures. The physical reason for this success is that, as in the Planck calculation, at low temperatures only a few oscillators possess enough energy to begin oscillating. At higher temperatures, there is enough energy available for all the oscillators to become active: all $3N$ oscillators contribute, and the heat capacity approaches its classical value.

The overall temperature dependence predicted by the Einstein formula is plotted in Fig. 11.3: the general shape of the curve is satisfactory, but the numerical agreement is in fact quite poor. The poor fit arises from Einstein's assumption that all the atoms oscillate with the same frequency whereas in fact, they oscillate with a range of frequencies. This complication is taken into account by averaging over all the frequencies present, the final result being the **Debye formula**. The details of this modification, which, as Fig. 11.3 shows, gives improved agreement with experiment, need not distract us at this stage from the main conclusion, which is that quantization must be introduced in order to explain the thermal properties of solids.

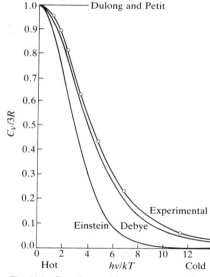

Fig. 11.3 Experimental low-temperature heat capacities and theoretical predictions. The Dulong and Petit law predicts no falling off at low temperatures; the Einstein formula (eqn 14) predicts the general temperature dependence fairly well, but is everywhere too low. Debye's modification of Einstein's calculation gives very good agreement with experiment. For copper, $hv/kT = 2$ corresponds to about 170 K, and so the detection of deviations from Dulong and Petit's Law had to await advances in low-temperature physics.

The photoelectric effect

Evidence for quantization also comes from the measurement of the energies of electrons produced by the **photoelectric effect**, the ejection of electrons from metals when they are exposed to ultraviolet radiation. The characteristics of the photoelectric effect are as follows:

(1) No electrons are ejected, regardless of the intensity of the radiation, unless its frequency exceeds a threshold value characteristic of the metal.
(2) The kinetic energy of the ejected electrons is linearly proportional to the frequency of the incident radiation but independent of its intensity.
(3) Even at low light intensities, electrons are ejected immediately if the frequency is above threshold.

These observations strongly suggest that the photoelectric effect depends on an electron being ejected if it is involved in a collision with a particle-like projectile carrying enough energy to knock it out of the metal. If we suppose that the projectile is a photon of energy $h\nu$, where ν is the frequency of the radiation, the conservation of energy requires that the kinetic energy of the electron should obey

$$\tfrac{1}{2}m_e v^2 = h\nu - \Phi \tag{15}$$

where Φ is the **work function** of the metal, the energy required to remove an electron (the analogue of ionization energy for atoms). If $h\nu < \Phi$, photoejection cannot occur because the photon brings insufficient energy: this accounts for observation (1). The expression predicts that the kinetic energy of an ejected electron should be proportional to the frequency, in agreement with observation (2). It therefore provides another route to the measurement of h, since a plot of kinetic energy of the photoelectron should give a straight line of slope h. When a photon collides with an electron, it gives up all its energy, and so we should expect electrons to appear as soon as the collisions begin, provided they have sufficient energy: this agrees with observation (3).

A **corpuscular theory** of light, one that regards light as consisting of particles, seems to be at variance with its well-established wave properties, such as diffraction. If the wave theory of light is to be discarded, or at least modified, we need further evidence. The next topic provides it.

The Compton effect

The **Compton effect**, which was discovered by the American physicist Arthur Compton in 1923, is the observation that when X-rays are scattered from electrons, their wavelength is slightly increased. According to classical physics, the electron is expected to be accelerated by the electric field of the light, and many different wavelengths are to be expected in the scattered ray. What in fact is observed is that the wavelength is increased by a single, definite amount that depends only on the angle through which the light is scattered and is independent of the wavelength of the incident radiation:

$$\delta\lambda = \lambda_C(1 - \cos\theta) \qquad \lambda_C = 2.43 \text{ pm} \tag{16}$$

The constant λ_C is called the **Compton wavelength** of the electron. The

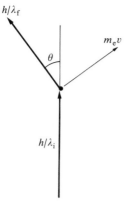

Fig. 11.4 The directions and momenta involved in the Compton effect. λ_i is the wavelength of the incident radiation, λ_f that of the scattered radiation. The electron acquires a momentum $m_e v$.

maximum wavelength shift, which occurs for $\theta = \pi$ (scattering back towards the light source) is only 4.86 pm, whatever the incident wavelength.

The photon theory of light explains the observations neatly if we suppose that, as well as having an energy, a photon of light of frequency v and wavelength λ also has a linear momentum given by

$$p = \frac{hv}{c} \tag{17a}$$

$$p = \frac{h}{\lambda} \tag{17b}$$

If the photon has a momentum, its scattering can be treated as a collision between a particle of momentum h/λ and another of mass m_e (Fig. 11.4). By requiring both energy and linear momentum to be conserved in the collision, the wavelength shift is predicted to be that given in eqn 16 with

$$\lambda_C = \frac{h}{m_e c} \tag{18}$$

Inserting the fundamental constants gives $\lambda_C = 2.426$ pm, in excellent agreement with experiment.

The diffraction of electrons

The photoelectric effect and the Compton effect both show that light has the attributes of particles. Although contrary to the long-established wave theory of light, a similar view had been held before, but discarded. No significant scientist, however, had taken the view that matter is wavelike. Nevertheless, experiments carried out in 1925 forced people to even that conclusion. The crucial experiment was performed by the American physicists Clinton Davisson and Lester Germer, who observed the diffraction of electrons by a crystal (Fig. 11.5). Their experiment was a lucky accident, because a chance rise of temperature caused their polycrystalline sample to anneal, and the ordered planes of atoms then acted as a diffraction grating.

The Davisson–Germer experiment, which has since been repeated with other particles (including molecular hydrogen), shows clearly that particles have wavelike properties. We have also seen that waves have particle properties. Thus we are brought to the heart of modern physics. When examined on an atomic scale, the concepts of particle and wave melt together, particles taking on the characteristics of waves, and waves the characteristics of particles.

Some progress towards coordinating these properties was made by Louis de Broglie when, in 1924, he suggested that any particle, not only photons, travelling with a momentum p should have (in some sense) a wavelength given by the **de Broglie relation**:

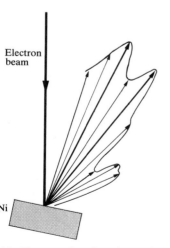

Electron beam

Ni

Fig. 11.5 The scattering of an electron beam from a nickel crystal shows a variation of intensity characteristic of a diffraction experiment in which waves interfere constructively and destructively in different directions.

$$\lambda = \frac{h}{p} \tag{19}$$

This expression was confirmed for particles by the Davisson–Germer experiment and for photons by the Compton effect. We shall build on it, and understand it more, in the next section.

Example 11.6: *Estimating the de Broglie wavelength*

Estimate the wavelength of electrons that have been accelerated from rest through a potential difference of 1.00 kV.

Answer. The energy acquired by an electron is $e \times \Delta\phi$, where e is its charge and $\Delta\phi$ is the potential difference. At the end of the period of acceleration all the acquired energy is in the form of kinetic energy, and we can write

$$\tfrac{1}{2}m_e v^2 = e\Delta\phi$$

Hence, the momentum of the electron is

$$p = m_e v = (2m_e e\Delta\phi)^{1/2}$$

and the de Broglie wavelength is

$$\lambda = \frac{h}{(2m_e e\Delta\phi)^{1/2}}$$

Substituting the data and the fundamental constants (from inside the front cover) gives

$$\lambda = \frac{6.626 \times 10^{-34}\,\text{J s}}{(2 \times 9.109 \times 10^{-31}\,\text{kg} \times 1.602 \times 10^{-19}\,\text{C} \times 1.00 \times 10^{3}\,\text{V})^{1/2}}$$

$$= 3.88 \times 10^{-11}\,\text{m}$$

Comment. We have used $1\,\text{C V} = 1\,\text{J}$ and $1\,\text{J} = 1\,\text{kg m}^2\,\text{s}^{-2}$. The wavelength of 38.8 pm is comparable to typical bond lengths in molecules (about 100 pm). Electrons accelerated in this way are used in the technique of electron diffraction (Section 21.10) for the determination of molecular structure.

Exercise. Calculate the wavelength of an electron in a 10 GeV particle accelerator ($1\,\text{GeV} = 10^9\,\text{eV}$).   [12 fm]

Atomic and molecular spectra

The most directly compelling evidence for the quantization of energy comes from the observation of the frequencies of radiation absorbed and emitted by atoms and molecules.

Fig. 11.6 The spectrum of light emitted by excited mercury atoms consists of radiation at a series of discrete frequencies (called 'spectral lines').

A typical atomic spectrum is shown in Fig. 11.6, and a typical molecular spectrum is shown in Fig. 11.7. The obvious feature of both is that radiation is emitted or absorbed at a series of discrete frequencies. This can be understood if the energy of the atoms or molecules is also confined to discrete values, for then energy can be discarded or absorbed only in

Fig. 11.7 When a molecule changes its state, it does so by absorbing light at definite frequencies. This suggests that it can possess only discrete energies, not an arbitrary energy. This is a part of the ultraviolet absorption spectrum of ScF (provided by Dr R. F. Barrow).

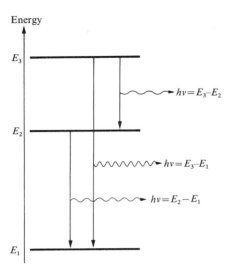

Energy

Fig. 11.8 Spectral lines can be accounted for if we assume that a molecule emits a photon as it changes between discrete energy levels. Note that high-frequency radiation is emitted when the energy change is large.

discrete amounts (Fig. 11.8). Then, if the energy of an atom decreases by ΔE, the energy is carried away as a photon of frequency $v = \Delta E/h$, and a line appears in the spectrum.

Classical mechanics utterly failed in its attempts to account for the appearance of spectra, just as it failed to account for the other experiments described above. Such total failure showed that the basic concepts of classical mechanics were false. A new mechanics, **quantum mechanics**, had to be devised to take its place.

The dynamics of microscopic systems

We shall take the de Broglie relation $p = h/\lambda$ as our starting point, and abandon the classical concept of particles moving on trajectories. From now on, we suppose that the position of a particle is distributed through space like the amplitude of a wave. In order to describe this distribution, we introduce the concept of **wavefunction** ψ in place of the trajectory, and then set up a scheme for calculating and interpreting ψ.

11.3 The Schrödinger equation

In 1926, the Austrian physicist Erwin Schrödinger proposed an equation for finding the wavefunction of any system. The **Schrödinger equation** for a particle of mass m moving in one dimension with energy E is

$$-\frac{\hbar^2}{2m}\frac{d^2\psi}{dx^2} + V\psi = E\psi \tag{20}$$

V is the potential energy of the particle; $\hbar$ (which is read h-cross or h-bar) is a convenient modification of Planck's constant:

$$\hbar = \frac{h}{2\pi} = 1.055 \times 10^{-34}\,\text{J s}$$

Various ways of expressing this equation, of incorporating the time-dependence of the wavefunction, and of extending it to more dimensions, are collected in Box 11.1.

Box 11.1 The Schrödinger equation

For one-dimensional systems:

$$-\frac{\hbar^2}{2m}\frac{d^2\psi}{dx^2} + V\psi = E\psi$$

or:

$$\frac{d^2\psi}{dx^2} + \frac{2m}{\hbar^2}(E - V)\psi = 0$$

V is the potential energy of the particle. For a free particle $V = 0$ (or some constant) and for a harmonic oscillator $V = \frac{1}{2}kx^2$. For three-dimensional systems:

$$-\frac{\hbar^2}{2m}\nabla^2\psi + V\psi = E\psi$$

where ∇^2 ('del squared') is

$$\nabla^2 = \frac{\partial^2}{\partial x^2} + \frac{\partial^2}{\partial y^2} + \frac{\partial^2}{\partial z^2}$$

In systems with spherical symmetry:

$$\nabla^2 = \frac{\partial^2}{\partial r^2} + \frac{2}{r}\frac{\partial}{\partial r} + \frac{1}{r^2}\Lambda^2$$

where

$$\Lambda^2 = \frac{1}{\sin^2\theta}\frac{\partial^2}{\partial\phi^2} + \frac{1}{\sin\theta}\frac{\partial}{\partial\theta}\sin\theta\frac{\partial}{\partial\theta}$$

In the general case, the Schrödinger equation is written

$$H\psi = E\psi$$

where H is the hamiltonian operator for the system:

$$H = -\frac{\hbar^2}{2m}\nabla^2 + V$$

For the evolution of a system with time, we solve the time-dependent Schrödinger equation

$$H\psi = i\hbar\frac{\partial\psi}{\partial t}$$

The justification of the Schrödinger equation

The form of the Schrödinger equation can be justified to a certain extent by the following remarks. Consider first the case of motion in a region where the potential energy is zero. Then

$$-\frac{\hbar^2}{2m}\frac{d^2\psi}{dx^2} = E\psi \tag{21a}$$

and a solution is

$$\psi = e^{ikx} = \cos kx + i \sin kx \qquad k = \left(\frac{2mE}{\hbar^2}\right)^{1/2} \tag{21b}$$

$\cos kx$ (or $\sin kx$) is a wave of wavelength $\lambda = 2\pi/k$, as can be seen by comparing $\cos kx$ with the standard form of a harmonic wave, $\cos(2\pi x/\lambda)$. The energy of the particle is entirely kinetic (because $V = 0$ everywhere), and so $E = p^2/2m$; but since the energy is related to k by $E = k^2\hbar^2/2m$, it follows that

$$p = k\hbar \tag{22}$$

Therefore, the linear momentum is related to the wavelength of the wavefunction by

$$p = \frac{2\pi}{\lambda} \times \frac{h}{2\pi} = \frac{h}{\lambda}$$

which is de Broglie's relation. Therefore, in the case of freely moving particles, the Schrödinger equation has led to an experimentally verified conclusion.

The kinetic energy and the wavefunction

If the potential energy is uniform but non-zero, the Schrödinger equation is

$$-\frac{\hbar^2}{2m}\frac{d^2\psi}{dx^2} = (E - V)\psi \tag{23a}$$

The solutions are the same as in eqn 21b, but with

$$E - V = \frac{k^2\hbar^2}{2m} \tag{23b}$$

Now the relation $\lambda = 2\pi/k$ leads to

$$\lambda = \frac{h}{\{2m(E - V)\}^{1/2}} \tag{23c}$$

This equation shows that the greater the difference between the total energy and the potential energy, the smaller the wavelength of the wavefunction. In other words, the greater the kinetic energy, the smaller the wavelength. A stationary particle, one with zero kinetic energy, has infinite wavelength,

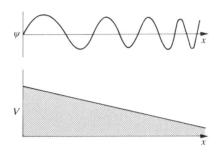

Fig. 11.9 The wavefunction of a particle in a potential decreasing towards the right and hence subjected to a constant force to the right. Only the real part of the wavefunction is shown, the imaginary part is similar, but displaced to the right.

which means that its wavefunction has the same value everywhere. That is, for a particle at rest, $\psi = $ constant.

A more general aspect of the wavefunction, one that remains well-defined even when it is is not possible to speak in terms of wavelength, is its *curvature*, which for our purposes we interpret as the second derivative $d^2\psi/dx^2$. When the wavefunction is sharply curved (when it has a short wavelength), the kinetic energy is large. When the wavefunction is not sharply curved (when its wavelength is long), the kinetic energy is low.

The association of sharp curvature with high kinetic energy will turn out to be a valuable clue to the interpretation of wavefunctions and a guide to guessing their shapes. For example, suppose we need to know the wavefunction for a particle with a potential energy that decreases with increasing x, as in the lower half of Fig. 11.9. Since the difference $E - V$ increases from left to right, the wavefunction must become more sharply curved as x increases: its wavelength decreases as its kinetic energy increases. We can therefore guess that the wavefunction will look like the function drawn in the upper half of the illustration, and more detailed calculation confirms this to be so.

11.4 The interpretation of the wavefunction

The next step in the argument is to interpret ψ. We shall see that the interpretation we adopt implies that although there are an infinite number of different solutions of the Schrödinger equation, only some of the solutions are acceptable physically. In particular, physically acceptable solutions exist only for certain values of E. That is, the interpretation of the wavefunction implies the quantization of energy.

The Born interpretation

The interpretation of ψ is based on a suggestion made by Max Born. He made use of an analogy with the wave theory of light, in which the square of the amplitude of an electromagnetic wave is interpreted as its intensity and therefore (in quantum terms) as the number of photons present. The **Born interpretation** is that the square of the wavefunction (or $\psi^*\psi$ if ψ is complex) is proportional to the probability of finding the particle at each point in space. Specifically, for a one-dimensional system:

> If the amplitude of the wavefunction of a particle is ψ at some point x, then the probability of finding the particle between x and $x + dx$ is proportional to $\psi^*\psi \, dx$.

Thus $\psi^*\psi$ is a **probability density** (since it must be multiplied by the length of the infinitesimal region dx to obtain the probability); ψ itself is called a **probability amplitude**. For a particle free to move in three dimensions (e.g. an electron near a nucleus in an atom), the wavefunction depends on the point r with coordinates x, y, and z, and the interpretation of $\psi(r)$ is then (Fig. 11.10):

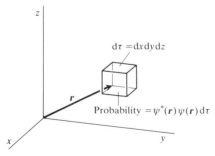

Fig. 11.10 The Born interpretation of the wavefunction in three-dimensional space implies that the probability of finding the particle in the volume element $d\tau = dx \, dy \, dz$ at some location r is proportional to the product of $d\tau$ and the value of $\psi^*\psi$ at that location.

> If the amplitude of the wavefunction of a particle is ψ at some point r, then the probability of finding the particle in an infinitesimal volume $d\tau = dx dy dy$ at the point r is proportional to $\psi^*\psi \, d\tau$.

Example 11.7: *Interpreting a wavefunction*

The wavefunction of an electron in the lowest energy state of a hydrogen atom is $\psi = e^{-r/a_0}$, with $a_0 = 52.9$ pm and r the distance from the nucleus. (Notice that the wavefunction depends only on this distance, not the angular position.) Calculate the relative probabilities of finding the electron inside a small volume of magnitude 1.0 pm³ located at (a) the nucleus, (b) a distance a_0 from the nucleus.

Answer. The probability is proportional to $\psi^2\,d\tau$ evaluated at the location in question. The volume is so small (on the scale of the atom) that we can ignore the variation of ψ within it and write

$$\text{Probability} \propto \int_{\text{Volume}} \psi^2\,d\tau \approx \psi^2 \times 1.0\,\text{pm}^3$$

with ψ^2 evaluated at the point in question. (a) At the nucleus, $r = 0$, and so there $\psi^2 = 1.0$ and

$$\text{Probability} \propto 1.0\,\text{pm}^3$$

(b) At a distance $r = a_0$ in an arbitrary (but definite) direction, $\psi^2 = e^{-2} = 0.14$ and

$$\text{Probability} \propto 0.14 \times 1.0\,\text{pm}^3$$

Therefore, the ratio of probabilities is $1.0/0.14 = 7.1$.

Comment. Note that it is more probable (by a factor of 7) that the electron will be found at the nucleus than in the same volume element located at a distance a_0 from the nucleus.

Exercise. The wavefunction for the lowest energy orbital in the ion He^+ is $\psi = e^{-2r/a_0}$. Repeat the calculation for this ion. Any comment?

[55; more compact wavefunction]

Normalization

If ψ is a solution of the Schrödinger equation, then so is $N\psi$, where N is any constant. (This is confirmed by noting that ψ occurs in every term in the equation, so any constant factor can be cancelled.) Hence we can always find a factor, the **normalization constant**, such that the proportionality of the Born interpretation becomes an equality.

We find the normalization factor by noting that for the normalized wavefunction $N\psi$, the probability that a particle is in the region dx is *equal* to $(N\psi^*)(N\psi)\,dx$. Furthermore, the sum over all space of these individual probabilities must be 1 (the probability of the particle being somewhere in the system is 1). Expressed mathematically:

$$N^2 \int \psi^*\psi\,dx = 1$$

where the integral is over all the space accessible to the particle. It follows that

$$N = \left\{ \frac{1}{\int \psi^*\psi\,dx} \right\}^{1/2} \tag{24}$$

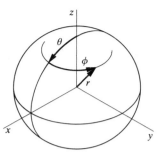

Fig. 11.11 Spherical polar coordinates. The radius ranges from 0 to ∞; the colatitude θ ranges from 0 (North pole) to π (South pole), and the azimuth ϕ ranges from 0 to 2π.

Therefore, by evaluating the integral, we can find the normalization constant N. This procedure is called **normalizing the wavefunction**. From now on, unless we state otherwise, we always use normalized wavefunctions; that is, from now on we assume that ψ already includes a factor which ensures that

$$\int \psi^* \psi \, dx = 1$$

In three dimensions, the wavefunction is normalized if

$$\int \psi^* \psi \, dx\,dy\,dz = 1$$

or, succinctly, if

$$\int \psi^* \psi \, d\tau = 1 \tag{25}$$

For systems with spherical symmetry, it is convenient to work in **spherical polar coordinates** (Fig. 11.11), in which

$$x = r \sin \theta \cos \phi \qquad y = r \sin \theta \sin \phi \qquad z = r \cos \theta$$

Then the volume element is

$$d\tau = r^2 \sin \theta \, dr \, d\theta \, d\phi \tag{26}$$

with the radius r ranging from 0 to ∞, the **colatitude** θ ranging from 0 to π, and the **azimuth** ϕ from 0 to 2π.

Example 11.8: *Normalizing a wavefunction*

Normalize the wavefunction used for the hydrogen atom in Example 11.7.

Answer. Evaluate N from eqn 25:

$$\int \psi^* \psi \, d\tau = N^2 \int_0^\infty r^2 e^{-2r/a_0} \, dr \int_0^\pi \sin \theta \, d\theta \int_0^{2\pi} d\phi$$

$$= N^2 \frac{a_0^3}{4} \times 2 \times 2\pi = \pi a_0^3 N^2$$

Therefore, in order for this integral to equal 1,

$$N = \left(\frac{1}{\pi a_0^3} \right)^{1/2}$$

and the normalized wavefunction is

$$\psi = \left(\frac{1}{\pi a_0^3} \right)^{1/2} e^{-r/a_0}$$

Comment. If Example 11.7 is now repeated, we can obtain the actual probabilities of finding the electron in the volume element at each location, not just their relative values. This gives (a) 2.2×10^{-6} and (b) 3.1×10^{-7}.

Exercise. Normalize the wavefunction given in the exercise of Example 11.7.

$$[N = (8/\pi a_0^3)^{1/2}]$$

Quantization

The Born interpretation puts severe restrictions on wavefunctions. The principal constraint is that ψ must not be infinite anywhere.[2] If it were, the integral in eqn 24 would be infinite and the normalization constant would be zero. The normalized function would be zero everywhere, except where it is infinite, which would be absurd. The requirement that ψ is finite everwhere rules out many possible solutions of the Schrödinger equation, because many mathematically acceptable solutions curl up to infinity. We shall see examples shortly.

The requirement that ψ is finite everywhere is not the only restriction implied by the Born interpretation. We could imagine (and will shortly meet) a function that gave rise to more than one value of $\psi^*\psi$ at a single point. The Born interpretation implies that such functions are unacceptable, because it would be absurd to have more than one probability that a particle is at some point. This restriction is expressed by saying that the wavefunction must be **single-valued**.

The Schrödinger equation itself also implies some mathematical restrictions on the type of functions that will occur. Since it is a second-order differential equation, the second derivative of ψ must be well-defined if the equation is to be applicable everywhere. We can take the second derivative of a function only if it is continuous (so there are no sharp steps in it, Fig. 11.12) and if its first derivative, its slope, is continuous (so there are no kinks, Fig. 11.12).[3] Therefore, wavefunctions must be continuous and have continuous first derivatives.

At this stage we see that ψ must (a) be continuous, (b) have a continuous slope, (c) be single-valued, and (d) be finite almost everywhere. These are such severe restrictions that acceptable solutions of the Schrödinger equation do not in general exist for arbitrary values of the energy E. In other words, a particle may possess only certain energies, for otherwise its wavefunction would be physically unacceptable. That is, the energy of a particle is quantized. We can find the acceptable energies by solving the Schrödinger equation for motion of various kinds, and selecting the solutions that conform to the restrictions listed above. That is the work we do in the next chapter.

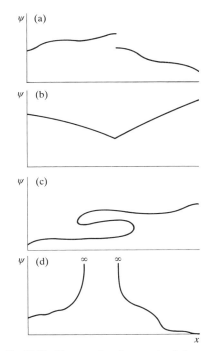

Fig. 11.12 The wavefunction must satisfy stringent conditions for it to be acceptable. (a) Unacceptable because it is not continuous, (b) unacceptable because its slope is discontinuous, (c) unacceptable because it is not single-valued, (d) unacceptable because it is infinite over a finite region.

Quantum mechanical principles

Much of quantum theory can be illustrated in a simple way by considering its application to a free particle in one dimension. We have already written down a solution of the Schrödinger equation for free translational motion

---

[2] Infinitely sharp spikes are acceptable so long as they have zero width. The true constraint is that the wavefunction must not be infinite over any finite region. In elementary quantum mechanics the simpler restriction, to finite ψ, is sufficient.

[3] Nothing, or almost nothing, is ever perfectly true. There are cases, and we shall meet them, where acceptable wavefunctions have kinks. These arise when the potential energy has peculiar properties, such as rising abruptly to infinity. When the potential energy is smoothly well-behaved and finite, the slope of the wavefunction must be continuous; if the potential energy becomes infinite, then the slope of the wavefunction need not be continuous. There are only two cases of this behaviour in elementary quantum mechanics, and the peculiarity will be mentioned when we meet them.

(eqn 21b). That solution is a special case of the general solution

$$\psi = Ae^{ikx} + Be^{-ikx} \qquad E = \frac{k^2\hbar^2}{2m} \qquad (27)$$

We have seen that the wavefunction with a given value of k corresponds to a particle with linear momentum $p = k\hbar$. One of the questions we must now address is the significance of the coefficients A and B. We shall make progress by setting up quantum mechanics in a more general and powerful way than we have presented it so far.

11.5 Operators and observables

Consider the Schrödinger equation (eqn 20) rewritten in the succinct form

$$H\psi = E\psi \qquad (28a)$$

with

$$H = -\frac{\hbar^2}{2m}\frac{d^2}{dx^2} + V \qquad (28b)$$

H is an **operator**, something that operates on the function ψ. In this case, it takes the second derivative of ψ and (after multiplication by $-\hbar^2/2m$), adds it to the result of multiplying ψ by V. The operator H plays a special role in quantum mechanics, and is called the **hamiltonian operator** after the 19th century mathematician William Hamilton, who developed a form of classical mechanics that could readily be converted into quantum mechanics.

Eigenvalues and eigenfunctions

The Schrödinger equation in eqn 28a is an **eigenvalue equation**, an equation of the form

(Operator)(function) = (Numerical factor) × (same function)

In symbols, denoting the function by f, the operator by $\hat{\Omega}$, and the number by ω:

$$\hat{\Omega}f = \omega f \qquad (29)$$

The numerical factor ω is called the **eigenvalue** of the operator $\hat{\Omega}$. In eqn 28a, the eigenvalue is the energy. The function f (which must be the same on each side in an eigenvalue equation) is the **eigenfunction** corresponding to that eigenvalue. In eqn 28a, the eigenfunction is the wavefunction corresponding to the energy E.

Example 11.9: *Identifying an eigenfunction*

Show that e^{ax} is an eigenfunction of the operator d/dx, and find the corresponding eigenvalue. Show that e^{ax^2} is not an eigenfunction of d/dx.

Answer. We must operate on the function with the operator and check whether the result is a numerical factor times the *original* function. For $\hat{\Omega} = d/dx$ and $f = e^{ax}$:

$$\hat{\Omega}f = \frac{de^{ax}}{dx} = ae^{ax} = af$$

Therefore e^{ax} is an eigenfunction of d/dx, and its eigenvalue is a. For $f = e^{ax^2}$,

$$\hat{\Omega}f = \frac{de^{ax^2}}{dx} = 2axe^{ax^2} = 2a\{xe^{ax^2}\}$$

which is not an eigenfunction equation.

Comment. Much of quantum mechanics boils down to looking for functions that are eigenfunctions of a given operator, especially of the hamiltonian operator for the energy.

Exercise. Is $\cos ax$ an eigenfunction of (a) d/dx, (b) d^2/dx^2? [(a) No, (b) yes]

The importance of eigenvalue equations is that the pattern

$$(\text{Energy operator})(\text{wavefunction}) = (\text{energy})(\text{wavefunction})$$

exemplified by the Schrödinger equation itself is repeated for other properties, which in quantum mechanics are called **observables**. In general we can write

$$(\text{Operator})(\text{wavefunction}) = (\text{observable})(\text{wavefunction})$$

and symbolically

$$\hat{\Omega}\psi = \omega\psi \tag{30}$$

where $\hat{\Omega}$ is the operator (e.g. the hamiltonian H) corresponding to the observable ω (e.g. the energy E). Therefore, if we know both the wavefunction ψ and the operator corresponding to the observable of interest, we can predict the outcome of an observation of that property (e.g. an atom's energy) by picking out the factor ω in the corresponding eigenvalue equation.

Operators

The first step is to find the operator corresponding to a given observable. The form of the operator for **linear momentum** is one of the postulates of quantum mechanics:

$$\hat{p} = \frac{\hbar}{i}\frac{d}{dx} \tag{31a}$$

That is, to find the linear momentum of a particle, we differentiate the wavefunction with respect to x, and then pick out the momentum p from the eigenvalue equation

$$\frac{\hbar}{i}\frac{d\psi}{dx} = p\psi$$

The operator for position is also one of the basic postulates of quantum mechanics: it is simply multiplication by the coordinate x:

$$\hat{x} = x \times \qquad (31b)$$

For example, suppose we select $B = 0$ for one of the free-particle wavefunctions given in eqn 27, then $\psi = A e^{ikx}$ and

$$\frac{\hbar}{i} \frac{d\psi}{dx} = \frac{\hbar}{i} A \frac{d e^{ikx}}{dx} = \frac{\hbar}{i} A \times i k e^{ikx} = \hbar k A e^{ikx} = k\hbar\psi$$

Hence $p = k\hbar$, as we already knew. Now suppose instead that we chose the wavefunction with $A = 0$. By the same reasoning, $p = -k\hbar$, which shows that a particle described by the wavefunction e^{-ikx} has the same magnitude of momentum (and the same kinetic energy) as before but directed towards $-x$ (the momentum is a vector quantity, and the sign gives the direction).

11.6 Superpositions and expectation values

Suppose now that the wavefunction of the free particle has $A = B$. What is the linear momentum of the particle? We quickly run into trouble if we use the operator technique. The wavefunction is

$$\psi = A(e^{ikx} + e^{-ikx}) = 2A \cos kx$$

which is a perfectly respectable wavefunction. However, when we operate with $\hat{p}$, we find

$$\frac{\hbar}{i} \frac{d\psi}{dx} = \frac{2A\hbar}{i} \frac{d}{dx} \cos kx = -\frac{2kA\hbar}{i} \sin kx$$

which is not an eigenvalue equation because the function on the right is different from the original one.

Linear superpositions of wavefunctions

When the wavefunction of a particle is not an eigenfunction of an operator, the property is indefinite. However, in the current example the momentum is not completely indefinite because the cosine wavefunction is a **linear superposition**, or sum, of e^{ikx} and e^{-ikx}, and these, as we have seen, individually correspond to definite momentum states. Symbolically we can write the superposition as

$$\psi = \psi_\rightarrow + \psi_\leftarrow$$

The interpretation of this wavefunction is that if the momentum of the particle is measured, its *magnitude* will be found to be $k\hbar$ (since that is the value for each component of the wavefunction), but half the measurements will show that it is moving to the right, and half the measurements will show that it is moving to the left.

The same interpretation applies to any wavefunction written as a superposition. Thus, suppose the wavefunction is known to be a sum of many different linear momentum eigenfunctions and written in the form

$$\psi = c_1\psi_1 + c_2\psi_2 + \ldots = \sum_n c_n\psi_n \qquad (32)$$

where the cs are numerical coefficients and the various ψ_n correspond to different momentum states. Then quantum mechanics tells us the following:

(1) When the momentum is measured, *one* of the values corresponding to the ψ_n that contribute to the superposition will be found.

(2) Which of these possible values will be found is unpredictable, but the probability of measuring a particular value in a series of observations is proportional to the square of its coefficient in the superposition (to c^*c if c is complex).

(3) The *average* value of a large number of observations is given by the **expectation value** $\langle \Omega \rangle$ of the observable Ω:

$$\langle \Omega \rangle = \int \psi^* \hat{\Omega} \psi \, d\tau \qquad (33)$$

(This formula is valid only for normalized wavefunctions.)

Example 11.10: *Calculating an expectation value*

Calculate the average value of the distance of an electron from the nucleus in the hydrogen atom.

Answer. The normalized wavefunction was obtained in Example 11.8. The operator corresponding to its distance from the nucleus is multiplication by r (eqn 31b). Therefore, the average value is given by the expectation value

$$\langle r \rangle = \int \psi^* r \psi \, d\tau$$

which we evaluate using spherical polar coordinates. Using the normalized function in Example 11.8, gives

$$\langle r \rangle = \frac{1}{\pi a_0^3} \int_0^\infty r^3 e^{-2r/a_0} \, dr \int_0^\pi \sin \theta \, d\theta \int_0^{2\pi} d\phi$$

$$= \frac{1}{\pi a_0^3} \times \frac{3! \, a_0^4}{2^4} \times 2 \times 2\pi = \tfrac{3}{2} a_0$$

We have used the integral

$$\int_0^\infty x^n e^{-ax} \, dx = \frac{n!}{a^{n+1}}$$

which is very useful in calculations of this kind. Since $a_0 = 52.9$ pm (Example 11.5), $\langle r \rangle = 79$ pm.

Comment. The result that $\langle r \rangle = 79$ pm means that if a very large number of measurements of the distance of the electron from the nucleus are made, then the *mean* value will be 79 pm. However, each different observation will give a different individual result, because the wavefunction is not an eigenfunction of the operator corresponding to r.

Exercise. Evaluate the root mean square distance $\langle r^2 \rangle^{1/2}$ of the electron from the nucleus in the hydrogen atom.   [$3^{1/2} a_0$]

The uncertainty principle

We have seen that if the particle's wavefunction is $A e^{ikx}$, then it corresponds to a definite state of linear momentum, namely travelling to the right with momentum $k\hbar$. But we might also ask for the *position* of the particle when it

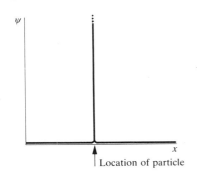

Fig. 11.13 The wavefunction for a particle at a well-defined location is a sharply spiked function which has zero amplitude everywhere except at the particle's position.

is in this state. The Born interpretation instructs us to answer this question by forming the probability density $\psi^*\psi$. In this case:

$$\psi^*\psi = (Ae^{ikx})^*(Ae^{ikx}) = A^2(e^{-ikx})(e^{ikx}) = A^2$$

This probability density is a constant, A^2, independent of x. Therefore, the particle has an equal probability of being found anywhere. In other words, if the momentum is specified precisely, it is impossible to predict the location of the particle. This is one half of **Heisenberg's uncertainty principle**, one of the most celebrated results of quantum mechanics:

> It is impossible to specify simultaneously, with arbitrary precision, both the momentum and the position of a particle.

Before discussing the principle further, we must establish the other half: that if we know the position of a particle exactly, we can say nothing about its momentum. The argument draws on the idea of expressing a wavefunction as a superposition of eigenfunctions, and runs as follows. If we know that the particle is at a definite location, then its wavefunction must be large there and zero everywhere else (Fig. 11.13). Such a wavefunction can be created by adding together a large number of harmonic (sine and cosine) functions, or, what is equivalent, a number of e^{ikx} functions. In other words, we can create a sharply localized wavefunction by superimposing wavefunctions corresponding to many different linear momenta. The superposition of a few harmonic functions gives a broad, ill-defined wavefunction (Fig. 11.14a), but as the number used increases, the wavefunction becomes sharper because of the more complete interference between the positive and negative regions of the components (Fig. 11.14b). When an infinite number of components are used, the wavefunction is a sharp, infinitely narrow spike (Fig. 11.14c), which corresponds to perfect localization of the particle. Now the particle is perfectly localized, but at the expense of discarding all information about its momentum. This is because, as we saw above, a measurement of the momentum will give a result corresponding to any one of the infinite number of waves in the superposition, and which one it will give is unpredictable. Hence, if we know the location of the particle, its momentum is unpredictable.

Werner Heisenberg arrived at a quantitative version of this result by considering the expectation values of position and momentum. For our purposes, only his result is important. The quantitative version of the **position–momentum uncertainty relation** is

$$\Delta p\, \Delta q \geq \tfrac{1}{2}\hbar \qquad (34)$$

Δp is the 'uncertainty' in the linear momentum (strictly, it is the root mean square deviation of the momentum from its mean value) and Δq is the uncertainty in position (the r.m.s. deviation of the position from the mean position, essentially the half-width of the superposition in Fig. 11.14). The p and q that appear in eqn 34 refer to the same direction in space. Therefore, whereas position on the x axis and momentum parallel to it are restricted by the uncertainty relation, simultaneous location of position on x and motion along y or z are not restricted.

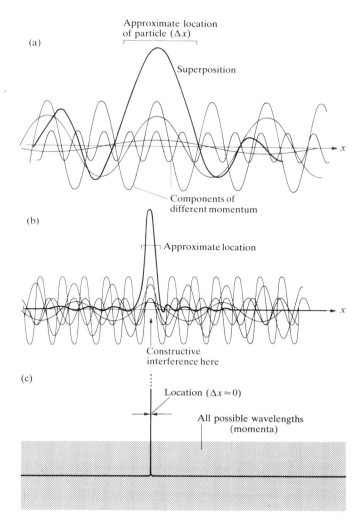

(a)

Approximate location
of particle (Δx)

Superposition

x

Components of
different momentum

(b)

Approximate location

x

Constructive
interference here

(c)

Location ($\Delta x \approx 0$)

All possible wavelengths
(momenta)

Fig. 11.14 (a) The wavefunction for a particle
with an ill-defined location can be regarded as
the sum (superposition) of several
wavefunctions of definite wavelength which
interfere constructively in one place but
destructively elsewhere. (b) As more waves
are used in the superposition, the location
becomes more precise at the expense of
uncertainty in the particle's momentum. (c) An
infinite number of waves are needed to
construct the wavefunction of a perfectly
localized particle.

Example 11.11: *Using the uncertainty principle*

The speed of a 1.0 g projectile is known to within $10^{-6}\,\mathrm{m\,s^{-1}}$. Calculate the minimum uncertainty in its position.

Answer. Estimate Δp from $m\Delta v$, where Δv is the uncertainty in the speed; then use eqn 34 to estimate the minimum uncertainty in position, Δq. The minimum uncertainty in position is

$$\Delta q = \frac{\hbar}{2m\Delta v} = 5 \times 10^{-26}\,\mathrm{m}$$

Comment. The uncertainty is completely negligible for all practical purposes. However, when the mass is that of an electron, then the same uncertainty in speed implies an uncertainty in position far larger than the diameter of an atom, and so the concept of a trajectory, the simultaneous possession of a precise position and momentum, is untenable.

Exercise. Estimate the minimum uncertainty in the speed of an electron in a hydrogen atom (taking its diameter as $2a_0$). [500 km s^{-1}]

The Heisenberg uncertainty relation applies to a number of pairs of observables called **complementary observables**, which are defined in terms of the properties of their operators.[4] Other than position and momentum, they include properties related to angular momentum (which we meet in the next chapter). With the discovery that some observables are complementary we are at the heart of the difference between classical and quantum mechanics. Classical mechanics supposed, falsely as we now know, that the position and momentum of a particle could be specified simultaneously with arbitrary precision: that, after all, is what is meant by a trajectory, as we saw in Section 11.1. However, quantum mechanics shows that position and momentum are complementary, and that we have to make a choice: we can specify position at the expense of momentum, or momentum at the expense of position.

The realization that some observables are complementary allows us to make considerable progress with the calculation of atomic and molecular properties; but it does away with some of classical physics' most cherished concepts.

Further reading

Historical background

T. S. Kuhn, *Black-body theory and the quantum discontinuity*, 1894–1912. Oxford University Press (1978).

M. Jammer, *The conceptual development of quantum mechanics*. McGraw-Hill, New York (1966).

Principles and concepts

P. W. Atkins, *Quantization* (lecture cassette and workbook). Royal Society of Chemistry, London (1981).

P. W. Atkins, *Quanta: a handbook of concepts*. Oxford University Press (1974).

P. W. Atkins, *Molecular quantum mechanics* (2nd edn). Oxford University Press (1983).

R. P. Feynman, R. B. Leighton, and M. Sands, *Lectures in physics*. W. H. Freeman & Co., San Francisco (1963).

---

[4] Specifically, two observables Ω_1 and Ω_2 are complementary if their operators satisfy the relation

$$\hat{\Omega}_1\hat{\Omega}_2 \neq \hat{\Omega}_2\hat{\Omega}_1$$

Thus, position x and linear momentum along x are complementary because (noting that operators always operate on a wavefunction)

$$\hat{x}\hat{p}\psi = x \times \frac{\hbar}{i}\frac{d\psi}{dx}$$

but

$$\hat{p}\hat{x}\psi = \frac{\hbar}{i}\frac{d(x\psi)}{dx} = \frac{\hbar}{i}\psi + x \times \frac{\hbar}{i}\frac{d\psi}{dx}$$

which is not the same.

Exercises

11.1 Calculate the power radiated by a 2.0 m × 3.0 m section of the surface of a hot body at 1500 K.

11.2 The power delivered to a photodetector that collects 8.0×10^7 photons in 3.8 ms from monochromatic light is $0.72\ \mu W$. What is the frequency of the radiation?

11.3 A diffraction experiment requires the use of electrons of wavelength 0.45 nm. Calculate the velocity of the electrons.

11.4 Calculate the linear momentum of photons of wavelength (a) 750 nm, (b) 70 pm, (c) 19 m.

11.5 The energy required for the ionization of a certain atom is 3.44×10^{-18} J. The absorption of a photon of unknown wavelength ionizes the atom and ejects an electron with velocity 1.03×10^6 m s^{-1}. Calculate the wavelength of the incident radiation.

11.6 In an experiment on the Compton scattering of X-rays by electrons, the incident beam has wavelength 70.78 pm. Calculate the wavelength of the X-rays that are scattered through an angle of 70°.

11.7 The speed of a certain proton is 4.5×10^5 m s^{-1}. If the uncertainty in its momentum is to be reduced to 0.0100 per cent, what uncertainty in its location must be tolerated?

11.8 A particle of mass 6.65×10^{-27} kg is confined to an infinite square well of width L. The energy of the level with $n = 3$ is 2.00×10^{-24} J. Calculate the width of the box given that $E_n = n^2 h^2 / 8mL^2$.

11.9 Calculate the location of a point in a box of length L at which the probability of a particle being found is 25 per cent of its maximum probability when $n = 1$ given that its wavefunction is $\psi_n = (2/L)^{1/2} \sin(n\pi x/L)$.

11.10 Calculate the spacing between the levels with $n = 4$ and $n = 5$ of a deuterium atom in a one-dimensional box of length 5.0 nm.

11.11 Calculate the energy per photon and the energy per mole of photons for radiation of wavelength (a) 600 nm (red), (b) 550 nm (yellow), (c) 400 nm (blue), (d) 200 nm (ultraviolet), (e) 150 pm (X-ray), (f) 1.00 cm (microwave).

11.12 Calculate the linear momenta of the photons specified in Exercise 11.11.

11.13 Calculate the speed to which a stationary H atom would be accelerated by each of the photons in Exercise 11.11.

11.14 A glow-worm of mass 5.0 g emits red light (650 nm) with a power of 0.10 W entirely in the backward direction. To what speed will it have accelerated after 10 years if released into free space (and assumed to live)?

11.15 A sodium lamp emits yellow light (550 nm). How many photons does it emit each second if its power is (a) 1.0 W, (b) 100 W?

11.16 The peak of the sun's emission occurs at about 480 nm; estimate the temperature of its surface.

11.17 The work function for metallic caesium is 2.14 eV. Calculate the kinetic energy and the speed of the electrons ejected by light of wavelength (a) 700 nm, (b) 300 nm.

11.18 By how much does the wavelength of radiation change when it scatters from (a) a free electron, (b) a free proton, and is deflected through 90°?

11.19 Calculate the size of the quantum involved in the excitation of (a) an electronic motion of period 10^{-15} s, (b) a molecular vibration of period 10^{-14} s, (c) a pendulum of period 1 s. Express the results in J and in kJ mol^{-1}.

11.20 Calculate the de Broglie wavelength of (a) a mass of 1.0 g travelling at 1.0 cm s^{-1}, (b) the same, travelling at 100 km s^{-1}, (c) an He atom travelling at 1000 m s^{-1} (a typical speed at room temperature).

11.21 Calculate the de Broglie wavelength of an electron accelerated from rest through a potential difference of (a) 100 V, (b) 1.0 kV, (c) 100 kV.

11.22 Calculate the minimum uncertainty in the speed of a ball of mass 500 g that is known to be within 1.0 μm of a certain point on a bat. What is the minimum uncertainty in the position of a bullet of mass 5.0 g that is known to have a speed somewhere between 350.00001 m s^{-1} and 350.00000 m s^{-1}?

11.23 An electron is confined to a linear region with a length of the same order as the diameter of an atom (ca. 100 pm). Calculate the minimum uncertainties in its linear momentum and speed.

11.24 In an X-ray photoelectron experiment, a photon of wavelength 150 pm ejects an electron from the inner shell of an atom and it emerges with a speed of 2.14×10^7 m s^{-1}. Calculate the binding energy of the electron.

Problems
Numerical problems

11.1 The Planck distribution (eqn 12) gives the energy in the wavelength range dλ at the wavelength λ. Calculate the energy density in the range 650 nm to 655 nm inside a cavity of volume 100 cm^3 when its temperature is (a) 25°C, (b) 3000°C.

11.2 The wavelength of the emission maximum from a small pinhole in an electrically heated container was determined at a series of temperatures, and the results are given below.

Deduce a value for Planck's constant.

θ/°C	1000	1500	2000	2500	3000	3500
λ_{max}/nm	2181	1600	1240	1035	878	763

11.3 Write a computer program to evaluate the Planck distribution at any temperature and wavelength or frequency, and add to it a routine for evaluating integrals for the energy density of the radiation between any two wavelengths. Use it to calculate the total energy density in the visible region (600 nm to 350 nm) for a black body at (a) 100°C, (b) 500°C, (c) 700°C. What are the classical values at these temperatures?

11.4 The Einstein frequency is often expressed in terms of an equivalent temperature θ_E, where $\theta_E = h\nu/k$. Confirm that θ_E has the dimensions of temperature, and express the criterion for the validity of the high-temperature form of the Einstein equation in terms of it. Evaluate θ_E for (a) diamond, for which $\nu = 4.65 \times 10^{13}$ Hz and (b) for copper, for which $\nu = 7.15 \times 10^{12}$ Hz. What fraction of the Dulong and Petit value of the heat capacity does each substance reach at 25°C?

11.5 The ground-state wavefunction for a particle confined to a one-dimensional box of length L is

$$\psi = \left(\frac{2}{L}\right)^{1/2} \sin\frac{\pi x}{L}$$

Suppose the box is 10.0 nm long. Calculate the probability that the particle is (a) between $x = 4.95$ nm and 5.05 nm, (b) between $x = 1.95$ nm and 2.05 nm, (c) between $x = 9.90$ and 10.00 nm, (d) in the right half of the box, (e) in the central third of the box.

11.6 The ground state wavefunction of a hydrogen atom is

$$\psi = \left(\frac{1}{\pi a_0^3}\right)^{1/2} e^{-r/a_0}$$

where $a_0 = 53$ pm. Calculate the probability that the electron will be found somewhere within a small sphere of radius 1.0 pm centred on the nucleus. Now suppose that the same sphere is relocated at $r = a_0$. What is the probability that the electron is inside it?

Theoretical problems

11.7 Derive the Wien law, that $\lambda_{max}T$ is a constant, from the Planck distribution, and find an expression for the constant.

11.8 Derive the Compton formula (eqn 16) using the following procedure. When the electron is at rest it has an energy $m_e c^2$. When it is in motion with linear momentum p, its energy is $(p^2 c^2 + m_e^2 c^4)^{1/2}$. Suppose that the photon of wavelength λ_i strikes an electron and is scattered with a new wavelength λ_f and emerges at an angle θ, and that the electron, which is stationary initially, moves off with a momentum p' at an angle θ' to the incoming photon. Set up the three conservation expressions (for energy, momentum along the line of approach and momentum perpendicular to

it), eliminate θ', then eliminate p, and hence arrive at an expression for $\delta\lambda$.

11.9 Normalize the following wavefunctions: (a) $\sin n\pi x/L$ in the range $0 \le x \le L$, (b) a constant in the range $-L \le x \le L$, (c) $e^{-r/a}$ in three-dimensional space, (d) $xe^{-r/2a}$ in three-dimensional space. Hint: The volume element in three dimensions is $d\tau = r^2\, dr \sin\theta\, d\theta\, d\phi$, with $0 \le r \le \infty$, $0 \le \theta \le \pi$, and $0 \le \phi \le 2\pi$. A useful integral is

$$\int_0^\infty x^n e^{-ax}\, dx = \frac{n!}{a^{n+1}}$$

11.10 Two (unnormalized) excited state wavefunctions of the H atom are

$$\text{(a)} \quad \psi = \left(2 - \frac{r}{a_0}\right)e^{-r/2a_0}$$

$$\text{(b)} \quad \psi = r\sin\theta\cos\phi\, e^{-r/2a_0}$$

Normalize both functions to 1.

11.11 Identify which of the following functions are eigenfunctions of the operator d/dx: (a) e^{ikx}, (b) $\cos kx$, (c) k, (d) kx, (e) $e^{-\alpha x^2}$. Give the corresponding eigenvalue where appropriate.

11.12 Which of the functions in Problem 11.11 are (a) also eigenfunctions of d^2/dx^2 and (b) only eigenfunctions of d^2/dx^2? Give the eigenvalues where appropriate.

11.13 A particle is in a state described by the wavefunction

$$\psi = \cos\chi e^{ikx} + \sin\chi e^{-ikx}$$

where χ is a parameter. What is the probability that the particle will be found with a linear momentum (a) $+k\hbar$, (b) $-k\hbar$? (c) What form would the wavefunction have if it were 90 per cent certain that the particle had linear momentum $+k\hbar$?

11.14 Evaluate the kinetic energy of the particle with wavefunction given in Problem 11.13.

11.15 The expectation value of momentum is evaluated using the expression in eqn 33. Calculate the average linear momentum of a particle described by the following wavefunctions: (a) e^{ikx}, (b) $\cos kx$, (c) $e^{-\alpha x^2}$, each one in the range from $-\infty$ to $+\infty$.

11.16 Evaluate the expectation values of r and r^2 for a hydrogen atom with wavefunctions given in Problem 11.10.

11.17 Calculate the mean potential energy of an electron in the ground state of a hydrogenic atom.

11.18 Write a computer program for constructing superpositions of cosine functions like those drawn in Fig. 11.14 and explore how the particle a wavefunction describes becomes more localized as more components are included. Include routines that determine the probability that a given momentum will be observed. If you plot the superposition (which you should), set $x = 0$ at the centre of the screen and build the superposition there. Include a routine that includes the evaluation of the root mean square location of the packet, $\langle x^2 \rangle^{1/2}$.

Quantum theory: techniques and applications

12

Check-list of key ideas

1. The *boundary conditions* that must be satisfied by the wavefunctions of a *particle in a box* and the resulting quantized *energy levels* (eqn 3).

2. The properties of the wavefunctions of a particle in a box (Section 12.1) and the *zero-point energy*.

3. The use of *quantum numbers* for labelling a state and specifying an observable (Section 12.1).

4. The *correspondence principle* and the emergence of classical behaviour at high quantum numbers (Section 12.1).

5. The technique of *separation of variables* for separating the Schrödinger equation in several variables into separate equations (Section 12.2) for each variable.

6. The wavefunctions and energies of a *particle in a two-dimensional square well* (eqn 5) and the *degeneracy* of the levels.

7. The *tunnelling* of a particle into classically forbidden regions (Section 12.3) and the calculation of tunnelling probabilities (Examples 12.4 and 12.5).

8. The quantum mechanical description of *vibrational motion* (Section 12.4) and the wavefunctions and energy levels of a *harmonic oscillator* (eqns 9 and 10).

9. The mean potential energy and the mean kinetic energy of a harmonic oscillator as a special case of the *virial theorem* (eqn 12).

10. The calculation of the extent to which a harmonic oscillator may be found in classically forbidden regions of extension and contraction (Example 12.8).

11. The quantum mechanical description of *rotational motion in two dimensions* and the role of *cyclic boundary conditions* (Section 12.6).

12. The *quantized energy levels* of a particle moving on a circle (eqn 15) and the *quantization of angular momentum* about an axis (eqn 16).

13. The *vector representation* of angular momentum (Fig. 12.12).

14. The quantum mechanical description of *rotational motion in three dimensions* (Section 12.7).

15. The *quantized energy levels* of a particle free to rotate in three dimensions (eqn 20) and the *spherical harmonic* wavefunctions (Section 12.7).

16. The *quantization of angular momentum* and *space quantization* (Section 12.7), and the representation of angular momentum in terms of the *vector model* (Fig. 12.17).

17. The *Stern–Gerlach experiment* and the property of electron spin (Section 12.8).

The three basic modes of motion—translation, vibration, and rotation—all play an important role in chemistry because they are ways in which molecules can store energy. For example, gas molecules undergo translational motion, and their kinetic energy is a contribution to the total internal energy of a sample. Molecules can also store energy as rotational kinetic energy, and transitions between their rotational energy levels are responsible for rotational spectra. Molecules can also store energy in their vibrations, and transitions between their vibrational energy levels give rise to vibrational spectra.

In this chapter we see how the concepts of quantum mechanics introduced in Chapter 11 can be developed into a powerful set of techniques for dealing with these types of motion. In later chapters we shall see how these calculations are used to account for atomic structure, molecular structure, and spectroscopy.

Translational motion

We outlined the quantum mechanical description of free motion in Section 11.3. We saw there that the Schrödinger equation is

$$-\frac{\hbar^2}{2m}\frac{d^2\psi}{dx^2} = E\psi$$

and that the general solutions are

$$\psi = Ae^{ikx} + Be^{-ikx} \qquad E = \frac{k^2\hbar^2}{2m}$$

In this case, all values of k, and therefore all values of the energy E, are permitted: the energy of a free particle is not quantized. For $B = 0$ the wavefunction describes a particle with linear momentum $p = k\hbar$ towards positive x. For $A = 0$ the wavefunction describes a particle with the same momentum travelling towards negative x. In either state the position of the particle is completely unpredictable (Section 11.6) because the momentum is completely certain.

12.1 The particle in a box

In this section we consider the problem of a **particle in a box**, in which a particle of mass m is confined between two walls at $x = 0$ and $x = L$. In the **infinite square well**, the potential energy of the particle is zero inside the box but rises abruptly to infinity at the walls (Fig. 12.1). This is an

Fig. 12.1 A particle in a one-dimensional region with impenetrable walls. Its potential energy is zero between $x = 0$ and $x = L$, and rises abruptly to infinity as soon as it touches the walls.

idealization of a gas molecule free to move in a one-dimensional container; later we shall generalize the calculation to a particle in two- and three-dimensional containers.

The Schrödinger equation

The Schrödinger equation for the region between the walls (where $V = 0$) is

$$-\frac{\hbar^2}{2m}\frac{d^2\psi}{dx^2} = E\psi \qquad (1a)$$

This equation is the same for the free particle, so the general solutions are also the same. It is convenient[1] to write them as

$$\psi = A\sin kx + B\cos kx \qquad E = \frac{k^2\hbar^2}{2m} \qquad (1b)$$

The best way of dealing with the regions where the potential energy is infinite is to suppose that V is large but not infinite, and then to allow V to become infinite later. The Schrödinger equation is then

$$-\frac{\hbar^2}{2m}\frac{d^2\psi}{dx^2} + V\psi = E\psi \qquad (2a)$$

which we can rearrange into

$$\frac{d^2\psi}{dx^2} = \frac{2m}{\hbar^2}(V - E)\psi \qquad (2b)$$

It will be helpful to bear in mind that if the second derivative is positive the wavefunction looks like $\smile$; if it is negative, it looks like $\frown$.

The boundary conditions

Suppose that $\psi > 0$ just within the material of the walls (for x slightly less than 0 or greater than L). Then since the right-hand side of eqn 2b is positive (because V is so large that it certainly exceeds E, and ψ is positive), the curvature of ψ is positive. Therefore, ψ curls off rapidly towards infinite values within the walls. This makes it unacceptable according to the criteria set out in Section 11.4. If $\psi < 0$ just inside the material of the walls, the second derivative is negative, and so ψ falls very rapidly to negatively infinite values, which also makes it unacceptable. Thus, the wavefunction can be neither positive nor negative just inside the material of the walls. Hence it must be zero there, and this requirement is increasingly stringent as V approaches infinity. When V is actually infinite (in the infinite square well), the only acceptable wavefunctions are those with $\psi = 0$ everywhere inside the material of the walls. Since the wavefunction must be continuous everywhere, it follows that ψ must be zero *at* the walls themselves (at $x = 0$ and $x = L$). Hence there are two **boundary conditions** on the wavefunction,

---

[1] We use $e^{i\theta} = \cos\theta + i\sin\theta$ and $e^{-i\theta} = \cos\theta - i\sin\theta$, and absorb all numerical factors into the coefficients A and B.

or conditions that the function must satisfy at stated locations. In this case they are

$$\psi = 0 \text{ at } x = 0 \text{ and at } x = L$$

The acceptable solutions

At this stage we know that the wavefunction has the general form given in eqn 1b but with the additional requirement that it must satisfy the two boundary conditions set out above.

Consider the wall at $x = 0$. According to eqn 1b, $\psi(0) = B$ (because $\sin 0 = 0$ and $\cos 0 = 1$). But the boundary condition there is that $\psi(0) = 0$, which requires $B = 0$. It follows that the wavefunction must be of the form

$$\psi = A \sin kx$$

The amplitude at the other wall (at $x = L$) is

$$\psi = A \sin kL$$

which must also be zero. Taking $A = 0$ would give $\psi = 0$ for all x, which would conflict with the Born interpretation (the particle must be somewhere). Therefore, kL must be chosen so that $\sin kL = 0$, which is satisfied by

$$kL = n\pi \quad n = 1, 2, \ldots$$

($n = 0$ is ruled out because it implies $k = 0$ and $\psi = 0$ everywhere, which is unacceptable, and negative values of n merely change the sign of $\sin n\pi$.) Since k and E are related by eqn 1b, it follows that the energy of the particle is limited to the values

$$E = \frac{n^2\hbar^2\pi^2}{2mL^2} = \frac{n^2h^2}{8mL^2} \quad n = 1, 2, \ldots$$

We see that the energy of the particle is quantized, and that the quantization arises from the boundary conditions that ψ must satisfy if it is to be an acceptable wavefunction.

Normalization

Before discussing the solution in more detail, we shall complete the derivation of the wavefunctions by finding the normalization constant (here written A). To do so, we look for the value of A that ensures that the integral of ψ^2 over all x is equal to 1:

$$1 = \int_{-\infty}^{\infty} \psi^*\psi \, dx = A^2 \int_0^L \sin^2 kx \, dx = \tfrac{1}{2}A^2L \quad \text{or} \quad A = \left(\frac{2}{L}\right)^{1/2}$$

Therefore, the complete solution to the problem is

$$\text{Energies:} \quad E_n = \frac{n^2h^2}{8mL^2} \quad n = 1, 2, \ldots \tag{3a}$$

$$\text{Wavefunctions:} \quad \psi_n = \left(\frac{2}{L}\right)^{1/2} \sin \frac{n\pi x}{L} \tag{3b}$$

We have labelled the energies and wavefunctions with the **quantum number** n. A quantum number is an integer (in some cases, as we shall see, a half-integer) that labels the state of the system. For a particle in a box there are an infinite number of acceptable solutions, and the quantum number *n* simply specifies the one of interest. As well as acting as a label a quantum number is used to calculate the energy corresponding to that state (through eqn 3a) and to write down the wavefunction explicitly (using eqn 3b).

The properties of the solutions

Figure 12.2 shows the shapes of some of the wavefunctions of a particle in a box. It is easy to see the origin of the quantization in pictorial terms: each wavefunction is a standing wave and, in order to fit into the cavity, successive functions must possess one more half-wavelength. Shortening the wavelength in order to fit another half-wavelength into the container implies sharpening the curvature of the wavefunction, and therefore increasing the kinetic energy of the particle it describes.

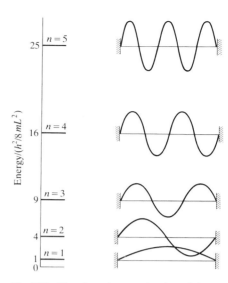

Fig. 12.2 The allowed energy levels and the corresponding (sine wave) wavefunctions for a particle in a box. Note that the energy levels increase as n^2, and so their separation increases as the quantum number increases. Each wavefunction is a standing wave, and successive functions possess one more half wave and a correspondingly shorter wavelength.

Example 12.1: *Deriving the energies of a particle in a box*

Derive the energy levels of a particle in a box from the de Broglie relation.

Answer. We assume that successive wavefunctions possess one more half-wavelength, and therefore that the permitted wavelengths satisfy

$$L = n \times \tfrac{1}{2}\lambda \quad n = 1, 2, \ldots$$

and therefore that

$$\lambda = \frac{2L}{n} \quad n = 1, 2, \ldots$$

According to the de Broglie relation, these wavelengths correspond to the momenta

$$p = \frac{h}{\lambda} = \frac{nh}{2L} \quad n = 1, 2, \ldots$$

The particle has only kinetic energy inside the box (where $V = 0$), so the permitted energies are

$$E = \frac{p^2}{2m} = \frac{n^2 h^2}{8mL^2} \quad n = 1, 2, \ldots$$

Exercise. What is the average value of the linear momentum of a particle in a box with quantum number *n*? $[\langle p \rangle = 0]$

The linear momentum of a particle in a box is not well defined because the wavefunction $\sin kx$ is a standing wave and not an eigenfunction of the linear momentum operator (Section 11.5). However, each wavefunction is a superposition of momentum eigenfunctions (because $\sin kx = (e^{ikx} - e^{-ikx})/2i$), so measurement of the linear momentum would give the value $k\hbar = nh/2L$ half the time and $-k\hbar = -nh/2L$ the other half. This is the quantum mechanical version of the classical picture that a particle in a box rattles from wall to wall, and travels to the right for half the time and to the left for the other half.

Because *n* cannot be zero, the lowest energy that the particle may possess is not zero (as would be allowed by classical mechanics) but $E_1 = h^2/8mL^2$.

This lowest, irremovable energy is called the **zero-point energy**. The physical origin of the zero-point energy can be explained in two ways. First, the uncertainty principle requires a particle to possess kinetic energy if it is confined to a finite region. This is because the particle's location is not completely indefinite, and so its momentum cannot be precisely zero. Alternatively, if the wavefunction is to be zero at the walls, but smooth, continuous, and not zero everywhere, then it must be curved, and curvature in a wavefunction implies the possession of kinetic energy.

The separation between adjacent energy levels is

$$\Delta E = E_{n+1} - E_n = (2n + 1)\frac{h^2}{8mL^2}$$

This separation decreases as the length of the container increases and is extremely small when the container is large; ΔE becomes zero when the walls are infinitely far apart. Atoms and molecules free to move in laboratory-sized vessels may therefore be treated as though their translational energy is not quantized.

Example 12.2: *Using the particle in a box solutions* (1)

An electron is confined to a molecule of length 1.0 nm (about five atoms long). What is (a) its minimum energy and (b) the minimum excitation energy from that state?

Answer. We use eqn 3 with $m = m_e$ and $n = 1$ for (a); for (b), the minimum excitation energy is $E_2 - E_1$. For $L = 1.0$ nm, $h^2/8m_eL^2 = 6.02 \times 10^{-20}$ J. Therefore, $E_1 = 6.0 \times 10^{-20}$ J (corresponding to 0.37 eV). The minimum excitation energy is

$$E_2 - E_1 = (2^2 - 1)\frac{h^2}{8m_eL^2} = 18 \times 10^{-20} \text{ J}$$

which corresponds to 1.1 eV.

Comment. The electron-in-a-box is a crude model of molecular structure; it can be used for estimating rough values of transition energies and hence used to predict the colours of dye and indicator molecules.

Exercise. Calculate the first excitation energy of a proton confined to a region roughly equal to the diameter of a nucleus (10^{-15} m). [600 MeV]

The distribution of the particle in a box is not uniform: the probability density at x is

$$\psi^2 = \frac{2}{L}\sin^2\frac{n\pi x}{L}$$

The nonuniformity is pronounced when n is small (Fig. 12.2) but ψ^2 becomes more uniform as n increases. The distribution at high quantum numbers reflects the classical result that a particle bouncing between the walls spends, on the average, equal times at all points. That the quantum result corresponds to the classical prediction at high quantum numbers is an aspect of the **correspondence principle**, which states that classical mechanics emerges from quantum mechanics as high quantum numbers are reached.

Example 12.3: *Using the particle in a box solutions* (2)

What is the probability of locating the electron between $x = 0$ and $x = 0.2$ nm in the box described in Example 12.2?

Answer. The distribution of the electron is given by eqn 3b with $n = 1$. The total probability of finding the electron in the specified region is the integral of $\psi_1^2\, dx$ over that region:

$$P = \frac{2}{L}\int_0^l \sin^2\frac{n\pi x}{L}\, dx = \frac{l}{L} - \frac{1}{2n\pi}\sin\frac{2n\pi l}{L}$$

with $n = 1$ and $l = 0.2$ nm. This gives $P = 0.05$, or a chance of 1 in 20 of finding the electron in that region.

Exercise. For the proton Exercise in Example 12.2, calculate the probability that in its ground state it will be found between $x = 0.25L$ and $x = 0.75L$.

[0.82]

12.2 Motion in two dimensions

We now consider the two-dimensional analogue of the particle in a box. Now the particle is confined to a rectangular surface of length L_1 in the x-direction and L_2 in the y-direction and the potential energy is zero everywhere except at the walls, where it is infinite. The Schrödinger equation is

$$-\frac{\hbar^2}{2m}\left(\frac{\partial^2\psi}{\partial x^2} + \frac{\partial^2\psi}{\partial y^2}\right) = E\psi \tag{4}$$

and ψ is a function of both x and y, which we write $\psi = \psi(x, y)$.

Separation of variables

In some cases, partial differential equations (differential equations in more than one variable, like eqn 4) can be solved very simply by the method called the **separation of variables**. This has the effect of dividing the complete equation into two or more ordinary differential equations, one for each variable. The method works in this case, as we can see by testing whether a solution of eqn 4 can be found by writing the wavefunction as a product of functions, one depending only on x and the other only on y:

$$\psi(x, y) = X(x)Y(y)$$

The first step in the procedure is to note that since

$$\frac{\partial^2\psi}{\partial x^2} = Y\frac{d^2X}{dx^2} \qquad \frac{\partial^2\psi}{\partial y^2} = X\frac{d^2Y}{dy^2}$$

the equation becomes

$$-\frac{\hbar^2}{2m}\left(Y\frac{d^2X}{dx^2} + X\frac{d^2Y}{dy^2}\right) = EXY$$

Then on dividing by XY we obtain

$$-\frac{\hbar^2}{2m}\left(\frac{X''}{X} + \frac{Y''}{Y}\right) = E$$

where $X'' = d^2X/dx^2$ and $Y'' = d^2Y/dy^2$.

Now we take the crucial step. Since X''/X is independent of y, if y is varied only the term Y''/Y can change. But the sum of these two terms is a constant; therefore, even the Y''/Y term cannot change. In other words, Y''/Y is a constant, which we denote E^Y. Similarly, X''/X is also a constant, E^X. Therefore, we can write

$$-\frac{\hbar^2 X''}{2mX} = E^X, \quad \text{or} \quad -\frac{\hbar^2}{2m}\frac{d^2 X}{dx^2} = E^X X$$

$$-\frac{\hbar^2 Y''}{2mY} = E^Y, \quad \text{or} \quad -\frac{\hbar^2}{2m}\frac{d^2 Y}{dy^2} = E^Y Y$$

with $E^X + E^Y = E$.

Each of the last two equations is the same as the one-dimensional square-well Schrödinger equation; hence we can adapt the results in eqn 3 without further calculation:

$$X_{n_1} = \left(\frac{2}{L_1}\right)^{1/2} \sin\frac{n_1 \pi x}{L_1}$$

$$Y_{n_2} = \left(\frac{2}{L_2}\right)^{1/2} \sin\frac{n_2 \pi y}{L_2}$$

Since $\psi = XY$ and $E = E^X + E^Y$, we obtain

$$\psi_{n_1,n_2} = \left(\frac{4}{L_1 L_2}\right)^{1/2} \sin\frac{n_1 \pi x}{L_1} \sin\frac{n_2 \pi y}{L_2} \tag{5a}$$

$$E_{n_1,n_2} = \left\{\left(\frac{n_1}{L_1}\right)^2 + \left(\frac{n_2}{L_2}\right)^2\right\}\frac{h^2}{8m} \tag{5b}$$

with the quantum numbers taking the values $n_1 = 1, 2, \ldots$ and $n_2 = 1, 2, \ldots$ independently. Some of these functions are plotted in Fig. 12.3; they are the two-dimensional versions of the wavefunctions shown in Fig. 12.2.

A particle in an actual three-dimensional box can be treated in the same way. The wavefunctions have another factor (for the z-dependence), and the energy has an additional term in n_3^2/L_3^2.

Degeneracy

An interesting feature of the solutions is obtained when the plane surface is square, when $L_1 = L$ and $L_2 = L$. Then

$$\psi_{n_1,n_2} = \frac{2}{L} \sin\frac{n_1 \pi x}{L} \sin\frac{n_2 \pi y}{L}$$

$$E_{n_1,n_2} = (n_1^2 + n_2^2)\frac{h^2}{8mL^2}$$

Consider the cases $n_1 = 1$, $n_2 = 2$, and $n_1 = 2$, $n_2 = 1$:

$$\psi_{1,2} = \frac{2}{L} \sin\frac{\pi x}{L} \sin\frac{2\pi y}{L} \qquad E_{1,2} = \frac{5h^2}{8mL^2}$$

$$\psi_{2,1} = \frac{2}{L} \sin\frac{2\pi x}{L} \sin\frac{\pi y}{L} \qquad E_{2,1} = \frac{5h^2}{8mL^2}$$

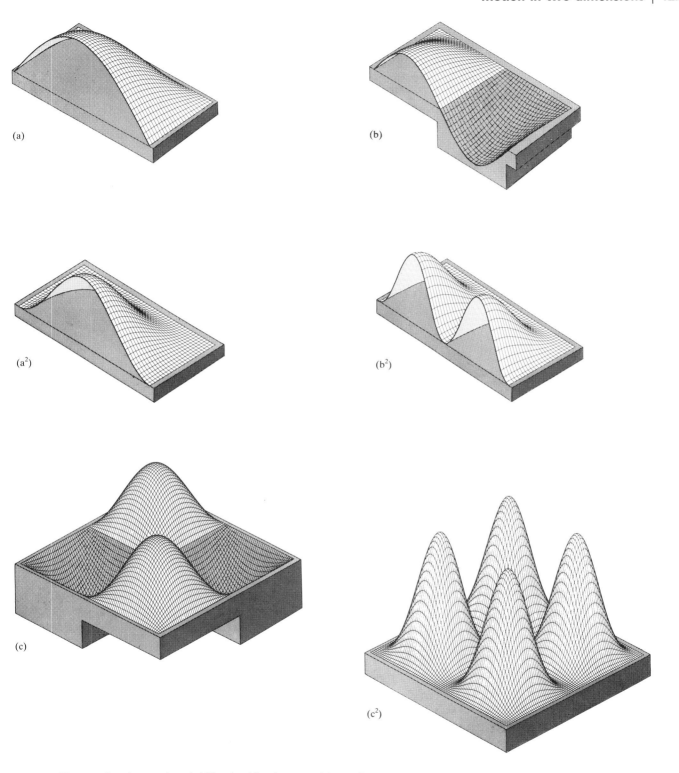

Fig. 12.3. The wavefunctions and probability densities for a particle confined to a rectangular surface. (a) $\psi_{1,1}$ section; (b) $\psi_{2,1}$ section (rotate by 90° for $\psi_{1,2}$). (c) $\psi_{2,2}$. The corresponding probability densities are labelled (a²), (b²), and (c²). Each section is half the total function.

(a)

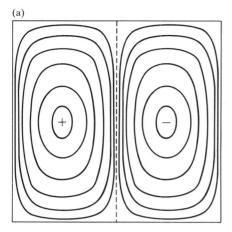

(b)

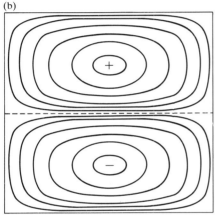

Fig. 12.4. It is easier to represent the functions in terms of contour diagrams. Here we show contour diagrams for (a) $\psi_{2,1}$ and (b) $\psi_{1,2}$ in a square square-well. Note that one can be converted into the other by a 90° rotation: we say that they are related by a 'symmetry transformation'. These two functions are also degenerate (i.e. have the same energy).

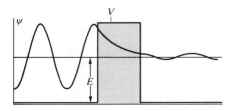

Fig. 12.5. A particle incident on a barrier from the left has an oscillating wave function, but inside the barrier it decays exponentially (for $E < V$). If the barrier is not too thick, the wavefunction is non-zero at its opposite face, and so oscillates there, which corresponds to the particle penetrating the barrier. (Only the real component is shown.)

We see that different wavefunctions correspond to the same energy, the condition called **degeneracy**. In this case, in which there are two degenerate wavefunctions, we say that the level with energy $5(h^2/8mL^2)$ is **doubly degenerate**.

The occurrence of degeneracy is related to the symmetry of the system. The two degenerate functions $\psi_{1,2}$ and $\psi_{2,1}$ are shown in Fig. 12.4: because the box is square, we see that we can convert one into the other simply by rotating the plane by 90°. Interconversion by rotation through 90° is not possible when the plane is not square, and $\psi_{1,2}$ and $\psi_{2,1}$ are then not degenerate. We shall see many examples of degeneracy in the pages that follow (e.g. in the hydrogen atom), and all of them can be traced to the symmetry properties of the system.

12.3 Tunnelling

If the potential energy of the particle does not rise to infinity when it is in the walls of the container, the wavefunction does not decay abruptly to zero. If the walls are thin (so the potential energy falls to zero again after a finite distance), the exponential decay of the wavefunction stops, and it begins to oscillate again like the wavefunctions inside the box (Fig. 12.5). Hence the particle might be found on the outside of a container even though according to classical mechanics it has insufficient energy to escape. Such leakage through classically forbidden zones is called **tunnelling**.

The decay of wavefunctions within barriers

We can use the Schrödinger equation to calculate the probability of tunnelling and its dependence on the mass of the particle. Inside a barrier (a region where $V > 0$), the Schrödinger equation is

$$\frac{d^2\psi}{dx^2} = \frac{2m(V-E)\psi}{\hbar^2}$$

Its general solutions are

$$\psi = Ae^{\kappa x} + Be^{-\kappa x} \qquad \kappa = \left\{ \frac{2m(V-E)}{\hbar^2} \right\}^{1/2} \qquad (6)$$

(Note that the exponentials are now real functions.) If V never fell back to zero, the first term would increase without limit as x increased, and in due course would approach infinity. The only way we can ensure that the wavefunction does not become infinite is to set $A = 0$. Therefore, inside a long barrier the wavefunction is

$$\psi = Be^{-\kappa x}$$

which decays exponentially towards zero as x increases.

Since the wavefunction decreases exponentially inside the wall, and does so with a rate that depends on $m^{1/2}$, particles of low mass are more able to tunnel through barriers than heavy ones. Tunnelling is very important for electrons, and moderately important for protons; for heavier particles it is less important. A number of effects in chemistry (e.g. the isotope-dependence of some reaction rates) depend on the ability of the proton to tunnel more readily than the deuteron.

Example 12.4: *Estimating a tunnelling probability*

Estimate the relative probabilities that a proton and a deuteron can tunnel through the same barrier of height 1.0 eV and length 10 pm when their energy is 0.9 eV.

Answer. The probability of appearing on the right of a long barrier is proportional to the square of the wavefunction there, and so from the equation above

$$\frac{P'}{P} = e^{-2(\kappa' - \kappa)l}$$

where P and κ refer to the proton and P' and κ' to the deuteron. Since $V - E = 1.602 \times 10^{-20}$ J, and the mass of a deuteron is approximately twice that of a proton,

$$2(\kappa' - \kappa)l = 2 \times (2^{1/2} - 1)$$

$$\times \left\{ \frac{2 \times 1.673 \times 10^{-27} \text{ kg} \times 1.602 \times 10^{-20} \text{ J}}{(1.055 \times 10^{-34} \text{ J s})^2} \right\}^{1/2} \times 1.0 \times 10^{-11} \text{ m}$$

$$= 0.57$$

Hence,

$$\frac{P'}{P} = e^{-0.57} = 0.57$$

Comment. The result shows that the tunnelling probability of a deuteron (in the system specified) is only about half that of a proton.

Exercise. Calculate the relative tunnelling probabilities when the barrier is twice as long, the other conditions being unchanged. [0.32]

Penetration probabilities

The kind of problem we can solve with the material developed so far is illustrated by the case of a projectile (such as an electron or a proton) fired at a barrier. Its potential energy increases sharply from zero to a finite value V when it enters the barrier, remains at that value for a distance L, and then falls to zero again (Fig. 12.6a). This is a model of what happens when particles are fired at an idealized metal foil or sheet of paper. We can ask for the proportion of incident particles that penetrate the barrier when their kinetic energy is less than V, when classically none can penetrate.

The strategy of the calculation (and of others like it) is as follows:

(1) Write down the Schrödinger equation for each zone of constant potential.

(2) Write down the general solutions for each zone using eqn 1b for the regions where $V < E$ and eqn 6 for regions where $V > E$.

(3) Find the coefficients by ensuring that (a) the wavefunction is continuous at each zone boundary, and (b) the first derivatives of the wavefunctions are also continuous at the zone boundaries.

The way this is carried through for the arrangement shown in Fig. 12.6 is illustrated in the following example.

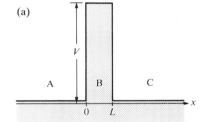

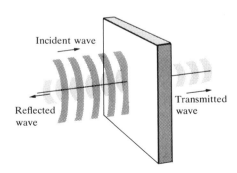

Fig. 12.6 (a) Shows the zones of constant potential energy used in the calculation of a tunnelling probability, and (b) represents the various components of the wavefunction: there is a strong incident wave (e^{ikx}, momentum to the right), a weaker reflected wave (e^{-ikx}), and a very weak transmitted wave (e^{ikx} in Zone C).

Example 12.5: *Calculating a penetration probability*

Deduce an expression for the probability that a particle of mass m and energy E will penetrate a potential energy barrier of height V (with $V > E$) and width L.

Answer. We follow the procedure set out above, taking the particle to be incident from the left. The three zones are shown in Fig. 12.6a. There are no further barriers to the right to reflect back particles, and so no particles have momentum to the left in that zone. The three general solutions are as follows:

$$\text{Zone A: } \psi_A = Ae^{ikx} + A'e^{-ikx} \qquad k = \left(\frac{2mE}{\hbar^2}\right)^{1/2}$$

$$\text{Zone B: } \psi_B = Be^{\kappa x} + B'e^{-\kappa x} \qquad \kappa = \left(\frac{2m(V - E)}{\hbar^2}\right)^{1/2}$$

$$\text{Zone C: } \psi_C = Ce^{ikx} + C'e^{-ikx} \qquad k = \left(\frac{2mE}{\hbar^2}\right)^{1/2}$$

Since there are no particles with negative momentum in zone C, we can set $C' = 0$. The probability of penetration is proportional to $|C|^2$, the probability of penetration relative to the incident probability, which is proportional to $|A|^2$, is

$$P = \frac{|C|^2}{|A|^2}$$

The boundary conditions are the continuity of the wavefunction at the boundaries of the zones:

$$\psi_A(0) = \psi_B(0) \qquad \psi_B(L) = \psi_C(L)$$

and the continuity of its slope at the boundary:

$$\psi'_A(0) = \psi'_B(0) \qquad \psi'_B(L) = \psi'_C(L)$$

with $\psi' = d\psi/dx$. These imply

$$A + A' = B + B' \qquad\qquad Be^{\kappa L} + B'e^{-\kappa L} = Ce^{ikL}$$

$$ikA - ikA' = \kappa B - \kappa B' \qquad \kappa Be^{\kappa L} - \kappa B'e^{-\kappa L} = ikCe^{ikL}$$

When we solve these four simultaneous equations for P, the result is eqn 7 below.

Comment. A 'trap', a region of negative V, has interesting properties as it becomes transparent for some values of E (see Problem 12.9). The classical analogy is the coating on a lens, which is chosen to have a refractive index and thickness that make it transparent to the incident light.

Exercise. Find an expression for P in the case when the potential energy in Zone C is $V \neq 0$

[Problem 12.7]

The example shows that the probability of penetration of the barrier is

$$P = \frac{1}{1 + G} \qquad G = \frac{(e^{L/D} - e^{-L/D})^2}{16\varepsilon(1 - \varepsilon)} \qquad (7)$$

where

$$D = \left\{\frac{\hbar^2}{2m(V - E)}\right\}^{1/2} \qquad \varepsilon = \frac{E}{V}$$

When the barrier is high and long, so $L/D \gg 1$, the first exponential dominates the second, and since then $G \gg 1$, $P \approx 1/G$. In this case

$$P = 4\varepsilon(1 - \varepsilon)e^{-2L/D}$$

We see that P depends exponentially on the square-root of the mass of the particle (as we anticipated) and on the length of the barrier. Figure 12.7 shows how the tunnelling probabilities of a proton and a deuteron depend on the incident energy for a barrier of height 5 eV and width 100 pm.

Vibrational motion

We saw in Section 11.1 that a particle undergoes harmonic motion if it experiences a restoring force that is proportional to its displacement:

$$F = -kx$$

k is the force constant. As explained in Example 11.2, a force of this form corresponds to a potential energy

$$V = \tfrac{1}{2}kx^2$$

The Schrödinger equation for the particle is therefore

$$-\frac{\hbar^2}{2m}\frac{d^2\psi}{dx^2} + \tfrac{1}{2}kx^2\psi = E\psi \tag{8a}$$

The appearance of this equation is simplified by introducing the following dimensionless quantities:

$$y = \frac{x}{\alpha} \qquad \varepsilon = \frac{E}{\tfrac{1}{2}\hbar\omega}$$

where

$$\alpha = \left(\frac{\hbar^2}{mk}\right)^{1/4} \qquad \omega = \left(\frac{k}{m}\right)^{1/2}$$

These transform the equation into

$$\frac{d^2\psi}{dy^2} + (\varepsilon - y^2)\psi = 0 \tag{8b}$$

12.4 The energy levels

It is helpful at the outset to identify the similarities between the harmonic oscillator and the particle in a box. As for the particle in a box, a particle undergoing harmonic motion is trapped in a symmetrical well in which the potential energy rises to large values (and ultimately to infinity) for sufficiently large displacements (Fig. 12.8). As boundary conditions must be satisfied, we should expect the energy of the particle to be quantized.

The solution of eqn 8 is outlined in *Further information 1* at the end of the chapter. The most important feature of the result is that the permitted energy levels of a harmonic oscillator are

$$E_v = (v + \tfrac{1}{2})\hbar\omega \quad \text{with} \quad v = 0, 1, 2, \ldots \quad \text{and} \quad \omega = \left(\frac{k}{m}\right)^{1/2} \tag{9}$$

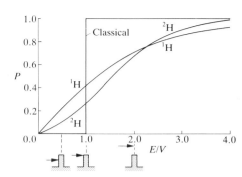

Fig. 12.7 The probability of transmission of a proton and a deuteron through a potential barrier for $E < V$, calculated using eqn 7. The diagrams beneath the horizontal axis indicate the relative values of E and V.

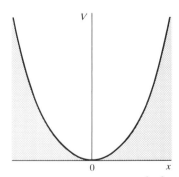

Fig. 12.8 The potential energy $V = \tfrac{1}{2}kx^2$ of harmonic oscillator. The narrowness of the curve depends on the force constant k. x is the displacement from equilibrium. V is a 'parabolic' potential energy.

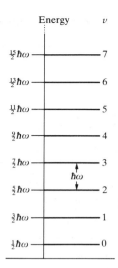

Energy v

$\frac{15}{2}\hbar\omega$ ———————— 7

$\frac{13}{2}\hbar\omega$ ———————— 6

$\frac{11}{2}\hbar\omega$ ———————— 5

$\frac{9}{2}\hbar\omega$ ———————— 4

$\frac{7}{2}\hbar\omega$ ———————— 3

$\frac{5}{2}\hbar\omega$ ———————— 2

$\frac{3}{2}\hbar\omega$ ———————— 1

$\frac{1}{2}\hbar\omega$ ———————— 0

Fig. 12.9 The first few energy levels of a harmonic oscillator. Notice that the levels have constant separation, and that there is a minimum energy of $\frac{1}{2}\hbar\omega$, the zero-point energy.

Hence, the separation between adjacent levels is

$$\Delta E = E_{v+1} - E_v = \hbar\omega$$

which is the same for all v. Therefore, the energy levels form a uniform ladder of spacing $\hbar\omega$ (Fig. 12.9).

The energy separation $\hbar\omega$ is negligibly small for macroscopic objects, but is of great importance for objects with mass similar to that of atoms. For instance, the force constant of a typical chemical bond is around $500\,\text{N m}^{-1}$, and since the mass of a proton is about $1.7 \times 10^{-27}\,\text{kg}$, the frequency is $\omega \approx 5 \times 10^{14}\,\text{s}^{-1}$ and the separation of adjacent levels is $\hbar\omega \approx 6 \times 10^{-20}\,\text{J}$. This energy separation corresponds to $30\,\text{kJ mol}^{-1}$, which is chemically significant. The excitation of a harmonic oscillator from one level to the one immediately above requires an energy $\Delta E = 6 \times 10^{-20}\,\text{J}$ (0.4 eV), and hence, if it is caused by a photon, requires radiation of frequency

$$v = \frac{\Delta E}{h} \approx 9 \times 10^{13}\,\text{Hz}$$

and therefore of wavelength

$$\lambda = \frac{c}{v} \approx 3000\,\text{nm}$$

Therefore, transitions between the vibrational energy levels of molecules require infrared radiation, as we shall describe in Chapter 16.

Since the smallest value of v is 0, an oscillator has a zero-point energy

$$E_0 = \tfrac{1}{2}\hbar\omega$$

For the typical molecular oscillator specified above, the zero-point energy is about $3 \times 10^{-20}\,\text{J}$. This corresponds to $0.2\,\text{eV}$, or $15\,\text{kJ mol}^{-1}$. The *mathematical* reason for the zero-point energy is that v cannot take the value $v = -\frac{1}{2}$, for if it did the wavefunction would be ill-behaved. The *physical* reason is the same as for the particle in a square well: the particle is confined, its position is not completely uncertain, and therefore its momentum, and hence its kinetic energy, cannot be exactly zero. We can picture this zero-point state as one in which the particle fluctuates incessantly around its equilibrium position; classical mechanics would allow the particle to be perfectly still.

12.5 The wavefunctions

We can expect the wavefunctions of a harmonic oscillator to resemble the particle in a box wavefunctions, but with two differences. First, their amplitudes fall towards zero more slowly at large displacements, because the potential energy climbs towards infinity only as x^2 and not abruptly. Second, as the kinetic energy of the particle depends on the displacement in a more complex way (on account of the variation of the potential energy), the curvature of the wavefunction also varies in a more complex way.

The form of the wavefunctions

As we show in *Further information 1*, the wavefunction for a harmonic oscillator in a state with quantum number v is

$$\psi_v = N_v H_v(y)\mathrm{e}^{-y^2/2} \qquad (10)$$

The factor H_v is a **Hermite polynomial**, some of which are listed in Table 12.1. For instance, since $H_0(y) = 1$, the wavefunction for the ground state (the lowest energy state) of the harmonic oscillator is

$$\psi_0 = N_0 e^{-y^2/2}$$

and the probability density is the bell-shaped **gaussian function**

$$\psi_0^2 = N_0^2 e^{-y^2}$$

This has its maximum amplitude at zero displacement (at $y = 0$), and so it captures the classical picture of the zero-point motion as arising from the fluctuation of the particle about its equilibrium position. The shapes of some of the other wavefunctions are also shown in Fig. 12.10.

Example 12.6: *Normalizing a harmonic oscillator wavefunction*

Find the normalization constant for the harmonic oscillator wavefunctions.

Answer. We must choose N_v such that

$$\int_{-\infty}^{\infty} \psi_v^2 \, dx = 1$$

We begin by expressing the integral as an integration over the dimensionless variable $y = x/\alpha$, implying that $dx = \alpha \, dy$, and then use the integration properties given in Table 12.1:

$$\alpha \int_{-\infty}^{\infty} \psi_v^2 \, dy = \alpha N_v^2 \int_{-\infty}^{\infty} H_v^2 e^{-y^2} \, dy = \alpha N_v^2 \pi^{1/2} 2^v v! = 1$$

Therefore,

$$N_v = \left(\frac{1}{\alpha \pi^{1/2} 2^v v!} \right)^{1/2}$$

Comment. The Hermite polynomials are one of a class of functions called 'orthogonal polynomials'. These have a wide range of important properties which allow a number of quantum mechanical calculations to be done with relative ease. See *Further reading* for a reference to their properties.

Exercise. Confirm, by explicit evaluation of the integral, that ψ_0 is normalized to 1. [Use $\int_{-\infty}^{\infty} e^{-z^2} \, dz = \pi^{1/2}$]

At high quantum numbers, harmonic oscillator wavefunctions have their largest amplitudes near the turning points of the classical motion (where $V = E$, so the kinetic energy is zero). Once again we see classical properties emerging in the correspondence limit of high quantum numbers, for the particle is most likely to be found at the turning points (where it travels most slowly) and is least likely to be found at zero displacement (where it travels with maximum velocity).

The properties of the oscillator

Once the wavefunctions are available, we can start calculating the properties of the harmonic oscillator. For instance, we can calculate the expectation values of an observable Ω by evaluating integrals of the type

$$\langle \Omega \rangle = \int \psi_v^* \hat{\Omega} \psi_v \, dx$$

Table 12.1. The Hermite polynomials $H_v(y)$

v	H_v
0	1
1	$2y$
2	$4y^2 - 2$
3	$8y^3 - 12y$
4	$16y^4 - 48y^2 + 12$
5	$32y^5 - 160y^3 + 120y$
6	$64y^6 - 480y^4 + 720y^2 - 120$

The Hermite polynomials (which continue up to infinite v) satisfy the equation

$$H_v'' - 2yH_v' + 2vH_v = 0$$

and the recursion relation

$$H_{v+1} = 2yH_v - 2vH_{v-1}$$

An important integral is

$$\int_{-\infty}^{\infty} H_v H_{v'} e^{-y^2} \, dy = \begin{cases} 0 & \text{if } v' \neq v \\ \pi^{1/2} 2^v v! & \text{if } v' = v \end{cases}$$

(a)
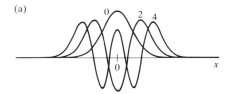

(b)
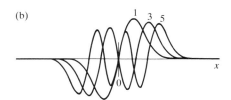

(c)
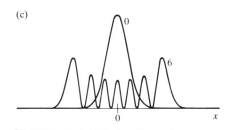

Fig. 12.10 Probability densities and wavefunctions for some states of a harmonic oscillator. (a) ψ for even v; (b) ψ for odd v; (c) ψ^2 for $v = 0$, 6. Notice that the most probable location of the oscillator lies closer to the turning points of the classical motion as the quantum number v increases.

When the explicit wavefunctions are substituted the integrals look fearsome, but the Hermite polynomials have many simplifying features. For instance, we show in the following example that the mean displacement $\langle x \rangle$ and the mean square displacement $\langle x^2 \rangle$ of the oscillator when it is in the state with quantum number v are

$$\langle x \rangle = 0 \qquad \langle x^2 \rangle = (v + \tfrac{1}{2})\left(\frac{\hbar^2}{mk}\right)^{1/2} \tag{11}$$

The result for $\langle x \rangle$ shows that the oscillator is as likely to be found on either side of $x = 0$ (like a classical oscillator). The result for $\langle x^2 \rangle$ shows that the mean square displacement increases with v. The increase is apparent from the probability densities in Fig. 12.10, and corresponds to the classical amplitude of swing increasing as the oscillator becomes more highly excited.

Example 12.7: *Calculating properties of the harmonic oscillator*

Calculate the mean displacement of the oscillator $\langle x \rangle$ from equilibrium when it is in a quantum state v.

Answer. We should calculate the expectation value of x using the normalized wavefunctions. The operator for position along x is multiplication by the value of x (Section 11.5).

$$\langle x \rangle = \int \psi_v^2 x \, dx = N_v^2 \int_{-\infty}^{\infty} (H_v e^{-y^2/2})^2 x \, dx$$

$$= \alpha^2 N_v^2 \int_{-\infty}^{\infty} H_v \, y H_v e^{-y^2} \, dy$$

Now use the recursion relation given in Table 12.1 to write

$$yH_v = vH_{v-1} + \tfrac{1}{2}H_{v+1}$$

which turns the integral into

$$\int_{-\infty}^{\infty} H_v \, y H_v e^{-y^2} \, dy = v \int_{-\infty}^{\infty} H_v H_{v-1} e^{-y^2} \, dy + \frac{1}{2} \int_{-\infty}^{\infty} H_v H_{v+1} e^{-y^2} \, dy$$

Both integrals are zero (Table 12.1), and so $\langle x \rangle = 0$.

Comment. The result is in fact obvious from the probability densities drawn in Fig. 12.10c: they are both symmetrical about $x = 0$, and so displacements to the right are as probable as displacements to the left. The reason for going through the calculation in detail even though the result is obvious is that the same technique is applicable to other observables for which the result is not obvious.

Exercise. Calculate the mean square displacement $\langle x^2 \rangle$ of the particle from its equilibrium position. (Use the recursion relation twice.) [eqn 11]

We can now calculate the mean potential energy of the oscillator very simply:

$$\langle V \rangle = \tfrac{1}{2}k\langle x^2 \rangle = \tfrac{1}{2}(v + \tfrac{1}{2})\hbar\left(\frac{k}{m}\right)^{1/2} = \tfrac{1}{2}(v + \tfrac{1}{2})\hbar\omega \tag{12a}$$

and as the total energy in the state v is $(v + \tfrac{1}{2})\hbar\omega$,

$$\langle V \rangle = \tfrac{1}{2}E_v \tag{12b}$$

Since the total energy is the sum of the potential and kinetic energies, it follows at once that the mean kinetic energy of the oscillator is the same:

$$\langle E_K \rangle = \tfrac{1}{2} E_v \qquad (12c)$$

The result that the mean potential and kinetic energies are equal (and therefore both equal to half the total energy) is a special case of the **virial theorem**:

If the potential energy of a particle has the form $V = ax^b$, then its mean potential and kinetic energies are related by

$$2\langle E_K \rangle = b \langle V \rangle \qquad (13)$$

In the case of a harmonic oscillator $b = 2$, and so $\langle E_K \rangle = \langle V \rangle$, as we have found. The virial theorem is a short cut to the result; and we shall use it again.

Example 12.8: *Calculating the probability of non-classical extensions*

Calculate the probability that a harmonic oscillator will be found extended into a classically forbidden region.

Answer. According to classical mechanics, the turning point x_{tp} of an oscillating particle occurs when its kinetic energy is zero, which is when its potential energy $\tfrac{1}{2}kx^2$ is equal to its total energy E. This occurs when

$$x_{tp}^2 = \frac{2E}{k} \quad \text{or} \quad x_{tp} = \pm \left(\frac{2E}{k}\right)^{1/2}$$

The probability of finding the particle beyond a displacement x_{tp} is the sum of the probabilities $\psi^2 \, dx$ of finding it in any of the intervals dx lying between x_{tp} and infinity:

$$P = \int_{x_{tp}}^{\infty} \psi^2 \, dx$$

The variable of integration is best expressed in terms of $x = \alpha y$, and then the turning point lies at

$$y_{tp} = x_{tp}/\alpha = (2v + 1)^{1/2}$$

For the ground state, $y_{tp} = 1$ and the probability is

$$P = \int_{x_{tp}}^{\infty} \psi_0^2 \, dx = \alpha N_0^2 \int_1^{\infty} e^{-y^2} \, dy = \frac{1}{\pi^{1/2}} \int_1^{\infty} e^{-y^2} \, dy$$

The integral is a special case of the error function, erf z, which is defined as follows:

$$\text{erf}(z) = 1 - \frac{2}{\pi^{1/2}} \int_z^{\infty} e^{-y^2} \, dy$$

and is tabulated (Table 12.2). In the present case

$$P = \tfrac{1}{2}(1 - \text{erf } 1) = \tfrac{1}{2}(1 - 0.843) = 0.079$$

Comment. In 7.9 per cent of a large number of observations, an oscillator in the state $v = 0$ (whatever its mass and the value of the force constant) will be found at a classically forbidden extension.

Exercise. Calculate the probability that the oscillator will have a classically forbidden compression when $v = 0$. \qquad [0.079]

Table 12.2. The error function

z	erf (z)
0	0
0.01	0.0113
0.05	0.0564
0.10	0.1125
0.50	0.5205
1.00	0.8427
1.50	0.9661
2.0	0.9953

The example shows that the oscillator may be found in classically forbidden regions, and that when $v = 0$ there is about 8 per cent chance of finding it with a classically forbidden extension. The probability of its being found with a non-classical extension decreases quickly with increasing v, and vanishes entirely as v approaches infinity, as we would expect from the correspondence principle. Since macroscopic oscillators (such as pendulums) are in states with very high quantum numbers, the probability that they will be found at a classically forbidden displacement is wholly negligible. Molecules, however, are normally in their vibrational ground states, and for them the probability is very significant.

Rotational motion

The treatment of rotational motion can be broken down into two parts. The first deals with motion of a particle on a ring, and the second with rotation in three dimensions.

12.6 Rotation in two dimensions

We consider a particle of mass m moving in a circular path of radius r. The total energy is equal to the kinetic energy, because the potential energy is zero everywhere. We can therefore write $E = p^2/2m$. According to classical mechanics (Section 11.1), the magnitude of the angular momentum is $J = pr$, and so the energy can be expressed as $J^2/2mr^2$. Since mr^2 is the **moment of inertia** I, it follows that

$$E = \frac{J^2}{2I}$$

We shall now see that not all the values of the angular momentum are permitted in quantum mechanics and therefore that the energy is quantized.

The qualitative origin of quantized rotation

Since $J = pr$, and $p = h/\lambda$, the magnitude of the angular momentum is

$$J = \frac{hr}{\lambda}$$

The smaller the wavelength of the particle on the ring, the greater its angular momentum. It follows that if we can see why the wavelength is restricted to discrete values, we shall understand why the angular momentum is restricted.

Suppose for the moment that λ can take an arbitrary value. In that case, the amplitude of the wavefunction depends on the azimuthal angle ϕ as shown in Fig. 12.11a. When ϕ increases beyond 2π, the wavefunction continues to change, but for an arbitrary wavelength it gives rise to a different amplitude at each point. Since different values are given on successive circuits, the wavefunction is not single-valued. As we saw in Section 11.4, a multiple-valued wavefunction is unacceptable.

An acceptable solution is obtained if the wavefunction reproduces itself on successive circuits, as in Fig. 12.11b. Then, since only some wavefunctions have this property, only some angular momenta are acceptable, and therefore only some energies exist. Hence, the energy of the particle is

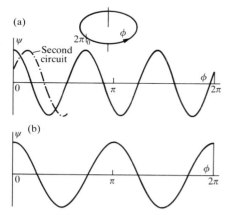

Fig. 12.11 Two solutions of the Schrödinger equation for a particle on a ring. The circumference has been opened out into a straight line; the points at $\phi = 0$ and 2π are identical. The solution in (a) is unacceptable because it is not single-valued. Moreover, on successive circuits it interferes destructively with itself, and does not survive. The solution in (b) is acceptable: it is single-valued, and on successive circuits it reproduces itself.

quantized. In particular, the wavelength must be a whole-number fraction of the circumference if its ends are to match after each circuit. That is,

$$\lambda = \frac{2\pi r}{n} \quad \text{with} \quad n = 0, 1, 2, \ldots$$

($n = 0$, which gives an infinite wavelength, corresponds to a uniform amplitude.) At this stage, therefore, it looks as though the angular momentum is limited to the values

$$J = \frac{nhr}{2\pi r} \quad \text{or} \quad J = n\hbar$$

(because $\hbar = h/2\pi$) and that the energy is limited to the values

$$E = \frac{J^2}{2I} = \frac{n^2\hbar^2}{2I} \quad \text{with} \quad n = 0, 1, 2, \ldots$$

A moment's thought, however, shows that this solution cannot be complete. The angular momentum can arise from motion in either direction, and so l, like p, ought to carry a sign, $J = +n\hbar$ indicating one sense of rotation and $J = -n\hbar$ indicating the other. We should therefore expect the quantum number for rotational motion to take both positive and negative values. How this comes about, and how it affects the expression for the energy, we can see by solving the Schrödinger equation explicitly.

The formal solution

The two-dimensional Schrödinger equation for a particle in a plane (with $V = 0$) is the same as in eqn 4:

$$-\frac{\hbar^2}{2m}\left(\frac{\partial^2\psi}{\partial x^2} + \frac{\partial^2\psi}{\partial y^2}\right) = E\psi$$

Instead of trying to solve this equation as it stands, it is better to transform it to polar coordinates, and to write

$$x = r\cos\phi \qquad y = r\sin\phi$$

because then the condition $r = $ constant is much easier to impose. (In general, it is always a good idea to use coordinates that reflect the full symmetry of the system.) Since r is constant, by standard manipulations we can write

$$\frac{\partial^2}{\partial x^2} + \frac{\partial^2}{\partial y^2} = \frac{1}{r^2}\frac{d^2}{d\phi^2}$$

and hence transform the Schrödinger equation into

$$-\frac{\hbar^2}{2mr^2}\frac{d^2\psi}{d\phi^2} = E\psi$$

The moment of inertia $I = mr^2$ has appeared automatically and the equation may be written

$$\frac{d^2\psi}{d\phi^2} = \frac{-2IE\psi}{\hbar^2} \tag{14a}$$

The normalized general solutions of the equation are

$$\psi_{m_l} = \left(\frac{1}{2\pi}\right)^{1/2} e^{im_l\phi} \qquad m_l = \pm\left(\frac{2IE}{\hbar^2}\right)^{1/2} \tag{14b}$$

(m_l is some dimensionless number at this stage; it will soon be promoted to the status of a quantum number; the origin of the notation m_l will be made clear later in the section.) We now select the acceptable solutions from among these general solutions by imposing the condition that the wavefunction should be single-valued. That is, the wavefunction ψ must satisfy the **cyclic boundary condition** and match at points separated by a complete revolution.

$$\psi(\phi + 2\pi) = \psi(\phi)$$

On substituting the general form into this condition, we find

$$\psi_{m_l}(\phi + 2\pi) = \left(\frac{1}{2\pi}\right)^{1/2} e^{im_l(\phi + 2\pi)} = \left(\frac{1}{2\pi}\right)^{1/2} e^{im_l\phi} e^{2\pi im_l}$$

$$= \psi_{m_l}(\phi) e^{2i\pi m_l}$$

As $e^{i\pi} = -1$, this is equivalent to

$$\psi_{m_l}(\phi + 2\pi) = (-1)^{2m_l} \psi_{m_l}(\phi)$$

Hence, $2m_l$ must be a positive or negative *even* integer, and therefore m_l must be an integer:

$$m_l = 0, \pm 1, \pm 2, \ldots$$

Quantization of rotation

We can collect these conclusions together as follows. In the first place, since the energy is related to the value of m_l by eqn 14b, and the cyclic boundary conditions restrict m_l to integral values, we conclude that the energy is quantized, and that the allowed values are given by

$$E_{m_l} = \frac{m_l^2 \hbar^2}{2I} \qquad m_l = 0, \pm 1, \pm 2, \ldots \tag{15}$$

The occurrence of m_l as its square means that the energy of rotation is independent of the sense of rotation (the sign of m_l), as we expect physically. Furthermore, although the result has been derived for the rotation of a single mass point, it also applies to any body of moment of inertia I constrained to rotate about one axis.

We can also conclude that the angular momentum is quantized. This is our first example of an observable other than the energy that is confined to discrete values. In the present case, the relation is established by comparing eqn 15 and the classical expression for the energy, which shows that the angular momentum is limited to the values

$$J_z = m_l \hbar \qquad m_l = 0, \pm 1, \pm 2, \ldots \tag{16}$$

(The subscript z is there to remind us that the angular momentum corresponds to motion about the z-axis.) The increasing angular momentum is associated with the increasing number of nodes in the wavefunction: the wavelength decreases stepwise as m_l increases, and so the momentum with which the particle travels round the ring increases.

In our discussion of free motion in one dimension, we saw that the opposite signs in the wavefunctions e^{ikx} and e^{-ikx} corresponded to opposite directions of travel, and that the linear momentum was given by the eigenvalue of the linear momentum operator. The same conclusions can be drawn here, but now we need the eigenvalues of the angular momentum operator. In classical mechanics the **orbital angular momentum** l_z about the z-axis is defined as

$$l_z = xp_y - yp_x$$

p_x is the component of linear motion parallel to the x-axis and p_y the component parallel to the y-axis. The operators for the two linear momentum components are differentiation with respect to x and y, so in quantum mechanics the operator for angular momentum about the z-axis is

$$\hat{l}_z = \frac{\hbar}{i}\left(x\frac{\partial}{\partial y} - y\frac{\partial}{\partial x}\right)$$

When expressed in terms of polar coordinates, this equation becomes

$$\hat{l}_z = \frac{\hbar}{i}\frac{\partial}{\partial \phi} \qquad (17)$$

With the angular momentum operator available, we can test the wavefunction in eqn 14b. Disregarding the normalization constant, we find

$$\hat{l}_z \psi_{m_l} = \frac{\hbar}{i}\frac{d e^{im_l\phi}}{d\phi} = im_l \times \frac{\hbar}{i} e^{im_l\phi} = m_l\hbar\psi_{m_l}$$

That is, ψ_{m_l} is an eigenfunction of the orbital angular momentum and corresponds to an angular momentum $m_l\hbar$. When m_l is positive, the angular momentum is positive (clockwise when seen from below); when m_l is negative, the angular momentum is negative (counter-clockwise when seen from below). This is the origin of the **vector representation** of the angular momentum, in which the magnitude is represented by the length of a vector and the direction of motion by its orientation (Fig. 12.12).

Finally, we can explore the question of the position of the particle when it is in a state of definite angular momentum. As usual, we form the probability density:

$$\psi_{m_l}^* \psi_{m_l} = \frac{1}{2\pi} \times e^{-im_l\phi} \times e^{im_l\phi} = \frac{1}{2\pi}$$

Since $\psi^*\psi$ is independent of ϕ, the probability of locating the particle at any point on the ring is also independent of ϕ. Hence the location of the particle is completely indefinite, and knowing the angular momentum precisely eliminates the possibility of specifying the particle's location. Angular momentum and angle are a pair of complementary observables (in the sense defined in Section 11.6), and the inability to specify them simultaneously to arbitrary precision is another example of the uncertainty principle.

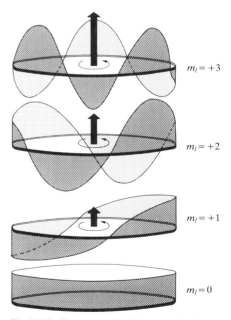

$m_l = +3$

$m_l = +2$

$m_l = +1$

$m_l = 0$

Fig. 12.12 The real parts of the wavefunctions of a particle on a ring. As shorter wavelengths are achieved, the angular momentum grows in steps of $\hbar$. This momentum can be represented by a vector of length m_l units along the z-axis. Note the sign convention (the right-hand screw rule) for the direction of motion.

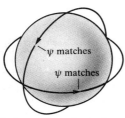

Fig. 12.13 A particle on the surface of a sphere must satisfy two cyclic boundary conditions, and this leads to two quantum numbers for its state of angular momentum.

12.7 Rotation in three dimensions

We now consider a mass point free to move over the surface of a sphere of radius r. The requirement that the wavefunction should match as a path is traced over the poles as well as round the equator (Fig. 12.13) introduces a second cyclic boundary condition and therefore a second quantum number. We shall use the results of this calculation when we come to describe the states of electrons in atoms (Chapter 13) and of rotating molecules (Chapter 16). The latter application arises from the fact that the rotation of a solid body of mass m can be represented by a single point of mass m rotating at a radius R_g, the **radius of gyration** of the body, which is defined so that $I = mR_g^2$.

The Schrödinger equation

The Schrödinger equation in three dimensions (Box 11.1) is

$$-\frac{\hbar^2}{2m}\nabla^2\psi + V\psi = E\psi \qquad (18)$$

where

$$\nabla^2\psi = \frac{1}{r}\frac{\partial^2(r\psi)}{\partial r^2} + \frac{1}{r^2}\Lambda^2\psi$$

and

$$\Lambda^2\psi = \frac{1}{\sin^2\theta}\frac{\partial^2\psi}{\partial\phi^2} + \frac{1}{\sin\theta}\frac{\partial}{\partial\theta}\left(\sin\theta\frac{\partial\psi}{\partial\theta}\right)$$

These equations look fearsome, but they can be simplified quite readily. Thus, since the particle is moving on the surface of a sphere, the radius is not a variable and the differentiations respect to r can be ignored. Moreover, on the surface of the sphere, V is a constant and may be set equal to zero. Therefore, eqn 18 simplifies to

$$-\frac{\hbar^2}{2mr^2}\Lambda^2\psi = E\psi$$

Notice that the moment of inertia $I = mr^2$ has appeared automatically. The equation may therefore be written

$$-\frac{\hbar^2}{2I}\Lambda^2\psi = E\psi$$

which rearranges into

$$\Lambda^2\psi = \frac{-2IE\psi}{\hbar^2} \qquad (19)$$

This equation is still a partial differential equation in the two angular variables θ and ϕ, but we show in *Further information 2* at the end of the chapter that it separates into two equations, one for θ and the other for ϕ, and that we can solve the two equations separately. That is, the wavefunction can be written

$$\psi = \Theta\Phi$$

where Θ is a function of θ alone and Φ is a function of ϕ alone.

The properties of the solutions

We show in *Further information 2* that the acceptable wavefunctions are specified by *two* quantum numbers l and m_l. (There are two cyclic boundary conditions to satisfy, one arising from the angle ϕ and the other from the angle θ.) These quantum numbers are restricted to the values

$$l = 0, 1, 2, \ldots$$

$$m_l = 0, \pm 1, \pm 2, \ldots, \pm l$$

Equivalently:

$$m_l = l, l - 1, l - 2, \ldots, -l$$

Note that l is positive and that, for a given value of l, there are $2l + 1$ permitted values of m_l. The normalized wavefunctions are usually denoted Y_{l,m_l} and are called the **spherical harmonics**. Some of the spherical harmonics are listed in Table 12.3, and their amplitudes at different points on the spherical surface are illustrated in Fig. 12.14.

The energy E of the particle is restricted to the values

$$E = l(l + 1) \frac{\hbar^2}{2I} \qquad l = 0, 1, 2, \ldots \tag{20}$$

We see that the energy is quantized, and that it is independent of m_l. Since there are $2l + 1$ different wavefunctions (one for each value of m_l) that correspond to the same energy, a level with quantum number l is $(2l + 1)$-fold degenerate.

Angular momentum

The energy of a rotating particle is related classically to its angular momentum by $E = J^2/2I$. Therefore, by comparing this equation with eqn 20, we can deduce that the magnitude of the angular momentum is quantized, and confined to the values

$$J = \{l(l + 1)\}^{1/2} \hbar \qquad l = 0, 1, 2, \ldots \tag{21a}$$

We have already seen that the angular momentum about the z-axis is quantized, and that it has the values

$$J_z = m_l \hbar \qquad m_l = 0, \pm 1, \pm 2, \ldots, \pm l \tag{21b}$$

The higher the value of l the larger the number of nodal lines (the positions at which $\psi = 0$) in the wavefunction. This reflects the fact that higher angular momentum implies higher kinetic energy, and therefore a more sharply buckled wavefunction. We can also see that the states corresponding to high angular momentum around the z-axis are those in which most nodes cut the equator: this indicates a high kinetic energy arising from motion parallel to the equator because the curvature is greatest in that direction.

Table 12.3. The spherical harmonics $Y_{l,m_l}(\theta, \phi)$

l	m_l	Y_{l,m_l}
0	0	$\left(\dfrac{1}{4\pi}\right)^{1/2}$
1	0	$\left(\dfrac{3}{4\pi}\right)^{1/2} \cos\theta$
	± 1	$\mp\left(\dfrac{3}{8\pi}\right)^{1/2} \sin\theta \, e^{\pm i\phi}$
2	0	$\left(\dfrac{5}{16\pi}\right)^{1/2} (3\cos^2\theta - 1)$
	± 1	$\mp\left(\dfrac{15}{8\pi}\right)^{1/2} \cos\theta \sin\theta \, e^{\pm i\phi}$
	± 2	$\left(\dfrac{15}{32\pi}\right)^{1/2} \sin^2\theta \, e^{\pm 2i\phi}$
3	0	$\left(\dfrac{7}{16\pi}\right)^{1/2} (5\cos^3\theta - 3\cos\theta)$
	± 1	$\mp\left(\dfrac{21}{64\pi}\right)^{1/2} (5\cos^2\theta - 1)\sin\theta \, e^{\pm i\phi}$
	± 2	$\left(\dfrac{105}{32\pi}\right)^{1/2} \sin^2\theta \cos\theta \, e^{\pm 2i\phi}$
	± 3	$\mp\left(\dfrac{35}{64\pi}\right)^{1/2} \sin^3\theta \, e^{\pm i\phi}$

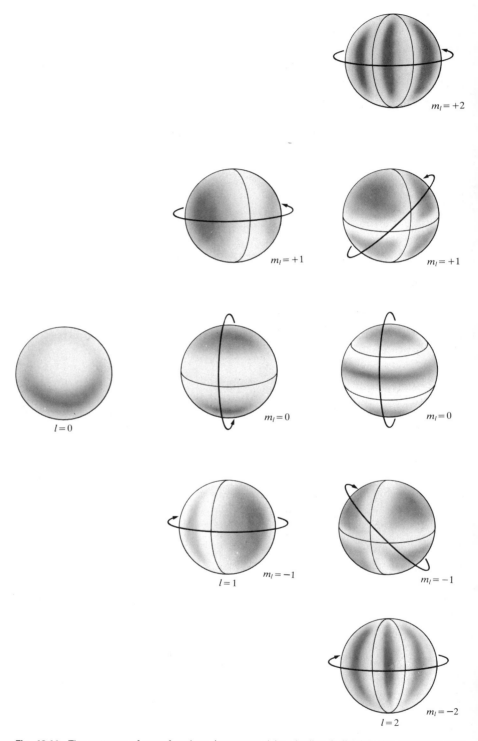

Fig. 12.14 The patterns of wavefunctions (represented by shading indicating the corresponding probability densities) for a particle on the surface of a sphere. Note that the number of nodes increases as the value of *l* increases, and that their orientation is determination by the value of m_l. The corresponding classical motion is indicated in each case. Only the real parts of the wavefunctions are shown.

Example 12.9: *Calculating the energy levels of a rotating molecule*

Calculate the energies of the first five rotational levels of H_2 ($I = 4.603 \times 10^{-48}$ kg m^2), the corresponding magnitudes of the angular momentum, and the number of different values of m_l in each case.

Answer. Use eqn 20 with $l = 0, 1, 2, 3, 4$. For the magnitude, use eqn 21a and express the results as multiples of $\hbar$. For a state of given l there are $2l + 1$ values of m_l in integral steps from $-l$ to $+l$. In the case of rotating molecules the angular momentum quantum number is normally denoted J. We need $\hbar^2/2I = 1.208 \times 10^{-21}$ J (corresponding to 0.727 kJ mol^{-1}). Draw up the following table; $|J|$ denotes the magnitude of the angular momentum.

J	0	1	2	3	4
$J(J+1)$	0	2	6	12	20
$E_J/(\text{kJ mol}^{-1})$	0	1.46	4.37	8.73	12.6
$|J|/\hbar$	0	1.41	2.45	3.46	4.47
Components	1	3	5	7	9

Comment. Note that the smaller the moment of inertia of the molecule, the greater the separation of the rotational energy levels. Since all components of a given J correspond to the same energy, the rotational level with quantum number J is $(2J + 1)$-fold degenerate.

Exercise. Repeat the calculation for a deuterium molecule (same bond length, approximately twice the mass).

The final point we shall make is that eqn 19 may be written in terms of l, so giving the differential equation satisfied by the spherical harmonics:

$$\Lambda^2 Y_{l,m_l} = -l(l+1)Y_{l,m_l} \tag{22}$$

This is a very useful result, which we shall draw on when discussing the hydrogen atom in the next chapter.

Space quantization

The result that m_l is confined to the discrete values $l, l-1, \ldots, -l$ for a given value of l means that the component of angular momentum about the z-axis may take only $2l + 1$ values. If the angular momentum is represented by a vector of length proportional to its magnitude (i.e. of length $\{l(l+1)\}^{1/2}$ units), then to represent correctly the value of the component of angular momentum, the vector must be oriented so that its projection on the z-axis is of length m_l (Fig. 12.15). In classical terms, this means that the plane of rotation of the particle can take only a discrete range of orientations (Fig. 12.15). The remarkable implication is that the *orientation* of a rotating body is quantized.

The quantum mechanical result that a rotating body may not take up an arbitrary orientation with respect to some specified axis (e.g. an axis defined by the direction of an externally applied electric or magnetic field) is called **space quantization**. It was confirmed by an experiment first performed by Otto Stern and Walther Gerlach in 1921, who shot a beam of silver atoms

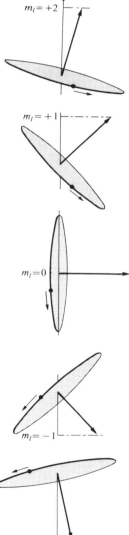

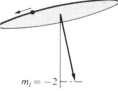

Fig. 12.15 The permitted orientations of angular momentum when $l = 2$. We shall see soon that this is too specific because the azimuthal orientation of the vector (its angle around z) is indeterminate.

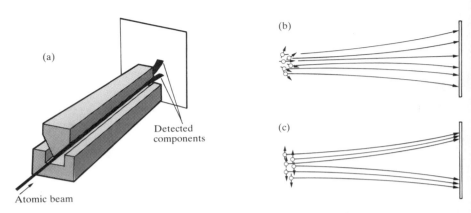

Fig. 12.16 (a) The experimental arrangement for the Stern–Gerlach experiment: the magnet provides an inhomogeneous field. (b) The classically expected result. (c) The observed outcome using silver atoms.

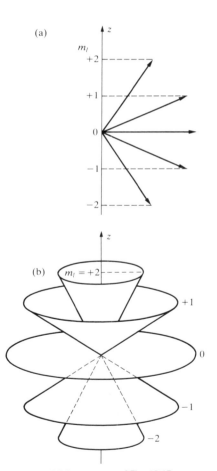

Fig. 12.17 (a) A summary of Fig. 12.15. However, since the azimuthal angle of the vector is indeterminate a better representation is as in (b), where each vector lies at an unspecified azimuth on its cone.

through an inhomogeneous magnetic field (Fig. 12.16a). The idea behind the experiment was that a rotating, charged body behaves like a magnet and interacts with the applied field. According to classical mechanics, since the orientation of the angular momentum can take any value, the associated magnet can take any orientation. Since the direction in which the magnet is driven by the inhomogeneous field depends on the orientation, it follows that a broad band of atoms is expected to emerge from the region where the magnetic field acts (Fig. 12.16b). According to quantum mechanics, since the angular momentum is quantized, the associated magnet lies in a number of discrete orientations, and so several sharp bands of atoms are expected (Fig. 12.16c).

In their first experiment, Stern and Gerlach appeared to confirm the classical prediction. However, the experiment is difficult because the atoms in the beam collide with each other and this blurs the bands. When the experiment was repeated with a beam of very low intensity (so that collisions were less frequent) they observed discrete bands, and so confirmed the quantum prediction.

The vector model

Throughout the preceding discussion, we have referred to the z-component of angular momentum, and have made no reference to the x- and y-components. The reason for this omission is that the uncertainty principle forbids the simultaneous, exact specification of more than one component. Therefore, if l_z is known, it is impossible to ascribe values to the other two components. It follows that the illustration in Fig. 12.15, which is summarized in Fig. 12.17a, gives a false impression of the state of the system, because it suggests definite values for the x- and y-components. A better picture must reflect the indeterminance of l_x and l_y if l_z is known.

The **vector model** of angular momentum uses pictures like that in Fig. 12.17b. The cones are drawn with side $\{l(l+1)\}^{1/2}$ units, and represent the magnitude of the angular momentum. Each cone has a definite projection (of m_l units) on the z-axis, representing the system's precise value of l_z. The l_x and l_y projections, however, are indefinite. The vector representing the state of angular momentum can be thought of as lying with its tip on any point on the mouth of the cone. (At this stage it should not be thought of as sweeping round the cone, that aspect of the model will be added later when we allow the picture to convey more information.)

The vector model of angular momentum, although only a pictorial representation of aspects of the quantum mechanical properties, turns out to be surprisingly useful when we turn to the structure and spectra of atoms.

12.8 Spin

When Stern and Gerlach used Ag atoms in their experiment, they observed *two* bands. This seems to conflict with one of the predictions of quantum mechanics, because an angular momentum l gives rise to $2l + 1$ orientations, which is equal to 2 only if $l = \frac{1}{2}$, contrary to the conclusion that l must be an integer. The conflict was resolved by the suggestion that the angular momentum they were observing was not due to the atom's orbital angular momentum (which arises from the motion of an electron around the atomic nucleus) but arose from the motion of the electron about its own axis. The internal angular momentum of the electron is called its **spin**.

The wavefunction of an electron spinning at a single point in space does not have to satisfy the same boundary conditions as those for a particle moving over the surface of a sphere, and so the quantum numbers are not subject to the same restrictions. In order to distinguish this spin angular momentum from orbital angular momentum we use the quantum number s (in place of l) and m_s for the projection on the z-axis. The magnitude of the spin angular momentum is $\{s(s + 1)\}^{1/2}\hbar$ and the component $m_s\hbar$ is restricted to the $2s + 1$ values

$$m_s = s, s - 1, s - 2, \ldots, -s$$

The detailed analysis of the spin of a particle is quite sophisticated, and shows that the property should not be taken to be an actual spinning motion. However, that picture can be very useful when used with care. For an electron it turns out that only one value of s is allowed, and $s = \frac{1}{2}$, corresponding to an angular momentum of magnitude $\frac{1}{2}\sqrt{3}\hbar = 0.866\hbar$. This spin angular momentum is an intrinsic property of the electron, like its rest mass and its charge, and every electron has exactly the same value. The spin may lie in either of $2s + 1 = 2$ different orientations (Fig. 12.18). One orientation corresponds to $m_s = +\frac{1}{2}$ (this is often called an α-electron and denoted ↑); the other corresponds to $m_s = -\frac{1}{2}$ (and is called a β-electron and denoted ↓).

The outcome of the Stern–Gerlach experiment can now be explained if we suppose that each silver atom possesses an angular momentum due to the spin of a single electron, because the two bands of atoms then correspond to the two spin orientations. The reason why the atoms behave like this will be explained in Chapter 15.

Like the electron, other elementary particles have characteristic spins. For example, protons and neutrons are spin-$\frac{1}{2}$ particles (i.e. $s = \frac{1}{2}$) and so invariably spin with angular momentum $\frac{1}{2}\sqrt{3}\hbar = 0.866\hbar$. Since the masses of a proton and a neutron are so much greater than the mass of an electron, yet they all have the same spin angular momentum, the classical picture would be of particles spinning much more slowly than an electron. Some elementary particles have $s = 1$, and so have an intrinsic angular momentum of magnitude $\sqrt{2}\hbar = 1.414\hbar$. Some mesons are spin-1 particles (as are some atomic nuclei), but for our purposes the most important spin-1 particle is the photon. We shall see the importance of photon spin in the next chapter.

The properties of angular momentum that we have developed are set out in Box 12.1. As mentioned there, when we use the quantum numbers l and

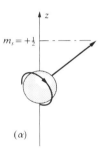

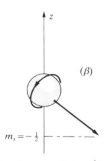

Fig. 12.18 An electron spin ($s = \frac{1}{2}$) can take only two orientations with respect to a specified axis. An α-electron is an electron with $m_s = +\frac{1}{2}$; a β-electron is an electron with $m_s = -\frac{1}{2}$.

Box 12.1 Angular momentum

The quantum numbers:
  Orbital angular momentum quantum number:
  $l = 0, 1, 2, \ldots$
  z-component quantum number: $m_l = 0, \pm 1, \ldots, \pm l$
  (m_l is also known as the magnetic quantum number)

Spin angular momentum quantum number: $s = \frac{1}{2}$
z-component quantum number: $m_s = \pm \frac{1}{2}$

In general, an angular momentum is denoted by the quantum numbers j and m_j. The magnitude of the angular momentum is equal to $\{j(j+1)\}^{1/2}\hbar$ and the z-component of angular momentum is equal to $m_j\hbar$.

For the total angular momentum of a composite system see Section 13.8.

---

m_l we shall mean orbital angular momentum; when we use s and m_s we shall mean spin; and when we use j and m_j we shall mean either (or, in some contexts to be described in Chapter 13, a combination of orbital and spin momenta).

Further information 1: the harmonic oscillator

The first step in solving the Schrödinger equation for harmonic oscillation (eqn 8) is to find the **asymptotic solutions**, the solutions that show how the wavefunctions behave as $x \to \pm\infty$. When x (and therefore y) is very large, the equation simplifies to

$$\frac{\mathrm{d}^2\psi}{\mathrm{d}y^2} - y^2\psi = 0$$

One approximate solution[2] is $\mathrm{e}^{y^2/2}$, but it is obviously unacceptable because it increases rapidly with the displacement, and approaches infinity. The other solution is $\mathrm{e}^{-y^2/2}$, which is well-behaved for large positive and negative displacements, and so is acceptable.

The next step is to suppose that the exact wavefunction has the form

$$\psi = f\mathrm{e}^{-y^2/2}$$

where f is a function that does not increase more rapidly than $\mathrm{e}^{-y^2/2}$ decays: then $f\mathrm{e}^{-y^2/2}$ remains acceptable at large displacements. We substitute this trial solution into the exact equation and find

$$f'' - 2yf' + (\varepsilon - 1)f = 0$$

where $f' = \mathrm{d}f/\mathrm{d}y$ and $f'' = \mathrm{d}^2f/\mathrm{d}y^2$. This differential equation, which is called 'Hermite's equation', is one that has been thoroughly studied by mathematicians, and its solutions are known.

Acceptable solutions of Hermite's equation (those not going to infinity more quickly than the factor $\mathrm{e}^{-y^2/2}$ falls to zero, so that the product is never infinite) exist only for ε equal to a positive, odd integer. Therefore we write $\varepsilon = 2v + 1$, with $v = 0, 1, 2, \ldots$. It follows from the relation between E and

---

[2] Substitution of this solution into the simplified equation gives

$$\frac{\mathrm{d}^2\psi}{\mathrm{d}y^2} = \mathrm{e}^{y^2/2} + y^2\mathrm{e}^{y^2/2}$$

but the first of the terms on the right is much smaller than the second (because we are interested only in the region where y is very large), and so may be ignored.

ε that the energy levels of the oscillator are

$$E_v = (2v + 1) \times \tfrac{1}{2}\hbar\omega$$

The shapes of the wavefunctions can be obtained once we know the solutions of Hermite's equation. The solutions for v a positive integer are the Hermite polynomials H_v, which are polynomials[3] that are generated by differentiating e^{-y^2} v times:

$$H_v(y) = (-1)^v e^{y^2}\left(\frac{d}{dy}\right)^v e^{-y^2}$$

The explicit forms of the first few polynomials can be found in Table 12.1 with some of their more useful properties.

Further information 2: rotational motion

To solve eqn 19 we use the separation of variables technique (Section 12.2) and write the wavefunction as the product

$$\psi = \Theta\Phi$$

where Θ is a function only of θ, and Φ is a function only of ϕ. Then

$$\Lambda^2\psi = \frac{1}{\sin^2\theta}\,\Theta\,\frac{d^2\Phi}{d\phi^2} + \Phi\,\frac{1}{\sin\theta}\,\frac{d}{d\theta}\sin\theta\,\frac{d\Theta}{d\theta}$$

$$= \frac{-2IE}{\hbar^2}\,\Theta\Phi$$

Division through by $\Theta\Phi$ gives

$$\frac{1}{\sin^2\theta}\,\frac{\Phi''}{\Phi} + \frac{1}{\Theta}\,\frac{1}{\sin\theta}\,\frac{d}{d\theta}(\Theta'\sin\theta) = \frac{-2IE}{\hbar^2}$$

where $\Theta' = d\Theta/d\theta$ and $\Phi' = d\Phi/d\phi$. We ensure that each term depends on only one variable by multiplying through by $\sin^2\theta$:

$$\frac{\Phi''}{\Phi} + \frac{\sin\theta}{\Theta}\,\frac{d}{d\theta}(\Theta'\sin\theta) = \frac{-2IE}{\hbar^2}\sin^2\theta$$

The first term depends only on ϕ, and the rest depend only on θ. Therefore, by the same argument as in Section 12.2, the first term must be equal to a constant, which we write $-m_l^2$ (because it will turn out to have the same significance as the m_l we have already encountered). The separation is therefore successful, and after some rearrangement the two ordinary differential equations we must solve are found to be

$$\Phi'' = -m_l^2\Phi$$

$$\sin\theta\,\frac{d}{d\theta}(\Theta'\sin\theta) + \left(\frac{2IE}{\hbar^2}\sin^2\theta - m_l^2\right)\Theta = 0$$

The first of this pair of equations is the same as for a particle on a ring, eqn 14a. The cyclic boundary conditions are also the same, and so the

[3] A polynomial in y has the form $a_0 + a_1 y + a_2 y^2 + \ldots + a_v y^v$ running to a finite number of terms. It is unlike $\cos y$, for example, which runs to an infinite number of terms when expressed in the same way.

acceptable solutions are the same. Hence, we can write

$$\Phi_{m_l} = \left(\frac{1}{2\pi}\right)^{1/2} e^{im_l\phi} \qquad m_l = 0, \pm 1, \pm 2, \ldots$$

However, since m_l also occurs in the equation for Θ, it may be the case that the boundary conditions that Θ must satisfy also limit the values that m_l may take.

The equation for θ is one that has been studied extensively by mathematicians. We turn it into a standard form by making the substitutions

$$\zeta = \cos\theta \qquad l(l+1) = \frac{2IE}{\hbar^2}$$

where l is a dimensionless number (shortly to be promoted to a quantum number). Since

$$\frac{d}{d\theta} = \left(\frac{d\zeta}{d\theta}\right)\frac{d}{d\zeta} = -\sin\theta\frac{d}{d\zeta}$$

$$\sin\theta = (1 - \cos^2\theta)^{1/2} = (1 - \zeta^2)^{1/2}$$

the equation transforms into

$$(1 - \zeta^2)\Theta'' - 2\zeta\Theta' + \left\{l(l+1) - \frac{m_l^2}{1 - \zeta^2}\right\}\Theta = 0$$

where now $\Theta' = d\Theta/d\zeta$. This is known as the **associated Legendre equation** (because it is associated with another equation known by Legendre's name). It has acceptable solutions (i.e., solutions that are single valued and do not become infinite anywhere) so long as two conditions are fulfilled:

(a) $l = 0, 1, 2, \ldots$ (b) $|m_l| < l$

That is, acceptable solutions exist only for *positive* integral l, and the absolute value of the number m_l must not exceed l (this is the second constraint on m_l mentioned above). When these conditions are satisfied, the functions Θ, which now must be denoted $\Theta_{l,|m_l|}$, are the **associated Legendre functions**. They can be written in terms of sine and cosine functions, and the first few are listed in Table 12.3 as components of the spherical harmonics, $Y = \Theta\Phi$.

Further reading

P. W. Atkins, *Molecular quantum mechanics* (2nd edn). Oxford University Press (1983).

A. I. M. Rae, *Quantum mechanics*. McGraw-Hill, London (1981).

D. A. McQuarrie, *Quantum chemistry*. University Science Books and Oxford University Press (1983).

I. N. Levine, *Quantum chemistry* (2nd edn). Allyn and Bacon, Boston (1974).

L. Pauling and E. B. Wilson, *Introduction to quantum mechanics*. McGraw-Hill, New York (1935).

A. S. Davydov, *Quantum mechanics* (2nd edn). Pergamon Press, Oxford (1976).

P. A. M. Dirac, *Principles of quantum mechanics* (4th edn). Clarendon Press, Oxford (1958).

R. P. Bell, *The tunnel effect in chemistry*. Chapman and Hall, New York (1980).

Exercises

12.1 Calculate the energy separations in J, kJ mol^{-1}, eV, and cm^{-1} between the levels (a) $n = 2$ and $n = 1$, (b) $n = 6$ and $n = 5$ of an electron in a box of length 1.0 nm.

12.2 Consider a particle in a cubic box. What is the degeneracy of the level that has an energy three times that of the lowest level?

12.3 Calculate the percentage change in a given energy level of a particle in a cubic box when the edge of the cube is decreased by 10 per cent in each direction.

12.4 Calculate the zero-point energy of a harmonic oscillator consisting of a particle of mass 2.33×10^{-26} kg and force constant 155 N m^{-1}.

12.5 For a harmonic oscillator consisting of a particle of mass 1.33×10^{-25} kg, the difference in adjacent energy levels is 4.82×10^{-21} J. Calculate the force constant of the oscillator.

12.6 Calculate the wavelength of a photon needed to excite a transition between neighbouring energy levels of a harmonic oscillator of mass equal to that of a proton and force constant 855 N m^{-1}.

12.7 Calculate the minimum excitation energies of (a) a pendulum of length 1.0 m on the surface of the earth, (b) the balance-wheel of a clockwork watch ($v = 5$ Hz), (c) the 33 kHz quartz crystal of a watch, and (d) the bond between two O atoms in O_2, for which $k = 1177$ N m^{-1}.

12.8 Refer to Exercise 12.6, and calculate the change in wavelength that would result from doubling the mass of the particle.

12.9 An Ar atom rotates in a circle about a fixed center with orbital angular momentum quantum number $l = 2$. If its energy of rotation is 2.47×10^{-23} J, calculate the distance of the mass from the centre of rotation.

12.10 A point mass moves on a spherical surface with $l = 1$. Calculate the magnitude of its angular momentum and the possible projections of the angular momentum on an arbitrary axis.

12.11 Calculate the probability that an electron will penetrate a potential barrier of height 2.00 eV and width 0.25 nm when its initial kinetic energy is 0.90 eV.

12.12 Draw scale vector diagrams to represent the states (a) $s = \frac{1}{2}$, $m_s = +\frac{1}{2}$, (b) $l = 1$, $m_l = +1$, (c) $l = 2$, $m_l = 0$.

12.13 Draw the vector diagram for all the permitted states of a particle with $l = 6$.

Problems

Numerical problems

12.1 Calculate the separation between the two lowest levels of an O_2 molecule in a one-dimensional container of length 5.0 cm. At what value of n does the energy of the molecule reach $\frac{1}{2}kT$ at 300 K, and what is the separation of this level from the one immediately below?

12.2 The mass to use in the expression for ω in a diatomic molecule is the 'reduced mass' $\mu = m_A m_B/(m_A + m_B)$, where m_A and m_B are the masses of the individual atoms. The following data on the infrared absorption wavenumbers (in cm^{-1}) of molecules is taken from *Spectra of diatomic molecules*, G. Herzberg, van Nostrand (1950):

H^{35}Cl	H^{81}Br	HI	CO	NO
2990	2650	2310	2170	1904

Calculate the force constants of the bonds and arrange them in order of increasing stiffness.

12.3 The rotation of an HI molecule can be pictured as the orbital motion of an H atom at a distance 160 pm from a stationary I atom. (This is quite a good picture; to be precise, both atoms rotate around their common centre of mass, which is very close to the I nucleus.) Suppose that the molecule rotates only in a plane. Calculate the energy needed to excite the molecule into rotation. What, apart from 0, is the minimum angular momentum of the molecule?

12.4 Calculate the energies of the first four rotational levels of HI free to rotate in three dimensions, using for its moment of inertia $I = \mu R^2$, with $\mu = m_H m_I/(m_H + m_I)$ and $R = 160$ pm.

Theoretical problems

12.5 Set up the Schrödinger equation for a particle of mass m in a three-dimensional square well with sides L_1, L_2, and L_3. Show that the wavefunction is defined by three quantum numbers and that the Schrödinger equation is separable. Find the energy levels, and specialize the result to a cubic box of side L.

12.6 The wavefunction inside a long barrier of height V is $\psi = Ne^{-\kappa x}$. Calculate (a) the probability that the particle is inside the barrier and (b) the average penetration depth of the particle into the barrier.

12.7 Calculate the penetration probability for the arrangement in Example 12.5, but for a potential that falls to a non-zero value V' on the right of the barrier.

12.8 Calculate the penetration probability for the arrangement in Example 12.5, but for particles with incident energy in excess of the barrier height.

12.9 Calculate the probability that a particle will be reflected back from a rectangular dip in an otherwise uniform poten-

345

tial. Evaluate the resulting expression for a dip of depth 5.0 eV and width 100 pm, and plot the penetration probability as a function of E/V for a proton and a deuteron.

12.10 Confirm that a function of the form e^{-gx^2} is a solution of the Schrödinger equation for the ground state of a harmonic oscillator and find an expression for g in terms of the mass and force constant of the oscillator.

12.11 Calculate the mean kinetic energy of a harmonic oscillator using the relations in Table 12.1.

12.12 Calculate the values of $\langle x^3 \rangle$ and $\langle x^4 \rangle$ for a harmonic oscillator using the relations in Table 12.1.

12.13 We shall see in Chapter 16 that the intensity of spectroscopic transitions between the vibrational states of a molecule are proportional to the square of the integral $\int \psi_{v'} x \psi_v \, dx$ over all space. Use the recursion relations in Table 12.1 to show that the only permitted transitions are those for which $v' = v \pm 1$ and evaluate the integral in these cases.

12.14 Use the virial theorem to obtain an expression for the relation between the mean kinetic and potential energies of an electron in a hydrogen atom.

12.15 Evaluate the z-component of the angular momentum and the kinetic energy of a particle on a ring that is described by the (unnormalized) wavefunctions (a) $e^{i\phi}$, (b) $e^{-2i\phi}$, (c) $\cos\phi$, and (d) $\cos\chi e^{i\phi} + \sin\chi e^{-i\phi}$.

12.16 A particle on a ring is described by the wavefunction

$$\psi = ae^{i\phi} + be^{2i\phi} + ce^{3i\phi}$$

Calculate (a) its mean angular momentum, (b) kinetic energy, (c) root-mean-square deviation of the angular momentum from its mean value.

12.17 Confirm that the spherical harmonics (a) $Y_{0,0}$, (b) $Y_{2,-1}$, and (c) $Y_{3,3}$ satisfy the Schrödinger equation for a particle free to rotate in three dimensions, and find its energy and angular momentum in each case.

12.18 Confirm that $Y_{3,3}$ is normalized. (The integration required is over the surface of a sphere.)

12.19 Derive an expression in terms of l and m_l for the half-angle of the apex of the cone used to represent an angular momentum according to the vector model. Evaluate the expression for an α spin. Show that the minimum possible angle for orbital angular momentum approaches 0 as $l \to \infty$.

Atomic structure and atomic spectra

13

Check-list of key ideas

1. The *spectrum of atomic hydrogen,* and the classification of the *spectral lines* into *series* (Section 13.1 and eqn 1).

2. The *Ritz combination principle* for expressing transitions using terms and the *Bohr frequency condition* (eqn 2).

3. The Schrödinger equation for *hydrogenic atoms,* the separation of the internal motion (eqn 4), and the separation of the *angular wavefunction* from the *radial wavefunction.*

4. The *energy levels* of hydrogenic atoms (eqn 8) in terms of the *principal quantum number.*

5. The measurement of the *ionization energy* of an atom from the *series limit* (Example 13.3).

6. The *shells* and *subshells* of atoms and the notion of *atomic orbitals* (Section 13.2).

7. The representation of atomic orbitals by their *boundary surfaces* and the *radial distribution function* (Section 13.2).

8. The form of *s orbitals, p orbitals,* and *d orbitals* (Section 13.2).

9. The *selection rules* of spectroscopic transitions (Section 13.3) and the representation of transitions on a *Grotrian diagram* (Fig. 13.13).

10. The *orbital approximation* for many-electron atoms and the *Pauli exclusion principle* (Section 13.4).

11. The concept of a *closed shell* and of *paired spins* (Section 13.4).

12. The effect of *penetration* and *shielding* on the *effective atomic number* of an atom and the relative order of orbitals in a many-electron atom (Section 13.4).

13. The *building-up principle* for predicting the ground-state configurations of many-electron atoms (Section 13.4).

14. *Hund's rule* and the role of *spin correlation* (Section 13.4).

15. The periodicity of atomic properties, particularly the *ionization energy* (Section 13.4).

16. The concept of *self-consistent field* orbitals of many-electron atoms (Section 13.5) and the *Hartree–Fock procedure.*

17. *Singlet and triplet states* of atoms (Section 13.6) and the origin

13

13

The structure and spectra of hydrogenic atoms

13.1 The structure of hydrogenic atoms

13.2 Atomic orbitals and their energies

13.3 Spectral transitions and selection rules

The structures of many-electron atoms

13.4 The orbital approximation

13.5 Self-consistent field orbitals

The spectra of complex atoms

13.6 Singlet and triplet states

13.7 Spin–orbit coupling

13.8 Term symbols and selection rules

13.9 The effect of magnetic fields

Further information 1: centre of mass coordinates

Further information 2: the Pauli principle

Further reading

Exercises

Problems

and effects of *spin–orbit coupling* and the *fine structure* of spectra (Section 13.7).

18. The use of the *Clebsch–Gordan series* to deduce the total angular momentum of a combined system (Section 13.8), and the use of *term symbols* to specify the state of an atom.

19. The *Russell–Saunders* and *jj-coupling* schemes for light and heavy atoms (Section 13.8), and the *selection rules* for transitions when Russell–Saunders coupling is appropriate (Section 13.8).

20. The *magnetogyric ratio* of an electron (eqn 14), the *Bohr magneton* (eqn 15), and the *g-factor* of the electron (eqn 17).

21. The energy of an electron in a magnetic field (eqn 17) and the normal and anomalous *Zeeman effects* (Section 13.9).

In this chapter we see how to use quantum mechanics to describe the **electronic structure** of an atom, the arrangement of electrons around its nucleus. The concepts we shall meet are of central importance for understanding the structures and reactions of atoms and molecules, and hence have extensive chemical applications. Right from the outset, we need to distinguish between **hydrogenic atoms** and **many-electron atoms**. Hydrogenic atoms are one-electron atoms and ions of general atomic number Z, and include H, He^+, Li^{2+}, and U^{91+}. Many-electron atoms are atoms and ions with more than one electron, and include all neutral atoms other than H. Hydrogenic atoms, and H in particular, are important because their structures can be discussed exactly. They also provide a set of concepts that are used to describe the structures of many-electron atoms and, as we shall see in the next chapter, the structures of molecules too.

One of the principal experimental techniques for determining the structures of atoms is spectroscopy, and in the course of the chapter we shall meet some of the principles involved. In this connection, we shall often refer to the **wavenumber** $\tilde{\nu}$ of the radiation emitted or absorbed by an atom. The wavenumber is related to the wavelength λ and frequency ν by

$$\tilde{\nu} = \frac{1}{\lambda} = \frac{\nu}{c}$$

where c is the speed of light. It is common in spectroscopy to express wavenumbers in the units cm^{-1}; a wavenumber of $1000\ cm^{-1}$ then signifies that there are 1000 complete wavelengths per centimeter. The record of wavenumbers (or frequencies and wavelengths) of the radiation emitted or absorbed by an atom (or a molecule) is called its **spectrum** (from the Greek word for appearance).

The structure and spectra of hydrogenic atoms

When an electric discharge is passed through gaseous hydrogen, the H_2 molecules are dissociated and the energetically excited H atoms that are produced emit light of discrete frequencies (Fig. 13.1). The first important contribution to understanding this spectrum was made by the Swiss schoolteacher Johann Balmer, who pointed out in 1885 that (in modern

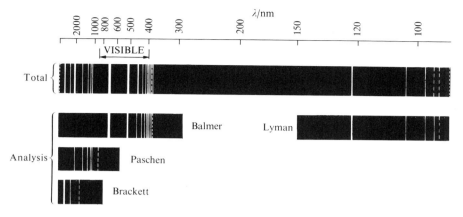

Fig. 13.1 The spectrum of atomic hydrogen. The spectrum is shown at the top, and is analyzed into its overlapping series below. Note that the Balmer series lies in the visible region.

terms) the wavenumbers of the light in the visible region fit the expression

$$\tilde{v} \propto \frac{1}{2^2} - \frac{1}{n^2} \qquad n = 3, 4, \ldots$$

The lines this formula describes are now called the **Balmer series**. When further lines were discovered in the ultraviolet, giving the **Lyman series**, and in the infrared, the **Paschen series**, the Swedish spectroscopist Johannes Rydberg noted (in 1890) that all could be fitted to

$$\tilde{v} = \mathscr{R}_H\left(\frac{1}{n_1^2} - \frac{1}{n_2^2}\right) \qquad \mathscr{R}_H = 109\,677 \text{ cm}^{-1} \tag{1}$$

with $n_1 = 1$ (Lyman), 2 (Balmer), and 3 (Paschen), and that in each case $n_2 = n_1 + 1$, $n_1 + 2$, ... The constant $\mathscr{R}_H$ is now called the **Rydberg constant** for the hydrogen atom.

Example 13.1: *Calculating the wavelengths of spectral lines*

Calculate the longest wavelength transition in the Lyman series of atomic hydrogen.

Answer. The Lyman series has wavenumbers

$$\tilde{v} = \mathscr{R}_H\left(1 - \frac{1}{n^2}\right) \qquad n = 2, 3, \ldots$$

The transition with the longest wavelength is the one with the smallest wavenumber, which is the transition with $n = 2$. Hence the wavenumber of this transition is

$$\tilde{v} = \frac{3}{4} \times \mathscr{R}_H = 82\,258 \text{ cm}^{-1}$$

The wavelength is the reciprocal of the wavenumber:

$$\lambda = \frac{1}{8.2258 \times 10^6 \text{ m}^{-1}} = 1.2157 \times 10^{-7} \text{ m}$$

Comment. The transition occurs at 121.57 nm, in the ultraviolet region of the spectrum.

Exercise. Calculate the shortest wavelength transition in the Paschen series.

[82.06 nm]

The form of eqn 1 strongly suggests that the wavenumber of each spectral line can be written as the difference of two **terms**, each of the form

$$T = \frac{\mathscr{R}_H}{n^2}$$

(For many-electron atoms the terms do not have such a simple form.) The concept of a spectroscopic term is the content of the **Ritz combination principle**, which states:

> The wavenumber of any spectral line is the difference of two terms.

We say that two terms T_1 and T_2 **combine** to produce a spectral line of wavenumber

$$\tilde{v} = T_1 - T_2$$

It is easy for us to explain the Ritz combination principle since, unlike the early spectroscopists, we know about the existence of photons of energy $h\nu$. The conservation of energy implies the **Bohr frequency condition**:

> When an atom changes its energy by ΔE, the difference is carried away as a photon of frequency ν, where
>
> $$\Delta E = h\nu \qquad (2)$$

Thus, if each term represents an energy hcT, the difference in energy when the atom undergoes a transition between two terms is

$$\Delta E = hcT_1 - hcT_2$$

and the frequency of the light emitted is given by

$$\nu = cT_1 - cT_2$$

This expression rearranges into the Ritz formula when expressed in terms of wavenumbers.

Since only certain wavenumbers are observed in a spectrum, only certain energy states of atoms are permitted. Our first tasks in this chapter are to determine the origin of this quantization, to find the permitted energy levels, and to account for the value of $\mathscr{R}_H$.

13.1 The structure of hydrogenic atoms

We can find a description of the structure of the hydrogen atom—and of any hydrogenic atom—by adopting Rutherford's nuclear model, in which an electron travels around a nucleus of charge Ze, and then solving the Schrödinger equation for that electron's wavefunction. At first sight this would seem to be a problem of considerable difficulty, because it involves two particles, the electron and the nucleus, and hence six coordinates.

However, physical intuition suggests that the full equation ought to separate into two equations, one for the motion of the atom as a whole through space, and the other for the motion of the electron around the nucleus.

We have already solved the first of these equations, because it corresponds to the free motion of a particle in space (Section 11.3). The second, the equation for the motion of the electron around the nucleus, still involves three coordinates of the electron relative to the nucleus. However, because the potential is spherically symmetrical, we can suspect that even that equation is separable. We shall see that one factor in the wavefunction is for the angular motion of the electron around a fixed point, which we have also already solved (Section 12.7). This suggests that the only new work concerns the radial dependence of the wavefunctions.

Our strategy, then, is to separate the relative motion of the electron and the nucleus from the motion of the atom as a whole. Then we separate the relative motion of the electron and the nucleus into angular and radial parts, and solve the latter.

The separation of internal motion

We can separate the internal motion from the motion of the atom as a whole by a series of straightforward steps that are given in detail in *Further information 1* at the end of the chapter. In outline, we begin by writing the Schrödinger equation for the nucleus and electron in the form

$$H\psi = E\psi$$

with

$$H = -\frac{\hbar^2}{2m_e}\nabla_e^2 - \frac{\hbar^2}{2m_N}\nabla_N^2 + V$$

where m_e is the mass of the electron, m_N that of the nucleus (a proton if the atom is H), and V is the Coulomb potential energy of an electron (of charge $-e$) at a distance r from a nucleus of charge Ze (with $Z = 1$ for H):

$$V = \frac{-Ze^2}{4\pi\varepsilon_0} \times \frac{1}{r} \qquad \varepsilon_0 = 8.854 \times 10^{-12}\,\text{J}^{-1}\,\text{C}^2\,\text{m}^{-1}$$

The laplacians ∇_e^2 and ∇_N^2 differentiate with respect to the coordinates of the electron and nucleus (so ∇_e^2 has terms like $\partial^2/\partial x_e^2$, etc, where x_e is the x-coordinate of the electron). Then, as shown in *Further information 1*, H transforms into

$$H = -\frac{\hbar^2}{2m}\nabla_{cm}^2 - \frac{\hbar^2}{2\mu}\nabla^2 + V$$

with m the total mass of the atom ($m = m_e + m_N$) and μ its **reduced mass**:

$$\frac{1}{\mu} = \frac{1}{m_e} + \frac{1}{m_N} \tag{3}$$

Since the laplacian ∇_{cm}^2 is expressed in terms of derivatives with respect to the coordinates of the centre of mass of the atom, the first term in the transformed H corresponds to the kinetic energy of the atom as a whole. The laplacian ∇^2 has derivatives with respect to the coordinates of the electron relative to the nucleus (Fig. 13.2). Hence it corresponds to the kinetic energy of motion within the atom.

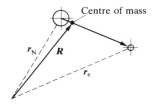

Fig. 13.2 Centre of mass coordinates for the hydrogen atom. Because the proton is so much more massive than the electron, the centre of mass of the atom lies very close to the proton (much closer than this diagram suggests).

Now we discard the first term in the hamiltonian. (In fact, as shown in *Further information 1*, what we really do is separate the variables and isolate the two types of motion.) This leaves the Schrödinger equation for the internal structure of the atom:

$$-\frac{\hbar^2}{2\mu}\nabla^2\psi + V\psi = E\psi \tag{4}$$

which is the equation we must solve.

Separation of the internal coordinates

The next step is to separate the angular variation of the wavefunction from its radial dependence. We do this by writing

$$\psi(r, \theta, \phi) = R(r)Y(\theta, \phi)$$

and testing whether the Schrödinger equation separates into two equations, one for R and the other for Y. The details of the separation are given in *Further information 1*. All we need here is to note that the separation is successful and that the two separated equations are

$$\Lambda^2 Y = -l(l+1)Y \tag{5a}$$

$$\frac{d^2(rR)}{dr^2} - \frac{2\mu}{\hbar^2}VrR = \frac{-2\mu ErR}{\hbar^2} \tag{5b}$$

where now

$$V = \frac{-Ze^2}{4\pi\varepsilon_0}\frac{1}{r} + \frac{l(l+1)\hbar^2}{2\mu r^2} \tag{5c}$$

Equation 5a, which is the **angular wave equation**, is the same as the one we solved for the angular momentum of a particle in three dimensions. The cyclic boundary conditions are also the same. Hence, the angular wavefunction Y is a spherical harmonic (Section 12.7) specified by the quantum numbers l and m_l, with

$$m_l = 0, \pm 1, \pm 2, \ldots \pm l$$

and l a positive integer. However, just as m_l came under a new constraint (that it lies between $-l$ and $+l$) when it appeared in an equation that involved l, so we shall see that l comes under a new constraint (that it cannot exceed a certain value) because it also appears in another equation (eqn 5b via eqn 5c).

The radial wave equation

Equation 5b is the **radial wave equation** for a hydrogenic atom. We can anticipate some features of its solution by examining the form of the potential energy V.

The first term in V is the Coulomb potential energy of the electron in the field of the nucleus. The second term in V stems from the centrifugal force that arises from the angular momentum of the electron around the nucleus. When $l = 0$, the electron has no angular momentum, and the potential energy is pure coulombic and attractive at all radii (Fig. 13.3). When $l \neq 0$, the centrifugal term gives a positive contribution to the effective potential energy. When the electron is close to the nucleus (r is small), this repulsive

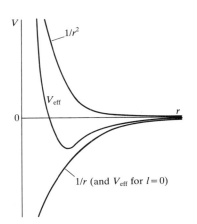

Fig. 13.3 The effective potential energy of an electron in the hydrogen atom. When it has zero angular momentum the effective potential is the pure Coulomb potential energy. When the electron has angular momentum, the centrifugal effect gives rise to a positive contribution which is very large close to the nucleus. We can expect the $l = 0$ and $l \neq 0$ wavefunctions to be very different near the nucleus.

term dominates the attractive coulombic component (Fig. 13.3) and the net effect is a repulsion of the electron from the nucleus. The two potential energies, the one for $l = 0$ and the one for $l \neq 0$, are qualitatively very different close to the nucleus, but they are similar at large distances because the centrifugal contribution tends to zero more rapidly than the coulombic contribution. Therefore, we can expect the solutions with $l = 0$ and $l \neq 0$ to be quite different near the nucleus but similar far away from it.

We shall not go through the technical steps of solving the radial equation. It is sufficient to know that acceptable solutions can be found only for the energies

$$E_n = \frac{-Z^2 \mu e^4}{32\pi^2 \varepsilon_0^2 \hbar^2} \times \frac{1}{n^2} \qquad n = 1, 2, \ldots$$

The corresponding wavefunctions are

$$R_{n,l} = \rho^l L_{n,l}(\rho) e^{-\rho/2}$$

with

$$n = 1, 2, \ldots \quad \text{and} \quad l = 0, 1, 2, \ldots, n - 1$$

The variable ρ that occurs in R is proportional to the radius r:

$$\rho = \frac{2Z}{n} \times \frac{\mu}{m_e} \times \frac{r}{a_0}$$

where a_0 is the **Bohr radius** (it first appeared in Bohr's model of the hydrogen atom as the radius of its lowest energy orbit):

$$a_0 = \frac{4\pi\varepsilon_0 \hbar^2}{m_e e^2} = 52.9 \text{ pm} \qquad (6)$$

For most atoms it is good enough to make the approximation that $\mu/m_e = 1$ (the actual value for H, the worst case, is 0.999 46), when the relation is simply

$$\rho = \frac{2Z}{n} \times \frac{r}{a_0}$$

The factor $L(\rho)$ that occurs in R is a polynomial in ρ, called an **associated Laguerre polynomial**. An example is the one with $n = 3$ and $l = 0$, which is

$$L_{3,0} = 6 - 6\rho + \rho^2$$

The first few associated Laguerre polynomials can be picked out from the functions listed in Table 13.1 and the full (normalized) radial dependence of the radial wavefunction R is illustrated in Fig. 13.4.

Example 13.2: *Locating the radial nodes*

Calculate the radii at which the radial wavefunction with $n = 3$ and $l = 0$ is zero in the hydrogen atom.

Answer. The radial wavefunction vanishes as $r \to \infty$ on account of the exponential factor in R. This is true whatever the value of n and l. However, it also vanishes where the polynomial factor $L = 0$, which is where

$$6 - 6\rho + \rho^2 = 0$$

The two roots of this quadratic equation are

$$\rho = 3 \pm \sqrt{3} = 4.732 \text{ and } 1.268$$

Since (when $n = 3$ and $Z = 1$)

$$r = \tfrac{3}{2}a_0\rho$$

The radial nodes are at $1.90a_0$ and $7.10a_0$ (101 pm and 375 pm).

Comment. The number of radial nodes in general is $n - l - 1$, not counting the one at infinity.

Exercise. At what distance from the nucleus is the radial node in the orbital with $n = 3$ and $l = 1$ in the hydrogen atom? [$6a_0$, 317 pm]

Table 13.1. Hydrogenic radial wavefunctions†

Orbital	n	l	$R_{n,l}$
$1s$	1	0	$2\left(\dfrac{Z}{a_0}\right)^{\frac{3}{2}} e^{-\frac{1}{2}\rho}$
$2s$	2	0	$\dfrac{1}{2\sqrt{2}}\left(\dfrac{Z}{a_0}\right)^{\frac{3}{2}} (2 - \rho)e^{-\frac{1}{2}\rho}$
$2p$	2	1	$\dfrac{1}{2\sqrt{6}}\left(\dfrac{Z}{a_0}\right)^{\frac{3}{2}} \rho e^{-\frac{1}{2}\rho}$
$3s$	3	0	$\dfrac{1}{9\sqrt{3}}\left(\dfrac{Z}{a_0}\right)^{\frac{3}{2}} (6 - 6\rho + \rho^2)e^{-\frac{1}{2}\rho}$
$3p$	3	1	$\dfrac{1}{9\sqrt{6}}\left(\dfrac{Z}{a_0}\right)^{\frac{3}{2}} (4 - \rho)\rho e^{-\frac{1}{2}\rho}$
$3d$	3	2	$\dfrac{1}{9\sqrt{30}}\left(\dfrac{Z}{a_0}\right)^{\frac{3}{2}} \rho^2 e^{-\frac{1}{2}\rho}$

† The full wavefunction is $\psi = RY$, where Y is given in Table 12.3. In the table, $\rho = 2Zr/na_0$.

(a)

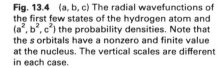

(b)

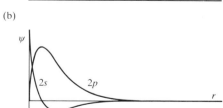

(c)

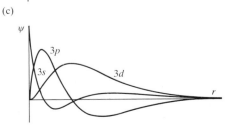

(a²)

(b²)

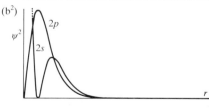

(c²)

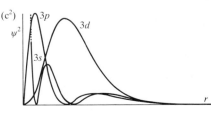

Fig. 13.4 (a, b, c) The radial wavefunctions of the first few states of the hydrogen atom and (a², b², c²) the probability densities. Note that the s orbitals have a nonzero and finite value at the nucleus. The vertical scales are different in each case.

Box 13.1 Hydrogenic atoms

The wavefunctions of hydrogenic atoms depend on three quantum numbers:

Principal quantum number: $n = 1, 2, 3, \ldots$
Angular momentum quantum number:

$$l = 0, 1, 2, \ldots, n-1$$

Magnetic quantum number: $m_l = l, l-1, l-2, \ldots, -l$

The energy is related to n by

$$E_n = \frac{-hc\mathcal{R}}{n^2} \qquad hc\mathcal{R} = \frac{Z^2 \mu e^4}{32\pi^2 \varepsilon_0^2 \hbar^2}$$

The magnitude of the orbital angular momentum of the electron is $\{l(l+1)\}^{1/2}$ and its component on an arbitrary axis is $m_l \hbar$. Each energy level is n^2-fold degenerate.

The wavefunctions are products of radial and angular components:

$$\psi = R(r)Y(\theta, \phi)$$

The angular wavefunctions Y are the spherical harmonics (Table 12.3) and the radial wavefunctions R are the associated Laguerre functions (Table 13.1).

The selection rules for spectroscopic transitions are

$$\Delta n \text{ unrestricted} \qquad \Delta l = \pm 1$$

13.2 Atomic orbitals and their energies

The wavefunctions of hydrogenic atoms are called **orbitals**, the name expressing something less definite than the 'orbits' of classical mechanics, and we shall use this term from now on. All the orbitals have the form

$$\psi_{n,l,m_l} = R_{n,l} Y_{l,m_l} \tag{7}$$

where R is one of the functions in Table 13.1 and Y one of the spherical harmonics listed in Table 12.3. Each orbital is defined by three quantum numbers (Box 13.1), n, l, and m_l. When an electron is described by one of these wavefunctions, we say that it **occupies** that orbital. Thus, an electron described by the wavefunction $\psi_{1,0,0}$ is said to occupy the orbital with $n = 1$, $l = 0$, and $m_l = 0$.

Two of the quantum numbers, l and m_l, come from the angular solutions, and relate to the angular momentum of the electron around the nucleus:

An electron in an orbital with quantum number l has an angular momentum of magnitude $\{l(l+1)\}^{1/2}\hbar$, with

$$l = 0, 1, 2, \ldots, n-1$$

An electron in an orbital with quantum number m_l has a z-component of angular momentum $m_l \hbar$, with

$$m_l = 0, \pm 1, \pm 2, \ldots, \pm l$$

The third quantum number, n, is called the **principal quantum number**. It can take the values

$$n = 1, 2, 3, \ldots$$

and determines the energy through

$$E_n = \frac{-hc\mathcal{R}}{n^2} \qquad hc\mathcal{R} = \frac{Z^2 \mu e^4}{32\pi^2 \varepsilon_0^2 \hbar^2} \tag{8}$$

It is a peculiarity of hydrogenic atoms that the energy is independent of l and m_l, and depends only on n. Therefore, all orbitals of a given n have the same energy—are degenerate—whatever their values of l and m_l.

The energy levels

All the energies in eqn 8 are negative. This means that we are dealing with the **bound states** of the atom, in which the energy of the electron is lower than when it is infinitely far away and at rest (which corresponds to the zero of energy). There are also solutions of the Schrödinger equation with positive energies. These correspond to **unbound states** of the electron, the states to which an electron is raised when it is ejected from the atom by a high-energy collision or photon. The energies of the unbound electron are not quantized, and form the **continuum states** of the atom.

In the case of hydrogen itself, $Z = 1$ and the bound state energies (the only ones we consider from now on) are given by

$$E_n = \frac{-hc\mathcal{R}_H}{n^2} \qquad hc\mathcal{R}_H = \frac{\mu e^4}{32\pi^2 \varepsilon_0^2 \hbar^2} \tag{9}$$

The levels are drawn in Fig. 13.5. The spectrum of atomic hydrogen can now be explained by supposing that it undergoes a transition from a state with principal quantum number n_2 (and energy $-hc\mathcal{R}_H/n_2^2$) to one with principal quantum number n_1 (and energy $-hc\mathcal{R}_H/n_1^2$), and discards the energy difference as a photon of energy $h\nu$ and frequency ν. Since energy is conserved overall, the frequency is given by

$$h\nu = \frac{hc\mathcal{R}_H}{n_1^2} - \frac{hc\mathcal{R}_H}{n_2^2} \quad \text{or} \quad \nu = c\mathcal{R}_H\left(\frac{1}{n_1^2} - \frac{1}{n_2^2}\right)$$

That is, since $\tilde{\nu} = \nu/c$,

$$\tilde{\nu} = \mathcal{R}_H\left(\frac{1}{n_1^2} - \frac{1}{n_2^2}\right)$$

which is the form required by experiment. When we insert the values of the fundamental constants into the expression for $\mathcal{R}_H$ we get almost exact agreement with experiment. The only discrepancies arise from the neglect of relativistic corrections, which the Schrödinger equation ignores.

Ionization energies

The **ionization energy** I of an atom is the minimum energy required to ionize the atom in its **ground state**, its lowest energy state. The ground state of hydrogen is the state with $n = 1$, which has energy

$$E_1 = -hc\mathcal{R}_H$$

The atom is ionized when the electron has been excited to the level $E = 0$, corresponding to $n = \infty$ (Fig. 13.5). Therefore, the energy that must be supplied is

$$I = hc\mathcal{R}_H = 2.179 \times 10^{-18}\ \text{J}$$

which corresponds to $13.60\ \text{eV}$.

The experimental determination of ionization energies of atoms depends on the detection of the **series limit**, the wavenumber at which the series

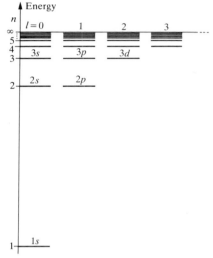

Fig. 13.5 The energy level diagram of the hydrogen atom. The values are relative to an infinitely separated, stationary electron and a proton. The degeneracy of the orbitals with the same n but different l is a special characteristic of a pure Coulomb potential.

terminates and becomes a continuum. Suppose that the upper state lies at an energy E_{upper}, then when the atom makes a transition to E_{lower} a photon is emitted of wavenumber $\tilde{\nu}$, where

$$hc\tilde{\nu} = E_{upper} - E_{lower}$$

Since the upper energy levels are given by an expression of the form $E_{upper} = -hc\mathcal{R}/n^2$, the emission lines will occur at

$$\tilde{\nu} = \frac{-\mathcal{R}}{n^2} - \frac{E_{lower}}{hc}$$

The separation between successive lines decreases as n increases (Fig. 13.5) and so the lines converge. Moreover, a plot of the wavenumbers against $1/n^2$ should give a straight line of slope $-\mathcal{R}$ and intercept $-E_{lower}/hc$. The value of $-E_{lower}$ is the ionization energy of the atom in that state (and is the ground state ionization energy when the lower state is the ground state).

Example 13.3: *Measuring an ionization energy spectroscopically*

The spectrum of atomic hydrogen shows lines at the wavenumbers 82 259, 97 492, 102 824, 105 292, 106 632, 107 440 cm^{-1}. Find (a) the ionization energy of the lower state, (b) the value of the Rydberg constant.

Answer. We need to plot wavenumber against $1/n^2$ with $n = 2, 3, \ldots$ (this assumes that the lower state is $n = 1$; if a straight line is not obtained, try $n = 2, 3, \ldots$, etc.). According to the equation above, the intercept at $1/n^2 = 0$ is E_{lower}/hc, or $-I/hc$, and the slope is $-\mathcal{R}_H$. Use a computer to make a least-squares fit of the data to get a result that reflects the precision of the data. The wavenumbers are plotted against $1/n^2$ in Fig. 13.6. The (least-squares) intercept lies at $-109\,679$ cm^{-1}, and so the ionization energy is $-hc$ times this quantity, or 2.1788×10^{-18} J (1312.1 kJ mol^{-1}). The slope is, in this instance, numerically the same, and so $\mathcal{R}_H = 109\,679$ cm^{-1}.

Exercise. The spectrum of atomic deuterium shows lines at the following wavenumbers: 15 238, 20 571, 23 039, 24 380 cm^{-1}. Determine (a) the ionization energy of the lower state, (b) the ionization energy of the ground state, (c) the mass of the deuteron.

[(a) 328.1 kJ mol^{-1}, (b) 1312.4 kJ mol^{-1}, (c) 3.4×10^{-27} kg]

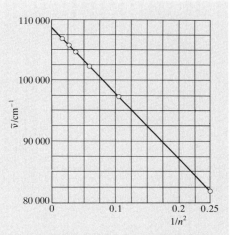

Fig. 13.6 The plot of the data in Example 13.3 used to determine the ionization energy of an atom (in this case, of H).

Shells and subshells

All the orbitals of a given value of n form a single **shell** of the atom. In a hydrogenic atom, all orbitals of given n, and therefore belonging to the same shell, have the same energy. It is common to refer to successive shells by the letters:

$$n = 1 \quad 2 \quad 3 \quad 4 \ldots$$
$$ K \quad L \quad M \quad N \ldots$$

Thus, all the orbitals of the shell with $n = 2$ form the L shell of the atom.

The orbitals with the same value of n but different values of l form the **subshells** of a given shell. These subshells are generally referred to by the letters s, p, ... using

$$l = 0 \quad 1 \quad 2 \quad 3 \quad 4 \ldots$$
$$s \quad p \quad d \quad f \quad g \ldots$$

(the letters then run alphabetically). Thus, the subshell with $l = 1$ of the shell with $n = 3$ is called the $3p$ subshell, and its three orbitals are called the $3p$ orbitals. An electron that occupies one of the $3p$ orbitals is called a **$3p$ electron**. Since l can range from 0 to $n - 1$, there are n subshells of a shell with principal quantum number n. Thus, when $n = 1$, there is only one subshell, the one with $l = 0$. When $n = 2$, there are two subshells, the $2s$ subshell (with $l = 0$) and the $2p$ subshell (with $l = 1$).

When $n = 1$ there is only one orbital, that with $l = 0$ and $m_l = 0$ (the only value of m_l permitted). When $n = 2$, there are four orbitals, one in the s subshell with $l = 0$ and $m_l = 0$, and three in the $l = 1$ subshell with $m_l = 0$, ± 1. When $n = 3$ there are nine (one with $l = 0$, three with $l = 1$, and five with $l = 2$):

	Number of orbitals					
n l:	0	1	2	3	4	Total number
	s	p	d	f	g	
1	1					1
2	1	3				4
3	1	3	5			9
4	1	3	5	7		16

In general, the number of orbitals in a shell of principal quantum number n is n^2, so in a hydrogenic atom each shell is n^2-fold degenerate.

s orbitals

The orbital corresponding to the ground state (for which the electron is most tightly bound to the nucleus), is the one with $n = 1$ (and therefore necessarily with $l = 0$ and $m_l = 0$, the only possible values of these quantum numbers when $n = 1$). From Table 13.1 we can write (for $Z = 1$):

$$\psi = \left(\frac{1}{\pi a_0^3}\right)^{1/2} e^{-r/a_0}$$

As this wavefunction is independent of angle it has the same value at all points of constant radius; that is, the $1s$ orbital is **spherically symmetrical**. However, ψ does depend on distance from the nucleus, and its amplitude decays exponentially from a maximum value of $(1/\pi a_0^3)^{1/2}$ at the nucleus (at $r = 0$). It follows that the most probable point at which the electron will be found is at the nucleus itself.

We can understand why the lowest energy wavefunction decays with distance at a particular rate by thinking about the potential and kinetic energy contributions to the total energy of the atom. The closer the electron is to the nucleus on average, the lower its average potential energy. This suggests that the lowest potential energy should be obtained with a sharply peaked wavefunction that has a large amplitude at the nucleus and is zero everywhere else. However, this implies a high kinetic energy, because such a wavefunction has a very sharp curvature close to the nucleus (Fig. 13.7a), and high curvature corresponds to high kinetic energy. The electron would

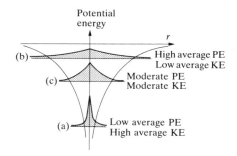

Fig. 13.7 The balance of kinetic and potential energies that accounts for the structure of the ground state of hydrogen (and similar atoms). (a) The sharply curved but localized orbital has high mean kinetic energy, but low potential energy; (b) the mean kinetic energy is low, but the potential energy is not very favourable; (c) the compromise of moderate kinetic energy and moderately favourable potential energy.

Electron densities

Boundary surfaces

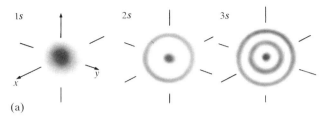

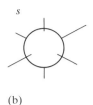

1s   2s   3s

s

(a)

(b)

Fig. 13.8 Representations of the first three s hydrogenic atomic orbitals in terms of (a) the electron densities (as represented by the density of shading) and (b) the boundary surfaces, within which there is a 90 per cent probability of finding the electron.

have very low kinetic energy is its wavefunction had only a very slight curvature (Fig. 13.7b). However, such a wavefunction spreads to great distances from the nucleus and the average potential energy of the electron will be less negative. The actual ground state is a compromise (Fig. 13.7c) between these two extremes: it spreads some way from the nucleus (so the potential energy is not as low as in the first example, but nor is it very high) and has a reasonably low curvature (so the kinetic energy is not very low, but nor is it as high as in the first example).

One way of depicting the probability of finding the electron in any volume element is to represent ψ^2 by the density of shading in a diagram (Fig. 13.8a). A simpler procedure, though, is to show only the **boundary surface**, the shape that captures about 90 per cent of the electron probability and shows the location of any angular nodes. For the 1s orbital, the boundary surface is a sphere (Fig. 13.8b).

All s orbitals are spherically symmetric, but differ in the number of radial nodes (the values of r at which $\psi = 0$). For instance, the 2s orbital has a radial node where $L_{2,0} = 0$:

$$L_{2,0} = 2 - \rho = 0 \quad \text{and} \quad r = \rho a_0 \text{ implies } r = 2a_0$$

Hence, the 2s orbital has a radial node at $2a_0$. Similarly, the 3s orbital has two nodes which are found by solving

$$L_{3,0} = 6 - 6\rho + \rho^2 = 0 \quad \text{with } r = \tfrac{3}{2}\rho a_0$$

One node is at $1.90a_0$ and the other at $7.10a_0$ (Example 13.2). The energies of the s orbitals increase (the electron becomes less tightly bound) as n increases because the average distance of the electron from the nucleus increases.

Example 13.4: *Calculating the mean radius of an orbital*

Use the hydrogenic orbitals to calculate the mean radius of (a) a 1s orbital and (b) a 2s orbital.

Answer. The mean value is the expectation value

$$\langle r \rangle = \int r \psi^2 \, d\tau$$

We therefore need to evaluate the integral $\int r\psi^2 \, d\tau$ using the wavefunctions given in Table 13.1 and $d\tau = r^2 \, dr \sin\theta \, d\theta \, d\phi$. The integration over the angular part gives unity in each case (because the spherical harmonics are normalized), and so the only work relates to the integration over r:

$$\langle r \rangle = \int_0^\infty r R^2 \times r^2 \, dr$$

Integrals of this kind are evaluated using

$$\int_0^\infty x^n e^{-ax} \, dx = \frac{n!}{a^{n+1}}$$

(a) For a 1s orbital,

$$R = 2\left(\frac{1}{a_0^3}\right)^{1/2} e^{-r/a_0}$$

and so

$$\langle r \rangle = \frac{4}{a_0^3} \int_0^\infty r^3 e^{-2r/a_0} \, dr = \tfrac{3}{2} a_0$$

(b) For a 2s orbital,

$$R = \left(\frac{1}{8a_0^3}\right)^{1/2} \left(2 - \frac{r}{a_0}\right) e^{-r/2a_0}$$

and so

$$\langle r \rangle = \frac{1}{8a_0^3} \int_0^\infty r^3 \left(2 - \frac{r}{a_0}\right)^2 e^{-r/a_0} \, dr = 6a_0$$

Comment. The general expression for the mean radius is

$$\langle r \rangle_{n,l,m_l} = n^2 \left\{ 1 + \frac{1}{2}\left(1 - \frac{l(l+1)}{n^2}\right) \right\} \times \frac{a_0}{Z}$$

Exercise. Evaluate the mean radius of a 3s orbital by integration, and of a 3p orbital using the general formula. $[27a_0/2; 25a_0/2]$

Radial distribution functions

The wavefunction tells us, through ψ^2, the probability of finding an electron in any region. We can imagine a probe with a volume $d\tau$ and sensitive to electrons, and which we can move around near the nucleus of a hydrogen atom. Since the probability density in the ground state of the atom is

$$\psi^2 \propto e^{-2r/a_0}$$

the reading from the detector decreases exponentially as the probe is moved out along any radius (Fig. 13.9).

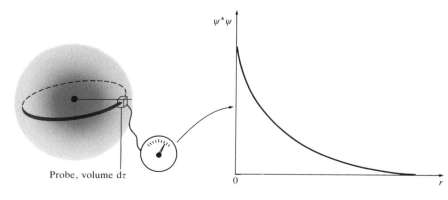

Fig. 13.9 A constant-volume electron-sensitive detector would show greatest reading at the nucleus, and a smaller reading elsewhere. The same reading would be obtained anywhere on a circle of given radius: the s orbital is isotropic.

Probe, volume $d\tau$

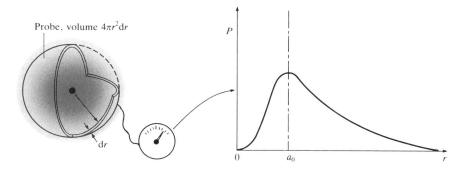

Probe, volume $4\pi r^2 dr$

dr

P

0 a_0 r

Fig. 13.10 The radial distribution function P gives the probability that the electron will be found anywhere in a shell of radius r. It is a maximum when r is equal to the Bohr radius a_0.

Now consider the probability of finding the electron *anywhere* on a spherical shell of thickness dr at a radius r. The sensitive volume of the probe is now the volume of the shell (Fig. 13.10), which is $4\pi r^2 \, dr$. The probability that the electron will be found in this shell is the probability density at the radius r multiplied by the volume of the probe:

$$\psi^2 \times 4\pi r^2 \, dr = P \, dr, \quad \text{with} \quad P = 4\pi r^2 \psi^2$$

The **radial distribution function** P is a probability density in the sense that when it is multiplied by dr, it gives the probability of finding the electron anywhere in a shell of thickness dr at the radius r.

For a $1s$ orbital,

$$P \propto r^2 e^{-2r/a_0}$$

Since r^2 increases with radius and ψ^2 decreases, P is zero at the nucleus (because $r^2 = 0$ there), zero at infinity (because the exponential is zero there), and passes through a maximum at an intermediate radius (Fig. 13.10). The maximum of P marks the most probable radius (as distinct from point) at which the electron will be found, and for a $1s$ orbital occurs at $r = a_0$, the Bohr radius. When we carry through the same calculation for the radial distribution function of the $2s$ orbital (by forming $4\pi r^2 \psi^2$), we find that the most probable radius at which the electron will be found is $5.2a_0 = 275$ pm, which reflects the expansion of the atom as its energy increases.

Example 13.5: *Calculating the most probable radius*

Calculate the most probable radius at which an electron will be found when it occupies a $1s$ orbital of a hydrogenic atom of atomic number Z, and tabulate the values for the atoms from H to Ne^{9+}.

Answer. We need to find the radius at which the radial distribution function of the hydrogenic $1s$ orbital has a maximum value (by solving $dP/dr = 0$). Since

$$\psi = \left(\frac{Z^3}{\pi a_0^3}\right)^{1/2} e^{-Zr/a_0}$$

$$P = 4\pi r^2 \psi^2 = \frac{4Z^3}{a_0^3} \times r^2 e^{-2Zr/a_0}$$

$$\frac{dP}{dr} = \frac{4Z^3}{a_0^3}\left(2r - \frac{2Zr^2}{a_0}\right)e^{-2Zr/a_0} = 0 \quad \text{at} \quad r = r^*$$

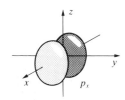

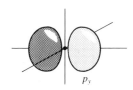

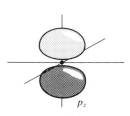

Fig. 13.11 The boundary surfaces of p orbitals. The nodal plane passes through the nucleus and separates the two lobes of each orbital. The shaded areas denote regions of opposite sign of the wavefunction.

Therefore,

$$r^* = \frac{a_0}{Z}$$

Then, with $a_0 = 52.9$ pm,

	H	He	Li	Be	B	C	N	O	F	Ne
r^*/pm	52.9	26.5	17.6	13.2	10.6	8.82	7.56	6.61	5.88	5.29

Comment. Notice how the $1s$ orbital is drawn towards the nucleus as the nuclear charge increases. At uranium the most probable radius is only 0.58 pm, almost 100 times closer than for hydrogen. (On a scale where $r^* = 10$ cm for H, $r^* = 1$ mm for U.) The electron then experiences strong accelerations, and relativistic effects are important.

Exercise. Find the most probable distance of a $2s$ electron from the nucleus in a hydrogenic atom. $[(3 + \sqrt{5})a_0/Z]$

p orbitals

A p electron (Fig. 13.11) has an angular momentum (of magnitude $\sqrt{2}\hbar$), and this momentum has a profound effect on the shape of the wavefunction close to the nucleus: p orbitals have zero amplitude at $r = 0$. This can be understood classically in terms of the centrifugal effect of the angular momentum, which flings the electron away from the nucleus. It is also what we expect from the form of the effective potential energy shown in Fig. 13.3, which rises to infinity as $r \rightarrow 0$ and excludes the wavefunction from the nucleus. The same centrifugal effect appears in all orbitals with $l > 0$ (such as the d orbitals and the f orbitals). All orbitals with $l > 0$ have zero amplitude at the nucleus, and consequently zero probability of finding the electron there. We can see from the form of the radial wavefunction that

$$R \propto \rho^l = 0 \quad \text{at} \quad \rho = 0 \quad \text{for any} \quad l > 0$$

The three $2p$ orbitals are distinguished by the three different values that m_l can take when $l = 1$. Since the quantum number m_l tells us the angular momentum around an axis, these different values of m_l denote orbitals in which the electron has different angular momenta around an arbitrary z-axis but the same *magnitude* of momentum (because l is the same for all three). The orbital with $m_l = 0$, for instance, has zero angular momentum around the z-axis. It has the form $f(r) \cos \theta$, and the electron density, which is proportional to $\cos^2 \theta$, has its maximum value on either side of the nucleus along the z axis (for $\theta = 0$ and 180°). For this reason, the orbital is also called a **p_z orbital**. The orbital amplitude is zero when $\theta = 90°$, so the xy-plane is a **nodal plane** of the orbital (Fig. 13.11). On this plane there is zero probability of finding an electron that occupies this orbital.

The orbitals with $m_l = \pm 1$ (which are proportional to $\sin \theta \, e^{\pm i\phi}$) do have angular momentum about the z-axis. The orbital with the factor $e^{+i\phi}$ corresponds to rotation in one direction and that with the factor $e^{-i\phi}$ corresponds to motion in the opposite direction. They have zero amplitude where $\theta = 0$ and 180° (along the z-axis) and maximum amplitude where $\theta = 90°$, which is in the xy-plane. To draw the functions it is usual to take the real linear combinations

$$f \sin \theta (e^{i\phi} + e^{-i\phi}) \propto f \sin \theta \cos \phi \propto xf$$
$$f \sin \theta (e^{i\phi} - e^{-i\phi}) \propto f \sin \theta \sin \phi \propto yf$$

which (when normalized) are called the p_x and p_y orbitals respectively. These combinations are standing waves with no net angular momentum around the z-axis, since they are composed of equal and opposite values of m_l. The p_x orbital has the same shape as a p_z orbital, but is directed along the x-axis (Fig. 13.11); the p_y orbital is similarly directed along the y-axis.

d orbitals

When $n = 3$, l can be 0, 1, or 2. This gives one 3s orbital, three 3p orbitals, and five 3d orbitals. The five d orbitals have $m_l = 2$, 1, 0, −1, −2 and correspond to five different angular momenta around the z-axis (but the same magnitude of angular momentum, since $l = 2$ in each case). As for the p orbitals, d orbitals with opposite values of m_l (and hence opposite senses of motion around the z axis) may be combined in pairs to give standing waves, and the boundary surfaces of the resulting shapes are shown in Fig. 13.12.

13.3 Spectral transitions and selection rules

We now know the energies of all the states of hydrogenic atoms:

$$E_{n,l,m_l} = \frac{-hc\mathcal{R}}{n^2}$$

When the electron undergoes a **transition**, a change of state, from an orbital with quantum numbers n_1, l_1, m_{l1} to another (lower-energy) orbital with quantum numbers n_2, l_2, m_{l2}, it undergoes a change of energy ΔE and discards the excess energy as a photon of electromagnetic radiation with a frequency v given by the Bohr frequency condition (eqn 2).

It is tempting to think that all possible transitions are permissible, and that the spectrum of the atom arises from the transition of an electron from any initial orbital to any other orbital. However, this is not so because a photon has unit intrinsic spin angular momentum. If a photon is generated by an electron undergoing a transition, the angular momentum of the electron must change to compensate for the angular momentum carried away by the photon as its spin. Thus an electron in a d orbital with $l = 2$ cannot make a transition into an s orbital with $l = 0$ because the photon cannot carry away enough angular momentum. Similarly, an s electron cannot make a transition to another s orbital, because then there is no change in the electron's angular momentum to make up for the angular momentum carried away by the photon. It follows that some spectral transitions are **allowed**, meaning that they can occur, while others are **forbidden**, meaning that they cannot occur.

A **selection rule** is a statement about which transitions are allowed. They are derived (for atoms) by identifying the transitions that conserve angular momentum when a photon is emitted or absorbed. The selection rules for hydrogenic atoms are

$$\Delta l = \pm 1 \qquad \Delta m_l = 0, \pm 1$$

The principal quantum number n can change by any amount consistent with the Δl for the transition because it does not relate directly to the angular momentum.

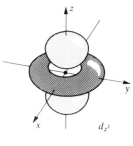

d_{z^2}

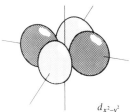

$d_{x^2-y^2}$

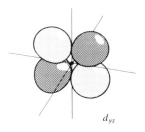

d_{yz}

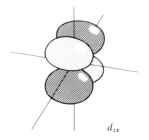

d_{zx}

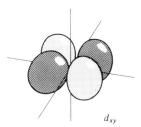

d_{xy}

Fig. 13.12 The boundary surfaces of d orbitals. Two nodal planes in each orbital intersect at the nucleus and separate the four lobes of each orbital. The grey and white areas denote regions of opposite sign of the wavefunction.

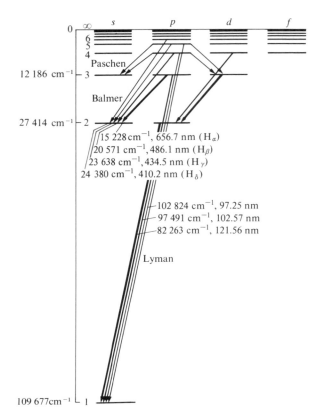

Fig. 13.13 A Grotrian diagram that summarizes the appearance and analysis of the spectrum of atomic hydrogen. The thicker the line, the more intense the transition.

Example 13.6: *Using the selection rules*

To what orbitals may a $4d$ electron make radiative transitions?

Answer. Since $l = 2$, the final orbital must have $l = 1$ or 3. Thus, an electron may make a transition from a $4d$ orbital to any np orbital (subject to $\Delta m_l = 0$ or ± 1) and to any nf orbital (subject to the same rule). However, it cannot undergo a transition to any other orbital, so a transition to any ns orbital or another nd orbital is forbidden.

Exercise. To what orbitals may a $4s$ electron make radiative transitions?

[np orbitals only]

The selection rules enable us to construct a **Grotrian diagram** (Fig. 13.13), which is a diagram that summarizes the energies of the states and the allowed transitions between them. The densities of the transition lines in the diagram denote their relative intensities in the spectrum. The intensities may also be calculated from the wavefunctions of the two states, but we shall not deal with this aspect here.

The structures of many-electron atoms

The Schrödinger equations for many-electron atoms are extremely complicated because all the electrons interact with each other. Even in the case of the He atom, with its two electrons, no analytical expression for the orbitals and energies can be given, and we are forced to make approximations. We shall adopt a simple approach based on what we already know about the structure of hydrogenic atoms. Later we shall see the kind of numerical computations that are currently used to obtain accurate wavefunctions and energies.

13.4 The orbital approximation

The actual wavefunction of a many-electron atom is a very complicated function of the coordinates of all the electrons, and we should write it $\psi(r_1, r_2, \ldots)$. However, in the **orbital approximation** we suppose that a reasonable first approximation to this exact wavefunction is obtained by thinking of each electron as occupying its 'own' orbital, and writing

$$\psi(r_1, r_2, \ldots) = \psi(r_1)\psi(r_2) \ldots$$

We can think of the individual orbitals as resembling the hydrogenic orbitals, but with nuclear charges that are modified by the presence of all the other electrons in the atom. This description is only approximate, but it is a useful model for discussing the chemical properties of atoms, and is the starting point for more sophisticated descriptions of atomic structure.

The helium atom

The orbital approximation allows us to express the electronic structure of an atom by reporting its **configuration**, the list of occupied orbitals (usually, but not necessarily, in its ground state). Thus, as the ground state of a hydrogenic atom consists of the single electron in a 1s orbital, we report its configuration as $1s^1$.

The He atom has two electrons. We can imagine forming the atom by adding the electrons in succession to the orbitals of the bare nucleus (of charge $2e$). The first electron occupies a 1s hydrogenic orbital, but since $Z = 2$ it is more compact than in H itself. The second electron joins the first in the 1s orbital, and so the electron configuration of the ground state of He is $1s^2$.

The Pauli principle

Lithium, with $Z = 3$, has three electrons. The first two occupy a 1s orbital drawn even more closely than in He around the more highly charged nucleus. The third electron, however, does not join the first two in the 1s orbital since that configuration is forbidden by the **Pauli exclusion principle**:

No more than two electrons may occupy any given orbital, and if two do occupy one orbital, their spins must be paired.

Electrons with **paired spins**, which we denote ↑↓, have zero net spin angular momentum since the spin angular momentum of one electron is cancelled by the spin of the other. This remarkable principle is the key to the structure of complex atoms, to chemical periodicity, and to molecular structure. It was proposed by Wolfgang Pauli in 1924 when he was trying to account for the absence of some lines in the spectrum of helium. Later he was able to derive a very general form of the principle from theoretical considerations. The Pauli exclusion principle in fact applies to any pair of identical **fermions**, or particles with half-integral spin. Thus it applies to protons, neutrons, and ^{13}C nuclei (all of which have spin $\frac{1}{2}$) and to ^{35}Cl nuclei (which have spin $\frac{3}{2}$). It does not apply to identical **bosons**, or particles with integral spin, which include photons (spin 1) and ^{12}C nuclei (spin 0). A deeper form of the principle is given in *Further information 2* at the end of the chapter.

In the case of Li ($Z = 3$), the third electron cannot enter the 1s orbital because that orbital is already full: we say the K shell is **complete** and that the two electrons form a **closed shell**. Since the same closed shell occurs in the He atom, we denote it [He]. The third electron is excluded from the K shell and must occupy the next available orbital, which is one with $n = 2$ and hence belonging to the L shell. However, we now have to decide whether the next available orbital is the 2s orbital or a 2p orbital, and therefore whether the lowest energy configuration of the atom is $[He]2s^1$ or $[He]2p^1$.

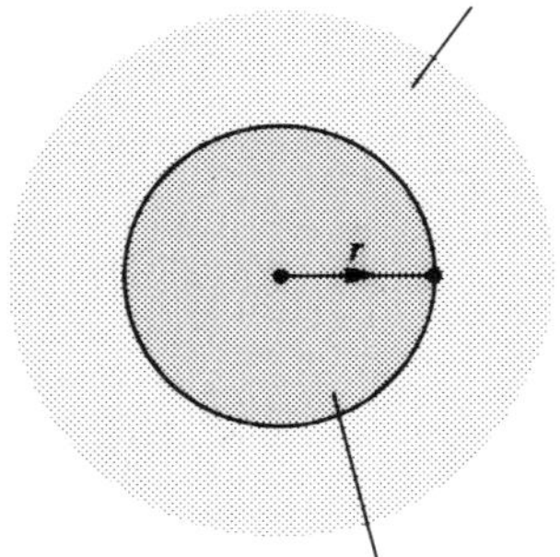

Penetration and shielding

Unlike in hydrogenic atoms, the 2s and 2p orbitals (and, in general, all subshells of a given shell) are not degenerate in many-electron atoms. For reasons we shall now explain, s electrons generally lie lower in energy than p electrons of a given shell, and p electrons lie lower than d electrons.

An electron in a many-electron atom experiences a coulombic repulsion from all the other electrons present. If it is at an average distance r from the nucleus, it experiences an average repulsion that can be represented by a point negative charge located at the nucleus and equal in magnitude to the total charge of the electrons within a sphere of radius r (Fig. 13.14). The effect of this point negative charge is to reduce the full charge of the nucleus from Ze to $Z_{eff}e$, where Z_{eff} is the **effective atomic number**. We say that the electron experiences a **shielded** nuclear charge, and that the true atomic number is reduced to Z_{eff} by an amount called the **shielding constant** σ:

$$Z_{eff} = Z - \sigma$$

The electrons do not actually 'block' the full coulombic attraction of the nucleus: the effective charge is simply a way of expressing the *net* outcome of the nuclear attraction and the electronic repulsions in terms of a single equivalent charge at the center of the atom.

The effective atomic number is different for s and p electrons because they have different radial wavefunctions (Fig. 13.15). An s electron has a

No net effect
of these electrons

Net effect of these
electrons is equivalent
to a point negative
charge on the nucleus

Fig. 13.14 An electron at a distance r from the nucleus experiences a coulombic repulsion from all the electrons within a sphere of radius r and which is equivalent to a point negative charge located on the nucleus. The negative charge reduces the effective nuclear charge of the nucleus from Ze to $Z_{eff}e$.

Fig. 13.15 An electron in an s orbital (here a 3s orbital) is more likely to be found close to the nucleus than an electron in a p orbital of the same shell. Hence it experiences less shielding and is more tightly bound.

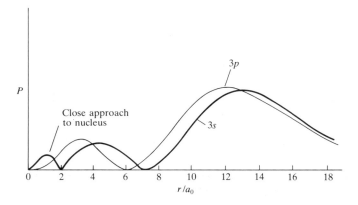

greater **penetration** through inner shells than a p electron in the sense that it is more likely to be found close to the nucleus than a p electron of the same shell (the p orbital, remember, has a node at the nucleus). Since only electrons inside the sphere defined by the location of the electron (in effect, the core electrons) contribute to shielding, an s electron experiences less shielding than a p electron and therefore experiences a larger Z_{eff}. Consequently, by the combined effects of penetration and shielding, an s electron is more tightly bound than a p electron of the same shell. Similarly, a d electron penetrates less than a p electron of the same shell, and therefore experiences more shielding and an even smaller Z_{eff}.

Effective atomic numbers for different types of electrons in atoms have been calculated from their wavefunctions (which are obtained by numerical solution of the Schrödinger equation for the atom). Some values are given in Table 13.2. We see that in general s electrons do experience higher effective atomic numbers than p electrons, although there are some discrepancies. We shall return to this point shortly.

The consequence of penetration and shielding is that the energies of subshells in a many-electron atom in general lie in the order

$$s < p < d$$

The individual orbitals of a given subshell (such as the three p orbitals of the p subshell) remain degenerate since they all have the same radial characteristics and so experience the same effective nuclear charge.

We can now complete the Li story. Since the shell with $n = 2$ consists of two nondegenerate subshells, with the $2s$ orbital lower in energy than the three $2p$ orbitals, the third electron occupies the $2s$ orbital. This results in the ground state configuration $1s^2 2s^1$, with the central nucleus surrounded by a complete helium-like shell of two $1s$ electrons, and around that a more diffuse $2s$ electron. The electrons in the outermost shell of an atom in its ground state are called the **valence electrons** since they are largely responsible for the chemical bonds that the atom forms. Thus, the valence electron in Li is a $2s$ electron and its other two electrons belong to its core.

The building-up principle

The extension of the procedure used for H, He, and Li to other atoms is called the **building-up principle**, or the **Aufbau principle**, from the German word for building up. The aim of the building-up principle is to find an order of occupation of the hydrogenic orbitals that leads to the ground state configuration of the neutral atom.

We imagine the bare nucleus of atomic number Z, and then feed into the orbitals Z electrons in succession. The order of occupation is

$$1s \quad 2s \quad 2p \quad 3s \quad 3p \quad 4s \quad 3d \quad 4p \quad 5s \quad 4d \quad 5p \quad 6s$$

and each orbital may accommodate up to two electrons. This order of occupation is *approximately* the order of energies of the individual orbitals, since in general the lower the energy of the orbital, the lower the total energy of the atom as a whole when that orbital is occupied. However there are complicating effects arising from electron–electron repulsions that are important when the orbitals have very similar energies (such as the $4s$ and $3d$ orbitals near Ca and Sr) and we must take special care then.

Table 13.2. Effective atomic numbers

	Z_{eff}
He $1s$	1.69
Na $1s$	10.6
$2s$	6.85
$2p$	6.85
$3s$	2.20

We feed the Z electrons in succession into the orbitals subject to the demand of the exclusion principle that no more than two can occupy any single oribtal. Since an s subshell consists of only one orbital, up to two electrons may occupy it. Since the p subshell consists of three orbitals, it can accommodate up to 6 electrons; the d subshell, which consists of five orbitals, can accommodate up to 10 electrons.

As an example, consider the carbon atom. Since $Z = 6$ for carbon, we must accommodate 6 electrons. Two enter and fill the $1s$ orbital, two enter and fill the $2s$ orbital, leaving two electrons to occupy the orbitals of the $2p$ subshell. Hence its ground configuration is $1s^2 2s^2 2p^2$, or more succinctly [He]$2s^2 2p^2$, with [He] the helium-like $1s^2$ core. However, we can be more precise. On electrostatic grounds, we can expect the last two electrons to occupy different $2p$ orbitals since they will then be further apart on average and repel each other less than if they were in the same orbital. Thus one electron can be thought of as occupying the $2p_x$ orbital and the other the $2p_y$ orbital (although the x, y, z designation is arbitrary), and the lowest energy configuration of the atom is [He]$2s^2 2p_x^1 2p_y^1$. The same rule applies whenever degenerate orbitals of a subshell are available for occupation. Another rule of the building-up principle is therefore:

Electrons occupy different orbitals of a given subshell before doubly occupying any one of them.

Thus nitrogen ($Z = 7$) has the configuration [He]$2s^2 2p_x^1 2p_y^1 2p_z^1$, and only when we get to oxygen ($Z = 8$) is a $2p$ orbital doubly occupied, giving [He]$2s^2 2p_x^2 2p_y^1 2p_z^1$.

An additional point arises when electrons occupy orbitals singly, for there is then no requirement that their spins should be paired. We need to know whether the lowest energy is achieved when the electron spins are the same (both α, for instance, denoted ↑↑, if there are two electrons in question, as in C) or when they are paired (↑↓). This is resolved by **Hund's rule**:

An atom in its ground state adopts a configuration with the greatest number of unpaired electrons.

The explanation of Hund's rule is complicated, but it reflects the quantum mechanical property of **spin correlation**, that electrons with parallel spins have a tendency to stay well apart and hence repel each other less.[1]

We can now conclude that in the ground state of the carbon atom, the two $2p$ electrons have the same spin, that all three $2p$ electrons in the N atom have the same spin, and that the two $2p$ electrons in different orbitals in the O atom have the same spin (the two in the $2p_x$ orbital are necessarily paired).

Neon, with $Z = 10$, has the configuration [He]$2s^2 2p^6$, which completes the L shell. This closed-shell configuration is denoted [Ne], and acts as a core for subsequent elements. The next electron must enter the $3s$ orbital and begin a new shell, and so an Na atom, with $Z = 11$, has the configuration [Ne]$3s^1$. Like lithium with the configuration [He]$2s^1$, sodium has a single s electron outside a complete core.

---

[1] The effect of spin correlation is to allow the atom to shrink slightly, so the electron–nucleus interaction is improved when the spins are parallel.

This analysis has brought us to the origin of chemical periodicity. The L shell is completed by eight electrons, and so the element with $Z = 3$ (Li) should have similar properties to the element with $Z = 11$ (Na). Likewise, Be ($Z = 4$) should be similar to $Z = 12$ (Mg), and so on up to the noble gases He ($Z = 2$), Ne ($Z = 10$), and Ar ($Z = 18$).

Argon has complete $3s$ and $3p$ subshells, and as the $3d$ orbitals are high in energy it counts as having a closed-shell configuration. Indeed, the $3d$ orbitals are so high in energy that the next electron (for K) occupies the $4s$ orbital and the K atom resembles an Na atom. The same is true of a Ca atom, which has the configuration $[Ar]4s^2$. However, at this point, the $3d$ orbitals become comparable in energy to the $4s$ orbitals, and they start to be filled.

Ten electrons can be accommodated in the five $3d$ orbitals, which accounts for the electron configurations of scandium to zinc. However, the building-up principle has less clear-cut predictions about the ground state configurations of these elements because electron–electron repulsions are comparable to the energy difference between the $4s$ and $3d$ orbitals, and a simple analysis no longer works. At gallium, the energy of the $3d$ orbitals has fallen so far below those of the $4s$ and $4p$ orbitals that they (the $3d$ orbitals) can be largely ignored, and the building-up principle can be used in the same way as in preceding periods. Now the $4s$ and $4p$ subshells constitute the valence shell, and the period terminates with krypton. Since 18 electrons have intervened since argon, this period is the first **long period** of the periodic table. The existence of the d-block elements (the 'transition metals') reflects the stepwise occupation of the $3d$ orbitals, and the subtle shades of energy differences along this series gives rise to the rich complexity of inorganic d-metal chemistry. A similar intrusion of the f orbitals in Periods 6 and 7 accounts for the existence of the f block of the periodic table (the lanthanides and actinides).

We derive the configurations of cations of elements in the s, p, and d blocks of the periodic table by removing electrons from the ground-state configuration of the neutral atom in a specific order. First, we remove p electrons (if any are present), then s electrons, and then as many d electrons as are necessary to achieve the stated charge. For instance, since the configuration of Fe is $[Ar]3d^64s^2$, the Fe^{3+} cation has the configuration $[Ar]3d^5$. We derive the configurations of anions simply by continuing the building-up procedure and adding electrons to the neutral atom until the configuration of the next noble gas has been reached. Thus, the configuration of the O^{2-} ion is achieved by adding two electrons to $[He]2s^22p^4$, giving $[He]2s^22p^6$, the analogue of the configuration of Ar.

The periodicity of ionization energies

The minimum energy necessary to remove an electron from a many-electron atom is its **first ionization energy** I_1. The **second ionization energy** I_2 is the minimum energy needed to remove a second electron (from the singly-charged cation). The variation of the first ionization energy through the periodic table is shown in Fig. 13.16 and some numerical values are given in Table 2.4.

Lithium has a low first ionization energy: its outermost electron is well-shielded from the nucleus by the core ($Z_{eff} = 1.3$ compared with $Z = 3$) and it is easily removed. Beryllium has a higher nuclear charge than lithium,

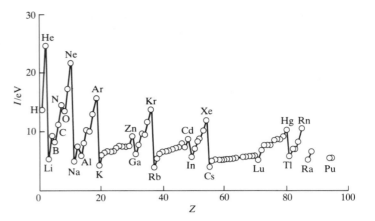

Fig. 13.16 The first ionization energies of the elements plotted against atomic number.

and its outermost electron (one of the two 2s electrons) is more difficult to remove: its ionization energy is larger. The ionization energy decreases between beryllium and boron because in the latter the outermost electron occupies a 2p orbital and is less strongly bound than if it had been a 2s electron. The ionization energy increases between boron and carbon because the latter's outermost electron is also 2p and the nuclear charge has increased. Nitrogen has a still higher ionization energy because of the further increase in nuclear charge.

There is now a kink in the curve which reduces the ionization energy of oxygen below what would be expected by simple extrapolation. This is because at oxygen a 2p orbital must become doubly occupied, and the electron-electron repulsions are increased above what would be expected by simple extrapolation along the row. (The kink is less pronounced in the next row, between phosphorus and sulphur, because their orbitals are more diffuse.) The values for oxygen, fluorine, and neon fall roughly on the same line, the increase of their ionization energies reflecting the increasing attraction of the nucleus for the outermost electrons.

The outermost electron in sodium is 3s. It is far from the nucleus, and the latter's charge is shielded by the compact, complete neon-like core. As a result, the ionization energy of sodium is substantially lower than that of neon. The periodic cycle starts again along this row, and the variation of the ionization energy can be traced to similar reasons.

13.5 Self-consistent field orbitals

The central difficulty of the Schrödinger equation is the presence of the electron–electron interaction terms:

$$V = \frac{e^2}{4\pi\varepsilon_0} \sum_{\substack{i,j \\ \text{pairs}}} \frac{1}{r_{ij}}$$

r_{ij} is the separation of electrons i and j, and the sum is over all pairs of electrons. It is hopeless to expect to find analytical solutions of a Schrödinger equation that has this as part of its potential energy term, but computational techniques are available that give very detailed and reliable numerical solutions for the wavefunctions and energies. The techniques were originally introduced by Douglas Hartree (before computers were available) and then modified by Vladimir Fock to take into account the Pauli principle correctly. In broad outline, the **Hartree–Fock procedure** is as follows.

Imagine that we have a rough idea of the structure of the atom. In the Na atom, for instance, the orbital approximation suggests the configuration $1s^2 2s^2 2p^6 3s^1$ with the orbitals approximated by hydrogenic atomic orbitals. Now consider the $3s$ electron. A Schrödinger equation can be written for this electron by ascribing to it a potential energy that arises from the nuclear attraction and the average electronic repulsion from the other electrons in their approximate orbitals. This equation has the form

$$\frac{-\hbar^2}{2m_e} \nabla^2 \psi_{3s} - \frac{Ze^2}{4\pi\varepsilon_0 r} \psi_{3s} + V_{ee}\psi_{3s} = E\psi_{3s} \tag{10}$$

where V_{ee}, which depends on the wavefunctions of all the other electrons, is the average repulsion term. The equation may be solved for ψ_{3s} (by numerical integration), and the solution obtained will be different from the solution guessed initially.

The procedure is then repeated for another orbital, such as $2p$. The Schrödinger equation is written in a form like eqn 10 but with the improved $3s$ orbital used in setting up the electron–electron repulsion term. The equation is then solved, giving an improved version of $2p$. This procedure is repeated for the $2s$ and $1s$ orbitals, each time using the improved orbitals found at the earlier stage. Then the whole procedure is repeated using the improved orbitals, and a second improved set of orbitals is obtained. The recycling continues until the orbitals and energies obtained are insignificantly different from those used at the start of the latest cycle. The solutions are then **self-consistent** and accepted as solutions of the problem.

Some of the self-consistent field (SCF) Hartree–Fock (HF) atomic orbitals (AO) for sodium are shown in Fig. 13.17. They show the grouping of electron density into shells, as was anticipated by the early chemists, and the differences of penetration as discussed above. These SCF calculations therefore support the qualitative discussions that are used to explain chemical periodicity. They also considerably extend that discussion by providing detailed wavefunctions and precise energies.

The spectra of complex atoms

The spectra of atoms rapidly become very complicated as the number of electrons increases, but there are some important and moderately simple features. The general idea is straightforward: lines in the spectrum (in either emission or absorption) occur when the atom undergoes a change of state with a change of energy ΔE, and emits or absorbs a photon of frequency $\nu = \Delta E/h$ and wavenumber $\tilde{\nu} = \Delta E/hc$. Hence, we can expect the spectrum to give information about the energies of electrons in atoms. However, the actual energy levels are not given solely by the energies of the orbitals, because the electrons interact with each other in various ways, and there are contributions to the energy in addition to those we have already considered.

13.6 Singlet and triplet states

Suppose we were interested in the energy levels of a He atom, with its two electrons. We know that the ground configuration is $1s^2$ and can anticipate

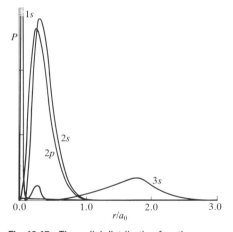

Fig. 13.17 The radial distribution functions for the orbitals of Na based on SCF calculations. Note the shell-like structure, with the $3s$ orbital outside the inner K and L shells.

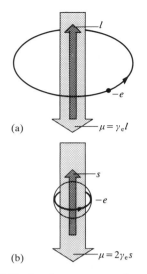

(a)

(b)

Fig. 13.18 Angular momentum gives rise to a magnetic moment (μ). In the case of an electron, (a) the magnetic moment is antiparallel to the orbital angular momentum, but proportional to it. In the case of spin angular momentum (b), there is a factor 2, which increases the magnetic moment (as explained in Section 13.9).

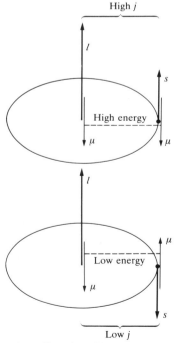

Fig. 13.19 The spin–orbit coupling is a magnetic interaction between spin and orbital magnetic moments. When the angular momenta are parallel, as in (a), the magnetic moments are aligned unfavourably; when they are opposed, as in (b), the interaction is favourable. This is the cause of the splitting of a configuration into levels.

that an excited configuration will be one in which one of the electrons has been promoted into a $2s$ orbital, giving the configuration $1s^12s^1$. The two electrons need not be paired because they occupy different orbitals. According to Hund's rule, the state of the atom with the spins parallel ($\uparrow\uparrow$) lies lower in energy than the state in which they are paired ($\uparrow\downarrow$). Both states are permissible, and can contribute to the spectrum of the atom.

The two spin arrangements differ in their overall spin angular momentum. In the paired case, the two spin momenta cancel each other, and there is zero net spin. For reasons we shall clarify shortly, this paired-spin arrangement is called a **singlet** state. In the parallel case, the two spins add together to give a nonzero total spin and the resulting state is called a **triplet**.

The fact that the ($\uparrow\uparrow$) arrangement of spins in the $1s^12s^1$ configuration of the He atom lies lower in energy than the ($\uparrow\downarrow$) arrangement can now be expressed by saying that the triplet ($\uparrow\uparrow$) state of the $1s^12s^1$ configuration of He lies lower in energy than the singlet ($\uparrow\downarrow$) state. This is a general conclusion that applies to other atoms (and molecules), and for states arising from the same configuration, the triplet state generally lies lower than the singlet. The origin of the energy difference lies in the effect of spin correlation, as we saw in the case of Hund's rule for ground-state configurations. Since the coulombic interaction between electrons in an atom is strong, the difference in energies between singlet and triplet states of the same configuration can be very large. The two states of $1s^12s^1$ He, for instance, differ by 6421 cm^{-1} (corresponding to 77 kJ mol^{-1} or 0.80 eV).

13.7 Spin–orbit coupling

The difference in energy between singlet and triplet states stems from the coulombic interaction in combination with the quantum mechanical effect of spin correlation. Another effect stems from *magnetic* interactions of the electrons in an atom. This interaction is weak for light atoms, but for heavy atoms may be comparable to the coulombic interaction and have a marked effect on the energy levels of atoms.

Because an electron has spin angular momentum, and because moving charges generate magnetic fields, an electron has a magnetic moment (Fig. 13.18) that arises from its spin. Similarly, an electron with orbital angular momentum is in effect a circulating current, and possesses a magnetic moment that arises from its orbital momentum. The interaction of the spin and orbital magnetic moments is called **spin–orbit coupling**. The strength of the coupling, and its effect on the energy levels of the atom, depends on the relative orientations of the spin and orbital magnetic moments, and therefore on the relative orientations of the two angular momenta (Fig. 13.19).

The total angular momentum of an electron

Another way of expressing the dependence of the spin–orbit interaction on the relative orientation of the spin and orbital momenta is to say that it depends on the *total* angular momentum of the electron, the vector sum of its spin and orbital momenta. Thus, when the spin and orbital angular momenta are parallel, the total angular momentum is high; when the two angular momenta are opposed, the total angular momentum is low.

The total angular momentum of a spinning, orbiting electron is quantized. It is described by the quantum numbers j and m_j, with the permitted values of j either $l + \frac{1}{2}$ (when the two angular momenta are in the same direction) or $l - \frac{1}{2}$ (when they are opposed, Fig. 13.20). The different values of j that can arise for a given configuration are called the **levels** of that configuration. For $l = 0$, the only permitted value is $j = s$ (the total angular momentum is the same as the spin angular momentum since there is no other angular momentum in the atom).

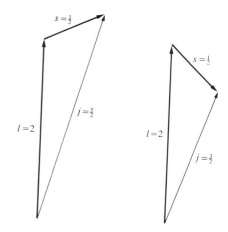

Fig. 13.20 The coupling of the spin and orbital angular momenta of a d electron ($l = 2$) gives two possible values of j.

Example 13.7: *Identifying the total angular momentum of an electron*

Calculate the values of the total angular momentum quantum number that may arise from (a) a d electron with spin, (b) an s electron with spin.

Answer. In each case, we must identify the value of l and then the two possible values of j. (a) For a d electron, $l = 2$ and there are two levels in a configuration with a single d electron, one with $j = \frac{5}{2}$ and the other with $j = \frac{3}{2}$. (b) For an s electron $l = 0$, so only one level is possible, and $j = \frac{1}{2}$.

Exercise. Calculate the values of j for a p electron and an f electron.

$$[p: \tfrac{3}{2}, \tfrac{1}{2}; f: \tfrac{7}{2}, \tfrac{5}{2}]$$

The dependence of the spin–orbit interaction on the value of j is expressed in terms of the **spin–orbit coupling constant** A (in cm^{-1}). A quantum-mechanical calculation leads to the result that the energies of the levels with quantum numbers s, l, and j are given by

$$E_{l,s,j} = \tfrac{1}{2}hcA\{j(j+1) - l(l+1) - s(s+1)\} \qquad (11)$$

Hence, the different levels of a configuration are split by the spin–orbit coupling and lie at different energies. That being so, we should expect to be able to detect spin–orbit coupling by examining the spectrum of the atom.

Example 13.8: *Calculating the spin–orbit interaction*

The spin–orbit interaction constant in the excited $[\mathrm{Ar}]5p^1$ configuration of K is 12.5 cm^{-1}. Calculate the splitting of the levels of this configuration.

Answer. The configuration has $l = 1$ and $s = \frac{1}{2}$; therefore, the two levels are $j = \frac{3}{2}$ and $j = \frac{1}{2}$. From eqn 11 the difference in their energies is

$$\Delta E = \tfrac{1}{2}hcA(\tfrac{3}{2} \times \tfrac{5}{2} - \tfrac{1}{2} \times \tfrac{3}{2}) = \tfrac{3}{2}hcA$$

Therefore, the splitting in wavenumbers is $\frac{3}{2} \times 12.5$ cm^{-1} = 18.8 cm^{-1}.

Exercise. The $[\mathrm{Ar}]3d^1$ of the same atom has $A = 0.93$ cm^{-1}. Calculate the interval between the two levels.

$$[2.3 \text{ cm}^{-1}]$$

The fine structure of spectra

We can see the effect of spin–orbit coupling on the states of the atom and its spectrum by considering the alkali metals, which consist of a single valence electron outside a closed core. As a good first approximation we can ignore the core electrons (which have no net orbital angular momentum) and concentrate on the valence electron alone.

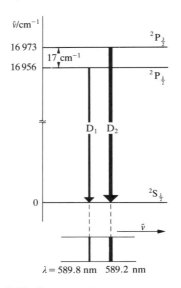

$\bar{v}/\text{cm}^{-1}$

16 973

16 956

$17\,\text{cm}^{-1}$

$^2P_{\frac{3}{2}}$

$^2P_{\frac{1}{2}}$

D_1 D_2

$^2S_{\frac{1}{2}}$

0

$\bar{v}$

$\lambda = 589.8\,\text{nm}$ 589.2 nm

Fig. 13.21 The energy-level diagram for the formation of the D lines of sodium. The splitting of the spectral lines (by 17 cm^{-1}) reflects the splitting of the levels.

In the ground state, the valence electron is an s electron, and the $[X]s^1$ configuration (where $[X]$ is a noble gas configuration) has only a single level, $j = \frac{1}{2}$. Since $l = 0$, the spin–orbit coupling energy is zero (as is confirmed by setting $j = s$ and $l = 0$ in eqn 11). If the electron is excited into a p orbital, it acquires orbital angular momentum (from the incoming photon), and occupies an orbital with $l = 1$; now the permitted values of j are $\frac{3}{2}$ and $\frac{1}{2}$ and the $[X]p^1$ configuration has two levels. The level with $j = \frac{3}{2}$ corresponds to a state in which the angular momenta (and their magnetic moments) are parallel, and the $j = \frac{1}{2}$ level is a state in which the two momenta (and their magnetic moments) are opposed. The former is a higher-energy arrangement, and so the level with $j = \frac{3}{2}$ lies above the level with $j = \frac{1}{2}$. Therefore, the $[X]p^1$ configuration splits into two levels (Fig. 13.21) with $j = \frac{3}{2}$ above $j = \frac{1}{2}$. When the excited atom undergoes a transition and the p electron falls into a lower s orbital, two spectral lines are observed, depending on which of the two levels of the $[X]p^1$ configuration was occupied initially. This splitting is called the **fine structure** of the spectrum.

Fine structure can be clearly seen in the emission spectrum from sodium vapour excited by an electric discharge (for example, in one kind of street lighting). The yellow line at 589 nm (17 000 cm^{-1}) is actually a doublet—a pair of lines—being composed of one line at 589.76 nm (16 956 cm^{-1}) and the other at 589.16 nm (16 973 cm^{-1}). The transitions (Fig. 13.21) are from the $j = \frac{3}{2}$ and $\frac{1}{2}$ levels of the $[\text{Ne}]3p^1$ configuration to the ground configuration $[\text{Ne}]3s^1$. Therefore, in Na, the spin–orbit coupling affects the energies by about 17 cm^{-1}.

Example 13.9: *Analysing a spectrum for the spin–orbit coupling constant*

The spectrum of atomic sodium shows that D lines are analyzed in Fig. 13.21. They lie at 16 956.2 cm^{-1} and 16 973.4 cm^{-1}. Calculate the spin–orbit coupling constant for the upper configuration of the Na atom.

Answer. The splitting of the lines is equal to the splitting of the $j = \frac{3}{2}$ and $\frac{1}{2}$ levels of the excited configuration, which we can express in terms of A using eqn 11. The two levels are split by

$$\Delta\bar{v} = \tfrac{1}{2}A\{\tfrac{3}{2}(\tfrac{3}{2}+1) - \tfrac{1}{2}(\tfrac{1}{2}+1)\} = \tfrac{3}{2}A$$

Since the experimental value is 17.2 cm^{-1}, it follows that

$$A = \tfrac{2}{3} \times 17.2\,\text{cm}^{-1} = 11.4\,\text{cm}^{-1}$$

Comment. The same calculation repeated for the other alkali metal atoms gives Li: 0.23 cm^{-1}, K: 38.5 cm^{-1}, Rb: 158 cm^{-1}, Cs: 370 cm^{-1}.

Exercise. The configuration $\ldots 4p^65d^1$ of rubidium has two levels at 25 700.56 cm^{-1} and 25 703.52 cm^{-1} above the ground configuration. What is the spin-orbit coupling constant in this excited state?   [1.18 cm^{-1}]

The strength of the coupling

The strength of the spin–orbit coupling depends on the nuclear charge. We can understand why this is so by imagining that we are riding on the orbiting electron. We then see a charged nucleus apparently orbiting around us, like the sun rising and setting, and as a result we find ourselves at the centre of a ring of current. The greater the nuclear charge the greater this current, and therefore the stronger the magnetic field we detect. Since the spin magnetic moment of the electron interacts with this orbital magnetic field, the greater the nuclear charge, the stronger the spin–orbit interaction.

The coupling increases sharply with atomic number (as Z^4), and whereas it is only small in H (giving rise to shifts of energy levels of no more than about $0.4 \, \text{cm}^{-1}$), in heavy atoms like Pb it is very large (giving shifts of the order of thousands of cm^{-1}).

13.8 Term symbols and selection rules

We have used expressions such as 'the $j = \frac{3}{2}$ level of a configuration'. A **term symbol**, which is a symbol looking like $^2P_{3/2}$ or $^3D_2$, conveys this information much more succinctly. It also enables us to extend the discussion from a configuration in which there is only one electron of interest (such as the valence electron in the alkali metal atoms) to cases where several electrons must be considered simultaneously, as in the alkaline earth metal atoms.

A term symbol gives three pieces of information:
(1) The letter (e.g. P or D in the examples) indicates the *total* orbital angular momentum.
(2) The left superscript in the term symbol (e.g. the 2 in $^2P_{3/2}$) gives the **multiplicity** of the term.
(3) The right subscript on the term symbol (e.g. the $\frac{3}{2}$ in $^2P_{3/2}$) is the value of the **total angular momentum quantum number J**.

We shall now say what each of these means; the contributions to the energies which we are about to discuss are summarized in Fig. 13.22.

The total orbital momentum

When several electrons are present, it is necessary to judge how their individual orbital angular momenta add together or oppose each other. The **total orbital angular momentum**[2] L is obtained by coupling the individual orbital angular momenta using the **Clebsch–Gordan series**:

$$L = l_1 + l_2, l_1 + l_2 - 1, \ldots, |l_1 - l_2| \tag{12}$$

The maximum value $L = l_1 + l_2$ is obtained when the two orbital angular momenta are in the same direction; the lowest value $|l_1 - l_2|$ is obtained when they are in opposite directions. The intermediate values represent possible intermediate relative orientations of the two momenta (Fig. 13.23). For two p electrons (for which $l_1 = l_2 = 1$), $L = 2$, 1, 0. The code for converting the value of l into a letter is the same as for the $s, p, d, f, \ldots$ designation of orbitals:

$$
\begin{array}{cccccc}
L: & 0 & 1 & 2 & 3 & 4 & \ldots \\
& S & P & D & F & G & \ldots
\end{array}
$$

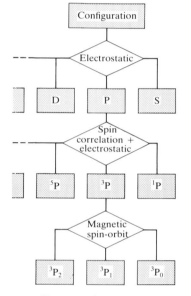

Fig. 13.22 The types of interaction that are responsible for the various kinds of splitting of energy levels in atoms. For light atoms, magnetic interactions are small, but in heavy atoms may dominate the electrostatic interactions.

---

[2] The total orbital angular momentum is quantized, and like the other momenta we have encountered, its magnitude is given by the value of L, and is $\{L(L+1)\}^{1/2}\hbar$. It also has $2L + 1$ orientations distinguished by the quantum number M_L, which can take the values L, $L - 1, \ldots, -L$. Similar remarks apply to the total spin S (its orientations are denoted M_S), and the total angular momentum J (with its orientations M_J).

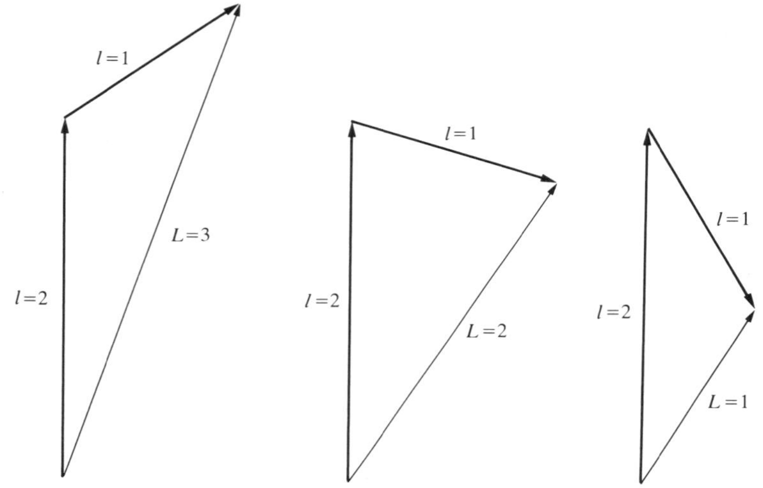

Fig. 13.23 The total orbital angular momenta of a *p* electron and a *d* electron correspond to $L = 3, 2$, and 1 and reflect the different relative orientations of the two momenta.

Hence, a p^2 configuration can give rise to D, P, and S terms; they differ in energy on account of the different electrostatic interactions between the electrons arising from their different orbital occupations.

A closed shell has zero orbital angular momentum because all the individual orbital angular momenta sum to zero. Therefore, when working out term symbols, we need consider only the electrons of the unfilled shell. In the case of a single electron outside a closed shell, the value of L is the same as the value of l, and so the configuration [Ne]$3s^1$ has only an S term.

Example 13.10: *Deriving the total orbital angular momentum of a configuration*

Find the terms that can arise from the configurations (a) d^2, (b) p^3.

Answer. We use the Clebsch–Gordan series and begin by finding the minimum value of L (so that we know where the series terminates). (a) Minimum value: $|l_1 - l_2| = |2 - 2| = 0$. Therefore,

$$L = 2 + 2, 2 + 2 - 1, \ldots, 0 = 4, 3, 2, 1, 0$$

corresponding to G, F, D, P, S terms respectively. In (b), we use two series in succession: first we couple two electrons, and then we couple the third to each combined state. (b) First coupling: Minimum value: $|1 - 1| = 0$. Therefore,

$$L' = 1 + 1, 1 + 1 - 1, \ldots, 0 = 2, 1, 0$$

Now couple l_3 with $L' = 2$, to give $L = 3, 2, 1$

$\qquad\qquad$ with $L' = 1$, to give $L = 2, 1, 0$

$\qquad\qquad$ with $L' = 0$, to give $L = 1$

The overall result is

$$L = 3, 2, 2, 1, 1, 1, 0$$

giving one F, two D, three P, and one S terms.

Exercise. Repeat the question for the configurations (a) f^1d^1 and (b) d^3.

$\qquad\qquad\qquad$ [(a) H, G, F, D, P; (b) I, 2H, 3G, 4F, 5D, 3P, S]

The multiplicity

When there are several electrons to be taken into account, we must assess their **total spin angular momentum** S. Once again, we use the Clebsch–Gordan series to decide on the value of S, noting that each electron has $s = \frac{1}{2}$, which gives (Fig. 13.24)

$$S = 1, 0$$

If there are three electrons, the total spin angular momentum is obtained by coupling the third spin to each of the values of S for the first two spins:

$$S = \tfrac{3}{2}, \tfrac{1}{2} \quad \text{and} \quad S = \tfrac{1}{2}$$

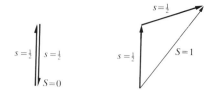

Fig. 13.24 For two electrons (which have $s = \frac{1}{2}$), only two total spin states are permitted ($S = 0, 1$). The state with $S = 0$ can have only one value of M_S ($M_S = 0$) and is a singlet; the state with $S = 1$ can have any of three values of M_S ($+1, 0, -1$) and is a triplet.

The **multiplicity** of a term is the value of $2S + 1$. When $S = 0$ (as for a closed shell) the electrons are all paired and there is no net spin: this gives a singlet term, such as ${}^1\mathrm{S}$. (Take care not to confuse the italic S of the spin and the roman S of the term symbol.) A single electron has $S = s = \frac{1}{2}$, and so a configuration such as $[\mathrm{Ne}]3s^1$ can give rise to a doublet term, ${}^2\mathrm{S}$. The configuration $[\mathrm{Ne}]3p^1$ likewise is a doublet, ${}^2\mathrm{P}$. When there are two unpaired electrons, $S = 1$ and so $2S + 1 = 3$, giving a triplet term, such as ${}^3\mathrm{D}$. We discussed the relative energies of singlets and triplets in Section 13.4 and saw that their energies differ on account of the different effects of spin correlation.

The total angular momentum

As we have seen, the total angular momentum quantum number j tells us the relative orientation of the spin and orbital angular momenta of a single electron. The **total angular momentum** J does the same for several electrons. If there is a single electron outside a closed shell, then $J = j$, with j either $l + \frac{1}{2}$ or $l - \frac{1}{2}$. The $[\mathrm{Ne}]3s^1$ configuration has $j = \frac{1}{2}$ (because $l = 0$ and $s = \frac{1}{2}$), and so the ${}^2\mathrm{S}$ term has a single level, ${}^2\mathrm{S}_{1/2}$. The $[\mathrm{Ne}]3p^1$ configuration has $l = 1$, and therefore $j = \frac{3}{2}$ and $\frac{1}{2}$; the ${}^2\mathrm{P}$ term therefore has two levels, ${}^2\mathrm{P}_{3/2}$ and ${}^2\mathrm{P}_{1/2}$, and these lie at different energies on account of the magnetic spin–orbit interaction (Fig. 13.19).

If there are several electrons outside a closed shell we have to consider the coupling of all the spins and all the orbital angular momenta. This complicated problem can be simplified when the spin–orbit coupling is weak (for atoms of low atomic number) for then we can use the **Russell–Saunders coupling scheme**. The Russell–Saunders scheme is based on the view that if spin–orbit coupling is weak, then it is effective only when all the orbital momenta are operating cooperatively. We therefore imagine that all the orbital angular momenta of the electrons couple to give some total L, and that all the spins are similarly coupled to give some total S. Only at this stage do we imagine the two kinds of momenta coupling through the spin–orbit interaction to give a total J. The permitted values of J are given by the Clebsch–Gordan series

$$J = L + S, \; L + S - 1, \ldots, |L - S| \tag{13}$$

For example, in the case of the ${}^3\mathrm{D}$ term of the configuration $[\mathrm{Ne}]2p^1 3p^1$, the permitted values of J are 3, 2, 1 (because ${}^3\mathrm{D}$ has $L = 2$ and $S = 1$), and so the term has three levels, ${}^3\mathrm{D}_3$, ${}^3\mathrm{D}_2$, and ${}^3\mathrm{D}_1$.

When $L \geq S$, the multiplicity is equal to the number of levels, as in ^2P having the two levels $^2P_{3/2}$ and $^2P_{1/2}$ and ^3D the three levels $^3D_3$, $^3D_2$, and $^3D_1$. However, this is not the case when $L < S$: ^2S, for example, has only the single level $^2S_{1/2}$.

Example 13.11: *Deriving term symbols*

Write the term symbols for the ground configurations of Na and F, and the excited configuration $1s^2 2s^2 2p^1 3p^1$ of C.

Answer. Write the configurations but ignore inner closed shells. Couple the orbital momenta to find L and the spins to find S. Couple L and S to find J. Express the term as $^{2S+1}\{L\}_J$, where $\{L\}$ is the appropriate letter. For F, treat the single gap in $2p^6$ as a single particle. For Na the configuration is $[Ne]3s^1$, and we consider the single $3s$ electron. Since $L = l = 0$ and $S = s = \frac{1}{2}$, it is possible for $J = j = s = \frac{1}{2}$ only. Hence the terms symbol is $^2S_{1/2}$. For F the configuration is $[He]2s^2 2p^5$, which we can treat as $[Ne]2p^{-1}$. Hence $L = l = 1$, and $S = s = \frac{1}{2}$. Two values of $J = j$ are allowed: $J = \frac{3}{2}, \frac{1}{2}$. Hence the term symbols for the two levels are $^2P_{3/2}$, $^2P_{1/2}$. For C the configuration is effectively $2p^1 3p^1$. This is a two-electron problem, and $l_1 = l_2 = 1$, $s_1 = s_2 = \frac{1}{2}$. It follows that $L = 2, 1, 0$ and $S = 1, 0$. The terms are therefore ^3D and ^1D, ^3P and ^1P, and ^3S and ^1S. For ^3D, $L = 2$ and $S = 1$; hence $J = 3, 2, 1$ and the levels are $^3D_3$, $^3D_2$, and $^3D_1$. For ^1D, $L = 2$ and $S = 0$, so that the single level is $^1D_2$. The triplet of levels of ^3P is $^3P_2$, $^3P_1$, and $^3P_0$, and the singlet is $^1P_1$. For the ^3S term there is only a single level, $^3S_1$ (because $J = 1$ only), and the singlet term is $^1S_0$.

Comment. The reason why we have treated an excited configuration of carbon is that in the ground configuration, $2p^2$, the Pauli principle forbids some terms, and deciding which survive (^1D, ^3P, ^1S) is quite complicated.

Exercise. Write down the terms arising from the configurations (a) $2s^1 2p^1$, (b) $2p^1 3d^1$.
[(a) $^3P_2$, $^3P_1$, $^3P_0$, $^1P_1$; (b) $^3F_4$, $^3F_3$, $^3F_2$, $^1F_3$, $^3D_3$, $^3D_2$, $^3D_1$, $^1D_2$, $^3P_2$, $^3P_1$, $^3P_0$, $^1P_1$]

Russell–Saunders coupling fails when the spin–orbit coupling is large (in heavy atoms). In that case, the individual spin and orbital momenta of the electrons are coupled into individual j values; then these momenta are combined into a grand total J. This is called **jj-coupling**. For example, in a p^2 configuration, the individual values of j are $\frac{3}{2}$ and $\frac{1}{2}$ for each electron. If the spin and the orbital angular momentum of each electron are coupled together strongly, it is best to consider each electron as a particle with angular momentum $j = \frac{3}{2}$ or $\frac{1}{2}$. These individual total momenta then couple as follows:

$$j_1 = \tfrac{3}{2} \text{ and } j_2 = \tfrac{3}{2} \text{ give } J = 3, 2, 1, 0$$
$$j_1 = \tfrac{3}{2} \text{ and } j_2 = \tfrac{1}{2} \text{ give } J = 2, 1$$
$$j_1 = \tfrac{1}{2} \text{ and } j_2 = \tfrac{3}{2} \text{ give } J = 2, 1$$
$$j_1 = \tfrac{1}{2} \text{ and } j_2 = \tfrac{1}{2} \text{ give } J = 1, 0$$

For heavy atoms, in which *jj*-coupling is appropriate, it is best to discuss their energies using these quantum numbers.

Selection rules

Any state of the atom, and any spectral transition, can be specified using term symbols. For example, the transitions giving rise to the yellow sodium doublet (Fig. 13.21) are

$$3p^{1} {}^{2}P_{3/2} \rightarrow 3s^{1} {}^{2}S_{1/2} \quad \text{and} \quad 3p^{1} {}^{2}P_{1/2} \rightarrow 3s^{1} {}^{2}S_{1/2}$$

The configuration need not always be specified, and if it is clear what we are talking about, we could specify the transitions as simply

$$^{2}P_{3/2} \rightarrow {}^{2}S_{1/2} \quad \text{and} \quad {}^{2}P_{1/2} \rightarrow {}^{2}S_{1/2}$$

Note that the upper term precedes the lower. The corresponding absorptions would therefore be denoted

$$^{2}P_{3/2} \leftarrow {}^{2}S_{1/2} \quad \text{and} \quad {}^{2}P_{1/2} \leftarrow {}^{2}S_{1/2}$$

We have seen that selection rules arise from the conservation of angular momentum during a transition and from the fact that a photon has a spin of 1. They can therefore be expressed in terms of the term symbols, because the latter carry information about angular momentum. A detailed analysis leads to the following rules:

$$\Delta S = 0 \quad \Delta L = 0, \pm 1 \quad \text{with} \quad \Delta l = \pm 1$$
$$\Delta J = 0, \pm 1 \text{ but } J = 0 \text{ cannot combine with } J = 0$$

The rule about ΔS (no change of overall spin) stems from the fact that the light does not affect the spin directly. The rule about ΔL and Δl expresses the fact that the orbital angular momentum of an individual electron must change (so $\Delta l = \pm 1$), but whether or not this results in an overall change of orbital momentum depends on the coupling.

The selection rules given above apply when Russell–Saunders coupling is valid (in light atoms). If we insist on labelling the terms of heavy atoms with symbols like ^{3}D, then we shall find that the selection rules progressively fail as the atomic number increases because the quantum numbers S and L become ill defined as jj-coupling becomes more appropriate. For this reason, transitions between singlet and triplet states (for which $\Delta S = \pm 1$), while forbidden in light atoms, are allowed in heavy atoms.

13.9 The effect of magnetic fields

Since orbital and spin angular momenta give rise to magnetic moments, it can be expected that the application of a magnetic field should modify an atom's spectrum. We shall first establish how the energies of an atom depend on the strength of an external field and then see how the spectrum is affected.

The magnetic moment of an electron

The orbital angular momentum of an electron around the z-axis (which we shall now take as the direction of the applied field) is $m_l \hbar$. Since the component of magnetic moment on the z-axis μ_z is proportional to the angular momentum around that axis, we can write

$$\mu_z = \gamma_e m_l \hbar \tag{14a}$$

where γ_e is a constant called the **magnetogyric ratio** of the electron. If the magnetic moment is treated as arising from the circulation of an electron of charge $-e$, standard electromagnetic theory gives

$$\gamma_e = -\frac{e}{2m_e} \qquad (14b)$$

The negative sign (reflecting the sign of the electron's charge) shows that the orbital magnetic moment of the electron is antiparallel to its orbital angular momentum. It follows that the possible values of μ_z are

$$\mu_z = -\frac{e\hbar}{2m_e} \times m_l = -\mu_B m_l \qquad (15a)$$

where

$$\mu_B = \frac{e\hbar}{2m_e} = 9.273 \times 10^{-24}\,\mathrm{J\,T^{-1}} \qquad (15b)$$

is the **Bohr magneton**, which is often regarded as the fundamental quantum of magnetic moment.

The energy of a magnetic moment in a magnetic field B is[3]

$$E = -\mu_z B \qquad (16a)$$

Therefore, in the presence of a magnetic field, an electron in a state with quantum number m_l has an additional contribution to its energy given by

$$E = \mu_B m_l B \qquad (16b)$$

The same expression, but with m_l replaced by M_L, applies when the orbital magnetic moment arises from several electrons.

A p electron has $l = 1$ and $m_l = 0, \pm 1$. In the absence of a magnetic field, these three states are degenerate. When a field is present, the degeneracy is removed: the state with $m_l = +1$ moves up in energy by an amount $\mu_B B$, the state with $m_l = 0$ is unchanged, and the state with $m_l = -1$ moves down by an amount $\mu_B B$:

$$E_{+1} = \mu_B B \qquad E_0 = 0 \qquad E_{-1} = -\mu_B B$$

The different energies arising from an interaction with an external field are sometimes represented on the vector model by picturing the vectors as **precessing**, or sweeping round their cones (Fig. 13.25), with the rate of precession proportional to the energy of the state.

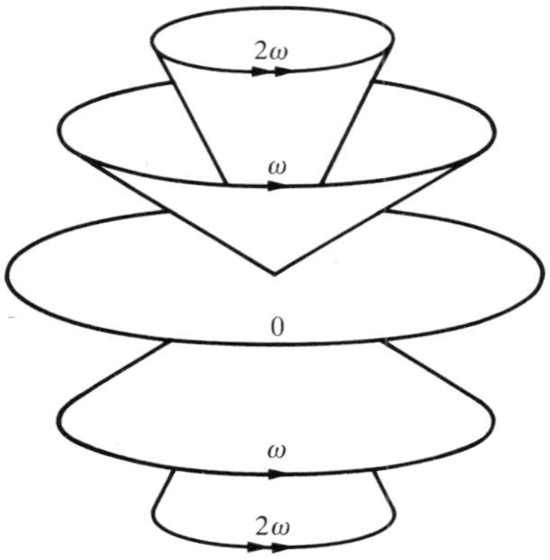

The spin magnetic moment of an electron is also proportional to its angular momentum. However, it is not given by $\gamma_e m_s \hbar$ but by about twice this value:

$$\mu_z = g_e \gamma_e m_s \hbar \qquad g_e = 2.0023 \qquad (17a)$$

The extra factor g_e, the **g-factor** of the electron, is derived from the Dirac equation, the relativistic version of the Schrödinger equation. The energy of

Fig. 13.25 The different energies of the m_l states in a magnetic field are represented by different rates of precession of the vectors representing the angular momentum. This diagram is for $l = 2$.

---

[3] This is a result from standard magnetic theory. B is actually the magnetic induction, and is measured in tesla T; gauss, G, are also still widely used: $1\,\mathrm{T} = 10^4\,\mathrm{G}$.

an electron in a state m_s in a magnetic field B is

$$E = -g_e \gamma_e m_s \hbar B = g_e \mu_B m_s B \qquad (17b)$$

The same expression, but with m_s replaced by M_S, applies to the magnetic moment arising from the spin of several electrons.

The Zeeman effect

The **Zeeman effect** is the modification of an atomic spectrum by the application of a strong magnetic field. In particular, the **normal Zeeman effect** is the observation of three lines in the spectrum where, in the absence of the field, there was only one (Fig. 13.26). The splitting is in fact extremely small: a field of 2 T (20 kG) is needed to produce a splitting of about $1 \, \text{cm}^{-1}$, which should be compared with typical optical transition wavenumbers of $20\,000 \, \text{cm}^{-1}$ and more.

Much more common than the normal Zeeman effect is the **anomalous Zeeman effect**, in which the original line splits into more than three components. The origin of this complexity is the anomalous magnetic moment of electron spin, and when spin is present the spin and orbital magnetic moments interact with the field in a more complicated way than in its absence.

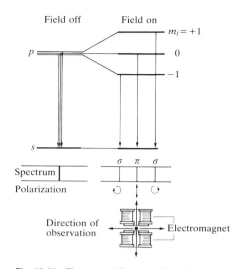

Fig. 13.26 The normal Zeeman effect. On the left, when the field is off, a single spectral line is observed. When the field is on, the line splits into three, with different polarizations. The circularly-polarized lines are called the 'σ-lines'; the plane-polarized lines are called 'π-lines'. Which line is observed depends on the orientation of the observer.

Further information 1: centre of mass coordinates

We want to show that the Schrödinger equation for hydrogen atoms can be expressed in terms of the separation of the particles, r, and the location of the centre of mass, R. We shall do the transformation of the derivatives with respect to x-coordinate: the others follow in the same way.

The location of the centre of mass is at

$$X = \frac{m_e}{m} x_e + \frac{m_N}{m} x_N$$

with m the total mass. The separation of the particles is

$$x = x_e - x_N$$

Since we can now write

$$\frac{\partial}{\partial x_e} = \frac{\partial X}{\partial x_e} \times \frac{\partial}{\partial X} + \frac{\partial x}{\partial x_e} \times \frac{\partial}{\partial x} = \frac{m_e}{m} \times \frac{\partial}{\partial X} + \frac{\partial}{\partial x}$$

$$\frac{\partial}{\partial x_N} = \frac{m_N}{m} \times \frac{\partial}{\partial X} - \frac{\partial}{\partial x}$$

the x-part of the sum of the two Laplacians becomes

$$\frac{1}{m_e} \frac{\partial^2}{\partial x_e^2} + \frac{1}{m_N} \frac{\partial^2}{\partial x_N^2} = \frac{1}{m} \frac{\partial^2}{\partial X^2} + \left(\frac{1}{m_e} + \frac{1}{m_N} \right) \frac{\partial^2}{\partial x^2}$$

The y- and z-components can be dealt with similarly, and so

$$\frac{1}{m_e} \nabla_e^2 + \frac{1}{m_N} \nabla_N^2 = \frac{1}{m} \nabla_{cm}^2 + \frac{1}{\mu} \nabla^2$$

with

$$\frac{1}{\mu} = \frac{1}{m_e} + \frac{1}{m_N}$$

as used in the text.

Now write the overall wavefunction ψ as

$$\psi = \psi(\boldsymbol{R})\psi(\boldsymbol{r})$$

and carry through the separation of variables procedure. The total equation separates into an equation for the free motion of a particle of mass m and the motion of a particle of mass μ around the centre of mass:

$$-\frac{\hbar^2}{2m_e}\nabla_e^2\psi - \frac{\hbar^2}{2m_N}\nabla_N^2\psi - \frac{Ze^2\psi}{4\pi\varepsilon_0 r}$$

$$= -\frac{\hbar^2\psi(\boldsymbol{r})}{2m}\nabla_{cm}^2\psi(\boldsymbol{R}) - \frac{\hbar^2\psi(\boldsymbol{R})}{2\mu}\nabla^2\psi(\boldsymbol{r}) - \frac{Ze^2\psi(\boldsymbol{r})\psi(\boldsymbol{R})}{4\pi\varepsilon_0 r}$$

$$= E\psi(\boldsymbol{r})\psi(\boldsymbol{R})$$

Division by $\psi(\boldsymbol{r})\psi(\boldsymbol{R})$ gives

$$-\frac{\hbar^2}{2m}\frac{\nabla_{cm}^2\psi(\boldsymbol{R})}{\psi(\boldsymbol{R})} - \frac{\hbar^2}{2\mu}\frac{\nabla^2\psi(\boldsymbol{r})}{\psi(\boldsymbol{r})} - \frac{Ze^2}{4\pi\varepsilon_0 r} = E$$

The first term is independent of r and the second and third are independent of $\boldsymbol{R}$; hence the equation is separable.

Further information 2: the Pauli principle

The Pauli exclusion principle is a special case of a general principle called the **Pauli principle**:

> When the labels of any two identical fermions are exchanged, the total wavefunction changes sign. When the labels of any two identical bosons are exchanged, the total wavefunction retains the same sign.

As remarked in the text, fermions are particles with half-integral spins, and include protons, neutrons, and some atomic nuclei. Bosons include photons ($s = 1$) and a number of atomic nuclei and atoms. For the time being we need consider only the form as stated above. By 'total wavefunction' we mean the entire wavefunction, including the spin of the particles.

Consider the wavefunction for two electrons $\psi(1, 2)$, which is a function of six variables, three the coordinates of electron 1 and three the coordinates of electron 2. The Pauli principle implies it is a fact of nature (which has its roots in the theory of relativity) that the wavefunction must change sign if we interchange the labels 1 and 2 wherever they occur in the function:

$$\psi(1, 2) = -\psi(2, 1)$$

The connection of this general form of the principle with the exclusion principle can be illustrated by the following argument, which has three stages.

An electron we might label 1 in a hydrogenic atom has a wavefunction that is the solution of the Schrödinger equation

$$H_1\psi(1) = E_1\psi(1)$$

An electron we might label 2 in a hydrogenic atom has a wavefunction that is a solution of

$$H_2\psi(2) = E_2\psi(2)$$

When both electrons are present in the same atom, the hamiltonian is

$$H = H_1 + H_2 + V(1, 2) \qquad V(1, 2) = \frac{e^2}{4\pi\varepsilon_0 r_{12}}$$

V is the potential energy of repulsion between the electrons. The wavefunction for the pair of electrons is the solution of

$$H\psi = E\psi$$

As a first approximation we can ignore their repulsion, then the total hamiltonian is $H = H_1 + H_2$, and the solution of the corresponding Schrödinger equation is the product, $\psi(1)\psi(2)$, of the individual wavefunctions:

$$\begin{aligned} H\psi &= (H_1 + H_2)\psi(1)\psi(2) = \{H_1\psi(1)\}\psi(2) + \psi(1)\{H_2\psi(2)\} \\ &= \{E_1\psi(1)\}\psi(2) + \psi(1)\{E_2\psi(2)\} \\ &= (E_1 + E_2)\psi(1)\psi(2) \\ &= E\psi, \text{ with } E = E_1 + E_2 \end{aligned}$$

That is, the product $\psi(1)\psi(2)$ is an eigenfunction of the (simplified) hamiltonian H.

To apply the Pauli principle, we must deal with the total wavefunction, the wavefunction including spin. There are four states for two spins:

both α, denoted $\alpha(1)\alpha(2)$
both β, denoted $\beta(1)\beta(2)$
one α, the other β, denoted $\alpha(1)\beta(2)$ or $\beta(1)\alpha(2)$

Since we cannot tell which electron is α and which is β, in the last case it is appropriate to express the spin states as the linear combinations

$$\sigma_+(1, 2) = \alpha(1)\beta(2) + \beta(1)\alpha(2) \qquad \sigma_-(1, 2) = \alpha(1)\beta(2) - \beta(1)\alpha(2)$$

because these allow one spin to be α and the other β with equal probability (Fig. 13.27). The total wavefunction of the system is therefore the product of the orbital part $\psi(1)\psi(2)$ and one of the four spins states:

$$\psi(1)\psi(2)\alpha(1)\alpha(2)$$
$$\psi(1)\psi(2)\beta(1)\beta(2)$$
$$\psi(1)\psi(2)\sigma_+(1, 2)$$
$$\psi(1)\psi(2)\sigma_-(1, 2)$$

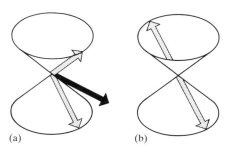

(a)   (b)

Fig. 13.27 (a) The two spins described by the + wavefunction are aligned so as to result in a net resultant angular momentum. (b) The two spins described by − are aligned so that their angular momenta cancel. They are perfectly antiparallel.

Now we take the Pauli principle into account. It says that if a wavefunction is to be acceptable (for electrons), it must change sign when the electrons are exchanged. In each case, exchanging the labels 1 and 2 converts the factor $\psi(1)\psi(2)$ into $\psi(2)\psi(1)$, which is the same, because the order of multiplying the functions does not change the value of the product. The same is true of $\alpha(1)\alpha(2)$ and $\beta(1)\beta(2)$. Therefore, the first two overall products are not allowed, because they do not change sign. The combination $\sigma_+(1, 2)$ changes to

$$\sigma_+(2, 1) = \alpha(2)\beta(1) + \beta(2)\alpha(1) = \sigma_+(1, 2)$$

because it is simply the original written in a different order. The third overall product is therefore also disallowed. Finally, consider $\sigma_-(1, 2)$:

$$\sigma_-(2, 1) = \alpha(2)\beta(1) - \beta(2)\alpha(1)$$
$$= -\{\alpha(1)\beta(2) - \beta(1)\alpha(2)\}$$
$$= -\sigma_-(1, 2)$$

and so it does change sign (it is 'antisymmetric'). Therefore the product $\psi(1)\psi(2)\sigma_-(1, 2)$ also changes sign under particle exchange, and therefore it is acceptable.

Now we see that only one of the four possible states are allowed by the Pauli principle, and the one that survives has one α and one β spin. This is the content of the Pauli exclusion principle. The exclusion principle is irrelevant when the orbitals occupied by the electrons are different, and both electrons may then have (but need not have) the same spin. Nevertheless, even then the overall wavefunction must still be anti-symmetric overall, and must still satisfy the Pauli principle itself.

Further reading

W. G. Richards and P. R. Scott, *Structure and spectra of atoms*. Wiley, London (1976).

G. Herzberg, *Atomic spectra and atomic structure*. Dover, New York (1944).

P. W. Atkins, *Molecular quantum mechanics* (2nd edn). Oxford University Press (1983).

M. Karplus and R. N. Porter, *Atoms and molecules*. Benjamin, New York (1970).

C. Candler, *Atomic spectra and the vector model*. Hilger and Watts, London (1964).

H. G. Kuhn, *Atomic spectra* (2nd edn). Longman, London (1969).

E. U. Condon and H. Odabaşi, *Atomic structure*. Cambridge University Press (1980).

Chemical consequences of atomic structure

R. J. Puddephatt and P. K. Monaghan, *The periodic table of the elements*. Oxford University Press (1986).

D. F. Shriver, P. W. Atkins, and C. H. Langford, *Inorganic chemistry*. Oxford University Press and W. H. Freeman & Co. (1990).

K. D. Sen, Electronegativity. In *Structure and bonding* **6,** Springer (1987).

Data

C. E. Moore, *Atomic energy levels*. NBS-Circ. 467, Washington, (1949, 1952, and 1958).

I. R. Stringanov and N. S. Sventitskii, *Tables of spectral lines of neutral and ionized atoms*. Plenum, New York (1968).

S. Bashkin and J. O. Stonor, Jr, *Atomic energy levels and Grotrian diagrams*. North-Holland, Amsterdam (1975 *et seq.*)

Exercises

13.1 Calculate the wavelength of the line with $n = 4$ in the Balmer series of the spectrum of atomic hydrogen.

13.2 The frequency of one of the lines in the Paschen series of the spectrum of atomic hydrogen is 2.7415×10^{15} Hz.

Calculate the principal quantum number of the upper state in the transition.

13.3 One of the terms of the H atom is at 27 414 cm^{-1}. What is the value of the term with which it combines to produce light of wavelength 486.1 nm?

13.4 When 58.4 nm ultraviolet radiation from a helium lamp is directed on to a sample of krypton, electrons are ejected at 1.59×10^6 m s^{-1}. Calculate the ionization energy of krypton.

13.5 By differentiation of the 2s radial wavefunction, show that it has a minimum in its amplitude and locate it.

13.6 Locate the radial nodes in the 3s orbital of an H atom.

13.7 What is the orbital angular momentum of an electron in the orbitals (a) 1s, (b) 3s, (c) 3d, (d) 2p, (e) 3p? Give the numbers of angular and radial nodes in each case.

13.8 Calculate the permitted values of j for (a) a d electron, (b) an f electron.

13.9 An electron in two different levels of an atom is known to have $j = \frac{3}{2}$ and $\frac{5}{2}$. What is its orbital angular momentum quantum number of the electron?

13.10 What are the allowed total angular momentum quantum numbers of a composite system in which $j_1 = 5$ and $j_2 = 3$?

13.11 State the orbital degeneracy of the levels in the hydrogen that have energy (a) $-hc\mathcal{R}_H$, (b) $-hc\mathcal{R}_H/9$, and (c) $-hc\mathcal{R}_H/25$.

13.12 What information does the term symbol $^1D_2$ provide about the angular momentum of an atom?

13.13 At what radius does the probability of finding an electron at a point in the H atom fall to 50 per cent of its maximum value?

13.14 At what radius in the H atom does the radial distribution function of the ground state have (a) 50 per cent, (b) 75 per cent of its maximum value.

13.15 Which of the following transitions are allowed in the normal electronic emission spectrum of an atom: (a) $2s \rightarrow 1s$, (b) $2p \rightarrow 1s$, (c) $3d \rightarrow 2p$, (d) $5d \rightarrow 2s$, (e) $5p \rightarrow 3s$?

13.16 How many electrons can occupy the following subshells: (a) 1s, (b) 3p, (c) 3d, and (d) 6g?

13.17 Give the electron configurations of the ground states of the first 18 elements in the periodic table.

13.18 Suppose that an atom has (a) 2, (b) 3, (c) 5 electrons in different orbitals. What are the possible values of the total spin quantum number S? What is the multiplicity in each case?

13.19 What values of J may occur in the terms (a) ^1S, (b) ^2P, (c) ^3P, (d) ^3D, (e) ^4D? How many states (distinguished by the quantum number M_J) belong to each level?

13.20 Give the possible term symbols for (a) Li [He]$2s^1$, (b) Na [Ne]$3p^1$, (c) Sc [Ar]$3d^14s^2$, and (d) Br [Ar]$3d^{10}4s^24p^5$.

13.21 The energy of an electron increases by 2.23×10^{-22} J when a magnetic field of 12.0 T is applied. What is the value of m_l of this electron?

13.22 Calculate the magnetic induction B required to produce a splitting of 1.0 cm^{-1} between the states of a P term.

Problems

Numerical problems

13.1 The 'Humphreys series' is another group of lines in the spectrum of atomic hydrogen. It begins at 12 368 nm and has been traced to 3281.4 nm. What are the transitions involved? What are the wavelengths of the intermediate transitions?

13.2 A series of lines in the spectrum of atomic hydrogen lies at 656.46 nm, 486.27 nm, 434.17 nm, and 410.29 nm. What is the wavelength of the next line in the series? What is the ionization energy of the atom when it is in the lower state of the transitions?

13.3 The Li^{2+} ion is hydrogenic and has a Lyman series at 740 747 cm^{-1}, 877 924 cm^{-1}, 925 933 cm^{-1}, and beyond. Show that the energy levels are of the form $\mathcal{R}/n^2$ and find the value of R for this ion. Go on to predict the wavenumbers of the two longest-wavelength transitions of the Balmer series of the ion and find the ionization energy of the ion.

13.4 A series of lines in the spectrum of neutral Li atoms arise from combinations of $1s^22p^1$ ^2P with $1s^2nd^1$ ^2D and occur at 610.36 nm, 460.29 nm, and 413.23 nm. The d orbitals are hydrogenic. It is known that the ^2P term lies at 670.8 nm

above the ground state, which is $1s^22s^1$ ^2S. Calculate the ionization energy of the ground-state atom.

13.5 The characteristic emission from K atoms when heated is purple and lies at 770 nm. On close inspection, the line is found to have two closely spaced components, one at 766.70 nm and the other at 770.11 nm. Account for this observation, and deduce what information you can.

13.6 Calculate the mass of the deuteron given that the first line in the Lyman series of H lies at 82 259.098 cm^{-1} whereas that of D lies at 82 281.476 cm^{-1}. Calculate the ratio of the ionization energies of H and D.

13.7 Positronium consists of an electron and a positron (same mass, opposite charge) orbiting round their common centre of mass. The broad features of the spectrum are therefore expected to be hydrogen-like, the differences arising largely from the mass differences. Predict the wavenumbers of the first three lines of the Balmer series of positronium. What is the ionization energy of the ground state of positronium?

13.8 In 1976 it was mistakenly believed that the first of the 'superheavy' elements had been discovered in a sample of mica. Its atomic number was believed to be 126. What is the most probable distance of the innermost electrons from the nucleus of an atom of this element? (In such elements, relativistic effects are very important, but ignore them here.)

Theoretical problems

13.9 Is an electron further from the nucleus on average when it is in a 2s orbital or a 2p orbital?

13.10 What is the most probable point (not radius) that a 2p electron will be found in the hydrogen atom?

13.11 One of the most famous of the obsolete theories of the hydrogen atom was proposed by Bohr. It has been replaced by quantum mechanics, but by a remarkable coin- cidence (not the first one where the Coulomb potential is concerned), the energies it predicts agree exactly with those obtained from the Schrödinger equation. In the Bohr atom, an electron travels in a circle around the nucleus. The coulombic force of attraction ($Ze^2/4\pi\varepsilon_0 r^2$) is balanced by the centrifugal effect of the orbital motion. Bohr proposed that the angular momentum is limited to integral values of $\hbar$. When the two forces are balanced, the atom remains in a 'stationary state' until it makes a spectral transition. Calculate the energies of a hydrogenic atom using the Bohr model.

13.12 The Bohr model of the atom is specified in Problem 13.11. What features of it are untenable according to quantum mechanics? How does the Bohr ground state differ from the actual ground state? Is there an experimental distinction between the Bohr and quantum mechanical models of the ground state?

Molecular structure

14

Check-list of key ideas

1. The *Born–Oppenheimer approximation* for the separation of electronic and nuclear motion (Section 14.1).

2. The approximation of molecular orbitals as *linear combinations of atomic orbitals* (Section 14.2).

3. The formation of *bonding orbitals* and *antibonding orbitals* in the hydrogen molecule-ion (eqns 4 and 6) and the corresponding electron distributions.

4. The role of *constructive interference* and *destructive interference* between atomic orbitals in bond formation (Section 14.2).

5. The use of a *molecular orbital energy level diagram* to derive the electron configurations of molecules (Sections 14.3 and 14.4).

6. The definition of *bond order* (eqn 8).

7. The formation of σ *orbitals* and π *orbitals* in homonuclear diatomic molecules (Section 14.4) and the assessment of overlap in terms of the *overlap integral* (eqn 9).

8. The ground state electron configurations of *homonuclear diatomic molecules* and their bond orders (Section 14.4).

9. The *notation* used to specify molecular orbitals, including the use of g and u to denote orbital *parity* (Section 14.5).

10. The *term symbols* used to classify overall molecular states (Section 14.5).

11. The ground state electron configurations of *heteronuclear diatomic molecules* and the *polarity* of covalent bonds (Section 14.6).

12. The *variation principle* for calculating the optimum orbitals and energies (Section 14.6) and the form and solution of the *secular equations* (eqn 14) using a *secular determinant* (eqn 15).

13. The use of ionization energies to judge the compositions and energies of molecular orbitals (Example 14.7).

14. The use of *hybrid orbitals* in the description of bond formation (Section 14.7).

15. The use of a *Walsh diagram* to show the dependence of orbital energy on molecular shape, and to judge the shapes of polyatomic molecules (Section 14.8).

16. The analysis of bonding in polyatomic molecules in terms of *hybrid orbitals* (Section 14.9) and the relation between bond angle and composition of hybrids (eqn 19).

17. The description of the *carbon–carbon double bond* (Section 14.9) and its *torsional rigidity*.

18. *The Hückel approximation* for the description of delocalized bonding, the identification of *frontier orbitals,* and the calculation of the π-*electron binding energy* (Section 14.10).

19. The molecular orbital description of *aromatic stability* (Section 14.10).

20. The classification of substances according to their electrical conductivity and the *band theory* of solids (Section 14.11).

21. The *band gap* in solids and the *Fermi–Dirac distribution* (eqn 22) for the population of bands.

22. The classification of *semiconductors* as *p-type* and *n-type* and the role of dopants (Section 14.11).

Now we turn to **valence theory**, the theory of the origin of the strengths, numbers, and three-dimensional arrangement of chemical bonds between atoms. The quantum mechanical description of chemical bonding has become highly developed through the use of computers, and it is now possible to consider the structures of molecules of almost any complexity. We shall concentrate on the quantum mechanical description of the **covalent bond**, which was identified by G. N. Lewis (in 1916, before quantum mechanics was established fully) as an electron pair shared between two neighbouring atoms.

In Chapter 13 we took the hydrogen atom as the primitive species for discussing atomic structure, and based our discussion of complex atoms on what we learned from it. In this chapter we use the simplest molecule of all, the hydrogen molecule-ion H_2^+, to introduce the essential features of bonding, and then use it as a guide to the structures of more complex systems.

The hydrogen molecule-ion

The hydrogen molecule-ion consists of two protons and one electron. However, right at the outset we shall make an approximation that greatly simplifies its treatment and establishes a framework for the language of 'bond lengths' and 'molecular structure' that is used throughout chemistry.

14.1 The Born–Oppenheimer approximation

The **Born–Oppenheimer approximation** supposes that the nuclei, being so much heavier than the electron, move relatively slowly and may be treated as stationary as the electron moves around them. We can therefore choose the nuclei to have a definite separation R and solve the Schrödinger equation for the electron alone, which is far easier than trying to solve the complete Schrödinger equation by treating all three particles on an equal footing. The approximation is quite good for ground-state molecules, for calculations suggest that the nuclei in H_2^+ move through only 1 pm while the electron speeds through about 1000 pm, and so the error of assuming that the nuclei are stationary is small. In molecules other than H_2^+ the nuclei are even heavier, and the approximation is generally (but not always) better. Exceptions to its validity include certain excited states of polyatomic molecules and the ground states of cations; both types of species are important when considering photoelectron spectroscopy (Section 17.9) and mass spectroscopy.

The Born–Oppenheimer approximation reduces the full problem to a single-particle Schrödinger equation for an electron in the field of two stationary protons at a separation R. The potential energy of the electron is

$$V = \frac{-e^2}{4\pi\varepsilon_0}\left(\frac{1}{r_A} + \frac{1}{r_B}\right) \qquad (1)$$

where r_A and r_B are the electron's distances from the nuclei A and B. This expression is used in the one-particle Schrödinger equation

$$-\frac{\hbar^2}{2m_e}\nabla^2\psi + V\psi = E\psi \qquad (2)$$

and exact solutions can be obtained. The total energy of the molecule at the selected separation R is then the sum of the eigenvalue E and the nucleus–nucleus repulsion

$$V_{\text{nuc–nuc}} = \frac{+e^2}{4\pi\varepsilon_0} \times \frac{1}{R} \qquad (3)$$

The Born–Oppenheimer approximation allows us to decide on a particular internuclear separation, and to solve the Schrödinger equation for the electron distribution. Then we can choose a different separation and repeat the calculation, and so on. In this way we can calculate how the energy of the molecule—with the nuclear kinetic energy ignored—varies with bond length (and, in more complex molecules, with angles too), and obtain the **molecular potential energy curve** (Fig. 14.1). We can then identify the equilibrium bond length of the molecule with the lowest point on this curve.

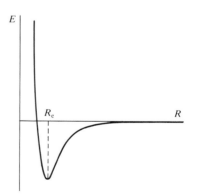

Fig. 14.1 A molecular potential energy curve. The equilibrium bond length R_e corresponds to the energy minimum.

14.2 The molecular orbital approximation

The one-electron wavefunctions obtained by solving the Schrödinger equation are called **molecular orbitals**. A molecular orbital ψ gives, through the value of ψ^2, the distribution of the electron in the molecule. A molecular orbital is like an atomic orbital, but spreads throughout the molecule.

Exact, analytical molecular orbitals may be obtained for H_2^+ (within the Born–Oppenheimer approximation), but they are very complicated functions and do not give much insight into the form of the orbitals and the contributions to the energy. Therefore, we shall adopt a simpler procedure that, while more approximate, gives more insight.

Linear combinations of atomic orbitals

The approximation we shall adopt is based on the fact that when the electron is very close to nucleus A, the term $1/r_A$ in V is very much bigger than $1/r_B$. That being so, the potential energy in eqn 1 reduces to

$$V \approx \frac{-e^2}{4\pi\varepsilon_0} \times \frac{1}{r_A}$$

The Schrödinger equation for the electron in the molecule is then the same as that for an isolated H atom, and its lowest energy solution is a $1s$ orbital on A, which we write $\psi_{1s}(A)$. Thus, close to A, the molecular orbital resembles an atomic $1s$ orbital. Likewise, close to B the molecular orbital resembles a $1s$ orbital on B, $\psi_{1s}(B)$. This discussion suggests that we can approximate the overall wavefunction ψ as a sum of the two atomic orbitals:

$$\psi = N\{\psi_{1s}(A) + \psi_{1s}(B)\} \tag{4}$$

where N is a normalization factor. In accord with the preceding discussion, when the electron is close to A its distance from B is large, $\psi_{1s}(B)$ is small, and therefore the wavefunction is almost pure $\psi_{1s}(A)$. Similarly, ψ is almost pure $\psi_{1s}(B)$ close to B.

The technical term for a sum of the kind in eqn 4 is a *linear combination of atomic orbitals* (LCAO), and we shall use that name from now on. An approximate molecular orbital formed from a linear combination of atomic orbitals is called an **LCAO-MO**. A molecular orbital that has cylindrical symmetry around the internuclear axis, such as the one we are discussing, is called a **σ orbital** (because it resembles an s orbital when viewed along the axis). Since the σ orbital in eqn 4 is formed from $1s$ orbitals, it is more fully described as a $1s\sigma$ orbital.

Example 14.1: *Normalizing a molecular orbital*

Normalize the molecular orbital in eqn 4.

Answer. We need to find the factor N in

$$\psi = N\{\psi_{1s}(A) + \psi_{1s}(B)\}$$

which ensures that

$$\int \psi^2 \, d\tau = 1$$

When we substitute the wavefunction, we find

$$N^2\left\{ \int \psi_{1s}(A)^2 \, d\tau + \int \psi_{1s}(B)^2 \, d\tau + 2 \int \psi_{1s}(A)\psi_{1s}(B) \, d\tau \right\} = N^2\{1 + 1 + 2S\}$$

since the atomic orbitals are normalized and where

$$S = \int \psi_{1s}(A)\psi_{1s}(B) \, d\tau$$

Therefore, the normalization factor is

$$N = \frac{1}{2^{1/2}(1+S)^{1/2}}$$

Comment. We shall shortly meet S as the 'overlap integral'. In H_2^+ its value is about 0.59, so $N = 0.56$.

Exercise. Normalize an orbital of the form $\psi_{1s}(A) - \psi_{1s}(B)$.
$$[N = 1/2^{1/2}(1 - S)^{1/2} = 1.10 \text{ in } H_2^+]$$

We should keep in mind the approximations we have made so far. The Born–Oppenheimer approximation separates the electronic and nuclear motions and allows us to talk in terms of the molecular orbitals of the electron in the field of the stationary nuclei. The LCAO approximation goes one step further and approximates a molecular orbital as a sum of atomic orbitals. It allows us to use atomic orbitals to discuss the distribution of electrons in molecules.

σ orbitals

According to the Born interpretation, the probability density of the electron in H_2^+ is proportional to the square of its wavefunction. The probability density of the LCAO-MO $1s\sigma$ orbital in eqn 4 is

$$\psi^2 = N^2\{\psi_{1s}(A) + \psi_{1s}(B)\}^2$$
$$= N^2\{\psi_{1s}(A)^2 + \psi_{1s}(B)^2 + 2\psi_{1s}(A)\psi_{1s}(B)\} \tag{5}$$

From Example 14.1 we know that $N^2 = 0.31$. Since $\psi_{1s}(A)$ is the function

$$\psi_{1s}(A) = \left(\frac{1}{\pi a_0^3}\right)^{1/2} e^{-r_A/a_0}$$

with r_A the distance of the electron from A, and similarly for $\psi_{1s}(B)$, it is easy to evaluate ψ and the probability density at any point. The result is shown in Fig. 14.2.

An important feature of eqn 5 becomes apparent when we examine the probability density in the internuclear region, where both atomic orbitals have similar amplitudes. According to eqn 5, the total probability density is proportional to the sum of

(a) $\psi_{1s}(A)^2$, the probability density if the electron were confined to the orbital on A.

(b) $\psi_{1s}(B)^2$, the probability density if the electron were confined to the orbital on B.

(c) $2\psi_{1s}(A)\psi_{1s}(B)$, an extra contribution to the density.

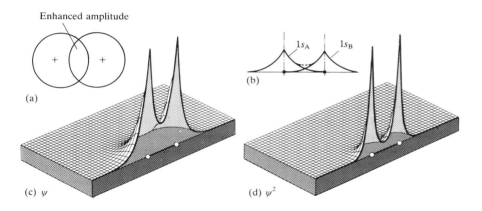

(a) Enhanced amplitude

(b) $1s_A$ $1s_B$

(c) ψ

(d) ψ^2

Fig. 14.2 (a) The orbital overlap responsible for bonding in the hydrogen molecule-ion and (b) the constructive interference in the internuclear region. (c) The orbital amplitude in a plane containing the two nuclei, and (d) the corresponding electron density. Diagram (b) is a side view of (c).

The last contribution, the **overlap density**, is crucial, since it represents an enhancement of the probability of finding the electron in the internuclear region above what it would be if it were confined to one of the two atoms. That is, because the electron is free to move from one nucleus to the other, the electron density in the internuclear region is increased. The enhancement can be traced to the **constructive interference** of the two atomic orbitals: each has a positive amplitude in the internuclear region, and so the total amplitude is greater there than if the electron were confined to a single atomic orbital.

We shall constantly use the result that electrons accumulate in regions where atomic orbitals overlap and interfere constructively. The accumulation of electron density between the nuclei puts the electron in a position where it interacts strongly with both nuclei. Hence the energy of the molecule is lower than that of the separate atoms, where each electron can interact strongly with only one nucleus.[1]

Bonding orbitals

The $1s\sigma$ orbital is an example of a **bonding orbital**, an orbital which, if occupied, contributes to a lowering of the energy of a molecule. An electron that occupies a σ orbital is called a **σ electron**, and if that is the only electron present in the molecule (as in the ground state of H_2^+), we report the configuration of the molecule as $1s\sigma^1$.

The energy of the $1s\sigma$ orbital decreases as R decreases from large values because electron density accumulates in the internuclear region as the two atomic orbitals increasingly overlap. However, at small separations, there is too little space between the nuclei for significant accumulation of electron density there. In addition, the nucleus–nucleus repulsion $V_{\text{nuc–nuc}}$ becomes large. As a result, the total energy rises at short distances, and there is a minimum in the potential energy curve. The internuclear separation at the minimum of the curve is called the **equilibrium bond length** R_e and the depth of the minimum is called the **bond dissociation energy** D_e.[2] Calculations on H_2^+ give $R_e = 130$ pm and $D_e = 1.77$ eV (171 kJ mol^{-1}); the experimental values are 106 pm and 2.6 eV, and so this simple LCAO-MO description of the molecule, while inaccurate, is not absurdly wrong.

---

[1] Unfortunately, this neat explanation is probably incorrect in the case of H_2^+ (at least). This is because shifting an electron away from a nucleus into the internuclear region raises its potential energy. The modern explanation is more subtle, and does not emerge from the simple LCAO treatment given here. It seems that at the same time as the electron shifts into the internuclear region, the atomic orbitals shrink. This orbital shrinkage improves the electron–nucleus attraction more than it is damaged by the migration to the internuclear region, and so there is a net lowering of potential energy. The kinetic energy of the electron is also modified, but it is dominated by the potential energy.

Even though the details are obscure for more complex molecules, it is generally found that bonding occurs when electrons accumulate between nuclei even though the actual cause of the bonding may be an accompanying shrinkage of the orbitals. Although orbital shrinkage reduces orbital overlap, it increases the electron–nucleus attraction and results in a new lowering of the energy of the molecule. Therefore, throughout the following discussion we ascribe the strength of chemical bonds to the accumulation of electron density in the internuclear region and leave open the question whether in molecules more complicated than H_2^+ the true source of energy lowering is that accumulation itself or some indirect but related effect.

[2] In Chapter 16 we shall see that the measured dissociation energy, the true dissociation energy D_0 is always smaller than D_e because the vibration of the molecule must be taken into account. This is explained in Section 16.7; in H_2^+ the two values differ by 0.3 eV, 30 kJ mol^{-1}.

Antibonding orbitals

The argument that led to the expression of ψ as the *sum* of two atomic orbitals is equally well satisfied by writing a molecular orbital as the *difference*

$$\psi' = N\{\psi_{1s}(A) - \psi_{1s}(B)\} \tag{6}$$

since, in this case too, the molecular orbital resembles one or other of the atomic orbitals close to the two nuclei. However, this linear combination corresponds to a higher energy than the orbital in eqn 4, and is in fact a good approximation to the next-higher exact solutions of the Schrödinger equation for H_2^+.

Since ψ' is cylindrically symmetrical around the internuclear axis it is also a σ orbital, and since it is formed from $1s$ orbitals it too is a $1s\sigma$ orbital. We distinguish it from the bonding $1s\sigma$ orbital by denoting it $1s\sigma^*$ (the asterisk is a label here, not complex conjugation). The normalization factor is given in the exercise of Example 14.1, and for H_2^+ is 1.10.

We can see from eqn 6 that the $1s\sigma^*$ orbital has a node where $\psi_{1s}(A)$ and $\psi_{1s}(B)$ cancel. Consequently there is zero probability of finding the electron

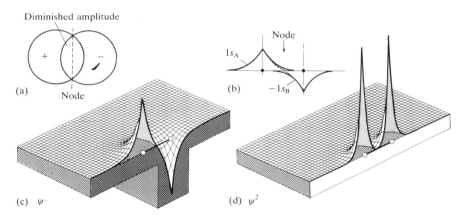

(a)   (b)   (c) ψ (d) ψ^2

Fig. 14.3 (a) The orbital overlap responsible for antibonding in the hydrogen molecule-ion and (b) the destructive interference in the internuclear region. (c) The orbital amplitude in a plane containing the two nuclei, and (d) the corresponding electron density. Diagram (b) is a side view of (c).

half-way between the nuclei if it occupies this orbital. We express this reduction in amplitude by saying that because the atomic orbitals are combined with opposite signs, they interfere *destructively* where they overlap (Fig. 14.3). In terms of the probability density,

$$\begin{aligned}
\psi'^2 &= N^2\{\psi_{1s}(A) - \psi_{1s}(B)\}^2 \\
&= N^2\{\psi_{1s}(A)^2 + \psi_{1s}(B)^2 - 2\psi_{1s}(A)\psi_{1s}(B)\}
\end{aligned} \tag{7}$$

and the third term reduces the probability of finding the electron between the nuclei relative to its value if the electron were confined to one of the atomic orbitals.

The $1s\sigma^*$ orbital is an example of an **antibonding orbital**, an orbital that, if occupied, raises the energy of the molecule relative to the separated atoms. Antibonding orbitals are labelled with a *. An antibonding electron destabilizes the molecule relative to the separated atoms (Fig. 14.4). This is partly because, since it is excluded from the internuclear region, the

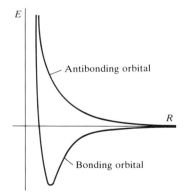

Fig. 14.4 The molecular potential energy curve for the hydrogen molecule-ion showing the variation of the energy of the molecule as the bond length is changed and the electron is in either the bonding or the antibonding orbital.

electron is distributed largely outside the bonding region. In effect, the electron pulls the nuclei apart from outside. The combined effect of the electron distribution and the internuclear repulsion results in an antibonding orbital being more strongly antibonding than the corresponding bonding orbital is bonding.

The structures of diatomic molecules

In Chapter 13 we used the hydrogenic atomic orbitals and the building-up principle to deduce the ground electronic configurations of many-electron atoms. We can do the same for many-electron diatomic molecules (such as H_2 with 2 electrons and Br_2 with 70), but using the H_2^+ molecular orbitals instead. As we shall illustrate in the following sections, first with H_2 and then with heavier molecules, the general procedure is to construct molecular orbitals by combining the atomic orbitals supplied by the atoms. The electrons supplied by the atoms are then accommodated in the orbitals so as to achieve the lowest overall energy subject to the constraint of the Pauli exclusion principle that no more than two electrons may occupy a single orbital (and then must be paired). As in the case of atoms, if several degenerate molecular orbitals are available, we add the electrons to each *individual* orbital before doubly occupying any one orbital (because that reduces electron–electron repulsions). We also take note of Hund's rule (Section 13.4), that if electrons do occupy different degenerate orbitals, they do so with parallel spins.

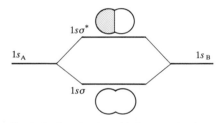

Fig. 14.5 A molecular orbital energy level diagram for orbitals constructed from (1s,1s)-overlap, the separation of the levels corresponding to the equilibrium bond length.

14.3 The hydrogen and helium molecules

We shall illustrate the general procedure by considering H_2, the simplest many-electron diatomic molecule. First, we need to build the molecular orbitals. Since each H atom of H_2 contributes a $1s$ orbital (as in H_2^+), we can form the $1s\sigma$ and $1s\sigma^*$ orbitals from them, as we have seen already. At the experimental internuclear separation these orbitals will have the energies shown in Fig. 14.5, which is called a **molecular orbital energy level diagram**. Note that from two atomic orbitals we can build two molecular orbitals. This is a special case of the general rule that from N atomic orbitals we can build N molecular orbitals.

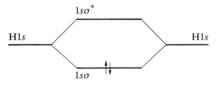

Fig. 14.6 The ground electronic configuration of H_2 is obtained by accommodating the two electrons in the lowest available orbital (the bonding orbital).

Electron configurations

There are two electrons to accommodate, and both can enter $1s\sigma$ by pairing their spins. The ground state configuration is therefore $1s\sigma^2$ (Fig. 14.6) and the atoms are joined by a bond consisting of an electron pair in a bonding σ orbital. This approach shows that an electron pair—the focus of Lewis's account of chemical bonding—represents the maximum number of electrons that can enter a bonding molecular orbital.

The same argument shows why He does not form diatomic molecules. Each He atom contributes a $1s$ orbital, and so the same two molecular orbitals can be constructed. (They differ in detail from those in H_2 because the He$1s$ orbitals are more compact, but the general shape is the same, and we can use the same energy level diagram, Fig. 14.5, in the discussion.) There are four electrons to accommodate. Two can enter the $1s\sigma$ orbital, but then it is full, and the next two must enter the $1s\sigma^*$ orbital (Fig. 14.7).

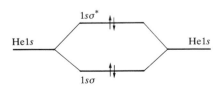

Fig. 14.7 The ground electronic configuration of the four-electron molecule He$_2$ has two bonding electrons and two antibonding electrons. It has a higher energy than the separated atoms, and so He$_2$ is unstable.

The ground electronic configuration of He_2 is therefore

$$He_2\ 1s\sigma^2 1s\sigma^{*2}$$

We see that there is one bond and one antibond. Since an antibond is slightly more antibonding than a bond is bonding, the He_2 molecule has a higher energy than the separated atoms, and so does not form.

Example 14.2: *Judging the stability of diatomic molecules*

Suppose that only the valence *s* orbitals contribute to molecular orbitals, and decide whether Li_2 and Be_2 are likely to exist.

Answer. Each molecular orbital is built from $2s$ atomic orbitals, which give one bonding and one antibonding combination. Each Li atom supplies one valence electron, which fills the $2s\sigma$ orbital to give the configuration $2s\sigma^2$, which is bonding. Each Be atom provides two valence electrons, which fill the bonding and antibonding combinations, resulting in no net bond.

Comment. In fact, as we shall see, Be_2 does exist, because the $2p$ orbitals contribute to the orbitals and provide another bonding orbital.

Exercise. Is LiH likely to exist if the Li atom uses only its $2s$ orbital for bonding? [Yes, $(Li2s, H1s)\sigma^2$]

Bond order

A measure of the net bonding in a diatomic molecule is its **bond order** b, defined as

$$b = \tfrac{1}{2}(n - n^*) \tag{8}$$

n is the number of electrons in bonding orbitals and n^* the number in antibonding orbitals. Thus each electron pair in a bonding orbital increases the bond order by 1 and each pair in an antibonding orbital decreases it by 1. For H_2, $b = 1$, corresponding to a single bond, H—H, between the two atoms. In He_2, $b = 0$, and there is no bond.

As we shall see, the bond order is a useful parameter for discussing the characteristics of bonds, since it correlates with bond length, and the greater the bond order between atoms of a given pair of elements the shorter the bond. It also correlates with bond strength, and the greater the bond order the greater the strength.

14.4 Period-2 diatomic molecules

We now see how the concepts we have introduced apply to **homonuclear diatomic molecules** in general, which are diatomic molecules formed from identical atoms, such as N_2 and Cl_2. In line with the building-up procedure, we first consider the molecular orbitals that may be formed and do not (at this stage) trouble about how many electrons are available. The atomic orbitals available are the **core orbitals**, those of the inner, closed shells, the **valence orbitals**, those of the valence shell, and the **virtual orbitals**, those of the atom that are unoccupied in its ground state. In elementary treatments (but not in the sophisticated treatments using computers), the core orbitals are ignored as being too compact to have significant overlap with orbitals on other atoms. The virtual orbitals are ignored on the grounds that they are too high in energy to participate in bonding. Therefore we form molecular orbitals using only the valence orbitals.

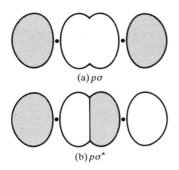

(a) pσ

(b) pσ*

Fig. 14.8 (a) The interference leading to the formation of a 2pσ bonding orbital and (b) the corresponding antibonding orbital.

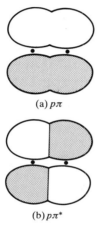

(a) pπ

(b) pπ*

Fig. 14.9 (a) The interference leading to the formation of a 2pπ bonding orbital and (b) the corresponding antibonding orbital.

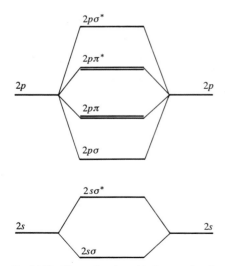

Fig. 14.10 The molecular orbital energy level diagram for homonuclear diatomic molecules. As remarked in the text, this diagram should be used for O_2 and F_2.

In Period 2, the valence orbitals are $2s$ and $2p$. The $2s$ orbitals on the two atoms overlap to give a bonding $2s\sigma$ orbital and an antibonding $2s\sigma^*$ orbital, in exactly the same way as we have already seen for $1s$ orbitals. The new feature of these molecules however, is that p orbitals are also available for bonding. The two $2p_z$ orbitals directed along the internuclear axis overlap strongly, and may do so either constructively or destructively, to give a bonding or antibonding $2p\sigma$ orbital respectively (Fig. 14.8).

π orbitals

Now consider the $2p_x$ and $2p_y$ orbitals of each atom, which are perpendicular to the internuclear axis and may overlap broadside-on. This overlap may be constructive or destructive, and results in a bonding or an antibonding **π orbital**. The notation π is the analogue of p in atoms, for when viewed along the axis of the molecule, a π orbital looks like a p orbital (Fig. 14.9). the two $2p_x$ orbitals overlap to give a bonding and antibonding $2p_x\pi$ orbital, and the two $2p_y$ orbitals overlap to give two $2p_y\pi$ orbitals. The $2p_x\pi$ and $2p_y\pi$ orbitals have the same energy, as do their antibonding partners.

In some cases, $2p\pi$ orbitals are less strongly bonding than $2p\sigma$ orbitals because their maximum overlap occurs off-axis, away from the optimum bonding region. This suggests that the molecular orbital energy level diagram ought to be as shown in Fig. 14.10. However, there is no guarantee that this order of energies should prevail, and it is found experimentally (by spectroscopy) and by detailed calculation that the order varies along Period 2 (Fig. 14.11). The order shown in Fig. 14.12 is therefore appropriate as far as N_2, and Fig. 14.10 applies for O_2 and F_2. We shall see that the relative order is controlled by the separation of the $2s$ and $2p$ orbitals in the atoms,

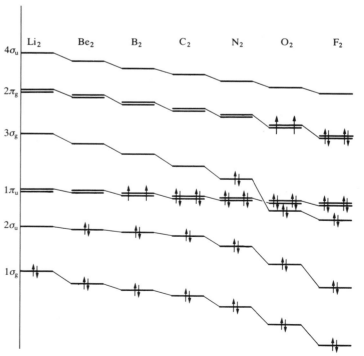

Fig. 14.11 The variation of the orbital energies of Period-2 homonuclear diatomics.

which increases across the period. The consequent switch in order occurs at about N_2.

s,p overlap

The origin of the variation in relative energy levels lies in the ability of both $2s$ and $2p$ orbitals to contribute to the same molecular orbitals. So far, we have considered s,s overlap and p,p overlap as occurring distinctly, and have constructed distinct $2s\sigma$ and $2p\sigma$ orbitals. One reason for this is that strong bonds arise from the overlap of atomic orbitals having similar energies (a feature we shall confirm shortly). In homonuclear diatomic molecules, the energies of the $2s$ orbitals of the two atoms are identical, as are those of the $2p$ orbitals, and so the principal contributions to molecular orbitals will come from their respective overlaps. However, we should really write a σ orbital as a linear combination of *all* possible atomic orbitals that have the correct symmetry to contribute, and express it as

$$\psi = c_{2s}(A)\psi_{2s}(A) + c_{2p_z}(A)\psi_{2p_z}(A) + c_{2s}(B)\psi_{2s}(B) + c_{2p_z}(B)\psi_{2p_z}(B)$$

A $2s$ and a $2p_z$ orbital can both contribute to a σ orbital because they both have cylindrical symmetry around the internuclear axis. In other words, s and p_z orbitals on different atoms have nonzero overlap (Fig. 14.13a) and can participate in orbital formation with each other. On the other hand, $2s,2p_x$ overlap makes no contribution to bonding because the effect of the constructive overlap in one region (Fig. 14.13b) is exactly cancelled by the effect of the destructive overlap in another, and so there is no net overlap and no net contribution to bonding.

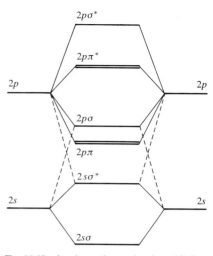

Fig. 14.12 An alternative molecular orbital energy level diagram for homonuclear diatomic molecules. As remarked in the text, this diagram should be used for diatomics as far as N_2.

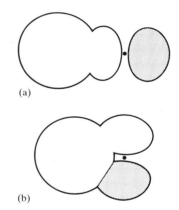

Fig. 14.13 Overlapping s- and p-orbitals. (a) End-on overlap leads to non-zero overlap and to the formation of an axially symmetric bond. (b) Broadside overlap leads to no net accumulation of electron density.

Example 14.3: *Assessing the contribution of d orbitals*

Can d orbitals contribute to σ and π orbitals in diatomic molecules?

Answer. We need to assess the symmetry of d orbitals with respect to the internuclear z axis. A d_{z^2} orbital has cylindrical symmetry around z and so can contribute to σ orbitals. The d_{xz} and d_{yz} orbitals have π symmetry with respect to the axis (Fig. 14.14), and so can contribute to π orbitals.

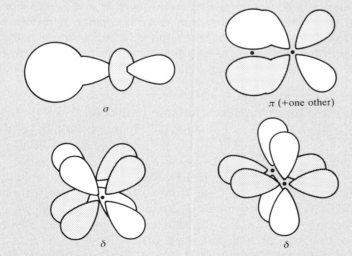

Fig. 14.14 The types of molecular orbital to which d orbitals can contribute. The σ and π combinations can be formed with s, p, and d orbitals of the appropriate symmetry but the δ orbitals arise from the overlap of d orbitals on different atoms.

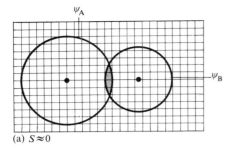

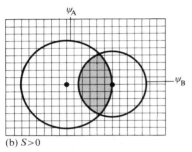

(a) $S \approx 0$

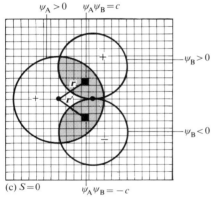

(b) $S > 0$

(c) $S = 0$

Fig. 14.15 A schematic representation of the contributions to the overlap integral. (a) $S \approx 0$ because the orbitals are far apart and their product is always small. (b) S is large (but less than 1) because $\psi(A)\psi(B)$ is large over a substantial region. (c) $S = 0$ because the positive region of overlap is exactly cancelled by the negative region.

Exercise. Sketch the 'δ orbitals' that may be formed by the remaining two d orbitals (and which contribute to bonding in some d-metal cluster compounds). [Fig. 14.14]

The overlap integral

The extent to which two orbitals overlap is measured by the **overlap integral** S:

$$S = \int \psi(A)^* \psi(B) \, d\tau \qquad (9)$$

If the atomic orbital $\psi(A)$ on A is small wherever the orbital $\psi(B)$ on B is large, or vice versa, the product of their amplitudes is everywhere small and the integral—the sum of these products—is small. If $\psi(A)$ and $\psi(B)$ are simultaneously large in some region of space, S may be large. This is illustrated in Fig. 14.15a and b. If the two normalized atomic orbitals are identical (e.g. $1s$ orbitals on the same nucleus), $S = 1$. In some cases, simple formulas can be given for overlap integrals. For example, for two hydrogenic $1s$ orbitals on nuclei separated by a distance R,

$$S = \left\{ 1 + \frac{ZR}{a_0} + \frac{1}{3}\left(\frac{ZR}{a_0}\right)^2 \right\} e^{-ZR/a_0} \qquad (10)$$

$S = 0.59$ for two $1s$ orbitals at the equilibrium bond length in H_2^+, which is an unusually large value. Typical values for orbitals with $n = 2$ are in the range 0.2 to 0.3.

Now consider the arrangement in Fig. 14.15c in which an s orbital overlaps a p orbital of a different atom. At some point $\mathbf{r}$ the product $\psi(A)^* \psi(B)$ may be large. However, there is a point $\mathbf{r}'$ where $\psi(A)^* \psi(B)$ has exactly the same magnitude but an opposite sign. When the integral is evaluated, these two contributions are added together and cancel. For every point in the upper half of the diagram, there is a point in the lower that cancels, and so $S = 0$. Therefore, there is no net overlap between the s and p orbitals in this arrangement. Orbitals for which $S = 0$ are called **orthogonal**; two different atomic orbitals on the same atom are always orthogonal.

The structures of homonuclear diatomic molecules

We show the general layout of the valence-shell atomic orbitals of Period-2 atoms on the left and right of the molecular orbital energy level diagram in Fig. 14.16. The lines in the middle are an indication of the energies of the molecular orbitals that can be formed by overlap of atomic orbitals: from the eight valence shell orbitals (four from each atom), we can form eight molecular orbitals. With the orbitals established, we can deduce the ground configurations of the molecules by adding the appropriate number of electrons to the orbitals and following the building-up rules. Charged species (such as the peroxide ion, O_2^{2-}) need either more or fewer electrons than the neutral molecules.

We shall illustrate the procedure with N_2, which has ten valence electrons. Two electrons pair, enter, and fill the $2s\sigma$ orbital; the next two enter and fill the $2s\sigma^*$ orbital. Six electrons remain. There are two $2p\pi$ orbitals, and so four electrons can be accommodated in them. The last two enter the $2p\sigma$ orbital. The ground state configuration of N_2 is therefore

$$N_2 \quad 2s\sigma^2 2s\sigma^{*2} 2p\pi^4 2p\sigma^2$$

and the bond order is

$$b = \tfrac{1}{2}(8-2) = 3$$

This bond order accords with the Lewis structure of the molecule ($N\!\equiv\!N$) and is consistent with its high dissociation energy (945 kJ mol^{-1}).

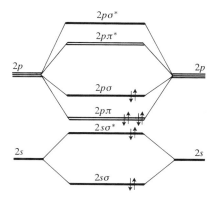

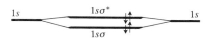

Fig. 14.16 The molecular orbital energy level diagram used for the discussion of the homonuclear diatomics of the second row. The electron configuration shown is that of N_2.

Example 14.4: *Writing the electron configuration of a diatomic molecule*

Write the ground state electron configuration of O_2 using N_2 as a guide.

Answer. The O_2 molecule ($2Z = 16$) has two more electrons than N_2, and 12 of the 16 are in its valence shell. The first 10 recreate the N_2 configuration (with a reversal of the order of the $2p\sigma$ and $2p\pi$ orbitals); the last two must enter the $2p\pi^*$ orbitals. Its configuration and bond order are therefore

$$O_2 \quad 2s\sigma^2 2s\sigma^{*2} 2p\sigma^2 2p\pi^4 2p\pi^{*2} \quad b = 2$$

Comment. The bond order accords with the classical view that oxygen has a double bond and may be denoted $O\!=\!O$. We see below that two outermost electrons are unpaired; to emphasize this feature the structure is sometimes written $O\!\cdot\!\cdot\!O$.

Exercise. Write the electron configuration of F_2 and deduce its bond order.

$$[2s\sigma^2 2s\sigma^{*2} 2p\sigma^2 2p\pi^4 2p\pi^{*4}, \ b = 1]$$

According to the building-up principle, the two $2p\pi^*$ electrons in O_2 will occupy different orbitals: one will enter $2p_x\pi^*$ and the other will enter $2p_y\pi^*$. Since they are in different orbitals, they will have parallel spins. Therefore, we can predict that an O_2 molecule will have a net spin angular momentum $S = 1$ and, in the language introduced in Section 13.6, be in a triplet state. Since electron spin is the source of a magnetic moment, we can go on to predict that oxygen should be paramagnetic,[3] which is in fact the case.

An F_2 molecule has two more electrons than an O_2 molecule and the configuration and bond order

$$F_2 \quad 2s\sigma^2 2s\sigma^{*2} 2p\sigma^2 2p\pi^4 2p\pi^{*4} \quad b = 1$$

We conclude that F_2 is a singly-bonded molecule, in agreement with its Lewis structure F—F. The low bond order is consistent with its low dissociation energy (154 kJ mol^{-1}). Ne_2 has two further electrons:

$$Ne_2 \quad 2s\sigma^2 2s\sigma^{*2} 2p\sigma^2 2p\pi^4 2p\pi^{*4} 2p\sigma^{*2} \quad b = 0$$

The zero bond order agrees with the monatomic nature of Ne.

---

[3] A paramagnetic substance tends to move into a magnetic field; a diamagnetic substance tends to move out of one. Paramagnetism, the rarer property, arises when the molecules have unpaired electron spins. Both properties are discussed in more detail in Section 22.10.

Example 14.5: *Judging the relative bond strengths of molecules and ions*

Judge whether N_2^+ is likely to have a larger or smaller dissociation energy than N_2.

Answer. Since the molecule with the larger bond order is likely to have the larger dissociation energy, we should compare their electronic configurations, and assess their bond orders. From Fig. 14.16:

$$N_2 \quad 2s\sigma^2 2s\sigma^{*2} 2p\pi^4 2p\sigma^2 \quad b = 3$$
$$N_2^+ \quad 2s\sigma^2 2s\sigma^{*2} 2p\pi^4 2p\sigma^1 \quad b = 2.5$$

Since the cation has the smaller bond order, we expect it to have the smaller dissociation energy.

Comment. The experimental dissociation energies are 945 kJ mol^{-1} for N_2 and 842 kJ mol^{-1} for N_2^+.

Exercise. Which can be expected to have the higher dissociation energy, F_2 or F_2^+? $\quad [F_2^+]$

14.5 More about notation

Parity

The molecular orbitals of homonuclear diatomic molecules are labelled with a subscript g or u that specifies their **parity**, their behaviour under inversion. To decide on the parity, we consider any point in a homonuclear diatomic, and note the sign of the orbital. Then we travel through the centre of the molecule (the 'centre of inversion') and go to the corresponding point on the other side (this process is the 'operation of inversion'). If the orbital has the same sign, it has even parity and is denoted g (from *gerade*, the German for even). If the orbital has opposite sign, it has odd parity and is denoted u (from *ungerade*, uneven). The parity designation applies only to homonuclear diatomic molecules, since heteronuclear diatomic molecules (such as HCl) do not have a centre of inversion.

We see from Fig. 14.17 that a bonding σ orbital has even parity; so we write it σ_g; a σ^* orbital has odd parity and is written σ_u. A bonding π orbital has odd parity and is denoted π_u and a π^* orbital has even parity, denoted π_g. Later we shall see the importance of the parity designation in a statement of spectroscopic selection rules.

Term symbols

The term symbols of molecules (the analogues of the symbols ^2P, etc, for atoms) are constructed in a similar way to those in atoms, but now we must pay attention to the orbital angular momentum about the internuclear axis. The *total* angular momentum of all the electrons around the axis is denoted by the symbols $\Sigma, \Pi, \Delta, \ldots$ corresponding to the S, P, D, $\ldots$ of atoms.

A single electron in a σ orbital has zero orbital angular momentum: the orbital is cylindrically symmetrical and has no angular nodes. The term symbol for H_2^+ is therefore Σ. As in atoms, we use a superscript with the value of $2S + 1$ to denote the multiplicity of the term. In this case, since there is only one electron, $S = \frac{1}{2}$ and the term symbol is $^2\Sigma$, a doublet term.

Centre of inversion

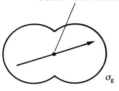

σ_g

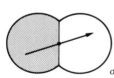

σ_u

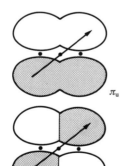

π_u

π_g

Fig. 14.17 The parity of an orbital is even (g) if its amplitude is unchanged under inversion in the centre of symmetry of the molecule, but odd (u) if the amplitude changes sign. Heteronuclear diatomic molecules do not have a centre of inversion, and so the g, u classification is irrelevant.

The overall parity of the term is added as a right subscript, and (if there are several electrons) is calculated using

$$g \times g = g \qquad u \times u = g \qquad u \times g = u$$

For H_2^+, the parity of the only occupied orbital is g, so the term itself is also g, and in full dress is $^2\Sigma_g$. The term symbol is $^1\Sigma_g$ for any closed-shell homonuclear diatomic because the spin is zero (all electrons paired), there is no orbital angular momentum from a closed shell, and the overall parity is g.

A π electron has one unit of orbital angular momentum about the internuclear axis, and if it is the only electron outside a closed shell, gives rise to a Π term. If there are two π electrons (as in O_2) the term symbol may be either Σ (if the electrons are orbiting in opposite directions) or Δ (if they are orbiting in the same direction). For O_2 it is known that Σ lies lower in energy than Δ, and so its ground term is Σ. We have already seen that O_2 is a triplet, and the overall parity is

$$\text{(closed shell)} \times g \times g = g$$

The term symbol is therefore $^3\Sigma_g$.

Finally, we add another superscript to a Σ term symbol to denote the behaviour of the molecular wavefunction under reflection in a plane containing the nuclei (Fig. 14.18). For O_2, one electron is $2p_x\pi$ (which changes sign under reflection in the xy-plane) and the other is $2p_y\pi$ (which does not); the overall reflection symmetry is therefore

$$\text{(closed shell)} \times (+) \times (-) = (-)$$

and the full term symbol is $^3\Sigma_g^-$. The need for all this dressing of a basic symbol will become apparent when we deal with the spectroscopic selection rules in Chapter 17.

14.6 Heteronuclear diatomic molecules

A **heteronuclear diatomic molecule** is a diatomic molecule formed from atoms of two different elements, such as CO and HCl. The electron distribution in the covalent bond between the atoms is not symmetrical because it is energetically favourable for the electron pair to be found closer to one atom than the other. This imbalance results in a **polar bond**, which is a covalent bond in which the electron pair is shared unequally by the two atoms. The bond in HF, for instance, is polar, with the electron pair closer to the F atom. The accumulation of the electron pair near the F atom results in that atom having a net negative charge, which is called a **partial negative charge** and denoted $\delta-$. There is a compensating **partial positive charge** $\delta+$ on the H atom.

Polar covalent bonds

A polar covalent bond consists of two electrons in an orbital of the form

$$\psi = c_A\psi(A) + c_B\psi(B)$$

with unequal coefficients. The proportion of the atomic orbital $\psi(A)$ in the bond is c_A^2 and that of $\psi(B)$ is c_B^2. A nonpolar bond has $c_A^2 = c_B^2$, and a pure ionic bond has one coefficient zero (so that A^+B^- would have $c_A = 0$ and

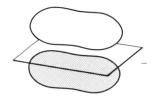

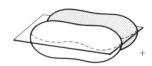

Fig. 14.18 The $\pm$ symmetry refers to the symmetry of an orbital when it is reflected in a plane containing the two nuclei.

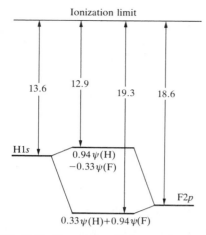

Fig. 14.19 The atomic orbital energy levels of H and F atoms and the molecular orbitals they form. The bonding orbital has predominantly (88 per cent) F atom character and the antibonding orbital has predominantly (88 per cent) H atom character.

$c_B = 1$). The atomic orbital with the lower energy makes the larger contribution to the bonding molecular orbital. The opposite is true of the antibonding orbital, for which the dominant component comes from the atomic orbital with higher energy.

These points can be illustrated by considering HF, and judging the energies of the orbitals that contribute to the molecular orbital by noting the ionization energies of the atoms. The general form of the molecular orbitals is

$$\psi = c_H \psi(H) + c_F \psi(F)$$

where $\psi(H)$ is an H1s orbital and $\psi(F)$ is an F2p orbital. The H1s orbital lies at 13.6 eV below the zero of energy (the separated proton and electron) and the F2p orbital lies at 18.6 eV below the zero of energy (Fig. 14.19). Hence, the bonding σ orbital in HF is mainly F2p and the antibonding σ orbital is mainly H1s orbital in character. The two electrons in the bonding orbital are most likely to be found in the F2p orbital, so there is a partial negative charge on the F atom and a partial positive charge on the H atom.

The variation principle

A systematic way of finding the coefficients in the linear combinations used to build molecular orbitals is provided by the **variation principle**:

> If an arbitrary wavefunction is used to calculate the energy, then the value calculated is never less than the true energy.

The arbitrary wavefunction is called the **trial wavefunction**. The principle implies that if we vary the coefficients in the trial wavefunction until we achieve the lowest energy, then those coefficients will be the best. We might get a lower energy if we use a more complicated wavefunction (for example, by taking a linear combination of several atomic orbitals on each atom), but we shall have the optimum molecular orbital that can be built from the given set of atomic orbitals.

The method can be illustrated by the trial wavefunction

$$\psi = c_A \psi(A) + c_B \psi(B)$$

This function is real but not normalized (because the coefficients can take arbitrary values), so in the following we cannot assume that $\int \psi^2 \, d\tau = 1$. The energy of the orbital is the expectation value of the energy operator (the hamiltonian, H, Section 11.5):

$$E = \frac{\int \psi H \psi \, d\tau}{\int \psi^2 \, d\tau} \tag{11}$$

We must search for values of the coefficients in the trial function that minimize the value of E. This is a standard problem in calculus, and is solved by finding the coefficients for which

$$\frac{\partial E}{\partial c_A} = 0 \quad \text{and} \quad \frac{\partial E}{\partial c_B} = 0$$

The secular equations

The first step is to express the two integrals in terms of the coefficients. The denominator is

$$\int \psi^2 \, d\tau = \int \{c_A \psi(A) + c_B \psi(B)\}^2 \, d\tau$$

$$= c_A^2 \int \psi(A)^2 \, d\tau + c_B^2 \int \psi(B)^2 \, d\tau + 2c_A c_B \int \psi(A) \psi(B) \, d\tau$$

$$= c_A^2 + c_B^2 + 2c_A c_B S \qquad (12)$$

since the individual atomic orbitals are normalized and the third integral is the overlap integral S (eqn 9). The numerator is

$$\int \psi H \psi \, d\tau = \int \{c_A \psi(A) + c_B \psi(B)\} H \{c_A \psi(A) + c_B \psi(B)\} \, d\tau$$

$$= c_A^2 \int \psi(A) H \psi(A) \, d\tau + c_B^2 \int \psi(B) H \psi(B) \, d\tau$$

$$+ 2c_A c_B \int \psi(A) H \psi(B) \, d\tau$$

There are some complicated integrals in this expression, but we can denote them by the constants

$$\alpha_A = \int \psi(A) H \psi(A) \, d\tau \qquad \alpha_B = \int \psi(B) H \psi(B) \, d\tau \qquad \beta = \int \psi(A) H \psi(B) \, d\tau$$

Then

$$\int \psi H \psi \, d\tau = c_A^2 \alpha_A + c_B^2 \alpha_B + 2c_A c_B \beta \qquad (13)$$

α is called a **Coulomb integral**. It is negative, and can be interpreted as the energy of the electron when it occupies $\psi(A)$ (for α_A) or $\psi(B)$ (for α_B). In a homonuclear diatomic molecule, $\alpha_A = \alpha_B$. β is called a **resonance integral** (for classical reasons). It vanishes when the orbitals do not overlap, and at equilibrium bond lengths it is normally negative.

The complete expression for E is

$$E = \frac{c_A^2 \alpha_A + c_B^2 \alpha_B + 2c_A c_B \beta}{c_A^2 + c_B^2 + 2c_A c_B S}$$

Its minimum is found by differentiation with respect to the two coefficients. This involves elementary but slightly tedious work, the end result being the two **secular equations**[4]

$$(\alpha_A - E)c_A + (\beta - ES)c_B = 0$$
$$(\beta - ES)c_A + (\alpha_B - E)c_B = 0 \qquad (14)$$

---

[4] 'Secular' is derived from the Latin word for age or generation. The term comes via astronomy, where the same equations appear in connection with slowly accumulating modifications of planetary orbits.

As for any set of simultaneous equations, these two have a solution if the **secular determinant**, the determinant of the coefficients, is zero; that is, if

$$\begin{vmatrix} \alpha_A - E & \beta - ES \\ \beta - ES & \alpha_B - E \end{vmatrix} = 0 \tag{15}$$

This determinant expands to a quadratic equation in E (see the example below). Its two roots give the energies of the bonding and antibonding molecular orbitals formed from the atomic orbitals and, according to the variation principle, these are the best energies for the given set of atomic orbitals.

The values of the coefficients in the linear combination are obtained by solving the secular equations using the two energies: the lower energy gives the coefficients for the bonding molecular orbital, the upper energy the coefficients for the antibonding molecular orbital. The secular equations give expressions for the ratio of the coefficients in each case, and so we need a further equation in order to find their individual values. This is obtained by demanding that the best wavefunction should be normalized, which means that, at this final stage, we must also ensure (from eqn 12) that

$$\int \psi^2 \, d\tau = c_A^2 + c_B^2 + 2c_A c_B S = 1$$

Example 14.6: *Finding the roots of a secular determinant*

Find the energies E of the bonding and antibonding orbitals of a homonuclear diatomic molecule by solving eqn 15.

Answer. A 2×2 determinant expands as follows:

$$\begin{vmatrix} A & B \\ D & C \end{vmatrix} = A \times C - B \times D$$

when we apply this rule to eqn 15 with $\alpha_A = \alpha_B = \alpha$ we get

$$\begin{vmatrix} \alpha - E & \beta - ES \\ \beta - ES & \alpha - E \end{vmatrix} = (\alpha - E)^2 - (\beta - ES)^2 = 0$$

The solutions of this equation are

$$E_+ = \frac{\alpha + \beta}{1 + S} \qquad E_- = \frac{\alpha - \beta}{1 - S}$$

Exercise. Find the coefficients corresponding to these two energies.

[See below]

Two simple cases

The complete solutions of the secular equations are very cumbersome even for 2×2 determinants, but there are two cases where the roots can be written down very simply.

We saw in Example 14.6 that when the two atoms are the same, and we can write $\alpha_A = \alpha_B = \alpha$, the solutions are

$$E_+ = \frac{\alpha + \beta}{1 + S} \qquad c_A = \left\{\frac{1}{2(1 + S)}\right\}^{1/2} \qquad c_B = c_A$$

$$E_- = \frac{\alpha - \beta}{1 - S} \qquad c_A = \left\{\frac{1}{2(1 - S)}\right\}^{1/2} \qquad c_B = -c_A$$

In this case, the best bonding function has the form

$$\psi_+ = \left\{\frac{1}{2(1 + S)}\right\}^{1/2} \{\psi(A) + \psi(B)\}$$

and the corresponding antibonding function is

$$\psi_- = \left\{\frac{1}{2(1 - S)}\right\}^{1/2} \{\psi(A) - \psi(B)\}$$

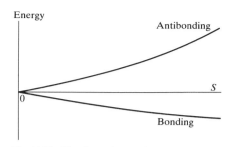

Fig. 14.20 The dependence of the bonding and antibonding orbital energies on the overlap, calculated using the full solutions of the secular equations and making the assumption that the resonance integral is proportional to the overlap integral.

in agreement with the discussion of homonuclear diatomics we have already given. However, we can now explore how the energies of the orbitals change with the resonance integral β and the overlap integral S. The easiest way of doing so is to suppose that β is directly proportional to S, and to write

$$\beta = \kappa S$$

where κ is a constant. Figure 14.20 shows how the bonding and antibonding effects fall sharply with decreasing overlap, which justifies the neglect of $1s$, $1s$ bonding in the Period-2 elements, for the overlap is so slight. However, care must be exercised when setting β proportional to S, for whereas the approximation is safe for overlap of s and p orbitals, it sometimes fails for d orbitals. For instance, in d-metal clusters, β is large but S is small.

When it is justifiable to neglect overlap, the secular determinant is

$$\begin{vmatrix} \alpha_A - E & \beta \\ \beta & \alpha_B - E \end{vmatrix} = 0$$

and its solutions can be expressed in terms of the parameter ζ, with

$$\tan 2\zeta = \frac{2\beta}{\alpha_A - \alpha_B} \tag{16a}$$

The solutions are

$$\begin{aligned} E = \alpha_A - \beta \cot \zeta \qquad & \psi = -\sin \zeta\, \psi(A) + \cos \zeta\, \psi(B) \\ E = \alpha_B + \beta \cot \zeta \qquad & \psi = \cos \zeta\, \psi(A) + \sin \zeta\, \psi(B) \end{aligned} \tag{16b}$$

An important feature revealed by these solutions is that as the difference in energy $\alpha_A - \alpha_B$ between the two atomic orbitals increases, the value of ζ decreases.[5] When the energy difference is large the energies of the

[5] Since $\tan x \approx x$ and $\cot x \approx 1/x$ when $x \ll 1$, when $\alpha_A - \alpha_B$ is large, $\zeta \approx \beta/(\alpha_A - \alpha_B)$ and $\beta \cot \zeta \approx \alpha_A - \alpha_B$. Then the energies of the two molecular orbitals are

$$E = \alpha_A - \beta \cot \zeta \approx \alpha_B \quad \text{and} \quad E = \alpha_B + \beta \cot \zeta \approx \alpha_A$$

since $\sin x \approx x$ and $\cos x \approx 1$ when $x \ll 1$, the orbitals are respectively almost pure $\psi(B)$ and almost pure $\psi(A)$.

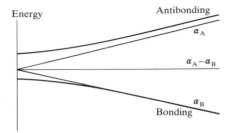

Fig. 14.21 The dependence of the bonding and antibonding orbital energies on the energy separation of the contributing atomic orbitals calculated using $\beta = \kappa S$.

molecular orbitals differ by only a small amount from those of the atomic orbitals, which implies in turn that the bonding and antibonding effects are small. That is, the strongest bonding and antibonding effects are obtained when the two contributing orbitals have closely similar energies. This is illustrated for the full secular determinant (without neglecting overlap) in Fig. 14.21.

In HF, the ionization energies from the valence orbitals are as follows:

$$H1s:\ 13.6\ eV \qquad F2s:\ 40.2\ eV \qquad F2p:\ 18.6\ eV$$

Since the $F2p$ and $H1s$ orbitals are much closer in energy than the $F2s$ and $H1s$ orbitals, to a first approximation we can neglect the contribution of the $F2s$ orbital. More generally, the difference in energy is the justification for neglecting the contribution to bonding of core orbitals, since these differ in energy markedly from the valence orbitals. The core orbitals of one atom have a similar energy to the core orbitals of the other atom; but core–core interaction is largely negligible because, as we have seen, the overlap between them (and hence the value of β) is so small.

Example 14.7: *Calculating the molecular orbitals of HF*

Calculate the wavefunctions and energies of the σ orbitals in the HF molecule, taking $\beta = -2.0\ eV$.

Answer. We need to know the values of the Coulomb integrals α_H and α_F. Since they represent the energies of the $H1s$ and $F2p$ electrons respectively, we can estimate that they are approximately equal to (the negative of) the ionization energies of the atoms, and use $\alpha_H \approx -13.6\ eV$, $\alpha_F \approx -18.6\ eV$. Using these numbers gives $\tan 2\zeta = -0.800$, so $\zeta = -19.33°$. Then

$$E = -19.3\ eV \qquad \psi = 0.33\psi(H) + 0.94\psi(F)$$
$$E = -12.9\ eV \qquad \psi = 0.94\psi(H) - 0.33\psi(F)$$

Comment. Notice how the lower energy orbital (the one with energy $-19.3\ eV$) has a composition that is more $F2p$ orbital than $H1s$, and that the opposite is true of the higher energy, antibonding orbital (Fig. 14.19).

Exercise. The ionization energy of Cl is $13.1\ eV$; find the form and energies of the σ orbitals in the HCl molecule using $\beta \approx -2.0\ eV$.

$$[E = -11.3\ eV,\ \psi = -0.66\psi(H) + 0.75\psi(Cl);$$
$$E = -15.4\ eV,\ \psi = 0.75\psi(H) + 0.66\psi(Cl)]$$

Semi-empirical and *ab initio* methods

We can now see how the LCAO coefficients for heteronuclear diatomics are found: the secular equations are solved for the energies, and those energies are used to obtain the optimum coefficients. There is still the problem of knowing the values of the Coulomb and resonance integrals. One approach has been to estimate them from spectroscopic information (such as ionization energies, as in Example 14.7). This combination of empirical data and quantum mechanical calculation has given rise to the **semi-empirical methods** of molecular structure calculation. The modern tendency, however, particularly for small molecules but increasingly for bigger ones too, is to calculate the integrals from first principles. These **ab initio methods** demand extensive numerical computation, and for this reason theoretical chemists are among the heaviest users of computers.

14.7 Hybridization

When dealing with heteronuclear diatomics, there is often no clear-cut distinction between the energies of the atomic orbitals, and therefore no precise criterion about which ones should be combined. As we see in Fig. 14.22 in connection with LiH, although Li2s lies closer to H1s, the Li2p orbitals are not far away and cannot be ignored: the ionization energies are

$$\text{H1}s: \ 13.6\,\text{eV} \qquad \text{Li2}s: \ 5.4\,\text{eV} \qquad \text{Li2}p: \ 3.7\,\text{eV}$$

While we could approximate the σ orbitals by a linear combination of H1s and Li2s orbitals (the two orbitals closest in energy), it is more satisfactory to include the Li2p orbitals too, and to write the molecular orbitals as

$$\psi = c_{2s}\psi_{2s}(\text{Li}) + c_{2p}\psi_{2p}(\text{Li}) + c_{1s}\psi_{1s}(\text{H})$$

A variational calculation then gives

$$\psi = 0.323\,\psi_{2s}(\text{Li}) + 0.231\,\psi_{2p}(\text{Li}) + 0.685\,\psi_{1s}(\text{H})$$

for the lowest-energy orbital of this form. We see that the H1s orbital makes the greatest contribution (which is consistent with its lying lowest in energy of the three), and that the Li2p orbital does indeed make a substantial contribution.

By including the p orbitals we seem to have lost an attractive feature of LCAO-molecular orbital theory, for it now seems impossible to think of an Li–H bond as being formed from the overlap of two orbitals, one on Li and the other on H. That is indeed the stance of molecular orbital theory, which treats each orbital as a combination of many atomic orbitals, and does not seek to analyze individual bonds as formed by pairwise overlap of the orbitals of adjacent atoms. However, there are occasions when it is helpful to regard two atoms as bonded together by the overlap of an orbital on each of them, and it is possible to recover this description from the molecular orbital treatment. It should be borne in mind, though, that the procedure we are about to describe is unnecessary and can be distracting. It is generally better to adopt the pure molecular orbital view that the bonding influence of electrons is distributed over many atoms in orbitals composed of all atomic orbitals of the appropriate symmetry.[6]

We can regard the bonding molecular orbital in LiH as arising from the overlap of $\psi_{1s}(\text{H})$ and a **hybrid orbital** $\phi(\text{Li})$ on Li, a mixture of the 2s and 2p orbitals on Li. The molecular orbital can be expressed alternatively as

$$\psi = 0.323\{\psi_{2s}(\text{Li}) + 0.715\,\psi_{2p}(\text{Li})\} + 0.685\,\psi_{1s}(\text{H})$$
$$= 0.323\,\phi(\text{Li}) + 0.685\,\psi_{1s}(\text{H})$$

where

$$\phi(\text{Li}) = \psi_{2s}(\text{Li}) + 0.715\,\psi_{2p}(\text{Li})$$

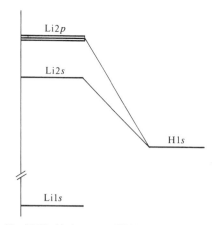

Fig. 14.22 Hydrogen and lithium atomic energy levels: H1s overlaps with both Li2s and Li2p, and the resulting orbital can be regarded as arising from the overlap of H1s with a (Li2s, Li2p)-hybrid orbital. L1s is a core orbital, and plays only a minor role in the bonding.

[6] The description of bonding in terms of the overlap of orbitals belonging to pairs of atoms is the basis of the 'valence bond' description of molecules. This approach to bonding was a serious competitor to molecular orbital theory in the 1930s, but fell into disfavour. However, there are some signs of a revival in interest in valence bond theory and it is important to be aware that it exists. For instance, one of the theories of high-temperature superconductors makes use of its concepts.

407

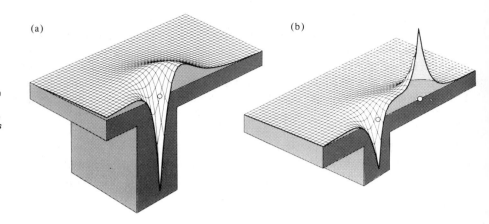

Fig. 14.23 (a) A cross-section through the (Li2s, Li2p)-hybrid showing the accumulation of amplitude on one side of the nucleus. (b) The H1s-orbital overlaps the hybrid strongly, and a stronger bond is formed than with Li2s alone.

Through hybridization we retain the simplicity of the basic LCAO–MO picture, but at the expense of a more complicated atomic contribution. Hybridization is unnecessary in the sense that it arises only from a wish to find a formally simple description of a bond: there is no mathematical reason why it should be introduced, and no compelling physical reason.

The shape of the Li hybrid is shown in Fig. 14.23, and we see that the bulk of its amplitude lies in the internuclear region. This shape arises from the interference between the Li2s and Li2p atomic orbitals, because they add on one side of the nucleus (where their amplitudes have the same sign) but partially cancel on the other side (where their signs are opposite). We can begin to see why the inclusion of the Li2p orbital in the LCAO leads to a lowering of the energy of the bonding orbital (and hence, by the variation principle, a better description of the molecule): the inclusion of the Li2p orbital gives a hybrid which has a better overlap with H1s than a 2s orbital has alone. Exactly the same enhancement is present in the 'three orbital' description of the bond, but hybridization focuses our attention on it.

The structures of polyatomic molecules

The bonds in polyatomic molecules are built in the same way as in diatomic molecules, the only difference being that we use more atomic orbitals to construct the molecular orbitals, and these molecular orbitals spread over the entire molecule. In general, a molecular orbital has the form

$$\psi = \sum_i c_i \psi_i \qquad (17)$$

where the ψ_i are the atomic orbitals. In H_2O, for instance, the atomic orbitals are the two H1s orbitals, the O2s orbital, and the three O2p orbitals (if we consider only the valence shell). From these six orbitals we can construct six molecular orbitals which spread over the three atoms. The molecular orbitals differ in energy, the lowest energy, most strongly bonding orbitals having the least number of nodes between adjacent atoms, and the highest energy, most strongly antibonding orbitals having the greatest numbers of nodes between neighbouring atoms.

The principal difference between diatomic and polyatomic molecules lies in the greater range of shapes that are possible: a diatomic molecule is necessarily linear, but a triatomic molecule, for instance, may be linear or angular with a certain bond angle. In principle, the shape of a polyatomic molecule—the specification of all its bond lengths and bond angles—can be predicted by calculating the total energy of the molecule for a variety of conformations, and then identifying the one that leads to the lowest energy. However, better insight into the features that control molecular geometry can be obtained by analysing the orbitals and their energies in a more pictorial fashion. We shall illustrate what is involved by considering H_2O, which has an experimental bond angle of 104°.

14.8 Walsh diagrams

The molecular orbitals of H_2O (and of H_2X molecules in general) have the form

$$\psi = c_1\psi(H_A) + c_2\psi(H_B) + c_3\psi_{2s}(O) + c_4\psi_{2p_x}(O) + c_5\psi_{2p_y}(O) + c_6\psi_{2p_z}(O)$$

There are six such orbitals (because they are built from six atomic orbitals) and eight valence electrons to accommodate in them. The simplest way of proceeding is to consider two conformations of the molecule, the linear 180° molecule and the angular 90° molecule, and then to identify how the molecular orbitals of one shape turn into the molecular orbitals of the other as the bond angle changes from 180° to 90°. The procedure results in the construction of a **Walsh diagram**, a diagram showing the variation of orbital energy with molecular geometry.

The Walsh diagram for H_2X molecules

We can identify the general form of the molecular orbitals we need to consider in the angular H_2O molecule by deciding which linear combinations have the correct symmetry for orbital formation. At this stage we shall proceed intuitively; in Chapter 15 we shall see how to select the combinations by calculation.

In the hypothetical linear HOH molecule, the molecular orbitals (Fig. 14.24) are classified as either σ or π and have the form

$$\sigma_g: \quad \psi = c_1\psi_{2s}(O) + c_2\{\psi(H_A) + \psi(H_B)\} \quad \text{(two orbitals)}$$

$$\pi_u: \quad \psi = \psi_{2p_x}(O) \text{ and } \psi_{2p_z}(O) \quad \text{(one orbital each)}$$

$$\sigma_u: \quad \psi = c_1\psi_{2p_y}(O) + c_2\{\psi(H_A) - \psi(H_B)\} \quad \text{(two orbitals)}$$

We have added the parity labels, but they no longer tell us which is bonding and antibonding. Thus, there are two σ_g orbitals, one bonding (with the two coefficients the same sign) and the other antibonding (with the coefficients of opposite sign). There are no orbitals of π symmetry on the H atoms, so the $O2p_x$ and $O2p_z$ orbitals do not form bonding and antibonding molecular orbitals. They are examples of **non-bonding orbitals**, orbitals which if occupied neither raise not lower the energy of the molecule. The coefficients in the molecular orbitals may be found in the normal way, by setting up and solving the secular determinants using estimates of the Coulomb and resonance integrals, and the energies of the orbitals are shown on the left of the diagram in Fig. 14.25.

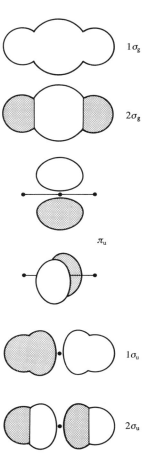

Fig. 14.24 The molecular orbitals that can be constructed from the H1s, O2s, and O2p atomic orbitals in a hypothetical linear H_2O molecule.

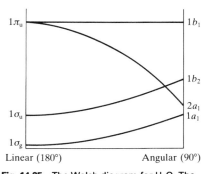

Fig. 14.25 The Walsh diagram for H_2O. The energies of the linear molecule are shown on the left (see Fig. 14.24 for their compositions) and those of the 90° molecule are shown on the right (see Fig. 14.26). The actual molecule has a bond angle of 104°.

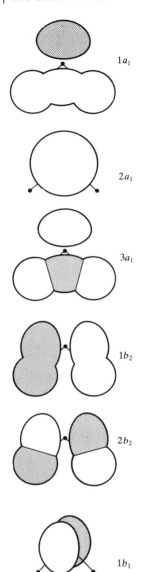

Fig. 14.26 The molecular orbitals that can be constructed from the H1s, O2s, and O2p atomic orbitals in a hypothetical 90° H_2O molecule.

In the hypothetical 90° angular molecule, the molecular orbitals are formed from the following groupings of atomic orbitals (Fig. 14.26):

$$a_1: \quad \psi = c_1\psi_{2s}(O) + c_2\psi_{2p_z}(O) + c_3\{\psi(H_A) + \psi(H_B)\}$$

$$b_1: \quad \psi = \psi_{2p_x}(O)$$

$$b_2: \quad \psi = c_1\psi_{2p_y}(O) + c_2\{\psi(H_A) - \psi(H_B)\}$$

We can no longer classify the orbitals as σ and π because those labels apply only when there is an axis of symmetry; the labels used here will be explained in Chapter 15 (as will be the choice of the orbitals from which each molecular orbital is built). There are three a_1 orbitals (because we are combining three atomic orbitals), two b_2 orbitals, and one (non-bonding) b_1 orbital. The central feature of molecular orbital theory is the formation of molecular orbitals from all the atomic orbitals available that have the same symmetry, and the linear combinations listed above can be regarded as a grouping of the atomic orbitals into different symmetry classes. This grouping is the subject of Chapter 15.

The lowest energy orbital in 90° H_2O is the one labeled $1a_1$, which is built from the overlap of the O2p_z orbital with the $\psi(A) + \psi(B)$ combination of H1s orbitals. As the bond angle changes to 180° the two H1s orbitals overlap less but the contribution to the molecular orbital of the O2s orbital increases. In the 180° molecule, O2s is the only contribution from the O atom to the $1\sigma_g$ orbital (Fig. 14.24). The replacement of O2p_z by O2s lowers the energy of the orbital. The energy of the $1b_2$ orbital is also lowered because the adverse H–H overlap decreases and the H1s orbitals move into a better position for overlap with the O2p_y orbital. The biggest change occurs for the $2a_1$ orbital. It is a pure O2s-orbital in the 90° molecule, but correlates with a pure O2p_z orbital in the 180° molecule. Hence, it shows a steep rise in energy as the bond angle increases. The $1b_1$ orbital is a non-bonding O2p orbital perpendicular to the molecular plane in the 90° molecule and remains non-bonding in the linear molecule. Hence, its energy barely changes with angle.

The shape of the H_2O molecule

The principal feature that determines whether or not the H_2O molecule is bent is whether the $2a_1$ orbital is occupied. This is the orbital that has considerable O2s character in the bent molecule but not in the linear molecule. Hence, a lower total energy is achieved if, when it is occupied, the molecule is bent. The shape adopted by an H_2O molecule therefore depends on the number of electrons that occupy the orbitals.

Example 14.8: *Using a Walsh diagram to predict a shape*

Predict the shape of the H_2O molecule from the Walsh diagram.

Answer. We choose an intermediate bond angle along the horizontal axis of the H_2O diagram in Fig. 14.25 and accommodate eight electrons. The resulting configuration is $1a_1^2 1b_2^2 2a_1^2 1b_1^2$. The $2a_1$ orbital is occupied, so we expect the nonlinear molecule to be more stable than the linear.

Exercise. Predict the shape of the BeH_2 molecule. [Linear]

14.9 Orthogonality and hybridization

In the molecular orbital description of H_2O, the three atoms are bound together by electrons occupying **delocalized molecular orbitals**, or orbitals that spread over all the atoms in the molecule, and the binding influence of an electron pair is spread throughout the molecule. Thus, two electrons can contribute to the binding together of more than two atoms. This is a major departure from the conventional interpretation of Lewis's theory, which took the electron pair lying between a pair of atoms as the principal concept: in that theory, one electron pair could bind one pair of atoms. In molecular orbital theory, one electron pair can contribute to the bonding of an indefinitely large number of atoms, but its contribution is weak if the orbital spreads over many atoms.

We saw in the discussion of LiH that the concept of a 'bond' as formed by the overlap of two orbitals could be recreated by hybridization. The same is true of polyatomic molecules, and for some purposes it is helpful to think of bonds being formed by the overlap of pairs of orbitals. For instance, whereas in H_2O the molecular orbital theory leads to electron pairs spreading into both bonding regions, we can recreate the electron distribution by forming two hybrid orbitals on the O atom that point towards the two H atoms, and then allow these hybrids to overlap the H1s orbitals.

Orthogonality and similarity

The question we address in this section is how to combine s and p orbitals on a single atom into identical hybrids with a specific angle between them. Our aim is to construct hybrids that are (a) equivalent but (b) spatially distinct. This translates into forming hybrids of the s, p_x, and p_y orbitals (where the xy-plane is the plane in which the hybrid orbitals are to lie) that (a) have the same composition (in the sense that they have the same proportions of s and p character) and (b) are **orthogonal** in the sense that one hybrid orbital has zero overlap with another (see the discussion of eqn 9):

$$S = \int \psi_A^* \psi_B \, d\tau = 0$$

Example 14.9: *Judging the orthogonality of orbitals*

Confirm that a (real) $2p_x$ orbital is orthogonal to a (real) $2p_y$ orbital of the same atom.

Answer. We need to evaluate the integral

$$S = \int \psi_{2p_x} \psi_{2p_y} \, d\tau$$

and can do so by noting that

$$\psi_{2p_x} = xf(r) \qquad \psi_{2p_y} = yf(r)$$

Then, using polar coordinates

$$\psi_{2p_x} = rf(r) \sin \theta \cos \phi \qquad \psi_{2p_y} = rf(r) \sin \theta \sin \phi$$

all we need show is that the integral over either angle vanishes. The two orbitals differ in their dependence on ϕ, so we begin with that integration:

$$S \propto \int_0^{2\pi} \cos \phi \sin \phi \, d\phi = 0$$

> **Comment.** All the spherical harmonics are mutually orthogonal, and so all hydrogenic atomic orbitals are mutually orthogonal so long as they belong to the same atom.
>
> **Exercise.** Show that $3d_{xy}$ is orthogonal to $2p_x$.

Two O2p orbitals are mutually orthogonal and point at 90° to each other. When O2s character is mixed into them in the process of hybridization, they are no longer orthogonal (they are similar to the extent of having O2s character). Only if the hybrids bend further away from each other, so reducing their spatial resemblance, can they become distinct. Therefore, we can anticipate that the greater the angle between the hybrids, the greater their s orbital content.

The construction of directed hybrids

Suppose we want to construct a p orbital that points along a line making an angle $\frac{1}{2}\Phi$ to the x-axis (Fig. 14.27). The combination of p orbitals required is[7]

$$p = p_x \cos \tfrac{1}{2}\Phi + p_y \sin \tfrac{1}{2}\Phi \tag{18a}$$

The equivalent orbital pointing along $-\frac{1}{2}\Phi$ is

$$p' = p_x \cos \tfrac{1}{2}\Phi - p_y \sin \tfrac{1}{2}\Phi \tag{18b}$$

Although the p_x and p_y orbitals are mutually orthogonal and normalized, these two combinations are not orthogonal:

$$\int pp' \, d\tau = \int (p_x \cos \tfrac{1}{2}\Phi + p_y \sin \tfrac{1}{2}\Phi)(p_x \cos \tfrac{1}{2}\Phi - p_y \sin \tfrac{1}{2}\Phi) \, d\tau$$

$$= \cos^2 \tfrac{1}{2}\Phi \int p_x^2 \, d\tau - \sin^2 \tfrac{1}{2}\Phi \int p_y^2 \, d\tau$$

$$= \cos^2 \tfrac{1}{2}\Phi - \sin^2 \tfrac{1}{2}\Phi = \cos \Phi$$

The integral vanishes only if $\Phi = 90°$.

Now we add s-character. The same proportion must be added to both (so that the hybrids remain equivalent). We therefore write

$$h = as + bp \qquad h' = as + bp'$$

We can find the coefficients a and b by requiring the hybrids to satisfy two conditions. One is the requirement that the hybrids are normalized:

$$\int h^2 \, d\tau = a^2 \int s^2 \, d\tau + b^2 \int p^2 \, d\tau + 2ab \int sp \, d\tau = a^2 + b^2 = 1$$

(The s and p orbitals are individually normalized and mutually orthogonal.)

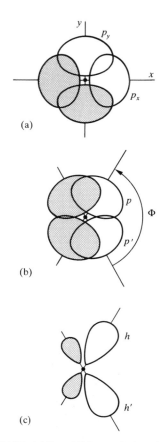

Fig. 14.27 (a) the orbitals p and p' can be expressed as linear combinations of p_x and p_y using eqn 18, but the combinations are not orthogonal. (b) The orthogonal hybrids, obtained by mixing s character into p and p'.

---

[7] A p_x orbital is proportional to $x = r \cos \phi$ and a p_y orbital is proportional to $y = r \sin \phi$. A p orbital directed along the line at $\frac{1}{2}\Phi$ is proportional to $\cos(\phi - \frac{1}{2}\Phi)$ because it is like a p_x orbital but rotated through an angle $\frac{1}{2}\Phi$. Now note that

$$p(\tfrac{1}{2}\Phi) \propto \cos(\phi - \tfrac{1}{2}\Phi) \propto \cos \phi \cos \tfrac{1}{2}\Phi + \sin \phi \sin \tfrac{1}{2}\Phi$$
$$\propto p_x \cos \tfrac{1}{2}\Phi + p_y \sin \tfrac{1}{2}\Phi$$

The second requirement is for the hybrids to be distinct; that is, orthogonal:

$$\int hh' \, d\tau = \int (as + bp)(as + bp') \, d\tau$$

$$= a^2 \int s^2 \, d\tau + b^2 \int pp' \, d\tau + ab \int (sp + sp') \, d\tau$$

$$= a^2 + b^2 \int pp' \, d\tau = a^2 + b^2 \cos \Phi$$

On combining these two results we get

$$a^2 = \frac{\cos \Phi}{\cos \Phi - 1} \qquad (19)$$

This expression is plotted in Fig. 14.28. It shows that as the angle between the hybrids, Φ, increases from 90° to 180°, their s character increases from $a^2 = 0$ (pure p) to $a^2 = 0.5$ (a 50:50 mixture of s and p).

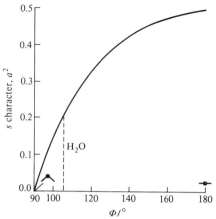

Fig. 14.28 The dependence of s character on the angle between two equivalent hybrids calculated using eqn 19. The p character is $1 - a^2$.

Example 14.10: *Calculating the composition of hybridized orbitals*

Calculate the hybridization of the O–H bonds in H_2O, which has a bond angle of 104°.

Answer. We can calculate the s character of a hybrid from eqn 19 and its p character from $1 - a^2$. Hybrids are often denoted $s^{a^2} p^{b^2}$. When $\phi = 104°$, $\cos \Phi = -0.24$; hence $a^2 = 0.20$. The p character of the hybrid is therefore 0.80, and it can be denoted $s^{0.20} p^{0.80}$.

Comment. Note that hybrids need not be of integral composition (such as sp^3). The coefficients are the square roots of 0.20 and 0.80, and so the hybrid wavefunctions are of the form $0.45s + 0.89p$, where p is a p orbital directed from O to H.

Exercise. What is the hybridization for a 120° bond angle?

$[s^{1/3} p^{2/3}, \text{ or } sp^2 \text{ in terms of ratios}]$

Symmetrical hybrids

The linear combination of three atomic orbitals leads to three orthogonal hybrids (and in general N orbitals give N hybrids). The third hybrid is $a's + b'p'$, where p' is an orbital directed along $-x$ (so $p' = -p_x$, Fig. 14.29). Normalization requires that

$$a'^2 + b'^2 = 1$$

and the orthogonality of this hybrid to the other two requires

$$aa' + bb' \cos \tfrac{1}{2}\Phi = 0$$

After a little rearrangement, these conditions lead to

$$a'^2 = \frac{1 + \cos \Phi}{1 - \cos \Phi} \qquad (20)$$

for the s character of this nonbonding, lone pair orbital. For H_2O,

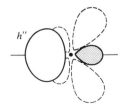

Fig. 14.29 Three orthogonal atomic orbitals hybridize to give three orthogonal hybrid orbitals. This is the third; its composition is given by eqn 20.

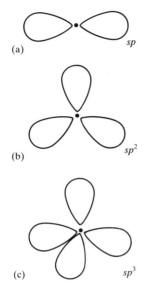

Fig. 14.30 Three important types of symmetrical equivalent hybrid orbitals: (a) linear (180°) *sp* hybrids, (b) trigonal planar (120°) *sp²* hybrids, and (c) tetrahedral (109°) *sp³* hybrids.

Table 14.1. Hybrid orbitals

Coordination number	Shape	Hybridization
2	Linear	sp
3	Trigonal planar	sp^2
4	Tetrahedral	sp^3
5	Bipyramidal	sp^3d
6	Octahedral	sp^3d^2

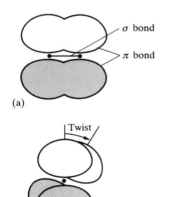

Fig. 14.31 (a) The structure of a double bond consists of a σ bond plus a π bond. (b) The bond is torsionally rigid because the (p,p)-overlap is reduced as the groups are twisted out of the plane.

$a'^2 = 0.60$, implying 60 per cent s character and a hybrid orbital of the form $0.77s + 0.63p$.

In a number of important molecules (e.g. NH_3 and CH_4), there are more than two equivalent bonds. For instance, three equivalent hybrids may be formed from the basis $(2s, 2p_x, 2p_y)$, which is just a special case of the H_2O calculation with $\Phi = 120°$. We find three hybrids of the form $s + \sqrt{2}p$ directed towards the corners of an equilateral triangle (Fig. 14.30). The s and p orbitals are present in the proportion $1:2$, and so the orbitals are called sp^2 hybrids. The basis $(2s, 2p_x, 2p_y, 2p_z)$ leads to four equivalent hybrids of composition $s + \sqrt{3}p$. These sp^3 hybrids point towards the corners of a regular tetrahedron.

Numerous other forms of hybridization may occur. Some hybridizations giving symmetrical arrangements are listed in Table 14.1.

The carbon–carbon double bond

The structure of ethene (ethylene, $CH_2\!\!=\!\!CH_2$) is easy to understand in the light of what has already been described. The two carbon atoms are both approximately sp^2 hybridized, giving three almost equivalent trigonal-planar orbitals and a single p orbital perpendicular to the plane. Two H atoms form $(Csp^2, H1s)\sigma$ bonds to each atom, and the two CH_2 groups link through a $(Csp^2, Csp^2)\sigma$ bond (Fig. 14.31a). The latter draws the $2p$ orbitals together: they overlap and form a π bond. The double bond of unsaturated compounds is this $\sigma^2\pi^2$ configuration (all double bonds, in fact, consist of a σ bond and a π bond). The **torsional rigidity** of a double bond, its resistance to twisting, arises from the reduction of the $(C2p, C2p)$ overlap and the rise of energy (the loss of bonding stabilization) that occurs as one CH_2 group is rotated relative to the other (Fig. 14.31b). The reactivity of a carbon–carbon double bond reflects the fact that the energy is lowered if the single π bond is replaced by two σ bonds, one for each atom.

Delocalized systems

Molecular orbital theory takes large molecules and extended aggregates of atoms, such as solid materials, into its stride. We shall consider two examples: **conjugated molecules**, in which there is an alternation of single and double bonds along a chain of carbon atoms, and solids.

14.10 The Hückel approximation

The π molecular orbital energy level diagrams of conjugated molecules can be constructed using a set of approximations suggested by Erich Hückel. In his approach, the π orbitals are treated separately from the σ orbitals, and the latter form a rigid framework that determines the shape of the molecule. All the C atoms are treated identically, so all Coulomb integrals α are set equal. For example, in ethene, we take the σ bonds as fixed, and concentrate on finding the energies of the single π bond and its companion antibond. In butadiene $(CH_2\!\!=\!\!CH\!\!-\!\!CH\!\!=\!\!CH_2)$, the σ framework $H_2C\!\!-\!\!CH\!\!-\!\!CH\!\!-\!\!CH_2$ is taken as fixed, and we concentrate on finding the π orbitals spreading across the four C atoms.

The secular determinant

We express the π orbitals as LCAOs of the C$2p$ orbitals. In ethene we would write

$$\psi = c_A \psi(A) + c_B \psi(B)$$

and in butadiene

$$\psi = c_A \psi(A) + c_B \psi(B) + c_C \psi(C) + c_D \psi(D)$$

where the $\psi(A)$ is a C$2p$ orbital on atom A, and so on. Next, the optimum coefficients and energies are found by the variation principle as explained in Section 14.6. That is, we have to solve the secular determinant, which in the case of ethene is eqn 15 with $\alpha_A = \alpha_B = \alpha$. The determinant for butadiene is similar, but more atoms contribute and, being at various distances from each other, they have different overlap and resonance integrals:

Ethene:

$$\begin{vmatrix} \alpha - E & \beta - ES \\ \beta - ES & \alpha - E \end{vmatrix} = 0$$

Butadiene:

$$\begin{vmatrix} \alpha - E & \beta_{AB} - ES_{AB} & \beta_{AC} - ES_{AC} & \beta_{AD} - ES_{AD} \\ \beta_{BA} - ES_{BA} & \alpha - E & \beta_{BC} - ES_{BC} & \beta_{BD} - ES_{BD} \\ \beta_{CA} - ES_{CA} & \beta_{CB} - ES_{CB} & \alpha - E & \beta_{CD} - ES_{CD} \\ \beta_{DA} - ES_{DA} & \beta_{DB} - ES_{DB} & \beta_{DC} - ES_{DC} & \alpha - E \end{vmatrix} = 0$$

The roots of the ethene determinant can be found very easily (they are the same as those in Example 14.6). However, for elementary calculations, the roots of the butadiene determinant are obviously going to prove difficult to find. In a modern computation all the resonance integrals and overlap integrals would be computed, but a rough idea of the molecular orbital energy level diagram can be obtained very readily if we make the following additional **Hückel approximations**:

All overlap integrals are set equal to zero.
All resonance integrals between non-neighbours are set equal to zero.
All remaining resonance integrals are set equal (to β).

These are obviously very severe approximations, but they let us calculate at least a general picture of the molecular orbital energy levels with very little work.

Ethene and frontier orbitals

For ethene, the Hückel approximations lead to

$$\begin{vmatrix} \alpha - E & \beta \\ \beta & \alpha - E \end{vmatrix} = 0$$

and its roots are

$$E_\pm = \alpha \pm \beta$$

The + sign corresponds to the bonding combination (β is negative) and the − sign corresponds to the antibonding combination (Fig. 14.32). The building-up principle then leads to the configuration $1\pi^2$, because each carbon atom supplies one electron to the π system, in agreement with the

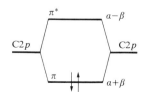

Fig. 14.32 The Hückel molecular orbital energy levels of ethene. Two electrons occupy the lower π orbital.

previous qualitative discussion. However, it goes beyond, because we have an estimate of the bond energy (β), and can see that an excited state of the molecule, when an electron is excited into the π^* orbital, lies about 2β above the ground state. (β is often left as a parameter; a rough value for $(C2p,C2p)\pi$-overlap unoccupied-bonds is about $-75\,\text{kJ}\,\text{mol}^{-1}$, $-0.8\,\text{eV}$.) The **highest occupied molecular orbital**, the HOMO, is the π orbital; the **lowest unoccupied molecular orbital**, the LUMO, is the π^* orbital. These two orbitals jointly form the **frontier orbitals** of the molecule. The frontier orbitals are important because they are largely responsible for the chemical and spectroscopic properties of the molecule.

Butadiene and π-electron binding energy

For butadiene, the approximations result in the determinant

$$\begin{vmatrix} \alpha - E & \beta & 0 & 0 \\ \beta & \alpha - E & \beta & 0 \\ 0 & \beta & \alpha - E & \beta \\ 0 & 0 & \beta & \alpha - E \end{vmatrix} = 0$$

Example 14.11: *Finding the roots of a determinant*

Find the roots of the butadiene secular determinant.

Answer. The determinant is expanded in a series of moves like the 2×2 determinant treated in Example 14.6:

$$\begin{vmatrix} \alpha - E & \beta & 0 & 0 \\ \beta & \alpha - E & \beta & 0 \\ 0 & \beta & \alpha - E & \beta \\ 0 & 0 & \beta & \alpha - E \end{vmatrix}$$

$$= (\alpha - E)\begin{vmatrix} \alpha - E & \beta & 0 \\ \beta & \alpha - E & \beta \\ 0 & \beta & \alpha - E \end{vmatrix} - \beta\begin{vmatrix} \beta & \beta & 0 \\ 0 & \alpha - E & \beta \\ 0 & \beta & \alpha - E \end{vmatrix}$$

$$= (\alpha - E)(\alpha - E)\begin{vmatrix} \alpha - E & \beta \\ \beta & \alpha - E \end{vmatrix} - (\alpha - E)\beta\begin{vmatrix} \beta & \beta \\ 0 & \alpha - E \end{vmatrix}$$

$$- \beta^2\begin{vmatrix} \alpha - E & \beta \\ \beta & \alpha - E \end{vmatrix} + \beta^2\begin{vmatrix} 0 & \beta \\ 0 & \alpha - E \end{vmatrix}$$

$$= (\alpha - E)^4 - (\alpha - E)^2\beta^2 - (\alpha - E)^2\beta^2 - (\alpha - E)^2\beta^2 + \beta^4$$

$$= (\alpha - E)^4 - 3(\alpha - E)^2\beta^2 + \beta^4$$

$$= 0$$

The expanded determinant has the form of a quadratic equation

$$x^2 - 3x + 1 = 0 \quad \text{with} \quad x = \left(\frac{\alpha - E}{\beta}\right)^2$$

and the roots are $x = 2.62$ and 0.38. Therefore, the energies of the four LCAO–MOs are

$$E = \alpha \pm 1.62\beta, \qquad \alpha \pm 0.62\beta$$

Exercise. Write down and expand the secular determinant for cyclobutadiene.

[See Example 14.12]

We show in the example that the energies of the four LCAO–MOs are

$$E = \alpha \pm 1.62\beta, \ \alpha \pm 0.62$$

These are drawn in Fig. 14.33. There are four electrons to accommodate, and so the ground state configuration is $1\pi^2 2\pi^2$. The frontier orbitals of butadiene are the 2π orbital (the HOMO, which is bonding) and the $3\pi^*$ orbital (the LUMO, which is antibonding).

There is an important point that emerges when we calculate the total **π-electron binding energy** E_π, the sum of the energies of each π electron, and compare it with what we find in ethene. In ethene the total energy is

$$E_\pi = 2(\alpha + \beta) = 2\alpha + 2\beta$$

In butadiene it is

$$E_\pi = 2(\alpha + 1.62\beta) + 2(\alpha + 0.62\beta) = 4\alpha + 4.48\beta$$

Therefore, the energy of the molecule lies lower by 0.48β (about $-36\,\text{kJ mol}^{-1}$) than the sum of two individual π bonds. This extra stabilization of a conjugated system is called the **delocalization energy**.

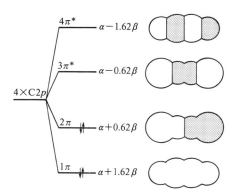

Fig. 14.33 The Hückel molecular orbital energy levels of butadiene and the top view of the corresponding π orbitals. The four p electrons (one supplied by each C) occupy the two lower π orbitals. Note that the orbitals are delocalized.

Example 14.12: *Estimating the delocalization energy*

Use the Hückel approximation to find the energies of the π orbitals of cyclobutadiene, and estimate the delocalization energy.

Answer. We need to set up the secular determinant using the same basis as for butadiene, but noting that A and D are now neighbours:

$$\begin{vmatrix} \alpha - E & \beta & 0 & \beta \\ \beta & \alpha - E & \beta & 0 \\ 0 & \beta & \alpha - E & \beta \\ \beta & 0 & \beta & \alpha - E \end{vmatrix} = 0$$

The determinant expands to

$$z^2(z^2 - 4) = 0 \qquad z = \frac{\alpha - E}{\beta}$$

Since the solutions are

$$z = 0, 0, -2, +2$$

the energies of the orbitals are

$$E = \alpha + 2\beta, \ \alpha, \ \alpha, \ \alpha - 2\beta$$

Four electrons must be accommodated. Two occupy the lowest orbital (of energy $\alpha + 2\beta$), and two occupy the doubly degenerate orbitals (of energy α). The total energy, disregarding electron–electron repulsions, is therefore $4\alpha + 4\beta$. Two isolated π bonds would have an energy $4\alpha + 4\beta$; therefore, in this case, the delocalization energy is zero.

Exercise. Repeat the calculation for benzene. [Next subsection]

Benzene and aromatic stability

The most notable example of delocalization conferring extra stability is benzene and the aromatic molecules based on its structure. Its six C atoms

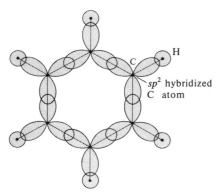

C

H

sp² hybridized C atom

Fig. 14.34 The σ framework of benzene is formed by the overlap of C*sp²*-hybrids, which fit without strain into a hexagonal arrangement.

are sp^2 hybridized with a single perpendicular $2p$ orbital. One H atom is bonded by $(Csp^2, H1s)$ overlap to each C carbon, and the remaining hybrids overlap to give a regular hexagon of atoms (Fig. 14.34). The internal angle of a regular hexagon is 120° and so the sp^2 hybridization is ideally suited for forming σ bonds. We see that benzene's hexagonal shape permits strain-free σ bonding.

The six C$2p$ orbitals overlap to give six π orbitals that spread all round the ring. Their energies are calculated within the Hückel approximation by solving the secular determinant

$$\begin{vmatrix} \alpha - E & \beta & 0 & 0 & 0 & \beta \\ \beta & \alpha - E & \beta & 0 & 0 & 0 \\ 0 & \beta & \alpha - E & \beta & 0 & 0 \\ 0 & 0 & \beta & \alpha - E & \beta & 0 \\ 0 & 0 & 0 & \beta & \alpha - E & \beta \\ \beta & 0 & 0 & 0 & \beta & \alpha - E \end{vmatrix} = 0$$

When this determinant is expanded in the same way as in Examples 14.11 and 14.12, the roots are found to be simply

$$E = \alpha \pm 2\beta, \ \alpha \pm \beta, \ \alpha \pm \beta$$

as shown in Fig. 14.35. The orbitals there have been given special labels that we explain in Chapter 15.

We now apply the building-up principle. There are six electrons to accommodate (one from each C atom), and so the three lowest orbitals (a_{2u} and the doubly-degenerate pair e_{1g}) are fully occupied, giving the ground state configuration $a_{2u}^2 e_{1g}^4$. A significant point is that the only molecular orbitals occupied are those with net bonding character.

The π-electron energy of benzene is

$$E_\pi = 2(\alpha + 2\beta) + 4(\alpha + \beta) = 6\alpha + 8\beta$$

If we ignored delocalization and thought of the molecule as having three

Fig. 14.35 The Hückel orbitals of benzene and the corresponding energy levels. The labels are explained in Chapter 15. The bonding and antibonding character of the delocalized orbitals reflects the numbers of nodes between the atoms. In the ground state, only the bonding orbitals are occupied.

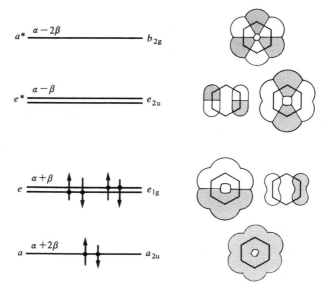

a^* $\alpha - 2\beta$ b_{2g}

e^* $\alpha - \beta$ e_{2u}

e $\alpha + \beta$ e_{1g}

a $\alpha + 2\beta$ a_{2u}

isolated π bonds, then it would be ascribed a π-electron energy of only $3(2\alpha + 2\beta) = 6\alpha + 6\beta$. The delocalization energy is therefore $2\beta \approx -150$ kJ mol^{-1}, which is considerably more than for butadiene.

In terms of molecular orbital theory, aromatic stability can be traced to two main contributions. First, the shape of the regular hexagon is ideal for the formation of strong σ bonds: the σ framework is relaxed and without strain. Second, the π orbitals are such as to be able to accommodate all the electrons in bonding orbitals, and the delocalization energy is large.

14.11 The band theory of solids

The extreme case of delocalization is a solid, in which atom after atom lies in a three-dimensional array and takes part in bonding spreading throughout the sample. We shall distinguish two types of solid according to the variation of their electrical conductivity with temperature. A **metallic conductor** is a substance with a conductivity that decreases as the temperature is raised. A **semiconductor** is a substance with a conductivity that increases as the temperature is raised. A semiconductor generally has a lower conductivity than a metallic conductor, but the magnitude of the conductivity is not the criterion of the distinction. It is conventional to classify semiconductors with very low electrical conductivities as **insulators**. We shall use the latter term, but it should be appreciated that it is one of convenience rather than one of fundamental significance.

We shall consider a **one-dimensional solid** initially, which consists of a single, infinitely long line of atoms, each one having one s orbital available for forming molecular orbitals. We can construct the LCAO–MOs of the solid by adding atoms to a line and then find the electronic structure using the building-up principle.

The formation of bands

One atom contributes one s orbital at a certain energy (Fig. 14.36a). When a second atom is brought up it overlaps the first, and forms a bonding and

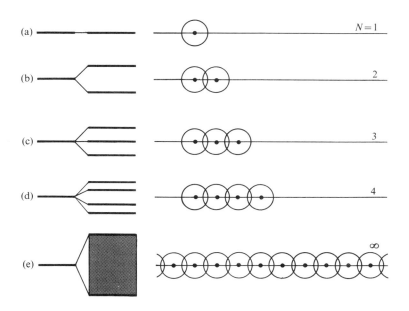

Fig. 14.36 The formation of a band of N molecular orbitals by successive addition of N atoms to a line. Note that the band remains of finite width, and although it looks continuous, it consists of N different orbitals.

antibonding orbital (Fig. 14.36b). The third atom overlaps its nearest neighbour (and only slightly the next-nearest), and from these three atomic orbitals, three molecular orbitals are formed (Fig. 14.36c). The fourth atom leads to the formation of a fourth molecular orbital (Fig. 14.36d). At this stage we can begin to see that the general effect of bringing up successive atoms is to spread the range of energies covered by the molecular orbitals, and also to fill in the range of energies with more and more orbitals (one more for each atom). When N atoms have been added to the line, there are N molecular orbitals covering a band of finite width, and the Hückel secular determinant is

$$
\begin{vmatrix}
\alpha - E & \beta & 0 & 0 & 0 & & 0 \\
\beta & \alpha - E & \beta & 0 & 0 & \cdots & 0 \\
0 & \beta & \alpha - E & \beta & 0 & \cdots & 0 \\
0 & 0 & \beta & \alpha - E & \beta & \cdots & 0 \\
\vdots & \vdots & \vdots & \vdots & \vdots & \cdots & \vdots \\
0 & 0 & 0 & 0 & 0 & & \alpha - E
\end{vmatrix} = 0
$$

where β is now the (s, s) resonance integral. The theory of determinants applied to such a symmetrical example as this (technically a 'tridiagonal determinant') leads to the following expression for the roots:

$$
E_k = \alpha + 2\beta \cos \frac{k\pi}{N + 1} \qquad k = 1, 2, \ldots, N \tag{21}
$$

When N is infinitely large, the difference between neighbouring energy levels (the energies corresponding to k and $k + 1$) is infinitely small, but the band still has finite width:

$$
E_N - E_1 \to 4\beta \quad \text{as} \quad N \to \infty
$$

We can think of this band as consisting of N different molecular orbitals, the lowest-energy orbital ($k = 1$) being fully bonding, and the highest-energy orbital ($k = N$) being fully antibonding between adjacent atoms (Fig. 14.37).

The band formed from overlap of s orbitals is called the **s band**. If the atoms have p orbitals available, the same procedure leads to a **p band** (Fig. 14.37). If the atomic p orbitals lie higher in energy than the s orbitals, then the p band lies higher than the s band, and there may be a **band gap**.

The occupation of orbitals at $T = 0$

Now consider the electronic structure of a solid formed from atoms each able to contribute one electron (e.g. the alkali metals). There are N atomic orbitals and therefore N molecular orbitals squashed into an apparently continuous band. There are N electrons to accommodate. Since the orbitals in the bands are so close together, electrons can be excited out of them by thermal motion of the atoms. This is a complication we can avoid by considering the solid at $T = 0$ when there is no such motion, and all the electrons occupy the lowest available orbitals.

At $T = 0$, only the lowest $\frac{1}{2}N$ molecular orbitals are occupied (Fig. 14.38), and the HOMO is called the **Fermi level**. However, unlike in the discrete

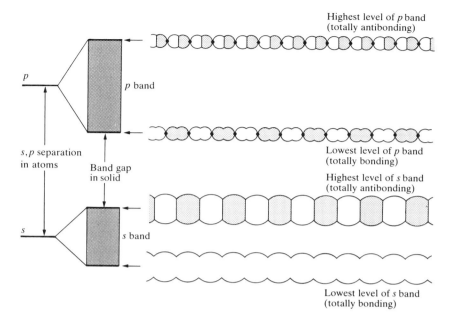

Highest level of p band
(totally antibonding)

p band

s,p separation
in atoms

Band gap
in solid

s band

Lowest level of p band
(totally bonding)

Highest level of s band
(totally antibonding)

Lowest level of s band
(totally bonding)

Fig. 14.37 The overlap of s orbitals gives rise to an s band, and the overlap of p orbitals gives rise to a p band. In this case the s and p orbitals of the atoms are so widely spaced that there is a band gap. In many cases the separation is less , and the bands overlap.

molecules we have considered so far, there are empty orbitals very close in energy to the Fermi level, and so it requires hardly any energy to excite the uppermost electrons. Some of the electrons are therefore very mobile, and give rise to electrical conductivity.

The occupation of orbitals at $T > 0$

At temperatures above absolute zero, there is no sharp distinction between occupied and unoccupied orbitals in a band since electrons can be excited by the thermal motion of the atoms. The population P of the orbitals is given by the **Fermi–Dirac distribution**, a version of the Boltzmann distribution that takes into account the effect of the Pauli principle:

$$P = \frac{1}{e^{(E - E_F)/kT} + 1} \tag{22a}$$

E_F is the **Fermi energy**, the energy of the level for which $P = \frac{1}{2}$. The shape of the Fermi–Dirac distribution is shown in Fig. 14.39. For energies well above the Fermi energy, the 1 in the denominator can be neglected, and

$$P \approx e^{-(E - E_F)/kT} \tag{22b}$$

The population now resembles a Boltzmann distribution, decaying exponentially with increasing energy. The higher the temperature, the longer the exponential tail of the Fermi–Dirac distribution.

The electrical conductivity of a metallic solid *decreases* with increasing temperature even though more electrons are excited into empty orbitals. This apparent paradox is resolved by noting that the increase in temperature causes more vigorous thermal motion of the atoms, so collisions between the moving electrons and an atom are more likely. That is, the electrons are scattered out of their paths through the solid, and are less efficient at transporting charge.

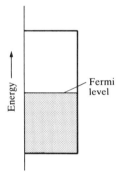

Fig. 14.38 When N electrons occupy a band of N orbitals, it is only half full and the electrons near the Fermi level (the top of the filled levels) are mobile.

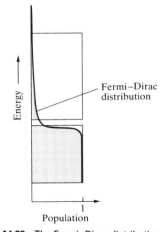

Fig. 14.39 The Fermi–Dirac distribution, which gives the population of the levels at a temperature T. The high-energy tail decays exponentially towards zero.

421

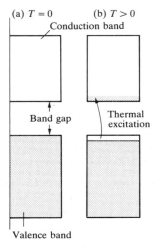

Fig. 14.40 (a) When 2N electrons are present, the band is full and the material is an insulator at T = 0. (b) At temperatures above T = 0, electrons populate the levels of the upper 'conduction' band at the expense of the filled 'valence' band and the solid is a semiconductor.

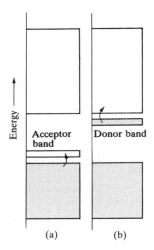

Fig. 14.41 (a) A dopant with fewer electrons than its host can form a narrow band that accepts electrons from the valence band. The holes in the band are mobile, and the substance is a p-type semiconductor. (b) A dopant with more electrons than its host forms a narrow band that can supply electrons to the conduction band. The electrons it supplies are mobile, and the substance is an n-type semiconductor.

Insulators and semiconductors

When each atom provides two electrons, the $2N$ electrons fill the N orbitals of the s band. The Fermi level now lies at the top of the band (at $T = 0$), and there is a gap before the next band begins (Fig. 14.40a). As the temperature is increased, the tail of the Fermi–Dirac distribution extends across the gap, and electrons populate the empty orbitals of the upper band (Fig. 14.40b). They are now mobile, and the solid is an electric conductor. In fact, it is a semiconductor, because the electrical conductivity depends on the number of electrons that are promoted across the gap, and that increases as the temperature is raised. If the gap is large, though, very few electrons will be promoted at ordinary temperatures, and the conductivity will remain close to zero, giving an insulator. Thus, the conventional distinction between an insulator and a semiconductor is related to the size of the band gap and is not absolute like the distinction between a metal (incomplete bands at $T = 0$) and a semiconductor (full bands at $T = 0$).

Another method of increasing the number of charge carriers and enhancing the semiconductivity of a solid is to implant foreign atoms into an otherwise pure material. If these dopants can trap electrons, they withdraw electrons from the filled band, leaving holes which allow the remaining electrons to move (Fig. 14.41a). This gives rise to **p-type semiconductivity**, the p indicating that the holes are relatively positive to the electrons in the band. Alternatively, a dopant might carry excess electrons (e.g. P atoms introduced into germanium), and these additional electrons occupy otherwise empty bands, giving **n-type semiconductivity** (Fig. 14.41b), where n denotes the negative charge of the carriers. The preparation of doped but otherwise ultrapure materials was described in Section 8.6.

Further reading

Principles of molecular structure

C. A. Coulson (revised by R. McWeeny), *The shape and structure of molecules* (2nd edn). Oxford University Press (1982).

R. L. De Kock and H. B. Gray, *Chemical structure and bonding*. Benjamin Cummings, Menlo Park (1980).

J. N. Murrell, S. F. A. Kettle, and J. M. Tedder, *Valence theory*. Wiley, New York (1965).

P. W. Atkins, *Molecular quantum mechanics* (2nd edn). Oxford University Press (1983).

D. F. Shriver, P. W. Atkins, and C. H. Langford, *Inorganic chemistry*. Oxford University Press and W. H. Freeman & Co (1990).

R. McWeeny, *Coulson's valence*. Oxford University Press (1979).

L. Pauling, *The nature of the chemical bond*. Cornell University Press (1960).

Calculations

A. Hinchliffe, *Computational quantum chemistry*. Wiley, New York (1988).

M. Karplus and R. N. Porter, *Atoms and molecules*. Van Nostrand, New York (1970).

W. L. Jorgensen and L. Salem, *The organic chemist's book of orbitals*. Academic Press, New York (1973).

E. Steiner, *Molecular wavefunctions*. Cambridge University Press (1976).

H. Shaefer III, *Quantum chemistry (the development of ab initio methods in molecular electronic structure theory)*. Clarendon Press, Oxford (1984).

Solids

P. A. Cox, *The electronic structure and chemistry of solids*. Oxford University Press (1987).

A. R. West, *Solid state chemistry*. Wiley, New York (1984).

Exercises

14.1 Give the ground state electron configurations and bond orders of (a) Li_2, (b) Be_2, and (c) C_2.

14.2 Give the ground state electron configurations of (a) H_2^-, (b) N_2, and (c) O_2.

14.3 Give the ground state electron configuration of (a) CO, (b) NO, and (c) CN^-.

14.4 From the ground state electron configurations of B_2 and C_2, predict which molecule should have the greater bond dissociation energy.

14.5 Which of the molecules N_2, NO, O_2, C_2, F_2, and CN would you expect to be stabilized by (a) the addition of an electron to form AB^-, (b) the removal of an electron to form AB^+?

14.6 Sketch the molecular orbital energy level diagrams for (a) CO, (b) XeF and deduce their ground-state electron configurations. Is XeF likely to have a shorter bond length than XeF^+?

14.7 Where it is appropriate, give the parity of (a) π^* in F_2, (b) σ^* in NO, (c) δ in Tl_2, (d) δ^* in Fe_2.

14.8 Give the parities of the six π molecular orbitals of benzene.

14.9 The term symbol for the ground state of the N_2^+ ion is $^2\Sigma_g$. What is the total spin and total orbital angular momentum of the molecule? Show that the term symbol agrees with the electron configuration that would be predicted using the building-up principle.

14.10 One of the excited states of the C_2 molecule has the valence electron configuration $2s\sigma_g^2 2s\sigma_u^{*2} 2p\pi_u^3 2p\pi_g^1$. Give the multiplicity and parity of the term.

14.11 Use the electron configurations of NO and N_2 to predict which is likely to have the shorter bond length.

14.12 One of the excited states of H_2 is $^3\Pi_u$, and can be considered to be formed from one H atom in its ground state and the other in an excited state. Give the electron configuration of the molecule.

14.13 Show that the sp^2 hybrid orbital $(s + \sqrt{2}p)/3^{1/2}$ is normalized to 1 if the s and p orbitals are normalized to 1.

14.14 Normalize the molecular orbital $\psi_s(A) + \lambda\psi_s(B)$ in terms of the parameter λ and the overlap integral S.

14.15 Confirm that the bonding and antibonding combinations $\psi_s(A) \pm \psi_s(B)$ are mutually orthogonal in the sense that their mutual overlap is zero.

14.16 Estimate the proportion of S $3s$ orbital character in the bonding orbitals of H_2S, for which the bond angle is 92.2°.

14.17 Which of the following triatomic molecules and ions are expected to be linear: (a) CO_2, (b) NO_2, (c) NO_2^+, (d) NO_2^-, (e) SO_2, (f) H_2O, (g) H_2O^{2+}? Give reasons in each case.

14.18 Construct the molecular orbital energy level diagrams of (a) ethene (ethylene) and (b) ethyne (acetylene) on the basis that the molecules are formed from the appropriately hybridized CH_2 or CH fragments.

14.19 Write down the secular determinants for (a) linear H_3, (b) cyclic H_3 within the Hückel approximation.

14.20 Predict the elecronic configurations of (a) the benzene anion, (b) the benzene cation. Estimate the π-bond energy in each case.

Problems

Numerical problems

14.1 Show that a wave $\cos kx$ centred on A (so that x is measured from A) interferes with a similar wave $\cos k'x$ centred on B (with x measured from B) a distance R away, then constructive interference occurs in the intermediate region when $k = k' = \pi/2R$ and destructive interference if $kR = \frac{1}{2}\pi$ and $k'R = \frac{3}{2}\pi$.

14.2 The overlap integral between two H1s orbitals on

nuclei separated by a distance R is

$$S = \left\{1 + \frac{R}{a_0} + \frac{1}{3}\left(\frac{R}{a_0}\right)^2\right\}e^{-R/a_0}$$

Plot this function for $0 \leq R < \infty$.

14.3 Before doing the calculation below, sketch how the overlap between an s orbital and a σp orbital can be expected to depend on their separation. The overlap integral between an H1s orbital and a H2p orbital on nuclei separated by a distance R is

$$S = \frac{R}{2a_0}\left\{1 + \frac{R}{a_0} + \frac{1}{3}\left(\frac{R}{a_0}\right)^2\right\}e^{-R/a_0}$$

Plot this function, and find the separation for which the overlap is a maximum.

14.4 Calculate the total amplitude of the normalized bonding and antibonding LCAO–MOs that may be formed from two H1s orbitals at a separation of 106 pm. Plot the two amplitudes for positions along the molecular axis both inside and outside the internuclear region.

14.5 Repeat the calculation in Problem 14.4 but plot the probability densities of the two orbitals. Then form the 'difference density', the difference between ψ^2 and $\frac{1}{2}\{\psi_s(A)^2 + \psi_s(B)^2\}$.

14.6 Imagine a small electron-sensitive probe of volume 1.00 pm^3 inserted into an H$_2^+$ molecule ion in its ground state. Calculate the probability that it will register the presence of an electron at the following positions: (a) at nucleus A, (b) at nucleus B, (c) half way between A and B, (d) at a point 20 pm along the bond from A and 10 pm perpendicularly. Do the same for the molecule ion the instant after the electron has been excited into the antibonding LCAO–MO.

14.7 The energy of H$_2^+$ with internuclear separation R is given by the expression

$$E = E_H - \frac{V_1 + V_2}{1 + S} + \frac{e^2}{4\pi\varepsilon_0 R}$$

where E_H is the energy of an isolated H atom, V_1 is the attractive potential energy between the electron centred on one nucleus and the charge of the other nucleus; V_2 is the attraction between the overlap density and one of the nuclei; S is the overlap integral. The values are given below. Plot the molecular potential energy curve and find the bond dissociation energy (in eV) and the equilibrium bond length.

R/a_0	0	1	2	3	4
$V_1/\bar{R}_H$	1.000	0.729	0.473	0.330	0.250
$V_2/\bar{R}_H$	1.000	0.736	0.406	0.199	0.092
S	1.000	0.858	0.587	0.349	0.189

where $\bar{R}_H = 27.3$ eV, $a_0 = 52.9$ pm, and $E_H = -\frac{1}{2}\bar{R}_H$.

14.8 The same data as in Problem 14.7 may be used to calculate the molecular potential energy curve for the anti-

bonding orbital, which is given by

$$E = E_H - \frac{V_1 - V_2}{1 - S} + \frac{e^2}{4\pi\varepsilon_0 R}$$

Plot the curve.

14.9 In the 'free electron molecular orbital' (FEMO) theory, the electrons in a conjugated molecule are treated as independent particles in a box of length L. Sketch the form of the two occupied orbitals in butadiene predicted by this model and predict the minimum excitation energy of the molecule. The tetraene CH$_2$=CHCH=CHCH=CHCH=CH$_2$ can be treated as a box of length 8R, where $R \approx 140$ pm (as in this case, an extra half bond-length is often added at each end of the box). Calculate the minimum excitation energy of the molecule and sketch the HOMO and LUMO. Estimate the colour a sample of the compound is likely to appear in white light.

Theoretical problems

14.10 An sp^2 hybrid orbital that lies in the xy plane and makes an angle of 120° to the x-axis has the form

$$\psi = \frac{1}{3^{1/2}}\left(s - \frac{1}{2^{1/2}}p_x + \frac{3^{1/2}}{2^{1/2}}p_y\right)$$

Use hydrogenic atomic orbitals to write the explicit form of the hybrid orbital. Show that it has its maximum amplitude in the direction specified.

14.11 Use the expressions in Problems 14.7 and 14.8 to show that the antibonding orbital is more antibonding than the bonding orbital is bonding at all internuclear separations.

14.12 Derive the expressions used in Problem 14.7 using the normalized LCAO–MOs for the H$_2^+$ molecule-ion. Proceed by evaluating the expectation value of the hamiltonian for the ion. Make use of the fact that $\psi_s(A)$ and $\psi_s(B)$ each individually satisfy the Schrödinger equation for an isolated H atom.

14.13 Construct the Walsh diagram for an AH$_3$ molecule, and use it to predict the shapes of (a) NH$_3$, (b) CH$_3^+$.

14.14 Take as a trial function for the ground state of the hydrogen atom (a) e^{-kr}, (b) e$^{-kr^2}$ and use the variation principle to find the optimum value of k in each case. Identify the better wavefunction. The only part of the laplacian that need be considered is the part that involves radial derivatives:

$$\nabla^2\psi = \frac{1}{r}\frac{d^2(r\psi)}{dr^2}$$

Symmetry: its description and consequences

15

Check-list of key ideas

1. The significance of *symmetry operations* and of the corresponding *symmetry elements* (Section 15.1).

2. The *classification* of molecules into *point groups* according to the symmetry elements they possess (Section 15.2).

3. The identification of *polar molecules* and *chiral molecules* from their point groups (Section 15.3).

4. The *representation* of symmetry operations by matrices (Section 15.5).

5. The *character* of symmetry operations and the recognition of *classes* of operation (Section 15.5).

6. A representation as a *direct sum* of *irreducible representations* (Section 15.6).

7. The *character table* of a point group and the recognition of the *symmetry species* of the irreducible representations of the group (Section 15.6).

8. The transformation properties of x, y, and z and of functions that can be built from them (Section 15.7).

9. The identification of the symmetry properties of functions and products of functions that guarantees that their integral must be zero (Section 15.8 and Examples 15.7 and 15.8).

10. Using symmetry to decide if orbitals have zero overlap (Section 15.8 and Example 15.9).

11. The deduction of *selection rules* from the point group of the molecule (Section 15.8 and Example 15.10).

12. The construction of *symmetry-adapted linear combinations* of atomic orbitals for the construction of molecular orbitals (Section 15.9).

In this chapter we sharpen the concept of 'shape' into a precise definition of 'symmetry', and show that it may be discussed systematically. We shall see how to classify any molecule according to its symmetry, and how to use this classification to discuss molecular properties without the need for detailed calculation. We shall see how to treat systematically many of the topics that we have mentioned in previous chapters. We shall see, for instance, how to select linear combinations of atomic orbitals that match the symmetry of the nuclear framework, and we remarked in Chapter 14 that this grouping of orbitals is one of the central features of molecular orbital theory. We shall see how to judge from its symmetry classification alone whether a molecule is polar or chiral. Through symmetry we shall be able to derive the selection rules that govern the intensities of spectroscopic transitions, not only of the electronic transitions that are responsible for the colours of substances, but also of the rotational and vibrational transitions that give rise to their microwave and infrared spectra.

The detailed discussion of symmetry is called **group theory**. Much of group theory is a systematic summary of common sense about the symmetries of objects. However, because group theory is systematic, its rules can be applied in a straightforward, mechanical way, and in some cases it gives unexpected results. In most cases the theory gives a simple, direct method for arriving at useful conclusions with the minimum of calculation, and this is the aspect we stress here.

The symmetry elements of objects

Some objects are 'more symmetrical' than others. A sphere is more symmetrical than a cube because it looks the same after it has been rotated through any angle about any diameter. A cube looks the same only if it is rotated through 90°, 180°, or 270° about an axis passing through the centres of any of its opposite faces (Fig. 15.1), or by 120° or 240° about an axis passing through any of its opposite corners. Similarly the NH_3 molecule is 'more symmetrical' than the H_2O molecule because it looks the same after rotations of 120° or 240° about the axis shown in Fig. 15.2, whereas H_2O looks the same only after a rotation of 180°.

An action that leaves an object looking the same after it has been carried out is called a **symmetry operation**. There is a corresponding **symmetry element** for each symmetry operation; this is the point, line, or plane with respect to which the symmetry operation is performed. We shall use the same symbol for the symmetry operation and the corresponding element, but always say which is intended if it is important to distinguish between them. We shall see that we can classify molecules by identifying all their symmetry elements, and grouping together those with identical elements. This procedure puts a sphere into a different group from a cube, and NH_3 into a different group from H_2O.

15.1 Operations and elements

There are five kinds of symmetry operation (and five kinds of symmetry element) that leave at least a single point unchanged and hence give rise to the **point groups**. When we consider crystals (Chapter 21), we shall refer to the symmetries arising from translation through space. These more extensive groups are called **space groups**.

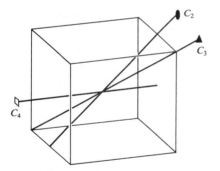

Fig. 15.1 Some of the symmetry elements of a cube. The twofold, threefold, and fourfold axes are labelled with the conventional symbols.

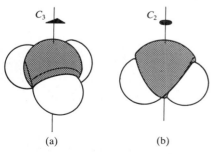

Fig. 15.2 (a) The NH_3 molecule has a threefold (C_3) axis and (b) the H_2O molecule has a twofold (C_2) axis.

The **identity** E consists of doing nothing; the corresponding element is the entire object. Since every molecule is indistinguishable from itself if nothing is done to it, every object possesses at least the identity element. One reason for including it is that some molecules (e.g. CHClBrF) have only this symmetry element; another reason is technical and connected with the formulation of group theory.

An **n-fold rotation** (the operation) about an **n-fold axis of symmetry** C_n (the corresponding element) is a rotation through $360°/n$. An H_2O molecule has a two-fold axis, C_2. An NH_3 molecule has a three-fold axis C_3 with which are associated two operations, one being $120°$ rotation in a clockwise sense and the other $120°$ rotation in a counter-clockwise sense. (There is only one two-fold rotation associated with a C_2 axis because clockwise and counter-clockwise $180°$ rotations are identical.) A cube has three C_4 axes, four C_3 axes, and six C_2 axes. However, even this high symmetry is capped by a sphere, which possesses an infinite number of symmetry axes (along any diameter) of all possible orders of n. If a molecule possesses several rotation axes, the one (or more) with the greatest value of n is called the **principal axis**.

A **reflection** (the operation) in a **plane of symmetry** or a **mirror plane** σ (the element) may be either parallel or perpendicular to a principal axis of a molecule. If the plane is parallel to the principal axis, it is called **vertical** and denoted σ_v. An H_2O molecule has two vertical planes of symmetry (Fig. 15.3) and an NH_3 molecule has three. When the plane of symmetry is perpendicular to the principal axis it is called **horizontal** and denoted σ_h. The C_6H_6 molecule has a C_6 principal axis and a horizontal mirror plane (as well as several other elements). A vertical mirror plane that bisects the angle between two C_2 axes (Fig. 15.4) is called a **dihedral plane** and denoted σ_d.

In an **inversion** (the operation) through a **centre of symmetry** i (the element) we imagine taking each point in a molecule, moving it to its centre, and then moving it out the same distance on the other side. Neither an H_2O molecule nor an NH_3 molecule has a centre of inversion, but both the sphere and the cube do have one. A C_6H_6 molecule does have a centre of inversion, as does a regular octahedron (Fig. 15.5); a regular tetrahedron and a CH_4 molecule do not.

An **improper rotation** or a **rotary-reflection** (the operation) about an **axis of improper rotation** or a **rotary-reflection axis** S_n consists of an n-fold rotation axis together with a horizontal reflection. A CH_4 molecule (Fig. 15.6) has three S_4 axes.

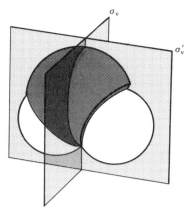

Fig. 15.3   The H_2O molecule has two mirror planes. They are both vertical (i.e. contain the principal axis), and so are denoted σ_v and σ_v'.

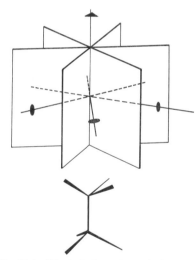

Fig. 15.4   Dihedral mirror planes (σ_d) bisect the C_2 axes perpendicular to the principal axis. The staggered conformation of ethane is an example.

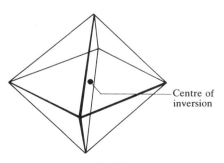

Centre of inversion

Fig. 15.5

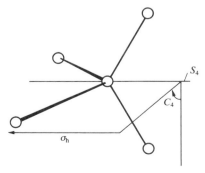

σ_h

Fig. 15.6

Fig. 15.5   A regular octahedron has a centre of inversion; a regular tetrahedron does not.

Fig. 15.6   The CH_4 molecule has a fourfold rotary-reflection axis (S_4): the molecule is indistinguishable after a $90°$ rotation followed by a reflection across the horizontal plane.

15.2 The symmetry classification of molecules

To classify molecules according to their symmetries, we list their symmetry elements and group together those with the same list. This procedure puts CH_4 and CCl_4, which both have the same symmetry as a regular tetrahedron, into the same group, and H_2O into another group.

The name of the group to which a molecule belongs is determined by the symmetry elements it possesses. There are two systems of notation (Box 15.1). The **Schoenflies system** is more common for the discussion of individual molecules, and the **Hermann–Mauguin system**, or **International system**, is used almost exclusively in the discussion of crystal symmetry.

Box 15.1 The notation for point groups

In the **international system** for point groups, which is also called the **Hermann–Mauguin system**, a number n denotes the presence of an n-fold axis and a letter m denotes a mirror plane. A diagonal line / indicates that the mirror plane is perpendicular to the symmetry axis. It is important to distinguish symmetry elements of the same type but of different classes, as in $4/mmm$ when there are the three classes of reflection σ_h, σ_v, and σ_d. A bar over a symbol indicates that that element is combined with an inversion.

The table on the right translates the **Schoenflies system** into the international system. The only groups listed are the crystallographic point groups (Section 21.1).

C_i	$\bar{1}$								
C_s	m								
C_1	1	C_2	2	C_3	3	C_4	4	C_6	6
		C_{2v}	$2mm$	C_{3v}	$3m$	C_{4v}	$4mm$	C_{6v}	$6mm$
		C_{2h}	$2/m$	C_{3h}	$\bar{6}$	C_{4h}	$4/m$	C_{6h}	$6/m$
		D_2	222	D_3	32	D_4	422	D_6	622
		D_{2h}	mmm	D_{3h}	$\bar{6}2m$	D_{4h}	$4/mmm$	D_{6h}	$6/mmm$
		D_{2d}	$\bar{4}2m$	D_{3d}	$\bar{3}m$	S_4	$\bar{4}$	S_6	$\bar{3}$
T	23	T_d	$\bar{4}3m$	T_h	$m3$				
O	432	O_h	$m3m$						

The group D_2 (222) is sometimes denoted V and called the *Vierer group* (the 'group of four').

The groups C_1, C_i, C_s

A molecule belongs to C_1 if it has no element other than the identity (e.g. CBrClFI, Fig. 15.7a), to C_i if it has the identity and the inversion (e.g. *meso*-tartaric acid, Fig. 15.7b), and to C_s if it has the identity and a plane of reflection (e.g. the quinoline molecule, Fig. 15.7c).

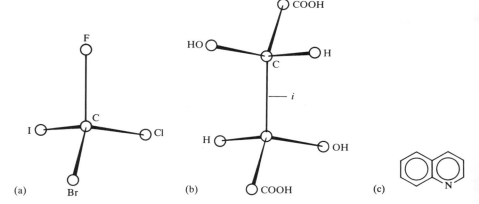

Fig. 15.7 Examples of molecules belonging to the groups (a) C_1, (b) C_i, and (c) C_s.

The groups C_n, C_{nv}, and C_{nh}

A molecule belongs to the group C_n if it possesses an n-fold axis. (Note that C_n is now playing a triple role: as the label of a symmetry element, a symmetry operation, and a group.) The H_2O_2 molecule (Fig. 15.8) has the elements E and C_2 and so belongs to the group C_2.

If in addition to the identity and a C_n axis a molecule has n vertical mirror planes σ_v, it belongs to the group C_{nv}. The H_2O molecule, for example, has the symmetry elements E, C_2, and $2\sigma_v$, and so belongs to the group C_{2v}. The NH_3 molecule has the elements E, C_3, and $3\sigma_v$, and so belongs to the group C_{3v}. A heteronuclear diatomic molecule such as HCl belongs to the group $C_{\infty v}$ because all rotations around the axis and reflections across it are symmetry operations. $C_{\infty v}$ is also the group of the linear OCS molecule and of a cone.

Objects that in addition to the identity and an n-fold principal axis also have a horizontal mirror plane σ_h belong to the groups C_{nh}. An example is *trans*-CHCl=CHCl (Fig. 15.9), which has the elements E, C_2, and σ_h, and so belongs to the group C_{2h}. Sometimes the presence of a symmetry element is implied by others: in this case C_2 and σ_h jointly imply the presence of a centre of inversion (Fig. 15.10).

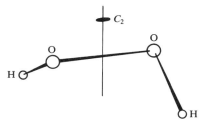

Fig. 15.8 The H_2O_2 molecule belongs to the group C_2 when it is in the conformation shown.

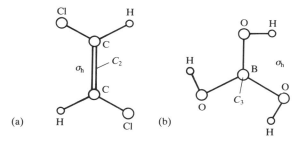

(a)   (b)

Fig. 15.9 (a) *Trans*-dichloroethene belongs to C_{2h} and (b) B(OH)$_3$ belongs to C_{3h}.

The groups D_n, D_{nh}, and D_{nd}

A molecule that has an n-fold principal axis and n two-fold axes perpendicular to C_n (Fig. 15.11) belongs to the group D_n. Molecules belong to D_{nh} if they also possess a horizontal mirror plane (Fig. 15.12). The planar trigonal BF_3 molecule has the elements E, C_3, $3C_2$, and σ_h (with one C_2 axis along each B—F bond), and so belongs to D_{3h}. The C_6H_6 molecule has the elements E, C_6, $6C_2$, and σ_h together with some others that these imply, and so it belongs to D_{6h}. All homonuclear diatomic molecules, such as N_2, belong to the group $D_{\infty h}$ since all rotations around the axis are symmetry operations, as are end-to-end rotation and end-to-end reflection; $D_{\infty h}$ is also the group of the linear OCO and HC≡CH molecules and of a uniform cylinder.

A molecule belongs to the group D_{nd} if it has the elements of D_n and, in addition, n dihedral mirror planes σ_d. The twisted, 90° allene shown in Fig. 15.13a belongs to D_{2d}, and the staggered form of ethane (Fig. 15.13b) belongs to D_{3d}.

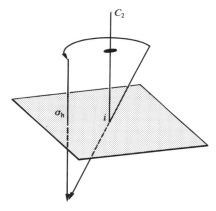

Fig. 15.10 The presence of a twofold axis and a horizontal mirror plane jointly imply the presence of an inversion centre in the molecule.

The groups S_n

Molecules that have not been classified into one of the groups we have mentioned so far, but which possess *one* S_n axis, belong to the group S_n. An example is shown in Fig. 15.14. Molecules belonging to S_n with $n > 4$ are rare. Note that the group S_2 is the same as C_i, so such a molecule will already have been classified as C_i.

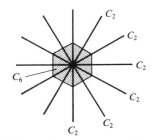

Fig. 15.11 A molecule with n twofold rotation axes perpendicular to an n-fold rotation axis belongs to the group D_n.

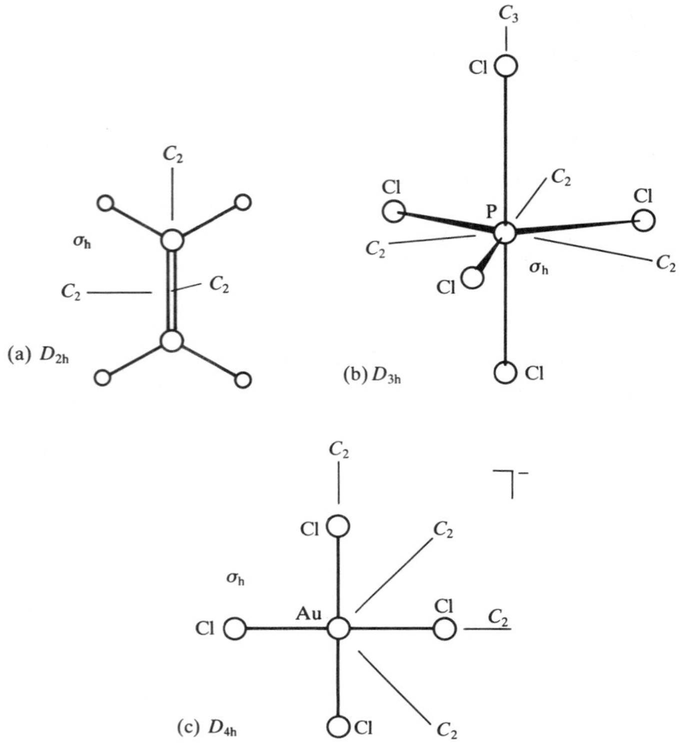

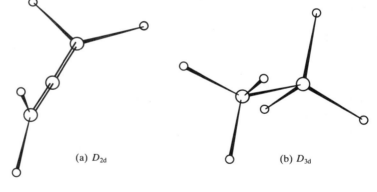

Fig. 15.12 Three examples of molecules belonging to D_{nh}. (a) C_2H_4 belongs to D_{2h}, (b) PCl_5, which has different axial and equatorial bond lengths, belongs to D_{3h}, and (c) the square-planar complex $[AuCl_4]^-$ belongs to D_{4h}.

(a) D_{2h}

(b) D_{3h}

(c) D_{4h}

Fig. 15.13 (a) The 90° allene molecule belongs to the group D_{2d} and (b) the staggered conformation of ethane belongs to D_{3d}.

(a) D_{2d}

(b) D_{3d}

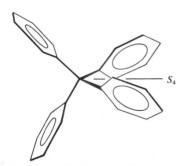

Fig. 15.14 Tetraphenylmethane is an example of a molecule that belongs to the group S_4.

The cubic groups

A number of very important molecules (e.g. CH_4) possess more than one principal axis. They all belong to the **cubic groups**, and in particular to the **tetrahedral groups** T, T_d, and T_h or to the **octahedral groups** O, O_h (Fig. 15.15). A few icosahedral (twenty-faced) molecules, belonging to the **icosahedral group** are also known: they include some of the boranes. T_d and O_h are the groups of the regular tetrahedron (e.g. CH_4) and the regular octahedron (e.g. SF_6) respectively. If the object possesses the rotational symmetry of the tetrahedron or the octahedron, but none of their planes of reflection, then it belongs to the simpler groups T or O. The group T_h is based on T but also contains a centre of inversion.

The full rotation group, R_3

The full rotation group consists of an infinite number of rotation axes with all possible values of n. A sphere and an atom belong to R_3, but no molecule does. Exploring the consequences of R_3 is a very important way of applying symmetry arguments to atoms and is an alternative approach to the theory of orbital angular momentum.

Example 15.1: *Identifying a point group of a molecule*

Identify the point group to which the sandwich molecule ruthenocene (two eclipsed cyclopentadiene rings) belongs.

Answer. The identification of a molecule's group is simplified by referring to the flow diagram in Box 15.2 and the shapes shown in Fig. 15.16. The path we trace through the flow diagram in Box 15.2 is shown by a dotted line; it ends at D_{nh}. Since the molecule has a fivefold axis, it belongs to the group D_{5h}.

Comment. If the rings were staggered, the horizontal reflection plane would be absent, but dihedral planes would be present.

Exercise. Classify (a) the conformation with staggered rings and (b) benzene.

[(a) D_{5d}, (b) D_{6h}]

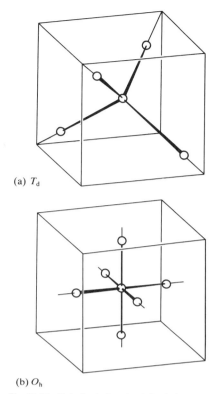

(a) T_d

(b) O_h

Fig. 15.15 Tetrahedral and octahedral molecules are best drawn in a way that shows their relation to a cube: they belong to one of the cubic groups. (a) CCl_4 belongs to T_d and (b) SF_6 belongs to O_h.

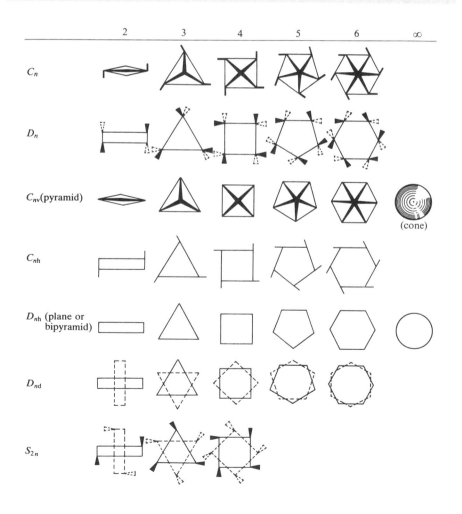

Fig. 15.16 A summary of the shapes corresponding to different point groups. The group to which a molecule belongs can often be identified from this diagram without going through the formal procedure in Box 15.2.

Box 15.2 The determination of a point group

To arrive at the point group of a given molecule, work through the following table. The use of a subgroup (i.e. not travelling to the end of a route) is permissible but gives less complete information. The dotted line shows the path used in Example 15.1.

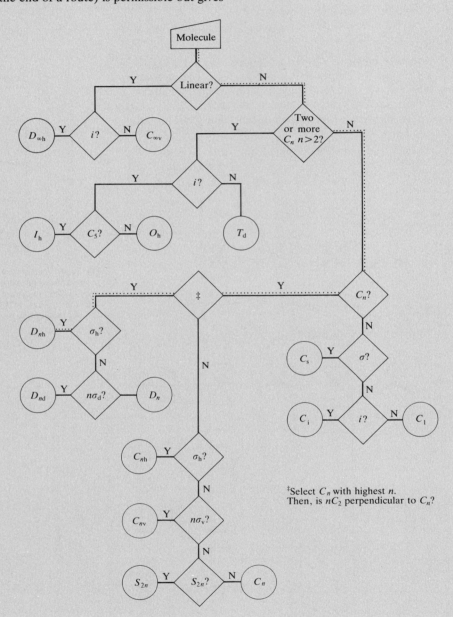

‡Select C_n with highest n.
Then, is nC_2 perpendicular to C_n?

15.3 Some immediate consequences of symmetry

We can make some statements about the properties of a molecule as soon as we have identified its point group.

Polarity

A **polar molecule** is one with a permanent electric dipole moment (HCl, O_3, and NH_3 are examples). If the molecule belongs to the group C_n with $n > 1$, it cannot possess a charge distribution corresponding to a dipole moment perpendicular to the axis. However, as the group makes no reference to operations relating the two ends of the molecule, a charge distribution may exist that results in a dipole along the axis (Fig. 15.17). The same remarks apply to C_{nv}, so molecules belonging to any of the C_{nv} groups may be polar. In all the other groups, such as C_{3h}, D, etc, there are symmetry operations that take one end of the molecule into the other. Therefore, as well as having no dipole perpendicular to the axis, such molecules can have none along the axis, for otherwise these additional operations would not be symmetry operations.

We can conclude that only molecules belonging to the groups C_n, C_{nv}, and C_s may have an electric dipole moment, and in the case of C_n and C_{nv} that that dipole must lie along the rotation axis. Thus O_3, which is angular and belongs to the group C_{2v}, may be polar, but CO_2, which is linear and belongs to the group $D_{\infty h}$, is not.

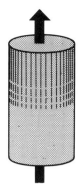

Fig. 15.17 A molecule with a C_n axis cannot have a dipole perpendicular to the axis, but it may have one parallel to the axis.

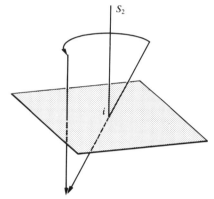

Fig. 15.18 Some elements are implied by the other elements in a group. Any molecule containing an inversion also possesses at least an S_2 element because i and S_2 are equivalent.

Chirality

A **chiral molecule** (from the Greek word for hand) is a molecule that cannot be superimposed on its mirror image. Chiral molecules are **optically active** in the sense that they rotate the plane of polarized light (a property discussed in more detail in Section 22.2). A chiral molecule and its mirror-image partner constitute an **enantiomeric pair**.

A molecule may be chiral only if it does not possess an axis of improper rotation, S_n. However, such an axis may be present under a different name and be implied by other symmetry elements that are present. For example, molecules belonging to the groups C_{nh} include S_n implicitly because they possess both C_n and σ_h. Any molecule containing a centre of inversion i also possesses an S_2 axis because i is equivalent to C_2 in conjunction with σ_h, which is S_2 (Fig. 15.18). It follows that all molecules with centres of inversion are non-chiral and hence optically inactive.

A molecule may be chiral if it does not have a centre of inversion or a mirror plane, which is the case with the amino acid alanine $CH_3CH(NH_2)COOH$ but not with glycine $CH_2(NH_2)COOH$. However, a molecule may be non-chiral even though it does not have a centre of inversion. For example, an S_4 molecule (Fig. 15.19) is non-chiral and optically inactive, for though it lacks i it does have an S_4 axis.

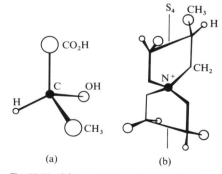

Fig. 15.19 (a) An optically active molecule without a centre of inversion. (b) Although this molecule has no centre of inversion (i.e. no S_2 axis) it is not optically active because it has an S_4 axis.

Groups, representations, and characters

Many properties may be analysed once the point group of a molecule is known, but to extract them we often need to use numerical aspects of group theory. We introduce these here, but we shall do no more than skim the surface of this very subtle and powerful subject.

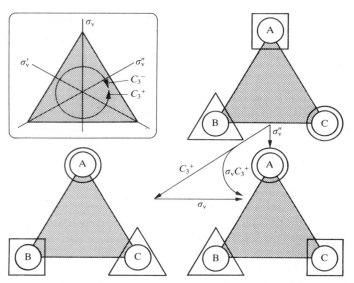

Fig. 15.20 The symmetry operations of the group C_{3v} are shown in the box, and the equivalence $\sigma_v C_3^+ = \sigma_v''$ is constructed by considering the effect of successive operations.

15.4 Group multiplication

Consider NH_3, which belongs to C_{3v} and is symmetrical under the operations E, C_3^+, C_3^-, σ_v, σ_v', and σ_v'' (Fig. 15.20). C_3^+ is a counter-clockwise $120°$ rotation as seen from above, C_3^- the corresponding clockwise rotation.[1] It should be obvious that the operation C_3^+ followed by C_3^- is the identity. We can express this symbolically by writing

$$C_3^- C_3^+ = E$$

Similarly, two successive counter-clockwise rotations by $120°$ are equivalent to one clockwise rotation:

$$C_3^+ C_3^+ = C_3^-$$

We can see from Fig. 15.20 that C_3^+ followed by σ_v is equivalent to σ_v'', and so we can write

$$\sigma_v C_3^+ = \sigma_v''$$

Note that in working out these relations, all the operations refer to some *fixed* arrangement of symmetry elements. That is, the planes and axes remain where they were first drawn on the page, and are unaffected by the performance of an operation. Note, too, that the second operation is written to the left of the first, so in the last example σ_v is carried out after C_3^+.

---

[1] This sign convention may seem odd, but it matches the convention used for angular momentum, when a clockwise rotation as seen from below (i.e. counter-clockwise as seen from above) is associated with positive values of m_l.

The table of all such combinations is called the **group multiplication table**, and for C_{3v} is as follows:

Second	First					
	E	C_3^+	C_3^-	σ_v	σ_v'	σ_v''
E	E	C_3^+	C_3^-	σ_v	σ_v'	σ_v''
C_3^+	C_3^+	C_3^-	E	σ_v'	σ_v''	σ_v
C_3^-	C_3^-	E	C_3^+	σ_v''	σ_v	σ_v'
σ_v	σ_v	σ_v''	σ_v'	E	C_3^-	C_3^+
σ_v'	σ_v'	σ_v	σ_v''	C_3^+	E	C_3^-
σ_v''	σ_v''	σ_v'	σ_v	C_3^-	C_3^+	E

A glance at the group multiplication table shows that the outcome of successive symmetry operations is always equivalent to a single symmetry operation of the group, which is called the **group property**. The group property is the main feature of the structure of groups: a set of operations form a group if they satisfy the group property together with some other mild conditions (Box 15.3). All symmetry operations on molecules satisfy the conditions in Box 15.3, which is why the theory of the symmetry of molecules is called group theory.

Box 15.3   The definition of a group

A group G of **order** h is a set of h **elements** (such as the symmetry operations of a molecule)

$$G = \{g_1, g_2, \ldots, g_h\}$$

together with a rule of combination that gives the symbol $g_i g_j$ a meaning (such as the symmetry operation g_j followed by the symmetry operation g_i) and which satisfy the following criteria:

(1) G includes the **identity** E, the element for which $Eg_i = g_i E = g_i$ for all the elements of the group

(2) G includes the **inverse** (g_i^{-1}) of each element, the element for which

$$g_i g_i^{-1} = g_i^{-1} g_i = E$$

(3) The rule of combination is **associative**, so

$$g_i(g_j g_k) = (g_i g_j)g_k$$

(4) The elements satisfy the **group property** that the combination of any pair of elements is itself an element:

$$g_i g_j = g_k$$

The definition of a group does not require the elements to **commute**:

$$g_i g_j = g_j g_i \text{ is not required}$$

However, if all the elements of a group do commute, the group is called **Abelian**.

15.5   The representation of transformations

Expressions like $C_3^- C_3^+ = E$ are symbolic ways of writing what happens when various physical operations are carried out in succession. However, it is possible to give them an actual algebraic significance, which means that we can deal with numbers instead of abstract symbols for operations. By dealing with numbers we shall arrive at precise conclusions.

The remainder of this section establishes the language and sets out the background of group theory. It makes use of the properties of matrices, which are reviewed in the *Further information* section at the end of the chapter. The rules needed for applying group theory are set out in Sections

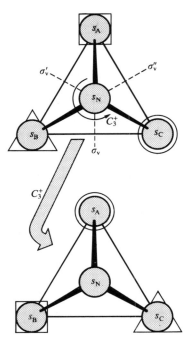

Fig. 15.21 The basis used for the discussion of the transformation properties of a C_{3v} molecule. We use the convention that the basis is always written in the order $\{\square, \triangle, \bigcirc\}$ and that the operations move the shapes on the diagram without affecting the N, A, B, C labels. The interpretation of the effect of C_3^+ is shown.

15.8 and 15.9. Most of them can be used simply as recipes without needing to work through the background material presented in this section.

Matrix representatives

Consider a C_{3v} molecule (such as NH_3) with s orbitals on each atom (Fig. 15.21) and think about what happens to these functions under a symmetry operation. Under σ_v, the change

$$(s_N, s_A, s_C, s_B) \leftarrow (s_N, s_A, s_B, s_C)$$

takes place. We can express this transformation using matrix multiplication:

$$(s_N, s_A, s_C, s_B) = (s_N, s_A, s_B, s_C)\begin{pmatrix} 1 & 0 & 0 & 0 \\ 0 & 1 & 0 & 0 \\ 0 & 0 & 0 & 1 \\ 0 & 0 & 1 & 0 \end{pmatrix} = (s_N, s_A, s_B, s_C)\boldsymbol{D}(\sigma_v)$$

The matrix $\boldsymbol{D}(\sigma_v)$ is called a **representative** of the operation σ_v.

We can use the same technique to find matrices that reproduce the other symmetry operations. For instance, C_3^+ has the effect

$$(s_N, s_B, s_C, s_A) \leftarrow (s_N, s_A, s_B, s_C)$$

and we can express the transformation as

$$(s_N, s_B, s_C, s_A) = (s_N, s_A, s_B, s_C)\begin{pmatrix} 1 & 0 & 0 & 0 \\ 0 & 0 & 0 & 1 \\ 0 & 1 & 0 & 0 \\ 0 & 0 & 1 & 0 \end{pmatrix} = (s_N, s_A, s_B, s_C)\boldsymbol{D}(C_3^+)$$

The operation σ_v'', which causes

$$(s_N, s_C, s_B, s_A) \leftarrow (s_N, s_A, s_B, s_C)$$

is represented by the matrix multiplication

$$(s_N, s_C, s_B, s_A) = (s_N, s_A, s_B, s_C)\begin{pmatrix} 1 & 0 & 0 & 0 \\ 0 & 0 & 0 & 1 \\ 0 & 0 & 1 & 0 \\ 0 & 1 & 0 & 0 \end{pmatrix} = (s_N, s_A, s_B, s_C)\boldsymbol{D}(\sigma_v'')$$

Since the identity leaves (s_N, s_A, s_B, s_C) unchanged,

$$\boldsymbol{D}(E) = \begin{pmatrix} 1 & 0 & 0 & 0 \\ 0 & 1 & 0 & 0 \\ 0 & 0 & 1 & 0 \\ 0 & 0 & 0 & 1 \end{pmatrix}$$

Example 15.2: *Finding a matrix representative of an operation*

Consider the four H1*s* orbitals of CH_4. Find matrix representatives for the operations C_3^+ and S_4^+.

Answer. CH_4 belongs to the group T_d. The axis corresponding to C_3^+ runs along a C—H bond (e.g. C—H_D), and so it rotates the other three H atoms into each other in a counterclockwise sense seen from above. A good plan is to put different shaped receptacles on each atom location (the shapes in Fig. 15.22) and to allow the operations to move them (the letters remaining stationary): the row vector representing the basis is then determined by writing the appropriate letter in the receptacle always written in the same order ($\square, \diamond, \triangle, \square$). The effect of $C_3^+(D)$ is

$$C_3^+(H_A, H_B, H_C, H_D) = (H_C, H_A, H_B, H_D)$$

$$= (H_A, H_B, H_C, H_D)\begin{pmatrix} 0 & 1 & 0 & 0 \\ 0 & 0 & 1 & 0 \\ 1 & 0 & 0 & 0 \\ 0 & 0 & 0 & 1 \end{pmatrix} = (H_A, H_B, H_C, H_D)\boldsymbol{D}(C_3^+)$$

S_4^+ rotates counterclockwise by 90° about a bisector of a CH_2 angle (e.g. H_ACH_C) and then reflects across the perpendicular plane:

$$S_4^+(H_A, H_B, H_C, H_D) = (H_B, H_C, H_D, H_A)$$

$$= (H_A, H_B, H_C, H_D)\begin{pmatrix} 0 & 0 & 0 & 1 \\ 1 & 0 & 0 & 0 \\ 0 & 1 & 0 & 0 \\ 0 & 0 & 1 & 0 \end{pmatrix} = (H_A, H_B, H_C, H_D)\boldsymbol{D}(S_4^+)$$

Comment. The representation depends on the basis selected: in this case it is four-dimensional because the basis has four members.

Exercise. Find the representatives for $S_4^-(CD)$, where the four-fold axis bisects the H_CCH_D angle, and $C_3^+(B)$.

$$\boldsymbol{D}(S_4^-) = \begin{pmatrix} 0 & 0 & 1 & 0 \\ 0 & 0 & 0 & 1 \\ 0 & 1 & 0 & 0 \\ 1 & 0 & 0 & 0 \end{pmatrix}, \qquad \boldsymbol{D}(C_3^+) = \begin{pmatrix} 0 & 0 & 1 & 0 \\ 0 & 1 & 0 & 0 \\ 0 & 0 & 0 & 1 \\ 1 & 0 & 0 & 0 \end{pmatrix}$$

Matrix representations

A very important property of the $\boldsymbol{D}$ matrices may now be identified. Using the rules of matrix multiplication gives

$$\boldsymbol{D}(\sigma_v)\boldsymbol{D}(C_3^+) = \begin{pmatrix} 1 & 0 & 0 & 0 \\ 0 & 1 & 0 & 0 \\ 0 & 0 & 0 & 1 \\ 0 & 0 & 1 & 0 \end{pmatrix}\begin{pmatrix} 1 & 0 & 0 & 0 \\ 0 & 0 & 0 & 1 \\ 0 & 1 & 0 & 0 \\ 0 & 0 & 1 & 0 \end{pmatrix} = \begin{pmatrix} 1 & 0 & 0 & 0 \\ 0 & 0 & 0 & 1 \\ 0 & 0 & 1 & 0 \\ 0 & 1 & 0 & 0 \end{pmatrix} = \boldsymbol{D}(\sigma_v'')$$

The importance of this result is that its structure

$$\boldsymbol{D}(\sigma_v)\boldsymbol{D}(C_3^+) = \boldsymbol{D}(\sigma_v'')$$

has exactly the same form as the group multiplication rule

$$\sigma_v C_3^+ = \sigma_v''$$

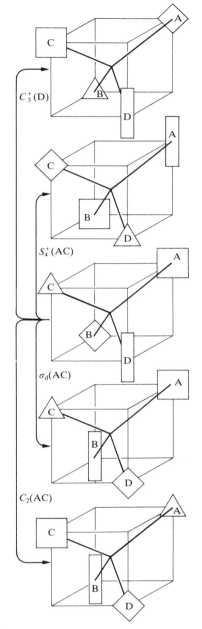

Fig. 15.22 The symmetry transformations used in Example 15.2. Note that we are using the convention defined in Fig. 15.21.

Whichever operations of a group we chose, we find that the matrix representatives multiply in an analogous way. We say that the multiplications are **homomorphous**. Therefore, the whole of the group multiplication table is reproduced by the matrix multiplication of the representatives. The set of six matrices is called a **matrix representation** of the C_{3v} group for the particular **basis** we have chosen, which in this case consists of the four orbitals s_N, s_A, s_B, and s_C. We denote the representation by the symbol Γ, and since it is four-dimensional, more specifically as $\Gamma^{(4)}$. The discovery of a matrix representation of the group means that we have found a link between the symbolic manipulations of the operations and algebraic manipulations involving numbers.

Example 15.3: *Verifying the matrix representation of a group*

Confirm that the matrix representatives found in Example 15.2 satisfy the group multiplication property.

Answer. Identify $C_3^+ S_4^+$ by assessing the effect on the basis of successive symmetry operations, and confirm the homomorphism by multiplying the representatives. The effect of the joint operation $S_4^+ C_3^+$ is

$$S_4^+ C_3^+ (H_A, H_B, H_C, H_D) = (H_D, H_B, H_C, H_A) = \sigma(H_A, H_B, H_C, H_D)$$

where the plane of the operation σ bisects $H_A C H_D$. The product of the representatives is

$$\begin{pmatrix} 0 & 0 & 0 & 1 \\ 1 & 0 & 0 & 0 \\ 0 & 1 & 0 & 0 \\ 0 & 0 & 1 & 0 \end{pmatrix} \begin{pmatrix} 0 & 1 & 0 & 0 \\ 0 & 0 & 1 & 0 \\ 1 & 0 & 0 & 0 \\ 0 & 0 & 0 & 1 \end{pmatrix} = \begin{pmatrix} 0 & 0 & 0 & 1 \\ 0 & 1 & 0 & 0 \\ 0 & 0 & 1 & 0 \\ 1 & 0 & 0 & 0 \end{pmatrix}$$

which is the representative of σ.

Exercise. Confirm that the representatives for $S_4^-(CD)$ and $C_3^+(B)$ multiply homomorphously with the elements of the group. $[S_4^- C_3^+ = S_4^+(AC)]$

The character of symmetry operations

In common parlance, the rotations C_3^+ and C_3^- of the group C_{3v} have the same 'character' but differ in direction. Likewise, the three reflections have the same character, but are different from the rotations. We say that the two rotations belong to the same **class** of operation, and that the three reflections form another class. This notion of class can be given precise, numerical significance.

Inspection of the matrix representation of C_{3v} for the s-orbital basis of Fig. 15.21 shows a remarkable fact. If we form the **trace** of each matrix, that is, sum the diagonal elements of each matrix, we get the following numbers:

	$D(E)$	$D(C_3^+)$	$D(C_3^-)$	$D(\sigma_v)$	$D(\sigma_v')$	$D(\sigma_v'')$
Trace:	4	1	1	2	2	2

We see that the matrices representing operations of the same class have identical diagonal sums. We call the trace of the representative the **character** χ of the operation; hence we can conclude that symmetry operations in the same class have the same character.

Example 15.4: *Calculating the character of an operation*

Calculate the characters of the operations C_3^+, S_4^+, and σ in the basis used in Example 15.3.

Answer. We refer to the representatives calculated in the example, and sum their diagonal elements.

$$\chi(C_3^+) = 0+0+0+1 = 1$$
$$\chi(S_4^+) = 0+0+0+0 = 0$$
$$\chi(\sigma) = 0+1+1+0 = 2$$

Comment. A quick rule for determining the character without first having to set up the matrix representation is to count 1 each time a basis function is left unchanged by the operation, because only these functions give a non-zero entry on the diagonal of the matrix representative. In some cases there is a sign change, $(\ldots -f \ldots) \leftarrow (\ldots f \ldots)$; then -1 occurs on the diagonal, and so count -1. The character of the identity is always equal to the dimension of the basis since each function contributes 1 to the trace.

Exercise. Calculate the character of (a) C_2 and (b) S_4^- in the same basis.

[(a) 0, (b) 0]

The character of an operation depends on the basis. For example, if instead of considering the four s orbitals, we used only s_N as the basis, then since each operation results in $s_N \leftarrow s_N$, which may be written $s_N = s_N \times \mathbf{1}$, with $\mathbf{1}$ a 1×1 matrix, the characters of the operations would be

	$D(E)$	$D(C_3^+)$	$D(C_3^-)$	$D(\sigma_v)$	$D(\sigma_v')$	$D(\sigma_v'')$
$\chi =$	1	1	1	1	1	1

We shall write this representation $\Gamma^{(1)}$ because it is one-dimensional. It is still true that the characters of the operations of the same class are equal; however, we also see that the characters of different classes may be the same. Furthermore, it is obvious that because $1 \times 1 = 1$, the matrices do reproduce the group multiplication table. However, as they do so in a trivial and uninformative way, the representation in which 1 represents each element is called the **unfaithful representation** of the group.

15.6 Irreducible representations

The unfaithful representation of the group, although apparently trivial, is a representation, and should not be discarded as being of no importance. In the next few sections, in fact, we shall see that it is the most important representation for many chemical applications.

The direct sum

The representatives in the basis (s_N, s_A, s_B, s_C) are four-dimensional (i.e. they are 4×4 matrices), but inspection shows that they are all of the form

$$\begin{pmatrix} 1 & 0 & 0 & 0 \\ 0 & & & \\ 0 & & & \\ 0 & & & \end{pmatrix}$$

and that the symmetry operations never mix s_N with the other three basis functions. This suggests that the basis can be cut into two parts, one consisting of s_N alone and the other of (s_A, s_B, s_C). The s_N orbital is a basis for the unfaithful representation, as we have seen, and we now see that the other three orbitals are a basis for a three-dimensional representation $\Gamma^{(3)}$ consisting of the following matrices:

$$
\begin{array}{cccccc}
D(E) & D(C_3^+) & D(C_3^-) & D(\sigma_v) & D(\sigma_v') & D(\sigma_v'') \\
\begin{pmatrix} 1 & 0 & 0 \\ 0 & 1 & 0 \\ 0 & 0 & 1 \end{pmatrix} &
\begin{pmatrix} 0 & 0 & 1 \\ 1 & 0 & 0 \\ 0 & 1 & 0 \end{pmatrix} &
\begin{pmatrix} 0 & 1 & 0 \\ 0 & 0 & 1 \\ 1 & 0 & 0 \end{pmatrix} &
\begin{pmatrix} 1 & 0 & 0 \\ 0 & 0 & 1 \\ 0 & 1 & 0 \end{pmatrix} &
\begin{pmatrix} 0 & 1 & 0 \\ 1 & 0 & 0 \\ 0 & 0 & 1 \end{pmatrix} &
\begin{pmatrix} 0 & 0 & 1 \\ 0 & 1 & 0 \\ 1 & 0 & 0 \end{pmatrix} \\
\chi = \quad 3 & 0 & 0 & 1 & 1 & 1
\end{array}
$$

The characters still satisfy the rule about symmetry operations of the same class. The matrices of $\Gamma^{(3)}$ are the same as those of the four-dimensional representation, except for the loss of the first row and column. We say that the original four-dimensional representation has been **reduced** to the **direct sum** of a one-dimensional representation **spanned** by s_N and a three-dimensional representation spanned by (s_A, s_B, s_C). This reduction is consistent with the common sense view that the central orbital plays a role different from the other three. The reduction is denoted symbolically by

$$\Gamma^{(4)} = \Gamma^{(1)} + \Gamma^{(3)}$$

The reduction of a representation

The one-dimensional representation $\Gamma^{(1)}$, which consists of the six 1×1 matrices 1, 1, 1, 1, 1, 1, obviously cannot be reduced any further, and is called an **irreducible representation** (or 'irrep') of the group.

We can demonstrate that $\Gamma^{(3)}$ is reducible by switching attention from s_A, s_B, and s_C to the linear combinations

$$s_1 = s_A + s_B + s_C \qquad s_2 = 2s_A - s_B - s_C \qquad s_3 = s_B - s_C$$

These are sketched in Fig. 15.23 (their form is justified later). Even at this stage it is clear that, because of the presence of the node in s_2 and s_3, these two have different symmetry from s_1. The decomposition

$$\Gamma^{(3)} = \Gamma^{(1)} + \Gamma^{(2)}$$

is beginning to emerge.

The representatives in the new basis can be constructed from the old. For example, since under σ_v,

$$(s_A, s_C, s_B) \leftarrow (s_A, s_B, s_C)$$

it follows (by applying these transformations to the linear combinations) that

$$(s_1, s_2, -s_3) \leftarrow (s_1, s_2, s_3)$$

The transformation is achieved by writing

$$(s_1, s_2, -s_3) = (s_1, s_2, s_3)\begin{pmatrix} 1 & 0 & 0 \\ 0 & 1 & 0 \\ 0 & 0 & -1 \end{pmatrix}$$

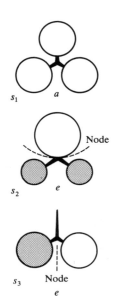

s_1 a

Node

s_2 e

s_3 Node

e

Fig. 15.23 Three linear combinations of three s orbitals in a C_{3v} molecule. The symmetry species they span (see later) have been added.

which gives us the representative $D(\sigma_v)$ in the new basis. The representative of C_3^+ takes a little more calculation, but depends on the transformation

$$(s_B, s_C, s_A) \leftarrow (s_A, s_B, s_C)$$

We know how the individual s_Q transform (with Q = A, B, and C), and so by substitution in the expressions defining the new basis gives the transformation of the s_n (with $n = 1$, 2, and 3):

$$(s_1, -\tfrac{1}{2}s_2 + \tfrac{3}{2}s_3, -\tfrac{1}{2}s_2 - \tfrac{1}{2}s_3) \leftarrow (s_1, s_2, s_3)$$

This transformation is reproduced by

$$(s_1, -\tfrac{1}{2}s_2 + \tfrac{3}{2}s_3, -\tfrac{1}{2}s_2 - \tfrac{1}{2}s_3) = (s_1, s_2, s_3)\begin{pmatrix} 1 & 0 & 0 \\ 0 & -\tfrac{1}{2} & -\tfrac{1}{2} \\ 0 & \tfrac{3}{2} & -\tfrac{1}{2} \end{pmatrix}$$

so giving $D(C_3^+)$. The complete representation and its characters may be found similarly, and are

$$
\begin{array}{ccc}
D(E) & D(C_3^+) & D(C_3^-) \\
\begin{pmatrix} 1 & 0 & 0 \\ 0 & 1 & 0 \\ 0 & 0 & 1 \end{pmatrix} & \begin{pmatrix} 1 & 0 & 0 \\ 0 & -\tfrac{1}{2} & -\tfrac{1}{2} \\ 0 & \tfrac{3}{2} & -\tfrac{1}{2} \end{pmatrix} & \begin{pmatrix} 1 & 0 & 0 \\ 0 & -\tfrac{1}{2} & \tfrac{1}{2} \\ 0 & -\tfrac{3}{2} & -\tfrac{1}{2} \end{pmatrix} \\
\chi = \quad 3 & 0 & 0
\end{array}
$$

$$
\begin{array}{ccc}
D(\sigma_v) & D(\sigma_v') & D(\sigma_v'') \\
\begin{pmatrix} 1 & 0 & 0 \\ 0 & 1 & 0 \\ 0 & 0 & -1 \end{pmatrix} & \begin{pmatrix} 1 & 0 & 0 \\ 0 & -\tfrac{1}{2} & \tfrac{1}{2} \\ 0 & \tfrac{3}{2} & \tfrac{1}{2} \end{pmatrix} & \begin{pmatrix} 1 & 0 & 0 \\ 0 & -\tfrac{1}{2} & -\tfrac{1}{2} \\ 0 & -\tfrac{3}{2} & \tfrac{1}{2} \end{pmatrix} \\
\chi = \quad 1 & 1 & 1
\end{array}
$$

The new representatives are all in **block diagonal form**

$$\begin{pmatrix} 1 & 0 & 0 \\ 0 & \blacksquare & \\ 0 & & \end{pmatrix}$$

and the s_1 combination is not mixed with the other two by any operation of the group. We have therefore achieved the reduction of $\Gamma^{(3)}$ to $\Gamma^{(1)} + \Gamma^{(2)}$, with s_1 a basis for the same one-dimensional irreducible representation $(1, 1, 1, 1, 1, 1)$ as before and (s_2, s_3) a basis for a two-dimensional representation $\Gamma^{(2)}$:

$$
\begin{array}{cccccc}
D(E) & D(C_3^+) & D(C_3^-) & D(\sigma_v) & D(\sigma_v') & D(\sigma_v'') \\
\begin{pmatrix} 1 & 0 \\ 0 & 1 \end{pmatrix} & \begin{pmatrix} -\tfrac{1}{2} & -\tfrac{1}{2} \\ \tfrac{3}{2} & -\tfrac{1}{2} \end{pmatrix} & \begin{pmatrix} -\tfrac{1}{2} & \tfrac{1}{2} \\ -\tfrac{3}{2} & -\tfrac{1}{2} \end{pmatrix} & \begin{pmatrix} 1 & 0 \\ 0 & -1 \end{pmatrix} & \begin{pmatrix} -\tfrac{1}{2} & \tfrac{1}{2} \\ \tfrac{3}{2} & \tfrac{1}{2} \end{pmatrix} & \begin{pmatrix} -\tfrac{1}{2} & -\tfrac{1}{2} \\ -\tfrac{3}{2} & \tfrac{1}{2} \end{pmatrix} \\
\chi = \quad 2 & -1 & -1 & 0 & 0 & 0
\end{array}
$$

It is easy to check that these matrices are a representation by multiplying pairs together and seeing that they reproduce the original group multiplication table.

The irreducibility of some representations

No linear combination of s_2 and s_3 exists that reduces $\Gamma^{(2)}$ to two one-dimensional representations, and so the $\Gamma^{(2)}$ representation is irreducible. We can conclude that while s_N and s_1 'have the same symmetry'—technically, are the bases for the same irreducible representation of the group—the pair s_2, s_3 'have different symmetry'—they span a different irreducible representation and must be treated as a pair (because that representation is two-dimensional). These features agree with what common sense tells us by inspection of the diagrams in Fig. 15.23.

But how do we know that $\Gamma^{(2)}$ is irreducible? This information, together with a great deal more, is included in one of the principal tools of group theory, the 'character tables' of the point groups.

Character tables

The **character table** of a group is the list of the characters of all its irreducible representations. The C_{3v} character table is shown below.

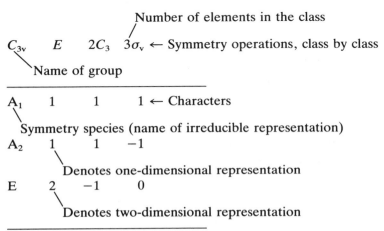

The columns in the table are labelled by the symmetry operations: it is not necessary to show the character for every individual operation because all those in the same class have the same value. The number of operations in each class is given (e.g. the 2 in $2C_3$, showing that there are two threefold rotations). The total number of operations in the group is called its **order**: the order of the C_{3v} group is 6.

The column on the left labels the **symmetry species** of the irreducible representations of the group. An A or B signifies a one-dimensional representation; A is used when the character is $+1$ under a principal rotation and B when it is -1. There are two species of one-dimensional representation in C_{3v}, and both have $+1$ for the character of the principal rotation; they are therefore labelled A_1 and A_2. A_1 and A_2 are distinguished by the character under a vertical reflection, and in some cases by their behaviour under a C_2 rotation perpendicular to the principal axis. An E denotes a two-dimensional representation, and T denotes a three-dimensional representation (there is none in C_{3v}).

The characters of the $\Gamma^{(2)}$ representation spanned by (s_2, s_3) are those of E. Since E is an irreducible representation, we know at once that $\Gamma^{(2)}$ is irreducible too.

Perhaps the most surprising feature of the character table is that there are so few allowed symmetry species. That the three given exhaust all possibilities is confirmed by an elegant theorem in group theory which states that

Number of symmetry species = Number of classes

In C_{3v} there are three classes (three columns in the character table), and so there are only three species of irreducible representation.

Although we have introduced these points through the group C_{3v}, they are entirely general, and the characters of all possible species of irreducible representation of any group may be listed. A selection of these very important tables is given at the end of the *Data section*.

15.7 Transformations of other bases

We shall now show that the three p orbitals on the central atom are a basis for another three-dimensional (reducible) representation. Each orbital has the form

$$p_x = xf \qquad p_y = yf \qquad p_z = zf$$

where f is a function of distance from the nucleus. Since the distance of a point from the nucleus does not change when the molecule is rotated and reflected, we can ignore f and use the basis (x, y, z).

Under the reflection σ_v in C_{3v} (Fig. 15.24),

$$(-x, y, z) \leftarrow (x, y, z)$$

That is, the p_x orbital changes sign. (Note that the coordinate system is an unchanging background on which these transformations are played out.) We can express the transformation as a matrix equation:

$$(-x, y, z) = (x, y, z)\begin{pmatrix} -1 & 0 & 0 \\ 0 & 1 & 0 \\ 0 & 0 & 1 \end{pmatrix} = (x, y, z)\boldsymbol{D}(\sigma_v)$$

Similarly, under C_3^+,

$$(-\tfrac{1}{2}x + \tfrac{1}{2}\sqrt{3}y, -\tfrac{1}{2}\sqrt{3}x - \tfrac{1}{2}y, z) \leftarrow (x, y, z)$$

which can be expressed as

$$(-\tfrac{1}{2}x + \tfrac{1}{2}\sqrt{3}y, -\tfrac{1}{2}\sqrt{3}x - \tfrac{1}{2}y, z) = (x, y, z)\begin{pmatrix} -\tfrac{1}{2} & -\tfrac{1}{2}\sqrt{3} & 0 \\ \tfrac{1}{2}\sqrt{3} & -\tfrac{1}{2} & 0 \\ 0 & 0 & 1 \end{pmatrix} = (x, y, z)\boldsymbol{D}(C_3^+)$$

All the representatives can be compiled in this way, and the complete representation is

$$\begin{array}{ccc}
\boldsymbol{D}(E) & \boldsymbol{D}(C_3^+) & \boldsymbol{D}(C_3^-) \\
\begin{pmatrix} 1 & 0 & 0 \\ 0 & 1 & 0 \\ 0 & 0 & 1 \end{pmatrix} &
\begin{pmatrix} -\tfrac{1}{2} & -\tfrac{1}{2}\sqrt{3} & 0 \\ \tfrac{1}{2}\sqrt{3} & -\tfrac{1}{2} & 0 \\ 0 & 0 & 1 \end{pmatrix} &
\begin{pmatrix} -\tfrac{1}{2} & \tfrac{1}{2}\sqrt{3} & 0 \\ -\tfrac{1}{2}\sqrt{3} & -\tfrac{1}{2} & 0 \\ 0 & 0 & 1 \end{pmatrix}
\end{array}$$

$$\chi = \qquad 3 \qquad\qquad 0 \qquad\qquad 0$$

$$\begin{array}{ccc}
\boldsymbol{D}(\sigma_v) & \boldsymbol{D}(\sigma_v') & \boldsymbol{D}(\sigma_v'') \\
\begin{pmatrix} -1 & 0 & 0 \\ 0 & 1 & 0 \\ 0 & 0 & 1 \end{pmatrix} &
\begin{pmatrix} \tfrac{1}{2} & -\tfrac{1}{2}\sqrt{3} & 0 \\ \tfrac{1}{2}\sqrt{3} & -\tfrac{1}{2} & 0 \\ 0 & 0 & 1 \end{pmatrix} &
\begin{pmatrix} \tfrac{1}{2} & \tfrac{1}{2}\sqrt{3} & 0 \\ -\tfrac{1}{2}\sqrt{3} & -\tfrac{1}{2} & 0 \\ 0 & 0 & 1 \end{pmatrix}
\end{array}$$

$$\chi = \qquad 1 \qquad\qquad 1 \qquad\qquad 1$$

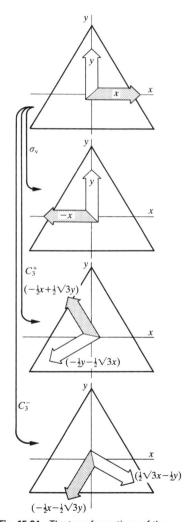

Fig. 15.24 The transformations of the functions x and y under the operations of the group C_{3v}. Notice that they take place against a fixed coordinate system (just as the N, A, B, C are fixed in Fig. 15.21).

The representation is reducible because all the matrices have block-diagonal form and the parts relating to z may be sliced off. The characters of the matrices for the remaining functions in the basis (x, y) are

$$2 \quad -1 \quad -1 \quad 0 \quad 0 \quad 0$$

By comparing this with the C_{3v} character table it is clear that (x, y) spans an irreducible representation of symmetry species E.

Example 15.5: *Finding the symmetry species of a representation*

Find the symmetry species of the irreducible representations spanned by (x, y, z) in the group C_{2v}.

Answer. We need to establish the effect of the operations E, C_2, σ_v, and σ_v' on the three functions. To do so, we write the matrix representation in the basis, and identify the symmetry species from the characters. The four transformations are

$$E: (x, y, z) \leftarrow (x, y, z) \qquad C_2: (-x, -y, z) \leftarrow (x, y, z)$$
$$\sigma_v: (x, -y, z) \leftarrow (x, y, z) \qquad \sigma_v': (-x, y, z) \leftarrow (x, y, z)$$

The matrix representation is therefore

$$
\begin{array}{cccc}
\boldsymbol{D}(E) & \boldsymbol{D}(C_2) & \boldsymbol{D}(\sigma_v) & \boldsymbol{D}(\sigma_v') \\
\begin{pmatrix} 1 & 0 & 0 \\ 0 & 1 & 0 \\ 0 & 0 & 1 \end{pmatrix} &
\begin{pmatrix} -1 & 0 & 0 \\ 0 & -1 & 0 \\ 0 & 0 & 1 \end{pmatrix} &
\begin{pmatrix} 1 & 0 & 0 \\ 0 & -1 & 0 \\ 0 & 0 & 1 \end{pmatrix} &
\begin{pmatrix} -1 & 0 & 0 \\ 0 & 1 & 0 \\ 0 & 0 & 1 \end{pmatrix}
\end{array}
$$

This representation is in block-diagonal form, and may be decomposed into the following one-dimensional irreducible representations

$$
\begin{array}{ccccc}
x: & 1 & -1 & 1 & -1 \\
y: & 1 & -1 & -1 & 1 \\
z: & 1 & 1 & 1 & 1
\end{array}
$$

The characters of the representatives are the numbers themselves (because they are 1×1 matrices), and so the symmetry species of the irreducible representations spanned by x, y, and z are B_1, B_2, and A_1 respectively.

Exercise. What symmetry species does xy span in C_{2v}? [A_2]

The conclusion of the analysis given above is that in C_{3v}, z spans an irreducible representation of symmetry species A_1 and (x, y) spans one of symmetry species E. Information like this will turn out to be so important that the symmetry species of the irreducible representations spanned by x, y, and z are normally reported in the character table. The same technique may be applied to the **quadratic forms** x^2, xy, xz, $\ldots$, z^2 which represent the shapes of the d orbitals, and the symmetry species of the irreducible representations they span are also usually listed. A complete character table therefore looks something like the following:

C_{3v}	E	$2C_3$	$3\sigma_v$				
A_1	1	1	1	z	$x^2 + y^2 + z^2$	$2z^2 - x^2 - y^2$	
A_2	1	1	-1				R_z
E	2	-1	0	(x, y)	(xz, yz)	$(xy, x^2 - y^2)$	(R_x, R_y)

The R_x, etc, in the final column denote rotations, and their listing shows how they transform under the operations of the group. The transformation properties of rotations can be deduced from those of angular momentum; for instance, R_z transforms as $l_z = xp_y - yp_x$ (Section 12.6), and the linear momentum components transform like x, y, z.

Example 15.6: *Deducing the transformation properties of a rotation*

Decide how a rotation around the z-axis transforms in the group C_{2v}.

Answer. We need to consider the transformation properties of the angular momentum operator

$$l_z = xp_y - yp_x$$

Under a C_2 rotation both x and y and p_x and p_y change sign, so l_z is unchanged. We know, therefore, that the rotation is the basis for either A_1 or A_2. Under a reflection in the yz-plane, x and p_x both change sign but neither y nor p_y does; therefore l_z does change sign, and its character is -1. It has the same character for reflection in the xz-plane. Hence, from the C_{2v} character table we can identify the irreducible representation spanned by rotation around the z-axis as A_2.

Exercise. Identify the irreducible representation spanned by a rotation around the x-axis in C_{2v}. [B_2]

Using character tables

Although the characters do not carry all the information contained in the representatives (they are, after all, only the diagonal sums), they do contain enough to make them of central importance in chemistry. One of the reasons for the importance of character tables is that they let us say, almost at a glance, whether an integral is zero. Integrals of interest to chemists include overlap integrals and 'transition dipole moments', which, as we shall see, govern the intensities of spectroscopic transitions. Thus group theory enables us to make powerful statements about chemical bonding and the appearance of spectra.

15.8 Vanishing integrals

Suppose we had to evaluate the integral

$$I = \int f_1 f_2 \, d\tau \tag{1}$$

where f_1 and f_2 are wavefunctions. For example, f_1 might be an atomic orbital on one atom and f_2 an atomic orbital on another atom, in which case I would be their overlap integral S. If we knew that the integral is zero, we could say at once that a molecular orbital does not result from (f_1, f_2) overlap in that molecule.

Integrals over the product of two functions

The key point in dealing with I is that the value of any integral, and of an overlap integral in particular, is independent of the orientation of the

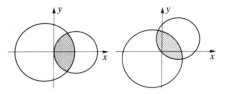

Fig. 15.25 The value of an integral I (e.g. an area) is independent of the coordinate system used to evaluate it. That is, I is a basis of a representation of symmetry species A_1.

molecule (Fig. 15.25). In group theoretical language we express this by saying that I is unchanged by any symmetry operation of the molecule, and that each operation brings about the trivial transformation

$$I \leftarrow I$$

Since the volume element $d\tau$ is unchanged by a symmetry operation, it follows that the integral is non-zero only if the product $f_1 f_2$ is unchanged by any symmetry operation of the molecular point group. If the product of functions changed sign under a symmetry operation, the integral would be the sum of equal and opposite contributions, and hence would be zero. It follows that the only contribution to a non-zero integral comes from functions for which under any symmetry operation of the molecular point group

$$f_1 f_2 \leftarrow f_1 f_2$$

and hence for which the representatives, and the characters, are all equal to 1. Therefore, for I not to be zero, the integrand $f_1 f_2$ must be a basis for the totally symmetric A_1 irreducible representation of the molecular point group.

We use the following procedure[2] to deduce the symmetry species spanned by the product $f_1 f_2$ and hence to see whether it does indeed span A_1.

(1) Decide on the symmetry species of the functions by reference to the character table, and write the characters in two rows in the same order as in the table.

For example, if f_1 is the s_N orbital in NH_3 and f_2 is the linear combination s_3 (Fig. 15.23), then since s_N spans A_1 and s_3 is a member of the basis spanning E, we write

$$f_1: \quad 1 \quad 1 \quad 1$$
$$f_2: \quad 2 \quad -1 \quad 0$$

(2) Multiply the numbers in each column, writing the results in the same order.

For the NH_3 calculation,

$$f_1 f_2: \quad 2 \quad -1 \quad 0$$

(3) Inspect the row so produced, and see if it can be expressed as a sum of characters of the irreducible representations of the group. If this sum does not contain A_1, the integral must be zero.

---

[2] The procedure here and in the rest of the chapter is based on a very important theorem in group theory known as the 'little orthogonality theorem' which states that

$$\sum_C g(C)\chi^{(\Gamma)}(C)\chi^{(\Gamma')}(C) = 0$$

where the sum is over the classes of the group (the columns in the character table), g is the number of elements in each class, and Γ and Γ' are two *different* irreducible representations.

In C_{3v}, for instance, the character of each operation can always be expressed as

$$\chi = c_1\chi(A_1) + c_2\chi(A_2) + c_3\chi(E)$$

and the integral must be zero if $c_1 = 0$. In the present example, the characters $2, -1, 0$ are those of E alone, and so the integral must be zero. Inspection of the form of the functions (Fig. 15.23) shows why this is so: s_3 has a node running through s_N. Had we taken $f_1 = s_N$ and $f_2 = s_1$ instead, then since each spans A_1 with characters $1, 1, 1$,

$$f_1: \quad 1 \quad 1 \quad 1$$

$$f_2: \quad 1 \quad 1 \quad 1$$

$$f_1 f_2: \quad 1 \quad 1 \quad 1$$

The characters of the product are those of A_1 itself. Therefore, s_1 and s_N may have non-zero overlap.

It is important to note that group theory is specific about when an integral must be zero, but integrals that it allows to be non-zero may be zero for reasons unrelated to symmetry. For example, the N—H distance may be so great that the s_1, s_N overlap integral is zero simply because the orbitals are so far apart.

Example 15.7: *Deciding if an integral must be zero* (1)

May the integral of the function $f = xy$ be non-zero when evaluated over a region the shape of an equilateral triangle centred on the origin?

Answer. An equilateral triangle has the point-group symmetry C_{3v}. If we refer to the character table of the group, we see that xy is a member of a basis that spans the irreducible representation E. Therefore, its integral must be zero, since the integrand has no component that spans A_1.

Comment. In this case we have used a special case of the argument given above in which the integrand is expressed as a single function. We can regard f_2 as present, but as equal to 1.

Exercise. Can the function $x^2 + y^2$ have a non-zero integral when integrated over a regular pentagon centred on the origin? [Yes, spans A_1]

Integrals over the product of three functions

The same technique may be used to decide whether integrals of the form

$$I = \int f_1 f_2 f_3 \, d\tau \qquad (2)$$

necessarily disappear. In this case the triple product $f_1 f_2 f_3$ must contain a component that spans A_1. To test whether this is so, the characters of all three functions are multiplied together in the same way as in the rules set out above.

Example 15.8: *Deciding if an integral must be zero* (2)

Does the integral $\int (3d_{z^2}) x (3d_{xy}) \, d\tau$ vanish in a tetrahedral molecule?

Answer. We must refer to the T_d character table (at the end of the *Data section*) and find the characters of the irreducible representations spanned by $3z^2 - r^2$ (the form of the d_{z^2} orbital), x, and xy; then we can use the procedure set out above (with one more row of multiplication). Note that $3z^2 - r^2 = 2z^2 - x^2 - y^2$. We then draw up the following table:

	E	$8C_3$	$3C_2$	$6S_4$	$6\sigma_d$	
$f_3 = d_{xy}$	3	0	-1	-1	1	(T_2)
$f_2 = x$	3	0	-1	-1	1	(T_2)
$f_1 = d_{z^2}$	2	-1	2	0	0	(E)
$f_1 f_2 f_3$	18	0	2	0	0	

The characters are the sum of $A_1 + A_2 + 2E + 2T_1 + 2T_2$. Therefore, the integral is not necessarily zero.

Comment. Closer inspection of the integral (e.g. using the representation itself, not just the characters) show that the integral must in fact vanish. This is a warning that arguments based on the character tables, since they carry only incomplete information in general, show only when an integral is necessarily zero. A second point is that the decomposition of the result to discover if A_1 is included is sometimes a lengthy job. The following procedure[3] can always be used if the answer is not obvious:

(1) Multiply the characters $(18, 0, \ldots)$ by the number of elements in each class $(1, 8, \ldots)$.
(2) Add together the numbers this produces $(18 + 0 + \ldots = 24)$.
(3) Divide the sum by the order of the group (the number of elements, 24).

This gives the number of times A_1 appears in the reducible representation (1).

Exercise. Does the integral $\int (2p_x)(2p_y)(2p_z) \, d\tau$ necessarily vanish in an octahedral environment? [No; spans A_1]

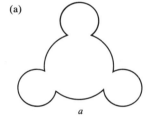

(a)

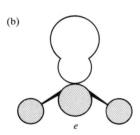

(b)

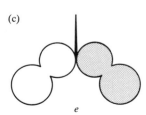

(c)

Fig. 15.26 Orbitals of the same symmetry species may have non-vanishing overlap. This diagram illustrates the three bonding orbitals that may be constructed from (N2s, H1s) and (N2p, H1s) overlap in a C_{3v} molecule, and their symmetry labels. (There are also three antibonding orbitals of the same species.)

Orbitals with non-zero overlap

The rules just given let us decide which atomic orbitals may have non-zero overlap in a molecule. We have seen that s_N may have non-zero overlap with s_1 (the combination $1s_A + 1s_B + 1s_C$), and so (s_N, s_1) overlap bonding and antibonding molecular orbitals can form (Fig. 15.26a). The general rule is that:

Only orbitals of the same symmetry species may have non-zero overlap, and so only orbitals of the same symmetry species form bonding and antibonding combinations.

[3] The procedure is based on an orthogonality theorem like that given in the previous footnote. The number of times c_1 that the irreducible representation A_1 appears in a representation Γ with characters $\chi^{(\Gamma)}$ is:

$$c_1 = \frac{1}{h} \sum_C g(C) \chi^{(\Gamma)}(C)$$

where h is the order of the group and g is the number of elements in each class C.

It should be recalled from Chapter 14 that the selection of atomic orbitals that had mutual non-zero overlap is the central and initial step in the construction of molecular orbitals by the LCAO procedure. We are therefore at the point of contact between group theory and the material introduced in that chapter. The molecular orbitals formed from a particular set of atomic orbitals with non-zero overlap are labelled with the lower-case letter corresponding to the symmetry species. Thus, the (s_N, s_1)-overlap orbitals are called a_1 orbitals (and a_1^* if we wish to emphasize that they are antibonding).

The s_2 and s_3 linear combinations span an irreducible representation of symmetry species E. Does the N atom have orbitals that have non-zero overlap with them (and give rise to an e orbital)? Intuition (Fig. 15.26b and c) suggests that N$2p_x$ and N$2p_y$ should be suitable. We can confirm this conclusion by noting that the character table shows that in C_{3v}, x and y jointly span an irreducible representation of symmetry species E. Therefore, N$2p_x$ and N$2p_y$ also span E, and so may have non-zero overlap with s_2 and s_3 (verify this by multiplying the characters: $E \times E = A_1 + A_2 + E$ in C_{3v}). The two e orbitals that result are shown in Fig. 15.26 (there are also two antibonding e orbitals).

The power of the method can be illustrated by exploring whether any d orbitals on the central atom can take part in bonding. The d orbitals have the forms

$$d_{z^2} = (3z^2 - r^2)f \qquad d_{x^2-y^2} = (x^2 - y^2)f$$

$$d_{xy} = xyf \qquad d_{yz} = yzf \qquad d_{zx} = zxf$$

Their symmetries can be taken from the character tables by noting how the quadratic forms xy, yz, etc, transform. Reference to the C_{3v} table shows that d_{z^2} has A_1 symmetry and that the pairs $(d_{x^2-y^2}, d_{xy})$ and (d_{yz}, d_{zx}) each transform as E. It follows that molecular orbitals may be formed by s_1, d_{z^2} overlap and by overlap of the s_2, s_3 combinations with the E d orbitals. Whether or not the d orbitals are in fact important is a question group theory cannot answer because that depends on energy considerations, not symmetry.

Although we have illustrated the technique with the group C_{3v}, it is entirely general, and the importance of knowing which s, p, and d orbitals overlap is one of the reasons why the transformation properties of x, xz, etc, are listed in the character tables.

Example 15.9: *Determining which orbitals can contribute to bonding*

The four H$1s$ orbitals of methane span $A_1 + T_2$. With which of the C atom orbitals can they overlap? What if the C atom had d orbitals available?

Answer. We need to refer to the T_d character table (at the end of the *Data section*) and look for s, p, and d orbitals spanning A_1 or T_2. An s orbital spans A_1, and so it may have non-zero overlap with the A_1 combination of H$1s$ orbitals. The C$2p$ orbitals span T_2, and so they may have non-zero overlap with the T_2 combination. The d_{xy}, d_{yz}, and d_{zx} orbitals span T_2, and so may overlap the same combination. Neither of the other two d orbitals span A_1 (they span E), and so they remain nonbonding orbitals.

Comment. It follows that in methane, there are (C2s,H1s)-overlap a_1 orbitals and (C2p,H1s)-overlap t_2 orbitals. The C3d orbitals might contribute to the latter. The lowest energy configuration is probably $a_1^2 t_2^6$, with all bonding orbitals occupied.

Exercise. Consider the octahedral SF_6 molecule, with the bonding arising from overlap of S orbitals and a 2p orbital on each F directed towards the central S. The latter span $A_{1g} + E_g + T_{1u}$. What S orbitals have non-zero overlap? Suggest what the ground configuration is likely to be.

$$[3s(A_{1g}), 3p(T_{1u}), 3d(E_g); a_{1g}^2 t_{1u}^6 e_g^4]$$

Selection rules

In Chapters 16 and 17 we shall see that the intensity of a spectral line arising from a molecular transition between some initial state with wavefunction ψ_i and a final state with wavefunction ψ_f depends on the (electric) **transition dipole moment** μ. The z-component of this vector is defined through

$$\mu_z = -e \int \psi_f^* z \psi_i \, d\tau \qquad (3)$$

where $-e$ is the charge of the electron. Stating the conditions for this quantity (and the x- and y-components) to be zero amounts to specifying the **selection rules** for the transition, the statement of which transitions are possible. The transition moment has the form of the integral in eqn 2, and so, once we know the symmetry species of the states, we can use group theory to decide which transitions have zero transition dipole moment and therefore cannot occur.

As an example, we investigate whether an a_1 electron in H_2O (which belongs to C_{2v}) can make an electric dipole transition to a b_1 orbital (Fig. 15.27). We must examine all three components of the transition dipole, and take f_2 in eqn 2 as x, y, and z in turn. Reference to the C_{2v} character table shows that these transform as B_1, B_2, and A_1 respectively. The three calculations run as follows:

	x-component				y-component				z-component				
f_1:	1	1	1	1	1	1	1	1	1	1	1	1	(A_1)
$f_2(x, y, \text{or } z)$:	1	-1	1	-1	1	-1	-1	1	1	1	1	1	
f_3:	1	-1	1	-1	1	-1	1	-1	1	-1	1	-1	(B_1)
$f_1 f_2 f_3$:	1	1	1	1	1	1	-1	-1	1	-1	1	-1	
		A_1				A_2				B_1			

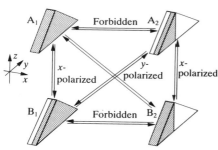

Fig. 15.27 The polarizations of the allowed transitions in a C_{2v} molecule. The shading indicates the structure of the orbitals of the specified symmetry species.

Only the first product (with $f_2 = x$) spans A_1, and so only the x-component of the transition dipole may be non-zero. Therefore, we conclude that the electric dipole transition $a_1 \leftarrow b_1$ (and $a_1 \rightarrow b_1$) is allowed. We can go on to state that the radiation emitted (or absorbed) is **x-polarized** and has its electric field vector in the x-direction, because that form of radiation couples with the x-component of a transition dipole.

Example 15.10: *Deducing a selection rule*

Is $p_x \leftarrow p_y$ an allowed transition in a tetrahedral molecule?

Answer. We must decide whether the product $p_y q p_x$, with $q = x$, y, or z, spans A_1 using the T_d character table. The procedure works out as follows:

$$
\begin{array}{lrrrrrl}
f_3(p_y): & 3 & 0 & -1 & 1 & -1 & (T_2) \\
f_2(q): & 3 & 0 & -1 & 1 & -1 & (T_2) \\
f_1(p_x): & 3 & 0 & -1 & 1 & -1 & (T_2) \\
f_1 f_2 f_3: & 27 & 0 & -1 & 1 & -1 &
\end{array}
$$

A_1 occurs (once) in this set of characters, and so $p_x \rightarrow p_y$ is allowed.

Comment. Closer analysis (using the representatives) shows that only $q = z$ gives a non-zero integral, and so the transition is z-polarized.

Exercise. What are the allowed transitions, and their polarizations, of a b_1 electron in a C_{4v} molecule?

$$[b_1 \rightarrow b_1(z); b_1 \rightarrow e(x, y)]$$

15.9 Symmetry-adapted linear combinations

So far we have only asserted the forms of the linear combinations (such as s_1, etc) that act as bases for irreducible representations. Group theory also provides machinery that takes an arbitrary basis (s_A, etc) as input and generates combinations of the specified symmetry. Because these combinations are adapted to the symmetry of the molecule, they are called **symmetry-adapted linear combinations** (SALC). Symmetry-adapted linear combinations are the building blocks of LCAO molecular orbitals, for they include combinations such as the $\psi_{1s}(A) \pm \psi_{1s}(B)$ used to construct molecular orbitals in H_2O (Section 14.8) and some of the complex examples that we have seen since then. The selection of symmetry-adapted linear combinations of atomic orbitals is the first step in any molecular orbital treatment of molecules, and is central, for instance, to the construction and analysis of Walsh diagrams and to the description of d-metal complexes.

The technique for building symmetry-adapted linear combinations is derived using the full power of group theory. We shall not show the derivation, which is very lengthy, but present the main conclusions as a set of rules:[4]

(1) Construct a table showing the effect of each operation on each orbital of the original basis.

For example, from the (s_N, s_A, s_B, s_C) basis in NH_3 we form the following table:

---

[4] Once again, it is possible to express the rules given here in a succinct formula derived from group theory. In this case, to form an orbital of symmetry species Γ we form $P\psi$, where

$$P\psi = \frac{1}{h} \sum_R \chi^{(\Gamma)}(R) R\psi$$

where R is an operation of the group. The implementation of this formula is described in the following remarks. Note that the actual operations occur in the formula, not the classes as in the earlier expressions.

Original basis:	s_N	s_A	s_B	s_C
Under E	s_N	s_A	s_B	s_C
C_3^+	s_N	s_B	s_C	s_A
C_3^-	s_N	s_C	s_A	s_B
σ_v	s_N	s_A	s_C	s_B
σ_v'	s_N	s_B	s_A	s_C
σ_v''	s_N	s_C	s_B	s_A

(2) To generate the combination of a specified symmetry species, take each column in turn and:
  (i) Multiply each member of the column by the character of the corresponding operation.
  (ii) Add together all the orbitals in each column with the factors as determined in (i).
(iii) Divide the sum by the order of the group.

In our example, in order to generate the A_1 combination we take the characters for A_1 $(1, 1, 1, 1, 1, 1)$, and so rules (i) and (ii) lead to

$$\psi \propto s_N + s_N + \ldots = 6s_N$$

The order of the group (the number of elements) is 6, and so the combination of A_1 symmetry that can be generated from s_N is s_N itself. Applying the same technique to the column under s_A gives

$$\psi = \tfrac{1}{6}(s_A + s_B + s_C + s_A + s_B + s_C) = \tfrac{1}{3}(s_A + s_B + s_C)$$

The same combination is built from the other two columns, and so they give no further information. The combination we have just formed is the s_1 combination we used before (apart from the numerical factor).

We now form the overall molecular orbital by forming a linear combination of all the symmetry-adapted linear combinations of the specified symmetry species. In this case, therefore, the a_1 molecular orbital is

$$\psi = c_N s_N + c_1 s_1$$

This is as far as group theory can take us. The coefficients must be found by solving the Schrödinger equation because they do not come directly from the symmetry of the system.

Suppose we try to generate a symmetry-adapted linear combination of species A_2 despite the fact that the previous work has shown that there is no such combination. The characters for A_2 are 1, 1, 1, -1, -1, -1. The column under s_N generates zero, and so do the other three. Therefore, we find that we generate no combination of A_2 symmetry.

When we try to generate the E symmetry-adapted combinations we run into a problem because, for representations of dimension 2 or more, where the characters are the sums of the diagonal elements, the rules generate sums of the symmetry-adapted combinations. (A more detailed rule based on the representatives themselves gives the individual combinations.) This can be seen as follows. The E characters are 2, -1, -1, 0, 0, 0, and so the column under s_N gives

$$\psi = 2s_N - s_N - s_N + 0 + 0 + 0 = 0$$

The other columns give

$$\tfrac{1}{6}(2s_A - s_B - s_C) \quad \tfrac{1}{6}(2s_B - s_A - s_C) \quad \tfrac{1}{6}(2s_C - s_B - s_A)$$

However, any one can be expressed as a sum of the other two (they are not linearly independent). The difference of the second and third gives $\tfrac{1}{2}(s_B - s_C)$, and this and the first, $\tfrac{1}{6}(2s_A - s_B - s_C)$, are the two linearly independent symmetry-adapted combinations we have used in the discussion of e orbitals.

The following chapters will show many more examples of how the systematic use of symmetry using the techniques of group theory can greatly simplify the analysis of molecular structure and spectra.

Further information: matrices

Matrices are arrays of numbers with special rules for combining them together. We shall consider $n \times n$ square matrices M with n^2 elements M_{rc}:

$$M = \begin{pmatrix} M_{11} & M_{12} & \cdots & M_{1n} \\ M_{21} & M_{22} & \cdots & M_{2n} \\ & & & \\ M_{n1} & M_{n2} & \cdots & M_{nn} \end{pmatrix}$$

M_{rc} is the element in row r, column c (r,c is a map reference). Two matrices M and N are equal (written $M = N$) only if all corresponding elements are equal: $M_{rc} = N_{rc}$ for all r,c.

The addition of two matrices,

$$M + N = P$$

is defined through

$$P_{rc} = M_{rc} + N_{rc}$$

That is, add corresponding elements. For example,

$$M = \begin{pmatrix} 1 & 2 \\ 3 & 4 \end{pmatrix} \quad N = \begin{pmatrix} 5 & 6 \\ 7 & 8 \end{pmatrix} \quad M + N = \begin{pmatrix} 1+5 & 2+6 \\ 3+7 & 4+8 \end{pmatrix} = \begin{pmatrix} 6 & 8 \\ 10 & 12 \end{pmatrix}$$

The multiplication of two matrices is written

$$MN = P$$

and defined through

$$P_{rc} = \sum_q M_{rq} N_{qc}$$

This rule can be remembered in terms of the following diagram:

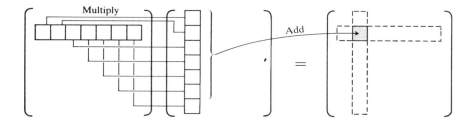

For example, using the same matrices as above,

$$MN = \begin{pmatrix} 1 & 2 \\ 3 & 4 \end{pmatrix}\begin{pmatrix} 5 & 6 \\ 7 & 8 \end{pmatrix} = \begin{pmatrix} 1\times 5 + 2\times 7 & 1\times 6 + 2\times 8 \\ 3\times 5 + 4\times 7 & 3\times 6 + 4\times 8 \end{pmatrix} = \begin{pmatrix} 19 & 22 \\ 43 & 50 \end{pmatrix}$$

In this case $NM \neq MN$, and so matrix multiplication is 'noncommutative' and depends on the order of multiplication in general.

Several types of matrix have special names or properties. Among them are the following:

A **diagonal matrix** is one in which all $M_{rc} = 0$ unless $r = c$. For example,

$$\begin{pmatrix} 1 & 0 \\ 0 & 5 \end{pmatrix} \text{ is diagonal}$$

$$\begin{pmatrix} 0 & 1 \\ 5 & 0 \end{pmatrix} \text{ is not diagonal}$$

The **unit matrix 1** diagonal with all diagonal elements equal to 1. Thus, the 2×2 unit matrix is

$$\mathbf{1} = \begin{pmatrix} 1 & 0 \\ 0 & 1 \end{pmatrix}$$

The **transpose** of a matrix M^{T} is related to M by interchange of rows and columns:

$$M = \begin{pmatrix} 1 & 2 \\ 3 & 4 \end{pmatrix} \qquad M^{\mathrm{T}} = \begin{pmatrix} 1 & 3 \\ 2 & 4 \end{pmatrix}$$

The **inverse** of a matrix M^{-1} satisfies

$$MM^{-1} = M^{-1}M = \mathbf{1}$$

The inverse can be constructed as follows:

(1) Calculate the **determinant**, det M, of the matrix M:

$$M = \begin{pmatrix} 1 & 2 \\ 3 & 4 \end{pmatrix} \qquad \det M = \begin{vmatrix} 1 & 2 \\ 3 & 4 \end{vmatrix} = 1\times 4 - 2\times 3 = -2$$

If det $M = 0$ the matrix is **singular** and M^{-1} does not exist (just as 0^{-1} is not defined in ordinary arithmetic).

(2) Form the transpose of M:

$$M^{\mathrm{T}} = \begin{pmatrix} 1 & 3 \\ 2 & 4 \end{pmatrix}$$

(3) Form the matrix M' of **cofactors** of M^{T}, where a cofactor of an element is the determinant formed from the matrix with row r, column c struck out and multiplied by $(-1)^{r+c}$:

$$M' = \begin{pmatrix} 4 & -2 \\ -3 & 1 \end{pmatrix}$$

(4) The inverse is then

$$M^{-1} = \frac{M'}{\det M}$$

$$M^{-1} = -\frac{1}{2}\begin{pmatrix} 4 & -2 \\ -3 & 1 \end{pmatrix} = \begin{pmatrix} -2 & 1 \\ \frac{3}{2} & -\frac{1}{2} \end{pmatrix}$$

An important application of matrices (apart from their role as representatives of symmetry operations) is in the solution of simultaneous equations. A set of equations

$$M_{11}x_1 + M_{12}x_2 + \ldots + M_{1n}x_n = c_1$$
$$M_{21}x_1 + M_{22}x_2 + \ldots + M_{2n}x_n = c_2$$
$$\cdots\cdots\cdots$$
$$M_{n1}x_1 + M_{n2}x_2 + \ldots + M_{nn}x_n = c_n$$

can be expressed in the compact form

$$Mx = c$$

where M is the matrix of coefficients and x and c are the $1 \times n$ matrices (or 'column vectors')

$$x = \begin{pmatrix} x_1 \\ x_2 \\ \vdots \\ x_n \end{pmatrix} \qquad c = \begin{pmatrix} c_1 \\ c_2 \\ \vdots \\ c_n \end{pmatrix}$$

Then, on multiplying both sides of the matrix equation by M^{-1} we obtain

$$x = M^{-1}c$$

Hence, solving the set of equations amounts to finding the inverse of the matrix of coefficients.

Further reading

F. A. Cotton, *Chemical applications of group theory*. Wiley, New York (1971).

S. F. A. Kettle, *Symmetry and structure*. Wiley, New York (1985).

D. C. Harris and M. D. Bertolucci, *Symmetry and spectroscopy*. Oxford University Press (1978).

B. E. Douglas and C. Hollingsworth, *Symmetry in bonding and structure*. Academic Press, New York (1985).

P. W. Atkins, *Molecular quantum mechanics* (2nd edn). Oxford University Press (1983).

D. M. Bishop, *Group theory and spectroscopy*. Oxford University Press (1973).

P. W. Atkins, M. S. Child, and C. S. G. Phillips, *Tables for group theory*. Oxford University Press (1970).

Exercises

15.1 The point group D_5 has four symmetry species of irreducible representation. How many classes of symmetry operations does it contain?

15.2 The CH_3Cl molecule belongs to the point group C_{3v}. List the symmetry elements of the group and locate them in the molecule.

15.3 Which of the following molecules may be polar? (a) pyridine (C_{2v}), (b) nitroethane (C_s), (c) gas-phase $HgBr_2$ ($D_{\infty h}$), (d) $B_3N_3H_6$ (D_{3h}), (e) CH_3Cl (C_{3v}), (f) $HW_2(CO)_{10}$ (D_{4h}), (g) $SnCl_4$ (T_d).

15.4 Use symmetry properties to determine whether or not the integral $\int p_x z p_z \, d\tau$ is necessarily zero in a molecule with symmetry C_{4v}.

15.5 Show that the transition $A_1 \rightarrow A_2$ is forbidden for electric dipole transitions in a C_{3v} molecule.

15.6 Show that a set of five hybrid orbitals in a molecule of C_{4v} symmetry and the characters

E	$2C_4$	C_2	$2\sigma_v$	$2\sigma_d$
5	1	1	3	1

may have the composition p^1d^4.

15.7 Show that the function xy has symmetry species B_2 in the group C_{4v}.

15.8 Molecules belonging to the point groups D_{2h}, C_{3h}, T_h, and T_d cannot be chiral. Which elements of these groups rule out chirality?

15.9 The group D_2 consists of the elements E, C_2, C_2', and C_2'', where the three twofold rotations are around mutually perpendicular axes. Construct the group multiplication table.

15.10 Identify the point groups to which the following objects belong: (a) a sphere, (b) an isosceles triangle, (c) an equilateral triangle, (d) an unsharpened cylindrical pencil, (e) a sharpened cylindrical pencil, (f) a three-bladed propellor, (g) a four-legged table, (h) yourself (approximately).

15.11 List the symmetry elements of the following molecules and name the point groups to which they belong: (a) NO_2, (b) N_2O, (c) $CHCl_3$, (d) $CH_2{=}CH_2$, (e) *cis*-$CHCl{=}CHCl$, (f) *trans*-$CHCl{=}CHCl$.

15.12 List the symmetry elements of the following molecules and name the point groups to which they belong: (a) naphthalene, (b) anthracene, (c) the three dichlorobenzenes.

15.13 Which of the molecules in Exercises 15.11 and 15.12 can be (a) polar, (b) chiral?

15.14 Consider the C_{2v} molecule NO_2. The combination $p_x(A) - p_x(B)$ of the two O atoms (with x perpendicular to the plane) spans A_2. Is there any orbital of the central N atom that can have a non-zero overlap with that combination of O orbitals? What would be the case in SO_2, where $3d$ orbitals might be available?

15.15 The ground state of NO_2 is A_1 in the group C_{2v}. To what excited states may it be excited by electric-dipole transitions, and what polarization of light is it necessary to use?

15.16 The ClO_2 molecule (which belongs to the group C_{2v}) was trapped in a solid. Its ground state is known to be B_1. Light polarized parallel to the y axis (parallel to the OO separation) excited the molecule to an upper state. What is the symmetry of that state?

15.17 What states of (a) benzene, (b) naphthalene may be reached by electric dipole transitions from their (totally symmetrical) ground states?

15.18 Confirm that the z component of orbital angular momentum is a basis for an irreducible representation of A_2 symmetry in C_{3v}.

15.19 Write $f_1 = \sin\theta$ and $f_2 = \cos\theta$, and show by symmetry arguments using the group C_s that the integral of their product over a symmetrical range around the origin is zero.

Problems

15.1 List the symmetry elements of the following molecules and name the point groups to which they belong: (a) staggered CH_3CH_3, (b) chair and boat cyclohexane, (c) B_2H_6, (d) $[Co(en)_3]^{3+}$ where en is ethylenediamine (ignore its detailed structure), (e) crown-shaped S_8. Which of these molecules can be (i) polar, (ii) chiral?

15.2 The group C_{2h} consists of the elements E, C_2, σ_h, i. Construct the group multiplication table and find an example of a molecule that belongs to the group.

15.3 The group D_{2h} has a C_2 axis perpendicular to the principal axis and a horizontal mirror plane. Show that the group must therefore have a centre of inversion.

15.4 Consider the H_2O molecule, which belongs to the group C_{2v}. Take as a basis the two H1s orbitals and the four valence orbital of the O atom and set up the 6×6 matrices that represent the group in this basis. Confirm by explicit matrix multiplication that the group multiplications (a) $C_2\sigma_v = \sigma_v'$ and (b) $\sigma_v\sigma_v' = C_2$. Confirm, by calculating the traces of the matrices, (a) that the representation is reducible, and (b) that the basis spans $3A_1 + B_1 + 2B_2$.

15.5 The (one-dimensional) matrices $\mathbf{D}(C_3) = 1$ and $\mathbf{D}(C_2) = 1$, and $\mathbf{D}(C_3) = 1$ and $\mathbf{D}(C_2) = -1$ both represent the group multiplication $C_3C_2 = C_6$ in the group C_{6v} with $\mathbf{D}(C_6) = +1$ and -1 respectively. Use the character table to

confirm these remarks. What are the representatives of σ_v and σ_d in each case?

15.6 What irreducible representations do the four H1s orbitals of CH_4 span? Are there s and p orbitals of the central C atom that may form molecular orbitals with them? Could d orbitals, even if they were present on the C atom, play a role in orbital formation in CH_4?

15.7 Suppose that a methane molecule became distorted to (a) C_{3v} symmetry by the lengthening of one bond, (b) C_{2v} symmetry, by a kind of scissors action in which one bond angle opened and another closed slightly. Would more d orbitals become available for bonding?

15.8 The algebraic forms of the f orbitals are a radial function multiplied by one of the factors

(a) $z(5z^2 - 3r^2)$, (b) $y(5y^2 - 3r^2)$, (c) $x(5x^2 - 3r^2)$
(d) $z(x^2 - y^2)$, (e) $y(x^2 - z^2)$, (f) $x(z^2 - y^2)$, (g) xyz

Identify the irreducible representations spanned by these orbitals in (a) C_{2v}, (b) C_{3v}, (c) T_d, (d) O_h. Consider a lanthanide ion at the center of (a) a tetrahedral complex, (b) an octahedral complex. What sets of orbitals do the seven f orbitals split into?

15.9 Which of the following transitions is allowed in (a) a tetrahedral complex, (b) an octahedral complex: (i) $d_{xy} \rightarrow$ d_{z^2}, (ii) $d_{xy} \rightarrow f_{xyz}$?

15.10 Does the product xyz necessarily vanish when integrated over (a) a cube, (b) a tetrahedron, (c) a hexagonal prism, each centred on the origin?

15.11 Treat the naphthalene molecule as belonging to the group C_{2v} with the C_2 axis perpendicular to the plane. Classify the irreducible representations spanned by the carbon $2p_z$ orbitals and find their symmetry-adapted linear combinations.

15.12 The NO_2 molecule belongs to the group C_{2v}, with the C_2 axis bisecting the ONO angle. Taking as a basis the N2s, N2p, and O2p orbitals, identify the irreducible representations they span, and construct the symmetry-adapted linear combinations.

15.13 Construct the symmetry-adapted linear combinations of C2p_z orbitals for benzene, and use them to calculate the Hückel secular determinant. This procedure leads to equations that are much easier to solve than using the original orbitals, and show that the Hückel orbitals are those specified in Section 14.10.

16 Rotational and vibrational spectra

Check-list of key ideas

1. The classification of spectra as *emission, absorption,* and *Raman* and the experimental techniques used for their study (Section 16.1).

2. The general principles of *Fourier transform spectroscopy* (Section 16.1).

3. The *Einstein transition probabilities,* and the coefficients of *stimulated absorption and emission* and of *spontaneous emission* (Section 16.2).

4. The *transition dipole moment* (eqn 10), the *gross selection rule,* and the *specific selection rules* of transitions (Section 16.2).

5. The contributions to spectral *linewidths* (Section 16.3), particularly *Doppler broadening* (eqn 11) and *lifetime broadening* (eqn 12).

6. The principles of *Lamb-dip spectroscopy* (Section 16.3).

7. The *rotational energy levels* of *rigid rotors* (Section 16.4) in terms of the *moments of inertia* of molecules and their *rotational constants* (eqns 14 to 16).

8. The *Stark effect* on the rotational energies of polar molecules in electric fields (eqn 17).

9. The effect of *centrifugal distortion* on the rotational energy levels of molecules (eqn 18).

10. The *selection rules* and transition moments of pure rotational transitions and the contribution of state populations to the intensities of spectral lines (Section 16.5).

11. The *electric polarizabilities* of molecules and their contribution to the detection of rotational Raman transitions (Section 16.6).

12. The *Stokes* and *anti-Stokes lines* in a Raman spectrum (eqn 21).

13. The *harmonic approximation* for the description of the vibrations of molecules (Section 16.7) and the *vibrational terms* of a diatomic molecule (eqn 22).

14. The *anharmonicity* of molecular vibrations, its description in terms of the *Morse potential* (eqn 23), and its effect on the vibrational spectrum (Section 16.7).

15. The *selection rules* for vibrational transitions and the appearance of vibrational spectra (Section 16.8).

16. The use of the *Birge–Sponer extrapolation* to determine dissociation energies (Section 16.8 and Example 16.9).

17. The formation of *P, Q, and R branches* in *vibration–rotation spectra* (Section 16.9).

18. The *vibrational Raman spectra* of diatomic molecules (Section 16.10).

19. The number of vibrations of polyatomic molecules (eqn 28) and their description in terms of *normal modes* (Section 16.11).

20. The *symmetry analysis* of normal modes and their *infrared and Raman activities* (Section 16.11 and Examples 16.10 to 16.12).

21. The appearance and analysis of the infrared spectra of polyatomic molecules (Section 16.12).

22. The *exclusion rule* for centrosymmetric molecules (Section 16.13) and information from the *depolarization ratios* of Raman transitions about the symmetries of normal modes (Section 16.13).

The origin of spectral lines in molecular spectroscopy is the emission or absorption of a photon when the energy of a molecule changes. The difference from atomic spectroscopy is that the energy of a molecule can change not only as a result of electronic transitions but also because it can undergo transitions between its rotational and vibrational states. Molecular spectra are therefore more complex than atomic spectra. However, they also contain information relating to more properties, and their analysis leads to values of bond strengths, lengths, and angles. They also provide a way of measuring a variety of molecular properties, particularly electric dipole moments.

A pure rotational spectrum (one in which only the rotational state of the molecule changes) can be observed, but vibrational spectra of gaseous samples show features that arise from simultaneous rotational transitions. Similarly, electronic spectra (Chapter 17) show features arising from simultaneous vibrational and rotational transitions. The simplest way of dealing with these complexities is to tackle each type of transition in turn, and then to see how simultaneous changes affect the appearance of the spectrum.

General features of spectroscopy

All types of spectra have some features in common, and we examine these first.

16.1 Experimental techniques

In **emission spectroscopy**, a molecule undergoes a transition from a state of high energy E_1 to a state of lower energy E_2 and emits the excess energy as a photon. In **absorption spectroscopy**, the net absorption[1] of nearly

---

[1] We say *net* absorption, since it will become clear that when a sample is irradiated, both absorption and emission at a given frequency are stimulated, and the detector measures the difference, the net absorption.

monochromatic incident radiation is monitored as it is swept over a range of frequencies. The energy hv of the photon emitted or absorbed, and therefore the frequency v of the radiation emitted or absorbed, is given by the same Bohr frequency condition as we met for atoms:

$$hv = E_1 - E_2 \tag{1}$$

This relation is often expressed in terms of the **vacuum wavelength** λ (usually in nanometers)

$$\lambda = \frac{c}{v} \tag{2a}$$

or the **vacuum wavenumber** $\tilde{v}$

$$\tilde{v} = \frac{v}{c} \tag{2b}$$

The units of the latter are almost always chosen as cm^{-1}. Figure 16.1 summarizes the frequencies, wavelengths, and wavenumbers of the various regions of the electromagnetic spectrum.

Emission and absorption spectroscopy give the same information about energy level separations, but practical considerations generally determine which technique is employed. In practice, emission spectroscopy, if it is used at all, is used only for optical and ultraviolet spectroscopy; absorption spectroscopy is much more widely used, and we shall concentrate on it.

All absorption spectrometers consist of a source of radiation, a sample cell, usually a dispersing element, and a detector, and the characteristics of each component depend on the region of the electromagnetic spectrum being considered. Most spectrometers also include a monochromator.

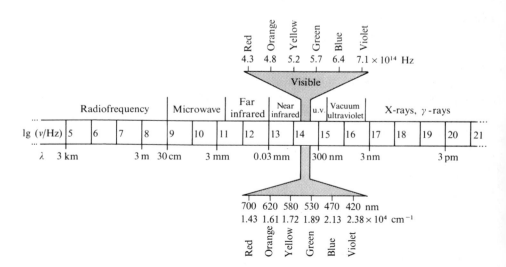

Fig. 16.1 The electromagnetic spectrum and the classification of the spectral regions.

Sources of radiation

The source generally produces radiation spanning a range of frequencies, but in a few cases it generates monochromatic—single frequency—radiation. One such case is the **klystron**, an electronic device used to generate microwaves. Lasers—which we discuss in more detail in the next chapter—give monochromatic radiation that can often be tuned over a range of frequencies. For the far infrared, the source is a mercury arc inside a quartz envelope, most of the radiation being generated by the hot quartz. A **Nernst filament** is used to generate radiation in the near infrared. This consists of a heated ceramic filament containing rare-earth oxides, which emits radiation closely resembling that of a true black body. For the visible region of the spectrum, a tungsten/iodine lamp is used, which gives out intense white light. A discharge through deuterium gas or xenon in quartz is still widely used for the near ultraviolet.

For certain applications, **synchrotron radiation** from a **synchrotron storage ring** is appropriate. A synchrotron ring consists of an electron beam (actually a series of closely spaced packets) travelling in a circle of several metres in diameter. Since accelerated charges emit radiation, and electrons travelling in a circle are being constantly accelerated by the forces that constrain them to their path, the beam emits radiation. The emitted radiation spans a wide range of frequencies, up to and including the far ultraviolet, and in all except the microwave region is much more intense than can be obtained by most conventional sources. The disadvantage of the source is that it is so large and costly that it is essentially a national facility, and not a laboratory commonplace.

The dispersing element and Fourier spectroscopy

In all but specialized techniques using monochromatic microwave radiation, spectrometers include a component for separating the frequencies of the radiation so that the variation of the absorption with frequency can be monitored. In conventional spectrometers, this component is a **dispersing element** that separates different frequencies into different spatial directions.

The simplest dispersing element is a glass or quartz prism (Fig. 16.2), which utilizes the variation of refractive index with frequency of the incident radiation. High-frequency radiation generally (but not always) results in a higher refractive index than low-frequency radiation, and therefore undergoes a greater deflection when passing through the prism. Problems of absorption by the prism can be avoided by replacing it by a **diffraction grating**. A diffraction grating consists of a glass or ceramic plate into which fine grooves have been cut about 1000 nm apart (comparable to the wavelength of visible light) and covered with a reflective aluminium coating. The grating causes interference between waves reflected from its surface, and constructive interference occurs at specific angles that depend on the frequency of the radiation being used. By shaping the grooves appropriately, the process called **blazing** (Fig. 16.3), the intensity of the interference pattern can be enhanced.

Fourier transform techniques

Modern spectrometers, particularly those operating in the infrared, now almost always use **Fourier transform techniques** of spectral detection and

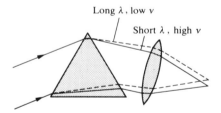

Fig. 16.2 One simple dispersing element is a prism, which separates frequencies spatially by making use of the higher refractive index of high-frequency radiation. The shortest wavelengths for which a glass prism can be used is about 400 nm, but quartz can be used down to 180 nm.

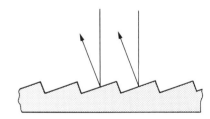

Fig. 16.3 A diffraction grating is 'blazed' as shown here in order to enhance the intensity of the diffracted radiation in each direction.

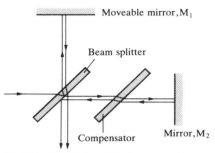

Fig. 16.4 A Michelson interferometer. The beam-splitting element divides the incident beam into two beams with a path difference that depends on the location of the mirror M_1. The compensator ensures that both beams pass through the same thickness of material.

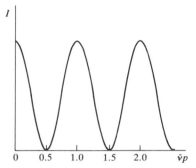

Fig. 16.5 An interferogram produced as the path length p is changed in the interferometer shown in Fig. 16.4 and radiation of wavelength λ is present.

analysis. The heart of a Fourier transform spectrometer is a **Michelson interferometer**, which is a device for analyzing the frequencies present in a composite signal. The total signal from a sample is the analogue of a chord played on a piano, and the Fourier transformation of the signal is equivalent to the separation of the chord into its individual notes, the spectrum. The advantage of this procedure is the greater sensitivity that stems from the fact that the detector monitors the entire spectrum simultaneously rather than one frequency at a time.

The Michelson interferometer works by splitting the beam from the sample into two (Fig. 16.4) and introducing a varying path difference p into one of them. When the two components recombine, there is a phase difference between them, and they interfere either constructively or destructively depending on the extra path that one has taken. The detected signal oscillates as the two components alternately come into and out of phase as p is changed (Fig. 16.5), and if the radiation has wavenumber $\tilde{v}$, the detected signal varies with p as

$$I(p) = I(\tilde{v}) \cos 2\pi\tilde{v}p$$

Hence, the interferometer converts the presence of a particular component in the signal into a variation in intensity of the radiation reaching the detector. An actual signal consists of radiation spanning a large number of wavenumbers, and the total intensity at the detector is the sum of all their oscillating intensities:

$$I(p) = \int_0^\infty I(\tilde{v}) \cos 2\pi\tilde{v}p \, d\tilde{v}$$

Example 16.1: *Calculating a Fourier transform spectrum*

Draw the total signal that would be expected for a signal consisting of three wavenumbers, one of $1000\,\text{cm}^{-1}$ of intensity I, another of wavenumber $1100\,\text{cm}^{-1}$ of intensity $2I$, and a third of wavenumber $1150\,\text{cm}^{-1}$ of intensity $\tfrac{1}{2}I$.

Answer. The only contributions to the integral defining $I(p)$ are at the three specified wavenumbers, so the integral consists of the three terms

$$I(p) = I\cos 2\pi \times \frac{1000p}{\text{cm}} + 2I\cos 2\pi \times \frac{1100p}{\text{cm}} + \tfrac{1}{2}I\cos 2\pi \times \frac{1150p}{\text{cm}}$$

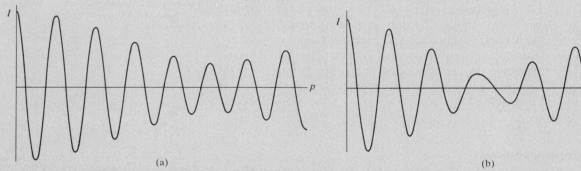

(a)   (b)

Fig. 16.6 (a) The interferogram obtained under the circumstances described in Example 16.1 when three components are present in the radiation and (b) the interferogram when five components are present.

The plot of $I(p)/I$ against p in the range $p = 0$ to $0.1\ \text{cm}$ is shown in Fig. 16.6a, and shows the signal that would be detected.

Comment. It is very easy to program calculations like these for a computer, and to explore how the signal changes as different components are included in the spectrum.

Exercise. Include two further components in the spectrum, one at $950\ \text{cm}^{-1}$ of intensity $2I$ and another at $975\ \text{cm}^{-1}$ of intensity $0.75I$. [Fig. 16.6b]

The problem is to find $I(\bar{v})$, the variation of intensity with wavenumber, which is the spectrum we require. This step is a standard technique of mathematics, and is the 'Fourier transformation' step from which this form of spectroscopy takes its name. Specifically:

$$I(\bar{v}) = \int_0^\infty I(p) \cos 2\pi \bar{v} p \ \mathrm{d}p$$

Thus, as we illustrate schematically in Fig. 16.7, we take the signal $I(p)$ for each path difference, multiply the intensity at each value of p by the value of $\cos 2\pi \bar{v} p$, and add all the products together. When the result is plotted against wavenumber, we get the absorption spectrum (Fig. 16.7b). In practice, the Fourier analysis step is carried out in a computer built into the spectrometer.

Detectors

The third component of a spectrometer is the **detector**, the device that converts incident radiation into an electric current for the appropriate signal processing or plotting. Radiation-sensitive semiconductor devices are increasingly dominating this role in the spectrometer. However, in the optical and ultraviolet region, photographic recording or a **photomultiplier** are widely used. In the latter device, an incident photon ejects an electron from a photosensitive surface, the electron is accelerated by a potential difference, and ejects a shower of electrons where it strikes a screen. These

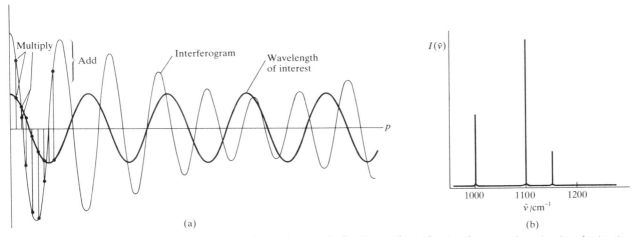

(a)   (b)

Fig. 16.7 (a) A schematic diagram illustrating the principle of the extraction of a Fourier transform of an interferogram. At each value of p the observed spectrum is multiplied by the amplitude of the cosine function at the wavelength of interest. The sum of the products (the integration) is zero if the harmonic is not present in the signal but is non-zero if it is present. (b) When the procedure is repeated for all wavelengths, the function obtained is the absorption spectrum of the sample.

electrons are accelerated, and each one releases a further shower on impact with another screen. Thus the impact of the initial photon is converted into a cascade of electrons, which is converted into a current in an external circuit.

Although semiconductor detectors are increasingly being used in the infrared, thermocouples are still widely used. A thermocouple detector consists typically of a blackened gold foil to which are attached thermo-electric alloys. A **thermistor bolometer** is essentially a resistance ther-mometer, and is typically formed from a mixture of oxides deposited on quartz. In each case the radiation is chopped by a shutter that rotates in the beam so that an alternating signal is obtained from the detector (which is easier to amplify than a steady signal). A microwave detector is typically a **crystal diode** consisting of a tungsten tip in contact with a semiconductor, such as germanium, silicon, or gallium arsenide.

The sample

The highest resolution is obtained when the sample is gaseous and of such low pressure that collisions between the molecules are infrequent. Gaseous samples are essential for rotational (microwave) spectroscopy, for only then can molecules rotate freely. To achieve sufficient absorption, the path lengths of gaseous samples must be very long, of the order of metres; long path lengths are achieved by multiple passage of the beam between two parallel mirrors at each end of the sample cavity.

For infrared spectroscopy, the sample is typically a liquid held between windows of sodium chloride (which is transparent down to $700\,cm^{-1}$) or potassium bromide (down to $400\,cm^{-1}$). Other ways of preparing the sample include grinding into a paste with 'Nujol', a hydrocarbon oil, or pressing it into a solid disk, perhaps with powdered potassium bromide.

Raman spectroscopy

In **Raman spectroscopy** the energy levels of molecules are explored by examining the frequencies present in the radiation scattered by molecules. In a typical experiment, a monochromatic incident beam—typically in the visible region of the spectrum—is passed through the sample and the radiation scattered perpendicular to the beam is monitored. Some of the incident photons collide with the molecules, give up some of their energy, and emerge with a lower energy. These scattered photons constitute the lower-frequency **Stokes radiation** from the sample. Other incident photons may collect energy from the molecules (if they are already excited), and emerge as higher-frequency **anti-Stokes radiation**. The shifts in frequency of the scattered radiation from the incident radiation are quite small, and the latter must be very monochromatic if the shifts are to be observed. Moreover, the intensity of scattered radiation is low, and so very intense incident beams are needed. Lasers are ideal in both respects, and have entirely displaced the mercury arcs used originally. Detection is usually with a photomultiplier.

16.2 The intensities of spectral lines

A glance at the spectra (whether emission, absorption, or Raman) illustrated in this chapter and the next shows that their lines occur with a variety of intensities. We shall also see that some lines that might be expected to occur do not appear at all. To account for these features, we must see how the intensities of spectral lines depend on the numbers of molecules in various states and how strongly individual transitions are able to interact with the electromagnetic field and generate or absorb photons.

The Einstein transition probabilities

Einstein considered the question of the rates of transitions between two levels in the presence of an electromagnetic field, and wrote the **transition rate**[2] w from the lower to the upper state as

$$w = B\rho \tag{3}$$

B is the **Einstein coefficient of stimulated absorption** and ρ is the energy density of radiation at the frequency of the transition. If the molecule is exposed to black-body radiation from a source of temperature T, ρ is given by the Planck distribution (eqn 12 of Section 11.2 expressed in terms of frequency[3]):

$$\rho = \frac{8\pi h \nu^3}{c^3} \frac{1}{e^{h\nu/kT} - 1} \tag{4}$$

The coefficient B depends only on the wavefunctions of the states involved in the transition, and we describe it in more detail later. For the time being we can treat B as an empirical parameter that characterizes the transition. The total **rate of absorption** W, the number of molecules undergoing excitation, is the transition rate of a single molecule multiplied by the number of molecules N in the lower state:

$$W = Nw \tag{5}$$

Einstein considered that the radiation was also able to induce the molecule in the upper state to undergo a transition to the lower state, and hence to generate a photon of frequency ν. Thus, he wrote the rate of this **stimulated emission** as

$$w' = B'\rho \tag{6}$$

where B' is the **Einstein coefficient of stimulated emission**. However, he

---

[2] Specifically, w is the rate of change of probability of the molecule being found in the upper state:

$$w = \frac{dP}{dt}$$

[3] The energy density is expressed in terms of frequency by writing

$$d\mathcal{U} = \rho(\nu)\,d\nu$$

in place of

$$d\mathcal{U} = \rho(\lambda)\,d\lambda$$

The relation between the ρ used here, which is $\rho(\nu)$, and $\rho(\lambda)$ in eqn 12 of Section 12.2 is then obtained using

$$d\nu = \frac{-c\,d\lambda}{\lambda^2}$$

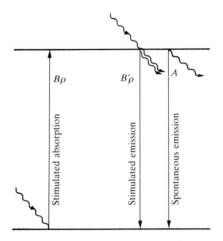

Fig. 16.8 The processes that account for absorption and emission of radiation and the attainment of thermal equilibrium. The excited state can return to the lower state spontaneously as well as by a process stimulated by radiation already present at the transition frequency.

realized that this was not the only means by which the excited state could generate radiation and return to the lower state. If it were, the total rate of return would be $N'w'$, and at thermal equilibrium, when the rate of absorption is equal to the rate of emission, we would be able to write

$$NB\rho = N'B'\rho$$

which rearranges to

$$\frac{N'}{N} = \frac{B}{B'}$$

However, it is known from very general grounds (which will be derived in Chapter 19), that the populations of the two states are given by the Boltzmann distribution:

$$\frac{N'}{N} = e^{-h\nu/kT} \quad h\nu = E' - E$$

The ratio of populations should therefore be temperature-dependent, but the ratio of the Einstein coefficients is independent of the temperature.

To resolve this inconsistency, Einstein proposed that the upper state was able to discard energy by **spontaneous emission** at a rate that is independent of the intensity of radiation already present (Fig. 16.8). He therefore wrote the total rate of transition to a lower state as

$$w' = A + B'\rho \tag{7}$$

A is the **Einstein coefficient of spontaneous emission**. The overall rate of emission is

$$W' = N'(A + B'\rho) \tag{8}$$

At thermal equilibrium, the rates of emission and absorption are equal, and so

$$NB\rho = N'(A + B'\rho)$$

Since the radiation intensity no longer cancels, and is temperature dependent, the ratio of populations is no longer inconsistent with the Boltzmann distribution. More specifically, we can arrange the last expression into

$$\rho = \frac{N'A}{NB - N'B'} = \frac{A}{B}\frac{1}{\dfrac{N}{N'} - \dfrac{B'}{B}}$$

$$= \frac{A}{B}\frac{1}{e^{h\nu/kT} - \dfrac{B'}{B}}$$

We have used the Boltzmann expression in the last line. This result is encouraging, since it has the same form as the Planck distribution (eqn 4), which is known to describe the radiation density at thermal equilibrium. Indeed, when we compare the two expressions for ρ, we can conclude that

$$B' = B \tag{9a}$$

$$A = \frac{8\pi h\nu^3}{c^3} \times B \tag{9b}$$

That is, the coefficients of stimulated absorption and emission are equal, and the relative importance of spontaneous emission grows as the cube of the frequency of the transition. The strong growth of the relative importance of spontaneous emission with increasing frequency is a very important conclusion, as we shall see when we consider the operation of lasers in the next chapter. The equality of B and B' implies that if two states happen to have equal populations, then the rate of stimulated emission is exactly equal to the rate of stimulated absorption, and there is then no net absorption.

The population of states

At low frequencies, such as those involved in rotational and vibrational transitions considered in this chapter, spontaneous emission can be largely ignored and the intensities of the transition discussed in terms of the coefficients of stimulated emission and absorption. Then the net rate of absorption is given by

$$W_{\text{net}} = NB\rho - N'B'\rho = (N - N')B\rho$$

and is proportional to the population difference of the two states involved in the transition. If the sample is at thermal equilibrium at a temperature T we may use the Boltzmann distribution to write the population difference as

$$N - N' = N\left(1 - \frac{N'}{N}\right)$$
$$= N(1 - e^{-h\nu/kT})$$

Therefore, the intensity of net absorption is proportional to the population N of the lower state as well as to the difference in population of the upper and lower states.

Example 16.2: *Estimating relative transition intensities*

Estimate the relative intensities at 25°C of absorptions originating in the ground state and the first excited state when the energy levels involved are separated by (a) $10\,000\ \text{cm}^{-1}$, (b) $1000\ \text{cm}^{-1}$, and (c) $1\ \text{cm}^{-1}$.

Answer. At 25°C, $kT/hc = 207\ \text{cm}^{-1}$, so the population differences between adjacent states is determined by the factor

$$1 - e^{-h\bar{\nu}c/kT} = \begin{cases} \text{(a)}\ 1 - e^{-48} = 1.000 \\ \text{(b)}\ 1 - e^{-4.8} = 0.992 \\ \text{(c)}\ 1 - e^{-0.0048} = 0.0048 \end{cases}$$

We draw two conclusions. The first is that the population of the upper state is negligible in (a) and (b), the only significant absorption is from the lower state. The second is that in these two cases, the stimulated emission from the upper state is also negligible, and we need consider only stimulated absorption when assessing the intensity of a transition. However, for (c) we can draw neither conclusion. Since in this case adjacent states are almost equally populated, transitions can originate with significant intensity from many states, and stimulated emission from upper states makes a significant contribution to the net absorption intensity.

Exercise. Repeat the analysis for a temperature of 1500 K.

[(a) As before, upper-state populations significant for (b) and (c)]

It follows from the last equation that the relative intensities of two lines corresponding to transitions originating from two different states should be proportional to the relative populations of the two initial states. Since the first electronically excited state of a molecule is usually of the order of $10^4\,\mathrm{cm}^{-1}$ above the ground state, it is not populated at room temperature (see Example 16.2). Therefore, an electronic absorption spectrum is normally due entirely to transitions originating from the ground electronic state. Vibrational energy levels are separated by around 10^2 to $10^3\,\mathrm{cm}^{-1}$, and so the principal transitions are also normally those from the ground vibrational state, and stimulated emission makes a negligible contribution to the net absorption. In contrast, rotational energy levels are separated by only 1 to $10^2\,\mathrm{cm}^{-1}$, and many states are occupied even at room temperature; consequently, rotational transitions occur from a wide range of initial states, not only the lowest, and stimulated emission from the occupied higher states is important.

Molecules are often prepared in short-lived excited states as a result of chemical reaction, electric discharge, or photolysis. In these cases the populations may be quite different from those at thermal equilibrium, and the spectra—if they can be taken quickly enough—then arise from transitions from all the populated levels.

Selection rules and transition moments

We met the concept of a 'selection rule' in Section 13.3 as a rule that determines whether a transition is forbidden or allowed. Selection rules also apply to molecular spectra, and the form they take depends on the type of transition. The underlying classical idea is that, for the molecule to be able to interact with the electromagnetic field and absorb or create a photon of frequency ν, it must possess, at least transiently, a dipole oscillating at that frequency. For emission and absorption spectra (we treat Raman spectra later) this transient dipole is expressed quantum mechanically in terms of the **transition dipole moment**, and for a transition between states with wavefunctions ψ_i and ψ_f is defined as

$$\boldsymbol{\mu}_{\mathrm{fi}} = -e \int \psi_f^* \boldsymbol{r} \psi_i \, \mathrm{d}\tau \qquad (10\mathrm{a})$$

where $\boldsymbol{r}$ is the location of the electron. The coefficient of stimulated absorption (and emission), and therefore the intensity of the transition, is proportional to the square of the transition dipole moment, and a detailed analysis gives

$$B = \frac{|\mu_{\mathrm{fi}}|^2}{6\varepsilon_0 \hbar^2} \qquad (10\mathrm{b})$$

and so only if the transition moment is non-zero does the transition contribute to the spectrum. We see that, to identify the selection rules, we must establish the conditions for which $\mu_{\mathrm{fi}} \neq 0$.

A **gross selection rule** specifies the general features a molecule must have if it is to have a spectrum of a given kind. For rotational transitions, the transition moment is zero unless the molecule has a permanent electric dipole. That is, the molecule must be polar. The classical basis of this rule is that a polar molecule appears to possess a fluctuating dipole when rotating (Fig. 16.9) but a nonpolar molecule does not. The permanent dipole can be regarded as a handle with which the molecule stirs the electromagnetic field into oscillation (and vice versa for absorption).

The transition moment is zero in a vibrational transition unless the electric dipole moment of the molecule changes during the vibration. The classical basis of this rule is that the molecule can shake the electromagnetic field into oscillation if its dipole changes as it vibrates (Fig. 16.10). The molecule need not have a permanent dipole: the rule requires only a *change* in dipole moment, possibly from zero. Some vibrations do not affect the molecule's dipole moment (e.g. the stretching motion of a homonuclear diatomic molecule), and so they neither absorb nor generate radiation: such vibrations are said to be **inactive** in the infrared.

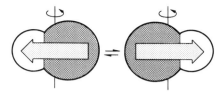

Fig. 16.9 To a stationary observer, a rotating polar molecule looks like an oscillating dipole which can stir the electromagnetic field into oscillation. This picture is the classical origin of the gross selection rule for rotational transitions.

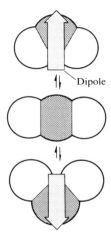

Fig. 16.10 The oscillation of a molecule, even if it is nonpolar, may result in an oscillating dipole that can interact with the electromagnetic field.

Example 16.3: *Using the gross selection rules*

State which of the following molecules have (a) rotational absorption spectra, (b) vibrational absorption spectra: N_2, CO_2, OCS, H_2O, $CH_2{=}CH_2$, C_6H_6.

Answer. (a) Molecules that give rise to rotational spectra have a permanent dipole moment; therefore, select the polar molecules. Only OCS and H_2O are polar, and so only these two give rise to a rotational absorption spectrum. (b) Molecules that give rise to vibrational spectra have dipole moments that change during the course of a vibration. Therefore, judge whether a distortion of the molecule can change its dipole moment (including changing it from zero). All the molecules except N_2 possess at least one vibrational mode that results in a change of a dipole moment, and so all except N_2 can show a vibrational absorption spectrum.

Comment. Not all the modes of complex molecules are vibrationally active. For example, the symmetric stretch of CO_2, in which the O—C—O bonds stretch and contract symmetrically is inactive because it leaves the dipole moment unchanged (at zero),

Exercise. Repeat the question for H_2, NO, N_2O, CH_4.
[(a) NO, N_2O; (b) NO, N_2O, CH_4]

A more detailed study of the transition moment leads to the **specific selection rules** that express the allowed transitions in terms of the changes in quantum numbers (as in the rule $\Delta l = \pm 1$ for atoms). Specific selection rules can often be interpreted in terms of the changes of angular momentum when a photon (with its intrinsic spin angular momentum $s = 1$) enters or leaves a molecule, and we shall discuss them once we have set up the quantum numbers needed to describe rotation and vibration.

16.3 Linewidths

Spectral lines are not infinitely narrow, and in condensed media may spread over several thousand cm^{-1}.

Doppler broadening

One important broadening process in gaseous samples is the **Doppler effect**, in which radiation is shifted in frequency when the source is moving towards or away from the observer. When a source emitting radiation of frequency v

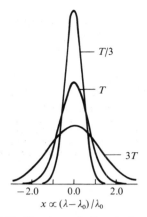

Fig. 16.11 The shape of a Doppler-broadened spectral line reflects the Maxwell distribution of speeds in the sample at the temperature of the experiment. Notice that the line broadens as the temperature is increased. The width at half height, $\delta\lambda$, is given by eqn 11.

recedes with a speed v, the observer detects radiation of frequency

$$\nu' = \frac{\nu}{1 + v/c}$$

where c is the speed of the radiation (the speed of light for electromagnetic radiation, the speed of sound for sound waves). A source approaching the observer appears to be emitting radiation of frequency

$$\nu' = \frac{\nu}{1 - v/c}$$

Molecules reach high speeds in all directions in a gas, and a static observer detects the corresponding Doppler shifted range of frequencies. Some molecules approach the observer, some move away; some move quickly, others slowly. The detected spectral 'line' is the absorption or emission profile arising from all the resulting Doppler shifts. The profile reflects the Maxwell distribution of molecular speeds parallel to the line of sight (see Section 24.2), which is a bell-shaped Gaussian curve (of the form e^{-x^2}). The Doppler line shape is therefore also a Gaussian curve (Fig. 16.11), and calculation shows that when the temperature is T and the mass of the molecule is m, the width of the line at half-height is

$$\delta\nu = \frac{2\nu}{c} \times \left(\frac{2kT}{m} \ln 2\right)^{1/2} \tag{11a}$$

In terms of the wavelength,

$$\delta\lambda = \frac{2\lambda}{c} \times \left(\frac{2kT}{m} \ln 2\right)^{1/2} \tag{11b}$$

Example 16.4: *Using the linewidth to measure a temperature*

The sun emits a spectral line at 677.4 nm which has been identified as arising from a transition in highly ionized ^{57}Fe. Its width at half-height is 5.3 pm. What is the temperature of the sun's surface?

Answer. Equation 11b rearranges to

$$T = \left(\frac{mc^2}{8k \ln 2}\right)\left(\frac{\delta\lambda}{\lambda}\right)^2 = 1.949 \times 10^{12} \text{ K} \times M/(\text{g mol}^{-1}) \times \left(\frac{\delta\lambda}{\lambda}\right)^2$$

where M is the molar mass of the emitter. Since $M = 57 \text{ g mol}^{-1}$, and $\delta\lambda/\lambda = 7.82 \times 10^{-6}$, we find $T = 6.8 \times 10^3$ K.

Comment. Temperatures may also be judged using Wien's Law (eqn 8 of Section 11.2) and looking for the maximum in the intensity of output of the sun (or star) regarded as a black body. Fitting the entire Planck distribution (eqn 4) to the observed intensity profile is another approach.

Exercise. What is the Doppler linewidth of the spectrum of an interstellar NH_3 molecule close to 240 GHz at 10 K? [50 Hz]

Doppler broadening increases with temperature because the molecules acquire a wider range of speeds. Therefore, to obtain spectra of maximum sharpness, it is best to work with cold samples.

Lamb-dip spectroscopy

A novel approach to the elimination of Doppler broadening has become available with the advent of lasers and their extremely high monochromaticity and of radiofrequency techniques with precise frequency control. The precise location of absorption frequencies in this way is called **Lamb-dip spectroscopy**, which is named after its discoverer.

When an intense, monochromatic beam with a frequency slightly higher than that of the absorption maximum passes through a gaseous sample, only the molecules that happen to be moving away from the source at some precise speed absorb radiation. If the beam is then reflected back through the sample (Fig. 16.12), more radiation is absorbed, but this time by the molecules that happen to be moving at the same precise speed but away from the mirror. The detector therefore observes a double dose of absorption. However, when the incident radiation is at the absorption peak, only those molecules moving perpendicular to the line of the beam (and therefore having no Doppler shift) absorb on the reflected path. Because some of those molecules were excited on the first passage, fewer are available to absorb the light on its second passage, and so a less intense absorption is observed. This appears as a dip, the **Lamb dip**, in the absorption curve, and its position gives a very precise location of the transition frequency.

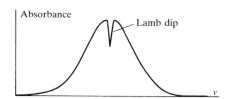

Fig. 16.12 A Lamb dip. The origin is explained in the text: Lamb-dip spectroscopy enables the positions of the centres of absorption lines to be pinpointed very precisely even if there is Doppler broadening.

Lifetime broadening

It is found that spectral lines are still not infinitely sharp even when Doppler broadening has been largely eliminated, either by working at low temperatures or by Lamb-dip spectroscopy.

When the Schrödinger equation is solved for a system that is changing with time, it is found that it is impossible to specify the energy levels exactly. If on average a system survives in a state for a time τ, the **lifetime** of the state, its energy levels are blurred to an extent of order δE, where

$$\delta E \approx \frac{\hbar}{\tau} \qquad (12a)$$

Equation 12a is reminiscent of the Heisenberg uncertainty principle (eqn 33 of Section 11.6), and although the connection is tenuous, lifetime broadening is often called 'uncertainty broadening'. Expressing the energy spread in wavenumbers through $\delta E = hc\delta\tilde{\nu}$ and using the values of the fundamental constants gives the practical form of the relation as

$$\delta\tilde{\nu} \approx \frac{5.3 \text{ cm}^{-1}}{\tau/\text{ps}} \qquad (12b)$$

No excited state has an infinite lifetime; therefore, all states are subject to some lifetime broadening, and the shorter the lifetimes of the states involved in a transition, the broader the spectral lines.

Three processes are principally responsible for the finite lifetimes of excited states. The dominant one is **collisional deactivation**, which arises from collisions between molecules or with the walls of the container. If the collisional lifetime is τ_{col}, the resulting collisional linewidth is $\delta E_{col} \approx \hbar/\tau_{col}$.

Table 16.1. Moments of inertia†

1. *Diatomics*

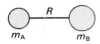

$$I = \frac{m_A m_B}{m} R^2 = \mu R^2$$

2. *Linear rotors*

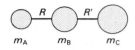

$$I = m_A R^2 + m_C R'^2$$
$$- \frac{(m_A R - m_C R')^2}{m}$$

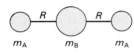

$$I = 2m_A R^2$$

3. *Symmetric rotors*

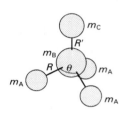

$$I_\parallel = 2m_A R^2 (1 - \cos \theta)$$
$$I_\perp = m_A R^2 (1 - \cos \theta)$$
$$+ \frac{m_A}{m}(m_B + m_C)R^2(1 + 2\cos\theta)$$
$$+ \frac{m_C R'}{m}\{(3m_A + m_B)R'$$
$$+ 6m_A R[\tfrac{1}{3}(1 + 2\cos\theta)]^{1/2}\}$$

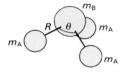

$$I_\parallel = 2m_A R^2 (1 - \cos \theta)$$
$$I_\perp = m_A R^2 (1 - \cos \theta)$$
$$+ \frac{m_A m_B}{m} R^2 (1 + 2\cos\theta)$$

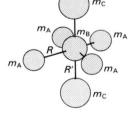

$$I_\parallel = 4m_A R^2$$
$$I_\perp = 2m_A R^2 + 2m_C R'^2$$

4. *Spherical rotors*

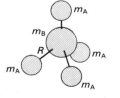

$$I = \tfrac{8}{3}m_A R^2$$

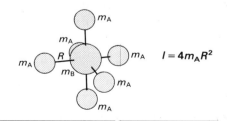

$$I = 4m_A R^2$$

† In each case *m* is the total mass of the molecule.

The collisional lifetime can be lengthened, and the broadening minimized, by working at low pressures.

The rate of spontaneous emission cannot be changed. Hence it is a natural limit to the lifetime of an excited state, and the resulting lifetime broadening is the **natural linewidth** of the transition. The natural linewidth is an intrinsic property of the transition, and cannot be changed by modifying the conditions.

Natural linewidths depend strongly on the transition frequency (they increase with A and therefore as v^3), and so low-frequency transitions (such as the microwave transitions of rotational spectroscopy) have very small natural linewidths, and collisional and Doppler line-broadening processes are dominant. The natural lifetimes of electronic transitions are very much shorter than for vibrational and rotational transitions, and so the natural linewidths of electronic transitions are much greater than those of vibrational and rotational transitions. For example, a typical electronic excited state natural lifetime is about 10^{-8} s (10^4 ps), corresponding to a natural width of about 5×10^{-4} cm^{-1} (15 MHz). A typical rotational natural lifetime is about 10^3 s, corresponding to a natural linewidth of only 5×10^{-15} cm^{-1} (10^{-4} Hz).

Pure rotational spectra

The general strategy that we shall adopt for discussing molecular spectra and the information they contain is to find expressions for the energy levels of molecules and then to calculate the transition frequencies by applying the selection rules. We then predict the appearance of the spectrum by taking into account the populations of the states. In this section we illustrate the strategy by considering the rotational energy levels and selection rules for molecules.

The key molecular parameter we shall need is the **moment of inertia** I of the molecule about various axes. The moment of inertia is defined as the mass of each atom multiplied by the square of its distance from the rotational axis (Fig. 16.13):

$$I = \sum_i m_i x_i^2$$

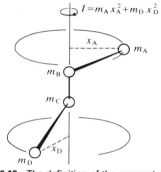

Fig. **16.13** The definition of the moment of inertia, the sum of each mass multiplied by the square of its perpendicular distance from the axis of rotation. Neither B nor C contributes to the moment about the axis shown.

The moment of inertia depends on the masses of the atoms present and the molecular geometry, so we can suspect (and later shall see explicitly) that rotational spectroscopy will give information about bond lengths and bond angles. The explicit expressions for the moments of inertia of some symmetrical molecules are given in Table 16.1. The convention is to label the moments of inertia I_A, I_B, and I_C, with $I_C \geq I_B \geq I_A$.

Example 16.5: *Calculating the moment of inertia of a molecule*

Calculate the moment of inertia of an H_2O molecule around its two-fold axis (the bisector of the HOH angle.

Answer. The moment of inertia is the sum of the masses multiplied by the squares of their distances from the axis of rotation:

$$I = \sum_i m_i x_i^2 = m_H x^2 + 0 + m_H x^2$$
$$= 2m_H x^2$$

If the bond angle of the molecule is denoted 2ϕ and the bond length is R, trigonometry gives

$$x = R \sin \phi$$

and therefore

$$I = 2m_H R^2 \sin^2 \phi$$

For H_2O, with bond angle 104.5° and bond length 95.7 pm, we obtain

$$I = 2 \times 1.67 \times 10^{-27} \text{ kg} \times (9.57 \times 10^{-11} \text{ m})^2 \times \sin^2 52.3°$$
$$= 1.91 \times 10^{-47} \text{ kg m}^2$$

Comment. The mass of the O atom makes no contribution to the moment of inertia for this mode of rotation as it is immobile while the H atoms rotate around it.

Exercise. Calculate the moment of inertia of a $CHCl_3$ molecule around its threefold axis. The C—Cl bond length is 177 pm and the HCCl angle is 142°.

$$[2.1 \times 10^{-45} \text{ kg m}^2]$$

We shall suppose initially that molecules are **rigid rotors** that do not distort under the stress of rotation. **Spherical rotors** are molecules with all three moments of inertia equal (such as CH_4); **symmetric rotors** have two equal moments of inertia (such as NH_3). **Linear rotors** have one moment of inertia (the one about the axis) equal to zero (such as CO_2 and HCl). In group theoretical language, a spherical rotor is a molecule belonging to a cubic point group and a symmetric rotor is one with at least a threefold axis of symmetry. All diatomic molecules are linear rotors. The energy levels of **asymmetric rotors**, which have three different moments of inertia (e.g. H_2O) are complicated and we shall not consider them.

16.4   The rotational energy levels

The rotational energy levels of a molecule may be obtained by solving the Schrödinger equation. Fortunately there is a much less onerous short-cut which depends on noting the classical expression for the energy of a rotating body, expressing it in terms of the angular momentum, and then importing the quantum mechanical properties of angular momentum into the equations.

The energy of a body rotating about some axis x is

$$E = \tfrac{1}{2}I_x \omega_x^2$$

where ω_x is the angular velocity (in rad s^{-1}) about the axis and I_x is the moment of inertia about the axis. A body free to rotate about three axes has an energy

$$E = \tfrac{1}{2}I_x \omega_x^2 + \tfrac{1}{2}I_y \omega_y^2 + \tfrac{1}{2}I_z \omega_z^2$$

Since the classical angular momentum about x is

$$J_x = I_x \omega_x$$

with similar expressions for the other directions, it follows that

$$E = \frac{J_x^2}{2I_x} + \frac{J_y^2}{2I_y} + \frac{J_z^2}{2I_z} \qquad (13)$$

This is the key equation. We described the quantum-mechanical properties of angular momentum in Section 12.7 (see Box 12.1), and we can now make use of them in conjunction with this equation to obtain the rotational energy levels.

Spherical rotors

When all three moments of inertia are equal to some value I, as in CH_4 and SF_6, the classical expression for the energy is

$$E = \frac{1}{2I}(J_x^2 + J_y^2 + J_z^2) = \frac{J^2}{2I}$$

where J is the magnitude of the angular momentum. We can immediately find the quantum expression by making the replacement

$$J^2 \rightarrow J(J+1)\hbar^2 \quad \text{with} \quad J = 0, 1, 2, \ldots$$

Therefore, the energy of a spherical rotor is confined to the values

$$E_J = J(J+1)\frac{\hbar^2}{2I} \quad \text{with} \quad J = 0, 1, 2, \ldots$$

The energy is normally expressed in terms of the **rotational constant** B of the molecule, where

$$hcB = \frac{\hbar^2}{2I}$$

The rotational constant as defined by this equation has the dimensions of a wavenumber[4] and is normally expressed in cm^{-1}. The energy of a rotational state is normally reported as the **rotational term** $F(J)$, its value expressed as a wavenumber, by division by hc:

$$F(J) = BJ(J+1) \qquad B = \frac{\hbar}{4\pi cI} \tag{14}$$

The separation of adjacent terms is

$$F(J) - F(J-1) = 2BJ$$

Since the separation decreases as I increases, we see that large molecules have closely spaced rotational energy levels. We can estimate the magnitude of the separation by considering CCl_4: from the bond lengths and masses of the atoms we find $I = 4.85 \times 10^{-45}$ kg m^2, and hence $B = 0.0577$ cm^{-1}.

Symmetric rotors

In symmetric rotors $I_x = I_y \neq I_z$ (as in CH_3Cl, NH_3, and C_6H_6) and z is the **figure axis** (the principal axis) of the molecule. We shall write $I_z = I_{\parallel}$ and $I_x = I_y = I_{\perp}$. If $I_{\parallel} > I_{\perp}$ the rotor is **oblate** (like a pancake, such as C_6H_6); if

---

[4] The definition of B as a wavenumber is convenient when we come to vibration-rotation spectra. However, for pure rotational spectroscopy it is more common to define B as a frequency and to report it in MHz or GHz. The appropriate definition is then

$$B = \frac{\hbar}{4\pi I}$$

$I_\parallel < I_\perp$ it is **prolate** (like a cigar, such as PCl_5). The classical expression for the energy becomes

$$E = \frac{J_x^2 + J_y^2}{2I_\perp} + \frac{J_z^2}{2I_\parallel}$$

We can rewrite this in terms of $J^2 = J_x^2 + J_y^2 + J_z^2$ by adding and subtracting $J_z^2/2I_\perp$:

$$E = \frac{J_x^2 + J_y^2 + J_z^2}{2I_\perp} + \frac{J_z^2}{2I_\parallel} - \frac{J_z^2}{2I_\perp}$$

$$= \frac{J^2}{2I_\perp} + \left(\frac{1}{2I_\parallel} - \frac{1}{2I_\perp}\right)J_z^2$$

Now we generate the quantum expression by replacing J^2 by $J(J+1)\hbar^2$, where J is the angular momentum quantum number. We also know from the quantum theory of angular momentum (Box 12.1) that the component of angular momentum about any axis is restricted to the values

$$J_z = K\hbar \qquad K = 0, \pm1, \ldots, \pm J$$

(K is the quantum number used to signify a component on the figure axis; M is reserved for a component on a laboratory axis.) Consequently we also replace J_z^2 by $K^2\hbar^2$. The rotational terms are therefore

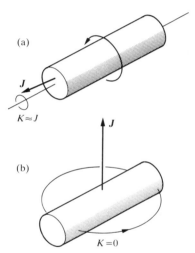

(a)

$K \approx J$

(b)

$K = 0$

Fig. 16.14 The significance of the quantum number K. (a) When $|K|$ is close to its maximum value, J, most of the molecular rotation is around the figure axis. (b) When $K = 0$ the molecule has no angular momentum about its figure axis: it is undergoing end-over-end rotation.

$$F(J, K) = BJ(J+1) + (A - B)K^2 \qquad (15a)$$

$$J = 0, 1, 2, \ldots; \qquad K = 0, \pm1, \ldots, \pm J$$

$$A = \frac{\hbar}{4\pi c I_\parallel} \qquad B = \frac{\hbar}{4\pi c I_\perp} \qquad (15b)$$

Equation 15a matches what we should expect for the dependence of the energy levels on the two moments of inertia of the molecule. When $K = 0$, there is no component of angular momentum about the figure axis (Fig. 16.14) and the energy levels depend only on $I_\perp$. When $K = \pm J$, almost all the angular momentum arises from rotation around the figure axis, and the energy levels are determined largely by $I_\parallel$. The sign of K does not affect the energy because opposite values of K correspond to opposite senses of rotation, and the energy does not depend on the sense of rotation.

Example 16.6: *Calculating the rotational energy levels of a molecule*

The $^{14}NH_3$ molecule is a symmetric rotor with bond length 101.2 pm and HNH bond angle 106.7°. Calculate its rotational terms.

Answer. We need to calculate the rotational constants A and B using the expressions for moments of inertia given in Table 16.1. Substitution of $m_A = 1.0078$ u, $m_B = 14.0031$ u, $R = 101.2$ pm, and $\theta = 106.7°$ into the second of the symmetric rotor expressions in Table 16.1 gives

$$I_\parallel = 4.4128 \times 10^{-47} \text{ kg m}^2 \qquad I_\perp = 2.8059 \times 10^{-47} \text{ kg m}^2$$

Hence, $A = 6.344$ cm^{-1} and $B = 9.977$ cm^{-1}. It follows from eqn 15a that

$$F(J, K)/\text{cm}^{-1} = 9.977J(J+1) - 3.633K^2$$

Comment. For $J = 1$, the energy needed for the molecule to rotate mainly about its figure axis ($K = J$) is equivalent to $16.32\ \text{cm}^{-1}$, but end-over-end rotation ($K = 0$) corresponds to $19.95\ \text{cm}^{-1}$.

Exercise. The $CH_3^{35}Cl$ molecule has a C—Cl bond length of 178 pm, a C—H bond length of 111 pm, and an HCH angle of 110.5°. Calculate its rotational energy levels.

$$[F(J, K)/\text{cm}^{-1} = 0.444J(J + 1) + 4.58K^2]$$

Linear rotors

For a linear rotor (such as CO_2, HCl, and C_2H_2) in which the atoms are regarded as mass points, the rotation occurs only about an axis perpendicular to the line of atoms and there is zero angular momentum around the line. Therefore we can use eqn 15a with $K \equiv 0$. The rotational terms are therefore

$$F(J) = BJ(J + 1) \qquad J = 0, 1, 2, \ldots \qquad (16)$$

Degeneracies and the Stark effect

The symmetric rotor has an energy that depends on J and K, and each level except those with $K = 0$ is doubly degenerate (the K, $-K$ degeneracy). However, we must not forget that an angular momentum has a component on a laboratory axis. This component is quantized, and its permitted values are $M_J \hbar$ with $M_J = 0, \pm 1, \ldots, \pm J$ for $2J + 1$ values in all. The quantum number M_J does not appear in the expression for the energy, but it is still necessary for a complete specification of the state of the rotor. Therefore, a symmetric rotor level is $2(2J + 1)$-fold degenerate for $K \neq 0$ and $(2J + 1)$-fold degenerate if $K = 0$:

$$g(J, K) = 2(2J + 1) \qquad g(J, 0) = 2J + 1$$

where g denotes the degeneracy. A linear rotor has K fixed at 0, but the angular momentum may still have $2J + 1$ components on the laboratory axis; hence,

$$g(J) = 2J + 1$$

A spherical rotor is the limit of a symmetric rotor with A vanishingly different from B: K may still take any one of $2J + 1$ values, but the energy is independent of which value it takes. Therefore, as well as having a $(2J + 1)$-fold degeneracy arising from the orientation in space, it also has a $(2J + 1)$-fold degeneracy arising from the orientation with respect to an axis selected in the molecule. The overall degeneracy of a spherical rotor with quantum number J is therefore

$$g(J) = (2J + 1)^2$$

This degeneracy increases very rapidly: when $J = 10$, for instance, there are 441 states of the same energy.

The M_J-degeneracy is removed when an electric field is applied to a polar molecule (e.g. HCl or NH_3) because now the energy of the molecule depends on its orientation in space. The splitting of states by an electric field is called the **Stark effect**. For a linear rotor in an electric field $\mathscr{E}$ the energy is given by

$$E_{J,M_J} = hcBJ(J+1) + \frac{\mu^2 \mathscr{E}^2 \{J(J+1) - 3M_J^2\}}{2hcBJ(J+1)(2J-1)(2J+3)} \tag{17}$$

Note that the energy depends on the square of the permanent electric dipole moment μ, and so the measurement of the Stark effect is a way of measuring this property.

Centrifugal distortion

The atoms of rotating molecules are subject to centrifugal forces that tend to distort the molecular geometry and change the moments of inertia. For a diatomic molecule, the centrifugal distortion stretches the bond, and therefore increases the moment of inertia; as a result, the energy levels are less far apart than the rigid-rotor expressions predict. The effect is usually taken into account by subtracting a term from the energy and writing

$$F(J) = BJ(J+1) - D_J J^2(J+1)^2 \tag{18a}$$

D_J is the **centrifugal distortion constant**: it is large when the bond is easily stretched. The centrifugal distortion constant of a diatomic molecule is related to the vibrational wavenumber of the bond $\tilde{v}$ (which, as we shall see later, is a measure of its stiffness):

$$D_J = \frac{4B^3}{\tilde{v}^2} \tag{18b}$$

Hence the observation of the convergence of the rotational levels as J increases can be interpreted in terms of the rigidity (specifically, the force constant) of the bond.

16.5 Rotational transitions

Typical values of B for small molecules are in the region of 1 to $10\ \mathrm{cm}^{-1}$ (e.g. $0.356\ \mathrm{cm}^{-1}$ for NF_3 and $10.59\ \mathrm{cm}^{-1}$ for HCl), and so the transitions lie in the microwave region of the spectrum. The transitions are detected by monitoring the net absorption of microwave radiation generated either by a klystron or, in modern instruments, by a 'backward wave oscillator', which is tunable over a wide range of frequencies. For technical reasons related to the detection system, it is desirable to modulate the energy levels (that is, vary them in an oscillatory manner) so that the absorption intensity, and therefore the detected signal, oscillates: it is easier to amplify an alternating signal than a steady one. The oscillation is achieved by **Stark modulation**, in which an alternating electric field (of strength of the order of $10^2\ \mathrm{V\ cm}^{-1}$ and frequency between 50 and $100\ \mathrm{kHz}$) is applied to the sample to modulate the energies of the rotational states.

If a *constant* electric field is applied to the sample, the levels shift to an extent determined by the magnitude of the molecular dipole moment (eqn 17), and so the modification in the spectrum brought about by a known field can be used to measure the dipole moment. Since microwave frequencies can be measured with very high precision, rotational spectroscopy is one of the most precise spectroscopic techniques.

Rotational selection rules

We have already seen (Section 16.2) that the gross selection rule for the observation of a pure rotational spectrum is that it must have a permanent electric dipole moment. Consequently, homonuclear diatomic molecules and symmetrical ($D_{\infty h}$) linear molecules such as CO_2 are rotationally inactive. Spherical rotors cannot have electric dipole moments unless they become distorted by rotation, and so they are also inactive except in special cases. An example of a spherical rotor that does become sufficiently distorted for it to acquire a dipole moment is SiH_4, which has a dipole moment of about 8.3 μD by virtue of its rotation (for comparison, HCl has a dipole moment of 1.1 D). The pure rotational spectrum of SiH_4 has been detected by using long path lengths (10 m) through high pressure (4 atm) samples.

The specific selection rules are found by evaluating the transition dipole moment between the states. For a linear molecule, the transition moment vanishes unless the following conditions are fulfilled:

$$\Delta J = \pm 1 \qquad \Delta M_J = 0, \pm 1$$

The change in J matches what we already know about the role of the conservation of angular momentum when a photon is emitted or absorbed. When these conditions are fulfilled, the total $J + 1 \leftarrow J$ transition intensity (the intensity summed over all the values of M_J that contribute to the line) is proportional to

$$|\mu_{J+1,J}|^2 = \frac{\mu^2(J+1)}{2J+1} \rightarrow \tfrac{1}{2}\mu^2 \text{ for } J \gg 1$$

where μ is the permanent electric dipole moment of the molecule.

The only extension needed for the discussion of symmetric rotors is a selection rule for K. If a symmetric rotor has a dipole it must lie parallel to the figure axis, as in NF_3. Such a molecule cannot be accelerated into different states of rotation around the figure axis by the absorption of radiation, so $\Delta K = 0$.

When these selection rules are applied to the expressions for the energy levels, it follows that the wavenumbers of the allowed $J + 1 \leftarrow J$ absorptions are

$$\tilde{\nu} = 2B(J+1) \qquad J = 0, 1, 2, \ldots \tag{19}$$

Example 16.7: *Predicting the appearance of a rotational spectrum*

Predict the form of the rotational spectrum of NH_3.

Answer. We calculated the energy levels in Example 16.6. The NH_3 molecule is a polar symmetric rotor, and so the selection rules $\Delta J = \pm 1$ and $\Delta K = 0$ apply. For absorption, $\Delta J = +1$ and we can use eqn 19. Intrinsic intensities are proportional to $|\mu_{J+1,J}|^2$ (we take populations into account later). Since $B = 9.977 \text{ cm}^{-1}$, we can draw up the following table for the $J + 1 \leftarrow J$ transitions.

$J =$	0	1	2	3	$\ldots$
ν/cm^{-1}	19.95	39.91	59.86	79.82	
$|\mu_{J+1,J}|^2/\mu^2$	1	0.67	0.60	0.57	

The line spacing is 19.95 cm^{-1}.

Comment. The intensities are the intrinsic intensities of the lines; that is, they do not take into account the different populations of the initial rotational levels. A complication that we have ignored is the modification of the numbers of possible rotational states by the Pauli principle, which allows only some of the states to be occupied.

Exercise. Repeat the problem for $CH_3^{35}Cl$ (see Example 16.6 for details).
[Lines of separation 0.888 cm^{-1}]

The appearance of rotational spectra

The form of the spectrum predicted by eqn 19 is shown in Fig. 16.15. The most significant feature is that it consists of a series of lines with wavenumbers $2B$, $4B$, $6B$, ... and separation $2B$. The intensities increase with increasing J and pass through a maximum before tailing off as J becomes large. It should be recalled from Section 16.2 that the observed absorption is the *net* outcome of the stimulated absorption less the

Fig. 16.15 The rotational energy levels of a linear rotor, the transitions allowed by the selection rule $\Delta J = \pm 1$, and a typical pure rotational absorption spectrum. The intensities reflect the populations of the initial level in each case and the strengths of the transition dipole moments.

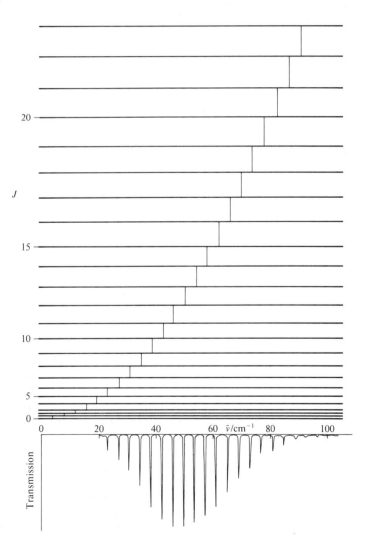

stimulated emission, and that the intensity of each transition depends on the value of J. Hence the value of J corresponding to the most intense line is not quite the same as the value of J for the most highly populated level.

The measurement of the line spacing gives B, and hence the moment of inertia perpendicular to the figure axis. Since the masses of the atoms are known, it is a simple matter to deduce the bond length of a diatomic molecule. However, in the case of a polyatomic molecule such as OCS or NH_3, the analysis gives only a single quantity $I_\perp$ and we cannot deduce both bond lengths (in OCS) or the bond length and bond angle (in NH_3). This difficulty can be overcome by using isotopically substituted molecules, such as ABC and A'BC; then, by assuming that $R(A{-}B) = R(A'{-}B)$, both A—B and B—C bond lengths can be extracted from the two moments of inertia. A famous example of this procedure is the study of OCS, and the actual calculation is worked through in Problem 16.9.

16.6 Rotational Raman spectra

The gross selection rule for rotational Raman transitions is that the molecule must be anisotropically polarizable. We shall begin by explaining what this means.

The distortion of a molecule in an electric field is determined by its **polarizability** α (we deal with polarizabilities in detail in Chapter 22). More precisely, if the strength of the field is $\mathscr{E}$, the molecule acquires a dipole moment

$$\mu = \alpha\mathscr{E} \qquad (20)$$

in addition to any dipole moment it may have in the absence of the field. We see that the greater the polarizability, the greater the dipole induced by a given field. A Xe atom, for example, has a greater polarizability than a He atom because its outer electrons are less tightly under the control of the more distant central nucleus and are more easily displaced by an externally applied field.

An atom is isotropically polarizable. That is, the same distortion is induced whatever the direction of the applied field. The polarizability of a spherical rotor is also isotropic. However, nonspherical rotors have polarizabilities that do depend on the direction of the field and hence are anisotropically polarizable (Fig. 16.16). The electron distribution in H_2, for example, is more distorted when the field is applied parallel to the bond than when it is applied perpendicular to it, and we write $\alpha_\parallel > \alpha_\perp$.

All linear molecules and diatomics (whether homonuclear or hetero-nuclear) have anisotropic polarizabilities and so are rotationally Raman active. This is one reason for the importance of rotational Raman spectroscopy: it enables us to examine many of the molecules that are inaccessible to pure rotational microwave spectroscopy. CH_4 and SF_6, however, being spherical rotors, are rotationally Raman inactive as well as rotationally inactive.

The specific rotational Raman selection rules are

$$\Delta J = \begin{cases} 0, \pm 2 & \text{linear rotor} \\ 0, \pm 1, \pm 2 \quad \Delta K = 0 & \text{symmetric rotor} \end{cases}$$

The $\Delta J = 0$ transitions do not lead to a shift of the scattered photon's

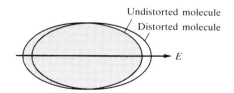

Undistorted molecule
Distorted molecule
E

Fig. 16.16 The electric polarizability of a molecule is a measure of its ability to distort in response to an applied field. For most molecules the polarizability is anisotropic.

frequency in pure rotational Raman spectroscopy, and contribute to the unshifted **Rayleigh scattered** light.[5]

We can predict the form of the Raman spectrum of a linear rotor (Fig. 16.18) by applying the selection rule $\Delta J = \pm 2$ to the rotational energy levels. When the molecule makes a transition with $\Delta J = +2$, the scattered radiation leaves it in a higher rotational state, and so the wavenumber of the incident radiation, initially $\tilde{v}_i$, is decreased. These transitions account for the **Stokes lines** in the spectrum:

$$\tilde{v}(J+2 \leftarrow J) = \tilde{v}_i - \{F(J+2) - F(J)\}$$

$$= \tilde{v}_i - 2B(2J+3) \tag{21a}$$

The Stokes lines appear to low frequency of the incident light and at displacements $6B, 10B, 14B, \ldots$ from $\tilde{v}_i$ for $J = 0, 1, 2, \ldots$ When the molecule makes a transition with $\Delta J = -2$, the scattered photon emerges with increased energy. These transitions account for the **anti-Stokes lines** of the spectrum:

$$\tilde{v}(J \rightarrow J-2) = \tilde{v}_i + \{F(J) - F(J-2)\}$$

$$= \tilde{v}_i + 2B(2J-1) \tag{21b}$$

The anti-Stokes lines occur at displacements of $6B, 10B, 14B, \ldots$ (for $J = 2, 3, \ldots$; $J = 2$ is the lowest state that can contribute under the selection rule $\Delta J = -2$) to high frequency of the incident radiation. The separation of the lines in both the Stokes and the anti-Stokes regions is $4B$, and so from its measurement $I_\perp$ can be determined and then used to find the bond lengths exactly as in the case of microwave spectroscopy.

---

[5] The classical origin of the 2 in the selection rule is as follows. In an electric field $\mathscr{E}$, a molecule acquires a dipole moment of magnitude $\alpha\mathscr{E}$, where α is the polarizability. If the electric field is that of a light wave of frequency ω_i, the induced dipole moment is time dependent and has the form

$$\mu = \alpha\mathscr{E} = \alpha\mathscr{E}_i \cos \omega_i t$$

If the molecule is rotating, its polarizability in the direction of the field is also time dependent (if it is anisotropic), and we can write

$$\alpha = \alpha_0 + \Delta\alpha \cos 2\omega_R t$$

the 2 appearing because the polarizability returns to its initial value twice each revolution (Fig. 16.17). Substituting this expression into the expression for the induced dipole moment gives

$$\mu = (\alpha_0 + \Delta\alpha \cos 2\omega_R t) \times (\mathscr{E}_i \cos \omega_i t)$$

$$= \alpha_0 \mathscr{E}_i \cos \omega_i t + \mathscr{E}_i \Delta\alpha \cos 2\omega_R t \cos \omega_i t$$

$$= \alpha_0 \mathscr{E}_i \cos \omega_i t + \tfrac{1}{2}\mathscr{E}_i \Delta\alpha \{\cos (\omega_i + 2\omega_R)t + \cos (\omega_i - 2\omega_R t)\}$$

This shows that the induced dipole has a component oscillating at the incident light frequency (so that it radiates Rayleigh radiation), and that it also has two components at $\omega_i \pm 2\omega_R$ which give rise to the shifted Raman lines. Note that these lines appear only if $\Delta\alpha \neq 0$, hence the need for anisotropy in the polarizability.

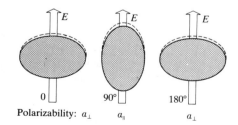

Polarizability: $a_\perp$ $a_\parallel$ $a_\perp$

Fig. 16.17 The distortion induced in a molecule by an applied electric field returns to its initial value after a rotation of only 180° (i.e. twice a revolution). This is the origin of the $\Delta J = \pm 2$ selection rule in rotational Raman spectroscopy.

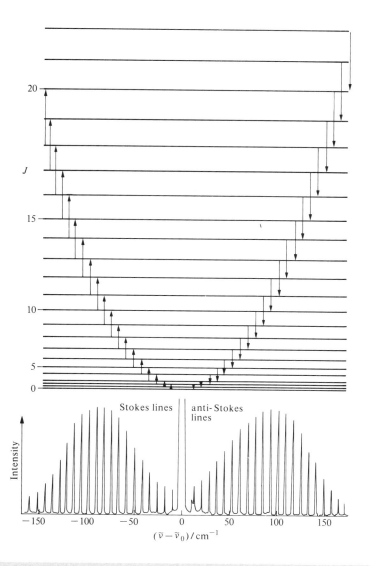

Fig. 16.18 The rotational energy levels of a linear rotor and the transitions allowed by the $\Delta J = \pm 2$ Raman selection rules. The form of a typical rotational Raman spectrum is also shown.

Example 16.8: *Predicting the form of a Raman spectrum*

Predict the form of the rotational Raman spectrum of $^{14}N_2$, for which $B = 1.99\ \text{cm}^{-1}$ when it is irradiated with monochromatic 336.732 nm laser light.

Answer. The molecule is rotationally Raman active because end-over-end rotation modulates its polarizability as viewed by a stationary observer. The Stokes and anti-Stokes lines are given by the expressions above. Since $\lambda_i = 336.732\ \text{nm}$ corresponds to $\tilde{\nu}_i = 29\,697.2\ \text{cm}^{-1}$, eqns 21a and 21b give the following line positions:

$J =$	0	1	2	3
Stokes:				
$\tilde{\nu}/\text{cm}^{-1}$	29 685.3	29 677.3	29 669.3	29 661.4
λ/nm	336.868	336.958	337.048	337.139
Anti-Stokes:				
$\tilde{\nu}/\text{cm}^{-1}$			29 709.1	29 717.1
λ/nm			336.597	336.507

Comment. There will be a strong central line at 336.732 nm accompanied on either side by lines of increasing and then decreasing intensity (as a result of transition moment and population effects). The spread of the entire spectrum is very small (about 300 cm^{-1} at room temperature), and so the incident light must be highly monochromatic.

Exercise. Repeat the calculation for the rotational Raman spectrum of NH_3 ($B = 9.977$ cm^{-1}).

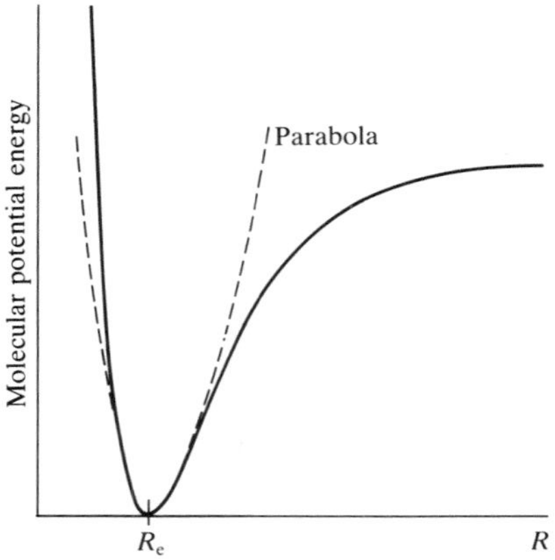

The vibrations of diatomic molecules

In this section, we adopt the same strategy of finding expressions for the energy levels, establishing the selection rules, and then discussing the form of the spectrum. We shall also see how the simultaneous excitation of rotation modifies the appearance of the spectrum.

16.7 Molecular vibrations.

We shall base our discussion on Fig. 16.19, which shows a typical potential energy curve (Section 14.1) of a diatomic molecule.

The harmonic approximation

In regions close to R_e (at the minimum of the curve) the potential energy can be approximated by a parabola, and we can write

$$V = \tfrac{1}{2}k(R - R_e)^2$$

where k is the **force constant** of the bond. The steeper the walls of the potential, the greater the force constant. This is another way of saying the stiffer the bond, the higher the force constant. The Schrödinger equation for the motion of the two atoms of masses m_1 and m_2 with this potential energy is

$$-\frac{\hbar^2}{2\mu}\frac{d^2\psi}{dx^2} + V\psi = E\psi$$

where μ is the **reduced mass**:

$$\frac{1}{\mu} = \frac{1}{m_1} + \frac{1}{m_2}$$

This equation is derived in the same way as in the *Further information* section of Chapter 14, where the separation of variables procedure was used to separate the relative motion of the atoms from the motion of the molecule as a whole.

The Schrödinger equation we have obtained is that for a particle of mass μ undergoing harmonic motion. Therefore, we can use the results of Section 12.4 directly, and immediately write down the permitted vibrational energy levels:

$$E_v = (v + \tfrac{1}{2})\hbar\omega \qquad \omega = \left(\frac{k}{\mu}\right)^{1/2} \qquad v = 0, 1, 2, \ldots$$

The **vibrational terms** of a molecule, the energies of its vibrational states

Fig. 16.19 The molecular potential energy curve can be approximated by a parabola near the bottom of the well. The parabolic potential leads to harmonic oscillations. At high excitation energies the parabolic approximation is poor (the true potential is less confining), and is totally wrong near the dissociation limit.

expressed in wavenumbers, are denoted by G, and so

$$G(v) = (v + \tfrac{1}{2})\tilde{v} \qquad \tilde{v} = \frac{\omega}{2\pi c} \tag{22}$$

The vibrational wavefunctions are the same as in Section 12.5.

It is important to note that the vibrational terms depend on the *reduced mass* of the molecule, not its total mass, which is physically reasonable. If atom 1 were as heavy as a brick wall, we would find $\mu \approx m_2$, the mass of the lighter atom, and the vibration would be that of a light atom relative to that of a stationary wall (this is approximately the case in HI, for example, where the I atom barely moves and $\mu \approx m_H$). In the case of a homonuclear diatomic molecule, for which $m_1 = m_2$, the reduced mass is half the total mass: $\mu = \tfrac{1}{2}m$.

Anharmonicity

The vibrational terms in eqn 22 are only approximate because they are based on the parabolic approximation to the actual potential energy curve. At high vibrational excitations the swing of the atoms (more precisely, the spread of the vibrational wavefunction) allows the molecule to explore regions of the curve where the parabolic approximation is poor. The motion then becomes **anharmonic** since the force is no longer proportional to the displacement. In particular, because the actual curve is less confining than a parabola (Fig. 16.19) we can anticipate that the energy levels become less widely spaced at high excitations.

One approach to the calculation of the energy levels over a wider range is to use a function that resembles the true potential energy more closely. The **Morse potential energy** is

$$V = D_e\{1 - e^{-a(R-R_e)}\}^2 \tag{23a}$$

where D_e is the depth of the potential minimum and

$$a = \left(\frac{\mu}{2D_e}\right)^{1/2}\omega \tag{23b}$$

Equation 23a is plotted in Fig. 16.20. Near the well minimum it resembles a parabola (as can be checked by expanding the exponential as far as the first term), but unlike a parabola it allows for dissociation at high energies. The Schrödinger equation can be solved for this potential and the permitted energy levels are

$$G(v) = (v + \tfrac{1}{2})\tilde{v} - (v + \tfrac{1}{2})^2 x_e \tilde{v} \qquad x_e = \frac{a^2\hbar}{2\mu\omega} \tag{24}$$

x_e is called the **anharmonicity constant**. The number of vibrational levels of a Morse oscillator is finite, and $v = 1, 2, \ldots, v_{max}$, as shown in Fig. 16.20. The second term in eqn 24 subtracts from the first, and gives rise to the convergence of the levels at high quantum numbers.

Although the Morse oscillator is quite useful theoretically, in practice the

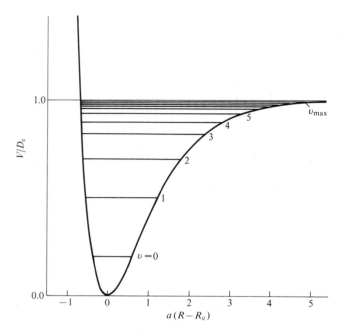

Fig. 16.20 The Morse potential energy curve reproduces the general shape of a molecular potential energy. The corresponding Schrödinger equation can be solved, and the values of the energies obtained. The number of bound levels is finite.

more general expression

$$G(v) = (v + \tfrac{1}{2})\tilde{v} - (v + \tfrac{1}{2})^2 x_e \tilde{v} + (v + \tfrac{1}{2})^3 y_e \tilde{v} + \ldots$$

where x_e and y_e are empirical constants, is used to fit the experimental data and to find the dissociation energy of the molecule.

16.8 The vibrational spectra of diatomic molecules

We have seen that the gross selection rule for vibrational transitions is that the electric dipole moment of the molecule must change in the course of a vibration. Homonuclear diatomic molecules are therefore inactive, because their dipole moments remain zero however long the bond, but heteronuclear diatomic molecules are infrared active.

The spectra of heteronuclear diatomic molecules

The specific selection rule is obtained from an analysis of the expression for the transition moment (and the properties of integrals over Hermite polynomials, Table 12.1), and is

$$\Delta v = \pm 1$$

It follows that the difference between terms of the allowed transitions, which is denoted $\Delta G_{v+1/2}$ for the transition $v + 1 \leftarrow v$, is

$$\Delta G_{v+1/2} = G(v + 1) - G(v) = \tilde{v} \qquad (25a)$$

in the harmonic approximation, and

$$\Delta G_{v+1/2} = \tilde{v} - 2(v + 1)x_e \tilde{v} + \ldots \qquad (25b)$$

if anharmonicity is taken into account. The latter equation shows that the lines converge as v increases. In the harmonic approximation all lines lie at the same wavenumber $\tilde{v}$.

HCl has a force constant of $516\,\text{N m}^{-1}$, a reasonably typical value. The reduced mass of $^1\text{H}^{35}\text{Cl}$ is $1.63 \times 10^{-27}\,\text{kg}$ (note that this is very close to the mass of the hydrogen atom, $1.67 \times 10^{-27}\,\text{kg}$, and so the Cl atom is like a brick wall). These values imply

$$\omega = 5.63 \times 10^{14}\,\text{s}^{-1} \qquad v = 8.95 \times 10^{13}\,\text{Hz}$$
$$\tilde{v} = 2990\,\text{cm}^{-1} \qquad \lambda = 3.35\,\mu\text{m}$$

The radiation lies in the infrared region of the spectrum, and so vibrational spectroscopy is an infrared technique.

At room temperature $kT/hc \approx 200\,\text{cm}^{-1}$, and the Boltzmann distribution implies that almost all the molecules will be in their vibrational ground states initially. Hence, the dominant spectral transition will be $1 \leftarrow 0$. As a result, the spectrum is expected to consist of a single absorption line. If the sample molecules are formed in a vibrationally excited state, such as when vibrationally 'hot' HF is formed in the reaction $\text{H}_2 + \text{F}_2 \rightarrow 2\text{HF}^*$, the transitions $5 \rightarrow 4$, $4 \rightarrow 3$, etc. may also appear (in emission). In the harmonic approximation, all these lines lie at the same frequency, and the spectrum is a single line. However, the presence of anharmonicity causes the transition to lie at slightly different frequencies, and so several lines are observed.

Anharmonicity also accounts for the appearance of additional weak absorption lines corresponding to the transitions $2 \leftarrow 0$, $3 \leftarrow 0, \ldots$, even though these second, third, ... **harmonics** are forbidden by the selection rule $\Delta v = \pm 1$. The reason is that the selection rule is derived using harmonic oscillator wavefunctions, and these are only approximately valid when anharmonicity is present. Therefore, the selection rule is also only an approximation. For an anharmonic oscillator, all values of Δv are allowed, but $\Delta v > 1$ is allowed only weakly if the anharmonicity is slight. The second harmonic, for example, gives rise to an absorption at

$$G(v + 2) - G(v) = 2\tilde{v} - 2(2v + 3)x_e\tilde{v} + \ldots$$

The Birge–Sponer extrapolation

When several vibrational transitions are detectable, a graphical technique called **Birge–Sponer extrapolation** may be used to determine the dissociation energy D_0 of the bond. The basis of this method is that the sum of successive energy separations $\Delta G_{v+1/2}$ from the zero-point level to the dissociation limit is the dissociation energy:

$$D_0 = \Delta G_{1/2} + \Delta G_{1+1/2} \ldots = \sum_v \Delta G_{v+1/2} \tag{26}$$

just as the height of the ladder is the sum of the separation of its rungs. The construction in Fig. 16.21 shows that the area under the plot of $\Delta G_{v+1/2}$ against v is equal to the sum, and therefore to D_0. The successive terms decrease linearly when only the x_e anharmonicity constant is taken into account and the inaccessible part of the spectrum can be estimated by linear extrapolation. Most actual plots differ from the linear plot as shown in the illustration, so the value of D_0 obtained in this way is usually an overestimate of the true value. The depth of the potential well, D_e, differs from D_0 by the

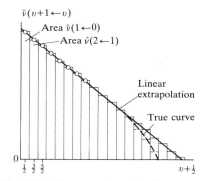

Fig. 16.21 The area under a plot of energy difference against vibrational quantum number is equal to the dissociation energy of the molecule. The assumption that the differences approach zero linearly is the basis of the Birge–Sponer extrapolation.

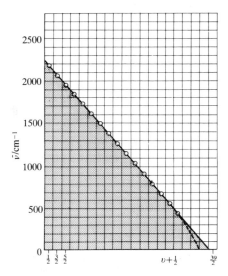

Fig. 16.22 The Birge–Sponor plot used in Example 16.9. The area is obtained simply by counting the squares beneath the line.

zero-point energy:

$$D_e = D_0 + \tfrac{1}{2}(1 - \tfrac{1}{2}x_e)\tilde{\nu}$$

Example 16.9: *Using a Birge–Sponer extrapolation*

The observed vibrational energy level separations of H_2^+ lie at the following values for $1 \leftarrow 0$, $2 \leftarrow 1$, ... respectively (in cm^{-1}): 2191, 2064, 1941, 1821, 1705, 1591, 1479, 1368, 1257, 1145, 1033, 918, 800, 677, 548, 411. Determine the dissociation energy of the molecule.

Answer. We need to plot the separations against υ, extrapolate linearly to the point cutting the υ axis, and then measure the area under the curve. The points are plotted in Fig. 16.22; the full line is the Birge–Sponer linear extrapolation. The area under the curve (count the squares) is 214. Each square corresponds to $100\ cm^{-1}$ (refer to the scale of the vertical axis); hence the dissociation energy is $21\,400\ cm^{-1}$ (corresponding to $256\ kJ\ mol^{-1}$).

Exercise. The vibrational levels of HgH converge rapidly, and successive separations are 1203.7, 965.6, 632.4, and $172\ cm^{-1}$. Estimate the dissociation energy.   [$40\ kJ\ mol^{-1}$]

16.9 Vibration–rotation spectra

At high resolution, each line of the vibrational spectrum of a gas-phase heteronuclear diatomic molecule is found to consist of a large number of closely spaced components (Fig. 16.23). For this reason, molecular spectra are often called **band spectra**. The separation between the components is of the order of $1\ cm^{-1}$, which suggests that the structure is due to rotational transitions accompanying the vibrational transition. A rotational change should be expected because classically we can think of the transition as leading to a sudden increase or decrease in the instantaneous bond length. Just as ice-skaters rotate more rapidly when they bring their arms in, and more slowly when they throw them out, so the molecular rotation is either accelerated or retarded. A detailed analysis of the quantum mechanics of the process shows that the rotational quantum number J changes by ± 1 during a vibrational transition. If the molecule possesses angular momentum about its axis, as in the case of the electronic orbital angular momentum of the $^2\Pi$ molecule NO, the selection rules also allow $\Delta J = 0$.

The appearance of the vibration–rotation spectrum of a diatomic mole-

Fig. 16.23 A high-resolution vibration–rotation spectrum of HCl. The lines appear in pairs because $H^{35}Cl$ and $H^{37}Cl$ both contribute (their abundance ratio is 3:1). There is no Q branch, because $\Delta J = 0$ is forbidden for this molecule.

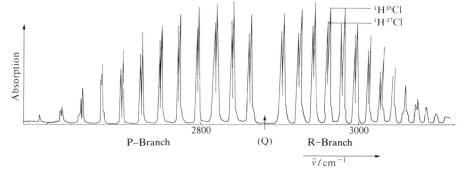

cule can be discussed in terms of the combined vibration–rotation terms S:

$$S(v, J) = G(v) + F(J)$$

If we ignore anharmonicity and centrifugal distortion,

$$S(v, J) = (v + \tfrac{1}{2})\tilde{\nu} + BJ(J + 1)$$

In a more detailed treatment B is allowed to depend on the vibrational state because as v increases the molecule swells slightly and the moment of inertia changes. However, these features lengthen the algebra without introducing anything new, and so we shall continue with the simple expression.

When the vibrational transition $v + 1 \leftarrow v$ occurs, J changes by ± 1 and in some cases by 0 (when $\Delta J = 0$ is allowed). The absorptions then fall into groups called the **P**, **Q**, and **R branches** of the spectrum. The P branch consists of all transitions with $\Delta J = -1$:

$$\tilde{\nu}_P(J) = \tilde{\nu} - 2BJ \tag{27a}$$

This branch (Figs. 16.23 and 16.24), consists of lines at $\tilde{\nu} - 2B$, $\tilde{\nu} - 4B, \ldots$ with an intensity distribution reflecting both the populations of the rotational levels and the $J - 1 \leftarrow J$ transition moment.

The Q branch consists of all lines with $\Delta J = 0$, and its wavenumbers are all

$$\tilde{\nu}_Q(J) = \tilde{\nu} \tag{27b}$$

for all values of J. This branch, when it is allowed, forms a single line at the vibrational transition wavenumber. In practice, since the rotational constants of the two vibrational levels are slightly different, the Q branch appears as a cluster of closely spaced lines. In Fig. 16.23 there is a gap at the expected location of the Q branch because it is forbidden in HCl.

The R branch consists of lines with $\Delta J = +1$:

$$\tilde{\nu}_R(J) = \tilde{\nu} + 2B(J + 1) \tag{27c}$$

This branch consists of lines displaced from $\tilde{\nu}$ to high wavenumber by $2B$, $4B, \ldots$ (Fig. 16.24).

The separation between the lines in the P and R branches of a vibrational transition gives the value of B, and so the bond length can be deduced without needing to record a pure rotation microwave spectrum (although the latter is more accurate).

16.10 Vibrational Raman spectra of diatomic molecules

The gross selection rule for vibrational Raman transitions is that the polarizability should change as the molecule vibrates. Both homonuclear and heteronuclear diatomic molecules swell and contract during a vibration, and the control of the nuclei over the electrons, and hence the molecular polarizability, changes too. Both types of diatomic molecule are therefore vibrationally Raman active.

The specific selection rule is $\Delta v = \pm 1$. The lines to high frequency of the incident light, the anti-Stokes lines, are those for which $\Delta v = -1$. They are

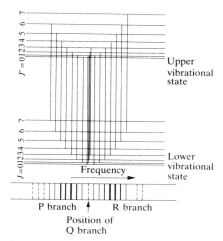

Fig. 16.24 The formation of P, Q, and R branches in a vibration–rotation spectrum. The intensities reflect the populations of the initial rotational levels.

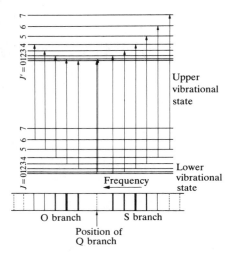

Upper vibrational state

Lower vibrational state

Frequency

O branch

S branch

Position of Q branch

Fig. 16.25 The formation of O, Q, and S branches in a vibration–rotation Raman spectrum of a linear rotor. Note that the frequency scale runs in the opposite direction to that in Fig. 16.24, because the higher-energy transitions (on the right) extract more energy from the incident beam and leave it at lower frequency.

Table 16.2. Properties of diatomic molecules.

	$\bar{v}/cm^{-1}$	B/cm^{-1}	$k/(N\,m^{-1})$
$^1H_2$	4400	60.86	575
$^1H^{35}Cl$	2991	10.59	516
$^1H^{127}I$	2308	6.61	314
$^{35}Cl_2$	560	0.244	323

usually weak because very few molecules are in an excited vibrational state initially. The lines to low frequency, the Stokes lines, correspond to $\Delta v = +1$. Superimposed on these is a branch structure arising from the simultaneous rotational transitions that accompany the vibrational excitation (Fig. 16.25). The selection rules are $\Delta J = 0, \pm 2$ (as in pure rotational Raman spectroscopy), and give rise to the **O branch** ($\Delta J = -2$), the **Q branch** ($\Delta J = 0$), and the **S branch** ($\Delta J = +2$).

The information available from vibrational Raman spectra adds to that from infrared spectroscopy because homonuclear diatomics can also be studied. The spectra can be interpreted in terms of the force constants, dissociation energies, and bond lengths, and some of the information obtained is included in Table 16.2.

The vibrations of polyatomic molecules

In a diatomic molecule there is only one mode of vibration, the bond stretch. In a polyatomic molecule there are several modes because bonds may stretch and angles may bend.

16.11 Normal modes

We begin by calculating the total number of vibrational modes of a polyatomic molecule consisting of N atoms. We then see that we can choose combinations of atomic motion that gives the simplest description of the vibrations of the molecule.

The number of vibrational modes

The total number of coordinates needed to specify the locations of all N atoms is $3N$. Each atom may change its location by varying one of its coordinates, and so the total number of displacements available is $3N$. We can group these displacements together in a physically sensible way. For example, three coordinates are needed to specify the location of the centre of mass of the molecule, and so three of the displacements correspond to the translational motion of the molecule as a whole. The remaining $3N - 3$ are non-translational 'internal' modes of the molecule.

Two angles are needed to specify the orientation of a linear molecule in space: in effect, we need to give only the latitude and longitude of the direction in which the molecular axis is pointing (Fig. 16.26a). However,

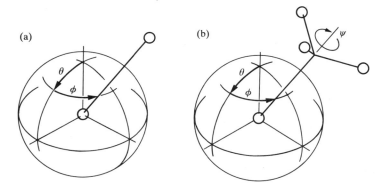

(a)

(b)

Fig. 16.26 The specification of the centre of mass of a molecule uses up three degrees of freedom. (a) The orientation of a linear molecule requires the specification of two angles. (b) The orientation of a non-linear molecule requires the specification of three angles.

three angles are needed for a non-linear molecule because we also need to specify the orientation of the molecule around the direction defined by the latitude and longitude (Fig. 16.26b). Therefore 2 (linear) or 3 (non-linear) of the $3N - 3$ internal displacements are rotational. This leaves $3N - 5$ (linear) or $3N - 6$ (non-linear) displacements of the atoms relative to each other: these are the vibrational modes. It follows that the number of modes of vibration N_{vib} is

$$N_{vib} = \begin{cases} 3N - 5 \text{ for linear molecules} \\ 3N - 6 \text{ for non-linear molecules} \end{cases} \tag{28}$$

For example, H_2O is a 3-atom non-linear molecule, and has three modes of vibration (and three modes of rotation); CO_2 is a 3-atom linear molecule, and has four modes of vibration (and only two modes of rotation). Even a middle-sized molecule such as naphthalene ($C_{10}H_8$) has 48 distinct modes of vibration.

Combinations of displacements

The next step is to find the best description of the modes. One choice for the four modes of CO_2, for example, might be the ones in Fig. 16.27a. This illustration shows the stretching of one bond (the mode ν_L), the stretching of the other (ν_R), and the two perpendicular bending modes (ν_2). The description, while permissible, has a disadvantage: when one C—O vibration is excited, the motion of the C atom sets the other C—O in motion, and so energy flows backwards and forwards between ν_L and ν_R.

The description of the vibrational motion is much simpler if linear combinations of ν_L and ν_R are taken. For example, one combination is ν_1 in Fig. 16.27b: this is the **symmetric stretch**, and in it the C atom is buffeted simultaneously from each side, and the motion continues indefinitely. Another mode is ν_3, the **antisymmetric stretch**, in which the two O atoms always move out of phase (in opposite directions). Both modes are independent in the sense that if one is excited, then it does not excite the other. They are two of the **normal modes** of the molecule, its independent, collective vibrational displacements. The two other normal modes are the bending modes ν_2. In general, a normal mode is an independent, synchronous motion of atoms or groups of atoms that may be excited without leading to the excitation of any other normal mode.

The four normal modes of CO_2, and the N_{vib} normal modes of polyatomics in general, are the key to the description of molecular vibrations. Each normal mode behaves like an independent harmonic oscillator (if anharmonicities are neglected), and so each has a series of terms

$$G_Q(v) = (v + \tfrac{1}{2})\bar{\nu}_Q$$

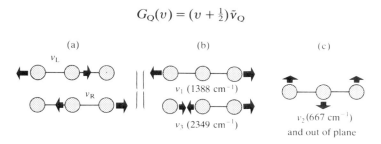

(a)   (b)   (c)

ν_L

ν_R

ν_1 (1388 cm^{-1})

ν_3 (2349 cm^{-1})

ν_2 (667 cm^{-1}) and out of plane

Fig. 16.27 Alternative descriptions of the vibrations of CO_2. (a) The stretching modes are not independent, and if one C–O is excited the other begins to vibrate. (b) The symmetric and antisymmetric stretches are independent, and one can be excited without affecting the other: they are normal modes. (c) The two perpendicular bending motions are normal modes.

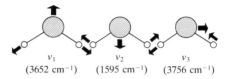

v_1 v_2 v_3
(3652 cm^{-1})   (1595 cm^{-1})   (3756 cm^{-1})

Fig. 16.28 The normal modes of H$_2$O. The mode v_2 is predominantly bending, and occurs at lower wavenumber than the other two.

$\bar{v}_Q$ is the wavenumber of mode Q and depends on the force constant k_Q for the mode and on the reduced mass μ_Q of the mode, with

$$\bar{v}_Q = \frac{\omega_Q}{2\pi c} \quad \text{and} \quad \omega_Q = \left(\frac{k_Q}{\mu_Q}\right)^{1/2}$$

The reduced mass of the mode is a measure of the mass that is swung about by the vibration. For example, in the symmetric stretch of CO_2 the C atom is stationary, and the reduced mass depends on the masses of only the O atoms. In the antisymmetric stretch and in the bends, all three atoms move, and so all contribute to the reduced mass. The three normal modes of H$_2$O are shown in Fig. 16.28: note that the predominantly bending mode (v_2) has a lower frequency than the others, which are predominantly stretching modes. One point that must be appreciated is that only in special cases (such as the CO_2 molecule) are the normal modes purely stretches or purely bends. In general a normal mode is a composite motion of simultaneous stretching and bending of bonds.

The symmetry species of normal modes

One of the most powerful ways of dealing with normal modes, especially of complex molecules, is to classify them according to their symmetries.

The procedure begins by deciding on the symmetry species of the irreducible representations spanned by all the $3N$ displacements of the atoms, using the characters of the molecular point group. We find these characters (as explained in Example 15.4) by counting 1 if the displacement is unchanged under a symmetry operation, -1 if it changes sign, and 0 if it is changed into some other displacement. Next, we subtract the symmetry species of the translations. These span the same symmetry species as x, y, and z, so they can be obtained from the right-hand columns of the character table. Finally, we subtract the symmetry species of the rotations, which are also given in the character table.

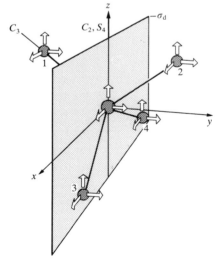

Fig. 16.29 The atomic displacements of CH$_4$ and the symmetry elements used to calculate the characters. Since all operations of the same class have the same character, there is no need to consider more than one element of each class.

Example 16.10: *Identifying the symmetry species of a normal mode*

Establish the symmetry species of the normal mode vibrations of CH$_4$.

Answer. There are $3 \times 5 = 15$ modes of motion, of which $3 \times 5 - 6 = 9$ are vibrations. Refer to Fig. 16.29. Under E, no displacement coordinates are changed, and so the character is 15. Under C_3, no displacements are left unchanged, and so the character is 0. Under a C_2 rotation the z-displacement of the central atom is left unchanged, while its x- and y-components both change sign. Therefore $\chi(C_2) = 1 - 1 - 1 + 0 + 0 + \ldots = -1$. Under S_4, the z-displacement of the central atom is reversed, and so $\chi(S_4) = -1$. Under σ_d, the z-displacement of C, H$_3$, and H$_4$ are left unchanged, three of the H displacements are left unchanged and three are reversed; hence $\chi(\sigma_d) = 3 + 3 - 3 = 3$. The characters are therefore 15 0 -1 -1 3, corresponding to $A_1 + E + T_1 + 3T_2$. The translations span T_2; the rotations span T_1. Hence the vibrations span $A_1 + E + 2T_2$.

Comment. We shall soon see that symmetry analysis gives a quick way of deciding which modes are active.

Exercise. Establish the symmetry species of the normal modes of H$_2$O.

           [$2A_1 + B_2$]

16.12 The vibrational spectra of polyatomic molecules

The gross selection rule for infrared activity is that the motion corresponding to a normal mode should be accompanied by a change of dipole moment. Deciding whether this is so can sometimes be done by inspection. For example, the symmetric stretch of CO_2 leaves the dipole moment unchanged (at zero), and so this mode is infrared inactive. The anti-symmetric stretch, however, changes the dipole moment because the molecule becomes unsymmetrical as it vibrates, and so this mode is infrared active. Since in this case the dipole moment change is parallel to the figure axis, the transitions arising from this mode are classified as **parallel bands** in the spectrum. Both bending modes are infrared active: they are accompanied by a changing dipole perpendicular to the figure axis, and so transitions involving them lead to a **perpendicular band** in the spectrum.

Symmetry and normal mode activity

It is best to use group theory to judge the activities of more complex modes of vibration. This is easily done by checking the character table of the molecular point group for the symmetry species of the irreducible representations spanned by x, y, and z, for these are also the symmetry species of the components of the electric dipole moment. Then the rule to apply is as follows:

If the symmetry species of a normal mode is the same as any of the symmetry species of x, y, or z, the mode is infrared active.

Example 16.11: *Identifying infrared active modes*

Which modes of CH_4 are infrared active?

Answer. Refer to the T_d character table to establish the symmetry species of x, y, and z; it is T_2. We found in Example 16.10 that the symmetry species of the normal modes are $A_1 + E + 2T_2$. Therefore, only the T_2 modes are infrared active.

Comment. The distortions accompanying the T_2 modes lead to a changing dipole moment. The A_1 mode, which is inactive, is the symmetrical 'breathing' mode of the molecule.

Exercise. Which of the normal modes of H_2O are infrared active?

[All three]

The appearance of the spectrum

The active modes are subject to the specific selection rule $\Delta v_Q = \pm 1$, and so the wavenumber of the **fundamental transition** (the first harmonic) of each active mode is $\bar{v}_Q$. From the analysis of the spectrum, a picture may be constructed of the stiffness of various parts of the molecule: that is, we can establish its **force field**, the set of force constants corresponding to all the displacements of the atoms.

Superimposed on this simple scheme are the complications arising from anharmonicities and the effects of molecular rotation. Very often the sample is a liquid or a solid, and the molecules are unable to rotate freely. In a liquid, for example, a molecule may be able to rotate through only a few degrees before it is struck by another, and so it changes its rotational state frequently. This random changing of orientation is called **tumbling**.

Since the lifetimes of rotational states in liquids are very short, the rotational energies are ill-defined. Collisions occur at a rate of about $10^{13}\,\mathrm{s}^{-1}$, and even allowing for only a 10 per cent success rate in knocking the molecule into another rotational state, a lifetime broadening (eqn 12) of more than $1\,\mathrm{cm}^{-1}$ can easily result. The rotational structure of the vibrational spectrum is blurred by this effect, and so the infrared spectrum of molecules in condensed phases usually consist of broad lines spanning the entire range of the resolved gas-phase spectrum, and showing no branch structure.

One very important application of infrared spectroscopy to condensed phase samples, and for which the blurring of the rotational structure by random collisions is a welcome simplification, is to chemical analysis. The vibrational spectra of different groups in a molecule give rise to absorptions at characteristic frequencies. Their intensities are also approximately transferable between molecules. Consequently, the molecules in a sample can often be identified by examining its infrared spectrum and accounting for all the bands by referring to a table of characteristic frequencies and intensities (Table 16.3 and Fig. 16.30).

Table 16.3. Typical vibrational wavenumbers, $\tilde{\nu}/\mathrm{cm}^{-1}$

C—H stretch	2850–2960
C—H bend	1340–1465
C—C stretch	700–1250
C=C stretch	1620–1680

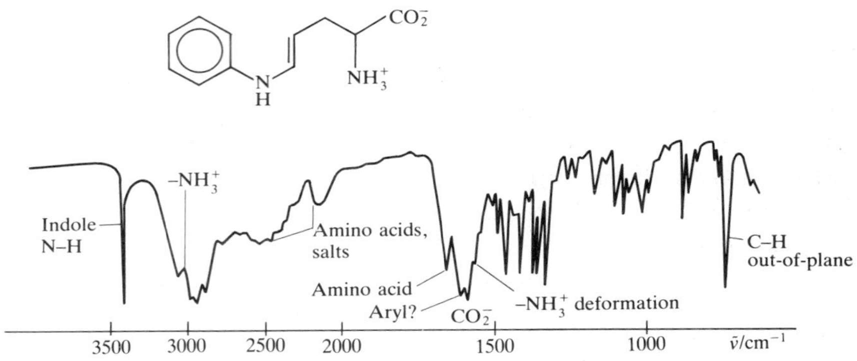

Fig. 16.30 The infrared absorption spectrum of an amino acid and a partial assignment.

16.13 Vibrational Raman spectra of polyatomic molecules

The normal modes of vibration of molecules are Raman active if they are accompanied by a changing polarizability. It is quite difficult to judge by inspection when this is so. The symmetric stretch of CO_2, for example, alternately swells and contracts the molecule: this motion changes its polarizability, and so the mode is Raman active. The other modes of CO_2 leave the polarizability unchanged, and so they are Raman inactive.

Symmetry aspects of Raman transitions

Group theory provides an explicit recipe for judging the Raman activity of a normal mode. In this case, the symmetry species of the quadratic forms (x^2, xy, etc) listed in the character table are noted (they transform in the same way as the polarizability), and then we use the following rule:

If the symmetry species of a normal mode is the same as the symmetry species of a quadratic form, the mode is Raman active.

Example 16.12: *Identifying Raman-active normal modes*

Which of the vibrations of CH_4 are Raman active?

Answer. Refer to the T_d character table. We found in Example 16.10 that the symmetry species of the normal modes are $A_1 + E + 2T_2$. Since the quadratic forms span $A_1 + E + T_2$, all the normal modes are Raman active.

Comment. All totally symmetric vibrations, whatever the point group of the molecule, are Raman active (and polarized, see below).

Exercise. Which of the vibrational modes of H_2O are Raman active?

[All three]

The **exclusion rule** also helps us to decide which modes are active:

If the molecule has a centre of symmetry, no modes can be both infrared and Raman active.

(A mode may be inactive in both.) Since we can often judge intuitively when a mode changes the molecular dipole moment, we can use this rule to identify modes that are not Raman active. The rule applies to CO_2 but to neither H_2O nor CH_4 because they have no centre of symmetry.

Depolarization

The assignment of Raman lines to particular vibrational modes is aided by noting the state of polarization of the scattered light. The **depolarization ratio** ρ of a line is the ratio of the intensities of the scattered light with a polarization parallel and perpendicular to the plane of polarization of the incident radiation (Fig. 16.31):

$$\rho = \frac{I_\perp}{I_\parallel}$$

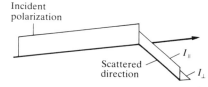

Fig. 16.31 The definition of the planes used for the specification of the depolarization ratio ρ in Raman scattering.

If the emergent light is not polarized, both intensities are the same and $\rho = 1$; if the light retains its initial polarization, $I_\perp = 0$ and so $\rho = 0$. We classify a line as **depolarized** if it has ρ close to 1 and as **polarized** if it has ρ close to zero. A general rule is that totally symmetrical vibrations give rise to polarized Raman lines in which the incident polarization is largely preserved. Vibrations that are not totally symmetrical give rise to depolarized lines since the incident radiation can give rise to radiation in the perpendicular direction too. This means that if we observe the Raman spectrum with a polarizing filter (a 'half-wave plate') first parallel and then perpendicular to the polarization of the incident beam, the intensity of the polarized lines will appear significantly reduced and hence these lines can be ascribed to symmetrical vibrations.

Applications

One application of vibrational Raman spectroscopy is to the determination of the structures of symmetrical molecules such as XeF_4 and SF_6. Another application makes use of the fact that the intensity characteristics of Raman transitions, which depend on molecular polarizabilities, are more readily transferred from molecule to molecule than the intensities of infrared spectra, which depend on dipole moments and are more sensitive to the

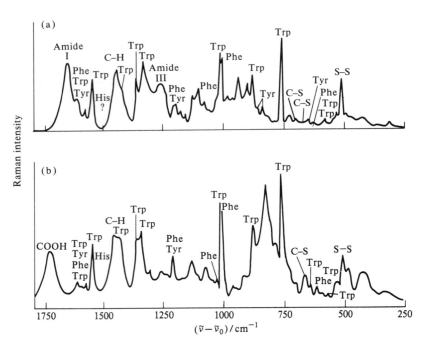

Fig. 16.32 The vibrational Raman spectrum of lysozyme in water and the superposition of the Raman spectra of the constituent amino acids. (From *Raman spectroscopy*, D. A. Long. Copyright 1977, McGraw-Hill Inc. Used with the permission of the McGraw-Hill Book Company.)

other groups present in a molecule and to the solvent. Hence, Raman spectra are useful in the identification of organic and inorganic species in solution. An example of the technique is shown in Fig. 16.32, which shows the vibrational Raman spectrum of an aqueous solution of lysozyme and, for comparison, a superposition of the Raman spectra of the constituent amino acids. The differences are indications of the effects of conformation, environment, and specific interactions (such as S–S linking) in the enzyme molecule.

Further reading

Principles

C. N. Banwell, *Fundamentals of molecular spectroscopy.* McGraw-Hill, New York (1972).

D. H. Whiffen, *Spectroscopy.* Longman, London (1972).

P. W. Atkins, *Molecular quantum mechanics* (2nd edn). Oxford University Press (1983).

J. M. Hollas, *Modern spectroscopy.* Wiley, New York (1987).

J. M. Hollas, *High resolution spectroscopy.* Butterworth, London (1982).

Microwave spectroscopy

T. M. Sugden and C. N. Kenney, *Microwave spectroscopy of gases.* Van Nostrand, London (1965).

W. Gordy and R. L. Cook, *Microwave molecular spectra.* Wiley, New York (1984).

H. W. Kroto, *Molecular rotation spectra.* Wiley, New York (1975).

W. H. Flygare, *Microwave spectroscopy.* In *Techniques in chemistry*, IIIA, p. 439 (ed. A. Weissberger and B. W. Rossiter). Wiley, New York (1972).

C. H. Townes and A. L Schawlow, *Microwave spectroscopy.* McGraw-Hill, New York (1955).

Infrared spectroscopy

D. H. Anderson and N. B. Woodall, *Infrared spectroscopy*. In *Techniques in chemistry*, IIIB, p. 1 (ed. A. Weissberger and B. W. Rossiter) Wiley, New York (1972).

P. Gans, *Vibrating molecules*. Chapman and Hall, London (1971).

L. J. Bellamy, *The infrared spectra of complex molecules*. Chapman and Hall, London (1980).

G. Herzberg, *Infrared and Raman spectra of polyatomic molecules*. Van Nostrand, New York (1945).

E. B. Wilson, J. C. Decius, and P. C. Cross, *Molecular vibrations*. McGraw-Hill, New York (1955).

Raman spectroscopy

J. R. Durig and W. C. Harris, *Raman spectroscopy*. In *Techniques in chemistry*, IIIB, p. 85 (ed. A. Weissberger and B. W. Rossiter) Wiley, New York (1972).

D. A. Long, *Raman spectroscopy*. McGraw-Hill, New York (1977).

Applications

D. H. Williams and I. Fleming, *Spectroscopic methods in organic chemistry*. McGraw-Hill, New York (1980).

R. Drago, *Physical methods in chemistry*. Saunders, Philadelphia (1977).

E. A. V. Ebsworth, D. W. H. Rankin, and S. Cradock, *Structural methods in inorganic chemistry*. Blackwell Scientific, Oxford (1987).

Exercises

16.1 Calculate the reduced masses of (a) $^1H^{35}Cl$, (b) $^2H^{35}Cl$, and (c) $^1H^{37}Cl$ and identify the atom that makes the greater contribution.

16.2 The bond length of $^{79}Br^{81}Br$ is 228 pm; calculate the moment of inertia of the molecule.

16.3 The rotational constant of $^{127}I^{35}Cl$ is 0.1142 cm^{-1}. Calculate the bond length of the molecule.

16.4 The wavenumber of the incident radiation in a Raman spectrometer is 20 487 cm^{-1}. What is the wavenumber of the scattered Stokes radiation for the $J = 2 \leftarrow 0$ transition of $^{14}N^{14}N$?

16.5 Infrared absorption by $^1H^{81}Br$ gives rise to an R branch from $v = 0$. What is the wavenumber of the line originating from the rotational state with $J = 2$? (Refer to Table 16.2.)

16.6 Calculate the percentage difference in the fundamental vibration wavenumber of $^{23}Na^{35}Cl$ and $^{23}Na^{37}Cl$ on the assumption that their force constants are the same.

16.7 The wavenumber of the fundamental vibrational transition of $^{35}Cl_2$ is 564.9 cm^{-1}. Calculate the force constant of the bond.

16.8 For $^{127}I^{35}Cl$, $\bar{v} = 384.3$ cm^{-1} and $x_e\bar{v} = 1.5$ cm^{-1}. Calculate the wavenumber of the pure fundamental ($\Delta v = 1$) vibrational transition with the highest wavenumber and that of the next highest.

16.9 The bond dissociation energy of $^{127}I^{35}Cl$ is 2.153 eV. Use the information in Exercise 16.8 to calculate the depth of the molecular potential energy curve of this molecule.

16.10 The molecule CH_2Cl_2 belongs to the point group C_{2v}. The displacements of the atoms span $5A_1 + 2A_2 + 4B_1 + 4B_2$. What are the symmetries of the normal modes of vibration?

16.11 Which of the following molecules may show a pure rotational microwave absorption spectrum: (a) H_2, (b) HCl, (c) CH_4, (d) CH_3Cl, (e) CH_2Cl_2, (f) H_2O, (g) H_2O_2, (h) NH_3?

16.12 Which of the following molecules may show infrared absorption spectra: (a) H_2, (b) HCl, (c) CO_2, (d) H_2O, (e) CH_3CH_3, (f) CH_4, (g) CH_3Cl, (h) N_2?

16.13 Which of the following molecules may show a pure rotational Raman spectrum: (a) H_2, (b) HCl, (c) CH_4, (d) CH_3Cl, (e) CH_2Cl_2, (f) CH_3CH_3, (g) SF_6?

16.14 What is the Doppler-shifted wavelength of a red (660 nm) traffic light approached at 50 m.p.h.? At what speed would it appear green (520 nm)?

16.15 A spectral line of $^{48}Ti^{8+}$ in a distant star was found to be shifted from 654.2 nm to 706.5 nm and to be broadened to 61.8 pm. What is the speed of recession and the surface temperature of the star?

16.16 Estimate the lifetime of a state that gives rise to a line of width (a) 0.1 cm^{-1}, (b) 1 cm^{-1}, (c) 100 MHz.

16.17 A molecule in a liquid undergoes about 1×10^{13} collisions in each second. Suppose that (a) every collision is effective in deactivating the molecule vibrationally and (b) that one collision in 100 is effective. Calculate the width (in cm^{-1}) of vibrational transitions in the molecule.

16.18 Calculate the relative numbers of Cl_2 molecules ($\tilde{v} = 559.7\,cm^{-1}$) in the ground and first excited vibrational states at (a) 298 K, (b) 500 K.

16.19 The pure rotational spectrum of $^1H^{131}I$ consists of a series of lines separated by $13.10\,cm^{-1}$. Calculate the bond length of the molecule.

16.20 The hydrogen halides have the following fundamental vibrational wavenumbers:

	HF	$H^{35}Cl$	$H^{81}Br$	$H^{127}I$
$\tilde{v}/cm^{-1}$	4143.3	2988.9	2649.7	2309.5

Calculate the force constants of the hydrogen–halogen bonds.

16.21 From the data in Exercise 16.20, predict the fundamental vibrational wavenumbers of the deuterium halides.

16.22 The first five vibrational energy levels of HCl are at 1481.86, 4367.50, 7149.04, 9826.48, and $12\,399.8\,cm^{-1}$. Calculate the dissociation energy of the molecule in cm^{-1} and eV.

16.23 The rotational Raman spectrum of $^{35}Cl_2$ shows a series of Stokes lines separated by $0.9752\,cm^{-1}$ and a similar series of anti-Stokes lines. Calculate the bond length of the molecule.

16.24 Which of the three vibrations of an AB_2 molecule are infrared or Raman active when it is (a) non-linear, (b) linear?

16.25 Consider the vibrational mode that corresponds to the uniform expansion of the benzene ring. Is it (a) Raman, (b) infrared active?

Problems

Numerical problems

16.1 Calculate the Doppler width (as a fraction of the transition wavelength) for any kind of transition in (a) HCl, (b) ICl at 25°C. What would be the widths of the rotational and vibrational transitions in these molecules (in MHz and cm^{-1} respectively), given $B(ICl) = 0.1142\,cm^{-1}$, $\tilde{v}(ICl) = 384\,cm^{-1}$, and the information in Table 16.2.

16.2 The number of collisions that a molecule undergoes per unit time in a gas of pressure p is

$$z = 4\sigma \left(\frac{kT}{\pi m}\right)^{1/2} \times \frac{p}{kT}$$

where σ is the collision cross-section. Find an expression for the collision-limited lifetime of an excited state assuming that every collision is effective. Estimate the width of a rotational transition in HCl ($\sigma = 0.30\,nm^2$) at 25°C and 1.0 atm. To what value must the pressure of the gas be reduced in order to ensure that collision broadening is less important than Doppler broadening?

16.3 The rotational constant of NH_3 is equivalent to 298 GHz. Compute the separation of the pure rotational spectrum lines in GHz and cm^{-1}, and show that the value of B is consistent with an N—H bond length of 101.4 pm and a bond angle of 106° 47′.

16.4 The vibrational energy levels of NaI lie at wavenumbers 142.81, 427.31, 710.31, and $991.81\,cm^{-1}$. Show that they fit the expression $(v + \frac{1}{2})\tilde{v} - (v + \frac{1}{2})^2 x_e \tilde{v}$ and deduce the force constant, zero-point energy, and dissociation energy of the molecule.

16.5 A space probe was designed to look for CO in the atmosphere of Saturn, and it was decided to use a microwave technique from an orbiting satellite. Given the bond length of the molecule as 112.82 pm, at what frequencies will the first four transitions of $^{12}C^{16}O$ lie? What precision is needed in

order to distinguish the 1–0 transition in the spectra of $^{12}C^{16}O$ and $^{13}C^{16}O$ in order to determine the relative abundances of the two carbon isotopes?

16.6 Rotational absorption lines from $^1H^{35}Cl$ gas were found at the following wavenumbers (R. L. Hausler and R. A. Oetjen, *J. chem. Phys.*, **21**, 1340 (1953)): 83.32, 104.13, 124.73, 145.37, 165.89, 186.23, 206.60, 226.86 cm^{-1}. Calculate the moment of inertia and the bond length of the molecule. Predict the positions of the corresponding lines in $^2H^{35}Cl$.

16.7 Is the bond length in HCl the same as that in DCl? The data (in cm^{-1}) from the rotational structure of the infrared spectrum of the two molecules are as follows (I. M. Mills, H. W. Thompson, and R. L. Williams, *Proc. R. Soc.*, **A218**, 29 (1953); J. Pickworth and H. W. Thompson, *Proc. R. Soc.*, **A218**, 37 (1953)):

J	0	1	2	3
$^1H^{35}Cl$(R branch)	2906.25	2925.92	2944.99	2963.35
$^1H^{35}Cl$(P branch)		2865.14	2843.63	2821.59
$^2H^{35}Cl$(R branch)	2101.60	2111.94	2122.05	2131.91
$^2H^{35}Cl$(P branch)		2080.26	2069.24	2058.02

J	4	5	6
$^1H^{35}Cl$(R branch)	2981.05	2998.05	3014.50
$^1H^{35}Cl$(P branch)	2799.00	2775.77	2752.01
$^2H^{35}Cl$(R branch)	2141.53	2150.93	2160.06
$^2H^{35}Cl$(P branch)	2046.58	2034.95	2023.12

16.8 Thermodynamic considerations suggest that the copper monohalides CuX should exist mainly as polymers in the gas phase, and indeed it proved difficult to obtain the monomers in sufficient abundance to detect spectroscopically. This difficulty was overcome by flowing the halogen gas over copper heated to 1100 K (E. L. Manson, F. C. de Lucia, and W. Gordy, *J. chem. Phys.*, **63**, 2724 (1975)). For CuBr the

$J = 13$–14, 14–15, and 15–16 transitions occurred at 84 421.34, 90 449.25, and 96 476.72 MHz respectively. Calculate the rotational constant and bond length of CuBr.

16.9 The microwave spectrum of $^{16}O^{12}CS$ (C. H. Townes, A. N. Holden, and F. R. Merritt, *Phys. Rev.*, **74**, 1113 (1948)) gave absorption lines (in GHz) as follows:

J	1	2	3	4
^{32}S	24.325 92	36.488 82	48.651 64	60.814 08
^{34}S	23.732 33		47.462 40	

Use the expressions for moments of inertia in Table 16.1 and assume that the bond lengths are unchanged by substitution, calculate the CO and CS bond lengths in OCS.

16.10 The HCl molecule is quite well described by the Morse potential with $D_e = 5.33$ eV, $\tilde{v} = 2989.7$ cm^{-1}, and $x_e\tilde{v} = 52.05$ cm^{-1}. Assuming that the potential is unchanged on deuteration, predict the dissociation energies (D_0) of (a) HCl, (b) DCl.

16.11 The Morse potential (eqn 23) is very useful as a simple representation of the actual molecular potential energy. When RbH was studied it was found that $\tilde{v} = 936.8$ cm^{-1} and $x_e\tilde{v} = 14.15$ cm^{-1}. Plot the potential energy curve from 50 pm to 800 pm around $R_e = 236.7$ pm. Then go on to explore how the rotation of a molecule may weaken its bond by allowing for the kinetic energy of rotation of a molecule and plotting

$$V^* = V + hcBJ(J + 1) \qquad B = \frac{\hbar}{4\pi c\mu R^2}$$

Plot these curves on the same diagram for $J = 40$, 80, and 100, and observe how the dissociation energy is affected by the rotation. (Taking $B = 3.020$ cm^{-1} at the equilibrium bond length will greatly simplify the calculation.)

Theoretical problems

16.12 Show that the moment of inertia of a diatomic molecule composed of atoms of masses m_A and m_B and bond length R is equal to μR^2, where μ is the reduced mass of the molecule.

16.13 Derive an expression for the value of J corresponding to the most highly populated rotational energy level of a diatomic rotor at a temperature T remembering that the degeneracy of each level is $2J + 1$. Evaluate the expression for ICl (for which $B = 0.1142$ cm^{-1}) at 25°C. Repeat the problem for the most highly populated level of a spherical rotor, taking note of the fact that each level is $(2J + 1)^2$-fold degenerate. Evaluate the expression for CH$_4$ (for which $B = 5.24$ cm^{-1}) at 25°C.

16.14 Derive expressions for the P, Q, and R branches of a diatomic rotor without making the assumption that the rotational constants are the same in the lower and upper vibrational states. Lines in the P branch of $^1H^{35}Cl$ were

observed at 2865.1, 2843.6, and 2821.6 cm^{-1} for $J = 1$, 2, and 3, and in the R branch at 2906.2, 2925.9, 2945.0, and 2963.3 cm^{-1} for $J = 0$, 1, 2, 3 (I. M. Mills, H. W. Thompson, and R. L. Williams, *Proc. R. Soc.*, **A218**, 29 (1953)). Calculate the force constant of the bond and the bond lengths of the upper and lower vibrational states.

16.15 The set of relations known as 'Kraitchman's equations' relate the change in rotational constant (or moments of inertia) to the positions at which isotopic substitution is made. Show that when an isotope is substituted at a distance z along the axis from the original centre of mass of a symmetric rotor, then the change in rotational constant B is given by

$$(z/\text{pm})^2 = 1.685\,90 \times 10^5 \frac{\Delta B/\text{cm}^{-1}}{(B/\text{cm}^{-1})(B'/\text{cm}^{-1})(\Delta M/\text{g mol}^{-1})}$$

where B is the rotational constant before substitution, B' that after substitution, and

$$\Delta M = \frac{M(M' - M)}{M'}$$

where M and M' are the initial and final molar masses of the molecule. The microwave spectra of various isotopic species of ClTeF$_5$ show rigid symmetric rotor behaviour (A. C. Legon, *J. chem. Soc. Faraday Trans.*, II, **29** (1973)). Four of the F atoms lie in a square, the Te atom lies just above the plane of the square they form, the fifth F atom lies beneath this plane and the Cl atom lies above it. Use Kraitchman's equation to deduce the Te–Cl bond length on the assumption that all the Te–F bond lengths are identical. The relevant data are that the 11–10 transition occurs at 30 711.18 MHz in $^{35}Cl^{126}TeF_5$, at 30 713.24 MHz in $^{35}Cl^{125}TeF_5$, and at 29 990.54 MHz in $^{37}Cl^{126}TeF_5$. Use $M(^{126}Te) = 125.0331$ g mol^{-1} and $M(^{125}Te) = 124.0443$ g mol^{-1}.

16.16 The moments of inertia of the linear mercury(II) halides are very large, and so the O and S branches of their vibrational Raman spectra show little rotational structure. Nevertheless, the peaks of both branches can be identified and have been used to measure the rotational constants of the molecules (R. J. H. Clark and D. M. Rippon, *J. chem. Soc. Faraday Trans.*, II, **69**, 1496 (1973)). Show, from a knowledge of the value of J corresponding to the intensity maximum, that the separation of the peaks of the O and S branches is given by the 'Placzek–Teller relation'

$$\delta\tilde{v} = \left(\frac{32BkT}{hc}\right)^{1/2}$$

The following widths were obtained at the temperatures stated:

	HgCl$_2$	HgBr$_2$	HgI$_2$
$\theta/°C$	282	292	292
$\delta\tilde{v}/\text{cm}^{-1}$	23.8	15.2	11.4

Calculate the bond lengths in the three molecules.

17 Electronic transitions

Check-list of key ideas

1. The *Beer–Lambert law* for the reduction in intensity of light passing through an absorbing medium and the definition of the *molar absorption coefficient* (eqn 1).

2. The intensity of absorption in terms of the *oscillator strength* (eqn 3a) and its relation to the transition dipole moment (eqn 3b).

3. The use of the *Franck–Condon principle* to account for the vibrational structure of electronic transitions and the concept of a *vertical transition* (Section 17.2).

4. The *Laporte selection rule* and the *vibronic character* of *d–d transitions* in complexes (Section 17.3).

5. *Charge-transfer transitions* and π^*,π and π^*,n transitions (Section 17.3).

6. The mechanisms of *fluorescence* and *phosphorescence* and the characteristics of a *fluorescence spectrum* (Section 17.4).

7. The mechanisms of *intersystem crossing* (Section 17.4) and *internal conversion* leading to *predissociation* (Section 17.5).

8. The principles of *laser action*, including *population inversion*, *pumping*, and the difference between *three-level* and *four-level* lasers (Section 17.6).

9. The characteristics of laser radiation and the formation of pulses by *Q-switching* and *mode locking* (Section 17.6).

10. Examples of practical lasers, including *solid-state lasers, gas lasers, ion lasers, chemical lasers, excimer lasers, dye lasers,* and *semiconductor lasers* (Section 17.7).

11. The applications of lasers in chemistry, particularly to *multiphoton spectroscopy, laser Raman spectroscopy,* and to precision *state-selection* and *fast reactions* (Section 17.8).

12. The techniques of *ultraviolet photoelectron spectroscopy* and *X-ray photoelectron spectroscopy* (Section 17.9 to 17.11).

Table 17.1. Colour, frequency, and energy of light

Colour	λ/nm	$\nu/10^{14}$ Hz	E/kJ mol^{-1}
Infrared	>1000	<3.0	<120
Red	700	4.3	170
Yellow	580	5.2	210
Blue	470	6.4	250
Ultraviolet	<300	>10	>400

The energies needed to change the electron distributions of molecules are of the order of several electronvolts (1 eV is about 8000 cm^{-1}). Consequently, the photons emitted or absorbed when such changes occur lie in the visible and ultraviolet regions of the spectrum, which spreads from about 14 000 cm^{-1} for red light to 21 000 cm^{-1} for blue, and on to 50 000 cm^{-1} for ultraviolet radiation (Table 17.1). In some cases the relocation of electrons may be so extensive that it results in the breaking of a bond and the dissociation of the molecule.

The characteristics of electronic transitions

The nuclei in a molecule are subjected to different forces after an electronic transition has occurred, and the molecule may respond by bursting into vibration. The resulting vibrational structure of electronic transitions in a gaseous sample can be resolved, but in a liquid or solid the lines usually merge together and result in broad, almost featureless bands (Fig. 17.1). Superimposed on the vibrational transitions that accompany the electronic transition is an additional branch structure that arises from rotational transitions. The electronic spectra of gaseous samples are therefore very complicated, but rich in information.

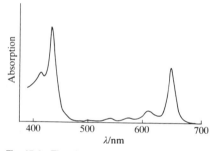

Fig. 17.1 The absorption spectrum of chlorophyll in the visible region. Note that it absorbs in the red and blue regions, and that green light is not absorbed.

17.1 Measures of intensity

We shall first establish how the intensity of an electronic transition is reported. Among other things, this will enable us to judge whether an electron moves over a large distance during the transition or whether its distribution remains largely unchanged.

The Beer–Lambert law

The reduction in intensity dI that occurs when light passes through a layer of thickness dx containing an absorbing species J at a molar concentration [J] is proportional to the thickness of the layer, the concentration, and the incident intensity I (because the rate of stimulated absorption is proportional to the intensity, Section 16.2). We can therefore write

$$dI = -\kappa[J]I\,dx$$

where κ is the proportionality coefficient, or equivalently

$$d\ln I = -\kappa[J]\,dx$$

These expressions apply to each successive layer into which the sample can be regarded as being divided. Therefore, to obtain the intensity I' that emerges from a sample of thickness l when the incident intensity is I, we sum all the successive changes:

$$\int_{\ln I}^{\ln I'} d \ln I = -\kappa \int_0^l [J] \, dx$$

If the concentration is uniform, $[J]$ is independent of x, and the expression integrates to the **Beer–Lambert law**:

$$I' = I e^{-\kappa [J] l} \qquad (1a)$$

Hence, the intensity decreases exponentially with the sample thickness and the concentration.

The Beer–Lambert law is often expressed as

$$I' = I \, 10^{-\varepsilon [J] l} \qquad (1b)$$

where

$$\kappa = \varepsilon \ln 10 = 2.303 \varepsilon$$

or as

$$\lg \frac{I'}{I} = -\varepsilon [J] l \qquad (1c)$$

ε is the **molar absorption coefficient** (formerly the 'extinction coefficient') of the species at the stated frequency, and depends on the frequency of the incident light. Its units are those of $1/(\text{concentration} \times \text{length})$, and it is normally convenient[1] to express it in $\text{M}^{-1} \, \text{cm}^{-1}$. The dimensionless product $A = \varepsilon [J] l$ is called the **absorbance** (formerly the 'optical density') of the sample, and I'/I is the **transmittance** T.

Example 17.1: *Calculating the molar absorption coefficient*

Light of wavelength 256 nm passes through a 1.0-mm cell containing a 0.050 M C_6H_6 solution. The light intensity is reduced to 16 per cent of its initial value. Calculate the absorbance and the molar absorption coefficient of the sample. What would be the transmittance through a 2.0-mm cell?

Answer. For ε we use eqn 1c with $I' = 0.16 I$:

$$\varepsilon = \frac{-1}{[C_6H_6] l} \lg \frac{I'}{I} = \frac{-\lg 0.16}{0.050 \, \text{M} \times 0.10 \, \text{cm}} = 160 \, \text{M}^{-1} \, \text{cm}^{-1}$$

The absorbance is therefore

$$A = \varepsilon [C_6H_6] l = 160 \, \text{M}^{-1} \text{cm}^{-1} \times 0.050 \, \text{M} \times 0.10 \, \text{cm} = 0.80$$

The transmittance through a 2.0-mm sample is given by

$$T = 10^{-\varepsilon [C_6H_6] l} = 10^{-160 \times 0.050 \times 0.20} = 0.025$$

That is, the emergent light is reduced to 2.5 per cent of its incident intensity.

Comment. The molar absorption coefficient may be expressed as $160 \, \text{cm}^2 \, \text{mmol}^{-1}$, and one benzene molecule effectively acts as a screen of cross-sectional area $0.27 \, \text{pm}^2$ to light of this wavelength.

---

[1] Alternative units are $\text{cm}^2 \, \text{mol}^{-1}$ (or $\text{cm}^2 \, \text{mmol}^{-1}$, with $1 \, \text{M}^{-1} \, \text{cm}^{-1} = 1 \, \text{cm}^2 \, \text{mmol}^{-1}$). These units bring out the point that ε is a molar cross-section for absorption, and the greater the cross-section of the molecule for absorption, the greater the attenuation of the intensity of the beam.

Exercise. The transmittance of a 0.10 M $CuSO_4(aq)$ solution at 600 nm was measured as 0.30 in a 5.0-mm cell. Calculate the molar absorption coefficient of $Cu^{2+}(aq)$ at that wavelength and the absorbance of the solution. What is the transmittance through a 1.0-mm cell?   [10 M^{-1} cm^{-1}, $A = 0.50$, $T = 0.79$]

The maximum value of the molar absorption coefficient ε_{max} is an indication of the intensity of a transition. Typical values for strong transitions are of the order 10^4–10^5 M^{-1} cm^{-1} (10^4–10^5 cm^2 $mmol^{-1}$), indicating that in a 0.01 M solution the intensity of light (of the appropriate frequency) falls to 10 per cent of its initial value after passing through about 0.1 mm of solution. Some values are listed in Table 17.2.

Table 17.2. Absorption characteristics of some groups and molecules

Group	$\tilde{v}_{max}/cm^{-1}$	λ_{max}/nm	$\varepsilon_{max}/M^{-1}cm^{-1}$
$C=C(\pi^*,\pi)$	61 000	163	15 000
	57 300	174	5 500
$C=O(\pi^*,n)$	36 000	280	10–20
$H_2O(\pi^*,n)$	60 000	167	7 000

The total intensity
Absorption bands generally spread over a range of frequencies, and so quoting the absorption coefficient at a single frequency might not give a true indication of the intensity of a transition. The **integrated absorption coefficient** $\mathscr{A}$ is the sum of the absorption coefficients over the entire band

$$\mathscr{A} = \int \varepsilon(v)\,dv \tag{2}$$

and it does measure the total strength of a transition. Its dimensions are 1/(concentration × length × time), the 1/time coming from the frequency integration. $\mathscr{A}$ is normally expressed in either M^{-1} cm^{-1} s^{-1} or cm^2 $mmol^{-1}$ s^{-1}.

The integrated absorption coefficient is sometimes expressed in terms of the dimensionless **oscillator strength** f, which is defined as

$$f = \left(\frac{4m_e c\varepsilon_0}{N_A e^2} \ln 10\right) \times \mathscr{A} \tag{3a}$$

where ε_0 is the vacuum permittivity.[2] The practical form of this relation is

$$f = 1.44 \times 10^{-19} \times \mathscr{A}/(cm^2\,mmol^{-1}\,s^{-1})$$

Very intense transitions have $f \approx 1$.

[2] The odd collection of constants in the definition of f ensure (a) that it is dimensionless and (b) that $f = 1$ for a three-dimensional harmonic oscillator.

Example 17.2: *Calculating an oscillator strength*

A transition in benzene has $\varepsilon \approx 160 \text{ cm}^2 \text{ mmol}^{-1}$ (Example 17.1) and spreads over about 4000 cm^{-1}. Estimate its oscillator strength.

Answer. We can approximate the area of the absorption band as (height) × (width) after expressing the width as a frequency using $v = c\tilde{v}$:

$$\mathcal{A} \approx (160 \text{ cm}^2 \text{ mmol}^{-1}) \times (4000 \text{ cm}^{-1}) \times (2.998 \times 10^{10} \text{ cm s}^{-1})$$
$$= 1.9 \times 10^{16} \text{ cm}^2 \text{ mmol}^{-1} \text{ s}^{-1}$$

The oscillator strength is therefore

$$f = (1.44 \times 10^{-19}) \times (1.9 \times 10^{16}) = 2.8 \times 10^{-3}$$

Comment. We see that $f \ll 1$, which indicates that the transition is forbidden. In a more accurate treatment, the area of the absorption curve would be measured as precisely as possible, not estimated in this crude manner.

Exercise. The absorption band of $Cu^{2+}(aq)$ considered in Example 17.1 spreads over about 3000 cm^{-1}; estimate its oscillator strength. $[1.3 \times 10^{-4}]$

The advantage of introducing the oscillator strength is that it has a direct connection with the wavefunctions of the initial and final states involved in the transition. The transition dipole between the initial and final states is defined in the usual way as

$$\boldsymbol{\mu}_{fi} = \int \psi_f^* \boldsymbol{\mu} \psi_i \, d\tau \qquad \boldsymbol{\mu} = -er$$

and is a measure of the dipole moment associated with the shift of charge that occurs when the electron redistribution takes place. The oscillator strength is related to this moment by

$$f = \frac{8\pi^2 m_e v}{3he^2} \times |\boldsymbol{\mu}_{fi}|^2 \qquad (3b)$$

Therefore we may estimate the intensity of a transition if we know the wavefunctions. This is an important link between theory and experiment.

Example 17.3: *Estimating a transition dipole moment*

Calculate the transition dipole moment for the benzene transition discussed in Examples 17.1 and 17.2.

Answer. Since the oscillator strength is 2.8×10^{-3} (Example 17.2), we can use it in eqn 3b. For v we use the frequency corresponding to the wavelength at the maximum of the band, and use the factor

$$\frac{3he^2}{8\pi^2 m_e} = 7.095 \times 10^{-43} \text{ m}^2 \text{ s}^{-1} \text{ C}^2$$

Since $v = 1.17 \times 10^{15} \text{ Hz}$, we find $|\boldsymbol{\mu}_{fi}|^2 = 1.7 \times 10^{-60} \text{ C}^2 \text{ m}^2$, so $|\boldsymbol{\mu}_{fi}| = 1.3 \times 10^{-30} \text{ C m}$.

Comment. The transition moment corresponds to 0.39 D. The transition can be visualized as corresponding to the movement of about 10 per cent of an electron through a distance about equal to the radius of an atom.

Exercise. Estimate the value of the transition dipole moment for the $Cu^{2+}(aq)$ transition treated in the Examples 17.1 and 17.2. $[0.13 \text{ D}]$

Selection rules

The expression for the transition dipole moment is used to derive the selection rules for electronic transitions. If the dipole moment is zero for a transition, the oscillator strength of that transition is zero, and the band has zero intensity.

The best way of judging whether the transition dipole moment is zero is to use group theory. We saw in Section 15.8 that the transition dipole moment must be zero unless the integral is totally symmetric under the operations of the molecular point group. That section should be reviewed now, particularly Example 15.10. We also saw in that example that the nonvanishing component of the transition dipole tells us the polarization of the light emitted or absorbed in the transition.

17.2 The vibrational structure

The width of electronic absorption bands in liquid samples can be traced to their unresolved vibrational structure. This structure, which can be resolved in gases and weakly interacting solvents, arises from the vibrational transitions that accompany electronic excitation.

The Franck–Condon principle

The details of the appearance of the vibrational structure of a band is explained by the **Franck–Condon principle**:

> Because the nuclei are so much more massive than the electrons, an electronic transition takes place faster than the nuclei can respond.

As a result of the transition, electron density is rapidly built up in new regions of the molecule and removed from others, and the initially stationary nuclei suddenly experience a new force field. They respond to the new force by beginning to vibrate, and swing backwards and forwards from their original separation (which was maintained during the rapid electronic excitation). The stationary equilibrium separation of the nuclei in the initial electronic state therefore becomes a stationary **turning point** (the point of a vibration when the nuclei are at an end point of their swing) in the final electronic state (Fig. 17.2).

The quantum mechanical version of the Franck–Condon principle refines this picture to the point of letting us calculate the intensities of the transitions to different vibrational levels of the electronically excited molecule. Before the absorption, the molecule is in the ground vibrational state of its ground electronic state (Fig. 17.3). The form of the vibrational wavefunction shows that the most probable location of the nuclei is at their equilibrium separation R_e. Consequently the electronic transition is most likely to take place when the nuclei have this separation. When the transition occurs, the molecule is excited to the state represented by the upper curve. According to the Franck–Condon principle, the nuclear framework remains constant during this excitation, and so we may imagine the transition as being up the vertical lines in Fig. 17.2 and Fig. 17.3. The vertical line is the origin of the expression **vertical transition**, which is used to denote an electronic transition that occurs without change of nuclear geometry.

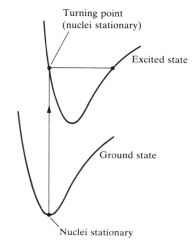

Fig. 17.2 According to the Franck–Condon principle, the most intense transition is from the ground vibrational state to the vibrational state lying vertically above it. Transitions to other vibrational levels also occur, but with lower intensity.

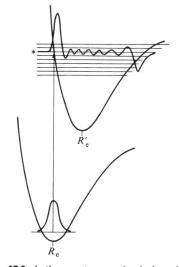

Fig. 17.3 In the quantum mechanical version of the Franck–Condon principle, the molecule undergoes a transition to the upper vibrational state that most closely resembles the vibrational wavefunction of the vibrational ground state of the lower electronic state. The two wavefunctions shown here have the greatest overlap integral of all the vibrational states of the upper electronic state and hence are most closely similar.

505

The vertical transition cuts through several vibrational levels of the upper electronic state. The level marked * is the one in which the nuclei are most probably at the same initial separation R_e (because the vibrational wavefunction has maximum amplitude there), and so this is the most probable level for the termination of the transition. However, it is not the only final vibrational level because several nearby levels have an appreciable probability of the nuclei being at the separation R_e. Therefore, transitions occur to all the vibrational levels in this region, but most intensely to the level with a vibrational wavefunction that peaks most strongly near R_e.

The vibrational structure of the spectrum depends on the relative displacement of the two potential energy curves, and a long **progression** of vibrations (a lot of vibrational structure) is stimulated if the two states are appreciably displaced. The upper curve is usually displaced to longer equilibrium bond lengths because excited states usually have more anti-bonding character than ground states.

The separation of the vibrational lines of an electronic absorption spectrum depends on the vibrational energies of the upper electronic state. Hence electronic absorption spectra may be used to assess the force fields and dissociation energies of electronically excited molecules (e.g. via a Birge–Sponer plot, Section 16.8).

Franck–Condon factors

The quantitative form of the Franck–Condon principle is derived from the expression for the transition dipole moment. The wavefunctions of the initial and final states are the products of the respective electronic and vibrational wavefunctions. That is, the total wavefunction for an electronic state ψ_ε and vibrational state ψ_v is $\psi_\varepsilon(r)\psi_v(R)$, where r stands for the electronic coordinates and R for the nuclear coordinates. The transition dipole moment for the excitation $\varepsilon', v' \leftarrow \varepsilon, v$ is therefore approximately

$$\boldsymbol{\mu} = -e \int \psi_{\varepsilon'}^*(r)\psi_{v'}^*(R) r \psi_\varepsilon(r)\psi_v(R) \, d\tau_{\text{elec}} \, d\tau_{\text{nuc}}$$

$$= -e \int \psi_{\varepsilon'}^*(r) r \psi_\varepsilon(r) \, d\tau_{\text{elec}} \int \psi_{v'}^*(R)\psi_v(R) \, d\tau_{\text{nuc}}$$

The second factor is the one to concentrate on:

$$S_{v',v} = \int \psi_{v'}^*(R)\psi_v(R) \, d\tau_{\text{nuc}} \tag{4}$$

We should recognize it as an overlap integral between the initial and final vibrational state wavefunctions.

Since the transition intensity from a vibrational level v of the ground electronic state to a vibrational level v' of the excited electronic state is proportional to the square of the transition dipole moment, that intensity is proportional to $S_{v',v}^2$. This quantity is known as the **Franck–Condon factor** for the $\varepsilon', v' \leftarrow \varepsilon, v$ transition. It follows that the greater the vibrational overlap the greater the absorption intensity of that particular simultaneous electronic and vibrational transition.

Example 17.4: *Calculating a Franck–Condon factor*

Consider the transition from one electronic state to another, their bond lengths being R_e and R'_e and their force constants equal. Calculate the Franck–Condon factor for the 0–0 transition and show that the vibronic transition is most intense when the bond lengths are equal.

Answer. We need to calculate $S_{0,0}$, the overlap integral of the two ground state harmonic oscillator wavefunctions (Table 12.1). We therefore use the wavefunctions

$$\psi_0 = \left(\frac{1}{\alpha\pi^{1/2}}\right)^{1/2} e^{-\frac{1}{2}y^2} \qquad \psi'_0 = \left(\frac{1}{\alpha\pi^{1/2}}\right)^{1/2} e^{-\frac{1}{2}y'^2}$$

where $y = (R - R_e)/\alpha$ and $y' = (R - R'_e)/\alpha$, with $\alpha^2 = \hbar(mk)^{1/2}$ (Section 12.2). The overlap integral is

$$S_{0,0} = \int_{-\infty}^{\infty} \psi'_0(R)\psi_0(R)\, dR = \left(\frac{1}{\alpha\pi^{1/2}}\right)\int_{-\infty}^{\infty} e^{-\frac{1}{2}(y^2 + y'^2)}\, dR$$

It is not difficult to manipulate this expression into one proportional to an integral over e^{-R^2}, the value of which is $\pi^{1/2}$. Therefore, the overlap integral is

$$S_{0,0} = e^{-(R_e - R'_e)^2/4\alpha}$$

and the intensity is proportional to $S_{0,0}^2$. This is unity when $R_e = R'_e$ and decreases as the equilibrium bond lengths diverge.

Comment. For Br_2, $R_e = 228$ pm and there is an upper state with $R'_e = 266$ pm. Taking the vibrational wavenumber as $250\ cm^{-1}$, gives $S_{0,0}^2 = 5 \times 10^{-10}$, and so the intensity of the 0–0 transition is only 5×10^{-10} of what it would have been if the potential curves had been directly above each other.

Exercise. Suppose the vibrational wavefunctions can be approximated by rectangular functions of width W and W', centred on the equilibrium bond lengths. Find the corresponding Franck–Condon factors when the centres are coincident and $W' < W$.

[W'^2/W^2]

17.3 Specific types of transitions

The absorption of a photon can often be traced to the excitation of specific electrons or electrons that belong to a small group of atoms. For example, when a carbonyl group is present, an absorption at about 290 nm is normally observed, although its precise location depends on the nature of the rest of the molecule. Groups with characteristic optical absorptions are called **chromophores** (from the Greek for 'colour bringer'), and their presence often accounts for the colours of substances.

d–d transitions

All the *d* orbitals of a given shell are degenerate in a free atom. In a *d*-metal complex, where the immediate environment of the atom is no longer spherical, the *d* orbitals are no longer all degenerate, and electrons can absorb energy by making transitions between them. In an octahedral complex, such as $[Ti(OH_2)_6]^{3+}$, the five *d* orbitals of the central atom are split into two sets (Fig. 17.4), a triply-degenerate set labelled t_{2g} and a doubly-degenerate set labelled e_g. The three t_{2g} orbitals lie below the two e_g orbitals, and the separation in energy is denoted Δ_O. The *d* orbitals also

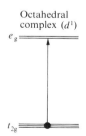

Fig. 17.4 In an octahedral complex the *d* orbitals are split into two groups by their interactions with the ligands. In a *d–d* transition of a d^1 complex, a t_{2g} electron is excited into an e_g orbital.

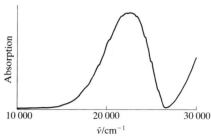

Fig. 17.5 The electronic absorption spectrum of $[Ti(OH_2)_6]^{3+}$ in aqueous solution. The transition responsible is the $e_g \leftarrow t_{2g}$ transition illustrated in Fig. 17.4.

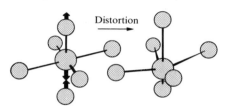

Fig. 17.6 A d–d transition is parity-forbidden because it corresponds to a g–g transition. However, a vibration of the molecule can destroy the inversion symmetry of the molecule so that the g,u classification no longer applies. This gives rise to a vibronically allowed transition.

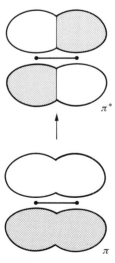

Fig. 17.7. The C=C double bond acts as a chromophore. One of its important transitions is the $\pi^* \leftarrow \pi$ transition illustrated here, in which an electron is promoted from a π orbital to the corresponding antibonding orbital.

divide into two sets in a tetrahedral complex, but in this case the e orbitals lie below the t_2 orbitals and their separation is written Δ_T. Neither separation is large, so transitions between the two sets of orbitals typically occur in the visible region of the spectrum. The transitions are responsible for many of the colours that are so characteristic of d-metal complexes. As an example, the spectrum of $[Ti(OH_2)_6]^{3+}$ near $20\,000\ cm^{-1}$ ($500\ nm$) is shown in Fig. 17.5, and can be ascribed to the promotion of its single d electron from a t_{2g} orbital to an e_g orbital. The wavenumber of the absorption maximum suggests that $\Delta_O \approx 20\,000\ cm^{-1}$ for this complex, which corresponds to about 2.5 eV.

Vibronic transitions

A major remaining problem is that d–d transitions are forbidden in octahedral complexes, for the **Laporte selection rule** for centrosymmetric complexes (those with a centre of inversion) and atoms states that:

> The only allowed transitions are those accompanied by a change of parity.

That is, u $\leftrightarrow$ g and g $\leftrightarrow$ u transitions are allowed, but g $\leftrightarrow$ g and u $\leftrightarrow$ u transitions are forbidden.[3]

A forbidden g $\leftrightarrow$ g transition can become allowed if the centre of symmetry is eliminated by an asymmetrical vibration, such as the one shown in Fig. 17.6. In a noncentrosymmetric molecule, d–d transitions are not parity-forbidden, and so the $e_g \leftarrow t_{2g}$ transition becomes weakly allowed (with $f \approx 10^{-4}$) if the complex vibrates asymmetrically. An electronic transition that derives its intensity from a vibration of a molecule is called a **vibronic transition**.

Charge-transfer transitions

A complex may absorb light as a result of the transfer of an electron from the ligands into the d orbitals of the central atom, or vice versa. In such **charge-transfer transitions** the electron moves through a considerable distance, which means that the transition dipole moment may be large and, because they are not parity-forbidden, the absorption is correspondingly intense. This mode of chromophore activity is shown by the permanganate ion, MnO_4^-, and accounts for its intense purple colour ($\lambda_{max} = 530\ nm$, $f \approx 0.03$).

π^*,π and π^*,n transitions

In the case of the C=C double bond, absorption lifts a π electron into an antibonding π^* orbital (Fig. 17.7). The chromophore activity is therefore due to a $\pi^* \leftarrow \pi$ **transition** (which is normally read 'π to π-star transition'). Its energy is around 7 eV for an unconjugated double bond, which

---

[3] The group theoretical argument is as follows. The transition dipole moment in eqn 3b vanishes unless the integral is a basis for the totally symmetric symmetry species of the group. This means that in a centrosymmetric complex it must have g parity (in O_h it needs to be A_{1g}, but the g symmetry is important for this argument). The three components of the dipole moment operator transform like x, y, and z and are all u. Therefore, for a d–d transition (a g–g transition), the overall parity of the transition dipole is g × u × g = u, and so it must be zero. Likewise, for a u–u transition, the overall parity is u × u × u = u, and it must also vanish. Hence, transitions without a change of parity are forbidden.

corresponds to an absorption at 180 nm (in the ultraviolet). When the double bond is part of a conjugated chain, the energies of the molecular orbitals lie closer together and the $\pi^* \leftarrow \pi$ transition shifts into the visible. We shall see an important example of such a transition in a moment.

The transition responsible for absorption in carbonyl compounds can be traced to the lone pairs of electrons on the O atom. One of these electrons may be excited into an empty π^* orbital of the carbonyl group (Fig. 17.8), which gives rise to a $\pi^* \leftarrow n$ **transition** (an 'n to π-star transition'). Typical absorption energies are about 4 eV (290 nm). Because $\pi^* \leftarrow n$ transitions in carbonyls are symmetry forbidden, the absorptions are weak (f around 2×10^{-4} to 6×10^{-4}).

An important example of $\pi^* \leftarrow \pi$ and $\pi^* \leftarrow n$ transitions is provided by the photochemical mechanism of vision. The retina of the eye contains 'visual purple', which is a protein in combination with 11-*cis*-retinal (**1**). The 11-*cis*-retinal acts as a chromophore, and is the primary receptor for photons entering the eye. A solution of 11-*cis*-retinal absorbs at about 380 nm, but in combination with the protein (a link which might involve the elimination of the terminal carbonyl) the absorption maximum shifts to about 500 nm and tails into the blue. The conjugated double bonds are responsible for the ability of the molecule to absorb over the entire visible region, but they also play another important role. In its electronically excited state the conjugated chain can isomerize, one half of the chain being able to twist about an excited C=C bond and forming all-*trans*-retinal (**2**). On account of its different shape, the new isomer cannot fit into the protein. The primary step in vision therefore appears to be photon absorption followed by isomerization: the uncoiling of the molecule then triggers a nerve impulse to the brain.

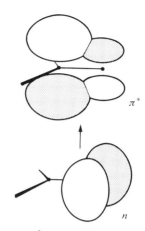

Fig. 17.8 The $>$C=O group acts as a chromophore primarily on account of the excitation of a non-bonding O lone-pair electron to an antibonding C—O π^* orbital.

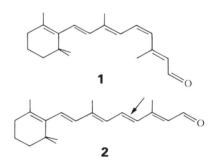

The fates of electronically excited states

The energy of an electronically excited state may be lost in a variety of ways. A **radiative decay** process is one in which a molecule discards its excitation energy as a photon. A more common fate is **nonradiative decay**, in which the excess energy is transferred into the vibration, rotation, and translation of the surrounding molecules. This **thermal degradation** converts the excitation energy into thermal motion of the environment (i.e. to 'heat'). An excited molecule may also take part in a chemical reaction, as we shall discuss in Part 3.

17.4 Fluorescence and phosphorescence

In this section we consider radiative decay by **spontaneous emission**, transitions that are independent of the number of photons present. In **fluorescence**, the spontaneously emitted radiation ceases immediately the exciting radiation is extinguished. In **phosphorescence**, the spontaneous emission may persist for long periods (even hours, but characteristically seconds or fractions of seconds). The difference suggests that fluorescence is an immediate conversion of absorbed light into re-emitted energy and that phosphorescence involves the storage of energy in a reservoir from which it slowly leaks.

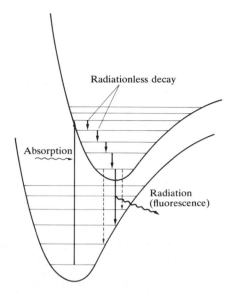

Fig. 17.9 The sequence of steps leading to fluorescence. After the initial absorption the upper vibrational states undergo radiationless decay by giving up energy to the surroundings. A radiative transition then occurs from the ground state of the upper electronic state. (In practice the separation of the curves is 10 to 100 times greater than the separation of the vibrational levels.)

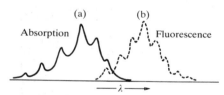

Fig. 17.10 The absorption spectrum (a) shows a vibrational structure characteristic of the upper state. The fluorescence spectrum (b) shows a structure characteristic of the lower state; it is also displaced to lower frequencies (but the 0–0 transitions are coincident) and resembles a mirror image of the absorption.

Fluorescence

Figure 17.9 shows the sequence of steps involved in fluorescence. The initial absorption takes the molecule to an excited electronic state, and if the absorption spectrum were monitored it would look like the one shown in Fig. 17.10a. The excited molecule is subjected to collisions with the surrounding molecules, and as it gives up energy it steps down the ladder of vibrational levels. The surrounding molecules, however, might be unable to accept the larger energy difference needed to lower the molecule to the ground electronic state. It might therefore survive long enough to undergo spontaneous emission, and emit the remaining excess energy as radiation. The downward electronic transition is vertical (in accord with the Franck–Condon principle) and the **fluorescence spectrum** (Fig. 17.10b) has a vibrational structure characteristic of the lower electronic state. The 0–0 absorption and downward fluorescence transitions are not always exactly coincident because the solvent may interact differently with the solute in the ground and excited states. Since the solvent molecules do not have time to rearrange during the transition, the absorption occurs in an environment characteristic of the solvated ground state but the fluorescence occurs in an environment characteristic of the solvated excited state.

Fluorescence occurs at a lower frequency than the incident because the radiation is emitted after some vibrational energy has been discarded into the surroundings. The vivid oranges and greens of fluorescent dyes are an everyday manifestation of this effect: they absorb in the ultraviolet and blue, and fluoresce in the visible. The mechanism also suggests that the intensity of the fluorescence ought to depend on the ability of the solvent molecules to accept the electronic and vibrational quanta. It is indeed found that a solvent composed of molecules with widely spaced vibrational levels (such as water) may be able to accept the large quantum of electronic energy and so extinguish, or 'quench', the fluorescence.

Phosphorescence

Figure 17.11 shows the sequence of events leading to phosphorescence. The first steps are the same as in fluorescence, but the presence of a triplet excited state plays a decisive role. (We first encountered triplet states in Section 13.6: they are states in which two electrons have parallel spins.)

The singlet and triplet excited states share a common geometry at the point where their potential energy curves intersect. Hence, if there is a mechanism for unpairing two electron spins, the molecule may undergo **intersystem crossing** and become a triplet state. We saw in the discussion of atomic spectra (Section 13.6) that singlet–triplet transitions may occur in the presence of spin–orbit coupling, and the same is true in molecules. We can expect intersystem crossing to be important when a molecule contains a moderately heavy atom (such as S), because then the spin–orbit coupling is large.

If an excited molecule crosses into a triplet state, it continues to deposit energy into the surroundings and to step down the vibrational ladder. However, it is now stepping down the triplet's ladder, and at the lowest vibrational energy level it is trapped. The solvent cannot extract the final, large quantum of electronic excitation energy, and the molecule cannot radiate its energy because return to the ground state is spin-forbidden. The radiative transition, however, is not totally forbidden because the spin–orbit coupling that was responsible for the intersystem crossing also breaks the selection rule. The molecules are therefore able to emit weakly, and the emission may continue long after the original excited state was formed.

The mechanism accounts for the observation that the excitation energy seems to get trapped in a slowly leaking reservoir. It also suggests (as is confirmed experimentally) that phosphorescence should be most intense from solid samples: energy transfer is then less efficient and the intersystem crossing has time to occur as the singlet excited state steps slowly past the intersection point. The mechanism also suggests that the phosphorescence efficiency should depend on the presence of a moderately heavy atom (with strong spin–orbit coupling), which is in fact the case. The confirmation of the mechanism is the experimental observation (using the sensitive resonance techniques we describe in the next chapter) that the sample is paramagnetic while the reservoir state, with its unpaired electron spins, is populated.

Internal conversion, intersystem crossing, and the various types of radiative transitions that can occur in molecules are often represented on a **Jablonski diagram** of the type shown in Fig. 17.12.

17.5 Dissociation and predissociation

Another fate for an electronically excited molecule is dissociation (Fig. 17.13). The onset of dissociation can be detected in an absorption spectrum by seeing that the vibrational structure of a band terminates at a certain energy. Absorption occurs in a continuous band above this

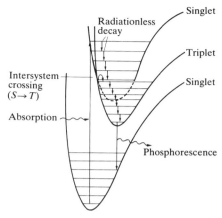

Fig. 17.11 The sequence of steps leading to phosphorescence. The important step is the intersystem crossing, the switch from singlet to triplet state brought about by spin–orbit coupling. The triplet state acts as a slowly radiating reservoir because the return to the ground state is spin-forbidden.

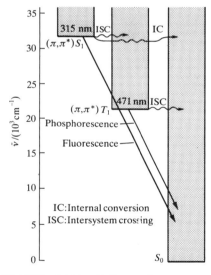

Fig. 17.12 A Jablonski diagram (for naphthalene) is a simplified portrayal of the relative positions of the electronic energy levels of a molecule. Vibrational levels of states of a given electronic state lie above each other, but the relative horizontal locations of the columns bear no relation to the nuclear separations in the states. The ground vibrational states of each electronic state are correctly located vertically but the other vibrational states are shown only schematically.

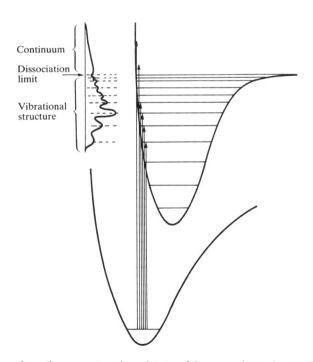

Fig. 17.13 When absorption occurs to unbound states of the upper electronic state, the molecule dissociates and the absorption is a continuum. Below the dissociation limit the electronic spectrum shows a normal vibrational structure.

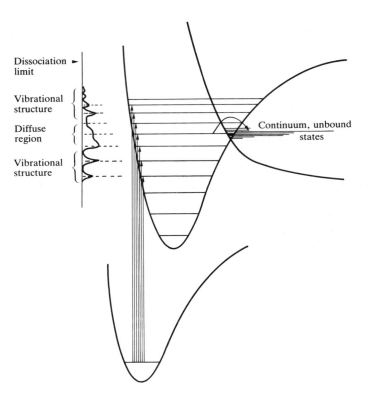

Fig. 17.14 When a dissociative state crosses a bound state, as in the upper part of the illustration, molecules excited to levels near the crossing may dissociate. This leads to predissociation, and is detected in the spectrum as a loss of vibrational structure that resumes at higher frequencies.

dissociation limit because the final state is an unquantized translational motion of the fragments. Locating the dissociation limit is a valuable way of determining the bond dissociation energy.

In some cases, the vibrational structure disappears but resumes at higher photon energies. This **predissociation** can be interpreted in terms of the potential energy curves shown in Fig. 17.14. When a molecule is excited to a vibrational level, its electrons may undergo a reorganization that results in it undergoing an **internal conversion**, a conversion to another state of the same multiplicity. An internal conversion occurs most readily at the point of intersection of the two molecular potential energy curves, because there the nuclear geometries of the two states are the same. The state into which the molecule converts may be dissociative, so the states near the intersection have a finite lifetime and hence are imprecisely defined. As a result, the absorption spectrum is blurred in the vicinity of the intersection. When the incoming photon brings enough energy to excite the molecule to a vibrational level high above the intersection, the internal conversion does not occur (the nuclei are unlikely to have the same geometry). Consequently, the levels resume their well-defined, vibrational character with correspondingly well-defined energies and the line structure resumes on the high-frequency side of the blurred region.

Lasers

Lasers have transformed chemistry as much as they have the everyday world. In this section, we see some of the principles of their operation, and then explore their applications in chemistry. Lasers lie very much on the frontier of physics and chemistry, since their operation depends on details of optics and, in some cases, of solid-state processes. We shall concentrate on the more chemical aspects of their operation, particularly the materials from which they are made and the events taking place within them, and largely ignore the engineering aspects of their design. Similarly, when we discuss their applications, we shall limit ourselves to purely chemical applications, such as spectroscopy and photochemistry.

17.6 General principles of laser action

The word **laser** is an acronym formed from **l**ight **a**mplification by **s**timulated **e**mission of **r**adiation. As this name suggests, it is a process that depends on stimulated emission as distinct from the spontaneous emission processes characteristic of fluorescence and phosphorescence. In stimulated emission (Section 16.2) an excited state is stimulated to emit a photon by the presence of radiation of the same frequency, and the more photons there are present, the greater the probability of the emission. The essential feature of laser action is positive-feedback and the strong **gain**, or growth of intensity, that results: the more photons present of the appropriate frequency, the more photons of that frequency will be stimulated to form.

The theory of stimulated emission and stimulated absorption described in Section 16.2 shows that, for a given intensity of illumination, the probability of any individual molecule undergoing a transition between two states is exactly the same in emission as it is in absorption. Therefore, if there is a greater number of molecules in the lower-energy state, there will be a net absorption of the incident radiation. However, if there are more molecules in the upper state, illumination of the sample will result in a net *emission* of radiation, and the intensity of the incident light will be enhanced (Fig. 17.15). This enhancement is the light amplification by stimulated emission of radiation that a laser achieves.

Population inversion

One requirement of laser action is the existence of a **metastable** excited state, one with a long enough lifetime for it to participate in stimulated emission. Another requirement is the existence of a greater population in the metastable state than in the lower state where the transition terminates. Since at thermal equilibrium the opposite is true (by the Boltzmann distribution), we see that it is necessary to achieve a **population inversion** in which there are more molecules in the upper state than in the lower.

One way of achieving population inversion is illustrated in Fig. 17.16. The inversion is achieved indirectly through an intermediate state I. Thus, the molecule is excited to I, which then gives up some of its energy non-radiatively and changes into a lower state A; the laser transition is the return of A to the ground state X. Since three levels are involved overall, this arrangement leads to a **three-level laser**. In practice, I is a range of states, all of which can convert to the upper of the two laser states A. The $I \leftarrow X$ transition is stimulated with an intense flash of light in the process called **pumping**. In some cases the pumping flash is achieved with an electric discharge through xenon or with the light of another laser. The conversion of I to A should be rapid, and the laser transitions from A to X should be relatively slow.

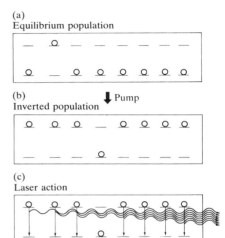

Fig. 17.15 A schematic illustration of the steps leading to laser action. (a) The Boltzmann population of states, with more atoms in the ground state. (b) When the initial state absorbs, the populations are inverted (the atoms are pumped to the excited state). (c) A cascade of radiation then occurs, as one emitted photon stimulates another atom to emit, and so on. The radiation is coherent (phases in step).

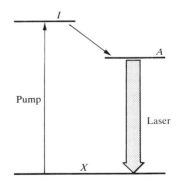

Fig. 17.16 The transitions involved in a three-level laser. The pumping pulse populates the intermediate state I, which in turn populates the laser state A. The laser transition is the stimulated emission $A \rightarrow X$.

513

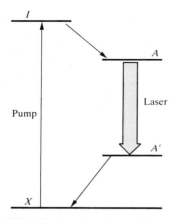

Fig. 17.17. The transitions involved in a four-level laser. Since the laser transition terminates in an excited state (A'), the population inversion between A and A' is much easier to achieve than in a three-level laser.

The disadvantage of the three-level arrangement is that it is difficult to achieve the population inversion, since so many ground-state molecules must be converted to the excited state by the pumping action. The **four-level laser** simplifies this task by having the laser transition terminate in a state A' other than the ground state (Fig. 17.17). Since A' is unpopulated initially, *any* population in A corresponds to a population inversion, and we can expect laser action if A is sufficiently metastable. Moreover, this population inversion can be maintained if the $A' \rightarrow X$ transitions are rapid, for these will deplete any population in A' that stems from the laser transition and keep that state relatively empty.

Cavity and mode characteristics

The laser medium is confined to a cavity that ensures that only certain types of photon are generated abundantly. (By type of photon we mean frequency, direction of travel, and state of polarization.) The cavity is essentially a region between two mirrors, which reflect the light back and forth. The only wavelengths that can be sustained by the cavity satisfy

$$N \times \frac{\lambda}{2} = L \tag{5a}$$

where N is an integer and L is the length of the cavity. That is, only an integral number of half-wavelengths fit into the cavity, all other waves undergoing destructive interference with themselves. The frequency difference between neighbouring values of N is

$$\Delta v = \frac{c}{2L} \tag{5b}$$

(We are ignoring the refractive index of the medium.) In addition, not all wavelengths that can be sustained by the cavity are amplified by the laser medium (many fall outside the range of frequencies of the laser transitions), so only a few contribute to the laser radiation. These are the **longitudinal modes** of the laser.

Photons with the correct wavelength for the resonant modes of the cavity and the correct frequency to stimulate the laser transition, are highly amplified. One photon might be generated spontaneously, and travel through the medium. It stimulates the emission of another photon, which in turn stimulates more (Fig. 17.15). The cascade of energy builds up rapidly, and soon the cavity is an intense reservoir of radiation at all the longitudinal modes that it can sustain. Some of this light can be withdrawn if one of the mirrors is half-reflecting.

The resonant modes of the cavity have various natural characteristics, and to some extent may be selected. In the first place, only photons that are travelling strictly parallel to the axis of the cavity undergo more than a couple of reflections, and so only they are amplified, all others simply vanishing into the surroundings. Hence, laser light is generally highly **collimated**, or forming a non-divergent beam. It may also be **polarized**, with its electric vector in a particular plane (or in some other state of polarization), by including a polarizing filter into the cavity. Polarization is also achieved in certain lasers by the shape of the cell containing the laser

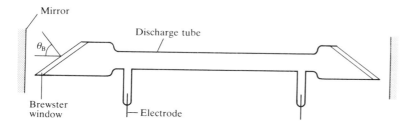

Fig. 17.18 The Brewster window at the end of a laser leads to polarized emission. If light strikes a transparent material of refractive index n, no light polarized parallel to the plane of incidence is reflected if the light strikes at an angle θ_B given by $\tan \theta_B = n$. For glass with $n = 1.5$, the Brewster angle is $\theta_B \approx 57°$. Some light with perpendicular polarization is reflected, so the gain is greater for the light of parallel polarization, which therefore dominates the output.

medium. Thus, to cut down reflections that result where a refractive index changes abruptly (as at the walls of a glass container), the walls are arranged to make a special angle, known as the 'Brewster angle', to the direction of propagation of the light (Fig. 17.18). The Brewster windows are more transparent to light of one polarization than the other. Consequently the gain is greater for that polarization, and the laser emits polarized light.

Although laser radiation is highly monochromatic, it is not perfectly so, since many longitudinal modes, each one with a different wavelength, are present in a resonant cavity. However, if it is desired to obtain strictly monochromatic radiation, a dispersing element, such as a diffraction grating, may be included in the path of the light, and the optics arranged so that only one particular wavelength becomes amplified.

Laser light is also **coherent** in the sense that the electromagnetic waves are all in step. We have to distinguish between two types of coherence. In **spatial coherence** we are concerned with the waves being in step across the cross-section of the beam emerging from the cavity. In **temporal coherence** we are concerned with the extent to which the waves remain in step along the beam. The latter is normally expressed in terms of a **coherence length** l_C, and is related to the range of wavelengths $\Delta\lambda$ present in the beam:

$$l_C = \frac{\lambda^2}{2\Delta\lambda} \qquad (6)$$

If the beam were perfectly monochromatic, with strictly one wavelength present, $\Delta\lambda$ would be zero, and the waves would remain in step for infinite distances. If many wavelengths are present, the waves get out of step in a short distance and the coherence length is small. A typical light bulb gives out light with a coherence length of only about 400 nm; a He–Ne laser with $\Delta\lambda$ about 2 pm has a coherence length of 10 cm or so.

Q-switching

A laser can generate light for as long as the population inversion is maintained. When heat is easily dissipated, the laser may act continuously, for the population of the upper level can be replenished by pumping. In some cases, practical considerations govern whether or not continuous pumping is feasible, as we shall see when we consider some particular lasers. In other cases, especially when overheating is a problem, the laser can be operated only in pulses, perhaps of microsecond or millisecond duration, so that the medium has a chance to cool or the lower state discard its population.

It is sometimes desirable to have pulses of radiation rather than a continuous output, with a lot of power concentrated into a brief pulse. One way of achieving pulses is by **Q-switching**, the modification of the resonance characteristics of the laser cavity. (The name comes from the 'Q-factor' used as a measure of the quality of a resonance cavity in microwave engineering.)

Example 17.5: *Relating the power and energy of a laser*

A laser rated at 0.10 J can generate radiation in 3.0 ns pulses. What is the power output per pulse?

Answer. The power output P is the energy per unit time, and is expressed in watts (1 W = 1 J s^{-1}). Therefore, from the data,

$$P = \frac{0.10\,\text{J}}{3.0 \times 10^{-9}\,\text{s}} = 3.3 \times 10^{7}\,\text{J s}^{-1}$$

That is, the pulses deliver 33 MW of power.

Comment. If the power consumption of the laser is 100 W, the 0.10 J required for each pulse can be supplied in 0.10 ms assuming 100 per cent conversion efficiency.

Exercise. Calculate the power output of a laser in which a 2.0 J pulse can be delivered in 1.0 ns.

[2.0 GW]

The aim of Q-switching is to achieve a healthy population inversion in the absence of the resonant cavity, then to plunge the population-inverted medium into a cavity, and hence to obtain a sudden pulse of radiation. The switching may be achieved by damaging the resonance characteristics of the cavity in some way while the pumping pulse is active, and then suddenly restoring them (Fig. 17.19). Q-switching can give pulses of about 10 ns duration.

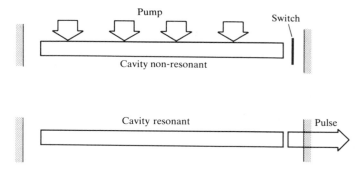

Fig. 17.19 The principle of Q-switching. The excited state is populated while the cavity is nonresonant. Then the resonance characteristics are suddenly restored, and the stimulated emission emerges in a giant pulse.

There are several ways of Q-switching a laser. The earliest technique was to rotate one of the cavity mirrors so that it periodically came back into correct alignment with its partner. A faster method is to use a **Pockels cell**, which is an electro-optical device based on the ability of crystals of ammonium dihydrogenphosphate to convert plane-polarized light to circularly polarized light when a potential difference is applied. If a Pockels cell is made part of a laser cavity, its action and the change of polarization that occurs when light is reflected from a mirror converts light polarized in one plane into reflected light polarized in the perpendicular plane (Fig. 17.20). As a result, the reflected light does not stimulate more emission. However, if the cell is suddenly turned off, the polarization effect is extinguished and all the energy stored in the cavity can emerge as an intense pulse of stimulated radiation. An alternative technique is to use a **saturable dye**, a dye that loses its power to absorb when many of its molecules have been excited by intense radiation. It then suddenly becomes transparent, and the cavity becomes resonant.

Mode locking

The technique of **mode locking** can produce pulses of picosecond duration and less. We have seen that a laser radiates at a number of different frequencies, depending on the precise details of the resonance characteristics of the cavity and in particular on the number of half-wavelengths of radiation that can be trapped between the mirrors (the cavity modes). The longitudinal modes differ in frequency by multiples of $c/2L$ (eqn 5b). Normally, they have random phases relative to each other. However, it is possible to lock their phases together so that they interfere with each other. Then constructive interference occurs at a series of sharp peaks separated by regions of destructive interference, and the power of the laser is obtained in picosecond bursts (Fig. 17.21). The sharpness of the peaks depends on the range of modes superimposed, and the wider the range, the narrower the pulses.

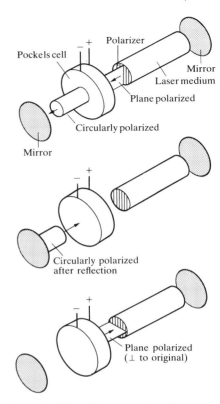

Fig. 17.20 When light passes through a Pockels cell which is 'on', its plane of polarization is rotated and so the laser cavity is non-resonant (its Q-factor, a measure of its resonant quality, is reduced). When the cell is turned off, no change of polarization occurs, and the cavity is resonant. The laser action then takes place in a single pulse. This is one method of Q-switching.

Example 17.6: *Demonstrating the effect of mode locking*

Show that if N modes with frequencies differing by $c/2L$ are superimposed, they give rise to a series of peaks separated by $2L/c$.

Answer. Each wave has the form

$$\psi_n = E_0 e^{2\pi i(\nu + nc/2L)t}$$

where ν is the lowest frequency and $n = 0, 1, \ldots, N-1$. The total amplitude of the wave is

$$\psi = \sum_n \psi_n = E_0 e^{2\pi i \nu t} \sum_0^{N-1} e^{i\pi nct/L}$$

The sum we require is a geometrical progression:

$$\sum_0^{N-1} e^{i\pi nct/L} = 1 + e^{i\pi ct/L} + e^{2i\pi ct/L} + \ldots + e^{(N-1)i\pi ct/L}$$

$$= \frac{\sin(N\pi ct/2L)}{\sin(\pi ct/2L)} \times e^{(N-1)i\pi ct/2L}$$

The intensity of the radiation is equal to the square modulus of the total amplitude, so

$$I = \psi^*\psi = E_0^2 \frac{\sin^2(N\pi ct/2L)}{\sin^2(\pi ct/2L)}$$

The function I is shown in Fig. 17.22. We see that it is a series of peaks with maxima separated by $t = 2L/c$, the round-trip transit time of the light in the cavity. The time between adjacent minima is $2t/N$.

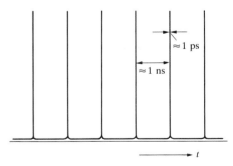

Fig. 17.21 The output of a mode-locked laser consists of a stream of very narrow pulses separated by an interval equal to the time it takes for light to make a round trip inside the cavity.

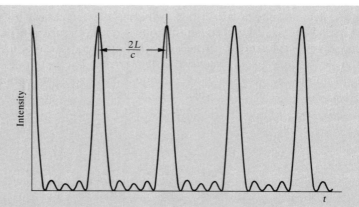

Fig. 17.22 The function derived above showing in more detail the structure of the pulses generated by a mode-locked laser.

Comment. In a laser with a cavity of length 30 cm, the peaks will be separated by 2 ns. If 10^3 modes contribute, the width of the pulses will be 4 ps.

Exercise. Show that the intensity is the sum of the individual mode intensities if there are random phase relations between the modes.

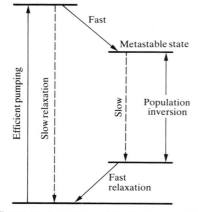

Fig. 17.23 A summary of the features needed for efficient laser action.

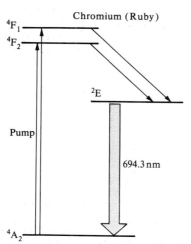

Fig. 17.24 The transitions involved in the ruby laser. The laser medium, ruby, consists of Al_2O_3 doped with Cr^{3+} ions.

Mode locking is obtained by varying the loss of the cavity periodically at the frequency $c/2L$. The modulation can be pictured as the opening of a shutter in time with the round-trip travel time of the photons in the cavity, so that only photons making the journey in that time are amplified. The modulation can be achieved by linking a prism in the cavity to a transducer driven by a radiofrequency source at a frequency $c/2L$. The transducer sets up standing-wave vibrations in the prism and modulates the loss it introduces into the cavity. Mode locking may also be accomplished passively, by including a saturable dye. This procedure makes use of the fact that the gain is very sensitive to amplification, and once a particular frequency begins to grow it can quickly dominate. If a saturable dye is included in the cavity, a spontaneous fluctuation in intensity—a bunching of photons—may result in its becoming transparent, and the bunch can pass through and travel to the far end of the cavity, amplifying as it goes. The dye immediately shuts down again (if it is well chosen), but opens when the intense pulse returns from the mirror at the far end and saturates it. In this way, that particular bunch of photons may grow to considerable intensity since it alone is stimulating emission in the cavity.

17.7 Practical lasers

Figure 17.23 summarizes the requirements for an efficient laser. In practice, the requirements can be satisfied using a variety of different systems, and in this section we review some that are commonly available.

Solid-state lasers

A **solid-state laser** is one in which the active medium is in the form of a single crystal or a glass. The first successful laser, the **ruby laser** built by Theodore Maiman in 1960, is an example (Fig. 17.24). Ruby is Al_2O_3 containing a small proportion of Cr^{3+} ions; the normal green of Cr^{3+} is modified to red by the distortion of the local crystal field stemming from the replacement of an Al^{3+} ion by a slightly larger Cr^{3+} ion. Ruby is a three-level laser, and the ground state, which is also the lower level of the laser transition, is $^4A_2$ with three unpaired spins on each Cr^{3+} ion. The population inversion results from pumping a majority of the Cr^{3+} ions into an excited state using an intense flash from another source, followed by a radiationless transition to another excited state. The pumping flash need not be monochromatic because the upper level actually consists of several states spanning a band of frequencies. The transition from the lowest of the excited states to the ground state ($^2E \rightarrow {}^4A_2$) is the laser transition, and gives rise to red 694 nm radiation. The population inversion is very difficult to sustain continuously, and in practice the ruby laser is pulsed. Typical pulses might consist of 2-J pulses persisting for 10 ns, corresponding to a power of 0.2 GW.

The **neodymium laser** is an example of a four-level laser (Fig. 17.25). In one form it consists of Nd^{3+} ions at low concentration in yttrium aluminium garnet (YAG, specifically $Y_3Al_5O_{12}$), and is then known as a Nd–YAG laser. A cheaper medium is glass, but glass is a poorer thermal conductor than YAG and then the laser must be pulsed. A neodymium laser operates at a number of wavelengths in the infrared, the band at 1064 nm being most common. The transition at 1064 nm is very efficient and the laser is capable of substantial power output. The power is great enough for **frequency doubling** to be used efficiently. Frequency doubling is a technique in which the laser beam is converted to radiation with a multiple of its initial frequency as it passes though a transparent material. A frequency-doubled Nd–YAG laser has a wavelength 532 nm, which corresponds to green light.

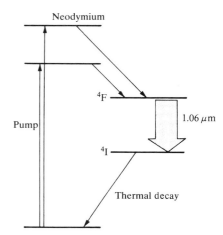

Fig. 17.25 The transitions involved in the neodymium laser. The laser action takes place between two excited states, and the population inversion is easier to achieve.

Example 17.7: *Accounting for multiphoton phenomena*

Show that if a substance responds nonlinearly to incident radiation of frequency ω, it may act as the source of radiation of twice the incident frequency.

Answer. The incident electric field E induces an electric dipole μ, and allowing for nonlinear response, we can write

$$\mu = \alpha E + \beta E^2 + \dots$$

The nonlinear terms can be expanded as follows if we suppose that the incident electric field is $E_0 \cos \omega t$:

$$\beta E^2 = \beta E_0^2 \cos^2 \omega t = \tfrac{1}{2}\beta E_0^2(1 + \cos 2\omega t)$$

Hence, the nonlinear term contributes an induced electric dipole that oscillates at the frequency 2ω and which can act as a source of radiation of that frequency.

Exercise. Show that if a substance responds nonlinearly to two sources of radiation, one of frequency ω_1 and the other of frequency ω_2, then it may give rise to radiation of the sum and difference of the two frequencies.
$$[\beta E^2 \propto \cos(\omega_1 + \omega_2)t + \cos(\omega_1 - \omega_2)t]$$

Gas lasers

Gas lasers are widely used, and since they can be cooled by a rapid flow of the gas through the cavity they can be used to generate high powers. The pumping is normally achieved using a gas that is different from the gas responsible for the laser emission itself.

In the **helium–neon laser** (Fig. 17.26) the active medium is a mixture of

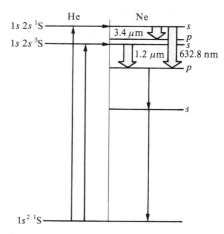

Fig. 17.26 The transitions involved in the helium–neon laser. The pumping (of the neon) depends on a coincidental matching of the helium and neon energy separations, so that excited He atoms can transfer their excess energy to Ne atoms during a collision.

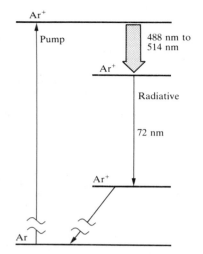

Fig. 17.27 The transitions involved in an argon-ion laser.

Fig. 17.28 The construction of an argon-ion laser. An electric discharge results in the formation of Ar^+ and Ar^{2+} ions in excited states and are neutralized by the electrodes inside the cavity. High currents are passed through the laser plasma, and in some designs, as shown here, it is held away from the walls by magnetic fields from electromagnets that surround the cavity.

helium and neon in a mole ratio of about $1:5$. The initial step is the excitation of a He atom to the metastable $1s^1 2s^1$ configuration using an electric discharge (the collisions of electrons and ions cause transitions that are not restricted by electric dipole selection rules). The excitation energy of this transition happens to match an excitation energy of neon, and during an He–Ne collision efficient transfer of energy may occur, leading to the production of highly excited, metastable Ne atoms with unpopulated intermediate states. Laser action generating 633 nm radiation (among about 100 other lines) then occurs.

The **argon-ion laser** (Fig. 17.27), one of a number of 'ion lasers', consists of argon at about 1 Torr, through which is passed an electric discharge. The discharge results in the formation of Ar^+ and Ar^{2+} ions in excited states, which undergo a laser transition to a lower state. These ions then revert to their ground states by emitting hard ultraviolet radiation (at 72 nm), and are then neutralized by a series of electrodes in the laser cavity (Fig. 17.28). One of the design problems is to find materials that can withstand this damaging residual radiation. There are many lines in the laser transition since the excited ions may make transitions to many lower states, but two

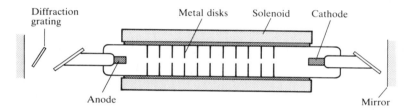

strong emissions from Ar^+ are at 488 nm (blue) and 514 nm (green): other transitions occur elsewhere in the visible region, in the infrared, and in the ultraviolet. The **krypton-ion laser** works similarly. It is less efficient, but gives a wider range of wavelengths, the most intense being at 647 nm (red), but it can also generate a yellow line. Both lasers are widely used in laser light shows (for this application Ar and Kr are often used simultaneously in the same cavity) as well as laboratory sources of high-power radiation.

The **carbon dioxide laser** (Fig. 17.29) works on a slightly different principle, since its radiation (between $9.2\,\mu m$ and $10.8\,\mu m$, with the strongest emission at $10.6\,\mu m$, in the infrared) arises from vibrational transitions. Most of the working gas is nitrogen, which becomes vibrationally excited by electronic and ionic collisions in an electric discharge. The vibrational levels happen to coincide with the ladder of antisymmetric stretching (ν_3) vibrational energy levels of CO_2, which pick up the energy during a collision. Laser action then occurs from the lowest excited level of ν_3 to the lowest excited level of the symmetric stretch (ν_1), which has remained unpopulated during the collisions. Some helium is included in the gas to help remove energy from this state and maintain the population inversion.

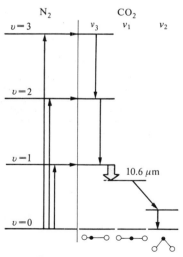

Fig. 17.29 The transitions involved in the carbon dioxide laser. The pumping also depends on the coincidental matching of energy separations; in this case the vibrationally excited N_2 molecules have excess energies that correspond to a vibrational excitation of the ν_3 (antisymmetric stretch) of CO_2. The laser transition is from $\nu_3 = 1$ to $\nu_1 = 1$.

In the **nitrogen laser**, the efficiency of the stimulated transition (at 337 nm, in the ultraviolet) is so great that a single passage of a pulse of radiation is enough to generate laser radiation and mirrors are unnecessary: such lasers are said to be **superradiant**.

Chemical and excimer lasers

Chemical reactions may also be used to generate molecules with nonequilibrium, inverted populations. For example, the photolysis of Cl_2 leads to the formation of Cl atoms which attack H_2 molecules in the mixture and produce HCl and H. The latter then attacks Cl_2 to produce vibrationally excited ('hot') HCl molecules. Since the newly formed HCl molecules have nonequilibrium vibrational populations, laser action can result as they return to lower states. Such processes are remarkable examples of the direct conversion of chemical energy into coherent electromagnetic radiation.

Chemical lasers are currently under consideration as sources of intense radiation for the Strategic Defense Initiative ('Star Wars'). The reaction that fuels the laser in this case is between H_2 and F atoms, which produces hot HF that can produce infrared radiation in the range 2.6 to 3.0 μm; the production of DF from D_2 and F_2 produces radiation of slightly longer wavelength, 3.6 to 4.0 μm. For deployment in space, the F atoms could be present as F_2, and the reaction is between H_2 and F_2 directly, the laser emission taking place just downstream of the mixing chamber (Fig. 17.30). An alternative source of F atoms is an electric discharge through SF_6. Although the population inversion in HF does not survive for very long, the ground-state molecules are swept rapidly out of the laser zone and replaced by excited molecules.

The population inversion needed for laser action is achieved in a more underhand way in **excimer lasers**, for in these (as we shall see) the lower state does not effectively exist. This odd situation is achieved by forming an **excimer**, a combination of two atoms that survives only in an excited state and which dissociates as soon as the excitation energy has been discarded. An example of an excimer laser is a mixture of xenon, chlorine, and neon (which acts as a buffer gas). An electric discharge through the mixture produces excited Cl atoms, which attach to the Xe atoms to give the excimer XeCl*. The excimer survives for about 10 ns, which is time for it to participate in laser action at 308 nm (in the ultraviolet). As soon as XeCl* has discarded a photon, the atoms separate because the molecular potential energy curve of the ground state is dissociative, and the ground state of the excimer cannot become populated (Fig. 17.31). The KrF* excimer laser is another example: it produces radiation at 249 nm.

Dye lasers

A solid state laser and a gas laser operate at discrete frequencies, and although the frequency required may be selected by suitable optics, the laser cannot be tuned continuously. The tuning problem is overcome by using a **dye laser**, which has such broad spectral characteristics (because the solvent broadens the vibrational structure of the transitions into bands, as we have seen). Hence, it is possible to scan the wavelength continuously (by rotating the diffraction grating in the cavity) and achieve laser action at any chosen wavelength. A commonly used dye is Rhodamine 6G in methanol (Fig. 17.32). As the gain is very high, only a short length of the optical path need be through the dye. The excited states of the active medium, the dye, are sustained by another laser, and the dye solution is flowed through the laser cavity (Fig. 17.33).

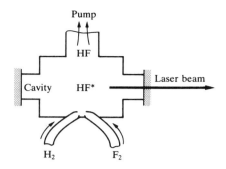

Fig. 17.30 The arrangement used to achieve laser action in a chemical hydrogen/fluorine laser. The emission is from vibrationally excited HF molecules produced in the reaction.

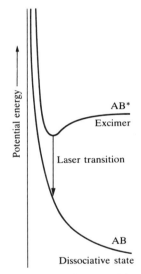

Fig. 17.31 The molecular potential energy curves for an excimer. The species can survive only as an excited state, since on discarding its energy it enters the lower, dissociative state. Since only the upper state can exist, there is never any population in the lower state.

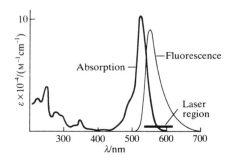

Fig. 17.32 The optical absorption spectrum of the dye Rhodamine 6G and the region used for laser action.

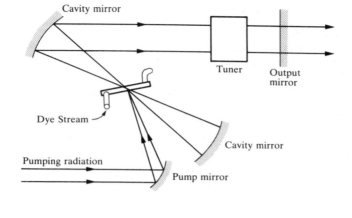

Fig. 17.33 The configuration used for laser dye action. The dye is flowed through the cell inside the laser cavity. The flow helps to keep it cool.

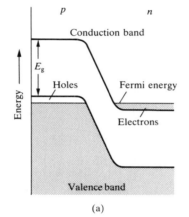

(a)

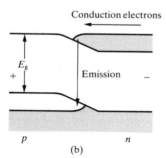

(b)

Fig. 17.34 The structure of a diode junction (a) without bias and (b) with bias.

Light-emitting diodes and semiconductor lasers

We have seen (in Section 14.11) that a semiconductor is classified as 'n-type' if its conduction band is partly populated and as 'p-type' if its valence band has a small number of holes. In this section we need to consider the properties of a **p–n junction**, the interface of the two types of semiconductor.

The band structure at the junction is shown in Fig. 17.34. When a 'forward bias' is applied to the junction, in the sense that electrons are supplied through an external circuit to the n side of the junction, the electrons in the conduction band of the n-type semiconductor fall into the holes in the valence band of the p-type semiconductor. As they fall, they emit energy. In silicon semiconductors this energy is largely in the form of heat because the transition can occur only if the electron transfers linear momentum to the lattice, and the device becomes warm. However, in some materials, most notably gallium arsenide, GaAs, the transition can occur without the lattice needing to be involved (because the electron has the same linear momentum in the ground state as in the upper state), and the energy is emitted as light. Practical **light-emitting diodes** of this kind are widely used in electronic displays. Gallium arsenide itself emits infrared light, but the band gap is widened by incorporating phosphorus, and a material of composition approximately $GaAs_{0.6}P_{0.4}$ emits light in the red region of the spectrum.

A light emitting diode is not a laser, since no resonance cavity and stimulated emission are involved. However, it is easy (in principle) to employ the light emission of electron–hole recombination as the basis of laser action. The population inversion can be sustained by sweeping away the electrons that fall into the holes of the p-type semiconductor, and a resonance cavity can be formed by using the high refractive index of the semiconducting material and cleaving single crystals so that the light is trapped by the sudden variation of refractive index. One widely used material is $Ga_{1-x}Al_xAs$, which produces infrared laser radiation and is widely used in compact-disc players.

17.8 Applications of lasers in chemistry

Laser radiation has five striking characteristics (Table 17.3). Each of them (sometimes in combination with the others) opens up interesting oppor-

Table 17.3. Characteristics of laser radiation and their chemical applications.

Characteristic	Advantages	Applications
High power density	Multiphoton processes	Nonlinear spectroscopy Saturation spectroscopy
	Low detector noise	Improved sensitivity
	High scattering intensity	Raman spectroscopy
Monochromatic	High resolution	Spectroscopy
	State selection	Isotope separation
		Photochemically precise
		State-to-state reaction dynamics
Collimated beam	Long path lengths	Sensitivity
	Forward-scattering observable	Nonlinear Raman spectroscopy
Coherent	Interference between separate beams	CARS
Pulsed	Precise timing of excitation	Fast reactions
		Relaxation
		Energy transfer

tunities in spectroscopy, giving rise to **laser spectroscopy**, and in photochemistry, giving rise to **laser photochemistry**.

Spectroscopy at high photon fluxes

The high spectral power density of a laser—the high intensity of the radiation it produces at well-defined frequencies—is an aid to conventional spectroscopy. Thus, it reduces the problem of detector noise and the interfering effects of background radiation. The high intensity is particularly advantageous in Raman spectroscopy, which until the introduction of lasers was plagued by the low intensity of the scattered radiation (which could be overcome only by using long exposures) and by interference from background scattering (which obscured the signal).

The high power of an incident laser beam also enhances the intensity of the radiation caused by fluorescence, since higher populations of the excited state may be achieved. Higher populations enhance the intensity of fluorescence and simplified the observation of fluorescence spectra. It therefore makes more substances open to study by that technique.

The large number of photons in an incident beam generated by a laser also gives rise to a qualitatively different branch of spectroscopy, since the photon density is so high that more than one photon may be absorbed by a single molecule and give rise to **multiphoton processes**. One application of multiphoton processes is that states inaccessible by conventional one-photon spectroscopy become observable, since the overall transition occurs with no change of parity. Thus, in one-photon spectroscopy, only g ↔ u transitions are observable; in two-photon spectroscopy, the overall outcome of absorbing two photons is a g ↔ g or a u ↔ u transition.

High powers and monochromatic beams make possible the technique of **saturation spectroscopy**, which permits the very precise location of absorption maxima. Saturation spectroscopy makes use of a technique very much like that involved in Lamb-dip spectroscopy (Section 16.3). As illustrated in Fig. 17.35, the output of a tunable laser is divided into an intense saturating beam and a less intense probe beam that pass through the

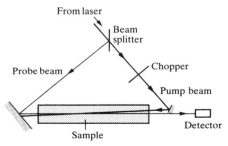

Fig. 17.35 The configuration of laser radiation used for saturation spectroscopy.

sample cavity in nearly opposite directions. The chopped saturating beam periodically excites molecules that are Doppler-shifted to its frequency. The probe beam gives a modulated signal at the detector, but only if it is interacting with the same Doppler-shifted molecules despite the fact that it is coming from an opposite direction. Since those molecules must be ones that are not moving parallel to the beams, the technique selects molecules that have essentially zero Doppler shift and hence gives very high resolution.

Collimated beams

The collimated beams generated by most kinds of lasers are also useful in spectroscopy for a number of reasons. One is that they permit the use of very long path lengths through samples, since the ray has only a small cross-section even after many reflections. A well-defined beam also implies that the detector can be designed to collect only the radiation that has passed through a sample, and can be screened much more effectively against stray scattered light. Moreover, with a collimated beam, the interaction zone in Raman spectroscopy is much more well-defined than in conventional spectroscopy, so the optics of the spectrometer can be optimized.

As with the high power output of lasers, the availability of a non-divergent beam makes possible a qualitatively different kind of spectroscopy. The beam is so well-defined that it is possible to observe Raman transitions very close to the direction of propagation of the incident beam (rather than perpendicular to it). This configuration is employed in the technique called **stimulated Raman spectroscopy**, where the Stokes and anti-Stokes radiation in the forward direction are powerful enough to undergo more scattering. Hence, the scattered beam gives up or acquires more quanta of energy from the molecules in the sample, so resulting in lines of frequency $v_i \pm 2v_M$, $v_i \pm 3v_M$, and so on, where v_i is the frequency of the incident radiation and v_M the frequency of a molecular excitation.

Laser Raman spectroscopy

Raman spectroscopy is one of the techniques that has been most revitalized by the introduction of lasers. We have already commented on the enhancements of the technique that stem from the high powers and collimation of the incident beam. Its monochromaticity is also a great advantage, since it is now possible to observe scattered light that differs by only fractions of cm^{-1} from the incident radiation. Such high resolution is particularly useful for observing the rotational structure of Raman lines, since rotational transitions are of the order of cm^{-1}. Monochromaticity also allows observations to be made very close to absorption frequencies, giving rise to the technique of **resonance Raman spectroscopy**.

As before, the laser opens up qualitatively different varieties of the technique, not merely enhancements of the established technique. For instance, the intensity of Raman transitions may be enhanced by **coherent anti-Stokes Raman spectroscopy** (CARS). The technique relies on the fact that if two coherent laser beams of frequencies v_1 and v_2 pass through a sample, they may mix together and give rise to radiation of several different frequencies, one of which is

$$v' = v_1 + v_1 - v_2$$

If v_1 is kept fixed, we can also expect it to give rise to Raman lines, and the anti-Stokes lines in the spectrum will include frequencies

$$v'' = v_1 + v_M$$

We now see that it may be possible to vary v_2 so that $v_1 - v_2$ coincides with v_M, in which case v' and v'' will be the same. When this occurs, the radiation in v' can stimulate the production of the Raman line v'', and it is observed with much higher intensity than in the absence of this resonance. A further advantage of CARS is that it can be used to study Raman transitions in the presence of competing incoherent background radiation, and so can be used to observe the Raman spectra of species in flames.

Precision-specified transitions

The monochromatic character of laser radiation is a very powerful characteristic, since it allows us to excite specific states with very high precision. One consequence of state-specificity for photochemistry is that the illumination of a sample may be photochemically precise and hence efficient in stimulating a reaction, since its frequency can be tuned exactly to an absorption. This is in contrast to radiation from a conventional broad-band source, much of which is not absorbed and is therefore wasted.

The specific excitation of a particular excited state of a molecule may greatly enhance the rate of a reaction even at low temperatures. The rate of a reaction is generally increased by raising the temperature, since the energy of the various modes of motion of the molecule is enhanced. However, this enhancement increases the energy of all the modes, even those that do not contribute appreciably to the reaction rate. With a laser we can excite the kinetically significant mode, so rate enhancement is achieved most efficiently. An example is the reaction

$$BCl_3 + C_6H_6 \rightarrow C_6H_5-BCl_2 + HCl$$

which normally proceeds only above 600°C in the presence of a catalyst; exposure to 10.6 μm CO_2 radiation results in the formation of products at room temperature without a catalyst. The commercial potential of this procedure is considerable (provided laser photons can be produced sufficiently cheaply), since heat-sensitive compounds, such as pharmaceuticals, may be made at lower temperatures than global energy enhancement by heating would require.

A related application is the study of **state-to-state** reaction dynamics, in which a specific state of a reactant molecule is excited and we monitor not only the rate at which it forms products but also the states in which they are produced. This gives highly detailed information about the deployment of energy in chemical reactions (Section 28.7).

Isotope separation

The precision state-selectivity of lasers is also of considerable potential for **laser isotope separation**. Isotope separation is possible because two species that differ only in their isotopic composition have slightly different energy levels and hence slightly different absorption frequencies. An example of the use of lasers to separate isotopes is the photoionization of uranium vapour, in which the incident laser is tuned to a photoionization of ^{235}U but not ^{238}U. The ^{235}U atoms in the atomic beam are ionized, and are attracted to a negative electrode, and may be collected.

An alternative procedure is to make use of the different vibrational frequencies of isotopic molecules. Thus SF_6 absorbs infrared radiation of slightly different wavenumber depending on whether the S atom is ^{34}S or ^{32}S. The multiphoton vibrationally excited SF_6 molecules may then undergo reaction with some other reagent, and hence give a product that is rich in one isotope. This procedure has been employed successfully to separate isotopes of B, N, O, and, most efficiently, H.

Pulsed techniques

The ability of lasers to produce pulses of very short duration is particularly useful in chemistry when we want to follow processes in time. Q-switched lasers produce nanosecond pulses, which are generally fast enough to study reactions with rates controlled by the speed with which reactants can move through a fluid medium. However, when we want to study the rates at which energy is converted from one mode to another within a molecule, we need the shorter timescale of picosecond pulses. These are available from mode-locked lasers, and modern techniques have reduced timescales of pulses to the femtosecond region (1 fs $= 10^{-15}$ s), the shortest pulse currently reported being about 4 fs, corresponding to a packet of electromagnetic radiation of only a few wavelengths long. We shall see some of the information obtained from this **femtosecond spectroscopy** in Section 28.5. Pulse techniques are used to study ultrafast dynamical processes such as energy transfer and conversion from one mode of motion to another. They are used to study relaxation of a disturbed set of level populations to thermal equilibrium, and, of particular importance in chemistry, to study the rates of fast reactions.

Photoelectron spectroscopy

The technique of **photoelectron spectroscopy** (PES) measures the ionization energies of molecules when electrons are ejected from different orbitals, and uses the information to infer the orbital energies. The technique is also used to study solids, and in Chapter 29 we shall see the important information that it gives about species at or on surfaces.

17.9 The technique

Since energy is conserved when a photon ionizes a sample, the energy of the incident photon $h\nu$ must be equal to the sum of the ionization energy I of the sample and the kinetic energy of the **photoelectron**, the ejected electron (Fig. 17.36):

$$h\nu = \tfrac{1}{2}m_e v^2 + I$$

This basic equation (which is like the one used for the photoelectric effect, Section 11.2) can be refined in two ways. First, a photoelectron may originate from a number of different orbitals, and each one requires a different ionization energy. Hence, a series of different kinetic energies of the photoelectrons will be obtained, each one satisfying

$$h\nu = \tfrac{1}{2}m_e v^2 + I_i \qquad (7a)$$

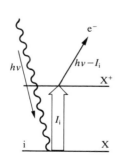

Fig. 17.36 An incoming photon carries an energy $h\nu$; an energy I_i is needed to remove an electron from an orbital ψ_i, and the difference appears as the kinetic energy of the electron.

where I_i is the ionization energy for ejection of an electron from an orbital ψ_i. Therefore, by measuring the kinetic energies of the photoelectrons, and knowing v, these ionization energies can be determined. Photoelectron spectra are interpreted in terms of an approximation called **Koopmans' theorem** which states that the ionization energy I_i is equal to the orbital energy of the ejected electron (formally: $I_i = -\varepsilon_i$). That is, we can identify the ionization energy with the energy of the orbital from which it is ejected. However, the theorem is only an approximation because it ignores the fact that the remaining electrons rearrange their distributions when ionization occurs.

The second refinement is that the ejection of an electron may leave an ion in a vibrationally excited state. Then not all the excess energy of the photon appears as kinetic energy of the photoelectron, and we should write

$$h\nu = \tfrac{1}{2}m_e v^2 + I_i + E_{vib}^+ \tag{7b}$$

where E_{vib}^+ is the energy used to excite the ion into vibration. Each vibrational quantum that is excited leads to a different kinetic energy of the photoelectron, and gives rise to the **fine structure** in the photoelectron spectrum.

Since the ionization energies of molecules are several electronvolts even for valence electrons, it is essential to work in at least the ultraviolet region of the spectrum and with wavelengths of less than about 200 nm. A great deal of work has been done with light generated by a discharge through helium: the He(I) line ($1s^1 2p^1 \rightarrow 1s^2$) lies at 58.43 nm, corresponding to a photon energy of 21.22 eV. Its use gives rise to the technique of **ultraviolet photoelectron spectroscopy** (UPS). If the core electrons are being studied, then photons of even higher energy are needed to expel them: X-rays are used, and the technique is denoted XPS. A modern version of PES makes use of synchrotron radiation (Section 16.1) which may be continuously tuned between UV and X-ray energies. The additional information that stems from the variation of the photoejection probability with wavelength is a valuable guide to the identity of the element and the orbital from which photoionization occurs.

Example 17.8: *Interpreting a photoelectron spectrum*

Photoelectrons ejected from N_2 with He(I) radiation had kinetic energies of 5.63 eV. What is their ionization energy?

Answer. We use eqn 7a and $1\,eV = 8065.5\ cm^{-1}$. He(I) radiation of wavelength 58.43 nm has wavenumber $1.711 \times 10^5\ cm^{-1}$ and therefore corresponds to an energy of 21.22 eV. Then, from eqn 7a,

$$21.22\ eV = 5.63\ eV + I_i$$

so $I_i = 15.59$ eV.

Comment. This ionization energy (corresponding to 1504 kJ mol^{-1}) is the energy needed to remove an electron from the HOMO of the N_2 molecule, the $2p\sigma_g$ bonding orbitals (Fig. 14.12).

Exercise. Under the same circumstances, 4.53 eV electrons are also detected. To what ionization energy does that correspond? Suggest an origin.

[16.7 eV, $2p\pi_u$]

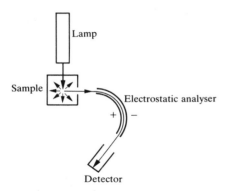

Fig. 17.37 A photoelectron spectrometer consists of a source of ionizing radiation (such as a helium discharge lamp for UPS and an X-ray source for XPS), an electrostatic analyser, and an electron detector. The deflection of the electron path caused by the analyzer depends on their speed.

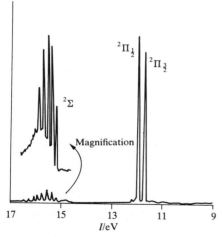

Fig. 17.38 The photoelectron spectrum of HBr. The lowest ionization energy band corresponds to the ionization of a Br lone-pair electron. The higher ionization energy band corresponds to the ionization of a bonding electron. The structure on the latter is due to the vibrational excitation of HBr$^+$ that results from the ionization.

The kinetic energies of the photoelectrons are measured using an electrostatic deflector (Fig. 17.37) which produces different deflections in the paths of the photoelectrons as they pass between charged plates. As the field strength is increased, electrons of different speeds, and therefore kinetic energies, reach the detector. Thus the electron flux can be recorded and plotted against kinetic energy to obtain the photoelectron spectrum. Present techniques give energy resolutions of about 5 meV, corresponding to about 40 cm^{-1}.

17.10 Ultraviolet photoelectron spectroscopy

A typical photoelectron spectrum (of HBr) is shown in Fig. 17.38. If we disregard the fine structure, we see that the HBr lines fall into two main groups. The least tightly bound electrons (with the lowest ionization energies and hence highest kinetic energies when ejected) are those in the nonbonding lone pairs of Br. The next ionization energy lies at 15.2 eV, and corresponds to the removal of an electron from the H—Br σ bond.

The HBr spectrum in Fig. 17.38 shows that ejection of a σ electron is accompanied by a long vibrational progression. The Franck–Condon principle would account for this if ejection were accompanied by an appreciable change of equilibrium bond length between HBr and HBr$^+$ because the ion is formed in a bond-compressed state, which is consistent with the important bonding effect of the σ electrons. The lack of much vibrational structure in the other band is consistent with the nonbonding role of the Br$2p\pi$ lone-pair electrons, for the equilibrium bond length is little changed when one is removed.

Example 17.9: *Interpreting a UV photoelectron spectrum*

The highest kinetic-energy electrons in the spectrum of H_2O using 21.22 eV He(I) radiation are at about 9 eV and show a large vibrational spacing of 0.41 eV. The symmetric stretching mode of the neutral H_2O molecule lies at 3652 cm^{-1}. What conclusions can be drawn about the nature of the orbital from which the electron is ejected?

Answer. Since 0.41 eV corresponds to 3310 cm^{-1}, which is similar to the 3652 cm^{-1} of the non-ionized molecule, we can suspect that the electron is ejected from an orbital that has little influence on the bonding in the molecule. That is, photoejection is from a largely non-bonding orbital.

Exercise. In the same spectrum of H_2O, the band near 7.0 eV shows a long vibrational series with spacing 0.125 eV. The bending mode of H_2O lies at 1596 cm^{-1}. What conclusions can you draw about the characteristics of the orbital occupied by the photoelectron?

[The electron contributed to H—H bonding]

17.11 X-ray photoelectron spectroscopy

In X-ray PES, the energy of the incident photon is so great that electrons are ejected from inner cores of atoms. As a first approximation, we would not expect core ionization energies to be sensitive to the bonds between atoms because they are too tightly bound to be greatly affected by the changes that accompany bond formation. This insensitivity turns out to be largely true, and the core ionization energies are characteristic of the individual atom rather than the overall molecule. Consequently, X-ray PES gives lines characteristic of the elements present in a compound or alloy. For instance, the K-shell ionization energies of the second row elements are

Li	Be	B	C	N	O	F	
50	110	190	280	400	530	690	eV

Detection of one of these values (and values corresponding to ejection from other inner shells) indicates the presence of the corresponding element. Figure 17.39, for example, shows the result of an X-ray photoelectron analysis of lunar rock brought back on an Apollo mission. This chemical analysis application is responsible for the alternative name **electron spectroscopy for chemical analysis**, or ESCA. The technique is mainly limited to the study of surface layers (as we shall explore in Chapter 29) because, even though X-rays may penetrate into the bulk sample, the ejected electrons cannot escape except from within a few nanometres of the surface. Despite (or because of) this limitation, the technique is very useful for studying the surface state of heterogeneous catalysts, the differences between surface and bulk structures, and the processes that can cause damage to high-temperature superconductors and semiconductor wafers.

While it is largely true that core ionization energies are unaffected by bond formation, it is not entirely true, and small but detectable shifts can be detected and interpreted in terms of the environments of the atoms. For example, the azide ion N_3^- gives the spectrum shown in Fig. 17.40. Although the spectrum lies in the region of 400 eV (and hence is typical of N1s electrons), it has a doublet structure with splitting 6 eV. This can be understood by noting that the structure of the ion is N=N=N, with charge distribution $(-, +, -)$. Hence the presence of the negative charges on the terminal atoms lowers the core ionization energies, while the positive charge on the central atom raises it. This results in two lines in the spectrum with intensities in the ratio 2:1. Observations like this can be used to obtain valuable information about the presence of chemically inequivalent atoms of the same element.

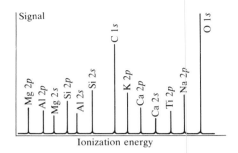

Fig. 17.39 The X-ray photoelectron spectrum of a sample of moon dust, with the assignment. (Adapted from *Physical methods and molecular structure*, The Open University Press (1977).)

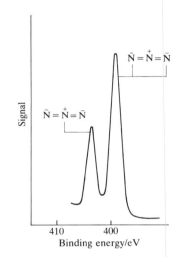

Fig. 17.40 The photoelectron spectrum of solid NaN_3 excited by Al K-radiation showing the region of N core ionization and the assignment. (K. Siegbahn et al, *Science* **176**, 245 (1972).)

Further reading

For references to books on general aspects of spectroscopy, see Chapter 16.

Ultraviolet and visible spectra

H. H. Jaffé and M. Orchin, *Theory and applications of ultraviolet spectroscopy*. Wiley, New York (1962).

C. N. R. Rao, *Ultraviolet and visible spectroscopy*. Butterworth, London (1967).

F. Grum, Visible and ultraviolet spectroscopy. In *Techniques of chemistry*, IIIB, p. 207 (ed. A. Weissberger and B. W. Rossiter). Wiley–Interscience, New York (1972).

E. A. V. Ebsworth, D. W. H. Rankin, and S. Cradock, *Structural methods in inorganic chemistry*. Blackwell Scientific, Oxford (1987).

G. Herzberg, *Spectra of diatomic molecules*. Van Nostrand, New York (1950).

G. Herzberg, *Electronic spectra and electronic structure of polyatomic molecules*. Van Nostrand, New York (1966).

Processes in excited states

N. Wotherspoon, G. K. Oster, and G. Oster, Determination of fluorescence and phosphorescence. In *Techniques of chemistry*, IIIB, p. 429 (ed. A. Weissberger and B. W. Rossiter). Wiley–Interscience, New York (1972).

A. G. Gaydon, *Dissociation energies*. Chapman and Hall, London (1952).

J. G. Calvert and J. N. Pitts, *Photochemistry*. Wiley, New York (1966).

R. P. Wayne, *Principles and applications of photochemistry*. Oxford University Press (1988).

Lasers

J. Hecht, *Understanding lasers*. Howard Sams & Co., Indianapolis (1988).

J. Wilson and J. F. B. Hawkes, *Lasers: principles and applications*. Prentice Hall, Englewood Cliffs (1987).

A. E. Siegman, *Lasers*. University Science Books, Mill Valley and Oxford University Press (1986).

M. J. Weber, *Handbook of laser science and technology*. CRC Press, Boca Raton (1982).

M. Hollas, *High resolution spectroscopy*. Butterworth, London (1982).

T. R. Evans (ed.), Applications of lasers to chemical problems. In *Techniques of chemistry*, XVII, Wiley–Interscience, New York (1982).

W. Demtröder, *Laser spectroscopy*. Springer, Berlin (1988).

G. R. Fleming, *Chemical applications of ultrafast spectroscopy*. Oxford University Press (1986).

Photoelectron spectra

J. H. D. Eland, *Photoelectron spectra*. Open University Press, Milton Keynes (1977).

A. D. Baker and D. Betteridge, *Photoelectron spectroscopy*. Pergamon Press, Oxford (1977).

D. W. Turner, C. Baker, A. D. Baker, and C. R. Brundle, *Molecular photoelectron spectroscopy*. Wiley–Interscience, New York (1970).

C. R. Brundle and A. D. Baker (ed.), *Electron spectroscopy: theory, techniques, and applications*, Vols. 1 and 2, Academic Press, London (1977).

Exercises

17.1 The molar absorption coefficient of a substance dissolved in hexane is known to be $855 \, M^{-1} \, cm^{-1}$ at 270 nm. Calculate the percentage reduction in intensity when light of that wavelength passes through 2.5 mm of a solution of concentration $3.25 \times 10^{-3} \, M$.

17.2 When light of wavelength 400 nm passes through 3.5 mm of a solution of an absorbing substance at a concentration $6.67 \times 10^{-4} \, M$, the transmission is 65.5 per cent. Calculate the molar absorption coefficient of the solute at this wavelength and express the answer in $cm^2 \, mol^{-1}$.

17.3 The molar absorption coefficient of a solute at 540 nm is $286 \, M^{-1} \, cm^{-1}$. When light of that wavelength passes through a 6.5 mm cell containing a solution of the solute, 46.5 per cent of the light was absorbed. What is the concentration of the solution?

17.4 The absorption associated with a particular transition begins at 230 nm, peaks sharply at 260 nm, and ends at 290 nm. The maximum value of the molar absorption coefficient is $1.21 \times 10^4 \, M^{-1} \, cm^{-1}$. Estimate the oscillator strength of the transition assuming a triangular lineshape.

17.5 The transition moment calculated from the initial and final state wavefunctions of a certain transition at $35\,000 \, cm^{-1}$ is $2.65 \times 10^{-30} \, C \, m$. Calculate the oscillator strength of the transition.

17.6 The oscillator strengths of a series of transitions were calculated as (a) 2.9×10^{-2}, (b) 0.75, (c) 6.2×10^{-5}, (d) 3.2×10^{-9}, and (e) 0.91. Classify the transitions as strong, weak, or forbidden.

17.7 The two compounds 2,3-dimethyl-2-butene and 2,5-dimethyl-2,4-hexadiene are to be distinguished by their ultraviolet absorption spectra. The maximum absorption in one compound occurs at 192 nm and in the other at 243 nm. Match the maxima to the compounds and justify the assignment.

17.8 The compound $CH_3CH{=}CHCHO$ has a strong absorption in the ultraviolet at $46\,950\ cm^{-1}$ and a weak absorption at $30\,000\ cm^{-1}$. Justify these features in terms of the structure of the compound.

17.9 The photoionization of H_2 by 21 eV photons produces H_2^+. Explain why the intensity of the $v = 2 \leftarrow 0$ transition is stronger than that of the $0 \leftarrow 0$ transition.

17.10 The following data were obtained for the absorption by Br_2 in carbon tetrachloride using a 2.0 mm cell. Calculate the molar absorption coefficient (ε) of bromine at the wavelength employed:

$[Br_2]/M$	0.0010	0.0050	0.0100	0.0500
$T/\%$	81.4	35.6	12.7	3.0×10^{-3}

17.11 A 2.0-mm cell was filled with a solution of benzene in a nonabsorbing solvent. The concentration of the benzene was 0.010 M and the wavelength of the radiation was 256 nm (where there is a maximum in the absorption). Calculate the molar absorption coefficient of benzene at this wavelength given that the transmission was 48 per cent. What will the transmittance be in a 4.0-mm cell at the same wavelength?

17.12 A swimmer enters a gloomier world (in one sense) on diving to greater depths. Given that the mean molar absorption coefficient of sea water in the visible region is $6.2 \times 10^{-5}\ M^{-1}\ cm^{-1}$, calculate the depth at which a diver will experience (a) half the surface intensity of light, (b) one-tenth the surface intensity.

17.13 The absorption bands of many molecules in solution have widths at half-height of about $5000\ cm^{-1}$. Estimate the integrated absorption coefficients and the oscillator strengths of bands for which (a) $\varepsilon_{max} \approx 1 \times 10^4\ M^{-1}\ cm^{-1}$, (b) $\varepsilon_{max} \approx 5 \times 10^2\ M^{-1}\ cm^{-1}$.

Problems

Numerical problems

17.1 In a certain laboratory, the precision of the equipment used to measure light intensity is 2 per cent. The only available sample cells have a thickness of 1.5 mm. What is the limit of precision in determing the concentrations with this equipment if the molar absorption coefficient of the solute is $275\ M^{-1}\ cm^{-1}$?

17.2 In many cases it is possible to assume that an absorption band has a gaussian lineshape (one proportional to e^{-x^2}) centred on the band maximum. Assume such a line shape, and show that

$$\mathscr{A} = 1.0645c\varepsilon_{max}\Delta\bar{v}_{1/2}$$

where $\Delta\bar{v}_{1/2}$ is the width at half-height. Use this result to obtain the corresponding expression for the oscillator strength f. The absorption spectrum of azoethane ($CH_3CH_2N_2$) between $24\,000\ cm^{-1}$ and $34\,000\ cm^{-1}$ is shown in Fig. 17.41. First, estimate $\mathscr{A}$ and f for the band by assuming that it is gaussian. Then integrate the absorption band graphically.

(The latter can be done either by ruling and counting squares, or by tracing the line shape on to paper and weighing.) Is the transition forbidden or allowed?

17.3 A more complex absorption spectrum, that of biacetyl ($CH_3COCOCH_3$) is shown in Fig. 17.42. Estimate the oscillator strengths of both principal transitions and decide whether they are forbidden or allowed. Notice that the spectrum is plotted on a wavelength scale and therefore that extra work must be done before the oscillator strengths can be obtained. There are three approaches. One is to estimate the half-width from the wavelength data and to use the gaussian line shape expression derived above. The second is to transfer the spectrum to a wavenumber scale and to carry out a graphical integration. The third is to find an expression for f in terms of wavelength integration and to use the spectrum directly. The molar absorption coefficients of the two peaks are $11\ M^{-1}\ cm^{-1}$ (at 280 nm) and $18\ M^{-1}\ cm^{-1}$ (at 430 nm).

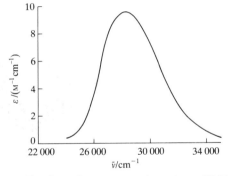

Fig. 17.41 The absorption spectrum of azoethane, $CH_3CH_2N_2$.

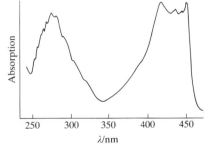

Fig. 17.42 The electronic absorption spectrum of biacetyl, $CH_3COCOCH_3$, in the ultraviolet and blue regions. Note the blurred vibrational structure of the absorption bands.

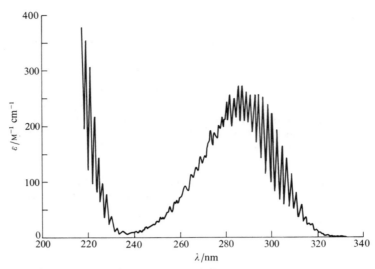

Fig. 17.43 The gas-phase absorption spectrum of SO_2.

17.4 A lot of information about the energy levels and wavefunctions of small inorganic molecules can be obtained from their ultraviolet spectra. An example of a spectrum with considerable vibrational structure, that of gaseous SO_2 at 25°C, is shown in Fig. 17.43. Calculate the oscillator strength of the transition. Is it forbidden or allowed? What electronic states are accessible from the A_1 ground state of this C_{2v} molecule by electric dipole transitions?

17.5 Consider an indicator InH in which InH and In$^-$ absorb in different parts of the spectrum and that the absorbances of the acid and conjugate base forms are A and A'' respectively. When a mixture of the two forms is present, the absorbance is A'. Show that the degree of ionization α of the indicator is given by

$$\alpha = \frac{A' - A}{A'' - A}$$

Hence show that the pH of the solution can be expressed in terms of the acidity constant K_a of InH and the absorbance A'. Sketch the form of the pH-dependence of A'/A for a colourless conjugate base form, taking $pK_a = 4$. The data below are for the molar absorption coefficient of p-nitrophenol in water (A. I. Biggs, *Trans. Faraday Soc.*, **50**, 800 (1954)). Calculate the pK_a of the phenol.

pH	4	5	6	7
$\varepsilon/(10^3\,\text{M}^{-1}\,\text{cm}^{-1})$	8.33	8.33	7.64	4.16
$\varepsilon/(10^3\,\text{M}^{-1}\,\text{cm}^{-1})$			1.66	8.16

pH	8	9	10	
$\varepsilon/(10^3\,\text{M}^{-1}\,\text{cm}^{-1})$	0.42			at 317 nm
$\varepsilon/(10^3\,\text{M}^{-1}\,\text{cm}^{-1})$	17.50	18.33	18.33	at 407 nm

17.6 A transition of particular importance in O_2 gives rise to the 'Schumann–Runge band' in the ultraviolet region. The wavenumbers (in cm^{-1}) of transitions from the ground state to the vibrational levels of the first excited state ($^3\Sigma_u^-$) are 50 062.6, 50 725.4, 51 369.0, 51 988.6, 52 579.0, 53 143.4, 53 679.6, 54 177.0, 54 641.8, 55 078.2, 55 460.0, 55 803.1, 56 107.3, 56 360.3, 56 570.6. What is the dissociation energy of the upper electronic state? (Use a Birge–Sponer plot). The same excited state is known to dissociate into one ground state O atom and one excited state O atom with an energy 190 kJ mol^{-1} above the ground state. (This excited atom is responsible for a great deal of photochemical mischief in the atmosphere.) Ground state O_2 dissociates into two ground state atoms. Use this information to calculate the dissociation energy of ground-state O_2 from the Schumann–Runge data.

17.7 The photoelectron spectra of N_2 and CO are shown in Fig. 17.44. Ascribe the lines to the ionization processes involved and classify the orbitals from which the electrons are ejected as bonding, nonbonding, or antibonding in the light of the extent of vibrational structure in the band. Analyze the bands near 4 eV in terms of the vibrational energy levels of the ions.

17.8 The photoelectron spectrum of NO can be described as follows (D. W. Turner, in *Physical methods in advanced inorganic chemistry*, (ed. H. A. O. Hill and P. Day), Wiley (1968)). Using He 584 pm (21.22 eV) radiation there is a single strong peak at kinetic energy 4.69 eV and a long series of 24 lines starting at 5.56 eV and ending at 2.2 eV. A shorter series of six lines begins at 12.0 eV and ends at 10.7 eV. Account for this spectrum.

Theoretical problems

17.9 Use a group theoretical argument to decide which of the following transitions are electric-dipole allowed: (a) the π^*,π transition in ethene, (b) the π^*,n transition in a carbonyl group in a C_{2v} environment.

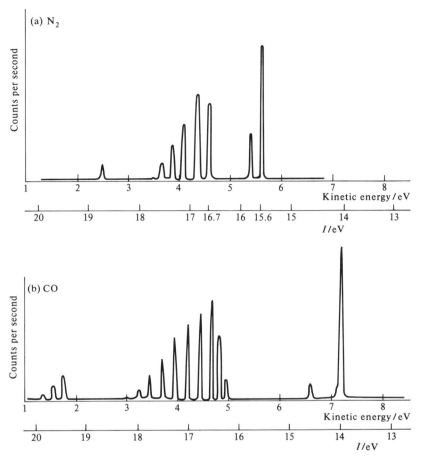

Fig. 17.44 The photoelectron spectrum of (a) N_2 and (b) CO.

17.10 The line marked A in Fig. 17.45 is the fluorescence spectrum of benzophenone in solid solution in ethanol at low temperatures observed when the sample is illuminated with 360 nm light. What can be said about the vibrational energy levels of the carbonyl group in (a) its ground electronic state

and (b) its excited electronic state? When naphthalene is illuminated with 360 nm light it does not absorb, but the line marked B in the illustration is the phosphorescence spectrum of a solid solution of a mixture of naphthalene and benzophenone in ethanol. Now a component of fluorescence from naphthalene can be detected. Account for this observation.

17.11 The fluorescence spectrum of anthracene vapour shows a series of peaks of increasing intensity with individual maxima at 440 nm, 410 nm, 390 nm, and 370 nm followed by a sharp cut-off at shorter wavelengths. The absorption spectrum rises sharply from zero to a maximum at 360 nm with a tail of peaks of lessening intensity at 345 nm, 330 nm, and 305 nm. Account for these observations.

17.12 The oscillator strength can be calculated from the molecular orbitals of the ground and excited states involved in the transition. Consider an electron in a one-dimensional square well of length L, and calculate the oscillator strengths of the transitions $n+1 \leftarrow n$ and $n+2 \leftarrow n$. Begin by cal-

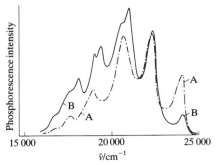

Fig. 17.45 The phosphorescence spectra of naphthalene and benzophenone.

3

culating the transition dipole moment

$$\mu = -e \int_0^L \psi_n \cdot x \psi_n \, dx$$

and calculate f from eqn 3. (Only the x component of the transition dipole moment is non-zero.) Estimate the excitation energy of β-carotene (**3**) using the free-electron molecular orbital theory of its structure and $R = 140$ pm (see Problem 14.9 for a review of what is involved). Suggest what colour β-carotene will appear in white light and estimate the molar absorption coefficient. Estimate the thickness of a 0.0010-M sample that will give 50 per cent absorption of incident light.

17.13 Suppose that you are a colour chemist and had been asked to intensify the colour of a dye without changing the type of compound, and that the dye in question was a polyene. Would you choose to lengthen or to shorten the chain? Would the modification to the length shift the apparent colour of the dye towards the red or the blue?

17.14 Consider an electron in an atom to be oscillating harmonically (that was an early model of atomic structure). The wavefunctions for such an electron are those in Table 12.1. Show that the oscillator strength for the transition of

this electron from its ground state is exactly $\frac{1}{3}$.

17.15 Estimate the oscillator strength of a charge-transfer transition modelled as the migration of an electron from a hydrogen $1s$ orbital on one atom to another hydrogen $1s$ orbital on an atom a distance R away. Approximate the transition moment by $-eRS$ where S is the overlap integral of the two orbitals. Plot the oscillator strength as a function of R using the value of S given in eqn 10 of Section 14.4. At what separation is the intensity greatest? Why does the intensity fall to zero as R approaches 0 and infinity?

17.16 When pyridine is added to a solution of iodine in carbon tetrachloride, the band at 520 nm shifts towards 450 nm but the absorption at 490 nm remains unchanged. The point at 490 nm is called an 'isosbestic point' of the mixture. Use the technique established in Problem 17.5, and especially the relation between α and A', to show that the isosbestic point should always be present when two absorbing species are in equilibrium.

17.17 Spin angular momentum is conserved when a molecule dissociates into atoms. What atom multiplicities are permitted when (a) an O_2 molecule, (b) an N_2 molecule dissociates into atoms?

Magnetic resonance

Check-list of key ideas

1. The *magnetic moments* of nuclei, their energies in magnetic fields (eqn 1), and the *resonance condition* (eqn 3).

2. The technique of *nuclear magnetic resonance* (Section 18.1).

3. The *shielding constant* and the *chemical shift* (Section 18.2), and the *δ-scale* of chemical shifts (eqn 4).

4. The origin of chemical shifts, including the *neighbouring group effect* and *ring currents* (Section 18.2).

5. The *fine structure* of NMR spectra, the *spin–spin coupling constant,* and the patterns expected from particular groups of nuclei (Section 18.3).

6. The origin of the spin–spin coupling, including the *polarization mechanism* and the *Fermi contact interaction* (Section 18.3).

7. The concepts of *chemically equivalent nuclei* and *magnetically equivalent nuclei* (Section 18.3).

8. *First-order* and *second-order* spectra, *dilute spins* and *abundant spins,* and *proton decoupling* (Section 18.3).

9. The *magnetization vector,* its *precession,* and the effect of a radiofrequency field (Section 18.4).

10. The effect of a *90° pulse* and the *free-induction decay* signal (Section 18.4).

11. *Fourier transforms* and the relation between *time–domain spectra* and *frequency–domain spectra* (eqn 8 and Example 18.5).

12. *Spin relaxation* and the significance of the *longitudinal relaxation time* and the *transverse relaxation time* (Section 18.5).

13. The effect of *inhomogeneities* and of *exchange processes* on the appearance of spectra (Section 18.5).

14. The measurement of relaxation times by *inversion recovery* and *spin echoes* (Section 18.5).

15. The *electron spin resonance* technique and the electron *g-factor* (Section 18.7).

16. The *hyperfine structure* of ESR spectra, its interpretation, and its origin (Section 18.8).

When two pendulums are joined by the same slightly flexible support and one is set in motion, the other is forced into oscillation by the motion of the common axle, and energy flows between the two. The energy transfer occurs most efficiently when the frequencies of the two oscillators are identical. The condition of strong effective coupling when the frequencies are identical is called **resonance**, and the excitation energy is said to **resonate** between the coupled oscillators.

Resonance is the basis of a number of everyday phenomena, including the response of radios to the weak oscillations of the electromagnetic field generated by a distant transmitter. In this chapter we explore some spectroscopic applications that when originally developed (and in some cases still) depend on matching a set of energy levels to a source of monochromatic radiation and observing the strong absorption that occurs at resonance.

Nuclear magnetic resonance

Many nuclei possess spin angular momentum. A nucleus with spin quantum number I (which may be an integer or a half-integer) may take $2I + 1$ different orientations relative to an arbitrary axis that are distinguished by the quantum number m_I:

$$m_I = I, I - 1, \ldots, -I$$

A proton has $I = \frac{1}{2}$ and may adopt either of two orientations; a ^{14}N nucleus has $I = 1$ and may adopt any of three orientations. For much of this chapter we shall consider **spin-$\frac{1}{2}$ nuclei**, those with $I = \frac{1}{2}$; as well as protons, spin-$\frac{1}{2}$ nuclei include ^{13}C, ^{19}F, and ^{31}P nuclei. As for electrons, the state with $m_I = +\frac{1}{2}(\uparrow)$ is denoted α and that with $m_I = -\frac{1}{2}(\downarrow)$ is denoted β. It is worth bearing in mind that two very common nuclei, ^{12}C and ^{16}O, have zero spin and hence are invisible in magnetic resonance.

18.1   The energies of nuclei in magnetic fields

A nucleus with non-zero spin has a magnetic moment. The component μ_z of this moment on the z axis depends on the orientation of the spin, and we write

$$\mu_z = \gamma \hbar m_I$$

where γ is the **magnetogyric ratio** of the nucleus. Theories of nuclear structure are not yet sufficiently advanced for γ to be calculated reliably, and it is treated as an empirical factor. The magnetic moment is often expressed in terms of the **nuclear g-factor** g_I and the **nuclear magneton** μ_N using

$$\gamma \hbar = g_I \mu_N \qquad \mu_N = \frac{e\hbar}{2m_p} = 5.051 \times 10^{-27} \text{ J T}^{-1}$$

where m_p is the mass of the proton. Nuclear g-factors are numbers of the order of 1 (Table 18.1): positive values of g denote a magnetic moment that is parallel to the spin, negative values that the magnetic moment and spin are antiparallel. The nuclear magneton is about 2000 times smaller than the Bohr magneton, and so nuclear magnetic moments are about 2000 times weaker than the electron spin magnetic moment.

Table 18.1. Nuclear spin properties

Nuclide	Natural abundance	Spin I	g-factor g_I
^{1}n		$\frac{1}{2}$	−3.826
^{1}H	99.98	$\frac{1}{2}$	5.586
^{2}H	0.02	1	0.857
^{13}C	1.11	$\frac{1}{2}$	1.405
^{14}N	99.64	1	0.404

The basic resonance experiment

Each value of m_I corresponds to a different orientation of the nuclear magnetic moment. In a magnetic field B the $2I + 1$ orientations of the nucleus have different energies, which are given by

$$E_{m_I} = -\mu_z B = -\gamma \hbar B m_I \qquad (1)$$

These energies are often expressed in terms of the **Larmor frequency** ω

$$\omega = \gamma B \qquad E_{m_I} = -m_I \hbar \omega \qquad (2)$$

The higher the magnetic field, the greater the Larmor frequency. A field of 12 T corresponds to Larmor frequencies of about 500 MHz for protons.

The energy separation of the two states of spin-$\frac{1}{2}$ nuclei (Fig. 18.1) is

$$\Delta E = E_\beta - E_\alpha = \tfrac{1}{2}\gamma \hbar B - (-\tfrac{1}{2}\gamma \hbar B) = \gamma \hbar B$$

For most nuclei γ is positive (and $g_I > 0$). Therefore, the β state lies above the α state, and there are slightly more α spins than β spins. If the sample is bathed in radiation of frequency ν, the energy separations come into resonance with the radiation when the frequency satisfies the **resonance condition**:

$$h\nu = \gamma \hbar B \qquad (3)$$

At resonance there is strong coupling between the nuclear spins and the radiation, and strong absorption occurs as the spins make the transition $\beta \leftarrow \alpha$. At 12 T, protons come into resonance at about 500 MHz.

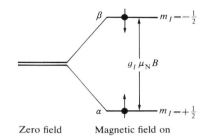

Fig. 18.1 The nuclear spin energy levels of a spin-$\frac{1}{2}$ nucleus (e.g. ^{1}H or ^{13}C) in a magnetic field. Resonance occurs when the energy separation of the levels matches the energy of the photons in the electromagnetic field.

The technique

In its simplest form, **nuclear magnetic resonance** (NMR) is the study of the properties of molecules containing magnetic nuclei by applying a magnetic field and observing the frequency at which they come into resonance with an electromagnetic field. When applied to proton spins, the technique is called **proton magnetic resonance** (^{1}H–NMR). In the early days of the technique only protons (which have relatively large magnetic moments) could be studied, but now a wide variety of nuclei (especially ^{13}C and ^{31}P) are investigated routinely.

An NMR spectrometer consists of a magnet that can produce a uniform, intense field and the appropriate sources of radiofrequency electromagnetic radiation. In simple instruments the magnetic field is provided by an electromagnet; for serious work, a superconducting magnet capable of producing fields of the order of 10 T and more is used. The use of high magnetic fields has two advantages. One is that the field increases the population difference between the two spin states, for according to the Boltzmann distribution

$$\frac{N_\beta}{N_\alpha} = e^{-\Delta E/kT}$$

and the ratio differs more from 1 as ΔE increases, and hence a stronger net absorption is obtained. Secondly, a high field simplifies the appearance of certain spectra, as we shall see.

The sample is placed in the cylindrically wound magnet, and is rotated at about 15 Hz. The spinning helps to remove magnetic inhomogeneities and ensure that all the magnetic nuclei experience the same field. Although a superconducting magnet operates at the temperature of liquid helium (4 K), the sample itself is normally at room temperature.

18.2 The chemical shift

Nuclear magnetic moments interact with the *local* magnetic field. The local field may differ from the applied field because the latter induces electronic orbital angular momentum which gives rise to a small additional magnetic field δB. This additional field is proportional to the applied field, and it is conventional to express it as

$$\delta B = -\sigma B$$

where σ is the **shielding constant** (which may be positive or negative). The ability of the applied field to induce orbital angular momentum depends on the details of the electronic structure near the magnetic nucleus of interest, and nuclei in different chemical groups have different shielding constants.

The δ scale of chemical shifts

Since the total local field is

$$B_{\text{loc}} = B + \delta B = (1 - \sigma)B$$

the Larmor frequency is

$$\omega = \gamma B_{\text{loc}} = (1 - \sigma)\gamma B$$

and is different for nuclei in different environments. Hence, different nuclei, even of the same element, come into resonance at different frequencies.

The **chemical shift** of a nucleus is the difference between its resonance frequency and that of a reference standard. The standard for protons is the proton resonance in tetramethylsilane ($Si(CH_3)_4$, commonly referred to as TMS), which bristles with protons and dissolves without reaction in many samples. Other references are used for other nuclei. For ^{13}C, the reference frequency is the ^{13}C resonance in TMS and for ^{31}P it is the ^{31}P resonance in 85 per cent $H_3PO_4(aq)$. The separation of the resonance of a particular group of nuclei from the standard increases with the strength of the applied magnetic field because the induced field is proportional to the applied field, and the stronger the latter, the greater the shift.

Chemical shifts are reported on the **δ scale**, which is defined as

$$\delta = \frac{\nu - \nu^{\circ}}{\nu^{\circ}} \times 10^6 \qquad (4a)$$

where ν° is the resonance frequency of the standard. The advantage of the δ-scale is that shifts reported on it are *independent* of the applied field (since both numerator and denominator are proportional to the applied field). The resonance frequencies themselves, however, do depend on the applied field through

$$\nu - \nu^{\circ} = \nu^{\circ}\delta \times 10^{-6} \qquad (4b)$$

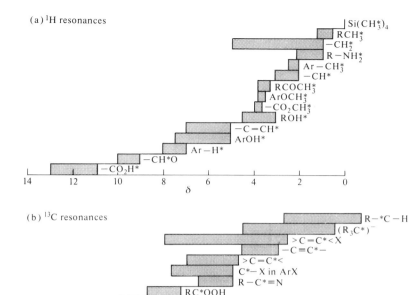

(a) ^{1}H resonances

δ

(b) ^{13}C resonances

δ

Fig. 18.2 The range of typical chemical shifts for (a) ^{1}H resonances and (b) ^{13}C resonances.

A positive δ indicates that the resonance frequency of the group of nuclei in question is higher than that of the standard. Hence $\delta > 0$ indicates that the local magnetic field is stronger than that experienced by the nuclei in the standard under the same conditions. Some typical chemical shifts are given in Fig. 18.2.

Example 18.1: *Using the chemical shift*

At what frequency shift from TMS would a group of nuclei with $\delta = 1.00$ resonate in a spectrometer operating at 500 MHz?

Answer. From eqn 4b, the shift in resonance from the standard (TMS) is

$$\nu - \nu^\circ = 500 \text{ MHz} \times 1.00 \times 10^{-6} = 500 \text{ Hz}$$

Comment. In a spectrometer operating at 100 MHz, the shift would be only 100 Hz.

Exercise. What is the shift of the resonance from TMS of a group of nuclei with $\delta = 3.50$ and an operating frequency of 350 MHz?   [1.23 kHz]

Resonance of different groups of nuclei

The existence of a chemical shift explains the general features of the spectrum of ethanol shown in Fig. 18.3. The CH_3 protons form one group of nuclei with $\delta = 1$. The two CH_2 protons are in a different part of the molecule, experience a different local magnetic field, and resonate at $\delta = 3$. Finally, the OH proton is in another environment, and has a chemical shift of $\delta = 4$.

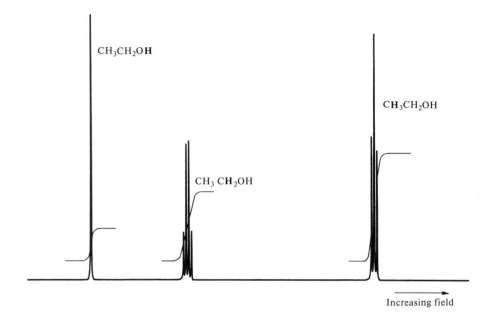

Fig. 18.3 An NMR spectrum of ethanol. The bold letters denote the protons giving rise to the resonance peak, and the step-like curve is the integrated signal.

We can use the relative intensities of the signal (the areas under the absorption lines) to help distinguish which group of lines corresponds to which chemical group, and spectrometers can integrate the absorption automatically (Fig. 18.3). In ethanol the group intensities are in the ratio $3:2:1$ because there are three CH_3 protons, two CH_2 protons, and one OH proton in each molecule. Counting the number of magnetic nuclei as well as noting their chemical shifts is valuable analytically since it helps us identify the compound present in a sample.

The origin of shielding constants

If $\sigma > 0$, the chemical shift is negative (to low frequencies) and we say that the nucleus is **shielded**.[1] If $\sigma < 0$, the chemical shift is positive and we say that the nucleus is **deshielded**. For some purposes it is convenient to regard σ as the sum of a positive **diamagnetic contribution** σ_d and a negative **paramagnetic contribution** σ_p, and to write

$$\sigma = \sigma_d + \sigma_p$$

Then σ is positive if the diamagnetic contribution dominates and is negative if the paramagnetic contribution dominates.

The diamagnetic contribution arises from the ability of the applied field to generate orbital motion of the electrons in the molecule. The resulting circulation of charge generates a magnetic field that opposes the applied field and reduces the frequency needed for resonance. The magnitude of σ_d depends on the electron density near the nucleus and can be calculated from

---

[1] The shielding constant σ and the chemical shift δ indicate similar properties, but whereas the shielding constant is an indication of the shielding or antishielding of a proton in an absolute sense, the value of δ indicates the shielding relative to the standard (TMS).

the **Lamb formula**:

$$\sigma_d = \frac{e^2 \mu_0}{3m_e} \int_0^\infty r\rho(r)\,\mathrm{d}r \qquad (5)$$

where ρ is the electron density.

Example 18.2: *Using the Lamb formula*

Calculate the shielding constant for a hydrogen atom.

Answer. The electron density at a distance r from the nucleus of this one-electron atom is equal to ψ^2, where ψ is the $1s$ orbital of the H atom (Section 13.2). Since

$$\psi = \left(\frac{1}{\pi a_0^3}\right)^{1/2} e^{-r/a_0}$$

we obtain (using the integral given in Example 13.4):

$$\sigma_d = \frac{e^2 \mu_0}{3\pi m_e a_0^3} \int_0^\infty r e^{-2r/a_0}\,\mathrm{d}r = \frac{e^2 \mu_0}{3\pi m_e a_0^3} \times \frac{a_0^2}{2^2}$$

$$= \frac{e^2 \mu_0}{12\pi m_e a_0}$$

With the values of the fundamental constants inside the front cover, this expression evaluates to 1.78×10^{-5}.

Exercise. Calculate σ_d for a hydrogenic atom with atomic number Z.

$$[Z\sigma_d(\mathrm{H})]$$

The paramagnetic contribution σ_p arises from the ability of the applied field to force the electrons to circulate through the molecule by taking advantage of the availability of orbitals that are unoccupied in the ground state. It is zero in free atoms and around the axes of linear molecules (such as $HC\equiv CH$) where the electrons can circulate freely. Proton shielding constants are often dominated by the diamagnetic contribution, but this is not the case for the nuclei of other elements.

One important contribution to the shielding constant arises from the currents induced in nearby groups of atoms. For this **neighbouring group effect** the division of σ into diamagnetic and paramagnetic contributions is no longer appropriate, since either kind of current can shield or deshield the nucleus depending on the relative location of the nucleus to the neighbouring group.[2] For example, the protons in $HC\equiv CH$ are shielded by the currents induced in the triple bond, but a proton that lies perpendicular to the bond (as part of a larger molecule) is deshielded (Fig. 18.4a). The opposite is true for protons near a $C=C$ double bond (Fig. 18.4b). The difference can be traced to the fact that the applied field can induce only a

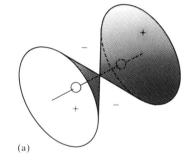

(a)

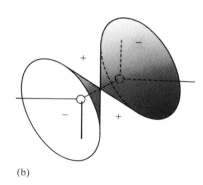

(b)

Fig. 18.4 The neighbouring group effect in NMR. (a) The protons in $HC\equiv CH$ are shielded by the currents induced in the triple bond, but a proton perpendicular to the bond is deshielded. (b) The opposite is true for protons near a $C=C$ double bond because the applied field can induce a paramagnetic current parallel to the axis of a double bond.

---

[2] The shielding constant depends on the difference in the magnetic susceptibilities χ parallel and perpendicular to the group, the angle θ that the vector to the magnetic nucleus makes to the axis of symmetry of the group, and the separation r of the two:

$$\sigma = \frac{1}{4\pi r^3}(\chi_\parallel - \chi_\perp)(1 - 3\cos^2\theta)$$

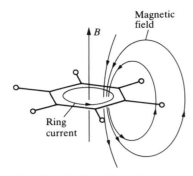

Fig. 18.5 The shielding effects of the ring current induced in the benzene ring by the applied field.

diamagnetic current parallel to the axis of HC≡CH (where the electrons can circulate freely) but it can also induce a paramagnetic current parallel to the axis of a double bond (where the electrons cannot circulate freely because the bond is not cylindrically symmetrical).

The neighbouring group effect depends on the magnetic anisotropy of the group, with the field able to induce currents of different strengths in different directions. A prime example of this anisotropy is the benzene ring. The strong anisotropy is ascribed to the ability of the field to induce a **ring current**, a circulation of electrons around the ring, when it is applied perpendicular to the molecular plane. Protons in the plane are deshielded but any that happen to lie above or below the plane are shielded (Fig. 18.5).

18.3 The fine structure

The splitting of resonances into individual lines in Fig. 18.3 is called the **fine structure** of the spectrum. It arises because each magnetic nucleus contributes to the local field experienced by the other nuclei and modifies their resonance frequencies. The strength of the interaction is expressed in terms of the **spin–spin coupling constant** J and reported in hertz (Hz). Spin–spin coupling constants are independent of the strength of the applied field since they do not depend on the latter's ability to generate local fields.

Patterns of coupling

We shall consider first a molecule that contains two spin-$\frac{1}{2}$ nuclei A and X. Suppose that X is α; then A will precess at a certain frequency as a result of the combined effect of the external field, the shielding constant, and the spin–spin interaction of the nucleus A with X. The spin–spin coupling will result in one line in the spectrum of A being shifted by $\frac{1}{2}J$ from the frequency it would have in the absence of coupling. If X is β, A will precess at a frequency shifted by $-\frac{1}{2}J$. Therefore, instead of a single line from A, we get a doublet of lines separated by a frequency J (Fig. 18.6). The same splitting occurs in the X resonance: instead of a single line it is a doublet with splitting J (the same value as for the splitting of A).

A subtle point is that the A resonance in an A_nX species (such as an A_2X or A_3X species) is also a doublet with splitting J. As we shall explain below, a group of equivalent nuclei resonates like a single nucleus, the only difference being that the intensity of absorption is greater (Fig. 18.7).

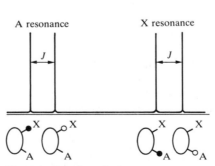

Fig. 18.6 The effect of spin–spin coupling on an AX spectrum. Each resonance is split into two lines separated by J. Full circles indicate α spins, open circles indicate β spins.

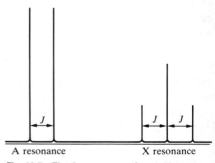

Fig. 18.7 The A resonance of an A_2X species is also a doublet, since the two equivalent nuclei behave like a single A nucleus, but the overall absorption is more intense than for an AX species. The fine structure of the X resonance is explained later.

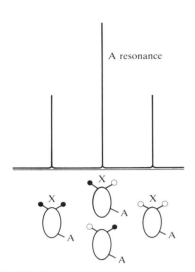

Fig. 18.8 The origin of the 1:2:1 triplet in the A resonance of an AX_2 species. The two X nuclei may have the $2^2 = 4$ spin arrangements $(\alpha\alpha)$; $(\alpha\beta)$; $(\beta\alpha)$; $(\beta\beta)$. The middle two arrangements are responsible for the coincident resonances of A.

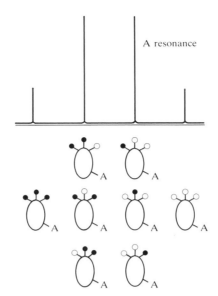

Fig. 18.9 The origin of the 1:3:3:1 quartet in the A resonance of an AX_3 species. There are $2^3 = 8$ arrangements of the spins of the three X nuclei, and their effects on the A nucleus gives rise to four groups of resonances.

If there is another X nucleus in the molecule with the same chemical shift as the first X (giving an AX_2 species), the resonance of A is split into a doublet by one X, and each line of the doublet is split again by the same amount by the second X. This splitting results in three lines in the intensity ratio 1:2:1 (Fig. 18.8) because the central frequency can be obtained in two ways. As in the A_2X case discussed above, the X resonance of the AX_2 species is split into a doublet by A.

Three equivalent X nuclei (an AX_3 species) split the resonance of A into four lines of intensity ratio 1:3:3:1 (Fig. 18.9). The X resonance is still a doublet. In general, N equivalent spin-$\frac{1}{2}$ nuclei split the resonance of a nearby spin or group of equivalent spins into $N + 1$ lines with an intensity distribution given by **Pascal's triangle** (as shown in the margin). Subsequent rows of the triangle are formed by adding together the two adjacent numbers in the line above.

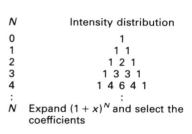

N	Intensity distribution
0	1
1	1 1
2	1 2 1
3	1 3 3 1
4	1 4 6 4 1
⋮	⋮
N	Expand $(1 + x)^N$ and select the coefficients

Example 18.3: *Accounting for the fine structure in a spectrum*

Account for the fine structure in the ^1H–NMR spectrum of the C—H protons of ethanol.

Answer. The three protons of the CH_3 group split the single resonance of the CH_2 protons into a 1:3:3:1 quartet with a splitting J. Likewise, the two protons of the CH_2 group split the single resonance of the CH_3 protons into a 1:2:1 triplet. Each of these lines is split into a doublet to a small extent by the OH proton.

Exercise. What fine-structure can be expected for the protons in NH_4^+?

[1:1:1 triplet from N]

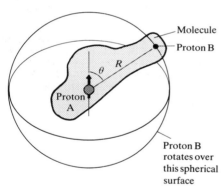

1

The size of coupling constants

The spin–spin coupling constant of two nuclei joined by N bonds is normally denoted $^N J$, with subscripts for the types of nuclei involved. Thus, $^1 J_{CH}$ is the coupling constant for a proton joined directly to a ^{13}C atom, and $^2 J_{CH}$ is the coupling constant when the same two nuclei are separated by two bonds (as in $^{13}C—C—H$). A typical value of $^1 J_{CH}$ is between 10^2 and 10^3 Hz; $^2 J_{CH}$ is about 10 times less, between about 10 and 10^2 Hz. Both $^3 J$ and $^4 J$ give detectable effects in a spectrum, but couplings over larger numbers of bonds can generally be ignored. One of the longest couplings that has been detected is $^9 J_{HH} = 0.4$ Hz in $CH_3C{\equiv}CC{\equiv}CC{\equiv}CCH_2OH$.

The value of J varies with the angle between the bonds. Thus, a $^3 J$ coupling constant depends on the angle ϕ (**1**) as

$$J = A \cos 2\phi + B \cos \phi + C$$

with A, B, and C empirical constants. J also depends on the hybridization of the C atom, as the following values indicate:

	sp	sp^2	sp^3
$^1 J_{CH}$:	250	160	125 Hz

The spin-spin coupling constant may be positive or negative. A positive J signifies that the nuclei have a higher energy when their spins are parallel; a negative sign signifies that their energy is higher when they are antiparallel.

The origin of spin–spin coupling

Spin–spin coupling is a very subtle phenomenon, and it is better to treat J as an empirical parameter than to use calculated values. However, it is possible to get some insight into its origins, if not its precise magnitude or always reliably its sign, by considering the magnetic interactions within molecules.

A nucleus with spin projection m_I gives rise to a magnetic field with z component B_{nuc} at a distance R, where

$$B_{nuc} = \frac{-g_I \mu_N \mu_0}{4 \pi R^3} (1 - 3 \cos^2 \theta) m_I \tag{6}$$

The angle θ is defined in Fig. 18.10 (μ_0 is the vacuum permeability). The magnitude of this field is about 0.1 mT when $R = 0.3$ nm, corresponding to $J \approx 10^4$ Hz, and is of the order of magnitude of the splitting observed in solid samples. However, in a liquid the angle θ sweeps over all values as the molecule tumbles, and $1 - 3 \cos^2 \theta$ averages to zero.[3] Hence the direct dipolar interaction between spins cannot account for the fine structure of the spectra of rapidly tumbling molecules. The direct interaction does make an important contribution to the spectra of solid samples and to the spectra of molecules that tumble only slowly in solution, such as biological and synthetic macromolecules (Section 23.5).

Spin–spin coupling in molecules in solution can be explained in terms of the **polarization mechanism**, in which the interaction is transmitted through the electrons of the bonds. The simplest case to consider is that of $^1 J_{XY}$ where X and Y are spin-$\frac{1}{2}$ nuclei joined by an electron–pair bond (Fig.

Fig. 18.10 The dipole–dipole interaction between two nuclei depends on the angle θ. When the molecule rotates, θ sweeps through all values. As a result, $1 - 3 \cos^2 \theta$ averages to zero, and the dipole–dipole interaction averages to zero.

---

[3] The volume element in polar coordinates is proportional to $\sin \theta \, d\theta$ and θ ranges from 0 to π. Therefore the average value of B_{nuc} for a tumbling molecule is proportional to

$$\int_0^\pi (1 - 3 \cos^2 \theta) \sin \theta \, d\theta = 0$$

18.11). The coupling mechanism depends on the fact that in some atoms it is favourable for the nucleus and a nearby electron spin to be parallel (both α or both β), but in others it is favourable for them to be antiparallel (one α and the other β). The electron–nucleus coupling is magnetic in origin, and may be either a dipolar interaction between the magnetic moments of the electron and nuclear spins or a **Fermi contact interaction**. The latter depends on the very close approach of an electron to the nucleus and hence can occur only if the electron occupies an s orbital. We shall suppose that it is energetically favourable for an electron spin and a nuclear spin to be parallel.

If the X nucleus is α, an α electron of the bonding pair will tend to be found nearby (since that is energetically favourable). The second electron in the bond, which must have β spin if the other is α, will therefore be found mainly at the far end of the bond, near the Y nucleus. Since it is energetically favourable for the Y nucleus to be parallel to an electron spin, the β spin of the Y nucleus (the upper state in a magnetic field) has a lower energy than in the absence of X and the α spin of Y (its lower state) has a higher energy. Since the two levels of Y are closer together, the resonance frequency of Y is lower than in the absence of X_α. If the spin of X is β, the opposite is true. Now the upper (β) state of Y has a higher energy and its lower state (α) a lower energy. The resonance frequency of Y is now increased by the presence of X_β. The overall result is a splitting of the Y resonance into two components by the spin of X.

To account for the value of $^2J_{XY}$, as in X–C–Y, we need a mechanism that can transmit the spin alignments through the central C atom (which may be ^{12}C with no nuclear spin of its own). In this case (Fig. 18.12), the X_α nucleus polarizes the electrons in its bond, and the β electron is closer to the C nucleus. Since the more favourable arrangement of two electrons on the same atom is with their spins parallel (Hund's rule, Section 13.6), the more favourable arrangement is for the β electron of the neighbouring bond to be close to the C nucleus. Consequently, the α electron of that bond is more likely to be found close to the Y nucleus, and therefore that nucleus will have a lower energy if it is α. Hence, by the same argument as before, the upper and lower states of Y are now further apart in the presence of X_α, and Y has a higher resonance frequency. Likewise, a lower Y resonance frequency is obtained if X is β.

The coupling of nuclear spin to electron spin by the Fermi contact interaction is most important for proton spins, but it is not necessarily the most important mechanism for other nuclei. These may also interact by a dipolar mechanism with the electron magnetic moments and with their orbital motion, and there is no simple way of specifying whether J will be positive or negative.

Equivalent nuclei

A group of nuclei are **chemically equivalent** if they are related by a symmetry operation of the molecule and have the same chemical shifts. They are **magnetically equivalent** if, as well as being chemically equivalent, they also have identical spin–spin interactions.

The difference between chemical and magnetic equivalence is illustrated by CH_2F_2 and $H_2C=CF_2$, in both of which the protons are chemically equivalent. However, although the protons in CH_2F_2 are magnetically equivalent, those in $CH_2=CF_2$ are not. One proton in the latter has

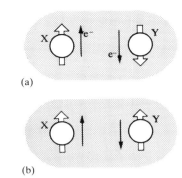

Fig. 18.11 The polarization mechanism for spin–spin coupling. The arrangement in (a) has slightly lower energy than that in (b), and so there is an effective coupling between the spins of magnitude J. In this case J is positive.

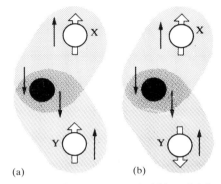

Fig. 18.12 The arrangement in (a) has slightly lower energy than that in (b) and so there is an effective coupling between the proton spins; in this case the spin–spin coupling constant is negative.

spin-coupling interactions with *cis* and *trans* F nuclei ($I = \frac{1}{2}$), which might be α and β respectively. However, the second proton in the same molecule will be *cis* to the β F nucleus and *trans* to the α F nucleus. In CH_2F_2 both protons are equally distant from the two F nuclei, so there is no distinction between them. Strictly speaking, the CH_3 protons in ethanol (and other compounds) are magnetically inequivalent on account of their different interactions with the CH_2 protons in the next group. However, they are in practice made magnetically equivalent by the rapid rotation of the CH_3 group, which averages out any differences.

An important feature of chemically equivalent magnetic nuclei is that, although they do couple together, the coupling has no effect on the appearance of the spectrum. The reason for the invisibility of the coupling is that all allowed nuclear spin transitions are *collective* reorientations of groups of equivalent nuclear spins that do not change the *relative* orientations of the spins within the group (Fig. 18.13). Since the relative orientations of nuclear spins are not changed in any transition, the magnitude of the coupling between them is undetectable. Hence, an isolated CH_3 group gives a single, unsplit line because all the allowed transitions of the group of three protons occur without change of their relative orientations.

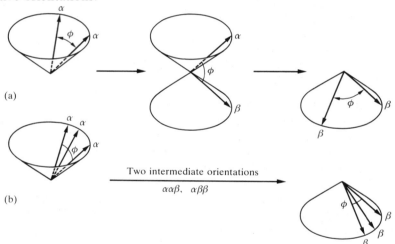

Fig. 18.13 (a) A group of two equivalent nuclei realigns as a group, without change of angle between the spins, when a resonant absorption occurs. Hence it behaves like a single nucleus and the spin–spin coupling between the individual spins of the group is undetectable. (b) Three equivalent nuclei also realign as a group without change of their relative orientations.

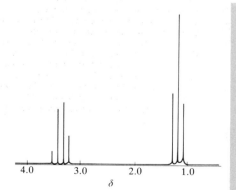

Fig. 18.14 The NMR spectrum referred to in Example 18.4. The field increases to the right.

Example 18.4: *Interpreting an NMR spectrum*

Suggest an interpretation of the 60 MHz ^{1}H–NMR spectrum in Fig. 18.14.

Answer. We need to look for groups with characteristic chemical shifts (Fig. 18.3) and account for the fine structure as was done for ethanol. The resonance at $\delta = 3.4$ corresponds to $>CH_2$ in an ether; that at $\delta = 1.2$ corresponds to CH_3 in CH_3CH_2. The fine structure of the $>CH_2$ group (a $1:3:3:1$ quartet) is characteristic of splitting caused by CH_3; the fine structure of the CH_3 resonance is characteristic of splitting caused by CH_2. The fine structure constant is $J = \pm 60$ Hz (the same for each group). The compound is probably $(CH_3CH_2)_2O$.

Exercise. What changes in the spectrum would be observed on recording it at 300 MHz?   [Groups of lines 5 times further apart in frequency (but the same δ values); no change in spin–spin splitting]

Second-order spectra

NMR spectra are usually much more complex than the foregoing simple analysis suggests. We have described the extreme case in which the differences in chemical shifts are much greater than the spin–spin coupling constants. In such cases it is simple to identify groups of chemically equivalent nuclei and to think of the groups as reorientating relative to each other. The spectra that result are called **first-order spectra**.

Nuclei cannot be allocated to definite groups when the differences in their chemical shifts are comparable to their spin–spin coupling interactions. The complicated spectra that are then obtained are called **second-order spectra**, and are much more difficult to analyse (Fig. 18.15). Since the difference in resonance frequencies increases with field, but spin–spin coupling constants are independent of it, a second-order spectrum may become simpler (and first-order) at high fields because groups of nuclei become identifiable again.

A clue to the type of analysis that is appropriate is given by the notation for the types of spins involved. Letters near each other in the alphabet are used to denote nuclei with similar chemical shifts. Letters far apart are used to denote nuclei with large differences in chemical shift. Thus an AX spin system consists of two nuclei with a large chemical shift difference and a first-order spectrum. An AB system, on the other hand, gives a second-order spectrum. An AX system may have widely different chemical shifts because A and X are nuclei of different elements (such as ^{13}C and ^1H), in which case they form a **heteronuclear spin system**. AX may also denote a **homonuclear spin system** in which the nuclei are of the same element but are in markedly different environments.

Dilute and abundant spins: spin decoupling

^{13}C is a **dilute-spin species** in the sense that it is unlikely that more than one ^{13}C nucleus will be found in any given molecule (provided the sample has not been enriched with that isotope; the natural abundance of ^{13}C is only 1.1 per cent). Hence, it is not normally necessary to take into account ^{13}C–^{13}C spin–spin coupling within a molecule.

Protons are **abundant-spin species** in the sense that a molecule is likely to contain many of them. If we were observing a ^{13}C-NMR spectrum, we would obtain a very complex spectrum on account of the coupling of the one ^{13}C nucleus with all the protons that are present. In order to avoid this difficulty ^{13}C-NMR spectra are normally observed using the technique of **proton decoupling**. Thus, if the CH_3 protons of ethanol are irradiated with a second, strong, resonant radiofrequency source they undergo rapid spin reorientations and the ^{13}C nucleus senses an average orientation. As a result, its resonance is a single line and not a $1:3:3:1$ quartet. Proton decoupling has the additional advantage of enhancing sensitivity, because the intensity is concentrated into a single transition instead of being spread over several. A disadvantage, however, is that the signal strength is no longer proportional to the number of nuclei, and so signal integration is unhelpful.

Pulse techniques in NMR

Modern methods of detecting the energy separation ΔE between nuclear spin states are more sophisticated than simply looking for the frequency at

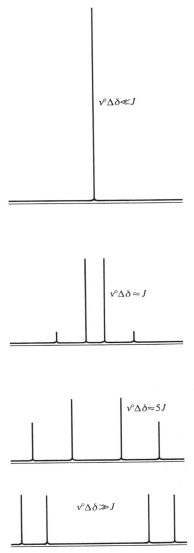

Fig. 18.15 NMR spectra are complicated when spin–spin coupling constants and chemical shifts are comparable. This diagram shows how the resonance of a two-spin system changes as J increases relative to the difference $\Delta\delta$ of chemical shifts. At the top $(\Delta\delta)v° \ll J$ and the spectrum is that of an A_2 species. At the bottom $(\Delta\delta)v° \gg J$ and the (first-order) spectrum is that of an AX species. The intermediate illustrations are for $(\Delta\delta)v° \approx J$ and are the second-order spectra typical of an AB species.

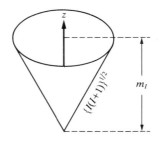

Fig. 18.16 The vector model of angular momentum for a single spin-$\frac{1}{2}$ nucleus. The angle around the z-axis is indeterminate.

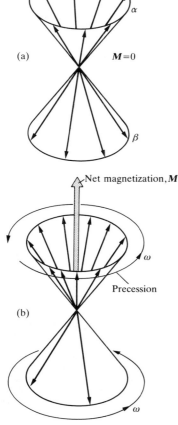

Fig. 18.17 The magnetization of a sample of spin-$\frac{1}{2}$ nuclei is the resultant of all their magnetic moments. (a) In the absence of an externally applied field, there are equal numbers of α and β spins at random angles around the z-axis (the field direction) and the magnetization is zero. (b) In the presence of a field, the spins precess around their cones and there are also slightly more α spins than β spins. As a result, there is a net magnetization along the z-axis.

which resonance occurs. One of the best analogies that has been suggested to illustrate the difference between the old and new ways of observing an NMR spectrum is that of detecting the spectrum of vibrations of a bell. We could stimulate the bell with a gentle vibration at a gradually increasing frequency, and note the frequencies at which it resonates with the stimulation. A lot of time would be spent getting zero response when the frequency was between the bell's vibrational modes. However, if we were simply to hit the bell with a hammer, we would immediately obtain a tone composed of all the frequencies that the bell can produce. The equivalent in NMR is to stimulate the nuclei to occupy their upper spin state, and then to monitor the radiation they emit as they return to the ground state.

We need to understand how the equivalent of the hammer blow is delivered and how the signal is monitored and interpreted. These features are generally expressed in terms of the vector model of angular momentum introduced in Section 12.7.

18.4 The magnetization vector

We consider a sample composed of many identical nuclei with spin quantum number $I = \frac{1}{2}$. As we saw in Section 12.7, an angular momentum can be represented by a vector of length $\{I(I + 1)\}^{1/2}$ units with a component of length m_I units along the z axis. Since the uncertainty principle does not allow us to specify the x and y components of the angular momentum, all we know is that the vector lies somewhere on a cone around the z axis. For $I = \frac{1}{2}$, the length of the vector is $\frac{1}{2}\sqrt{3}$ and it makes an angle of 55° to the z axis (Fig. 18.16).

In the absence of a magnetic field, the sample consists of equal numbers of α and β nuclear spins with their vectors lying at random angles ϕ on the cones (Fig. 18.17a). The angles ϕ are unpredictable, and at this stage we picture the spin vectors as stationary. The **magnetization M** of the sample, its net nuclear magnetic moment, is zero.

The effect of the static field

Two changes occur in the magnetization when a magnetic field is present. First, the energies of the two orientations change, the α spins moving to low energy and the β spins to high energy (provided $g_I > 0$). At 10 T, the Larmor frequency for protons is $2.7 \times 10^9\,\text{s}^{-1}$, so we can now picture the individual vectors as **precessing**, or sweeping round their cones, at this rate (which corresponds to 430 MHz, 4.3×10^8 cycles per second). As the field is increased the Larmor frequency increases and the precessional motion becomes faster. Secondly, the populations of the two spin states—the numbers of α and β spins—change, and there will be more α spins than β spins. Since the β state lies at an energy $\hbar\omega$ above the α state, at a temperature T the populations are related by

$$\frac{N_\beta}{N_\alpha} = e^{-\hbar\omega/kT} = 1 - \frac{\hbar\omega}{kT} + \cdots$$

Because $\hbar\omega/kT \approx 7 \times 10^{-5}$ for protons at room temperature and 10 T, there is only a tiny imbalance of populations, and it is even smaller for other nuclei with their smaller magnetogyric ratios. However, despite its smallness, the imbalance means that there is a net magnetization that we can represent by a vector M pointing in the z direction and with a length proportional to the population difference (Fig. 18.17b).

The effect of the radiofrequency field

We now consider the effect of a circularly polarized radiofrequency field in the xy plane. We have considered the *electric* component of this field in the other forms of spectroscopy that we have treated. However, in this chapter we consider only the *magnetic* component, for it is this component that interacts with the nuclear magnetic moment. The strength of the oscillating magnetic field is B_1.

Suppose we choose the frequency of the oscillating field to be equal to the Larmor frequency of the spins (about 430 MHz in a 10 T magnetic field). This is equivalent to choosing the resonance condition in the conventional experiment. The nuclei now experience a *steady* B_1 field (Fig. 18.18) because the rotating magnetic field is in step with the precessing spins. Under the influence of this steady field, the magnetization vector begins to precess around its direction. If we apply the B_1 field in a pulse of a certain duration, the magnetization precesses into the xy plane (Fig. 18.19) and we

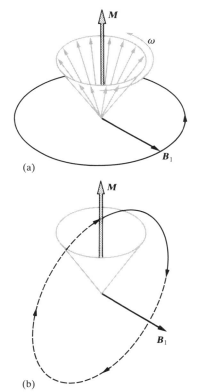

(a)

(b)

Fig. 18.18 (a) In a resonance experiment, a circularly polarized radiofrequency magnetic field B_1 is applied in the xy-plane. (b) If we step into a frame rotating at the Larmor frequency, the radiofrequency field appears to be stationary if its frequency is the same as the Larmor frequency. When the two frequencies coincide, the magnetization vector of the sample begins to rotate around the direction of the B_1 field.

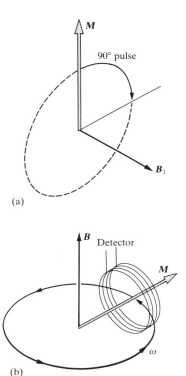

(a)

(b)

Fig. 18.19 (a) If the radiofrequency field is applied for a certain time, the magnetization vector is rotated into the xy-plane. (b) To an external observer, the vector is rotating at the Larmor frequency, and can induce a signal in the coils.

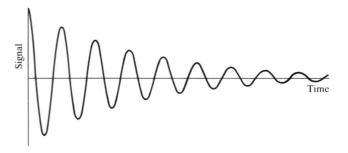

Fig. 18.20 A simple free-induction decay of a sample of spins with single resonance frequency.

say that we have applied a **90° pulse** (or a '$\pi/2$ pulse'). The duration of the pulse depends on the strength of the B_1 field, but is typically of the order of milliseconds. To a stationary external observer (a radiofrequency coil), the magnetization vector is now rotating in the xy plane at the Larmor frequency (at about 430 MHz). The rotating magnetization induces a 430 MHz signal in the coil, which can be amplified and processed.

As time passes, the individual spins move out of step (partly because they are precessing at slightly different rates, as we shall explain later) and so the magnetization vector shrinks exponentially with a time constant T_2 and induces an ever weaker signal in the detector coil. The form of the signal that we can expect is therefore the oscillating-decaying **free-induction decay** (FID) shown in Fig. 18.20, and the y-component of the magnetization varies as

$$M_y(t) = M_0 \cos \omega t \, e^{-t/T_2} \qquad (7)$$

We have considered the effect of a pulse of precise frequency. However, virtually the same effects are obtained if we expose the sample to a 90° pulse of radiation that spans a range of frequencies.[4] Although all the different B_1 fields are not equally efficient at rotating the magnetization vector into the xy plane, they do not differ very much and we can be confident that the magnetization will end up largely in the plane. Once there, it rotates at the Larmor frequency, and generates a free-induction decay signal. Note that we do not need to know this frequency initially: the broad-band pulse is the analogue of the hammer blow on the bell, and the detected signal is a sign that a particular frequency (in our case, a nuclear spin energy level separation) is present.

Time- and frequency-domain spectra

We can think of the magnetization vector of a homonuclear AX spin system as consisting of two parts, one formed by the A spins and the other by the X spins. When the 90° pulse is applied, both magnetization vetors are rotated into the xy plane. However, because the A and X nuclei precess at different frequencies they induce two signals in the detector coils, and the overall FID curve may resemble that in Fig. 18.21. The composite FID curve is the

---

[4] In fact, any pulse must cover a range of frequencies: we can have a monochromatic wave only if it continues for ever. For exactly the same reason as there is an inverse relation between the width of a spectral line and the lifetime of a state (Section 16.3), the length of the pulse τ and its frequency spread $\Delta \nu$ are related by $\Delta \nu \approx 1/2\pi\tau$. A pulse lasting 1 μs has a frequency spread of nearly 200 kHz.

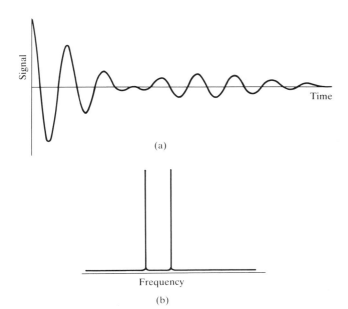

(a)

(b)

Fig. 18.21 (a) A free induction decay signal of a sample of AX species and (b) its analysis into its frequency components.

analogue of the struck bell emitting a rich tone composed of all the frequencies at which it can vibrate.

The problem we must address is how to recover the resonance frequencies present in a free-induction decay. We encountered a similar problem when discussing Fourier-transform infrared spectra in Section 16.1, where all the vibrational frequencies were detected at once. We use the same technique here. We know that the FID curve is a sum of oscillating functions, so the problem is to analyse it into its harmonic components.

The analysis of the FID curve is achieved by the standard mathematical technique of Fourier transformation. We start by noting that the signal $S(t)$ in the **time domain**, the total FID curve, is the sum (more precisely, the integral) over all the contributing frequencies

$$S(t) = \int_{-\infty}^{\infty} I(v)e^{-2\pi i v t}\, dv \qquad (8a)$$

└─Signal oscillating with frequency v

└─Intensity of the contribution of the frequency v

└─Sum over all possible frequencies

We need $I(v)$, the spectrum in the **frequency domain**; it is obtained by evaluating the integral

$$I(v) = 2\,\mathrm{re} \int_{0}^{\infty} S(t)e^{2\pi i v t}\, dt \qquad (8b)$$

The integration is carried out at a series of frequencies v on a computer that is an integral part of the spectrometer. When the signal in Fig. 18.21a is transformed in this way, we get the frequency-domain spectrum shown in Fig. 18.21b. One line represents the Larmor frequency of the A nuclei and the other that of the X nuclei.

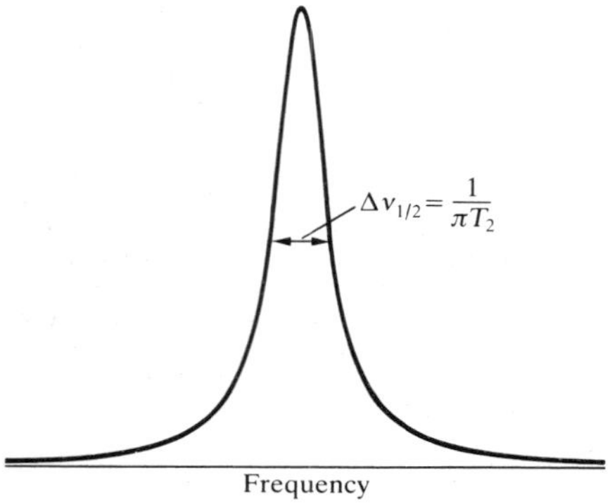

Fig. 18.22 A Lorentzian line typical of NMR spectral lines.

Example 18.5: *Performing a Fourier transformation*

Suppose that the signal is proportional to the magnetization $M_y(t)$ given in eqn 7. What is the form of the frequency-domain spectrum?

Answer. We must evaluate $I(\nu)$; the best way of proceeding is to use the relation

$$\cos \omega t = \tfrac{1}{2}(e^{i\omega t} + e^{-i\omega t})$$

because then all the integrals are of exponentials. The transformation we require is

$$I(\nu) \propto 2 \, \mathrm{re} \int_0^\infty M_y(t) e^{2\pi i \nu t} \, dt$$

$$\propto \mathrm{re} \, M_0 \left\{ \int_0^\infty e^{(2\pi i \nu + i\omega - 1/T_2)t} \, dt + \int_0^\infty e^{(2\pi i \nu - i\omega - 1/T_2)t} \, dt \right\}$$

$$\propto \mathrm{re} \, M_0 \left\{ \frac{1}{(2\pi\nu + \omega)i - 1/T_2} + \frac{1}{(2\pi\nu - \omega)i - 1/T_2} \right\}$$

At typical operating frequencies, both ν and $\omega/2\pi$ are close to 10^8 Hz, so the first term is of the order of 10^{-8} s. However, since $2\pi\nu - \omega$ is close to zero, the second term is of the order of T_2, which may be about 1 s. Therefore, we neglect the first term, and obtain

$$I(\nu) \propto M_0 \, \mathrm{re} \, \frac{1}{(2\pi\nu - \omega)i - 1/T_2} = M_0 \frac{T_2}{1 + (2\pi\nu - \omega)^2 T_2^2}$$

This curve has its maximum value at $\nu = \omega/2\pi$, corresponding to resonance when the radiofrequency field matches the Larmor frequency.

Comment. The absorption curve obtained is called a 'Lorentzian line'; one is plotted in Fig. 18.22; it is the typical shape of an NMR absorption.

Exercise. Find the relation between the width of the absorption at half-height and T_2.

$$[\Delta\nu_{1/2} = 1/\pi T_2]$$

The FID curve in Fig. 18.23 is obtained from a sample of ethanol. The frequency-domain spectrum obtained from it by Fourier transformation is the one that we have already discussed (Fig. 18.3). We can now see why the

Fig. 18.23 A free induction decay signal of a sample of ethanol. Its Fourier transform is the frequency-domain spectrum shown in Fig. 18.3.

FID curve in Fig. 18.23 is so complex: it arises from the precession of a magnetization vector that is composed of eight components, each with a characteristic frequency.

18.5 Linewidths and rate processes

The linewidths of NMR spectra, in common with other spectroscopic techniques, provide information about the rates of processes relating to the molecules in the sample. We have seen that the free-induction decay signal decays with time, which implies that the component of the magnetization vector in the xy plane must be shrinking. In this section we see some of the processes involved.

Spin relaxation

There are two reasons why the component of the magnetization vector in the xy plane shrinks. Both reflect the fact that the nuclear spins are not in thermal equilibrium with their surroundings (for then M lies parallel to z) and returning to equilibrium is the process called **spin relaxation**.

At thermal equilibrium the spins have a Boltzmann distribution, with more α spins than β spins; however, a magnetization vector in the xy plane immediately after a 90° pulse has equal numbers of α and β spins. The populations revert to their thermal equilibrium values exponentially, and as they do so the z component of magnetization reverts to its equilibrium value M_z with a time constant called the **longitudinal relaxation time** T_1 (Fig. 18.24):

$$M_z(t) - M_z \propto e^{-t/T_1}$$

Since this relaxation process involves giving up energy to the surroundings (the 'lattice') as β spins revert to α spins, the time constant T_1 is also called the **spin–lattice relaxation time**. Spin–lattice relaxation is caused by fluctuating local magnetic fields arising from the motion of the molecules. These fluctuations can stimulate the spins to change from β to α and hence to relax towards the thermal equilibrium population.

A second aspect of spin relaxation is the fanning-out of the spins in the xy plane if they precess at different rates (Fig. 18.25). The magnetization vector is large when all the spins are bunched together immediately after a 90° pulse. However, this orderly bunching of spins is not at equilibrium, and even if there were no spin–lattice relaxation, we would expect the individual spins to spread out until they were uniformly distributed with all possible angles around the z-axis, when the component of magnetization vector in the plane would be zero. The randomization of the spin directions occurs exponentially with a time constant called the **transverse relaxation time** T_2:

$$M_y(t) \propto e^{-t/T_2}$$

Since the relaxation involves the relative orientation of the spins, T_2 is also known as the **spin–spin relaxation time**.

We saw in Example 18.5 that if the y component of magnetization decays with a time constant T_2, the spectral line is broadened and its width at half-height becomes

$$\Delta v_{1/2} = \frac{1}{\pi T_2} \tag{9}$$

Fig. 18.24 In longitudinal relaxation the spins relax back towards their thermal equilibrium populations. In (a) we see the precessional cones representing spin-$\frac{1}{2}$ angular momenta, and they do not have their thermal equilibrium populations (there are more β-spins than α-spins). The time constant for return to the Boltzmann distribution in (b) is T_1.

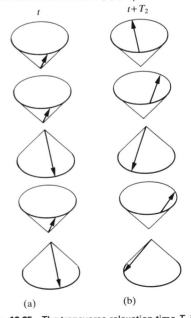

Fig. 18.25 The transverse relaxation time T_2 is the time it takes for the phases of the spins to become randomized (another condition for equilibrium) and to change from the orderly arrangement shown in (a) to the disorderly arrangement in (b).

Typical values of T_2 in proton NMR are of the order of seconds, and so linewidths of around 0.1 Hz can be anticipated, in broad agreement with observation. In mobile liquids, $T_2 \approx T_1$.

So far, we have assumed that the equipment, and in particular the magnet, are perfect, and that the differences in Larmor frequencies arise solely from interactions within the sample. In practice, the magnet is not perfect, and the field is different at different locations in the sample. The inhomogeneity broadens the resonance, and in most cases this **inhomogeneous broadening** of the lines dominates the broadening that we have discussed so far. It is common to express the extent of inhomogeneous broadening in terms of an **effective transverse relaxation** time T_2^* using a relation like eqn 9 but writing

$$T_2^* = \frac{1}{\pi \Delta v_{1/2}} \tag{10}$$

where $\Delta v_{1/2}$ is the *observed* width at half-height.

Example 18.6: *Estimating the effective relaxation time*

A line in the ^{1}H–NMR spectrum of GeH_4 had a width of 10 Hz. What is the effective transverse relaxation time?

Answer. We use eqn 10, with $1\,Hz = 1\,s^{-1}$:

$$T_2^* = \frac{1}{\pi \times 10\ s^{-1}} = 32\ ms$$

Exercise. A spectrum gave a linewidth of 0.30 Hz. What is the effective transverse relaxation time?

[1.1 s]

Exchange processes

The appearance of an NMR spectrum is changed if magnetic nuclei can jump rapidly between different environments. Consider a fluxional molecule that can jump between conformations, such as the boat–chair inversion of a substituted cyclohexane (Fig. 18.26). In one conformation the substituent is axial, in another it is equatorial. The chemical shifts of the axial and equatorial ring protons are different, and so they are in effect jumping between two magnetic environments. When the inversion rate is slow, the spectrum shows two sets of lines, one from molecules with an axial substituent and one from molecules with an equatorial substituent. When the inversion is fast, the spectrum shows a single line at the mean of the two chemical shifts.

At intermediate inversion rates, the line is very broad. This broadening occurs when the lifetime τ of a conformation is so short that the lifetime broadening (eqn 12 of Section 16.3) is comparable to the difference of

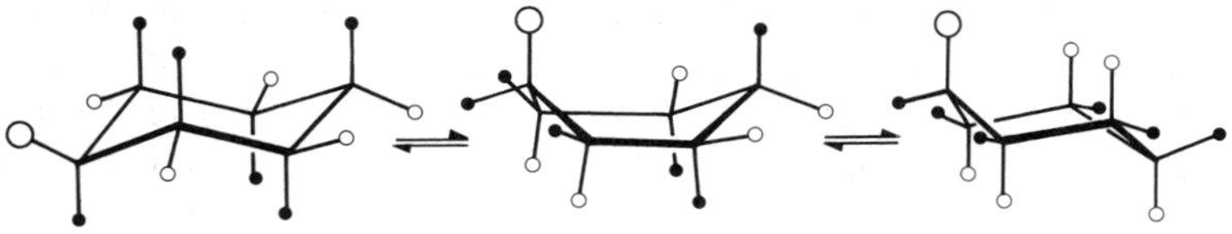

Fig. 18.26 When a substituted cyclohexane molecule undergoes inversion, the axial and equatorial protons are interchanged and jump between magnetically distinct environments.

chemical shifts. That is, the collapse of structure occurs when

$$\tau < \frac{1}{2\pi\delta\nu} \qquad (11)$$

For example, if the chemical shifts differ by 100 Hz, the spectrum collapses into a single line when the conformation lifetime is less than about 2 ms.

Example 18.7: *Interpreting lifetime broadening*

The NO group in N,N-dimethylnitrosamine, $(CH_3)_2N$—NO, rotates and, as a result, the magnetic environments of the two CH_3 groups are interchanged. In a 60 MHz spectrometer the two CH_3 resonances are separated by 39 Hz. At what rate of interconversion will the resonance collapse to a single line?

Answer. We can use eqn 11 for the average lifetimes of the tautomers with $\delta\nu = 39$ Hz. This gives $\tau < 4.1$ ms. The rate of interconversion is the inverse of the average lifetime, so the signal will collapse to a single line when the interconversion rate exceeds $250 \, s^{-1}$.

Comment. The dependence of the rate of collapse on the temperature is used to determine the energy barrier to the interconversion.

Exercise. What would the minimum lifetime need to be in a 300 MHz spectrometer for a single line to be observed?   [0.8 ms]

A similar explanation accounts for the loss of structure in solvents able to exchange protons with the sample. For example, hydroxyl protons are able to exchange with water protons. When this **chemical exchange** occurs, a molecule ROH with an α-spin proton (we write this ROH_α) rapidly converts to ROH_β and then perhaps to ROH_α again because the protons provided by the solvent molecules in successive exchanges have random spin orientations. Therefore, instead of seeing a spectrum composed of contributions from both ROH_α and ROH_β molecules (that is, a spectrum showing the doublet structure due to the OH proton) we see a single, unsplit line at the mean position (as in Fig. 18.3). The effect is observed when the lifetime of a molecule due to this chemical exchange is so short that the lifetime broadening is greater than the doublet splitting. Since this splitting is often very small (about 1 Hz), a proton must remain attached to the same molecule for longer than about 0.1 s if the splitting is to be observable. In water the exchange rate is much faster than that, and so alcohols show no splitting from the OH protons. In very dry alcohol the exchange rate may be slow enough for splitting to be detected.

The measurement of T_1

The longitudinal relaxation time can be measured by the **inversion recovery technique**. The first step is to apply a **180° pulse** to the sample. A 180° pulse is achieved by applying the B_1 field for twice as long as for a 90° pulse, so that the magnetization vector precesses through 180° and points in the $-z$ direction (Fig. 18.27). No signal can be seen at this stage because there is no component of magnetization in the xy plane where the detection coils are sensitive. The β spins begin to relax back into α spins, and the

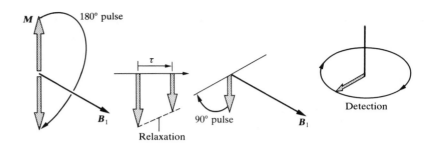

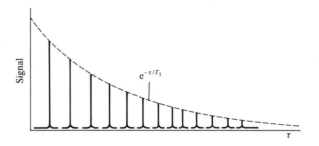

Fig. 18.27 The result of applying a 180° pulse to the magnetization in the rotating frame and the effect of a subsequent 90° pulse. The longer the interval between the two pulses, the smaller the amplitude of the frequency–domain spectrum since spin–lattice relaxation has occurred.

magnetization vector shrinks exponentially back towards its thermal equilibrium value M_z. After an interval τ, a 90° pulse is applied that rotates the magnetization into the xy plane, where it starts to generate a free-induction decay signal. The frequency–domain spectrum is then obtained by Fourier transformation.

The intensity of the spectrum obtained in this way depends on the length of the magnetization vector that is rotated into the xy plane. The length of the vector decreases exponentially as the interval between the two pulses is increased and so the intensity of the spectrum also decreases exponentially with increasing τ. We can therefore measure T_1 by fitting an exponential curve to the series of spectra obtained after different values of τ.

Spin echoes

The measurement of T_2 (as distinct from T_2^*) depends on being able to eliminate the effects of inhomogeneous broadening. The cunning required is at the root of some of the most important advances that have been made in NMR since its introduction.

A **spin echo** is a magnetic analogue of an audible echo: a pulse of magnetization is formed, allowed to spread, is reflected, and is then detected as another pulse a short time later. The sequence of events is illustrated in Fig. 18.28. First, a 90° pulse is applied to the sample in the x direction, and the magnetization rotates into the xy plane. The spins now begin to fan out because they have different Larmor frequencies, with some going faster than the mean and others going more slowly, and the signal decays with a time-constant T_2^*. We shall consider the overall magnetization as being made up of a number of different magnetizations, each one of which arises from a **spin packet** of nuclei with closely similar precession frequencies.

A 180° pulse is applied in the y direction after an interval τ. The pulse rotates the magnetization vectors of the spin packets, moving the magnetization vectors of the fast packets into the angles occupied by the slow packets, and vice versa (Fig. 18.28). The vectors continue to precess, but after an interval τ they will all be found to lie in the y direction again. The regrouping is called **refocusing**, and when the spins are aligned along y a signal will be observed, which then decays again with a time constant T_2^*.

The important feature of the technique is that the size of the echo is independent of any local fields that remain constant during the two τ intervals: if a spin packet is 'fast' because it happens to be composed of spins in a region of the sample that experiences higher than average fields, it remains fast throughout both intervals, and what it gains on the first interval it makes up on the second. Hence, the size of the echo is independent of inhomogeneities in the magnetic field, for these remain constant. The true transverse relaxation arises from fields that fluctuate on a molecular time scale, and there is no guarantee that an individual 'fast' spin will remain 'fast' in the refocussing phase: the spins *within* the packets therefore spread with a time constant T_2. Hence, the effects of the true relaxation are not refocused, and the size of the echo decays with the time constant T_2.

In an actual experiment, the **Carr–Purcell–Meiboom–Gill sequence** of pulses (the CPMG sequence) is used. This sequence consists of a series of 180° pulses after the initial 90° pulse:

$$90°(x) - [\tau - 180°(y) - \tau - \text{Echo}] - [\tau - 180°(y) - \tau - \text{Echo}]$$
$$- [\tau - 180°(y) - \tau - \text{Echo}] - \cdots$$

The amplitudes of the echoes die away with a time constant T_2 as spin–spin relaxation occurs, and the value of T_2 is obtained by fitting an exponential curve to their decay (Fig. 18.29).

18.6 Two-dimensional NMR

An NMR spectrum contains a great deal of information and, if many protons are present, is very complex. Even a first-order spectrum is complex, since the fine structure of different groups of lines can overlap. The complexity would be reduced if we could use two axes to display the data, with resonances belonging to different groups lying at different locations on the second axis. This separation is essentially what is achieved in **two-dimensional NMR**.

We have seen that a spin–echo experiment refocuses spins that are in a constant environment. Hence, if two spins are in environments with different chemical shifts, they will be refocused and a single line will be obtained. That is, we can eliminate chemical shifts from a spectrum. Since we saw earlier that we can also remove the effects of spin–spin coupling by decoupling techniques, we can separate the two contributions to the spectrum. In practice, a clever choice of pulses and Fourier transformation techniques makes it possible to display spin coupling in one dimension and the chemical shifts in another, and so greatly simplify the appearance of a spectrum. The type of spectrum that is obtained is shown in Fig. 18.30 and the techniques themselves are described in the books listed in *Further reading* at the end of the chapter. A two-dimensional plot of the spectrum enriches the information that can be obtained from one-dimensional spectroscopy and makes it possible to interpret spectra that would have been far too complicated to analyze by conventional techniques.

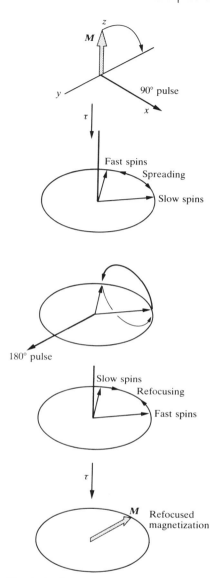

Fig. 18.28 The sequence of pulses leading to the observation of a spin echo (see text).

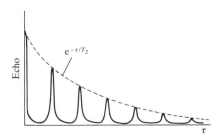

Fig. 18.29 The exponential decay of the spin echoes in a CPMG pulse sequence gives the transverse relaxation time.

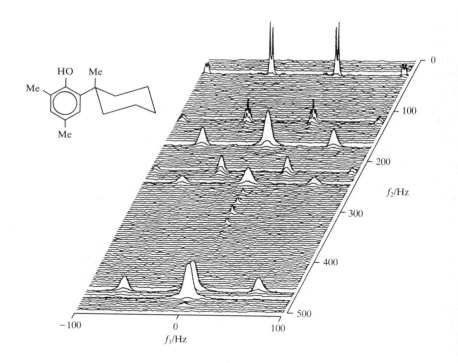

Fig. 18.30 A typical two-dimensional ^{13}C–NMR spectrum. The horizontal axis gives the ^{13}C—H coupling constants and the vertical axis gives the ^{13}C chemical shifts. [Reproduced with permission from M. H. Levitt and R. Freeman. *J. Magn. Reson.,* **34,** 675 (1979).]

Electron spin resonance

The energy levels of an electron spin in a magnetic field B (Fig. 18.31) are

$$E_{m_s} = g_e \mu_B m_s B \qquad m_s = \pm\tfrac{1}{2} \qquad (12)$$

Equation 12 shows that the energy of an α electron ($m_s = +\tfrac{1}{2}$) increases and the energy of a β electron ($m_s = -\tfrac{1}{2}$) decreases as the field is increased, and that the separation of the levels is

$$\Delta E = E_\alpha - E_\beta = g_e \mu_B B$$

The resonance condition is therefore

$$h\nu = g_e \mu_B B \qquad (13)$$

and when this condition is fulfilled strong absorption of radiation occurs. **Electron spin resonance** (ESR; or electron paramagnetic resonance, EPR), is the study of molecules containing unpaired electrons by observing the magnetic fields at which they come into resonance with monochromatic radiation. Magnetic fields of about 0.3 T (the value used in most commercial ESR spectrometers) correspond to resonance with an electromagnetic field of frequency 10 GHz (10^{10} Hz) and wavelength 3 cm. Since 3 cm radiation falls in the X-band of the microwave region of the electromagnetic spectrum, ESR is a microwave technique.

The layout of an ESR spectrometer is shown in Fig. 18.32. It consists of a microwave source (a klystron), a cavity in which the sample is inserted in a glass or quartz container, a microwave detector, and an electromagnet with a field that can be varied in the region of 0.3 T. The ESR spectrum is obtained by monitoring the microwave absorption as the field is changed,

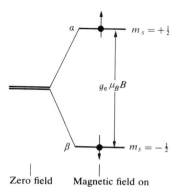

Fig. 18.31 Electron spin levels in a magnetic field. Note that the β state is the lower. Resonance is achieved when the frequency of the incident radiation matches the frequency corresponding to the energy separation.

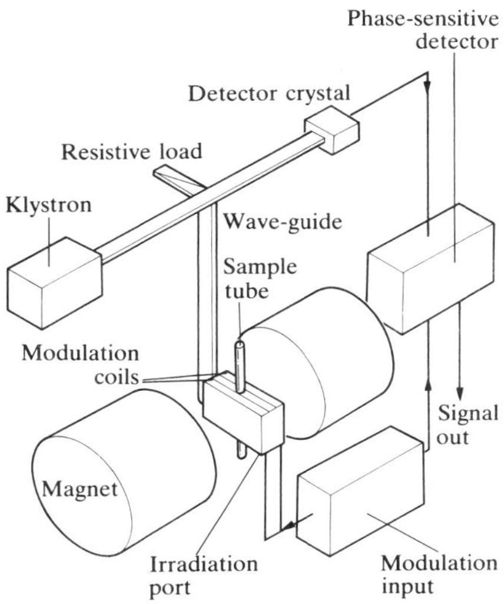

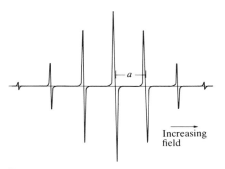

Fig. 18.32 The layout of an ESR spectrometer. A typical magnetic field is 0.3 T, which requires 9 GHz (3 cm) microwaves for resonance.

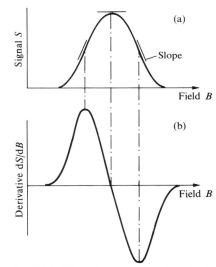

Fig. 18.33 The ESR spectrum of the benzene radical anion, $C_6H_6^-$, in fluid solution. *a* is the hyperfine splitting of the spectrum; the centre of the spectrum is determined by the *g*-value of the radial.

and a typical spectrum (of the benzene radical-anion, $C_6H_6^-$) is shown in Fig. 18.33. The peculiar appearance of the spectrum—which is in fact the first-derivative of the absorption—arises from the detection technique, which is sensitive to the slope of the absorption curve (Fig. 18.34).

The sample molecules must possess unpaired spins, and so ESR is less widely applicable than NMR. It is used to study radicals formed during chemical reactions or by radiation, many *d*-metal complexes, and molecules in triplet states (such as those involved in phosphorescence, Section 17.4). It is insensitive to normal, spin-paired molecules. The sample may be a gas, a liquid, or a solid, but the free rotation of molecules in the gas phase gives rise to complications.

Fig. 18.34 When phase-sensitive detection is used, the signal is the first-derivative of the absorption. (a) The absorption, (b) the signal, the slope of the absorption signal at each point.

18.7 The *g*-factor

As in NMR, the spin magnetic moment interacts with the local magnetic field, and so the resonance condition should be written

$$h\nu = g_e\mu_B B_{loc} = g_e\mu_B(1 - \sigma)B$$

However, in ESR it is conventional to write $g = (1 - \sigma)g_e$, where g is the **g-factor** of the radical or complex. Then the resonance condition is

$$h\nu = g\mu_B B \qquad (14)$$

Example 18.8: *Measuring a g-factor*

The centre of the ESR spectrum of the methyl radical occurred at 329.4 mT in a spectrometer operating at 9.233 GHz. What is the g factor?

Answer. We use eqn 14; a convenient factor is

$$\frac{h}{\mu_B} = 7.144\,48 \times 10^{-11}\,\text{T s} = 71.4448\,\text{mT GHz}^{-1}$$

Therefore, from the data,

$$g = 71.44\,\text{mT GHz}^{-1} \times \frac{9.233\,\text{GHz}}{329.4\,\text{mT}} = 2.0026$$

Comment. Many organic radicals have g-factors close to 2.0026; inorganic radicals have g-factors typically in the range 1.9–2.1; d-metal complexes have g-factors in a wider range (e.g. 0–4).

Exercise. At what field would the methyl radical come into resonance in a spectrometer operating at 9.468 GHz?   [337.8 mT]

The deviation of g from $g_e = 2.0023$ depends on the ability of the applied field to induce local electron currents in the radical, and therefore its value gives some information about electronic structure. However, since g-values differ very little from g_e in many radicals (e.g. 2.003 for H, 1.999 for NO_2, 2.01 for ClO_2), its main use in chemical applications is to aid the identification of the species present in a sample.

18.8 Hyperfine structure

The most important feature of ESR spectra is their **hyperfine structure**, the splitting of individual resonance lines into components. The source of the splitting is the magnetic field arising from nuclear magnetic moments within the radical.

The effects of nuclear spin

Consider the effect on the ESR spectrum of a single H nucleus located somewhere in a radical. The proton spin is a source of magnetic field, and depending on the orientation of the nuclear spin, the field it gives rise to adds to or subtracts from the applied field. The total local field is therefore

$$B_{loc} = B + am_I \quad m_I = \pm\tfrac{1}{2}$$

where a is the **hyperfine coupling constant**. Half the radicals in a sample have $m_I = +\tfrac{1}{2}$, and so half resonate when the applied field satisfies the condition

$$h\nu = g\mu_B(B + \tfrac{1}{2}a) \quad \text{or} \quad B = \frac{h\nu}{g\mu_B} - \tfrac{1}{2}a$$

and the other half (which have $m_I = -\tfrac{1}{2}$) resonate when

$$h\nu = g\mu_B(B - \tfrac{1}{2}a) \quad \text{or} \quad B = \frac{h\nu}{g\mu_B} + \tfrac{1}{2}a$$

Therefore, instead of a single line, the spectrum shows two lines of half the original intensity separated by a and centred on the field determined by g (Fig. 18.35).

If the radical contains an N atom ($I = 1$), its ESR spectrum is split into three lines of equal intensity because the ^{14}N nucleus has three possible spin orientations, and each spin orientation is possessed by one third of all the radicals in the sample. In general, a spin-I nucleus splits the spectrum into $2I + 1$ hyperfine lines of equal intensity.

When there are several magnetic nuclei present in the radical, each one contributes to the hyperfine structure. In the case of equivalent protons (for example, the two CH_2 protons in the radical $CH_3CH_2\cdot$) some of the hyperfine lines are coincident. It is not hard to show that if the radical contains N equivalent protons, there are $N + 1$ hyperfine lines with a binomial intensity distribution (i.e. the intensity distribution given by

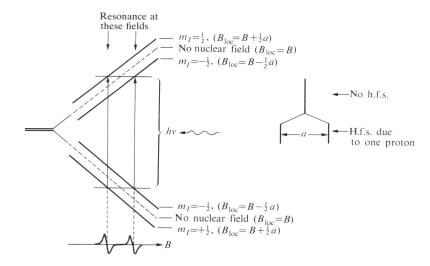

Resonance at these fields

$m_I = \frac{1}{2}$, $(B_{loc} = B + \frac{1}{2}a)$
No nuclear field $(B_{loc} = B)$
$m_I = -\frac{1}{2}$, $(B_{loc} = B - \frac{1}{2}a)$

$h\nu$

No h.f.s.

H.f.s. due to one proton

$m_I = -\frac{1}{2}$, $(B_{loc} = B - \frac{1}{2}a)$
No nuclear field $(B_{loc} = B)$
$m_I = +\frac{1}{2}$, $(B_{loc} = B + \frac{1}{2}a)$

B

Fig. 18.35 The hyperfine interaction between an electron and a spin-$\frac{1}{2}$ nucleus results in four energy levels in place of the original two. As a result, the spectrum consists of two lines (of equal intensity) instead of one. The intensity distribution can be summarized by a simple stick diagram.

Pascal's triangle). The spectrum of the benzene radical anion in Fig. 18.33, which has seven lines with intensity ratio

$$1:6:15:20:15:6:1$$

is consistent with a radical containing six equivalent protons.

Example 18.9: *Predicting the hyperfine structure of an ESR spectrum*

A radical contains one ^{14}N nucleus ($I = 1$) with hyperfine constant 1.03 mT and two equivalent protons ($I = \frac{1}{2}$) with hyperfine constant 0.35 mT. Predict the form of the ESR spectrum.

Answer. The ^{14}N nucleus gives three hyperfine lines of equal intensity separated by 1.03 mT. Each line is split into doublets of spacing 0.35 mT by the first proton, and each line of these doublets is split into doublets with the same 0.35 mT splitting (Fig. 18.36a). The central lines of each split doublet coincide, and so the proton splitting gives $1:2:1$ triplets of internal splitting 0.35 mT. Therefore, the spectrum consists of three equivalent $1:2:1$ triplets.

Comment. Often it is quicker to realize that a group of equivalent protons give characteristic hyperfine patterns (two giving a $1:2:1$ triplet, in this case), and to superimpose the patterns directly.

Exercise. Predict the form of the ESR spectrum of a radical containing three equivalent ^{14}N nuclei. [Fig. 18.36b]

The hyperfine structure of an ESR spectrum is a kind of fingerprint that helps to identify the radicals present in a sample. Moreover, since the magnitude of the splitting depends on the distribution of the unpaired electron near the magnetic nuclei present, the spectrum can be used to map the molecular orbital occupied by the unpaired electron. For example, since the hyperfine splitting in $C_6H_6^-$ is 0.375 mT, and one proton is close to a C atom with one-sixth the unpaired electron spin density (because the electron is spread uniformly around the ring), the hyperfine splitting caused by a proton in the electron spin entirely confined to a single adjacent C atom should be 6×0.375 mT = 2.25 mT. If in another aromatic radical we find a hyperfine splitting constant a, the spin density there can be calculated from

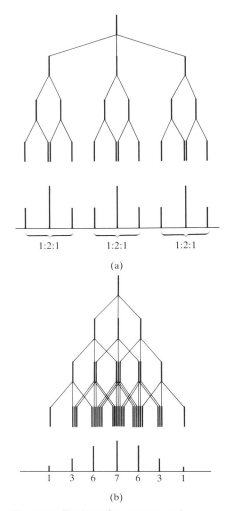

1:2:1   1:2:1   1:2:1

(a)

1   3   6   7   6   3   1

(b)

Fig. 18.36 The hyperfine structures of radicals containing (a) one ^{14}N nucleus and two equivalent protons and (b) three equivalent ^{14}N nuclei.

the **McConnell equation**:

$$a = Q\rho \qquad Q = 2.25\,\text{mT} \tag{15}$$

where ρ is the unpaired electron spin density on a C atom and a is the hyperfine splitting observed for the H atom to which it is attached.

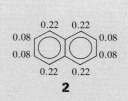

2

0.193   0.097

0.048

3

Example 18.10: *Mapping the unpaired spin density*

The hyperfine structure of the ESR spectrum of (naphthalene)$^-$ can be interpreted as arising from two groups of four equivalent protons. Those at the α positions in the ring have $a = 0.490\,\text{mT}$ and for those in the β positions $a = 0.183\,\text{mT}$. Map the unpaired spin density round the ring.

Answer. We use the McConnell equation, finding $\rho = 0.22$ for the α positions and $\rho = 0.08$ for the β positions (**2**).

Comment. Notice how the unpaired electron density accumulates in the α positions. The same value of Q may be used for the approximate mapping of spin density in heterocyclic radicals.

Exercise. The spin density in (anthracene)$^-$ is shown in (**3**). Predict the form of its ESR spectrum.
[A 1:2:1 triplet of splitting 0.43 mT split into a 1:5:6:5:1 quintet of splitting 0.22 mT, split into a 1:4:6:4:1 quintet of splitting 0.11 mT, 375 lines in all]

The origin of the hyperfine interaction

The hyperfine interaction is an interaction between the magnetic moments of the unpaired electron and the nuclei. There are two contributions to the interaction.

An electron in a p orbital does not approach the nucleus very closely, and so it experiences a field that appears to arise from a point magnetic dipole. The resulting interaction is called the **dipole–dipole interaction**, and its magnitude is given by an expression like that in eqn 6. A characteristic of this type of interaction is that it is anisotropic: that is, its magnitude (and sign) depends on the orientation of the radical with respect to the applied field. Furthermore, just as in the case of NMR, the dipole–dipole interaction averages to zero when the radical is free to tumble. Therefore, it is observed only for radicals trapped in solids.

An s electron is spherically distributed around a nucleus and so has zero average dipole–dipole interaction with it even in a solid sample. However, since an s electron has a non-zero probability of being at the nucleus, it is incorrect to treat the interaction as one between two *point* dipoles. An s electron has a Fermi contact interaction with the nucleus, which as we saw earlier is a magnetic interaction that occurs when the point dipole approximation fails. The contact interaction is isotropic (that is, independent of the radical's orientation), and consequently is shown even by rapidly tumbling molecules in fluids (so long as the spin density has some s character).

The dipole–dipole interactions of p electrons and the Fermi contact interaction of s electrons can be quite large. For example, an N2p electron experiences an average field of about 3.4 mT from the ^{14}N nucleus. An H1s

electron experiences a field of about $50\,mT$ as a result of its Fermi contact interaction with the proton. More values are listed in Table 18.2. The magnitudes of the contact interactions in radicals can be interpreted in terms of the s orbital character of the molecular orbital occupied by the unpaired electron, and the dipole–dipole interaction can be interpreted in terms of the p character. This gives information about the composition of the orbital, and especially the hybridization of the atomic orbitals (see Problem 18.6).

We still have the source of the hyperfine structure of the $C_6H_6^-$ anion and other aromatic radical anions to explain. The sample is fluid, and so the hyperfine structure cannot be due to the dipole–dipole interaction. Moreover, the protons lie in the nodal plane of the π orbital occupied by the unpaired electron, and so the structure cannot be due to a Fermi contact interaction. The explanation lies in a **polarization mechanism** similar to the one responsible for spin–spin coupling in NMR. There is a magnetic interaction between a proton and the σ electrons which results in one of the electrons tending to be found with a greater probability nearby (Fig. 18.37). The other electron is therefore more likely to be close to the C atom at the other end of the bond. The unpaired electron on the C atom has a lower energy if it is parallel to that electron (Hund's rule favours parallel electrons on atoms), and so the unpaired electron can detect the spin of the proton indirectly. Calculation using this model leads to a hyperfine interaction in agreement with the observed value of $2.8\,mT$.

Further reading

Nuclear magnetic resonance

R. K. Harris, *Nuclear magnetic resonance spectroscopy*. Longman, London (1986).

A. E. Derome, *Modern NMR techniques for chemistry research*. Pergamon, Oxford (1987).

J. K. M. Sanders and B. K. Hunter, *Modern NMR spectroscopy*. Oxford University Press (1987).

R. Freeman, *A handbook of nuclear magnetic resonance spectroscopy*. Longman, London (1987).

Electron spin resonance

J. E. Wertz and J. R. Bolton, *Electron spin resonance: elementary theory and practical applications*. McGraw-Hill, New York (1972).

M. C. R. Symons, *Chemical and biochemical aspects of electron spin resonance spectroscopy*. Van Nostrand Reinhold, New York (1978).

E. A. V. Ebsworth, D. W. H. Rankin, and S. Cradock, *Structural methods in inorganic chemistry*. Blackwell Scientific, Oxford (1987).

Table 18.2. Hyperfine coupling constants for atoms, a/mT

Nuclide	Isotropic coupling	Anisotropic coupling
^1H	50.8 (1s)	
^2H	7.8 (1s)	
^{14}N	55.2 (2s)	4.8 (2p)
^{19}F	1720 (2s)	108.4 (2p)

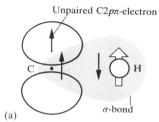

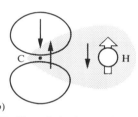

Fig. 18.37 The polarization mechanism for the hyperfine interaction in π-electron radicals. The arrangement in (a) is lower in energy than that in (b), and so there is an effective coupling between the unpaired electron and the proton.

Exercises

18.1 ^{32}S has a nuclear spin of $\frac{3}{2}$ and a nuclear g factor of 0.4289. Calculate the energies of the nuclear spin states in a magnetic field of $7.500\,T$.

18.2 Calculate the energy difference between the lowest and highest nuclear spin states of a ^{14}N nucleus ($I = 1$, $g = 0.4036$) in a $15.00\,T$ magnetic field.

18.3 Calculate the magnetic field needed to satisfy the resonance condition for unshielded protons in a $150.0\,MHz$ radiofrequency field.

18.4 Use Table 18.1 to predict the magnetic fields at which ^1H, ^2H, ^{13}C, ^{14}N, ^{19}F, and ^{31}P come into resonance at (a) $60\,MHz$, (b) $300\,MHz$.

18.5 Calculate the relative population differences ($\delta N/N$) for protons in fields of (a) 0.3 T, (b) 1.5 T, and (c) 10 T at 25°C.

18.6 The chemical shift of the CH_3 protons in acetaldehyde (ethanal) is $\delta = 2.20$ and that of the CHO proton is 9.80. What is the difference in local magnetic field between the two regions of the molecule when the applied field is (a) 1.5 T, (b) 7.0 T?

18.7 Using the information in Exercise 18.6, state the splitting (in Hz) between the methyl and aldehydic proton resonances in a spectrometer operating at (a) 60 MHz, (b) 350 MHz.

18.8 Sketch the appearance of the ^1H–NMR spectrum of acetaldehyde using $J = 2.90$ Hz and the data in Exercise 18.6 in a spectrometer operating at (a) 60 MHz, (b) 350 MHz.

18.9 Two groups of protons are made equivalent by the isomerization of a fluxional molecule. At low temperatures, where the interconversion is slow, one group resonates at $\delta = 4.0$ and the other at $\delta = 5.2$. At what rate of interconversion will the two signals merge in a spectrometer operating at 60 MHz?

18.10 Sketch the form of the ^{19}F–NMR spectra of a natural sample of $^{10}BF_4^-$ and $^{11}BF_4^-$.

18.11 Sketch the form of an $A_3M_2X_4$ spectrum, where A, M, and X are protons with distinctly different chemical shifts and $J_{AM} > J_{AX} > J_{MX}$.

18.12 Which of the following molecules have sets of nuclei that are chemically but not magnetically equivalent? (a) CH_3CH_3, (b) $CH_2{=}CH_2$, (c) $CH_2{=}C{=}CF_2$.

18.13 What magnetic field would be required in order to use an ESR X-band spectrometer (9 GHz) to observe ^1H–NMR and a 60 MHz spectrometer to observe ESR?

18.14 Some commercial ESR spectrometers use 8 mm microwave radiation. What magnetic field is needed to satisfy the resonance condition?

18.15 Commercial ESR spectrometers can detect as few as 10^{10} spins per cm^3. What is the molar concentration of doublet radicals in this limit?

18.16 Suppose a sample is known to contain 2.5×10^{14} electron spins. Calculate the population difference between the α and β spin states when the applied field is 0.3 T and the temperature 25°C.

18.17 The centre of the ESR spectrum of atomic hydrogen lies at 329.12 mT in a spectrometer operating at 9.2231 GHz. What is the electronic g-factor of the atom?

18.18 In a spectrometer operating at 9.032 GHz, one of the two hyperfine lines of the H atom occurred at 357.3 mT and the other at 306.6 mT. What is the hyperfine constant for the atom?

18.19 A radical containing two equivalent protons shows a three-line spectrum with an intensity distribution 1:2:1. The lines occur at 330.2 mT, 332.5 mT, and 334.8 mT. What is the hyperfine coupling constant for each proton? What is the g-factor of the radical given that the spectrometer is operating at 9.319 GHz?

18.20 A radical containing two inequivalent protons with hyperfine constants 2.0 mT and 2.6 mT gives a spectrum centred on 332.5 mT. At what fields do the hyperfine lines occur and what are their relative intensities?

18.21 Predict the intensity distribution in the hyperfine lines of the ESR spectra of (a) CH_3, (b) CD_3.

18.22 The benzene radical anion has $g = 2.0025$. At what field should you search for resonance in a spectrometer operating at (a) 9.302 GHz, (b) 33.67 GHz?

18.23 The mean g-factor of ClO_2 is 2.0102. What is the additional local field when the applied field is (a) 0.340 T, (b) 1.23 T?

18.24 The fluxional lifetime of a molecule is 200 ms. At 100 MHz, the difference between the two forms is reflected in a separation of 90.0 Hz between the two resonances. Determine whether or not the two lines will be merged or resolved in the observed spectrum.

18.25 The ESR spectrum of a radical with a single magnetic nucleus is split into four lines of equal intensity. What is the spin of the nucleus?

18.26 Sketch the form of the hyperfine structures of radicals XH_2 and XD_2, where the nucleus X has $I = \frac{5}{2}$.

Problems

Numerical problems

18.1 A scientist investigates the possibility of neutron spin resonance, and has available a commercial NMR spectrometer operating at 60 MHz. What field is required for resonance? What is the relative population difference at room temperature? Which is the lower spin state of the neutron?

18.2 Two groups of protons have $\delta = 4.0$ and $\delta = 5.2$ and are interconverted by a conformational change of a fluxional molecule. In a 60 MHz spectrometer the spectrum collapsed into a single line at 280 K but at 300 MHz the collapse did not occur until the temperature had been raised to 300 K. What is the activation energy of the interconversion?

18.3 The non-linear NO_2 molecule has a single unpaired electron and can be trapped in a solid matrix or prepared inside a nitrite crystal by radiation damage of NO_2^- ions. When the applied field is parallel to the OO direction the centre of the spectrum lies at 333.64 mT in a spectrometer operating at 9.302 GHz. When the field lies along the bisector of the ONO angle, the resonance lies at 331.94 mT. What are the g-factors in the two orientations.

18.4 The hyperfine coupling constant in CH_3 is 2.3 mT. Use the information in Table 18.2 to predict the splitting between the hyperfine lines of the spectrum of CD_3. What are the overall widths of the hyperfine spectra in each case?

18.5 The p-dinitrobenzene radical anion can be prepared by reduction of p-dinitrobenzene. The radical anion has two equivalent N nuclei $(I = 1)$ and four equivalent protons. Predict the form of the ESR spectrum using $a(N) = 0.148$ mT and $a(H) = 0.112$ mT.

18.6 When an electron occupies a 2s orbital on an N atom it has a hyperfine interaction of 55.2 mT with the nucleus. The spectrum of NO_2 shows an isotropic hyperfine interaction of 5.7 mT. For what proportion of its time is the unpaired electron of NO_2 occupying a 2s orbital? The hyperfine coupling constant for an electron in a 2p orbital of an N atom is 3.4 mT. In NO_2 the anisotropic part of the hyperfine coupling is 1.3 mT. What proportion of its time does the unpaired electron spend in the 2p orbital of the N atom in NO_2? What is the total probability that the electron will be found on (a) the N atoms, (b) the O atoms? What is the hybridization ratio of the N atom? Does the hybridization support the view that NO_2 is angular? Use the discussion of the hybridization ratio in Section 14.9 to interpret the hybridization as a bond angle. (The experimental bond angle is 134°.)

18.7 The hyperfine coupling constants observed in the radical anions (**4**, **5**, and **6**) are shown (in mT). Use the value for the benzene radical anion to map the probability of finding the unpaired electron in the π orbital on each C atom.

4   **5**   **6**

18.8 The pyridyl radical (**7**) and its carboxy derivative (**8**) have the proton hyperfine coupling constants shown (H. Zemel and R. W. Fessenden, *J. Phys. Chem.* **79**, 1419 (1975)). Calculate the electron spin density around the ring.

7   **8**

18.9 The motion of nitroxide radicals trapped in a solid clathrate was examined at low temperature, and the hyperfine coupling constant to the N atom was found to depend on the orientation of the radical to the applied field (A. A. McConnell, D. D. MacNicol, and A. L. Porte, *J. Chem. Soc.*

A, 3516 (1971)). With the field perpendicular to the bond the coupling constant varies from 113.1 MHz to 11.2 MHz; when the field is parallel to the bond, the coupling constant is 14.1 MHz. When the temperature was raised to 115 K, motion about the parallel axis began to influence the appearance of the spectrum. How rapidly does the molecule rotate in the clathrate cage at that temperature?

Theoretical problems

18.10 The z-component of the magnetic field at a distance R from a magnetic moment parallel to the z-axis is given by eqn 6. In a solid, a proton at a distance R from another can experience such a field and the measurement of the splitting it causes in the spectrum can be used to calculate R. In gypsum, for instance, the splitting in the H_2O resonance can be interpreted in terms of a magnetic field of 0.715 mT generated by one proton and experienced by the other. What is the separation of the protons in the H_2O molecule.

18.11 In a liquid, the dipolar magnetic field averages to zero: show this by evaluating the average of the field given above. (Hint: the volume element is $\sin\theta \, d\theta \, d\phi$ in polar coordinates.)

18.12 In a liquid crystal, a molecule might not rotate freely in all directions and the dipolar interaction might not average to zero. Suppose a molecule is trapped so that, although the vector separating two protons may rotate freely around the z-axis, the colatitude may vary only between 0 and θ'. Average the dipolar field over this restricted range of orientations and confirm that the average vanishes when θ' is equal to π (corresponding to free rotation over a sphere). What is the average value of the local dipolar field for the H_2O molecule in Problem 18.10 if it is dissolved in a liquid crystal that enables it to rotate up to $\theta' = 30°$?

18.13 The shape of a spectral line $I(\omega)$ is related to the free induction decay signal $G(t)$ by

$$I(\omega) = A \, \text{re} \int_0^\infty G(t) e^{i\omega t} \, dt$$

where A is a constant and 're' means take the real part of what follows. Calculate the line shape corresponding to an oscillating, decaying function

$$G(t) = \cos \omega_0 t \, e^{-t/\tau}$$

18.14 In the language of Problem 18.13, show that if

$$G(t) = (a \cos \omega_1 t + b \cos \omega_2 t) e^{-t/\tau}$$

then the spectrum consists of two lines with intensities proportional to a and b and located at $\omega = \omega_1$ and $\omega = \omega_2$ respectively.

18.15 Write a computer program to construct $G(t)$ for spectra with absorption lines at arbitrary frequencies, and explore the different free induction decay signals that are associated with different numbers of protons and patterns of coupling.

19

Statistical thermodynamics: the concepts

Check-list of key ideas

1. The concept of a *configuration* of a system and of its *weight* (Section 19.1 and eqn 1).

2. The use of *Stirling's approximation* to find the most probable configuration and hence to deduce the *Boltzmann distribution* (eqn 6) for the most probable distribution of populations.

3. The definition of the *molecular partition function* (eqn 7) and its interpretation as a measure of the number of thermally accessible states (Section 19.2).

4. The calculation of the molecular partition function for a uniform ladder of states (eqn 9) and for translational motion (eqn 10).

5. The relation between the *internal energy* of non-interacting molecules and the molecular partition function (eqn 11), and the internal energy of a perfect gas (eqn 12).

6. The calculation of *heat capacity* from the molecular partition function (Example 19.5).

7. The justification of the *Boltzmann formula* for the entropy (eqn 13) and the relation between the *statistical entropy* and the molecular partition function (eqn 15).

8. The definition and significance of the *canonical ensemble* of systems composed of interacting molecules (Section 19.5).

9. The *canonical distribution* and the definition of the *canonical partition function* (eqn 16).

10. The *internal energy* (eqn 17) and the *entropy* (eqn 20) in terms of the canonical partition function.

11. The relation between the canonical and molecular partition functions for non-interacting molecules (eqn 21).

12. The *Sackur–Tetrode equation* for the entropy of a monatomic gas (eqn 23).

The preceding chapters of this part of the text have shown how the energy levels of molecules can be calculated, measured spectroscopically, and related to their sizes and shapes. The next major step is to see how a knowledge of these energy levels can be used to account for the properties of matter in bulk. To do so, we now introduce the concepts of **statistical thermodynamics**, the link between molecular properties and bulk thermodynamics.

The crucial step in going from the quantum mechanics of individual molecules to the thermodynamics of bulk samples is to recognize that the latter deals with *average* behaviour. For example, the pressure of a gas is the average force per unit area exerted by its particles, and there is no need to specify which particles are striking the wall at any instant. Nor is it necessary to consider the fluctuations in the pressure as different numbers of particles happen to collide with the wall at different moments. The amplitude of these fluctuations is very small: it is highly improbable that there will be a sudden lull in the number of collisions, or a sudden storm.

This chapter introduces statistical thermodynamics in two stages. The first, the derivation of the Boltzmann distribution for individual particles, is of restricted applicability, but it has the advantage of taking us directly to a result of central importance in a straightforward and elementary way: we can *use* statistical thermodynamics once we have deduced the Boltzmann distribution. Then (in Section 19.5) we elaborate the arguments a little (but not much) so as to be able to cope with systems composed of interacting particles.

The distribution of molecular states

We shall consider a system composed of N molecules. Although the total energy is constant at E, it is not possible to be definite about how that energy is distributed over the molecules. Collisions take place, and result in the ceaseless redistribution of energy not only between the molecules but also among their different modes of motion. The closest we can come to a description of the distribution of energy is the statement of the **population** of a state, the average number of molecules that occupy it, and to say that on average there are n_i molecules in a state of energy ε_i. This population remains almost constant, but the identities of the molecules in each state may change at every collision.

The problem we address in this section is the calculation of these populations for any type of molecule in any mode of motion at any temperature. The only restriction is that the molecules should be independent in the sense that the total energy of the system is a sum of their individual energies. We are discounting (at this stage) the possibility that in a real system a contribution to the total energy may arise from interactions between molecules. We shall also accept the **principle of equal *a priori* probabilities**,[1] the assumption that all possibilities for the distribution of energy are equally probable. That is, we assume that vibrational states of a certain energy, for instance, are as likely to be populated as rotational states of the same energy.

---

[1] *A priori* means in this context loosely 'as far as one knows'. We have no reason to presume otherwise than that all states are equally likely to be occupied whatever their nature.

19.1 Configurations and weights

Any individual molecule may exist in states with energies $\varepsilon_0, \varepsilon_1, \ldots$. We shall almost always take ε_0, the lowest state, as the zero of energy ($\varepsilon_0 = 0$), and measure all other energies relative to it. As a result, we shall have to add a constant to the calculated energy of the system in order to obtain the internal energy U based on some other energy origin. For example, we must add the total zero-point energy of any oscillators in the sample.

Instantaneous configurations

At any instant there will be n_0 molecules in the state with energy ε_0, n_1 with ε_1, and so on. The specification of the set of populations $n_0, n_1, \ldots$ in the form $\{n_0, n_1, \ldots\}$ gives the instantaneous **configuration** of the system.

The instantaneous configuration of the system fluctuates with time because the instantaneous populations change. We can picture a large number of different instantaneous configurations. One, for example, might be $\{N, 0, 0, \ldots\}$, corresponding to every molecule being in its ground state. Another might be $\{N-2, 2, 0, \ldots\}$, in which two molecules are in a higher state.[2] The latter configuration is intrinsically more likely to be found because it can be achieved in more ways: $\{N, 0, 0, \ldots\}$ can be achieved in only one way, but $\{N-2, 2, \ldots\}$ can be achieved in $\frac{1}{2}N(N-1)$ different ways.[3] If the system were free to fluctuate between the two configurations (as a result of collisions), it would almost always be found in the second, more likely state (especially if N were large). In other words, a system free to switch between the two would show properties characteristic almost exclusively of the second.

A general configuration $\{n_0, n_1, \ldots\}$ can be achieved in W different ways. W is called the **weight** of the configuration, and is given by

$$W = \frac{N!}{n_0! \, n_1! \, n_2! \ldots} \tag{1}$$

(Bear in mind that $0! = 1$.) This expression is a generalization of $W = \frac{1}{2}N(N-1)$, and reduces to it for the configuration $\{N-2, 2, 0, \ldots\}$. Equation 1 comes from elementary probability theory, for W is the number of distinguishable ways in which N objects can be sorted into bins with n_i in bin i (the objects are now molecules, and the bins are their available states).

Example 19.1: *Calculating the weight of a configuration*

Calculate the number of ways of distributing 20 identical objects into five boxes, with the arrangement $1, 0, 3, 5, 10, 1$.

---

[2] At this stage of the argument, we are ignoring the requirement that the total energy of the system should be constant: the energy of the second configuration is greater than that of the first. We impose the constraint of constant total energy later.

[3] One candidate for entering the upper state can be selected in N ways. There are then $N-1$ candidates for the second choice, corresponding to $N(N-1)$ choices overall. However, we should not distinguish the choice (Jack, Jill) from the choice (Jill, Jack), and so only half the choices lead to distinguishable configurations.

Answer. The configuration of the system is $\{1, 0, 3, 5, 10, 1\}$ with $N = 20$; therefore the weight is

$$W = \frac{20!}{1! \times 0! \times 3! \times 5! \times 10! \times 1!} = 9.31 \times 10^8$$

Exercise. Calculate the weight of the configuration in which 20 objects are distributed in the arrangement 0, 1, 5, 0, 8, 0, 3, 2, 0, 1. $\qquad [4.19 \times 10^{10}]$

We shall find it more convenient to deal with $\ln W$ rather than W itself, and so will need

$$\ln W = \ln \frac{N!}{n_0! \, n_1! \ldots} = \ln N! - \ln (n_0! \, n_1! \ldots)$$

$$= \ln N! - \sum_i \ln n_i!$$

One reason for introducing $\ln W$ is that we can simplify the factorials in this expression with **Stirling's approximation** in the form[4]

$$\ln x! \approx x \ln x - x \qquad (2)$$

for then

$$\ln W = N \ln N - N - \sum_i (n_i \ln n_i - n_i)$$

$$= N \ln N - \sum_i n_i \ln n_i \qquad (3)$$

because the sum of the n_i is equal to N.

The dominating configuration

We have seen that the configuration $\{N-2, 2, 0, \ldots\}$ dominates $\{N, 0, 0, \ldots\}$, and it should be easy to believe that there may be other configurations that greatly dominate both. Is there, in fact, a configuration with so great a weight that it overwhelms all the rest in importance to such an extent that the system will almost always be found in it? If that configuration exists, then the properties of the system will be characteristic of it.

---

[4] A simple justification of Stirling's approximation is to write

$$\ln x! = \ln 1 + \ln 2 + \ldots + \ln x = \sum_{i=1}^{x} \ln y_i$$

$$\approx \int_0^x \ln y \, dy = (y \ln y - y)\big|_0^x = x \ln x - x$$

where the approximation depends on x being large. The precise form of the approximation is

$$x! \approx (2\pi)^{1/2} x^{x+1/2} e^{-x}$$

and is reliable when x is large (greater than about 10). We are dealing with far larger values of x, and the simplified version in eqn 2 is normally adequate.

We can look for the dominating configuration by looking for the values of n_i that lead to a maximum value of W. Since W is a function of all the n_i, we can do this by varying the n_i and looking for the values that correspond to $dW = 0$ (just as in the search for the maximum of any function). However, there are two difficulties with this procedure.

The first is that the only permitted configurations are those corresponding to the specified, constant, total energy of the system. This requirement rules out many configurations, because $\{N, 0, 0, \ldots\}$ and $\{N-2, 2, 0, \ldots\}$ correspond to the different energies $N\varepsilon_0$ and $(N-2)\varepsilon_0 + 2\varepsilon_1$ respectively, and so both cannot be achieved by the same isolated system. It follows that in looking for the configuration with the greatest weight, we must ensure that the configuration also satisfies

$$\sum_i n_i \varepsilon_i = E$$

where E is the total energy of the system.

The second constraint is that, because the total number of molecules present is also fixed (at N), we cannot arbitrarily vary all the populations simultaneously. Thus, increasing the population of one state by 1 demands that the population of another must be reduced by 1. Therefore, the search for the maximum value of W is also subject to

$$\sum_i n_i = N$$

The most probable configuration

We are looking for the set of numbers $n_0, n_1, \ldots$ for which W has its maximum value. It is equivalent, and it turns out to be simpler, to find the criterion for which $\ln W$ is a maximum.

Since $\ln W$ depends on all the n_i, when a configuration changes and the n_i change to $n_i + dn_i$, $\ln W$ changes to $\ln W + d \ln W$, where

$$d \ln W = \sum_i \left(\frac{\partial \ln W}{\partial n_i} \right) dn_i \tag{4}$$

At a maximum, $d \ln W = 0$. However, when the n_i change they do so subject to the constraints

$$\sum_i \varepsilon_i \, dn_i = 0 \quad \text{and} \quad \sum_i dn_i = 0$$

These two equations prevent us from solving the equation for $\ln W$ simply by setting all $(\partial \ln W / \partial n_i) = 0$ because the dn_i in eqn 4 are not all independent.

The way to take constraints into account was devised by the French mathematician Lagrange and is called the **method of undetermined multipliers**. The technique is described in the *Further information* section at the end of the chapter. All we need here is the rule that a constraint should be multiplied by a constant and then added to the main variation equation (eqn 4). The variables are then treated as though they are all independent, and the constants are evaluated at the end of the calculation.

We employ the technique as follows. The two constraints are multiplied by the constants $-\beta$ and α respectively and then added to eqn 4:

$$\text{d} \ln W = \sum_i \left(\frac{\partial \ln W}{\partial n_i}\right) \text{d}n_i + \alpha \sum_i \text{d}n_i - \beta \sum_i \varepsilon_i \, \text{d}n_i$$

$$= \sum_i \left\{\left(\frac{\partial \ln W}{\partial n_i}\right) + \alpha - \beta \varepsilon_i\right\} \text{d}n_i$$

(We have chosen one coefficient to be negative for future convenience; it will then turn out that $\beta = 1/kT$.) All the $\text{d}n_i$ are now treated as independent. Hence the only way of satisfying $\text{d} \ln W = 0$ is to require that for each i

$$\left(\frac{\partial \ln W}{\partial n_i}\right) + \alpha - \beta \varepsilon_i = 0 \qquad (5)$$

when the n_i have their most probable values.

The expression for $\ln W$ is given in eqn 3. Differentiation of it with respect to n_i gives[5]

$$\frac{\partial \ln W}{\partial n_i} = -\ln n_i - 1$$

Since the 1 in $\ln n_i + 1$ may be neglected (because $\ln n_i$ is so large), eqn 5 becomes

$$-\ln n_i + \alpha - \beta \varepsilon_i = 0$$

The most probable population of the state of energy ε_i is therefore

$$n_i = e^{\alpha - \beta \varepsilon_i}$$

The Boltzmann distribution

The final step is to evaluate the constants α and β. Since

$$N = \sum_j n_j = e^{\alpha} \sum_j e^{-\beta \varepsilon_j}$$

(we are free to label the states with j instead of i, and it will prove

---

[5] The argument runs as follows:

$$\frac{\partial \ln W}{\partial n_i} = -\sum_j \left(\frac{\partial (n_j \ln n_j)}{\partial n_i}\right) = -\sum_j \left\{\left(\frac{\partial n_j}{\partial n_i}\right) \ln n_j + n_j \left(\frac{\partial \ln n_j}{\partial n_i}\right)\right\}$$

$$= -\sum_j \left(\frac{\partial n_j}{\partial n_i}\right)(\ln n_j + 1) = -(\ln n_i + 1)$$

because

$$\frac{\partial \ln n_j}{\partial n_i} = \frac{1}{n_j}\frac{\partial n_j}{\partial n_i}$$

and for $j \neq i$, n_j is independent of n_i, so

$$\left(\frac{\partial n_j}{\partial n_i}\right) = 0$$

and for $j = i$,

$$\left(\frac{\partial n_j}{\partial n_i}\right) = \left(\frac{\partial n_j}{\partial n_j}\right) = 1$$

convenient to do so at this point) it follows that

$$e^{\alpha} = \frac{N}{\sum\limits_{j} e^{-\beta\varepsilon_j}}$$

which leads to the **Boltzmann distribution**:

$$n_i = \frac{Ne^{-\beta\varepsilon_i}}{\sum\limits_{j} e^{-\beta\varepsilon_j}} \tag{6a}$$

We shall confirm shortly that $\beta = 1/kT$, where T is the thermodynamic temperature. We shall also find it convenient to express the Boltzmann distribution in terms of the fraction $p_i = n_i/N$ of molecules in the state i:

$$p_i = \frac{e^{-\beta\varepsilon_i}}{\sum\limits_{j} e^{-\beta\varepsilon_j}} \tag{6b}$$

19.2 The molecular partition function

From now on we shall write the Boltzmann distribution as

$$p_i = \frac{e^{-\beta\varepsilon_i}}{q} \tag{6c}$$

where q is the **molecular partition function**:

$$q = \sum_{j} e^{-\beta\varepsilon_j} \qquad \beta = \frac{1}{kT} \tag{7}$$

The sum in q is sometimes expressed slightly differently. It may happen that several states have the same energy, and so give the same contribution to the sum. If, for example, g_j states have the same energy ε_j (so the energy level is g_j-fold degenerate), we could write

$$q = \sum_{j} g_j e^{-\beta\varepsilon_j}$$

where the sum is now over energy *levels,* not individual states. We shall normally employ the sum-over-states form.

Example 19.2: *Writing a partition function*

Write an expression for the partition function of a linear rotor (such as HCl).

Answer. We know from Section 16.4 that the energy levels of a linear rotor are $hcBJ(J+1)$ and that each J level consists of $2J+1$ degenerate states. Therefore,

$$q = \sum_{J} (2J+1)e^{-\beta hcBJ(J+1)}$$

Comment. The sum can be evaluated numerically by supplying the value of B (from spectroscopy or calculation) and the temperature. The expression is confined to unsymmetrical linear rotors (e.g. HCl, not CO_2) for reasons explained in Section 20.2.

Exercise. Write down the partition function for a two-level system, the lower state being non-degenerate, and the upper state (at an energy ε) doubly degenerate.
$$[q = 1 + 2e^{-\beta\varepsilon}]$$

An interpretation of the partition function

We can get some insight into the partition function by considering how it depends on the temperature. When T is close to zero, the parameter $\beta = 1/kT$ is close to infinity. Then every term except one in the sum defining q is zero because each one has the form e^{-x} with $x \to \infty$. The exception is the term with $\varepsilon_0 \equiv 0$ (or the g_0 terms if the ground state is g_0-fold degenerate), because then $\varepsilon_0/kT = 0$ whatever the temperature. As there is only one surviving term when $T = 0$, and its value is g_0, it follows that

$$\text{As } T \to 0, q \to g_0$$

At the other extreme, consider the case when T is so large that for each term in the sum $\varepsilon_j/kT \approx 0$. Since $e^{-x} = 1$ when $x = 0$, each term in the sum contributes 1. It follows that the sum is equal to the number of molecular states, which in general is infinite:

$$\text{As } T \to \infty, q \to \infty$$

In some idealized cases the molecule may have only a finite number of states; then the upper limit of q is equal to the number of states. For example, if we were considering only the spin energy levels of a radical, there would be only two states ($m_s = \pm\frac{1}{2}$).

We see that the molecular partition function gives an indication of the average number of states that are thermally accessible to a molecule at the temperature of the system. At $T = 0$, only the ground level is accessible and $q = g_0$. At very high temperatures, virtually all states are accessible, and q is correspondingly large.

The partition function for a uniform ladder of energy levels

As an explicit example, consider a molecule with a set of equally spaced non-degenerate energy levels (Fig. 19.1). (These can be thought of as the vibrational energy levels of a diatomic molecule in the harmonic approximation.) If the separation between levels is ε, the partition function is[6]

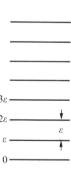

Fig. 19.1 The equally-spaced infinite array of energy levels used in the calculation of the partition function. A harmonic oscillator has the same spectrum of levels.

$$q = 1 + e^{-\beta\varepsilon} + e^{-2\beta\varepsilon} + \ldots$$
$$= 1 + e^{-\beta\varepsilon} + (e^{-\beta\varepsilon})^2 + \ldots$$

[6] The sum of a geometric series
$$S = a + ar + ar^2 + \ldots + ar^{n-1}$$

for $r < 1$ is

$$S = \frac{a(1 - r^n)}{1 - r}$$

For an infinite number of terms, $r^n = 0$. The two versions we shall use frequently are

$$1 - x + x^2 - \ldots = \frac{1}{1+x} \quad \text{and} \quad 1 + x + x^2 + \ldots = \frac{1}{1-x}$$

and hence

$$q = \frac{1}{1 - e^{-\beta\varepsilon}} \qquad (8)$$

It follows that the proportion of molecules in the state with energy ε_i is almost certain to be

$$p_i = (1 - e^{-\beta\varepsilon})e^{-\beta\varepsilon_i} \qquad (9)$$

The variation of p_i with temperature is shown in Fig. 19.2. We see that at very low temperatures q is close to 1, and only the lowest state is significantly populated. As the temperature is raised, the population breaks out of the lowest state, and the upper states become progressively more highly populated. At the same time the partition function rises from unity, and its value gives an indication of the range of states populated. The name 'partition function' reflects the sense in which q measures how the total number of molecules is distributed—partitioned—over the available states.

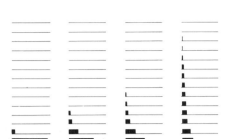

$\beta\varepsilon = 3.0$ $\beta\varepsilon = 1.0$ $\beta\varepsilon = 0.7$ $\beta\varepsilon = 0.3$
$q = 1.05$ $q = 1.58$ $q = 1.99$ $q = 3.86$

Fig. 19.2 The populations of the energy levels of the system shown in Fig. 19.1 at different temperatures, and the corresponding values of the partition function calculated using eqn 9. Note that $\beta = 1/kT$.

Example 19.3: *Using the partition function to calculate a population*

Calculate the proportion of I_2 molecules in their ground, first excited, and second excited vibrational states at 25°C. The vibrational wavenumber is $\tilde{v} = 214.6 \text{ cm}^{-1}$.

Answer. Vibrational energy levels have a constant separation (in the harmonic approximation, Section 16.7), and so the partition function is given by eqn 8 and the populations by eqn 9. Identify i with the quantum number v, and calculate p_v for $v = 0, 1,$ and 2. The following quantity is useful in this and similar calculations:

$$\frac{k\mathcal{T}}{hc} = 207.223 \text{ cm}^{-1}$$

(We have used the notation $\mathcal{T} = 298.15 \text{ K}$ introduced in Part 1.) Since

$$\beta\varepsilon = \frac{214.6 \text{ cm}^{-1}}{207.223 \text{ cm}^{-1}} = 1.036$$

the populations are

$$p_v = (1 - e^{-\beta\varepsilon})e^{-v\beta\varepsilon} = 0.645 \, e^{-1.036v}$$

Therefore, $p_0 = 0.645$, $p_1 = 0.229$, and $p_2 = 0.081$.

Comment. The I—I bond is not stiff and the atoms are heavy: as a result, the vibrational energy separations are small and at room temperature several vibrational levels are significantly populated. The value of the partition function, $q = 1.55$, reflects this small but significant spread.

Exercise. At what temperature would the $v = 1$ level have (a) half the population of the ground state, (b) the same population as the ground state?
[(a) 445 K, (b) infinite]

The translational partition function

A second example of the calculation of q is that for a molecule of mass m free to move in a container of volume V. In this case we need to make an approximation in order to obtain a formula.

We take the length of the container in the x direction to be X (and similarly Y, Z for the other two directions, with $XYZ = V$). The ε_i in the expression for q is the translational energy of the molecule and i is the label of its state. Since ε_i is the sum of the energies of translation in all three directions,

$$\varepsilon_i = \varepsilon_{n_1}^{(X)} + \varepsilon_{n_2}^{(Y)} + \varepsilon_{n_3}^{(Z)}$$

where n_1, n_2, and n_3 are the quantum numbers for motion in the x, y, and z directions respectively. Therefore, because the exponential of a sum of terms is equal to the product of the individual exponentials,[7] the partition function factorizes:

$$q = \sum_{n_1,n_2,n_3} e^{-\beta\varepsilon_{n_1}^{(X)} - \beta\varepsilon_{n_2}^{(Y)} - \beta\varepsilon_{n_3}^{(Z)}}$$

$$= \left(\sum_{n_1} e^{-\beta\varepsilon_{n_1}^{(X)}}\right)\left(\sum_{n_2} e^{-\beta\varepsilon_{n_2}^{(Y)}}\right)\left(\sum_{n_3} e^{-\beta\varepsilon_{n_3}^{(Z)}}\right)$$

$$= q_X q_Y q_Z$$

We calculated the energy levels of a molecule of mass m in a container of length X in Section 12.1 (eqn 3):

$$E_n = \frac{n^2 h^2}{8mX^2} \qquad n = 1, 2, 3, \ldots$$

(For simplicity, we are omitting the subscript on the quantum number n_1.) The lowest level has energy $h^2/8mX^2$, and so the energies relative to that level are

$$\varepsilon_n = (n^2 - 1)\varepsilon \qquad \varepsilon = \frac{h^2}{8mX^2}$$

The sum we must evaluate is therefore

$$q_X = \sum_{n=1}^{\infty} e^{-(n^2-1)\beta\varepsilon}$$

The last expression is exact. However, we cannot evaluate the sum explicitly (other than by doing it numerically), so we need to make an approximation. The translational energy levels are very close together in a container the size of a typical laboratory vessel; therefore, the sum can be replaced by an integral:

$$q_X = \int_1^{\infty} e^{-(n^2-1)\beta\varepsilon} dn$$

Extending the lower limit to $n = 0$ and replacing $n^2 - 1$ by n^2 introduces

[7] That is,

$$e^{a+b+c} = e^a e^b e^c$$

negligible error but turns the integral into standard form. We make the substitution $x^2 = n^2 \beta \varepsilon$, implying $dn = dx/(\beta \varepsilon)^{1/2}$, and therefore that

$$q_x = \left(\frac{1}{\beta \varepsilon}\right)^{1/2} \int_0^\infty e^{-x^2} dx = \left(\frac{1}{\beta \varepsilon}\right)^{1/2} \times \frac{\pi^{1/2}}{2}$$

$$= \left(\frac{2\pi m}{h^2 \beta}\right)^{1/2} X$$

We now have the partition function for translational motion in the x direction. The only change for the other two directions is to replace X by Y or Z. Hence the partition function for motion in three dimensions is

$$q = \left(\frac{2\pi m}{h^2 \beta}\right)^{3/2} \times XYZ$$

Since $XYZ = V$, the volume of the container,

$$q = \frac{V}{\Lambda^3} \qquad \Lambda = h\left(\frac{\beta}{2\pi m}\right)^{1/2} \qquad (10)$$

The quantity Λ has the dimensions of length and is called the **thermal wavelength** of the molecule.

Example 19.4: *Calculating a translational partition function*

Calculate the translational partition function of an H_2 molecule confined to a $100\ cm^3$ vessel at 25°C.

Answer. We simply substitute the fundamental constants and the data into eqn 10, using $kT = 4.116 \times 10^{-21}\ J$ and $m = 2.016\ u$:

$$\Lambda = 6.626 \times 10^{-34}\ J\ s \times \left(\frac{1}{2\pi \times 4.116 \times 10^{-21}\ J \times 2.016 \times 1.6605 \times 10^{-27}\ kg}\right)^{1/2}$$

$$= 7.12 \times 10^{-11}\ m$$

Therefore,

$$q = \frac{100 \times 10^{-6}\ m^3}{(7.12 \times 10^{-11}\ m)^3} = 2.8 \times 10^{26}$$

Comment. We see that about 10^{26} quantum states are thermally accessible, even at room temperature and for this light molecule. Many states are occupied if the thermal wavelength (which in this case is 71.2 pm) is small compared with the linear dimensions of the container.

Exercise. Calculate q for the D_2 molecule under the same conditions.

[$q = 7.9 \times 10^{26}$, $2^{3/2}$ times larger]

The internal energy and the entropy

The importance of the molecular partition function is that it contains all the information needed to calculate the thermodynamic properties of a system of independent molecules at thermal equilibrium.

19.3 The internal energy

We shall begin to unfold the importance of q by showing that we can derive the internal energy of the system from it. Then we shall show that we can also derive the entropy from q.

The relation between U and q

The total energy of the system is

$$E = \sum_i n_i \varepsilon_i$$

Since the most probable configuration is so dominating, we can use the Boltzmann expression for the populations and write

$$E = \frac{N}{q} \sum_i \varepsilon_i e^{-\beta \varepsilon_i}$$

This expression can be manipulated into a form involving only q. Thus, since

$$\frac{\mathrm{d}e^{-\beta \varepsilon_i}}{\mathrm{d}\beta} = -\varepsilon_i e^{-\beta \varepsilon_i}$$

we can write

$$E = \frac{-N}{q} \frac{\mathrm{d}}{\mathrm{d}\beta} \sum_i e^{-\beta \varepsilon_i} = \frac{-N}{q} \frac{\mathrm{d}q}{\mathrm{d}\beta}$$

There are several points in relation to this result. Since $\varepsilon_0 = 0$ (because we are measuring all energies from the lowest available level), E is the value of the internal energy relative to its value at $T = 0$. Therefore, to obtain the conventional internal energy U, we must add the internal energy at $T = 0$:

$$U = U(0) + E$$

Secondly, since the partition function may depend on variables other than the temperature (e.g. the volume), the derivative with respect to β is really a partial derivative with these other variables held constant. The final expression relating the molecular partition function to the thermodynamic internal energy of a system of independent molecules is therefore

$$U - U(0) = \frac{-N}{q} \left(\frac{\partial q}{\partial \beta} \right)_V \tag{11a}$$

An equivalent form is obtained using $\mathrm{d}x/x = \mathrm{d}\ln x$:

$$U - U(0) = -N \left(\frac{\partial \ln q}{\partial \beta} \right)_V \tag{11b}$$

These two equations confirm that we need know only the partition function in order to calculate the internal energy. In due course we shall see that all thermodynamic functions can be calculated once we know q (and its generalization for interacting molecules). In this respect, q plays a role in statistical thermodynamics very similar to that played by the wavefunction in quantum mechanics: q is a kind of thermal wavefunction.[8]

The internal energy of a perfect gas

As an illustration of how eqn 11 may be used, we shall see how to calculate the internal energy of a perfect monatomic gas from the translational partition function in eqn 10. First, we calculate

$$\left(\frac{\partial q}{\partial \beta}\right)_V = V\frac{\mathrm{d}}{\mathrm{d}\beta}\frac{1}{\Lambda^3} = -\frac{3V}{\Lambda^4}\frac{\mathrm{d}\Lambda}{\mathrm{d}\beta}$$

$$= -\frac{3V}{2\beta\Lambda^3}$$

because

$$\frac{\mathrm{d}\Lambda}{\mathrm{d}\beta} = \frac{\Lambda}{2\beta}$$

Then, by eqn 11a,

$$U - U(0) = -N \times \frac{\Lambda^3}{V} \times \left(\frac{-3V}{2\beta\Lambda^3}\right) = \frac{3N}{2\beta} \tag{12}$$

The confirmation that $\beta = 1/kT$

We can use eqn 12 to confirm that $\beta = 1/kT$. We saw in Section 11.1 that, according to the equipartition theorem, the total kinetic energy of molecules free to move in three dimensions is

$$E = \tfrac{3}{2}nRT$$

Therefore, by equating this result with the value of $U - U(0)$ just calculated, we can conclude that

$$\beta = \frac{N}{nRT} = \frac{nN_A}{nN_A kT} = \frac{1}{kT}$$

Example 19.5: *Calculating the heat capacity from the partition function*

Calculate the constant-volume heat capacity of a monatomic gas.

Answer. By definition,

$$C_V = \left(\frac{\partial U}{\partial T}\right)_V$$

Hence, from the expression just derived for U,

$$C_V = \tfrac{3}{2}nR$$

The molar heat capacity is therefore

$$C_V = \tfrac{3}{2}R = 12.5\ \mathrm{J\,K^{-1}\,mol^{-1}}$$

[8] The analogy can be seen very clearly by writing the expression for the mean energy per molecule, $\varepsilon = E/N$, as $\varepsilon q = -\partial q/\partial\beta$ and noticing the striking resemblance to the time-dependent Schrödinger equation $H\psi = -(\hbar/i)\partial\psi/\partial t$ (Box 11.1). In the sense that we can go from one to the other by replacing q by ψ and β by $it/\hbar$, there may be pleasure to be had from noticing that inverse temperature is imaginary time.

Comment. This value agrees exactly with experimental data on monatomic gases at normal pressures. In more complex molecules, other modes of motion contribute (Section 20.4). Differentiation with respect to T can often more conveniently be treated in terms of differentiation with respect to β using

$$\frac{d}{dT} = \frac{d\beta}{dT} \times \frac{d}{d\beta} = \frac{-1}{kT^2}\frac{d}{d\beta}$$

Exercise. Calculate the heat capacity of N two-level systems of the kind described in the Exercise of Example 19.2. $\quad [C = 2Nkf^2, f = e^{-\beta\varepsilon/2}/(1 + e^{-\beta\varepsilon})]$

19.4 The statistical entropy

If it is true that the partition function contains all thermodynamic information, it must be possible to use it to calculate the entropy as well as the internal energy. Since we know (Section 4.2) that entropy is related to the dispersal of energy and that the partition function is a measure of the number of states that are thermally accessible, we can be confident that the two are related.

The Boltzmann formula for the entropy

We can arrive at a relation between the entropy and the partition function by noting that a change in the internal energy

$$U - U(0) = \sum_i n_i\varepsilon_i$$

may arise from either a modification of the energy levels of a system (when ε_i changes to $\varepsilon_i + d\varepsilon_i$) or from a modification of the populations (when n_i changes to $n_i + dn_i$). The most general change is therefore

$$dU = dU(0) + \sum_i n_i\,d\varepsilon_i + \sum_i \varepsilon_i\,dn_i$$

Since the energy levels do not change when a system is heated at constant volume (Fig. 19.3), in the absence of all changes other than heating

$$dU = \sum_i \varepsilon_i\,dn_i$$

We know from thermodynamics (and specifically from eqn 1 of Chapter 5) that under the same conditions

$$dU = dq_{rev} = T\,dS$$

Therefore,

$$dS = \frac{1}{T}\sum_i \varepsilon_i\,dn_i = k\beta\sum_i \varepsilon_i\,dn_i$$

We also know that for changes in the most probable configuration (the only one we need consider)

$$\left(\frac{\partial \ln W}{\partial n_i}\right) + \alpha - \beta\varepsilon_i = 0$$

(this is eqn 5 of this chapter), and so substituting for $\beta\varepsilon_i$ in the expression

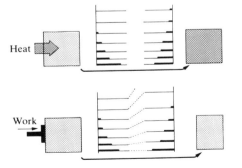

Fig. 19.3 (a) When a system is heated, the energy levels are unchanged but their populations are changed. (b) When work is done on a system, the energy levels are changed. The levels in this case are the particle-in-a-box energy levels of Chapter 12: they depend on the size of the container and move apart as its volume is decreased.

for $\mathrm{d}S$ using this relation gives

$$\mathrm{d}S = k \sum_i \left(\frac{\partial \ln W}{\partial n_i}\right) \mathrm{d}n_i + k\alpha \sum_i \mathrm{d}n_i$$

But the sum over the $\mathrm{d}n_i$ is zero, because the number of molecules is constant. Hence

$$\mathrm{d}S = k \sum_i \left(\frac{\partial \ln W}{\partial n_i}\right) \mathrm{d}n_i = k\,\mathrm{d}\ln W$$

This expression strongly suggests that we should define the **statistical entropy** as

$$S = k \ln W \tag{13}$$

where W is the weight of the most probable configuration of the system. Equation 13 is the **Boltzmann formula** for the entropy.

The statistical entropy behaves in exactly the same way as the thermodynamic entropy. Thus, as the temperature is lowered, the value of W, and hence of S, decreases because fewer configurations are compatible with the total energy. In the limit $T \to 0$, $W = 1$, so $\ln W = 0$, because only one configuration (every molecule in the lowest level) is compatible with $\varepsilon = 0$. It follows that $S \to 0$ as $T \to 0$, which is compatible with the Third Law of thermodynamics, that the entropies of all perfect crystals approach the same value as $T \to 0$ (Section 4.5).

The entropy and the partition function

We can relate the expression for the entropy to the partition function, using the expression for $\ln W$ in eqn 3. The first stage is to write

$$S = k \sum_i (n_i \ln N - n_i \ln n_i)$$

$$= -Nk \sum_i \frac{n_i}{N} \ln \frac{n_i}{N}$$

$$= -Nk \sum_i p_i \ln p_i \tag{14}$$

It follows from eqn 6c that

$$\ln p_i = -\beta \varepsilon_i - \ln q$$

and therefore that

$$S = -Nk\left(-\beta \sum_i p_i \varepsilon_i - \sum_i p_i \ln q\right)$$

$$= k\beta\{U - U(0)\} + Nk \ln q$$

because the sum over the p_i is unity, and the sum over $N p_i \varepsilon_i$ is equal to $U - U(0)$. We have already established that $\beta = 1/kT$, and so we find the following very important expression for the statistical entropy:

$$S = \frac{U - U(0)}{T} + Nk \ln q \tag{15}$$

Example 19.6: *Calculating the entropy of a collection of oscillators*

Calculate the entropy of a collection of N independent harmonic oscillators, and evaluate it using vibrational data for I_2 at 25°C.

Answer. The molecular partition function is given in eqn 8 for a molecule with evenly spaced energy levels (as in a harmonic oscillator):

$$q = \frac{1}{1 - e^{-\beta\varepsilon}}$$

The internal energy is obtained using eqn 11a:

$$U - U(0) = \frac{-N}{q}\left(\frac{\partial q}{\partial \beta}\right)_V = \frac{N\varepsilon e^{-\beta\varepsilon}}{1 - e^{-\beta\varepsilon}}$$

Data for I_2 are given in Example 19.3. The entropy is then obtained by substituting this expression in the preceding one. For I_2 at 25°C, $q = 1.550$ (Example 19.3). Therefore, since 214.6 cm^{-1} corresponds to $2.567 \text{ kJ mol}^{-1}$, and setting $N = nN_A$, $U - U(0) = 1.412n \text{ kJ mol}^{-1}$. Then, since $Nk \ln q = 3.644n \text{ J K}^{-1} \text{mol}^{-1}$, we find $S_m = 8.38 \text{ J K}^{-1} \text{mol}^{-1}$.

Comment. We shall see in Section 20.2 that the total entropy of a molecule may be ascribed to its different modes of motion, and the value we have just calculated will turn out to be the contribution arising from the vibrations of the I_2 molecule.

Exercise. Evaluate for $\beta\varepsilon = 1$ the molar entropy of the two-level system introduced in the exercise in Example 19.2 and treated in subsequent exercises.
$$[8.1 \text{ J K}^{-1} \text{mol}^{-1}]$$

The canonical partition function

In this section we see how to generalize our conclusions to include systems composed of interacting molecules and molecules free to exchange positions with each other.

19.5 The canonical ensemble

The crucial new concept we need when treating systems of interacting particles is the **ensemble**. Like so many scientific terms, the term has basically its normal meaning of 'collection', but sharpened and refined into a precise significance.

The concept of ensemble

In order to set up an ensemble, we take a closed system of specified volume, composition, and temperature, and think of it as replicated $\mathbb{N}$ times (Fig. 19.4). All the identical closed systems are regarded as being in thermal contact with each other, and hence they can exchange energy with each other. The total energy of all the systems is $\mathbb{E}$ and, because they are in thermal equilibrium with each other, they all have the same temperature T.

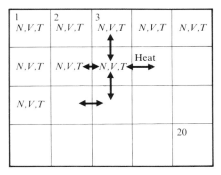

Fig. 19.4 A representation of the canonical ensemble, in this case for $\mathbb{N} = 20$. The individual replications of the actual system all have the same composition and volume. They are all in mutual thermal contact, and so all have the same temperature. Energy may be transferred between them as heat, and so they do not all have the same energy. The total energy ($\mathbb{E}$) of all 20 replications is a constant because the ensemble is isolated overall.

This imaginary collection of replications of the actual system is called the **canonical ensemble**.[9]

The important point about an ensemble is that it is a collection of *imaginary* replications of the system, and so we are free to let the number of members be as large as we like; when appropriate, we can let $\mathbb{N}$ become infinite.[10] The number of members of the ensemble in a state with energy E_i is denoted $\mathfrak{m}_i$, and we can speak of the configuration of the ensemble (by analogy with the configuration of the system used in Section 19.1) and its weight $\mathbb{W}$.

Dominating configurations

Just as in Section 19.1, some of the configurations of the ensemble will be very much more probable than others. For instance, it is very unlikely that the whole of the energy $\mathbb{E}$ will accumulate in one system (giving the configuration $\{\mathbb{N}, 0, 0, \dots\}$). By analogy with the earlier discussion, we can anticipate that there will be a dominating configuration, and that we can evaluate the thermodynamic properties by taking the average over the ensemble using that single, most probable, composition. In the thermodynamic limit of $\mathbb{N} \to \infty$, this dominating configuration is overwhelmingly the most probable, and it dominates the properties of the system virtually completely.

The quantitative discussion follows the argument in Section 19.1 with the modification that N and n_i are replaced by $\mathbb{N}$ and $\mathfrak{m}_i$. The weight of a configuration $\{\mathfrak{m}_0, \mathfrak{m}_1, \dots\}$ is

$$\mathbb{W} = \frac{\mathbb{N}!}{\mathfrak{m}_0! \, \mathfrak{m}_1! \dots}$$

and the configuration of greatest weight subject to the constraints that the total energy of the ensemble is constant and that the total number of members is fixed at $\mathbb{N}$ is given by the **canonical distribution**

$$\frac{\mathfrak{m}_i}{\mathbb{N}} = \frac{e^{-\beta E_i}}{Q} \qquad Q = \sum_i e^{-\beta E_i} \qquad (16)$$

Q is called the **canonical partition function**.

Fluctuations from the most probable

The shape of the canonical distribution is only *apparently* an exponentially decreasing function of the energy of the system.

---

[9] The word 'canon' means 'according to a rule'. There are two other important ensembles. In the 'microcanonical ensemble' the condition of constant temperature is replaced by the requirement that all the systems should have exactly the same energy: each system is individually isolated. In the 'grand canonical ensemble' the volume and temperature of each system is the same, but they are open. This means that matter can be imagined as able to pass between the systems; the composition of each one may fluctuate, but now the chemical potential is the same in each system. We can summarize these ensembles as follows:

Microcanonical: N, V, E common
Canonical: N, V, T common
Grand canonical: μ, V, T common

[10] Note that $\mathbb{N}$ is unrelated to N, the number of molecules in the actual system; $\mathbb{N}$ is the number of imaginary replications of that system.

We must appreciate that eqn 16 gives the probability of occurrence of members in a single state i (of the entire system) of energy E_i. There may in fact be numerous states with almost identical energies. For example, in a gas the identities of the molecules moving slowly or quickly can change without necessarily affecting the total energy. The **density of states**, the number of states per unit energy range, is a very sharply increasing function of energy. It follows that the probability of a member of an ensemble having a specified energy (as distinct from a specified state) is given by eqn 16, a sharply decreasing function, multiplied by a sharply increasing function (Fig. 19.5). Therefore, the overall distribution is a sharply peaked function near some energy: most members of the ensemble have an energy very close to the mean value.

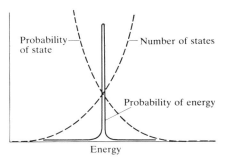

Fig. 19.5 To construct the form of the distribution of members of the canonical ensemble in terms of their energies, we multiply the probability that any one is in a state of given energy, eqn 16, by the number of states corresponding to that energy (a steeply rising function). The product is a sharply peaked function at the mean energy, which shows that almost all the members of the ensemble have that energy.

19.6 The thermodynamic information in the partition function

Like the molecular partition function, the canonical partition function carries all the thermodynamic information about a system. However, Q is more general than q because it is not based on the assumption that the molecules are independent. We can therefore use Q to discuss the properties of condensed phases and real gases where molecular interactions are important.

The internal energy

If the total energy of the ensemble is $\mathbb{E}$, and there are $\mathbb{N}$ members, the average energy of any one member is the *arithmetic mean* over the entire ensemble

$$E = \frac{\mathbb{E}}{\mathbb{N}}$$

We use this quantity to calculate the internal energy of the system in the limit of $\mathbb{N}$ (and $\mathbb{E}$) approaching infinity:

$$U = U(0) + E$$

$$= U(0) + \frac{\mathbb{E}}{\mathbb{N}} \tag{17}$$

The fraction of members of the ensemble in a state i with energy E_i is given by eqn 16 as

$$\mathbb{p}_i = \frac{e^{-\beta E_i}}{Q}$$

and the internal energy is given by

$$U - U(0) = \frac{\mathbb{E}}{\mathbb{N}} = \frac{1}{\mathbb{N}} \sum_i \mathbb{n}_i E_i = \frac{1}{Q} \sum_i E_i e^{-\beta E_i}$$

By the same argument that led to eqn 11,

$$U - U(0) = -\frac{1}{Q}\left(\frac{\partial Q}{\partial \beta}\right)_V = -\left(\frac{\partial \ln Q}{\partial \beta}\right)_V \tag{18}$$

The entropy

The total weight $\mathbb{W}$ of a configuration of the ensemble is the product of the average weight W of each member of the ensemble,

$$\mathbb{W} = W^{\mathbb{N}}$$

Hence, we can calculate S from

$$S = k \ln \mathbb{W} = k \ln \mathbb{W}^{1/\mathbb{N}}$$

$$= \frac{k}{\mathbb{N}} \ln \mathbb{W} \tag{19}$$

By the same argument that we used in Section 19.4,

$$S = \frac{U - U(0)}{T} + k \ln Q \tag{20}$$

19.7 Independent molecules

We shall now see how to recover the molecular partition function from the more general canonical partition function when the molecules are independent. Since the total energy of a collection of N independent molecules is the sum of the energies of the molecules, we can write the total energy of a state j of the system as

$$E_j = \varepsilon_j(1) + \varepsilon_j(2) + \ldots + \varepsilon_j(N)$$

$\varepsilon_j(1)$ is the energy of molecule 1 when the system is in the state j, $\varepsilon_j(2)$ the energy of molecule 2 when the system is in the same state j, and so on. The canonical partition function is then

$$Q = \sum_j e^{-\beta \varepsilon_j(1) - \beta \varepsilon_j(2) - \ldots - \beta \varepsilon_j(N)}$$

The sum over the states of the system can be achieved by letting each molecule enter all its own individual states (although we meet an important restriction shortly). Therefore, instead of summing over the states j of the system, we can sum over all the *individual* states j of molecule 1, all the states j of molecule 2, and so on. This leads to

$$Q = \left(\sum_j e^{-\beta \varepsilon_j} \right) \left(\sum_j e^{-\beta \varepsilon_j} \right) \ldots \left(\sum_j e^{-\beta \varepsilon_j} \right)$$

$$= \left(\sum_j e^{-\beta \varepsilon_j} \right)^N$$

We immediately recognize the molecular partition function on the right, and so it is tempting to write $Q = q^N$.

Indistinguishable molecules

Unfortunately, the derivation fails in a very important case. If all the molecules are identical and free to move through space, we cannot distinguish them. Suppose that molecule 1 is in some state a, molecule 2 in b, and molecule 3 in c, then one member of the ensemble has an energy

$E = \varepsilon_a + \varepsilon_b + \varepsilon_c$. This member, however, is indistinguishable from one formed by putting molecule 1 in state b, molecule 2 in state c, and molecule 3 in state a, or some other permutation (there are six in all, and $N!$ in general). In the case of indistinguishable molecules, it follows that we have counted too many states in going from the sum over system states to the sum over molecular states, and so writing $Q = q^N$ overestimates Q. The detailed argument is quite involved, but at all except very low temperatures it turns out that the correction factor is $1/N!$. Therefore:

$$\text{For distinguishable molecules: } Q = q^N \qquad (21a)$$

$$\text{For indistinguishable molecules: } Q = \frac{q^N}{N!} \qquad (21b)$$

For molecules to be indistinguishable, they must be of the same kind: an Ar atom is never indistinguishable from a Ne atom. Their identity, however, is not the only criterion. If identical molecules are in a crystal lattice, each one can be 'named' with a set of coordinates. Identical molecules in a lattice are therefore distinguishable, and we use eqn 21a. On the other hand, identical molecules in a gas are free to move to different locations, and there is no way of keeping track of the identity of a given molecule; we therefore use eqn 21b.

The entropy of a monatomic gas

For a gas of independent molecules, Q may be replaced by $q^N/N!$, with the result that eqn 20 becomes

$$S = \frac{U - U(0)}{T} + kN \ln q - k \ln N!$$

Since the number of molecules ($N = nN_A$) in a typical sample is large, we can use Stirling's approximation (and $R = kN_A$):

$$S = \frac{U - U(0)}{T} + nR \ln q - k(N \ln N - N)$$

$$= \frac{U - U(0)}{T} + nR(\ln q - \ln N + 1) \qquad (22)$$

For a gas of atoms, the only motion is translation, and the partition function is $q = V/\Lambda^3$, eqn 10. The internal energy is given by eqn 12, and so the entropy is

$$S = \tfrac{3}{2}nR + nR\left(\ln \frac{V}{\Lambda^3} - \ln nN_A + 1\right)$$

$$= nR\left(\ln e^{3/2} + \ln \frac{V}{\Lambda^3} - \ln nN_A + \ln e\right)$$

That is,

$$S = nR \ln\left(\frac{e^{5/2}V}{nN_A\Lambda^3}\right) \qquad \Lambda = \left(\frac{h^2}{2\pi mkT}\right)^{1/2} \qquad (23a)$$

585

Equation 23a is the **Sackur-Tetrode equation** for the entropy of a monatomic perfect gas. Since the gas is perfect, $V = nRT/p$, and an alternative form is

$$S = nR \ln\left(\frac{e^{5/2}kT}{p\Lambda^3}\right) \tag{23b}$$

Example 19.7: *Using the Sackur–Tetrode equation*

Calculate the standard molar entropy of gaseous argon at 25°C.

Answer. The standard molar entropy is obtained by dividing eqn 23b by n and setting p equal to $p^{\ominus}$:

$$S_m^{\ominus} = R \ln\left(\frac{e^{5/2}kT}{p^{\ominus}\Lambda^3}\right)$$

Since $m = 39.95\,u$ for Ar, at 25°C (by the same kind of calculation as in Example 19.4), its thermal wavelength is $\Lambda = 16.0\,pm$. Therefore,

$$S_m^{\ominus} = R \ln \frac{e^{5/2} \times 4.12 \times 10^{-21}\,J}{10^5\,N\,m^{-2} \times (1.60 \times 10^{-11}\,m)^3}$$

$$= 18.62\,R = 155\,J\,K^{-1}\,mol^{-1}$$

Comment. We can anticipate, on the basis of the number of accessible states for a lighter molecule, that the standard molar entropy of Ne is likely to be smaller than for Ar, and its actual value is $17.60\,R$.

Exercise. Calculate the translational contribution to the standard molar entropy of H_2 at 25°C. 　　　　　　　　　　　　　　　　[$14.2\,R$]

We can now make another connection with the material in Part 1: the Sackur–Tetrode equation implies that when a perfect gas expands isothermally from V_i to V_f, its entropy changes by

$$\Delta S = nR \ln aV_f - nR \ln aV_i = nR \ln \frac{V_f}{V_i}$$

where aV is the collection of quantities inside the logarithm of eqn 23a. This is exactly the expression we obtained using classical thermodynamics (Section 4.4).

Further information: undetermined multipliers

Suppose we need to find the maximum (or minimum) value of some function f that depends on several variables $x_1, x_2, \ldots, x_n$. When the variables undergo a small change from x_i to $x_i + \delta x_i$ the function changes from f to $f + \delta f$, where

$$\delta f = \sum_i \left(\frac{\partial f}{\partial x_i}\right)\delta x_i$$

At a minimum or maximum, $\delta f = 0$, and so then

$$\sum_i \left(\frac{\partial f}{\partial x_i}\right)\delta x_i = 0 \tag{A1}$$

If the x_i were all independent, all the δx_i would be arbitrary, and this equation could be solved by setting each $(\partial f / \partial x_i) = 0$ individually. When the x_i are not all independent the δx_i are not all arbitrary, and the simple solution is no longer valid. We proceed as follows.

Let the constraint connecting the variables be an equation of the form $g = 0$. For example, in this chapter, one constraint was $n_0 + n_1 + \ldots = N$, which can be written

$$g = 0 \quad \text{with} \quad g = (n_0 + n_1 + \ldots) - N$$

The constraint $g = 0$ is always valid, and so g remains unchanged when the x_i are varied:

$$\delta g = \sum_i \left(\frac{\partial g}{\partial x_i}\right) \delta x_i = 0$$

Since δg is zero, we can multiply it by a parameter λ and add it to eqn A1:

$$\sum_{i=1}^{n} \left\{ \left(\frac{\partial f}{\partial x_i}\right) + \lambda \left(\frac{\partial g}{\partial x_i}\right) \right\} \delta x_i = 0 \qquad \text{(A2)}$$

Equation A2 can be solved for one of the δx, δx_n for instance, in terms of all the other δx_i. All those other δx_i ($i = 1, 2, \ldots, n-1$) are independent, because there is only one constraint on the system. But here is the trick: λ is arbitrary; therefore we can choose it so that the coefficient of δx_n in eqn A2 is zero. That is, we choose λ so that

$$\left(\frac{\partial f}{\partial x_n}\right) + \lambda \left(\frac{\partial g}{\partial x_n}\right) = 0 \qquad \text{(A3)}$$

Then eqn A2 becomes

$$\sum_{i=1}^{n-1} \left\{ \left(\frac{\partial f}{\partial x_i}\right) + \lambda \left(\frac{\partial g}{\partial x_i}\right) \right\} \delta x_i = 0$$

Now the $n-1$ variations δx_i *are* independent, and so the solution of this equation is

$$\left(\frac{\partial f}{\partial x_i}\right) + \lambda \left(\frac{\partial g}{\partial x_i}\right) = 0 \qquad i = 1, 2, \ldots, n-1$$

But eqn A3 has exactly the same form, and so the maximum or minimum of f can be found by solving

$$\left(\frac{\partial f}{\partial x_i}\right) + \lambda \left(\frac{\partial g}{\partial x_i}\right) = 0 \qquad i = 1, 2, \ldots, n$$

The use of this approach was illustrated in the text for two constraints and therefore two undetermined multipliers λ_1 and λ_2 (α and $-\beta$).

The multipliers λ cannot always remain undetermined. One approach is to solve eqn A3 instead of incorporating it into the minimization scheme. In this chapter we used the alternative procedure of keeping λ undetermined until a property was calculated for which the value was already known. Thus, we found that $\beta = 1/kT$ by calculating the internal energy of a perfect gas.

Further reading

P. W. Atkins, *The second law*. Scientific American Books, New York (1984).

H. A. Bent, *The second law*. Oxford University Press (1965).

B. J. McClelland, *Statistical thermodynamics*. Wiley, New York (1973).

D. A. McQuarrie, *Statistical thermodynamics*. Harper & Row, New York (1976).

N. Davidson, *Statistical thermodynamics*. McGraw-Hill, New York (1962).

T. L. Hill, *An introduction to statistical thermodynamics*. Addison-Wesley, Reading (1960).

D. Chandler, *Introduction to statistical mechanics*. Oxford University Press (1987).

C. Hecht, *Statistical thermodynamics and kinetic theory*. W. H. Freeman & Co., New York (1990).

Exercises

19.1 What are the relative populations of the states of a two-level system when the temperature is infinite?

19.2 Calculate the translational partition function at (a) 300 K and (b) 400 K of a molecule of molar mass 120 g mol^{-1} in a container of volume 2.00 cm^3.

19.3 Calculate (a) the thermal wavelength, (b) the translational partition function of Ar in a cubic box of side 1.00 cm at (i) 300 K and (ii) 3000 K.

19.4 Calculate the ratio of the translational partition functions of D_2 and H_2 at the same temperature and volume.

19.5 A certain atom has a threefold degenerate ground level, a non-degenerate electronically excited level at 3500 cm^{-1}, and a threefold degenerate level at 4700 cm^{-1}. Calculate the partition function of these electronic states at 1900 K.

19.6 Calculate the electronic contribution to the molar internal energy at 1900 K for a sample composed of the atoms specified in Exercise 19.5.

19.7 A certain molecule has a non-degenerate excited state lying at 540 cm^{-1} above the non-degenerate ground state. At what temperature will 10 per cent of the molecules be in the upper state if these are the only states we need consider?

19.8 Test Stirling's approximation, that $\ln x! \approx x \ln x - x$, for $x = 5$, 10, and 15 by comparing its predictions with the exact values. Repeat the comparison for the more exact form of the approximation

$$\ln x! = (x + \tfrac{1}{2}) \ln x - x + \tfrac{1}{2} \ln 2\pi$$

19.9 An electron spin can adopt either of two orientations in a magnetic field, and its energies are $\pm \mu_B B$ where μ_B is the Bohr magneton. Deduce an expression for the partition function and mean energy of the electron and plot the functions against B. Calculate the relative populations of the spin states at (a) 4 K, (b) 298 K when $B = 1.0 \text{ T}$.

19.10 Write down an expression for (a) the partition function, (b) the mean energy of a ^{14}N nucleus ($I = 1$) in a magnetic field.

19.11 Calculate the standard molar entropy of neon gas at (a) 200 K, (b) 298.15 K.

19.12 Calculate the vibrational contribution to the entropy of Cl_2 at 500 K given that the wavenumber of the vibration is 560 cm^{-1}.

19.13 Identify the systems for which it is essential to include a factor of $1/N!$ on going from Q to q: (a) a sample of helium gas, (b) a sample of carbon monoxide gas, (c) a solid sample of carbon monoxide, (d) water vapour, (e) ice.

19.14 Confirm that the Sackur–Tetrode equation accounts for (a) the pressure dependence and (b) the temperature dependence of the entropy of a monatomic perfect gas as derived from thermodynamics.

Problems

Numerical problems

19.1 A certain atom has a doubly degenerate ground level pair and an upper level of four degenerate states at 450 cm^{-1} above the ground level. In an atomic beam study of the atoms it was observed that 30 per cent of the atoms were in the upper level, and the translational temperature of the beam was 300 K. Are the electronic states of the atoms in thermal equilibrium with the translational states?

19.2 Explore the conditions under which the 'integral' approximation for the translational partition function is not valid by considering the translational partition function of Ar in a cubic box of side 1.00 cm. Estimate the temperature at which, according to the integral approximation, $q = 10$ and evaluate the exact partition function at that temperature.

19.3 (a) Calculate the electronic partition function of the Te atom at (i) 298 K, (ii) 5000 K by direct summation using the following data:

Term	Degeneracy	Wavenumber/cm^{-1}
Ground	5	0
1	1	4707
2	3	4751
3	5	10 559

(b) What proportion of the Te atoms are in the ground term and in the term labelled 2 at the two temperatures? (c) Calculate the electronic contribution to the standard molar entropy of gaseous Te atoms.

19.4 The NO molecule has a doubly degenerate excited level 121.1 cm^{-1} above the doubly degenerate ground term. Calculate and plot the electronic partition function of NO from $T = 0$ to 1000 K. Evaluate (a) the term populations and (b) the electronic contribution to the molar internal energy at 300 K. Calculate the electronic contribution to the molar entropy of the NO molecule at 300 K and 500 K.

19.5 Calculate, by explicit summation, the vibrational partition function and the vibrational contribution to the molar internal energy of I_2 molecules at (a) 100 K, (b) 298 K given that its vibrational energy levels lie at the following wavenumbers above the zero-point energy level: 0, 213.30, 425.39, 636.27, 845.93, 1054.38 cm^{-1}. What proportion of I_2 molecules are in the ground and first two excited levels at the two temperatures? Calculate the vibrational contribution to the molar entropy of I_2 at the two temperatures.

Theoretical problems

19.6 A sample consisting of five molecules has a total energy 5ε. Each molecule is able to occupy states of energy $j\varepsilon$ with $j = 0, 1, 2, \ldots$. (a) Calculate the weight of the configuration in which the energy is equally shared by the molecules. (b) Draw up a table with columns headed by the energy of the states and write beneath them all configurations that are consistent with the total energy. Calculate the weights of each configuration and identify the most probable configurations.

19.7 A sample of nine molecules is numerically tractable but on the verge of being thermodynamically significant. Draw up

a table of configurations for $N = 9$, total energy 9ε in a system with energy levels $j\varepsilon$ (as in the Problem 19.6). Before evaluating the weights of the configurations, guess (by looking for the most 'exponential' distribution of populations) which of the configurations will turn out to be the most probable. Go on to calculate the weights and identify the most probable configuration.

19.8 The most probable configuration is characterized by a parameter we know as the 'temperature'. The temperatures of the system specified in Problem 19.7 must be such as to give a mean value of ε for the energy of each molecule and a total energy $N\varepsilon$ for the system. (a) Show that the temperature can be obtained by plotting p_j against j, where p_j is the (most probable) fraction of molecules in the state with energy $j\varepsilon$. Apply the procedure to the system in Problem 19.7. What is the temperature of the system when ε corresponds to 50 cm^{-1}? (b) Choose configurations other than the most probable, and show that the same procedure gives a worse straight line, indicating that a temperature is not well-defined for them.

19.9 Consider a system with energy levels $\varepsilon_j = j\varepsilon$ and N molecules. (a) Show that if the mean energy per molecule is $a\varepsilon$, then the temperature is given by

$$\beta = \frac{1}{\varepsilon} \ln\left(1 + \frac{1}{a}\right)$$

Evaluate the temperature for a system in which the mean energy is ε, taking ε equivalent to 50 cm^{-1}. (b) Calculate the molecular partition function q for the system when its mean energy is $a\varepsilon$. (c) Show that the entropy of the system is

$$S/k = (1 + a) \ln(1 + a) - a \ln a$$

and evaluate this expression for a mean energy ε.

19.10 Suppose by some means we contrive to invert the population of a two-level system in the sense that the upper and lower levels have the populations of the lower and upper levels respectively at the system in thermal equilibrium at a temperature T. Show that the relative populations are still given by a Boltzmann-like expression but with a temperature $-T$. Under what circumstances is it possible to speak of negative temperatures of an evenly-spaced three-level system?

20

Statistical thermodynamics: the machinery

Check-list of key ideas

1. The calculation of the *Helmholtz function* and the *pressure* from the partition function (eqns 2 and 3).

2. The calculation of the *enthalpy* and the *Gibbs function* from the partition function (eqns 4 and 5).

3. The definition of the *molar partition function* (eqn 6).

4. The calculation of the *translational partition function* of a gas-phase molecule (eqn 7).

5. The calculation of the *rotational partition function* of a gas-phase molecule (eqn 8).

6. The significance of the *rotational temperature* of a molecule and its use for judging when the high temperature form of the rotational partition function is valid (eqn 9).

7. The significance and calculation of the *symmetry number* of a molecule (Section 20.2).

8. The calculation of the *vibrational partition function* of a gas-phase molecule (eqn 10) and its high temperature form (eqn 11).

9. The significance of the *vibrational temperature* of a molecule and its use for judging when the high temperature form of the vibrational partition function is valid (eqn 11).

10. The *electronic contribution* to the partition function and the *overall partition function* (eqn 12).

11. The calculation of the *Giauque function* from spectroscopic data (Example 20.7).

12. The calculation of the *mean energy* of each mode of motion of a molecule (Section 20.3) and the justification of the *equipartition theorem*.

13. The calculation of the contribution of different modes of motion to the *heat capacity* of a sample from the partition function (eqn 18).

14. The definition and interpretation of the *residual entropy* of substances that are disorderly at $T = 0$ (eqn 19).

15. The calculation of the *equilibrium constant* of a gas-phase reaction in terms of the partition functions of the reactants and products (eqn 22).

The partition function is the bridge between thermodynamics, spectroscopy, and quantum mechanics. Once it is known, we can calculate thermodynamic functions, heat capacities, entropies, and equilibrium constants, and in the process discover new ways of thinking about their significance.

Fundamental relations

We do two things in this section. First, we show how all the thermodynamic functions can be obtained once we know the partition function. Then we show how to calculate the molecular partition function, and through that the thermodynamic functions, from spectroscopic data.

20.1 The thermodynamic functions

We have already derived (in Section 19.6) the two formulas for calculating the internal energy and the entropy of a system from its partition function Q:

$$U - U(0) = -\left(\frac{\partial \ln Q}{\partial \beta}\right)_V \tag{1a}$$

$$S = \frac{U - U(0)}{T} + k \ln Q \tag{1b}$$

If we are dealing with independent molecules we can go on to make the substitutions $Q = q^N$ (if the molecules are distinguishable) or $Q = q^N/N!$ (if they are indistinguishable). All the thermodynamic functions are related to U and S, and so we have a route to their calculation. We shall now show how the relations can be built by drawing on some of the results established in Part 1; for convenience, the expressions we shall derive are collected in Table 20.1.

The Helmholtz function

The Helmholtz function A is given by $A = U - TS$. Since this definition implies that $A(0) = U(0)$, substitution for U and S using eqn 1 leads to the very simple expression

$$A - A(0) = -kT \ln Q \tag{2}$$

The pressure

Once we know the Helmholtz function we can derive an expression for the pressure. Thus, since

$$p = -\left(\frac{\partial A}{\partial V}\right)_T$$

we can relate the pressure p to the partition function through

$$p = kT\left(\frac{\partial \ln Q}{\partial V}\right)_T \tag{3}$$

This relation is entirely general, and may be used for any type of substance, including real gases and liquids.

Since eqn 3 relates the pressure to the volume and temperature (because

Table 20.1. Statistical thermodynamic relations

In terms of the canonical partition function Q

$$U - U(0) = -\left(\frac{\partial \ln Q}{\partial \beta}\right)_V$$

$$S = \frac{U - U(0)}{T} + k \ln Q$$

$$A - A(0) = -kT \ln Q$$

$$p = kT\left(\frac{\partial \ln Q}{\partial V}\right)_T$$

$$H - H(0) = -\left(\frac{\partial \ln Q}{\partial \beta}\right)_V + kTV\left(\frac{\partial \ln Q}{\partial V}\right)_T$$

$$G - G(0) = -kT \ln Q + kTV\left(\frac{\partial \ln Q}{\partial V}\right)_T$$

For indistinguishable, independent particles $Q = q^N/N!$

$$U - U(0) = -N\left(\frac{\partial \ln q}{\partial \beta}\right)_V$$

$$S = \frac{U - U(0)}{T} + nR(\ln q - \ln N + 1)$$

$$G - G(0) = -nRT \ln \frac{q_m}{N_A}$$

where q_m is the molar partition function.

For distinguishable, independent particles $Q = q^N$

$$U - U(0) = -N\left(\frac{\partial \ln q}{\partial \beta}\right)_V$$

$$S = \frac{U - U(0)}{T} + nR \ln q$$

$$G - G(0) = -nRT \ln q$$

In general, for indistinguishable independent particles,

$$Q = \frac{(q_{external}q_{internal})^N}{N!} = \frac{(q_{external})^N}{N!} \times (q_{internal})^N$$

The thermodynamic functions are then the sums of internal and external (translational) contributions.

these properties occur in Q), it is a very important route to the equations of state of real gases in terms of intermolecular forces (which can be built into Q, Section 22.6).

Example 20.1: *Deriving an equation of state*

Derive the perfect gas law from the canonical partition function.

Answer. As a perfect gas consists of independent molecules, we use $Q = q^N/N!$ in eqn 3, with $q = V/\Lambda^3$:

$$p = \frac{kT}{Q}\left(\frac{\partial Q}{\partial V}\right)_T = \frac{NkT}{q}\left(\frac{\partial q}{\partial V}\right)_T$$

$$= nRT \times \frac{\Lambda^3}{V} \times \frac{1}{\Lambda^3} = \frac{nRT}{V}$$

as required.

Comment. This calculation can be regarded as yet another way of deducing that $\beta = 1/kT$.

Exercise. Derive the equation of state of a sample for which $Q = q^N f/N!$, where f depends on the volume. $\qquad [p = nRT/V + kT(\partial \ln f/\partial V)_T]$

The enthalpy

Since U and p have now both been related to Q, the enthalpy, H, can also be related to it through $H = U + pV$, which gives

$$H - H(0) = -\left(\frac{\partial \ln Q}{\partial \beta}\right)_V + kTV\left(\frac{\partial \ln Q}{\partial V}\right)_T \tag{4}$$

Example 20.2: *Calculating the enthalpy*

Calculate the enthalpy of a perfect monatomic gas at a temperature T.

Answer. The procedure is to substitute $Q = q^N/N!$ into eqn 4. However, there is a short cut, because most of the work has already been done for this calculation. Thus, in Section 19.3 we showed that

$$U - U(0) = \tfrac{3}{2}nRT$$

and we have just confirmed that $pV = nRT$. Hence,

$$H - H(0) = U - U(0) + pV$$
$$= \tfrac{3}{2}nRT + nRT = \tfrac{5}{2}nRT$$

Exercise. Repeat the calculation by explicit use of eqn 4.

The Gibbs function

One of the most important thermodynamic functions for chemistry is the Gibbs function $G = A + pV$. We can now express this function in terms of the partition function because we know the expressions for A and p:

$$G - G(0) = -kT \ln Q + kTV\left(\frac{\partial \ln Q}{\partial V}\right)_T \tag{5}$$

$G(0)$ is the Gibbs function at $T = 0$.

The expression for G takes a simple form for a perfect gas since we can replace pV by nRT:

$$G - G(0) = -kT \ln Q + nRT$$

Furthermore, since $Q = q^N/N!$,

$$G - G(0) = -NkT \ln q + kT \ln N! + nRT$$
$$= -nRT \ln q + kT(N \ln N - N) + nRT$$
$$= -nRT \ln \frac{q}{N} \tag{6a}$$

with $N = nN_A$. It will turn out to be convenient to define the **molar partition function** q_m (with units mol^{-1}) as $q_m = q/n$. Then

$$G - G(0) = -nRT \ln \frac{q_m}{N_A} \tag{6b}$$

20.2 The partition function

The energy of a molecule is the sum of contributions from its different modes of motion:

$$\varepsilon_j = \varepsilon_j^T + \varepsilon_j^R + \varepsilon_j^V + \varepsilon_j^E$$

where T denotes translation, R rotation, V vibration, and E electronic contributions. This separation is only approximate (except for translation) because the modes are not completely independent, but in most cases it is satisfactory. The separation of the electronic and vibrational motions, for example, is justified by the Born–Oppenheimer approximation (Section 14.1), and the separation of the vibrational and rotational modes is valid if we treat the molecule as a rigid rotor.

Given that the energy separates in this way, the partition function factorizes into contributions from each mode:

$$q = \sum_j e^{-\beta \varepsilon_j^T - \beta \varepsilon_j^R - \beta \varepsilon_j^V - \beta \varepsilon_j^E}$$

$$= \left(\sum_j e^{-\beta \varepsilon_j^T} \right) \left(\sum_j e^{-\beta \varepsilon_j^R} \right) \left(\sum_j e^{-\beta \varepsilon_j^V} \right) \left(\sum_j e^{-\beta \varepsilon_j^E} \right)$$

$$= q^T q^R q^V q^E$$

This factorization means that we can investigate each contribution separately.

The translational contribution

We derived the translational partition function of a molecule of mass m in a container of volume V in Section 19.2:

$$q^T = \frac{V}{\Lambda^3} \qquad \Lambda = \left(\frac{h^2 \beta}{2\pi m} \right)^{1/2} \tag{7}$$

Notice that $q^T \to \infty$ as $T \to \infty$ because an infinite number of states become accessible. Even at room temperature $q^T \approx 2 \times 10^{28}$ for an O_2 molecule in a $100 \, cm^3$ vessel.

The size of the thermal wavelength Λ lets us judge whether the approximations that led to the expression for q^T are valid. The approximations are valid if many states are occupied, which requires V/Λ^3 to be large. That will be so if Λ is small compared with the linear dimensions of the container. $\Lambda = 71 \, pm$ for H_2 at 25°C, which is far smaller than any conventional container is likely to be (but comparable to pores in zeolites or cavities in clathrates). For a more typical molecule, such as O_2, $\Lambda = 18 \, pm$.

The rotational contribution

As we saw in Example 19.2, the partition function of a linear rotor is

$$q^R = \sum_J (2J + 1)e^{-\beta E(J)} \qquad E(J) = hcBJ(J+1) \tag{8}$$

The direct method of calculating q^R is to substitute the experimental values of the rotational energy levels into this expression and to sum the series numerically.

Example 20.3: *Evaluating the rotational partition function explicitly*

Evaluate the rotational partition function of $^1H^{35}Cl$ at 25°C, for which $B = 10.591 \text{ cm}^{-1}$.

Answer. We use eqn 8, and evaluate it term by term using $kT/hc = 207.22 \text{ cm}^{-1}$, so $hcB/kT = 0.05111$.

J	0	1	2	3	4	. . .	10
$J(J+1)$	0	2	6	12	20	. . .	110
$x = 2J+1$	1	3	5	7	9	. . .	21
$y = e^{-0.05111J(J+1)}$	1	0.903	0.736	0.542	0.360	. . .	0.004
xy	1	2.71	3.68	3.79	3.24	. . .	0.08

The sum required by eqn 8 is equal to 19.9; hence $q^R = 19.9$ at this temperature. Taking J up to 16 gives 19.902.

Comment. Notice that about ten J-levels are significantly populated (but the number of populated states is larger on account of the $(2J+1)$-fold degeneracy of each level). We shall shortly encounter the approximation that $q^R \approx kT/hcB$. In the present case this gives $q^R = 19.6$, in good agreement with the exact value, and with much less work. Calculations based on eqn 8 are easily programmed for a microcomputer.

Exercise. Evaluate the rotational partition function for HCl at 0°C. [18.26]

At room temperature $kT/hc \approx 200 \text{ cm}^{-1}$. The rotational constants of many molecules are around 1 cm^{-1} and often smaller:

	CO_2	I_2	HI	HCl
B/cm^{-1}	0.39	0.037	6.5	10.6

(H_2, for which $B = 60.9 \text{ cm}^{-1}$, is an exception) and many levels are populated. When this is the case, the sum may be approximated by an integral:

$$q^R = \int_0^\infty (2J+1)e^{-\beta hcBJ(J+1)} \, dJ$$

Although this integral looks complicated, we can evaluate it without much effort by noticing that it can also be written as

$$q^R = \left(\frac{-1}{\beta hcB}\right) \int_0^\infty \frac{d}{dJ} e^{-\beta hcBJ(J+1)} dJ$$

Then, since the integral of a derivative of a function is the function itself,

$$q^R = \left(\frac{-1}{\beta hcB}\right) e^{-\beta hcBJ(J+1)} \Big|_0^\infty = \frac{1}{\beta hcB}$$

The approximate form of the rotational partition function of a linear molecule is therefore

$$q^R = \frac{kT}{hcB} = \frac{2IkT}{\hbar^2}$$

A similar calculation for a nonlinear molecule with rotational constants A, B, and C gives

$$q^R = \left(\frac{kT}{hc}\right)^{3/2}\left(\frac{\pi}{ABC}\right)^{1/2}$$

However, before using these expressions, read on (to eqn 9).

A useful way of expressing the temperature above which the approximation is valid is to introduce the **rotational temperature** θ_R as

$$k\theta_R = hcB$$

Then 'high temperature' means $T \gg \theta_R$. Some typical values are:

	I_2	CO_2	HI	HCl	CH_4	H_2
θ_R/K	0.053	0.56	7.5	9.4	15	88

The value for H_2 is abnormally high and we must be careful with the approximation for this molecule. The general conclusion at this stage is that molecules with large moments of inertia (small rotational constants) have large rotational partition functions. This reflects the closeness (compared with kT) of the rotational levels in large, heavy molecules, and the large number of them that are populated at normal temperatures.

We must take care not to include too many rotational states in the sum (see *Further information* at the end of the chapter for the quantum mechanical origin of this remark). In the case of a homonuclear diatomic molecule or a symmetrical linear molecule (such as CO_2 or $HC\equiv CH$), a rotation through 180° results in an indistinguishable state of the molecule. Hence, the number of thermally accessible states is only half the number that can be occupied by a heteronuclear diatomic molecule, where rotation through 180° does result in a distinguishable state. Therefore, for a symmetrical linear molecule,

$$q^R = \frac{kT}{2hcB}$$

We can combine the equations for symmetrical and non-symmetrical molecules into one by introducing the **symmetry number** σ, which is the number of indistinguishable orientations of the molecule:

$$q^R = \frac{kT}{\sigma hcB} \tag{9a}$$

For a heteronuclear diatomic molecule $\sigma = 1$; for a homonuclear diatomic molecule or a symmetrical ($D_{\infty h}$) linear molecule $\sigma = 2$. The same problem arises for other types of symmetrical molecule, and for a nonlinear molecule we write

$$q^R = \frac{1}{\sigma}\left(\frac{kT}{hc}\right)^{3/2}\left(\frac{\pi}{ABC}\right)^{1/2} \tag{9b}$$

Some typical values of the symmetry numbers required are

	H_2O	NH_3	CH_4	C_6H_6
σ:	2	3	12	12

The value $\sigma(H_2O) = 2$ reflects the fact that a 180° rotation about its C_2 axis interchanges two indistinguishable atoms; in NH_3, there are three indistinguishable orientations around its C_3 axis. For CH_4, any of three 120° rotations about any of its four C—H bonds leaves the molecule in an indistinguishable state, and so the symmetry number is $3 \times 4 = 12$. For benzene, any of six orientations around its C_6 axis leaves it apparently unchanged, as does a rotation of 180° around any of six C_2 axes in the plane of the molecule.

A more formal way of arriving at the value of σ is to note that σ is the order of the **rotational subgroup** of the molecule, the point group of the molecule with all but the identity and the rotations removed. The rotational subgroup of H_2O is $\{E, C_2\}$, and so $\sigma = 2$. The rotational subgroup of NH_3 is $\{E, 2C_3\}$, and so $\sigma = 3$. This recipe makes it easy to find the symmetry numbers for more complicated molecules. The rotational subgroup of CH_4 is obtained from the T character table in Table 17.1 as $\{E, 8C_3, 3C_2\}$, and so $\sigma = 12$. For benzene, the rotational subgroup of D_{6h} is $\{E, 2C_6, 2C_3, C_2, 3C_2', 3C_2''\}$, and so $\sigma = 12$.

Example 20.4: *Estimating a rotational partition function*

Estimate the rotational partition function of ethene at 25°C given that $A = 4.828 \text{ cm}^{-1}$, $B = 1.0012 \text{ cm}^{-1}$, and $C = 0.8282 \text{ cm}^{-1}$.

Answer. We use eqn 9b with $kT/hc = 207.223 \text{ cm}^{-1}$. From the D_{2h} character table, the rotational subgroup of the molecule is of order 4 ($E, C_{2x}, C_{2y}, C_{2z}$) and so $\sigma = 4$. Therefore, since $ABC = 4.0033 \text{ cm}^{-3}$, $q^R = 661$.

Comment. Ethene is quite a big molecule, the energy levels are close together (compared with kT), and so many are populated at room temperature.

Exercise. Evaluate the rotational partition function of pyridine, C_5H_5N, at room temperature ($A = 0.2014 \text{ cm}^{-1}$, $B = 0.1936 \text{ cm}^{-1}$, $C = 0.0987 \text{ cm}^{-1}$).

$$[4.3 \times 10^4]$$

The vibrational contribution

We can evaluate the vibrational partition function of a molecule by substituting the measured vibrational energy levels into the exponentials appearing in the definition and summing them numerically. In a polyatomic molecule each normal mode (Section 16.11) has its own partition function (so long as the anharmonicities are so small that the modes are independent), and the overall vibrational partition function is the product of these individual ones:

$$q^V = q^V(1)q^V(2) \ldots$$

$q^V(K)$ is the partition function for the Kth normal mode and is calculated by direct summation of the observed spectroscopic levels.

Example 20.5: *Estimating the vibrational partition function*

Given that a typical value of the vibrational partition function of one normal mode is about 1.1, estimate the overall vibrational partition function of a nonlinear molecule containing ten atoms.

Answer. The overall vibrational partition function will be of the order of $(1.1)^n$ where n is the number of normal modes. We saw in Section 16.11 that $n = 3N - 6$ for a nonlinear molecule containing N atoms; therefore in the present case $n = 24$. The overall vibrational partition function is therefore approximately

$$q^V \approx (1.1)^{24} = 9.9$$

Comment. Even though vibrational modes are not appreciably excited, there may be so many modes in a molecule that overall their excitation is significant.

Exercise. Repeat the calculation for a linear molecule containing ten atoms.

[10.8]

So long as the vibrational excitation is not too great, the harmonic approximation may be made, and the vibrational energy levels written as

$$E(v) = (v + \tfrac{1}{2})hc\tilde{v} \qquad v = 0, 1, 2, \ldots$$

If we measure energies from the zero-point level (our general rule), the permitted values are $\varepsilon_v = vhc\tilde{v}$. Then the partition function is

$$q^V = \sum_v e^{-\beta vhc\tilde{v}} = \sum_v (e^{-\beta hc\tilde{v}})^v$$

We met this series in Section 19.2 (which is no acccident: the ladder-like array of levels in Fig. 19.1 is exactly the same as that of a harmonic oscillator). The series can be summed in the same way, and gives

$$q^V = \frac{1}{1 - e^{-\beta hc\tilde{v}}} \qquad (10)$$

In a polyatomic molecule, each normal node gives rise to a partition function of this form.

Example 20.6: *Calculating a vibrational partition function*

The wavenumbers of the three normal modes of H_2O are $3656.7 \, \text{cm}^{-1}$, $1594.8 \, \text{cm}^{-1}$, and $3755.8 \, \text{cm}^{-1}$. Evaluate the vibrational partition function at 1500 K.

Answer. We use eqn 10 for each mode, and then form the product. At 1500 K, $kT/hc = 1042.5 \, \text{cm}^{-1}$.

Mode:	1	2	3
$\tilde{v}/\text{cm}^{-1}$	3656.7	1594.8	3755.8
$hc\tilde{v}/kT$	3.508	1.530	3.603
q^V	1.031	1.276	1.028

The overall vibrational partition function is therefore

$$q = 1.031 \times 1.276 \times 1.028 = 1.35$$

Comment. The vibrations of H_2O are at such high wavenumber that even at 1500 K most of the molecules are in their vibrational ground state.

Exercise. Repeat the calculation for CO_2, where the vibrational wavenumbers are $1388 \, \text{cm}^{-1}$, $667.4 \, \text{cm}^{-1}$, and $2349 \, \text{cm}^{-1}$, $\tilde{v}_2$ being the doubly-degenerate bending mode.

[6.71]

In many molecules the vibrational wavenumbers are so great that $\beta hc\tilde{v} > 1$. For example, the lowest vibrational wavenumber of CH_4 is 1306 cm^{-1}, and so $\beta hc\tilde{v} = 6$ at room temperature. C—H stretches normally lie in the range 2850 to 2960 cm^{-1}, and so for them $\beta hc\tilde{v} = 14$. In these cases $e^{-\beta hc\tilde{v}}$ may be neglected in the denominator of q^V (for example, $e^{-6} = 0.002$), and the vibrational partition function for a single mode is simply $q^V \approx 1$ (implying that only the zero-point level is significantly occupied).

At temperatures so high that $kT \gg hc\tilde{v}$, the partition function may be approximated by expanding the exponential:

$$q^V = \frac{1}{1 - (1 - \beta hc\tilde{v} + \ldots)} \approx \frac{1}{\beta hc\tilde{v}}$$

and hence

$$q = \frac{kT}{hc\tilde{v}} \tag{11}$$

We can express the temperatures for which this expansion is valid in terms of the **characteristic vibrational temperature** θ_V, which is defined through

$$k\theta_V = hc\tilde{v}$$

In terms of the vibrational temperature, 'high temperature' means $T \gg \theta_V$. Some typical values are

	I_2	F_2	HCl	H_2
$\tilde{v}$/cm^{-1}	215	892	2990	4400
θ_V/K	309	1280	4300	6330

The value for H_2 is abnormally high because the atoms are so light and the vibrational frequency consequently high.

The electronic contribution

Electronic energy separations from the ground state are usually very large, and so for most cases $q^E = 1$. An important exception arises in the case of atoms and molecules having electronically degenerate ground states, in which case

$$q^E = g^E$$

where g^E is the degeneracy of the electronic ground state. The alkali metal atoms, for example, have doubly-degenerate ground states (corresponding to the two orientations of their electron spin), and so $q^E = 2$.

Some atoms and molecules have low-lying electronically excited states (at high enough temperatures, all atoms and molecules have thermally accessible excited states). One case is NO, which has the configuration $\ldots \pi^1$ (the molecule has one electron more than N_2). The orbital angular momentum may take two orientations with respect to the molecular axis, and the spin angular momentum may also take two, giving four states in all (Fig. 20.1). The energy of the two states in which the orbital and spin momenta are parallel (giving the $^2\Pi_{3/2}$ term) is slightly greater than that of the two other states in which they are antiparallel (giving the $^2\Pi_{1/2}$ term). The separation, which arises from spin-orbit coupling (Section 13.7), is only 121 cm^{-1}, and so at normal temperatures, all four states are thermally

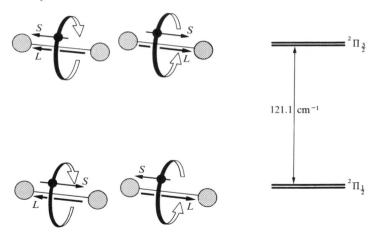

Fig. 20.1 The doubly-degenerate ground electronic level of NO (with the spin and orbital angular momentum around the axis in opposite directions) and the doubly-degenerate first excited level (with the spin and orbital momenta parallel). The upper level is thermally accessible at room temperature.

accessible, but the doubly degenerate lower term is more heavily populated. Denoting the energies of the two levels as $E_{1/2} = 0$ and $E_{3/2} = \varepsilon$, leads to

$$q^{E} = \sum_{j} g_{j}e^{-\beta\varepsilon_{j}} = 2 + 2e^{-\beta\varepsilon}$$

At $T = 0$, $q^{E} = 2$, because only the doubly degenerate ground state is accessible. At very high temperatures, $q^{E} \to 4$ because all four states are accessible. At 25°C, $q^{E} = 3.1$.

The overall partition function

The partition functions for each mode of motion of a molecule are collected in Table 20.2. The overall partition function is the product of each contribution. For a diatomic molecule

$$q = g^{E} \times \frac{V}{\Lambda^{3}} \times \frac{1}{\sigma\beta hcB} \times \frac{1}{1 - e^{-\beta hc\tilde{\nu}}} \tag{12}$$

Overall partition functions obtained in this way are approximate because they assume that the rotational levels are very close and that the vibrational levels are harmonic. These approximations are avoided by evaluating the sums explicitly using spectroscopic data.

Example 20.7: *Calculating the Giauque function from spectroscopic data*

The Giauque function $\Phi_{0} = \{G_{m}^{\ominus} - H_{m}^{\ominus}\}/T$ was introduced in Section 9.4. Calculate its value for $H_{2}O(g)$ at 1500 K given that $A = 27.8778 \text{ cm}^{-1}$, $B = 14.5092 \text{ cm}^{-1}$, and $C = 9.2869 \text{ cm}^{-1}$ and the information in Example 20.6.

Answer. From eqn 6b, with $G(0) = H(0)$,

$$\Phi = -R \ln \frac{q_{m}^{\ominus}}{N_{A}}$$

and we can use the expressions in Table 20.2 for $q_{m}^{\ominus}$. The vibrational partition function was calculated in Example 20.6. Use the expressions in Table 20.2 for the other contributions. For this C_{2v} molecule, $\sigma = 2$. Since $m = 18.015$ u, $q_{m}^{\ominus T} = 1.706 \times 10^{8}$. For the vibrational contribution we have already found that $q^{V} = 1.353$. For the rotational contribution, $q^{R} = 486.7$. The overall molar partition function is therefore $q_{m}^{\ominus}/N_{A} = 1.123 \times 10^{11}$; and so

$$\Phi = -R \ln (1.123 \times 10^{11}) = -211.5 \text{ J K}^{-1} \text{mol}^{-1}$$

Using statistical thermodynamics

Any thermodynamic quantity can now be calculated from a knowledge of the energy levels of molecules; we have merged thermodynamics and spectroscopy. In this section we indicate how to do the calculations for four important properties.

20.3 Mean energies

It is often useful to know the mean energy of various modes of motion. When the molecular partition function can be factorized into contributions

Table 20.2. Contributions to the molecular partition function†

Translation

$$q = \frac{V}{\Lambda^3} \qquad \Lambda = \left(\frac{h^2\beta}{2\pi m}\right)^{1/2}$$

$$\Lambda/\mathrm{pm} = \frac{1749}{(T/\mathrm{K})^{1/2}(M/\mathrm{g\,mol^{-1}})^{1/2}}$$

$$\frac{q_m^\ominus}{N_A} = \frac{kT}{p^\ominus\Lambda^3} = 2.561 \times 10^{-2}(T/\mathrm{K})^{5/2}(M/\mathrm{g\,mol^{-1}})^{3/2}$$

Rotation
  (a) *Linear molecules*

$$q = \frac{1}{\sigma hcB\beta} = \frac{0.6950}{\sigma} \times \frac{T/\mathrm{K}}{(B/\mathrm{cm^{-1}})}$$

  (b) *Nonlinear molecules*

$$q = \frac{1}{\sigma}\left(\frac{1}{hc\beta}\right)^{3/2}\left(\frac{\pi}{ABC}\right)^{1/2}$$

$$= \frac{1.0270}{\sigma} \times \frac{(T/\mathrm{K})^{3/2}}{(ABC/\mathrm{cm^{-3}})^{1/2}}$$

Vibration

$$q = \frac{1}{1 - e^{-hc\tilde{v}\beta}} = \frac{1}{1 - e^{-a}} \qquad a = \frac{1.4388(\tilde{v}/\mathrm{cm^{-1}})}{T/\mathrm{K}}$$

Electronic

$$q = g$$

where g is the degeneracy of the electronic ground state (when that is the only accessible level); at high temperatures, evaluate q explicitly.

† $\beta = 1/kT$. It is often useful to note that

$$\frac{hc}{k} = 1.438\,79\ \mathrm{cm\,K}$$

See also inside front cover for further information.

from each mode, we can write

$$U - U(0) = \frac{-N}{q}\left(\frac{\partial q}{\partial \beta}\right)_V = \frac{-N}{q^T q^R q^V q^E}\left(\frac{\partial q^T q^R q^V q^E}{\partial \beta}\right)_V$$

$$= N\left\{-\frac{1}{q^T}\left(\frac{\partial q^T}{\partial \beta}\right)_V - \frac{1}{q^R}\left(\frac{\partial q^R}{\partial \beta}\right)_V - \frac{1}{q^V}\left(\frac{\partial q^V}{\partial \beta}\right)_V - \frac{1}{q^E}\left(\frac{\partial q^E}{\partial \beta}\right)_V\right\}$$

Since the internal energy is the sum of the mean energies of all the modes, we can also write

$$U - U(0) = N\{\langle \varepsilon^T\rangle + \langle \varepsilon^R\rangle + \langle \varepsilon^V\rangle + \langle \varepsilon^E\rangle\}$$

where $\langle \ldots \rangle$ denotes a mean value. It follows that the mean energy of each mode is

$$\langle \varepsilon^M\rangle = -\frac{1}{q^M}\left(\frac{\partial q^M}{\partial \beta}\right)_V \qquad M = T, R, V, \text{ and } E \qquad (13)$$

The mean translational energy

To see a pattern emerging, we consider first a one-dimensional system of length X, for which $q^T = X/\Lambda$. Then

$$\langle \varepsilon^T\rangle = \frac{-\Lambda}{X}\left(\frac{\partial}{\partial \beta}\frac{X}{\Lambda}\right)_V = \frac{1}{2\beta} = \tfrac{1}{2}kT$$

For a molecule free to move in three dimensions, the analogous calculation leads to

$$\langle \varepsilon^T\rangle = \tfrac{3}{2}kT$$

Both conclusions are in agreement with the classical equipartition theorem (Section 11.1), that the mean energy of each quadratic contribution to the energy is $\tfrac{1}{2}kT$. Furthermore, the fact that the mean energy is independent of the size of the container is consistent with the thermodynamic result that internal energy of a perfect gas is independent of its volume (Section 5.1).

The mean rotational energy

The mean rotational energy of a linear molecule is obtained from the partition function given in eqn 8:

$$q^R = \sum_J (2J + 1)e^{-\beta \varepsilon J(J+1)} \qquad \varepsilon = hcB$$

When the temperature is low ($T < \theta_R$), we must sum the series term by term, which gives

$$q^R = 1 + 3e^{-2\beta\varepsilon} + 5e^{-6\beta\varepsilon} + \ldots$$

Hence

$$\langle \varepsilon^R\rangle = \frac{\varepsilon(6e^{-2\beta\varepsilon} + 30e^{-6\beta\varepsilon} + \ldots)}{1 + 3e^{-2\beta\varepsilon} + 5e^{-6\beta\varepsilon} + \ldots}$$

At high temperatures ($T \gg \theta_R$), q^R is given by eqn 9a and

$$\langle \varepsilon^R\rangle = -\frac{1}{q^R}\left(\frac{\partial q^R}{\partial \beta}\right)_V = \frac{1}{\beta} = kT$$

The high-temperature result is also in agreement with the equipartition theorem, since the classical expression for the energy of a linear rotor is

$$E_K = \tfrac{1}{2}I\omega_x^2 + \tfrac{1}{2}I\omega_y^2$$

(there is no z component because there is no rotation around the line of atoms), and the mean energy is $2 \times \tfrac{1}{2}kT = kT$.

The mean vibrational energy

The vibrational partition function is given in eqn 10. Since q^V is independent of the volume,

$$\frac{dq^V}{d\beta} = \frac{d}{d\beta}\frac{1}{1 - e^{-\beta hc\tilde{v}}} = \frac{hc\tilde{v}e^{-\beta hc\tilde{v}}}{(1 - e^{-\beta hc\tilde{v}})^2}$$

and we obtain

$$\langle \varepsilon^V \rangle = \frac{hc\tilde{v}}{e^{\beta hc\tilde{v}} - 1} \tag{14}$$

This formula is exact, apart from the zero-point energy of $\tfrac{1}{2}hc\tilde{v}$ (which can be added to the right-hand side if required).

Example 20.8: *Deriving the high-temperature form of an expression*

Derive an expression for the mean energy of a harmonic oscillator when the temperature is high.

Answer. By 'high temperature' in this vibrational context we mean $T \gg \theta_V$, or $\beta hc\tilde{v} \ll 1$. The exponential functions can therefore be expanded and all but the leading terms discarded:

$$\langle \varepsilon^V \rangle = \frac{hc\tilde{v}}{(1 + \beta hc\tilde{v} + \ldots) - 1} \approx \frac{1}{\beta} = kT$$

Comment. A mean energy of kT is in agreement with the value predicted by the classical equipartition theorem, since the energy of a one-dimensional oscillator is

$$E = \tfrac{1}{2}mv_x^2 + \tfrac{1}{2}kx^2$$

and the mean energy of each term is $\tfrac{1}{2}kT$.

Exercise. Suppose that we falsely believe that $h = 0$ (as in classical physics). Calculate the mean energy of a harmonic oscillator at a temperature T. [kT]

20.4 Heat capacities

The constant-volume heat capacity is

$$C_V = \left(\frac{\partial U}{\partial T}\right)_V$$

Since

$$\frac{d}{dT} = \frac{d\beta}{dT}\frac{d}{d\beta} = -\frac{1}{kT^2}\frac{d}{d\beta} = -k\beta^2\frac{d}{d\beta}$$

an alternative form is

$$C_V = -k\beta^2 \left(\frac{\partial U}{\partial \beta}\right)_V \tag{15}$$

Because the internal energy of a perfect gas is a sum of contributions, the heat capacity is also a sum:

$$C_V = C_V^T + C_V^R + C_V^V + C_V^E$$

and the contribution of each mode is

$$C_V^M = N\left(\frac{\partial \langle \varepsilon^M \rangle}{\partial T}\right)_V = -Nk\beta^2 \left(\frac{\partial \langle \varepsilon^M \rangle}{\partial \beta}\right)_V \qquad M = T, R, V, \text{ or } E \tag{16}$$

The individual contributions

The temperature is always high enough (so long as the gas is above its condensation temperature) for the mean translational energy to be $\frac{3}{2}kT$, the equipartition value. Therefore, the molar constant volume heat capacity is

$$C_V^T = N_A \frac{d(\frac{3}{2}kT)}{dT} = \frac{3}{2}R$$

Translation is the only mode of motion for a monatomic gas, so for such a gas

$$C_V^T = \tfrac{3}{2}R = 12.47 \text{ J K}^{-1}\text{mol}^{-1}$$

This is a very reliable result: helium, for example, has this value over a range of 2000 K. Since the molar constant pressure heat capacity of a perfect gas is related to C_V by

$$C_p = C_V + R$$

we can also conclude at once that for a monatomic perfect gas

$$C_p = \tfrac{5}{2}R = 20.78 \text{ J K}^{-1}\text{mol}^{-1}$$

and that

$$\gamma = \frac{C_p}{C_V} = \tfrac{5}{3}$$

When the temperature is high enough for the rotations of the molecules to be highly excited (when $T \gg \theta_R$), we can use the equipartition value $\langle \varepsilon^R \rangle = kT$ (for a linear rotor) to obtain

$$C_V^R = R$$

For nonlinear molecules, the mean rotational energy rises to $\frac{3}{2}kT$, and so the molar rotational heat capacity rises to $\frac{3}{2}R$ when $T \gg \theta_R$. When the temperature is very low, only the lowest rotational state is occupied, and the rotations do not contribute to the heat capacity ($C_V^R = 0$). At intermediate temperatures we can calculate the rotational heat capacity by differentiating the equation for $\langle \varepsilon^R \rangle$ obtained earlier. The resulting (untidy) expression, which is plotted in Fig. 20.2, shows that the contribution rises from zero (when $T = 0$) to the equipartition value (when $T \gg \theta_R$). Since the translational contribution is always present, we can expect the molar heat

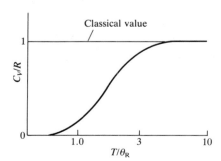

Fig. 20.2 The temperature dependence of the rotational contribution to the heat capacity of a linear molecule. The rotational temperature is $\theta_R = hcB/k$; the quantum value is close to the classical value for $T > 3\theta_R$.

capacity of a gas of diatomic molecules $(C_V^T + C_V^R)$ to rise from $12.5\,\mathrm{J\,K^{-1}\,mol^{-1}}$ to $20.8\,\mathrm{J\,K^{-1}\,mol^{-1}}$ as the temperature is increased above θ_R.

Molecular vibrations contribute to the heat capacity, but only when the temperature is high enough for them to be significantly excited. The equipartition mean energy is kT per mode, and so the maximum contribution to the molar heat capacity is R. However, it is very unusual for the vibrations to be so highly excited that equipartition is valid, and it is more appropriate to use the full expression for the vibrational heat capacity, which is obtained by differentiating eqn 14:

$$C_V^V = Rf^2 \qquad f = \frac{\theta_V}{T}\left\{\frac{e^{-\theta_V/2T}}{1-e^{-\theta_V/T}}\right\} \qquad (17)$$

$\theta_V = hc\bar{\nu}/k$ is the vibrational temperature; in Chapter 11 we met θ_V as the Einstein temperature. The curve in Fig. 20.3 shows how C_V^V depends on temperature: note that it is close to the equipartition value when $T \approx \theta_V$.

The overall heat capacity

The total heat capacity of a system is the sum of each contribution (Fig. 20.4). When equipartition is valid (when the temperature is well above the characteristic temperature of the mode, $T \gg \theta_M$) we can estimate the heat capacity by counting the numbers of modes that are active.

In gases, all three translational modes are always active, and contribute $\frac{3}{2}R$ to the molar heat capacity. If we denote the number of active rotational modes by ν_R^* (so that for most molecules at normal temperatures $\nu_R^* = 2$ for linear molecules, and $\nu_R^* = 3$ for nonlinear molecules) then the rotational contribution is $\frac{1}{2}\nu_R^* R$. If the temperature is high enough for ν_V^* vibrational modes to be active, the vibrational contribution to the molar heat capacity is $\nu_V^* R$. In most case $\nu_V^* = 0$. Therefore, the total molar heat capacity is

$$C_V = \tfrac{1}{2}(3 + \nu_R^* + 2\nu_V^*)R \qquad (18)$$

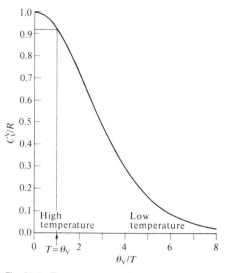

Fig. 20.3 The temperature dependence of the vibrational heat capacity calculated using eqn 17. Note that temperature decreases from left to right.

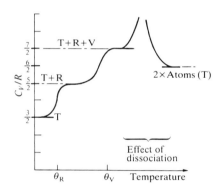

Fig. 20.4 The general temperature dependence of the heat capacity of diatomic molecules is as shown here. Each mode becomes active when its characteristic temperature is exceeded. The heat capacity becomes very large when the molecule dissociates because the energy is used to cause dissociation and not to raise the temperature. Then it falls back to the translation-only value of the atoms.

Example 20.9: *Estimating the molar heat capacity of a gas*

Estimate the molar constant-volume heat capacity of water vapour at 100°C.

Answer. We need to assess whether the rotational and vibrational modes are active by computing their characteristic temperatures from the data in Example 20.6 and the rotational constants 27.9, 14.5, and 9.3 cm^{-1} (to do so, we use $hc/k = 1.439$ cm K). The characteristic temperatures (in round numbers) of the vibrations are 5300 K, 2300 K, and 5400 K and are therefore inactive at 373 K. The three rotational modes have characteristic temperatures 40 K, 21 K, and 13 K, and so are fully active, like the three translational modes. The translational contribution is $\frac{3}{2}R = 12.5\,\mathrm{J\,K^{-1}\,mol^{-1}}$. Fully active rotations contribute a further $12.5\,\mathrm{J\,K^{-1}\,mol^{-1}}$. Therefore a value close to $25\,\mathrm{J\,K^{-1}\,mol^{-1}}$ is predicted.

Comment. The experimental value is $26.1\,\mathrm{J\,K^{-1}\,mol^{-1}}$. The discrepancy is probably due to deviations from perfect gas behaviour.

Exercise. Estimate the molar heat capacity of gaseous I_2 at 25°C ($B = 0.037$ cm^{-1}).
[21 J K^{-1} mol^{-1}]

20.5 Residual entropies

Entropies may be calculated from spectroscopic data; they may also be measured experimentally (Section 4.5). In many cases there is good agreement, but in some the experimental entropy is less than the calculated value. One possibility is that the experimental determination failed to take a phase transition into account. Another is that some disorder is present in the solid even at $T = 0$. The entropy at $T = 0$ is then greater than zero, and is called the **residual entropy**.

We can understand the origin and magnitude of the residual entropy by considering a crystal composed of AB molecules, where A and B are similar atoms (such as CO, with its very small electric dipole moment). There may be so little energy difference between ...AB AB AB AB..., ...AB BA BA AB..., and other random arrangements that the molecules adopt either orientation arbitrarily in the solid. We can readily calculate the entropy arising from residual disorder using the Boltzmann formula $S = k \ln W$.

Suppose that two orientations are equally probable, and that the sample consists of N molecules. Since the same energy can be achieved in 2^N different ways (because each molecule can take either of two orientations), the total number of ways of achieving the same energy is $W = 2^N$. It follows that

$$S = k \ln 2^N = Nk \ln 2 = nR \ln 2$$

We can therefore expect a residual molar entropy of $R \ln 2 = 5.8 \, \text{J K}^{-1} \text{mol}^{-1}$ for solids composed of molecules that can adopt either of two orientations at $T = 0$.

If s orientations are possible, the residual molar entropy will be

$$S_m = R \ln s \qquad (19)$$

The $FClO_3$ molecule can adopt four orientations with about the same energy, and the calculated residual entropy of $R \ln 4 = 11.5 \, \text{J K}^{-1} \text{mol}^{-1}$ is in good agreement with the experimental value ($10.1 \, \text{J K}^{-1} \text{mol}^{-1}$). For CO the measured residual entropy is $5 \, \text{J K}^{-1} \text{mol}^{-1}$, which is close to $R \ln 2$, the value expected for a random structure of the form ...CO CO OC CO OC OC....

The residual entropy of ice is $3.4 \, \text{J K}^{-1} \text{mol}^{-1}$. This value can be explained in terms of the hydrogen-bonded structure of the solid. Each O atom is surrounded tetrahedrally by four H atoms, two of which are attached by short σ bonds, the other two being attached by long hydrogen bonds. The randomness lies in which two of the four bonds are short, and an approximate analysis leads to $S_m(0) \approx R \ln \frac{3}{2} = 3.4 \, \text{J K}^{-1} \text{mol}^{-1}$, in good agreement with the experimental value.

20.6 Equilibrium constants

The Gibbs function of a system of independent molecules is given by eqn 6b as

$$G - G(0) = -nRT \ln \frac{q_m}{N_A}$$

where q_m is the molar partition function q/n. The equilibrium constant K_p is related to the standard Gibbs function of reaction by

$$\Delta G^{\ominus} = -RT \ln K_p$$

To calculate the equilibrium constant, we must combine these two equations.

The relation between K_p and the partition function

We need the *standard* molar Gibbs function $G^{\ominus}/n$ of each component. To find it, we need the value of the molar partition function when $p = p^{\ominus}$ (where $p^{\ominus} = 1$ bar): we denote this $q_m^{\ominus}$. Since only the translational component depends on the pressure (through $V_m = V/n$) we can find $q_m^{\ominus}$ simply by evaluating q_m with $V_m = RT/p^{\ominus}$. For a component J it follows that

$$G_{J,m}^{\ominus} - G_{J,m}^{\ominus}(0) = -RT \ln \frac{q_J^{\ominus}}{N_A} \qquad (20)$$

where $q_J^{\ominus}$ is the standard molar partition function of species J. (We shall omit the additional 'molar' subscript on $q_{J,m}$; whenever q has a substance label, it should be interpreted as a molar partition function.)

We shall consider the reaction

$$0 = \sum_J \nu_J S_J$$

The standard molar Gibbs function of the reaction is

$$\Delta G^{\ominus} = \sum_J \nu_J G_{J,m}^{\ominus} = \sum_J \nu_J G_{J,m}^{\ominus}(0) - RT \sum_J \nu_J \ln \frac{q_J^{\ominus}}{N_A}$$

Since for a perfect gas $G(0) = U(0)$,[1] the first term on the right is

$$\sum_J \nu_J G_{J,m}^{\ominus}(0) = \Delta G^{\ominus}(0) = \Delta U^{\ominus}(0)$$

$\Delta U^{\ominus}(0)$ is the difference in (molar) energy between the zero-point levels of the reactants and products (weighted by the stoichiometric coefficients), and we normally write it ΔE_0:

$$\Delta E_0 = \sum_J \nu_J U_J(0) \qquad (21)$$

Example 20.10: *Calculating the reaction energy difference*

Calculate ΔE_0 for the reaction

$$N_2(g) + 3H_2(g) \rightarrow 2NH_3(g)$$

---

[1] This follows from

$$G = H - TS = U + pV - TS$$
$$= U + nRT - TS = U \text{ at } T = 0$$

Answer. We write the reaction in the form

$$0 = 2NH_3(g) - N_2(g) - 3H_2(g)$$

and from eqn 21 obtain

$$\Delta E_0 = 2U(NH_3, 0) - U(N_2, 0) - 3U(H_2, 0)$$

If the internal energies are measured relative to those of the free atoms, we can replace them by the (mean) bond dissociation energies D:

$$\Delta E_0 = 2 \times 3 \times D(N-H) - D(N\equiv N) - 3 \times D(H-H)$$

Exercise. Express ΔE_0 for the reaction $2H_2(g) + O_2(g) \rightarrow 2H_2O(g)$.

$$[2D(O-H) - 2D(H-H) - D(O\!=\!O)]$$

For the last part of the calculation we write

$$\Delta G^{\ominus} = \Delta E_0 - RT \sum_J \ln \left(\frac{q_J^{\ominus}}{N_A}\right)^{\nu_J}$$

$$= -RT \left\{ \frac{-\Delta E_0}{RT} + \ln \prod_J \left(\frac{q_J^{\ominus}}{N_A}\right)^{\nu_J} \right\}$$

We can now pick out an expression for K_p by comparing this equation with $\Delta G^{\ominus} = -RT \ln K_p$:

$$\ln K_p = \frac{-\Delta E_0}{RT} + \ln \prod_J \left(\frac{q_J^{\ominus}}{N_A}\right)^{\nu_J}$$

Hence:

$$K_p = \prod_J \left(\frac{q_J^{\ominus}}{N_A}\right)^{\nu_J} \times e^{-\Delta E_0/RT} \qquad (22)$$

A dissociation equilibrium

We shall illustrate the application of eqn 22 to an equilibrium in which a diatomic molecule X_2 dissociates into its atoms: $X_2(g) \rightleftharpoons 2X(g)$.

Since the forward reaction is

$$0 = 2X(g) - X_2(g) \qquad \nu(X) = 2,\ \nu(X_2) = -1$$

the equilibrium constant is

$$K_p = \prod_J \left(\frac{p_J}{p^{\ominus}}\right)^{\nu_J} = \left(\frac{p(X)}{p^{\ominus}}\right)^2 \left(\frac{p(X_2)}{p^{\ominus}}\right)^{-1} = \frac{p(X)^2}{p(X_2)p^{\ominus}}$$

In terms of partition functions

$$K_p = \left(\frac{q^{\ominus}(X)}{N_A}\right)^2 \left(\frac{q^{\ominus}(X_2)}{N_A}\right)^{-1} e^{-\Delta E_0/RT}$$

$$= \frac{q^{\ominus}(X)^2 e^{-\Delta E_0/RT}}{q^{\ominus}(X_2)N_A}$$

ΔE_0 is the (molar) dissociation energy D_0 of the molecule

$$\Delta E_0 = 2U(X) - U(X_2) \quad \text{at} \quad T = 0$$

$$= D(X_2)$$

The atomic molar partition functions relate only to their translational motion and any electronic degeneracy:

$$q^{\ominus}(X) = \frac{g(X)V_{m}^{\ominus}}{\Lambda(X)^3} \qquad V_{m}^{\ominus} = \frac{RT}{p^{\ominus}}$$

The diatomic molecule also has rotational and vibrational modes of motion and so its molar partition function is

$$q^{\ominus}(X_2) = \frac{g(X_2)V_{m}^{\ominus}q^R(X_2)q^V(X_2)}{\Lambda(X_2)^3}$$

It follows that the equilibrium constant is

$$K_p = \frac{g(X)^2}{g(X_2)} \times \frac{\Lambda(X_2)^3 V_{m}^{\ominus}}{\Lambda(X)^6 N_A} \times \frac{1}{q^R(X_2)q^V(X_2)} \times e^{-D_0/RT}$$

Example 20.11: *Evaluating an equilibrium constant*

Evaluate K_p for $Na_2(g) \rightleftharpoons 2Na(g)$ at 1000 K from the following data: $B = 0.1547$ cm^{-1}, $\tilde{v} = 159.2$ cm^{-1}, $D_0 = 70.4$ kJ mol^{-1} (0.73 eV). The Na atoms have doublet ground terms.

Answer. We can evaluate the partition functions using the expressions in Table 20.2:

$$\Lambda(Na_2) = 8.14 \text{ pm} \qquad \Lambda(Na) = 11.5 \text{ pm}$$

$$q^R(Na_2) = 2246 \qquad q^V(Na_2) = 4.885$$

$$g(Na) = 2 \qquad g(Na_2) = 1$$

Then, with $V_{m}^{\ominus} = 8.206 \times 10^{-2}$ m^3 mol^{-1}, the equation above evaluates to

$$K_p = \frac{4 \times (8.14 \text{ pm})^3 \times (8.206 \times 10^{-2} \text{ m}^3 \text{ mol}^{-1})}{(11.5 \text{ pm})^6 \times (6.022 \times 10^{23} \text{ mol}^{-1})} \times \frac{1}{2246 \times 4.885} \times e^{-8.47}$$

$$= 2.42$$

Comment. Note that the procedure leads to K_p; for conversion to K_c we use $[J] = p(J)/RT$.

Exercise. Evaluate K_p at 1500 K. [52]

Contributions to the equilibrium constant

We are now in a position to appreciate the physical basis of equilibrium constants. To show what is involved, we consider a simple $A \rightleftharpoons B$ equilibrium.

In Fig. 20.5 one set of levels belongs to A, and the other to B. The populations are given by the Boltzmann distribution, and are independent of whether any given level happens to belong to A or to B. We can therefore imagine a single Boltzmann distribution spreading, without distinction, over the two sets of energy levels. If the spacings of A and B are similar (as in Fig. 20.5), and B lies above A, the diagram indicates that A will dominate in the equilibrium mixture. However, if B has a high density of states (as in Fig. 20.6), then even though its zero-point energy lies above that of A, it might still dominate at equilibrium.

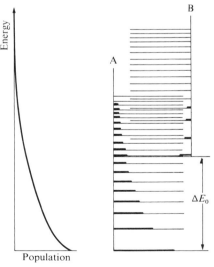

Fig. 20.5 The array of A and B energy levels. At equilibrium all are accessible, and the equilibrium composition of the system reflects the overall Boltzmann distribution of populations. As ΔE_0 increases, A becomes dominant.

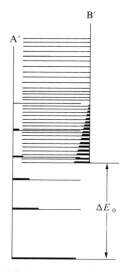

Fig. 20.6 It is important to take into account the densities of states of the molecules. Even though B might lie well above A in energy (i.e. ΔE_0 is large and positive), B might have so many states that its total population dominates in the mixture. In classical thermodynamic terms, we have to take entropies into account as well as enthalpies when considering equilibria.

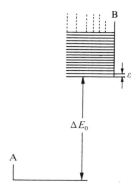

Fig. 20.7 The model used in the text for exploring the effects of energy separations and densities of states on equilibria. B can dominate so long as ΔE_0 is not too large.

We can in fact show that an analysis of diagrams like those in Figs. 20.5 and 20.6 leads to an expression for the equilibrium constant in terms of partition functions. The population of a state i of the composite (A + B) system is

$$n_i = \frac{Ne^{-\beta\varepsilon_i}}{q}$$

where N is the total number of molecules. The total number of type A molecules is the sum of these populations taken over the states belonging to A (these we label a with energies ε_a), and the total number of B is the sum over the states belonging to B (which we label b with energies ε_b'; we shall explain the prime on ε_b' in a moment):

$$N_A = \sum_a n_a = \frac{N}{q}\sum_a e^{-\beta\varepsilon_a} \qquad N_B = \sum_b n_b = \frac{N}{q}\sum_b e^{-\beta\varepsilon_b'}$$

Since the sum over the states of A is its partition function,

$$N_A = \frac{Nq(A)}{q}$$

The sum over the states of B is also a partition function, but the energies are measured from the ground state of the total system, which is the ground state of A. However, since $\varepsilon_b' = \varepsilon_b + \Delta\varepsilon_0$, where $\Delta\varepsilon_0$ is the separation of zero-point energies (Fig. 20.7),

$$N_B = \frac{N}{q}\sum_b e^{-\beta(\varepsilon_b+\Delta\varepsilon_0)} = \frac{Nq(B)}{q}e^{-\Delta E_0/RT}$$

(The switch from $\Delta\varepsilon_0/k$ to $\Delta E_0/R$ is the conversion of molecular energies to molar energies.)

The equilibrium constant of the A $\rightleftharpoons$ B reaction is proportional to the ratio of the numbers of the two types of molecule:

$$K_p = \frac{N_B}{N_A} = \frac{q(B)}{q(A)}e^{-\Delta E_0/RT} \tag{23}$$

This is the same[2] as the expression given by eqn 22.

The content of eqn 23 can be seen most clearly by exaggerating the molecular features that contribute to it. We shall suppose that A has only a single accessible level, which implies that $q(A) = 1$, and that B has a large number of evenly, closely spaced levels (Fig. 20.7). Its partition function is

$$q(B) \approx \frac{kT}{h\nu}$$

In this model system, the A $\rightleftharpoons$ B equilibrium constant is

$$K_p = \frac{kTe^{-\Delta E_0/RT}}{h\nu}$$

---

[2] For an A $\rightleftharpoons$ B equilibrium, the V factors in the partition functions cancel, and so the appearance of q in place of $q^\ominus$ has no effect. In the case of a more general reaction, the conversion from q to $q^\ominus$ comes about at the stage of converting the pressures that occur in K_p into numbers.

When ΔE_0 is very large, the exponential term dominates and $K_p \approx 0$, which implies that very little B is present at equilibrium. When ΔE_0 is small but still positive, K_p can exceed unity. This reflects the predominance of B at equilibrium on account of its high density of states. At low temperatures $K_p \approx 0$, and the system consists entirely of A. At high temperatures the exponential approaches 1 and the pre-exponential factor is large. Hence B becomes dominant. We see that in this endothermic reaction (endothermic because B lies above A), a rise in temperature favours B, because its states become accessible. This behaviour is what we saw, from the outside, in Chapter 9.

Further information: nuclear statistics and rotational states

In this section we shall demonstrate that certain molecules cannot occupy all the available rotational states and that, as a consequence, the partition function must be reduced by a factor (the symmetry number σ). To do so, we draw on two fundamental features of nature. One is that particles are classified as fermions if they have half-integral spin (like electrons, protons, and ^{13}C nuclei) and as bosons if they have integral spins, including zero (like deuterons, photons, and ^{16}O nuclei). The second point is that the wavefunctions of systems of identical particles must satisfy the Pauli principle (see *Further information 2*, Chapter 13):

> When two identical particles are interchanged, the total wavefunction ψ changes to $-\psi$ if they are fermions and to $+\psi$ if they are bosons.

To apply the Pauli principle we also need to know that when a molecule is rotated by 180° its rotational wavefunction is unchanged except for a factor $(-1)^J$. We have seen similar behaviour in atomic orbitals, where the relevant factor is $(-1)^l$: an s orbital is unchanged by the rotation, a p orbital changes sign, and a d orbital is unchanged.

The Pauli principle forbids the occurrence of some rotational states of molecules that interchange identical nuclei when they rotate, such as when the two ^{16}O nuclei of CO_2 are interchanged by a 180° rotation. Since the two ^{16}O nuclei are bosons, the wavefunction of CO_2 must be unchanged in a 180° rotation; hence only even values of J are allowed, for only then does $(-1)^J = +1$. No CO_2 molecules (where O is ^{16}O) can exist in a rotational state with J odd.

The problem is more complicated when the nuclei have non-zero spin, and we shall show what is involved by considering the $^1H_2$ molecule. The ^1H are identical fermions, so the only rotational states allowed are those that have wavefunctions that change sign under a 180° rotation. However, the Pauli principle applies to the *total* wavefunction, and we need to see how the rotation affects the nuclear spin states as well as the rotational wavefunction. The wavefunctions representing the states of the two spin-$\frac{1}{2}$ nuclei are

$$\alpha(1)\alpha(2), \qquad \alpha(1)\beta(2) + \beta(1)\alpha(2), \qquad \beta(1)\beta(2), \qquad \alpha(1)\beta(2) - \beta(1)\alpha(2)$$

The first three are symmetric under interchange of the nuclei (i.e. they do not change sign when 1 and 2 are exchanged), whereas the last is antisymmetric and does change sign.

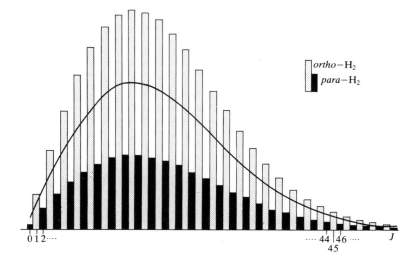

Fig. 20.8 The values of the individual terms $(2J + 1)e^{-hcB\beta J(J+1)}$ contributing to the mean partition function of a 3:1 mixture of *ortho*- and *para*-H_2. The partition function is the sum of all these terms. At high temperatures the sum is approximately equal to the sum of the terms over all values of J, each with a weight of $\frac{1}{2}$. This is the sum of the contributions indicated by the curve.

We now have two factors in the overall H_2 wavefunction that may change when the molecule turns through 180°. However, the Pauli principle instructs us that only totally antisymmetric wavefunctions are allowed (for spin-$\frac{1}{2}$ molecules and other fermions). Therefore we must take the symmetric (even J) rotational wavefunctions with the antisymmetric spin function and the antisymmetric rotational functions (odd J) with any of the three symmetric spin functions. This means that there are two kinds of molecular hydrogen:

> *para*-hydrogen: paired nuclear spins ($\uparrow\downarrow$) $J = 0, 2, 4, \ldots$
>
> *ortho*-hydrogen: parallel nuclear spins ($\uparrow\uparrow$) $J = 1, 3, 5, \ldots$

There are three states of o-H_2 to each value of J (because there are three parallel spin states of the nuclei).

Now we are ready to set up the rotational partition function. In order to do so, we have to take note of the fact that 'ordinary' molecular hydrogen is a mixture of one part p-H_2 (with only its even-J rotational states occupied) and three parts o-H_2 (with only its odd-J rotational states occupied). Therefore, the average partition function per molecule is

$$q^{\mathrm{R}} = \frac{1}{4}\left\{ \sum_{\text{even } J} (2J + 1)e^{-\beta hcBJ(J+1)} + 3 \sum_{\text{odd } J} (2J + 1)e^{-\beta hcBJ(J+1)} \right\}$$

and the odd-J states are more heavily weighted than the even-J states. This is illustrated in Fig. 20.8. From that illustration we see that we would obtain approximately the same answer for the partition function (the sum of all the populations) if each J term contributed half its normal value to the sum. That is, the last equation can be approximated as

$$q^{\mathrm{R}} = \tfrac{1}{2} \sum_{J} (2J + 1)e^{-\beta hcBJ(J+1)}$$

and this approximation is very good when many terms contribute (at high temperatures).

The same type of argument may be used when rotations through angles other than 180° interchange equivalent nuclei (as in CH_4). In each case, in the high temperature limit, the true rotational partition function is a factor $1/\sigma$ smaller than calculated by summing over all the states regardless of the Pauli principle.

Further reading

B. J. McClelland, *Statistical thermodynamics*. Wiley, New York (1973).

D. A. McQuarrie, *Statistical thermodynamics*. Harper & Row, New York (1976).

N. Davidson, *Statistical thermodynamics*. McGraw-Hill, New York (1962).

T. L. Hill, *An introduction to statistical thermodynamics*. Addison-Wesley, Reading (1960).

D. Chandler, *Introduction to statistical mechanics*. Oxford University Press (1987).

D. Wu and D. Chandler, *Solutions manual for 'Introduction to statistical mechanics'*. Oxford University Press (1988).

Exercises

20.1 Use the equipartition theorem to estimate the constant volume heat capacity of (a) I_2, (b) CH_4, (c) C_6H_6 in the gas phase at 25°C.

20.2 Estimate the rotational partition function of HCl at (a) 25°C and (b) 250°C.

20.3 Give the symmetry number for each of the following molecules: (a) CO, (b) O_2, (c) H_2S, and (d) SiH_4, (e) $CHCl_3$.

20.4 Calculate the rotational partition function of H_2O at 298 K from its rotational constants 27.878 cm^{-1}, 14.509 cm^{-1}, and 9.287 cm^{-1}. Above what temperature is the high-temperature approximation valid?

20.5 Calculate the rotational partition function of CH_4 (a) by direct summation of the energy levels at 298 K and 500 K, and (b) by the high temperature approximation. Take $B = 5.28$ cm^{-1}.

20.6 The bond length of O_2 is 120.75 pm. Use the high-temperature approximation to calculate the rotational partition function of the molecule at 300 K.

20.7 The NOF molecule is an asymmetric rotor with rotational constants 3.1752 cm^{-1}, 0.3951 cm^{-1}, and 0.3505 cm^{-1}. Calculate the rotational partition function of the molecule at (a) 25°C, (b) 100°C.

20.8 Plot the molar heat capacity of a collection of harmonic oscillators as a function of T/θ_V and predict the vibrational heat capacity of acetylene at (a) 298 K, (b) 500 K. The normal modes occur at wavenumbers 612(2), 729(2), 1974, 3287, and 3374 cm^{-1}.

20.9 Derive expressions for the partition function, internal energy, and heat capacity of a system that consists of two levels separated by an energy ε. Plot the temperature dependence in each case as a function of ε/kT.

20.10 The CO_2 molecule is linear, and its vibrational wavenumbers are 1388.2 cm^{-1}, 667.4 cm^{-1}, and 2349.2 cm^{-1}, the second being doubly degenerate and the others non-degenerate. The rotational constant of the molecule is 0.3092 cm^{-1}. Calculate the rotational and vibrational contributions to the molar Gibbs function at 298 K.

20.11 The ground level of Cl is $^2P_{3/2}$ and $^2P_{1/2}$ level lies 881 cm^{-1} above it. Calculate the electronic contribution to the heat capacity of Cl atoms at (a) 500 K and (b) 900 K.

20.12 The first electronically excited state of O_2 is $^1\Delta_g$ and lies 7918.1 cm^{-1} above the ground state, which is $^3\Sigma_g^-$. Calculate the electronic contribution to the molar Gibbs function of O_2 at 400 K.

20.13 The ground state of the Co^{2+} ion in $CoSO_4\cdot7H_2O$ may be regarded as $^4T_{9/2}$. The entropy of the solid at temperatures below 1 K is derived almost entirely from the electron spin. Estimate the molar entropy of the solid at these temperatures.

20.14 Estimate the constant pressure molar heat capacity of gaseous ammonia at 25°C from the equipartition theorem. Suggest reasons for the difference of the value calculated from the experimental value, which is 35.06 J K^{-1} mol^{-1}.

20.15 Calculate the residual molar entropy of a solid in which the the molecules can adopt (a) three, (b) five, (c) six orientations of equal energy at $T = 0$.

20.16 Suppose that the hexagonal molecule $C_6H_nF_{6-n}$ has a residual entropy on account of the similarity of the H and F atoms. Calculate the residual entropy for each value of n.

20.17 Calculate the standard molar entropy of $N_2(g)$ at 298 K from its rotational constant $B = 1.9987$ cm^{-1} and its vibrational wavenumber $\bar{v} = 2358$ cm^{-1}. The thermochemical value is 192.1 J K^{-1} mol^{-1}. What does this suggest about the solid at $T = 0$?

20.18 Calculate the equilibrium constant of the reaction $I_2(g) \rightleftharpoons 2I(g)$ at 1000 K from the following data for I_2:

$$\bar{v} = 214.36 \text{ cm}^{-1}, \qquad B = 0.0373 \text{ cm}^{-1}, \qquad D_e = 1.5422 \text{ eV}$$

The ground state of the I atoms is $^2P_{3/2}$, implying fourfold degeneracy.

20.19 Calculate the value of K_p at 298 K for the isotopic exchange reaction

$$2\,^{79}Br^{81}Br \rightleftharpoons\, ^{79}Br^{79}Br +\, ^{81}Br^{81}Br$$

The Br_2 molecule has a non-degenerate ground state, with no other electronic states nearby. Base the calculation on the wavenumber of the vibration of $^{79}Br^{81}Br$, which is 323.33 cm^{-1}.

Problems

Numerical problems

20.1 Calculate the range of vibrational wavenumbers for which the vibrational partition function at 500 K differs from 1 by no more than 0.10 per cent.

20.2 The NO molecule has a doubly degenerate electronic ground state and a doubly degenerate excited state at 121.1 cm^{-1}. Calculate the electronic contribution to the molar heat capacity of the molecule at (a) 50 K, (b) 298 K, and (c) 500 K.

20.3 Explore whether a magnetic field can influence the heat capacity of the molecule by calculating the electronic contribution to the heat capacity of the NO_2 molecule in a magnetic field. Estimate the total constant volume heat capacity using equipartition, and calculate the percentage change in heat capacity brought about by a 5.0 T magnetic field at (a) 50 K, (b) 298 K.

20.4 The energy levels of a CH_3 group attached to a larger fragment are given by the expression for a particle on a ring so long as it is rotating freely. What is the high-temperature contribution to the heat capacity and the entropy of such a freely rotating group at 25°C? The moment of inertia is 5.341×10^{-47} kg m^2.

20.5 Calculate the temperature dependence of the heat capacity of p-H_2 (in which only rotational states with even values of J are populated) at low temperatures on the basis that its rotational levels $J = 0$ and $J = 2$ constitute a system that resembles a two-level system except for the degeneracy of the upper level. Use $B = 60.864$ cm^{-1} and sketch the heat capacity curve. The experimental heat capacity of p-H_2 does in fact show a peak at low temperatures.

20.6 Calculate the rotational entropy of C_6H_6 that is free to rotate in three dimensions at 362 K given its moments of inertia $I_{\parallel} = 2.93 \times 10^{-38}$ g cm^2 and $I_{\perp} = 1.46 \times 10^{-38}$ g cm^2. Calculate the change in rotational entropy when the molecule attaches to a surface and can rotate only about its C_6 axis. In an experiment to study the adsorption of hydrocarbons on graphite at 362 K (D. Dollimore, G. R. Heal, and D. R. Martin, *J. Chem. Soc. Faraday Trans.* I, 1784 (1973)), a change in entropy of -111 J K^{-1} mol^{-1} was observed when there was only a small surface coverage but -52 J K^{-1} mol^{-1} was measured when surface coverage was complete. Propose a model for the adsorption of benzene on graphite, and test it quantitatively given that each molecule occupies an area of 4.08×10^{-2} nm^2.

20.7 The pure rotational microwave spectrum of HCl has absorption lines at the following wavenumbers (in cm^{-1}): 21.19, 42.37, 63.56, 84.75, 105.93, 127.12, 148.31, 169.49, 190.68, 211.87, 233.06, 254.24, 275.43, 296.62, 317.80, 338.99, 360.18, 381.36, 402.55, 423.74, 444.92, 466.11, 487.30, 508.48. Calculate the rotational partition function at 25°C by direct summation.

20.8 Calculate and plot as a function of temperature, in the range 300 K to 1000 K, the equilibrium constant K_p for

$$CD_4(g) + HCl(g) \rightleftharpoons CHD_3(g) + DCl(g)$$

using the following data (numbers in parentheses are degeneracies):

$\tilde{v}(CHD_3)/cm^{-1} = 2993(1), 2142(1), 1003(3), 1291(2), 1036(2)$

$\tilde{v}(CD_4)/cm^{-1} = 2109(1), 1092(2), 2259(3), 996(3)$

$\tilde{v}(HCl)/cm^{-1} = 2991 \qquad \tilde{v}(DCl)/cm^{-1} = 2145$

$B(HCl)/cm^{-1} = 10.59 \qquad B(DCl)/cm^{-1} = 5.445$

$B(CHD_3)/cm^{-1} = 3.28 \qquad A(CHD_3)/cm^{-1} = 2.63$

$B(CD_4)/cm^{-1} = 2.63$

20.9 The exchange of deuterium between acid and water is an important type of equilibrium, and we can examine it using spectroscopic data on the molecules. Calculate the equilibrium constant at (a) 298 K and (b) 800 K for the gas phase exchange reaction $H_2O + DCl \rightleftharpoons HDO + HCl$ from the following data:

$\tilde{v}(H_2O)/cm^{-1} = 3656.7, 1594.8, 3755.8$

$\tilde{v}(HDO)/cm^{-1} = 2726.7, 1402.2, 3707.5$

$B(H_2O)/cm^{-1} = 27.88, 14.51, 9.29$

$B(HDO)/cm^{-1} = 23.38, 9.102, 6.417$

$B(HCl)/cm^{-1} = 10.59 \qquad B(DCl)/cm^{-1} = 5.449$

$\tilde{v}(HCl)/cm^{-1} = 2991 \qquad \tilde{v}(DCl)/cm^{-1} = 2154$

20.10 Calculate the Giauque function Φ_0 at 1000 K for (a) H_2, (b) N_2, (c) NH_3 using data assembled from Table 16.2 and Exercise 20.17. Go on to calculate the value of K_p from Φ_0 for the reaction $N_2(g) + 3H_2(g) \rightleftharpoons 2NH_3(g)$. Use $A = 6.34$ cm^{-1}, $B = 9.44$ cm^{-1}, and $\tilde{v}(NH_3)/cm^{-1} = 3336.7(1), 950.4(1), 3443.8(2), 1626.8(2)$.

20.11 A rich source of data on atomic energy levels is C. E. Moore's *Atomic energy levels,* NBS Circular No. 476 (1949). Use the data in the table below (which has been reproduced from Moore's collection) to calculate Φ_0 for Na(g) from 200 K to 5000 K. Calculations of this kind are ideal for programming on a computer or programmable calculator, especially when many electronic states are accessible at the temperature of interest. If you do not have ready access to either, calculate the value of Φ_0 at 3000 K.

20.12 Sodium boils at 1163 K and the vapour consists of atoms and diatomic molecules in equilibrium. Calculate the equilibrium constant for the dimerization at this temperature and the proportion of dimers in the vapour at the normal boiling point. The value of Φ_0 for Na(g) may be taken from the work in Problem 20.11; for Na$_2$ use $B = 0.1547$ cm^{-1}, $\tilde{\nu} = 159$ cm^{-1} and $D_0 = 70.4$ kJ mol^{-1}.

Na I

Config.	Desig.	J	Level/cm^{-1}	Interval
3s	3s ^{2}S	$\frac{1}{2}$	0.000	
3p	3p ^{2}P°	$\frac{1}{2}$	16 956.183	17.1963
		$1\frac{1}{2}$	16 973.379	
4s	4s ^{2}S	$\frac{1}{2}$	25 739.86	
3d	3d ^{2}D	$2\frac{1}{2}$	29 172.855	−0.0494
		$1\frac{1}{2}$	29 172.904	
4p	4p ^{2}P°	$\frac{1}{2}$	30 266.88	5.63
		$1\frac{1}{2}$	30 272.51	
5s	5s ^{2}S	$\frac{1}{2}$	33 200.696	
4d	4d ^{2}D	$2\frac{1}{2}$	34 548.754	−0.0346
		$1\frac{1}{2}$	34 548.789	
4f	4f ^{2}F°	$\begin{Bmatrix}2\frac{1}{2}\\3\frac{1}{2}\end{Bmatrix}$	34 588.6	
5p	5p ^{2}P°	$\frac{1}{2}$	35 040.27	2.52
		$1\frac{1}{2}$	35 042.79	
6s	6s ^{2}S	$\frac{1}{2}$	36.372.647	
5d	5d ^{2}D	$2\frac{1}{2}$	37 036.781	−0.0230
		$1\frac{1}{2}$	37 036.805	
5f	5f ^{2}F°	$\begin{Bmatrix}2\frac{1}{2}\\3\frac{1}{2}\end{Bmatrix}$	37 057.6	
5g	5g ^{2}G	$\begin{Bmatrix}3\frac{1}{2}\\4\frac{1}{2}\end{Bmatrix}$	37 060.2	
6p	6p ^{2}P°	$\frac{1}{2}$	37 296.51	1.25
		$1\frac{1}{2}$	37 297.76	
7s	7s ^{2}S	$\frac{1}{2}$	38 012.074	
6d	6d ^{2}D	$2\frac{1}{2}$	38 387.287	−0.0124
		$1\frac{1}{2}$	38 387.300	

Theoretical problems

20.13 Derive the Sackur–Tetrode equation for a monatomic gas confined to a two-dimensional surface, and hence derive an expression for the standard molar entropy at condensation to form a mobile surface film.

20.14 Derive expressions for the internal energy, heat capacity, entropy, Helmholtz function, Gibbs function, and the Giauque function Φ_0 of a harmonic oscillator. Express the results in terms of the vibrational temperature θ_V and plot graphs of each property against θ_V/T. Use the graphs to estimate the vibrational contribution to the Giauque function of ammonia at 1000 K, for which the vibrational wavenumbers (and degeneracies) are 3336.7(1), 950.4(1), 3443.8(2), 1626.8(2) cm^{-1}.

20.15 Although expressions like $\varepsilon = -\mathrm{d}\ln q/\mathrm{d}\beta$ are useful for formal manipulations in statistical thermodynamics, and for expressing thermodynamic functions in neat formulas, they are sometimes more trouble than they are worth in practical applications. When presented with a table of energy levels, it is often much more convenient to evaluate the following sums directly:

$$q = \sum_j e^{-\beta\varepsilon_j} \qquad \dot{q} = \sum_j (\varepsilon_j\beta)e^{-\beta\varepsilon_j} \qquad \ddot{q} = \sum_j (\varepsilon_j\beta)^2 e^{-\beta\varepsilon_j}$$

(a) Derive expressions for the internal energy, heat capacity, and entropy in terms of these three functions. (b) Apply the technique to the calculation of the electronic contribution to the enthalpy, heat capacity, and the Giauque function Φ_0 for magnesium vapour at 5000 K using the following data:

Term:	^{1}S	^{3}P$_0$	^{3}P$_1$	^{3}P$_2$	^{1}P$_1$	^{3}S
Degeneracy:	1	1	3	5	3	3
$\tilde{\nu}/$cm^{-1}	0	21 850	21 870	21 911	35 051	41 197

20.16 Determine whether a magnetic field can influence the value of an equilibrium constant. Consider the equilibrium $I_2(g) \rightleftharpoons 2I(g)$ at 1000 K, and calculate the ratio of equilibrium constants $K_p(B)/K_p$, where $K_p(B)$ is the equilibrium constant when a magnetic field B is present and removes the degeneracy of the four states of the ^{2}P$_{3/2}$ level. Data on the species are given in Exercise 20.18. The electronic g value of the atoms is $\frac{4}{3}$. Calculate the field required to change the equilibrium constant by 1 per cent.

20.17 The heat capacity of a gas determines the speed of sound in it through the formula

$$c_s = \left(\frac{\gamma RT}{M}\right)^{1/2}$$

where $\gamma = C_p/C_V$ and M is the molar mass of the gas molecules. Deduce an expression for the speed of sound in a perfect gas of (a) diatomic, (b) linear triatomic, (c) nonlinear triatomic molecules at high temperatures (with translation and rotation active). Estimate the speed of sound in air at 25°C.

Diffraction methods

Check-list of key ideas

1. The significance of an *asymmetric unit* and a *space lattice* in the description of the structures of crystals (Section 21.1).

2. The *unit cell* of a crystal lattice and their classification into *crystal systems* on the basis of their *essential symmetries* (Section 21.1).

3. The *Bravais lattices* and their classification into different types (Fig. 21.3).

4. The description of lattice planes using *Miller indices* (Section 21.2) and the calculation of their separation (eqn 2).

5. The principles of *X-ray diffraction*, and the interpretation of a *reflection* using the *Bragg law* (eqn 2).

6. The *Debye–Scherrer method* using a powdered sample (Section 21.4).

7. *Indexing* the reflections in the Debye–Scherrer method and identifying the symmetry of the unit cell (Example 21.3).

8. The origin of *systematic absences* (Section 21.4) and the intensity of scattering in terms of the *scattering factors* of atoms (eqn 4).

9. The *Bragg technique* of X-ray diffraction using a single crystal and a *four-circle diffractometer* (Section 21.5).

10. The interpretation of X-ray diffraction in terms of the *structure factor* (eqn 6 and Example 21.5) and the use of structure factors in the *Fourier synthesis* of the electron density (eqn 7 and Example 21.6).

11. The *phase problem* and the *Patterson synthesis* and the *direct method* for its solution (Section 21.5).

12. *Close-packed structures* of identical spheres, their possible *polytypes,* and the *coordination number* and *packing fraction* of the lattice (Section 21.6).

13. The *caesium chloride structure* (Fig. 21.24) and the *rock salt structure* (Fig. 21.25) encountered for some ionic solids.

14. The *radius-ratio rule* for predicting structures and the definition of *ionic radius* (Section 21.7).

15. The determination of *absolute configurations* using *anomalous scattering* (Section 21.8).

16. The principles of *neutron diffraction*, particularly for the detection of the location of hydrogen atoms and the role of *magnetic scattering* (Section 21.9).

17. The principles of *electron diffraction* (Section 21.10) and the interpretation of the scattering in terms of the *Wierl equation* (eqn 8).

A characteristic property of waves is that they interfere with each other, giving a greater displacement where their displacements add and a smaller one where they subtract (Fig. 21.1). Since the intensity of electromagnetic radiation is proportional to the square of the amplitude of the waves, the regions of constructive and destructive interference show up as regions of enhanced and diminished intensities. The phenomenon of **diffraction** is the interference that is caused by an object in the path of waves, and the pattern of varying intensity that results is called the **diffraction pattern**. Diffraction occurs when the dimensions of the diffracting object are comparable to the wavelength of the radiation.

The diffraction of waves by atoms and molecules is utilized in several powerful techniques for the determination of the structures of molecules and solids. X-rays have wavelengths comparable to bond lengths in molecules and the spacing of atoms in crystals (about 100 pm), and so they are diffracted by them. By analysing the diffraction pattern, it is possible to draw up a detailed picture of the location of atoms even of such complex molecules as proteins. Electrons moving at about $20\,000\,\mathrm{km\,s^{-1}}$ (after acceleration through about 4 kV) have wavelengths of 40 pm, and may also be diffracted by molecules. Neutrons generated in a nuclear reactor, and then slowed to thermal velocities, have similar wavelengths and may also be used for diffraction studies.

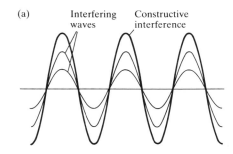

(a) Interfering waves — Constructive interference

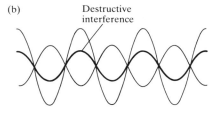

(b) Destructive interference

Fig. 21.1 When two waves are in the same region of space they interfere. Depending on their relative phase, they may interfere (a) constructively, to give an enhanced amplitude, or (b) destructively, to give a smaller amplitude.

Crystal structure

Early in the history of modern science it was suggested that the regular external form of crystals implied an internal regularity. In this section we shall see how atoms lie together in crystals.

21.1 Lattices and unit cells

We need to distinguish the individual entities from which a crystal is built from the patterns these entities form. That is, we need to distinguish the **asymmetric unit**, the atom, ion, or molecule (or, indeed, part of a molecule or a group of molecules) from which the crystal is built, from the **space lattice**, the pattern formed by points representing the locations of the asymmetric units. The space lattice is, in effect, an abstract scaffolding for the crystal structure.

More formally, the space lattice is a three-dimensional, infinite array of points, each of which is surrounded in an identical way by its neighbours, and which defines the basic structure of the crystal. In some cases there may be an asymmetric unit at each lattice point, but that is not necessary. For instance, each lattice point might be at the centre of a group of three asymmetric units. The **crystal structure** itself is obtained by associating with

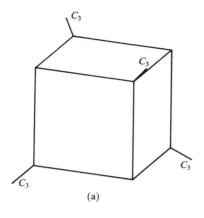

(a)

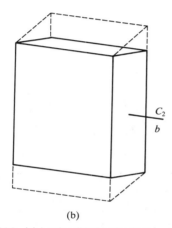

(b)

Fig. 21.2 (a) A unit cell belonging to the cubic system has four three-fold axes arranged tetrahedrally. (b) A unit cell belonging to the monoclinic system has one C_2 axis (along b).

Table 21.1. The seven crystal systems

The systems	Essential symmetries
Triclinic	None
Monoclinic	One C_2 axis
Orthorhombic	Three perpendicular C_2 axes
Rhombohedral	One C_3 axis
Tetragonal	One C_4 axis
Hexagonal	One C_6 axis
Cubic	Four C_3 axes in a tetrahedral arrangement

each lattice point an identical assembly of asymmetric units. The asymmetric units might be as small as an atom of neon or as large as a virus.

The **unit cell** is the fundamental unit from which the entire crystal may be constructed by purely translational displacements (like bricks in a wall). An infinite number of different unit cells can describe the same lattice, but we normally choose the one with sides that have the shortest lengths and that are most nearly perpendicular to each other.

Unit cells are classified into one of seven **crystal systems** in terms of the *rotational* symmetry elements they possess. The cubic system, for example, has four threefold axes in a tetrahedral array (Fig. 21.2a); the monoclinic system has one twofold axis (Fig. 21.2b). The **essential symmetries**, the elements that must be present for the unit cell to belong to a particular system, are listed in Table 21.1.

In three dimensions there are fourteen distinct types of unit cell that can stack together to give fourteen types of lattice called **Bravais lattices** (Fig. 21.3). Unit cells with lattice points only at the corners are called **primitive** and denoted P. A **body-centred unit cell** (I) also has a lattice point at its centre. A **face-centred unit cell** (F) has lattice points at its corners and also at the centres of its six faces. A **side-centred unit cell** (A, B, or C) has lattice points at its corners and at the centres of two opposite faces. As we remarked above, there is no need for an asymmetric unit actually to lie at one of these lattice points, but in many cases it is both possible and convenient to regard them as the location of an atom, ion, or molecule.

21.2 The identification of lattice planes

The spacing of the planes defined by the lattice points in a crystal is an important quantitative aspect of its structure and its investigation by diffraction techniques. However, there are many different sets of planes (Fig. 21.4), and we need to be able to label them. Since two-dimensional lattices are easier to visualize than three-dimensional lattices, we shall introduce the concepts involved by referring to them initially, and then extend the conclusions by analogy to three dimensions.

The Miller indices

Consider the two-dimensional rectangular lattice formed from a unit cell of sides a, b (Fig. 21.4). We can distinguish the four sets of planes in the illustration by the distances along the axes where they are intersected by a representative member of each set. One way of labelling the planes would therefore be to denote each set by the smallest intersection distances. For example, we could denote the four sets in the illustration as $(1a, 1b)$, $(\frac{1}{2}a, \frac{1}{3}b)$, $(-1a, 1b)$, and $(\infty a, 1b)$. If we agree to quote distances along the axes in terms of the lengths of the unit cell, we can label the planes more simply as $(1, 1)$, $(\frac{1}{2}, \frac{1}{3})$, $(-1, 1)$, and $(\infty, 1)$. If the lattice in Fig. 21.4 is the top view of a three-dimensional rectangular lattice in which the unit cell has a length c in the z direction, all four sets of planes intersect the z axis at infinity, and so the full labels are $(1, 1, \infty)$, $(\frac{1}{2}, \frac{1}{3}, \infty)$, $(-1, 1, \infty)$, and $(\infty, 1, \infty)$.

The presence of ∞ in the labels is inconvenient, and can be eliminated by taking the reciprocals of the labels (this turns out to have further advantages, as we shall see). The **Miller indices** are the reciprocals of the

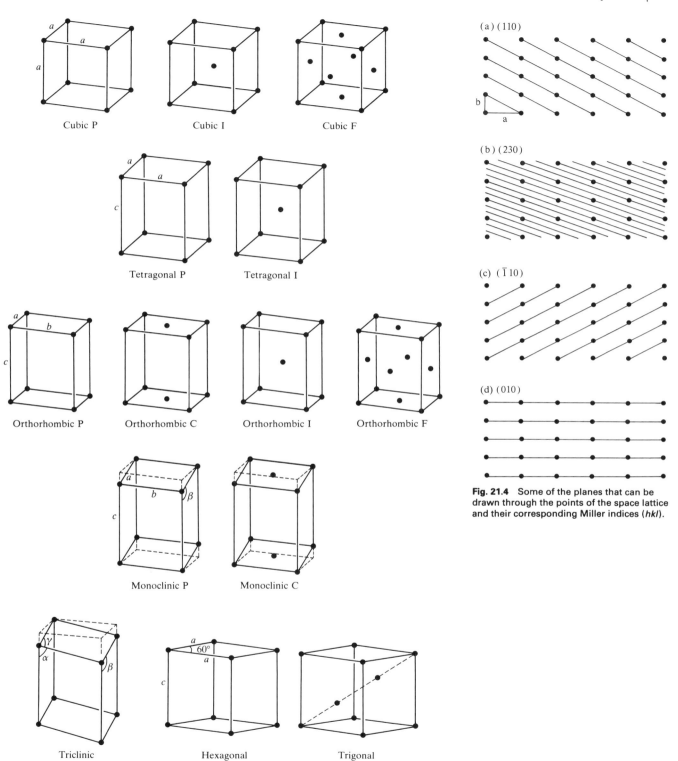

Fig. 21.3 The fourteen Bravais lattices. The points are lattice points, and are not necessarily occupied by atoms. P denotes a primitive unit cell, I a body-centred unit cell, F a face-centred unit cell, and C (or A or B) a cell with lattice points on two opposite faces.

Fig. 21.4 Some of the planes that can be drawn through the points of the space lattice and their corresponding Miller indices (*hkl*).

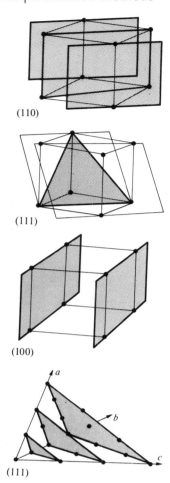

(110)

(111)

(100)

(111)

Fig. 21.5 Some representative planes in three dimensions and their Miller indices. Note that a 0 indicates that a plane is parallel to the corresponding axis, and that the indexing may also be used for unit cells with non-orthogonal axes.

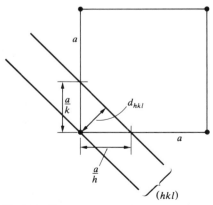

Fig. 21.6 The separation of the planes (hkl) is written d_{hkl}. This illustration shows the values of d_{hkl} for a square lattice.

numbers in the parentheses with fractions cleared. For example, the $(1, 1, \infty)$ planes in Fig. 21.4a are the (110) planes in the Miller notation. Similarly, the $(\frac{1}{2}, \frac{1}{3}, \infty)$ planes in Fig. 21.4b are the (230) planes in the Miller system. Negative indices are written with a bar over the number, and Fig. 21.4c shows the ($\bar{1}$10) planes. Figure 21.5 shows some planes in three dimensions, including an example of a lattice with non-orthogonal axes. It is helpful to keep in mind that when $h = 0$ the planes intersect the a axis at infinity, and so the (0kl) planes are parallel to the a axis. Similarly, the (h0l) planes are parallel to b and the (hk0) planes are parallel to c.

The separation of planes

The Miller indices are very useful for expressing the separation of planes. The separation of the (hk0) planes in the square lattice shown in Fig. 21.6 is given by

$$\frac{1}{d_{hk0}^2} = \frac{1}{a^2}(h^2 + k^2) \quad \text{or} \quad d_{hk0} = \frac{a}{(h^2 + k^2)^{1/2}}$$

By extension to three dimensions, the separation of the (hkl) planes of a cubic lattice is given by

$$\frac{1}{d_{hkl}^2} = \frac{1}{a^2}(h^2 + k^2 + l^2) \quad \text{or} \quad d_{hkl} = \frac{a}{(h^2 + k^2 + l^2)^{1/2}}$$

The corresponding expression for a general orthorhombic lattice (a rectangular lattice built from a unit cell with sides of arbitrary length) is the generalization of the last equation:

$$\frac{1}{d_{hkl}^2} = \frac{h^2}{a^2} + \frac{k^2}{b^2} + \frac{l^2}{c^2} \tag{1}$$

Example 21.1: *Using the Miller indices*

Calculate the separation of (a) the (123) planes and (b) the (246) planes of an orthorhombic cell with $a = 0.82$ nm, $b = 0.94$ nm, and $c = 0.75$ nm.

Answer. For the first part, we simply substitute the information into eqn 1:

$$\frac{1}{d_{123}^2} = \frac{1^2}{(0.82 \text{ nm})^2} + \frac{2^2}{(0.94 \text{ nm})^2} + \frac{3^2}{(0.75 \text{ nm})^2}$$

$$= 22._0 \text{ nm}^{-2}$$

Hence, $d_{123} = 0.21$ nm. For the second part, instead of repeating the calculation, we note that if all three Miller indices are multiplied by 2,

$$\frac{1}{d_{2h,2k,2l}^2} = \frac{(2h)^2}{a^2} + \frac{(2k)^2}{b^2} + \frac{(2l)^2}{c^2} = 2^2\left(\frac{h^2}{a^2} + \frac{k^2}{b^2} + \frac{l^2}{c^2}\right)$$

$$= \frac{2^2}{d_{hkl}^2}$$

That is,

$$d_{2h,2k,2l} = \frac{1}{2}d_{hkl}$$

Hence, in the present case, $d_{246} = 0.11$ nm.

Comment. In general, increasing the indices uniformly by a factor n decreases the separation of the planes by n.

Exercise. Calculate the separation of the (133) and (399) planes in the same lattice.

[0.19 nm, 0.063 nm]

X-ray diffraction

X-rays—which are electromagnetic radiation with wavelengths of about 100 pm—are produced by bombarding a metal with high-energy electrons. The electrons decelerate as they plunge into the metal and generate radiation with a continuous range of wavelengths called **Bremsstrahlung** (*Bremse* is German for brake, *Strahlung* for ray). Superimposed on the continuum are a few high-intensity, sharp peaks. These peaks arise from the interaction of the incoming electrons with the electrons in the inner shells of the atoms. A collision expels an electron, and an electron of higher energy drops into the vacancy, emitting the excess energy as an X-ray photon.

21.3 The Bragg Law

In 1913, shortly after X-rays had been discovered by Wilhelm Röntgen, Max von Laue suggested that they might be diffracted when passed through a crystal, for their wavelengths are comparable to the separation of lattice planes. Laue's suggestion was confirmed almost immediately by Walter Friedrich and Paul Knipping, and has grown since then into a technique of extraordinary power.

The earliest approach to the analysis of diffraction patterns produced by crystals was to regard a lattice plane as a mirror, and to model a crystal as stacks of reflecting lattice planes of separation d (Fig. 21.7). The model makes it easy to calculate the angle the crystal must make to the incoming beam of X-rays for constructive interference to occur. It has also given rise to the name **reflection** to denote an intense spot arising from constructive interference.

The path-length difference of the two rays shown in Fig. 21.7 is

$$AB + BC = 2d \sin \theta$$

where θ is the **glancing angle**. For many glancing angles the path-length difference is not an integral number of wavelengths, and the waves interfere destructively. However, when the path-length difference is an integral number of wavelengths ($AB + BC = n\lambda$), the reflected waves are in phase and interfere constructively. It follows that a bright reflection should be observed when the glancing angle satisfies the **Bragg Law**

$$n\lambda = 2d \sin \theta \tag{2a}$$

Reflections with $n = 2, 3, \ldots$ are called second order, third order, and so on; they correspond to path-length differences of $2, 3, \ldots$ wavelengths. In modern work it is normal to absorb the n into d, to write the Bragg law as

$$\lambda = 2d \sin \theta \tag{2b}$$

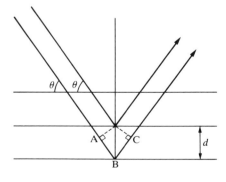

Fig. 21.7 The conventional derivation of the Bragg Law treats each lattice plane as reflecting the incident radiation. The path lengths differ by AB + BC, which depends on the glancing angle θ. Constructive interference (a 'reflection') occurs when AB + BC is equal to an integral number of wavelengths.

and to regard the nth order reflection as arising from the (nh, nk, nl) planes (see *Comment* in Example 21.1).

The primary use of the Bragg law is in the determination of the spacing between the layers in the lattice, for once the angle θ corresponding to a reflection has been determined, d may readily be calculated.

Example 21.2: *Using the Bragg Law*

A reflection from the (111) planes of a cubic crystal was observed at a glancing angle of 11.2° when Cu Kα X-rays of wavelength 154 pm were used. What is the length of the side of the unit cell?

Answer. According to the Bragg Law, the (111) planes responsible for the diffraction have separation

$$d_{111} = \frac{\lambda}{2 \sin \theta} = \frac{154 \text{ pm}}{2 \times \sin 11.2°}$$

$$= 396 \text{ pm}$$

The separation of the (111) planes of a cubic lattice of side a is given by eqn 1 as

$$\frac{1}{d_{111}^2} = \frac{1^2}{a^2} + \frac{1^2}{a^2} + \frac{1^2}{a^2} = \frac{3}{a^2}$$

Therefore,

$$a = 3^{1/2} \times d_{111} = 687 \text{ pm}$$

Exercise. Calculate the angle at which the same lattice will give a reflection from the (123) planes. [24.8°]

21.4 The powder method

Laue's original method consisted of passing a broad-band beam of X-rays into a single crystal, and recording the diffraction pattern photographically. The idea behind the approach was that a crystal might not be suitably orientated to act as a diffraction grating for a single wavelength, but whatever its orientation the Bragg law would be satisfied for at least one of the wavelengths if a range was used. There is currently a resurgence of interest in this approach because synchrotron radiation (Section 16.1) spans a range of X-ray wavelengths.

The Debye–Scherrer method

An alternative technique to Laue's was developed by Peter Debye and Paul Scherrer and independently by Albert Hull, who used monochromatic radiation and a powdered sample.

When the sample is a powder, some of the crystallites will always be orientated so as to satisfy the Bragg condition. For example, some of them will be oriented so their (111) planes, of spacing, d_{111}, give rise to diffracted intensity at the glancing angle θ (Fig. 21.8). The (111) planes of some other crystallites may be at the angle θ to the beam, but at an arbitrary angle about the line of its approach. It follows that the diffracted beams lie on a cone around the incident beam of half-angle 2θ. Other crystallites will be

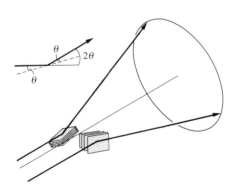

Fig. 21.8 The same set of planes in two microcrystallites with different orientations around the direction of the incident beam give diffracted rays that lie on a cone. The full powder diffraction pattern is formed by cones corresponding to reflections from all the sets of (hkl) planes that satisfy the Bragg Law. (A reflection at a glancing angle θ gives rise to a reflection at an angle 2θ to the direction of the incident beam; see inset.)

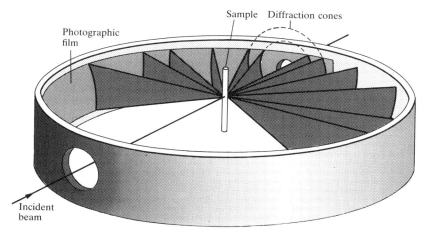

Fig. 21.9 In the Debye–Scherrer method, a monochromatic X-ray beam is diffracted by a powder sample. The crystallites give rise to cones of intensity, which are detected by a photographic film wrapped round the circumference of the camera.

oriented with different planes satisfying the Bragg Law and give rise to a cone of diffracted intensity with a different half-angle. In principle, each set of (*hkl*) planes gives rise to a diffraction cone, because some of the randomly orientated crystallites will have the correct angle to diffract the incident beam.

The original Debye–Scherrer method is illustrated in Fig. 21.9. The sample is in a capillary tube, which is rotated to ensure that the crystallites are randomly orientated. The diffraction cones are photographed as arcs of circles where they cut the strip of film, and some typical patterns are shown in Fig. 21.10. In modern diffractometers the sample is spread on a flat plate and the diffraction pattern is monitored electronically. The major application is now for qualitative analysis because the diffraction pattern is a kind of fingerprint and may be recognizable. The technique is also used for the initial determination of the dimensions and symmetries of unit cells.

Indexing the reflections

The angle θ can be measured from the location of the diffraction intensity, and if the values of *h*, *k*, and *l* are known the value of d_{hkl} can be deduced from the Bragg law. The crux of the technique is the **indexing** of the reflection, or ascribing the indexes *hkl* to it.

Some types of unit cell give characteristic and easily recognizable patterns of lines. For example, in a cubic lattice of unit cell dimension *a* the spacing is given by eqn 1, and so the angle at which the (*hkl*) planes diffract is given by

$$\sin \theta_{hkl} = \frac{\lambda (h^2 + k^2 + l^2)^{1/2}}{2a} \tag{3}$$

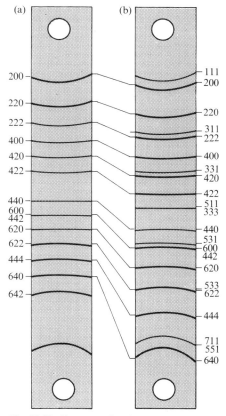

Fig. 21.10 X-ray powder photographs of (a) KCl, (b) NaCl and the indexed reflections. The smaller number of lines in (a) is a consequence of the similarity of the K^+ and Cl^- scattering factors, as discussed in the text. Note that the film covers one half of the circumference of the powder camera, the lines at the top corresponding to the largest diffraction angles.

The reflections are then predicted by substituting the values of *h*, *k*, and *l*:

(*hkl*):	(100)	(110)	(111)	(200)	(210)	(211)	(220)	(300)	(221)	(310)
$h^2 + k^2 + l^2$:	1	2	3	4	5	6	8	9	9	10

etc. Notice that 7 (and 15, . . .) is missing because the sum of the squares of three integers cannot equal 7 (or 15, . . .). Therefore the pattern has omissions that are characteristic of the cubic P lattice.

Example 21.3: *Identifying the unit cell*

A powder diffraction photograph of a substance gave lines at the following distances (in mm) from the centre spot when 70.8 pm Mo X-rays were used in a camera of radius 5.73 cm:

13.2,   18.4,   22.8,   26.2,   29.4,   32.2,   37.2,   39.6,   41.8,   43.8,   46.0

Identify the unit cell and determine its dimensions.

Answer. From eqn 3 we write

$$\sin^2 \theta = A(h^2 + k^2 + l^2) \qquad A = \left(\frac{\lambda}{2d}\right)^2$$

Therefore, we should convert distances to $\sin^2 \theta$, find the common factor A, and find $h^2 + k^2 + l^2$. The angle θ in radians is related to the distances D of the reflection line from the centre of the pattern by $\theta = D/2R$. Convert to degrees by multiplication by $360°/2\pi$. Since $R = 57.3$ mm, we have

$$\theta/° = \frac{\frac{1}{2}D}{57.3 \text{ mm}} \times \frac{360°}{2\pi} = \frac{1}{2}D/\text{mm}$$

and so the conversion is very simple. We draw up the following table:

D/mm	13.2	18.4	22.8	26.2	29.4	32.2	37.2	39.6	41.8	43.8	46.0
$\theta/°$	6.60	9.20	11.4	13.1	14.7	16.1	18.6	19.8	20.9	21.9	23.0
$100 \sin^2 \theta$	1.32	2.56	3.91	5.24	6.44	7.69	10.2	11.5	12.7	13.9	15.3

The common divisor is 1.32/100. Divide through to identify $h^2 + k^2 + l^2$:

$h^2 + k^2 + l^2$	1	2	3	4	5	6	8	9	10	11	12

The corresponding indexes are

(100) (110) (111) (200) (210) (211) (220) (300) (310) (311) (222)

We have now indexed the lines. Note the absence between 6 and 8, which indicates a primitive cubic (cubic P) cell. From $(\lambda/2a)^2 = 0.0132$, we find $a = 308$ pm.

Comment. Later we shall see that additional information comes from the intensities of the lines. Note that a cunning choice of R saves a lot of work.

Exercise. In the same camera, another cubic crystal gave reflections at the following distances (in mm): 2.3, 3.2, 4.0, 4.6, 5.1, 5.6. Identify the cell (refer to Fig. 21.10) and its dimensions.   [Cubic I; 550 pm]

Systematic absences

The powder diffraction patterns from KCl and NaCl are surprisingly different (Fig. 21.10) despite the fact that the two crystals have the same structure. The reason is that the K^+ and Cl^- ions have the same electron configurations (both have 18 electrons), and act as almost identical scatterers of X-rays. Therefore, although the crystal contains two kinds of ions, the X-rays scatter as though the crystal is composed of identical ions. Because an Na^+ ion has only ten electrons, its scattering power is less than that of Cl^-, and the diffraction pattern shows the presence of two kinds of ion in the crystal.

We can discuss the problem qualitatively by treating the NaCl crystal as two mutually interpenetrating face-centred cubic lattices, one of Na^+ ions and the other of Cl^- ions. Some reflections from the Na^+ ions are out of phase with the Cl^- reflections, and the two tend to cancel. For other orientations the two sets of reflections are in-phase, and intense lines appear. Cancellation is complete only if the ions have identical scattering power (as is nearly true for KCl, the precise extent of the similarity depending on the scattering angle), but in NaCl the cancellation is incomplete and an alternation of lines, some intense and others faint, appears and accounts for the general appearance of the NaCl diffraction pattern shown in Fig. 21.10.

We can discuss the type of diffraction pattern to expect by considering the two-dimensional lattice shown in Fig. 21.11a. The lattice is composed of A and B atoms with scattering powers measured by their **scattering factors** f_A and f_B. These factors are related to the electron density distribution in the atom, $\rho(r)$, by

$$f = 4\pi \int_0^\infty \rho \frac{\sin kr}{kr} r^2 \, dr \qquad k = \frac{4\pi}{\lambda} \sin \theta \qquad (4)$$

The scattering factor of an atom is the ratio of its scattering intensity to that of a single electron situated at the atom's centre. The scattering factor is largest in the forward direction and smaller for directions away from the forward direction. Any analysis of the intensities of reflections must take this dependence on direction into account.

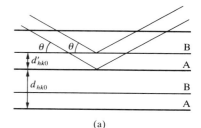

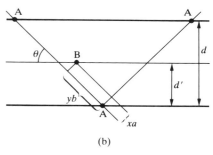

Fig. 21.11 Diffraction from a crystal containing two kinds of atom. (a) The interference between the waves scattered from the B atom (of scattering factor f_B) and of the A atoms (of scattering factor f_A) depends on the relative locations of B in the unit cell. (b) The trigonometric relations needed to show that $d' = (hx + ky)d$.

Example 21.4: *Calculating a scattering factor*

Calculate the scattering factor of Na^+, K^+, and Cl^- in the forward direction.

Answer. For forward scattering $\theta = 0$ and $k = 0$. However, because k occurs in the numerator and denominator in the integrand, we must take the limit using $\sin kr \to kr$ as $kr \to 0$:

$$\frac{\sin kr}{kr} \to \frac{kr}{kr} = 1$$

The scattering factor in the forward direction is therefore

$$f = 4\pi \int_0^\infty \rho r^2 \, dr$$

The integral over the electron density ρ multiplied by the volume element $4\pi r^2 \, dr$ is simply the total number of electrons N_e in the atom. Hence, in the forward direction

$$f = N_e$$

Therefore, the scattering factors of Na^+, K^+, and Cl^- are 10, 18, and 18 respectively.

Exercise. Show that scattering factor is greatest in the forward direction.
$$[(\sin kr)/kr < 1 \text{ for } \theta > 0]$$

We shall now calculate the diffraction pattern to expect when two kinds of atom are present in the unit cell. We shall suppose that in the unit cell there is an A atom at the origin and a B atom at (xa, yb, zc). When the incident beam is at a glancing angle θ with respect to the $(hk0)$ planes of A atoms, it is also at that same angle with respect to the $(hk0)$ planes of B atoms (Fig. 21.11a). Since the A and B planes are separated by d'_{hk0}, the path-length difference between waves from the B and A planes is $2d'_{hk0}\sin\theta$. However, since θ is a diffraction angle,

$$\sin\theta = \frac{\lambda}{2d_{hk0}}$$

Consequently, the A,B path-length difference is $d'_{hk0}\lambda/d_{hk0}$. Since (see Fig. 21.11)

$$d'_{hk0} = (hx + ky)d_{hk0}$$

the path-length difference is $(hx + ky)\lambda$. It follows that the A,B phase difference ϕ is

$$\phi = (hx + ky)\lambda \times \frac{2\pi}{\lambda} = 2\pi(hx + ky)$$

and in three dimensions

$$\phi = 2\pi(hx + ky + lz) \tag{5}$$

The A and B reflections interfere destructively with each other when the phase difference is π (180°). If the atoms have the same scattering power the total intensity is then zero. For example, if the unit cells are cubic I (Fig. 21.3) with a B atom at $x = y = z = \frac{1}{2}$, the A,B phase difference is $\pi(h + k + l)$. Therefore, all reflections corresponding to odd values of $h + k + l$ vanish because the waves are displaced in phase by 180°. Hence the diffraction pattern for a cubic I lattice can be constructed from that for the cubic P lattice (a lattice without the atom at the centre) by striking out all reflections with odd values of $h + k + l$. Recognition of these **systematic absences** in a powder spectrum (Fig. 21.12) immediately indicates a cubic I lattice.

If the displacement of the waves scattered from A is f_A at the detector, that of the waves scattered from B is $f_B e^{i\phi}$ with ϕ the phase difference given above. The total displacement at the detector is therefore

$$F = f_A + f_B e^{i\phi}$$

Because the intensity is proportional to the square modulus of the amplitude of the wave, the intensity at the detector is

$$I_{hkl} \propto F^*F = (f_A + f_B e^{-i\phi})(f_A + f_B e^{i\phi})$$

This expression expands to

$$I_{hkl} \propto f_A^2 + f_B^2 + f_A f_B(e^{i\phi} + e^{-i\phi})$$

$$\propto f_A^2 + f_B^2 + 2f_A f_B \cos\phi$$

The cosine term either adds to or subtracts from $f_A^2 + f_B^2$ depending on the value of ϕ, which in turn depends on h, k, and l (through eqn 5). Hence, there is a variation in the intensities of the lines with different hkl (as well as the smooth variation of intensity that arises from the angle-dependence of the scattering factors), which is exactly what is observed for NaCl.

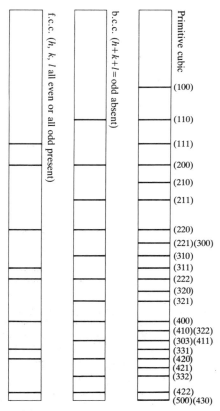

Fig. 21.12 The powder photographs and the systematic absences of the cubic I and cubic F unit cells. Comparison of the observed photograph with patterns like these enable the unit cell to be identified. The locations of the lines give the cell dimensions.

f.c.c. (h, k, l all even or all odd present)

b.c.c. ($h+k+l$ = odd absent)

Primitive cubic

(100)
(110)
(111)
(200)
(210)
(211)
(220)
(221)(300)
(310)
(311)
(222)
(320)
(321)
(400)
(410)(322)
(303)(411)
(331)
(420)
(421)
(332)
(422)
(500)(430)

21.5 Single-crystal X-ray diffraction

The method developed by the Braggs (William and his son Lawrence, who later jointly won the Nobel Prize) is the foundation of almost all modern work in X-ray crystallography. They used a single crystal and a monochromatic beam of X-rays, and rotated the crystal until a reflection was detected. There are many different sets of planes in a crystal, and so there are many angles at which a reflection occurs. The complete set of data consists of the list of angles at which reflections are observed and their intensities.

The technique

The preliminary investigation of a crystal, to identify the symmetry of its unit cell and its dimensions, makes use of a photographic technique. One method is to use the **oscillation camera**, in which the crystal (which typically might be of side 0.1 mm) is set on a mount called a 'goniometer head' and oscillated through about 10°. The diffraction pattern is recorded on a cylindrical film surrounding the sample. The problem of overlapping reflections is overcome by the **Weissenberg technique** in which a screen is placed in front of the film so that only one set of reflections is exposed, and gearing the film to the oscillating camera head so that it moves parallel to the axis of oscillation of the crystal. Although the Weissenberg technique greatly simplifies the indexing problem, the photographs are severely distorted. A refinement that overcomes this difficulty is to have a different coupling between the motions of the crystal and the screened film, and in the **precession camera** technique the pattern is undistorted and can be indexed quite readily.

Computing techniques are now available that lead not only to automatic indexing but also to the automated determination of the shape, symmetry, and size of the unit cell. The most sophisticated of the current techniques uses the **four-circle diffractometer** (Fig. 21.13). The crystal is set in an

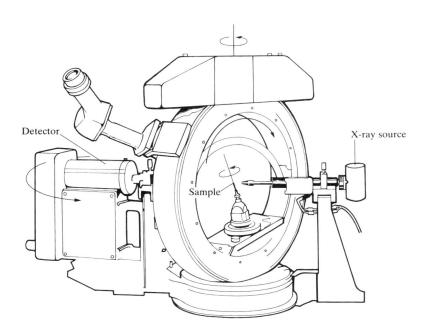

Detector

Sample

X-ray source

Fig. 21.13 A four-circle diffractometer. The settings of the orientations of the components is controlled by computer; each (*hkl*) reflection is monitored in turn, and their intensities are recorded.

arbitrary orientation in the goniometer head. Since its unit cell dimensions will already have been determined (e.g. using a precession camera) the settings of the diffractometer's four angles that are needed to observe any particular (hkl)-reflection can be computed. Setting the angles requires knowledge of the orientation of the crystal, but that can be determined by finding reflections that can be identified from the diffraction photographs taken earlier. The computer controls the settings, and moves the diffractometer to each one in turn. At each setting, the diffraction intensity is measured (using some kind of crystal detector or photomultiplier) and background intensities are assessed by making measurements at slightly different settings.

Structure factors

The problem we now address is how to interpret the data from a four-circle diffractometer in terms of the structure of the crystal. To do so, we must go beyond the simple Bragg law.

If the unit cell contains several atoms with scattering factors f_i and coordinates ($x_i a$, $y_i b$, $z_i c$), the overall amplitude of a wave diffracted by the (hkl) planes is a generalization of the expression $F = f_A + f_B e^{i\phi}$ that we obtained earlier:

$$F_{hkl} = \sum_i f_i e^{2\pi i(hx_i + ky_i + lz_i)} \tag{6}$$

The sum is over all the atoms in the unit cell. F_{hkl} is called the **structure factor**. The intensity of the (hkl)-reflection (the intensity of the reflection from the set of (hkl) planes) is proportional to $|F_{hkl}|^2$.

Example 21.5: *Calculating a structure factor*

Calculate the structure factors for a NaCl unit cell in Fig. 21.14.

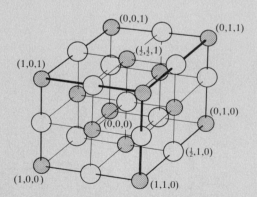

Fig. 21.14 The location of the atoms for the structure factor calculation. The tinted circles are Na$^+$, the open circles are Cl$^-$.

Answer. The unit cell has Na$^+$ ions at $(0, 0, 0)$, $(0, 1, 0)$, $(0, 1, 1)$, etc. and Cl$^-$ ions at $(0, \frac{1}{2}, 0)$, $(\frac{1}{2}, 1, 0)$, $(\frac{1}{2}, \frac{1}{2}, \frac{1}{2})$, etc. We shall write f^+ for the Na$^+$ scattering factor and f^- for the Cl$^-$ scattering factor. Note that ions on faces are shared between two cells (use $\frac{1}{2}f$), those on edges by four (use $\frac{1}{4}f$), and those at corners by eight (use $\frac{1}{8}f$). Then, from eqn 6, and summing over the coordinates of all 27 atoms in the illustration:

$$F_{hkl} = f^+ \{ \tfrac{1}{8} + \tfrac{1}{8}e^{2\pi il} + \ldots + \tfrac{1}{8}e^{2\pi i(h+k+l)} + \ldots + \tfrac{1}{2}e^{2\pi i(\frac{1}{2}h+\frac{1}{2}k+l)} \}$$
$$+ f^- \{ e^{2\pi i(\frac{1}{2}h+\frac{1}{2}k+\frac{1}{2}l)} + \tfrac{1}{4}e^{2\pi i(\frac{1}{2}h)} + \ldots + \tfrac{1}{4}e^{2\pi i(\frac{1}{2}h+l)} \}$$

Now use

$$e^{2\pi ih} = e^{2\pi ik} = e^{2\pi il} = 1$$

because h, k, and l are all integers and $e^{2\pi i} = 1$. The expression then simplifies considerably, and becomes

$$F_{hkl} = f^+ \{ 1 + \cos(h+k)\pi + \cos(h+l)\pi + \cos(k+l)\pi \}$$
$$+ f^- \{ \cos(h+k+l)\pi + \cos h\pi + \cos k\pi + \cos l\pi \}$$

Then, since $\cos h\pi = (-1)^h$,

$$F_{hkl} = f^+ \{ 1 + (-1)^{h+k} + (-1)^{h+l} + (-1)^{k+l} \}$$
$$+ f^- \{ (-1)^{h+k+l} + (-1)^h + (-1)^k + (-1)^l \}$$

Now note that:

If h, k, l are all even,

$$F_{hkl} = f^+ \{ 1+1+1+1 \} + f^- \{ 1+1+1+1 \} = 4(f^+ + f^-)$$

If h, k, l are all odd,

$$F_{hkl} = 4(f^+ - f^-)$$

If one index is odd and two are even, or vice versa,

$$F_{hkl} = 0$$

Comment. Note that the hkl all-odd reflections are less intense than the hkl all-even. For $f^+ = f^-$, which is the case for identical atoms in a simple cubic arrangement, the hkl all-odd have zero intensity, corresponding to the systematic absences of simple cubic unit cells.

Exercise. Deduce the rule for the systematic absences of a cubic I lattice.

$$[h+k+l \text{ odd}]$$

The electron density

The structure of the unit cell is contained in the structure factor because it depends on the atoms present (through f_i) and on their locations (through $hx_i + ky_i + lz_i$ and the angle dependence of f). Given that we know the structure factors for all the reflections (from their intensities), we can construct the electron density $\rho(r)$ in the unit cell using the relation

$$\rho(r) = \sum_{hkl} F_{hkl} e^{-2\pi i(hx+ky+lz)} \qquad (7)$$

Equation 7 is called a **Fourier synthesis** of the electron density.

Example 21.6: *Calculating an electron density by Fourier synthesis*

Consider the $(h00)$ planes of a crystal extending indefinitely in the x direction. In an X-ray analysis the structure factors were found as follows:

h:	0	1	2	3	4	5	6	7	8	9	10	11	12	13	14	15
F_h:	16	−10	2	−1	7	−10	8	−3	2	−3	6	−5	3	−2	2	−3

(and $F_{-h} = F_h$). Construct a plot of the electron density along the x axis of the unit cell.

Answer. Since $F_{-h} = F_h$,

$$\rho(x) = F_0 + 2 \sum_h F_h \cos(2\pi hx)$$

and we evaluate the sum for $x = 0, 0.1, 0.2, \ldots 1.0$ (or more finely if a computer is available). The results are plotted in Fig. 21.15 (the full line).

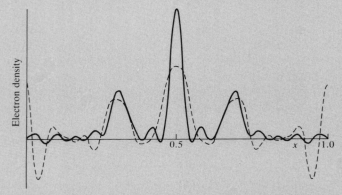

Fig. 21.15 The plot of the electron density calculated in Examples 21.6 (full line) and 21.7 (broken line). Note how a different choice of phases for the structure factors leads to a markedly different structure (and, in the second case, to an unacceptable negative electron density in two regions).

Comment. The positions of three atoms can be discerned very readily. The more terms there are included, the more accurate the density plot. Terms corresponding to high values of h (short wavelength cosine terms in the sum) account for the finer details of the electron density; low values of h account for the broad features.

Exercise. Write a short computer program to evalute Fourier sums, and then experiment with different structure factors (including changing signs as well as amplitudes).

The phase problem

From the measured I_{hkl} we get the F_{hkl}, and then evaluate eqn 7 to find ρ. Unfortunately, I_{hkl} is proportional to the square modulus $|F_{hkl}|^2$, and so we cannot say whether we should use $+|F_{hkl}|$ or $-|F_{hkl}|$ in the sum. In fact, the difficulty is more severe, because if we write F_{hkl} as the complex number $|F_{hkl}|e^{i\alpha}$, where α is the phase of F_{hkl} and $|F_{hkl}|$ is its magnitude, the intensity lets us determine $|F_{hkl}|$ but tells us nothing of its phase. This ambiguity is called the **phase problem**. Some way must be found to assign phases to the structure factors, for otherwise the sum for ρ could not be evaluated and the method would be useless. Phases can take any value for structures without a centre of symmetry, but must be 0 or 180° if the unit cell is centro-symmetric.

Example 21.7: *Illustrating the phase problem*

Repeat Example 21.6, but suppose that all the structure factors after $h = 6$ have the same positive phase.

Answer. We repeat the evaluation of the 16-term sum, but with positive values of F for $h > 6$. The resulting electron density is plotted in Fig. 21.15 with a broken line.

Comment. This exercise illustrates the importance of the phase problem: different phase choices give quite different electron densities.

Exercise. Calculate the electron density using structure factors with signs that alternate as h increases.

The phase problem can be overcome to some extent by a variety of methods, one of the most important of which has been the **Patterson synthesis**, in which the intensities I_{hkl} are used in place of the structure factors themselves in eqn 7 and the phases do not appear in the sum. A Patterson synthesis produces a map of *separations* of atoms in the unit cell. For example, if the unit cell has the structure shown in Fig. 21.16a, the Patterson synthesis would be the map shown in Fig. 21.16b, where the location of each spot relative to the origin gives the orientations and separations of all the pairs of atoms in the original structure.

If some atoms are heavy, they dominate the scattering (because their scattering factors are large, of the order of their atomic number) and their locations may be deduced quite readily. In the technique known as **isomorphous replacement**, such atoms are introduced artificially into a complex molecule without affecting its structure significantly. Then the principal features of its structure are established before the locations of the lighter atoms are explored.

Modern structural analyses also make extensive use of the **direct method**. The direct method is based on statistical procedures and depends on the possibility of treating the atoms in a unit cell as being virtually randomly distributed (from the radiation's point of view), and then to use statistical techniques to compute the probabilities that the phases have a particular value. It is possible to deduce relations between some structure factors and sums (and sums of squares) of others, which have the effect of constraining the phases to particular values (with about 99.7 per cent probability).

Since the statistical relations can be built into computer programs, the determination of a crystal structure from a collection of observed intensities has become almost completely automated. Unfortunately, the reliability of the direct approach decreases as $1/N^{1/2}$, where N is the number of atoms per unit cell, and so it is unsuitable when the molecules are large ($N > 100$).

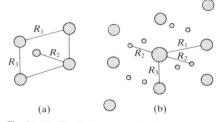

(a)   (b)

Fig. 21.16 The Patterson synthesis corresponding to the pattern in (a) is the pattern in (b). The distance and orientation of each spot from the origin gives the orientation and separation of one atom–atom separation in (a).

The overall procedure

The X-ray technique depends on the availability of a single crystal of the sample—although only a small one—which has led to a great deal of effort to obtain single crystals of large, biologically important molecules, such as

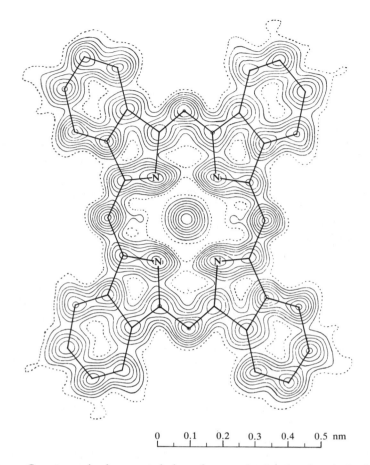

Fig. 21.17 The electron density contours of nickel phthalocyanine computed by Fourier synthesis from the observed structure factors (J. M. Robertson, *Organic crystals and molecules,* Cornell University Press (1953)). The conventional structure has been superimposed. Notice that in this early example of the technique the hydrogen atoms do not show.

```
0    0.1   0.2   0.3   0.4   0.5  nm
```

proteins. Once a single crystal has been obtained, the indexing and determination of the unit cell characteristics are carried out with a precession camera, and then the intensities are measured on a four-circle diffractometer and interpreted as structure factors with signs assigned by the direct method. (With a modern four-circle diffractometer, it is no longer necessary to determine the unit cell first.) The electron density can then be calculated, and drawn out as a contour diagram, such as that in Fig. 21.17. We shall take up the subject again in Chapter 23 when we consider the structures of macromolecules.

Various refinement techniques may be employed. For example, the atoms vibrate about their mean positions, which blurs their location. The magnitude of the blurring can be calculated, and it can be eliminated from the electron density map.

The remarkable and striking results obtained from X-ray diffraction suggest that it is an ideal technique for the determination of structures. This would be true were it not for a number of limitations. The first is its restriction to the solid state. We need to remember that in their natural environment biological molecules might unwind significantly, and that in the solid the packing of molecules might impose constraints on their stereo-chemistry. NMR is often more suitable because it can be used to study fluid samples, and the development of special resonance techniques (Chapter 23) has enabled enzymes to be studied in their natural environment. Another limitation (which makes NMR complementary) is the poor response of X-ray diffraction to the presence of hydrogen atoms (because they have so few scattering electrons). However, the sensitivity of modern instruments is such that the detection of hydrogen atoms has now become largely routine.

Information from X-ray analysis

The bonding within a solid may be of various kinds. Simplest of all (in principle) are metals, where electrons are delocalized over arrays of identical cations and bind the whole together into a rigid but malleable structure. In many cases their crystal structures are dictated largely by the way spherical metal cations can pack together into an orderly array.

In covalent solids, covalent bonds in a definite spatial orientation link the atoms in a network extending through the crystal. The stereochemical demands of valence, which have only a small effect on the structures of many metals, now override the geometrical problem of packing spheres together, and elaborate and extensive structures may be formed. A famous example of a covalent solid is diamond (Fig. 21.18), in which each sp^3-hybridized carbon atom is bonded tetrahedrally to its four neighbours. Covalent solids are often hard and unreactive.

Molecular solids, which are the subject of the overwhelming majority of modern structural determinations, are bonded together by van der Waals interactions (which we treat in detail in the next chapter). The observed crystal structure is nature's solution to the problem of condensing objects of various shapes into an aggregate of minimum energy (actually, for temperatures above zero, of minimum Gibbs function). Predicting the structure is very difficult and rarely possible. The problem is made more complicated by the role of hydrogen bonds, which in some cases dominate the crystal structure, as in ice (Fig. 21.19), but in others (e.g. phenol) distort a structure that is determined largely by other van der Waals interactions.

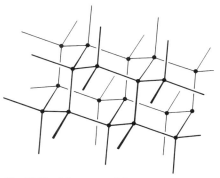

Fig. 21.18 A fragment of the structure of diamond. Each C atom is tetrahedrally bonded to four neighbours. This framework-like structure results in a rigid crystal with a high thermal conductivity.

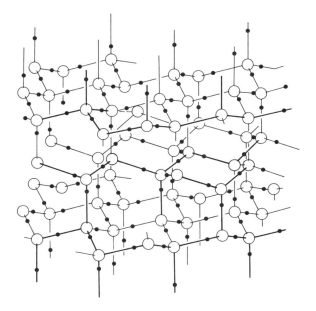

Fig. 21.19 A fragment of the crystal structure of ice (ice-I). Each O atom is at the centre of a tetrahedron of four O atoms at a distance of 276 pm. The central O atom is attached by two short O—H bonds to two H atoms and by two long hydrogen bonds to H atoms of two of the neighbouring molecules. Overall, the structure consists of planes of hexagonal puckered rings of H_2O molecules (like the chair form of cyclohexane).

21.6 The packing of identical spheres: metal crystals

Most pure metals crystallize in one of three simple forms, two of which can be explained in terms of organizing spheres into the closest possible packing.

Close packing

A **close-packed layer** of identical spheres, one with maximum utilization of space, can be formed as shown in Fig. 21.20a. A close-packed three-dimensional structure is formed by stacking such two-dimensionally close-packed layers on top of each other. However, this stacking can be done in different ways and can result in close-packed **polytypes**, or structures that are identical in two dimensions (the close-packed planes) but differ in the third dimension.

We can form a second close-packed layer by placing spheres in the depressions of the first layer (Fig. 21.20b). The third layer may be added in either of two ways. In one, the spheres are placed so that they reproduce the first layer (Fig. 21.21), to give an ABA pattern of layers. Alternatively, the spheres may be placed over the gaps in the first layer (Fig. 21.22), so giving an ABC pattern.

Two polytypes are formed if the two stacking patterns are repeated in the vertical direction. If the ABA pattern is repeated, to give the sequence of layers ABABAB..., the spheres are **hexagonally close-packed** (hcp). Alternatively, if the ABC pattern is repeated, to give the sequence ABCABC..., the spheres are **cubic close-packed** (ccp). The ccp structure gives rise to face-centred unit cells, and so may also be denoted cubic F (or

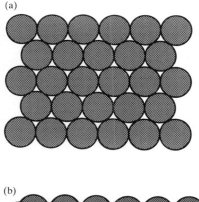

(a)

(b)

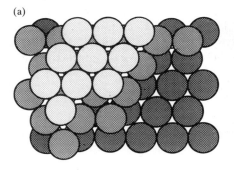

(a)

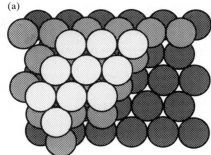

(a)

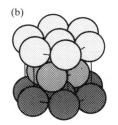

(b)

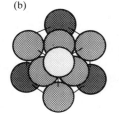

(b)

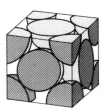

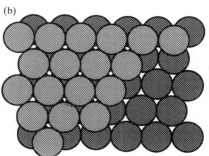

Fig. 21.20 The close-packing of identical spheres. (a) The first layer of close-packed spheres. (b) The second layer of close-packed spheres occupies the dips of the first layer. The two layers are the AB component of the structure.

Fig. 21.21 The third layer of close-packed spheres might occupy the dips lying directly above the spheres in the first layer, resulting in an ABA structure (a) which corresponds to hexagonal close-packing (b). This hcp structure is possessed by the elements Be, Cd, Co, He, Mg, Ti, and Zn.

Fig. 21.22 Alternatively, the third layer might lie in the dips that are not above the spheres in the first layer, resulting in an ABC structure (a) which correspond to cubic close-packing (b). This ccp (or fcc) structure is possessed by the elements Ag, Al, Ar, Au, Ca, Cu, Ne, Ni, Pb, Pt, and Xe.

fcc, for face-centred cubic). It is also possible to have ABCABAB ... structures and even random sequences; however, the hcp and ccp polytypes are the most important, and we shall deal only with them. Some of the elements possessing these structures are listed in the captions of Figs. 21.21 and 21.22.

The compactness of the ccp and hcp structures is indicated by their **coordination number**, the number of atoms immediately surrounding any selected atom, which is 12 in both cases. Another measure of their compactness is the **packing fraction**, the fraction of space occupied by the spheres, which is 0.740. That is, in a close-packed solid of identical spheres, 26.0 per cent of the volume is empty space. The fact that many metals are close-packed accounts for one of their common characteristics, their high densities.

Example 21.8: *Calculating the packing fraction*

Calculate the packing fraction of an hcp lattice.

Answer. Refer to Fig. 21.23. The area of the base of the unit cell (the tinted parallelogram) is $3^{1/2}R \times 2R$. The next layer of atoms lies above the point marked B, which is at a distance $R/\cos 30° = 2R/3^{1/2}$ from the centre of the adjacent ion. The overlaying atom has a centre a distance $2R$ from the ions just mentioned, and so is at a height $\{(2R)^2 - (2R/3^{1/2})^2\}^{1/2} = 2(\frac{2}{3})^{1/2}R$. The height of the unit cell is twice this value, or $4(\frac{2}{3})^{1/2}R$. The volume of the unit cell is (area) × (height) = $(2 \times 3^{1/2})R^2 \times 4(\frac{2}{3})^{1/2}R = 8 \times 2^{1/2}R^3$. There are the equivalent of two complete atoms per unit cell, and so the volume occupied per atom is $4 \times 2^{1/2}R^3$. The volume of a spherical atom is $\frac{4}{3}\pi R^3$. Therefore, the packing fraction, the fraction of 'full' space is $(\frac{4}{3}\pi R^3)/(4 \times 2^{1/2}R^3) = \pi/(3 \times 2^{1/2}) = 0.740$.

Comment. A ccp structure has the same packing fraction. The packing fraction of a cubic I (bcc) structure is 0.68 and that of a cubic P structure is 0.52, so both are less closely packed than the hcp and ccp structures.

Exercise. Calculate the packing fraction of a cubic I solid composed of identical spheres. [0.68]

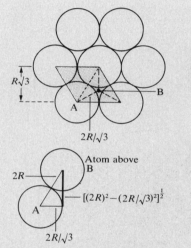

Fig. 21.23 The calculation of the packing fraction of an hcp unit cell.

Less closely packed structures

A number of common metals adopt structures that are less than close-packed, which suggests that specific covalent bonding between neighbouring atoms is beginning to influence the structure and impose a specific geometrical arrangement. One such arrangement results in a cubic I (bcc, for body-centred cubic) lattice, with one sphere at the centre of a cube formed by eight others. The bcc structure is adopted by a number of common metals, including Ba, Cs, Cr, Fe, K, and W. The coordination number is only 8, and its packing fraction is only 0.68, showing that only about two-thirds of the available space is actually occupied.

21.7 Ionic crystals

When we model ionic crystals by stacks of spheres we must allow for the fact that they have different radii (generally with the cations smaller than the anions) and different charges. The coordination number of an ionic

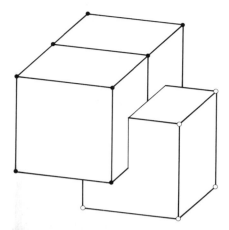

Fig. 21.24 The caesium chloride structure consists of two interpenetrating simple cubic lattices, one of cations and the other of anions, so that each cube of ions of one kind has a counter-ion at its centre.

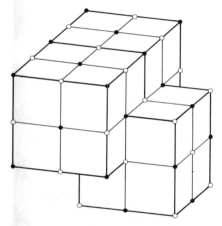

Fig. 21.25 The rock salt (NaCl) structure consists of two mutually interpenetrating slightly expanded face-centred cubic lattices.

Table 21.2. Ionic radii, R/pm

Na⁺	102(6*), 116(8)
K⁺	138(6), 151(8)
F⁻	128(2), 131(4)
Cl⁻	181 (Close packing)

* Coordination number.

lattice is the number of nearest neighbours of opposite charge, and cations and anions may have different environments in the same crystal.

Even if, by chance, the ions have the same size, the problems of ensuring that the unit cells are electrically neutral makes it impossible to achieve 12-coordinate close-packed structures (which is a reason why ionic solids are generally less dense than metals). The best that can be done is the 8-coordination of the **caesium chloride structure** (Fig. 21.24) in which each cation is surrounded by eight anions and each anion is surrounded by eight cations. In the caesium chloride structure, an ion of one charge occupies the body-centred position of a cubic unit cell with eight counter ions at its corners. The structure is adopted by CsCl itself and by CaS, CsCN (with some distortion), and CuZn.

Eight-coordinate packing cannot be achieved when the radii of the ions differ more than in CsCl. One common structure adopted is the six-coordinated **rock salt structure** (Fig. 21.25) typified by NaCl in which each cation is surrounded by six anions and each anion is surrounded by six cations. The rock salt structure can be interpreted as the interpenetration of two slightly expanded cubic F (fcc) lattices, one of cations and the other of anions. It is the structure of NaCl itself and of many other MX compounds, including KBr, AgCl, MgO, and ScN.

The switch from the caesium chloride to the rock salt structure occurs (sometimes) in accord with the **radius-ratio rule**, which is based on the value of the **radius ratio**

$$\rho = \frac{r(\text{smaller})}{r(\text{larger})}$$

where the two radii are those of the larger and smaller ions in the crystal. The radius-ratio rule states that the caesium chloride structure should be expected when

$$\rho > 3^{1/2} - 1 = 0.732$$

and that the rock salt structure should be expected when

$$2^{1/2} - 1 = 0.414 < \rho < 0.732$$

For $\rho < 0.414$, the most efficient packing leads to 4-coordination of the type exhibited by the **sphalerite** (or zinc blende) form of ZnS. The radius-ratio rule, which is based on geometrical considerations of sphere packing, is fairly well supported by observation. The deviation of a structure from the prediction is often taken to be an indication of a shift from ionic towards covalent bonding.

The **ionic radii** that are used to calculate ρ, and wherever else it is important to know the sizes of ions, are derived from the internuclear separation of adjacent ions in a crystal. However, we need to apportion the total distance between the two ions by defining the radius of one ion and reporting all others on that basis. One scale that is widely used is based on the value 140 pm for the radius of the O^{2-} ion (Table 21.2); however, other scales are also available (such as one based on F^- for discussing halides), and it is essential not to mix values from different scales. Since ionic radii are so arbitrary, predictions based on them (such as with the radius-ratio rule) must be viewed cautiously.

21.8   Absolute configurations

Although it has long been possible to separate enantiomers (mirror-image chiral isomers, Section 15.3), it was not until X-ray crystallography was developed that the absolute stereochemical configuration of an isomer could be determined. It is now possible to state, for example, that D-tartaric acid (**1**) is the isomer responsible for rotating light clockwise (i.e. it is the (+) isomer) and that L-tartaric acid (**2**) is the (−) isomer. The X-ray method is not trivial, because a crystal of one isomer gives an almost identical intensity pattern to that of its enantiomer. The information about the absolute configuration is contained in the phase of the diffracted radiation, and its extraction is based on a technique introduced by J. M. Bijvoet.

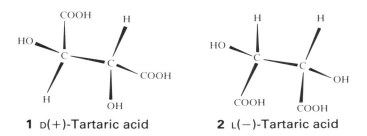

1 D(+)-Tartaric acid   **2** L(−)-Tartaric acid

Consider first the diagrams in Fig. 21.26a, which represent an idealized crystal and its mirror image. Each plane of atoms gives rise to a scattered wave, and the superpositions are shown. Note that the two superpositions have the same amplitude, but differ in phase. The diffraction pattern therefore has the same intensity for each enantiomer and, at this stage, cannot be used to distinguish them.

The essence of the original Bijvoet method was to incorporate a heavy atom (such as Rb) into the molecule. Such an atom causes an extra phase shift in the scattered X-ray because (as a simple way of picturing it) the X-rays tend to excite the atom, and get delayed in the process. The effect is called **anomalous scattering**. If the layer marked A in the crystal contains the anomalous scatterers, the scattered waves are as shown in Fig. 21.26b. The essential point is that the superpositions now differ slightly in amplitude, not just phase, and so the diffracted intensities are slightly different in each case. Therefore, the enantiomers can in fact be distinguished because the scattering intensities differ.

Modern diffractometers are so sensitive that the incorporation of a heavy atom is no longer strictly necessary. It is now possible to detect the small intensity variations arising from the light atoms normally present, but the procedure is much easier and more reliable if some moderately heavy atoms (such as S or Cl) are present.

Neutron and electron diffraction

A neutron generated in a reactor and slowed to thermal velocities by repeated collisions with a moderator (such as graphite) until it is travelling at about $4 \, km \, s^{-1}$ has a wavelength of about 100 pm. Since 100 pm is comparable to X-ray wavelengths, similar diffraction phenomena can be expected.

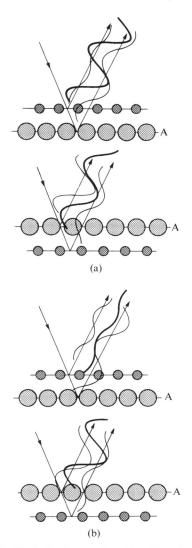

Fig. 21.26   The two versions (D and L) of the two layers of atoms represent enantiomers. The interference between their scattered waves results in composite waves (heavy line) that differ in phase (a), but the absolute phase cannot be determined, and the intensities of the reflections are identical. If, however, A modifies the phase of the waves it scatters, as represented by the light lines in (b), then the resultant superpositions differ in amplitude as well as phase, the reflections have different intensities, and the absolute configuration can be established.

Example 21.9: *Calculating the wavelength of thermal neutrons*

Calculate the wavelength of neutrons that have reached thermal equilibrium with their surroundings at 100°C.

Answer. From the equipartition theorem we know that the mean translational kinetic energy of a neutron at a temperature T travelling in the x direction is $E_K = \frac{1}{2}kT$. The kinetic energy is also equal to $p^2/2m$, where p is the momentum of the neutron and m is its mass. Hence,

$$p = (mkT)^{1/2}$$

The de Broglie relation then gives the neutron's wavelength as

$$\lambda = \frac{h}{p} = \frac{h}{(mkT)^{1/2}}$$

Therefore, at 100°C,

$$\lambda = \frac{6.626 \times 10^{-34}\,\text{J s}}{(1.675 \times 10^{-27}\,\text{kg} \times 1.381 \times 10^{-23}\,\text{J K}^{-1} \times 373\,\text{K})^{1/2}}$$

$$= 226\,\text{pm}$$

Exercise. Calculate the temperature needed for the wavelength of the neutrons to be 100 pm. $[1.6 \times 10^3\,°\text{C}]$

Electrons can be accelerated to precisely controlled energies by a known potential difference. When accelerated through 10 keV they acquire a wavelength of 12 pm, which makes them suitable for structural studies too.

21.9 Neutron diffraction

The scattering of X-rays is caused by the oscillations an incoming wave generates in the electrons of atoms. The scattering of neutrons is a *nuclear* phenomenon. Neutrons pass through the electronic structures of atoms and interact with the nuclei through the 'strong forces' that are responsible for binding nucleons together. As a result, the intensity with which neutrons are scattered is independent of the number of electrons and is not dominated by the heavy atoms present in a molecule (in contrast to X-rays, which are scattered more by electron-rich heavy atoms). Neutron diffraction therefore shows up the positions of hydrogen nuclei much more clearly than do X-rays.

The difference in sensitivity to hydrogen nuclei can have a pronounced effect on the measurement of C—H bond lengths. Because X-rays respond to accumulations of electrons, the weak peaks in an X-ray diffraction map represent the locations of the bulk of the electron density in the bonds, and these may be shifted towards the C atom. For example, X-ray measurements on sucrose give $R(\text{C—H}) = 96$ pm; neutron measurements, which respond to the location of the nuclei, give $R(\text{C—H}) = 109.5$ pm. The O—H bond lengths in sucrose show similar differences, being 79 pm by X-rays but 97 pm by neutrons.

Another attribute of neutrons that distinguishes them from X-ray photons is their possession of a magnetic moment due to their spin. This magnetic moment can couple to the magnetic fields of ions in a crystal (if they have unpaired electrons) and modify the diffraction pattern. A simple example of **magnetic scattering** is provided by metallic chromium. The lattice is cubic I (bcc), and the diffraction pattern using X-rays has systematic absences. These absences are not observed when neutrons are used because the structure is such that atoms at the body-centre location have magnetic moments opposite to those at the corners, and the structure is better regarded as consisting of two interpenetrating lattices (Fig. 21.27) of magnetically different Cr atoms. Therefore, although the atoms are identical as far as X-rays are concerned, they are different from the viewpoint of neutrons, and diffraction intensity is observed at the predicted systematic X-ray absences. Neutron diffraction is especially important for investigating these **magnetically ordered** lattices.

21.10 Electron diffraction

Electrons are scattered strongly by their interaction with the charges of electrons and nuclei, and so cannot be used to study the interiors of solid samples. However, they can be used to study molecules in the gas phase, on surfaces, and in thin films.

A typical electron diffraction apparatus is illustrated in Fig. 21.28. Electrons are emitted from the hot filament on the left and accelerated through a potential gradient. They then pass through the stream of gas, and on to a fluorescent screen. (The modifications used when studying surfaces are described in Section 29.2.)

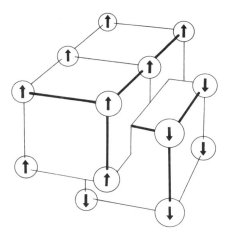

Fig. 21.27 If the spins of atoms at lattice points are orderly, as in this antiferromagnetic material, where the spins of one set of atoms are aligned antiparallel to those of the other set, then neutron diffraction detects two interpenetrating simple cubic lattices on account of the magnetic interaction of the neutron with the atoms, but X-ray diffraction would see only a single bcc lattice.

Example 21.10: *Calculating the wavelength of electrons*

Calculate the wavelength of electrons that have been accelerated from rest through a potential difference of 50 kV.

Answer. The kinetic energy of an electron accelerated through a potential difference V is eV; hence, by equating the kinetic energy to $p^2/2m_e$, its linear momentum is

$$p = (2m_e eV)^{1/2}$$

From the de Broglie relation it follows that its wavelength is

$$\lambda = \frac{h}{(2m_e eV)^{1/2}}$$

Therefore, with $V = 50$ kV,

$$\lambda = \frac{6.626 \times 10^{-34}\,\text{J s}}{(2 \times 9.110 \times 10^{-31}\,\text{kg} \times 1.602 \times 10^{-19}\,\text{C} \times 5.0 \times 10^3\,\text{V})^{1/2}}$$

$$= 17\,\text{pm}$$

Exercise. Calculate the potential difference needed to produce electrons of wavelength 100 pm.   [150 V]

The gaseous sample presents all possible orientations of atom–atom separations to the electron beam, and the resulting diffraction pattern is like an X-ray powder photograph. The pattern consists of a series of concentric undulations on a background with an intensity that decreases steadily with increasing scattering angle (Fig. 21.29a). The undulations are due to the

Diffraction methods

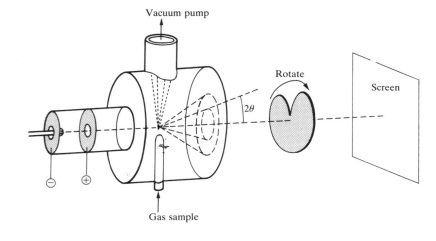

Fig. 21.28 The layout of an electron diffraction apparatus. The diffraction pattern is photographed from the fluorescent screen. A rotating heart-shaped sector emphasizes the scattering from the nuclear positions and suppresses the smoothly varying background due to scattering from the continuous electron distribution in the molecules.

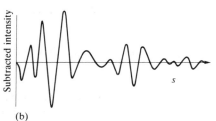

Fig. 21.29 (a) The scattering intensity consists of a smoothly varying background with undulations superimposed. (b) The undulations are emphasized if a sector is rotated in front of the screen, and then the densitometer trace taken from the photograph plotted against $s = (4\pi/\lambda) \sin \frac{1}{2}\theta$.

sharply defined scattering from the nuclear positions, and the background is due to scattering from the less well-defined continuous electron density distribution in the molecule. One way of eliminating the unwanted background is to insert a rotating heart-shaped disk in front of the screen: the disk exposes the outer parts more than the inner, and helps to emphasize the undulations (Fig. 21.29b).

The scattering from a pair of nuclei separated by a distance R_{ij} and orientated at a definite angle to the incident beam can be calculated. The overall diffraction pattern is then calculated by allowing for all possible orientations of this pair of atoms. In other words, we integrate over orientations. When the molecule consists of a number of atoms, we sum over the contribution from all pairs, and the total intensity has an angular variation given by the **Wierl equation**

$$I(\theta) = \sum_{i,j} f_i f_j \frac{\sin sR_{ij}}{sR_{ij}} \qquad s = \frac{4\pi}{\lambda} \sin \frac{1}{2}\theta \qquad (8)$$

where λ is the wavelength of the electrons in the beam and θ is the scattering angle. The electronic scattering factors f are a measure of the intensity of the electron scattering powers of the atoms.

The electron diffraction pattern gives the distances between all possible pairs of atoms in the molecule (not just those bonded together). When there are only a few atoms, the peaks can be analysed reasonably quickly, and the analysis proceeds by assuming a geometry and calculating the intensity pattern using the Wierl equation. The best fit is then taken as the actual molecular geometry.

Further reading

Crystals

A. F. Wells, *The third dimension in chemistry*. Clarendon Press, Oxford (1956, reissued 1968).

F. C. Phillips, *An introduction to crystallography*. Longman, London (1979).

M. J. Buerger, *Introduction to crystal geometry*. McGraw-Hill, New York (1971).

A. F. Wells, *Structural inorganic chemistry* (5th edn). Clarendon Press, Oxford (1984).

R. W. G. Wycoff, *Crystal structure* (5 sections and supplements). Wiley-Interscience, New York (1959).

X-ray diffraction

P. J. Wheatley, *The determination of molecular structure* (2nd edn). Clarendon Press, Oxford (1968).

J. P. Glusker and K. N. Trueblood. *Crystal structure analysis* (2nd edn). Oxford University Press (1985).

W. N. Lipscomb and R. A. Jacobson, X-ray crystal structure analysis. In *Techniques of chemistry* (ed. A. Weissberger and B. W. Rossiter) IIID, Wiley-Interscience, New York (1972).

M. M. Woolfson, *An introduction to X-ray crystallography*, Cambridge University Press (1970).

J. D. Dunitz, *X-ray analysis and the structure of organic molecules*. Cornell University Press, Ithaca (1979).

M. F. C. Ladd and R. A. Palmer, *Structure determination by X-ray crystallography*. Plenum, New York (1985).

Neutron and electron diffraction

G. E. Bacon, *Neutron diffraction*. Oxford University Press (1975).

L. S. Bartell, Electron diffraction by gases. In *Techniques of chemistry* (ed. A. Weissberger and B. W. Rossiter) IIID, p. 125, Wiley-Interscience, New York (1972).

T. B. Rymer, *Electron diffraction*. Chapman and Hall, London (1970).

Exercises

21.1 Draw an array of points as a rectangular lattice based on unit cells of side a and b, and mark the planes with Miller indices (10), (01), (11), (12), (23), (41), (4$\bar{1}$).

21.2 Repeat Exercise 21.1 for a lattice in which the a and b axes make $60°$ to each other.

21.3 In a certain unit cell, planes cut through the crystal axes at $(2a, 3b, c)$, (a, b, c), $(6a, 3b, 3c)$, $(2a, -3b, -3c)$. Identify the Miller indices of the planes.

21.4 Draw an orthorhombic unit cell and mark on it the (100), (010), (001), (011), (101), and (111) planes.

21.5 Draw a triclinic unit cell and mark on it the (100), (010), (001), (011), (101), and (111) planes.

21.6 Calculate the separations of the planes (111), (211), and (100) in a crystal in which the cubic unit cell has side 432 pm.

21.7 The glancing angle of a Bragg reflection from a set of crystal planes separated by 99.3 pm is $20.85°$. Calculate the wavelength of the X-rays.

21.8 Copper Kα radiation consists of two components of wavelengths 154.433 pm and 154.051 pm. Calculate the separation of the diffraction lines arising from the two components in a powder diffraction pattern recorded in a camera of radius 5.74 cm from planes of separation 77.8 pm.

21.9 Rb_3TlF_6 has a tetragonal unit cell with dimensions $a = 651$ pm and $c = 934$ pm. Calculate the volume of the unit cell.

21.10 The orthorhombic unit cell of $NiSO_4$ has the dimensions $a = 634$ pm, $b = 784$ pm, and $c = 516$ pm, and the density of the solid is estimated as $3.9\ \mathrm{g\,cm^{-3}}$. Determine the number of formula units per unit cell and calculate a more precise value of the density.

21.11 The unit cells of $SbCl_3$ are orthorhombic with dimensions $a = 812$ pm, $b = 947$ pm, and $c = 637$ pm. Calculate the spacing of the (411) planes.

21.12 The orthorhombic unit cell of Li_3AlAs_2 has dimensions $a = 1198$ pm, $b = 1211$ pm, and $c = 1186$ pm. Calculate the distance from the origin to each of the intercepts of the (523) plane on the three coordinate axes.

21.13 A substance known to have a cubic unit cell gives reflections with Cu Kα radiation (wavelength 154 pm) at the glancing angles $19.4°$, $22.5°$, $32.6°$, and $39.4°$. The reflection at $32.6°$ is known to be due to the (220) planes. Index the other reflections.

21.14 Potassium nitrate crystals have orthorhombic unit cells of dimensions $a = 542$ pm, $b = 917$ pm, and $c = 645$ pm. Calculate the glancing angles for the (100), (010), and (111) reflections using Cu Kα radiation (154 pm).

21.15 Copper(I) chloride forms cubic crystals with four molecules per unit cell. The only reflection present in a powder photograph are those with either all even indices or all odd indices. What is the symmetry of the unit cell?

21.16 A powder diffraction photograph from tungsten shows lines which index as (110), (200), (211), (220), (310), (222), (321), (400), ... Identify the symmetry of the unit cell.

21.17 The coordinates, in units of a, of the atoms in a simple cubic lattice are $(0, 0, 0)$, $(0, 1, 0)$, $(0, 0, 1)$, $(0, 1, 1)$, $(1, 0, 0)$, $(1, 1, 0)$, $(1, 0, 1)$, and $(1, 1, 1)$. Calculate the structure factors F_{hkl} when all the atoms are identical.

21.18 The coordinates of the four I atoms in the unit cell of KIO_4 are $(0, 0, 0)$, $(0, \frac{1}{2}, \frac{1}{4})$, $(\frac{1}{2}, \frac{1}{2}, \frac{1}{2})$, $(\frac{1}{2}, 0, \frac{3}{4})$. By calculating the phase of the I reflection in the structure factor, show that the I atoms contribute no net intensity to the (114) reflection.

21.19 Calculate the packing fraction for close-packed cylinders.

21.20 The separation of (100) planes of lithium metal is 350 pm and its density is $0.53\ \mathrm{g\,cm^{-3}}$. Is the structure of lithium fcc or bcc?

21.21 Copper crystallizes in an fcc structure with unit cells of side 361 pm. Predict the appearance of the powder diffraction pattern using 154 pm radiation. What is the density of copper?

21.22 In a Patterson synthesis, the spots correspond to the lengths and directions of the vectors joining the atoms in a unit cell. Sketch the pattern that would be obtained for (a) a trigonal-planar isolated BF_3 molecule, (b) the C atoms in an isolated benzene molecule.

21.23 A diffraction experiment requires a neutron beam with wavelength 70 pm. Calculate (a) the average kinetic energy and (b) the effective temperature of the beam.

21.24 What velocity should neutrons have if they are to have wavelength 50 pm?

21.25 Calculate the wavelength of neutrons that have reached thermal equilibrium by collision with a moderator at 300 K.

21.26 What accelerating potential difference must be applied to electrons to generate a beam with wavelength 18 pm?

21.27 Calculate the wavelengths of electrons that have been accelerated through (a) 1.0 kV, (b) 10 kV, (c) 40 kV.

Problems

Numerical problems

21.1 In the early days of X-ray crystallography there was an urgent need to know the wavelengths of X-rays. One technique was to measure the diffraction angle from a mechanically ruled grating. Another method was to estimate the separation of lattice planes from the measured density of a crystal. The density of NaCl is $2.17\ \mathrm{g\,cm^{-3}}$ and the (100) reflection using Pd Kα radiation occurred at 6.0°. Calculate the wavelength of the X-rays.

21.2 A powder diffraction pattern of KCl in a camera of radius 28.7 mm gave lines at the following distances from the centre spot on the film when 154 pm X-rays were used: 14.5, 20.6, 25.4, 29.6, 33.4, 37.1, 44.0, 47.5, 50.9, 54.4, 58.2, 62.1, 66.4, 78.1 mm. Index the lines and calculate the size of the cubic unit cell.

21.3 The powder diffraction patterns of (a) tungsten, (b) copper are shown in Fig. 21.30. Both were obtained with 154 pm X-rays and the scales are marked. Identify the unit cell in each case, and calculate the lattice spacing. Estimate the metallic radii of W and Cu. The camera radius is 28.7 mm.

21.4 Mercury(II) chloride is orthorhombic, and with 154 pm radiation the (100), (010), and (001) reflections occur at 7° 25′, 3° 28′, and 10° 13′ respectively. The density of the solid is $5.42\ \mathrm{g\,cm^{-3}}$. Calculate the dimensions of the unit cell and the number of $HgCl_2$ formula units that it contains.

21.5 Genuine pearls consist of concentric layers of calcite ($CaCO_3$) crystals in which the trigonal axes are oriented along the radii. The nucleus of a cultured pearl is a piece of mother-of-pearl that has been worked into a sphere on a lathe. The oyster then deposits concentric layers of calcite on the central seed. Suggest an X-ray method for distinguishing between real and cultured pearls.

21.6 In their book *X-rays and crystal structures* (which begins 'It is now two years since Dr. Laue conceived the idea ... ') the Braggs give a number of simple examples of X-ray analysis. For instance, they report that the reflection from (100) planes in KCl occurs at 5° 23′, but for NaCl it occurs at 6° 0′ for X-rays of the same wavelength. If the side of the NaCl unit cell is 564 pm, what is the side of the KCl unit cell? The densities of KCl and NaCl are $1.99\ \mathrm{g\,cm^{-3}}$ and $2.17\ \mathrm{g\,cm^{-3}}$ respectively. Do these values support the X-ray analysis?

21.7 The volume of a monoclinic unit cell is $abc \sin \beta$.

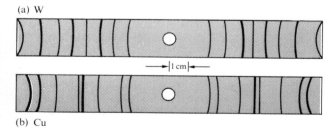

(a) W

←|1 cm|←

(b) Cu

Fig. 21.30 The powder diffraction patterns of (a) tungsten, (b) copper.

Naphthalene has a monoclinic unit cell with two molecules per cell and sides in the ratio $1.377:1:1.436$. The angle β is $122° 49'$ and the density of the solid is 1.152 g cm^{-3}. Calculate the dimensions of the cell.

21.8 The density of LiF is 2.601 g cm^{-3}. The (111) reflection occurs at $8° 44'$ when 70.8 pm Mo X-rays are used. Calculate Avogadro's constant from the data and the fact that there are four LiF units per unit cell.

21.9 The only example of a metal that crystallizes with a primitive cubic structure is polonium, and its unit cell is of side 334.5 pm. Predict (a) the form of the powder diffraction pattern using 154 pm X-rays and (b) the density of the metal. Suppose it makes a transition to an fcc structure but retains the same metallic radius and calculate its density.

21.10 Calculate the coefficient of thermal expansion of diamond given that the (111) reflection shifts from $22° 2' 25''$ to $21° 57' 59''$ on heating a crystal from 100 K to 300 K and 154.0562 pm X-rays are used.

21.11 Use the Wierl equation to predict the appearance of the electron diffraction pattern of CCl_4 with an (as yet) undetermined C—Cl bond length but of known tetrahedral symmetry. Take $f_{Cl} = 17f$ and $f_C = 6f$ and note that $R(Cl, Cl) = \left(\frac{8}{3}\right)^{1/2} R(C, Cl)$. Plot I/f^2 against $x = sR(C, Cl)$. In an actual experiment using 10.0 eV electrons the positions of the maxima occurred at $3° 10'$, $5° 22'$, and $7° 54'$ and minima occurred at $1° 46'$, $4° 6'$, $6° 40'$, and $9° 10'$. What is the C—Cl bond length in CCl_4?

Theoretical problems

21.12 Show that the separation of the (hkl) planes in an orthorhombic crystal with sides a, b, and c is given by

$$\frac{1}{d_{hkl}^2} = \frac{h^2}{a^2} + \frac{k^2}{b^2} + \frac{l^2}{c^2}$$

21.13 Show that the volume of a triclinic unit cell of sides a, b, and c and angles α, β, and γ is

$$V = abc(1 - \cos^2 \alpha - \cos^2 \beta - \cos^2 \gamma + 2 \cos \alpha \cos \beta \cos \gamma)^{1/2}$$

Use this expression to derive expressions for monoclinic and orthorhombic unit cells. For the derivation, it may be helpful to use the result from vector analysis that $V = \boldsymbol{a} \cdot \boldsymbol{b} \wedge \boldsymbol{c}$ and to calculate V^2 initially.

21.14 Calculate the packing fractions of (a) a primitive cubic lattice, (b) a bcc unit cell, (c) an fcc unit cell.

21.15 The coordinates, in units of a, of the A atoms, with scattering factor f_A, in a cubic lattice are $(0, 0, 0)$, $(0, 1, 0)$, $(0, 0, 1)$, $(0, 1, 1)$, $(1, 0, 0)$, $(1, 1, 0)$, $(1, 0, 1)$, and $(1, 1, 1)$. There is also a B atom, with scattering factor f_B, at $(\frac{1}{2}, \frac{1}{2}, \frac{1}{2})$. Calculate the structure factors F_{hkl} and predict the form of the powder diffraction pattern when (a) $f_A = f$, $f_B = 0$, (b) $f_B = \frac{1}{2}f_A$, and (c) $f_A = f_B = f$.

22

The electric and magnetic properties of molecules

Check-list of key ideas

1. The definition of the *electric dipole moment* and the approximate additivity of dipole moments in polyatomic molecules (Section 22.1).

2. The *polarizability* and *polarizability volume* of a molecule and its contribution to the *polarization* of a medium (Section 22.1).

3. The *polarization* of a medium as the mean dipole moment per unit volume (Section 22.1).

4. The variation of the polarizability with frequency and the *orientation, distortion,* and *electronic contributions* (Section 22.1).

5. The *Debye equation* for the contribution of the permanent dipole moment and the polarizability to the polarization (eqn 5a).

6. The *Clausius–Mossotti equation* for the polarization when the permanent dipole does not contribute (eqn 5b).

7. The *molar polarization* (eqn 6a) and the *molar refractivity* (eqn 9) of a substance.

8. The *refractive index* of a substance, its *dispersion* (Section 22.2), and its calculation from molecular contributions (eqn 10).

9. Circular birefringence, the origin of *optical activity,* and *optical rotatory dispersion* (Section 22.2).

10. The *circular dichroism* of chiral molecules (Section 22.2).

11. The origin of *van der Waals forces* between molecules (Section 22.3).

12. The *potential energy* of interaction between a charge and an electric dipole (eqn 11), between two dipoles (eqn 12), and between higher *multipoles*.

13. The *electric field* developed by an electric dipole (eqn 13).

14. The average *dipole/dipole interaction* of rotating polar molecules (eqn 16).

15. The *dipole/induced-dipole interaction* between a polar molecule and a polarizable molecule (eqn 17).

16. The *induced-dipole/induced-dipole interaction* between molecules and the *London dispersion interaction* (eqn 18).

17. The contribution of *three-body interactions* and the *Axilrod–Teller formula* (Section 22.3).

18. The *hard-sphere potential* and the *Lennard–Jones potential* (eqn 20).

19. The study of molecular interactions in *molecular beams* and the significance of the *differential scattering cross-section* and the *impact parameter* (Section 22.5).

20. The phenomena of *glory scattering* and *rainbow scattering* (Section 22.5).

21. The contribution of the intermolecular potential to the *virial coefficients* of a real gas and its calculation in terms of the *configuration integral* and the *Mayer f-function* (eqn 22).

22. The description of liquids in terms of the *pair distribution function* and calculations using the *Monte Carlo method* and *molecular dynamics* (Section 22.7).

23. The *mesophase* and the classification of *liquid crystals* (Section 22.8).

24. Molecular motion in liquids and the temperature dependence of the *viscosity* (Section 22.9).

25. The *magnetic susceptibility* of a substance and its classification as *paramagnetic* and *diamagnetic* (Section 22.10).

26. The *Curie Law* of magnetism (eqn 26).

27. The contribution of electron spin to the magnetic susceptibility and the transition to *ferromagnetic* and *antiferromagnetic* phases (Section 22.11).

28. The contribution of *induced magnetic moments* to the magnetic susceptibility and the origin of *temperature-independent paramagnetism* (Section 22.12).

In this chapter we examine some of the electric and magnetic properties of molecules and interpret them in terms of electronic structure. The properties we consider include the electric dipole moments and polarizabilities of molecules and some related properties that include refractive index, optical activity, and intermolecular forces. We also discuss the analogous magnetic properties, particularly the magnetizabilities and the magnetic susceptibilities of molecules.

Electric properties

An **electric dipole** consists of two charges q and $-q$ separated by a distance l, and is represented by a vector μ that points from the negative charge to the positive charge. The magnitude of this vector (**1**) is ql and is called the **electric dipole moment** μ. An electric dipole is represented by the arrow $\leftrightarrow$ added to the Lewis structure for the molecule, with the + marking the positive end, as in $\overset{\longleftarrow}{\text{H}—\text{Cl}}$ (the direction of $\leftrightarrow$ is opposite to that of the

1

vector $\boldsymbol{\mu}$). Dipole moments are generally reported in **debye** D, where

$$1\,D = 3.336 \times 10^{-30}\,C\,m$$

The debye is named after Peter Debye, a pioneer in the study of dipole moments of molecules. The dipole moment of a pair of charges e and $-e$ separated by 100 pm (1 Å) is $1.6 \times 10^{-29}\,C\,m$, corresponding to 4.8 D. Dipole moments of small molecules are typically about 1 D.

22.1 Permanent and induced electric dipole moments

A **polar molecule** is a molecule with a permanent electric dipole moment that arises from the partial charges on atoms linked by polar bonds (Section 14.6). Non-polar molecules may acquire a dipole moment in an electric field on account of the distortion the field causes in their electronic distributions and nuclear positions. Similarly, polar molecules may have their existing dipole moments modified by the applied field.

Permanent and induced dipole moments are important in chemistry through their role in intermolecular forces (as we describe later) and in their contribution to the ability of a substance to act as a solvent for ionic solids. The latter ability stems from the fact that one end of a dipole may be coulombically attracted to an ion of opposite charge and hence contribute an exothermic term to the enthalpy of solution (Fig. 22.1).

The average electric dipole moment per unit volume of a sample is called its **polarization** P. The polarization of a fluid sample is zero in the absence of an applied field because the molecules adopt random orientations and the average dipole moment is zero. In the presence of a field the dipoles are partially aligned and there is an additional contribution from the dipole moment induced by the field. Hence, the polarization of a medium in the presence of an applied field is non-zero.

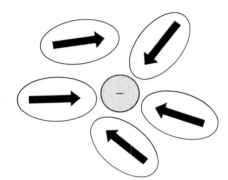

Fig. 22.1 An ion in a polar solvent is solvated as a result of its interaction with the dipoles of the solvent molecules.

Polar molecules

We explained in Section 16.4 how the Stark effect can be used to measure electric dipole moments of molecules for which a rotational spectrum can be observed. When microwave spectroscopy cannot be used (because the sample is not volatile, decomposes, or consists of complex molecules), the dipole moment may be obtained by measurements on a liquid or solid bulk sample using a method that we explain later. In the following we refer to the sample as a **dielectric**, by which we mean a polarizable, non-conducting medium.

All heteronuclear diatomic molecules are polar, and typical values are 1.08 D for HCl and 0.42 for HI (Table 22.1). A very approximate relation between the dipole moment and the difference in electronegativities of the two atoms $\Delta\chi$ is

$$\mu/D = \Delta\chi$$

Table 22.1. Dipole moments (μ) and polarizability volumes (α')

	μ/D	$\alpha'/(10^{-24}\,cm^3)$
CCl_4	0	10.5
H_2	0	0.819
H_2O	1.85	1.48
HCl	1.08	2.63

The more electronegative atom is normally the negative end of the dipole, but there are exceptions, particularly when antibonding orbitals are occupied.[1] Thus the dipole moment of CO is very small (0.12 D) but the

[1] We remarked in Chapter 14 that the major contribution to an antibonding orbital is made by the atomic orbitals of the less electronegative atom. Hence, if that orbital is occupied there may be so much electron density of the less electronegative atom that that atom has a partial negative charge.

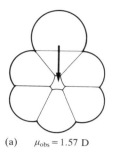

(a) $\mu_{obs} = 1.57$ D

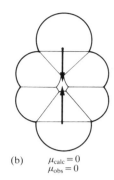

(b) $\mu_{calc} = 0$
        $\mu_{obs} = 0$

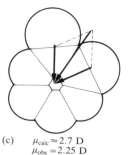

(c) $\mu_{calc} \approx 2.7$ D
        $\mu_{obs} = 2.25$ D

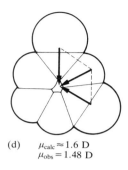

(d) $\mu_{calc} \approx 1.6$ D
        $\mu_{obs} = 1.48$ D

Fig. 22.2 The dipole moments of the dichlorobenzene isomers can be obtained approximately by vectorial addition of two chlorobenzene dipole moments (1.57 D).

negative end of the dipole is on the C atom even though the O atom is more electronegative.

Whether or not a polyatomic molecule is polar depends on its symmetry. The criteria were discussed in Section 15.3, where we saw that a molecule is non-polar if it belongs to a D point group or to one of the cubic point groups. We also saw that the dipole moment of a molecule with a symmetry axis cannot lie perpendicular to that axis. The symmetry criterion is more important than the question of whether or not the atoms in the molecule are the same. Thus the homonuclear triatomic molecule O_3 (which is angular, with C_{2v} symmetry) is polar because the electron density on the central O atom is different from that on the two outer O atoms. The heteronuclear triatomic molecule CO_2 (which is linear, with $D_{\infty h}$ symmetry) is nonpolar because the two bond dipoles point in opposite directions and cancel.

To some extent it is possible to resolve the dipole moment of a polyatomic molecule into contributions of various components (Fig. 22.2). Thus, p-dichlorobenzene is non-polar because of the cancellation of the two equal but opposing moments. o-Dichlorobenzene has a dipole moment which is approximately the resultant of two chlorobenzene dipole moments arranged at 60°. The technique of vector addition can be applied with fair success to other series of related molecules, and the resultant of two dipole moments that make an angle θ to each other (**2**) is

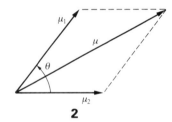

2

$$\mu = (\mu_1^2 + \mu_2^2 + 2\mu_1\mu_2 \cos \theta)^{1/2}$$

If the two dipole moments are equal, this equation simplifies to

$$\mu = 2\mu_1 \cos \tfrac{1}{2}\theta$$

Induced dipole moments

As long as the applied electric field is not too strong the magnitude of the induced dipole moment μ^* is proportional to the field $\mathscr{E}$, and we write

$$\mu^* = \alpha\mathscr{E} \qquad (1)$$

where the constant α is the **polarizability** of the molecule. When the field is strong (as in laser beams) the induced moment also depends on $\mathscr{E}^2$ and higher powers of the field. We then write

$$\mu^* = \alpha\mathscr{E} + \tfrac{1}{2}\beta\mathscr{E}^2 + \dots \qquad (2)$$

where β is called the **hyperpolarizability**.

The polarizability has the units $J^{-1}C^2 m^2$ when the field strength is expressed in $V m^{-1}$ and μ is in $C m$. Since that collection of units is awkward, α is usually converted to a **polarizability volume** α' using the relation

$$\alpha' = \frac{\alpha}{4\pi\varepsilon_0} \tag{3}$$

The units of $4\pi\varepsilon_0$ are $J^{-1}C^2 m^{-1}$, so α' has the dimensions of volume (hence its name). Values are usually quoted in cm^3 (and $Å^3$, where $1 Å^3 = 10^{-24} cm^3$).[2] The larger the polarizability volume, the larger the polarizability of the molecule.

Some experimental polarizability volumes of molecules are given in Table 22.1. The values reflect the strengths with which the nuclear charges control the electron distributions and prevent their distortion by the applied field. If the molecule has few electrons, they are tightly controlled by the nuclear charges and the polarizability of the molecule is low. If the molecule contains large atoms with electrons some distance from the nucleus, the nuclear control is less, the electron distribution is flabbier, and the polarizability is greater.

For all molecules other than those belonging to one of the cubic groups, the polarizability depends on the orientation of the molecule with respect to the field. We saw in Section 16.6 that the anisotropy of the polarizability of a molecule was essential to the observation of a rotational Raman spectrum. The values reported in Table 22.1 are mean values.

Polarization at high frequencies

When the applied field changes direction periodically, the permanent dipole moments reorientate and follow the field. However, when the frequency of the field is high, the molecules cannot change direction fast enough to follow it and then make no contribution to the polarization of the sample. Since a molecule takes about 1 ps to rotate in a fluid, the change occurs when measurements are made at frequencies greater than about 10^{11} Hz (in the microwave region). We say that the **orientation polarization**, the polarization arising from the permanent dipole moments, is lost at high frequencies.

The next contribution to the polarization to be lost as the frequency is taken higher is the **distortion polarization**, the polarization that arises from the distortion of the positions of the nuclei by the applied field. The molecule is bent and stretched by the applied field, and as it does so its molecular dipole changes. The time it takes for a molecule to bend is approximately the time it takes for it to vibrate, and so the distortion polarization disappears when the frequency of the radiation is increased through the infrared. The disappearance occurs in stages, each one occurring as the incident frequency rises above the frequency of a particular mode of vibration.

At even higher frequencies, in the visible region, only the electrons are

---

[2] When using older compilations of data, it is useful to note that polarizability volumes have the same numerical values as the 'polarizabilities' reported using c.g.s. electrical units, and so the tabulated values previously called 'polarizabilities' can be used directly.

light enough to respond to the rapidly changing direction of the applied field. The polarization that remains is now due entirely to the distortion of the electron distribution, and the surviving contribution to the molecular polarizability is called the **electronic polarizability**.

The relative permittivity

When two charges q_1 and q_2 are separated by a distance r in a medium of permittivity ε, the potential energy of their interaction is

$$V = \frac{q_1 q_2}{4\pi \varepsilon r} \tag{4a}$$

In a vacuum the permittivity is ε_0, a fundamental constant with the value $8.854 \times 10^{-12}\,\text{C}^2\,\text{J}^{-1}\,\text{m}^{-1}$. The permittivity is larger than ε_0 when the medium is not a vacuum, and is normally expressed in terms of the **relative permittivity** ε_r (which is also called the dielectric constant) of the medium:

$$\varepsilon_r = \frac{\varepsilon}{\varepsilon_0} \tag{4b}$$

The relative permittivity of a substance is easily measured by comparing the capacitance of a capacitor with and without the sample present (C and C_0 respectively) and using

$$\varepsilon_r = \frac{C}{C_0}$$

Example 22.1: *Determining the relative permittivity of a substance*

The capacitance of an empty cell is 5.01 pF; when it was filled with a sample of camphor at 25°C its capacitance was 57.1 pF. Calculate the relative permittivity of camphor at 25°C.

Answer. Since the relative permittivity of dry air is very close to 1 (actually, 1.000 536) we obtain immediately that

$$\varepsilon_r = \frac{C}{C_0} = \frac{57.1\,\text{pF}}{5.01\,\text{pF}} = 11.4$$

Comment. This trivial example introduces a substance we consider in more detail later. F denotes the unit of capacitance called the 'farad':

$$1\,\text{F} = 1\,\text{C}\,\text{V}^{-1}$$

Exercise. The same cell when filled with ethanol at 25°C had a capacitance of 121.7 pF. Calculate the relative permittivity of ethanol at this temperature.

[24.3]

The magnitude of ε_r can have a very significant effect on the strength of the interactions between ions in solution. For instance, water has $\varepsilon_r = 78$ at 25°C, and reduces the long-range interionic coulombic interaction energy by nearly two orders of magnitude from its vacuum value. We examined some of the consequences of this reduction in Chapter 10.

The relative permittivity of a substance is large if its molecules are polar and highly polarizable. The quantitative relation between the relative permittivity and the electric properties of the molecules is expressed by the **Debye equation**:

$$\frac{\varepsilon_r - 1}{\varepsilon_r + 2} = \frac{\mathcal{N}}{3\varepsilon_0}\left(\alpha + \frac{\mu^2}{3kT}\right) \tag{5a}$$

The term in $\mu^2/3kT$ stems from the thermal averaging of the electric dipole moment in the presence of the applied field, and its derivation is described in the *Further information* section at the end of the chapter. $\mathcal{N}$ is the number density of molecules, the number per unit volume. The same expression, but without the permanent dipole moment contribution, is called the **Clausius–Mossotti equation**:

$$\frac{\varepsilon_r - 1}{\varepsilon_r + 2} = \frac{\mathcal{N}\alpha}{3\varepsilon_0} \tag{5b}$$

The Clausius–Mossotti equation is used when there is no contribution from permanent electric dipole moments to the polarization, either because the molecules are non-polar or because the frequency of the applied field is high.

Equations 5a and 5b are normally written in terms of the density ρ of the sample using $\mathcal{N} = \rho N_A/M$, where M is the molar mass of the molecules, and the **molar polarization** P_m:

$$P_m = \frac{N_A}{3\varepsilon_0}\left(\alpha + \frac{\mu^2}{3kT}\right) \tag{6a}$$

Then

$$\frac{\varepsilon_r - 1}{\varepsilon_r + 2} = \frac{\rho P_m}{M} \tag{6b}$$

We see from eqn 6 that we can measure the polarizability and permanent dipole moment of the molecules in a sample by measuring ε_r at a series of temperatures, calculating P_m, and plotting it against $1/T$. The slope of the graph is $N_A\mu^2/9\varepsilon_0 k$ and its intercept at $1/T = 0$ is $N_A\alpha/3\varepsilon_0$.

Example 22.2: *Determining dipole moment and polarizability*

A series of experiments on camphor were made in the same sample cell as in Example 22.1 but at different temperatures. Use the following data to find the dipole moment and the polarizability volume of the molecule.

$\theta/°C$	0	20	40	60	80	100	120	140	160	200
$\rho/(\text{g cm}^{-3})$	0.99	0.99	0.99	0.99	0.99	0.99	0.97	0.96	0.95	0.91
C/pF	62.6	57.1	54.1	50.1	47.6	44.6	40.6	38.1	35.6	31.1

Answer. We must calculate ε_r at each temperature and form $(\varepsilon_r - 1)/(\varepsilon_r + 2)$. Multiply by M/ρ to form P_m (via eqn 6). For camphor, $M = 152.23 \text{ g mol}^{-1}$. Plot P_m against $1/T$. The intercept at $1/T = 0$ is $N_A\alpha/3\varepsilon_0 = (4\pi N_A/3)\alpha'$; the slope is $N_A\mu^2/9\varepsilon_0 k$.

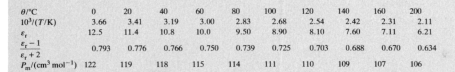

$\theta/°C$	0	20	40	60	80	100	120	140	160	200
$10^3/(T/K)$	3.66	3.41	3.19	3.00	2.83	2.68	2.54	2.42	2.31	2.11
ε_r	12.5	11.4	10.8	10.0	9.50	8.90	8.10	7.60	7.11	6.21
$\dfrac{\varepsilon_r - 1}{\varepsilon_r + 2}$	0.793	0.776	0.766	0.750	0.739	0.725	0.703	0.688	0.670	0.634
$P_m/(\text{cm}^3 \text{ mol}^{-1})$	122	119	118	115	114	111	110	109	107	106

The points are plotted in Fig. 22.3. The intercept lies at 82.7, and so $\alpha' = 3.3 \times 10^{-23} \text{ cm}^3$. The slope is 10.9, so $\mu = 4.46 \times 10^{-30} \text{ C m}$, corresponding to 1.34 D.

Comment. Since the Debye equation describes molecules that are free to rotate, the data show that camphor, which does not melt until 175°C, is rotating even in the solid. It is an approximately spherical molecule.

Exercise. The relative permittivity of chlorobenzene is 5.71 at 20°C and 5.62 at 25°C. Assuming a constant density (1.11 g cm^{-3}), estimate its polarizability volume and dipole moment.
$[1.4 \times 10^{-23} \text{ cm}^3 \text{ mol}^{-1}; 1.2 \text{ D}]$

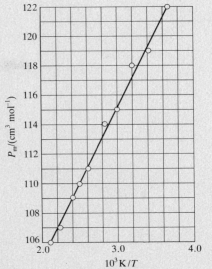

Fig. 22.3 The plot of $P_m/(\text{cm}^3 \text{ mol}^{-1})$ against $10^3/(T/K)$ used for the determination of the polarizability and dipole moment of camphor.

22.2 Refractive index

The **refractive index** n_r of a medium is the ratio of the speed of light in a vacuum c to its speed v in the medium:

$$n_r = \frac{c}{v} \tag{7}$$

It follows from the Maxwell equations, which describe the properties of electromagnetic radiation, that the refractive index at a frequency v is related to the relative permittivity at that frequency by

$$n_r = \varepsilon_r^{1/2} \tag{8}$$

The molar polarizability, and hence the molecular polarizability, can therefore be measured at frequencies typical of visible light (about 10^{15} to 10^{16} Hz) by measuring the refractive index of the sample (Table 22.2) and using the Clausius–Mossotti equation.

Table 22.2. Refractive indices (at different wavelengths of light) relative to air at 20°C

	434 nm	589 nm	656 nm
$C_6H_6(l)$	1.524	1.501	1.497
$CS_2(l)$	1.675	1.628	1.618
$H_2O(l)$	1.340	1.333	1.331
$KI(s)$	1.704	1.666	1.658

The molar refractivity

The Clausius–Mossotti equation is normally expressed in terms of the **molar refractivity** R_m:

$$R_m = \frac{M}{\rho} \times \frac{n_r^2 - 1}{n_r^2 + 2} \tag{9}$$

Then, from eqn 6:

$$\alpha = \frac{3\varepsilon_0 R_m}{N_A} \quad \text{or} \quad \alpha' = \frac{3R_m}{4\pi N_A}$$

Example 22.3: *Calculating the polarizability from the refractive index*

The refractive index of water at 20°C is 1.3330 for light of wavelength 589 nm and its density is 0.9982 g cm^{-3}. Calculate the polarizability volume of the molecule at this frequency.

Answer. Since

$$\frac{n_r^2 - 1}{n_r^2 + 2} = 0.2057$$

$R_m = 3.699 \text{ cm}^3 \text{ mol}^{-1}$. Therefore, $\alpha' = 1.467 \times 10^{-24} \text{ cm}^3$ (1.467 Å³).

Comment. The same calculation repeated for 434 nm light, for which the refractive index is 1.3404, gives $1.501 \times 10^{-24} \text{ cm}^3$. The H_2O molecule is more polarizable at the higher frequency because the photons of incoming light carry more energy and are better able to distort the electron distribution.

Exercise. Calculate the polarizability volume of ethanol at the frequency corresponding to the sodium D lines, given that its refractive index is 1.361 at 20°C and its density is 0.789 g cm^{-3}. $[5.12 \times 10^{-24} \text{ cm}^3]$

Table 22.3. Molar refractivities at 589 nm, $R_m/(\text{cm}^3 \text{ mol}^{-1})$

C—H	1.65	Na⁺	0.46
C—C	1.20	K⁺	2.12
C=C	2.79	F⁻	2.65
C=O	3.34	Cl⁻	9.30

To a fair approximation, molar refractivities are the sums of contributions from the individual groups making up a molecule or solid. Specifically, **bond refractivities** are contributions that can be ascribed to the bonds in covalent molecules, and **ion refractivities** are contributions that can be ascribed to ions in ionic solids. Some values are given in Table 22.3. Once we have calculated the molar refractivity by adding together the various contributions, we can deduce the refractive index of the sample by rearranging eqn 9 into

$$n_r^2 = \frac{V_m + 2R_m}{V_m - R_m} \qquad V_m = \frac{M}{\rho} \qquad (10)$$

Example 22.4: *Estimating the refractive index from refractivity data*

Estimate the refractive index of acetic acid for sodium-D light. Its density is 1.046 g cm^{-3}.

Answer. We use the bond refractivity information in Table 22.3, taking the CH_3COOH molecule to consist of $3(C—H) + (C—C) + (C=O) + (C—O) + (O—H)$:

$$R_m/(\text{cm}^3 \text{ mol}^{-1}) = 3 \times 1.65 + 1.20 + 3.34 + 1.41 + 1.85 = 12.75$$

$$V_m = \frac{M}{\rho} = \frac{60.05 \text{ g mol}^{-1}}{1.046 \text{ g cm}^{-3}} = 57.41 \text{ cm}^3 \text{ mol}^{-1}$$

Therefore, by substitution in eqn 10,

$$n_r = \left(\frac{57.41 + 2 \times 12.75}{57.41 - 12.75}\right)^{1/2} = 1.36$$

Comment. The experimental value is 1.37, and so in this case the agreement is very good.

Exercise. Estimate the refractive index for ethanol for sodium-D light ($\rho = 0.789 \text{ g cm}^{-3}$). [1.34]

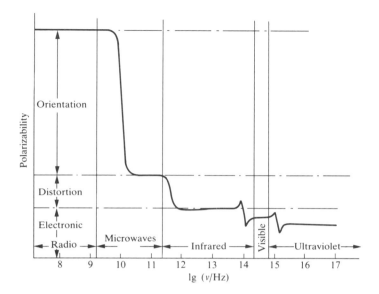

Fig. 22.4 The general form of the dependence of the polarizability on the frequency of the applied field. Note the considerable reduction in polarizability when the field is reversing direction so rapidly that the polar molecules cannot reorientate quickly enough to follow it.

The refractive index is related to the molecular polarizability because the propagation of light through a medium can be imagined as occurring by the incident light inducing an oscillating dipole moment, which then radiates light of the same frequency. The newly formed radiation is delayed in phase relative to the incident radiation, so the light propagates more slowly through the medium than through a vacuum. Since photons of high-frequency light carry more energy than those of low-frequency light, they can distort the electronic distributions of the molecules in their path more effectively. Therefore we can expect the polarizabilities of molecules, and hence the refractive index, to increase with increasing frequency. The dependence on frequency is the origin of the dispersion of white light by a prism: the refractive index is greater for blue light than for red, and therefore the blue rays are bent more than the red. **Dispersion** is a name carried over from this phenomenon to mean the variation of the refractive index, or of any property, with frequency. Figure 22.4 shows the typical dispersion of the polarization of a sample.

Optical activity

An optically active substance rotates the plane of polarization of plane polarized light. The effect arises from the difference of the refractive indices for right and left circularly polarized light, n_R and n_L respectively.[3] A sample in which these two refractive indices are different is said to be **circularly birefringent**.

That optical rotation is a consequence of circular birefringence may be seen by reference to Fig. 22.5. Before entering the medium the beam is plane polarized at an angle $\theta = 0$ to some axis. This plane-polarized light may be regarded as a superposition of two oppositely rotating circularly polarized components). On entering the medium one component propagates faster than the other if their refractive indices are different, and if the

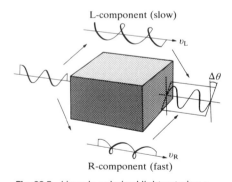

Fig. 22.5 Linearly polarized light entering a sample (from the left) can be regarded as the superposition of two counter-rotating circularly polarized components with a definite phase relation. If one component propagates more rapidly than the other in the medium, when they emerge the phase relation is changed, and the resultant is plane-polarized light rotated through an angle $\Delta\theta$ to its original orientation.

[3] We use the normal convention that in right-handed circularly polarized light the electric vector rotates clockwise as seen by an observer facing the oncoming beam (Fig. 22.5).

sample is of length l the difference in the times of passage is

$$\Delta t = \frac{l}{v_R} - \frac{l}{v_L}$$

where v_R and v_L are the speeds of the two components. In terms of the refractive indices the difference is

$$\Delta t = \frac{(n_R - n_L)l}{c}$$

The phase difference between the two components when they emerge from the sample is therefore

$$\Delta \theta = c\Delta t \times \frac{2\pi}{\lambda} = (n_R - n_L) \times \frac{2\pi l}{\lambda}$$

where λ is the wavelength of the light. Since the two rotating electric vectors are in phase at a different angle when they leave the sample, their superposition gives rise to a plane-polarized beam rotated through an angle $\Delta \theta$ relative to the plane of the incoming beam. Hence, the angle of optical rotation is proportional to the difference in refractive index $n_R - n_L$.

To explain why the refractive indices depend on the handedness of the light, we must examine why the electronic polarizabilities depend on the handedness. The full theory of optical rotation is complicated, but one interpretation is that if a molecule has a helical structure (including, if the molecule is small, a structure that can be regarded as being a fragment of a helix), its polarizability depends on whether or not the electric field of the incident radiation rotates in the same sense as the helix. Molecules having a helical structure are chiral, which is the criterion for optical activity discussed in Section 15.3.

The variation of the angle of optical rotation with the frequency of light is called **optical rotatory dispersion** (ORD). It arises from the individual dispersions of the polarizabilities (and refractive indices) for left- and right-circularly polarized light, and can be used to investigate the stereochemistry of molecules.

Associated with the differences in the two refractive indices (the circular birefringence of the medium) is a difference in absorption intensities I_R and I_L for right and left circularly polarized light. This difference is known as **circular dichroism** (CD) and the **CD spectrum** of a sample is a plot of the variation of $I_R - I_L$ with frequency of the radiation. CD spectroscopy is particularly useful for determining the absolute configurations of d-metal complexes since similar complexes with similar geometries give CD spectra with similar features.

Intermolecular forces

The term **van der Waals forces** denotes the interactions between closed-shell molecules. The attractive contributions to these forces include the interactions between the partial electric charges of polar molecules. Van der Waals forces also include the repulsive interactions that are responsible for the prevention of the complete collapse of matter to nuclear densities. The repulsive interactions arise from the Pauli principle and the exclusion of electrons from regions of space where the orbitals of closed-shell species overlap. In this section we consider the attractive van der Waals forces between molecules, and see how they are related to the electrical properties treated in Section 22.1.

22.3 Interactions between dipoles

Most of the work in this section will be based in one way or the other on the coulombic potential energy of interaction between two charges (eqn 4a). It is easy to adapt this expression to give the interaction between a charge and a dipole and to extend it to the interaction between two dipoles.

The potential energy of interaction

The total potential energy of interaction between the dipole $\mu_1 = q_1 l$ and the point charge q_2 fixed in the orientation shown in Fig. 22.6 is the sum of a repulsion and an attraction:

$$4\pi\varepsilon_0 V = -\frac{q_1 q_2}{r - \frac{1}{2}l} + \frac{q_1 q_2}{r + \frac{1}{2}l}$$

Since $l \ll r$, we can simplify this expression by writing

$$4\pi\varepsilon_0 V = \frac{-q_1 q_2}{r}\left(\frac{1}{1-x} - \frac{1}{1+x}\right)$$

where $x = l/2r$ and then expanding the terms in x:

$$4\pi\varepsilon_0 V = \frac{-q_1 q_2}{r}\left\{(1 + x + x^2 + \ldots) - (1 - x + x^2 - \ldots)\right\}$$

$$= \frac{-q_1 q_2}{r} \times 2x = \frac{-q_1 q_2 l}{r^2}$$

Therefore, since $\mu_1 = q_1 l$, the potential energy of interaction is

$$V = \frac{-\mu_1 q_2}{4\pi\varepsilon_0} \times \frac{1}{r^2} \qquad (11)$$

The interaction energy decreases more rapidly than that between two point charges because, from the viewpoint of the single charge q_2, the partial charges of the dipole seem to merge and cancel as the distance r increases.

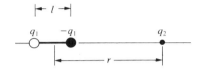

Fig. 22.6 The potential energy of interaction between a dipole and a point charge is the sum of the repulsion of like charges and the attraction of opposite charges.

Example 22.5: *Calculating the interaction energy of two dipoles*

Calculate the potential energy of interaction of two dipoles in the arrangement shown in Fig. 22.7a when their separation is r.

Answer. We proceed in exactly the same way, but now the total interaction energy is the sum of four pairwise terms:

$$4\pi\varepsilon_0 V = \frac{q_1 q_2}{r + l} - \frac{q_1 q_2}{r} - \frac{q_1 q_2}{r} + \frac{q_1 q_2}{r - l}$$

$$= \frac{q_1 q_2}{r}\left(\frac{1}{1+x} - 2 + \frac{1}{1-x}\right)$$

with $x = l/r$. As before, we expand the two terms in x and retain only the first

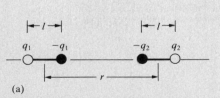

Fig. 22.7 The potential energy of interaction between two dipoles is the sum of the repulsions of like charges and the attractions of opposite charges. (a) A collinear arrangement of dipoles treated in Example 22.5 and (b) the more general arrangement treated in the exercise.

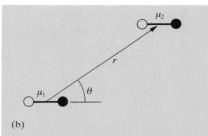

(b)

Fig. 22.7(b) The more general arrangement treated in the exercise.

surviving term. This results in

$$4\pi\varepsilon_0 V = \frac{q_1 q_2}{r} \times 2x^2$$

Therefore, since $\mu_1 = q_1 l$ and $\mu_2 = q_2 l$, the potential energy of interaction in the alignment shown in the illustration is

$$V = \frac{2\mu_1 \mu_2}{4\pi\varepsilon_0} \times \frac{1}{r^3} \qquad (12)$$

Comment. Notice that the interaction energy decreases more rapidly than for the previous case: now both interacting entities seem neutral to each other at large separations.

Exercise. Find an expression for the potential energy when the dipoles are in the arrangement shown in Fig. 22.7b.
$$\left[V = \frac{\mu_1 \mu_2}{4\pi\varepsilon_0} \times \frac{1}{r^3} \times (1 - 3\cos^2\theta) \right]$$

Table 22.4. Multipole interaction potential energies

Interaction Type	Distance dependence of potential energy	Typical energy kJ mol^{-1}	Comment
Ion–ion	$1/r$	250	Only between ions
Ion–dipole	$1/r^2$	15	
Dipole–dipole	$1/r^3$	2	Between stationary polar molecules
	$1/r^6$	0.3	Between rotating polar molecules
London (dispersion)	$1/r^6$	2	Between all types of molecules

The energy of a hydrogen bond A–H $\cdots$ B is typically 20 kJ mol^{-1} and occurs on contact for A, B = N, O, or F.

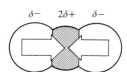

$\delta-$ $2\delta+$ $\delta-$

Fig. 22.8 Although CO_2 has no permanent dipole moment, it does have a permanent quadrupole moment, which can be regarded as two dipole moments back-to-back. These make an important contribution to the intermolecular forces, but are shorter range than dipole–dipole interactions.

The various expressions that arise for the interaction of charges and dipoles are summarized in Table 22.4. It is quite easy to extend the formulas given there to obtain expressions for the energy of interaction of higher multipoles, such as **quadrupoles**, which consist of four charges that sum to zero but have no net dipole moment (as for CO_2 molecules, Fig. 22.8), and **octupoles**, which consist of eight charges that sum to zero but have neither a dipole moment nor a quadrupole moment (as for CH_4 molecules). The feature to remember is that the interaction energy falls off more rapidly the higher the order of the multipole. Thus, for the interaction of an n-pole with an m-pole, the potential energy varies with distance as

$$V \propto \frac{1}{r^{n+m-1}}$$

The reason for the faster decrease with distance is the same as before: the array of charges seems to blend into neutrality more rapidly with distance the higher the number of individual charges contributing to the multipole.

The electric field

Much the same kind of argument can be used to establish the distance-dependence of the strength of the electric field generated by a dipole. We shall need this expression when we calculate the dipole moment induced in one molecule by another.

The starting point for the calculation is the strength of the electric field[4] generated by a point electric charge:

$$\mathscr{E} = \frac{q_1}{4\pi\varepsilon_0 r^2}$$ (13)

The field generated by a dipole is the difference between the fields generated by each partial charge, and can be calculated by combining the contributions from the individual charges in much the same way as the potential energy. For the arrangement shown in Fig. 22.9 this gives

$$\mathscr{E} = \frac{\mu_2}{4\pi\varepsilon_0} \times \frac{2}{r^3}$$ (14)

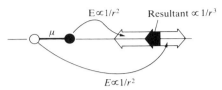

Fig. 22.9 The electric field of a dipole is the sum of the opposing fields from the positive and negative charges, each of which is proportional to $1/r^2$. The difference is proportional to $1/r^3$.

As usual, the electric field of a multipole (in this case a dipole) decreases more rapidly with distance than a monopole (a point charge).

Dipole/dipole interactions

The potential energy of interaction between two polar molecules is a complicated function of the angle between them. However, when the two dipoles are parallel, the energy is simply

$$V = \frac{\mu_1\mu_2 f}{4\pi\varepsilon_0 r^3} \qquad f = 1 - 3\cos^2\theta$$ (15)

This expression applies to polar molecules in a fixed orientation in a solid.

In a fluid of freely rotating molecules, the interaction between dipoles averages to zero because f changes sign as θ changes and the average value $\langle f \rangle$ is zero. Physically, the like partial charges of two freely rotating molecules are close together as much as the two opposite charges, and the attraction is cancelled by the repulsion. However, since the potential energy depends on the orientation, the molecules do not in fact rotate completely freely even in a gas, since the lower energy orientations are marginally favoured, and so there is a non-zero interaction between polar molecules (Fig. 22.10).

The detailed calculation of the interaction energy is quite complicated, but the form of the final answer can be constructed quite simply. First, we note that the average interaction energy of two polar molecules rotating at a fixed separation r is given by

$$\langle V \rangle = \frac{\mu_1\mu_2}{4\pi\varepsilon_0 r^3} \langle f \times P \rangle$$

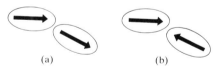

(a) (b)

Fig. 22.10 When a pair of molecules can adopt all relative orientations with equal probability, the favourable orientations (a) and the unfavourable ones (b) cancel, and the average interaction is zero. In an actual fluid, the interactions in (a) slightly predominate.

where P is a weighting factor in the averaging, and is equal to the probability that a particular orientation θ will be adopted. This probability is given by the Boltzmann distribution

$$P \propto e^{-E/kT}$$

---

[4] The electric field is actually a vector and we cannot simply add and subtract magnitudes without taking into account the directions of the fields. In the cases we consider this will not be a complication because the two charges of the dipoles will be collinear and give rise to fields in the same direction. Be careful, though, with more general arrangements of charges.

where E is the potential energy of interaction of the two dipoles in that orientation. That is,

$$P \propto e^{-V/kT} \qquad V = \frac{\mu_1 \mu_2 f}{4\pi\varepsilon_0 r^3}$$

Since the potential energy of interaction of the two dipoles is very small compared with the energy of thermal motion, we can make use of the fact that $V \ll kT$, expand the exponential function in P, and retain only the first two terms:

$$P \propto 1 - \frac{V}{kT} + \dots$$

The average interaction energy is therefore

$$\langle V \rangle \propto \frac{\mu_1 \mu_2}{4\pi\varepsilon_0 r^3} \left\{ \langle f \rangle - \frac{\mu_1 \mu_2 \langle f^2 \rangle}{4\pi\varepsilon_0 kT r^3} + \dots \right\}$$

The average value of f is zero, so the first term vanishes. However, the average value of f^2 is non-zero because f^2 is always positive, so we can write

$$\langle V \rangle \propto -\frac{\mu_1^2 \mu_2^2 \langle f^2 \rangle}{(4\pi\varepsilon_0)^2 kT r^6}$$

The average value $\langle f^2 \rangle$ is a number which we can expect to be close to 1 (because f^2 ranges from 0 to 4) and in fact turns out to be $\frac{2}{3}$ when the calculation is carried through in detail. Therefore, the final result is that

$$V = \frac{-C}{r^6} \qquad C = \frac{2}{3kT} \left(\frac{\mu_1 \mu_2}{4\pi\varepsilon_0} \right)^2 \qquad (16)$$

The important features of eqn 16 are the dependence of the average interaction energy on the inverse sixth power of the separation and its inverse dependence on the temperature. The latter reflects the way that the greater thermal motion overcomes the mutual orientating effects of the dipoles at higher temperatures.

At 25°C the average interaction energy for pairs of molecules with $\mu = 1\,\text{D}$ is about $-1.4\,\text{kJ mol}^{-1}$ when the separation is 0.3 nm. This energy should be compared with the average molar kinetic energy of $\frac{3}{2}RT = 3.7\,\text{kJ mol}^{-1}$ at the same temperature: the two are not very dissimilar but they are both much less than the energies involved in the making and breaking of chemical bonds.

Dipole/induced-dipole interactions

A polar molecule with dipole moment μ_1 can induce a dipole μ_2^* in a polarizable molecule (which may itself be either polar or non-polar). The induced dipole interacts with the permanent dipole of the first molecule, and the two are attracted together.

We can establish the *form* of the potential energy of interaction in a fairly straightforward calculation that shows the essentials of what is involved without being bogged down in the details of the actual calculation. Thus, if we assume that the dipoles—the permanent and the induced dipole—

already exist, their energy of interaction is given by eqn 15 with $\theta = 0$:

$$V = -\frac{2\mu_1\mu_2^*}{4\pi\varepsilon_0 r^3}$$

However, the induced dipole moment depends on the field generated by the polar molecule, and hence on the separation of the two molecules. Since we can write

$$\mu_2^* = \alpha_2\mathscr{E}$$

where α_2 is the polarizability of the molecule and $\mathscr{E}$ is the field generated by the polar molecule, the potential energy is

$$V = -\frac{2\mu_1\alpha_2\mathscr{E}}{4\pi\varepsilon_0 r^3}$$

The electric field generated by the polar molecule is given by eqn 14, so

$$V = -\frac{2\mu_1\alpha_2}{4\pi\varepsilon_0 r^3} \times \frac{2\mu_1}{4\pi\varepsilon_0 r^3}$$

$$= -\frac{4\mu_1^2\alpha_2}{(4\pi\varepsilon_0)^2} \times \frac{1}{r^6}$$

As the induced dipole follows the direction of the inducing dipole (Fig. 22.11), the thermal motion does not average the interaction to zero because both dipoles remain aligned however fast the molecules tumble. The average interaction energy when the separation of the molecules is r is

$$V = \frac{-C}{r^6} \qquad C \approx \frac{\mu_1^2\alpha_2'}{4\pi\varepsilon_0} \tag{17}$$

where α_2' is the polarizability volume of molecule 2 and μ_1 the permanent dipole moment of molecule 1.

The dipole/induced-dipole interaction energy is independent of the temperature (because we do not need to use the Boltzmann distribution for the averaging) but, like the dipole/dipole interaction, it depends on $1/r^6$. For a molecule with $\mu = 1$ D (such as HCl) near a molecule of polarizability volume $\alpha' = 10 \times 10^{-24}$ cm^3 (such as benzene, Table 22.1) the average interaction energy is about -0.8 kJ mol^{-1} when the separation is 0.3 nm.

Induced-dipole/induced-dipole interactions

Non-polar molecules can interact even though neither has a permanent dipole moment. The evidence for the existence of interactions between them is the formation of condensed phases of non-polar substances, such as the condensation of hydrogen and argon.

The interaction between non-polar molecules arises from the *instantaneous* transient dipoles that all molecules possess as a result of changes in the instantaneous positions of electrons (Fig. 22.12). Suppose the electrons in one molecule flicker into an arrangement that gives it an instantaneous dipole moment μ_1^*. This dipole polarizes the other molecule, and induces in it an instantaneous dipole moment μ_2^*. The two dipoles attract each other and the potential energy of the pair is lowered. Although

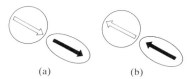

Fig. 22.11 A polar molecule (dark arrow) can induce a dipole (light arrow) in a non-polar molecule, and the latter's orientation follows the former's, so that the interaction does not average to zero.

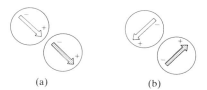

Fig. 22.12 In the dispersion interaction, an instantaneous dipole (shaded arrow) on one molecule induces a dipole (light arrow) on another molecule, and the two dipoles then interact to lower the energy. The two instantaneous dipoles are correlated, and although they occur in different orientations at different instants, the interaction does not average to zero.

the first molecule will go on to change the size and direction of its dipole, the second will follow it; that is, the two dipoles are 'correlated' in direction. Because of this correlation, the attraction between the two instantaneous dipoles does not average to zero and gives rise to an **induced-dipole/induced-dipole interaction**. This interaction is also called the **dispersion interaction** and the **London interaction** for Fritz London, who first described it. Polar molecules also interact by a dispersion interaction: such molecules also possess instantaneous dipoles, the only difference being that the time average of each fluctuating dipole does not vanish, but corresponds to the permanent dipole.

The strength of the dispersion interaction depends on the polarizability of the first molecule because the instantaneous dipole moment μ_1^* depends on the looseness of the nuclear charge's control over the outer electrons. It also depends on the polarizability of the second molecule, for that polarizability determines how readily a dipole can be induced by another molecule. The actual calculation of the dispersion interaction is quite involved, but a reasonable approximation to the interaction energy is given by the **London formula**:

$$V = \frac{-C}{r^6} \qquad C = \tfrac{3}{2}\alpha_1'\alpha_2'\frac{I_1 I_2}{I_1 + I_2} \tag{18}$$

where I_1 and I_2 are the ionization energies of the two molecules (Table 2.4). Once again, the interaction turns out to be proportional to the inverse sixth power of the separation.

In the case of two CH_4 molecules we can substitute $\alpha' = 2.6 \times 10^{-24}\ cm^3$ and $I \approx 7\ eV$ to obtain $V = -5\ kJ\ mol^{-1}$ for $r = 0.3\ nm$. A very rough check on this figure is the enthalpy of vaporization of methane, which is $8.2\ kJ\ mol^{-1}$. (The comparison is very rough because the enthalpy of vaporization is a many-body quantity.)

The total attractive interaction

The total attractive interaction energy between rotating molecules is the sum of the three contributions discussed above. (Only the dispersion interaction contributes if both molecules are non-polar.) All three vary as the inverse sixth power of the separation, and so we may write

$$V = \frac{-C_6}{r^6} \tag{19}$$

where C_6 is a coefficient that depends on the identity of the molecules.

Although attractive interactions between molecules are often expressed in the form of eqn 19, we must remember that this equation has only limited validity. First, we have taken into account only dipolar interactions, for they have the longest range and are dominant if the average separation of the molecules is large. However, in a complete treatment we should also consider quadrupolar and higher-order multipolar interactions, particularly if the molecules do not have dipole moments. Secondly, the expressions have been derived by assuming that the molecules can rotate reasonably freely. That is not the case in a solid, and then the dipole–dipole interaction is proportional to $1/r^3$ because the Boltzmann averaging procedure is irrelevant when the molecules are trapped into a fixed orientation.

A different kind of limitation is that eqn 19 relates to the interactions of *pairs* of molecules, and there is no reason to suppose that the energy of interaction of three (or more) molecules is the sum of the pairwise interaction energies. The total dispersion energy of three closed-shell atoms, for instance, is given by the **Axilrod–Teller formula**:

$$V = -\frac{C_6}{r_{AB}^6} - \frac{C_6}{r_{BC}^6} - \frac{C_6}{r_{CA}^6} + \frac{C'}{(r_{AB}r_{BC}r_{CA})^3}$$

where

$$C' = a(3 \cos \theta_A \cos \theta_B \cos \theta_C + 1)$$

a is a constant (which is approximately equal to $\frac{3}{4}\alpha' C_6$) and the angles θ are the internal angles of the triangle formed by the three atoms (**3**). The term in C' (which represents the non-additivity of the pairwise interactions) is negative for a linear arrangement of atoms (so that arrangement is stabilized) and positive for an equilateral triangular cluster. It is found that the three-body term contributes about 10 per cent of the total interaction energy in liquid argon.

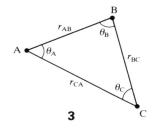

3

22.4 Repulsive and total interactions

When molecules are squeezed together, the nuclear and electronic repulsions and the rising electronic kinetic energy begin to dominate the attractive forces. The repulsions increase steeply with decreasing separation (Fig. 22.13) in a way that can be deduced only by very extensive, complicated molecular structure calculations of the kind described in Chapter 14.

In many cases, however, progress can be made by using a greatly simplified representation of the potential energy, where the details are ignored and the general features expressed by a few adjustable parameters. One such approximation is the **hard-sphere potential**, where it is assumed that the potential energy rises abruptly to infinity as soon as the particles come within a separation σ called the **collision diameter**:

$$V = \infty \text{ for } r \leq \sigma \qquad V = 0 \text{ for } r > \sigma$$

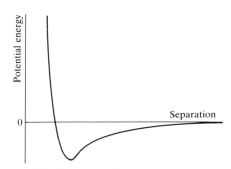

Fig. 22.13 The general form of an intermolecular potential energy curve. At long range the interaction is attractive, but at close range the repulsions dominate.

This very simple potential is surprisingly useful for assessing a number of properties, as we shall see.

Another widely used approximation is to express the short-range repulsive potential energy as inversely proportional to a high power of r:

$$V = \frac{C_n}{r^n}$$

Typically $n = 12$, in which case the repulsion dominates the $1/r^6$ attractions strongly at short separations because C_{12}/r^{12} is then much larger than C_6/r^6. The sum of the repulsive interaction and the attractive interaction given by eqn 19 is called the **Lennard-Jones (n, 6)-potential**:

$$V = \frac{C_n}{r^n} - \frac{C_6}{r^6} \qquad (20a)$$

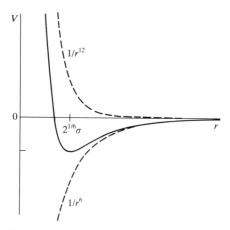

Fig. 22.14 The Lennard–Jones (12, 6)-potential, and the relation of the parameters to the features of the curve. The dotted lines are the two contributions.

Table 22.5. Lennard–Jones (12, 6) parameters.*

	$(\varepsilon/k)/\text{K}$	σ/pm
Ar	124	342
CCl_4	327	588
He	10.22	258
N_2	91.5	368

* ε is expressed as an effective temperature by division by Boltzmann's constant k.

The (12, 6)-potential is often written in the form

$$V = 4\varepsilon \left\{ \left(\frac{\sigma}{r} \right)^{12} - \left(\frac{\sigma}{r} \right)^{6} \right\} \qquad (20\text{b})$$

and is drawn in Fig. 22.14. The two parameters are now ε, the depth of the well, and σ, the separation at which $V = 0$. The well minimum occurs at $r_e = 2^{1/6}\sigma$. Some typical values are listed in Table 22.5. Although the (12, 6)-potential has been used in many calculations, there is plenty of evidence to show that $1/r^{12}$ is a very poor representation of the repulsive potential, and that an exponential form $e^{-r/\sigma}$ is greatly superior. An exponential function is more faithful to the exponential decay of atomic wavefunctions at large distances, and hence to the overlap that is responsible for repulsion.

22.5 Molecular interactions in beams

A notable advance in the experimental study of intermolecular forces has come from the development of **molecular beams**, which consist of a narrow stream of molecules in an otherwise empty vessel. The beam is directed towards other molecules, and the scattering that occurs on impact is related to the intermolecular interactions.

The basic principles

The basic arrangement for a molecular beam experiment is shown in Fig. 22.15. The slotted disks make up the **velocity selector**. They rotate in the path of the beam and allow only those molecules having a certain speed to pass. There are also more sophisticated devices for generating molecules with a desired velocity. Among the most important are **supersonic nozzles**, in which the molecules stream out of the source with a very narrow velocity distribution and are then stripped down to a beam by passing through a skimmer shaped like an inverted cone (Fig. 22.16). Other types of selection are also possible: for example, electric fields may be used to deflect polar molecules and to obtain a beam of aligned molecules. The target gas may be either a bulk sample or another molecular beam. The latter **crossed beam** technique gives a lot of information because the states of both the target and projectile molecules may be controlled. The intensity of the incident beam is measured by the **incident beam flux** I, which is the number of particles per unit area per unit time.

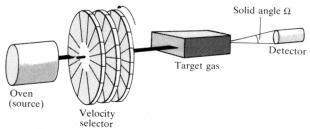

Fig. 22.15 The basic arrangement of a molecular beam apparatus. The atoms or molecules emerge from a heated source, and pass through the velocity selector, a train of rotating disks. The scattering occurs from the target gas (which might take the form of another beam), and the flux of particles entering the detector set at some angle is recorded.

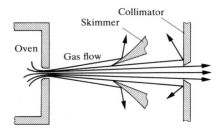

Fig. 22.16 A supersonic nozzle skims off some of the molecules of the beam and leads to a beam with well-defined velocity.

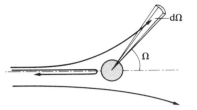

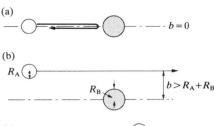

Fig. 22.17 The definition of the differential scattering cross-section dσ. The angles θ and φ are collectively denoted Ω, and the solid angle dΩ is equal to sin θ dθ dφ.

Fig. 22.18 The definition of the impact parameter b as the perpendicular separation of the initial paths of the particles.

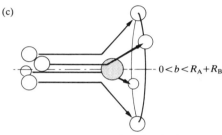

The detectors may consist of a chamber fitted with a sensitive pressure gauge or an ionization detector, in which the incoming molecule is first ionized and then detected electronically. The state of the scattered molecules may also be determined spectroscopically, and is of interest when the collisions change their vibrational or rotational states.

Fig. 22.19 The collisions of two hard spheres as the impact parameter changes. (a) $b = 0$, giving backward scattering; (b) $b > R_A + R_B$, giving forward scattering; (c) $0 < b < R_A + R_B$, leading to scattering into one direction on a ring of possibilities. (The shaded circle is taken to be so heavy that it remains virtually stationary.)

The experimental observations

The principal piece of experimental information is the fraction of the molecules in the incident beam that are scattered into a particular direction. The fraction is normally expressed in terms of dI, the number of molecules that are scattered per unit time into a cone that represents the area covered by the 'eye' of the detector (Fig. 22.17), and reported as the **differential scattering cross-section** d$σ$

$$d\sigma = \frac{dI}{I} \qquad (21)$$

The value of d$σ$ (which has the dimensions of area) depends on the **impact parameter** b (the initial perpendicular separation of the paths of the colliding molecules, Fig. 22.18), and the details of the intermolecular potential. The role of the impact parameter is most easily seen by considering the impact of two hard spheres (Fig. 22.19). If $b = 0$, the lighter projectile is on a trajectory that leads to a head-on collision, and so the only scattering intensity is detected when the detector is at $\theta = 180°$. If the impact parameter is so great that the spheres do not make contact $(b > R_A + R_B)$, there is no scattering and the scattering cross-section is zero at all angles. Glancing blows, with $0 < b < R_A + R_B$, lead to scattering intensity in cones around the forward direction (Fig. 22.19c).

Scattering effects

The scattering pattern of real molecules, which are not hard spheres, depends on the details of the intermolecular potential, including the anisotropy that is present when the molecules are non-spherical. The scattering also depends on the relative speed of approach of the two particles: a very fast particle might pass through the interaction region without much deflection, whereas a slower one on the same path might be temporarily captured and undergo considerable deflection (Fig. 22.20). The dependence of the scattering cross-section on the relative speed of approach should therefore give information about the strength and range of the intermolecular potential.

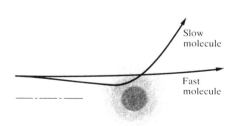

Fig. 22.20 The extent of scattering may depend on the relative speed of approach as well as the impact parameter. The dark zone represents the repulsive core; the fuzzy outer zone represents the long-range attractive potential.

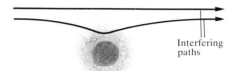

Fig. 22.21 Two paths leading to the same destination will interfere quantum mechanically; in this case they give rise to glory scattering in the forward direction.

A further point is that the outcome of collisions is determined by quantum, not classical, mechanics. The wave nature of the particles can be taken into account, at least to some extent, by drawing all classical trajectories that take the projectile particle from source to detector, and then considering the effects of interference between them.

Two quantum mechanical effects are of great importance. A particle with a certain impact parameter might approach the attractive region of the potential in such a way that it is deflected towards the repulsive core (Fig. 22.21), which then repels it out through the attractive region to continue its flight in the forward direction, just like particles with impact parameters so large that they are undeflected. The paths interfere, and the particle intensity in the forward direction is modified. The effect is called **glory scattering** because the same phenomenon accounts for the optical glory effect where a bright halo can sometimes be seen surrounding an illuminated object. (The coloured rings around the shadow of an aircraft cast on clouds by the sun, and often seen in flight, is an example of a glory.)

The second quantum effect we need consider is the observation of a strongly enhanced scattering in some non-forward direction. This effect is called **rainbow scattering** because the same mechanism accounts for the appearance of an optical rainbow. The origin of the phenomenon is illustrated in Fig. 22.22. As the impact parameter decreases, there comes a point where the scattering angle passes through a maximum and the interference between the paths results in a strongly scattered beam. The **rainbow angle** is the angle for which $d\theta/db = 0$ and the scattering is strong.

The detailed analysis of scattering data can be very complicated, but the main outcome should be clear: the intensity distribution of scattered particles can be related to the intermolecular potential, and a detailed picture can be built up of its radial and angular variation. One outcome is the value of C_6 in the van der Waals interaction and the testing of the Lennard–Jones potential (or some more elaborate version). It is found, for example, that although $1/r^6$ is quite a good representation of the attractive component of the potential, as remarked before the repulsive component is not all well described by $1/r^{12}$.

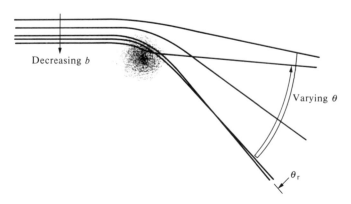

Fig. 22.22 The interference of paths leading to rainbow scattering. The rainbow angle θ_r is the maximum scattering angle as b is decreased, and the interference between the numerous paths at that angle modifies the scattering intensity markedly.

Gases and liquids

Intermolecular forces are responsible for the imperfections of real gases and for the formation of condensed phases of molecular compounds.

22.6 Gas imperfections

Real gases differ from perfect gases in their equations of state (Section 1.3), the existence of a non-zero Joule–Thomson coefficient (Section 3.2), and in various transport properties (which we leave until Chapter 24). The meeting place of theory and experiment is the virial equation of state

$$\frac{pV_m}{RT} = 1 + \frac{B}{V_m} + \frac{C}{V_m^2} + \ldots$$

where B is the second virial coefficient and C is the third virial coefficient; both coefficients depend on the temperature. The virial coefficients are consequences of the intermolecular forces, and so we have a route to the latter's measurement.

The virial coefficients and the potential

The pressure of a sample is related to the canonical partition function Q by eqn 3 of Section 20.1:

$$p = kT\left(\frac{\partial \ln Q}{\partial V}\right)_T$$

We saw in Section 20.1 that in the absence of interactions, the canonical partition function is related to the molecular partition function $q = V/\Lambda^3$ by $Q = q^N/N!$, so

$$Q = \frac{(V^N/N!)}{\Lambda^{3N}} \quad \text{where} \quad \Lambda^2 = \frac{h^2\beta}{2\pi m} \quad \text{and} \quad \beta = \frac{1}{kT}$$

This form of the partition function leads to

$$p = \frac{RT}{V_m}$$

corresponding to $B, C, \ldots$ all zero.

The total kinetic energy of a gas of molecules is the sum of their individual kinetic energies even in the presence of interactions between them. Therefore, even in a real gas the canonical partition function factorizes into a part arising from the kinetic energy, which is the same as for the perfect gas, and a new factor Z that depends on the intermolecular potentials. We shall therefore write

$$Q = \frac{Z}{\Lambda^{3N}}$$

Z is the **configuration integral**. For a perfect gas we expect

$$Z = \frac{V^N}{N!}$$

for then Q is the partition function of a collection of non-interacting particles.

For a real gas, Z is related to the total potential energy V_N of interaction of all the particles by

$$Z = \frac{1}{N!} \int e^{-\beta V_N} \, d\boldsymbol{r}_1 \, d\boldsymbol{r}_2 \ldots d\boldsymbol{r}_N$$

We can quickly verify that this expression gives the perfect gas partition function when there is no interaction, for then

$$V_N = 0 \quad \text{and} \quad e^{-\beta V_N} = 1$$

and

$$Z = \frac{1}{N!} \int d\boldsymbol{r}_1 \, d\boldsymbol{r}_2 \ldots d\boldsymbol{r}_N = \frac{V^N}{N!}$$

because

$$\int d\boldsymbol{r} = V$$

Hence

$$Q = \frac{1}{N!} \left(\frac{V}{\Lambda^3} \right)^N$$

the form it has for a perfect gas.

Pairwise interactions

When we consider only pairwise interactions (e.g. dispersion interactions between pairs of particles) and confine attention to low densities the configuration integral simplifies to

$$Z_2 = \tfrac{1}{2} \int e^{-\beta V_2} \, d\boldsymbol{r}_1 \, d\boldsymbol{r}_2 \tag{22a}$$

The second virial coefficient, which is a consequence of these pairwise interactions, then turns out to be

$$B = \frac{-N_A}{2V} \int f \, d\boldsymbol{r}_1 \, d\boldsymbol{r}_2 \qquad f = e^{-\beta V_2} - 1 \tag{22b}$$

f is the **Mayer f-function**: it goes to zero when the two particles are so far apart that $V_2 = 0$. When the intermolecular potential depends only on the separation r of the particles, as in the dispersion interaction of closed-shell atoms and spherical molecules, eqn 22b simplifies to

$$B = -2\pi N_A \int_0^\infty f r^2 \, dr \tag{22c}$$

The integral can be evaluated (usually numerically) by substituting one of the intermolecular potential expression discussed earlier. One reason why the Lennard–Jones potential has been so widely used is that eqn 22c may be evaluated analytically (but the result is quite complicated).

As an example, we consider the hard-sphere potential. Since

$$e^{-\beta V_2} = 0 \qquad f = -1 \quad \text{when} \quad r < \sigma$$

$$e^{-\beta V_2} = 1 \qquad f = 0 \quad \text{when} \quad r > \sigma$$

so the second virial coefficient is

$$B = 2\pi N_A \int_0^{\sigma} r^2 \, dr = \tfrac{2}{3}\pi N_A \sigma^3 \qquad (22d)$$

Long ago, in Section 1.4, we saw that the part of the van der Waals equation of state that represented the hard-sphere repulsive part of the intermolecular potential (i.e. b) contributed to B in exactly this way. This result raises the question as to whether a potential can be found which, when the virial coefficients are evaluated, gives the full van der Waals equation of state. Such a potential can be found: it consists of a hard-sphere repulsive core and a long-range, shallow attractive region (Problem 22.14).

Example 22.6: *Evaluating a virial coefficient*

Take the Lennard–Jones potential for Ar in the form of eqn 20b with $\varepsilon/k = 120$ K and $\sigma = 340$ pm, and evaluate the second virial coefficient at 298 K.

Answer. Evaluate the integral in eqn 22d numerically (the technical term is 'numerical quadrature'). One elementary approach is to use Simpson's rule (Fig. 22.23):

(1) Consider the range of integration as being covered by an even number of strips of width h.
(2) Form the sum

$$\tfrac{1}{3}h\{y_0 + 4(y_1 + y_3 + \ldots + y_{m-1}) + 2(y_2 + y_4 + \ldots + y_{m-2}) + y_m\}$$

Since the integrand rises from zero up to a maximum, then crosses through zero at 340 pm, and has a long tail out to several 1000 pm, it is sensible to do the integration in two ranges, one from $r = 0$ to 340 pm; the second from 340 pm out to about 5000 pm. The integration is readily programmed for a microcomputer. Integration in the first range gives

$$\frac{B}{2\pi N_A} = 1.130 \times 10^7 \text{ pm}^3$$

and in the second range

$$\frac{B}{2\pi N_A} = -0.516 \times 10^7 \text{ pm}^3$$

The complete integral is therefore

$$B = 0.613 \times 10^7 \text{ pm}^3 \times 2\pi N_A = 23.2 \text{ cm}^3 \text{ mol}^{-1}$$

Comment. The two domains of integration, and the opposite signs of their contributions to B, reflect regions where the repulsions and attractions are dominant respectively.

Exercise. Repeat the determination for 273 K. $\qquad [+22.0 \text{ cm}^3 \text{ mol}^{-1}]$

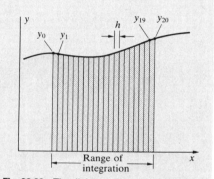

Fig. 22.23 The division of the region of integration into strips that is used for the application of Simpson's rule for numerical integration of an arbitrary function.

Once the second virial coefficient has been calculated, we can calculate the isothermal Joule–Thomson coefficient μ_T (Section 3.1) from the thermodynamic relation

$$\lim_{p \to 0} \mu_T = B - T\frac{dB}{dT}$$

and from the result calculate the Joule–Thomson coefficient itself using eqn 10 of Section 3.1.

The third virial coefficient C arises in part from three-particle contributions to the potential energy. We have seen that these are the sum of three pair-wise interactions and a correction term (e.g. the Axilrod–Teller contribution). Therefore, C consists of two parts, one arising from the pairwise terms, and another from the correction. By exploring the values of C (e.g. from p, V, T measurements or Joule–Thomson coefficient determinations and theoretical calculations), the non-additivity can be investigated.

22.7 The structures of liquids

The starting point for the discussion of solids is the well-ordered structure of perfect crystals. The starting point for the discussion of gases is the totally chaotic distribution of the molecules of a perfect gas. With liquids we are between these two extremes: there is some structure and some chaos.

The pair distribution function

The particles of a liquid are held together by intermolecular forces, but their kinetic energies are comparable to their potential energies. As a result, the whole structure is very mobile. The best description of the average locations of the particles comes from the radial distribution functions. The **pair distribution function** g is the most important, and we shall concentrate on it. The pair distribution function is defined so that $g(r)dr$ is the probability that a molecule will be found in the range dr at a distance r from another.

In a crystal, g is a periodic array of sharp spikes, representing the certainty (in the absence of defects and thermal motion) that particles lie at definite locations. This regularity continues out to large distances, and so we say that crystals have **long-range order**. When the crystal melts the long-range order is lost, and wherever we look at long distances from a given particle, there is equal probability of finding a second particle. Close to the first particle, though, there may be a remnant of order. Its nearest neighbours might still adopt approximately their original positions, and even if they are displaced by newcomers, the new particles might adopt their vacated positions. It may still be possible to detect a sphere of nearest

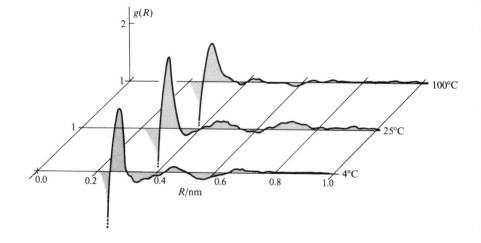

Fig. 22.24 The radial distribution function of the oxygen atoms in liquid water at three temperatures. Note the expansion as the temperature is raised. (A. H. Narten, M. D. Danford, and H. A. Levy, *Discuss. Faraday. Soc.* **43**, 97 (1967).)

neighbours at a distance r_1, and perhaps beyond them a sphere of next-nearest neighbours at r_2. The existence of this short-range order means that the PDF can be expected to oscillate close to $r = 0$, with a peak at r_1, a smaller peak at r_2, and perhaps some more structure beyond that.

The shape of the pair distribution function can be determined by X-ray diffraction, for g can be extracted from the diffuse diffraction pattern characteristic of liquid samples in much the same way as a crystal structure is obtained from X-ray diffraction of crystals. The shells of local structure shown in the example in Fig. 22.24 (for water) are unmistakable. Closer analysis shows that any given H_2O molecule is surrounded by other molecules at the corners of a tetrahedron, similar to the arrangement in ice (Fig. 21.19). The form of g at 100°C shows that the intermolecular forces (in this case, largely hydrogen bonds) are strong enough to affect the local structure right up to the boiling point.

The calculation of g

Since the pair distribution function can be calculated by making assumptions about the intermolecular forces, it can be used to test theories of liquid structure. However, even a fluid of hard spheres without attractive interactions (a box of ball-bearings) gives a function that oscillates near the origin (Fig. 22.25), and one of the factors influencing, and sometimes dominating, the structure of a liquid is the geometrical problem of stacking together reasonably hard spheres. The attractive part of the potential modifies this basic structure, but sometimes only quite weakly, by gathering and trapping particles into each other's vicinity. One of the reasons behind the difficulty of describing liquids theoretically is the importance of both the attractive and repulsive, hard-core, parts of the potential.

There are several ways of building the intermolecular potential into the calculation of g. Numerical methods take a box of between 100 and 1000 particles (the number increases as computers grow more powerful), and the rest of the liquid is simulated by surrounding the box with replications (Fig. 22.26). Then, whenever a particle leaves through a face, its image arrives through the opposite face so that the total number remains constant.

Monte Carlo methods

In the **Monte Carlo method** the particles in the box are moved by a random amount, and the total potential energy is calculated using one of the intermolecular potentials discussed earlier (including the important three-body terms). Whether or not this new configuration is accepted is then judged from the following criteria:

(1) if the potential energy is not greater than before the change, the configuration is accepted;
(2) if it is greater, then it is accepted in proportion to the value of $e^{-\beta \Delta V_N}$, where ΔV_N is the change in total potential energy of the N particles in the box.

This procedure ensures that the probability of occurrence of any configuration is proportional to the Boltzmann factor $e^{-\beta V_N}$. The configurations generated in this way can then be analysed for the value of g simply by counting the number of particles at a distance r from a selected one, and averaging the result over the whole collection.

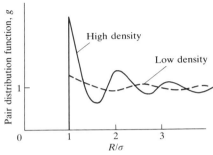

Fig. 22.25 The radial distribution function for a simulation of a liquid using impenetrable hard spheres (ball bearings).

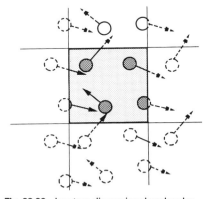

Fig. 22.26 In a two-dimensional molecular dynamics simulation that uses 'periodic boundary conditions', when one particle leaves the cell its mirror image enters through the opposite face.

Molecular dynamics

In the molecular dynamics approach, the history of an initial arrangement is followed by calculating the trajectories of all the particles under the influence of the intermolecular potentials. Newton's laws are used to predict where each particle will be after a short time interval (about 10^{-14} s, which is shorter than the time between collisions), and then the calculation is repeated for tens of thousands of such steps. The procedure gives a series of snapshots of the liquid and g can be calculated as before.

Once g is known, it can be used to calculate the thermodynamic properties of liquids. For example, the contribution of the pairwise additive intermolecular potential to the internal energy is given by the integral

$$U = \frac{2\pi N^2}{V} \int_0^\infty g V_2 r^2 \, \mathrm{d}r$$

That is, it is the average two-particle potential energy weighted by g, the probability that the pair of particles have each separation r. Likewise, the pressure is given by

$$\frac{pV}{nRT} = 1 - \frac{2\pi N}{VkT} \int_0^\infty g\left\{r\left(\frac{\mathrm{d}V_2}{\mathrm{d}r}\right)\right\} r^2 \, \mathrm{d}r \tag{23}$$

The quantity $r(\mathrm{d}V/\mathrm{d}r)$ is called the **virial** (hence the term 'virial equation of state').

22.8 Liquid crystals

A feature that makes calculations even more difficult is the possibility that molecules have strongly anisotropic interactions. An important extreme case of anisotropy gives rise to a **mesophase**, a phase intermediate between solid and liquid.

A mesophase may arise when molecules have highly anisotropic shapes, such as being long and thin, as in (**4**), or disk-like. When the solid melts, some aspects of the long-range order characteristic of the solid may be retained, and the new phase may be a **liquid crystal**, a substance having liquid-like imperfect long-range order but some crystal-like aspects of short-range order. One type of retained long-range order gives rise to a **smectic phase** (from the Greek word for soapy), in which the molecules align themselves in layers (Fig. 22.27a). Other materials, and some smectic liquid crystals at higher temperatures, lack the layered structure but retain a parallel alignment (Fig. 22.27b): this mesophase is the **nematic phase** (from the Greek for thread). In the **cholesteric phase** (from the Greek for bile solid) the molecules lie in sheets at angles that change slightly between each sheet. As a result, they form helical structures with a pitch that depends on the temperature. These cholesteric liquid crystals diffract light, and have colours that depend on the temperature. The strongly anisotropic optical properties of nematic liquid crystals, and their response to electric fields, is the basis of their use as data displays in calculators and watches.

22.9 Molecular motion in liquids

The motion of molecules in liquids can be studied theoretically using a molecular dynamics simulation, because the detailed information it gives

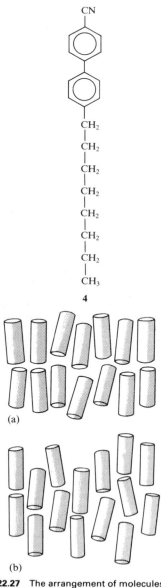

Fig. 22.27 The arrangement of molecules in (a) the smectic phase and (b) the nematic phase of a liquid crystal. They both have elements of order, but it is different in different directions.

includes the velocities of the particles. Their motion may be studied experimentally by a variety of methods. Relaxation time measurements in NMR and ESR (Section 18.5) can be interpreted in terms of the mobilities of the particles, and have been used to show that some types of molecule rotate in a series of small (about 5°) steps, but that others jump through about 1 radian (57°) in each step. Another important technique is **inelastic neutron scattering**, in which the energy neutrons collect or discard as they pass through a sample is interpreted in terms of the motion of its particles, and very detailed information can be obtained.

More mundane than these experiments are viscosity measurements (some values are given in Table 22.6). In order to move, a particle in a liquid must escape from its neighbours, and so it needs a minimum energy. Since the probability that a molecule has an energy E_a is proportional to $e^{-E_a/RT}$, the mobility of the molecules in the liquid should follow this Boltzmann temperature dependence. Since the viscosity η is inversely proportional to the mobility of the particles, we should expect that

$$\eta \propto e^{E_a/RT}$$

implying that the viscosity should decrease exponentially with increasing temperature. This is found to be the case, at least over reasonably small temperature ranges (Fig. 22.28). One problem is that the density of the liquid changes as it is heated, and makes a pronounced contribution to the viscosity. Thus, the temperature dependence of viscosity at constant volume, when the density is constant, is much less than that at constant pressure. The intermolecular potentials govern the magnitude of E_a, but the problem of calculating it is immensely difficult and still largely unsolved.

Magnetic properties

The magnetic and electric properties of molecules are analogous. For instance, some molecules possess permanent magnetic dipole moments, and an applied magnetic field can induce a magnetic moment.

22.10 Magnetic susceptibility

The analogue of the electric polarization P is the **magnetization** M, the magnetic dipole moment per unit volume. The magnetization in the presence of a field of strength H is proportional to H and we write

$$M = \chi H$$

where χ is the dimensionless **volume magnetic susceptibility**. A closely related quantity is the **molar magnetic susceptibility** χ_m:

$$\chi_m = \chi \times V_m \tag{24}$$

where V_m is the molar volume of the substance (we shall soon see why it is sensible to introduce this quantity). The **magnetic flux density** B is related to the applied field strength and the magnetization by

$$B = \mu_0(H + M) = \mu_0(1 + \chi)H$$

μ_0 is the **vacuum permeability**:

$$\mu_0 = 4\pi \times 10^{-7}\,\text{J}\,\text{C}^{-2}\,\text{m}^{-1}\,\text{s}^2$$

Table 22.6. Viscosities of liquids at 298 K, $\eta/(10^{-3}\,\text{kg}\,\text{m}^{-1}\,\text{s}^{-1})$

Benzene	0.601
Mercury	1.53
Pentane	0.224
Water	0.891

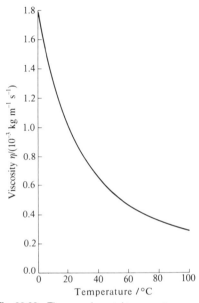

Fig. 22.28 The experimental temperature dependence of the viscosity of water. As the temperature is increased, more molecules are able to escape from the potential wells provided by their neighbours, and so the liquid becomes more fluid.

B can be thought of as the density of magnetic lines of force permeating the medium. This density is increased if M adds to H (when $\chi > 0$), but is decreased if M opposes H (when $\chi < 0$). Materials for which χ is positive are called **paramagnetic**; those for which χ is negative are called **diamagnetic**.

The molecules might possess permanent magnetic dipole moments of magnitude m, which contribute to the magnetization an amount proportional to $m^2/3kT$, as in the electric case (eqn 5). An applied field can also induce a magnetic moment to an extent determined by the **magnetizability** ξ of the molecules, and the magnetic analogue of eqn 5 is

$$\chi = \mathcal{N}\mu_0\left(\xi + \frac{m^2}{3kT}\right) \tag{25a}$$

We can now see why it is convenient to introduce χ_m, for the product of the number density $\mathcal{N}$ and the molar volume is Avogadro's constant N_A:

$$\mathcal{N} \times V_m = \frac{N}{V} \times V_m = \frac{nN_A}{nV_m} \times V_m = N_A$$

Hence

$$\chi_m = N_A\mu_0\left(\xi + \frac{m^2}{3kT}\right) \tag{25b}$$

and the density dependence of the susceptibility has been eliminated. The expression for χ_m is in agreement with the empirical **Curie Law**,

$$\chi_m = A + \frac{B}{T} \tag{26}$$

with $A = N_A\mu_0\xi$ and $B = N_A\mu_0 m^2/3k$.

The magnetic susceptibility may be measured with a **Gouy balance**, consisting of a sensitive balance from which the sample hangs in the form of a narrow cylinder (Fig. 22.29) and lies between the poles of a magnet. If the sample is paramagnetic, it is drawn into the field, and its apparent weight is greater than when the field is off. A diamagnetic sample tends to be expelled from the field and appears to weigh less when the field is turned on. The balance is normally calibrated against a sample of known susceptibility. Some experimental values obtained in this way are listed in Table 22.7; a typical paramagnetic volume susceptibility is about 10^{-3}, and a typical diamagnetic volume susceptibility is about $(-)10^{-5}$. The permanent magnetic moment can be extracted from susceptibility measurements by plotting χ against $1/T$.

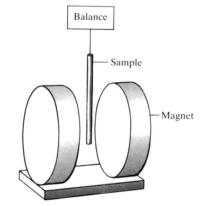

Fig. 22.29 The arrangement of the Gouy balance for measuring magnetic susceptibilities. A paramagnetic sample appears to weigh more and a diamagnetic sample appears to weigh less, when the magnetic field is on.

Table 22.7. Magnetic susceptibilities at 298 K

	$\chi/10^{-6}$	$\chi_m/(10^{-5}\,\text{cm}^3\,\text{mol}^{-1})$
$H_2O(l)$	-90	-160
$NaCl(s)$	-13.9	-38
$Cu(s)$	-9.6	-6.8
$CuSO_4 \cdot 5H_2O(s)$	$+176$	$+1920$

22.11 The permanent magnetic moment

The permanent magnetic moment of a molecule arises from any unpaired electron spins in the molecule. We saw in Section 13.9 that the magnetic moment of an electron is proportional to its spin, and the square of the magnitude of the magnetic moment is proportional to the square of the magnitude of the angular momentum, $s(s + 1)\hbar^2$:

$$m^2 = g_e^2\gamma_e^2\{s(s + 1)\hbar^2\}$$

g_e is the electron's g-factor ($g_e = 2.0023$) and

$$\gamma_e = -\frac{e}{2m_e}$$

In terms of the Bohr magneton

$$\mu_B = \frac{e\hbar}{2m_e}$$

$$m^2 = g_e^2 s(s+1)\mu_B^2$$

If there are several electron spins in each molecule they combine to a total spin S, and then $s(s+1)$ should be replaced by $S(S+1)$. It follows that the spin contribution to the magnetic susceptibility is

$$\chi = \frac{Ng_e^2 S(S+1)\mu_0\mu_B^2}{3kT} \tag{27}$$

This expression shows that χ is positive, and so the spin magnetic moments contribute to the paramagnetic susceptibilities of materials. The contribution decreases with increasing temperature because the thermal motion randomizes the spin orientations. In practice a contribution to the paramagnetism also arises from the orbital angular momenta of electrons: we have discussed the spin-only contribution.

Example 22.7: *Calculating the magnetic susceptibility*

Calculate the volume and molar paramagnetic susceptibilities of a sample of a complex salt with three unpaired electrons at 298 K, given that its density is 3.24 g cm^{-3} and its molar mass is 200 g mol^{-1}.

Answer. We use eqn 27 and convert to molar susceptibility by multiplying by the molar volume using $V_m = M/\rho$, which gives

$$\chi_m = \frac{N_A g_e^2 \mu_0 \mu_B^2 S(S+1)}{3kT}$$

A convenient quantity is

$$\frac{N_A g_e^2 \mu_0 \mu_B^2}{3k} = 6.3001 \text{ cm}^3 \text{ K}^{-1} \text{ mol}^{-1}$$

Consequently,

$$\chi_m = 6.3001 \times \frac{S(S+1)}{(T/K)} \text{ cm}^3 \text{ mol}^{-1}$$

and the numerical values can be obtained by substitution using $S = \frac{3}{2}$. Substituting the data gives

$$\chi_m = \frac{6.3001 \times (15/4)}{298} \text{ cm}^3 \text{ mol}^{-1} = 7.9 \times 10^{-2} \text{ cm}^3 \text{ mol}^{-1}$$

Then, from $V_m = 61.7 \text{ cm}^3 \text{ mol}^{-1}$, $\chi = 1.3 \times 10^{-3}$.

Comment. Note that χ is dimensionless and (in this case) positive, indicating paramagnetism. Note also that the density is not needed for the calculation of χ_m.

Exercise. Repeat the calculation for a complex with five unpaired spins, a density of 2.87 g cm^{-3}, and a molar mass of 322.4 g mol^{-1} at 273 K.

$$[\chi_m = 0.20 \text{ cm}^3 \text{ mol}^{-1}, \chi = 1.79 \times 10^{-3}]$$

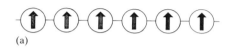

(a)

(b)

Fig. 22.30 In (a) a ferromagnetic material electron spins are locked into a parallel alignment over large domains; in (b) an antiferromagnetic material the electron spins are locked into an antiparallel arrangement.

At low temperatures some paramagnetic solids make a phase transition to a state in which large domains of spins align with parallel orientations. This cooperative alignment gives rise to a very strong magnetization and is called **ferromagnetism** (Fig. 22.30a). In other cases the cooperative effect leads to alternating spin orientations, and so they are locked into a low-magnetization arrangement (Fig. 22.30b) to give an **antiferromagnetic phase**. The ferromagnetic phase has a non-zero magnetization in the absence of an applied field, but the antiferromagnetic phase has a zero magnetization because the spin magnetic moments cancel. The ferromagnetic transition occurs at the **Curie temperature**, and the antiferromagnetic transition occurs at the **Néel temperature**.

22.12 Induced magnetic moments

Whereas an electric field polarizes a molecule by stretching it, a magnetic field magnetizes a molecule by twisting it. More precisely, an applied magnetic field induces the circulation of electronic currents. These currents give rise to a magnetic field which usually opposes the applied field, so the susceptibility is diamagnetic. In a few cases the induced field augments the applied field, and the susceptibility is then paramagnetic.

The great majority of molecules with no unpaired electron spins are diamagnetic. This is because the diamagnetic flow occurs within the orbitals of the molecule that are occupied in its ground state, whereas an orbital contribution to the paramagnetic susceptibility depends on currents being stimulated by forcing the electrons to pass through unoccupied, higher-energy orbitals. When orbital paramagnetism does occur it can be distinguished from spin paramagnetism by the fact that it is temperature independent: this is why it is called **temperature-independent paramagnetism** (TIP).

We can summarize these remarks as follows. All molecules have a diamagnetic component to their susceptibility, but it is dominated by spin paramagnetism if the molecules have unpaired electrons. In a few cases (where there are low-lying excited states) TIP is strong enough to make the molecules paramagnetic even though all their electrons are paired.

Further information: the Langevin function

When a field $\mathscr{E}$ is present, the energy of a molecule that makes an angle θ to the field is

$$E = -\mu \mathscr{E} \cos \theta \qquad (A1)$$

Some molecular orientations have lower energy than others, and the mean dipole moment $\langle \mu \rangle$ (and therefore the polarization) depends on a competition between the aligning influence of the field and the randomizing influence of thermal motion. The net moment of a fluid sample is evaluated using the Boltzmann distribution (Section 19.1, see below) for a sample at a temperature T, and is

$$\langle \mu \rangle = \mu \mathscr{L} \qquad \mathscr{L} = \frac{e^x + e^{-x}}{e^x - e^{-x}} - \frac{1}{x} \quad \text{with} \quad x = \frac{\mu \mathscr{E}}{kT} \qquad (A2)$$

$\mathscr{L}$ is called the **Langevin function**.

Under most circumstances x is very small (e.g. if $\mu = 1\,\mathrm{D}$ and $T = 300\,\mathrm{K}$, x exceeds 0.01 only if the field strength exceeds $100\,\mathrm{kV\,cm^{-1}}$, and most measurements are done at much lower strengths). When $x \ll 1$ the exponentials in the Langevin function can be expanded, and the largest term that survives is

$$\mathscr{L} = \tfrac{1}{3}x + \ldots$$

Therefore, the average molecular dipole moment is

$$\langle \mu \rangle = \frac{\mu^2 \mathscr{E}}{3kT}$$

If the number density of molecules (the number per unit volume) is $\mathscr{N}$, the mean dipole moment per unit volume in a field $\mathscr{E}$ is

$$P = \mathscr{N}\langle \mu \rangle = \frac{\mathscr{N}\mu^2 \mathscr{E}}{3kT} \tag{A3}$$

The Langevin function is derived by considering a molecule of dipole moment μ that makes an angle θ to the z axis (the direction of the field). This molecule gives a contribution $\mu \cos \theta$ to the average dipole moment of the sample. The probability that a dipole has the orientation θ is proportional to

$$P = \mathrm{e}^{-E/kT} = \mathrm{e}^{x \cos \theta} \qquad x = \frac{\mu \mathscr{E}}{kT}$$

where E is the energy at that orientation (eqn A1). Since the volume element is $\sin \theta \, \mathrm{d}\theta$, with θ ranging from 0 to π, the average value of $\mu \cos \theta$ is

$$\langle \mu \rangle = \frac{\int_0^\pi \mu \cos \theta P \sin \theta \, \mathrm{d}\theta}{\int_0^\pi P \sin \theta \, \mathrm{d}\theta} = \frac{\mu \int_{-1}^1 \cos \theta \, P \, \mathrm{d}\cos \theta}{\int_{-1}^1 P \, \mathrm{d}\cos \theta}$$

$$= \frac{\mu \int_{-1}^1 y P \, \mathrm{d}y}{\int_{-1}^1 P \, \mathrm{d}y}$$

The integral in the denominator is

$$\int_{-1}^1 P \, \mathrm{d}y = \int_{-1}^1 \mathrm{e}^{xy} \, \mathrm{d}y = \frac{\mathrm{e}^x - \mathrm{e}^{-x}}{x}$$

The integral in the numerator is similarly

$$\int_{-1}^1 y\mathrm{e}^{xy} \, \mathrm{d}y = \frac{\mathrm{e}^x + \mathrm{e}^{-x}}{x} - \frac{\mathrm{e}^x - \mathrm{e}^{-x}}{x^2}$$

It is now straightforward algebra to combine these two results and to obtain eqn A2.

Further reading

Dipole moments and polarizabilities

C. P. Smyth, Determination of dipole moments. In *Techniques of chemistry* (ed. A. Weissberger and B. W. Rossiter), IV, 397, Wiley-Interscience, New York (1972).

W. E. Vaughan, C. P. Smyth, and J. C. Powles, Determination of dielectric constant and loss. In *Techniques of chemistry* (ed. A. Weissberger and B. W. Rossiter), IV, 351, Wiley-Interscience, New York (1972).

A. I. McClellan, *Tables of experimental dipole moments*. W. H. Freeman & Co., San Francisco (1963).

Intermolecular forces

M. Rigby, E. B. Smith, W. A. Wakeham, and G. C. Maitland, *The forces between molecules*. Oxford University Press (1986).

G. C. Maitland, M. Rigby, E. B. Smith, and W. A. Wakeham, *Intermolecular forces: their origin and determination*. Clarendon Press, Oxford (1981).

M. A. D. Fluendy and K. P. Lawley, *Molecular beams in chemistry*. Chapman and Hall, London (1973).

U. Buck, Elastic scattering. In *Adv. chem. Phys.* **30**, 313 (1975).

Liquids

J. S. Rowlinson, The structure of liquids. In *Essays in chemistry* (ed. J. N. Bradley, R. D. Gillard, and R. F. Hudson), **1**, 1 (1970).

J. S. Rowlinson and F. L. Swinton, *Liquids and liquid mixtures*. Butterworths, London (1982).

J. N. Murrell and E. A. Boucher, *Properties of liquids and solutions*. Wiley-Interscience, New York (1982).

M. P. Allen and D. Tildesley, *Computer simulation of liquids*. Clarendon Press, Oxford (1986).

Magnetism

A. Earnshaw, *Introduction to magnetochemistry*. Academic Press, New York (1968).

W. E. Hatfield, Magnetic measurements. In *Solid-state chemistry: Techniques* (ed. A. K. Cheetham and P. Day), Clarendon Press, Oxford (1987).

L. N. Mulay, Instrumentation and techniques for measuring magnetic susceptibility. In *Techniques of chemistry* (ed. A. Weissberger and B. W. Rossiter) IV, 431, Wiley-Interscience, New York (1972).

Exercises

22.1 Calculate the capacitance of a condenser when the space between the plates is filled with a substance of relative permittivity 35.5. When the space is a vacuum, the capacitance is 6.2 pF.

22.2 The molar polarization of fluorobenzene vapour is proportional to T^{-1}, and is 70.62 cm^3 mol^{-1} at 351.0 K and 62.47 cm^3 mol^{-1} at 423.2 K. Calculate the polarizability and dipole moment of the molecule.

22.3 At 0°C, the molar polarization of liquid ClF$_3$ is 27.18 cm^3 mol^{-1} and its density is 1.89 g cm^{-3}. Calculate the relative permittivity of the liquid.

22.4 The ClF$_3$ molecule has five electron pairs around the central Cl atom, and so may have either three equatorial F atoms or two axial and one equatorial F atoms. Since the molecule is polar, which structure does it have?

22.5 Use the tabulated values of molar refractivity for 589 nm light to estimate the refractive index of diethyl ether. The density of the liquid is 0.715 g cm^{-3} at 20°C.

22.6 The refractive index of CH$_2$I$_2$ is 1.732 for 656 nm light. Its density at 20°C is 3.32 g cm^{-3}. Calculate the polarizability of the molecule at this wavelength.

22.7 The electric dipole moment of toluene (methylbenzene) is 0.4 D. Estimate the dipole moments of the three xylenes (dimethylbenzene). Which answer can you be sure about?

22.8 Calculate the resultant of two dipole moments of magnitude 1.5 D and 0.80 D that make an angle 109.5° to each other.

22.9 The polarizability volume of H_2O is 1.48×10^{-24} cm^3; calculate the dipole moment of the molecule (in addition to the permanent dipole moment) induced by an applied electric field of strength 1.0 kV cm^{-1}.

22.10 The polarizability volume of H_2O at optical frequencies is 1.5×10^{-24} cm^3: estimate the refractive index of water. The experimental value is 1.33; what may be the origin of the discrepancy?

22.11 The dipole moment of chlorobenzene is 1.57 D and its polarizability volume is 1.23×10^{-23} cm^3. Estimate its relative permittivity at 25°C, when its density is 1.173 g cm^{-3}.

22.12 Estimate the refractive index of crystals of $CaCl_2$,

NaCl, and solid Ar from the data in Table 22.3. The densities of the solids are 2.15, 2.16, and 1.42 g cm^{-3} respectively.

22.13 The Lennard-Jones $(12, 6)$ potential gives the potential energy of interaction between molecules. Given that the force is the negative slope of the potential, calculate the distance-dependence of the force acting between the molecules. What is the separation at which the force is zero?

22.14 The magnetic moment of $CrCl_3$ is $3.81\mu_B$. How many unpaired electrons does the Cr possess?

22.15 Calculate the molar susceptibility of benzene given that its volume susceptibility is -7.2×10^{-7} and its density 0.879 g cm^{-3} at 25°C.

22.16 Estimate the spin-only molar susceptibility of $CuSO_4 \cdot 5H_2O$ at 25°C.

Problems

Numerical problems

22.1 Allow an H_2O molecule ($\mu = 1.85$ D) to approach an anion. What is the favourable orientation of the molecule? Calculate the electric field (in V m^{-1}) experienced by the anion when the water dipole is (a) 1.0 nm, (b) 0.3 nm, (c) 30 nm from the ion.

22.2 An H_2O molecule is aligned by an external electric field of strength 1.0 kV m^{-1} and an Na^+ ion is brought up slowly from one side. At what separation is it energetically favourable for the H_2O molecule to flip over and point towards the approaching ion?

22.3 The relative permittivities of gaseous hydrogen halides fit the expression $\varepsilon_r = 1 + \Delta/\tau$ between 0°C and 300°C, the values of Δ being given below and $\tau = T/(273 \text{ K})$. Calculate the dipole moments and the polarizability volumes of the molecules.

$\theta/°C$	0	100	200	300
1000Δ: HCl	4.3	3.5	3.0	2.6
HBr	3.1	2.6	2.3	2.1
HI	2.3	2.2	2.1	2.1

The molar volume varies as $V_m = \tau V_m^{\circ}$ with $V_m^{\circ}(273.15 \text{ K}) = 2.24 \times 10^4$ cm^3 mol^{-1}.

22.4 The relative permittivity of chloroform was measured over a range of temperatures with the following results:

$\theta/°C$	−80	−70	−60	−40	−20	0	20
ε_r	3.1	3.1	7.0	6.5	6.0	5.5	5.0
$\rho/(\text{g cm}^{-3})$	1.65	1.64	1.64	1.61	1.57	1.53	1.50

The freezing point of chloroform is −64°C. Account for these results and calculate the dipole moment and polarizability volume of the molecule.

22.5 The relative permittivities of methanol (m.p. −95°C) corrected for density variation are given below. What mole-

cular information can be deduced from these values? Take $\rho = 0.791$ g cm^{-3} at 20°C.

$\theta/°C$	−185	−170	−150	−140	−110	−80	−50	−20	0	20
ε_r	3.2	3.6	4.0	5.1	67	57	49	42	38	34

22.6 In his classic book *Polar molecules*, Debye reports some early measurements of the polarizability of ammonia. From the selection below, find the dipole moment and the polarizability volume of the molecule.

T/K	292.2	309.0	333.0	387.0	413.0	446.0
$P_m/(\text{cm}^3 \text{ mol}^{-1})$	57.57	55.01	51.22	44.99	42.51	39.59

The refractive index of ammonia at 273 K and 1 atm is 1.000 379 (for yellow sodium light). Calculate the molar polarizability of the gas at this temperature and at 292.2 K. Combine the value calculated with the static molar polarizability at 292.2 K and deduce from this information alone the molecular dipole moment.

Theoretical problems

22.7 Calculate the potential energy of the interaction between two linear quadrupoles when they are collinear and separated by a distance r.

22.8 Show that in a gas (for which the refractive index is close to 1), the refractive index depends on the pressure as

$$n_r = 1 + \text{const} \times p$$

and find the constant of proportionality. Go on to show how to deduce the polarizability volume of a molecule from measurements of the refractive index of a gaseous sample.

22.9 The refractive index of benzene is constant (at 1.51) from 0.4 GHz up to 0.55 GHz (in the microwave region of the spectrum), but then shows a series of oscillations between

1.47 and 1.54. Throughout the same frequency range, methylbenzene shows a higher refractive index (about 1.55), the same oscillations as in benzene, and additional oscillations between 1.52 and 1.56 near 0.4 GHz. Account for these observations.

22.10 Acetic acid vapour contains a proportion of planar, hydrogen bonded dimers. The relative permittivity of pure liquid acetic acid is 7.14 at 290 K and increases with increasing temperature. Suggest an interpretation of the latter observation. What effect should isothermal dilution have on the relative permittivities of solutions of acetic acid in benzene?

22.11 Show that the mean interaction energy of N atoms of diameter d interacting with a potential energy of the form C_6/R^6 is given by $U = -2\pi N^2 C_6/3Vd^3$, where V is the volume in which the molecules are confined and all effects of clustering are ignored. Hence, find a connection between the van der Waals parameter a and C_6 from $n^2a/V^2 = (\partial U/\partial V)_T$.

22.12 The relation between the second virial coefficient B in terms of the intermolecular potential energy V is given by eqn 22. Suppose that the atoms of a gas have a distance of closest approach d and that outside this range they are attracted together by a $-C_6/r^6$ potential. Suppose also that when the atoms are not in contact the term V/kT is so small that the exponential can be expanded and all but the first term beyond 1 ignored. Find an expression for B in terms of C_6 and d.

22.13 Suppose the repulsive term in a Lennard-Jones (12, 6) potential is replaced by an exponential function of the form $\exp(-r/\sigma)$. Sketch the form of the potential energy and locate the distance at which it is a minimum.

22.14 Calculate the second virial coefficient when the intermolecular potential energy has the form

$$V = \infty \quad \text{for} \quad R < \sigma_1 \qquad V = -\varepsilon \quad \text{for} \quad \sigma_1 < R < \sigma_2$$
$$V = 0 \quad \text{for} \quad R > \sigma_2$$

Explore the relation of the parameters in this potential energy to the van der Waals parameters a and b when the potential well is shallow but long-ranged.

22.15 Write a computer program to evaluate eqn 22c as suggested in Example 22.6. Use the program to evaluate the second virial coefficient of argon at several temperatures using a Lennard–Jones potential. Plot the result and locate the Boyle temperature.

22.16 The 'cohesive energy density' is defined as $-U/V$, where U is the mean potential energy of attraction within the sample and V its volume. Show that the cohesive energy density is equal to $-\frac{1}{2}\mathcal{N}^2 \int V(r)\, d\tau$, where $\mathcal{N}$ is the number density of the molecules and $V(r)$ is their attractive potential energy and where the integration ranges from d to infinity and over all angles. Go on to show that the cohesive energy density of a uniform distribution of molecules that interact by a van der Waals attraction of the form $-C_6/r^6$ is equal to $(2\pi/3)(N_A^2/d^3M^2)\rho^2C_6$, where ρ is the density of the solid sample and M is the molar mass of the molecules.

22.17 Consider the collision between a hard-sphere molecule of radius R_1 and mass m_1 and an infinitely massive impenetrable sphere of radius R_2. Plot the scattering angle θ as a function of the impact parameter b. Carry out the calculation using simple geometrical considerations.

22.18 The dependence of the scattering characteristics of atoms on the energy of the collision can be modelled as follows. We suppose that the two colliding atoms behave as impenetrable spheres, as in Problem 22.17, but that the effective radius of the heavy atoms depends on the speed v of the light atom. Suppose its effective radius depends on v as $R_2 \exp(-v/v^*)$ where v^* is a constant. Take $R_1 = \frac{1}{2}R_2$ for simplicity and an impact parameter $b = \frac{1}{2}R_2$, and plot the scattering angle as a function of (a) speed, (b) kinetic energy of approach.

22.19 The magnetizability ξ and the volume and molar magnetic susceptibilities can be calculated from the wavefunctions of molecules. For instance, the magnetizability of a hydrogenic atom is

$$\xi = \frac{-e^2}{6m_e}\langle r^2\rangle$$

where $\langle r^2\rangle$ is the expectation (mean) value of r^2 in the atom. Calculate ξ and χ_m for the ground state of a hydrogen atom.

22.20 The NO molecule has thermally accessible electronically excited states. It also has an unpaired electron, and so may be expected to be paramagnetic. However, its ground state is not paramagnetic because the magnetic moment of the orbital motion of the unpaired electron almost exactly cancels the spin magnetic moment. The first excited state (at 121 cm^{-1}) is paramagnetic because the orbital magnetic moment adds to, rather than cancels, the spin magnetic moment. The upper state has a magnetic moment of $2\mu_B$. Since the upper state is thermally accessible, the paramagnetic susceptibility of NO shows a pronounced temperature dependence even near room temperature. Calculate the molar paramagnetic susceptibility of NO and plot it as a function of temperature.

Macromolecules

Check-list of key ideas

1. The use of *osmometry* for the determination of molar mass (Section 23.1).

2. The determination and significance of the *osmotic virial coefficient* (Section 23.1 and eqn 4).

3. The significance of the *Flory temperature* and a *θ solution* (Section 23.1).

4. The number-average molar mass of a *polydisperse* polymer (eqn 5).

5. The technique of *vapour-phase osmometry* (Section 23.1).

6. The role of *polyelectrolytes* in osmosis and the *Donnan equilibrium* established when a polyelectrolyte affects the distribution of ion concentrations (eqns 6 and 7).

7. The rate of *sedimentation* and the use of the *sedimentation constant* in the determination of molar mass (eqn 9).

8. The *sedimentation equilibrium* established in a gravitational field or in a centrifuge (Section 23.2) and its use for the determination of molar mass (eqn 14).

9. The determination of molar mass by *electrophoresis* and *gel filtration* (Section 23.2).

10. The definition of *intrinsic viscosity* (eqn 16) and its correlation with the molar mass of polymers (eqn 17).

11. The use of *light scattering* to determine the *mass-average molar mass* (eqn 19) of a macromolecule (Section 23.4).

12. The *turbidity* of a solution (eqn 20).

13. The differences between light scattering by small particles (eqn 18) and large particles (eqn 21).

14. The use of *shift reagents* and *broadening reagents* in magnetic resonance (Section 23.5).

15. The *primary*, *secondary*, *tertiary*, and *quaternary structure* of polypeptides.

16. The properties of *random coils* (Section 23.6), including the *random walk* for calculating the *radial distribution* (eqn 23) and the dimensions of a random coil (Example 23.7).

17. The *Corey–Pauling rules* for the secondary structure of polypeptides (Section 23.7) and the formation of the α *helix* and the β-*pleated sheet*.

18. The *conformational energy analysis* of polypeptides (Section 23.7).

19. The classification, preparation, and purification of *colloids* (Section 23.9).

20. The stability of colloids and the formation of *micelles* (Section 23.10).

21. The role of the *electrokinetic potential* and the *electric double layer* (Section 23.10) in the stability of colloids.

22. The definition of *surface excess* (eqn 28).

23. The *Gibbs surface tension equation* relating the change in surface tension to the surface excess (eqn 30) and to the accumulation at a surface of a surfactant (eqn 31).

There are macromolecules everywhere, inside us and outside us. Some are natural and include polysaccharides such as cellulose, polypeptides such as enzymes, and nucleic acids such as DNA. Others are synthetic, and include the polymers like nylon and polystyrene that are manufactured by stringing together and (in some cases) cross-linking smaller units known as **monomers**. Now that chemists are succeeding in synthesising polypeptides and nucleic acids, however, the distinction between natural and synthetic is becoming old-fashioned. Molecules and atoms sometimes swarm together under the influence of intermolecular forces, and the large conglomerates, which are known as **colloids**, behave like macromolecules in a number of respects. Varieties of colloids are intimately involved in the processes of the world. Life in all its forms, from its intrinsic nature to its technological interaction with its environment, is the chemistry of macromolecules.

Although the concepts of physical chemistry apply equally to macromolecules as well as to small molecules, macromolecules do give rise to special questions and problems. These include the determination of their sizes, the shapes and the lengths of polymer chains, and the large deviations from ideality of their solutions. We concentrate on these special characteristics here and treat the relation of the rate of formation of synthetic polymers to their physical properties in Chapter 27.

Size and shape

X-ray diffraction (Section 21.5) can reveal the position of almost every atom even in highly complex molecules. However, there are several reasons why other techniques must also be used. In the first place, the sample might be a mixture of polymers with different chain lengths and extents of cross-linking, in which case sharp X-ray images are unobtainable. Even if all the molecules in the sample are identical, it might prove impossible to obtain a single crystal. Furthermore, although the work on enzymes, proteins, and DNA has shown how immensely stimulating the data can be, the information is incomplete. For instance, what can be said about the shape of the molecule in its natural environment, a biological cell? What can be said about the response of its shape to changes in its environment? Shape and function go hand in hand, and it is essential to know how the shapes of biological macromolecules, which often carry both acidic and basic groups, respond to the pH of the medium. It is also useful to be able to follow the collapse of a macromolecule into a less orderly form: this **denaturation** is often accompanied by loss of function, but when it happens in a controlled way it is sometimes an essential step in the fulfilment of function, as in the replication of DNA.

23.1 Colligative properties

The classical methods of determining molar mass utilize colligative properties (Section 7.5). For macromolecules, where the number of molecules in solution may be very small even though the mass of the solute may be appreciable, only osmometry (Fig. 23.1) is sufficiently sensitive. We shall see that molar masses are often determined more by empirical comparisons than by absolute determinations.

Osmometry

The van't Hoff equation (eqn 19 of Section 7.5) for the osmotic pressure of an ideal solution resembles the perfect gas law:

$$\Pi = [P]RT$$

[P] is the molar concentration of the macromolecule P, and Π is the osmotic pressure. The extension of the equation to non-ideal solutions is written like the virial equation for real gases

$$\Pi = [P]RT(1 + B[P] + \ldots) \tag{1a}$$

and B is called the **osmotic virial coefficient**. Since [P] is related to the mass concentration c by $[P] = c/M$, where M is the molar mass of P, another version of this equation is

$$\frac{\Pi}{c} = \frac{RT}{M}\left(1 + \frac{Bc}{M} + \ldots\right) \tag{1b}$$

Therefore, by plotting Π/c against c and extrapolating to $c = 0$, we can obtain M from the intercept and B from the slope (Fig. 23.2). We illustrated the procedure in Example 7.10.

Macromolecules give strongly non-ideal solutions partly because, being so large, they displace a large quantity of solvent instead of replacing individual solvent molecules with negligible disturbance. In thermodynamic terms, the displacement of solvent molecules implies that the entropy change is important when a macromolecule dissolves.[1] Furthermore, its

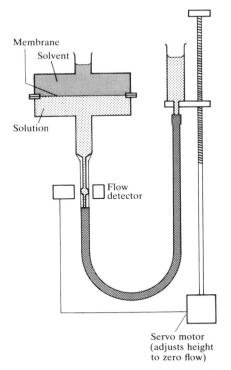

Fig. 23.1 One version of an osmometer used to determine the molar masses of macromolecules. The pressure on the solution is adjusted until there is no flow through the semipermeable membrane: its value is the osmotic pressure.

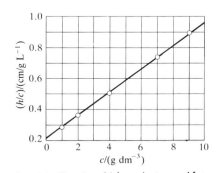

Fig. 23.2 The plot of h/c against c used for the determination of molar mass (from the intercept) and the osmotic virial coefficient (from the slope). h is proportional to Π.

[1] A good starting point for a discussion of the entropy of mixing is a generalization of the expression given in Section 7.2 for the ideal entropy of mixing. Instead of using eqn 8 of that section, which depends on the mole fractions of the components, it turns out to be better to use

$$\Delta S_{\text{mix}} = -R(x_A \ln v_A + x_B \ln v_B)$$

where v_J is the 'volume fraction' of component J. This reduces to the mole fraction expression when the molecular volumes of the two components are the same, but is a better starting point when they are markedly different.

great bulk means that a macromolecule is unable to move freely through the solution since it is excluded from the regions occupied by others.

The origin of the virial coefficient

The most straightforward thermodynamic justification of the osmotic virial expansion, and an interpretation of B, is obtained from the expression for the osmotic pressure in terms of the chemical potential. We saw in eqn 18 of Section 7.5 that

$$-RT \ln x_A = \int_p^{p+\Pi} V_m \, dp$$

where V_m is the molar volume of the pure solvent and x_A is its mole fraction. The more general form of this expression for non-ideal solutions has the solvent activity a_A in place of x_A:

$$-RT \ln a_A = \int_p^{p+\Pi} V_m \, dp$$

Since the integral is equal to ΠV_m if the solvent is incompressible, the osmotic pressure is given by

$$\Pi V_m = -RT \ln a_A \tag{2}$$

If the solution is ideal

$$\ln a_A = \ln x_A = \ln (1 - x_P)$$

$$\approx -x_P$$

where x_P is the mole fraction of the solute macromolecule and we have supposed that $x_P \ll 1$. This approximation leads to the van't Hoff equation. If the solution is non-ideal, we suppose that $-x_P$ is the first term in a series,[2] and write

$$\ln a_A = -x_P(1 + B'x_P + \ldots) \tag{3}$$

When we substitute the expansion in eqn 3 into eqn 2 we get the osmotic virial expansion.

The coefficient B arises largely from the effect of excluded volume. The same is true of a van der Waals gas, for which

$$B = b - \frac{a}{RT} \approx b$$

when the excluded volume effect is dominant. If we imagine a solution of a macromolecule being built by the successive addition of macromolecules to the solvent, each one being excluded by the ones that preceded it, the value

---

[2] The virial expansion in eqn 3 is not a trivial result, but it is confirmed by the *MacMillan–Meyer theory* of solutions of non-electrolytes. That is, it is not obvious that the expansion of $\ln a$ should have this form. A counterexample that we have already met is in the Debye–Hückel theory (Section 10.2), where we saw that for solutions of electrolytes the first term in the deviation of the activity coefficient from 1 is proportional to the *square root* of the concentration of the ions.

of B turns out to be

$$B = \tfrac{1}{2} N_A v_P \tag{4}$$

where v_P is the excluded volume due to a single molecule.

Example 23.1: *Estimating the volume of polymer molecules*

Use the information in Example 7.10 to estimate the volume of the polymer molecules regarded as impenetrable spheres.

Answer. The excluded volume of spherical molecules of volume v is $v_P = 8v$ because the minimum separation of the centers of two spheres is the sum of their radii. We can estimate v_P from the osmotic virial coefficient B using eqn 4, and can find B from the slope of the graph plotted in Fig. 7.12. To do so we use (as in Example 7.10):

$$\frac{h}{c} = \frac{RT}{\rho g M}\left(1 + \frac{Bc}{M}\right)$$

The intercept $RT/\rho g M$ was found in Example 7.10 to be 0.21. The slope of the straight line in Fig. 7.12, which is equal to $(RT/\rho g M) \times B/M$, is 0.073. It follows that

$$\frac{B}{M} = 0.35/(\mathrm{g\,L}^{-1})$$

so $B = 0.35 \times (123 \times 10^3\,\mathrm{g\,mol}^{-1})/(\mathrm{g\,L}^{-1}) = 43 \times 10^3\,\mathrm{L\,mol}^{-1}$. Therefore, from eqn 4,

$$v_P = \frac{2B}{N_A} = 1.4 \times 10^{-22}\,\mathrm{m}^3$$

From this value of v_P it follows that the molecular volume is approximately $1.8 \times 10^4\,\mathrm{nm}^3$.

Comment. The radius of the molecule is approximately 16 nm.

Exercise. Another sample in the same solvent resulted in the following heights of solution at the same temperature and the concentrations specified in Example 7.10: 0.22, 0.53, 1.39, 3.32, 5.02 cm. Calculate the molar mass and the molecular volume of the solute. $[14\,\mathrm{kg\,mol}^{-1}, 9 \times 10^{-23}\,\mathrm{m}^3]$

In broad terms, the excluded volume contributes to the excess entropy of solution (the entropy change in excess of the ideal value, Section 7.4), and the attractions and repulsions between macromolecules contribute to the excess enthalpy. For most solute/solvent systems there is a unique temperature (which is not always experimentally attainable) where these effects cancel and the solution is virtually ideal. This temperature (the analogue of the Boyle temperature for real gases) is called the **Flory temperature** θ, and at it the osmotic virial coefficient B is zero. As an example, for polystyrene in cyclohexane $\theta \approx 306\,\mathrm{K}$, the exact value depending on the molar mass of the polymer. A solution at its Flory temperature is called a θ **solution**. Since a θ solution behaves nearly ideally, its thermodynamic and structural properties are easier to describe even though the concentration is not low.

Number-average molar mass

A further complication affecting the interpretation of the results obtained from osmometry is that the sample may consist of molecules covering a range of molar masses. A pure protein is **monodisperse**, meaning that it has a single, definite molar mass. (There may be small variations, such as one amino acid replacing another, depending on the source of the sample.) A synthetic polymer is **polydisperse**, in the sense that a sample is a mixture of molecules with various chain lengths and molar masses. In the latter case osmometry gives the **number-average molar mass** $\langle M \rangle_N$:

$$\langle M \rangle_N = \frac{1}{N} \sum_i N_i M_i \tag{5}$$

where N_i is the number of molecules with molar mass M_i and there are N molecules in all. The number average is also used for the mean score in a test or height of a population. We shall meet more complicated averages shortly, but the reason why osmometry leads to the number-average molar mass is that osmosis is a colligative property: it depends on the number, not the identity, of the solute molecules.

Vapour-phase osmometry

Osmosis, like all colligative properties, is related to the lowering of vapour pressure that occurs when a solute is present in a solution. The vapour pressure is the pressure established when the rate of evaporation of the solvent matches the rate of return of the vaporized solvent molecules to the solution. Since the latter rate is unchanged when a solute is present, the lowering of vapour pressure corresponds to a reduction in the rate at which solvent molecules leave the solution. Furthermore, since vaporization is endothermic, it leads to a reduction of the temperature in an adiabatic system. If the system is not fully adiabatic, the temperature will fall until the rate at which energy leaks in as heat maintains a steady temperature.

Now we put these remarks together. It follows from them that the temperature of a vaporizing pure solvent will be different from that of a vaporizing solution, and that the temperature difference will be proportional to the mole fraction of the polymer in the solution (like any colligative property). Hence, by measuring the small temperature differential obtained when droplets of solvent and solution evaporate, the molar mass of the polymer can be inferred. This technique, which is called **vapour-phase osmometry**, gives the number-average molar mass and can be used for polymers of molar mass up to about 40 kg mol^{-1}.

Polyelectrolytes and dialysis

Some polymers are strings of acid groups, as in poly(acrylic acid) —$(CH_2CHCOOH)_n$—, or strings of bases, as in nylon, —$[NH(CH_2)_6$-$NHCO(CH_2)_4CO]_n$—; proteins have both acid and base groups. Macromolecules may therefore be **polyelectrolytes**, and, depending on their state of ionization, **polyanions** or **polycations**. A macromolecule with mixed cation and anion character is known as a **polyampholyte**.

One consequence of dealing with polyelectrolytes is that it is necessary to know the extent of ionization before osmotic data can be interpreted. For example, suppose the sodium salt of a polyelectrolyte is present in solution as ν Na$^+$ ions and a single polyanion P$^{\nu-}$, then it gives rise to $\nu + 1$ particles for each formula unit of salt that dissolves. If we guess that $\nu = 1$ when in fact $\nu = 10$, the estimate of the molar mass will be wrong by an order of magnitude. We can find a way out of this difficulty by considering another feature of charged macromolecules.

Suppose the solution of the polyelectrolyte $Na_\nu P$ also contains added NaCl, and that it is in contact through a semipermeable membrane (such as cellophane or a cell wall) with another salt solution. Furthermore, suppose the membrane is permeable to the solvent *and* to the salt ions, but not to the polyanion. This arrangement is one that actually occurs in living systems, where osmosis is an important feature of cell operation. The presence of the salt affects the osmotic pressure because the anions and cations cannot migrate through the membrane to an arbitrary extent. Electrical neutrality must be preserved on both sides: if an anion migrates, a cation must accompany it.

Suppose that $Na_\nu P$ is at a concentration [P] on one side of the membrane, and that NaCl is added to each side. On the left (L) there are $P^{\nu-}$, Na^+, and Cl^- ions. On the right (R) there are Na^+ and Cl^- ions. The condition for equilibrium is that the chemical potential of NaCl should be the same on both sides of the membrane, and so a net flow of Na^+ and Cl^- ions occurs until $\mu_L(NaCl) = \mu_R(NaCl)$. This equality occurs when

$$\mu^{\ominus}(NaCl) + RT \ln a_L(Na^+)a_L(Cl^-) = \mu^{\ominus}(NaCl) + RT \ln a_R(Na^+)a_R(Cl^-)$$

If we ignore activity coefficients, the two expressions are equal when

$$[Na^+]_L[Cl^-]_L = [Na^+]_R[Cl^-]_R$$

As the Na^+ ions are supplied by the polyelectrolyte as well as the added salt, the conditions for electrical neutrality are

$$[Na^+]_L = [Cl^-]_L + \nu[P]$$
$$[Na^+]_R = [Cl^-]_R$$

We can now combine these three conditions to obtain expressions for the differences in ion concentrations across the membrane:

$$[Na^+]_L - [Na^+]_R = \frac{\nu[P][Na^+]_L}{[Na^+]_L + [Na^+]_R} = \frac{\nu[P][Na^+]_L}{2[Cl^-] + \nu[P]} \qquad (6a)$$

$$[Cl^-]_L - [Cl^-]_R = \frac{-\nu[P][Cl^-]_L}{[Cl^-]_L + [Cl^-]_R} = \frac{-\nu[P][Cl^-]_L}{2[Cl^-]} \qquad (6b)$$

where

$$[Cl^-] = \tfrac{1}{2}([Cl^-]_L + [Cl^-]_R)$$

$[Cl^-]$ is the average concentration of Cl^- ions on each side of the membrane.

The final step is to note that the osmotic pressure depends on the *difference* in the numbers of solute particles on each side of the membrane. That being so, the van't Hoff equation

$$\Pi = RT[\text{Solute}]$$

becomes

$$\Pi = RT\{([P] + [Na^+]_L + [Cl^-]_L) - ([Na^+]_R + [Cl^-]_R)\}$$

$$= RT[P](1 + B[P]) \qquad B = \frac{\nu^2}{4[Cl^-] + \nu[P]} \qquad (7)$$

When the concentration of added salt is so great that $B[P] \ll 1$, eqn 7 reduces to

$$\Pi = RT[P]$$

a result *independent* of the value of v. Therefore, if we measure the osmotic pressure in the presence of high concentrations of salt, the molar mass may be obtained unambiguously.

A second point arises from eqn 6. There is often interest in the extent to which ions are bound to macromolecules, especially when a membrane (such as a cell wall) separates two regions. The equations show, however, that cations will dominate the anions in the compartment containing the polyanion (because the concentration difference is positive for Na^+ and negative for Cl^-) simply as a result of the equilibrium and electroneutrality conditions. The equilibrium distribution of ions in two compartments connected by a semipermeable membrane, in one of which there is a polyelectrolyte, is called a **Donnan equilibrium**.

Example 23.2: *Analysing the Donnan equilibrium*

Two equal volumes of $0.200\,M\,NaCl(aq)$ are separated by a membrane. A macromolecule of molar mass $55\,kg\,mol^{-1}$, which cannot pass through the membrane, is added as its sodium salt Na_6P to a concentration of $50\,g\,L^{-1}$ to the left-hand compartment. What are the equilibrium concentrations of Na^+ and Cl^- in each compartment?

Answer. We use eqn 6a to calculate the concentration differences and eqn 6b to calculate their sum as

$$[Na^+]_L + [Na^+]_R = [Cl^-]_L + [Cl^-]_R + v[P]$$

$$= 2[Cl^-] + v[P]$$

Then use $[Cl^-] = 0.200\,M$. Since $[P] = 9.1 \times 10^{-4}\,M$, eqn 6 gives

$$[Na^+]_L - [Na^+]_R = \frac{6 \times (9.1 \times 10^{-4}\,M) \times [Na^+]_L}{0.400\,M + 6 \times (9.1 \times 10^{-4}\,M)}$$

$$= 1.35 \times 10^{-2}[Na^+]_L$$

and the sum above gives

$$[Na^+]_L + [Na^+]_R = 0.400\,M + 6 \times (9.1 \times 10^{-4})\,M$$

$$= 0.405\,M$$

Solving these equations gives

$$[Na^+]_L = 0.204\,M \qquad [Na^+]_R = 0.201\,M$$

Then

$$[Cl^-]_R = [Na^+]_R = 0.201\,M \qquad [Cl^-]_L = [Na^+]_L - 6[P] = 0.199\,M$$

Comment. Note how the Na^+ accumulates slightly in the compartment containing the macromolecule.

Exercise. Repeat the calculation for $0.300\,M\,NaCl(aq)$ and polyelectrolyte $Na_{10}P$ of molar mass $33\,kg\,mol^{-1}$ at a concentration of $50\,g\,L^{-1}$.

$$[Na^+: 0.311\,M, 0.304\,M]$$

23.2 Sedimentation

In a gravitational field, heavy particles settle towards the foot of a column of solution by the process called **sedimentation**. The rate of sedimentation depends on the strength of the field and on the masses and shapes of the particles. Spherical molecules (and compact molecules in general) sediment faster than rod-like or extended molecules. For example, DNA helices sediment much faster when they are denatured to a random coil, and so sedimentation rates can be used to study denaturation. When the sample is at equilibrium, the particles are dispersed over a range of heights in accord with the Boltzmann distribution (because the gravitational field competes with the stirring effect of thermal motion). The spread of heights depends on the masses of the molecules, and so the equilibrium distribution is another way of determining molar mass.

Sedimentation is normally very slow, but it can be accelerated by replacing the gravitational field by a centrifugal field. The latter can be achieved in an ultracentrifuge, which is essentially a cylinder that can be rotated at high speed about its axis with a sample in a cell near its periphery (Fig. 23.3). Modern ultracentrifuges can produce accelerations equivalent to about 10^5 that of gravity ('$10^5 g$'). Initially the sample is uniform, but the 'top' (innermost) boundary of the solute moves outwards as sedimentation proceeds. The rate at which the boundary moves can be monitored by making use of the effect of the concentration on the refractive index of the sample. Since there is a sharp change of refractive index between the solution and the solvent left behind by the receding solute, the sample behaves like a prism and bends the light passing through it. The **Schlieren optical system** (which is also used to study air flow in wind tunnels and shock tubes) turns a refractive index gradient into a dark image (Fig. 23.4). Alternatively, in the **interference technique** the concentration profile is monitored through the effect of refractive index on the interference of two beams of light, one coming through the sample and the other through a blank (Fig. 23.5).

The rate of sedimentation

A solute particle of mass m has an effective mass $m_{eff} = bm$ on account of the buoyancy of the medium, with

$$b = 1 - \rho v_s \qquad (8)$$

ρ is the solution density, v_s is the solute's specific volume (its volume per unit mass), and mv_s is the mass of solvent displaced by the solute. (More precisely, v_s is the partial specific volume, in the sense described in Section 7.1.) The solute particles at a distance r from the axis of a rotor spinning at an angular velocity ω experience a centrifugal force of magnitude $m_{eff}r\omega^2$. The acceleration outwards is countered by a frictional force proportional to the speed s of the particles through the medium, and which is written fs, where f is the **frictional coefficient**. The particles therefore adopt a **drift speed**, a steady speed through the medium, which is found by equating the two forces $m_{eff}r\omega^2$ and fs. The forces are equal when

$$s = \frac{m_{eff}r\omega^2}{f} = \frac{bmr\omega^2}{f}$$

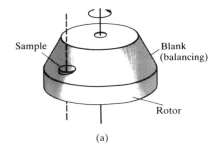

(a)

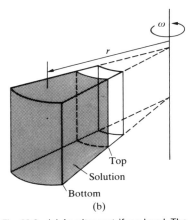

Top
Solution
Bottom
(b)

Fig. 23.3 (a) An ultracentrifuge head. The sample on one side is balanced by a blank diametrically opposite. (b) Detail of the sample cavity: the 'top' surface is the inner surface, and the centrifugal force causes sedimentation towards the outer surface; a particle at a radius r experiences a force of magnitude $mr\omega^2$.

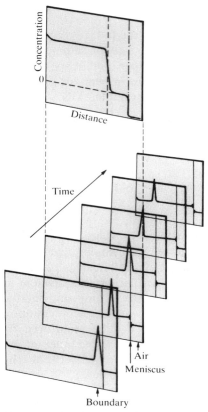

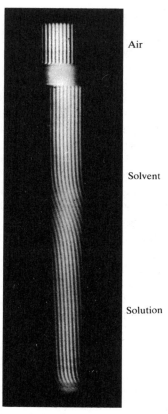

Fig. 23.4 A Schlieren photograph indicates regions where the refractive index is changing. This set of diagrams, corresponding to a series of times, shows the sedimentation of the solute; one has been interpreted in terms of the concentration profile through the cell.

Fig. 23.5. An interference photograph also shows the concentration profile, but in a different manner. The regions of air, solvent, and solution have been marked. (Provided by Professor D. Freifelder, from his *Physical biochemistry*, W. H. Freeman & Co (1982).)

The drift speed depends on the angular velocity and the radius, and it is convenient to focus on the **sedimentation constant**

$$S = \frac{s}{r\omega^2} \tag{9a}$$

Then, since the mass of an individual molecule is related to the molar mass through $m = M/N_A$,

$$S = \frac{bM}{fN_A} \tag{9b}$$

Example 23.3: *Determining a sedimentation constant*

The sedimentation of bovine serum albumin (BSA) was monitored at 25°C. The initial radius of the solute surface was 5.50 cm and during centrifugation at 56 850 r.p.m. it receded as follows:

t/s	0	500	1000	2000	3000	4000	5000
r/cm	5.50	5.55	5.60	5.70	5.80	5.91	6.01

Calculate the sedimentation constant.

Answer. From eqn 9a with $s = dr/dt$,

$$\frac{dr}{dt} = r\omega^2 S$$

which integrates to

$$\ln\frac{r}{r_0} = \omega^2 St$$

Therefore, a plot of $\ln(r/r_0)$ against t should be a straight line of slope $\omega^2 S$. Use $\omega = 2\pi\nu$, where ν is in cycles/second, and draw up the following table:

t/s	0	500	1000	2000	3000	4000	5000
$100\ln(r/r_0)$	0	0.90	1.80	3.57	5.31	7.19	8.87

The plot has slope 1.79×10^{-5}; and so $\omega^2 S = 1.79 \times 10^{-5}\,\text{s}^{-1}$. Since $\omega = 2\pi(56\,850/60)\,\text{s}^{-1} = 5.95 \times 10^3\,\text{s}^{-1}$, it follows that $S = 5.1 \times 10^{-13}\,\text{s}$.

Comment. We develop this result below. The unit $10^{-13}\,\text{s}$ is sometimes called a 'svedberg' and denoted Sv; in this case $S = 5.1$ Sv. Accurate results are obtained by extrapolating to zero concentration.

Exercise. Calculate the sedimentation constant given the following data (the other conditions being the same as above):

t/s	0	500	1000	2000	3000	4000	5000
r/cm	5.65	5.68	5.71	5.77	5.84	5.90	5.97

[3.08 Sv]

To make progress we need to know the frictional coefficient f. For a spherical particle of radius a in a solvent of viscosity η, and for solute molecules that are not small compared with the solvent molecules, f is given by **Stokes' relation**

$$f = 6\pi a\eta \tag{10}$$

Therefore, for spherical molecules,

$$S = \frac{bM}{6\pi a\eta N_A} \tag{11}$$

and S may be used to determine either M or a. If the molecules are not spherical we use the appropriate value of f given in Table 23.1. As always when dealing with macromolecules, the measurements must be extrapolated to zero concentration in order to avoid the complications that arise from the interference between bulky molecules.

At this stage it appears that we need to know the molecular radius a (and in general the frictional coefficient f) in order to obtain the molar mass from the value of S. Fortunately, this can be avoided by drawing on the **Stokes–Einstein relation** between f and the **diffusion coefficient** D:

$$f = \frac{kT}{D} \tag{12}$$

The diffusion coefficient measures the rate at which molecules spread down

Table 23.1. Frictional coefficients and molecular geometry*

Major axis		
Minor axis	Prolate	Oblate
2	1.04	1.04
4	1.18	1.17
6	1.31	1.28
8	1.43	1.37
10	1.54	1.46

* Entries are the ratio f/f_0 where $f_0 = 6\pi\eta c$ with $c = (ab^2)^{1/3}$ for prolate ellipsoids and $c = (a^2 b)^{1/3}$ for oblate ellipsoids; $2a$ is the major axis and $2b$ is the minor axis.

Table 23.2. Diffusion coefficients in water at 20°C

	$M/(\text{g mol}^{-1})$	$D/(\text{cm}^2\,\text{s}^{-1})$
Sucrose	342	4.59×10^{-6}
Lysozyme	14 100	1.04×10^{-6}
Haemoglobin	68 000	6.9×10^{-7}
Collagen	345 000	6.9×10^{-8}

a concentration gradient (Section 24.6). It can be measured by observing the rate at which a boundary spreads, or the rate at which a more concentrated solution diffuses into a less concentrated one. Some typical values are given in Table 23.2. The diffusion coefficient may also be measured using light scattering, as we shall see later. Then, from eqn 9b,

$$M = \frac{SRT}{bD} \tag{13}$$

This result is independent of the shape of the solute molecules. It follows that we can find the molar mass by combining measurements of sedimentation and diffusion rates (for S and D respectively).

Example 23.4: *Interpreting a sedimentation experiment*

Use the result from Example 23.3 in combination with the data below to calculate the molar mass of BSA. Estimate its axial ratio on the basis that it is a prolate ellipsoid. Take $D = 6.97 \times 10^{-7}\,\text{cm}^2\,\text{s}^{-1}$, $\rho = 1.0024\,\text{g cm}^{-3}$, $v_s = 0.734\,\text{cm}^3\,\text{g}^{-1}$, $\eta = 0.890 \times 10^{-3}\,\text{kg m}^{-1}\,\text{s}^{-1}$, and the temperature as 25°C.

Answer. First, we find f from eqn 12 and $f_0 = 6\pi c\eta$ from the assumption that the molecule is a sphere of radius c; we can then obtain c from v_s using $v = (4\pi/3)c^3$. Substitution into eqn 13 with $b = 1 - \rho v_s$ gives $M = 69\,\text{kg mol}^{-1}$. Then, using $f = kT/D = 5.91 \times 10^{-11}\,\text{kg s}^{-1}$ and

$$v = \frac{(0.734\,\text{cm}^3\,\text{g}^{-1}) \times (69\,000\,\text{g mol}^{-1})}{N_A} = 8.4 \times 10^{-26}\,\text{m}^3$$

$$c = (3v/4\pi)^{1/3} = 2.7 \times 10^{-9}\,\text{m}$$

$$f_0 = 6\pi c\eta = 4.5 \times 10^{-11}\,\text{kg s}^{-1}$$

Therefore, $f/f_0 = 1.3$. Reference to Table 23.1 shows that this ratio corresponds to an axial ratio of about 6.

Comment. The ellipsoid is like a small cigar, six times longer than it is broad. Extrapolation to zero concentration gives an axial ratio of 4.4.

Exercise. Use the result in the exercise of Example 23.3 to find the molar mass and the axial ratio of that macromolecule. Use the following additional data: $D = 5.89 \times 10^{-7}\,\text{cm}^2\,\text{s}^{-1}$, $\rho = 1.0024\,\text{g cm}^{-3}$, $v_s = 0.728\,\text{cm}^3\,\text{g}^{-1}$, $\eta = 0.890 \times 10^{-3}\,\text{kg m}^{-1}\,\text{s}^{-1}$, and the same temperature. $[48\,\text{kg mol}^{-1}, f/f_0 = 1.7]$

Sedimentation equilibria

The difficulty with using sedimentation rates to measure molar masses lies in the inaccuracies inherent in the determination of diffusion coefficients, such as the blurring of the boundary by convection currents. We can avoid this problem by allowing the system to reach equilibrium, for the transport property D is then no longer relevant.

Since the number of solute molecules with any given potential energy E is proportional to $e^{-E/kT}$, the ratio of the concentrations at different heights (or radii in a centrifuge) can be used to determine their masses. The kinetic energy of a molecule of mass m_{eff} arising from its circulation around the axis of the centrifuge is

$$E = \tfrac{1}{2}m_{\text{eff}}r^2\omega^2$$

when it is travelling in a circuit of radius r with angular velocity ω. Therefore, the ratio of the concentrations at radii r_1 and r_2 is

$$\frac{c(1)}{c(2)} = \frac{N(1)}{N(2)} = \frac{e^{-E(1)/kT}}{e^{-E(2)/kT}}$$

$$= e^{mb\omega^2(r_1^2 - r_2^2)/2kT}$$

so

$$M = \frac{2RT}{(r_2^2 - r_1^2)b\omega^2} \ln \frac{c(2)}{c(1)} \qquad (14)$$

The centrifuge is run more slowly in this technique than in the sedimentation rate method because it is no use having all the solute pressed in a thin film against the bottom of the cell.

Electrophoresis

Many macromolecules are charged, and move in an electric field: this motion is called **electrophoresis**. In **gel electrophoresis** the migration takes place through a cross-linked polyacrylamide gel. The mobilities of macromolecules depends on their masses and their shapes, and a constant drift speed is reached when the driving force $ez\mathscr{E}$ (where z is the charge number and $\mathscr{E}$ is the field strength) is matched by the viscous retarding force fs.

One way of avoiding the problem of knowing neither the hydrodynamic shape of the molecules nor their charge is to denature them in a controlled way. Sodium dodecylsulphate has been found to be very useful in this respect. It denatures proteins into rods by forming a complex with them, and so all proteins, whatever their initial shapes, are made rod-like. Moreover, most proteins have been found to bind a constant amount of the anion per unit mass, and so the charge per protein molecule is well regulated. The molar mass of the protein is determined by comparing its mobility in its rod-like complexed form with standard samples.

Gel filtration

Beads of porous polymeric material about 0.1 mm in diameter can capture molecules selectively, according to their size. Thus, if a solution is filtered through a column, the small molecules require a long **elution time**, or time to pass through a particular length of column, while the larger ones, which are not captured, pass through rapidly. The molar mass of a macromolecule may therefore be determined by observing its elution time in a column calibrated against standard samples. The range of molar masses that can be determined can be altered by selecting columns made from polymers with different degrees of cross-linking. The elution time depends on shape in a complicated way and the technique works best if the molecules are spherical.

23.3 Viscosity

The presence of a macromolecular solute increases the viscosity of a solution. The effect is large even at low concentrations, because big molecules affect the fluid flow over a long range. At low concentrations the viscosity of the solution η is related to the viscosity of the pure solvent η^*,

(a)

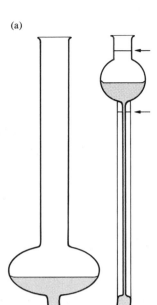

(b)

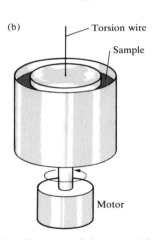

Torsion wire

Sample

Motor

Fig. 23.6 Two types of viscometer. (a) In the Ostwald viscometer the time for the liquid to drain between the two marks is recorded. (b) In the rotating drum viscometer the torque on the inner drum is observed when the outer container is rotated.

by

$$\eta = \eta^*(1 + [\eta]c + \dots) \tag{15}$$

The **intrinsic viscosity** $[\eta]$ is the analogue of a virial coefficient (and has the dimensions of 1/concentration). It follows from eqn 15 that

$$[\eta] = \lim_{c \to 0} \frac{(\eta/\eta^*) - 1}{c} \tag{16}$$

Viscosities are measured in several ways. In the 'Ostwald viscometer' shown in Fig. 23.6a, the time taken for the solution to flow through the capillary is noted, and compared with a standard sample. The method is well suited to the determination of $[\eta]$ because the ratio of the viscosities of the solution and the pure solvent is proportional to the drainage times t and t^* after correcting for different densities ρ and ρ^*:

$$\frac{\eta}{\eta^*} = \frac{t}{t^*} \times \frac{\rho}{\rho^*}$$

This ratio can be used directly in eqn 16. Viscometers in the form of rotating concentric cylinders are also used (Fig. 23.6b), and the torque on the inner cylinder is monitored while the outer one is rotated. Such 'rotating drum viscometers' have the advantage over the Ostwald type that the shear gradient between the cylinders is simpler than in the capillary, and effects of the kind we shall shortly describe can be studied more easily.

There are many complications in interpreting viscosity measurements. Much of (but not all) the work is based on empirical observations, and the determination of molar mass is usually based on comparisons with standard samples. Some regularities are observed that help in the determination. For example, it is found that θ solutions of macromolecules often fit

$$[\eta] = KM^a \tag{17}$$

where K and a are constants that depend on the solvent and type of macromolecule (Table 23.3). As an example, solutions of poly(γ-benzyl-L-glutamate) in its rod-like form has an intrinsic viscosity four times greater than when it is denatured and the rods collapse into random coils. Conversely, solutions of natural ribonuclease are less viscous than the denatured form: this suggests that the natural protein is more compact than when it is denatured.

Table 23.3. Intrinsic viscosity

Macromolecule	Solvent	$\theta/°C$	$K/(\text{cm}^3\,\text{g}^{-1})$	a
Polystyrene	Benzene	25	9.5×10^{-3}	0.74
Polyisobutylene	Benzene	23	8.3×10^{-2}	0.50
Various proteins	Guanidine hydrochloride + $HSCH_2CH_2OH$		7.2×10^{-3}	0.66

Example 23.5: *Using intrinsic viscosity to measure molar mass*

The viscosities of a series of solutions of polystyrene in toluene were measured at 25°C with the following results:

$c/(\text{g L}^{-1})$	0	2.0	4.0	6.0	8.0	10.0
$\eta/(10^{-4}\,\text{kg m}^{-1}\text{s}^{-1})$	5.58	6.15	6.74	7.35	7.98	8.64

Calculate the intrinsic viscosity and estimate the molar mass of the polymer using eqn 17 with $K = 3.80 \times 10^{-5}\,\text{L g}^{-1}$ and $a = 0.63$.

Answer. The intrinsic viscosity is defined in eqn 16; therefore, form this ratio and extrapolate to $c = 0$. To do so, we draw up the following table:

$c/(\text{g L}^{-1})$	0	2.0	4.0	6.0	8.0	10.0
η/η^*	1	1.102	1.208	1.317	1.430	1.549
$100[(\eta/\eta^*)-1]/(c/\text{g L}^{-1})$		5.11	5.20	5.28	5.38	5.49

The points are plotted in Fig. 23.7. The extrapolated intercept at $c = 0$ is 0.0504, and so $[\eta] = 0.0504\,\text{L g}^{-1}$. Therefore,

$$M = \left(\frac{[\eta]}{K}\right)^{1/a} = 90 \times 10^3\,\text{g mol}^{-1}$$

Comment. When $\eta \approx \eta^*$

$$\ln\frac{\eta}{\eta^*} = \ln\left(1 + \frac{\eta - \eta^*}{\eta^*}\right)$$

$$\approx \frac{\eta - \eta^*}{\eta^*} = \frac{\eta}{\eta^*} - 1$$

Hence, $[\eta]$ can also be defined as the limit of $(1/c)\ln(\eta/\eta^*)$ as $c \to 0$. The intercept is identified more precisely by plotting both functions.

Exercise. Evaluate the molar mass using the second plotting technique.

$$[90\,\text{kg mol}^{-1}]$$

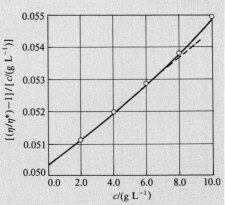

Fig. 23.7 The plot used for the determination of intrinsic viscosity, which is taken from the intercept at $c = 0$.

One complication in viscosity measurements is that in some cases it is found that the fluid is **non-Newtonian** in that its viscosity changes as the rate of flow increases. A decrease in viscosity with increasing rate of flow indicates the presence of long rod-like molecules that are orientated by the flow and hence slide past each other more freely. In some cases the stresses set up by the flow are so great that long molecules are broken up, with further consequences for the viscosity.

23.4 Light scattering

When light falls on an object, it drives the electrons into oscillation and they radiate. If the medium is perfectly homogeneous (e.g. a perfect crystal or a completely random collection of molecules homogeneous on the scale of the wavelength of the light, like a sample of water), all the secondary waves interfere destructively except in the original propagation direction. Therefore, an observer sees the beam only when looking towards the source along the initial direction. If the medium is inhomogeneous (e.g. an imperfect crystal or a solution containing foreign bodies, such as macromolecules in a solvent or smoke in air), radiation is scattered into other directions. A

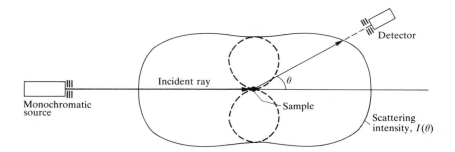

Fig. 23.8 Rayleigh scattering from a sample of point-like particles follows a $1 + \cos^2 \theta$ dependence (full line) when unpolarized light is used, but a $\sin^2 \theta$ dependence (broken line) when plane polarized light is used (e.g. when the source is a laser).

familiar example is light scattered by specks of dust in a sunbeam (and in advertisers' photographs of laser beams).

Light scattering by particles much smaller than the wavelength of the light is called **Rayleigh scattering**. The intensity of scattered radiation depends on $1/\lambda^4$, and short wavelength radiation is scattered more intensely than long. The blue of the sky arises from the predominant scattering of the blue component of white sunlight by the atoms and molecules of the atmosphere. The intensity also depends on the scattering angle θ, and is proportional to $1 + \cos^2 \theta$ when the light is unpolarized and to $\sin^2 \theta$ when it is polarized (Fig. 23.8). In practice it turns out to be easier to make observations in a non-forward direction. The intensity also depends on the strength of the interaction of the light with the molecules, the interaction being large when their polarizability is large.

When these remarks are combined into a quantitative theory, it turns out that the scattering intensity I at the angle θ is

$$I = A I_0 c M g(\theta) \qquad g(\theta) = \begin{cases} 1 + \cos^2 \theta \text{ for unpolarized light} \\ \sin^2 \theta \text{ for polarized light} \end{cases} \tag{18}$$

where I_0 is the incident intensity, c is the solute concentration, M its molar mass, and A a constant that depends on the refractive index of the solution, the wavelength, and the distance of the detector from the sample. Equation 18 is an 'ideal' result in the sense that it ignores the complications that arise from the interactions between solute particles, and in an actual experiment it is important to extrapolate to zero concentration.

The mass-average molar mass

The scattered intensity I is measured at several concentrations, and M is obtained from the limiting value of I/c. With a polydisperse sample this procedure leads to the **mass-average molar mass** $\langle M \rangle_\text{M}$:

$$\langle M \rangle_\text{M} = \frac{1}{m} \sum_i m_i \times M_i \tag{19a}$$

m_i is the total mass of molecules of molar mass M_i and m is the total mass of the sample. Light scattering gives the mass-average molar mass because the intensity of scattering is greater for larger particles, and so their contribution is emphasized.

Since $m_i = N_i M_i / N_A$, we can also express the mass average as

$$\langle M \rangle_M = \frac{\sum\limits_i N_i M_i^2}{\sum\limits_i N_i M_i} \qquad (19b)$$

Hence, the mass-average molar mass is proportional to the **mean square molar mass**. Sedimentation experiments give yet another average molar mass, the **Z-average molar mass**:

$$\langle M \rangle_Z = \frac{\sum\limits_i N_i M_i^3}{\sum\limits_i N_i M_i^2} \qquad (19c)$$

The Z-average molar mass is proportional to a mean cubic molar mass.

Example 23.6: *Calculating number and mass averages*

A sample of polymer consists of two components present in equal masses, one having $M = 30 \text{ kg mol}^{-1}$ and the other $M = 12 \text{ kg mol}^{-1}$. Calculate the mass-average and number-average molar masses.

Answer. We use eqn 19b for the mass average and eqn 5 for the number average. Since $m_1 = m_2$, the proportions by mass are each $\frac{1}{2}$, so

$$\langle M \rangle_M = \tfrac{1}{2}(M_1 + M_2) = 21 \text{ kg mol}^{-1}$$

We obtain the proportions by number using the molar masses:

$$\frac{N_i}{N} = \frac{(m_i / M_i)}{(m_1 / M_1) + (m_2 / M_2)}$$

So, because $m_1 = m_2$,

$$\frac{N_1}{N} = \frac{M_2}{M_1 + M_2} \qquad \frac{N_2}{N} = \frac{M_1}{M_1 + M_2}$$

From eqn 5,

$$\langle M \rangle_N = \frac{2M_1 M_2}{M_1 + M_2} = 17 \text{ kg mol}^{-1}$$

Comment. Note that the two averages are significantly different, their ratio being about 1.2.

Exercise. Evaluate the Z-average molar mass of the sample.   [25 kg mol^{-1}]

The example shows that while at first sight it might appear troublesome to have several types of average, the observation that they have different values gives additional information about the range of molar masses in the sample. In the determination of protein molar masses we expect the two averages to be the same because the sample is monodisperse (unless there has been degradation). In samples of synthetic polymers there is normally a range of molar masses and the two averages are expected to be different.

Typical synthetic materials have

$$\frac{\langle M \rangle_M}{\langle M \rangle_N} \approx 3$$

The term 'monodisperse' is conventionally applied to synthetic polymers in which this ratio is less than 1.1. One consequence of a narrow molar mass distribution is a higher crystallinity, and therefore density and melting point, of synthetic polymers. The spread of values is controlled by the choice of catalyst and reaction conditions (Section 27.4).

Turbidity

Since the solute scatters light away from the forward direction, the transmitted intensity is reduced. The intensity in the forward direction is given by a type of Beer–Lambert law (Section 17.1): if the incident intensity is I_0, then the intensity that survives after passing through a solution of length l is

$$I_l = I_0 e^{-\tau l} \tag{20}$$

where τ is the **turbidity**. For macromolecules of moderate size the turbidity is related to the concentration by

$$\frac{Hc}{\tau} = \frac{1}{M}(1 + 2Bc + \dots) \qquad H \propto \frac{dn}{dc}$$

where n_0 and n are the solvent and solution refractive indices respectively. Therefore, by plotting Hc/τ against c, the intercept gives the molar mass. Typical values of τ are $\tau \approx 10^{-5}\,\text{cm}^{-1}$ for pure transparent liquids, $10^{-3}\,\text{cm}^{-1}$ for polymers at 1 per cent concentration, and $10\,\text{cm}^{-1}$ for milk. For the first of these values, the sample would need to be 1 km long before the intensity drops to 1/e of its initial value as a result of scattering alone (absorption in fact dominates). The last value, loosely interpreted, means that we can see about 1 mm into a glass of milk.

Large-particle scattering

When the wavelength of light is comparable to the size of the scattering particles, scattering occurs at different sites of the same molecule (Fig. 23.9), and the interference between different rays is important.[3] As a result, the scattering intensity is distorted from the form characteristic of small-particle, Rayleigh scattering of light given in eqn 18. A measure of the distortion is the ratio

$$P = \frac{I(\text{Obs})}{I(\text{Rayleigh})}$$

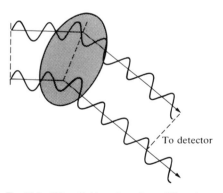

Fig. 23.9 When light scatters from different regions of the same molecule the emergent waves interfere and modify the Rayleigh scattering intensity distribution. The modification can be interpreted in terms of the shape of the macromolecules.

To detector

---

[3] The effect accounts for the appearance of clouds, which, although we see them by scattered light, look white, not blue like the sky. The reason is that the water molecules group together into droplets of a size comparable to the wavelength of light, and scatter cooperatively. Although blue light scatters more strongly, more molecules can contribute cooperatively when the wavelength is longer (as for red light), and so the net result is uniform scattering for all wavelengths: white light scatters as white light. This paper looks white for the same reason. Cigarette smoke is blue before it is inhaled, but brownish after it is exhaled because the particles aggregate in the lungs.

measured at several angles, where I(Obs) is the observed intensity and I(Rayleigh) that predicted for Rayleigh scattering.

If we treat the molecule as being composed of a collection of atoms i at distances R_i from some convenient origin, interference occurs between the light scattered by each pair (Fig. 23.9). The scattering from all the particles is then calculated by allowing for contributions from all possible orientations of each pair of atoms in each molecule. This description is very much like the one we used when discussing electron diffraction (Section 21.10), so we can expect the intensity pattern to be described by a kind of Wierl equation. This turns out to be so, and if there are N atoms in the macromolecule, and if all are assumed to have the same scattering power, then

$$P = \frac{1}{N^2} \sum_i \sum_j \frac{\sin sR_{ij}}{sR_{ij}} \qquad s = \frac{4\pi}{\lambda} \sin \tfrac{1}{2}\theta \qquad (21)$$

(Compare eqn 8 of Section 21.10.) R_{ij} is the separation of atoms i and j, and λ is the wavelength of the light. The observed intensity is equal to I(Rayleigh)P, with I(Rayleigh) given by eqn 18.

Small-particle scattering

When the molecule is much smaller than the wavelength of light, $sR_{ij} \ll 1$ (e.g. if $R = 5$ nm, and $\lambda = 500$ nm, all the sR_{ij} are about 0.1), and we can use the expansion

$$\sin sR_{ij} = sR_{ij} - \tfrac{1}{6}(sR_{ij})^3 + \ldots$$

Then

$$P = \frac{1}{N^2} \sum_i \sum_j \{1 - \tfrac{1}{6}(sR_{ij})^2 + \ldots\}$$

$$= 1 - \frac{s^2}{6N^2} \sum_{i,j} R_{ij}^2$$

The sum over the squares of the separations gives the **radius of gyration** R_g of the molecule:[4]

$$R_g^2 = \frac{1}{2N^2} \sum_{i,j} R_{ij}^2 \qquad (22)$$

The radius of gyration is the radius of a thin hollow spherical shell of the same mass and moment of inertia as the molecule. Then

$$P \approx 1 - \tfrac{1}{3}s^2 R_g^2 = 1 - \frac{16\pi^2 R_g^2}{3\lambda^2} \sin^2 \tfrac{1}{2}\theta$$

The last equation shows that an analysis of the deviation of the scattering from the Rayleigh form will give the value of R_g for the molecule in solution. This in turn can be interpreted in terms of the size of the molecule.

---

[4] In Problem 23.19 this definition is shown to be equivalent to another and more easily visualized one in the case of a chain of identical atoms: the radius of gyration is the root mean square distance of the atoms from the centre of mass.

Table 23.4. Radius of gyration

	$M/(\text{g mol}^{-1})$	R_g/nm
Serum albumin	66 000	2.98
Polystyrene	3.2×10^6	49*
DNA	4×10^6	117

* In a poor solvent.

For example, a solid sphere of radius R has $R_g = (3/5)^{1/2}R$, and a long thin rod of length L has $R_g = L/(2\sqrt{3})$. Once again, it must be emphasized that the analysis must be performed on data obtained by extrapolation to zero concentration. Some experimental values are listed in Table 23.4.

The use of laser light has led to further refinements in the application and interpretation of light scattering. There has been a shift of emphasis towards the investigation of the time-dependence of the positions of atoms and the orientation of macromolecules in solution. The properties can be studied by measuring the shift frequency that occurs when monochromatic light is scattered by a moving target in the technique called **dynamic light scattering**. In particular, laser light scattering can be used for the direct determination of the diffusional characteristics of macromolecules, and provides a fast, direct, and reliable method for the measurement of diffusion coefficients, even of macromolecules of low stability.

23.5 Magnetic resonance

NMR and ESR (Chapter 18) are both used to study macromolecules in solution to determine the locations of their magnetic nuclei and details of their motion.

Shift reagents

The data from magnetic resonance spectra are principally resonance frequencies and relaxation times. The resonance frequencies of NMR lines depend on the local magnetic fields at the protons or ^{13}C nuclei (or other magnetic nuclei) being examined. These local fields are modified if there are **shift reagents** present, which are ions or groups of atoms that possess a magnetic moment and that may have been deliberately incorporated into the molecule. The resonant nuclei at a distance R from the dipole experience an additional magnetic field with a strength that depends on distance as $1/R^3$, and so the distances of the nuclei from the ion may be determined. One complication is that in a tumbling molecule the shift gives distance and not direction, but this difficulty can be resolved by substituting an ion into different places and constructing a three-dimensional array of positions from separate experiments.

Shift reagents may be of several kinds. The technique was developed using lanthanide ions. However, it is not necessary to use paramagnetic species because the applied field can induce magnetic moments into sufficiently magnetizable groups (e.g. aromatic rings) that are already a part of the molecule, and these magnetized groups then act as sources of local field.

Broadening reagents

As explained in Chapter 18, the relaxation time depends on the strength and rate of fluctuation of the local field causing the relaxation. The relaxing efficiency of a magnetic dipole at a distance R from a resonant nucleus varies as the *square* of the field strength that it generates, and therefore it depends on $1/R^6$. Consequently, their locations can be deduced by studying relaxation times of nuclei in the presence of a paramagnetic **broadening reagent**. As for shift reagents, if this is done for a variety of substitutional sites, the network of separations can be used to build a map of positions. If two closely related lanthanide ions are used, one being a shift reagent (e.g.

Eu(III)) and the other a broadening reagent (e.g. Gd(III)), and which can reasonably be supposed to bind to the same sites, then since the former explores $1/R^3$ and the latter explores $1/R^6$, a detailed map of the locations of the magnetic nuclei can be constructed.

The paramagnetic centre itself can be examined by ESR. The electron spin relaxation times depend on the tumbling rate of the molecule. If it is supposed that the molecule tumbles like a sphere of radius a in a medium of viscosity η, then the **rotational correlation time** τ_R, the time it takes to rotate through about 1 rad (57°) in a large number of small steps, is given by the **Debye expression**:

$$\tau_R = \frac{4\pi a^3 \eta}{3kT} \tag{23}$$

For example, a sphere of radius 2 nm in water at room temperature has $\tau_R = 8$ ns, and the ESR technique is sensitive to processes on this time scale.

The use of **spin labels**, a paramagnetic nitroxide group, $-R(NO)R'$, attached to a molecule, has been particularly fruitful. They have been used to show, for example, that some molecules tumble at different rates around different axes. This **anisotropic rotational diffusion** can be detected from the shapes of ESR spectral lines and interpreted in terms of the shapes of the molecules. The same technique can be used to study the mobilities of groups within a macromolecule and to show that some groups can wave around loosely but others are sterically trapped into rigid conformations. We can thus explore not only the size and shape of a molecule, but also the flexibility of its structure.

Conformation and configuration

The **primary structure** of a macromolecule is the sequence of chemical segments making up the chain (or network if there is cross-linking). In the case of a synthetic polymer, virtually all the segments are identical, and it is sufficient to name the monomer used in the synthesis. Thus, the primary structure of polyethylene is the $-CH_2CH_2-$ unit, and the chemical constitution of the chain is specified by denoting it as $-(CH_2CH_2)_n-$.

The concept of primary structure ceases to be trivial in the case of biological macromolecules, for these are often chains of different molecules. Proteins, for example, are **polypeptides**, the name signifying chains formed from a number of different amino acids (about twenty occur naturally) strung together by the **peptide link**, $-CO-NH-$. The determination of the primary structure is then a highly complex problem of chemical analysis called **sequencing**. The **degradation** of a polymer is a disruption of its primary structure, when the chain breaks into shorter components.

The **secondary structure** of macromolecules refers to the (often local) spatial, well-characterized arrangement of the basic structural units. The secondary structure of an isolated molecule of polyethylene is a random coil, whereas that of a protein is a highly organized arrangement determined largely by hydrogen bonds, and taking the form of helices or sheets in various segments of the molecule. When the hydrogen bonds in a protein are destroyed (for instance, by heating, as when cooking an egg) the structure denatures into a random coil.

The difference between primary and secondary structure is closely related to the difference between the configuration and the conformation of a chain. The **configuration** refers to the structural features that can be changed only by breaking chemical bonds and forming new ones. Thus, the chains —A—B—C— and —A—C—B— have different configurations. The **conformation** of a chain refers to the spatial arrangement of its different parts, and one conformation can be changed into another by rotating one part of a chain round the bond joining it to another.

The **tertiary structure** of a protein refers to the overall three-dimensional structure of the molecule. For instance, many proteins have a helical secondary structure, but in many the helix is bent and distorted and the molecule has a globular tertiary structure. The **quaternary structure** is the manner in which some molecules are formed by the aggregation of others. Haemoglobin is a famous example: each molecule consists of four subunits of two types (the α and the β chains).

23.6 Random coils

As the first step in unravelling the various aspects of structure, we consider the most likely conformation of a chain of identical units that are incapable of forming hydrogen bonds or any other type of specific bond. Polyethylene is a simple example, but the general idea applies to a denatured protein. The simplest model is the **freely jointed chain**, in which any bond is free to make any angle with respect to the preceding one (Fig. 23.10a). The model is obviously an oversimplification, because a bond is actually constrained to a cone of angles around a direction defined by its neighbour (Fig. 23.10b).

The random coil is the least structured conformation of a polymer chain, and so corresponds to the state of maximum **conformational entropy**. Any stretching of the coil introduces order and reduces the entropy. Conversely, the formation of a random coil from a more extended form is a spontaneous process (so long as enthalpy contributions do not interfere). The elasticity of a **perfect elastomer**, a flexible polymer in which the internal energy is independent of the extension, may be discussed in these terms (see *Further information* at the end of the chapter). The random coil model is also a helpful starting point for estimating the orders of magnitude of the hydrodynamic properties (e.g. sedimentation rates) of polymers and denatured proteins in solution.

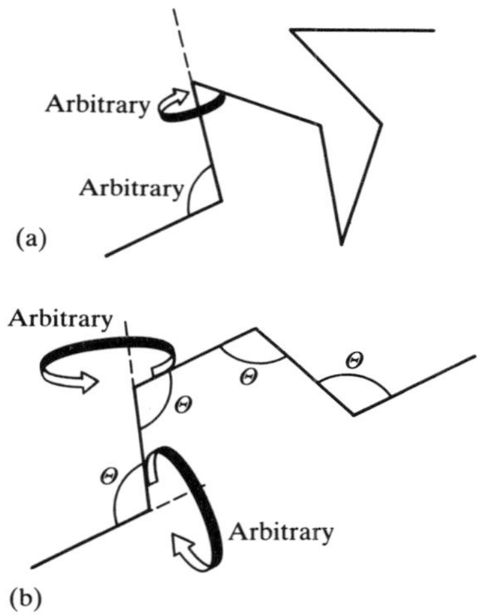

The radial distribution

A random coil is a version of the three-dimensional **random walk** (Chapter 25), with each bond representing a step taken in a random direction. For our present purposes all we need to know is that the probability of the ends of the chain lying in the range R to $R + dR$ is $f\,dR$, where

Arbitrary

Arbitrary

(a)

Arbitrary

Θ Θ

Θ

Θ

Arbitrary

(b)

Fig. 23.10 (a) A freely-jointed chain is like a three-dimensional random walk, each step being in an arbitrary direction but of the same length. (b) A better description is obtained by fixing the bond angle (e.g. at the tetrahedral angle) and allowing free rotation about a bond direction.

$$f = 4\pi \left(\frac{a}{\pi^{1/2}}\right)^3 R^2 e^{-a^2 R^2} \qquad a^2 = \frac{3}{2Nl^2} \tag{24}$$

N is the number of bonds and l is the bond length.[5] Equation 24 shows that in some coils (the proportion being given by the value of f with R large), the ends may be far apart, while in others their separation is small. An

[5] Here and elsewhere we are ignoring the fact that the chain cannot be longer than Nl. Although eqn 24 gives a non-zero probability for $R > Nl$, the values are so small that the errors in pretending that R can range up to infinity are negligible.

alternative interpretation is to regard each coil as writhing continually from one conformation to another, then $f \, dR$ is the probability that at any instant the chain will be found with the separation of its ends between R and $R + dR$.

Measures of size

There are several measures of the geometrical size of a random coil. The **contour length** R_c is the length of the macromolecule measured along its backbone from atom to atom (the total distance that the random walker walked). For a polymer of N C—C backbone bonds each of length l, the contour length is

$$R_c = Nl$$

The **root mean square separation** R_{rms} is a measure of the *average* separation of the ends of a random coil. It is the square root of the mean value of R^2, calculated by weighting each possible value of R^2 with the probability that R occurs:

$$R_{rms} = \left\{ \int_0^\infty R^2 f \, dR \right\}^{1/2} = N^{1/2} l \qquad (25)$$

We see that as the number of monomer units increases, the root mean square separation of its ends increases as $N^{1/2}$ (and its volume increases as $N^{3/2}$). Similarly, the radius of gyration of the coil (which is defined in eqn 21) is

$$R_g = \frac{N^{1/2} l}{6^{1/2}} \qquad (26)$$

The radius of gyration also increases as $N^{1/2}$.

Example 23.7: *Calculating the dimensions of a random coil*

Calculate the mean separation of the ends of a freely jointed polymer chain of N bonds of length l.

Answer. Use eqn 24 in the expression

$$\langle R^n \rangle = \int_0^\infty R^n f \, dR$$

with $n = 1$. The resulting integral is standard:

$$\int_0^\infty x^3 e^{-a^2 x^2} \, dx = \frac{1}{2a^4}$$

Therefore,

$$\langle R \rangle = 4\pi \left(\frac{a}{\pi^{1/2}} \right)^3 \int_0^\infty R^3 e^{-a^2 R^2} \, dR = \frac{2}{a\pi^{1/2}}$$

$$= \left(\frac{8}{3\pi} \right)^{1/2} N^{1/2} l$$

Comment. The result must be multipled by a factor when the chain is not freely jointed: see below.

Exercise. Evaluate the root mean square separation of the ends of the chain.
$$[N^{1/2} l]$$

Constrained chains

Before making use of these conclusions we must remove the obvious absurdity of allowing bond angles to take any value. This is simple in the case of long chains, for it is possible to take groups of neighbouring bonds, and to think about the direction of their resultant. Although individual bonds are constrained to a single cone of angle θ, the resultant of several lies in a random direction. By concentrating on groups rather than individuals, it turns out that for long chains the average values given above should be multiplied by

$$F = \left(\frac{1 - \cos\theta}{1 + \cos\theta}\right)^{1/2}$$

In the case of tetrahedral bonds, for which $\cos\theta = -\frac{1}{3}$ (i.e. $\theta = 109.5°$), $F = 2^{1/2}$. Therefore:

$$R_{rms} = 2^{1/2}N^{1/2}l \qquad R_g = \frac{N^{1/2}l}{3^{1/2}} \qquad (27)$$

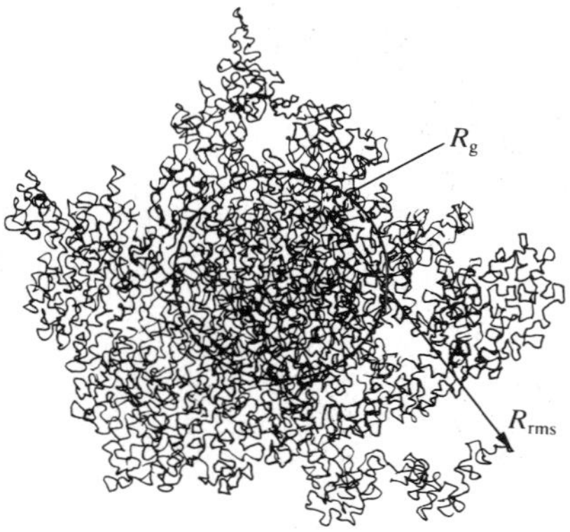

For example, in the case of a polyethylene chain with $M = 56\ \text{kg mol}^{-1}$, corresponding to $N = 4000$, since $l = 154\ \text{pm}$ for a C—C bond, we find $R_{rms} = 4.4\ \text{nm}$ and $R_g = 1.8\ \text{nm}$ (Fig. 23.11). This value of R_g means that, on average, the coils rotate like hollow spheres of radius 1.8 nm and mass equal to the molecular mass.

The model of a randomly coiled molecule is still an approximation even after the bond angles have been restricted because it does not take into account the impossibility of two or more atoms occupying the same place. Such self-avoidance tends to swell the coil, and so it is better to regard R_{rms} and R_g as lower bounds to the actual values. The model also ignores the role of the solvent: a poor solvent will tend to cause the coil to tighten so that solute–solvent contacts are minimized; a good solvent does the opposite.

23.7 Helices and sheets

Natural macromolecules need a precisely maintained conformation in order to function. The achievement of a specific conformation is the major remaining problem in protein synthesis, for although primary structures can be built, the product is inactive because the secondary structure cannot yet be produced.

The Corey–Pauling rules

The origin of the secondary structures of proteins is found in the rules formulated by Linus Pauling and Robert Corey. The essential feature is the stabilization of structures by hydrogen bonds involving the —CO—NH— peptide link. The latter can act both as a donor of the H atom (the NH part of the link) and as an acceptor (the CO part).

The **Corey–Pauling rules** are as follows:
(a) The atoms of the peptide link lie in a plane (Fig. 23.12).
(b) The N, H, and O atoms of a hydrogen bond lie in a straight line (with displacements of H tolerated up to not more than 30° from the N—O vector.

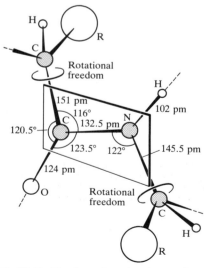

Fig. 23.11 A random coil in three dimensions. This one contains about 4000 units. The root mean square distance between the ends (R_{rms}) and the radius of gyration (R_g) are indicated.

Fig. 23.12 The dimensions that characterise the peptide link. The C—CO—NH—C atoms define a plane (the C—N bond has partial double-bond character), but there is rotational freedom around the C—CO and N—C bonds.

(c) All NH and CO groups are engaged in bonding.

The rules are satisfied by two structures. One, in which H-bonding occurs between peptide links of the same chain, is the *α* **helix**. The other in which H-bonding links different chains, is the *β*-**pleated sheet**; this form is the secondary structure of the protein fibroin, a principal constituent of silk.

The *α* helix is illustrated in Fig. 23.13. Each turn of the helix contains 3.6 amino acid residues, and so the period of the helix corresponds to 5 turns (18 residues). The pitch of a single turn is 544 pm. The N—H $\cdots$ O bonds lie parallel to the axis and link every fourth group (so residue i is linked to residues $i - 4$ and $i + 4$). There is freedom for the helix to be arranged as either a right- or a left-handed screw, but the overwhelming majority of natural polypeptides are right-handed on account of the preponderance of the L-configuration of the naturally occurring amino acids, as we explain below. The reason for their preponderance is uncertain, but it may be related to the symmetries of fundamental particles and the non-conservation of parity (the fact that this universe behaves differently from its hypothetical mirror image).

Conformational energy

The stabilities of different polypeptide geometries can be investigated by calculating the total potential energy of all the interactions between nonbonded atoms, and looking for a minimum. It turns out, in agreement with experience, that a right-handed *α* helix of L-amino acids is marginally more stable than a left-handed helix of the same acids.

The geometry of the chain can be specified by two angles, ϕ (the torsional angle for the N—C bond) and ψ (the torsional angle for the C—C bond). The illustration in Fig. 23.14 defines these angles,[6] and shows the all-*trans* form of the chain, in which all ϕ and ψ are 180°. A helix is obtained when all the ϕ are equal and when all the ψ are equal. For a right-handed *α* helix, both angles are positive. Since only two angles are needed to specify the conformation of a helix, and they range from $-180°$ to $+180°$, the potential energy of the entire molecule can be represented by a point on a plane in which one axis represents ϕ and the other ψ.

The potential energy of a given conformation (ϕ, ψ) can be calculated using the expressions developed in Section 22.4. For example, the interaction energy of two atoms separated by a distance R (which we know once ϕ

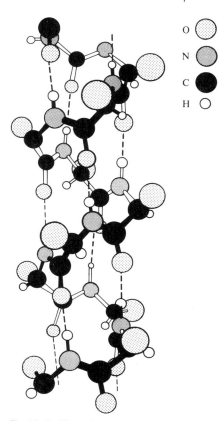

Fig. 23.13 The polypeptide *α* helix. There are 3.6 residues per turn, and a translation along the helix of 150 pm per residue, giving a pitch of 540 pm. The diameter (ignoring side chains) is about 600 pm.

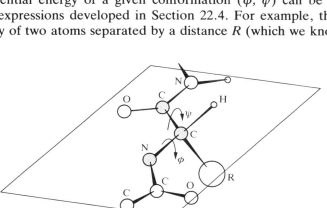

Fig. 23.14 The definition of the torsional angles ψ and ϕ between two peptide units. In this case (an *α*-L-polypeptide) the chain has been drawn in its all-trans form, with $\psi = \phi = 180°$.

[6] The sign convention is that a positive angle means that the front atom must be rotated clockwise to bring it into an eclipsed position relative to the rear atom.

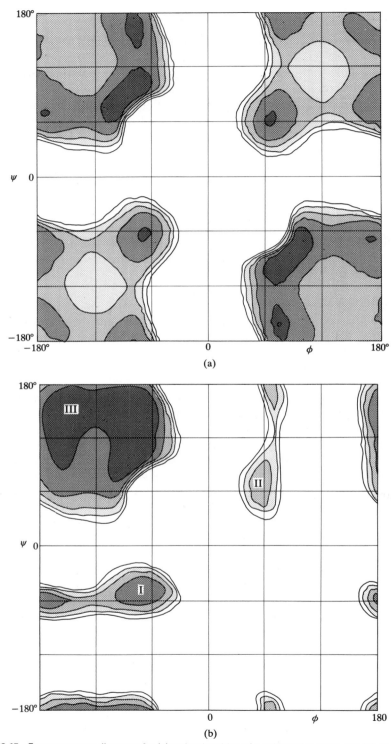

Fig. 23.15 Energy contour diagrams for (a) a glycyl residue of a polypeptide chain and (b) an alanine diagram of the potential energy. The glycyl diagram is symmetrical, but regions I and II in the alanine diagram correspond to right- and left-handed helices, are unsymmetrical, and the minimum in region I lies lower than that in region II. (After D. A. Brant and P. J. Flory, *J. mol. Biol.* 23, 47 (1967).)

and ψ are specified) can be given the Lennard–Jones $(12, 6)$ form. If the partial charges on the atoms (arising from ionic character in the bonds) are known, we can include a coulombic contribution of the form $1/R$. This is sometimes done by ascribing charges $-0.28e$ and $+0.28e$ to N and H respectively, and $-0.39e$ and $+0.39e$ to O and C respectively. There is also a torsional contribution arising from the barrier to internal rotation of one bond relative to another (just like the barrier to internal rotation in ethane), and which is normally expressed as

$$V = A(1 + \cos 3\phi) + B(1 + \cos 3\psi)$$

A and B are constants of the order of $1 \, kJ \, mol^{-1}$.

The potential energy contours for helical forms of polypeptide chains formed from the nonchiral amino acid glycine $(R = H)$ and the chiral molecule alanine are shown in Figs. 23.15a and 23.15b respectively. They were computed by summing all the contributions described above for each choice of angles, and then plotting contours of equal potential energy. The glycine map is symmetrical, with minima of equal depth at $\phi = -80°$, $\psi = +90°$ and at $\phi = +80°$, $\psi = -90°$. In contrast, the map for L-alanine is unsymmetrical, and there are three distinct low-energy conformations (marked I, II, III). The minima of regions I and II lie close to the angles typical of right- and left-handed α helices, but the former has a lower minimum, which is consistent with the formation of right-handed helices from the naturally occurring L-amino acids.

23.8 Higher-order structure

Helical polypeptide chains are folded into a tertiary structure if there are other bonding influences between the residues of the chain that are strong enough to overcome the interactions responsible for the secondary structure. The folding influences include —S—S— links, ionic interactions (which depend on the pH), and strong H-bonds (such as O—H $\cdots$ O—), and is illustrated by the structure of myoglobin (Fig. 23.16), the full structure (of 2600 atoms) having been determined by X-ray diffraction. The folding of the basic α helix caused by —S—S— links can be seen in the structure: about 77 per cent of the structure is α helix, the rest being involved in the folds.

Proteins with $M > 50 \, kg \, mol^{-1}$ are often found to be aggregates of two or more polypeptide chains. The possibility of such quaternary structure often confuses the determination of their molar masses, since different techniques might give values differing by factors of 2 or more. Haemoglobin, which consists of four myoglobin-like chains, is an example.

Protein denaturation can be caused by several means, and different aspects of structure may be affected. The permanent waving of hair, for example, is reorganization at the quaternary level. Hair is a form of the protein keratin, and its quaternary structure is thought to be a multiple helix, with the α helices bound together by —S—S— links and H-bonds (although there is some dispute about its precise structure). The process of permanent waving consists of disrupting these links, unravelling the keratin quaternary structure, and then reforming it into a more fashionable disposition. The 'permanence' is only temporary, however, because the structure of the newly formed hair is genetically controlled. Incidentally,

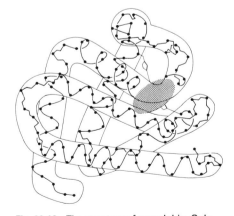

Fig. 23.16 The structure of myoglobin. Only the α-carbon atom positions are shown. The haem group, the oxygen binding group, is shown as a shaded region. (Based on M. F. Perutz, copyright *Scientific American*, 1964; with permission.)

normal hair grows at a rate that requires at least 10 twists of the keratin helix to be produced each second, and so very close inspection of the human scalp would show it to be literally writhing with activity.

Denaturation at the secondary level is brought about by agents that destroy hydrogen bonds. Thermal motion may be sufficient, in which case denaturation is a kind of intramolecular melting. When eggs are cooked the albumin is denatured irreversibly, and the protein collapses into a structure resembling a random coil. The **helix-coil transition** is sharp, like ordinary melting, because it is a cooperative process: when one hydrogen bond has been broken it is easier to break its neighbours, and then even easier to break theirs, and so on. The disruption cascades down the helix, and the transition occurs sharply. Denaturation may also be brought about chemically. For instance, a solvent that forms stronger H-bonds than those within the helix will compete successfully for the NH and CO groups. Acids and bases can cause denaturation by protonation or deprotonation of various groups.

Colloids

Colloids are dispersions of small particles of one material in another. 'Small' means something less than about 500 nm in diameter (about the wavelength of light). In general they are aggregates of numerous atoms or molecules, but are too small to be seen with an ordinary optical microscope. They pass through most filter papers, but can be detected by light-scattering, sedimentation, and osmosis.

23.9 Classification and preparation

The name given to the colloid depends on the two phases involved. **Sols** are dispersions of solids in liquids (such as clusters of gold atoms in water) or of solids in solids (such as ruby glass, which is a gold-in-glass sol, and achieves its colour by scattering). **Aerosols** are dispersions of liquids in gases (like fog and many sprays) and of solids in gases (such as smoke): the particles are often large enough to be seen with a microscope. **Emulsions** are dispersions of liquids in liquids (such as milk).

A further classification of colloids is as **lyophilic** (solvent attracting) and **lyophobic** (solvent repelling); in the case of water as solvent, the terms **hydrophilic** and **hydrophobic** are used instead. Lyophobic colloids include the metal sols. Lyophilic colloids generally have some chemical similarity to the solvent, such as —OH groups able to form H-bonds. A **gel** is a semi-rigid mass of a lyophilic sol in which all the dispersion medium has been absorbed by the sol particles.

The preparation of aerosols can be as simple as sneezing (which produces an aerosol). Laboratory and commercial methods make use of several techniques. Material (e.g. quartz) may be ground in the presence of the dispersion medium. Passing a heavy electric current through a cell may lead to the crumbling of an electrode into colloidal particles; arcing between electrodes immersed in the support medium also produces a colloid. Chemical precipitation sometimes results in a colloid. A precipitate (e.g. silver iodide) already formed may be dispersed by the addition of a **peptizing agent** (e.g. potassium iodide). Clays may be peptized by alkalis, the OH$^-$ ion being the active agent.

Emulsions are normally prepared by shaking the two components together, although some kind of **emulsifying agent** has to be used in order to stabilize the product. This emulsifier may be a soap (a long chain fatty acid), a surfactant (Section 23.11), or a lyophilic sol that forms a protective film around the dispersed phase. In milk, which is an emulsion of fats in water, the emulsifying agent is casein, a protein containing phosphate groups. That casein is not completely successful in stabilizing milk is apparent from the formation of cream on the surface: the dispersed fats coalesce into oily droplets which float to the surface. This may be prevented by ensuring that the emulsion is dispersed very finely initially: violent agitation with ultrasonics brings this about, the product being 'homogenized' milk.

Aerosols are formed when a spray of liquid is torn apart by a jet of gas. The dispersal is aided if a charge is applied to the liquid, for then the electrostatic repulsions blast the jet apart into droplets. This procedure may also be used to produce emulsions, for the charged liquid phase may be squirted into another liquid.

Colloids are often purified by dialysis. The aim is to remove much (but not all, for reasons explained later) of the ionic material that may have accompanied their formation. As in the discussion of the Donnan effect, a membrane (e.g. cellulose) is selected which is permeable to solvent and ions, but not to the colloidal particles. Dialysis is very slow, and is normally accelerated by applying an electric field and making use of the charge carried by many colloids; the technique is then called **electrodialysis**.

23.10 Surface, structure, and stability

The principal feature of colloids is the very great surface area of the dispersed phase in comparison with the same amount of ordinary material. For example, a 1 cm cube of material has a surface area of 6 cm^2, but when it is dispersed as 10^{18} little 10 nm cubes the total surface area is $6 \times 10^6 \text{ cm}^2$ (about the size of a tennis court). This dramatic increase in area means that surface effects are of dominating importance in colloid chemistry.

The stability of colloids

As a result of their great surface area, colloids are thermodynamically unstable with respect to the bulk: since $dG = \gamma \, d\sigma$ where γ is the surface tension, dG is negative for a decrease in area. The apparent stability must therefore be a consequence of the kinetics of collapse: colloids are kinetically, not thermodynamically, stable.

At first sight even the kinetic argument seems to fail: colloidal particles attract each other over large distances, and so there is a long-range force tending to collapse them down into a single blob. The reasoning behind this remark is as follows. The energy of attraction between two individual atoms, one in each colloidal particle, varies as their separation as $1/R_{ij}^6$ (Section 22.3). The *sum* of all these pairwise interactions between all the atoms in the two neighbouring molecules, however, decreases only as $1/R^2$, where R is the separation of the centres of the particles, which has a much longer range than the $1/R^6$ dependence characteristic of individual particles and small molecules.

Several factors oppose the long-range dispersion attraction. For example, there may be a protective film at the surface of the colloidal particles that stabilizes the interface and cannot be penetrated when two particles touch. Thus the surface atoms of a platinum sol in water react chemically and are turned into $—Pt(OH)_3H_3$, and this layer encases the particle like a shell. A fat can be emulsified by a soap because the long hydrocarbon tails penetrate the oil droplet but the $—CO_2^-$ head groups (or other hydrophilic groups in detergents) surround the surface, form hydrogen bonds with water, and give rise to a shell of negative charge that repels a possible approach from another similarly charged particle.

Micelle formation and the hydrophobic interaction

Soap molecules can group together as **micelles**, colloid-sized clusters of molecules, even in the absence of grease droplets, for their hydrophobic tails tend to congregate, and their hydrophilic heads provide protection (Fig. 23.17). Micelles form only above the **critical micelle concentration** (CMC) and above the **Kraft temperature**. Non-ionic surfactant molecules may cluster together in swarms of 1000 or more, but ionic species tend to be disrupted by the electrostatic repulsions between head groups and are normally limited to groups of between 10 and 100 molecules. The micelle population is often polydisperse, and the shape of the individual micelles varies with concentration. Although spherical micelles do occur, they are more commonly flattened spheres close to the CMC, and rod-like at higher concentrations. The interior of a micelle is like a droplet of oil, and magnetic resonance shows that the hydrocarbon tails are mobile, but slightly more restricted than in the bulk.

Micelles are important in industry and biology on account of their solubilizing function: matter can be transported by water after it has been dissolved in their hydrocarbon interiors. For this reason, micellar systems are used as detergents and drug carriers, and for organic synthesis, froth flotation, and petroleum recovery.

The thermodynamics of micelle formation shows that the enthalpy of formation in aqueous systems is probably positive (that is, that they are endothermic) with $\Delta H \approx 1–2\,kJ$ per mole of surfactant. That they do form above the CMC indicates that the entropy change accompanying their formation must then be positive, and measurements suggest a value of about $+140\,J\,K^{-1}\,mol^{-1}$ at room temperature. That the entropy change is positive even though the molecules are clustering together shows that there must be a contribution to the entropy from the solvent and that its molecules must be more free to move once the solute molecules have herded into small clusters. This is plausible, because each individual solute molecule is held in an organized solvent cage (Fig. 23.18), but once the micelle has formed the solvent molecules need form only a single (admittedly larger) cage. The increase in energy when hydrophobic groups cluster together and reduce their structural demands on the solvent is the origin of the **hydrophobic interaction** that tends to stabilize groupings of hydrophobic groups in biological macromolecules. The hydrophobic interaction is an example of an ordering process that is stabilized by a tendency toward greater disorder of the solvent.

The electric double layer

Apart from the physical stabilization of colloid particles, a major source of kinetic stability is the existence of an electric charge on their surfaces. On

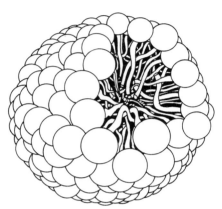

Fig. 23.17 A representation of a spherical micelle. The hydrophilic groups are represented by spheres, and the hydrophobic hydrocarbon chains are represented by the stalks: the latter are mobile.

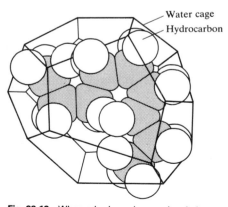

Water cage
Hydrocarbon

Fig. 23.18 When a hydrocarbon molecule is surrounded by water, the water molecules form a clathrate cage. As a result of this acquisition of structure, the entropy of the water decreases, and so the dispersal of the hydrocarbon into water is entropy-opposed; its coalescence is entropy-favoured.

account of this charge, ions of opposite charge tend to cluster nearby, and an ionic atmosphere is formed, just as was described for ions in Section 10.2.

Two regions of charge must be distinguished. First, there is a fairly immobile layer of ions that stick tightly to the surface of the colloidal particle, and which may include water molecules (if that is the dispersion medium). The radius of the sphere that captures this rigid layer is called the **radius of shear**, and is the major factor determining the mobility of the particles. The electric potential at the radius of shear relative to its value in the distant, bulk medium is called the **zeta potential** ζ or the **electrokinetic potential**. The charged unit attracts an oppositely charged ionic atmosphere. The inner shell of charge and the outer atmosphere is called the **electric double layer**. The structure of the atmosphere can be described in the same way as in the Debye–Hückel theory of ionic solutions, and we shall consider it in more detail in Chapter 30.

At high ionic strengths the atmosphere is dense and the potential falls to its bulk value within a short distance. In this case there is little electrostatic repulsion to hinder the close approach of two colloid particles. As a result **flocculation**, the coagulation of the colloid, readily occurs as a consequence of the van der Waals forces.

Since the ionic strength is increased by the addition of ions, particularly those of high charge type, such ions act as flocculating agents. This is the basis of the empirical **Schulze–Hardy rule**, that hydrophobic colloids are flocculated most efficiently by ions of opposite charge type and high charge number. The Al^{3+} ions in alum are very effective, and are used in stiptic pencils for inducing the congealing of blood. When river water containing colloidal clay flows into the sea, the brine induces coagulation and is a major cause of silting in estuaries.

Metal oxide sols tend to be positively charged whereas sulphur and the noble metals tend to be negatively charged. Naturally occurring macro-molecules also acquire a charge when dispersed in water, and an important feature of proteins and other natural macromolecules is that their overall charge depends on the pH of the medium. For instance, in acidic environments protons attach to basic groups, and the net charge of the macromolecule is positive; in basic media the net charge is negative as a result of proton loss. At the **isoelectric point** the pH is such that there is no net charge on the macromolecule.

Example 23.8: *Determining the isoelectric point*

The mobility of BSA in aqueous solution was monitored at several values of pH, and the data are listed below. What is the isoelectric point of the protein?

pH	4.20	4.56	5.20	5.65	6.30	7.00
Mobility/$(\mu m\,s^{-1})$	0.50	0.18	−0.25	−0.65	−0.90	−1.25

Answer. The macromolecule has zero electrophoretic mobility when it is uncharged. Therefore, the isoelectric point is the pH at which it does not migrate in an electric field. We should therefore plot mobility against pH and find by interpolation the pH of zero mobility. The data are plotted in Fig. 23.19. The mobility is zero at pH = 4.8; hence pH = 4.8 is the isoelectric point.

Comment. Sometimes the isoelectric point must be obtained by extrapolation because the macromolecule might not be stable over the whole pH range.

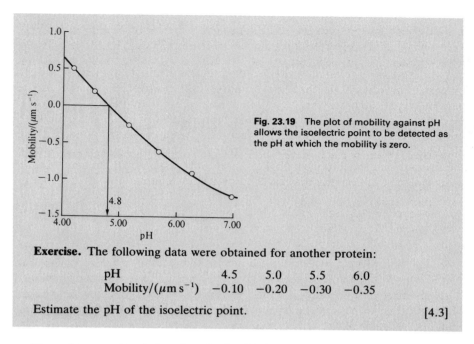

Fig. 23.19 The plot of mobility against pH allows the isoelectric point to be detected as the pH at which the mobility is zero.

Exercise. The following data were obtained for another protein:

pH	4.5	5.0	5.5	6.0
Mobility/$(\mu m\,s^{-1})$	−0.10	−0.20	−0.30	−0.35

Estimate the pH of the isoelectric point.   [4.3]

The primary role of the electric double layer is to confer kinetic stability. Colliding colloidal particles break through the double layer and coalesce only if the collision is sufficiently energetic to disrupt the layers of ions and solvating molecules, or if thermal motion has stirred away the surface accumulation of charge. This may happen at high temperatures, which is one reason why sols precipitate when they are heated. The protective role of the double layer is the reason why it is important not to remove all the ions when a colloid is being purified by dialysis, and why proteins coagulate most readily at their isoelectric point.

The presence of charge on colloidal particles and natural macromolecules also permits us to control their motion, such as in dialysis and electrophoresis. Apart from its application to the determination of molar mass (Section 23.2), electrophoresis has several analytical and technological applications. One analytical application is to the separation of different macromolecules, and a typical apparatus is illustrated in Fig. 23.20. Technical applications

Fig. 23.20 The layout of a simple electrophoresis apparatus. The sample is introduced into the trough in the gel, and the different components form separated bands under the influence of the potential difference.

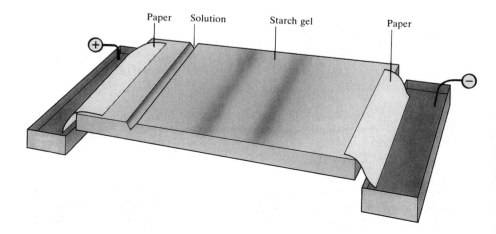

include silent ink-jet printers, the painting of objects by airborne charged paint droplets, and electrophoretic rubber forming by deposition of charged rubber molecules on anodes formed into the shape of the desired product (e.g. surgical gloves).

Surface tension and surfactants

A **surfactant** or **surface-active agent** is a species that is active at the interface between two phases, such as the interface between hydrophilic and hydrophobic phases. A surfactant accumulates at the interface, and modifies the surface tension.

23.11 The surface excess

Consider two phases α and β in contact; the system consists of several components J, each one present in an overall amount n_J. If the components were distributed uniformly through the phases right up to the interface, which is taken to be a plane of surface area σ, the total Gibbs function G would be the sum of the Gibbs function of each phase,

$$G = G(\alpha) + G(\beta)$$

But the components are not uniform, and the sum of the two Gibbs functions differs from G by an amount called the **surface Gibbs function** $G(\sigma)$:

$$G(\sigma) = G - \{G(\alpha) + G(\beta)\}$$

Similarly, if the bulk contains an amount $n_J(\alpha)$ of J in phase α and an amount $n_J(\beta)$ in phase β, the total amount of J differs from their sum by the amount

$$n_J(\sigma) = n_J - \{n_J(\alpha) + n_J(\beta)\}$$

We can express this excess as an amount per unit area of the surface by introducing the **surface excess** Γ_J:

$$\Gamma_J = \frac{n_J(\sigma)}{\sigma} \tag{28}$$

$n_J(\sigma)$ and Γ_J may be either positive (an accumulation of J at the interface) or negative (a deficiency at the surface).

A general change in G is brought about by changes in T, p, σ, and the n_J:

$$dG = -S\, dT + V\, dp + \gamma\, d\sigma + \sum_J \mu_J\, dn_J$$

When this relation is applied to G, $G(\alpha)$, and $G(\beta)$ we find

$$dG(\sigma) = -S(\sigma)\, dT + \gamma\, d\sigma + \sum_J \mu_J\, dn_J(\sigma) \tag{29a}$$

because at equilibrium the chemical potential of each component is the same in every phase, $\mu_J(\alpha) = \mu_J(\beta) = \mu_J(\sigma)$. Just as in the discussion of partial molar quantities (Section 7.1), eqn 29a integrates at constant temperature to

$$G(\sigma) = \gamma\sigma + \sum_{J} \mu_J n_J(\sigma) \qquad (29b)$$

23.12 The Gibbs surface tension equation

We are seeking a connection between the change in surface tension $d\gamma$ and the change in composition at the interface. Therefore, we use the argument which in Section 7.1 led to the Gibbs–Duhem equation (eqn 6 of that section) but this time we compare eqn 29a at constant temperature ($dT = 0$) with

$$dG(\sigma) = \gamma\, d\sigma + \sigma\, d\gamma + \sum_{J} n_J(\sigma)\, d\mu_J + \sum_{J} \mu_J\, dn_J(\sigma)$$

The comparison implies that, at constant temperature,

$$\sigma\, d\gamma + \sum_{J} n_J(\sigma)\, d\mu_J = 0$$

Division by σ then gives the **Gibbs surface tension equation**:

$$d\gamma = -\sum_{J} \Gamma_J\, d\mu_J \qquad (30)$$

which relates the change in surface tension and the chemical potentials of the substances present.

The effect of a surfactant

It is easy to derive a simpler form of the Gibbs equation for a surfactant S distributed between the two phases of the system by making the approximation that the 'oil' and 'water' phases are separated by a geometrically flat surface. This implies that only the surfactant accumulates at the surface, and so both Γ_{Oil} and Γ_{Water} are zero. Then the Gibbs equation becomes

$$d\gamma = -\Gamma_S\, d\mu_S$$

For dilute solutions,

$$d\mu_S = RT\, d \ln c$$

where c is the concentration of the surfactant. It follows that

$$d\gamma = \frac{-RT\Gamma_S\, dc}{c}$$

at constant temperature, or

$$\left(\frac{\partial\gamma}{\partial c}\right)_T = \frac{-RT\Gamma_S}{c} \qquad (31)$$

If the surfactant accumulates at the interface, its surface excess is positive and eqn 31 implies that $(\partial\gamma/\partial c)_T$ is then negative. That is, the surface tension *decreases* when a solute accumulates at a surface. Conversely, if the concentration dependence of γ is known, the surface excess may be predicted. The predictions have been tested by the simple (but technically elegant) procedure of slicing thin layers off the surfaces of solutions and analysing their compositions. Note that the result summarized by eqn 31 is based on the assumption that the solutions are ideal; there may be marked deviations at the concentrations of detergents used in practice.

The surface phase

We can round off this section, chapter, and part of the text by linking the properties of an ideal surface phase with the properties of perfect gases with which the text began. At low surfactant concentrations the surface tension can be expected to decrease linearly, and we can write

$$\gamma = \gamma^* - Kc$$

with K some constant. Then, eqn 31 leads to

$$\Gamma = \frac{Kc}{RT} = \frac{\gamma^* - \gamma}{RT}$$

We define the **surface pressure** π as

$$\pi = \gamma^* - \gamma$$

and using eqn 28 find that

$$\pi\sigma = n(\sigma)RT$$

This is the equation of state of a two-dimensional perfect gas. The excess solute at the interface of dilute, ideal solutions can therefore be pictured as a perfect gas confined to a two-dimensional surface.

Further information: the elasticity of rubber

Consider a one-dimensional freely-jointed polymer. The conformation can be expressed in terms of the number of bonds pointing to the right (N_R) and the number pointing to the left (N_L). The distance between the ends of the chain is $(N_R - N_L)l$, where l is the length of an individual bond. We write $n = N_R - N_L$ and the total number of bonds as $N = N_R + N_L$.

The number of ways of forming a chain with a given end-to-end distance nl is the number of ways of having N_R right-pointing and N_L left-pointing bonds, and is given by the binomial coefficient

$$W = \frac{N!}{N_L! \, N_R!} = \frac{N!}{\{\tfrac{1}{2}(N + n)\}! \, \{\tfrac{1}{2}(N - n)\}!}$$

The conformational entropy of the chain, $S = k \ln W$, is therefore

$$S/k = \ln N! - \ln N_R! - \ln N_L!$$

Since the factorials are large (except for large extensions), we can use Stirling's approximation in the form

$$\ln x! \approx \ln (2\pi)^{1/2} + (x + \tfrac{1}{2}) \ln x - x$$

to obtain

$$S/k = -\ln (2\pi)^{1/2} + (N + 1) \ln 2 + (N + \tfrac{1}{2}) \ln N$$
$$- \tfrac{1}{2} \ln \{(N + n)^{N+n+1}(N - n)^{N-n+1}\}$$

The most probable conformation of the chain is the one with the ends

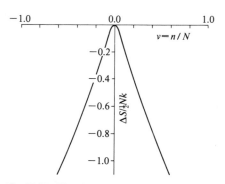

Fig. 23.21 The change in molar entropy of a perfect rubber as its extention changes. $v = 1$ corresponds to complete extension; $v = 0$, the conformation of highest entropy, corresponds to the random coil.

together $(n = 0)$, as may be confirmed by differentiation. Therefore, the maximum entropy is

$$S_{\text{max}}/k = -\ln (2\pi)^{1/2} + (N + 1) \ln 2 - \tfrac{1}{2} \ln N$$

The change in entropy when the chain is stretched from its most probable conformation until the distance between its ends is nl is therefore

$$\Delta S = S - S_{\text{max}} = -\tfrac{1}{2}kN \ln \{(1 + v)^{1+v}(1 - v)^{1-v}\} \qquad \text{(A1)}$$

with $v = n/N$. Equation A1 is plotted in Fig. 23.21. Note that it is negative for all extensions, and so we conclude that adiabatic contraction of the chain to its fully coiled state is spontaneous.

The work done on a piece of rubber when it is extended through a distance dx is $F\,dx$, where F is the restoring force. The First Law is therefore

$$dU = T\,dS - p\,dV + F\,dx$$

It follows that

$$\left(\frac{\partial U}{\partial x}\right)_{T,V} = T\left(\frac{\partial S}{\partial x}\right)_{T,V} + F$$

In a perfect rubber, as in a perfect gas, the internal energy is independent of the dimensions (at constant temperature), and so $(\partial U/\partial x)_{T,V} = 0$. The restoring force is therefore

$$F = -T\left(\frac{\partial S}{\partial x}\right)_{T,V}$$

If the statistical expression for the entropy is introduced into this equation (and we evade problems arising from the constant volume constraint by assuming that the sample contracts laterally when it is stretched), we obtain

$$F = \frac{-T}{l}\left(\frac{\partial S}{\partial n}\right)_{T,V} = \frac{-T}{Nl}\left(\frac{\partial S}{\partial v}\right)_{T,V}$$

$$= \frac{kT}{2l} \ln \frac{1 + v}{1 - v} \qquad \text{(A2)}$$

which is plotted in Fig. 23.22. At low extensions $(v \ll 1)$,

$$F \approx \frac{vkT}{l} = (N_{\text{R}} - N_{\text{L}})\frac{kT}{Nl}$$

and the sample obeys Hooke's Law (that the restoring force is proportional to the displacement), but it departs from it at higher extensions.

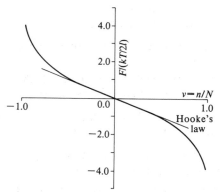

Fig. 23.22 The restoring force, f, of a one-dimensional perfect rubber. For small extensions, f is linearly proportional to the extension, in accord with Hooke's law.

Further reading

Biological macromolecules

K. E. van Holde, *Physical biochemistry*. Prentice–Hall, Englewood Cliffs (1971).
D. Freifelder, *Physical biochemistry*. W. H. Freeman & Co., San Francisco (1982).
A. G. Marshall, *Biophysical chemistry*. Wiley-Interscience, New York (1978).

C. R. Cantor and P. R. Schimmel, *Biophysical chemistry*. W. H. Freeman & Co., San Francisco (1980).
L. Stryer, *Biochemistry* (3rd edn). W. H. Freeman & Co., New York (1988).

Polymers

P. Flory, *Principles of polymer chemistry*. Cornell University Press (1953).
F. W. Billmeyer, *Textbook of polymer science*. Wiley-Interscience, New York (1984).
N. G. McCrum, C. P. Buckley, and C. B. Bucknall, *Principles of polymer engineering*. Oxford University Press (1988).

Colloids

D. Eagland, Colloids and interparticle forces. In *Contemporary Physics* **14,** 119 (1973): J. N. Israelichvili; *ibid.* **15,** 159 (1974).
A. W. Adamson, *Physical chemistry of surfaces*. Wiley, New York (1976).
M. J. Rosen, *Surfactants and interfacial phenomena*. Wiley-Interscience, New York (1978).
R. J. Hunter, *Foundations of colloid science,* Volume I. Clarendon Press, Oxford (1987).

Exercises

23.1 Calculate the number-average molar mass and the mass-average molar mass of a mixture of equal amounts of two polymers, one having $M = 62 \text{ kg mol}^{-1}$ and the other $M = 78 \text{ kg mol}^{-1}$.

23.2 The concentration dependence of the osmotic pressure of solutions of a macromolecule at 20°C was found to be as follows:

$c/(\text{g L}^{-1})$	1.21	2.72	5.08	6.60
Π/Pa	134	321	655	898

Find the molar mass of the macromolecule and the osmotic virial coefficient.

23.3 The osmotic pressures of solutions of polystyrene in toluene were measured at 25°C, with the following results:

$c/(\text{mg cm}^{-3})$	3.2	4.8	5.7	6.9	7.8
h/cm	3.11	6.22	8.40	11.73	14.90

h is the height of the solution of density 0.867 g cm^{-3} corresponding to the osmotic pressure. Obtain the molar mass of the polymer by plotting h/c against c.

23.4 The osmotic pressure of a fraction of polyvinylchloride in a ketone solvent was measured at 25°C. The density of the solvent (which is virtually equal to the density of the solution) was 0.798 g cm^{-3}. Calculate the molar mass and the virial coefficient B of the fraction from the following data:

$c/(\text{g}/100 \text{ cm}^3)$	0.200	0.400	0.600	0.088	1.000
h/cm	0.48	1.2	1.86	2.76	3.88

23.5 A polymer chain consists of 700 segments, each 0.90 nm long. If the chain were ideally flexible, what would be the r.m.s. separation of the ends of the chain?

23.6 The radius of gyration of a long chain molecule is found to be 7.3 nm. The chain consists of C—C links. Assume the chain is randomly coiled and estimate the number of links in the chain.

23.7 The concentration dependence of the viscosity of a polymer solution is found to be as follows:

$c/(\text{g L}^{-1})$	1.32	2.89	5.73	9.17
$\eta/(\text{g m}^{-1}\text{s}^{-1})$	1.08	1.20	1.42	1.73

The viscosity of the solvent is $0.985 \text{ g m}^{-1}\text{s}^{-1}$. What is the intrinsic viscosity of the polymer?

23.8 In a sedimentation experiment the position of the boundary as a function of time was found to be as follows:

$t/\text{minutes}$	15.5	29.1	36.4	58.2
r/cm	5.05	5.09	5.12	5.19

The rotational speed of the centrifuge was 45 000 r.p.m. Calculate the sedimentation constant of the solute.

23.9 In an ultracentrifuge experiment at 20°C on bovine serum albumin the following data were obtained: $\rho = 1.001 \text{ g cm}^{-3}$, $v_s = 1.112 \text{ cm}^3 \text{ g}^{-1}$, $\omega/2\pi = 322 \text{ Hz}$,

r/cm	5.0	5.1	5.2	5.3	5.4
$c/(\text{mg cm}^{-3})$	0.536	0.284	0.148	0.077	0.039

Evaluate M.

23.10 At 20°C the diffusion coefficient of a macromolecule is found to be $8.3 \times 10^{-7} \text{ cm}^2 \text{ s}^{-1}$. Its sedimentation constant is 3.2 Sv in a solution of density 1.06 g cm^{-3}. The specific volume of the macromolecule is $0.656 \text{ cm}^3 \text{ g}^{-1}$. Calculate the molar mass of the macromolecules.

23.11 A solution consists of solvent, 30 per cent by mass of a dimer with $M = 30 \text{ kg mol}^{-1}$ and its monomer. What average molar mass would be obtained from measurement of: (a) osmotic pressure, (b) light scattering?

23.12 A polyelectrolyte $Na_{20}P$ with $M = 100 \text{ kg mol}^{-1}$ at a concentration $1.00 \text{ g}/(100 \text{ cm}^3)$ was equilibrated in the presence of $0.0010 \text{ M NaCl}(aq)$ (i.e. $[Na^+]_R = 0.0010 \text{ M}$). What is $[Na^+]_L$ at equilibrium?

23.13 At the start of a membrane equilibrium experiment, the first compartment contains 1.00 L of solution with a concentration of 0.100 M of NaX, where X^- cannot pass through the membrane. The second compartment has 2.00 L of 0.030 M NaCl solution. Find the concentration of Cl^- ion in the first compartment after equilibrium is established.

23.14 The data from a sedimentation equilibrium experiment performed at 300 K on a macromolecular solute in aqueous solution show that a graph of $\ln c$ against r^2 is a straight line with a slope of 729 cm^{-2}. The rotational speed of the centrifuge was $50\,000 \text{ r.p.m.}$ The specific volume of the solute is $0.61 \text{ cm}^3 \text{ g}^{-1}$. Calculate the molar mass of the solute using $\rho = 0.997 \text{ g cm}^{-3}$.

23.15 Calculate the radial acceleration (as so many g) in a cell placed at 6.0 cm from the centre of rotation in an ultracentrifuge operating at $80\,000 \text{ r.p.m.}$

23.16 Evaluate the rotational correlation time for serum albumin in water at $25°C$ on the basis that it is a sphere of radius 3.0 nm. What is the value for a CCl_4 molecule in carbon tetrachloride at $25°C$? (Viscosity data in Table 22.6; take $a(CCl_4) \approx 250 \text{ pm}$.)

23.17 The surface tensions of aqueous salt solutions are normally greater than that of water itself. Does the salt accumulate at the surface?

Problems

Numerical problems

23.1 Calculate the speed of operation (in r.p.m.) of an ultracentrifuge needed to obtain a readily measurable concentration gradient in a sedimentation equilibrium experiment. Take that gradient to be a concentration at the bottom of the cell about five times greater than at the the top. Use $r_{top} = 5.0 \text{ cm}$, $r_{bottom} = 7.0 \text{ cm}$, $M \approx 10^5 \text{ g mol}^{-1}$, $\rho v_s \approx 0.75$, $T = 298 \text{ K}$.

23.2 At the start of a Donnan equilibrium experiment, the first compartment contains 2.00 L of solution which is 0.015 M in the polyelectrolyte Na_2P and 0.010 M in NaCl. The second compartment has 2.00 L of solution which is 0.0050 M in NaCl. What is the potential difference across the membrane arising from the Na^+ ion concentration difference at 300 K?

23.3 Investigation of the composition of the solutions used to study the osmotic pressure due to a polyelectrolyte with $v = 20$ showed that at equilibrium the concentrations corresponded to $[Cl^-] \approx 0.020 \text{ M}$. Calculate the osmotic virial coefficient for $v = 20$. Does it dominate the effect of excluded volume?

23.4 Sedimentation studies on haemoglobin in water gave a sedimentation constant $S = 4.5 \text{ Sv}$ at $20°C$. The diffusion coefficient is $6.3 \times 10^{-7} \text{ cm}^2 \text{ s}^{-1}$ at the same temperature. Calculate the molar mass of haemoglobin using $v_s = 0.75 \text{ cm}^3 \text{ g}^{-1}$ for its partial specific volume and $\rho = 0.998 \text{ g cm}^{-3}$ for the density of the solution. Estimate the effective radius of the haemoglobin molecule given that the viscosity of the solution is $1.00 \times 10^{-3} \text{ kg m}^{-1}\text{s}^{-1}$.

23.5 The diffusion coefficient for bovine serum albumin, a prolate ellipsoid, is $6.97 \times 10^{-7} \text{ cm}^2 \text{ s}^{-1}$ at $20°C$, its partial specific volume is $0.734 \text{ cm}^3 \text{ g}^{-1}$, and its sedimentation constant is 5.01 Sv in a solution of density 1.0023 g cm^{-3} and viscosity $1.00 \times 10^{-3} \text{ kg m}^{-1}\text{s}^{-1}$. Estimate its dimensions.

23.6 The rate of sedimentation of a recently isolated protein was monitored at $20°C$ and with a rotor speed of $50\,000 \text{ r.p.m.}$ The boundary receded as follows:

t/s	0	300	600	900	1200	1500	1800
r/cm	6.127	6.153	6.179	6.206	6.232	6.258	6.284

Calculate the sedimentation constant and the molar mass of the protein on the basis that its partial specific volume is $0.728 \text{ cm}^3 \text{ g}^{-1}$ and its diffusion coefficient is $7.62 \times 10^{-7} \text{ cm}^2 \text{ s}^{-1}$ at $20°C$, the density of the solution then being 0.9981 g cm^{-3}. Suggest a shape for the protein given that the viscosity of the solution is $1.00 \times 10^{-3} \text{ kg m}^{-1}\text{s}^{-1}$ at $20°C$.

23.7 The viscosities of solutions of polyisobutylene in benzene were measured at $24°C$ (the θ temperature for the system) with the following results:

$c/(g/100 \text{ cm}^3)$	0	0.2	0.4	0.6	0.8	1.0
$\eta/(10^{-3} \text{ kg m}^{-1}\text{s}^{-1})$	0.647	0.690	0.733	0.777	0.821	0.865

Use the information in Table 23.2 to deduce the molar mass of the polymer.

23.8 Evaluate the radius of gyration of (a) a solid sphere of radius a, (b) a long straight rod of radius a and length l. Show that in the case of a solid sphere of specific volume v_s,

$$R_g/\text{nm} \approx 0.056\,902 \times \{(v_s/\text{cm}^3\,\text{g}^{-1})(M/\text{g mol}^{-1})\}^{1/3}$$

Evaluate R_g for species with $M = 100 \text{ kg mol}^{-1}$, $v_s = 0.750 \text{ cm}^3 \text{ g}^{-1}$, and, in the case of the rod, a radius 0.5 nm.

23.9 Use the information below and the expression for R_g of a solid sphere derived in Problem 23.8, to classify the species below as globular or rod-like.

	$M/(\text{kg mol}^{-1})$	$v_s/(\text{cm}^3 \text{g}^{-1})$	Measured R_g/nm
Serum albumin	66×10^3	0.752	2.98
Bushy stunt virus	10.6×10^6	0.741	12.0
DNA	4×10^6	0.556	117.0

23.10 In formamide as solvent, poly(γ-benzyl- L-glutamate) is found by light scattering experiments to have a radius of gyration proportional to M; in contrast, polystyrene in butanone has R_g proportional to $M^{1/2}$. Present arguments to show that the first polymer is a rigid rod, while the second is a random coil.

23.11 The surface tensions of a series of aqueous solutions of a surfactant were measured at 20°C, with the following results.

$[A]/\text{M}$	0	0.10	0.20	0.30	0.40	0.50
$\gamma/(\text{mN m}^{-1})$	72.8	70.2	67.7	65.1	62.8	59.8

Calculate the surface excess concentration and the surface pressure π exerted by the surfactant, and investigate whether $\pi\sigma = nRT$ is satisfied.

23.12 The surface tensions of solutions of salts in water at molar concentration c can be expressed in the form $\gamma = \gamma^* + (c/M)\Delta\gamma$. The values of $\Delta\gamma$ at 20°C and near $c = 1 \text{ M}$ are as follows:

$$\Delta\gamma/(\text{mN m}^{-1}) = 1.4(\text{KCl}), 1.64(\text{NaCl}), 2.7(\text{Na}_2\text{CO}_3)$$

Calculate the surface excess concentrations when the bulk concentrations are 1.0 M.

Theoretical problems

23.13 A polymerization process produced a gaussian distribution of polymers in the sense that the proportion of molecules having a molar mass in the range M to $M + dM$ was proportional to $\exp\{-(M - \bar{M})^2/2\Gamma\} \, dM$. What is the number average molar mass when the distribution is narrow?

23.14 Calculate the excluded volume in terms of the molecular volume on the basis that the molecules are spheres of radius a. Evaluate the osmotic virial coefficient in the case of bushy stunt virus, $a \approx 14.0$ nm, and haemoglobin, $a \approx 3.2$ nm. Evaluate the percentage deviation of the osmotic pressures of $1.00 \text{ g}/(100 \text{ cm}^3)$ solutions of bushy stunt virus ($M \approx 1.07 \times 10^7 \text{ g mol}^{-1}$) and haemoglobin ($M \approx 66.5 \text{ kg mol}^{-1}$) from the ideal solution values.

23.15 The effective radius of a random coil a is related to its radius of gyration R_g by $a = \gamma R_g$, with $\gamma \approx 0.85$. Deduce an expression for the osmotic virial coefficient B in terms of the

number of chain units for (a) a freely jointed chain, (b) a chain with tetrahedral bond angles. Evaluate B for $l = 154$ pm and $N = 4000$. Estimate the osmotic virial coefficient B for a randomly coiled polyethylene chain of arbitrary M and evaluate it for $M = 56 \text{ kg mol}^{-1}$.

23.16 Show that the ratio $[\text{Na}^+]_L/[\text{Na}^+]_R$ in the Donnan equilibrium is equal to $x + (1 + x^2)$, where $x = v[\text{P}]/2[\text{Na}^+]_R$ and sketch the ratio as a function of the polyelectrolyte concentration.

23.17 Consider the thermodynamic description of stretching rubber. The observables are the tension t and length l (like p and V for gases). Since $dw = t \, dl$, the basic equation is $dU = T \, dS + t \, dl$. ($p \, dV$ terms are supposed negligible throughout.) If $G = U - TS - tl$, find expressions for dG and dA and deduce the Maxwell relations $(\partial S/\partial l)_T = -(\partial t/\partial T)_l$ and $(\partial S/\partial t)_T = (\partial l/\partial T)_t$. Go on to deduce the equation of state for rubber,

$$\left(\frac{\partial U}{\partial l}\right)_T = t - T\left(\frac{\partial t}{\partial T}\right)_l$$

23.18 On the assumption that the tension required to keep a sample at a constant length is proportional to the temperature ($t = aT$, the analogue of $p \propto T$), show that the tension can be ascribed to the dependence of the entropy on the length of the sample. Account for this result in terms of the molecular nature of the sample.

23.19 The radius of gyration is defined in eqn 22. Show that an equivalent definition is that R_g is the average root mean square distance of the atoms or groups (all assumed to be of the same mass); that is, that $R_g^2 = (1/N^2) \sum_j R_j^2$, where R_j is the distance of atom j from the centre of mass.

23.20 Use eqn 24 to deduce expressions for (a) the root mean square separation of the ends of the chain, (b) the mean separation of the ends, and (c) their most probable separation. (Integrals over Gaussians are given in Box 24.1.) Evaluate these three quantities for an $N = 4000$, $l = 154$ pm fully flexible chain.

23.21 Construct a two-dimensional random walk either using tables of random numbers (e.g. Abramowitz and Stegun, *Handbook of mathematical functions*) or using a random number generating program of a computer or calculator. Construct a walk of 50 and 100 steps. If there are many people working on the problem, investigate the mean and most probable separations in the plots by direct measurement. Do they vary as $N^{1/2}$?

23.22 Evaluate eqn 21 for $P(\theta)$ in the case of a long rigid rod of N identical scattering units with separation l, and assume that the units are so closely spaced that sums may be replaced by integrals. Plot $P(\theta)$ as a function of θ on polar graph paper in the case $L \approx \lambda$. The integral $\int_0^z (\sin x/x) \, dx$ is the 'sine-integral' $\text{Si}(z)$, and is tabulated in Abramowitz and Stegun, *Handbook of mathematical functions*.

Part 3

Change

In Part 3 of the text we consider the processes responsible for chemical change. We prepare the ground for a discussion of the rates of chemical reactions by considering the motion of molecules in gases and in liquids. Then we establish the precise meaning of reaction rate, and see how the overall rate—and the complex behaviour of some reactions— may be expressed in terms of elementary steps and the atomic events that take place when molecules meet. Some chemical reactions take place on surfaces, and we see how to describe heterogeneous catalysis. A special type of surface is that of an electrode, and we shall see how to describe and understand the rate at which electrons are transferred between a metal and a substance in solution.

The kinetic theory of gases

24

Check-list of key ideas

1. The assumptions of the *kinetic theory* of gases (Introduction).

2. The definition of *collision frequency* and *mean free path* (Introduction).

3. The calculation of the *pressure* from the kinetic theory (eqn 1).

4. The calculation of the *average* molecular speed of molecules in a gas (eqn 2).

5. The derivation of the *Maxwell–Boltzmann distribution* of molecular velocities (eqn 5) and of the *Maxwell distribution* of molecular speeds (eqn 6).

6. The calculation of the *most probable speed* (eqn 7a), the *mean speed* (eqn 7b), and the *root mean square speed* (eqn 7c) of molecules in a gas.

7. The definition of *collision diameter* and *collision cross-section* (Section 24.3) and the calculation of the *collision frequency* of a single molecule (eqn 9).

8. The calculation of the *collision density* of intermolecular collisions in a gas (eqns 10 and 11).

9. The calculation of the *mean free path* (eqn 12).

10. The calculation of the frequency of collisions with the walls of a container (eqn 13).

11. The *rate of effusion* of a gas through a hole in a container (Section 24.5).

12. The definition of a *transport coefficient* and the *flux* of a property, and the statement of *Fick's first law* of diffusion (Section 24.6).

13. The definition of the *diffusion coefficient* (eqn 15), the *coefficient of thermal conductivity* (eqn 16), and the *coefficient of viscosity* (eqn 17).

14. The calculation of the *diffusion coefficient* from the kinetic theory of gases (eqn 19) and its variation with pressure and temperature.

15. The calculation of the *coefficient of thermal conductivity* from the kinetic theory of gases (eqn 20) and its variation with pressure and temperature.

16. The calculation and measurement of the *viscosity* of a perfect gas (eqn 21) and its variation with pressure and temperature.

One of the most remarkable theories of matter is the kinetic theory of gases, for from a very simple model we can deduce an equation of state of a perfect gas and derive reasonable approximations to its physical properties. We shall present the theory as an example of model-building in science, where a model of a state of matter is proposed, expressed quantitatively, and finally compared with experiment. Apart from the equation of state that we shall derive, we shall also describe some properties of perfect gases that relate to the transfer of energy and matter from one location to another. This chapter therefore serves as an introduction to Part 3 of the text, where we deal with the rates of physical and chemical change.

The model and the basic calculations

The **kinetic theory** of gases is based on three assumptions:

(1) The gas consists of molecules of mass m and diameter d in ceaseless random motion.
(2) The size of the molecules is negligible (in the sense that their diameters are much smaller than the average distance travelled between collisions).
(3) The molecules do not interact, except that they make perfectly elastic collisions when the separation of their centres is equal to d.

An **elastic collision** is one in which the total translational kinetic energy of a pair of molecules is the same before and after a collision: no energy is transferred to their internal modes of motion. The assumption that the molecules do not interact implies that there is no potential energy of interaction between them. That is, the total energy of a sample is the sum of the kinetic energies of the molecules in flight. Thus the kinetic model of gases is a 'kinetic energy only' model of their properties.

The collisions ensure that the particles constantly change their speed and direction. The **collision frequency** z is the average number of collisions per unit time made by a single molecule. The **mean free path** λ is the average distance a molecule travels between collisions. Hence we can express assumption 2 more succinctly as $d \ll \lambda$. We should note that any mechanical property can be expressed as a combination of quantities with dimensions of mass, length, and time. Therefore, once m, λ, and $1/z$ have been calculated we should be able to find expressions for any mechanical property of a gas by forming suitable combinations.

24.1 The pressure of a gas

The kinetic theory accounts for the steady pressure exerted by a gas in terms of collisions the molecules make with the walls of the container. These collisions are so numerous that the walls experience a virtually constant force, and hence a steady pressure (for pressure is force per unit area).

The force of collision

Consider the system in Fig. 24.1. When a particle of mass m collides with the wall on the right its component of momentum parallel to the x axis changes from mv_x to $-mv_x$, its other components remaining unchanged. The momentum therefore changes by $2m|v_x|$ on each collision. The number of collisions in an interval Δt is equal to the number of particles able to reach the wall in that interval. Since a particle with velocity component v_x can travel a distance $|v_x|\Delta t$ in an interval Δt, all the particles within a distance $|v_x|\Delta t$ of the wall will strike it if they are travelling towards it. If the wall has area A, all the particles in a volume $A|v_x|\Delta t$ will reach the wall (if they are travelling towards it). If the number density, the number of particles per unit volume, is $\mathcal{N}$, the number in the volume $A|v_x|\Delta t$ is $\mathcal{N}A|v_x|\Delta t$.

On average, half the particles are moving to the right, and half are moving to the left. Therefore, the average number of collisions with the wall during the interval Δt is $\frac{1}{2}\mathcal{N}A|v_x|\Delta t$. The total momentum change in that interval is the product of this number and the change $2m|v_x|$:

$$\text{Momentum change} = \tfrac{1}{2}\mathcal{N}A\,|v_x|\,\Delta t \times 2m\,|v_x| = \mathcal{N}Amv_x^2\,\Delta t$$

The rate of change of momentum is this change of momentum divided by the interval Δt during which it occurs:

$$\text{Rate of change of momentum} = \mathcal{N}Amv_x^2$$

The rate of change of momentum is equal to the force (by Newton's second law of motion), and so the force exerted by the gas on the wall is also $m\mathcal{N}Av_x^2$. It follows that the pressure, the force per unit area, is

$$\text{Pressure} = \mathcal{N}mv_x^2$$

The average pressure

Not all the particles travel with the same velocity, and so the detected pressure p is the *average* (denoted $\langle \ldots \rangle$) of the quantity just calculated:

$$p = m\mathcal{N}\langle v_x^2 \rangle$$

The **root mean square speed** c of the particles is

$$c = \langle v^2 \rangle^{1/2} = (\langle v_x^2 \rangle + \langle v_y^2 \rangle + \langle v_z^2 \rangle)^{1/2}$$

However, since the particles are moving randomly (and there is no net flow in a particular direction), the average of v_x^2 is the same as the average of the analogous quantities in the y and z directions. Since $\langle v_x^2 \rangle$, $\langle v_y^2 \rangle$, and $\langle v_z^2 \rangle$ are all equal,

$$c = (3\langle v_x^2 \rangle)^{1/2} \text{ implying that } \langle v_x^2 \rangle = \tfrac{1}{3}c^2$$

Therefore,

$$p = \tfrac{1}{3}\mathcal{N}mc^2 \qquad\qquad (1)^\circ$$

Equation 1 is one of the key results of kinetic theory.[1]

---

[1] The convention used in Parts 1 and 2 is that a superscript $^\circ$ on an equation signifies a perfect gas; by extension it now also means a result stemming from kinetic theory.

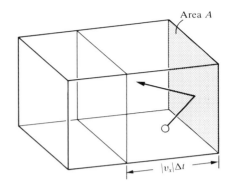

Fig. 24.1 The calculation of the pressure of a gas considers the total force exerted on the shaded wall by the colliding particles. In an interval Δt, all particles within the volume $A|v_x|\Delta t$ and moving towards the wall strike it and undergo a reversal of the x-component of their momentum.

The root mean square molecular speed

The number density $\mathcal{N}$ is equal to N/V, where N is the total number of particles present in the volume V. Then, since $N = nN_A$, where N_A is Avogadro's constant,

$$pV = \tfrac{1}{3}nN_A mc^2$$

Since a perfect gas satisfies the equation of state

$$pV = nRT = nN_A kT$$

where k is Boltzmann's constant, we can conclude at once that

$$c = \left(\frac{3kT}{m}\right)^{1/2} \tag{2a}°$$

Example 24.1: *Calculating the root mean square molecular speed*

Calculate the root mean square speed of CO_2 molecules at 298 K.

Answer. It will be useful to express eqn 2a in terms of the molar mass M by multiplying both k and m by Avogadro's constant:

$$c = \left(\frac{3RT}{M}\right)^{1/2} \tag{2b}°$$

Since the molar mass of CO_2 is 44.01 g mol^{-1}, we find

$$c = \left(\frac{3 \times 8.314\,\text{J K}^{-1}\,\text{mol}^{-1} \times 298\,\text{K}}{44.01 \times 10^{-3}\,\text{kg mol}^{-1}}\right)^{1/2} = 411\,\text{m s}^{-1}$$

Comment. The root mean square speed increases as the square root of the temperature and is inversely proportional to the square root of the molar mass. Thus doubling the temperature (e.g. from 300 K to 600 K) increases c by a factor of 1.4.

Exercise. Calculate the root mean square speed of He atoms at 298 K.

[1.36 km s^{-1}]

24.2 The distribution of molecular velocities

We have seen that the kinetic theory of gases enables us to calculate the mean values of properties. However, it also does much more, for it also enables us to discuss the spread of values around the mean and to calculate the fraction of molecules that have speeds in a particular range.

Some properties of mean values

To make progress we need to draw on some of the properties of mean values and their calculation. The properties we require are reviewed in the *Further information* section at the end of the chapter. In brief, the mean value of a property X that can have any of a continuous range of values (like a speed) is

$$\langle X \rangle = \int Xf(X)\,\mathrm{d}X \tag{3}$$

The function $f(X)$, which is called the **distribution** of the property X, gives the probability that the property lies in the range X to $X + dX$. For example, $f(v)$ is the distribution of speed v, and $f(v)\,dv$ is the probability that the speed lies in the range v to $v + dv$. A concrete example should make this clear.

Example 24.2: *Interpreting a distribution function*

The expression that we derive below for the distribution of speeds of molecules of molar mass M at a temperature T is

$$f(v) = 4\pi\left(\frac{M}{2\pi RT}\right)^{3/2} v^2 e^{-Mv^2/2RT}$$

What is the probability that a CO_2 molecule will be found travelling within $\pm 1\ \mathrm{m\ s^{-1}}$ of its root mean square speed at 25°C?

Answer. We need to calculate

$$P \approx f(v) \times \Delta v$$

with f evaluated at the root mean square speed ($411\ \mathrm{m\ s^{-1}}$, Example 24.1) and $\Delta v = 2\ \mathrm{m\ s^{-1}}$, the (almost infinitesimal) range of speeds of interest. For $M/RT = 1.766 \times 10^{-5}\ \mathrm{m^{-2}\ s^2}$, $v = 411\ \mathrm{m\ s^{-1}}$, and $Mv^2/2RT = 1.500$,

$$f = 4\pi \times (2.827 \times 10^{-6}\ \mathrm{m^{-2}\ s^2})^{3/2} \times (411\ \mathrm{m\ s^{-1}})^2 \times e^{-1.500}$$

$$= 2.25 \times 10^{-3}\ \mathrm{m^{-1}\ s}$$

Therefore,

$$P \approx 2.25 \times 10^{-3}\ \mathrm{m^{-1}\ s} \times 2.0\ \mathrm{m\ s^{-1}} = 4.5 \times 10^{-3}$$

That is, 1 molecule in 222 is likely to be found in the range.

Comment. The range is so narrow compared with the total spread at this temperature that we have assumed that f is constant over it. If the range is broad, the variation must be taken into account by integration.

Exercise. What is the probability that a He atom will be found within $\pm 10\ \mathrm{m\ s^{-1}}$ of its root mean square speed at 298 K?   [1 in 74]

The second property of mean values that we shall use (see *Further information*) concerns the probability that two properties have particular values simultaneously. If the probability that one property lies in the range X to $X + dX$ is $f(X)\,dX$, and the probability that an *independent* property Y lies in the range Y to $Y + dY$ is $f(Y)\,dY$, then the probability of X and Y lying simultaneously in these ranges is the product of the individual probabilities:

$$f(X,\ Y)\,dX\,dY = f(X)f(Y)\,dX\,dY$$

For example, the components of velocity v_x and v_y of a molecule in a gas are independent of each other, and the probability that the molecule has a particular value of v_x and simultaneously a particular value of v_y is the product of the individual probabilities. In such cases, the joint distribution $f(X,\ Y)$ is the product of the individual distributions:

$$f(X,\ Y) = f(X)f(Y) \tag{4}$$

The Maxwell–Boltzmann distribution

We now have enough background to find the distribution of the components of velocities of molecules in a perfect gas. The result we shall derive will prove to be useful when we are doing calculations relating to the rates of reactions and to the properties of molecular beams.

The three velocity components of a molecule v_x, v_y, and v_z are independent of each other, and so the probability $f(v_x, v_y, v_z)\,\mathrm{d}v_x\,\mathrm{d}v_y\,\mathrm{d}v_z$ that a molecule has a velocity with components in the range v_x to $v_x + \mathrm{d}v_x$, v_y to $v_y + \mathrm{d}v_y$, and v_z to $v_z + \mathrm{d}v_z$ is the product of the individual probabilities

$$f(v_x, v_y, v_z)\,\mathrm{d}v_x\,\mathrm{d}v_y\,\mathrm{d}v_z = f(v_x)f(v_y)f(v_z)\,\mathrm{d}v_x\,\mathrm{d}v_y\,\mathrm{d}v_z$$

The joint distribution is the product of the individual distributions:

$$f(v_x, v_y, v_z) = f(v_x)f(v_y)f(v_z)$$

We now assume that the probability of a molecule having a particular range of velocity components is independent of its direction of flight. That is, we assume that f depends on the speed v, where $v^2 = v_x^2 + v_y^2 + v_z^2$, but not the individual components. For example, the probability of a molecule having a velocity with components $(1.0\ \mathrm{km\ s^{-1}},\ 2.0\ \mathrm{km\ s^{-1}},\ 3.0\ \mathrm{km\ s^{-1}})$ and hence speed $3.7\ \mathrm{km\ s^{-1}}$ is the same as the probability of its having a velocity with components $(2.0\ \mathrm{km\ s^{-1}},\ 1.0\ \mathrm{km\ s^{-1}},\ 3.0\ \mathrm{km\ s^{-1}})$, or any other set corresponding to a speed of $3.7\ \mathrm{km\ s^{-1}}$. It follows that f depends only on $v_x^2 + v_y^2 + v_z^2$, and so we denote it $f(v_x^2 + v_y^2 + v_z^2)$. Then the last equation becomes

$$f(v_x^2 + v_y^2 + v_z^2) = f(v_x)f(v_y)f(v_z)$$

Only an exponential function satisfies a relation of this kind because

$$e^{a+b+c} = e^a e^b e^c$$

Consequently

$$f(v_x) = K e^{\pm \zeta v_x^2}$$

with K and ζ constants. The two constants are the same for $f(v_y)$ and $f(v_z)$ because the distributions are the same in each direction. Therefore,

$$f(v_x)f(v_y)f(v_z) = K^3 e^{\pm \zeta(v_x^2 + v_y^2 + v_z^2)} = f(v_x^2 + v_y^2 + v_z^2)$$

as required. We can resolve the $\pm \zeta$ ambiguity on physical grounds: the probability of extremely high velocities must be very small; therefore, the negative sign must be taken.

Next we determine K. Since the molecule must have some velocity in the range $-\infty < v_x < \infty$, the total probability of the x component of velocity being in that range is 1:

$$\int_{-\infty}^{\infty} f(v_x)\,\mathrm{d}v_x = 1$$

Substitution of the expression given above leads to

$$\int_{-\infty}^{\infty} f(v_x)\,\mathrm{d}v_x = K \int_{-\infty}^{\infty} e^{-\zeta v_x^2}\,\mathrm{d}v_x = K\left(\frac{\pi}{\zeta}\right)^{1/2}$$

Therefore, $K = (\zeta/\pi)^{1/2}$ and

$$f(v_x) = \left(\frac{\zeta}{\pi}\right)^{1/2} e^{-\zeta v_x^2}$$

Finally, we determine ζ by calculating a property that we already know. The mean value of v_x^2 is

$$\langle v_x^2 \rangle = \int_{-\infty}^{\infty} v_x^2 f(v_x)\, dv_x = \left(\frac{\zeta}{\pi}\right)^{1/2} \int_{-\infty}^{\infty} v_x^2 e^{-\zeta v_x^2}\, dv_x$$

The integral on the right is standard (Box 24.1) and is equal to $\frac{1}{2}(\pi/\zeta^3)^{1/2}$. It follows that

$$\langle v_x^2 \rangle = \frac{1}{2}\left(\frac{\zeta}{\pi}\right)^{1/2}\left(\frac{\pi}{\zeta^3}\right)^{1/2} = \frac{1}{2\zeta}$$

and that the mean square speed is $c^2 = 3/2\zeta$. However, we have already established a value for c (eqn 2a); hence we conclude that

$$\zeta = \frac{m}{2kT}$$

Therefore, the complete form of the velocity distribution is

$$f(v_x) = \left(\frac{m}{2\pi kT}\right)^{1/2} e^{-mv_x^2/2kT} \qquad (5)°$$

Equation 5 is known as the **Maxwell–Boltzmann distribution** of molecular velocities, representing both Maxwell's contribution (he derived it originally) and Boltzmann's (who proved it rigorously).

The Maxwell distribution of speeds

We can now derive the distribution of the speeds of the molecules irrespective of their direction of motion. The probability that a molecule has velocity components in the range v_x to $v_x + dv_x$, v_y to $v_y + dv_y$, and v_z to $v_z + dv_z$ is

$$f(v_x, v_y, v_z)\, dv_x\, dv_y\, dv_z = f(v_x)f(v_y)f(v_z)\, dv_x\, dv_y\, dv_z$$

$$= \left(\frac{m}{2\pi kT}\right)^{3/2} e^{-mv^2/2kT}\, dv_x\, dv_y\, dv_z$$

The probability $f(v)\, dv$ that the molecule has a speed in the range v to $v + dv$ is the sum of the probabilities that it lies in any of the volume elements $dv_x\, dv_y\, dv_z$ in a spherical shell of radius v (Fig. 24.2). The sum of the volume elements on the right-hand side of the last expression is the volume of this shell, which is $4\pi v^2\, dv$. Therefore,

$$f(v) = 4\pi\left(\frac{m}{2\pi kT}\right)^{3/2} v^2 e^{-mv^2/2kT} \qquad (6)°$$

which is the **Maxwell distribution of speeds**.

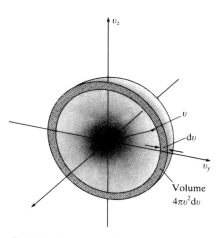

Fig. 24.2 The density of shading represents the probabilities of having each velocity. For the distribution of molecular speeds, the total probability of having a value in the shell of thickness dv and radius v is calculated.

Example 24.3: *Calculating the mean speed*

What is the mean speed of Cs atoms in an oven heated to 500°C? Integrals are listed in Box 24.1.

Answer. The mean speed $\bar{c}$ is obtained by evaluating the integral

$$\bar{c} = \int v f \, dv$$

with f the distribution in eqn 6 and substituting $M = 132.9 \text{ g mol}^{-1}$ and $T = 773 \text{ K}$:

$$\bar{c} = \int_0^\infty v f \, dv = 4\pi \left(\frac{m}{2\pi kT}\right)^{3/2} \int_0^\infty v^3 e^{-mv^2/2kT} \, dv$$

$$= 4\pi \left(\frac{m}{2\pi kT}\right)^{3/2} \times \tfrac{1}{2}\left(\frac{2kT}{m}\right)^2 = \left(\frac{8kT}{\pi m}\right)^{1/2}$$

$$= \left(\frac{8RT}{\pi M}\right)^{1/2}$$

Then, for $M = 132.9 \text{ g mol}^{-1}$, $\bar{c} = 351 \text{ m s}^{-1}$.

Exercise. Evaluate the root mean square speed of the atoms.
$$[(3RT/M)^{1/2}, 380 \text{ m s}^{-1}]$$

Figure 24.3, which summarizes the main features of the Maxwell distribution, shows that the distribution of speeds broadens as the temperature increases. It also shows that the **most probable speed** c^*, the speed corresponding to the maximum in the distribution (which is found by differentiating $f(v)$ with respect to v and identifying the speed at which the slope is zero) shifts to higher values:

$$c^* = \left(\frac{2kT}{m}\right)^{1/2} \tag{7a}°$$

The **mean speed** $\bar{c}$ is calculated from the Maxwell distribution as explained in Example 24.3:

$$\bar{c} = \left(\frac{8kT}{\pi m}\right)^{1/2} \tag{7b}°$$

Box 24.1 Integrals over Gaussian functions

Let

$$I_n = \int_0^\infty x^n e^{-ax^2} dx$$

Then:

n	0	1	2	3	4	5	6
I_n	$\frac{1}{2}\left(\frac{\pi}{a}\right)^{1/2}$	$\frac{1}{2a}$	$\frac{1}{4}\left(\frac{\pi}{a^3}\right)^{1/2}$	$\frac{1}{2a^2}$	$\frac{3}{8}\left(\frac{\pi}{a^5}\right)^{1/2}$	$\frac{1}{a^3}$	$\frac{15}{16}\left(\frac{\pi}{a^7}\right)^{1/2}$

The **error function** is the integral

$$\text{erf } z = \frac{2}{\pi^{1/2}} \int_0^z e^{-x^2} dx = 1 - \frac{2}{\pi^{1/2}} \int_z^\infty e^{-x^2} dx$$

It is listed in tables (such as M. Abramowitz and I. A. Stegun, *Handbook of mathematical functions*, Dover (1965)), and the following is a brief abstract:

z	0	0.200	0.400	0.600	0.800	1.000	1.200	1.400	1.600	1.800
erf z	0	0.223	0.428	0.604	0.742	0.843	0.910	0.952	0.976	0.989

See Table 12.2 for other values.

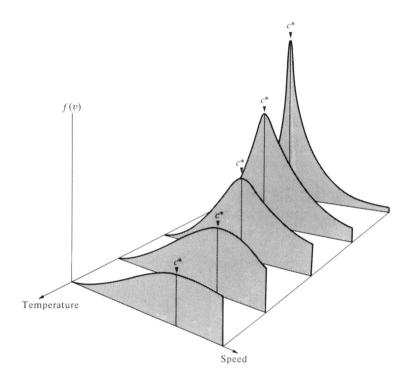

$f(v)$

Temperature

Speed

Fig. 24.3 The Maxwell distribution of speeds and its dependence on temperature calculated using eqn 6. Higher temperatures lie at the front: note the broadening of the distribution and the shift of the most probable speed at the temperature is increased.

It differs slightly from the root mean square speed

$$c = \left(\frac{3kT}{m}\right)^{1/2}$$
(7c)°

(In each case, we can replace m by the molar mass M if at the same time we replace k by R.) Numerically, $c = 1.225c^*$ and $\bar{c} = 1.128c^*$. Light molecules have higher mean speeds than heavy molecules (Table 24.1).

The Maxwell distribution has been verified experimentally. For example, molecular speeds can be measured directly with a velocity selector of the sort shown in Fig. 24.4 (the same kind of device as used with molecular beams, Section 22.5). The spinning disks have slots that permit the passage of only those molecules moving through them at the appropriate speed, and the number of molecules can be determined by collecting them at a detector. Indirect methods of measuring speeds make use of the Doppler effect on the wavelength of emitted light, as we discussed in Section 16.3.

Table 24.1. Mean speeds at 25°C, $\bar{c}/(\text{m s}^{-1})$

C_6H_6	284
CO_2	379
He	1256
N_2	475

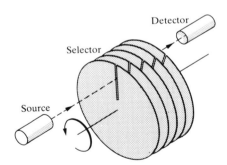

Fig. 24.4 The experimental determination of molecular speeds makes use of a series of rotating disks: the response of the detector is proportional to the number of particles having a speed that enables them to pass through the slots in the selector when it is rotating at a known rate.

Collisions

The time-scale of events in a gas, such as the rates of chemical reactions, are determined by the collision frequency, the number of collisions a molecule makes per unit time.

24.3 Intermolecular collisions

We count a 'hit' whenever the centres of two molecules come within some distance d of each other, where d, the **collision diameter**, is of the order of

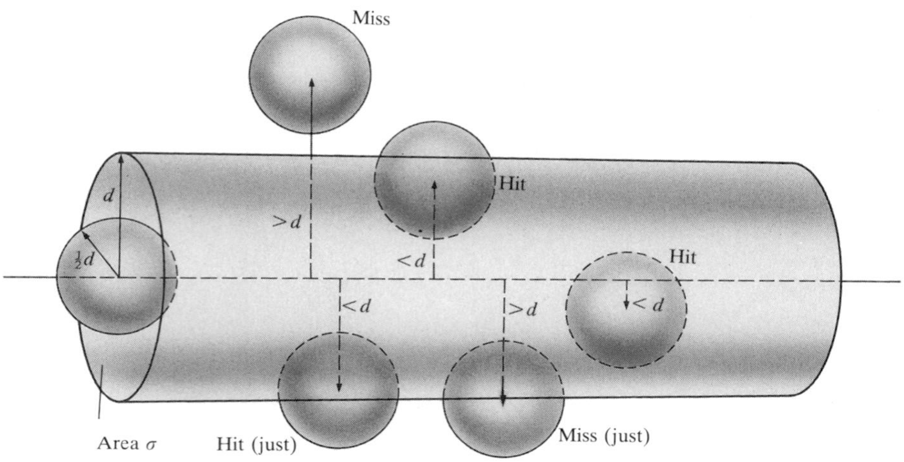

Fig. 24.5 The collision cross section σ and the collision tube. A 'hit' is scored when two particles come within a distance equal to their diameter.

Table 24.2. Collision cross-sections, σ/nm^2

C_6H_6	0.88
CO_2	0.52
He	0.21
N_2	0.43

the actual diameters of the molecules (for hard spheres d is the diameter). The simplest approach to calculating the frequency of such collisions is to freeze the positions of all the molecules except one. Then we note what happens as that mobile molecule travels through the gas with a mean speed $\bar{c}$ for a time Δt. In doing so it sweeps out a 'collision tube' of cross-sectional area $\sigma = \pi d^2$ and length $\bar{c}\Delta t$, and therefore of volume $\sigma\bar{c}\Delta t$ (Fig. 24.5). The area σ is called the **collision cross-section**. Some typical collision cross-sections are given in Table 24.2 (they are obtained by measuring various transport properties, as we shall see in Sections 24.7 to 24.9).

The collision frequency

The number of stationary molecules with centres inside the collision tube is given by the volume of the tube times the number density $\mathcal{N} = N/V$, and is $\mathcal{N}\sigma\bar{c}\Delta t$. The number of hits scored in the interval Δt is equal to this number, and so the collision frequency z, the number of collisions per unit time, is $\mathcal{N}\sigma\bar{c}$. However, the molecules are not stationary, and so we should use the *relative* speed of the colliding molecules. For collisions between different types of molecule, the mean relative speed is

$$\bar{c}_{Rel} = \left(\frac{8kT}{\pi\mu}\right)^{1/2} \qquad \mu = \frac{m_A m_B}{m_A + m_B} \tag{8}°$$

where μ is the reduced mass. (We saw in the *Further information 1* section of Chapter 13 that the reduced mass occurs when the relative motion of particles is considered.) For identical molecules $\mu = \frac{1}{2}m$, and so

$$\bar{c}_{Rel} = 2^{1/2}\bar{c}$$

Therefore, the collision frequency is

$$z = \frac{2^{1/2}\sigma\bar{c}N}{V} = \frac{2^{1/2}\sigma\bar{c}p}{kT} \tag{9}°$$

(since $N/V = p/kT$).

The collision density

The collision frequency z gives the number of collisions made by a single molecule. We can obtain the *total* collision frequency, the rate of collisions between all the molecules in the gas, by multiplying z by $\frac{1}{2}N$ (the factor $\frac{1}{2}$ ensures that the $A \ldots A'$ and $A' \ldots A$ collisions are counted as one). Therefore, the **collision density** Z, the total number of collisions per unit time per unit volume, is

$$Z_{AA} = \frac{\frac{1}{2}zN}{V} = \frac{\sigma \bar{c}}{2^{1/2}}\left(\frac{N}{V}\right)^2 \qquad (10a)°$$

The mean speed is given by eqn 7b, and the molar concentration of A is related to its number density by Avogadro's constant:

$$[A]N_A = \frac{N}{V}$$

Therefore,

$$Z_{AA} = \sigma\left(\frac{4kT}{\pi m}\right)^{1/2} N_A^2 [A]^2 \qquad (10b)°$$

Collision densities may be very large. For example, in nitrogen at room temperature and pressure, with $d = 280$ pm, $Z = 5 \times 10^{34}\,\text{s}^{-1}\,\text{m}^{-3}$.

Example 24.4: *Calculating the collision frequency*

We continue to explore the properties of Cs atoms in an oven at 500 °C. Calculate the number of collisions a single Cs atom makes per second. If the volume of the oven is 50.0 cm³, what is the total number of collisions inside the oven per second? The vapour pressure of Cs(*l*) at 500 °C is 80 Torr; the collision diameter of a Cs atom is 540 pm.

Answer. We use eqn 9 for the collision frequency of a single atom with $\bar{c}$ from Example 24.3 (351 m s⁻¹). The collision cross-section is

$$\sigma = \pi d^2 = 9.16 \times 10^{-19}\,\text{m}^2$$

and the pressure (80 Torr) is 1.07×10^4 Pa. Therefore,

$$z = \frac{2^{1/2}\sigma \bar{c} p}{kT} = 4.6 \times 10^8\,\text{s}^{-1}$$

From eqn 10a written in the form

$$Z = \frac{zp}{2kT}$$

we find $Z = 2.3 \times 10^{32}\,\text{s}^{-1}\,\text{m}^{-3}$. Hence, in a 50.0 cm³ container the total collision frequency is $1.1 \times 10^{28}\,\text{s}^{-1}$.

Comment. To study individual isolated molecules it is essential to reduce the collision frequency so the molecules spend appreciable periods in free flight. This involves working at very low pressures.

Exercise. What is the interval between collisions for a single CO_2 molecule at 298 K and 760 Torr?

$$[1.5 \times 10^{-10}\,\text{s}]$$

For collisions between different types of molecule, the mean relative speed is given by eqn 8 and the collision cross-section by $\sigma = \pi d^2$, but with $d = \frac{1}{2}(d_A + d_B)$. The number of collisions an A molecule makes per unit time with all the N' B molecules present is $\sigma \bar{c} N'/V$. There are N A molecules, and so the total number of A–B collisions per unit time is $(\sigma \bar{c} N'/V)N$. The collision density Z_{AB}, the total number of A–B collisions per unit time per unit volume, is

$$Z_{AB} = \sigma \left(\frac{8kT}{\pi\mu}\right)^{1/2} \times \frac{NN'}{V^2} = \sigma \left(\frac{8kT}{\pi\mu}\right)^{1/2} N_A^2 [A][B] \qquad (11)°$$

The mean free path

Once we have the collision frequency, we can calculate the mean free path λ, the average distance a molecule travels between collisions. If a molecule moving with a mean speed $\bar{c}$ collides with a frequency z, it spends a time $1/z$ in free flight between collisions, and therefore travels a distance $(1/z)\bar{c}$. Therefore, the mean free path is

$$\lambda = \frac{\bar{c}}{z} \qquad (12a)°$$

Note that since z is proportional to the pressure, λ is inversely proportional to the pressure. Doubling the pressure reduces the mean free path by half.

Example 24.5: *Calculating the mean free path*

Calculate the mean free path of Cs atoms under the conditions specified in Example 24.4.

Answer. First, we substitute the expression for z derived in Example 24.4 into eqn 12a, which gives

$$\lambda = \frac{kT}{2^{1/2} \sigma p} \qquad (12b)°$$

Then we use $p = 80$ Torr, which corresponds to 11 kPa, and $\sigma = 0.92$ nm². This gives $\lambda = 770$ nm.

Comment. The mean free path corresponds to about 1400 diameters, and so the gas is under conditions where the kinetic theory is likely to be reliable.

Exercise. Calculate the mean free path of CO_2 molecules at 298 K and 1.0 atm. [55 nm]

Since $pV = nRT$, eqn 12b is equivalent to

$$\lambda = \frac{V}{2^{1/2} \sigma N} = \frac{1}{2^{1/2} \sigma N_A [A]} \qquad (12c)°$$

This relation shows that, at constant volume, the mean free path in a sample of a gas is independent of the temperature. The *distance* between collisions is determined by the number of molecules present in the given volume, not the rate at which they travel.

24.4 Collisions with walls and surfaces

Consider a wall of area A perpendicular to the x axis. If a molecule has $v_x > 0$ it will strike the wall within an interval Δt if it lies within a distance $v_x \Delta t$ of it. (If $v_x < 0$, the molecule is moving away from the wall.) Therefore, all molecules in the volume $A v_x \Delta t$, and with positive velocities, will strike the wall in the interval Δt. The total average number of collisions in this interval is therefore the average of this quantity multiplied by the number density of molecules:

$$\text{Number of collisions} = \mathcal{N} A \, \Delta t \int_0^\infty v_x f(v_x) \, \mathrm{d}v_x$$

(Notice that the integration is over only positive velocities.) We can evaluate the integral easily using the velocity distribution:

$$\int_0^\infty v_x f(v_x) \, \mathrm{d}v_x = \left(\frac{m}{2\pi kT} \right)^{1/2} \int_0^\infty v_x \mathrm{e}^{-mv_x^2/2kT} \, \mathrm{d}v_x = \left(\frac{kT}{2\pi m} \right)^{1/2}$$

Therefore, the number of collisions per unit time per unit area is

$$Z_{\mathrm{W}} = \left(\frac{kT}{2\pi m} \right)^{1/2} \frac{N}{V} = \frac{\bar{c}N}{4V} \tag{13a}°$$

Then, since $N/V = p/kT$,

$$Z_{\mathrm{W}} = \frac{p\bar{c}}{4kT} = \frac{p}{(2\pi mkT)^{1/2}} \tag{13b}°$$

When $p = 1$ atm and $T = 300$ K, a container receives about 3×10^{23} collisions per second per cm^2. We shall see in Chapter 29 that a knowledge of how Z_{W} depends on the pressure and the mass of the molecule is very important for discussing processes occurring at solid surfaces.

24.5 The rate of effusion

When a gas at a pressure p and temperature T is separated from a vacuum by a very small hole, the rate of escape of its molecules is equal to the rate at which they strike the area of the hole (eqn 13b). Therefore, if the area of the hole is A_0, the number of molecules that escape per unit time is

$$Z_{\mathrm{W}} A_0 = \frac{p A_0}{(2\pi mkT)^{1/2}} \tag{14}°$$

The fact that the right-hand side is proportional to $1/m^{1/2}$ is the origin of **Graham's law of effusion**, that the rate of effusion is inversely proportional to the square-root of the molar mass.

Example 24.6: *Deducing the time-dependence of the pressure of an effusing gas*

Derive an expression that shows how the pressure of an effusing perfect gas varies with time.

Answer. The rate of change of pressure of a gas in a container at constant temperature is related to the rate of change of the number of molecules present by

$$\frac{dp}{dt} = \frac{d(nRT/V)}{dt} = \frac{RT}{V}\frac{dn}{dt} = \frac{RT}{N_A V}\frac{dN}{dt} = \frac{kT}{V}\frac{dN}{dt}$$

The rate of change of the number of molecules is equal to the collision frequency multiplied by the area of the hole:

$$\frac{dN}{dt} = -Z_W A_0 = \frac{-pA_0}{(2\pi mkT)^{1/2}}$$

Substituting this expression into the one above gives

$$\frac{dp}{p} = -\left(\frac{kT}{2\pi m}\right)^{1/2}\frac{A_0}{V}dt$$

which integrates to

$$p = p_0 e^{-t/\tau} \qquad \tau = \left(\frac{2\pi m}{kT}\right)^{1/2}\frac{V}{A_0}$$

Comment. The pressure decreases exponentially towards zero; the decrease is faster the higher the temperature, the bigger the hole, the lighter the molecules, and the smaller the volume of the container.

Exercise. Show that $t_{1/2}$, the time required for the pressure to decrease to half its initial value, is independent of the initial pressure. $[t_{1/2} = \tau \ln 2]$

The formula in eqn 14 is the basis of the **Knudsen method** for the determination of molar mass or, if that is known, the vapour pressure of the sample (e.g. of a solid). Thus, if the vapour pressure of a solid is p, and it is enclosed in a cavity with a small hole, the rate of loss of mass from the container is proportional to p. The method is reliable if the mean free path of the atoms is long compared with the diameter of the hole.

Example 24.7: *Calculating the vapour pressure from a mass loss*

When a 0.50 mm hole was opened in the oven used in previous examples, a mass loss of 385 mg was measured in the course of 100 s. Calculate the vapour pressure of liquid caesium at 500°C.

Answer. The mass loss Δm in an interval Δt is related to the collision frequency by

$$\Delta m = Z_W A_0 m \, \Delta t$$

where A_0 is the area of the hole and m the molecular mass. The vapour pressure is constant inside the oven because the liquid is present in equilibrium with its vapour. Since Z_W is related to the pressure by eqn 13, we can write

$$p = \left(\frac{2\pi RT}{M}\right)^{1/2} \times \frac{\Delta m}{A_0 \, \Delta t}$$

Since $M = 132.9 \text{ g mol}^{-1}$, substitution of the data gives $p = 11$ kPa (using $1 \text{ Pa} = 1 \text{ N m}^{-2} = 1 \text{ J m}^{-1}$), or 83 Torr.

Exercise. How long would it take 1.0 g of Cs atoms to effuse out of the oven?
[225 s]

Transport properties

A **transport property** of a substance is its ability to transfer matter, energy, or some other specified property from one place to another. For example, molecules (in gases, liquids, and solids) diffuse down a concentration gradient until the composition is uniform, and the rate of this diffusion is a transport property. The rate of thermal conduction, the transport of energy down a temperature gradient, is another transport property characteristic of the substance and its state. Electric conduction is the transport of charge (by ions or electrons) in a potential gradient, and the electric conductivity of the substance is a transport property. Viscosity, we shall see, is a measure of the rate at which linear momentum is transported through a fluid, and hence is another transport property.

We shall look at some of the general aspects of transport initially, and then calculate some of the transport properties using the kinetic theory of gases. We return to the subject in the next chapter, where we deal with transport in fluids in a more general way.

24.6 Flux

The rate of migration of a property is measured by its **flux** J, the amount of that property passing through unit area per unit time. If mass is flowing (as in diffusion), we speak of a mass flux of so many $kg\,m^{-2}\,s^{-1}$; if the property is energy (as in thermal conduction), we speak of the energy flux and express it in $J\,m^{-2}\,s^{-1}$, and so on.

Experimental observations on transport properties show that the flux of a property is usually proportional to the gradient of a related property of a system. For example, the flux of matter J_z diffusing parallel to some axis z is found to be proportional to the concentration gradient along that axis:

$$J_z(\text{matter}) \propto \frac{d\mathcal{N}}{dz}$$

The proportionality of the flux of matter to the concentration gradient is sometimes called **Fick's first law of diffusion**. Similarly, the rate of thermal diffusion (the flux of energy of thermal motion) is found to be proportional to the temperature gradient:

$$J_z(\text{energy}) \propto \frac{dT}{dz}$$

The flux J_z is the component of a vector. If $J_z > 0$, the flux is towards increasing z (to the right); if $J_z < 0$, the flux is towards the left. Matter flow occurs down a concentration gradient, and so if $d\mathcal{N}/dz < 0$ (implying that the concentration decreases to the right, Fig. 24.6), J_z is positive (flow to the right). Therefore the coefficient of proportionality in the matter flux expression must be negative, and we denote it $-D$, where the constant D is the **diffusion coefficient**. Hence

$$J_z(\text{matter}) = -D\frac{d\mathcal{N}}{dz} \qquad (15)$$

Energy of thermal motion flows down a temperature gradient, and the same

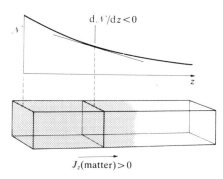

Fig. 24.6 The flux of particles down a concentration gradient. Fick's first law states that the flux of matter (the number of particles per unit area per unit time) is proportional to the density gradient at that point.

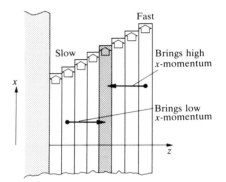

Fig. 24.7 The viscosity of a gas arises from the transport of linear momentum. In this illustration the liquid is undergoing laminar flow, and particles bring their initial momentum when they enter a new layer. If they arrive with high x-component of momentum they accelerate the layer; if with low x-component of momentum they retard it.

reasoning leads to

$$J_z(\text{energy}) = -\kappa \frac{dT}{dz} \tag{16}$$

where κ is the **coefficient of thermal conductivity**.

To see the connection between the flux of momentum and the viscosity, consider a fluid in a state of **Newtonian flow**, a flow that can be imagined as occurring as a series of laminas (Fig. 24.7). The layer next to the wall of the vessel is stationary, and the velocity of successive layers varies linearly with distance from a wall. Molecules continuously move between the laminas and bring with them the x-component of momentum they possessed in their original lamina. A lamina is retarded by molecules arriving from the left (from a more slowly moving lamina) because they have a low momentum in the x direction. A lamina is accelerated by molecules arriving from the right (from a more rapidly moving lamina). Since fast laminas are retarded and slow laminas are accelerated by the arriving molecules, the laminas tend towards a uniform velocity, and we interpret the retarding effect of the slow layers on the fast as the fluid's viscosity.

Since the effect depends on the transfer of x-momentum into the lamina of interest, the viscosity depends on the flux of x-momentum in the z-direction (Fig. 24.7). The flux of x-momentum is proportional to dv_x/dz because there is no net flux when all the laminas move at the same velocity. We can therefore write

$$J_z(\text{momentum along } x) \propto \frac{dv_x}{dz}$$

or

$$J_z = -\eta \frac{dv_x}{dz} \tag{17}$$

η is the **coefficient of viscosity** (or simply 'the viscosity').

Example 24.8: *Deducing the units of a transport coefficient*

Deduce the SI units of viscosity.

Answer. The units of η must be such as to make the units of the right of eqn 17 equal to those on the left. The flux of momentum has the dimensions of [momentum][area]$^{-1}$[time]$^{-1}$, which in SI units is

$$[\text{flux}] = (\text{kg m s}^{-1}) \times \text{m}^{-2} \times \text{s}^{-1} = \text{kg m}^{-1}\,\text{s}^{-2}$$

The dimensions of the velocity gradient are [velocity] × [length]$^{-1}$, which in SI units is

$$[\text{velocity gradient}] = (\text{m s}^{-1}) \times \text{m}^{-1} = \text{s}^{-1}$$

Therefore, to satisfy

$$\text{kg m}^{-1}\,\text{s}^{-2} = [\text{viscosity}] \times \text{s}^{-1}$$

we need

$$[\text{viscosity}] = \text{kg m}^{-1}\,\text{s}^{-1}$$

Comment. Viscosities are often reported in poise (P), which are defined as $1\,\text{P} = 10^{-1}\,\text{kg m}^{-1}\,\text{s}^{-1}$.

Exercise. What are the SI units of thermal conductivity? $\quad [\text{J K}^{-1}\,\text{m}^{-1}\,\text{s}^{-1}]$

24.7 Diffusion

We shall now show the origin of Fick's First Law in terms of kinetic theory and obtain an expression for the diffusion coefficient of a perfect gas. Our task is to show that the flux of molecules is proportional to their concentration gradient.

Consider the arrangement depicted in Fig. 24.8. On average, the molecules passing through the area A at $z = 0$ have travelled about one

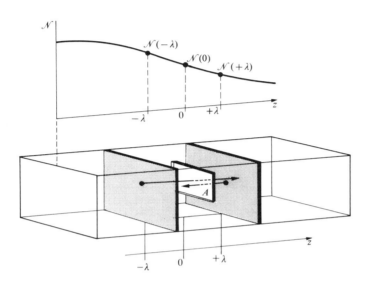

Fig. 24.8 The calculation of the rate of diffusion of a gas is based on the net flux through a plane of area A as a result of arrivals from (on average) a distance λ away, where λ is the mean free path.

mean free path λ. Therefore, the number density where they originated is $\mathcal{N}(z)$ evaluated at $z = -\lambda$. This number density is approximately

$$\mathcal{N}(-\lambda) = \mathcal{N}(0) - \lambda\left(\frac{d\mathcal{N}}{dz}\right)_0$$

where the subscript 0 indicates that the gradient should be evaluated at $z = 0$. Since (from eqn 13) the average number of impacts on the imaginary window from the left during an interval Δt is $\frac{1}{4}\mathcal{N}(-\lambda)\bar{c}A_0\,\Delta t$, the flux from left to right $J(\text{L} \to \text{R})$ arising from the supply of molecules on the left is

$$J(\text{L} \to \text{R}) = \tfrac{1}{4}\mathcal{N}(-\lambda)\bar{c}$$

There is also a flux of molecules from right to left. On average, the molecules making the journey have originated from $z = +\lambda$ where the number density is $\mathcal{N}(\lambda)$. Therefore,

$$J(\text{L} \leftarrow \text{R}) = \tfrac{1}{4}\mathcal{N}(\lambda)\bar{c}$$

The average number density at $z = +\lambda$ is approximately

$$\mathcal{N}(\lambda) = \mathcal{N}(0) + \lambda\left(\frac{d\mathcal{N}}{dz}\right)_0$$

The flow from the more concentrated region on the left dominates the

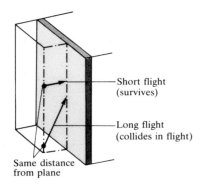

Fig. 24.9 One approximation ignored in the simple treatment is that some particles might make a long flight to the plane even though they are only a short perpendicular distance away, and therefore they have a higher chance of colliding during their journey.

backwash, and the net flux is

$$J_z = J(\text{L} \rightarrow \text{R}) - J(\text{L} \leftarrow \text{R})$$

$$= \frac{1}{4}\left\{\left[\mathcal{N}(0) - \lambda\left(\frac{d\mathcal{N}}{dz}\right)_0\right] - \left[\mathcal{N}(0) + \lambda\left(\frac{d\mathcal{N}}{dz}\right)_0\right]\right\}\bar{c}$$

$$= -\tfrac{1}{2}\lambda\bar{c}\left(\frac{d\mathcal{N}}{dz}\right)_0 \tag{18}°$$

Equation 18 shows that the flux is proportional to the gradient of concentration, in agreement with Fick's law.

At this stage it looks as though we can pick out a value of the diffusion coefficient by comparing eqn 18 with eqn 15:

$$D = \tfrac{1}{2}\lambda\bar{c}$$

It must be remembered, however, that the calculation is quite crude, and is little more than an assessment of the order of magnitude of D. One aspect that has not been taken into account is illustrated in Fig. 24.9 which shows that although a molecule may have begun its journey very close to the window, it could have a long flight before its gets there. Since the path is long, the molecule is likely to collide before reaching the window, and so it ought to be added to the graveyard of other molecules that have collided. Taking this effect into account involves a lot of work, but the end result is the appearance of a factor of $\tfrac{2}{3}$, representing the lower flux. The modification results in

$$D = \tfrac{1}{3}\lambda\bar{c} \tag{19}°$$

for the diffusion coefficient of a perfect gas (Table 24.3).

Since the mean free path λ shortens as the pressure is increased, D decreases with increasing pressure and the molecules diffuse more slowly. Since the mean speed $\bar{c}$ increases with the temperature, D also increases with temperature, and molecules in a hot sample diffuse more quickly than those in a cool sample (for a given concentration gradient). From eqns 12c and 2, at constant volume $D \propto T^{1/2}$. Since the mean free path decreases as the collision diameter of the molecules increases, the diffusion coefficient is greater for small molecules than for large.

Table 24.3. Transport properties of perfect gases

Property	Transported quantity	Simple kinetic theory	Units
Diffusion	Matter	$D = \tfrac{1}{3}\lambda\bar{c}$	$m^2\,s^{-1}$
Thermal conductivity	Energy	$\kappa = \tfrac{1}{3}\lambda\bar{c}C_V[X]$ $= \dfrac{\bar{c}C_V}{(3\sqrt{2})\sigma N_A}$	$J\,K^{-1}\,m^{-1}\,s^{-1}$
Viscosity	Momentum	$\eta = \tfrac{1}{3}\lambda\bar{c}m\mathcal{N}$ $= \dfrac{m\bar{c}}{(3\sqrt{2})\sigma}$	$kg\,m^{-1}\,s^{-1}$

24.8 Thermal conductivity

In thermal conduction the flux of interest is that of energy: our job now is to account for the proportionality of the flux to the temperature gradient, and then to find an expression for the coefficient of thermal conductivity κ. Some experimental values are given in Table 24.4. Thermal conductivities are made use of in the 'Pirani gauge', in which pressure is measured by monitoring the temperature of a heated wire, and in 'katharometer detectors' in chromatography (GLC), where changes of composition are detected similarly. In a katharometer, the change in thermal conductivity of the carrier gas around a heated platinum or tungsten wire is monitored and compared with a reference wire surrounded by the pure gas.

Suppose each molecule carries an average energy $\varepsilon = vkT$, where v is a number close to 1 and can be obtained from the equipartition theorem (Section 11.1). For monatomic particles, $v = \frac{3}{2}$. When one molecule passes through the imaginary window, it transports that energy on average. We shall suppose that the number density is uniform (so there is no mass diffusion) but that the temperature is not. On average, molecules arrive from the left after travelling a mean free path from a hotter region and therefore with a higher energy. They arrive from the right after travelling a mean free path from a cooler region. The energy fluxes in the two directions are therefore

$$J(\text{L} \to \text{R}) = \tfrac{1}{4}\bar{c}\mathcal{N}\varepsilon(-\lambda) \qquad \varepsilon(-\lambda) = vk\left\{T - \lambda\left(\frac{\mathrm{d}T}{\mathrm{d}z}\right)_0\right\}$$

$$J(\text{L} \leftarrow \text{R}) = \tfrac{1}{4}\bar{c}\mathcal{N}\varepsilon(\lambda) \qquad \varepsilon(\lambda) = vk\left\{T + \lambda\left(\frac{\mathrm{d}T}{\mathrm{d}z}\right)_0\right\}$$

and the net energy flux is

$$J_z = J(\text{L} \to \text{R}) - J(\text{L} \leftarrow \text{R}) = -\tfrac{1}{2}v\lambda\bar{c}k\mathcal{N}\left(\frac{\mathrm{d}T}{\mathrm{d}z}\right)_0$$

As before, we multiply by $\frac{2}{3}$ in order to take long flight paths into account, and so arrive at

$$J_z = -\tfrac{1}{3}v\lambda\bar{c}k\mathcal{N}\left(\frac{\mathrm{d}T}{\mathrm{d}z}\right)_0$$

The energy flux is proportional to the temperature gradient, as we wanted to show. Comparison of this equation with eqn 16 shows that

$$\kappa = \tfrac{1}{3}v\lambda\bar{c}k\mathcal{N} \tag{20a}°$$

Since for a perfect gas $C_V = vkN_A$, we can express this as

$$\kappa = \tfrac{1}{3}\lambda\bar{c}C_V[\text{A}] \tag{20b}°$$

where [A] is the molar concentration.[2]

[2] We have used

$$\mathcal{N} = N/V = nN_A/V = N_A[\text{A}]$$

with N_A Avogadro's constant and $[\text{A}] = n/V$.

Table 24.4 Transport properties of gases at 1 atm

	$\kappa/(\text{mJ cm}^{-2}\,\text{s}^{-1}$ $(\text{K cm}^{-1})^{-1})$ 273 K	$\eta/\mu\text{P}$ 273 K	293 K
Ar	0.163	210	223
CO_2	0.145	136	147
He	1.442	187	196
N_2	0.240	166	176

$1\,\mu\text{P} = 10^{-7}\,\text{kg m}^{-1}\,\text{s}^{-1}$

In either case, notice that because $\lambda \propto 1/[A]$ (eqn 12c), $\kappa \propto \bar{c}C_V$ and the thermal conductivity is independent of the pressure of the gas. The physical reason is that the thermal conductivity is large when many molecules are available to transport the energy (hence the proportionality of κ to $[A]$ in eqn 20), but the presence of many molecules limits their mean free path and they cannot carry the energy over a great distance. These two effects balance. The thermal conductivity is found experimentally to be independent of the pressure, except when the pressure is very low. At low pressures $\kappa \propto p$ because λ is then greater than the dimensions of the apparatus and the distance over which the energy is transported is determined by the size of the vessel and not by the other molecules present. The flux is still proportional to the number of carriers, but the length of the journey no longer depends on λ, and so $\kappa \propto [A] \propto p$.

Example 24.9: *Estimating the thermal conductivity of a gas*

Estimate the thermal conductivity of air at room temperature.

Answer. We use eqn 20, noting that the constant-volume heat capacity of a gas of diatomic molecules is $\frac{5}{2}R$ (Section 20.4) and that the molar mass of air is about 29 g mol^{-1}. Take $\sigma = 0.42 \text{ nm}^2$ for O_2 and N_2 from Table 24.2 and estimate $\bar{c}$ as 460 m s^{-1} from eqn 7b. Then

$$\kappa = \left(\frac{5}{6 \times 2^{1/2}}\right)\frac{\bar{c}k}{\sigma}$$

Substituting the values gives $\kappa = 8.9 \text{ mJ K}^{-1}\text{ m}^{-1}\text{ s}^{-1}$.

Comment. A more cumbersome but more useful form of the answer is $\kappa = 8.9 \times 10^{-2} \text{ mJ cm}^{-2}\text{ s}^{-1} \text{ (K cm}^{-1})^{-1}$. Then, in a temperature gradient of 1 K cm^{-1} the heat flux would be about $0.1 \text{ mJ cm}^{-2}\text{ s}^{-1}$.

Exercise. Would carbon dioxide be a better insulator under the same conditions? $[\kappa = 6 \times 10^{-2} \text{ mJ cm}^{-2}\text{ s}^{-1} \text{ (K cm}^{-1})^{-1}; \text{ yes}]$

24.9 Viscosity

The viscosity of a perfect gas

We have seen that viscosity is related to the flux of momentum. Molecules travelling from the right in Fig. 24.10 (from a fast layer to a slower) transport a momentum $mv_x(\lambda)$ to their new layer at $z = 0$, and those travelling from the left transport $mv_x(-\lambda)$ to it. If we assume that the density is uniform (an approximation), the number of impacts per unit area per unit time on the imaginary window is $\frac{1}{4}\mathcal{N}\bar{c}$. Those from the right on average carry a momentum

$$mv_x(\lambda) = mv_x(0) + m\lambda\left(\frac{dv_x}{dz}\right)_0$$

Those from the left bring a momentum

$$mv_x(-\lambda) = mv_x(0) - m\lambda\left(\frac{dv_x}{dz}\right)_0$$

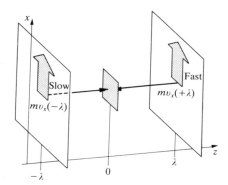

Fig. 24.10 The calculation of the viscosity of a gas examines the net *x*-component of momentum brought to a plane from faster and slower layers on average a mean free path away.

The net flux of x-momentum in the z-direction is therefore

$$J_z = \tfrac{1}{4}\mathcal{N}\bar{c}\left\{\left[mv_x(0) - m\lambda\left(\frac{dv_x}{dz}\right)_0\right] - \left[mv_x(0) + m\lambda\left(\frac{dv_x}{dz}\right)_0\right]\right\}$$

$$= -\tfrac{1}{2}\mathcal{N}m\lambda\bar{c}\left(\frac{dv_x}{dz}\right)_0$$

We see that the flux is proportional to the velocity gradient, as we wished to show. By comparing this expression with eqn 17, and multiplying by $\tfrac{2}{3}$ in the normal way, we can identify the viscosity as

$$\eta = \tfrac{1}{3}\mathcal{N}m\lambda\bar{c} = \tfrac{1}{3}m\lambda\bar{c}N_A[A] \qquad (21)°$$

As for thermal conductivity, the viscosity is independent of the pressure. Thus $\lambda \propto 1/p$ and $[A] \propto p$, implying that $\eta \propto \bar{c}$ independent of p. The physical reason is the same: more molecules are available to transport the momentum, but carry it less far on account of the shorter mean free path. Since $\bar{c} \propto T^{1/2}$, the viscosity is proportional to $T^{1/2}$. That is, the viscosity of a gas *increases* with temperature: the molecules travel more quickly, and the flux of momentum is greater.[3]

Measurement of gas viscosity

There are two main techniques for measuring viscosities of gases. One depends on the rate of damping of the torsional oscillations of a disk hanging in the gas, the time constant for the decay of the harmonic motion depending on the viscosity and the design of the apparatus. The other is based on **Poiseuille's formula** for the rate of flow of a fluid through a tube of radius r:

$$\frac{dV}{dt} = \frac{(p_1^2 - p_2^2)\pi r^4}{16l\eta p_0} \qquad (22)$$

where V is the volume flowing, p_1 and p_2 are the pressures at each end of the tube of length l, and p_0 is the pressure at which the volume is measured.

Such measurements confirm that the viscosities of gases are independent of pressure over a wide range. For instance, the results for argon from 10^{-3} atm to 10^2 atm are shown in Fig. 24.11, and we see that η is constant from about 0.01 atm to 50 atm. The measurements also confirm (to a lesser extent) the $T^{1/2}$ temperature dependence. The broken line in the illustration shows the calculated values using $\sigma = 22 \times 10^{-20}$ m², implying a collision diameter of 260 pm, in contrast to the van der Waals diameter of 335 pm obtained from the density of the solid. The agreement is not too bad, considering the simplicity of the model and the neglect of intermolecular forces.

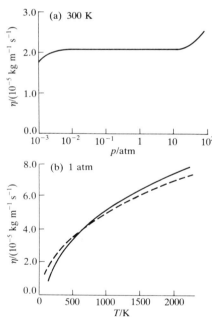

Fig. 24.11 The experimental results for (a) the pressure dependence of the viscosity of argon, and (b) its temperature dependence. The broken line in the latter is the calculated value. Fitting the observed and calculated curves is one way of determining the collision cross section.

---

[3] In a liquid the viscosity *decreases* with temperature because intermolecular interactions must be overcome (Section 22.9).

Example 24.10: *Measuring the viscosity using Poiseuille's method*

In a Poiseuille's flow experiment to measure the viscosity of air at 298 K, the sample was allowed to flow through a 100 cm tube of internal diameter 1.0 mm. The high-pressure end was at 765 Torr and the low-pressure end was at 760 Torr. The volume was measured at the latter pressure. In 100 s a volume of 90.2 cm³ passed through the tube. What is the viscosity of air at this temperature?

Answer. We use eqn 22, after converting pressures to Pa using 1 Torr = 133.3 Pa. The rate of flow is

$$\frac{dV}{dt} = \frac{90.2 \text{ cm}^3}{100 \text{ s}} = 9.02 \times 10^{-7} \text{ m}^3 \text{ s}^{-1}$$

Then, since

$$p_1^2 - p_2^2 = 1.355 \times 10^8 \text{ Pa}^2$$

and

$$\frac{\pi r^4}{16 l p_0} = 1.21 \times 10^{-19} \text{ N}^{-1} \text{ m}^5$$

from eqn 22, $\eta = 1.8 \times 10^{-5} \text{ kg m}^{-1} \text{ s}^{-1}$.

Comment. The kinetic theory expression (eqn 21) gives $\eta = 1.4 \times 10^{-5} \text{ kg m}^{-1} \text{ s}^{-1}$, and so the agreement is quite good. Viscosities are often expressed in centipoise (cP) or (for gases) micropoise (μP), the conversion being $1 \text{ cP} = 10^{-3} \text{ kg m}^{-1} \text{ s}^{-1}$: the viscosity of air is 180 μP.

Exercise. What volume would be collected if the pressure gradient were doubled, other conditions remaining constant? [180 cm³]

Some experimental results are given in Table 24.4, and the theoretical formulas are collected (together with some exact expressions from more detailed calculations) in Table 24.3.

Further information: mean values and distributions

Suppose we want to calculate the mean value $\langle X \rangle$ of a property X which may take any of the values $X_1, X_2, \ldots, X_Z$. These values are the possible **outcomes** of the observation. Moreover, suppose also that in a series of N measurements we find that X_1 occurs N_1 times, X_2 occurs N_2 times, and so on. Then the mean value is given by

$$\langle X \rangle = \frac{N_1 X_1 + N_2 X_2 + \ldots + N_Z X_Z}{N} = \sum_i \frac{N_i}{N} X_i$$

This formula can be expressed in terms of the **probability** P_i that an outcome X_i is obtained. Since $P_i = N_i / N$, we can write

$$\langle X \rangle = \sum_i P_i X_i \qquad \text{(A1)}$$

where the sum ranges over all possible outcomes.

Example 24.11: *Calculating a mean value*

The abundances of the isotopes of palladium and their masses are

Mass number:	102	104	105	106	108	110
Abundance/%	1.020	11.15	22.34	27.34	26.47	11.72
Mass/u:	101.906	103.904	104.905	105.903	107.904	109.905

Calculate the mean atomic mass.

Answer. The abundances are probabilities × 100%, so eqn A1 becomes

$$\langle m \rangle / u = 1.020 \times 10^{-2} \times 101.906 + 0.1115 \times 103.904 + \ldots$$

$$= 106.46$$

Exercise. Calculate the root mean square mass $\langle m^2 \rangle^{1/2}$ of the isotopes.

[106.45 u]

Now consider the case where the outcomes may take any of a continuous range of values, as in the height of a population or the speed of molecules in a gas. To cope with this problem, we express it in a way that resembles the discrete averaging procedure by dividing the continuous range of outcomes into segments (the boxes in Fig. 24.12). Then we count 1 each time a measurement has an outcome that falls *anywhere* in a given box. For example, consider the box of length ΔX at X. If in a series of 300 observations an outcome lying in this box is obtained in six of them, we write $N(X) = 6$. If the total number of observations is N, the probability that the outcome of any single observation is in the box lying between X and $X + \Delta X$ is $P(X) = N(X)/N$, which in this case is 1/50. The value of $N(X)$, and therefore of $P(X)$, is proportional to the length of the box at X (so long as X is small), and so we write

$$P(X) = f(X)\,\Delta X$$

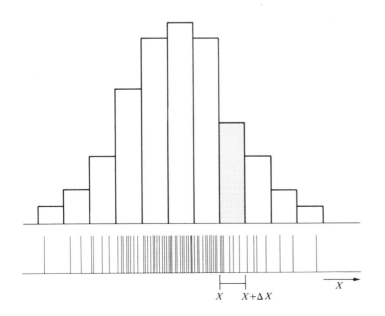

Fig. 24.12 A sequence of events: each line represents one observation of the corresponding value of *X*. An approximation to the average value is obtained by dividing the range of outcomes into packets of range *X* (one is shown), and lumping together all the observations that have an outcome in that range. In this case, six observations yield the outcome *X*.

743

The bunching procedure has turned the continuous problem into one resembling the discrete problem, and we can continue as we did there. The *approximate* average value of X is found by taking the value of X for each box, multiplying it by the probability that the observation has an outcome in that box, and then summing over all the boxes:

$$\langle X \rangle \approx \sum XP(X) = \sum Xf(X)\,\Delta X$$

The equality is only approximate because $f(X)$ might vary appreciably over the width of the box. However, it becomes *exact* if we allow each segment to become infinitesimal, for then $f(X)$ will be constant over the range. We therefore take the limit of the box width ΔX becoming infinitesimal ($\Delta X \to dX$) and simultaneously convert the discrete sum into an integral over all possible outcomes:

$$\langle X \rangle = \int Xf(X)\,dX \qquad (A2)$$

The function $f(X)$ is called the **distribution** of the property X. From its original definition $P(X) = f(X)\,\Delta X$, which becomes

$$dP(X) = f(X)\,dX \qquad (A3)$$

now that we are dealing with infinitesimal ranges, we see that f gives the probability that a property lies in the range X to $X + dX$. The interpretation of eqn A3 was illustrated in Example 24.2.

If two properties X and Y are independent of each other, the probability of an observation having both X_i and Y_j as outcomes is the product of the individual probabilities:

$$P(X_i, Y_j) = P(X_i)P(Y_j) \qquad (A4)$$

Example 24.12: *Calculating a joint probability*

Given that the probability of a person being a man in a given population is 0.495, and the probability of a person (man or woman) being left-handed is 0.110, what is the probability of selecting a left-handed man by random choice from a crowd?

Answer. If we suppose that the property of being left-handed is independent of sex, we can use eqn A4:

$$P(\text{left-handed, man}) = P(\text{left-handed}) \times P(\text{man})$$

$$= 0.110 \times 0.495 = 0.054$$

That is, 1 in 18 people is likely to be a left-handed man.

Comment. If left-handedness were a male characteristic, this calculation would be false: eqn A4 applies only to *uncorrelated* properties.

Exercise. What is the probability of selecting a right-handed woman from the same population? [1 in 2.2]

The same technique can be used for continuous properties. If the probability of X lying in the range X to $X + dX$ is $f(X)\,dX$, and the probability that an independent property Y lies in the range Y to $Y + dY$ is $f(Y)\,dY$, then the probability of X and Y lying simultaneously in these ranges is the product of the individual probabilities:

$$f(X, Y)\,dX\,dY = f(X)f(Y)\,dX\,dY$$

and the distribution is the product of the individual distributions:

$$f(X, Y) = f(X)f(Y) \qquad (A5)$$

We used this conclusion in the derivation of the Maxwell–Boltzmann distribution.

Further reading

R. D. Present, *Kinetic theory of gases*. McGraw-Hill, New York (1958).

D. Tabor, *Gases, liquids, and solids* (2nd edn). Cambridge University Press (1979).

A. J. Walton, *Three phases of matter*. Clarendon Press, Oxford (1983).

R. B. Bird, W. E. Stewart, and E. N. Lightfoot, *Transport phenomena*. Wiley, New York (1960).

J. O. Hirschfelder, C. F. Curtiss, and R. B. Bird, *The molecular theory of gases and liquids*. Wiley, New York (1954).

Exercises

24.1 Calculate the mean speed of (a) He atoms, (b) CH_4 molecules at (i) 77 K, (ii) 298 K, (iii) 1000 K.

24.2 At what pressure does the mean free path of argon at 25°C become comparable to the size of a 1 L vessel that contains it? Take $\sigma \approx 0.36\ \mathrm{nm}^2$.

24.3 At what pressure does the mean free path of argon at 25°C become comparable to the diameters of the atoms themselves?

24.4 At an altitude of 20 km the temperature is 217 K and the pressure 0.05 atm. What is the mean free path of N_2 molecules? ($\sigma \approx 0.43\ \mathrm{nm}^2$.)

24.5 How many collisions does a single Ar atom make in 1.0 s when the temperature is 25°C and the pressure is (a) 10 atm, (b) 1.0 atm, (c) 1.0 μatm?

24.6 Calculate the total number of collisions per second in 1.0 L of argon under the same conditions as in Exercise 24.5.

24.7 How many collisions per second does an N_2 molecule make at an altitude of 20 km? (See Exercise 24.4 for data.)

24.8 Calculate the collision density in air at 25°C and 1.0 atm (a) between O_2 molecules, (b) between O_2 and N_2 molecules. Take $d(O_2) \approx 357$ pm and $d(N_2) \approx 370$ pm.

24.9 How many collisions are made per second in 1.0 cm^3 of the interior of the envelope of a Dewar flask where the remaining air is at 1.2 Torr and 25°C?

24.10 Calculate the molar translational kinetic energy (in kJ mol^{-1}) of (a) H_2 molecules, (b) I_2 molecules in a gas at 300 K and 1.0 atm.

24.11 Calculate the mean free path of diatomic molecules in air using $\sigma = 0.43\ \mathrm{nm}^2$ at 25°C and (a) 10 atm, (b) 10^{-6} atm.

24.12 Use the Maxwell distribution of speeds to estimate the fraction of N_2 molecules at 500 K that have speeds in the range 290 to 300 m s^{-1}.

24.13 A solid surface with dimensions 2.5 mm × 3.0 mm is exposed to argon gas at 90 Pa and 500 K. How many collisions do the Ar atoms make with this surface in 15 s?

24.14 A 5.00 L container holds 1.50 mmol of N_2 and 3.00 mmol of H_2 at a total pressure of 2.00 kPa. Calculate the number of N_2–H_2 collisions that occur in the mixture in 1.0 ms.

24.15 How does the mean free path in a sample of a gas vary with temperature in a constant-volume container?

24.16 An effusion cell has a circular hole which is 2.50 mm in diameter. If the molar mass of the solid in the cell is 260 g mol^{-1} and its vapour pressure is 0.835 Pa at 400 K, by how much will the mass of the solid decrease in a period of 2 hours?

24.17 Calculate the flux of energy arising from a temperature gradient of 2.5 K m^{-1} in a sample of argon in which the mean temperature is 273 K.

24.18 Use the experimental value of the thermal conductivity of neon at 273 K to estimate the collision cross section of Ne atoms.

24.19 Use the experimental value of the coefficient of viscosity for neon (Table 24.4) to estimate the collision cross section of Ne atoms at 273 K.

24.20 Calculate the inlet pressure required to maintain a flow rate of $9.5 \times 10^5 \, \text{L h}^{-1}$ of nitrogen at 293 K flowing through a pipe of length 8.50 m and diameter 1.00 cm. The pressure of gas as it exits the tube is 1.00 bar. The volume of the gas is measured at that pressure.

24.21 Calculate the viscosity of air at (a) 273 K, (b) 298 K, (c) 1000 K. Take $\sigma \approx 0.40 \, \text{nm}^2$. (The experimental value is 173 μP at 273 K, 182 μP at 20°C, and 394 μP at 600°C.)

24.22 Calculate the thermal conductivities of (a) argon, (b)

helium at 300 K and 1.0 mbar. Each gas is confined in a cubic vessel of 10 cm side, one wall being at 310 K and the one opposite at 295 K. What is the rate of flow of energy as heat from wall to the other in each case?

24.23 The viscosity of carbon dioxide was measured by comparing its rate of flow through a long narrow tube (using Poiseuille's formula) with that of argon. For the same pressure differential, the same volume of carbon dioxide passed through the tube in 55 s as argon in 83 s. The viscosity of argon at 25°C is 208 μP; what is the viscosity of carbon dioxide? Estimate the molecular diameter of carbon dioxide.

24.24 Calculate the diffusion constant of argon at 25°C and (a) 10^{-6} atm, (b) 1 atm, (c) 100 atm. If a pressure gradient of $0.10 \, \text{atm cm}^{-1}$ is established in a pipe, what is the flow of gas due to diffusion?

Problems

Numerical problems

24.1 In an experiment to measure the speed of molecules by a rotating slotted-disk experiment, the apparatus consisted of five coaxial 5.0 cm diameter disks separated by 1.0 cm, the slots in their rims being displaced by 2.0° between neighbours. The relative intensities I of the detected beam of Kr atoms for two different temperatures and at a series of rotation rates were as follows:

ν/Hz	20	40	80	100	120
$I(40 \, \text{K})$	0.846	0.513	0.069	0.015	0.002
$I(100 \, \text{K})$	0.592	0.485	0.217	0.119	0.057

Find the distributions of molecular velocities $f(v_x)$ at these temperatures, and check that they conform to the theoretical prediction for a one-dimensional system.

24.2 Cars were timed by police radar as they pass in both directions below a bridge. Their velocites (m.p.h., numbers of cars in parentheses) to the east and west were as follows: 50 E (40), 55 E (62) 60 E (53), 65 E (12), 70 E (2); 50 W (38), 55 W (59), 60 W (50), 65 W (10), 70 W (2). What are (a) the mean velocity, (b) the mean speed, (c) the root mean square speed?

24.3 A population consists of people of the following heights (in feet and inches, numbers of individuals in parentheses): 5'5" (1), 5'6" (2), 5'7" (4), 5'8" (7), 5'9" (10), 5'10" (15), 5'11" (9), 6'0" (4), 6'1" (0), 6'2" (1). What is (a) the mean height, (b) the root mean square height of the population.

24.4 Calculate the ratio of the thermal conductivities of gaseous hydrogen at 300 K to gaseous hydrogen at 10 K? Be circumspect.

24.5 In a double-glazed window, the panes of glass are separated by 5.0 cm. What is the rate of transfer of heat by conduction from the warm room (25°C) to the cold exterior

(−10°C) through a window of area $1.0 \, \text{m}^2$? What power of heater is required to make good the loss of heat?

24.6 A manometer was connected to bulb containing carbon dioxide under slight pressure. The gas was allowed to escape through a small pinhole, and the time for the manometer reading to drop from 75 cm to 50 cm was 52 s. When the experiment was repeated using nitrogen (for which $M = 28.02 \, \text{g mol}^{-1}$) the same fall took place in 42 s. Calculate the molar mass of carbon dioxide.

24.7 A particular make of electric light bulb contains argon at 50 Torr and has a tungsten filament of radius 0.10 mm and length 5.0 cm. When operating, the gas close to the filament has a temperature of about 1000°C. How many collisions are made with the filament in each second?

24.8 A space vehicle of internal volume $3.0 \, \text{m}^3$ is struck by a meteor and hole 0.10 mm radius is formed. If the oxygen pressure within the vehicle is initially 0.80 atm and its temperature 298 K, how long will it take to fall to 0.70 atm?

24.9 In the Knudsen method for the determination of vapour pressure, a weighed amount of a sample is heated inside a container, in the wall of which there is a small hole. The mass loss over a period of time can be related to the vapour pressure at the temperature of the experiment. If Δm is the mass lost in an interval Δt through a circular hole of radius R, find an expression relating the vapour pressure p to Δm and Δt. A Knudsen cell was used to determine the vapour pressure of germanium at 1000°C. During an interval of 7200 s the mass loss through a hole of radius 0.50 mm amounted to 4.3×10^{-2} mg. What is the vapour pressure of germanium at 1000°C? Assume the gas to be monatomic.

24.10 In a study of the catalytic properties of a titanium surface it was necessary to maintain the surface free from contamination. Calculate the collision frequency per cm^2 of surface made by O_2 molecules at (a) 1.0 atm, (b) 1.0 μatm, (c) 1.0×10^{-10} atm and 300 K. Estimate the number of collisions made with a single surface atom in each second. The conclusions underline the importance of working at very low pressures in order to study the properties of uncontaminated surfaces. Take the nearest neighbour distance as 291 pm.

24.11 The nuclide ^{244}Bk (berkelium) decays by producing α particles, which capture electrons and form He atoms. Its half-life is 4.4 h. A 1.0 mg sample was placed in a 1.0 cm^3 container that was impermeable to α radiation, but there was also a hole of radius 2.0 μm in the wall. What is the pressure of helium, at 298 K, inside the container after (a) 1 h, (b) 10 h?

24.12 A spherical glass bulb of diameter 10 cm has a tube of radius 3.0 mm attached to it. While the bulb is maintained at 300 K the tube is at 77 K (it is immersed in liquid nitrogen). Initially the gas in the bulb is damp, the partial pressure of the water vapour being 1.0 Torr. Estimate the time required for condensation in the tube to lower the partial pressure to 10 μTorr if effusion is rate-determining.

24.13 An atomic beam is designed to function with (a) cadmium, (b) mercury. The source is an oven maintained at 380 K, there being a small slit of dimensions 1 cm $\times$ 10^{-3} cm. The vapour pressure of cadmium is 0.13 Pa and that of mercury is 152 kPa at this temperature. What is the atomic current (the number of atoms per unit time) in the beams?

24.14 The reaction $H_2 + I_2 \rightarrow 2HI$ depends on collisions between a variety of species in the reaction mixture, and we examine it in more detail in Chapter 26. Calculate the collision densities for the encounters (a) $H_2 + H_2$, (b) $I_2 + I_2$, (c) $H_2 + I_2$ for a gas with partial pressures of 0.50 atm for both species, the temperature being 400 K. Use $\sigma(H_2) \approx 0.27$ nm^2 and $\sigma(I_2) \approx 1.2$ nm^2.

24.15 Calculate the escape velocity (the minimum initial velocity that will take an object to infinity) from the surface of a planet of radius R? What is the value for (a) the earth, $R = 6.37 \times 10^6$ m, $g = 9.81$ m s^{-2}, (b) Mars, $R = 3.38 \times 10^6$ m, $m_{Mars}/m_{Earth} = 0.108$. At what temperatures do H_2, He, and O_2 molecules have mean speeds equal to their escape speeds?

Theoretical problems

24.16 The Maxwell–Boltzmann distribution was derived from arguments about probability, but it can also be derived from the Boltzmann distribution itself. Do so.

24.17 A specially constructed velocity-selector accepts a beam of molecules from an oven at a temperature T but blocks the passage of molecules with a speed greater than the mean. What is the mean speed of the emerging beam, relative to the initial value, treated as a one-dimensional problem?

24.18 What is the proportion of gas molecules having (a) more than, (b) less than the root mean square speed? (c) What are the proportions having speeds greater and smaller than the mean speed?

24.19 Calculate the fractions of molecules in a gas that have a speed in a range Δv at the speed nc^* relative to those in the same range at c^* itself? This calculation can be used to estimate the reaction of very energetic molecules (which is important for reactions). Evaluate the ratio for $n = 3$ and $n = 4$.

24.20 The rate of growth of droplets of lead from a vapour has been studied in the laboratories of an oil company (J. B. Homer and A. Prothero, *J. chem. Soc. Faraday Trans.* I, **69**, 673 (1973)). Virtually all the lead condensed within 0.5 ms of the initiation of the run, and the concentration in the gas phase was no more than about 3×10^{15} atoms/cm^3. Find an expression for the rate of growth of the radius of the spherical particles. Take $T = 935$ K and assume that every atom sticks to the growing surface when it collides with it.

25

Molecules in motion

Check-list of key ideas

1. The measurement of the *conductance* of an electrolyte solution (Section 25.1) and the definition of the *conductivity* and the *molar conductivity* (eqn 1).

2. The classification of electrolytes as *strong* or *weak* (Section 25.1).

3. The *Kohlrausch law* for the molar conductivities of strong electrolytes (eqn 2) and the *limiting molar conductivity*.

4. The *law of the independent migration of ions* (eqn 3).

5. The *degree of ionization* of a weak electrolyte (eqn 4b) and *Ostwald's dilution law* for the molar conductivity of the solution (eqn 5).

6. The *drift speed* of an ion in an electric field (eqn 6) and the variation of ionic conductivities with their effective hydrodynamic radii.

7. The *chain mechanism* for conduction by protons (Section 25.2).

8. The definition of the *mobility* of an ion (eqn 7) and its relation to the ionic conductivity (eqn 8).

9. The *transport number* of an ion (eqn 10) and its measurement by the *moving boundary* and *Hittorf* methods (Section 25.2).

10. The effect of *ion–ion interactions* on the conductivities of ions and the *Debye–Hückel–Onsager theory* (eqn 13).

11. The definition of *thermodynamic force* (eqn 14), its calculation (Example 25.6), and its use in the derivation of Fick's first law of diffusion.

12. The *Einstein relation* between the diffusion coefficient and the ionic mobility (eqn 16).

13. The *Nernst–Einstein equation* for the molar conductivity in terms of the diffusion coefficient (eqn 18).

14. The *Stokes–Einstein equation* relating the diffusion coefficient and the frictional constant (eqn 20).

15. The derivation of the *diffusion equation* (eqn 21) and the inclusion of *convection* (eqn 23).

16. The measurement of diffusion coefficients (Section 25.5).

17. The *net distance* that a molecule diffuses (eqn 26).

18. The interpretation of diffusion as a *random walk* (Section 25.7) and the derivation of the *Einstein–Smoluchowski equation* for the diffusion coefficient in terms of the step length and time of a random walk (eqn 28).

The general approach we describe in this chapter provides techniques for discussing the motion of all kinds of particles in all kinds of fluids. We shall set the scene by considering a simple type of motion, the migration of ions in the presence of an electric field. This motion is simple (as long as we do not enquire too closely into the details) because the ions drift steadily through the solvent towards one electrode or the other. We shall then see that the electric force acting on the ions can be regarded as a special case of a generalized force. The formulation of this generalized force will enable us to discuss the motion of ions and neutral molecules in greater detail and to describe their diffusion through solutions even in the absence of electric fields. We shall then be in a position to discuss (in the following chapters) the rates of chemical reactions, which depend on the migration of particles through media of various kinds.

Ion transport

The motion of ions in solution can be studied by measuring the electrical conductivities of electrolyte solutions. The migration of cations towards a negatively charged electrode and of anions towards a positively charged electrode carries charge through the solution, and is therefore a transport process of the kind considered in Chapter 24.

25.1 The conductivity of electrolyte solutions

The fundamental measurement used to study the motion of ions is that of the electrical resistance of the solution. The standard technique is to incorporate a cell of the type shown in Fig. 25.1 into one arm of a resistance

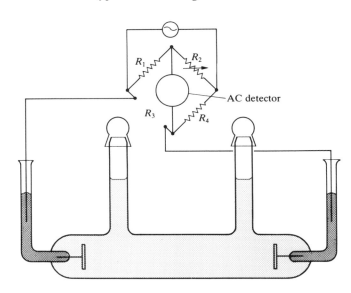

Fig. 25.1 The conductivity of an electrolyte solution is measured by making the cell one arm of a resistance bridge. The balance point is obtained when the resistances satisfy $R_1/R_2 = R_3/R_4$ (or, since the current is alternating, a similar relation for the impedances).

bridge and to search for the balance point, as explained in standard texts on electricity. The main complication is that alternating current must be used because a direct current would lead to electrolysis and to **polarization**, the modification of the composition of the layers of solution in contact with the electrodes. The use of alternating current (with a frequency of about 1 kHz) may avoid polarization because the change that occurs on one half of the cycle is undone during the second half (if the reverse reaction is kinetically feasible).

Conductance and conductivity

The **conductance** G of a solution is the inverse of its **resistance** R: the lower the resistance of a solution, the greater its conductance. Since resistance is expressed in ohm, Ω, the conductance of a sample is expressed in Ω^{-1}. The reciprocal ohm used to be called the mho, but its official designation is now the **siemens,** S, and $1\,S = 1\,\Omega^{-1}$.

The resistance of a sample increases with its length l and decreases with its cross-sectional area A. We therefore write

$$R = \rho \times \frac{l}{A}$$

The constant of proportionality ρ is called the **resistivity** of the sample. The **conductivity** κ is the inverse of the resistivity, and so

$$R = \frac{1}{\kappa} \times \frac{l}{A} \quad \text{or} \quad \kappa = \frac{l}{RA}$$

With the resistance in Ω and the dimensions in m, it follows that the units of κ are $S\,m^{-1}$ ($S\,cm^{-1}$ is sometimes more convenient).

It is unreliable to calculate the conductivity directly from the resistance of the sample and the cell dimensions l and A because the current distribution is complicated. In practice, the cell is calibrated using a sample of known conductivity κ^* (typically an aqueous solution of potassium chloride) and a **cell constant** C is determined from

$$\kappa^* = \frac{C}{R^*}$$

where R^* is the resistance of the standard. The dimensions of C are $[\text{length}]^{-1}$. If the sample has a resistance R in the same cell, its conductivity is

$$\kappa = \frac{C}{R}$$

The conductivity of a solution depends on the number of ions present, and it is normal to introduce the **molar conductivity** Λ_m, which is defined as

$$\Lambda_m = \frac{\kappa}{c} \tag{1}$$

where c is the molar concentration of the added electrolyte. The molar conductivity is normally expressed[1] in $S\,cm^2\,mol^{-1}$.

[1] The conductivity is normally available in $S\,cm^{-1}$ and the concentration in M, and so the practical relation is

$$\Lambda_m = \frac{100 \times \kappa/(S\,cm^{-1})\,S\,cm^2\,mol^{-1}}{c/M}$$

Example 25.1: *Calculating the molar conductivity of a solution*

The molar conductivity of $0.100\,\text{M}\ KCl(aq)$ at 298 K is $129\ \text{S cm}^2\ \text{mol}^{-1}$. The measured resistance in a conductivity cell was $28.44\ \Omega$. The resistance was $31.60\ \Omega$ when the same cell contained $0.0500\,\text{M}\ NaOH(aq)$. Calculate the molar conductivity of $NaOH(aq)$ at that concentration.

Answer. First, we determine the cell constant, and then use this constant and the given R to find the conductivity of the test solution. The conductivity of an $0.100\,\text{M}\ KCl(aq)$ solution is

$$\kappa = \Lambda_m c = 129\ \text{S cm}^2\ \text{mol}^{-1} \times 0.100\ \text{mol dm}^{-3}$$

$$= 1.29 \times 10^{-2}\ \text{S cm}^{-1}$$

and so

$$C = 1.29 \times 10^{-2}\ \text{S cm}^{-1} \times 28.44\ \Omega = 0.367\ \text{cm}^{-1}$$

Therefore, for $NaOH(aq)$,

$$\kappa = \frac{C}{R} = \frac{0.367\ \text{cm}^{-1}}{31.60\ \Omega} = 0.0116\ \text{S cm}^{-1}$$

The molar conductivity of $NaOH(aq)$ is therefore

$$\Lambda_m = \frac{0.0116\ \text{S cm}^{-1}}{0.0500\ \text{mol dm}^{-3}} = 232\ \text{S cm}^2\ \text{mol}^{-1}$$

Comment. It is good practice to calibrate cells with solutions having conductances close to the conductance of the test solution.

Exercise. The same cell was used to study $0.100\,\text{M}\ NH_4Cl(aq)$, and a resistance of $28.50\ \Omega$ was measured. Calculate the molar conductivity of $NH_4Cl(aq)$ at this concentration.

$$[129\ \text{S cm}^2\ \text{mol}^{-1}]$$

The molar conductivity of an electrolyte would be independent of concentration if κ were exactly proportional to the concentration of the electrolyte. However, in practice, the molar conductivity is found to vary with the concentration. One reason is that the number of ions in the solution might not be proportional to the concentration of the electrolyte. For instance, the concentration of ions in a solution of a weak acid depends on the concentration of the acid in a complicated way, and doubling the nominal concentration of the acid does not double the number of ions. Secondly, because ions interact with each other strongly, the conductivity of a solution is not exactly proportional to the number of ions present.

Measurement of the concentration dependence of molar conductivities shows that there are two classes of electrolyte. The characteristic of a **strong electrolyte** is that its molar conductivity decreases only slightly as its concentration is increased (Fig. 25.2). The characteristic of a **weak electrolyte** is that its molar conductivity is normal at concentrations close to zero, but falls sharply to low values as the concentration increases (Fig. 25.2). The classification depends on the solvent employed as well as the solute: lithium chloride, for example, is a strong electrolyte in water but a weak electrolyte in propanone.

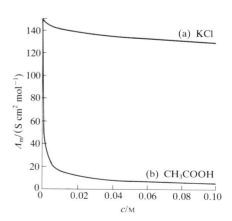

Fig. 25.2 The concentration dependence of the molar conductivities of (a) a typical strong electrolyte and (b) a typical weak electrolyte.

Strong electrolytes

Strong electrolytes are substances that are fully ionized in solution, and include ionic solids and strong acids. As a result of their complete ionization, the concentration of ions in solution is proportional to the concentration of electrolyte added.

In an extensive series of measurements during the 19th century, Friedrich Kohlrausch showed that at low concentrations the molar conductivities of strong electrolytes obey **Kohlrausch's law**, that the molar conductivity is proportional to the square root of the concentration:

$$\Lambda_m = \Lambda_m^\circ - \mathcal{K}c^{1/2} \tag{2}$$

Λ_m° is the **limiting molar conductivity**, the molar conductivity in the limit of zero concentration (when the ions do not interact with each other). $\mathcal{K}$ is a coefficient which is found to depend more on the stoichiometry of the electrolyte (i.e. whether it is of the form MA, or M_2A, etc) than on its specific identity.

Kohlrausch was also able to confirm that Λ_m° can be expressed as the sum of contributions from its individual ions. If the limiting molar conductivity of the cations is denoted λ_+ and that of the anions λ_-, his **law of the independent migration of ions** is that

$$\Lambda_m^\circ = \nu_+\lambda_+ + \nu_-\lambda_- \tag{3}$$

where ν_+ and ν_- are the numbers of cations and anions per formula unit of electrolyte (e.g. $\nu_+ = \nu_- = 1$ for HCl, NaCl, and $CuSO_4$, but $\nu_+ = 1$, $\nu_- = 2$ for $MgCl_2$). This simple result, which can be understood on the grounds that the ions behave independently in the limit of zero concentration, lets us predict the limiting molar conductivity of any strong electrolyte from the data in Table 25.1. For example, the limiting molar conductivity of $BaCl_2$ in water at 298 K is

$$\Lambda_m^\circ = (127.2 + 2 \times 76.3)\ \text{S cm}^2\ \text{mol}^{-1} = 279.8\ \text{S cm}^2\ \text{mol}^{-1}$$

Table 25.1. Limiting ionic conductances in water at 298 K, $\lambda/(\text{S cm}^2\ \text{mol}^{-1})$

H^+	349.6	OH^-	199.1
Na^+	50.1	Cl^-	76.3
K^+	73.5	Br^-	78.1
Zn^{2+}	105.6	SO_4^{2-}	160.0

Weak electrolytes

Weak electrolytes are substances that are not fully ionized in solution. They include weak Brønsted acids such as CH_3COOH and weak Brønsted bases such as NH_3. The marked concentration dependence of their conductivities arises from the displacement of the equilibrium

$$HA(aq) + H_2O(l) \rightleftharpoons H_3O^+(aq) + A^-(aq) \qquad K_a = \frac{a(H_3O^+)a(A^-)}{a(HA)}$$

towards products at low concentrations.

The conductivity depends on the number of ions in the solution, and therefore on the **degree of ionization** α of the electrolyte. The degree of ionization is defined so that, for the acid HA at a nominal concentration c, at equilibrium

$$[H_3O^+] = \alpha c \qquad [A^-] = \alpha c \qquad [HA] = (1-\alpha)c$$

If we ignore activity coefficients, the acidity constant is approximately

$$K_a = \frac{\alpha^2 c}{1 - \alpha} \tag{4a}$$

The degree of ionization when the concentration is c is therefore

$$\alpha = \frac{K_a}{2c} \left\{ \left(1 + \frac{4c}{K_a} \right)^{1/2} - 1 \right\} \tag{4b}$$

If the molar conductivity of the hypothetical fully ionized electrolyte is Λ_m', since only a fraction α is actually present as ions in the actual solution, the measured molar conductivity Λ_m is given by

$$\Lambda_m = \alpha \Lambda_m'$$

with α given by eqn 4b. When the concentration of ions in solution is low, we can approximate Λ_m' by its limiting value, and write

$$\Lambda_m = \alpha \Lambda_m^\circ \tag{4c}$$

Example 25.2: *Using conductivity measurements to determine pK$_a$*

The resistance of an 0.0100 M CH$_3$COOH(aq) solution was measured at 298 K in the same cell as in Example 25.1, and found to be 2220 Ω. Find the degree of ionization of the acid at this concentration and its pK$_a$.

Answer. We calculate Λ_m using the cell constant determined in Example 25.1 ($C = 0.367$ cm^{-1}), which gives $\Lambda_m = 16.5$ S cm^2 mol^{-1}. From the data in Table 25.1 we find $\Lambda_m^\circ = 390.5$ S cm^2 mol^{-1}. Therefore, $\alpha = 0.0423$. The acidity constant is then obtained by substituting this value of α into eqn 4a, and is 1.9×10^{-5}, implying that pK$_a$ = 4.72.

Comment. The thermodynamic value of pK$_a$ could be obtained by repeating the determination with different concentrations and extrapolating to zero concentration.

Exercise. The resistance of an 0.0250 M HCOOH(aq) solution was measured as 444 Ω under similar conditions; evaluate the pK$_a$ of the acid. [3.74]

Once we know K_a, we can use eqn 4b to predict the concentration dependence of the molar conductivity. The result agrees quite well with the experimental curve in Fig. 25.2. Alternatively, we can use the concentration dependence of Λ_m in measurements of the limiting molar conductance. It is quite easy to manipulate the expression for K_a into the form

$$\frac{1}{\alpha} = 1 + \frac{\alpha c}{K_a}$$

Then, using eqn 4c, we obtain **Ostwald's dilution law**:

$$\frac{1}{\Lambda_m} = \frac{1}{\Lambda_m^\circ} + \frac{\Lambda_m c}{K_a (\Lambda_m^\circ)^2} \tag{5}$$

Equation 5 suggests a procedure for determining the limiting molar conductivity of a solution, for it implies that if $1/\Lambda_m$ is plotted against $\Lambda_m c$, the intercept at $c = 0$ will be $1/\Lambda_m^\circ$.

25.2 The motion of ions

To interpret conductivity measurements we need to know why ions move at different rates, why they have different molar conductivities, and why the molar conductivities of strong electrolytes decrease with the square root of the concentration. The central concept we shall build on is that the greater the mobility of an ion in solution, the greater its contribution to the conductivity (for a given charge).

The drift speed

When two electrodes a distance l apart are at a potential difference $\Delta\phi$, the ions in the solution between them experience a uniform electric field of magnitude

$$E = \frac{\Delta\phi}{l}$$

In such a field an ion of charge[2] ze experiences a force of magnitude

$$\mathscr{F} = zeE = \frac{ze\,\Delta\phi}{l}$$

A cation responds by accelerating towards the negative electrode and an anion responds by accelerating towards the positive electrode. However, as an ion moves through the solvent it experiences a frictional retarding force $\mathscr{F}'$ that is proportional to its speed. If we assume that the Stokes formula (eqn 10 of Section 23.2) for a sphere of radius a and speed s applies even on a microscopic scale (and independent evidence from magnetic resonance suggests that it often gives at least the right order of magnitude), we can write this retarding force as

$$\mathscr{F}' = fs \qquad f = 6\pi\eta a$$

The two forces act in opposite directions, and the ions reach a terminal speed, the **drift speed**, when the accelerating force $\mathscr{F}$ is balanced by the viscous drag $\mathscr{F}'$. The net force is zero ($\mathscr{F} = \mathscr{F}'$) when

$$s = \frac{zeE}{f} \tag{6}$$

Since the drift speed governs the rate at which charge is transported, we might expect the conductivity to decrease with increasing solvent viscosity and ion size. Experiments confirm these predictions for bulky ions (such as R_4N^+ and RCO_2^-) but not for small ions. For example, the molar conductivities of the alkali metal ions *increase* from Li^+ to Cs^+ (Table 25.1) even though the ionic radii increase (Table 21.2). The paradox is resolved when we realize that the radius a in Stokes' formula is the **hydrodynamic radius** of the ion, its effective radius in the solution taking into account all the H_2O molecules it carries in its hydration sphere. Small ions give rise to

---

[2] In this chapter we shall ignore the sign of the charge number z.

stronger electric fields than large ones,[3] and so small ions are more extensively solvated than big ions. Thus an ion of small ionic radius may have a large hydrodynamic radius because it drags many solvent molecules through the solution as it migrates. The hydrating H_2O molecules are often very labile, however, and NMR and isotope studies have shown that the exchange between the coordination sphere of the ion and the bulk solvent is very rapid.

The proton, however, although it is very small, has a very high molar conductivity (Table 25.1)! The explanation is that the proton conducts by a mechanism that does not involve its actual motion through the solution. According to the **chain mechanism** of proton migration, instead of a single, highly solvated proton moving through the solution, there is an effective motion of a proton that involves the rearrangement of bonds in a group of water molecules (Fig. 25.3). The conductivity is governed by the rates at which the water molecules can rotate into orientations in which they can accept or donate protons and the rate at which the protons tunnel from one end of a hydrogen bond to another (from $O\text{—}H \cdots O$ to $O \cdots H\text{—}O$).

Ion mobilities

According to eqn 6, the drift speed of an ion is proportional to the strength of the applied field. We write

$$s = uE \tag{7}$$

where the coefficient of proportionality u is called the **mobility** of the ion.

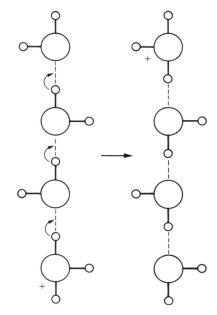

Fig. 25.3 The mechanism of conduction by water. Because protons can be transferred between molecules, the current is transported very rapidly down a chain by an effective, not an actual, migration of proton as each H_2O molecule passes one proton to its neighbour.

Example 25.3: *Estimating the mobility of an ion*

Estimate the mobility of an ion in water at 25°C.

Answer. Comparison of eqns 6 and 7 shows that

$$u = \frac{ze}{f} = \frac{ze}{6\pi\eta a}$$

For an order of magnitude estimate we can take $z = 1$ and a the radius of an ion such as Cs^+ (which might be typical of a smaller ion plus its hydration sphere), which is 170 pm. For the viscosity, we use $\eta = 1.0\,cP$ ($1.0 \times 10^{-3}\,kg\,m^{-1}\,s^{-1}$, Table 22.6). Then

$$u = \frac{1.602 \times 10^{-19}\,C}{6\pi \times (1.0 \times 10^{-3}\,kg\,m^{-1}\,s^{-1}) \times (1.7 \times 10^{-10}\,m)}$$

$$= 5 \times 10^{-8}\,C\,kg^{-1}\,s$$

Since $1\,C = 1\,J\,V^{-1}$ and $1\,J = 1\,kg\,m^2\,s^{-2}$,

$$u = 5 \times 10^{-8}\,m^2\,s^{-1}\,V^{-1}$$

Comment. Mobilities are often reported in $cm^2\,s^{-1}\,V^{-1}$, which in this case would be $5 \times 10^{-4}\,cm^2\,s^{-1}\,V^{-1}$.

Exercise. Estimate the hydrodynamic radius of SO_4^{2-} from the value of its mobility given in Table 25.2.   [205 pm]

---

[3] The electric field at the surface of a sphere of radius r is proportional to ze/r^2, and so the smaller the radius the stronger the field.

Table 25.2. Ionic mobilities in water at 298 K, $u/(10^{-4}\,cm^2\,s^{-1}\,V^{-1})$

H^+	36.23	OH^-	20.64
Na^+	5.19	Cl^-	7.91
K^+	7.62	Br^-	8.09
Zn^{2+}	5.47	SO_4^{2-}	8.29

Some ionic mobilities are given in Table 25.2 (we show below how they are related to conductivities and how they may be measured). Mobilities are typically in the region of $5 \times 10^{-4}\,cm^2\,s^{-1}\,V^{-1}$. This value means that when there is a potential difference of 1 V across a solution of length 1 cm (so $E = 1\,V\,cm^{-1}$), the drift speed is typically about $5 \times 10^{-4}\,cm\,s^{-1}$. That might seem slow, but not when expressed on a molecular scale, for it corresponds to an ion passing about 10^4 solvent molecules per second.

Mobility and conductivity

The usefulness of ionic mobilities is that they provide a link between measurable and theoretical quantities. As a first step we establish the relation between mobility and conductivity. To keep the calculation simple, we ignore signs in the following and concentrate on the magnitudes of quantities: the direction of flux can always be decided by common sense.

Consider a solution of a strong electrolyte at a molar concentration c. Let each formula unit give rise to v_+ cations of charge z_+e and v_- anions of charge z_-e. The concentration of each type of ion is therefore vc (with $v = v_+$ or v_-), and the number density of each type is vcN_A. The cations and anions travel with drift speeds s_+ and s_- respectively. The number of ions of one kind that pass through an imaginary window of area A (Fig. 25.4) during an interval Δt is equal to the number within the distance $s\Delta t$, and therefore to the number in the volume $s\Delta tA$. (The same sort of argument was used in Section 24.1 when we were discussing the pressure of a gas.) The number of ions in this volume is equal to $s\Delta tA \times vcN_A$. The flux through the window (the number of each type of ion passing through the window per unit area per unit time) is therefore

$$J(\text{ions}) = \frac{s\Delta tA \times vcN_A}{A\,\Delta t} = svcN_A$$

Each ion carries a charge ze, and so the flux of charge is

$$J(\text{charge}) = zsvceN_A = zsvcF$$

where $F = eN_A$ is Faraday's constant. Since $s = uE$, the flux is

$$J(\text{charge}) = zuvcFE$$

The current I through the window due to the ions we are considering is the charge flux times the area:

$$I = JA = zuvcFAE$$

Since the electric field is the potential gradient $\Delta\phi/l$,

$$I = \frac{zuvcFA\Delta\phi}{l}$$

Current and potential difference are related by **Ohm's law**, that

$$I = \frac{\Delta\phi}{R} = \frac{\kappa A\,\Delta\phi}{l}$$

When we compare the last two expressions we find that

$$\kappa = zuvcF \tag{8a}$$

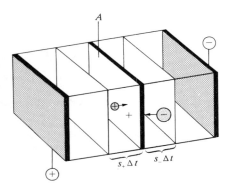

Fig. 25.4 In the calculation of the current, all the cations within a distance $s_+\,\Delta t$ (i.e. those in the volume $s_+A\,\Delta t$) will pass through the area A. Likewise, the anions in the corresponding volume the other side of the window will also contribute to the current.

Therefore, the molar conductivity of the ions is

$$\lambda = zuF \qquad (8b)$$

Equation 8 applies to the cations and the anions. Therefore, for the solution itself in the limit of zero concentration (when there are no interionic interactions),

$$\Lambda_m^\circ = (z_+ u_+ \nu_+ + z_- u_- \nu_-)F \qquad (9a)$$

For a symmetrical $z:z$ electrolyte (e.g. $CuSO_4$), eqn 9a simplifies to

$$\Lambda_m^\circ = z(u_+ + u_-)F \qquad (9b)$$

Example 25.4: *Estimating the limiting conductivity of a solution*

Estimate the typical limiting conductivity of a 1,1-aqueous solution at 25°C.

Answer. We estimate the typical ionic mobility in Example 25.3, and can use that value here for both the anions and the cations. With $z = 1$, eqn 9b gives

$$\Lambda_m^\circ = 2 \times (5 \times 10^{-8}\,m^2\,V^{-1}\,s^{-1}) \times (9.65 \times 10^4\,C\,mol^{-1})$$

$$= 1 \times 10^{-2}\,m^2\,C\,V^{-1}\,s^{-1}\,mol^{-1}$$

Since $1\,C\,s^{-1} = 1\,A$, and $1\,A\,V^{-1} = 1\,\Omega^{-1}$,

$$\Lambda_m^\circ = 1 \times 10^{-2}\,S\,m^2\,mol^{-1}$$

Comment. This answer converts to $1 \times 10^2\,S\,cm^2\,mol^{-1}$ for the value of a typical limiting molar conductivity, in accord with experiment. The value for KCl, for instance, is $150\,S\,cm^2\,mol^{-1}$.

Exercise. Estimate the typical molar conductivity of a 1,1-electrolyte methanol solution at 25°C, assuming that the hydrodynamic radii are the same as in water.
$$[2 \times 10^2\,S\,cm^2\,mol^{-1}]$$

Transport numbers

The **transport number** t is defined as the fraction of total current carried by the ions of a specified type. For a solution of two kinds of ion, the cation transport number is

$$t_+ = \frac{I^+}{I} \qquad (10)$$

where I^+ is the current carried by the cation and I is the total current through the solution. The anion transport number t_- is defined analogously in terms of I^-, the anion current. Since the total current I is the sum of the cation and anion currents,

$$t_+ + t_- = 1$$

The **limiting transport numbers** t° are defined in the same way but for the limit of zero concentration of the electrolyte solution. We shall consider only these limiting values from now on, for that avoids the problem of ionic interactions.

The current that can be ascribed to each type of ion is related to the mobility of the ion by $I \propto \kappa$, with κ given by eqn 8a. Hence the relation between $t°$ and u is

$$t° = \frac{zvu}{\sum\limits_{i} z_i v_i u_i} \tag{11a}$$

where the sum in the denominator is over the two kinds of ion. For a symmetrical electrolyte (in which the charge numbers and hence the v_i are the same for both ions), eqn 11a simplifies to

$$t° = \frac{u}{\sum\limits_{i} u_i} \tag{11b}$$

Alternatively, since the ionic conductivities are related to the mobilities by eqn 8b, it follows from eqn 11a that

$$t° = \frac{v\lambda}{\sum\limits_{i} v_i \lambda_i} = \frac{v\lambda}{\Lambda_m^°}$$

and hence, for each type of ion,

$$v\lambda = t°\Lambda_m^° \tag{12}$$

Consequently, since there are independent ways of measuring transport numbers of ions, we can determine the individual ionic conductivities and (through eqn 8) their mobilities.

The measurement of transport numbers

One of the most accurate methods for measuring transport numbers is the **moving boundary method**, in which the motion of a boundary between two ionic solutions having a common ion is observed as a current flows.

Let MX be the salt of interest and NX, the 'indicator', a salt giving a denser solution. The indicator solution occupies the lower part of a vertical tube (Fig. 25.5) and the MX solution, the 'leading solution', occupies the upper part with a sharp boundary between them. The indicator solution must be denser than the leading solution, and the mobility of the M ions must be greater than that of the N ions.[4] Thus, if any M ions diffuse into the lower solution, they will be pulled upwards more rapidly than the N ions around them, and the boundary will reform.

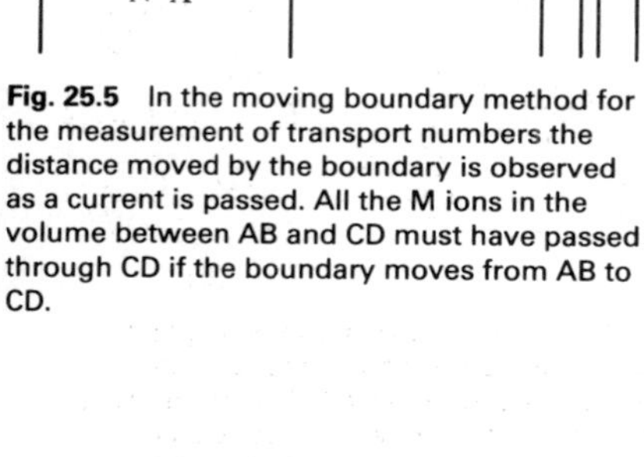

When a current I is passed for a time Δt, the boundary moves from AB to CD, and so all the M ions in the volume between AB and CD must have passed through CD. That number is cVN_A, and so the charge they transfer through the plane is $zcVeN_A$. However, the total charge transferred when a

Fig. 25.5 In the moving boundary method for the measurement of transport numbers the distance moved by the boundary is observed as a current is passed. All the M ions in the volume between AB and CD must have passed through CD if the boundary moves from AB to CD.

[4] One procedure is to add bromothymol blue indicator to a slightly alkaline solution of the ion of interest and to use a cadmium electrode at the lower end of the vertical tube. The electrode produces Cd^{2+} ions, which are slow moving and slightly acidic (the hydrated ion is a Brønsted acid), and the boundary is revealed by the colour change of the indicator.

current I flows for an interval Δt is $I\Delta t$. Therefore, the fraction due to the motion of the M ions, which is their transport number, is

$$t = \frac{zcVF}{I\Delta t}$$

Hence, by measuring the distance moved by the boundary in the time Δt (and calculating V from the distance moved) we can determine the transport number and hence the conductivity and mobility of the ions.

In the **Hittorf method** an electrolytic cell is divided into three compartments and an amount $I\Delta t$ of electricity is passed. An amount $I\Delta t/z_+F$ of cations are discharged at the cathode but $t_+(I\Delta t/z_+F)$ of cations migrate ino the cathode compartment. The net change in the amount of cations in that compartment is therefore

$$\frac{(t_+ - 1)I\Delta t}{z_+F} = \frac{-t_-I\Delta t}{z_+F}$$

Hence, by measuring the change of composition in the cathode compartment, we can deduce the anion transport number t_-. Likewise, the change in composition of the anode compartment is $-t_+(I\Delta t/z_-F)$, which gives the cation transport number t_+.

Finally, transport numbers may also be measured using galvanic cells. The potential E_t of a cell with transference having an electrode reversible with respect to anions, for example

$$\text{Ag} \,|\, \text{AgCl}(s) \,|\, \text{HCl}(m_1) \,|\, \text{HCl}(m_2) \,|\, \text{AgCl}(s) \,|\, \text{Ag}$$

is related to the potential of a cell with the same overall reaction but without transference E, such as

$$\text{Ag} \,|\, \text{AgCl}(s) \,|\, \text{HCl}(m_1) \,|\, \text{H}_2(g) \,|\, \text{Pt} \,|\, \text{H}_2(g) \,|\, \text{HCl}(m_2) \,|\, \text{AgCl}(s) \,|\, \text{Ag}$$

by

$$E_t = t_+ E$$

(The argument is similar to that for the Hittorf method; see Problem 25.18.) Therefore, comparison of the two cell potentials gives the transport number, in this case of H^+.

25.3 Conductivities and ion–ion interactions

The remaining problem is to account for the $c^{1/2}$ dependence of the Kohlrausch law (eqn 2). In Chapter 10 we saw something similar: the activity coefficients of ions also depend on $c^{1/2}$ and depend on their charge type rather than their specific identities. That $c^{1/2}$ dependence was explained in terms of the properties of the ionic atmosphere around each ion, and we can suspect that the same explanation applies here.

We need to modify the picture of an ionic atmosphere as a spherical haze of charge when an electric field is applied to a solution of ions and all the ions drift in a definite direction. In the first place, the ions forming the atmosphere do not adjust to the moving ion infinitely quickly, and the atmosphere is incompletely formed in front of the moving ion and incompletely decayed in its wake (Fig. 25.6). The overall effect is the displacement of the centre of charge of the atmosphere a short distance behind the moving ion. Since the two charges are opposite, the result is a

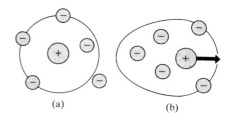

Fig. 25.6 (a) In the absence of an applied field, the ionic atmosphere is spherically symmetric, but (b) when a field is present it is distorted and the centres of negative and positive charge no longer coincide. This charge separation retards the motion of the central ion.

retardation of the moving ion. This reduction of the ions' mobility is called the **relaxation effect**. A confirmation of the picture is obtained by observing the conductivities of ions at high frequencies, which are greater than at low frequencies, because the atmosphere does not have time to follow the rapidly changing direction of motion of the ion, and its effect averages to zero.

The ionic atmosphere also introduces another effect on the motion of the ions. We have seen that the moving ion experiences a viscous drag. When the ionic atmosphere is present this drag is enhanced because the ionic atmosphere moves in an opposite direction to the central ion. The enhanced viscous drag, which is called the **electrophoretic effect**, reduces the mobility of the ions, and hence also reduces their conductivities.

The quantitative formulation of these effects is far from simple, but the **Debye–Hückel–Onsager theory** is an attempt to obtain quantitative expressions at about the same level of sophistication as the Debye–Hückel theory itself. The theory leads to a Kohlrausch-like expression in which

$$\mathcal{K} = A + B\Lambda_m^\circ \tag{13}$$

with

$$A = \frac{z^2 e F^2}{3\pi\eta}\left(\frac{2}{\varepsilon RT}\right)^{1/2} \qquad B = \frac{qz^3 e F}{24\pi\varepsilon RT}\left(\frac{2}{\varepsilon RT}\right)^{1/2}$$

where ε is the electric permittivity of the solvent and $q = 2 - \sqrt{2} = 0.586$ for a 1,1-electrolyte. The numerical values of A and B for 1,1-electrolytes in several solvents are listed in Table 25.3. The slopes of the conductivity curves are predicted to depend on the charge type of the electrolyte, as the Kohlrausch measurements require, and some comparisons between theory and experiment are shown in Fig. 25.7. The agreement is quite good at very low concentrations (less than about 10^{-3} M depending on the charge type).

Table 25.3. Debye–Hückel–Onsager coefficients for (1, 1)-electrolytes at 298 K

Solvent	$A/(\text{S cm}^2\ \text{mol}^{-1}\ \text{M}^{-1/2})$	$B/\text{M}^{-1/2}$
Methanol	156.1	0.923
Propanone	32.8	1.63
Water	60.20	0.229

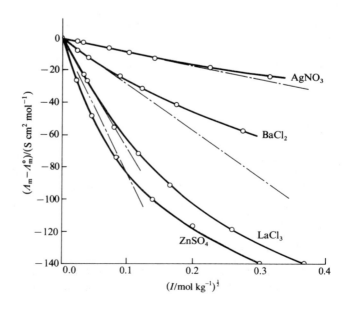

Fig. 25.7 The dependence of molar conductivities on the square root of the ionic strength, and comparison (broken lines) with the dependence predicted by the Debye–Hückel–Onsager theory.

Example 25.5: *Using the Kohlrausch law to measure a limiting conductivity*

The molar conductivity of aqueous potassium chloride at 298 K depends on concentration as reported below. Use the Kohlrausch law to find the limiting molar conductivity of the salt.

c/M	0.001	0.005	0.010	0.020
$\Lambda_m/(S\ cm^2\ mol^{-1})$	146.9	143.5	141.2	138.2

Answer. Equation 2 is valid only at very low concentrations, and so we take it to be the start of a more complicated expression, with the next term proportional to c:

$$\Lambda_m = \Lambda_m^\circ - (A + B\Lambda_m^\circ)c^{1/2} + xc$$

If we rearrange this expression to

$$\Lambda_m + (A + B\Lambda_m^\circ)c^{1/2} = \Lambda_m^\circ + xc$$

we see that a plot of the expression on the left against c should give a straight line with intercept Λ_m° at $c = 0$. For the coefficient of $c^{1/2}$ we use the values of A and B from Table 25.3, and make a preliminary estimate of $\Lambda_m^\circ = 149.8\ S\ cm^2\ mol^{-1}$ from the information in Table 25.1. Then we can write

$$\mathscr{K} = 60.20\ S\ cm^2\ mol^{-1}\ M^{-1/2} + (0.229\ M^{-1/2}) \times (149.8\ S\ cm^2\ mol^{-1})$$

$$= 94.50\ S\ cm^2\ mol^{-1}\ M^{-1/2}$$

We draw up the following table, with $X = \Lambda_m + (A + B\Lambda_m^\circ)c^{1/2}$:

c/M	0.001	0.005	0.010	0.020
$X/(S\ cm^2\ mol^{-1})$	149.9	150.2	150.7	151.6

These points are plotted in Fig. 25.8. The intercept is at $X = 149.8$, and so $\Lambda_m^\circ = 149.8\ S\ cm^2\ mol^{-1}$, in agreement with the preliminary value.

Comment. Had the agreement been poor, the new estimate of the limiting conductivity could have been used in repeats of the procedure until self-consistency was reached.

Exercise. The following results were obtained for NaI(aq) at 298 K:

c/M	0.001	0.005	0.010	0.020
$\Lambda_m/(S\ cm^2\ mol^{-1})$	124.2	121.2	119.2	116.6

Find the limiting molar conductivity of the salt. $[126.9\ S\ cm^2\ mol^{-1}]$

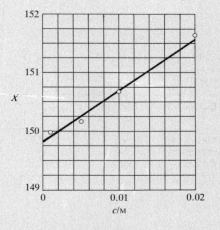

Fig. 25.8 The determination of the limiting molar conductance using an extrapolation based on the Debye–Hückel–Onsager theory.

Diffusion

We are now in a position to generalize the discussion of motion to cover the migration of neutral molecules and of ions in the absence of an applied electric field. We shall do this by expressing ion motion in a more general way, and will then discover that the same equations apply even when the charge on the particles is zero.

25.4 The thermodynamic view

We saw in Part 1 that, at constant temperature and pressure, the maximum work that can be done when a substance moves from a location where its chemical potential is μ to a location where it is $\mu + d\mu$ is $dw = d\mu$. In a system in which the chemical potential depends on the position x,

$$dw = d\mu = \left(\frac{\partial \mu}{\partial x}\right)_{p,T} dx$$

We also saw in Chapter 2 (Table 2.1) that in general work can always be expressed in terms of an opposing force (which here we shall write $\mathscr{F}$), and that

$$dw = -\mathscr{F}\, dx$$

By comparing these expressions, we see that the slope of the chemical potential can be interpreted as an effective force. We shall write this **thermodynamic force** as

$$\mathscr{F} = -\left(\frac{\partial \mu}{\partial x}\right)_{p,T} \tag{14}$$

There is not necessarily a real force pushing the particles down the slope of the chemical potential. As we shall see, the force may represent the spontaneous tendency of the molecules to disperse as a consequence of the Second Law and the hunt for maximum entropy.

The thermodynamic force of a concentration gradient

In a solution where the activity of the particles is a, the chemical potential is

$$\mu = \mu^{\ominus} + RT \ln a$$

If the solution is not uniform the activity depends on the position and we can write

$$\mathscr{F} = -RT\left(\frac{\partial \ln a}{\partial x}\right)_{p,T}$$

If the solution is ideal, a may be replaced by the concentration c, and then

$$\mathscr{F} = \frac{-RT}{c}\left(\frac{\partial c}{\partial x}\right)_{p,T} \tag{15}$$

because $(d/dx) \ln c = (1/c)\, dc/dx$.

Example 25.6: *Calculating the thermodynamic force*

Suppose that the concentration of a solute decays exponentially along the length of a container. Calculate the thermodynamic force on the solute at 25°C when the concentration decreases to half its value in every 10 cm.

Answer. The concentration varies with position as

$$c = c_0 e^{-x/\lambda}$$

where λ is the decay constant. Since the concentration falls to $\frac{1}{2}c_0$ when $x = 10$ cm, we can find λ from

$$\tfrac{1}{2} = e^{-(10\ \text{cm})/\lambda}$$

That is,

$$\lambda = \frac{10 \text{ cm}}{\ln 2} = 14.4 \text{ cm}$$

From the exponential form of the concentration it follows that

$$\frac{dc}{dx} = \frac{-c}{\lambda}$$

and therefore, from eqn 15, that

$$\mathscr{F} = \frac{RT}{\lambda}$$

Therefore,

$$\mathscr{F} = \frac{8.314 \text{ J K}^{-1} \text{ mol}^{-1} \times 298 \text{ K}}{14.4 \times 10^{-2} \text{ m}} = 17 \text{ kN mol}^{-1}$$

Comment. We have used $1 \text{ J} = 1 \text{ N m}$. Note that the thermodynamic force is a molar quantity.

Exercise. Calculate the thermodynamic force on the molecules of molar mass M in a vertical tube in a gravitational field on the surface of the earth and evaluate it for molecules of molar mass 100 g mol^{-1}. Comment on its magnitude relative to that just calculated.
[$\mathscr{F} = -Mg$, -0.98 N mol^{-1}; the force arising from the concentration gradient greatly dominates that arising from the gravitational gradient]

Fick's first law of diffusion

In Section 24.7 we showed that Fick's first law of diffusion (that the particle flux is proportional to the concentration gradient) could be deduced from the kinetic theory of gases. We shall now show that it can be deduced more generally.

We suppose that the flux of diffusing particles is motion in response to a thermodynamic force arising from a concentration gradient. The particles reach a steady drift speed s when the thermodynamic force $\mathscr{F}$ is matched by the viscous drag. This drift speed is proportional to the thermodynamic force, and we write $s \propto \mathscr{F}$. However, the particle flux J is proportional to the drift speed, and the thermodynamic force is proportional to the concentration gradient dc/dx. The chain of proportionalities ($J \propto s$, $s \propto \mathscr{F}$, and $\mathscr{F} \propto dc/dx$) implies that

$$J \propto \frac{dc}{dx}$$

which is the content of Fick's Law.

The Einstein relation

Fick's law for the particle flux in moles of molecules per unit area per unit time is

$$J = -D \frac{dc}{dx}$$

where D is the diffusion coefficient and dc/dx is the slope of the molar concentration.[5] The flux is related to the drift speeds by

$$J = sc$$

This relation follows from the argument that we have used several times before. Thus, all particles within a distance $s\Delta t$, and therefore in a volume $s\Delta tA$, can pass through a window of area A in an interval Δt. Hence, the number of moles that can pass through in that interval is $s\Delta tA \times c$, and so

$$sc = -D\frac{dc}{dx}$$

If now we express dc/dx in terms of $\mathscr{F}$ using eqn 15, we find

$$s = -\frac{D}{c}\frac{dc}{dx} = \frac{D}{RT} \times \mathscr{F}$$

Therefore, once we know the effective force and the diffusion coefficient D, we can calculate the drift speed of the particles (and vice versa) whatever the origin of the force.

There is one case where we already know the drift speed and the effective force acting on a particle: an ion in solution has a drift speed $s = uE$ when it experiences a force ezE, and zFE per mole, from an electric field of strength E. Therefore, substituting these known values into the equation above gives

$$uE = \frac{D}{RT} \times zFE \quad \text{or} \quad u = \frac{zFD}{RT}$$

This equation rearranges into the very important result known as the **Einstein relation** between the diffusion coefficient and the ionic mobility:

$$D = \frac{uRT}{zF} \tag{16}$$

If we insert the value $u = 5 \times 10^{-4}\ \text{cm}^2\,\text{s}^{-1}\,\text{V}^{-1}$ that we estimated in Example 25.3, we find $D \approx 1 \times 10^{-5}\ \text{cm}^2\,\text{s}^{-1}$ at 25°C as a typical value of the diffusion coefficient of an ion in water.

The Nernst–Einstein equation

We can develop the Einstein relation in two ways. In the first place, it can be used to find the relation between the molar conductivity of an electrolyte and the diffusion coefficients of its ions. First we write

$$\lambda = zuF = \frac{z^2DF^2}{RT} \tag{17}$$

for each ion. Then, from $\Lambda_m^\circ = \nu_+\lambda_+ + \nu_-\lambda_-$, the limiting molar conductivity

[5] This is derived from eqn 15 of Section 24.6 simply by dividing both sides by Avogadro's constant, which converts numbers into amounts (numbers of moles).

is

$$\Lambda_m^\circ = \frac{F^2}{RT}(v_+ z_+^2 D_+ + v_- z_-^2 D_-) \qquad (18)$$

Equation 18 is the **Nernst–Einstein equation**. One of its applications is to the determination of ionic diffusion coefficients from conductivity measurements; another is to the prediction of conductivities using models of ionic diffusion (see below).

The Stokes–Einstein equation

The Einstein relation can also be used to derive a relation between the diffusion coefficient for any solute and the viscosity of the solution. By combining the expression $s = uE$ (eqn 7) and $s = ezE/f$ (eqn 6) we obtain

$$u = \frac{ez}{f} \qquad (19)$$

as a relation between an ion's mobility and its hydrodynamic radius. However, since we have already deduced that $u = ezD/kT$, we obtain the **Stokes–Einstein equation** between the diffusion coefficient D and the frictional constant f:

$$D = \frac{kT}{f} \qquad (20)$$

An important feature of eqn 20 is that it is independent of the charge of the diffusing particle. Therefore, it also applies in the limit of vanishingly small charge; that is, it also applies to neutral molecules. Consequently, we may use viscosity measurements to estimate the diffusion coefficients for molecules in solution (and this relation was in fact used in Chapter 23 in the discussion of macromolecules). It must not be forgotten, however, that eqn 20 depends on the validity of Stokes's formula for the viscous drag (that the drag is proportional to the speed). Some diffusion coefficients are listed in Table 25.4 and the relations developed above are collected in Box 25.1.

Table 25.4. Diffusion coefficients at 298 K, $D/(10^{-5}\,\mathrm{cm}^2\,\mathrm{s}^{-1})$

H^+ in water	9.31
I_2 in hexane	4.05
Na^+ in water	1.33
Sucrose in water	0.521

Box 25.1 Transport properties in solution

Einstein–Smoluchowski equation relating the jump length λ and jump time τ:

$$D = \frac{\lambda^2}{2\tau}$$

Stokes–Einstein equation relating the diffusion coefficient D and the solution viscosity η:

$$D = \frac{kT}{f} \qquad f = 6\pi\eta a$$

a is the effective hydrodynamic radius of the spherical particle; values of f for other shapes are given in Table 23.1.

Einstein relation between the diffusion coefficient D and the ion mobility u:

$$D = \frac{ukT}{ze} = \frac{uRT}{zF}$$

z is the charge number of the ion.

Nernst–Einstein equation relating the diffusion coefficient D and the ionic conductivity λ:

$$\lambda = \frac{z^2 F^2 D}{RT}$$

Example 25.7: *Interpreting the mobility of an ion*

Evaluate the diffusion coefficient, limiting molar conductivity, and effective hydrodynamic radius of SO_4^{2-} in water at 298 K.

Answer. We take the mobility ($8.29 \times 10^{-4}\,cm^2\,s^{-1}\,V^{-1}$) from Table 25.2 and obtain D using eqn 16:

$$D = \frac{uRT}{zF} = 1.1 \times 10^{-5}\,cm^2\,s^{-1}$$

Then we use eqn 8 for the ionic conductivity:

$$\lambda = zuF = 160\,S\,cm^2\,mol^{-1}$$

Finally, we obtain a from eqn 20 and $f = 6\pi\eta a$ using 1.00 cP (or $1.00 \times 10^{-3}\,kg\,m^{-1}\,s^{-1}$) for the viscosity of water (Table 22.6):

$$a = \frac{kT}{6\pi\eta D} = 2.0 \times 10^{-10}\,m = 200\,pm$$

Comment. The bond length in SO_4^{2-} is 144 pm, and so the radius calculated here is compatible with a small degree of solvation.

Exercise. Repeat the calculation for the NH_4^+ ion.
$$[1.96 \times 10^{-5}\,cm^2\,s^{-1}, 73.6\,S\,cm^2\,mol^{-1}, 110\,pm]$$

Experimental support for the relations we have derived comes from conductivity measurements. In particular, **Walden's rule** is the empirical observation that the product $\eta\Lambda_m$ is very approximately constant for the same ions in different solvents. Since $\Lambda_m \propto D$, and we have just seen that $D \propto 1/\eta$, we do indeed predict that $\Lambda_m \propto 1/\eta$ as Walden's rule requires. The usefulness of the rule, however, is muddied by the role of solvation: different solvents solvate the same ions to different extents, and so both the hydrodynamic radius and the viscosity change with the solvent.

25.5 The diffusion equation

We now turn to the discussion of time-dependent diffusion processes, where we are interested in the spreading of inhomogeneities with time. One example is the temperature of a metal bar that has been suddenly heated at one end: if the source of heat is removed, then the bar gradually settles down into a state of uniform temperature. If the source of heat is maintained and the bar can radiate, it settles down into a steady state of non-uniform temperature. Another example (and one more relevant to chemistry) is the concentration distribution in a solvent to which a solute is added. We shall focus on the description of the diffusion of particles, but similar arguments apply to the diffusion of physical properties, such as temperature. Our aim is to obtain an equation for the rate of change of the concentration of particles in an inhomogeneous region.

The formulation of the equation

Consider a thin slab of cross-sectional area A and extending from x to $x + l$ (Fig. 25.9). Let the concentration at x be c at the time t. The number of moles of particles that enter the slab per unit time is JA, and so the increase in concentration inside the slab (which has volume Al) on the account of the

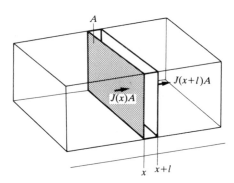

Fig. 25.9 The net flux in a region is the difference between the flux entering from the region of high concentration (on the left) and the flux leaving to the region of low concentration (on the right).

flux from the left is

$$\frac{dc}{dt} = \frac{JA}{Al} = \frac{J}{l}$$

There is also an outflow through the right-hand window. The flux through that window is J', and the change of concentration that results is

$$\frac{dc}{dt} = \frac{-J'A}{Al} = \frac{-J'}{l}$$

The net rate of change of concentration is therefore

$$\frac{dc}{dt} = \frac{J - J'}{l}$$

Each flux is proportional to the concentration gradient at the window. Using Fick's first law we write

$$J - J' = -D\frac{dc}{dx} + D\frac{dc'}{dx}$$

$$= -D\frac{dc}{dx} + D\frac{d}{dx}\left\{c + \left(\frac{dc}{dx}\right)l\right\}$$

$$= Dl\frac{d^2c}{dx^2}$$

When we substitute this relation into the expression for the rate of change of concentration in the slab, we get the **diffusion equation**:

$$\frac{\partial c}{\partial t} = D\frac{\partial^2 c}{\partial x^2} \tag{21}$$

(Equation 21 used to be called Fick's second law of diffusion; at this stage we have recognized that c is a function of both x and t and have written the derivatives as partial derivatives.)

The significance of the equation

The diffusion equation shows that the rate of change of concentration is proportional to the curvature (more precisely, the second derivative) of the concentration with respect to distance. If the concentration changes sharply from point to point (if the distribution is highly wrinkled) then the concentration changes rapidly with time. If the curvature is zero, then the concentration is constant in time. If the concentration decreases linearly with distance, then the concentration at any point is constant because the inflow of particles is exactly balanced by the outflow.

The diffusion equation can be regarded as a mathematical formulation of the intuitive notion that there is a natural tendency for the wrinkles in a distribution to disappear. More succinctly: Nature abhors a wrinkle.

Diffusion with convection

The transport of particles arising from the motion of a streaming fluid is called **convection**. If for the moment we ignore diffusion, the flux of particles through an area A in an interval Δt when the fluid is flowing at a velocity v can be calculated in the way we have used several times before (by counting the particles within a distance $v\Delta t$), and is

$$J = \frac{cAv\Delta t}{A\Delta t} = cv \qquad (22)$$

This J is called the **convective flux**. The rate of change of concentration in a slab of thickness l and area A is, by the same argument as before,

$$\frac{dc}{dt} = \frac{J - J'}{l} = \left\{ c - \left[c + \left(\frac{\partial c}{\partial x} \right) l \right] \right\} \frac{v}{l}$$

$$= -v \frac{\partial c}{\partial x}$$

(We have assumed that the velocity does not depend on the position.)

When both diffusion and convection are of similar importance the total change of concentration in a region is the sum of the two effects, and the **generalized diffusion equation** is

$$\frac{\partial c}{\partial t} = D \frac{\partial^2 c}{\partial x^2} - v \frac{\partial c}{\partial x} \qquad (23)$$

A further refinement, which is important in chemistry, is the possibility that the concentrations of particles may change as a result of reaction. When reactions are included in eqn 23 (as we describe in Section 28.2), we get a powerful differential equation for discussing the properties of reacting, diffusing, convecting systems and which is the basis of reactor design in chemical industry and of the utilization of resources in living cells.

Solutions of the equation

The diffusion equation is a second-order differential equation with respect to space and a first-order differential equation with respect to time. Therefore, we must specify two boundary conditions for the spatial dependence and a single initial condition for the time-dependence.

As an illustration, we consider the example of a solvent in which the solute is initially coated on one surface of the container (e.g. a layer of sugar on the bottom of a deep beaker of water). The single initial condition is that at $t = 0$ all N_0 particles are concentrated on the yz-plane (of area A) at $x = 0$. The two boundary conditions are (1) that the concentration must everywhere be finite and (2) that the total amount (number of moles) of particles present is n_0 at all times. The solution of the diffusion equation under these conditions is

$$c = \frac{n_0 e^{-x^2/4Dt}}{A(\pi Dt)^{1/2}} \qquad (24)$$

as may be verified by direct substitution. Figure 25.10 shows the shape of the concentration distribution at various times, and it is clear that the concentration spreads and tends to uniformity.

Another useful solution is for the case of a localized concentration of solute in a three-dimensional solvent (a sugar lump suspended in a large

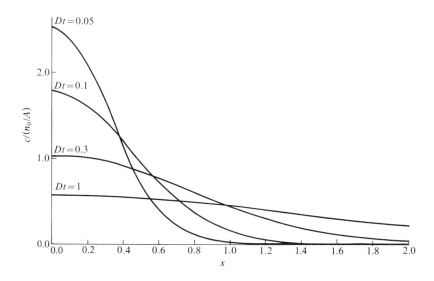

Fig. 25.10 The concentration profiles above a plane from which a solute is diffusing. These are plots of eqn 24. The units of Dt and x are arbitrary, but related so that Dt/x^2 is dimensionless. For example, if x is in cm, Dt would be in cm^2, and so for $D = 10^{-5}\,cm^2\,s^{-1}$, '$Dt = 0.1$' corresponds to $t = 10^4\,s$.

flask of water). The concentration of diffused solute is spherically symmetrical, and at a radius r is

$$c = \frac{n_0 e^{-r^2/4Dt}}{8(\pi Dt)^{3/2}} \qquad (25)$$

Other chemically (and physically) interesting arrangements can be treated, but the solutions are more cumbersome.

The measurement of diffusion coefficients

The solutions of the diffusion equation are useful for experimental determinations of diffusion coefficients.

In the **capillary technique**, a capillary tube, open at one end and containing a solution, is immersed in a well stirred larger quantity of solvent, and the change in concentration in the tube is measured at a series of times. The solute diffuses from the open end of the capillary at a rate that can be calculated by solving the diffusion equation with the appropriate boundary conditions, and so D may be determined.

In the **diaphragm technique**, the diffusion occurs through the capillary pores of a sintered glass diaphragm separating the well-stirred solution and solvent. The concentrations are monitored and then related to the solutions of the diffusion equation corresponding to this arrangement.

25.6 Diffusion probabilities

We can use the solutions of the diffusion equation to predict the concentration of particles (or the value of some other physical quantity, such as the temperature in a non-uniform system) at any location. We can also use them to calculate the net distance through which the particles diffuse in a given time.

Example 25.8: *Calculating the net distance of diffusion*

Calculate the net distance travelled on average by particles in a time t if they are diffusing with diffusion constant D.

Answer. The number of particles in a slab of thickness dx and area A at x, where the molar concentration is c, is $cAN_A dx$. The probability that any of the $N_0 = n_0 N_A$ particles is in the slab is therefore $cAN_A dx/N_0$. If the particle is in the slab, it has travelled a distance x from the origin. Therefore, the mean distance travelled by all the particles is the sum of each x weighted by the probability of its occurrence:

$$\langle x \rangle = \int_0^\infty \frac{xcAN_A}{N_0} \, dx = \frac{1}{(\pi Dt)^{1/2}} \int_0^\infty x e^{-x^2/4Dt} \, dx$$

$$= 2 \left(\frac{Dt}{\pi} \right)^{1/2}$$

If we use the Stokes–Einstein relation for the diffusion coefficient, the mean distance travelled by particles of radius a in a solvent of viscosity η is

$$\langle x \rangle = \left(\frac{2kT}{3\pi^2 \eta a} \right)^{1/2} t^{1/2}$$

Comment. The average distance of diffusion varies as the square root of the lapsed time.

Exercise. Derive an expression for the root mean square distance travelled by diffusing particles in a time t. $\qquad [\langle x^2 \rangle^{1/2} = (2Dt)^{1/2}]$

As shown in the example, the average distance travelled by diffusing particle in a time t (in a system like that illustrated in Fig. 25.10) is

$$\langle x \rangle = 2 \left(\frac{Dt}{\pi} \right)^{1/2} \tag{26a}$$

and the mean square distance travelled in the same time is

$$\langle x^2 \rangle = 2Dt \tag{26b}$$

The latter is a valuable measure of the spread of particles when they can diffuse in both directions from the origin (for then $\langle x \rangle = 0$ at all times). The root mean square distance travelled by particles with a typical diffusion coefficient ($D = 5 \times 10^{-6}$ cm^2 s^{-1}) is illustrated in Fig. 25.11, which shows how long it takes for diffusion to increase the net distance travelled on average to about 1 cm in an unstirred solution. The graph shows that diffusion is a very slow process (which is why solutions are stirred).

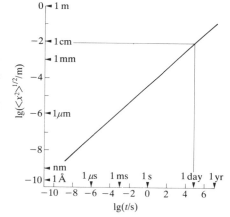

Fig. 25.11 The root mean square distance covered by particles with $D = 5 \times 10^{-6}$ cm^2 s^{-1}. Note the great slowness of diffusion.

25.7 The statistical view

An intuitive picture of diffusion is of the particles moving in a series of small steps and gradually migrating from their original positions. We shall explore this idea using a model in which the particles can jump through a distance λ in a time τ. The total distance travelled by a particle in a time t is therefore $t\lambda/\tau$. However, this does not mean that the particle will be found at that distance from the origin. The direction of each step may be different, and the net distance travelled must take the changing directions into account. If we simplify the discussion by allowing the particles to travel only along a straight line (the x-axis), and for each step (to the left or the right) to be through the same distance λ, we obtain the **one-dimensional random walk**. This process was first described in Section 23.6.

The random walk is described in the *Further information* section at the end of the chapter. We show there that the probability of being at a distance x from the origin after a time t is

$$P = \left(\frac{2\tau}{\pi t}\right)^{1/2} e^{-x^2\tau/2t\lambda^2}$$

(27)

Equation 27 has precisely the same form as eqn 24, the differences of detail arising from the fact that in the present calculation the particles can migrate in either direction from the origin. Moreover, they can be found only at discrete points separated by λ instead of being anywhere on a continuous line. The fact that the two expressions are so similar suggests that diffusion can indeed be interpreted as the outcome of a large number of steps in random directions.

We can now relate the coefficient D to the step length λ and the rate at which the jumps occur. Thus, by comparing the two exponents in eqn 24 and eqn 27 we can immediately write down the **Einstein–Smoluchowski equation**:

$$D = \frac{\lambda^2}{2\tau}$$

(28)

Example 25.9: *Using the Einstein–Smoluchowski equation*

Suppose that a SO_4^{2-} ion jumps through its own diameter each time it makes a move in an aqueous solution. How often does it change position in 1 s at 25°C?

Answer. From Example 25.7, we know that $D = 1.1 \times 10^{-5}\,\text{cm}^2\,\text{s}^{-1}$ and $a = 200$ pm. We therefore set $\lambda = 2a$ and find

$$\tau = \frac{\lambda^2}{2D} = 8 \times 10^{-11}\,\text{s}$$

Since τ is the time for one jump, in 1 s the ion makes about 10^{10} jumps.

Comment. If the jump length were taken to be the diameter of a water molecule (about 150 pm), the jump time would be about 10 ps.

Exercise. Repeat the calculation for the NH_4^+ ion.

[13 ps for 1 jump, 7.7×10^{10} jumps in 1 s]

The Einstein–Smoluchowski equation is the central connection between the microscopic details of particle motion and the macroscopic parameters relating to diffusion (e.g. the diffusion coefficient and, through the Stokes–Einstein relation, the viscosity). It also brings us back full circle to the properties of the perfect gas. For if we interpret λ/τ as $\bar{c}$, the mean speed of the molecules, and interpret λ as a mean free path, we can recognize in the Einstein–Smoluchowski equation exactly the same expression as we obtained from the kinetic theory of gases in Section 24.7. That is, the diffusion of a perfect gas is a random walk with an average step size equal to the mean free path.

Further information: the random walk

We picture the one-dimensional random walk as taking place in steps of length λ, each of duration τ. Our task is to calculate the probability that a particle will be found at a distance x from the origin after a time t. During that time it will have taken N steps, with $N = t/\tau$. If N_R of these are steps to the right, and N_L are steps to the left (with $N_L + N_R = N$), then the net distance travelled is $x = (N_R - N_L)\lambda$. That is, to arrive at x, we must ensure that

$$N_R = \tfrac{1}{2}(N + s) \quad \text{and} \quad N_L = \tfrac{1}{2}(N - s)$$

with $s = x/\lambda$. The probability of being at x after N steps of length λ is therefore the probability that, when N random steps are taken, the numbers to the right and the left are as given above.

The total number of different journeys for a walk of N steps is equal to 2^N, because each step can be in either of two directions. The number of journeys in which exactly N_R steps are taken to the right is equal to the number of ways of choosing N_R objects from N possibilities irrespective of the order:

$$W = \frac{N!}{N_R! \, (N - N_R)!} \tag{A1}$$

1

As a check, consider all journeys of 4 steps: there are 16 possible journeys (**1**). There are 6 ways of taking 2 steps to the right and two to the left, which tallies with the expression $4!/2!\,2! = 6$. The probability that the particle is back at the origin after four steps is therefore 6/16. The probability that it is out at $x = +4\lambda$ is only 1/16 because, in order to be there, all four steps must be to the right, and since $4!/4!\,0! = 1$ the chance of that occurring is 1/16.

Returning now to the general case, the probability of being at x after N steps is

$$P = \frac{\text{Number of journeys with } N_R \text{ steps to the right}}{\text{Total number of journeys}}$$

$$= \left\{ \frac{N!}{N_R! \, (N - N_R)!} \right\} \Big/ 2^N$$

$$= \left\{ \frac{N!}{[\tfrac{1}{2}(N + s)]! \, [\tfrac{1}{2}(N - s)]!\}} \right\} \Big/ 2^N \tag{A2}$$

At this stage the expression for the probability does not seem to resemble the gaussian function (eqn 24) obtained by solving the diffusion equation. However, consider the case when so much time has elapsed that the particles have taken many steps. Then the factorials may be expressed using Stirling's approximation

$$\ln N! \approx (N + \tfrac{1}{2}) \ln N - N + \ln (2\pi)^{1/2}$$

with similar expressions for the other factorials in eqn A2. Taking logarithms of eqn A2 then leads (after quite a lot of algebra) to

$$\ln P = \ln N! - \ln \{\tfrac{1}{2}(N+s)\}! - \ln \{\tfrac{1}{2}(N-s)\}! - N \ln 2$$

$$= \ln \left(\frac{2}{\pi N}\right)^{1/2} - \tfrac{1}{2}(N+s+1) \ln \left(1+\frac{s}{N}\right) - \tfrac{1}{2}(N-s+1) \ln \left(1-\frac{s}{N}\right)$$

So long as $s/N \ll 1$ (which is equivalent to x not being at great distances from the origin) we can use the approximation $\ln(1+z) \approx z$, and obtain

$$\ln P = \ln \left(\frac{2}{\pi N}\right)^{1/2} - \frac{s^2}{N}$$

which rearranges to

$$P = \left(\frac{2}{\pi N}\right)^{1/2} e^{-s^2/N}$$

and now looks very familiar. Finally, we replace s by $x/(\lambda\sqrt{2})$ and N by t/τ, and obtain

$$P = \left(\frac{2\tau}{\pi t}\right)^{1/2} e^{-x^2\tau/2t\lambda^2} \tag{A3}$$

which is the equation used in the text (eqn 27).

Further reading

Ion transport

T. Shedlovsky, Conductimetry. In *Techniques of chemistry* (ed. A. Weissberger and B. W. Rossiter) IIA, 163, Wiley-Interscience, New York (1971).

M. Spiro, Determination of transference numbers. In *Techniques of chemistry* (ed. A. Weissberger and B. W. Rossiter) IIA, 205, Wiley-Interscience, New York (1971).

R. A. Robinson and R. H. Stokes, *Electrolyte solutions*. Academic Press, New York and Butterworth, London (1959).

A. J. Bard and L. R. Faulkner, *Electrochemical methods: Fundamentals and applications*. Wiley, New York (1980).

Diffusion

P. J. Dunlop, B. J. Steel, and J. E. Lane, Experimental methods for studying diffusion in liquids, gases, and solids. In *Techniques of chemistry* (ed. A. Weissberger and B. W. Rossiter) IV, 205, Wiley-Interscience, New York (1972).

W. Jost, *Diffusion in solids, liquids and gases*. Academic Press, New York (1960).

J. Crank, *The mathematics of diffusion*. Clarendon Press, Oxford (1975).

Exercises

25.1 The molar conductivity of an electrolyte solution is known to be 135.5 S cm^2 mol^{-1} at 25°C and a concentration of 5.35×10^{-2} M. Calculate the conductivity of the solution.

25.2 A conductivity cell has plane parallel electrodes, each 2.2 cm × 2.2 cm in area, spaced 2.75 cm apart. When the cell is filled with an electrolytic solution, the resistance is found to be 351 Ω. What is the conductivity of the solution?

25.3 The molar conductivity of a strong electrolyte in water at 25°C was found to be 109.9 S cm^2 mol^{-1} for a concentration of 6.2×10^{-3} M and 106.1 S cm^2 mol^{-1} for a concentration of 1.50×10^{-2} M. What is the limiting molar conductivity of the electrolyte?

25.4 The mobility of the negative ion in a 1:1 electrolyte in aqueous solution at 25°C is measured as 6.85×10^{-8} m^2 s^{-1} V^{-1}. Calculate the molar ionic conductivity.

25.5 The mobility of the Rb^+ ion in aqueous solution is $7.92 \times 10^{-8} \, m^2 \, s^{-1} \, V^{-1}$ at 25°C. The potential difference between two electrodes placed in the solution is 35.0 V. If the electrodes are 8.00 mm apart, what is the drift speed of the Rb^+ ion?

25.6 What fraction of the total current is carried by Li^+ when current flows through an aqueous solution of LiBr at 25°C?

25.7 The limiting molar conductivities of KCl, KNO_3, and $AgNO_3$ are $149.9 \, S \, cm^2 \, mol^{-1}$, $145.0 \, S \, cm^2 \, mol^{-1}$ and $133.4 \, S \, cm^2 \, mol^{-1}$, respectively (all at 25°C). What is the limiting molar conductivity of AgCl at this temperature?

25.8 What is the molar conductivity, the conductivity, and the resistance (in a cell with constant $0.206 \, cm^{-1}$) of an $0.040 \, M$ solution of acetic acid at 25°C? Use $pK_a = 4.72$, $\gamma_{\pm} = 1$.

25.9 At 25°C the molar ionic conductivities of Li^+, Na^+, and K^+ are $38.7 \, S \, cm^2 \, mol^{-1}$, $50.1 \, S \, cm^2 \, mol^{-1}$, and $73.5 \, S \, cm^2 \, mol^{-1}$ respectively. What are their mobilities?

25.10 The diffusion constant of a certain globular protein in water at 20°C is $7.1 \times 10^{-11} \, m^2 \, s^{-1}$. The partial specific volume of the protein is $0.75 \, cm^3 \, g^{-1}$. The viscosity of water at 20°C is $1.00 \times 10^{-3} \, kg \, m^{-1} \, s^{-1}$. Estimate the molar mass of the protein.

25.11 The mobility of the NO_3^- ion in aqueous solution at 25°C is $7.40 \times 10^{-8} \, m^2 \, s^{-1} \, V^{-1}$. Calculate its diffusion constant in water at 25°C.

25.12 A macromolecule has a diffusion coefficient of $4.0 \times 10^{-11} \, m^2 \, s^{-1}$ in water at 20°C. The viscosity of water at that temperature is $1.00 \times 10^{-3} \, kg \, m^{-1} \, s^{-1}$. Assume that the macromolecule is spherical and determine its effective radius.

25.13 The diffusion coefficient of CCl_4 in heptane at 25°C is $3.17 \times 10^{-9} \, m^2 \, s^{-1}$. Estimate the time required for a CCl_4 molecule to have a root mean square displacement of 5.0 mm.

25.14 Estimate the effective radius of a sugar (sucrose) molecule in water 25°C given that its diffusion coefficient is $5.2 \times 10^{-6} \, cm^2 \, s^{-1}$ and that the viscosity of water is 1.00 cP.

25.15 The diffusion coefficient for molecular iodine in benzene is $2.13 \times 10^{-5} \, cm^2 \, s^{-1}$. How long does a molecule take to jump through about 300 pm (approximately the fundamental jump length for translational motion)?

25.16 What is the root mean square distance travelled by (a) an iodine molecule in benzene, (b) a sucrose molecule in water at 25°C in 1.0 s?

25.17 About how long, on average, does it take for the molecules in Exercise 25.16 to drift to a point (a) 1 mm, (b) 1 cm from their starting points?

Problems

Numerical problems

A $0.0200 \, M$ aqueous KCl solution has a molar conductivity of $138.3 \, S \, cm^2 \, mol^{-1}$ at 25°C, and in the cell its resistance was found to be $74.58 \, \Omega$. Calculate the cell constant $C = \kappa R$. This is 'the cell' of the problems that follow.

25.1 Conductivities are often measured by comparing the resistance of a cell filled with the sample to its resistance when filled with some standard solution, such as aqueous potassium chloride. The conductivity of water is $7.6 \times 10^{-4} \, S \, cm^{-1}$ at 25°C and the conductivity of $0.100 \, M$ aqueous KCl is $1.1639 \times 10^{-2} \, S \, cm^{-1}$. A cell had a resistance of $33.21 \, \Omega$ when filled with $0.100 \, M$ KCl solution and $300.0 \, \Omega$ when filled with $0.100 \, M$ acetic acid. What is the molar conductivity of acetic acid at that concentration and temperature?

25.2 The resistances of a series of aqueous NaCl solutions, formed by successive dilution of a sample, were measured in the cell. The following resistances were found:

c/M	0.0005	0.001	0.005	0.010	0.020	0.050
R/Ω	3314	1669	342.1	174.1	89.08	37.14

Check that the molar conductivity follows the Kohlrausch Law and find the limiting molar conductivity. Determine the coefficient $\mathcal{K}$. Use the value of $\mathcal{K}$ (which should depend only on the nature, not the identity of the ions) and the information that $\lambda(Na^+) = 50.1 \, S \, cm^2 \, mol^{-1}$ and $\lambda(I^-) = 76.8 \, S \, cm^2 \, mol^{-1}$ to predict (a) the molar conductivity, (b) the conductivity, (c) the resistance it would show in the cell, of a $0.010 \, M$ aqueous solution of NaI at 25°C.

25.3 After correction for the water conductivity, the conductivity of a saturated solution of AgCl in water at 25°C was found to be $1.887 \times 10^{-6} \, S \, cm^{-1}$. Use the results of Exercise 25.7 to find the solubility and the solubility product at this temperature.

25.4 The limiting molar conductivities of aqueous sodium acetate, hydrochloric acid, and sodium chloride are $91.0 \, S \, cm^2 \, mol^{-1}$, $425.0 \, S \, cm^2 \, mol^{-1}$, and $128.1 \, S \, cm^2 \, mol^{-1}$ respectively at 25°C. What is the limiting molar conductivity of acetic acid? The resistance of a $0.020 \, M$ solution of acetic acid in the cell was found to be $888 \, \Omega$. What is the degree of ionization of the acid at this concentration?

25.5 The resistances of aqueous acetic acid solutions were measured at 25°C in the cell with the following results.

c/M	0.0049	0.00099	0.00198	0.01581	0.06323	0.2529
R/Ω	6146	4210	2927	1004	497	253

Draw the appropriate graph to obtain values of pK_a.

25.6 What are the drift speeds of Li^+, Na^+, and K^+ in water when a potential difference of 10 V is applied across a 1.00 cm conductivity cell? How long would it take an ion to move from one electrode to the other? In conductivity measurements it is normal to use alternating current: what are the displacements of the ions in (a) cm, (b) solvent diameters, about 300 pm, during a half cycle of 1 kHz applied potential?

25.7 The mobilities of H^+ and Cl^- at 25°C in water are 3.623×10^{-3} cm^2 s^{-1} V^{-1} and 7.91×10^{-4} cm^2 s^{-1} V^{-1}, respectively. What proportion of the current is carried by the protons in 1.0 mM hydrochloric acid? What fraction do they carry when the NaCl is added to the acid so that it is 1.0 M in the salt? Note how concentration as well as mobility governs the transport of current.

25.8 In a moving boundary experiment on KCl the apparatus consisted of a tube of bore 4.146 mm, and it contained aqueous KCl at a concentration of 0.021 M. A steady current of 18.2 mA was passed, and the boundary advanced as follows:

$\Delta t/s$	200	400	600	800	1000
x/mm	64	128	192	254	318

Find the transport number of K^+, its mobility, and its ionic conductivity. Use $\Lambda_m^\circ(KCl) = 149.9$ S cm^2 mol^{-1}.

25.9 The proton possesses abnormal mobility in water, but does it behave normally in liquid ammonia? To investigate this question, a moving-boundary technique was used to determine the transport number of NH_4^+ in liquid ammonia (the analogue of H_3O^+ in liquid water) at −40°C (J. Baldwin, J. Evans, and J. B. Gill, *J. chem. Soc.* A, 3389 (1971)). A steady current of 5.000 mA was passed for 2500 s, during which time the boundary formed between mercury(II) iodide and ammonium iodide ammonical solutions moved 286.9 mm in a 0.01365 mol kg^{-1} solution and 92.03 mm in a 0.04255 mol kg^{-1} solution. Calculate the transport number of NH_4^+ at these molalities, and comment on the mobility of the proton in liquid ammonia. The bore of the tube is 4.146 mm and the density of liquid ammonia is 0.682 g cm^{-3}.

25.10 The conductivity of the purest water prepared is 5.5×10^{-8} S cm^{-1}. What would be the resistance measured for a sample of this water in the conductivity cell. What is the ionization constant of water? What are the values of pK_w and pH for pure water? (Take the mobilities of H^+ and OH^- from Table 25.2.)

25.11 A dilute solution of potassium permanganate in water at 25°C was prepared. The solution was in a horizontal 10 cm tube, and at first there was a linear gradation of intensity of the purple solution from the left (where the concentration was 0.100 M) to the right (where the concentration was 0.050 M). What is the magnitude and sign of the thermodynamic force acting on the solute (a) close to the left face of the container, (b) in the middle, (c) close to the right face. Give the force per mole and force per molecule in each case.

25.12 The diffusion coefficient for sucrose in water at 25°C is 5.2×10^{-6} cm^2 s^{-1}. Suppose that the permanganate in Problem 25.11 were replaced by sucrose and we had a way (what, for instance?) of monitoring its concentration. What is the drift velocity of the sugar at the three points (left, middle, right)? What effect on the concentration distribution do these different drift velocities bring about? Sketch the concentration distribution at several later times. What is the drift velocity after an infinite time?

25.13 Estimate the diffusion coefficients and the effective hydrodynamic radii of the alkali metal cations in water from their mobilities at 25°C. Estimate the approximate number of water molecules that are dragged along by the cations. Ionic radii are given in Table 21.1.

25.14 Nuclear magnetic resonance can be used to determine the mobility of molecules in liquids. A set of measurements on methane in carbon tetrachloride showed that its diffusion coefficient is 2.05×10^{-5} cm^2 s^{-1} at 0°C and 2.89×10^{-5} cm^2 s^{-1} at 25°C. Deduce what information you can about the mobility of methane in carbon tetrachloride.

25.15 A concentrated sucrose solution is poured into a 5 cm diameter cylinder. Take it as 10 g of sugar in 5.0 cm^3 of water. A further 1.0 L of water is then poured very carefully on top of the layer, without disturbing it. Ignore gravitational effects, and pay attention only to diffusional processes. Find the concentration at 5.0 cm above the lower layer after a lapse of (a) 10 s, (b) 1 year.

25.16 In a series of observations on the displacement of rubber latex spheres of radius 2.12×10^{-5} cm the mean square displacements after selected time intervals were on average as follows:

t/s	30	60	90	120
$10^8\langle x^2\rangle/cm^2$	88.2	113.5	128	144

These results were originally used to find the value of Avogadro's constant, but there are now better ways of determining it, and so the data can be used to find another quantity. Find the effective viscosity of water at the temperature of this experiment (25°C).

Theoretical problems

25.17 Show how the ratio of two transport numbers t', t'' for two cations in a mixture depends on their concentrations c', c'' and their mobilities u', u''.

25.18 Show that the zero-current potential of the cell

$$Ag \mid AgCl \mid HCl(m_1) \mid HCl(m_2) \mid AgCl \mid Ag$$

with transference E_t is related to its potential without transference E by $E_t = t_+ E$. By the same argument show that for electrodes reversible with respect to the cation, the cell potential is given by $E_t = t_- E$.

25.19 Confirm that eqn 24 is a solution of the diffusion equation with the correct initial value.

25.20 The diffusion equation is valid when many elementary steps are taken in the time interval of interest; but the random walk calculation lets us discuss distributions for short times as well as for long. Use eqn A2 to calculate the probability of being six paces from the origin (that is, at $x = 6\lambda$) after (a) four, (b) six, (c) twelve steps.

25.21 Write a program for calculating $P(x)$ in a one-dimensional random walk, and evaluate the probability of being at $x = n\lambda$ for $n = 6, 10, 14, \ldots, 60$. Compare the numerical value with the analytical value in the limit of a large number of steps. At what value of n is the discrepancy no more than 0.1 per cent?

The rates of chemical reactions

Check-list of key ideas

1. The *techniques* for following the concentrations of reactants and products (Section 26.1).

2. The definition of the *rate of reaction* (eqn 1).

3. The significance of a *rate law* and the *rate constant* of a reaction (Section 26.2).

4. The definition of *reaction order* and the *overall order* of a reaction (Section 26.2).

5. The *isolation method* for simplifying the determination of reaction order, the significance of a *pseudofirst-order reaction*, and the method of *initial rates* (eqn 4).

6. The *integrated rate laws* for first- and second-order reactions (eqns 5 and 6).

7. The time-dependence of *reactions approaching equilibrium* when the reverse reaction is significant (eqn 8a) and the relation between the *equilibrium constant* and the rate constants of forward and reverse reactions (eqn 8b and Example 26.5).

8. The definition of the *half-life* of a substance in a reaction (Section 26.5) and its relation to the initial concentration (eqn 12) and the reaction order (Example 26.6).

9. The *Arrhenius equation* (eqn 13) for the temperature dependence of the rate constant and the *Arrhenius parameters* of a reaction.

10. The use of *relaxation methods* and particularly *temperature-jump relaxation* to measure rate constants (Section 26.6).

11. The *reaction mechanism* in terms of *elementary reactions* and the classification of elementary reactions according to their *molecularity* (Section 26.7).

12. The rate laws for elementary *unimolecular* and *bimolecular reactions* (eqns 17 and 18).

13. The overall rate law stemming from *consecutive reactions* (Section 26.8) and the *rate-determining step* in a mechanism.

14. The *steady state approximation* for simplifying the analysis of a kinetic scheme (Section 26.8 and Example 26.10).

15. The assumption that a *pre-equilibrium* has been established and the approximate analysis of a kinetic scheme (Section 26.8).

16. The *Michaelis–Menten mechanism* of enzyme action (eqn 23).

17. The *Lindemann–Hinshelwood mechanism* of *unimolecular reactions* (Section 26.9) and improvements to the theory.

One reason for studying the rates of reactions is the practical importance of being able to predict how quickly a reaction mixture approaches equilibrium. The rate might depend on variables under our control, such as the pressure, the temperature, and the presence of a catalyst, and we may be able to optimize it by the appropriate choice of conditions. Another reason is that the study of reaction rates leads to an understanding of the **mechanisms** of reactions, their analysis into a sequence of elementary reactions. For example, we might discover that the reaction of hydrogen and bromine to form hydrogen bromide proceeds by the dissociation of Br_2, the attack of a Br atom on H_2, and several subsequent steps, and not by a single event in which an H_2 molecule encounters a Br_2 molecule and the atoms exchange partners to form two HBr molecules.

This chapter establishes the principles of chemical kinetics by showing how reaction rates may be measured and interpreted in simple cases. All the remaining chapters of this part of the text then develop this material in more detail and apply it to more complicated or more specialized cases.

Empirical chemical kinetics

The first stage in the kinetic analysis of reactions is to establish the stoichiometry of the reaction and identify any side reactions. The basic data of chemical kinetics are then the concentrations of the reactants and products at different times after a reaction has been initiated. Since the rates of chemical reactions are generally sensitive to the temperature (for reasons we explore later), the temperature of the reaction mixture must be held constant throughout the course of the reaction, for otherwise the observed rate would be a meaningless average of rates at different temperatures. This requirement puts severe demands on the design of an experiment. Gas phase reactions, for instance, are often carried out in a vessel held in contact with a substantial block of metal. Liquid phase reactions, including flow reactions, must be carried out in an efficient thermostat.

26.1 Experimental techniques

The method used to monitor the concentrations depends on the substances involved and the rapidity with which they change. Many reactions reach thermodynamic equilibrium over periods of minutes or hours, and several techniques may be used to follow the changing concentrations.

Monitoring the progress of a reaction

A reaction in which at least one component is a gas might result in an overall change in pressure in a constant-volume system, and so its progress may be followed by recording the variation of pressure with time. An example is the decomposition of nitrogen(V) oxide,

$$2N_2O_5(g) \rightarrow 4NO_2(g) + O_2(g)$$

For each mole of N_2O_5 molecules consumed, $\frac{5}{2}$ mol of gas molecules are formed, and so the total pressure increases as the reaction proceeds (if the volume is constant). A disadvantage of this method is that it is not specific: all the gas phase particles contribute to the pressure.

Example 26.1: *Monitoring the variation in pressure*

Predict how the total pressure varies during the gas-phase decomposition of N_2O_5.

Answer. Let the initial pressure be p_0 and the initial amount of N_2O_5 molecules present be n. When a fraction α of the N_2O_5 molecules have decomposed, the amounts of the components in the reaction mixture are:

	N_2O_5	NO_2	O_2	Total
Amount:	$n(1-\alpha)$	$2\alpha n$	$\frac{1}{2}\alpha n$	$n(1+\frac{3}{2}\alpha)$

The total pressure (at constant volume and temperature and assuming perfect gas behaviour) is proportional to the number of gas-phase molecules. When $\alpha = 0$ the pressure is p_0, and so at any stage the total pressure is

$$p = (1 + \tfrac{3}{2}\alpha)p_0$$

When the reaction is complete, the pressure will have risen to $\frac{5}{2}$ times its initial value.

Exercise. Repeat the calculation for the decomposition $2NOBr(g) \rightarrow 2NO(g) + Br_2(g)$.

$[p = (1 + \tfrac{1}{2}\alpha)p_0]$

Spectrophotometry, the measurement of the intensity of absorption in a particular spectral region, is widely applicable, and is especially useful when one substance (and only one) in the reaction mixture has a strong characteristic absorption in a conveniently accessible region of the spectrum. For example, the reaction

$$H_2(g) + Br_2(g) \rightarrow 2HBr(g)$$

can be followed by measuring the absorption of visible light by bromine. If the reaction changes the number or type of ions present in a solution it may be followed by monitoring the conductivity or the pH of the solution.

Other methods of determining composition include titration, mass spectrometry, gas chromatography, and magnetic resonance. Polarimetry, the observation of the optical activity of a reaction mixture, is occasionally applicable.

Application of the techniques

In **real-time analysis** the composition of the system is analysed while the reaction is in progress by withdrawing a small sample or monitoring the bulk. In the **quenching method** the reaction is stopped after it has been allowed to proceed for a certain time, and the composition is analysed at leisure. The quenching (of the entire mixture or of a sample drawn from it) can be achieved either by cooling suddenly, by adding the mixture to a large amount of solvent, or by rapid neutralization of an acid reagent. This

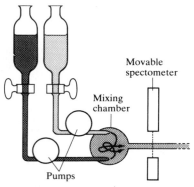

Fig. 26.1 The arrangement used in the flow technique for studying reaction rates. The reactants are pumped into the mixing chamber at a steady rate by the peristaltic pumps (i.e. pumps that squeeze the fluid through flexible tubes). The location of the spectrometer corresponds to different times after initiation.

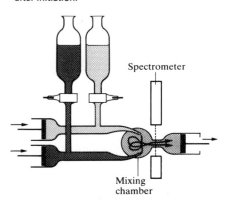

Fig. 26.2 In the stopped-flow technique the reagents are driven quickly into the mixing chamber and then the time-dependence of the concentrations is monitored.

method is suitable only for reactions that are slow enough for there to be little reaction during the time it takes to quench the mixture. In recent years notable advances have been made in the study of **fast reactions**, which we shall take to be reactions complete in less than about 1 s (and often very much less), and the present thrust of chemical kinetics is to ever shorter time-scales. With special laser techniques it is now possible to observe processes occurring in a few femtoseconds (1 fs = 10^{-15} s).

In the **flow method** the reactants are mixed as they flow together in a chamber (Fig. 26.1). The reaction continues as the thoroughly mixed solutions flow through the outlet tube, and observation of the composition at different positions along the tube (e.g. spectroscopically) is equivalent to the observation of the reaction mixture at different times after mixing.

The disadvantage of conventional flow techniques is that a large volume of reactant solution is necessary. This disadvantage is particularly important for fast reactions, because in order to spread the reaction over a length of tube the flow must be rapid. The **stopped-flow technique** (Fig. 26.2) avoids this disadvantage. The two solutions are mixed very rapidly by injecting them into a tangential mixing chamber designed to ensure that the flow is turbulent and that complete mixing occurs. Behind the mixing chamber there is an observation cell fitted with a plunger that moves back as the liquids flood in, but comes up against a stop. The reaction then continues in the thoroughly mixed solution and is monitored (usually spectrophotometrically). The suitability of the stopped-flow technique to the study of small samples means that it is appropriate for biochemical reactions, and it has been widely used to study the kinetics of enzyme action.

In **flash photolysis** the gaseous or liquid sample is exposed to a brief photolytic flash of light, and then the contents of the reaction chamber are monitored spectrophotometrically. Although discharge lamps can be used for flashes of about 10^{-5} s, most work is now done with lasers with flashes of about 10^{-9} s duration, but mode-locking (Section 17.6) has made picosecond pulses available (1 ps = 10^{-12} s), and many studies are now carried out on that timescale. The world record for pulse shortness at the time of writing stands at about 4 fs (1 fs = 10^{-15} s). Both emission and absorption spectroscopy may be used to monitor the reaction, and the spectra are observed electronically or photographically at a series of times following the flash.

26.2 The rates of reactions

The general outcome of experiments to measure reaction rates is the observation that they depend on the composition and the temperature of the reaction mixture. The next few sections look at these observations in more detail.

The definition of rate

Consider a reaction of the form

$$A + B \rightarrow C$$

in which at some instant the concentrations of the participants are [A], [B], and [C]. One measure of the rate of the reaction is the rate of formation or consumption of a specified species, for example, the **rate of consumption** of the reactant A is

$$v_A = -\frac{d[A]}{dt}$$

and the **rate of formation** of the product C is

$$v_C = \frac{d[C]}{dt}$$

Both rates are positive. In the present case the reaction stoichiometry implies that

$$\frac{d[C]}{dt} = -\frac{d[B]}{dt} = -\frac{d[A]}{dt}$$

because whenever a C molecule is formed, one A and one B molecule are destroyed. For a reaction with a more complicated stoichiometry, such as

$$A + 2B \rightarrow 3C + D$$

the relations between the various rates of formation and consumption are more complicated. In this case

$$\frac{d[D]}{dt} = \frac{1}{3}\frac{d[C]}{dt} = -\frac{d[A]}{dt} = -\frac{1}{2}\frac{d[B]}{dt}$$

The ambiguity in the definition of rate is avoided if we define the **rate of reaction** v as

$$v = \frac{1}{v_J}\frac{d[J]}{dt} \tag{1}$$

where v_J is the stoichiometric coefficient of substance J (with v_J negative for reactants and positive for products, Section 2.7). Now there is a single rate for the entire equation. A refinement of this approach is described in the *Further information* section at the end of the chapter.

Example 26.2: *Reporting rates of reaction*

The rate of formation of NO(g) in the reaction

$$2NOBr(g) \rightarrow 2NO(g) + Br_2(g)$$

was reported as $1.6 \times 10^{-4}\,\text{M s}^{-1}$. What is the rate of reaction, and the rate of consumption of NOBr?

Answer. The reaction is formally

$$0 = -2NOBr(g) + 2NO(g) + Br_2(g)$$

and so $v(NO) = +2$. Therefore, the rate of reaction is obtained from eqn 1 with $d[NO]/dt = 1.6 \times 10^{-4}\,\text{M s}^{-1}$:

$$v = \tfrac{1}{2} \times (1.6 \times 10^{-4}\,\text{M s}^{-1}) = 8.0 \times 10^{-5}\,\text{M s}^{-1}$$

Since $v(NOBr) = -2$, the rate of formation of NOBr is

$$\frac{d[NOBr]}{dt} = -2 \times (8.0 \times 10^{-5}\,\text{M s}^{-1}) = -1.6 \times 10^{-4}\,\text{M s}^{-1}$$

and so its rate of consumption is $1.6 \times 10^{-4}\,\text{M s}^{-1}$.

781

Comment. The rate of formation of a substance is the negative of the rate of consumption. Note that rates of reaction are normally expressed in $M\,s^{-1}$ or, equivalently in $mol\,dm^{-3}\,s^{-1}$.

Exercise. The rate of consumption of CH_3 radicals in the reaction $2CH_3(g) \rightarrow CH_3CH_3(g)$ was reported as $d[CH_3]/dt = -1.2\,M\,s^{-1}$ under particular conditions in a 5.0 litre vessel. What is (a) the rate of reaction and (b) the rate of formation of CH_3CH_3? [(a) $0.60\,M\,s^{-1}$, (b) $0.60\,M\,s^{-1}$]

Rate laws and rate constants

The measured rate of reaction is often found to be proportional to the concentrations of the reactants raised to some power. For example, it may be found that the rate is proportional to the concentrations of two reactants A and B, and that

$$v = k[A][B] \tag{2}$$

The coefficient k is called the **rate constant** and is independent of the concentrations (but dependent on the temperature). An experimentally determined equation of this kind is called the **rate law** of the reaction. More formally, a rate law is an equation that expresses the rate of reaction v as a function of the concentrations of all the species present, including the products.

Rate laws have two main applications. A practical application is that once we know the rate law and the rate constant we can predict the rate of reaction from the composition of the mixture. The theoretical application of a rate law is that it is a guide to the mechanism of the reaction, and any proposed mechanism must be consistent with the observed rate law. We shall see both these applications in operation in the following sections.

Reaction order

The **order** of a reaction with respect to some component is the power to which the concentration of that component is raised in the rate law. For example, a reaction with the rate law in eqn 2 is first order in A and also first order in B. The **overall order** of a reaction is the sum of the orders of all the components. The rate law in eqn 2 is therefore second order overall.

A reaction need not have an integral order, and many gas-phase reactions do not. For example, if a reaction is found to have the rate law

$$v = k[A]^{1/2}[B]$$

it is half order in A, first order in B, and three-halves order overall. When a rate law is not of the form $[A]^x[B]^y[C]^z \ldots$, the reaction does not have an order. Thus, the experimentally determined rate law for the gas-phase reaction $H_2 + Br_2 \rightarrow 2HBr$ is

$$v = \frac{k[H_2][Br_2]^{3/2}}{[Br_2] + k'[HBr]} \tag{3}$$

and although the reaction is first order in H_2, it has an indefinite order with respect to both Br_2 and HBr and overall (except under certain simplifying conditions, such as $[Br_2] \gg k'[HBr]$).

The rate law is arrived at experimentally, and cannot in general be inferred from the reaction equation. The reaction of hydrogen and bromine, for example, has a very simple stoichiometry, but its rate law (eqn 3) is very complicated. Similarly, for the thermal decomposition of nitrogen(V) oxide

$$2N_2O_5(g) \rightarrow 4NO_2(g) + O_2(g) \qquad v = k[N_2O_5]$$

and the reaction is first order. In some cases, however, the rate law does happen to reflect the reaction stoichiometry. This is the case with the oxidation of nitrogen(II) oxide, which under certain conditions has a third-order rate law:

$$2NO(g) + O_2(g) \rightarrow 2NO_2(g) \qquad v = k[NO]^2[O_2]$$

Some reactions obey a zeroth-order rate law, and therefore have a rate that is independent of the concentration of the reactant (so long as some is present). Thus, the catalytic decomposition of phosphine on hot tungsten at high pressures has the rate law

$$v = k$$

The PH_3 decomposes at a constant rate until it has entirely disappeared, when the reaction stops abruptly. Only heterogeneous reactions can have rate laws that are zeroth order overall.

These remarks point to three problems. First, we must see how to identify the rate law and obtain the rate constant from the experimental data. We shall concentrate on this aspect in this chapter. Second, we must see how to construct reaction mechanisms that are consistent with the rate law. We shall introduce the techniques for doing so in this chapter and develop them further in Chapter 27. Third, we must account for the values of the rate constants and explain their temperature dependence. We shall see a little of what is involved in this chapter, but leave the details until Chapter 28.

The determination of the rate law

The investigation of a reaction aims to determine the rate law and the rate constant, often at several temperatures. Ideally, the first step is to identify all the products, and to investigate whether any transient intermediates and side reactions are involved.

The determination of a rate law is simplified by the **isolation method** in which the concentrations of all the reactants except one are in large excess. If B is in large excess, for example, it is a good approximation to take its concentration as constant throughout the reaction. Then, although the true rate law might be second-order overall, and

$$v = k[A][B]$$

we can approximate [B] by [B]$_0$ and write

$$v = k'[A] \qquad k' = k[B]_0$$

which has the form of a first-order rate law. Since the true rate law has been forced into first-order form by assuming a constant B concentration, it is called a **pseudofirst-order rate law**. Had the rate law been more complicated, such as

$$v = \frac{k_1[A]^2[B]^{1/2}}{k_2 + k_3[B]}$$

the isolation technique with B in excess would result in

$$v = k[A]^2 \qquad k = \frac{k_1[B]_0^{1/2}}{k_2 + k_3[B]_0}$$

This is a pseudosecond-order rate law, and is much easier to analyse and identify than the complete law. The dependence of the rate on all the reactants may be found by isolating them in turn (by having all the other substances present in large excess), and so constructing a picture of the overall rate law.

In the method of **initial rates**, which is often used in conjunction with the isolation method, the rate is measured at the beginning of the reaction for several different initial concentrations of reactants. We shall suppose that the rate law for a reaction with A isolated is

$$v = k[A]^a$$

Then its initial rate v_0 is given by the initial values of the concentration of A

$$v_0 = k[A]_0^a$$

Taking logarithms:

$$\lg v_0 = \lg k + a \lg [A]_0 \tag{4}$$

For a series of initial concentrations, a plot of the logarithms of the initial rates against the logarithms of the initial concentrations of A should be a straight line with slope a.

Example 26.3: *Using the method of initial rates*

The recombination of I atoms in the gas phase in the presence of argon was investigated and the order of the reaction was determined by the method of initial rates. The initial rates of reaction of $2I(g) + Ar(g) \rightarrow I_2(g) + Ar(g)$ were as follows:

$[I]_0/(10^{-5}\,\text{M})$		1.0	2.0	4.0	6.0
$v/(\text{M s}^{-1})$	(a)	8.70×10^{-4}	3.48×10^{-3}	1.39×10^{-2}	3.13×10^{-2}
	(b)	4.35×10^{-3}	1.74×10^{-2}	6.96×10^{-2}	1.57×10^{-1}
	(c)	8.69×10^{-3}	3.47×10^{-2}	1.38×10^{-1}	3.13×10^{-1}

The Ar concentrations are (a) 1.0×10^{-3} M, (b) 5.0×10^{-3} M, and (c) 1.0×10^{-2} M. Find the orders of reaction with respect to the I and Ar atom concentrations and the rate constant.

Answer. In Fig. 26.3 we have plotted $\lg v_0$ against $\lg [I]_0$ for a given $[Ar]_0$ and against $\lg [Ar]_0$ for a given $[I]_0$. Intercepts at $\lg [I]_0 = 0$ give $\lg k + a \lg [Ar]_0$ and slopes give the orders. The slopes are 2 and 1 respectively, so the (initial) rate law is

$$v = k[I]_0^2[Ar]_0$$

This rate law signifies that the reaction is second order in I, first order in Ar, and third order overall. The intercepts correspond to $k = 9.9\,\text{M}^{-2}\,\text{s}^{-1}$.

Comment. The units of k come automatically from the calculation, and are always such as to convert the product of concentrations to concentration per unit time (e.g. M s^{-1}).

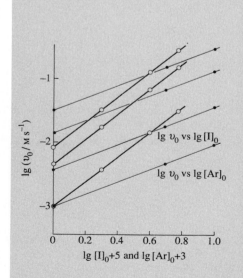

Fig. 26.3 The plot of $\lg v_0$ against $\lg [I]_0$ for a given $[Ar]_0$ and against $\lg [Ar]_0$ for a given $[I]_0$.

lg v_0 vs lg $[I]_0$

lg v_0 vs lg $[Ar]_0$

lg $[I]_0$+5 and lg $[Ar]_0$+3

Exercise. The initial rate of a reaction depended on concentration of a substance J as follows:

$[J]_0/(10^{-3}\,\text{M})$	5.0	8.2	17	30
$v_0/(10^{-7}\,\text{M}\,\text{s}^{-1})$	3.6	9.6	41	130

Find the order of the reaction with respect to J and the rate constant.

$$[2,\ 1.4 \times 10^{-2}\,\text{M}^{-1}\,\text{s}^{-1}]$$

26.3 Integrated rate laws

The method of initial rates might not reveal the full rate law, for in a complex reaction the products themselves might affect the rate. For example, products participate in the synthesis of HBr, since eqn 3 shows that the full rate law depends on the concentration of HBr, none of which is present initially. To avoid this difficulty, the rate law should be fitted to the data throughout the reaction. The fitting may be done, in simple cases at least, by using a proposed rate law to predict the concentration of any component at any time, and comparing it with the data. A law should also be tested by observing whether the addition of products or a change in the surface-to-volume ratio in the reaction chamber affects the rate.

Since rate laws are differential equations, we must integrate them if we want to find the concentrations as a function of time. Now that computers are so widely available, even the most complex rate laws may be integrated numerically. However, in a number of simple cases analytical solutions are easily obtained, and prove to be very useful. We shall examine a few of these simple cases here, and illustrate the computational approach in Chapter 27.

First-order reactions

The first-order rate law for the consumption of a reactant A is

$$-\frac{d[A]}{dt} = k[A] \tag{5a}$$

It rearranges to

$$\frac{d[A]}{[A]} = -k\,dt$$

which can be integrated directly. Since initially (at $t = 0$) the concentration of A is $[A]_0$, and at a later time t it is $[A]$, we write

$$\int_{[A]_0}^{[A]} \frac{d[A]}{[A]} = -\int_0^t k\,dt$$

and obtain

$$\ln\frac{[A]}{[A]_0} = -kt \tag{5b}$$

$$[A] = [A]_0 e^{-kt} \tag{5c}$$

These two equations are versions of an **integrated rate law**, the integrated form of the rate equation.

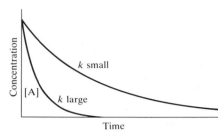

Fig. 26.4 The exponential decay of the reactant in a first-order reaction. The greater the rate constant, the more rapid the decay.

Table 26.1. Kinetic data for first-order reactions

Reactions	Phase	$\theta/°C$	k/s^{-1}	$t_{1/2}$
$2NO_5 \rightarrow 4NO_2 + O_2$	g	25	3.38×10^{-5}	2.85 h
$2N_2O_5 \rightarrow 4NO_2 + O_2$	$Br_2(l)$	25	4.27×10^{-5}	2.25 h
$C_2H_6 \rightarrow 2CH_3$	g	700	5.46×10^{-4}	21.2 min

Equation 5b shows that if $\ln [A]/[A]_0$ is plotted against t, then a first-order reaction will give a straight line. If the plot is straight, then the reaction is first order, and k may be obtained from the slope (which is equal to $-k$). Some rate constants determined in this way are given in Table 26.1. Equation 5c shows that in a first-order reaction the reactant concentration decreases exponentially with time with a rate determined by k (Fig. 26.4).

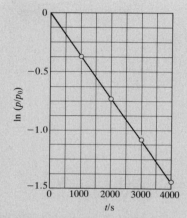

Fig. 26.5 The determination of the rate constant of a first-order reaction: a straight line is obtained when $\ln [A]$ (or $\ln p$) is plotted against t; the slope gives k.

Example 26.4: *Analysing a first-order reaction*

The variation in the partial pressure of azomethane with time was followed at 600 K, with the results given below. Confirm that the decomposition

$$CH_3N_2CH_3(g) \rightarrow CH_3CH_3(g) + N_2(g)$$

is first order in azomethane, and find the rate constant at this temperature.

t/s	0	1000	2000	3000	4000
$p/(10^{-2}\,\text{Torr})$	8.20	5.72	3.99	2.78	1.94

Answer. We plot $\ln (p/p_0)$ against t in Fig. 26.5. The plot is straight, confirming a first-order reaction, and its slope is -3.6×10^{-4}. Therefore, $k = 3.6 \times 10^{-4}\,\text{s}^{-1}$.

Exercise. The concentration of N_2O_5 in liquid bromine varied with time as follows:

t/s	0	200	400	600	1000
$[N_2O_5]/\text{M}$	0.110	0.073	0.048	0.032	0.014

Determine the order and rate constant. $\qquad [1, 2.1 \times 10^{-3}\,\text{s}^{-1}]$

Second-order reactions

If the rate law is

$$\frac{d[A]}{dt} = -k[A]^2 \tag{6a}$$

the integration required is

$$-\int_{[A]_0}^{[A]} \frac{d[A]}{[A]^2} = \int_0^t k \, dt$$

which evaluates to

$$\frac{1}{[A]} - \frac{1}{[A]_0} = kt \tag{6b}$$

and rearranges to

$$[A] = \frac{[A]_0}{1 + kt[A]_0} \tag{6c}$$

Equation 6b shows that in order to test for a second-order reaction we should plot $1/[A]$ against t and expect a straight line. If it is straight the reaction is second order in A and the slope of the line is equal to the rate constant. Some rate constants determined in this way are given in Table 26.2. Equation 6c lets us predict the concentration of A at any time after the start of the reaction. It shows that the concentration of A approaches zero more slowly than in a first-order reaction with the same initial rate (Fig. 26.6).

When a reaction is second order overall, but first order in each of two reactants A and B, the rate law is

$$\frac{d[A]}{dt} = -k[A][B] \tag{7a}$$

We cannot integrate this law until we know how the concentration of B is related to that of A, and that depends on the stoichiometry of the reaction.

For simplicity we consider

$$A + B \rightarrow \text{Products}$$

If the initial concentrations are $[A]_0$ and $[B]_0$, then when the concentration of A has fallen to $[A]_0 - x$, the concentration of B will have fallen to $[B]_0 - x$ (because each A that disappears entails the disappearance of one B). It follows that

$$-\frac{d[A]}{dt} = k([A]_0 - x)([B]_0 - x)$$

Then, since $d[A]/dt = -dx/dt$, the rate law is

$$\frac{dx}{dt} = k([A]_0 - x)([B]_0 - x)$$

Since $x = 0$ when $t = 0$,

$$kt = \int_0^x \frac{dx}{([A]_0 - x)([B]_0 - x)}$$

$$= \frac{-1}{[A]_0 - [B]_0} \int_0^x \left\{ \frac{1}{[A]_0 - x} - \frac{1}{[B]_0 - x} \right\} dx$$

$$= \frac{-1}{[A]_0 - [B]_0} \left\{ \ln \frac{[A]_0}{[A]_0 - x} - \ln \frac{[B]_0}{[B]_0 - x} \right\}$$

We can simplify this expression by combining the two logarithms and noting that $[A] = [A]_0 - x$ and $[B] = [B]_0 - x$; then

$$kt = \frac{1}{[A]_0 - [B]_0} \ln \frac{[A][B]_0}{[A]_0[B]} \tag{7b}$$

Table 26.2. Kinetic data for second-order reactions

Reaction	Phase	$\theta/°C$	$k/(\text{M}^{-1}\,\text{s}^{-1})$
2NOBr $\rightarrow$ 2NO + Br$_2$	g	10	0.80
2I $\rightarrow$ I$_2$	g	23	7×10^9
CH$_3$Cl + CH$_3$O$^-$ $\rightarrow$ CH$_3$OH(l)	20	2.29×10^{-6}

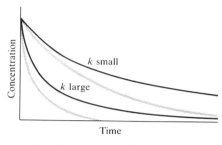

Fig. 26.6 The variation with time of the concentration of a reactant in a second-order reaction. The pale line is the corresponding decay in a first-order reaction with the same initial rate.

Similar calculations may be carried out to find the integrated rate laws for other orders, and some are listed in Box 26.1. However, since all the laws considered so far disregard the possibility that the reverse reaction is important, none of them describes the overall rate when the reaction is close to equilibrium. At that stage the products may be abundant and in principle the reverse reaction must be taken into account. In practice,

---

Box 26.1 Integrated rate laws

Order	Reaction	Rate law*	$t_{1/2}$
0	$A \to P$	$v = k$ $kt = x$ for $0 \le x \le [A]_0$	$\dfrac{[A]_0}{2k}$
1	$A \to P$	$v = k[A]$ $kt = \ln \dfrac{[A]_0}{[A]_0 - x}$	$\dfrac{\ln 2}{k}$
2	$A \to P$	$v = k[A]^2$ $kt = \dfrac{x}{[A]_0([A]_0 - x)}$	$\dfrac{1}{k[A]_0}$
	$A + B \to P$	$v = k[A][B]$ $kt = \dfrac{1}{[B]_0 - [A]_0} \ln \dfrac{[A]_0([B]_0 - x)}{([A]_0 - x)[B]_0}$	
	$A + 2B \to P$	$v = k[A][B]$ $kt = \dfrac{1}{[B]_0 - 2[A]_0} \ln \dfrac{[A]_0([B]_0 - 2x)}{([A]_0 - x)[B]_0}$	
	$A \to P$ with autocatalysis	$v = k[A][P]$ $kt = \dfrac{1}{[A]_0 + [P]_0} \ln \dfrac{[A]_0([P]_0 + x)}{([A]_0 - x)[P]_0}$	
3	$A + 2B \to P$	$v = k[A][B]^2$ $kt = \dfrac{2x}{(2[A]_0 - [B]_0)([B]_0 - 2x)[B]_0}$ $+ \dfrac{1}{(2[A]_0 - [B]_0)^2} \ln \dfrac{[A]_0([B]_0 - 2x)}{([A]_0 - x)[B]_0}$	
$n \ge 2$	$A \to P$	$v = k[A]^n$ $kt = \dfrac{1}{n-1} \left\{ \dfrac{1}{([A]_0 - x)^{n-1}} - \dfrac{1}{[A]_0^{n-1}} \right\}$	$\dfrac{2^{n-1} - 1}{(n-1)k[A]_0^{n-1}}$

*$x = [P]$, and $v = dx/dt$.

however, most kinetic studies are made on reactions that are far from equilibrium, and the reverse reactions are unimportant.

26.4 Reactions approaching equilibrium

We can explore the variation of the composition with time close to equilibrium by considering the reaction $A \rightleftharpoons B$ in which both forward and reverse reactions are first order

$$A \rightarrow B \qquad v = k[A]$$
$$B \rightarrow A \qquad v = k'[B]$$

and the total volume of the system is constant. The rate of change of $[A]$ has two contributions: it is depleted by the forward reaction at a rate $k[A]$ but is replenished by the reverse reaction at a rate $k'[B]$. The net rate of change is therefore

$$\frac{d[A]}{dt} = -k[A] + k'[B]$$

If the initial concentration of A is $[A]_0$, and there is no B present initially, at all times $[A] + [B] = [A]_0$. Therefore

$$\frac{d[A]}{dt} = -k[A] + k'([A]_0 - [A]) = -(k + k')[A] + k'[A]_0$$

The solution of this first-order differential equation is

$$[A] = \left\{ \frac{k' + ke^{-(k+k')t}}{k + k'} \right\}[A]_0 \qquad (8a)$$

The time dependence predicted by this equation is drawn in Fig. 26.7. As $t \rightarrow \infty$, the concentrations reach their equilibrium values:

$$[A] = \frac{k'[A]_0}{k + k'}$$

$$[B] = [A]_0 - [A] = \frac{k[A]_0}{k + k'}$$

It follows that the equilibrium constant of the reaction is

$$K = \frac{[B]}{[A]} = \frac{k}{k'} \qquad (8b)$$

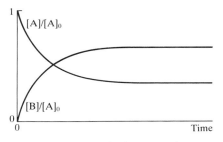

Fig. 26.7 The approach of concentrations to their equilibrium values as predicted by eqn 8a for a reaction $A \rightleftharpoons B$ that is first-order in each direction, and for which $k = 2k'$.

Equation 8b is a very important result, because it relates the thermodynamic quantity, the equilibrium constant, to quantities relating to rates. The practical importance of eqn 8b is that if one of the rate constants can be measured, then the other may be obtained if the equilibrium constant is known.

The same type of calculation may be made for other types of equilibria. For a reaction

$$A + B \rightleftharpoons C + D$$

that is second order in both directions, the rates of change of the concentration of A as a result of the forward and reverse reactions are

$$\text{Forward reaction:} \quad \frac{d[A]}{dt} = -k[A][B]$$

$$\text{Reverse reaction:} \quad \frac{d[A]}{dt} = k'[C][D]$$

$$\text{Net reaction:} \quad \frac{d[A]}{dt} = k'[C][D] - k[A][B]$$

At equilibrium the net rate of change is zero. Hence, at equilibrium,

$$k'[C][D] - k[A][B] = 0$$

and so

$$K = \frac{[C][D]}{[A][B]} = \frac{k}{k'}$$

as before.

Example 26.5: *Relating the equilibrium constant to the rate constants*

Show that, for a reaction that takes place in a sequence of reactions, the overall equilibrium constant is a product of ratios of rate constants.

Answer. We shall consider the reaction sequence

$$A + B \rightleftharpoons C + D \quad \frac{d[A]}{dt} = k_a'[C][D] - k_a[A][B]$$

$$C \rightleftharpoons E + F \quad \frac{d[C]}{dt} = k_b'[E][F] - k_b[C]$$

At equilibrium all the reactions are individually at equilibrium, and setting the net rates each equal to zero yields

$$\frac{[C][D]}{[A][B]} = \frac{k_a}{k_a'} \qquad \frac{[E][F]}{[C]} = \frac{k_b}{k_b'}$$

The overall reaction is

$$A + B \rightleftharpoons D + E + F$$

and its equilibrium constant is

$$K = \frac{[D][E][F]}{[A][B]} = \frac{[C][D][E][F]}{[A][B][C]}$$

$$= \frac{[C][D]}{[A][B]} \times \frac{[E][F]}{[C]} = \frac{k_a}{k_a'} \times \frac{k_b}{k_b'}$$

Comment. When an overall reaction is the sum of several steps, the equilibrium constant is a product of all the ratios:

$$K = \frac{k_a}{k_a'} \times \frac{k_b}{k_b'} \times \cdots$$

where the ks are the rate constants for the individual steps and the k's are for the corresponding reverse reaction step.

Exercise. Confirm the general equation for the reaction sequence $A + B \rightarrow 2C$, $C + D \rightarrow 2P$, $C + B \rightarrow D$ at equilibrium.

26.5 Half-lives

A useful indication of the rate of a chemical reaction is the **half-life** $t_{1/2}$ of a substance, the time it takes for its concentration to fall to half the initial value. Half-lives depend on the initial concentration of the substance in a characteristic way for reactions of different orders, and so their measurement is a guide to the reaction order.

The time for $[A]$ to decrease from $[A]_0$ to $\frac{1}{2}[A]_0$ in a first-order reaction is given by eqn 5 as

$$kt_{1/2} = -\ln \frac{\frac{1}{2}[A]_0}{[A]_0} = -\ln \frac{1}{2} = \ln 2$$

Hence

$$t_{1/2} = \frac{\ln 2}{k} = \frac{0.693}{k} \tag{9}$$

The main point to note about this result is that for a first-order reaction, the half-life of a reactant is independent of its initial concentration. Hence, if the concentration of A at some arbitrary stage of the reaction is $[A]$, it will have fallen to $\frac{1}{2}[A]$ after a further interval of $0.693/k$. Some half-lives are given in Table 26.1.

For a second-order reaction with eqn 6a as its rate law, we can use eqn 6b to find the half-life by substituting $t = t_{1/2}$ and $[A] = \frac{1}{2}[A]_0$. This substitution leads to

$$t_{1/2} = \frac{1}{k[A]_0} \tag{10}$$

Now the half-life depends on the initial concentration, and the greater the initial concentration the less time it takes to fall to half its value. Moreover, as the concentration of A decreases, the half-life lengthens, and it takes twice as long to fall from $\frac{1}{2}[A]_0$ to $\frac{1}{4}[A]_0$ as it did from $[A]_0$ to $\frac{1}{2}[A]_0$. This increase is consistent with the slower decline of the graph in Fig. 26.6 than for first-order reactions. For a general nth-order reaction with rate law

$$\frac{d[A]}{dt} = -k[A]^n \tag{11a}$$

the same procedures lead to the result that

$$t_{1/2} = \frac{2^{n-1} - 1}{(n-1)k[A]_0^{n-1}} \tag{11b}$$

We now have another way of checking for second-order kinetics: $t_{1/2}$ is measured for a series of different initial concentrations and plotted against $1/[A]_0$. A straight line confirms second-order kinetics, and the slope gives k. Alternatively, since $t_{1/2} \propto [A]_0^{1-n}$,

$$\lg t_{1/2} = \text{constant} + (1 - n) \lg [A]_0 \tag{12}$$

and so a plot of $\lg t_{1/2}$ against $\lg [A]_0$ will be a straight line of slope $1 - n$.

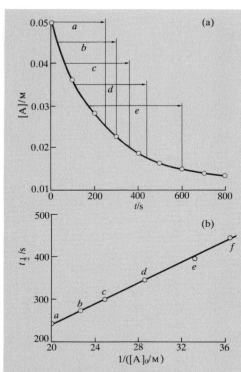

Fig. 26.8 The determination of the rate constant of a second-order reaction by the half-life method. (a) The original data; $a, b, \ldots, f$ are the half-lives for selected initial concentrations. (b) A straight line is obtained when the half lives are plotted against $1/[A]_0$; its slope is $1/k$.

Example 26.6: *Using the half-life to determine reaction order*

The results of the alkaline hydrolysis of ethyl nitrobenzoate (A) are reported below. Determine the order of reaction by the half-life method, and find the rate constant.

t/s	0	100	200	300	400	500	600	700	800
$[A]/(10^{-2}\,M)$	5.00	3.55	2.75	2.25	1.85	1.60	1.48	1.40	1.38

Answer. We need to plot $[A]/M$ against t/s (Fig. 26.8). To do so, we choose a sequence of times $a, b, c, \ldots$ to regard as 'initial' times, and note the corresponding 'initial' concentrations. Find the time for each concentration to fall to half its initial value, giving $t_{1/2}$, and if $t_{1/2}$ is independent of $[A]_0$ the reaction is first order:

	a	b	c	d	e	f
$[A]_0/(10^{-2}\,M)$	5.00	4.50	4.00	3.50	3.00	2.75
$t_{1/2}/s$	240	270	300	345	400	450

Clearly, $t_{1/2}$ is not independent of $[A]_0$, so we now try plotting $t_{1/2}$ against $1/[A]_0$, as in Fig. 26.8(b). This plot is a straight line, and so the reaction is second order. The slope is 12.5; Therefore

$$k = \frac{1}{12.5} M^{-1} s^{-1} = 8.0 \times 10^{-2} M^{-1} s^{-1}$$

Exercise. Use the same technique to determine the order and rate constant of the reaction given the following data:

t/s	0	100	200	300	400	500	600	700	800
$[A]/(10^{-3}\,M)$	4.40	3.55	2.98	2.57	2.26	2.01	1.81	1.65	1.52

$$[2, 0.54\,M^{-1}\,s^{-1}]$$

26.6 The temperature dependence of reaction rates

It is found that the rates of most reactions increase as the temperature is raised. Many reactions fall somewhere in the range spanned by the hydrolysis of methyl ethanoate (where the rate constant at 35°C is 1.82 times that at 25°C) and the hydrolysis of sucrose (where the factor is 4.13).

The Arrhenius parameters

An empirical observation is that many reactions have rate constants that follow the **Arrhenius equation**

$$\ln k = \ln A - \frac{E_a}{RT} \tag{13a}$$

That is, for many reactions it is found that a plot of $\ln k$ against $1/T$ gives a straight line. The Arrhenius equation is often written as

$$k = A e^{-E_a/RT} \tag{13b}$$

A is called the **pre-exponential factor** and E_a the **activation energy**. Collectively they are called the **Arrhenius parameters** of the reaction, and some experimental values are given in Table 26.3. Equation 13b is

Table 26.3. Arrhenius parameters

(1) First-order reactions	A/s^{-1}	$E_a/(kJ\,mol^{-1})$
$CH_3NC \rightarrow CH_3CN$	3.98×10^{13}	160
$2N_2O_5 \rightarrow 4NO_2 + O_2$	4.94×10^{13}	103.4
(2) Second-order	$A/(M^{-1}\,s^{-1})$	$E_a/(kJ\,mol^{-1})$
$OH + H_2 \rightarrow H_2O + H$	8×10^{10}	42
$NaC_2H_5O + CH_3I$ in ethanol	2.42×10^{11}	81.6

sometimes written in an alternative form that combines the two parameters:

$$k \propto e^{-\Delta G^{\ddagger}/RT} \quad \text{or} \quad -RT \ln k \propto \Delta G^{\ddagger} \qquad (13c)$$

where $\Delta G^{\ddagger}$ is called the **activation Gibbs function** (Section 28.6). In this form, the expression for the rate constant strongly resembles the formula for the equilibrium constant in terms of the standard reaction Gibbs function (eqn 5 of Section 9.1). We shall see the origin of this formal analogy in Chapter 28.

For the present chapter we shall regard the Arrhenius parameters as purely empirical parameters that enable us to discuss the variation of rate constants with temperature. However, it is worth anticipating the interpretation that we shall put on E_a in Section 28.1. There we shall see that the activation energy is the minimum energy that reactants must have in order to form products. For example, in a gas phase reaction there are numerous collisions each second, but only a tiny proportion of them are sufficiently energetic to lead to reaction. The fraction of collisions with a kinetic energy in excess of an energy E_a is given by the Boltzmann distribution as $e^{-E_a/RT}$, exactly as in eqn 13b. Hence, the exponential factor in eqn 13b can be interpreted as the fraction of collisions that have enough energy to lead to reaction.

The analogous interpretation of the pre-exponential factor is that it is a measure of the rate at which collisions occur irrespective of their energy (we shall see that it is proportional to the collision density in the gas). Hence the product of A and the exponential factor in eqn 13b gives the rate of *successful* collisions. We shall develop these remarks in Chapter 28 and see that they have their analogues for reactions that take place in liquids.

Example 26.7: *Determining the Arrhenius parameters*

The rate of the second-order decomposition of acetaldehyde (ethanal, CH_3CHO) was measured over the temperature range 700–1000 K, and the rate constants are reported below. Find the activation energy and the pre-exponential factor.

T/K	700	730	760	790	810	840	910	1000
$k/(M^{-1}\,s^{-1})$	0.011	0.035	0.105	0.343	0.789	2.17	20.0	145

Answer. We plot $\ln k$ against $1/T$ (Fig. 26.9). The least-squares best fit of the line is with slope -2.21×10^4 and intercept 27.0. Therefore, since the slope is $-E_a/R$ and the intercept at $1/T = 0$ is $\ln A$,

$$E_a = (2.21 \times 10^4\,K) \times (8.314\,J\,K^{-1}\,mol^{-1}) = 184\,kJ\,mol^{-1}$$

$$A = e^{27.0}\,M^{-1}\,s^{-1} = 5.3 \times 10^{11}\,M^{-1}\,s^{-1}$$

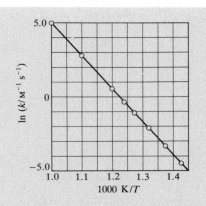

Fig. 26.9 The Arrhenius plot of $\ln k$ against $1/T$ for the decomposition of CH_3CHO, and the best straight line. The slope gives $-E_a/R$ and the intercept at $1/T = 0$ gives $\ln A$.

Comment. Note that A has the same units as k. For this reaction, the reaction rate doubles on increasing the temperature from 700 K to 716 K or from 900 K to 926 K, etc.

Exercise. Find A and E_a from the following data:

T/K	300	350	400	450	500
$k/(\text{M}^{-1}\,\text{s}^{-1})$	7.9×10^6	3.0×10^7	7.9×10^7	1.7×10^8	3.2×10^8

$$[8 \times 10^{10}\,\text{M}^{-1}\,\text{s}^{-1}, 23\ \text{kJ mol}^{-1}]$$

In some cases the temperature dependence is not Arrhenius-like. However, it is still possible to express the strength of the dependence by defining an activation energy as

$$E_a = RT^2\left(\frac{\partial \ln k}{\partial T}\right)_V \tag{14a}$$

This definition reduces to the earlier one (as the slope of an Arrhenius plot) for a temperature-independent activation energy. Thus, with $d(1/T) = -dT/T^2$ we can rearrange eqn 14a into

$$E_a = -R\left(\frac{\partial \ln k}{\partial (1/T)}\right)_V \tag{14b}$$

which integrates to eqn 13 if E_a is independent of temperature. However, the definition in eqn 14a is more general, because it allows E_a to be obtained from the slope (at the temperature of interest) of a plot of $\ln k$ against $1/T$ even if the Arrhenius plot is not a straight line. Equation 14a shows that the higher the activation energy, the stronger the temperature dependence of the rate constant. That is, a high activation energy signifies that the rate constant changes rapidly with temperature.

Temperature-jump relaxation methods

The term **relaxation** denotes the return of a system to equilibrium. It is used in chemical kinetics to indicate that an externally applied influence has shifted the equilibrium position of a reaction, normally very suddenly, and that the reaction is adjusting to the new equilibrium composition. We shall consider the response of reaction rates to a **temperature jump**, a sudden change in temperature.

We take as an example a simple $A \rightleftharpoons B$ equilibrium, which is first-order in each direction. At a certain temperature the rate of change of $[A]$ is

$$\frac{d[A]}{dt} = -k_a'[A] + k_b'[B]$$

At equilibrium under these conditions, $d[A]/dt = 0$, and the concentrations are $[A]_{eq}'$ and $[B]_{eq}'$. Therefore,

$$k_a'[A]_{eq}' = k_b'[B]_{eq}'$$

Now we increase the temperature suddenly. The rate constants change to k_a and k_b but the concentrations of A and B are still, momentarily, at their old equilibrium values. As the system is no longer at equilibrium, it readjusts to the new equilibrium concentrations, which are now given by

$$k_a[A]_{eq} = k_b[B]_{eq}$$

and it does so at a rate that depends on the new rate constants.

We write the deviation of $[A]$ from its new equilibrium value as x, so $[A] = x + [A]_{eq}$. At $t = 0$ (when the jump in conditions took place), $[A]_0 = [A]'_{eq}$ implying that $x_0 = [A]'_{eq} - [A]_{eq}$. The concentrations then change as follows:

$$\frac{d[A]}{dt} = -k_a(x + [A]_{eq}) + k_b(-x + [B]_{eq}) = -(k_a + k_b)x \qquad (15)$$

because the two terms involving the equilibrium concentrations cancel. Since $d[A]/dt = dx/dt$, eqn 15 is a first-order differential equation with the solution

$$x = x_0 e^{-t/\tau} \qquad \frac{1}{\tau} = k_b + k_a \qquad (16)$$

Equation 16 shows that the concentration of A (and of B) relaxes into the new equilibrium at a rate determined by the sum of the two new rate constants. Since the equilibrium constant under the new conditions is $K = k_a/k_b$ its value may be combined with the relaxation time measurement to find the individual k_a and k_b.

One way of achieving a temperature jump is to discharge an electric current through a sample made conducting by the addition of ions, but laser discharges can also be used. Temperature jumps of between 5 and 10 K can be achieved in about 10^{-7} s. An important application of the technique has been to the determination of the rate of the proton transfer reaction

$$H_3O^+(aq) + OH^-(aq) \rightarrow 2H_2O(l)$$

by monitoring the conductivity of the sample. A relaxation time of about 40 μs was measured at room temperature, corresponding to $k = 1.4 \times 10^{11} \, M^{-1} s^{-1}$, making it one of the fastest liquid-phase reactions known (but the reaction is faster in ice, where $k = 8.6 \times 10^{12} \, M^{-1} s^{-1}$).

Example 26.8: *Analysing a temperature-jump experiment*

The $H_2O(l) \rightleftharpoons H^+(aq) + OH^-(aq)$ equilibrium relaxes in 37 μs at 298 K and $pK_w = 14.01$. Calculate the rate constants for the forward and backward reactions.

Answer. We begin by deriving an expression for τ in terms of k_1 (forward, first-order reaction) and k_2 (reverse, second-order reaction). Proceed as above, but make the assumption that the deviation from equilibrium (x) is so small that terms in x^2 can be neglected. Relate k_1 and k_2 through the equilibrium constant, using $K = K_w/([H_2O]/M)$. The forward rate is $k_1[H_2O]$ and the reverse rate is $k_2[H^+][OH^-]$. The same procedure as before leads to

$$\frac{1}{\tau} = k_1 + k_2([H^+] + [OH^-])$$

The equilibrium condition is

$$k_1[H_2O]_e = k_2[H^+]_e[OH^-]_e$$

so that $K = k_1/(k_2 \text{M})$ with $K = (0.98 \times 10^{-14})/55.6 = 1.8 \times 10^{-16}$. Therefore,

$$\frac{1}{\tau} = k_2\{(K\,\text{M}) + [\text{H}^+]_e + [\text{OH}^-]_e\}$$

$$= k_2(K + K_w^{1/2} + K_w^{1/2})\,\text{M}$$

$$= 2.0 \times 10^{-7} \times k_2\,\text{M}$$

Hence,

$$k_2 = \frac{1}{(37 \times 10^{-6}\,\text{s}) \times (2.0 \times 10^{-7}\,\text{M})} = 1.4 \times 10^{11}\,\text{M}^{-1}\,\text{s}^{-1}$$

It follows that

$$k_1 = k_2 K\,\text{M} = 2.4 \times 10^{-5}\,\text{s}^{-1}$$

Comment. Notice how we keep track of units: K and K_w are dimensionless; k_2 is expressed in $\text{M}^{-1}\text{s}^{-1}$ and k_1 in s^{-1}.

Exercise. Derive an expression for the relaxation time of a concentration when the reaction $A + B \rightleftharpoons C + D$ is second order in both directions.
$$[1/\tau = k([A] + [B])_e + k'([C] + [D])_e]$$

The equilibrium constant of a reaction is temperature dependent so long as the reaction enthalpy is non-zero (Section 9.4), and so the temperature-jump method is widely applicable. The equilibrium composition also depends on pressure in some cases, and **pressure-jump techniques** may then be used.

Accounting for the rate laws

We now move on to the second stage of the analysis of kinetic data, their explanation in terms of a postulated reaction mechanism.

26.7 Elementary reactions

Most reactions occur in a sequence of steps called **elementary reactions**, each of which involves only one or two molecules. A typical elementary reaction may be

$$\text{H} + \text{Br}_2 \rightarrow \text{HBr} + \text{Br}$$

This equation signifies that a particular H atom attacks a particular Br_2 molecule to produce a molecule of HBr and a Br atom.[1] The **molecularity** of an elementary reaction is the number of molecules (or atoms or ions) coming together to react. In a **unimolecular reaction** a single molecule shakes itself apart or its atoms into a new arrangement, as in the isomerization of cyclopropane to propene:

$$\underset{\text{CH}_2-\text{CH}_2}{\overset{\text{CH}_2}{\triangle}} \rightarrow \text{CH}_3\text{CH}{=}\text{CH}_2$$

In a **bimolecular reaction** a pair of molecules collide and exchange energy,

[1] The chemical equation for an elementary reaction has a different significance from that for an overall reaction. It specifies the *individual event,* not merely the overall, net stoichiometry. Which meaning should be ascribed to the arrow $\rightarrow$ will always be clear from the context.

atoms, or groups of atoms, or undergo some other kind of change, as in the reaction between H and Br_2. It is most important to distinguish molecularity from order: the order is an empirical quantity, and obtained from the experimental rate law; the molecularity refers to an elementary reaction that has been postulated to be an individual step in some mechanism.

We can write down the rate law of an elementary reaction from its reaction equation. This can be done only in the case of an elementary reaction, not for reactions in general. Thus, the rate law of a unimolecular elementary reaction is first order in the reactant:

$$A \rightarrow Products \qquad \frac{d[A]}{dt} = -k[A] \qquad (17a)$$

A unimolecular reaction is first order because the number of A molecules that decay in a short interval is proportional to the number available to decay (ten times as many decay in the same interval when there are initially 1000 A molecules than when there are only 100 present). Therefore, the rate of decomposition of A is proportional to its concentration.

Similarly, the rate law of an elementary bimolecular reaction is second order:

$$A + B \rightarrow Products \qquad \frac{d[A]}{dt} = -k[A][B] \qquad (17b)$$

A bimolecular reaction is second order because its rate depends on the rate at which the reactants meet, which is proportional to their concentrations. Therefore, if we believe (or simply postulate) that a reaction is a single step, bimolecular process, we can write down the rate law (and then go on to test it). Bimolecular elementary reactions are believed to account for many homogeneous reactions, such as the dimerizations of alkenes and dienes and reactions such as

$$CH_3I + CH_3CH_2O^- \rightarrow CH_3OCH_2CH_3 + I^-$$

in alcohol solution. The mechanism of the last reaction is believed to be a single elementary step, which is consistent with the observed rate law

$$v = k[CH_3I][CH_3CH_2O^-]$$

The interpretation of a rate law is full of pitfalls, partly because a second-order rate law, for instance, can also result from a complex reaction scheme. We shall see below how to string simple steps together into a mechanism and how to arrive at the corresponding rate law. For the present we emphasize that if the reaction is an elementary bimolecular process, then it has second-order kinetics, but if the kinetics are second order, the reaction might be complex. The postulated mechanism can be explored only by detailed detective work on the system, and by investigating whether side products or intermediates appear during the course of the reaction. Detailed analysis of this kind was one of the ways, for example, in which the reaction $H_2 + I_2 \rightarrow 2HI$ was shown to proceed by a complex reaction after many years during which people had accepted on good, but insufficiently meticulous, evidence that it was a fine example of a simple bimolecular reaction in which atoms exchanged partners during a collision.

26.8 Consecutive elementary reactions

Some reactions proceed through the formation of an intermediate, as in the consecutive unimolecular reactions

$$A \xrightarrow{k_a} B \xrightarrow{k_b} C$$

An example is the decay of a radioactive family, such as

$$^{239}U \xrightarrow{23.5 \text{ min}} {}^{239}Np \xrightarrow{2.35 \text{ days}} {}^{239}Pu$$

(The times are half-lives.) We can discover the characteristics of this type of reaction by setting up the rate laws for the net rate of change of the concentration of each substance.

The variation of concentrations with time

The rate of unimolecular decomposition of A is

$$\frac{d[A]}{dt} = -k_a[A] \tag{18a}$$

and A is not replenished. The intermediate B is formed from A (at a rate $k_a[A]$) but decays to C (at a rate $k_b[B]$). Its net rate of formation is therefore

$$\frac{d[B]}{dt} = k_a[A] - k_b[B] \tag{18b}$$

C is formed by the unimolecular decay of B:

$$\frac{d[C]}{dt} = k_b[B] \tag{18c}$$

We suppose that initially only A is present, and that its concentration is $[A]_0$.

The first of the rate laws is an ordinary first-order decay, and so

$$[A] = [A]_0 e^{-k_a t} \tag{19a}$$

When this equation is substituted in eqn 18b, and we set $[B]_0 = 0$, the solution is

$$[B] = \frac{k_a}{k_b - k_a} \times (e^{-k_a t} - e^{-k_b t})[A]_0 \tag{19b}$$

Since at all times

$$[A] + [B] + [C] = [A]_0$$

it follows that

$$[C] = \left\{1 + \frac{k_a e^{-k_b t} - k_b e^{-k_a t}}{k_b - k_a}\right\}[A]_0 \qquad (19c)$$

The three concentrations are plotted in Fig. 26.10. We see that the intermediate's concentration rises to a maximum, and then falls to zero. The concentration of the final product C rises from zero and reaches $[A]_0$.

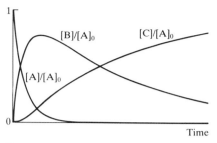

Fig. 26.10 The concentrations of A, B, and C in the consecutive reaction scheme A → B → C. The curves are plots of eqn 19 with $k_a = 10 k_b$. If B is the desired product, it is important to be able to predict when its concentration is greatest; see Example 26.9.

Example 26.9: *Analysing consecutive reactions*

Suppose that in an industrial batch process a substance A produces the desired product B which goes on to decay to a worthless product C, each stage of the reaction being first-order. At what time will product B be present in greatest concentration?

Answer. The time-dependence of the concentration of B is given by eqn 19b. We can find by differentiation the time at which it passes through a maximum. Thus

$$\frac{d[B]}{dt} = \frac{-k_a [A]_0 (k_a e^{-k_a t} - k_b e^{-k_b t})}{k_b - k_a}$$

and this is equal to zero when

$$k_a e^{-k_a t} = k_b e^{-k_b t}$$

Therefore, the time at which B has its maximum concentration is

$$t = \frac{1}{k_a - k_b} \ln \frac{k_a}{k_b}$$

Comment. For a given value of k_a, as k_b increases both the time at which $[B]$ is a maximum and the yield of B decrease.

Exercise. Calculate the maximum concentration of B and justify the last remark.

$$\left[\frac{[B]_{max}}{[A]_0} = \left(\frac{k_b}{k_a}\right)^c \text{ with } c = \frac{k_b}{k_a - k_b}\right]$$

The rate-determining step

Suppose now that $k_b \gg k_a$, then whenever a B molecule is formed it decays rapidly into C. Since

$$e^{-k_b t} \ll e^{-k_a t} \quad \text{and} \quad k_b - k_a \approx k_b$$

eqn 19c reduces to

$$[C] = (1 - e^{-k_a t})[A]_0$$

which shows that the formation of C depends on only the *smaller* of the two rate constants. That is, the rate of formation of C depends on the rate at which B is formed, not on the rate at which B changes into C. For this reason, the step A → B is called the **rate-determining step** of the reaction. Its existence has been likened to building a six-lane highway up to a single-lane bridge: the traffic flow is governed by the rate of crossing the bridge. Similar remarks apply to more complicated reaction mechanisms, and in general the rate-determining step is the one with the smallest rate constant.

The steady-state approximation

One feature of the calculation so far has probably not gone unnoticed: there is a considerable increase in mathematical complexity as soon as the reaction mechanism has more than a couple of steps. A mechanism involving many steps will be unsolvable analytically, and we shall need alternative methods of solution. One approach is to integrate the rate laws numerically (Chapter 27). An alternative, which continues to be widely used because it leads to convenient expressions, is to make an approximation.

The **steady-state approximation** assumes that during the major part of the reaction, the concentrations and the rates of change of all reaction intermediates are constant and small. The approximation can be justified by considering the rate law for the intermediate B in the consecutive reactions $A \rightarrow B \rightarrow C$ that we have been considering. We can combine eqns 19a and 19b to obtain

$$\frac{[B]}{[A]} = \frac{k_a(1 - e^{(k_a - k_b)t})}{k_b - k_a}$$

If $k_b \gg k_a$ this expression simplifies to

$$\frac{[B]}{[A]} = \frac{k_a(1 - e^{-k_b t})}{k_b}$$

If also the reaction has proceeded long enough for it to be true that $k_b t \gg 1$, the exponential is much smaller than 1 and

$$\frac{[B]}{[A]} = \frac{k_a}{k_b}$$

That is,

$$[B] = \frac{k_a}{k_b} \times [A] \ll [A]$$

and the concentration of the intermediate is only a small fraction of the concentration of A. Similarly, by differentiation of both sides of the last equation,

$$\frac{d[B]}{dt} = \frac{k_a}{k_b} \times \frac{d[A]}{dt} \ll \frac{d[A]}{dt}$$

and the rate of change for the concentration of the intermediate is much smaller than that of A. For this kinetic scheme at least, we can neglect the rate of change of the concentration of the intermediate in comparison with the rate of decomposition of the reactant. The steady-state approximation generalizes this conclusion, and states that for the major part of the reaction (after any induction period and before the reactants have become significantly depleted), the rates of change of the concentrations of all intermediates can be set equal to zero:

$$\frac{d[B]}{dt} \approx 0 \text{ for all intermediates}$$

The steady-state approximation greatly simplifies the discussion of reaction schemes. For example, when we apply it to the consecutive first-order mechanism, we set $d[B]/dt = 0$ in eqn 18b, which becomes

$$k_a[A] - k_b[B] = 0$$

Then

$$[B] = \frac{k_a[A]}{k_b}$$

On substituting this value of [B] into eqn 18c, that equation becomes

$$\frac{d[C]}{dt} = k_b[B] = k_a[A]$$

and we see that C is formed by a first-order decay of A, with a rate constant k_a, the rate-constant of the slower, rate-determining, step. We can write down the solution of this equation at once by substituting the solution for [A], eqn 19a, and integrating:

$$[C] = k_a[A]_0 \int_0^t e^{-k_a t}\, dt = (1 - e^{-k_a t})[A]_0$$

the same (approximate) result as before, but much more quickly obtained.

Example 26.10: *Using the steady-state approximation*

Account for the rate law for the decomposition of N_2O_5

$$2N_2O_5(g) \rightarrow 4NO_2(g) + O_2(g) \qquad v = k[N_2O_5]$$

on the basis of the mechanism

$$N_2O_5 \rightarrow NO_2 + NO_3 \qquad k_a$$

$$NO_2 + NO_3 \rightarrow N_2O_5 \qquad k_a'$$

$$NO_2 + NO_3 \rightarrow NO_2 + O_2 + NO \qquad k_b$$

$$NO + N_2O_5 \rightarrow 3NO_2 \qquad k_c$$

Answer. We first identify the intermediates (the substances that do not appear in the overall reaction) as NO and NO_3, and write expressions for their net rates of formation:

$$\frac{d[NO]}{dt} = k_b[NO_2][NO_3] - k_c[NO][N_2O_5]$$

$$\frac{d[NO_3]}{dt} = k_a[N_2O_5] - k_a'[NO_2][NO_3] - k_b[NO_2][NO_3]$$

According to the steady-state approximation, we set both rates equal to zero:

$$k_b[NO_2][NO_3] - k_c[NO][N_2O_5] = 0$$

$$k_a[N_2O_5] - k_a'[NO_2][NO_3] - k_b[NO_2][NO_3] = 0$$

The net rate of change of concentration of N_2O_5 is

$$\frac{d[N_2O_5]}{dt} = -k_a[N_2O_5] + k_a'[NO_2][NO_3] - k_c[NO][N_2O_5]$$

and replacing the concentrations of the intermediates using the equations above, gives

$$\frac{d[N_2O_5]}{dt} = \frac{-2k_ak_b[N_2O_5]}{k_a' + k_b}$$

Since $v(N_2O_5) = -2$, it follows that the reaction rate is

$$v = k[N_2O_5] \qquad k = \frac{k_ak_b}{k_a' + k_b}$$

in accord with observation.

Comment. The N_2O_5 decomposition is a problematic reaction, since its rate decreases more quickly than expected at low pressures. It is believed that this is due to changes in the value of the rate constants themselves (particularly k_a').

Exercise. Derive the rate law for the decomposition of ozone in the reaction $2O_3(g) \rightarrow 3O_2(g)$ on the basis of the (incomplete) mechanism

$$O_3 \rightarrow O_2 + O \quad k_a$$

$$O_2 + O \rightarrow O_3 \quad k_a'$$

$$O + O_3 \rightarrow 2O_2 \quad k_b$$

$$\left[v = \frac{k_ak_b[O_3]^2}{k_a'[O_2] + k_b[O_3]} \right]$$

Pre-equilibria

From a simple sequence of consecutive reactions we now turn to a slightly more complicated mechanism in which an intermediate reaches an equilibrium with the reactants:

$$A + B \underset{k_a'}{\overset{k_a}{\rightleftharpoons}} C \overset{k_b}{\longrightarrow} P \tag{20}$$

where C denotes the intermediate. This scheme involves a **pre-equilibrium**, and arises when the rates of formation of the intermediate and its decay back into reactants are much faster than its rate of formation of products. Since we assume that A, B, and C are in equilibrium, we can write

$$K = \frac{[C]}{[A][B]} \quad \text{with} \quad K = \frac{k_a}{k_a'}$$

In writing these equations, we are presuming that the rate of reaction of C to form P is too slow to affect the maintenance of the pre-equilibrium (see the example below). The rate of formation of P may now be written:

$$\frac{d[P]}{dt} = k_b[C] = k_bK[A][B]$$

This rate law has the form of a second-order rate law with a composite rate constant:

$$\frac{d[P]}{dt} = k[A][B] \qquad k = k_bK = \frac{k_ak_b}{k_a'} \tag{21}$$

Example 26.11: *Analysing a pre-equilibrium*

Repeat the pre-equilibrium calculation but without ignoring the fact that C is slowly leaking away as it forms P.

Answer. We begin by writing the net rates of formation of the substances:

$$\frac{d[P]}{dt} = k_b[C]$$

$$\frac{d[C]}{dt} = k_a[A][B] - k_a'[C] - k_b[C]$$

Now invoke the steady-state approximation for the intermediate C: the second of these two equations is set equal to 0, which solves to

$$[C] = \frac{k_a[A][B]}{k_a' + k_b}$$

When we substitute this into the expression for the rate of formation of P, we obtain

$$\frac{d[P]}{dt} = k[A][B] \qquad k = \frac{k_a k_b}{k_a' + k_b}$$

Comment. This expression reduces to that in eqn 21 when the rate constant for the decay of C into products is much smaller than that for its decay into reactants, $k_b \ll k_a'$.

Exercise. Show that the pre-equilibrium mechanism

$$2A \xrightleftharpoons{K} B \text{ followed by } B + C \xrightarrow{k_b} P$$

results in an overall third-order reaction. $\qquad [d[P]/dt = k_b K[A]^2[B]]$

Third-order reactions

An example will help to show how a pre-equilibrium assumption helps to elucidate a mechanism. The oxidation of nitrogen(II) oxide is found to be third-order overall:

$$2NO(g) + O_2(g) \rightarrow 2NO_2(g) \qquad \frac{d[NO_2]}{dt} = k[NO]^2[O_2] \qquad (22)$$

One explanation might be that the reaction is a single intermolecular simple step, but that requires the simultaneous collision of three particles, which occurs very infrequently. Furthermore, it is observed that the reaction rate *decreases* as the temperature is raised. This points to a complex reaction mechanism because simple reactions almost always go faster at higher temperatures.

A mechanism that accounts for the rate law and the temperature dependence is a pre-equilibrium

$$\text{(a)} \quad 2NO(g) \rightleftharpoons N_2O_2(g) \qquad K = \frac{[N_2O_2]}{[NO]^2}$$

followed by a simple bimolecular reaction

(b) $N_2O_2(g) + O_2(g) \rightarrow 2NO_2(g)$ $\dfrac{d[NO_2]}{dt} = k_b[N_2O_2][O_2]$

The reaction rate is obtained by combining the two equations into

$$\frac{d[NO_2]}{dt} = k_b K[NO]^2[O_2]$$

which has the observed overall third-order form.

The mechanism is consistent with the anomalous temperature dependence because K decreases with temperature (the dimerization of NO is exothermic, and from thermodynamics we know that K decreases as the temperature is raised, Section 9.4). Therefore, provided k_b does not increase more sharply, the rate constant $k = k_b K$ decreases as the temperature is increased and the effective activation energy of the reaction is negative.

Example 26.12: *Calculating an apparent activation energy*

Calculate the activation energy of a reaction with a pre-equilibrium step and show how it might be negative.

Answer. The rate constant of a reaction with pre-equilibrium is $k = k_b k_a / k_a'$, and we shall suppose that each individual step has an Arrhenius form with activation energies ε_a, ε_a', and ε_b. From the definition of activation energy, eqn 14a, we obtain

$$E_a = RT^2 \left(\frac{\partial \ln k}{\partial T}\right)_V = RT^2 \left(\frac{\partial}{\partial T} \ln \frac{k_b k_a}{k_a'}\right)_V$$

$$= RT^2 \left\{ \left(\frac{\partial \ln k_b}{\partial T}\right)_V + \left(\frac{\partial \ln k_a}{\partial T}\right)_V - \left(\frac{\partial \ln k_a'}{\partial T}\right)_V \right\}$$

$$= \varepsilon_b + \varepsilon_a - \varepsilon_a'$$

If the activation energy ε_a' is large enough, it may result in $E_a < 0$, and the reaction will form products more slowly at higher temperatures than low.

Comment. Since a reaction with a high activation energy is more responsive to temperature than one with a low activation energy, a way of understanding the conclusion is to say that the decay of the intermediate into reactants increases more quickly with temperature than either of the forward reactions.

The Michaelis–Menten mechanism

Another example of a reaction in which an intermediate is formed is the **Michaelis–Menten mechanism** of enzyme action. The rate of an enzyme catalyzed reaction in which a substrate S is converted into products P is found to depend on the concentration of the enzyme E even though the enzyme undergoes no net change. The proposed mechanism is

$$E + S \underset{k_a'}{\overset{k_a}{\rightleftharpoons}} (ES) \overset{k_b}{\longrightarrow} P + E \qquad (23)$$

(ES) denotes a bound state of the enzyme and its substrate. This mechanism has the same form as that treated in Example 26.11, so we can conclude at once that

$$[ES] = \frac{k_a[E][S]}{k_a' + k_b}$$

[E] and [S] are the concentrations of the free enzyme and free substrate, and if $[E]_0$ is the total concentration of enzyme,

$$[E] + [ES] = [E]_0$$

Since only a little enzyme is added, the free substrate concentration is almost the same as the total substrate concentration, and we can ignore the fact that [S] differs slightly from $[S]_{total}$. Therefore,

$$[ES] = \frac{k_a([E]_0 - [ES])[S]}{k_a' + k_b}$$

which rearranges to

$$[ES] = \frac{k_a[E]_0[S]}{k_a' + k_b + k_a[S]}$$

It follows that the rate of formation of product is

$$\frac{d[P]}{dt} = k[E]_0 \qquad k = \frac{k_b[S]}{K_M + [S]} \tag{24a}$$

where the **Michaelis constant** K_M is

$$K_M = \frac{k_b + k_a'}{k_a} \tag{24b}$$

According to eqn 24a, the rate of enzymolysis depends linearly on the enzyme concentration, but in a more complicated way on the concentration of substrate. Thus, when $[S] \gg K_M$, the rate law in eqn 24a reduces to

$$\frac{d[P]}{dt} = k_b[E]_0$$

and is zeroth-order in S. This means that the rate is constant because there is so much S present that it remains at effectively the same concentration even though products are being formed. Moreover, the rate of formation of products is a maximum, and $k_b[E]_0$ is called the **maximum velocity** of the enzymolysis; k_b itself is the **maximum turnover number**. When so little S is present that $[S] \ll K_M$, then the rate of formation of products is

$$\frac{d[P]}{dt} = \frac{k_b}{K_M}[E]_0[S]$$

Now the rate is proportional to [S] as well as to $[E]_0$.

It follows from eqn 24a that

$$\frac{1}{k} = \frac{1}{k_b} + \frac{K_M}{k_b[S]} \tag{25}$$

Hence, a plot of $1/k$ against $1/[S]$ will give the value of k_b (from the intercept at $1/[S] = 0$) and K_M (from the slope, K_M/k_b). However, the plot cannot give the values of the individual rate coefficients k_a and k_a' that appear in K_M. The stopped-flow technique can give the additional data needed, because the rate of formation of the enzyme–substrate complex can be found by monitoring its concentration after mixing enzyme and substrate. This gives k_a, and k_a' can then be found from the value of K_M.

26.9 Unimolecular reactions

A number of gas phase reactions follow first-order kinetics, as in the isomerization of cyclopropane mentioned earlier:

$$\overset{\displaystyle CH_2}{\underset{\displaystyle CH_2—CH_2}{\diagup\diagdown}} \to CH_3CH{=}CH_2 \qquad v = k[\text{cyclopropane}]$$

The problem with first-order rate laws is that presumably a molecule acquires enough energy to react as a result of its collisions with other molecules. However, collisions are simple bimolecular events, and so how can they result in a first-order rate law? First-order gas-phase reactions are widely called 'unimolecular reactions' since they also involve an elementary unimolecular step in which the reactant molecule changes into the product. However, this term must be used with caution, since the composite mechanism has bimolecular as well as unimolecular steps.

The Lindemann–Hinshelwood mechanism

The first successful explanation was provided by Frederick Lindemann in 1921 and then elaborated by Cyril Hinshelwood. In the **Lindemann–Hinshelwood mechanism** it is supposed that a reactant molecule A becomes energetically excited by collision with another A molecule:

$$A + A \to A^* + A \qquad \frac{d[A^*]}{dt} = k_a[A]^2 \qquad (26a)$$

The energized molecule might lose its excess energy by collision with another:

$$A^* + A \to 2A \qquad \frac{d[A^*]}{dt} = -k_a'[A^*][A] \qquad (26b)$$

Alternatively, the excited molecule might shake itself apart and form products P. That is, it might undergo the unimolecular decay

$$A^* \to P \qquad \frac{d[P]}{dt} = k_b[A^*] \qquad \frac{d[A^*]}{dt} = -k_b[A^*] \qquad (26c)$$

If the unimolecular step is slow enough to be the rate-determining step, the overall reaction will have first-order kinetics, as we require. We can demonstrate this explicitly by applying the steady-state approximation to the net rate of formation of A^*:

$$\frac{d[A^*]}{dt} = k_a[A]^2 - k_a'[A^*][A] - k_b[A^*] = 0$$

This equation solves to

$$[A^*] = \frac{k_a[A]^2}{k_b + k_a'[A]}$$

and so the rate law for the formation of P is

$$\frac{d[P]}{dt} = k_b[A^*] = \frac{k_a k_b[A]^2}{k_b + k_a'[A]} \qquad (27a)$$

At this stage the rate law is not first order. However, if the rate of deactivation by A^*, A collisions is much greater than the rate of unimolecular decay, in the sense that

$$k_a'[A^*][A] \gg k_b[A^*], \quad \text{or} \quad k_a'[A] \gg k_b$$

then we can neglect k_b in the denominator in eqn 27a and obtain

$$\frac{d[P]}{dt} = k[A] \qquad k = \frac{k_a k_b}{k_a'} \qquad (27b)$$

Equation 27b is a first-order rate law, as we set out to show.

The Lindemann–Hinshelwood mechanism can be tested because it predicts that as the concentration (and therefore the partial pressure) of A is reduced the reaction should switch to overall second-order kinetics. Thus, when $k_a'[A] \ll k_b$, the rate law in eqn 27a is approximately

$$\frac{d[P]}{dt} = k_a[A]^2 \qquad (27c)$$

The physical reason for the change of order is that at low pressures the rate-determining step is the bimolecular formation of A^*. If we write the full rate law in eqn 27a as

$$\frac{d[P]}{dt} = k[A] \qquad k = \frac{k_a k_b[A]}{k_b + k_a'[A]}$$

then the expression for the effective rate-constant can be rearranged to

$$\frac{1}{k} = \frac{1}{k_a[A]} + \frac{k_a'}{k_a k_b} \qquad (28)$$

Hence, a test of the theory is to plot $1/k$ against $1/[A]$, and to expect a straight line.

Whereas the Lindemann–Hinshelwood mechanism agrees in general with the switch in order of unimolecular reactions, it does not agree in detail. A typical graph of $1/k$ against $1/[A]$ is shown in Fig. 26.11. The graph has a pronounced curvature, corresponding to a larger value of k (smaller value

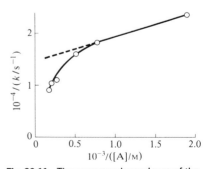

Fig. 26.11 The pressure-dependence of the unimolecular isomerization of *trans*-CHD=CHD showing a pronounced departure from the straight line predicted by eqn 28 based on the Lindemann–Hinshelwood mechanism.

of $1/k$) at high pressures (low $1/[A]$) than would be expected by extrapolation of the reasonably linear low pressure (high $1/[A]$) data.

Improvements to the theory

One of the reasons for the discrepancy is that the Lindemann–Hinshelwood mechanism fails to recognize that a specific excitation of the molecule may be required before reaction occurs. For example, in the isomerization of cyclopropane the crucial step is the breaking of one C—C bond, and occurs when that bond is highly vibrationally excited. In the collision leading to excitation, however, the excess energy is shared among all the bonds and the rotation of the molecule, and so the isomerization occurs only after the energy accumulates in the critical bond. The existence of this delay suggests that we should distinguish between the **energized molecule** A^*, where the excess energy is dispersed over many modes, and the **activated state** $A^\dagger$, where the exicitation is specific and the molecule is poised for reaction. The unimolecular part of the mechanism therefore ought to be modified to

$$A^* \rightarrow A^\dagger \rightarrow P$$

with a rate constant for each step. The values of the various rate constants are related to the numbers and frequencies of the vibrational modes available by the **Rice–Ramsperger–Kassel theory** (RRK theory) of unimolecular reactions, and by the more sophisticated **RRKM theory** (where M is Marcus), which also takes into account the effects of molecular rotation. Descriptions of these theories will be found in the books mentioned in *Further reading*.

Further information: reaction rates

The most elegant procedure for expressing the rate of a reaction is to use the extent of reaction ξ (Section 9.1). We express the chemical reaction in the general form introduced in Section 2.7:

$$0 = \sum_J v_J S_J \tag{A1}$$

When the extent of reaction increases by $d\xi$, the amount of component J changes by

$$dn_J = v_J\, d\xi$$

We now define the **rate of conversion** as the rate of change of the extent of reaction, and write

$$\dot{\xi} = \frac{d\xi}{dt}$$

Since $dn_J = v_J\, d\xi$, it follows that

$$\dot{\xi} = \frac{1}{v_J}\frac{dn_J}{dt} \tag{A2}$$

The same numerical value for the rate is obtained whichever component we select, and there is a single rate of conversion for a reaction (for a given composition and temperature of the reaction mixture).

The rate of conversion refers to the amounts of the substances present, not their concentrations. In principle, this is another advantages of using $\dot{\xi}$ because in systems where the volume changes during the reaction the concentration might change even though the numbers of molecules of each kind remain the same. However, in practice, most kinetic studies are made on systems at constant volume, and the measurement of rates in terms of concentrations is unambiguous and is the approach normally employed.

When the volume of the system is constant (at V) the rate of conversion is directly related to the **rate of reaction** v, which is defined as

$$v = \frac{1}{V}\frac{d\xi}{dt} \tag{A3}$$

The rate of reaction also has a single value for a given reaction (under specified conditions). The rate of reaction v should be carefully distinguished from the rates of formation and consumption v_J of individual substances, where

$$v_J = \frac{d[J]}{dt} \quad \text{if J is a product}$$

$$v_J = \frac{-d[J]}{dt} \quad \text{if J is a reactant}$$

The relation between these various rates is

$$v = \frac{1}{V v_J}\frac{dn_J}{dt} = \frac{1}{v_J}\frac{d[J]}{dt}$$

Therefore, for the rate of formation of J,

$$v_J = v_J \times v \tag{A4}$$

Since we can measure v_J experimentally by following the concentration of substance J, we can easily obtain the rate of reaction v and, if desired, the rate of conversion $\dot{\xi}$.

Further reading

K. J. Laidler, *Chemical kinetics*. Harper & Row, New York (1987).

J. Nicholas, *Chemical kinetics*. Harper & Row, New York (1976).

I. W. M. Smith, *Kinetics and dynamics of elementary gas reactions*. Butterworths, London (1980).

A. Maccoll, Homogeneous gas phase reactions. In *Techniques of chemistry* (ed. E. S. Lewis) VIA, 47, Wiley-Interscience, New York (1974).

J. F. Bunnett, Kinetics in solution. In *Techniques of chemistry* (ed. E. S. Lewis) VIA, 129, Wiley-Interscience, New York (1974).

C. H. Bamford and C. F. Tipper (ed.), *Comprehensive chemical kinetics*. Elsevier, Amsterdam (1969 *et seq.*)

Fast reactions

J. N. Bradley, *Fast reactions*. Clarendon Press, Oxford (1974).

G. Porter and M. A. West, Flash photolysis. In *Techniques of chemistry* (ed. G. G. Hammes) VIB, 367, Wiley-Interscience, New York (1974).

B. B. Chance, Rapid flow methods. In *Techniques of chemistry* (ed. G. G. Hammes) VIB, 5, Wiley-Interscience, New York (1974).

G. G. Hammes, Temperature-jump methods. In *Techniques of chemistry* (ed. G. G. Hammes) VIB, 147, Wiley-Interscience, New York (1974).

W. Knoche, Pressure-jump methods. In *Techniques of chemistry* (ed. G. G. Hammes) VIB, 187, Wiley-Interscience, New York (1974).

Data

S. W. Benson and H. E. O'Neal, *Kinetic data on gas phase unimolecular reactions.* NSRDS-NBS-21, US Department of Commerce, Washington (1970).

J. A. Kerr and S. J. Moss, *Handbook of bimolecular and termolecular gas reactions.* I and II, CRC Press, Boca Raton (1981).

J. A. Kerr and R. M. Drew, *Handbook of bimolecular and termolecular gas reactions.* IIIA and IIIB, CRC Press, Boca Raton (1987).

Exercises

26.1 The rate of the reaction $A + 2B \rightarrow 3C + D$ was reported as $1.0 \, M \, s^{-1}$. State the rates of formation and consumption of the participants.

26.2 The rate of formation of C in the reaction $2A + B \rightarrow 2C + 3D$ is $1.0 \, M \, s^{-1}$. State the reaction rate, and the rates of formation and consumption of A, B, and D.

26.3 The rate law for the reaction in Exercise 26.1 was found to be $v = k[A][B]$. What are the units of k? Express the rate law in terms of the rates of formation and consumption of (a) A, (b) C.

26.4 The rate law for the reaction in Exercise 26.2 was reported as $d[C]/dt = k[A][B][C]$. Express it in terms of the reaction rate; what are the units for k in each case?

26.5 The rate constant for the first-order decomposition of N_2O_5 in the reaction $2N_2O_5 \rightarrow 4NO_2 + O_2$ is $k = 3.38 \times 10^{-5} \, s^{-1}$ at 25°C. What is the half life of N_2O_5? What will be the partial pressure of N_2O_5, initially 500 Torr, (a) 100 s, (b) 100 min after initiation of the reaction?

26.6 If the rate laws are expressed with (a) concentrations in M, (b) pressures in atmospheres, what are the units of the second-order and third-order rate constants?

26.7 The half-life for the (first-order) radioactive decay of ^{14}C is 5730 y (it emits β rays with an energy of 0.16 MeV). An archaeological sample contained wood that had only 72 per cent of the ^{14}C found in living trees. What is its age?

26.8 One of the hazards of nuclear explosions is the generation of ^{90}Sr and its subsequent incorporation in place of calcium in bones. This nuclide emits β rays of energy 0.55 MeV, and has a half-life of 28.1 y. Suppose 1.00 μg was absorbed by a newly born child. How much will remain after (a) 18 y, (b) 70 y if none is lost metabolically?

26.9 The second-order rate constant for the reaction

$$CH_3COOC_2H_5(aq) + OH^-(aq)$$
$$\rightarrow CH_3CO_2^-(aq) + CH_3CH_2OH(aq)$$

is $0.11 \, M^{-1} \, s^{-1}$. What is the concentration of ester after (a) 10 s, (b) 10 min when ethyl acetate is added to sodium hydroxide so that the initial concentrations are $[NaOH] = 0.050 \, M$ and $[CH_3COOC_2H_5] = 0.100 \, M$?

26.10 A reaction $A + 2B \rightarrow P + Q$ obeys the rate law $v = k[A][B]$ with $k = 3.67 \times 10^{-3} \, M^{-1} \, s^{-1}$. At the start of the reaction, 0.255 mol of A is mixed with 0.605 mol of B in 1.70 L of solvent. Calculate the initial values of $d[A]/dt$, $d[B]/dt$, $d[P]/dt$, dn_B/dt, and the initial rate of reaction.

26.11 A substance decomposes according to the reaction $2A \rightarrow P$ with a second-order rate law and $k = 2.62 \times 10^{-3} \, M^{-1} \, s^{-1}$. What is the half life of A when $[A]_0 = 1.70 \, M$?

26.12 A reaction $2A \rightarrow P$ has a second order rate law with $k = 3.50 \times 10^{-4} \, M^{-1} \, s^{-1}$. Calculate the time required for the concentration of A to change from $0.260 \, M$ to $0.011 \, M$.

26.13 The composition of a liquid phase reaction $2A \rightarrow B$ was followed by a spectrophotometric method with the following results:

t/min	0	10	20	30	40	∞
$[B]/M$	0	0.089	0.153	0.200	0.230	0.312

Determine the order of the reaction and its rate constant.

26.14 The rate constant for the decomposition of a certain substance is $2.80 \times 10^{-3} \, M^{-1} \, s^{-1}$ at 30°C and $1.38 \times 10^{-2} \, M^{-1} \, s^{-1}$ at 50°C. Evaluate the Arrhenius parameters of the reaction.

26.15 The reaction $2H_2O_2(aq) \rightarrow 2H_2O(l) + O_2(g)$ is catalysed by Br^- ions. If the mechanism is

$$H_2O_2(aq) + Br^-(aq) \rightarrow H_2O(l) + BrO^-(aq) \qquad \text{(slow)}$$

$$BrO^-(aq) + H_2O_2(aq) \rightarrow H_2O(l) + O_2(g) + Br^-(aq) \qquad \text{(fast)}$$

give the order of the reaction with respect to the various participants.

26.16 The reaction mechanism

$$A_2 \rightleftharpoons 2A \quad \text{(fast)}$$

$$A + B \rightarrow P \quad \text{(slow)}$$

involves an intermediate A. Deduce the rate law for the reaction.

26.17 Consider the following mechanism for renaturation of a double helix from its strands A and B:

$$A + B \rightleftharpoons \text{unstable helix} \qquad \text{(fast)}$$

$$\text{Unstable helix} \rightarrow \text{stable double helix} \quad \text{(slow)}$$

Derive the rate equation for the formation of the double helix and express the rate constant of the renaturation reaction in terms of the rate constants of the individual steps.

26.18 Show that $t_{1/2} \propto 1/[A]_0^{n-1}$ for a reaction that is nth-order in A.

26.19 The enzyme-catalysed conversion of a substrate at 25°C has a Michaelis constant of 0.035 M. The rate of the reaction is $1.15 \times 10^{-3} \text{ M s}^{-1}$ when the substrate concentration is 0.110 M. What is the maximum velocity of this enzymolysis?

26.20 Find the condition for which the reaction rate of an enzymolysis that follows Michaelis–Menten kinetics is half its maximum value.

26.21 The effective rate constant for a gaseous reaction which has a Lindemann–Hinshelwood mechanism is $2.50 \times 10^{-4} \text{ s}^{-1}$ at 1.30 kPa (the partial pressure of the reactant) and $2.10 \times 10^{-5} \text{ s}^{-1}$ at 12 Pa. Calculate the rate constant for the activation step in the mechanism.

26.22 The pK_a of NH_3 is 9.25 at 25°C. The rate constant at 25°C for the reaction of NH_4^+ and OH^- to form aqueous NH_3 is $4.0 \times 10^{10} \text{ M}^{-1} \text{ s}^{-1}$. Calculate the rate constant for proton transfer to NH_3. What relaxation time would be observed if a temperature jump were applied to a solution of 0.15 M $NH_3(aq)$ at 25°C?

26.23 The equilibrium $A \rightleftharpoons B + C$ at 25°C is subjected to a temperature jump which slightly increases the concentrations of B and C. The measured relaxation time is 3.0 μs. The equilibrium constant for the system is 2.0×10^{-16} at 25°C, and the equilibrium concentrations of B and C at 25°C are both 2.0×10^{-4} M. Calculate the rate constants for the forward and reverse reactions.

26.24 Show that the reaction $A \rightleftharpoons B + C$, first-order forwards, second-order backwards, relaxes exponentially for small displacements from equilibrium. Find an expression for the relaxation time in terms of k_1 and k_2.

Problems

Numerical problems

26.1 In an experiment to investigate the stability of substituted allyl radicals (A. B. Trenwith, *J. chem. Soc. Faraday trans.* I, 1737 (1973)) the rate of formation of water in the reaction

$$CH_3CH(OH)CH = CH_2 \rightarrow H_2O + CH_2 = CHCH = CH_2$$

was monitored. At 810 K the results were as follows:

t/min	0.5	1.0	1.5	2.0	2.5	∞
V/cm^3	1.0	1.4	1.6	1.7	1.8	2.0

V is the volume of water produced. Find the order and rate coefficient. The C_4H_6 did not appear to form as rapidly as the water: suggest a reason.

26.2 The experiment described in Problem 26.1 was repeated at several temperatures and the following values of the rate constant were obtained:

T/K	773.5	786	797.5	810	810	824	834
k/units	1.63	2.95	4.19	8.13	8.19	14.9	22.2

What is the activation energy and pre-exponential factor A for the reaction? (Decide on the units and magnitude of k by referring to Problem 26.1.)

26.3 Cyclopropane isomerizes into propene when heated to 500°C in the gas phase. The extent of conversion for various initial pressures has been followed by gas chromatography by allowing the reaction to proceed for a time with various initial pressures:

p_0/Torr	200	200	400	400	600	600
t/s	100	200	100	200	100	200
p/Torr	186	173	373	347	559	520

where p_0 is the initial pressure and p is the final pressure of cyclopropane. What is the order and rate constant for the reaction under these conditions?

26.4 The radioactive decay of a family of nuclei was specified in Section 26.8. Calculate the abundance of the nuclides as a function of time, and plot the results as a graph.

26.5 The addition of hydrogen halides to alkenes has played a fundamental role in the investigation of organic reaction mechanisms. In one study (M. J. Haugh and D. R. Dalton, *J. Amer. chem. Soc.*, **97**, 5674 (1975)), high pressures of hydrogen chloride (up to 25 atm) and propene (up to 5 atm) were examined over a range of temperatures and the amount of 2-chloropropane formed was determined by NMR. Show that if the reaction $A + B \rightarrow P$ proceeds for a short time Δt, the concentration of product follows $[P]/[A] = k[A]^{m-1}[B]^n \Delta t$ if the reaction is mth-order in A and nth-order in B. In a series of runs the ratio of [chloropropane] to [propene] was independent of [propene] but the ratio of [chloropropane] to [HCl] for constant amounts of propene depended on [HCl]. For $\Delta t \approx 100$ h (which is short on the time scale of the reaction) the latter ratio rose from zero to 0.05, 0.03, 0.01 for $p(HCl) = 10$ atm, 7.5 atm, 5.0 atm. What are the orders of the reaction with respect to each reactant?

26.6 Show that the following mechanism can account for the rate law of the reaction in Problem 26.5:

$$2HCl \rightleftharpoons (HCl)_2 \qquad K_1$$

$$HCl + CH_3CH{=}CH_2 \rightleftharpoons complex \qquad K_2$$

$$(HCl)_2 + complex \rightarrow CH_3CHClCH_3 + 2HCl \quad k, \text{ slow}$$

What further tests could you apply to verify this mechanism?

26.7 In the experiments described in Problems 26.5 and 26.6 an inverse temperature dependence of the reaction rate was observed, the overall rate of reaction at 70°C being roughly one-third that at 19°C. Estimate the apparent activation energy and the activation energy of the rate determining step given that the enthalpies of the two equilibria are both of the order of -14 kJ mol^{-1}.

26.8 The second-order rate constants for the reaction of oxygen atoms with aromatic hydrocarbons have been measured (R. Atkinson and J. N. Pitts, *J. phys. Chem.*, **79**, 295 (1975)). In the reaction with benzene the rate constants are 1.44×10^7 at 300.3 K, 3.03×10^7 at 341.2 K, and 6.9×10^7 at 392.2 K, all in $M^{-1} s^{-1}$. Find the pre-exponential factor and activation energy of the reaction.

26.9 In a study of the autoxidation of hydroxylamine (M. N. Hughes, H. G. Nicklin, and K. Shrimanker, *J. chem. Soc.* A, 3845 (1971)), the rate constant k_{obs} in the rate equation $-d[NH_2OH]/dt = k_{obs}[NH_2OH][O_2]$ was found to have the following temperature-dependence:

$\theta/°C$	0	10	15	25	34.5
$10^4 k_{obs}/(M^{-1} s^{-1})$	0.237	0.680	1.02	2.64	5.90

Evaluate the activation energy of the reaction. This analysis is taken further in Problem 26.10.

26.10 The kinetics of autoxidation of hydroxylamine in the presence of ethylenediaminetetraacetic acid have been studied in the hydroxide ion concentration range 0.5–3.2 M. The reaction proceeds by the mechanism

$$NH_2OH + OH^- \rightarrow NH_2O^- + H_2O \quad \text{(fast)}$$

$$NH_2O^- + O_2 \rightarrow products \quad \text{(slow)}$$

with the rate laws

$$\frac{-d[(NH_2OH)]}{dt} = k_{obs}[(NH_2OH)][O_2]$$

$$\frac{-d[(NH_2OH)]}{dt} = k[NH_2O^-][O_2]$$

where $[(NH_2OH)]$ denotes the total hydroxylamine concentration, $[NH_2OH] + [NH_2O^-]$, and $k = k_{obs}/f$, where f is the fraction of hydroxylamine present as NH_2O^-. Show that a plot of $1/k_{obs}$ against $[H^+]$ should give a straight line, and that the acidity constant of NH_2OH can be determined from the slope. Find pK_a for the hydroxylamine from the following data:

$[OH^-]/M$	0.50	1.00	1.6	2.4
$10^4 k_{obs}/(M^{-1} s^{-1})$	2.15	2.83	3.32	3.54

26.11 In Problem 26.3 the isomerization of cyclopropane over a limited pressure range was examined. If the Lindemann mechanism of first-order reactions is to be tested we also need data at low pressures. These have been obtained (H. O. Pritchard, R. G. Sowden, and A. F. Trotman-Dickenson, *Proc. R. Soc.*, **A217**, 563 (1953)):

$p/$Torr	84.1	11.0	2.89	0.569	0.120	0.067
$10^4 k_{eff}/s^{-1}$	2.98	2.23	1.54	0.857	0.392	0.303

Test the Lindemann theory with these data.

26.12 The initial rate of O_2 production by the action of an enzyme on a substrate was measured for a range of substrate concentrations; the data are below. Evaluate the Michaelis constant for the reaction.

$[S]/M$	0.050	0.017	0.010	0.005	0.002
rate/(mm³ min⁻¹)	16.6	12.4	10.1	6.6	3.3

26.13 A certain second-order gas phase reaction has the form $2A \rightarrow B$. Find an expression for the time-dependence of the total pressure of the reacting system. Let the initial pressure, when no B is present, be p_0; then plot p/p_0 as a function of $x = p_0kt$. What time is needed for the pressure to fall half way toward its final value? The composition of a gas phase reaction of the form $2A \rightarrow B$ was monitored by measuring the total pressure as a function of time. The following are the results:

t/s	0	100	200	300	400
$p/$Torr	400	322	288	268	256

Find the order of the reaction and the rate constant.

Theoretical problems

26.14 The equilibrium $A \rightleftharpoons B$ is first-order in both directions. Derive an expression for the concentration of A as a function of time when the initial concentrations of A and B are $[A]_0$ and $[B]_0$. What is the final composition of the system?

26.15 Integrate the second-order rate law $v = k[A][B]$ for a reaction of stoichiometry $2A + 3B \rightarrow P$.

26.16 Derive the integrated form of a third-order rate law $v = k[A]^2[B]$ in which the stoichiometry is $2A + B \rightarrow P$ and the reactants are initially present in (a) their stoichiometric proportions, (b) with B present initially in twice the amount.

26.17 Derive an expression for the half life of the reaction in Problem 27.16 defined (a) as the time for A to halve its initial concentration, (b) defined as the time for B to halve its concentration, (c) defined as the time for the extent of reaction to reach $\xi = 0.5$ mol.

26.18 Show that the ratio $t_{1/2}/t_{3/4}$, where $t_{1/2}$ is the half-life and $t_{3/4}$ is the time for the concentration of A to decrease to $\frac{3}{4}$ of its initial value (implying $t_{3/4} < t_{1/2}$) can be written as a function of n alone, and so it can be used as a rapid assessment of the order of a reaction.

The kinetics of complex reactions

Check-list of key ideas

1. The structure of *chain reactions* in terms of *initiation*, *propagation*, *branching*, *retardation*, and *termination* (Section 27.1).

2. The *rate laws* of chain reactions using the steady-state assumption (Section 27.1).

3. The origin of *thermal explosions* and of *chain-branching explosions* (Section 27.2).

4. The processes that may follow *photoexcitation* (Box 27.1).

5. The *primary quantum yield* and the *overall quantum yield* of photochemical reactions (Section 27.3).

6. The formulation of *photochemical rate laws* (eqn 3).

7. The principles of *laser isotope separation* and *photosensitization* (Section 27.3).

8. The rate law governing *chain polymerization* and the characteristics of the molar mass distribution to which it leads (Section 27.4).

9. The *kinetic chain length* in a polymerization process (eqn 5) and the average molar mass (Example 27.3).

10. The rate law governing *stepwise polymerization* and the characteristics of the molar mass distribution to which it leads (Section 27.5).

11. The calculation of the *degree of polymerization* (Example 27.4) and the growth of the *number-average* and *mass-average molar masses* as polymerization proceeds (eqns 14 and 15).

12. The effect of *homogeneous catalysis* on the rate of reaction (Section 27.6).

13. The role of *autocatalysis* in chemical kinetics (Section 27.7 and Example 27.5).

14. The *Lotka–Volterra mechanism* of chemical oscillation (Section 27.8).

15. The *brusselator* and the *oregonator* as examples of chemical oscillation, and the property of *bistability* (Section 27.8).

Many reactions take place by mechanisms that involve several elementary steps, and some take place at a useful rate only if a catalyst is present. In this chapter we see how to develop the ideas introduced in Chapter 26 to deal with these special kinds of reactions. One very important example of both catalysis and a chain mechanism is the process of polymerization, and we shall see how the distribution of molar masses of polymers may be controlled kinetically. We shall also see that when certain reactions are far from equilibrium, the concentrations of the intermediates and products may oscillate with time.

Chain reactions

Many gas-phase reactions and liquid-phase polymerization reactions are **chain reactions**. In a chain reaction, an intermediate produced in one step generates a reactive intermediate in a subsequent step, then that intermediate generates another reactive intermediate, and so on.

27.1 The structure of chain reactions

The intermediates responsible for the propagation of a chain are called **chain carriers**. In a radical chain reaction the chain carriers are radicals. Ions may also propagate chains, and in nuclear fission the chain carriers are neutrons.

The classification of reaction steps

The first chain carriers are formed in the **initiation step** of the reaction. For example, Cl atoms are formed by the dissociation of Cl_2 molecules either as a result of vigorous intermolecular collisions in a thermolysis reaction or as a result of absorption of a photon in a photolysis reaction. The chain carriers produced in the initiation step attack other reactant molecules in the **propagation steps**, and each attack gives rise to a new chain carrier. An example is the attack of a methyl radical on ethane:

$$\cdot CH_3 + CH_3CH_3 \rightarrow CH_4 + \cdot CH_2CH_3$$

(The dot signifies the unpaired electron and marks the radical.) In some cases the attack results in the production of more than one chain carrier. An example of such a **branching step** is

$$\cdot O\cdot + H_2O \rightarrow HO\cdot + HO\cdot$$

where the attack of one O atom on an H_2O molecule forms two $\cdot OH$ radicals (an O atom has the configuration $[He]2s^2 2p_x^2 2p_y^1 2p_z^1$, with two unpaired electrons).

The chain carrier might attack a product molecule formed earlier in the reaction. Since this attack reduces the net rate of formation of product, it is called a **retardation step**. For example, in a photochemical reaction in which HBr is formed from H_2 and Br_2, an H atom might attack an HBr molecule, leading to H_2 and a Br atom:

$$\cdot H + HBr \rightarrow H_2 + \cdot Br$$

Retardation does not end the chain, because one radical ($\cdot H$) gives rise to another ($\cdot Br$), but it does deplete the concentration of the product.

Elementary reactions in which radicals combine and end the chain are called **termination steps**, as in

$$CH_3CH_2\cdot + \cdot CH_2CH_3 \rightarrow CH_3CH_2CH_2CH_3$$

In an **inhibition step** radicals are removed other than by chain termination, such as by reaction with the walls of the vessel or with foreign radicals:

$$CH_3CH_2\cdot + \cdot R \rightarrow CH_3CH_2R$$

The NO molecule has an unpaired electron and is a very efficient chain inhibitor. The observation that a reaction is quenched when NO is introduced is a good indication that a radical chain mechanism is in operation.

The rate laws of chain reactions

A chain reaction often leads to a complicated rate law (but not always). As a first example, consider the thermal reaction between H_2 and Br_2. The overall reaction and the observed rate law are

$$H_2(g) + Br_2(g) \rightarrow 2HBr(g) \qquad \frac{d[HBr]}{dt} = \frac{k[H_2][Br_2]^{3/2}}{[Br_2] + k'[HBr]}$$

The complexity of the rate law suggests that a complicated mechanism is involved, and the following radical chain mechanism has been proposed:

(a) Initiation:

$$Br_2 \rightarrow 2Br\cdot \qquad v = k_a[Br_2]$$

(At low pressures this elementary reaction is bimolecular and second order in Br_2.)

(b) Propagation:

$$Br\cdot + H_2 \rightarrow HBr + H\cdot \qquad v = k_b[Br][H_2]$$
$$H\cdot + Br_2 \rightarrow HBr + Br\cdot \qquad v = k_b'[H][Br_2]$$

(c) Retardation:

$$H\cdot + HBr \rightarrow H_2 + Br\cdot \qquad v = k_c[H][HBr]$$

(d) Termination:

$$Br\cdot + \cdot Br + M \rightarrow Br_2 + M \qquad v = k_d[Br]^2$$

(The third body M removes the energy of recombination; the constant concentration of M has been absorbed into the rate constant.) Other possible termination steps include the recombination of H atoms to form H_2 and the combination of H and Br atoms; however, it turns out that only Br atom recombination is important. The net rate of formation of the product HBr is

$$\frac{d[HBr]}{dt} = k_b[Br][H_2] + k_b'[H][Br_2] - k_c[H][HBr] \qquad (1)$$

There are now two ways to proceed. One is to solve the set of differential equations numerically, and a sketch of the procedures involved is given in *Further information 1* at the end of the chapter. Thus the concentrations of all the participants in the reaction may be computed and the rate constants determined by varying them until a best fit is obtained. An alternative procedure (the only one available until recently) is to look for approximate solutions and see if they agree with the empirical rate law. The latter route involves making the steady-state approximation for the concentrations of any intermediates ($\cdot$H and $\cdot$Br in the present case) and setting their net rates of change of concentration equal to zero.

Example 27.1: *Deriving the rate equation of a chain reaction*

Derive the rate law for the formation of HBr according to the mechanism given above.

Answer. We write down the net rates of formation of the intermediates ($\cdot$H and $\cdot$Br), and use the steady-state approximation:

$$\frac{d[H]}{dt} = k_b[Br][H_2] - k_b'[H][Br_2] - k_c[H][HBr] = 0$$

$$\frac{d[Br]}{dt} = 2k_a[Br_2] - k_b[Br][H_2] + k_b'[H][Br_2] + k_c[H][HBr] - 2k_d[Br]^2 = 0$$

The steady-state concentrations of the intermediates are therefore

$$[Br] = \left\{ \frac{k_a[Br_2]}{k_d} \right\}^{1/2}$$

$$[H] = \frac{k_b(k_a/k_d)^{1/2}[H_2][Br_2]^{1/2}}{k_b'[Br_2] + k_c[HBr]}$$

When we substitute these concentrations into eqn 1 we obtain

$$\frac{d[HBr]}{dt} = \frac{2k_b(k_a/k_d)^{1/2}[H_2][Br_2]^{3/2}}{[Br_2] + (k_c/k_b')[HBr]}$$

This equation is the same as the empirical law, and we can identify the two empirical rate constants as

$$k = 2k_b \left(\frac{k_a}{k_d} \right)^{1/2} \qquad k' = \frac{k_c}{k_b'}$$

Exercise. Deduce the rate law when the initiation step is bimolecular in Br_2.

$$\left[\frac{d[HBr]}{dt} = \frac{2k_b(k_a/k_d)^{1/2}[H_2][Br_2]^2}{[Br_2] + (k_c/k_b')[HBr]} \right]$$

The example shows that the observed rate law is reproduced by the mechanism. That used to be essentially the end of the calculation, but we can now make use of computers at this stage even though we did not use them to find the rate law. Thus, we can go on to integrate the approximate rate law numerically as explained in *Further information 1*, and hence predict the time dependence of the HBr concentration. Some typical results are plotted in Fig. 27.1: they show how the rate of formation of products is slow when the inhibition step is important.

Some chain reactions turn out to give rise to simple rate laws, which underlines yet again the care that must be taken when analysing kinetic data. For example, the dehydrogenation of ethane to ethene follows

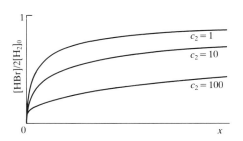

Fig. 27.1 The numerical integration of the HBr rate law, Example 27.1, can be used to explore how the concentration of HBr changes with time. These runs began with stoichiometric proportions of H_2 and Br_2; $x = c_1 t$; $c_1 = k/(2\sqrt{[H_2]_0})$, $c_2 = 2k' - 1$.

first-order kinetics:

$$CH_3CH_3(g) \rightarrow CH_2{=}CH_2(g) + H_2(g) \qquad v = k[CH_3CH_3]$$

and the 'obvious' interpretation is that the reaction occurs by a Lindemann–Hinshelwood mechanism (Section 26.9). Further experiments, however, show that methyl and ethyl radicals occur during the reaction, and so a radical chain mechanism is involved. It has been proposed that the following **Rice–Herzfeld mechanism** is operative:

(a) Initiation:

$$CH_3CH_3 \rightarrow H_3C{\cdot} + {\cdot}CH_3 \qquad v = k_a[CH_3CH_3]$$

(b) Propagation:

$$CH_3CH_3 + {\cdot}CH_3 \rightarrow CH_3CH_2{\cdot} + CH_4 \qquad v = k_b[CH_3][CH_3CH_3]$$
$$CH_3CH_2{\cdot} \rightarrow CH_2{=}CH_2 + H{\cdot} \qquad v = k_b'[CH_3CH_2]$$
$$CH_3CH_3 + {\cdot}H \rightarrow CH_3CH_2{\cdot} + H_2 \qquad v = k_b''[H][CH_3CH_3]$$

(c) Termination:

$$CH_3CH_2{\cdot} + {\cdot}H \rightarrow CH_3CH_3 \qquad v = k_c[H][CH_3CH_2]$$

When the steady-state approximation is used and certain conditions about the sizes of the rate constants are fulfilled (Problem 27.5), the rate law becomes

$$v = k[CH_3CH_3] \quad \text{with} \quad k = \left(\frac{k_a k_b' k_b''}{k_c}\right)^{1/2} \qquad (2)$$

in agreement with the observed first-order kinetics.

27.2 Explosions

A **thermal explosion** is due to the rapid increase of reaction rate with temperature. If the energy of an exothermic reaction cannot escape, the temperature of the reaction system rises, and the reaction goes faster. The acceleration of the rate results in a faster rise of temperature, and so the reaction goes even faster ... catastrophically fast. A **chain-branching explosion** may occur when there are chain branching steps in a reaction, for then the number of chain centres grows exponentially and the rate of reaction may cascade into an explosion.

An example of both types of explosion is provided by the reaction between hydrogen and oxygen:

$$2H_2(g) + O_2(g) \rightarrow 2H_2O(g)$$

Although the net reaction is very simple, the mechanism is very complex and has not yet been fully elucidated. It is known that a chain reaction is involved, and that the chain carriers include ${\cdot}H$, ${\cdot}O$, ${\cdot}OH$, and ${\cdot}O_2H$. Some steps are:

Initiation: $\qquad H_2 + O_2 \rightarrow {\cdot}O_2H + {\cdot}H$

Propagation:
$$H_2 + \cdot O_2H \rightarrow \cdot OH + H_2O$$
$$H_2 + \cdot OH \rightarrow \cdot H + H_2O$$
$$\cdot O_2 \cdot + \cdot H \rightarrow \cdot O \cdot + \cdot OH \quad \text{(branching)}$$
$$\cdot O \cdot + H_2 \rightarrow \cdot OH + \cdot H \quad \text{(branching)}$$

The last two branching steps can lead to a chain-branching explosion.

The occurrence of an explosion depends on the temperature and pressure of the system, and the explosion regions for the reaction are shown in Fig. 27.2. At very low pressures the system is outside the explosion region and the mixture reacts smoothly. At these pressures the chain carriers produced in the branching steps can reach the walls of the container where they combine (with an efficiency that depends on the composition of the walls). Increasing the pressure (along the broken line in the illustration) takes the system through the first explosion limit (if the temperature is greater than about 730 K). The mixture then explodes because the chain carriers react before reaching the walls and the branching reactions are explosively efficient. The reaction is smooth when the pressure is above the second explosion limit. The concentration of molecules in the gas is then so great that the radicals produced in the branching reaction combine in the body of the gas, and reactions such as $O_2 + \cdot H \rightarrow \cdot O_2H$ can occur. Recombination reactions like this are facilitated by three-body collisions because the third body (M) can remove the excess energy:

$$O_2 + \cdot H + M \rightarrow \cdot O_2H + M^*$$

At low pressures three-particle collisions are unimportant and recombination is much slower. At higher pressures, when three-particle collisions are important, the explosive propagation of the chain by the radicals produced in the branching step is partially quenched because $\cdot O_2H$ is formed in place of O_2: the reaction is essentially the reverse of a branching step. If the pressure is increased to above the **third explosion limit** the reaction rate increases so much that a thermal explosion occurs.

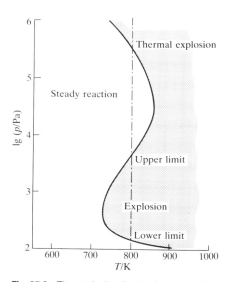

Fig. 27.2 The explosion limits of the $H_2 + O_2$ reaction. In the explosive regions the reaction proceeds explosively when heated homogeneously.

27.3 Photochemical reactions

Many reactions can be initiated by the absorption of light. The most important of all are the photochemical processes that capture the sun's radiant energy. Some of these reactions lead to the heating of the atmosphere during the daytime by absorption in the ultraviolet region as a result of reactions like those depicted in Fig. 27.3. Others include the absorption of red and blue light by chlorophyll and the subsequent use of the energy to bring about the synthesis of carbohydrates from carbon dioxide and water. Without photochemical processes the world would be simply a warm, sterile, rock. A summary of the processes that can occur following photochemical excitation is given in Box 27.1.

Quantum yield

A molecule acquires enough energy to react by absorbing photons. The **Stark–Einstein law** states that one photon is absorbed by each molecule responsible for the primary photochemical process. However, the law is not valid under all conditions, for in a laser beam the intensity is so great that a

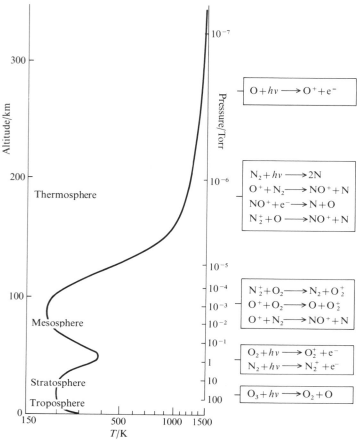

Fig. 27.3 The temperature profile through the atmosphere and some of the reactions that occur. The temperature peak at about 50 km is due to the absorption of solar radiation by the O_2 and N_2 ionization reactions.

Box 27.1 Photochemical processes

In the following, S denotes a singlet state and T a triplet state; A, B, and M are arbitrary.

Primary absorption: $S \rightarrow S^*$ followed by vibrational and rotational relaxation.

Physical processes:
  Fluorescence: $S^* \rightarrow S + h\nu$
  Collision-induced emission: $S^* + M \rightarrow S + M + h\nu$
  Stimulated emission: $S^* + h\nu \rightarrow S + 2h\nu$
  Intersystem crossing (ISC): $S^* \rightarrow T^*$
  Phosphorescence: $T^* \rightarrow S + h\nu$
  Internal conversion (IC): $S^* \rightarrow S'$
  Singlet electronic energy transfer: $S^* + S \rightarrow S + S^*$
  Energy pooling: $S^* + S^* \rightarrow S^{**} + S$

Triplet electronic energy transfer: $T^* + S \rightarrow S + T^*$
Triplet-triplet absorption: $T^* + h\nu \rightarrow T^{**}$

Ionization:
  Penning ionization: $A^* + B \rightarrow A + B^+ + e^-$
  Dissociative ionization: $A^* + B\text{—}C \rightarrow A\text{—}B^+ + C + e^-$
  Collisional ionization: $A^* + B \rightarrow A^+ + B + e^-$ (or B^-)
  Associative ionization: $A^* + B \rightarrow AB^+ + e^-$

Chemical processes
  Dissociation: $A\text{—}B^* \rightarrow A + B$
  Addition or insertion: $A^* + B \rightarrow AB$
  Abstraction or fragmentation: $A^* + B \rightarrow C + D$
  Isomerization: $A \rightarrow A'$
  Dissociative excitation: $A^* + C\text{—}D \rightarrow A + C^* + D$

single molecule may absorb several photons. We shall shortly see that a single molecule may absorb dozens of infrared photons from an intense laser beam.

Under normal conditions the Stark–Einstein law is valid, but it leaves open the possibility that, even though a reactant molecule absorbs a photon, it does not lead to products: there are many ways in which the excitation may be lost other than by dissociation or ionization. We therefore speak of the **primary quantum yield** ϕ, the number of reactant molecules producing specified primary products (atoms or ions, for instance) for each photon absorbed.

The primary product of photon absorption—a radical, a photoexcited molecule, or an ion—may be successful in initiating a process that leads to products. As a result of one successful initiation, many reactant molecules might be consumed. The **overall quantum yield** Φ is the number of reactant molecules that react for each photon absorbed. In the photolysis of HI, for example, the processes are

$$\text{HI} + h\nu \rightarrow \text{H} + \text{I}$$

$$\text{H} + \text{HI} \rightarrow \text{H}_2 + \text{I}$$

$$2\text{I} \rightarrow \text{I}_2$$

The overall quantum yield is 2 because the absorption of one photon leads to the destruction of two HI molecules. In a chain reaction Φ may be very large, and values of about 10^4 are common. In such cases the chain reaction acts as a chemical amplifier of the initial absorption step.

Example 27.2: *Using the quantum yield*

The overall quantum yield for the formation of ethene from 4-heptanone with 313 nm light is 0.21. How many molecules of the ketone per second, and what amount per second, are destroyed when the sample is irradiated with a 50 W, 313 nm source under conditions of total absorption?

Answer. Calculate the number of photons emitted by the lamp per second; all are absorbed (by assertion); the number of molecules destroyed per second is the number of photons absorbed multiplied by the quantum yield Φ. The energy of a 313 nm photon is

$$\frac{hc}{\lambda} = 6.35 \times 10^{-19}\,\text{J}$$

A 50 W source therefore generates photons at a rate

$$\frac{50\,\text{J s}^{-1}}{6.35 \times 10^{-19}\,\text{J}} = 7.9 \times 10^{19}\,\text{s}^{-1}$$

The number of 4-heptanone molecules destroyed per second is therefore 0.21 times this quantity, or $1.7 \times 10^{19}\,\text{s}^{-1}$, which corresponds to $2.8 \times 10^{-5}\,\text{mol s}^{-1}$.

Comment. The quantum yield depends on the wavelength of light used. One mole of photons is called an 'einstein'.

Exercise. The overall quantum yield for another reaction at 290 nm is 0.30. For what length of time must irradiation with a 100 W source continue in order to destroy 1.0 mol of molecules?   [3.8 h]

Photochemical rate laws

As an example of how to incorporate the photochemical activation step into a mechanism, consider the photochemical activation of the reaction

$$H_2(g) + Br_2(g) \rightarrow 2HBr(g)$$

In place of the first step in the thermal reaction we have

$$Br_2 \xrightarrow{h\nu} 2Br \qquad \upsilon = I_{abs}$$

where I_{abs} is the number of photons of the appropriate frequency absorbed per unit time per unit volume. It follows that I_{abs} should take the place of $k_a[Br_2]$ in the thermal reaction scheme, and so from Example 27.1 we can write

$$\frac{d[HBr]}{dt} = \frac{2k_b(1/k_d)^{1/2}[H_2][Br_2]I_{abs}^{1/2}}{[Br_2] + (k_c/k_b')[HBr]} \qquad (3)$$

Equation 3 predicts that the reaction rate should depend on the square root of the absorbed light intensity, which is confirmed experimentally.

Isotope separation

Laser radiation is so monochromatic that it may be absorbed by one molecule but not by an isotopically substituted version of the same molecule. The highly specific character of the absorption is utilized in **laser isotope separation**. In this technique, molecules containing different isotopes are irradiated with a weak monochromatic laser beam and simultaneously with an intense beam of infrared photons from a CO_2 laser. The absorption of the first few photons (from the weak beam) depends critically on the match between their frequency and the vibration-rotation energy separations of the molecule (which depend on the masses of the atoms, and hence differ between isotopic species). However, after the initial excitation has occurred, the photons from the intense laser may be absorbed, leading to the dissociation and reaction of the molecule and its fragments. Since the initial absorption is the doorway to the later dissociation, the isotopes can be separated. For example, irradiation of a sample of gaseous ICl with 605 nm light selectively excites the molecules containing ^{37}Cl atoms, and they react in collisions with bromobenzene molecules:

$$ICl \xrightarrow{h\nu} ICl^*, \quad \text{predominantly with } ^{37}Cl$$

$$C_6H_5Br + I^{37}Cl^* \rightarrow C_6H_5{}^{37}Cl + IBr$$

Photosensitization

The reactions of a molecule that does not absorb directly can be stimulated if another absorbing molecule is present because the latter may be able to transfer its energy during a collision. An example of this photosensitization is the reaction often used to generate atomic hydrogen, the irradiation of hydrogen gas containing a trace of mercury using 254 nm light from a mercury discharge lamp. The Hg atoms are excited by resonant absorption of the radiation, and then collide with H_2 molecules. Two reactions then take place:

$$Hg^* + H_2 \rightarrow Hg + 2H$$

$$Hg^* + H_2 \rightarrow HgH + H$$

The latter reaction, which is monitored by detecting the HgH spectrophoto-metrically, accounts for 67 per cent of the process (but 76 per cent when deuterium is used in place of hydrogen). It is the initiation step for other mercury photosensitized reactions, such as the synthesis of formaldehyde from carbon monoxide and hydrogen:

$$H + CO \rightarrow HCO$$

$$HCO + H_2 \rightarrow HCHO + H$$

$$2HCO \rightarrow HCHO + CO$$

Photosensitization also plays an important role in solution kinetics, and molecules containing the carbonyl chromophore (such as benzophenone, $C_6H_5COC_6H_5$) are often used to trap the incident light and to transfer it to some potentially reactive species.

Polymerization kinetics

In **chain polymerization** an activated monomer M attacks another mono-mer, links to it, then that unit attacks another monomer, and so on. The monomer is used up slowly through the reaction by linking to the growing chains. In chain polymerization high polymers are formed rapidly and, as we shall see in detail later, only the yield and not the molar mass of the polymer is increased by allowing long reaction times. In **step polymeriza-tion** any two monomers present in the reaction mixture can link together at any time, and growth is not confined to chains that are already forming. A consequence of the process is that monomers are removed early in the reaction and (as we shall see) the average molar mass of the product grows with time.

27.4 Chain polymerization

Chain polymerization results in the rapid growth of an individual polymer chain for each activated monomer. It commonly occurs by addition, often by a radical chain process, and examples include the addition polymeriza-tions of ethene, methyl methacrylate, and styrene.

The rate law for chain polymerization

There are three basic reaction steps in a chain polymerization process:
  (a) Initiation:

$$I \rightarrow 2R \cdot \qquad v = k_i[I]$$

$$M + \cdot R \rightarrow \cdot M_1 \quad \text{(fast)}$$

where I is the initiator and $\cdot M_1$ is a monomeric radical. We have shown a reaction in which a radical is produced, but in some polymerizations the initiation step leads to the formation of a cation or an anion chain carrier. The rate-determining step is the formation of the radicals $R \cdot$ by homolysis of the initiator, so the rate of initiation is equal to the v given above.

(b) Propagation:

$$M + \cdot M_1 \rightarrow \cdot M_2$$

$$M + \cdot M_2 \rightarrow \cdot M_3$$

$$\cdots\cdots\cdots\cdots$$

$$M + \cdot M_{n-1} \rightarrow \cdot M_n \qquad v = k_p[M][\cdot M]$$

Since this chain of reactions propagates quickly, the rate at which the total concentration of radicals grows is equal to the rate of the rate-determining initiation step, and we can write

$$\frac{d[\cdot M]}{dt} = 2\phi k_i[I]$$

ϕ is the yield of the initiation step, the fraction of radicals R· that successfully initiate a chain.

(c) Termination:

$$\cdot M_n + \cdot M_m \rightarrow M_{n+m}$$

If we suppose that the rate of termination is independent of the length of the chain, its rate law is

$$v = k_t[\cdot M]^2$$

and the rate of change of radical concentration by this process is

$$\frac{d[\cdot M]}{dt} = -2k_t[\cdot M]^2$$

In practice, other termination steps may intervene. Side reactions may also occur, such as **chain transfer**, in which the reaction

$$M + \cdot M_n \rightarrow \cdot M + M_n$$

initiates a new chain at the expense of the one currently growing.

The total radical concentration is approximately constant throughout the main part of the polymerization since the rate at which radicals are formed by initiation is approximately the same as the rate at which they are removed by termination. Hence, we can use the steady-state assumption to write

$$\frac{d[\cdot M]}{dt} = 2\phi k_i[I] - 2k_t[\cdot M]^2 = 0$$

The steady-state concentration of radical chains is therefore

$$[\cdot M] = \left\{ \frac{\phi k_i[I]}{k_t} \right\}^{1/2}$$

and the rate of propagation of the chains (the rate at which the monomer is consumed) is

$$\frac{d[M]}{dt} = -k_p[\cdot M][M]$$

Therefore,

$$\frac{d[M]}{dt} = -k_p \left(\frac{\phi k_i}{k_t} \right)^{1/2} [I]^{1/2}[M] \qquad (4)$$

That is, the overall rate of polymerization is proportional to the square root of the initiator concentration.

The kinetic chain length

The **kinetic chain length** v is a measure of the efficiency of the chain propagation mechanism. It is defined as the ratio of the number of monomer units consumed per active centre produced in the initiation step:

$$v = \frac{\text{number of monomer units consumed}}{\text{number of active centres produced}} \qquad (5)$$

The kinetic chain length is therefore equal to the ratio of the propagation and initiation rates:

$$v = \frac{\text{propagation rate}}{\text{initiation rate}}$$

Since the initiation rate is equal to the termination rate, we can write this expression as

$$v = \frac{k_p[M][M\cdot]}{2k_t[\cdot M]^2} = \frac{k_p[M]}{2k_t[\cdot M]}$$

When we substitute the steady-state expression for the radical concentration, we obtain

$$v = k[I]^{-1/2}[M] \qquad k = \frac{k_p}{2(\phi k_i k_t)^{1/2}} \qquad (6)$$

We see that the slower the initiation of the chain (the smaller the initiator concentration and the smaller the initiation rate constant), the greater the kinetic chain length.

Example 27.3: *Using the kinetic chain length*

Estimate the average number of units in a polymer produced by a chain mechanism in which termination occurs by combination of radicals.

Answer. Since termination occurs by the combination of two radicals, the number of monomers in a polymer molecule $\langle n \rangle$ produced by the reaction is the sum of the numbers in the two combining polymer chains. Since the latter are both equal to v (since by eqn 5 each initiation leads to a polymer of v monomer units), the number of monomers in a product molecule is

$$\langle n \rangle = 2v = 2k[I]^{-1/2}[M] \qquad (7)$$

with k given in eqn 6.

Comment. We see that the slower the initiation (as controlled by the rate constant k_i and the initiator concentration [I]), the longer the average chain length, and therefore the higher the molar mass of the polymer. Some of the consequences of molar mass for polymers were explored in Chapter 23: now we see how we can exercise kinetic control over them.

Exercise. Another termination mechanism is a disproportionation reaction of the form $M\cdot + \cdot M \rightarrow M + :M$. Calculate the polymer length.

$$[\langle n \rangle = k[I]^{-1/2}[M]]$$

27.5 Stepwise polymerization

Stepwise polymerization commonly proceeds by a **condensation reaction**, in which a small molecule (typically H_2O) is eliminated in each step. Stepwise polymerization is the mechanism of production of polyamides (nylon):

$$H_2N(CH_2)_6NH_2 + HOOC(CH_2)_4COOH$$
$$\rightarrow H_2N(CH_2)_6NHCO(CH_2)_4COOH + H_2O$$
$$\rightarrow H\text{---}[NH(CH_2)_6NHCO(CH_2)_4CO]_n\text{---}OH$$

Polyesters and polyurethanes are formed similarly (the latter without elimination). A polyester, for example, can be regarded as the outcome of the stepwise condensation of a hydroxyacid HO—M—COOH. We shall consider the formation of a polyester from such a monomer, and measure its progress in terms of the concentration of the —COOH groups (which we shall denote A) in the sample, since these gradually disappear as the condensation proceeds. Since the condensation reaction can occur between any appropriate monomers, chains of many different lengths can grow in the reaction mixture. We shall show how to use the reaction scheme to predict molar mass distributions.

The rate law of stepwise polymerization

The condensation can be expected to be overall second-order in the concentration of the —OH and —COOH (or A) groups:

$$\frac{d[A]}{dt} = -k[\text{---OH}][A]$$

However, since there is one —OH group for each —COOH group, this equation is the same as

$$\frac{d[A]}{dt} = -k[A]^2$$

We assume that the rate constant for the condensation is independent of the chain length, so k remains constant throughout the reaction. The solution of this rate law is given by eqn 6c of Section 26.3, and is

$$[A] = \frac{[A]_0}{1 + kt[A]_0} \tag{8}$$

The fraction p of —COOH groups that have condensed at time t is

$$p = \frac{[A]_0 - [A]}{[A]_0} \tag{9}$$

and so

$$p = \frac{kt[A]_0}{1 + kt[A]_0} \tag{10}$$

Example 27.4: *Calculating the degree of polymerization*

Find an expression for the growth with time of the degree of polymerization of a polymer formed by a stepwise process.

Answer. Suppose the current concentration of A groups is [A] and that initially their concentration was $[A]_0$. Since there is one A group per polymer molecule, the average number of monomers per polymer molecule will be

$$\langle n \rangle = \frac{[A]_0}{[A]} = \frac{1}{1-p}$$

For example, if there were initially 1000 A groups and there are now only 10, each polymer must be 100 units long on average. When we substitute the value for p given in eqn 10, we find

$$\frac{1}{1-p} = 1 + kt[A]_0$$

Therefore, the degree of polymerization grows as

$$\langle n \rangle = 1 + kt[A]_0$$

Comment. The length grows linearly with time, so the longer a stepwise polymerization proceeds, the higher the average molar mass of the product.

Exercise. Derive the expression for the time-dependence of p for a stepwise polymerization in which the reaction is acid-catalysed by the —COOH acid functional group. The rate law is $d[A]/dt = -k[A]^2[—OH]$.
$$[p = 1 - \alpha^{-1/2}, \ \alpha = 1 + 2kt[A]_0^2]$$

The statistics of polymerization

We can arrive at the same conclusion about the degree of polymerization as in Example 27.4 by considering the probability p that a group (e.g. —COOH) has formed a link to another molecule. At the start of the reaction $p = 0$; at the end $p = 1$.

The probability P_n that a polymer consists of n monomers (i.e. that it is an n-mer) is equal to the probability that it has $n - 1$ reacted —COOH groups and one unreacted —COOH group. The former probability is p^{n-1}; the latter probability is $1 - p$. Therefore, the total probability of finding an n-mer is

$$P_n = p^{n-1}(1-p) \tag{11}$$

Equation 11 is the key equation for predicting the molar masses of polymers from the kinetics of their formation. For example, we show in *Further information 2* at the end of the chapter that the degree of polymerization (the average number of monomers in a chain) is

$$\langle n \rangle = \frac{1}{1-p} \tag{12}$$

which is the same expression as that obtained from kinetics in Example 27.4 (Fig. 27.4). It is clear from the illustration that we need p very close to 1 in order to obtain polymers with high molar masses.

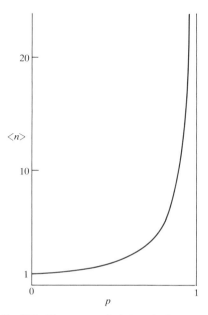

Fig. 27.4 The average chain length of a polymer as a function of the fraction of reacted monomers p. Note that p must be very close to 1 for the chains to be long.

Since p grows with time according to eqn 10, the degree of polymerization also grows with time:

$$\langle n \rangle = 1 + kt[A]_0 \tag{13}$$

Therefore the number-average molar mass increases with time as

$$\langle M \rangle_N = \langle n \rangle M_{mon} = (1 + kt[A]_0)M_{mon} \tag{14}$$

Equation 14 is the origin of the remark made earlier, that in stepwise polymerization (in contrast to chain polymerization), the molar mass of the product increases the longer the reaction proceeds.

We also show in *Further information 2* how to calculate the mass-average molar mass (Section 23.4) from p. The result obtained there is

$$\langle M \rangle_M = \frac{1+p}{1-p} \times M_{mon} \tag{15}$$

Hence, the ratio of the mass-average and number-average molar masses, which is a measure of the spread of molar masses in the sample (Section 23.4), is

$$\frac{\langle M \rangle_M}{\langle M \rangle_N} = 1 + p \tag{16}$$

The ratio increases towards 2 as p, and the average molar mass, increases with time.

Catalysis and oscillation

If the activation energy of a reaction is high, at normal temperatures only a small proportion of molecular encounters result in reaction. A catalyst lowers the activation energy of the reaction (often by providing an alternative path that avoids the slow, rate-determining step of the un-catalysed reaction), and results in a higher reaction rate at the same temperature. Catalysts can be very effective; for instance, the activation energy for the decomposition of hydrogen peroxide in solution is $76\ kJ\ mol^{-1}$, and the reaction is slow at room temperature. When a little bromide is added the activation energy falls to $57\ kJ\ mol^{-1}$, and the rate increases by a factor of 2000. Enzymes, biological catalysts, are very specific and can have a dramatic effect on the reaction they control. The activation energy for the acid hydrolysis of sucrose is $107\ kJ\ mol^{-1}$, but the enzyme saccharase reduces it to $36\ kJ\ mol^{-1}$, corresponding to an acceleration of the reaction by a factor of 10^{12} at blood temperature (310 K).

A **homogeneous catalyst** is a catalyst that is in the same phase as the reaction mixture (e.g. an acid added to an aqueous solution). A **heterogeneous catalyst** is in a different phase (e.g. a solid catalyst for a gas-phase reaction). We examine heterogeneous catalysis in Chapter 29 and consider only homogeneous catalysis here.

27.6 Homogeneous catalysis

Some idea of the mode of action of homogeneous catalysis can be obtained by examining the kinetics of the bromide catalysed decomposition of hydrogen peroxide:

$$2H_2O_2(aq) \rightarrow 2H_2O(aq) + O_2(g)$$

The reaction is believed to proceed through the following pre-equilibrium:

$$HOOH + H_3O^+ \rightleftharpoons HOOH_2^+ + H_2O \qquad K = \frac{[HOOH_2^+]}{[HOOH][H_3O^+]}$$

$$HOOH_2^+ + Br^- \rightarrow HOBr + H_2O \qquad v = k[HOOH_2^+][Br^-]$$

$$HOBr + HOOH \rightarrow H_3O^+ + O_2 + Br^- \quad (fast)$$

(In the pre-equilibrium, the almost constant H_2O concentration has been absorbed into K.) Since the second step is rate determining, the rate law of the overall reaction is predicted to be

$$\frac{d[O_2]}{dt} = kK[HOOH][H_3O^+][Br^-]$$

in agreement with the observed dependence of the rate on the Br^- concentration and the pH of the solution. The observed activation energy is the activation energy corresponding to the effective rate coefficient kK. In the absence of Br^- ions the reaction cannot proceed through the path set out above, and a different and much larger activation energy is observed.

Two important types of homogeneous catalysis are **acid catalysis** and **base catalysis**, and many organic reactions involve one or the other (and sometimes both). Brønsted acid catalysis is the transfer of a hydrogen ion to the substrate:

$$X + HA \rightarrow HX^+ + A^- \qquad HX^+ \text{ then reacts}$$

and is the primary process in the solvolysis of esters, keto-enol tautomerism, and the inversion of sucrose. Brønsted base catalysis is the transfer of a hydrogen ion from the substrate to a base:

$$XH + B \rightarrow X^- + BH^+ \quad X^- \text{ then reacts}$$

It is the primary step in the isomerization and halogenation of organic compounds, and in the Claisen and aldol reactions.

27.7 Autocatalysis

The phenomenon of **autocatalysis** is the catalysis of a reaction by the products. For example, in a reaction $A \rightarrow P$ it may be found that the rate law is

$$v = k[A][P] \qquad\qquad (17)$$

and so the reaction rate increases as products are formed. (The reaction gets started because there are usually other reaction routes for the formation of some P initially, which then takes part in the autocatalytic reaction proper.) An example of autocatalysis is provided by two steps in the **Belousov–Zhabotinskii reaction** (BZ reaction) which will figure in discussions later in the section:

$$BrO_3^- + HBrO_2 + H_3O^+ \rightarrow 2BrO_2 + 2H_2O$$

$$2BrO_2 + 2Ce(III) + 2H_3O^+ \rightarrow 2HBrO_2 + 2Ce(IV) + 2H_2O$$

The product $HBrO_2$ is a reactant in the first step.

Example 27.5: *Calculating concentrations in an autocatalytic reaction*

Set up and solve the rate equation for the $A \rightarrow P$ autocatalytic reaction in eqn 17.

Answer. Write $[A] = [A]_0 - x$, $[P] = [P]_0 + x$, when the rate law becomes

$$\frac{dx}{dt} = k([A]_0 - x)([P]_0 + x)$$

Integration by partial fractions, using

$$\frac{1}{([A]_0 - x)([P]_0 + x)} = \frac{1}{[A]_0 + [P]_0} \left\{ \frac{1}{[A]_0 - x} + \frac{1}{[P]_0 + x} \right\}$$

gives

$$\frac{1}{[A]_0 + [P]_0} \ln \frac{([P]_0 + x)[A]_0}{[P]_0([A]_0 - x)} = kt$$

Hence

$$\frac{x}{[P]_0} = \frac{e^{at} - 1}{1 + be^{at}}$$

with

$$a = ([A]_0 + [P]_0)k \quad \text{and} \quad b = \frac{[P]_0}{[A]_0}$$

Comment. The solution is plotted in Fig. 27.5. The rate of reaction is slow initially (little P present), then fast (when P and A are both present), and finally slow again (when A has disappeared).

Exercise. At what time is the reaction rate a maximum?

$$[t_{max} = -(1/ka) \ln b]$$

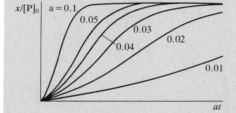

Fig. 27.5 The concentration of product during the autocatalysed $A \rightarrow P$ reaction discussed in the example (using $b = 0.1$).

The industrial importance of autocatalysis (which occurs in a number of reactions, such as oxidations) is that the rate of the reaction can be maximized by ensuring that the optimum concentrations of reactant and product are always present.

27.8 Oscillating reactions

One consequence of autocatalysis is the possibility that the concentrations of reactants, intermediates, and products will vary periodically either in space or in time (Fig. 27.6). Chemical oscillation is the analogue of electrical oscillation, with autocatalysis playing the role of positive feedback. Oscillating reactions are much more than a laboratory curiosity: they are known to occur in some industrial processes, and there are many examples in biochemical systems where a cell plays the role of a chemical reactor. Oscillating reactions, for example, maintain the rhythm of the heartbeat. They are also known to occur in the glycolytic cycle, in which

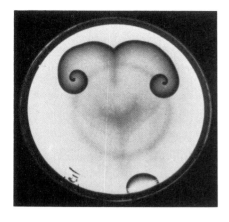
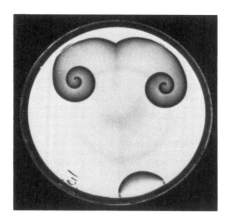
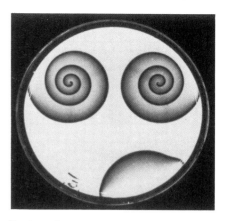

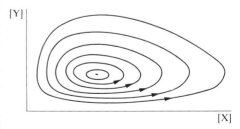

Fig. 27.6 Some reactions show oscillations in time; some show spatially periodic variations. This sequence of photographs shows the emergence of a spatial pattern.

one molecule of glucose is used to produce (through enzyme catalysed reactions involving ATP) two molecules of ATP. All the metabolites in the chain oscillate under some conditions, and do so with the same period but with different phases.

The Lotka–Volterra mechanism

We shall use a model autocatalytic reaction of a particularly simple form that illustrates how these oscillations may occur. The actual chemical examples that have been discovered so far have a different mechanism, as we shall see. The **Lotka–Volterra mechanism** is as follows:

$$
\begin{array}{lll}
\text{(a)} & A + X \rightarrow 2X & \dfrac{d[A]}{dt} = -k_a[A][X] \\[2mm]
\text{(b)} & X + Y \rightarrow 2Y & \dfrac{d[X]}{dt} = -k_b[X][Y] \\[2mm]
\text{(c)} & Y \rightarrow B & \dfrac{d[B]}{dt} = k_c[Y]
\end{array}
$$

Steps (a) and (b) are autocatalytic. The concentration of A is held constant by supplying it to the reaction vessel as needed. (B plays no part in the reaction once it has been produced, and so it is unnecessary to remove it; in practice, though, it would normally be removed.) These constraints leave [X] and [Y], the concentrations of the intermediates, as variables. Note that we are considering a steady-state condition, which is maintained by the flow of A into the reactor. This steady-state *condition* must not be be confused with the steady-state *approximation* made earlier: in the present case we shall solve the rate equations exactly for the variable concentrations of X and Y, but hold [A] at an arbitrary but constant value.

The Lotka–Volterra equations can be solved numerically using the techniques explained in *Further information 1*, and the results can be depicted in two ways. One way is to plot [X] and [Y] against time (Fig. 27.7). The same information can be displayed more succinctly by plotting one concentration against the other: this representation gives the series of closed curves shown in Fig. 27.8.

The periodic variation of the concentrations of the intermediates can be explained as follows. At some stage there may be only a little X present, but

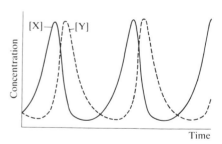

Fig. 27.7 The periodic variation of the concentrations of the intermediates X and Y in a Lotka–Volterra reaction. The system is in a steady-state, but not at equilibrium.

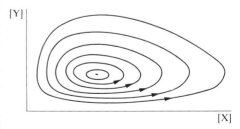

Fig. 27.8 An alternative representation of the periodic variation of [X] and [Y] is to plot one against the other. Then the system describes closed orbits, different ones being obtained with different starting conditions.

831

reaction (a) provides more, and the production of X autocatalyses the production of even more X. There is therefore a surge of X. However, as X is formed, reaction (b) can begin. It occurs slowly initially because [Y] is small, but autocatalysis leads to a surge of Y. This surge, though, removes X, and so reaction (a) slows, and less X is produced. Since less X is now available reaction (b) slows. As less Y becomes available to remove X, X has a chance to surge forward again, and so on.

The brusselator

Another very interesting set of equations is called the **brusselator** (because it is an oscillator that originated with Ilya Prigogine's group in Brussels):

$$(a) \quad A \rightarrow X \qquad\qquad \frac{d[X]}{dt} = k_a[A]$$

$$(b) \quad 2X + Y \rightarrow 3Y \qquad \frac{d[Y]}{dt} = k_b[X]^2[Y]$$

$$(c) \quad B + X \rightarrow Y + C \qquad \frac{d[Y]}{dt} = k_c[B][X]$$

$$(d) \quad X \rightarrow D \qquad\qquad \frac{d[X]}{dt} = -k_d[X]$$

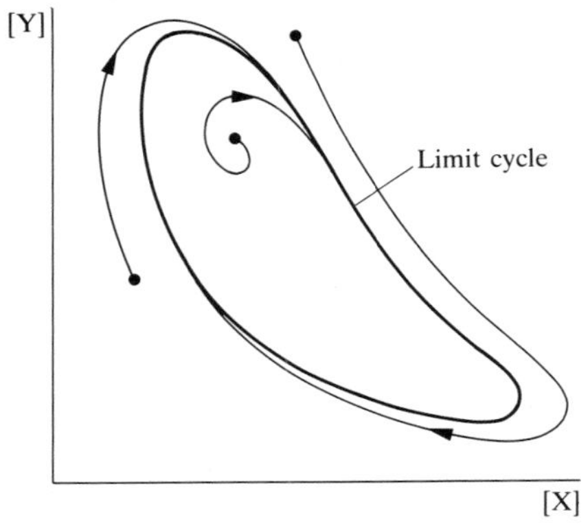

Since the reactants (A and B) are maintained at constant concentration, the two variables are the concentrations of X and Y. These two concentrations may be calculated by solving the rate equations numerically, and the results are plotted in Fig. 27.9. The interesting feature is that whatever the initial concentrations of X and Y, the system settles down into the same periodic variation of concentrations. The common trajectory is called a **limit cycle**, and its period depends on the values of the rate constants.

The oregonator

One of the first oscillating reactions to be reported and studied systematically was the BZ reaction mentioned earlier, which takes place in a mixture of potassium bromate, malonic acid, and a cerium(IV) salt in an acidic solution. The mechanism has been elucidated by Richard Noyes, and involves 18 elementary steps and 21 different chemical species. The main features of this awesomely complex mechanism can be reproduced by the following **oregonator** (so called because Noyes and his group work in Oregon), where A stands for $HBrO_2$, B for Br^-, and Z for Ce^{4+}:

$$(a) \quad A + Y \rightarrow X$$
$$(b) \quad X + Y \rightarrow C$$
$$(c) \quad B + X \rightarrow 2X + Z$$
$$(d) \quad 2X \rightarrow D$$
$$(e) \quad Z \rightarrow Y$$

A, B, C, D are held constant (in the model). The oscillations arise in a similar way to the brusselator, and can be traced to the autocatalysis in step 3 and the linkage between the reactions provided by the other steps.

Fig. 27.9 Some oscillating reactions approach a closed trajectory whatever their starting conditions. The closed trajectory is called a 'limit cycle'.

Bistability

Attempts have been made to discover the underlying causes of oscillation more deeply than simply recognizing the role of autocatalysis. It appears that three conditions must be fulfilled in order to obtain oscillations:

(1) The reactions must be far from equilibrium.
(2) The reactions must have autocatalytic steps.
(3) The system must be able to exist in two steady states.

The last criterion is called **bistability**, and is a property that takes us well beyond what is familiar from the equilibrium properties of systems. The property of bistability is a chemical version of supercooling, in which the temperature of a liquid may be lowered beneath its freezing point without it solidifying.

Consider a reaction in which there are two intermediates X and Y. If the concentration of Y is at some high value in a reactor, and X is added, then the concentration of Y might decrease as shown by the upper line in Fig. 27.10. If X is at some high value, then as Y is added the reaction might result in the slow increase of Y as shown by the lower line. However, in each case, a concentration may be reached at which the concentration will jump from one curve to the other (just as a supercooled liquid might suddenly solidify). The two curves represent the two stable states of the bistable system. Neither is an equilibrium state in the thermodynamic sense: they occur in steady states that are well removed from equilibrium, and the concentrations of X and Y represent the consequences of reactants continuously flowing into and of products flowing out of the reactor.

Now consider what happens when a third type of intermediate, Z, is present. Suppose Z reacts with both X and Y. In the absence of Z the flows of material might correspond to a certain state on the upper curve of Fig. 27.11. However, as Z reacts with Y to produce X the state of the system moves along the curve (to the right, as Y decreases and X increases) until the sudden transition occurs to the lower curve. Then Z reacts with X and produces Y, which means that the composition moves to the left along the lower curve. There comes a point, however, when the concentration of X has been reduced so much, and that of Y has risen so much, that there is a sudden transition to the upper curve, when the process begins again. It is the leaping from one stable state to another that we see as the sudden surge or depletion of the concentration of a species (Fig. 27.12).

By studying the regions of concentrations and the rate constants for the individual steps of a reaction it is now becoming possible to predict the occurrence of bistable chemical systems and to anticipate the occurrence of oscillations. We are still far, however, from being able to use these ideas to account for gene expression, the patterns on tigers and butterflies, and the oscillation of cool flames, in all of which it is thought these processes play a part.

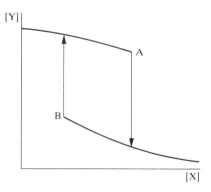

Fig. 27.10 A system showing bistability. As the concentration of X is increased (by adding it to a reactor) the concentration of Y decreases along the upper curve but at A drops sharply to a low value. If X is then decreased, the concentration increases along the lower curve, but rises sharply to a high value at B.

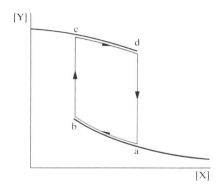

Fig. 27.11 Chemical oscillation in a bistable system occurs as a result of the effect of a third substance Z which can react with X to produce Y and with Y to produce X. As a result, the system switches periodically between the upper and lower curves.

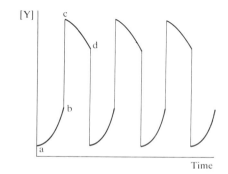

Fig. 27.12 The leaping from one branch to another of a bistable system appears to the observer as a periodic surge and depletion of concentration.

Further information 1: integration of rate equations

The general method of solving differential equations numerically is to approximate differentials by small but finite differences. Suppose we have the differential equation $df/dx = g(x)$, then instead of $df = g(x) \, dx$ we write

$f(x + \delta x) - f(x) = g(x)\delta x$, with δx a small increment of x. Then

$$f(x + \delta x) = f(x) + g(x)\delta x$$

This equation is exact when δx is infinitesimal. Now we just move along the x coordinate in steps of magnitude δx, using the known initial value of f for the first step, and evaluating $g\delta x$ at that initial value of x, which gives the value of f at $x + \delta x$; then we use this new value of f as the initial value for the next step together with the value of $g\delta x$, at the new value of x, to calculate f two steps away from the origin, and so on, until we have reached the value of x of interest. For example, to integrate the HBr rate law in Example 27.1 we write $[H_2] = a - x$, $[Br_2] = a - x$ (assuming equal concentrations initially), and $[HBr] = 2x$ (assuming that the intermediates are at such low concentrations that they can be neglected). Then, with $y = x/a$, the rate law becomes

$$\frac{dy}{dt} = \frac{c_1(1 - y)^{5/2}}{1 + c_2 y}$$

where $c_1 = \frac{1}{2}ka^{1/2}$ and $c_2 = 2k' - 1$. We integrate this expression by stepping along the t axis and calculating successive values of y.

The integration procedure is easy to implement on a microcomputer, and may be extended to the case where the values to use in one equation are determined by another (i.e. when the differential equations are coupled). A simple program would have the form:

Input step size
Input initial and final values of x
Evaluate f(initial x)
Loop from initial x to final x
    Evaluate $f(x + \text{step}) = f(x) + g(x) * \text{step}$
    Print or plot $f(x)$
Repeat loop

A set of coupled differential equations can be integrated similarly. Thus, if the two equations are

$$\frac{df}{dx} = af(x) + g(x)$$

$$\frac{dg}{dx} = bg(x) + f(x)$$

we express them as finite difference equations:

$$f(x + \delta x) = f(x) + \{af(x) + g(x)\}\delta x$$
$$g(x + \delta x) = g(x) + \{bg(x) + f(x)\}\delta x$$

and integrate them as before. There are, however, several problems with simple integration procedures. One is that it may be inaccurate to use the value of $g(x)$ at the start of the interval for which we are calculating $f(x + \delta x)$, and similarly for f in the equation for g. This source of inaccuracy is avoided by **predictor–corrector techniques** that make predictions of the later value of g and h on the basis of their recent values. A second problem is that one concentration may vary slowly but another might vary rapidly. Thus, while it might be tempting to use a large step size for the former, the latter demands that the step size must be small. Differential equations with widely varying rates are termed **stiff** and are not easy to integrate without special routines.

Further information 2: the statistics of stepwise polymerization

The key equation for the statistical analysis of polymers is eqn 11 for the probability P_n of finding an n-mer in terms of the probability p that a monomer has formed a link to another molecule:

$$P_n = p^{n-1}(1-p) \tag{A1}$$

We first confirm that the probability is correctly normalized by showing that the total probability of selecting *any* monomer with $1 \leq n \leq \infty$ is equal to 1:

$$\sum_{n=1}^{\infty} P_n = (1-p) \sum_{n=1}^{\infty} p^{n-1} = \frac{1-p}{p} \sum_{n=1}^{\infty} p^n$$

$$= \frac{1-p}{p} \left\{ \left(\sum_{n=0}^{\infty} p^n \right) - 1 \right\} = 1$$

We have used

$$\sum_{n=0}^{\infty} p^n = 1 + p + p^2 + \cdots = \frac{1}{1-p}$$

Next, we calculate the average chain length, the average value of n. The average is the weighted sum of the values of n that occur in the sample:

$$\langle n \rangle = \sum_{n=1}^{\infty} n P_n = (1-p) \sum_{n=1}^{\infty} n p^{n-1}$$

To evaluate the sum we use a trick first employed in connection with the evaluation of mean energies from the molecular partition function (Section 19.3):

$$\sum_{n=1}^{\infty} n p^{n-1} = \frac{\mathrm{d}}{\mathrm{d}p} \sum_{n=1}^{\infty} p^n = \frac{\mathrm{d}}{\mathrm{d}p} \sum_{n=0}^{\infty} p^n$$

$$= \frac{\mathrm{d}}{\mathrm{d}p} \frac{1}{1-p} = \frac{1}{(1-p)^2}$$

The change in the summation range is justified because $p^0 = 1$ and makes no contribution to the derivative. Therefore,

$$\langle n \rangle = \frac{1}{1-p}$$

which is the expression used in the text and drawn in Fig. 27.4.

The molar mass of a monomer is M_{mon} and the molar mass of an n-mer is $n M_{mon}$. The total mass of the sample is the molecular mass m of the monomer multiplied by the average polymer length:

$$m_{tot} = \frac{m}{1-p}$$

Since the mass of an n-mer is nm, and the probability that there is an n-mer in the sample is P_n, the mass-average molar mass is

$$\langle M \rangle_M = \sum_{n=1}^{\infty} \frac{nmP_n}{m_{tot}} \times nM_{mon} = (1-p)^2 \times M_{mon} \times \sum_{n=1}^{\infty} np^{n-1} \times n$$

$$= M_{mon}(1-p)^2 \sum_{n=0}^{\infty} n^2 p^{n-1} \quad (\text{because } n = 0 \text{ makes no contribution})$$

$$= \frac{(1-p)^2(1+p)}{(1-p)^3} M_{mon} = \frac{1+p}{1-p} \times M_{mon}$$

which is the expression used in the text (eqn 15).

Further reading

Chain reactions and polymerization kinetics

V. N. Kondratiev, Chain reactions. In *Comprehensive chemical kinetics,* Vol. 2 (ed. C. H. Bamford and C. F. Tipper), Elsevier, Amsterdam (1970).

F. G. R. Grimblett, *Kinetics of chemical chain reactions.* McGraw-Hill, New York (1970).

J. C. Bevington, Polymer chemistry. In *Photochemistry and reaction kinetics* (ed. P. G. Ashmore, F. S. Dainton, and T. M. Sugden). Cambridge University Press (1967).

F. W. Billmeyer, *Textbook of polymer science.* Wiley-Interscience, New York (1984).

Photochemistry

R. P. Wayne, *Principles and applications of photochemistry.* Oxford University Press (1988).

J. G. Calvert and J. N. Pitts, *Photochemistry.* Wiley, New York (1966).

J. S. Levine (ed.), *The photochemistry of atmospheres.* Academic Press, Orlando (1985).

Oscillating reactions

Self-organization in chemistry, a series of papers in *J. chem. Educ.,* **66**, 187–212 (1989).

I. R. Epstein, K. Kustin, P. De Kepper, and M. Orban, Oscillating chemical reactions. In *Scientific American,* **248**(3), 96 (1983).

K. G. Denbigh and J. C. R. Turner, *Chemical reactor theory.* Cambridge University Press (1984).

A. R. Peacocke, *The physical chemistry of biological organization.* Clarendon Press, Oxford (1983).

S. K. Scott, Oscillations in simple models of chemical systems. *Accounts chem. Res.,* **20**, 186 (1987).

Computational techniques

J. H. Noggle, *Physical chemistry on a microcomputer.* Little, Brown, & Co, Boston (1985).

Exercises

27.1 A saturated hydrocarbon undergoes pyrolysis at 800 K according to the mechanism:

$$A \underset{(2)}{\overset{(1)}{\rightleftharpoons}} CH_3 + R$$

$$R \underset{(4)}{\overset{(3)}{\rightleftharpoons}} B + H$$

$$H + A \xrightarrow{(5)} H_2 + G$$

$$G \underset{(7)}{\overset{(6)}{\rightleftharpoons}} H + M$$

$$G + G \xrightarrow{(8)} G_2$$

The stable products of the reaction are B, H_2, M, and G_2. Identify the chain initiation step, the chain propagation steps, and the chain termination steps.

27.2 Photolysis of $Cr(CO)_6$ in the presence of certain molecules M gives rise to the following reaction sequence:

$$Cr(CO)_6 + h\nu \rightarrow Cr(CO)_5 + CO \quad (1)$$

$$Cr(CO)_5 + CO \rightarrow Cr(CO)_6 \quad (2)$$

$$Cr(CO)_5 + M \rightarrow Cr(CO)_5M \quad (3)$$

$$Cr(CO)_5M \rightarrow Cr(CO)_5 + M \quad (4)$$

Suppose that the absorbed light intensity is so weak that $I \ll k_4[Cr(CO)_5M]$. Find the factor f in the equation

$$\frac{d[Cr(CO)_5M]}{dt} = -f[Cr(CO)_5M]$$

Show that a graph of $1/f$ against [M] should be a straight line.

27.3 Consider the following mechanism for the thermal decomposition of R_2:

$$R_2 \rightarrow 2R \quad (1)$$

$$R + R_2 \rightarrow P_B + R' \quad (2)$$

$$R' \rightarrow P_A + R \quad (3)$$

$$2R \rightarrow P_A + P_B \quad (4)$$

where R_2, P_A, P_B are stable hydrocarbons. R and R' are free radicals. Find the dependence of the rate of decomposition of R_2 on the concentration of R_2.

27.4 Refer to Fig. 27.2 and determine the pressure range for branching chain explosion in the hydrogen–oxygen reaction at 700 K, 800 K, and 900 K.

27.5 In a photochemical reaction $A \rightarrow 2B + C$, the quantum efficiency with 500 nm light is 2.1×10^2 mol einstein^{-1}. After exposure of 300 mmol of A to the light, 2.28 mmol of B are formed. How many photons were absorbed by A?

27.6 In an experiment to measure the quantum efficiency of a photochemical reaction, the absorbing substance was exposed to 490 nm light from a 100 W source for 45 minutes. The intensity of the transmitted light was 40 per cent of the intensity of the incident light. As a result of irradiation, 0.344 mol of the absorbing substance decomposed. Find the quantum efficiency.

27.7 The condensation reaction of acetone, $(CH_3)_2CO$, in aqueous solution is catalysed by bases, B, which react reversibly with acetone to form the carbanion $C_3H_5CO^-$. The carbanion then reacts with a molecule of acetone to give the product. A simplified version of the mechanism is

$$AH + B \rightarrow BH^+ + A^- \quad (1)$$

$$A^- + BH^+ \rightarrow AH + B \quad (2)$$

$$A^- + AH \rightarrow \text{product} \quad (3)$$

where AH stands for acetone and A^- its carbanion. Use the steady state approximation to find the concentration of the carbanion and derive the rate equation for the formation of the product.

27.8 Consider the acid-catalyzed reaction

$$HA + H^+ \underset{(2)}{\overset{(1)}{\rightleftharpoons}} HAH^+ \quad \text{(fast)}$$

$$HAH^+ + B \xrightarrow{(3)} BH^+ + AH \quad \text{(slow)}$$

Deduce the rate law and show that it can be made independent of the specific term $[H^+]$.

27.9 Consider the following chain mechanism:

$$AH \rightarrow A\cdot + H\cdot \quad (1)$$

$$A\cdot \rightarrow B\cdot + C \quad (2)$$

$$AH + B\cdot \rightarrow A\cdot + D \quad (3)$$

$$A\cdot + B\cdot \rightarrow P \quad (4)$$

Identify the initiation, propagation, and termination steps, and use the steady-state approximation to deduce that the decomposition of AH is first order in AH.

27.10 An autocatalytic reaction $A \rightarrow P$ is observed to have the rate law $d[P]/dt = k[A]^2[P]$. Solve the rate law for initial concentrations $[A]_0$ and $[P]_0$. Calculate the time at which the rate reaches a maximum.

27.11 Another reaction with the stoichiometry $A \rightarrow P$ has the rate law $d[P]/dt = k[A][P]^2$; integrate the rate law for initial concentrations $[A]_0$ and $[P]_0$. Calculate the time at which the rate reaches a maximum.

Problems

Numerical problems

27.1 The number of photons absorbed by a sample can be determined by a variety of methods, of which the classical one is chemical actinometry. The decomposition of oxalic acid $(COOH)_2$, in the presence of uranyl sulphate, $(UO_2)SO_4$, proceeds according to the sequence

$$UO_2^{2+} + h\nu \rightarrow (UO_2^{2+})^*$$

$$(UO_2^{2+})^* + (COOH)_2 \rightarrow UO_2^{2+} + H_2O + CO_2 + CO$$

with a quantum efficiency of 0.53 at the wavelength used. The amount of oxalic acid remaining after exposure can be determined by titration (with $KMnO_4$) and the extent of decomposition used to find the number of incident photons. In a particular experiment, the actinometry solution consisted of 5.232 g anhydrous oxalic acid, 25.0 cm³ water (together with the uranyl salt). After exposure for 300 s the remaining solution was titrated with 0.212 M $KMnO_4$, and 17.0 cm³ were required for complete oxidation of the remaining oxalic acid. What is the rate of absorption of photons at the wavelength of the experiment? Express the answer in photons/second and einstein/second.

27.2 When benzophenone is illuminated with ultraviolet light it is excited into a singlet state. This singlet changes rapidly into a triplet, which phosphoresces. Triethylamine acts as a quencher for the triplet. In an experiment in methanol as solvent, the phosphorescence intensity varied with amine concentration as shown below. A flash photolysis experiment had also shown that the half life of the fluorescence in the absence of quencher is 29 μs. What is the value of k_q?

$[Q]/M$	0.001	0.005	0.010
I_f/(arbitrary units)	0.41	0.25	0.16

27.3 Studies of combustion reactions depend on knowing the concentrations of H atoms and HO radicals. Measurements on a flow system using ESR for the detection of radicals gave information on the reactions

$$H + NO_2 \rightarrow OH + NO \qquad k = 2.9 \times 10^{10}\, M^{-1}\, s^{-1}$$

$$OH + OH \rightarrow H_2O + O \qquad k' = 1.55 \times 10^9\, M^{-1}\, s^{-1}$$

$$O + OH \rightarrow O_2 + H \qquad k'' = 1.1 \times 10^{10}\, M^{-1}\, s^{-1}$$

(J. N. Bradley, W. Hack, K. Hoyermann, and H. G. Wagner, *J. chem. Soc. Faraday Trans.* I, 1889 (1973)). Using initial H atom and NO_2 concentrations of 4.5×10^{-10} and 5.6×10^{-10} mol cm^{-3} respectively, compute and plot curves showing the O, O_2, and OH concentrations as a function of time in the range 0–10 ns.

27.4 In a flow study of the reaction between O atoms and Cl_2 (J. N. Bradley, D. A. Whytock, and T. A. Zaleski, *J. chem. Soc. Faraday Trans.* I, 1251 (1973)) at high chorine

pressures, plots of $\ln [O]_0/[O]$ against distances l along the flow tube, where $[O]_0$ is the oxygen concentration at zero chlorine pressure, gave straight lines. Given the flow velocity as 6.66 m s^{-1} and the data below, find the rate coefficient for the reaction $O + Cl_2 \rightarrow ClO + Cl$.

l/cm	0	2	4	6	8	10	12	14	16	18
$\ln [O]_0/[O]$	0.27	0.31	0.34	0.38	0.45	0.46	0.50	0.55	0.56	0.60

$[O]_0 = 3.3 \times 10^{-8}$ M, $[Cl_2] = 2.54 \times 10^{-7}$ M, $p = 1.70$ Torr.

Theoretical problems

27.5 The Rice–Herzfeld mechanism for the dehydrogenation of ethane is specified in Section 27.1, and it was noted there that it led to first-order kinetics. Confirm this remark, and find the approximations that lead to the rate law quoted there. How may the conditions be changed so that the reaction shows different orders?

27.6 The following mechanism has been proposed for the thermal decomposition of acetaldehyde (ethanal):

$$CH_3CHO \rightarrow \cdot CH_3 + CHO, \qquad k_a$$

$$\cdot CH_3 + CH_3CHO \rightarrow CH_4 + \cdot CH_2CHO, \qquad k_b$$

$$\cdot CH_2CHO \rightarrow CO + \cdot CH_3, \qquad k_c$$

$$\cdot CH_3 + \cdot CH_3 \rightarrow CH_3CH_3, \qquad k_d$$

Find an expression for the rate of formation of methane and the rate of disappearance of acetaldehyde.

27.7 The reaction $Cl_2 + CO \rightarrow COCl_2$ is believed to proceed by the following mechanism:

$$Cl_2 \rightleftharpoons 2Cl \qquad k_a \text{ forward,} \quad k_a' \text{ reverse}$$

$$Cl + CO \rightleftharpoons COCl \qquad k_b \text{ forward,} \quad k_b' \text{ reverse}$$

$$COCl + Cl_2 \rightarrow COCl_2 + Cl \quad k_c \text{ forward}$$

Deduce a rate law for the formation of $COCl_2$ on the assumption that the first two steps are fast. Integrate the rate law for equal initial concentrations of Cl_2 and CO. An improved rate law is obtained by assuming not that the first two reactions are fast, but by making the steady state approximation for Cl and COCl. Deduce a rate law for the formation of $COCl_2$ in this way. Finally, obtain an 'exact' version of the rate law for the formation of $COCl_2$ by numerical integration of the rate laws. Write a computer program for the mechanism, and use it to explore the time dependences of all the components in the reaction.

27.8 The derivation of eqn 10 assumed that the rates of termination and propagation were independent of the polymer chain length. Suppose that both reactions are slower as the chain length increases, and that k_t depends on the monomer concentration as $k_t = k_p^\circ(1 + a[M])$ and k_p depends on $k_p = k°(1 + b[M])$. Derive an expression for the rate of change of concentration of M.

27.9 Express the root mean square deviation of the molar mass of the condensation polymer in terms of p, where $\langle M^2 \rangle^{1/2} = \{\langle M^2 \rangle - \langle M \rangle^2\}^{1/2}$, and deduce its time-dependence.

27.10 Calculate the ratio of the mean cube molar mass to the mean square molar mass in terms of the fraction p.

27.11 Use the steady-state approximation with the following reaction scheme to calculate the concentration of H atoms in the hydrogen/oxygen reaction, and show that under some circumstances the concentration can become (explosively) high.

(a) $H_2 + O_2 \rightarrow 2OH$
(b) $H_2 + OH\cdot \rightarrow H_2O + H\cdot$
(c) $O_2 + H\cdot \rightarrow OH\cdot + \cdot O\cdot$
(d) $H_2 + \cdot O\cdot \rightarrow OH\cdot + H\cdot$
(e) $H\cdot \rightarrow P$

27.12 Derive an expression for the rate of disappearance of a species A in a photochemical reaction for which the mechanism is:

(a) initiation with light of intensity I, $A \rightarrow 2R\cdot$
(b) propagation, $A + R\cdot \rightarrow R\cdot + B$, rate constant k_p
(c) termination, $R\cdot + R\cdot \rightarrow R_2$, rate constant k_t.

Hence, show that rate measurements will give only a combination of k_p and k_t if a steady state is reached, but that both may be obtained if a steady state is not reached. The latter is the basis of the rotating sector method of studying photochemical reactions (see *Further reading*).

27.13 The photochemical chlorination of chloroform in the gas phase has been found to follow the rate law

$$\frac{d[CCl_4]}{dt} = k[Cl_2]^{1/2}I_a^{1/2}$$

Devise a mechanism that leads to this rate law when the chlorine pressure is high.

27.14 Conventional equilibrium considerations do not apply when a reaction is being driven by light absorption. Thus the steady-state concentration of products and reactants might differ significantly from equilibrium values. For instance, suppose the reaction $A \rightarrow B$ is driven by light absorption, and that its rate is I_a, but that the reverse reaction $B \rightarrow A$ is bimolecular and second-order with a rate $k[B]^2$. What is the stationary state concentration of B? Why does this 'photo-stationary state' differ from the equilibrium state?

27.15 When anthracene is dissolved in benzene and exposed to ultraviolet light it dimerizes. In concentrated solutions the dimerization occurs with high quantum efficiency (and little fluorescence), but in dilute solutions fluorescence deactivates the excited molecules and the dimerization proceeds with low quantum efficiency. Calculate the quantum efficiency as a function of concentration, basing the calculation on the mechanism $A + h\nu \rightarrow A^*$, $A^* + A \rightarrow A_2$, $A^* \rightarrow A + h\nu_f$, and account for the observations.

27.16 Write a program for the integration of the Lotka–Volterra equations and arrange for it to plot the concentration of Y against that of X. Explore the consequences of varying the starting concentrations for the integration. Identify the conditions (the concentrations of X and Y) corresponding to the steady state of the Lotka–Volterra equations (the point at the centre of the orbits).

27.17 Integrate the brusselator equations numerically, with an adaptation of the computer program devised for Problem 27.16, and plot [Y] against [X] for a selection of initial concentrations of X and Y: the solutions will show the existence of a limit cycle, Fig. 27.9.

27.18 Set up the overall rate equations for the concentrations of X and Y in terms of the oregonator and explore the periodic properties of the solutions.

28 Molecular reaction dynamics

Check-list of key ideas

1. The *collision theory* of bimolecular gas phase reaction rates and the significance of the *activation energy* and *pre-exponential factor* (Section 28.1).

2. The *reactive cross section* and the *steric factor* (Examples 28.1 and 28.2).

3. *Diffusion-controlled* and *activation-controlled* reactions in solution (Section 28.2).

4. The *diffusion-controlled rate constant* in terms of the solvent viscosity (eqn 7).

5. The formulation of the *material-balance equation* (eqn 8) and some of its solutions (Section 28.3).

6. The formulation of *activated complex theory* in terms of the *reaction coordinate* and the *transition state* (Section 28.4).

7. The derivation and application of the *Eyring equation* (eqn 15).

8. The *kinetic isotope effect* and its analysis in terms of activated complex theory (eqn 16).

9. The use of *ultrashort pulses* in *femtochemistry* and the direct observation of the activated complex (Section 28.5).

10. The description of rate constants in terms of *thermodynamic functions of activation* and the origin of *linear free energy relations* (Section 28.6).

11. The interpretation of collision theory in terms of the *entropy of activation* (Section 28.6).

12. The rate constant for reactions in solution and the *kinetic salt effect* (eqn 24).

13. The interpretation of reaction rates in terms of *potential energy surfaces* (Section 28.7).

14. The trajectories of reactants and products on *attractive and repulsive* potential surfaces and the distinction between *direct mode* and *complex mode* reactions (Section 28.8).

Now we are at the heart of chemistry. Here we examine the details of what happens to molecules at the climax of reactions. Massive changes of structure are taking place and energies the size of dissociation energies are being redistributed among bonds: old bonds are being ripped apart and new bonds are being formed.

As may be imagined, the calculation of rate constants from first principles is a very difficult undertaking. Nevertheless, like so many intricate problems, the broad features can be established quite simply, and only when we enquire more deeply do the complications emerge. In this chapter we look at three levels of approach to the calculation of a rate constant of an elementary bimolecular reaction. Although a great deal of information can be obtained from gas phase reactions, most reactions of interest take place in solution, and we shall also see to what extent their rates can be predicted.

Reactive encounters

In this section we consider two elementary approaches to the calculation of reaction rates. Both are based on the view that reactant molecules must meet, and that reaction will ensue only if they have a certain minimum energy. In the **collision theory** of bimolecular gas phase reactions, which we mentioned briefly in Section 26.6, products are formed only if the collision is sufficiently energetic. If the collision is insufficiently energetic, the colliding reactant molecules separate again. In solution, the reactant molecules may simply diffuse together and then acquire the necessary energy from their immediate surroundings while they are in contact.

28.1 Collision theory

We shall consider the simple bimolecular reaction

$$A + B \rightarrow P \qquad v = k_2[A][B]$$

where P denotes products, and aim to calculate the second-order rate constant k_2.

The basic calculation

In Section 24.3 we introduced the collision density Z_{AB}, the number of A, B collisions per unit volume per unit time. According to collision theory, we can write the rate of change in the number of A molecules per unit volume per unit time as the product of this collision density and the probability that the collision occurs with sufficient energy:

$$\frac{d\mathcal{N}_A}{dt} = -Z_{AB} \times f$$

f is the fraction of collisions that occur with a kinetic energy along the line of approach of the atoms in excess of some threshold value E_a. It is given by the Boltzmann distribution[1] as

$$f = e^{-E_a/RT}$$

The rate of change of the number of A particles per unit volume is equal to the product of the collision density and the fraction f. The rate of change of molar concentration of A is this rate divided by Avogadro's constant N_A. Therefore,

$$\frac{d[A]}{dt} = \frac{-Z_{AB}}{N_A} e^{-E_a/RT}$$

The collision density is given by eqn 11 of Section 24.3:

$$Z_{AB} = \sigma \left(\frac{8kT}{\pi\mu}\right)^{1/2} N_A^2 [A][B]$$

where σ is the collision cross-section. For hard spheres of radii R_A and R_B

$$\sigma = \pi(R_A + R_B)^2$$

and μ is the reduced mass of the colliding molecules:

$$\frac{1}{\mu} = \frac{1}{m_A} + \frac{1}{m_B}$$

Therefore,

$$\frac{d[A]}{dt} = -\sigma \left(\frac{8kT}{\pi\mu}\right)^{1/2} N_A e^{-E_a/RT} [A][B]$$

It follows that

$$k_2 = \sigma \left(\frac{8kT}{\pi\mu}\right)^{1/2} N_A e^{-E_a/RT} \tag{1}$$

Equation 1 has the Arrhenius form

$$k_2 = A e^{-E_a/RT}$$

so long as the exponential temperature dependence dominates the weak square-root temperature dependence of the pre-exponential factor.[2] It

---

[1] The Boltzmann distribution for translational motion in one dimension is

$$f(E) = \frac{1}{kT} e^{-E/kT}$$

where the factor $1/kT$ ensures that the distribution is normalized to 1. It follows that the fraction with an energy in excess of E_a is

$$f = \int_{E_a}^{\infty} f(E)\, dE = \frac{1}{kT} \int_{E_a}^{\infty} e^{-E/kT}\, dE = e^{-E_a/kT}$$

[2] From the general definition of activation energy in eqn 14 of Section 26.6,

$$RT^2 \left(\frac{\partial \ln k_a}{\partial T}\right)_V = E_a + \tfrac{1}{2}RT$$

and so the activation energy is weakly temperature dependent. Generally $E_a \gg \tfrac{1}{2}RT$, and the $\tfrac{1}{2}RT$ can be neglected.

Table 28.1 Arrhenius parameters for gas-phase reactions

	$A/(\text{M}^{-1}\,\text{s}^{-1})$		$E_a/(\text{kJ mol}^{-1})$	P
	Experiment	Theory		
$2\text{NOCl} \rightarrow 2\text{NO} + \text{Cl}_2$	9.4×10^9	5.9×10^{10}	102.0	0.16
$2\text{ClO} \rightarrow \text{Cl}_2 + \text{O}_2$	6.3×10^7	2.5×10^{10}	0.0	2.5×10^{-3}
$\text{H}_2 + \text{C}_2\text{H}_2 \rightarrow \text{C}_2\text{H}_6$	1.24×10^6	7.3×10^{11}	180	1.7×10^{-6}
$\text{K} + \text{Br}_2 \rightarrow \text{KBr} + \text{Br}$	1.0×10^{12}	2.1×10^{11}	0.0	4.8

follows that the activation energy E_a can be identified with the minimum kinetic energy along the line of approach that is needed for reaction, and that the pre-exponential factor is a measure of the rate at which collisions occur in the gas.

Comparison with experiment

The simplest procedure for calculating k_2 is to use for σ the values obtained for non-reactive collisions (e.g. typically those obtained from viscosity measurements) or from tables of molecular radii.

Table 28.1 compares some calculated values of the pre-exponential factor A with values obtained from Arrhenius plots of $\ln k$ against $1/T$ (Section 26.6). One of the reactions shows fair agreement between theory and experiment, but for others there are major discrepancies. In some cases the experimental values are orders of magnitude smaller than those calculated, which suggests that the collision energy is not the only criterion for reaction and that some other feature, such as the relative orientation of the colliding species, is important. Moreover, one reaction has a pre-exponential factor larger than theory, which seems to indicate that the reaction occurs more quickly than the particles collide!

The steric requirement

We can express the disagreement between experiment and theory by replacing σ in eqn 1 by the **reactive cross-section** σ^*. Sometimes it is convenient to express σ^* as a multiple of σ by introducing a **steric factor** P and writing

$$\sigma^* = P\sigma$$

Then the rate constant becomes

$$k_2 = \sigma^*\left(\frac{8kT}{\pi\mu}\right)^{1/2} N_A e^{-E_a/RT}$$

$$= \sigma P\left(\frac{8kT}{\pi\mu}\right)^{1/2} N_A e^{-E_a/RT} \tag{2}$$

P is normally found to be several orders of magnitude smaller than 1.

Example 28.1: *Estimating a steric factor* (1)

Estimate the steric factor for the reaction $H_2 + C_2H_4 \rightarrow C_2H_6$ at 628 K given that the pre-exponential factor is $1.24 \times 10^6 \, M^{-1} \, s^{-1}$.

Answer. The reduced mass is

$$\mu = \frac{mm'}{m + m'} = 3.12 \times 10^{-27} \, \text{kg}$$

since $M = 2.016 \, \text{g mol}^{-1}$ for H_2 and $M' = 28.05 \, \text{g mol}^{-1}$ for C_2H_4. Hence

$$\left(\frac{8kT}{\pi\mu}\right)^{1/2} = 2.66 \times 10^3 \, \text{m s}^{-1}$$

From Table 24.2, $\sigma(H_2) = 0.27 \, \text{nm}^2$ and $\sigma(C_2H_4) = 0.64 \, \text{nm}^2$, giving a mean collision cross-section of $\sigma = 0.46 \, \text{nm}^2$. Therefore,

$$A = \sigma P \left(\frac{8kT}{\pi\mu}\right)^{1/2} N_A = P \times 7.37 \times 10^{11} \, M^{-1} \, s^{-1}$$

Since experimentally $A = 1.24 \times 10^6 \, M^{-1} \, s^{-1}$, it follows that $P = 1.7 \times 10^{-6}$.

Comment. As a rough guide, the more complex the molecules, the smaller the value of P.

Exercise. It is found for the reaction $NO + Cl_2 \rightarrow NOCl + Cl$ that $A = 4.0 \times 10^9 \, M^{-1} \, s^{-1}$ at 298 K. Use $\sigma(NO) = 0.42 \, \text{nm}^2$ and $\sigma(Cl_2) = 0.93 \, \text{nm}^2$ to estimate the P factor.

[0.02]

Equation 2 displays very clearly the three contributions to the rate constant. If we write the rate as

$$k_2[A][B] = P \times Z_{AB} \times f$$

we see that the final factor (f) is the energy criterion; the second factor (Z_{AB}) is the transport property, which governs how often the particles come together; the first factor (P) takes care of the local properties of the reaction, such as the orientations necessary for reaction and the details of how close they must come.

An example of a reaction for which we can estimate the steric factor is

$$K + Br_2 \rightarrow KBr + Br$$

for which the experimental value of P is 4.8. In this reaction the distance of approach at which reaction can occur appears to be considerably larger than the distance needed for deflection of the path of the approaching particles in a non-reactive collision. It has been proposed that the reaction proceeds by a **harpoon mechanism**. This brilliant name is based on a model of the reaction which pictures the K atom as approaching the Br_2 molecule, and when the two are close enough an electron (the harpoon) flips across to the Br_2 molecule. In place of two neutral particles there are now two ions, and so there is a coulombic attraction between them: this attraction is the line on the harpoon. Under its influence the ions move together (the line is wound in), the reaction takes place, and $KBr + Br$ emerge. The harpoon extends the cross-section for the reactive encounter, and we greatly underestimate the reaction rate by taking for the collision cross-section the value for simple mechanical contact between $K + Br_2$.

Example 28.2: *Estimating a steric factor* (2)

Estimate the value of P for the harpoon mechanism by calculating the distance at which it is energetically favourable for the electron to leap from K to Br_2.

Answer. We should begin by identifying all the contributions to the energy of interaction between the colliding species. There are three contributions to the energy of the process $K + Br_2 \rightarrow K^+ + Br_2^-$. The first is the ionization energy I of K. The second is the electron affinity E_{ea} of Br_2. The third is the coulombic interaction energy between the ions when they have been formed: when their separation is R this energy is $-e^2/4\pi\varepsilon_0 R$. Therefore, the net change in energy when the transfer occurs at a separation R is

$$E = I - E_{ea} - \frac{e^2}{4\pi\varepsilon_0 R}$$

I is larger than E_{ea}, and so E becomes negative only when R has decreased to less than some critical value R^* given by

$$\frac{e^2}{4\pi\varepsilon_0 R^*} = I - E_{ea}$$

When the particles are at this separation the harpoon shoots across, so we can identify the reactive cross-section as

$$\sigma^* = \pi R^{*2}$$

This value of σ^* implies that the steric factor is

$$P = \frac{\sigma^*}{\sigma} = \frac{R^{*2}}{d^2} = \left\{ \frac{e^2}{4\pi\varepsilon_0 d(I - E_{ea})} \right\}^2$$

where $d = R(K) + R(Br_2)$. With $I = 420\ kJ\ mol^{-1}$, $E_{ea} \approx 250\ kJ\ mol^{-1}$, and $d = 400\ pm$, we find $P \approx 12$, which is consistent with the experimental value.

Exercise. Estimate the value of P for the harpoon reaction between Na and Cl_2 for which $d \approx 350\ pm$; take $E_{ea} \approx 230\ kJ\ mol^{-1}$. [6]

The example illustrates two points about steric factors. First, the concept of a steric factor is not wholly useless because in some cases its numerical value can be estimated. Second (and more pessimistically) most reactions are much more complex than $K + Br_2$, and we cannot expect to obtain P so easily. What we need is a more powerful theory that lets us calculate, and not merely guess, its value. We go some way to setting up that theory in Section 28.4.

28.2 Diffusion-controlled reactions

Encounters between reactants in solution occur in a very different way from those in gases. Particles have to jostle their way through the solvent, and so their encounter frequency is considerably less than in a gas. However, since a particle also migrates only slowly away from a location, two particles encountering each other stay near each other for much longer than in a gas, and may accumulate enough energy to react even though they do not have enough when they first meet. The activation energy of a reaction is a much more complicated quantity in solution than in a gas because the encounter pair is surrounded by solvent.

Classes of reaction

We can divide the complicated overall process into simpler parts by setting up a simple kinetic scheme. We suppose that the rate of formation of an **encounter pair** AB is second-order in the reactants A and B:

$$A + B \rightarrow AB \qquad v = k_d[A][B]$$

As we shall see, k_d (where the d signifies diffusion) is determined by the diffusional characteristics of A and B. The encounter pair can break up without reaction or it can go on to form products P. If we suppose that these are pseudofirst-order reactions (with the solvent perhaps playing a role), we can write

$$AB \rightarrow A + B \qquad v = k_d'[AB]$$

$$AB \rightarrow P \qquad v = k_a[AB]$$

We can now find the steady-state concentration of AB from the equation for the net rate of change of concentration of AB:

$$\frac{d[AB]}{dt} = k_d[A][B] - k_d'[AB] - k_a[AB] = 0$$

which solves to

$$[AB] = \frac{k_d[A][B]}{k_a + k_d'}$$

The overall rate law for the formation of products is therefore

$$\frac{d[P]}{dt} = k_a[AB] = k_2[A][B] \qquad k_2 = \frac{k_a k_d}{k_a + k_d'} \qquad (3a)$$

Two limits can now be distinguished. If the rate of separation of the unreacted encounter pair is much slower than the rate at which it forms products, then $k_d' \ll k_a$ and the effective rate constant is

$$k_2 \approx \frac{k_a k_d}{k_a} = k_d \qquad (3b)$$

In this **diffusion-controlled limit** the rate of reaction is governed by the rate at which the reactant particles diffuse through the medium. We shall see that a signal that a reaction is diffusion-controlled is that its rate constant is of the order of $10^9 \, \text{M}^{-1} \, \text{s}^{-1}$ or greater. Because the combination of radicals involves very little activation energy, radical and atom recombination reactions are often diffusion-controlled.

An **activation-controlled reaction** arises when a substantial activation energy is involved in the reaction of AB. Then $k_a \ll k_d'$ and

$$k_2 \approx \frac{k_a k_d}{k_d'} = k_a K \qquad (4)$$

where K is the equilibrium constant for $A + B \rightleftharpoons AB$. In this limit the reaction rate depends on the accumulation of energy in the encounter pair

Table 28.2 Arrhenius parameters for reactions in solution

	Solvent	$A/(\text{M}^{-1}\,\text{s}^{-1})$	$E_a/(\text{kJ mol}^{-1})$
$(CH_3)_3CCl$ solvolysis	Water	7.1×10^{16}	100
	Ethanol	3.0×10^{13}	112
	Chloroform	1.4×10^{4}	45
$CH_3CH_2Br + OH^-$	Ethanol	4.3×10^{11}	89.5

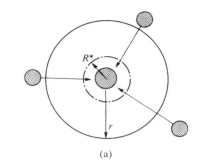

(a)

from the surrounding solvent. Some experimental data are given in Table 28.2.

Diffusion and reaction

We can calculate the rate of a diffusion-controlled reaction by considering the rate at which the reactants diffuse together. We begin with the case of a stationary A molecule in a solvent that contains mobile B molecules; later we allow the A molecules to move too.

Consider an imaginary sphere of radius r surrounding A (Fig. 28.1). The (molar) flux J of B through the surface is the amount (number of moles) of B molecules that pass through per unit area per unit time. Therefore, the total flow $\mathcal{J}$, the amount of B molecules per unit time passing through the entire surface of area $4\pi r^2$, is

$$\mathcal{J} = 4\pi r^2 J$$

We know from Fick's first law (Section 25.4) that the flux is proportional to the concentration gradient:

$$J = D_B \frac{d[B]}{dr}$$

[B] is the molar concentration of the B molecules and D_B is their diffusion constant in the medium. Therefore,

$$\mathcal{J} = 4\pi r^2 J = 4\pi r^2 D_B \frac{d[B]}{dr}$$

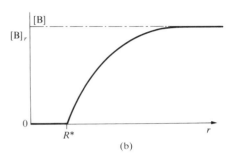

Fig. 28.1 (a) The flux through a spherical surface of B particles on their approach to a stationary A particle. The total flow through a surface is independent of r; reaction occurs when the particles are at a separation R^*. The corresponding concentration profile is shown in (b).

We now calculate the concentration of B at any distance from A by integrating this equation. We need two pieces of information. First, when r is infinite, the concentration of B is the same as in the bulk, which we denote [B]. Second, the total flow through a shell is the same whatever the radius, because no reaction occurs until A and B touch (Fig. 28.1). Therefore, under steady state conditions, $\mathcal{J}$ is a constant independent of r. It follows that

$$\int_{[B]_r}^{[B]} d[B] = \frac{\mathcal{J}}{4\pi D_B} \int_r^\infty \frac{dr}{r^2}$$

which evaluates to

$$[B]_r = [B] - \frac{\mathcal{J}}{4\pi D_B r}$$

We see that the concentration of B varies inversely with distance.

Now suppose that at some critical distance R^* the reactants touch, reaction occurs, and B is destroyed. The existence of this critical distance of approach implies that when $r = R^*$, $[B]_r = 0$. When this condition is substituted into the last equation we obtain an expression for the flow of B

847

towards A:

$$\mathscr{J} = 4\pi R^* D_B[B] \qquad (5)$$

This flow is the average amount of B molecules per unit time passing through any spherical surface centred on any A.

The rate of the diffusion-controlled reaction is equal to the average flow of B molecules to all the A molecules in the sample. If the bulk concentration of A is [A], the number of A molecules in the sample of volume V is $N_A[A]V$; the global flow of all B to all A is therefore $4\pi R^* D_B N_A[A][B]V$. Of course, it is unrealistic to suppose that all A are stationary; this feature is removed by replacing D_B by the sum of the diffusion coefficients of the two species and writing $D = D_A + D_B$. Then the rate of change of concentration of AB is

$$\frac{d[AB]}{dt} = 4\pi R^* D N_A[A][B]$$

By comparing this expression with eqn 3 we can identify the diffusion-controlled rate constant as

$$k_d = 4\pi R^* D N_A \qquad (6)$$

We can take eqn 6 further by incorporating the Stokes–Einstein equation (Section 25.4) relating the diffusion constant and the hydrodynamic radius R_A and R_B of each molecule in a medium of viscosity η

$$D_A = \frac{kT}{6\pi\eta R_A} \qquad D_B = \frac{kT}{6\pi\eta R_B}$$

Since these relations are approximate, we introduce little extra error if we write $R_A = R_B = \frac{1}{2}R^*$, which leads to

$$k_d = \frac{8RT}{3\eta} \qquad (7)$$

The radii have cancelled because although the diffusion constants are smaller when the radii are large, the reactive collision radius is larger and the particles need travel a shorter distance to meet. In this approximation the rate constant is independent of the identities of the reactants, and depends only on the temperature and the viscosity of the solvent.

Example 28.3: *Estimating the rate constant of a diffusion-controlled reaction*

Estimate the second-order rate constant for the recombination of I atoms in hexane at 298 K.

Answer. The viscosity of the solvent is 0.326 cP at this temperature (Table 22.6, and 1 cP = 10^{-3} kg m^{-1} s^{-1}). Then, from eqn 7,

$$k_d = \frac{6.61 \times 10^9 \, \text{M}^{-1} \, \text{s}^{-1}}{(\eta/\text{cP})} = 2.0 \times 10^{10} \, \text{M}^{-1} \, \text{s}^{-1}$$

Comment. The experimental value is $1.3 \times 10^{10} \, M^{-1} \, s^{-1}$, and so the agreement is very good considering the approximations involved.

Exercise. Estimate the rate in benzene at the same temperature.

$$[1.1 \times 10^{10} \, M^{-1} \, s^{-1}]$$

28.3 The material-balance equation

The diffusion of reactants plays an important role in many chemical processes, such as the diffusion of O_2 molecules into red blood corpuscles and of a gas towards a catalyst. We can have a glimpse of the kinds of calculations involved by considering the generalized diffusion equation, eqn 21 of Section 25.5, but generalized even more to take into account the possibility that the diffusing, convecting molecules are also reacting.

The formulation of the equation

Consider a small volume element in a chemical reactor (or a biological cell). During some small time interval Δt the total change of the number of molecules of some substance J is

Net change of number of J in the volume element

= Number entering − Number leaving

+ Number formed by reaction − Number consumed by reaction

The rate of change of the concentration is this expression divided by the volume and the time interval. We shall now show how to express the relations as a differential equation.

The net rate at which J molecules enter the region by diffusion and convection is given by eqn 23 of Section 25.5:

$$\frac{\partial [J]}{\partial t} = D \frac{\partial^2 [J]}{\partial x^2} - v \frac{\partial [J]}{\partial x}$$

The net rate of change due to chemical reaction is

$$\frac{\partial [J]}{\partial t} = -k[J]$$

if we suppose that J disappears by a pseudofirst-order reaction. Therefore, the mathematical form of the **material-balance equation** is

$$\frac{\partial [J]}{\partial t} = D \frac{\partial^2 [J]}{\partial x^2} - v \frac{\partial [J]}{\partial x} - k[J] \qquad (8)$$

A similar equation, with different reaction terms, applies to each species present.

Solutions of the equation

The material balance equation is a second-order partial differential equation, and it is far from easy to solve in general. Some idea of the difficulty of dealing with it can be seen by considering the special case in which there is no convective motion:

$$\frac{\partial [J]}{\partial t} = D \frac{\partial^2 [J]}{\partial x^2} - k[J] \qquad (9)$$

If the solution in the absence of reaction (i.e. $k = 0$ in this equation) is [J], then it is possible to show that the solution in the presence of reaction ($k \neq 0$) is

$$[J]^* = k \int_0^t [J] e^{-kt}\, dt + [J] e^{-kt} \qquad (10)$$

We have already met one solution of the diffusion equation in the absence of reaction, and eqn 24 of Section 25.5 is the solution for a system in which initially a layer of $n_0 N_A$ molecules are spread over a plane of area A:

$$[J] = \frac{n_0 e^{-x^2/4Dt}}{A(\pi Dt)^{1/2}}$$

When we substitute this expression into eqn 10 and evaluate the integral, we obtain the concentration of J as it diffuses away from its initial surface layer and undergoes reaction in the solution above (Fig. 28.2).

Even this example has led to a difficult equation to compute, and only in some special cases can the full material-balance equation be solved analytically. Most modern work on reactor design and cell kinetics now uses numerical methods to solve the equation, and detailed solutions for realistic environments can be obtained reasonably easily. Some of the interesting applications include the exploration of the spatial periodicity of auto-catalytic reactions mentioned in Section 27.8.

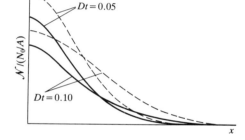

Fig. 28.2 The concentration profiles for a diffusing, reacting system (e.g. a column of solution) in which one reactant is initially in a layer at $x = 0$. In the absence of reaction (broken lines) the concentration profiles are the same as in Fig. 25.10. (Arbitrary values for D and k have been taken.)

Activated complex theory

We shall now consider a more detailed calculation of rate constants using some of the concepts of statistical thermodynamics developed in Chapter 20. The approach we describe, which is called **activated complex theory** (ACT), has the advantage that a quantity corresponding to the steric factor appears automatically, and P does not need to be grafted on to an equation in an *ad hoc* manner. Activated complex theory is an attempt to identify the principal features governing the size of a rate constant in terms of a model of the events that take place during the reaction.

28.4 The reaction coordinate and the transition state

The general features of how the potential energy of the reactants A and B change in the course of a simple bimolecular reaction are shown in Fig. 28.3. The horizontal axis of the diagram represents the course of the individual reaction event (a bimolecular collision in a gas phase reaction) and is called the **reaction coordinate**.

Initially only the reactants A and B are present. As the reaction event proceeds, A and B come into contact, distort, and begin to exchange or discard atoms. The potential energy rises to a maximum, and the cluster of atoms that corresponds to the region close to the maximum is called the

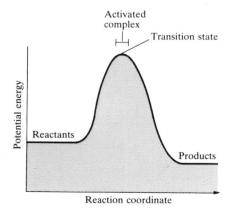

Fig. 28.3 A reaction profile. The horizontal axis is the reaction coordinate, and the vertical axis is potential energy. The activated complex is the region near the potential maximum, and the transition state corresponds to the maximum itself.

activated complex. Then the potential energy falls as the atoms rearrange in the cluster, and reaches a value characteristic of the products. The climax of the reaction is at the peak of the potential energy. Here two reactant molecules have come to such a degree of closeness and distortion that a small further distortion will send them in the direction of products. This crucial configuration is called the **transition state** of the reaction. Although some molecules entering the transition state might revert to reactants, if they pass through this configuration it is inevitable that products will emerge from the encounter.

As an example, consider the approach of an H atom to an F_2 molecule. For simplicity, we imagine the approach as occuring along the F—F bond direction. At great distances the potential energy is the sum of the potential energies of H and F_2. When H and F_2 are so close that their orbitals start to overlap, the F—F bond begins to stretch and a bond begins to form between H and the nearer F. The H atom comes closer, the F—F bond lengthens, the H—F bond shortens and strengthens, and the atoms enter the range of locations characteristic of the activated complex. There comes a stage when the activated complex has maximum potential energy and is poised at the transition state. An infinitesimal compression of the H—F bond and a stretch of F—F takes the complex through the transition state. Distances further along the reaction coordinate represent stages at which the H—F bond forms more fully and the F—F bond breaks. Motion along the reaction coordinate from left to right therefore represents the progress of H and F_2 through these configurations. Whether or not a colliding pair actually crosses the potential barrier depends on the kinetic energy the molecules have initially, because they must be able to climb the barrier and attain the transition state.

In an actual reaction, H atoms approach F_2 molecules from all angles and the specification of the reaction coordinate is a subtle problem. We shall therefore regard the coordinate simply as an indication of the distortions in the reactant molecules as the activated complex is formed, the critical transition state is reached, and the product molecules emerge. At the transition state, motion along the reaction coordinate corresponds to some complicated collective vibration-like motion of all the atoms in the complex.

28.5 The Eyring equation

Activated complex theory pictures a reaction between A and B as proceeding through the formation of an activated complex $C^{\ddagger}$ that falls apart by unimolecular decay into products P with a rate constant $k^{\ddagger}$:

$$C^{\ddagger} \rightarrow P \qquad v = k^{\ddagger}[C^{\ddagger}]$$

The concentration of the activated complex is likely to be proportional to the concentration of the reactants, and later we show explicitly that

$$[C^{\ddagger}] = K^{\ddagger}[A][B]$$

where $K^{\ddagger}$ is a proportionality constant (with dimensions 1/concentration). It follows that

$$v = k_2[A][B] \qquad k_2 = k^{\ddagger}K^{\ddagger} \tag{11}$$

and our task is to calculate the unimolecular rate constant $k^{\ddagger}$ and the proportionality constant $K^{\ddagger}$.

The rate of decay of the activated complex

The criterion for the activated complex turning into products is that it pass through the transition state. If its vibration-like motion along the reaction coordinate occurs with a frequency v, the frequency with which the cluster of atoms forming the complex approaches the transition state is also v. However, it is possible that not every oscillation along the reaction coordinate takes the complex through the transition state. For instance, the centrifugal effect of rotations might also be an important contribution to the break-up of the complex, and in some cases the complex might be rotating too slowly or about the wrong axis. Therefore, we suppose that the rate of passage of the complex through the transition state is proportional to the vibrational frequency along the reaction coordinate, and write

$$k^{\ddagger} = \kappa v \tag{12}$$

where κ is the **transmission coefficient**. In many cases $\kappa \approx 1$.

The concentration of the activated complex

The simplest procedure for estimating the concentration of the activated complex is to assume that there is a pre-equilibrium between the reactants and the complex,[3] and to write

$$A + B \rightleftharpoons C^{\ddagger} \qquad K_p = \frac{(p_C/p^{\ominus})}{(p_A/p^{\ominus})(p_B/p^{\ominus})} = \frac{p_C p^{\ominus}}{p_A p_B}$$

We can express the partial pressures p_J in terms of the concentrations using $p_J = RT[J]$, which gives

$$[C^{\ddagger}] = \frac{RTK_p}{p^{\ominus}}[A][B]$$

from which it follows that

$$K^{\ddagger} = \frac{RT}{p^{\ominus}} \times K_p \tag{13}$$

Our remaining task is to calculate the equilibrium constant K_p.

We saw in Section 20.6 how to calculate equilibrium constants from structural data. We can use eqn 22 of that section directly, which in this case gives

$$K_p = \frac{N_A q_C^{\ominus} e^{-\Delta E_0/RT}}{q_A^{\ominus} q_B^{\ominus}}$$

where ΔE_0 is the (molar) energy separation between the zero-point levels of

---

[3] In earlier editions a different line of argument was used: we recognized that almost nothing is known about the populations of the levels of the activated complex. Since nothing is known, the most honest approach is to assume that the populations depend on the energy and not on the identity of the level (i.e. whether it belongs to A, B, or $C^{\ddagger}$). This 'least-prejudiced' approach leads to the same results as here, but has the advantage that it avoids the presumption of equilibrium between the reactants and their activated complex. Its disadvantage is that it is slightly less direct. The Second Edition of this text should be consulted for details.

$C^{\ddagger}$ and $A + B$:

$$\Delta E_0 = E_0(C^{\ddagger}) - E_0(A) - E_0(B)$$

and the $q_J^{\ominus}$ are the standard molar partition functions as defined in Section 20.1.

In the final step of this part of the calculation we focus attention on the partition function of the activated complex. We have already assumed that a vibration of $C^{\ddagger}$ tips it through the transition state. The partition function for this vibration is

$$q = \frac{1}{1 - e^{-h\nu/kT}}$$

where ν is its frequency (the same frequency that determines $k^{\ddagger}$). This frequency is much lower than for an ordinary molecular vibration because the oscillation corresponds to the complex falling apart. Therefore, since $h\nu/kT \ll 1$, the exponential may be expanded and the partition function reduces to

$$q = \frac{1}{1 - \left(1 - \dfrac{h\nu}{kT} + \ldots\right)} \approx \frac{kT}{h\nu}$$

We can therefore write

$$q_C = \frac{kT}{h\nu} \times \bar{q}_C$$

where $\bar{q}$ denotes the partition function for all the other modes in the complex. The coefficient $K^{\ddagger}$ is therefore

$$K^{\ddagger} = \frac{kT}{h\nu} \times \bar{K} \tag{14a}$$

where

$$\bar{K} = \frac{RT}{p^{\ominus}} \times \bar{K}_p \tag{14b}$$

$$\bar{K}_p = \frac{N_A \bar{q}_C^{\ominus}}{q_A^{\ominus} q_B^{\ominus}} e^{-\Delta E_0/RT} \tag{14c}$$

with $\bar{K}_p$ a kind of equilibrium constant, but with one vibrational mode of $C^{\ddagger}$ discarded.

The rate constant

We can now combine all the parts of the calculation into

$$k_2 = k^{\ddagger} K^{\ddagger} = \kappa\nu \times \frac{kT}{h\nu} \times \bar{K}$$

Then, because the unknown frequency ν cancels, we obtain the **Eyring equation**:

$$k_2 = \kappa \frac{kT}{h} \times \bar{K} \tag{15}$$

Since $\bar{K}$ is given by eqn 14 in terms of the partition functions of A, B, and C‡, in principle we now have an explicit expression for calculating the second-order rate constant for a bimolecular reaction in terms of the molecular parameters for the reactants and the activated complex.

The partition functions for the reactants can normally be calculated quite readily, using either spectroscopic information about their energy levels or the approximate expressions set out in Box 20.2. The difficulty with the Eyring equation lies in the calculation of the partition function of the activated complex since C‡ cannot normally be investigated spectroscopically. We are normally forced to make assumptions about its size, shape, and structure. We shall illustrate what is involved for two simple cases.

The collision of structureless particles

As a first example, consider the case of two structureless gas phase particles A and B colliding to give an activated complex that resembles a diatomic molecule. Since the reactants are structureless 'atoms' the only contributions to their partition functions are the translational terms:

$$q_A^{\ominus} = \frac{V_m^{\ominus}}{\Lambda_A^3} \qquad q_B^{\ominus} = \frac{V_m^{\ominus}}{\Lambda_B^3}$$

with

$$\Lambda_J = \left(\frac{h^2}{2\pi m_J kT}\right)^{1/2} \qquad V_m^{\ominus} = \frac{RT}{p^{\ominus}}$$

The activated complex is a diatomic cluster of mass $m_C = m_A + m_B$ and moment of inertia I. It has one vibrational mode, but that corresponds to motion along the reaction coordinate and therefore does not appear in $\bar{q}_C$. It follows that the molar partition function of the activated complex is

$$\bar{q}_C^{\ominus} = \frac{2IkT}{\hbar^2} \times \frac{V_m^{\ominus}}{\Lambda_C^3}$$

and therefore that

$$k_2 = \kappa \times \frac{kT}{h} \times \frac{RT}{p^{\ominus}} \times \frac{N_A \Lambda_A^3 \Lambda_B^3}{\Lambda_C^3 V_m^{\ominus}} \times \frac{2IkT}{\hbar^2} \times e^{-\Delta E_0/RT}$$

$$= \kappa \frac{kT}{h} \times \left(\frac{\Lambda_A \Lambda_B}{\Lambda_C}\right)^3 \left(\frac{2IkT}{\hbar^2}\right) e^{-\Delta E_0/RT}$$

Since the moment of inertia of a diatomic molecule of bond length R is μR^2, where μ is the reduced mass, after some straightforward algebra we obtain

$$k_2 = \kappa N_A \left(\frac{8kT}{\pi \mu}\right)^{1/2} \pi R^2 e^{-\Delta E_0/RT}$$

Finally, by identifying the reactive cross-section σ^* as

$$\sigma^* = \kappa \pi R^2$$

we arrive at precisely the same expression as that obtained from simple collision theory (eqn 2).

Example 28.4: *Estimating the steric factor from the Eyring equation*

Obtain an order-of-magnitude estimate of the P factor for the reaction between two nonlinear molecules.

Answer. We use the Eyring equation twice, first for two structureless molecules, and then for two molecules with internal structure (and which can rotate and vibrate). For the order-of-magnitude estimate, assume that all the translational partition functions are the same. Likewise, we also assume that all rotational and all vibrational partition functions are the same. An N-atomic nonlinear molecule has three translational, three rotational, and $3N - 6$ vibrational modes. For no internal modes of the colliding molecules,

$$q_A^\ominus = \frac{V_m^\ominus}{\Lambda^3} \qquad q_B^\ominus = \frac{V_m^\ominus}{\Lambda^3} \qquad \bar{q}_C^\ominus = q_r^2 \frac{V_m^\ominus}{\Lambda^3}$$

Therefore, from eqn 15:

$$k_2 = \kappa \frac{kT}{h} \times \frac{RT}{p^\ominus} \times \frac{N_A q_r^2 \Lambda^6}{\Lambda^3 V_m^\ominus} \times e^{-\Delta E_0/RT}$$

$$= \kappa \frac{RT}{h} \times q_r^2 \Lambda^3 \times e^{-\Delta E_0/RT}$$

When all the modes are active,

$$q_A^\ominus = q_r^3 q_v^{3N-6} \times \frac{V_m^\ominus}{\Lambda^3} \qquad q_B^\ominus = q_r^3 q_v^{3N'-6} \times \frac{V_m^\ominus}{\Lambda^3}$$

$$\bar{q}_C^\ominus = q_r^3 q_v^{3(N+N')-7} \times \frac{V_m^\ominus}{\Lambda^3}$$

(We have discarded one vibrational mode from the activated complex.) It follows that

$$k_2 = \kappa \frac{kT}{h} \times \frac{RT}{p^\ominus} \times \frac{N_A q_r^3 q_v^{3(N+N')-7}}{q_r^6 q_v^{3(N+N')-12}} \times \frac{\Lambda^3}{V_m^\ominus} \times e^{-\Delta E_0/RT}$$

$$= \kappa \frac{RT}{h} \times \frac{q_v^5 \Lambda^3}{q_r^3} \times e^{-\Delta E_0/RT}$$

Comparison of the two expressions for k_2 gives

$$P = \left(\frac{q_v}{q_r}\right)^5$$

Since $q_v/q_r \approx 1/50$, $P \approx 3 \times 10^{-9}$.

Comment. The calculation suggests that, for reactions with similar activation energies, the rate constants for reactions between complex molecules in the gas phase should be much slower than reactions between simple molecules.

Exercise. Estimate the steric factor for two linear molecules that form a nonlinear activated complex.
$$[P = q_v/q_r \approx 1/50]$$

The kinetic isotope effect

As a second example, consider the effect of deuteration on a reaction in which the rate determining step is the breaking of a C—H bond. The reaction coordinate corresponds to the stretching of the C—H bond, and the potential energy profile is shown in Fig. 28.4. On deuteration, the

Fig. 28.4 Changes in the reaction profile when a bond undergoing cleavage is deuterated. The only significant change is to the zero-point energy of the reactants, which is lower for C–D than for C–H. As a result, the activation energy is greater for C–D than for C–H.

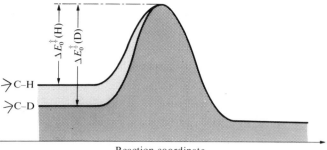

dominant change is the reduction of the zero-point energy of the bond (because the deuterium atom is heavier). The whole reaction profile is not lowered, however, because the relevant vibration in the activated complex has a very low force constant, and so there is little zero-point energy in either the proton or the deuterium forms of the complex.

We assume that the deuteration affects only the reaction coordinate, and hence that the partition functions for all the other internal modes remain unchanged. The translational partition functions are changed by deuteration, but the mass of the rest of the molecule is normally so great that the change is insignificant. The value of ΔE_0 changes on account of the change of zero-point energy, and

$$\Delta E_0(\text{C--D}) - \Delta E_0(\text{C--H}) = N_A \{ \tfrac{1}{2}\hbar\omega(\text{C--D}) - \tfrac{1}{2}\hbar\omega(\text{C--H}) \}$$

$$= \tfrac{1}{2} N_A \hbar k_f^{1/2} \left\{ \frac{1}{\mu_{\text{CH}}^{1/2}} - \frac{1}{\mu_{\text{CD}}^{1/2}} \right\}$$

where k_f is the force constant of the bond and μ the reduced mass. Since all the partition functions are the same (by assumption), the rate constants for the two species should be in the ratio

$$\frac{k(\text{C--D})}{k(\text{C--H})} = e^{-\lambda} \qquad \lambda = \frac{\hbar k_f^{1/2}}{2kT} \left\{ \frac{1}{\mu_{\text{CD}}^{1/2}} - \frac{1}{\mu_{\text{CH}}^{1/2}} \right\} \qquad (16)$$

and $\lambda < 0$ because $\mu_{\text{CD}} > \mu_{\text{CH}}$. This equation predicts that at room temperature C–H cleavage should be about seven times faster than C–D cleavage, other conditions being equal. The principal reason is the smaller activation energy for C–H scission on account of its greater zero-point energy.

The experimental observation of the activated complex

Until very recently there were no direct spectroscopic observations on activated complexes, for they have a very fleeting existence and often survive for only a picosecond or so. However, the development of femtosecond pulsed lasers (1 fs = 10^{-15} s) and their application to chemistry in the form of **femtochemistry** has opened up the activated complex to experimental observation.

In a typical femtosecond experiment, a femtosecond pulse is used to excite a molecule to a dissociative state, and a second femtosecond laser pulse is fired at a series of intervals after the dissociating pulse. The frequency of the second pulse is set at an absorption of one of the free fragmentation products, so its absorption is a measure of the abundance of the dissociation product. For example, when ICN is dissociated by the first pulse, the emergence of CN from the photoactivated state can be monitored by watching the growth of the free CN absorption (or, more commonly, its laser-induced fluorescence). In this way it has been found that the CN signal remains zero until the fragments have separated by about 600 pm, which takes about 205 fs.

Some sense of the progress that has been made in the study of the intimate mechanism of chemical reactions can be had by considering the femtochemistry of an analogue of the harpoon reaction introduced in Section 28.1. The decay of the ion pair $Na^{+}X^{-}$, where X is a halogen, has been studied by exciting it with a femtosecond pulse to an excited state that can be interpreted as a covalently bonded NaX molecule. The second probe pulse examines the system at an absorption frequency either of the free Na atom or at a frequency at which the atom absorbs when it is a part of the complex. The latter frequency depends on the Na–X distance, so an absorption (in practice, a laser-induced fluorescence) is obtained each time the vibration of the complex returns it to that separation.

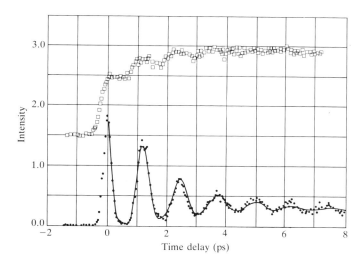

Fig. 28.5 Femtosecond spectroscopic results for the reaction in which NaI‡ separates into Na + I. The full circles are the absorption of the complex and the open squares the absorption of the free Na atoms (A. H. Zewail, *Science*, **242**, 1645 (1988)).

A typical set of results for NaI is shown in Fig. 28.5. The bound Na absorption intensity shows up as a series of pulses that recur in about 1 ps, showing that the complex vibrates with about that period. The decline in intensity shows the rate at which the complex can dissociate as the two atoms swing away from each other. The complex does not dissociate on every outward-going swing since there is a chance that the I atom can be harpooned again, in which case it fails to make good its escape. The free Na absorption also grows in an oscillating manner, showing the periodicity of the vibration of the complex that gives it a chance to dissociate each picosecond. The precise period of the oscillation in NaI is 1.25 ps, corresponding to a vibrational wavenumber of 27 cm^{-1} (recall that the activated complex theory assumes that such a vibration has a very low frequency). The complex survives for about ten oscillations. In contrast, the oscillation frequency of NaBr is similar, but the complex barely survives one period of vibration.

Femtosecond spectroscopy has also been used to examine the activated complex involved in bimolecular reactions. Thus, a molecular beam can be used to produce a **van der Waals molecule** in which two species are loosely bonded together by intermolecular forces. An example is the van der Waals molecule $IH\cdots OCO$ formed between HI and CO_2, where the interaction is by a kind of hydrogen bonding. The HI bond can be dissociated by a femtosecond pulse, and the H atom is ejected towards the O atom of the neighbouring CO_2 molecule to form HOCO. Hence, the van der Waals molecule is a source of the activated complex of the reaction

$$H + CO_2 \rightarrow [HOCO]^{\ddagger} \rightarrow HO + CO$$

The probe pulse is tuned to the OH radical, which enables the evolution of the activated complex $[HOCO]^{\ddagger}$ to be studied in real time.

28.6 Thermodynamic aspects

The statistical version of activated complex theory rapidly runs into difficulties because we rarely know anything about the structure of the activated complex. However, the concepts that it introduces, principally that of an equilibrium between the reactants and the activated complex, have motivated a more general approach in which the activation process is expressed in terms of thermodynamic functions.

Activation parameters

If we accept that $\bar{K}_p$ is an equilibrium constant (despite one mode of $C^{\ddagger}$ having been discarded), we can express it in terms of a **Gibbs function of activation** through

$$\Delta G^{\ddagger} = -RT \ln \bar{K}_p \qquad (17)$$

Then the rate constant becomes

$$k_2 = \kappa \frac{kT}{h} \times \frac{RT}{p^{\ominus}} e^{-\Delta G^{\ddagger}/RT}$$

Since $G = H - TS$, we can divide the Gibbs function of activation into an **entropy of activation** $\Delta S^{\ddagger}$ and an **enthalpy of activation** $\Delta H^{\ddagger}$ through

$$\Delta G^{\ddagger} = \Delta H^{\ddagger} - T \Delta S^{\ddagger} \qquad (18)$$

Gibbs functions, enthalpies, and entropies of activation are widely used to report experimental reaction rates, especially for organic reactions in solution. They are encountered when relationships between equilibrium constants and rates of reaction are explored using **correlation analysis** in which $\ln K$ (which is equal to $-\Delta G^{\ominus}/RT$) is plotted against $\ln k$ (which is proportional to $-\Delta G^{\ddagger}/RT$). In many cases the correlation is linear, signifying that as the reaction becomes thermodynamically more favourable, its rate constant increases (Fig. 28.6). This linear correlation is the origin of the alternative name **linear free energy relation** (LFER; see *Further reading*).

When eqn 14 is used in eqn 17 and κ is absorbed into the entropy term,

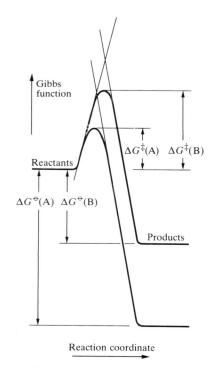

Fig. 28.6 For a related series of reactions, as the magnitude of the standard reaction Gibbs function increases, so the activation barrier decreases. The approximate correlation betwen $\Delta G^{\ddagger}$ and $\Delta G^{\ominus}$ is the origin of linear free energy relations.

we obtain

$$k_2 = Be^{\Delta S^{\ddagger}/R}e^{-\Delta H^{\ddagger}/RT} \qquad B = \frac{kT}{h} \times \frac{RT}{p^{\ominus}} \qquad (19)$$

We can express this equation in Arrhenius form by identifying the activation energy E_a using the definition in eqn 14 of Section 26.6. After some calculation, which is described in the *Further information* section at the end of the chapter, we find, for a bimolecular gas phase reaction,

$$\Delta H^{\ddagger} = E_a - 2RT \qquad (20a)$$

whereas for a reaction in solution (and for a unimolecular reaction in either phase)

$$\Delta H^{\ddagger} = E_a - RT \qquad (20b)$$

Consequently, we can write k_2 in the Arrhenius form $k_2 = Ae^{-E_a/RT}$ if we write (for a bimolecular gas phase reaction)

$$A = e^2 B \times e^{\Delta S^{\ddagger}/R} \qquad (21)$$

Hence, to calculate the entropy of activation from the pre-exponential factor in a bimolecular gas phase reaction, we use

$$\Delta S^{\ddagger} = R\left(\ln\frac{A}{B} - 2\right) \qquad (22)$$

Example 28.5: *Calculating the entropy of activation*

Calculate the activation Gibbs function, enthalpy, and entropy of the second-order hydrogenation of ethene at 355°C using information from Table 28.1.

Answer. Since $E_a = 180 \text{ kJ mol}^{-1}$, from eqn 20

$$\Delta H^{\ddagger} = E_a - 2RT = (180 - 2 \times 5.2) \text{ kJ mol}^{-1}$$

$$= +170 \text{ kJ mol}^{-1}$$

The practical form of eqn 22 is obtained using

$$\frac{A}{B} = \frac{5.7723 \times 10^{-10} \times A/(\text{M}^{-1}\text{s}^{-1})}{(T/\text{K})^2}$$

From the table, $A = 1.24 \times 10^6 \text{ M}^{-1}\text{s}^{-1}$; hence

$$\Delta S^{\ddagger} = R\left(\ln\frac{5.7723 \times 10^{-10} \times 1.24 \times 10^6}{628^2} - 2\right)$$

$$= -184 \text{ J K}^{-1} \text{ mol}^{-1}$$

Finally, from eqn 18,

$$\Delta G^{\ddagger} = \Delta H^{\ddagger} - T\Delta S^{\ddagger} = +286 \text{ kJ mol}^{-1}$$

Comment. Notice the large negative entropy of activation. The simple collision theory magnitude of A calculated in Example 28.1 corresponds to a much less negative value ($-73\,\mathrm{J\,K^{-1}\,mol^{-1}}$).

Exercise. Repeat the calculation for the $K + Br_2 \rightarrow KBr + Br$ reaction at 298 K.　　　　　　　　　　$[-2.5\,\mathrm{kJ\,mol^{-1}}, -58\,\mathrm{J\,K^{-1}\,mol^{-1}}, +15\,\mathrm{kJ\,mol^{-1}}]$

We can now make contact with the collision theory expression for the rate constant. In that theory, A is determined by the frequency of collisions in the gas. But collisions correspond to a decrease in entropy (because they correspond to molecules coming together and therefore to a reduction of randomness in the gas). Hence the negative value of $\Delta S^{\ddagger}$ reflects the occurrence of collisions. Furthermore, collisions with well-defined relative orientations correspond to an even greater reduction of entropy than is brought about by collisions in general, and so the entropy of activation should then be even more negative. This reduces the value of A, a feature taken into account in collision theory by the steric factor P.

Reactions between ions

The thermodynamic version of activated complex theory makes it relatively simple to discuss reactions in solution. The statistical theory is very complicated to apply because the solvent plays a role in the activated complex. In the thermodynamic approach we combine the rate law

$$\frac{d[P]}{dt} = k^{\ddagger}[C^{\ddagger}]$$

with the thermodynamic equilibrium constant

$$K = \frac{a_C}{a_A a_B} = K_{\gamma} \times \frac{[C^{\ddagger}]}{[A][B]}$$

K_{γ} is the ratio of activity coefficients,

$$K_{\gamma} = \frac{\gamma_C}{\gamma_A \gamma_B}$$

Then

$$\frac{d[P]}{dt} = k_2[A][B] \qquad k_2 = \frac{k^{\ddagger} K}{K_{\gamma}} \qquad (23a)$$

If k_2° is the rate constant when the activity coefficients are 1, we can write

$$k_2 = \frac{k_2^{\circ}}{K_{\gamma}} \qquad (23b)$$

At low concentrations the activity coefficients can be expressed in terms of the ionic strength I of the solution using the Debye–Hückel limiting law (Section 10.2, particularly eqn 5) in the form

$$\lg \gamma_J = -A z_J^2 I^{1/2}$$

Then

$$\lg k_2 = \lg k_2^\circ - A\{z_A^2 + z_B^2 - (z_A + z_B)^2\}I^{1/2}$$

and hence

$$\lg k_2 = \lg k_2^\circ + 2A z_A z_B I^{1/2} \qquad (24)$$

(The charges of A and B are z_A and z_B, and so the charge of the activated complex is $z_A + z_B$; the z_J are positive for cations and negative for anions.)

Equation 24 expresses the **kinetic salt effect**, the variation of the rate constant of a reaction between ions with the ionic strength of the solution (Fig. 28.7). If the ions have the same sign (as in a reaction between cations or between anions), increasing the ionic strength by the addition of inert ions increases the rate constant. The formation of a single, highly charged ionic complex from two less highly charged ions is favoured by a high ionic strength because the new ion has a denser ionic atmosphere. Conversely, ions of opposite charge react more slowly in solutions of high ionic strength. Now the charges cancel and the complex has a less favourable interaction with its atmosphere than the separated ions.

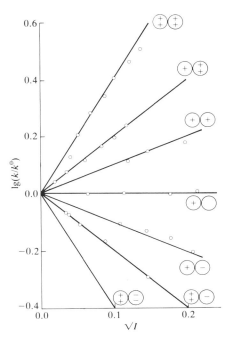

Fig. 28.7 Experimental tests of the kinetic salt effect for reactions in water at 298 K. The ion types are shown as spheres, and the slopes of the lines are those given by the Debye–Hückel limiting law and eqn 24.

Example 28.6: *Analysing the kinetic salt effect*

The rate constant for the base hydrolysis of $[CoBr(NH_3)_5]^{2+}$ depends on ionic strength as tabulated below. What can be deduced about the charge of the activated complex in the rate-determining step?

$I/(mol\ kg^{-1})$	0.005	0.010	0.015	0.020	0.025	0.030
k_2/k_2°	0.718	0.631	0.562	0.515	0.475	0.447

Answer. We plot $\lg k_2/k_2^\circ$ against $I^{1/2}$, and the slope will give us $1.02 z_A z_B$, from which we shall be able to infer the charges of the ions involved in the formation of the activated complex. Form the following table:

$I/(mol\ kg^{-1})$	0.005	0.010	0.015	0.020	0.025	0.030
$I^{1/2}/(mol\ kg^{-1})^{1/2}$	0.071	0.100	0.122	0.141	0.158	0.173
$\lg k_2/k_2^\circ$	−0.14	−0.20	−0.25	−0.29	−0.32	−0.35

These points are plotted in Fig. 28.8. The slope of the (least squares) line is −2.1, indicating that $z_A z_B = -2$. Since for the OH^- ion, $z_A = -1$, if that ion is involved in the formation of the activated complex, the charge number of the second ion is +2. This suggests that the pentamminebromocobalt(III) cation participates in the formation of the activated complex.

Comment. The rate constant is also influenced by the relative permittivity of the medium (see Problem 28.6).

Exercise. An ion of charge number +1 is known to be involved in the activated complex of a reaction. Deduce the charge number of the other ion from the following data:

$I/(mol\ kg^{-1})$	0.005	0.010	0.015	0.020	0.025	0.030
k_2/k_2°	0.98	0.97	0.97	0.96	0.96	0.95

[−1]

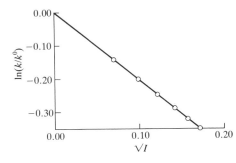

Fig. 28.8 The experimental ionic strength dependence of the rate constant of a hydrolysis reaction: the slope gives information about the charge types involved in the activated complex of the rate-determining step. See Example 28.6.

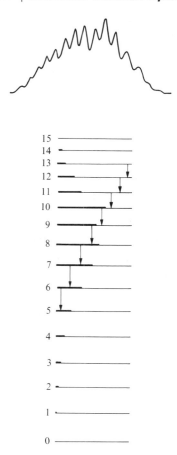

15
14
13
12
11
10
9
8
7
6
5
4
3
2
1
0

Fig. 28.9 Infrared chemiluminescence from CO produced in the reaction $O + CS \rightarrow CO + S$ arises from the non-equilibrium populations of the vibrational states of CO and the radiative relaxation to equilibrium.

The dynamics of molecular collisions

We now come to the third and most detailed level of our examination of the factors that govern the rates of reactions. Molecular beams allow us to study collisions between molecules in preselected states, and can be used to determine the states of the products of a reactive collision. Information of this kind is essential if a full picture of the reaction is to be built, because the rate constant is an average over events in which reactants in different initial states evolve into products in their final states.

Detailed experimental information comes from molecular beams, especially crossed molecular beams. The detector for the products of the collision of two beams can be moved to different angles, and so the angular distribution of the products can be determined. The detector can also distinguish between different energy states of the products. Therefore, since the molecules in the incoming beams can be prepared with different energies (e.g. different translational energies using rotating sectors and supersonic nozzles, and different vibrational energies using selective excitation with lasers) and different orientations (using electric fields), it is possible to study the dependence of the success of collisions on these variables and to study how they affect the properties of the outgoing product molecules. One method for examining the energy distribution in the products is **infrared chemiluminescence**, in which vibrationally excited molecules emit infrared radiation as they return to their ground states. By studying the intensities of the infrared emission spectrum, the populations of the vibrational states may be determined (Fig. 28.9).

28.7 Potential energy surfaces

One of the most important concepts for discussing beam results is the **potential energy surface** of a reaction, the potential energy as a function of the relative positions of all the atoms taking part in the reaction. For a collision between an H atom and an H_2 molecule, for instance, the potential energy surface is the plot of the potential energy for all relative locations of the three hydrogen nuclei. Detailed calculations show that the approach of an atom along the H—H axis requires less energy for reaction than any other approach, and so initially we confine our attention to a collinear approach. Two parameters are required to define the nuclear separations: one is the H_A—H_B separation R_{AB}, and the other is the H_B—H_C separation R_{BC}.

At the start of the encounter R_{AB} is infinite and R_{BC} is the H_2 equilibrium bond length. At the end of a successful reactive encounter R_{AB} is equal to the bond length and R_{BC} is infinite. The total energy of the three-atom system depends on their relative separations, and can be found by doing a molecular structure calculation using the Born–Oppenheimer approximation (Section 14.1) in which all atom locations are regarded as frozen in their instantaneous positions. The plot of the total energy of the system against each value of R_{AB} and R_{BC} gives the potential energy surface of this collinear reaction (Fig. 28.10). This surface is normally depicted as a contour diagram (Fig. 28.11).

When R_{AB} is very large, the variation in potential energy represented by the surface as R_{BC} changes are those of an isolated H_2 molecule as its bond length is altered. A section through the surface at $R_{AB} = \infty$, for example, is

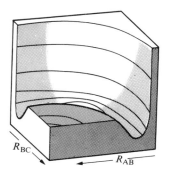

Fig. 28.10 The potential energy surface for the H + H_2 reaction when the atoms are constrained to be collinear.

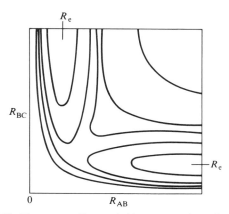

Fig. 28.11 The contour diagram (with contours of equal potential energy) corresponding to the surface in Fig. 28.10. R_e marks the equilibrium bond length of an H_2 molecule (strictly, it relates to the arrangement when the third atom is at infinity).

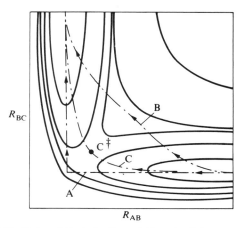

Fig. 28.12 Various trajectories through the potential energy surface shown in Fig. 28.11. A corresponds to a path in which R_{BC} is held constant as H_A approaches; B corresponds to a path in which R_{BC} lengthens at an early stage during the approach of H_A. C is the path along the floor of the potential valley.

the same as the H_2 potential energy curve like that drawn in Fig. 14.1. At the edge of the diagram where R_{BC} is very large, a section through the surface is the molecular potential energy curve of an isolated H_A—H_B molecule.

The actual path of the atoms in the course of the encounter depends on their total energy, the sum of their kinetic and potential energies. However, we can obtain an initial idea of the paths available to the system by considering the potential energy surface alone, and looking for paths that correspond to least potential energy. For example, consider the changes in potential energy as H_A approaches H_B—H_C. If the H_B—H_C bond length is constant during the initial approach of H_A, the potential energy of the H_3 cluster would rise along the path marked A in Fig. 28.12. We see that the potential energy rises to a high value as H_A is pushed into the molecule, and then decreases sharply as H_C breaks off and separates a great distance. An alternative reaction path can be imagined (B) in which the H_B—H_C bond length increases while H_A is still far away. It is clear that both paths, although feasible if the molecules have sufficient initial kinetic energy, take the three atoms to regions of high potential energy.

The path of least potential energy is the one marked C. It corresponds to R_{BC} lengthening as H_A approaches and begins to form a bond with H_B. The H_B—H_C bond relaxes at the demand of the incoming atom, and although the potential energy rises, it climbs only as far as the saddle-shaped region of the surface, to the **saddle point** marked $C^{\ddagger}$. The encounter of least potential energy is one in which the atoms take route C up the floor of the valley, through the saddle point, and down the floor of the other valley as H_C recedes and the new H_A—H_B bond achieves its equilibrium length. This path is the reaction coordinate we met in Section 28.4.

Motion through the surface

The type of question explored by molecular beam studies can be introduced by considering how the molecules move through their potential energy surface when they also possess kinetic energy. To travel successfully along C, the incoming molecules must possess enough kinetic energy to be able to climb to the saddle point. Therefore, the shape of the surface can be explored experimentally by changing the relative speed of approach (by selecting the beam velocity) and determining the kinetic energy at which reaction occurs. If the collision occurs with a high kinetic energy, the saddle point might be overshot and a path may be taken that leads to vibrational excitation of the product molecule (Fig. 28.13). By observing the degree of excitation of the products, details of the shape of the potential energy surface can be investigated.

The type of practical question that can be answered is whether it is better to smash the reactants together with a lot of translational kinetic energy or to ensure instead that they approach in highly excited vibrational states. For example, is trajectory C_2^* in Fig. 28.13, where the H_B-H_C molecule is initially vibrationally excited, more efficient at leading to reaction than the trajectory C_1^*, in which the total energy is the same but has a high translational kinetic energy? We see below something of the answers to questions like this.

The relation of the surface and the rate constant

A related question is how the information given by molecular beam studies is related to the value of the rate constant of the reaction. This is the most complex part of the analysis, because in an actual reaction the molecules collide with many different energies and in many different rotational and vibrational states. Each individual collision can be imagined as a trajectory

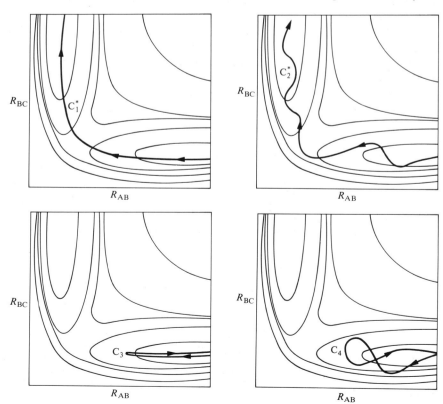

Fig. 28.13 Some successful (*) and unsuccessful encounters. C_1^* corresponds to the path along the foot of the valley; C_2^* corresponds to an approach of A to a vibrating BC molecule, and the formation of a vibrating AB molecule as C departs. C_3 corresponds to A approaching a non-vibrating BC molecule, but with insufficient translational kinetic energy; C_4 corresponds to A approaching a vibrating BC molecule, but still the energy, and the phase of the vibration, is insufficient for reaction.

on the potential energy surface. Some of these trajectories will be successful (C_1^* and C_2^* in Fig. 28.13) and some unsuccessful (C_3 and C_4) either because they lack sufficient energy or because the energy is distributed inappropriately. The rate of reaction is an average over all these trajectories, and so the calculation of the rate constant requires the calculation of many trajectories and then averaged in some way. One of the techniques for doing the average in a manner that is consistent with the Boltzmann distribution of populations over the states of the system is the Monte Carlo method, which was described in Section 22.7.

28.8 Some results from experiments and calculations

In this section we see how some of the questions just raised are answered, and how the study of collisions and the calculation of potential energy surfaces gives some insight into the course of reactions.

The direction of attack and separation

Figure 28.14 shows the results of a calculation of the potential energy as an H atom approaches an HI molecule from different angles, the H_2 bond being allowed to relax to the optimum length in each case. The potential energy barrier is least for collinear attack, as we assumed earlier. (But we must be aware that other lines of attack are feasible and contribute to the overall rate.) In contrast, Fig. 28.15 shows the potential energy changes that occur as a Cl atom approaches an H_2 molecule. The lowest barrier occurs for approaches within a cone of half-angle 30° surrounding the H atom. The relevance of this result to the calculation of the steric factor of collision theory should be noted: not every collision is successful, because not every one lies within the reactive cone.

If the collision is sticky, so that when the reactants collide they orbit around each other, the products can be expected to emerge in random directions because all memory of the approach direction has been lost. A rotation takes about 1 ps, and so if the collision is over in less than that time the complex will not have had time to rotate and the products will be thrown off in a specific direction. In the collision of K and I_2, for example, most of the products are thrown off in the forward direction.[4] This is consistent with the harpoon mechanism (Section 28.1) because the transition takes place at long range. In contrast, the collision of K with CH_3I leads to reaction only if the particles approach each other very closely. This is like K bumping into a brick wall, and the KI product bouncing out in the backward direction. The detection of this anisotropy in the angular distribution of products gives an indication of the distance and orientation of approach needed for reaction, as well as showing that the event is complete in less than 1 ps.

Attractive and repulsive surfaces

Some reactions are very sensitive to whether the energy has been predigested into a vibrational mode or left as the relative translational

[4] There is a subtlety here. In molecular beam work the remarks normally refer to directions in a centre-of-mass coordinate system. The origin of the coordinates is the centre of mass of the colliding reactants, and the collision takes place when the molecules are at the origin. The way in which centre-of-mass coordinates are constructed and the events in them interpreted involves too much detail for our present purposes, but we should bear in mind that 'forward' and 'backward' have unconventional meanings. The details are explained in the books in *Further reading*.

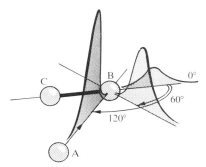

Fig. 28.14 An indication of the anisotropy of the potential energy changes as H approaches H_2 with different angles of attack. The collinear attack has the lowest potential barrier to reaction. The curves show the potential energy profiles along the reaction coordinate for each configuration.

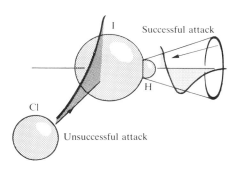

Fig. 28.15 Partial potential energy barriers for approach of Cl to HI. In this case successful encounters occur only when Cl approaches within a cone surrounding the H atom.

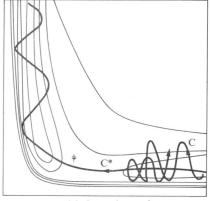

(a) Attractive surface

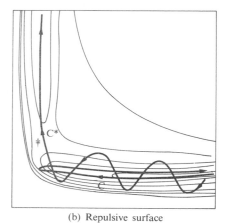

(b) Repulsive surface

Fig. 28.16 (a) An attractive potential energy surface. A successful encounter (*C**) involves high translational kinetic energy and results in a vibrationally excited product. (b) A repulsive potential energy surface. A successful encounter (*C**) involves initial vibrational excitation and the products have high translational kinetic energy. A surface that is attractive in one direction is repulsive in the reverse direction.

kinetic energy of the colliding molecules. For example, if two HI molecules are hurled together with more than twice the activation energy of the reaction, no reaction occurs if all the energy is translational. For $F + HCl \rightarrow Cl + HF$ it has been found that the reaction is about five times as efficient when the HCl is in its first vibrational excited state than when, although it has the same total energy, it is in its vibrational ground state.

The origin of these requirements can be found by examining the potential energy surface. Figure 28.16a shows an **attractive surface** in which the saddle point occurs early in the reaction coordinate. Figure 28.16b is a **repulsive surface** in which the saddle point occurs late. A surface that is attractive in one direction is repulsive in the reverse direction.

Consider first the attractive surface. If the original molecule is vibrationally excited, a collision with an incoming molecule takes the system along C. This path is bottled up in the region of the reactants, and does not take the system to the saddle point. If, however, the same amount of energy is present solely as translational kinetic energy, the system moves along C* and travels smoothly over the saddle point into products. We can therefore conclude that reactions with attractive potential energy surfaces proceed more efficiently if the energy is in relative translational motion. Moreover, the potential surface shows that once past the saddle point the trajectory runs up the steep wall of the product valley, and then rolls from side to side as it falls to the foot of the valley as the products separate. In other words, the products emerge in a vibrationally excited state.

Now consider the repulsive surface (Fig. 28.16b). On trajectory C the collisional energy is largely in translation. As the reactants approach, the potential energy rises. Their path takes them up the opposing face of the valley, and they are reflected back into the reactant region. This path corresponds to an unsuccessful encounter, even though the energy is sufficient for reaction. On C* some of the energy is in the vibration of the reactant molecule and the motion causes the trajectory to weave from side to side up the valley as it approaches the saddle point. This motion may be sufficient to tip the system round the corner to the saddle point and then on to products. In this case, the product molecule is expected to be in an unexcited vibrational state. It follows that reactions with repulsive potential surfaces can be expected to proceed more efficiently if the excess energy is present as vibrations. This is the case with the $H + Cl_2 \rightarrow HCl + Cl$ reaction, for instance.

Classical trajectories

A clear picture of the reaction event can be obtained using classical mechanics to calculate the trajectories of the atoms taking place in a reaction. Figure 28.17 shows the result of such a calculation of the positions of the three atoms in the reaction $H + H_2 \rightarrow H_2 + H$, the horizontal coordinate now being time and the vertical coordinate the separations. This illustration shows clearly the vibration of the original molecule and the approach of the attacking atom. The reaction itself, the switch of partners, takes place very rapidly and is an example of a **direct mode** process. Then the new molecule shakes, but quickly settles down to steady, harmonic vibration as the expelled atom departs. In contrast, Fig. 28.18 shows an example of a **complex mode** process, in which the activated complex survives for an extended period. The reaction in the illustration is the

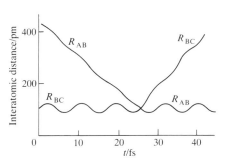

Fig. 28.17 The calculated trajectories for a reactive encounter between A and a vibrating BC molecule leading to the formation of a vibrating AB molecule. This 'direct mode' reaction is between H and H_2. (M. Karplus, R. N. Porter, and R. D. Sharma, *J. chem. Phys.* **43**, 3258 (1965).)

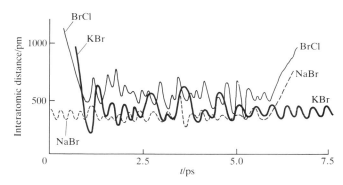

Fig. 28.18 An example of the trajectories calculated for a 'complex mode' reaction, KCl + NaBr → KBr + NaCl, in which the collision cluster has a long lifetime. (P. Brumer and M. Karplus, *Faraday Disc. chem. Soc.* **55**, 80 (1973).)

exchange reaction

$$KCl + NaBr \rightarrow KBr + NaCl$$

The tetraatomic activated complex survives for about 5 ps, during which time the atoms make about 15 oscillations before dissociating into products.

Although this kind of calculation gives a good sense of what happens during a reaction, its limitations must be kept in mind. In the first place, a real gas-phase reaction occurs with a wide variety of different speeds and angles of attack. In the second place, the motion of the atoms, electrons, and nuclei, is governed by quantum mechanics. The concept of trajectory then fades and is replaced by the unfolding of a wavefunction that represents initially the reactants and finally products. Nevertheless, recognizing these limitations should not be allowed to obscure the fact that recent advances in molecular reaction dynamics have given us a first glimpse of the processes going on at the core of reactions.

Further information: energies and entropies of activation

We establish here some of the relations between the Gibbs function, enthalpy, entropy, and energy of activation that were quoted in the text. The starting point is eqn 19 for the rate constant:

$$k_2 = Be^{\Delta S^{\ddagger}/R}e^{-\Delta H^{\ddagger}/RT}$$

The enthalpy of activation for a bimolecular reaction in the gas phase is related to the **internal energy of activation** $\Delta U^{\ddagger}$ by

$$\Delta H^{\ddagger} = \Delta U^{\ddagger} - RT \qquad (A1)$$

because two molecules form one activated complex. Hence,

$$k_2 = eBe^{\Delta S^{\ddagger}/R}e^{-\Delta U^{\ddagger}/RT}$$

For a reaction in solution, $\Delta H^{\ddagger}$ is negligibly different from $\Delta U^{\ddagger}$ and we can write

$$k_2 = Be^{\Delta S^{\ddagger}/R}e^{-\Delta U^{\ddagger}/RT}$$

A final preliminary piece of work is to define the **Helmholtz energy of activation** $\Delta A^{\ddagger}$ through

$$\Delta A^{\ddagger} = \Delta U^{\ddagger} - T\Delta S^{\ddagger} \qquad \text{(A2)}$$

for then we can write

$$k_2 = eBe^{-\Delta A^{\ddagger}/RT}$$

for a gaseous bimolecular reaction and a similar expression without the factor e for a reaction in solution (and for a unimolecular reaction).

The activation energy of a reaction is defined by eqn 14 of Section 26.6. Since it is natural to use ΔA when discussing changes at constant volume (and E_a is defined in terms of a partial derivative at constant volume), we write

$$E_a = RT^2 \left(\frac{\partial \ln k}{\partial T} \right)_V$$

$$= RT^2 \left(\frac{\partial \ln (Be)}{\partial T} \right)_V - T^2 \left(\frac{\partial (\Delta A^{\ddagger}/T)}{\partial T} \right)_V$$

$$= RT + T^2 \times \frac{\Delta U^{\ddagger}}{T^2} = RT + \Delta U^{\ddagger}$$

(In the first differentiation we use $B \propto T$ because $RT/p \propto V$ and the volume is constant.) The final steps follow from the constant-volume version of the Gibbs–Helmholtz equation (Section 5.2):

$$\frac{d(\Delta A/T)}{dT} = \frac{-\Delta U}{T^2}$$

It follows from eqn A1 that for a bimolecular gaseous reaction

$$E_a = \Delta H^{\ddagger} + 2RT$$

For a reaction in solution when $\Delta H^{\ddagger} \approx \Delta U^{\ddagger}$,

$$E_a = \Delta H^{\ddagger} + RT$$

Further reading

Activated complex theory

M. J. Pilling, *Reaction kinetics*. Clarendon Press, Oxford (1974).

K. J. Laidler, *Chemical kinetics*. Harper and Row, New York (1987).

I. W. M. Smith, *Kinetics and dynamics of elementary gas reactions*. Butterworth, London (1980).

J. Simons, *Energetic principles of chemical reactions*. Jones and Bartlett, Portola Valley (1983).

R. A. Marcus, Activated complex theory: current status, extensions, and applications. In *Techniques of chemistry* (ed. E. S. Lewis) VIA, p. 13. Wiley-Interscience, New York (1974).

Diffusion with reaction

K. G. Denbigh and J. C. R. Turner, *Chemical reactor theory* (3rd edn). Cambridge University Press (1984).

J. Crank, *The mathematics of diffusion*. Clarendon Press, Oxford (1975).

Time-resolved spectra

A. H. Zewail, Laser femtochemistry. In *Science,* **242**, 1645 (1988).

A. H. Zewail and R. B. Bernstein, Real-time laser femtochemistry. In *Chem. & Eng. News,* **66**, 24 (1988).

D. Husain, Fundamental collisional processes of alkali and alkaline earth atoms studied by spectroscopic methods. *J. chem. Soc. Faraday Trans.* 2, **85**, 85 (1989).

Molecular beams

M. A. D. Fluendy and K. P. Lawley, *Molecular beams in chemistry*. Chapman and Hall, London (1974).

J. E. Jordan, E. A. Mason, and I. Amdur, Molecular beams in chemistry. In *Techniques of chemistry* (ed. A. Weissberger and B. W. Rossiter) IIID, p. 365. Wiley-Interscience, New York (1972).

R. Grice, Reactive scattering. In *Adv. chem. Phys.,* **30**, 249 (1975).

R. D. Levine and R. B. Bernstein, *Molecular reaction dynamics*. Clarendon Press, Oxford (1987).

K. R. Lawley, Potential energy surfaces. In *Adv. chem. Phys.,* **42**, Wiley-Interscience, New York (1980).

R. B. Bernstein, *Chemical dynamics via molecular beam and laser techniques*. Clarendon Press, Oxford (1982).

D. M. Hirst, *Potential energy surfaces*. Taylor and Francis, London (1985).

Exercises

28.1 Calculate the collision frequency z and the collision density Z in (a) ammonia, $R \approx 190$ pm, (b) carbon monoxide, $R \approx 180$ pm at 25°C and 1 atm. What is the percentage increase when the temperature is raised by 10 K at constant volume?

28.2 Collision theory depends on knowing the fraction of molecular collisions having at least the kinetic energy E_a along the line of flight. What is this fraction when (a) $E_a = 10$ kJ mol^{-1}, (b) $E_a = 100$ kJ mol^{-1} at (1) 300 K and (2) 1000 K?

28.3 Calculate the percentage increase in the fractions in Exercise 28.2 when the temperature is raised by 10 K.

28.4 Calculate the magnitude of the diffusion-controlled rate constant at 298 K for a species in (a) water, (b) pentane, (c) decylbenzene. The viscosities are 1.00 g m^{-1} s^{-1}, 0.22 g m^{-1} s^{-1}, 3.36 g m^{-1} s^{-1} (1.00 cP, 0.22 cP, 3.36 cP), respectively.

28.5 The temperature dependence of the rate constant for a certain bimolecular reaction in the gas phase is given by the expression $k_2 = 3.72 \times 10^2 e^{-8600\text{K}/T}$ M^{-1} min^{-1} in the temperature range 10 to 90°C. The molar masses of the two reacting species are 16 and 100 g mol^{-1}. Use collision theory to calculate the reactive cross-section of the reaction at 25°C.

28.6 For the gaseous reaction A + B → P, the reactive cross-section obtained from the experimental value of the pre-exponential factor is 9.2×10^{-22} m^2. The collision cross-sections of A and B estimated from the transport properties are 0.95 and 0.65 nm^2 respectively. Calculate the P-factor for the reaction.

28.7 A and B, which are neutral species with diameters 588 pm and 1650 pm respectively, undergo a diffusion controlled reaction A + B → P in a solvent of viscosity 2.37×10^{-3} kg m^{-1} s^{-1} at 40°C. Calculate the initial rate d[P]/dt if the initial concentrations of A and B are 0.150 M and 0.330 M respectively.

28.8 The reaction of propylxanthate ion in acetic acid buffer solutions has the mechanism A$^-$ + H$^+$ → P. Near 30°C the rate constant is given by the empirical expression $k_2 = 2.05 \times 10^{13} e^{-8681\text{K}/T}$ M^{-1} s^{-1}. Evaluate the energy and entropy of activation at 30°C.

28.9 When the reaction in Exercise 28.8 occurs in a dioxane–water mixture which is 30 mass per cent dioxane, the rate constant fits $k_2 = 7.78 \times 10^{14} e^{-9134\text{K}/T}$ M^{-1} s^{-1} near 30°C. Calculate $\Delta G^{\ddagger}$ for the reaction at 30°C.

28.10 The gas phase association reaction between F$_2$ and IF$_5$ is first order in each of the reactants. The energy of activation for the reaction is 58.6 kJ mol^{-1}. At 65°C the rate constant is $7.84 \times 10^{-3} \text{ kPa}^{-1} \text{ s}^{-1}$. Calculate the Gibbs function and entropy of activation at 65°C.

28.11 Calculate the molar entropy of activation for a collision between two essentially structureless particles at 300 K, taking $M \approx 50 \text{ g mol}^{-1}$ and $\sigma \approx 0.4 \text{ nm}^2$.

28.12 The pre-exponential factor for the gas-phase decomposition of ozone at low pressures is $4.6 \times 10^{12} \text{ M}^{-1} \text{ s}^{-1}$ and its activation energy is 10.0 kJ mol^{-1}. What is (a) the entropy of activation, (b) the enthalpy of activation, (c) the Gibbs function of activation at 298 K?

28.13 The base-catalysed bromination of nitromethane-d_3 in water at room temperature (298 K) proceeds 4.3 times more slowly than the bromination of the undeuterated material. Account for this difference. Use $k_f(C—H) \approx 450 \text{ N m}^{-1}$.

28.14 Predict the order of magnitude of the isotope effect on the relative rates of displacement of (a) ^1H and ^3H, (b) ^{16}O and ^{18}O from an organic molecule. Will raising the temperature enhance the difference? Take $k_f(CH) \approx 450 \text{ N m}^{-1}$, $k_f(CO) \approx 1750 \text{ N m}^{-1}$.

28.15 The rate constant of the reaction

$$H_2O_2(aq) + I^-(aq) + H^+(aq) \rightarrow H_2O(l) + HIO(aq)$$

is sensitive to the ionic strength of the aqueous solution in which the reaction occurs. At 25°C and at an ionic strength of $0.0525 \text{ mol kg}^{-1}$, $k = 12.2 \text{ M}^{-2} \text{ min}^{-1}$. Use the Debye–Hückel limiting law to estimate the rate constant at zero ionic strength.

28.16 The rate constant of the reaction

$$I^-(aq) + H_2O_2(aq) \rightarrow H_2O(l) + IO^-(aq)$$

varies slowly with ionic strength, even though the Debye–Hückel limiting law predicts no effect. Use the following data from 25°C to find the dependence of $\lg k_r$ on the ionic strength:

$I/(\text{mol kg}^{-1})$	0.0207	0.0525	0.0925	0.1575
$k_r/(\text{M}^{-1} \text{min}^{-1})$	0.663	0.670	0.779	0.694

Evaluate the limiting value of k_r at zero ionic strength. What does the result suggest for the dependence of $\lg \gamma$ on ionic strength for a neutral molecule in an electrolyte solution?

28.17 Use the Debye–Hückel limiting law to show that changes in ionic strength can affect the rate of reaction catalysed by H^+ from the ionization of a weak acid. Consider the mechanism: $H^+(aq) + B(aq) \rightarrow P$, where H^+ comes from the ionization of the weak acid, HA. The weak acid has a fixed concentration. First show that $\lg [H^+]$, derived from the ionization of HA, depends on the activity coefficients of ions and thus depends on the ionic strength. Then find the relationship between $\lg$ (rate) and $\lg [H^+]$ to show that the rate also depends on the ionic strength.

Problems

Numerical problems

28.1 In the dimerization of methyl radicals at 25°C the experimental pre-exponential factor is $2.4 \times 10^{10} \text{ M}^{-1} \text{ s}^{-1}$. What is (a) the reactive cross-section, (b) the P-factor for the reaction if the C–H bond length is 154 pm?

28.2 Nitrogen dioxide reacts bimolecularly in the gas phase to give $2NO + O_2$. The temperature dependence of the second-order rate constant for the rate law $v = k[NO_2]^2$ is given below. What is the P-factor and the reactive cross-section for the reaction?

T/K	600	700	800	1000
$k/(\text{cm}^3 \text{mol}^{-1} \text{s}^{-1})$	4.6×10^2	9.7×10^3	1.3×10^5	3.1×10^6

Take $\sigma \approx 0.60 \text{ nm}^2$.

28.3 The diameter of the methyl radical is about 308 pm. What is the maximum rate constant in the expression $d[C_2H_6]/dt = k[CH_3]^2$ for second-order recombination of radicals at room temperature? 10 per cent of a sample of ethane at 298 K and 1.0 atm pressure is dissociated into methyl radicals. What is the minimum time for 90 per cent recombination?

28.4 The rates of thermolysis of a variety of *cis*- and *trans*-azoalkanes have been measured over a range of temperatures in order to settle a controversy concerning the mechanism of the reaction. In ethanol an unstable *cis*-azoalkane decomposed at a rate that was followed by observing the N_2 evolution, and this led to the rate constants listed below (P. S. Engel and D. J. Bishop, *J. Amer. chem. Soc.*, **97**, 6754 (1975)). Calculate the enthalpy, entropy, energy, and Gibbs function of activation at −20°C.

$\theta/°C$	−24.82	−20.73	−17.02	−13.00	−8.95
$10^4 \times k/\text{s}^{-1}$	1.22	2.31	4.39	8.50	14.3

28.5 In an experimental study of a bimolecular reaction in aqueous solution, the second-order rate constant was measured at 25°C and at a variety of ionic strengths and the results are tabulated below. It is known that a singly charged ion is involved in the rate-determining step. What is the charge on the other ion involved?

$I/(\text{mol kg}^{-1})$	0.0025	0.0037	0.0045	0.0065	0.0085
$k/(\text{M}^{-1} \text{s}^{-1})$	1.05	1.12	1.16	1.18	1.26

28.6 If the activated complex is formed from ions of charges $z'e$ and $z''e$, and there is some characteristic distance $R^{\ddagger}$ between them in the activated complex, then the Gibbs function of activation will contain a term proportional to $z'z''/R^{\ddagger}\varepsilon_r$ where ε_r is the relative permittivity of the solvent. Deduce the expression

$$\ln k_{\text{eff}} = \ln k_{\text{eff}} - \frac{z'z''B}{\varepsilon_r}$$

with $B = e^2/4\pi\varepsilon_0 R^{\ddagger}kT$, for the dependence of the rate constant on ε_r. (The problem does not involve questions of ionic strength.) The model can be tested using the following data. Bromophenol blue fades when OH^- is added, the reaction rate being controlled by the step that can be symbolized as $B^{2-} + OH^- \rightarrow BOH^{3-}$. The reaction between an azodicarbonate ion (A^{2-}) and H^+ has a rate-determining step that may be symbolized as $A^{2-} + H^+ \rightarrow AH^-$. Both reactions were carried out in solvents of different relative permittivities, and the results are below. Do they support the model?

Bromophenol blue reaction:

ε_r	60	65	70	75	79
$\lg k_{\text{eff}}$	−0.987	0.201	0.751	1.172	1.401

Azodicarbonate ion reaction:

ε_r	27	35	45	55	65	79
$\lg k_{\text{eff}}$	12.95	12.22	11.58	11.14	10.73	10.34

28.7 The total cross-sections for reactions between alkali metal atoms and halogen molecules are given in the table below (R. D. Levine and R. B. Bernstein, *Molecular reaction dynamics*, Clarendon Press, Oxford, p. 72 (1974)). Assess the data in terms of the harpoon mechanism.

σ^*/nm^2	Cl_2	Br_2	I_2
Na	1.24	1.16	0.97
K	1.54	1.51	1.27
Rb	1.90	1.97	1.67
Cs	1.96	2.04	1.95

Electron affinities are approximately 1.3 eV (Cl_2), 1.2 eV (Br_2), and 1.7 eV (I_2), and ionization energies are 5.1 eV (Na), 4.3 eV (K), 4.2 eV (Rb), and 3.9 eV (Cs).

Theoretical problems

28.8 Confirm that eqn 10 is a solution of eqn 9, where [J] is a solution of the same equation but with $k = 0$ and for the same initial conditions.

28.9 Evaluate [J]* numerically by writing a program for the numerical integration of eqn 10, and explore the effect of increasing reaction rate constant on the spatial distribution of J.

28.10 Estimate the orders of magnitude of the partition functions involved in the rate expression. State the order of magnitude of q_m^T/N_A, q^R, q^V, and q^E for typical molecules.

Check that in the collision of two structureless molecules the order of magnitude of the pre-exponential factor is of the same order as that predicted by collision theory. Go on to estimate the P-factor for a reaction in which $A + B \rightarrow P$, and A and B are nonlinear triatomic molecules.

28.11 The major difficulty in applying activated complex theory (and it must be admitted, in devising straightforward problems to illustrate it) is to decide on the structure of the activated complex and to ascribe appropriate bond strengths and lengths to it. The following exercise gives some familiarity with the difficulties involved, yet leads to a numerical result for a reaction of some interest. Consider the attack of H on D_2, which is one step in the $H_2 + D_2$ reaction. Suppose that the H approaches D_2 from the side and forms a complex in the form of an isosceles triangle. Take the H–D distance as 30 per cent greater than in H_2 (74 pm) and the D–D distance as 20 per cent greater than in H_2. Let the critical coordinate be the antisymmetric stretching vibration in which one H—D bond stretches as the other shortens. Let all the vibrations be at about $1000\ \text{cm}^{-1}$. Estimate k_2 for this reaction at 400 K using the experimental activation energy of about $35\ \text{kJ mol}^{-1}$.

28.12 Now change the model of the activated complex in Problem 28.11 and make it linear. Use the same estimated molecular bond lengths and vibrational frequencies to calculate k_{eff} for this choice of model.

28.13 Clearly, there is much scope for modifying the parameters of the models of the activated complex in the last pair of Problems. Write and run a program that allows you to vary the structure of the complex and the parameters in a plausible way, and look for a model (or more than one model) that gives a value of k close to the experimental value, $4 \times 10^5\ \text{M}^{-1}\text{s}^{-1}$.

28.14 The Eyring equation can also be applied to physical processes. As an example, consider the rate of diffusion of an atom stuck to the surface of a solid. Suppose that in order to move from one site to another it has to reach the top of the barrier where it can vibrate classically in the vertical direction and in one horizontal direction, but vibration along the other horizontal direction takes it into the neighbouring site. Find an expression for the rate of diffusion, and evaluate it for W atoms on a tungsten surface ($E_a = 60\ \text{kJ mol}^{-1}$). Suppose that the vibration frequencies at the transition state are (a) the same as, (b) one half the value for the adsorbed atom. What is the value of the diffusion coefficient D at 500 K? (Take the site separation as 316 pm and $\nu \approx 10^{11}\ \text{Hz}$.)

28.15 Suppose now that the adsorbed, migrating species treated in Problem 28.14 is a spherical molecule, and that it can rotate classically as well as vibrate at the top of the barrier, but that at the adsorption site itself it can only vibrate. What effect does this have on the diffusion constant? Take the molecule to be methane, for which $B = 5.24\ \text{cm}^{-1}$.

28.16 Show that the intensities of a molecular beam before and after passing through a chamber of length l containing inert scattering atoms are related by $I = I_0 e^{-\mathcal{N}\sigma l}$, where σ is the collision cross-section and $\mathcal{N}$ the number density of scattering atoms.

28.17 In a molecular beam experiment to measure collision cross-sections it was found that the intensity of a CsCl beam was reduced to 60 per cent of its intensity on passage through CH_2F_2 at $10\ \mu$Torr, but that when the target was Ar at the same pressure the intensity was reduced only 10 per cent. What are the relative cross-sections of the two types of collision? Why is one much larger than the other?

Processes at solid surfaces

29

Check-list of key ideas

1. The role of *defects* in the growth of surfaces and the self-propagating property of a *screw dislocation* (Section 29.1).

2. The importance of *high-vacuum techniques* when studying surfaces (Section 29.2).

3. The study of surface composition using *photoemission spectroscopy, electron energy-loss spectroscopy,* and *Auger spectroscopy* (Section 29.2).

4. The composition of surfaces from *low-energy electron diffraction, field emission microscopy,* and *field ionization microscopy* (Section 29.2).

5. The use of *scanning tunnelling microscopy* to show details of surface structure (Section 29.2).

6. The definition of *fractional coverage* and its determination.

7. The processes of *physisorption* and *chemisorption* and the experimental distinctions between them (Section 29.3).

8. The derivation of the *Langmuir isotherm* (eqn 2) and its modification when the adsorbate dissociates (eqn 3).

9. The determination of the *isosteric enthalpy of adsorption* from the adsorption isotherm (eqn 4 and Example 29.2).

10. The *Brunauer–Emmett–Teller isotherm* when multilayer adsorption may occur (eqn 5).

11. The *Temkin isotherm* (eqn 8) and the *Freundlich isotherm* (eqn 9).

12. The *rate of adsorption* in terms of the sticking probability (Section 29.5).

13. The *rate of desorption* in terms of the *desorption activation energy* (eqn 10) and the technique of *flash desorption spectroscopy* (Section 29.5).

14. The *mobility* of adsorbates on surfaces (Section 29.5).

15. The *Eley–Rideal mechanism* (eqn 12) and the *Langmuir–Hinshelwood mechanism* (eqn 14) of heterogeneous catalysis.

16. The significance of a *volcano curve* and examples of catalysis (Section 29.7).

Processes at surfaces govern most aspects of daily life, including life itself. Even if we limit our attention to solid surfaces the importance of the processes is hardly reduced. Processes at solid surfaces govern the viability of industry both constructively, as in catalysis, and destructively, as in corrosion.

A fundamental dynamical aspect of a surface is its role as a region where a solid grows, evaporates, and dissolves. We shall therefore begin with a brief look at the growth of solids and the structures of simple surfaces. Then we shall see how surfaces are contaminated by the deposition of foreign material, and how that contamination is responsible for heterogeneous catalysis. The special case of electrode surfaces is treated in the next chapter.

The growth and structure of surfaces

In this section we see how surfaces are extended and crystals grow, and begin to picture the structures of surfaces that are responsible for catalysis. The accumulation of particles at a surface is called **adsorption**. The substance that adsorbs is the **adsorbate** and the underlying material is the **adsorbent** or **substrate**. The reverse of adsorption is **desorption**.

29.1 Surface growth

A simple picture of a perfect crystal surface is as a cobbled street. A gas molecule that collides with the surface can be imagined as a ball bouncing erratically over the cobbles. The molecule loses energy as it bounces, but it is likely to escape from the surface before it has lost enough kinetic energy to be trapped. The same is true, to some extent, of an ionic crystal in contact with a solution. There is little energy advantage for an ion in solution to discard some of its solvating molecules and stick at an exposed position on the surface. However, the actual changes in energy and entropy (collectively, the Gibbs function) are difficult to assess because an ion interacts with other ions below the surface of the solid and discards its solvating molecules as it attaches to the surface.

The role of defects

The picture changes when the surface has **defects**, for then there are ridges of incomplete layers of atoms or ions. A typical type of surface defect is a **step** between two otherwise flat layers of atoms called **terraces** (Fig. 29.1). A step defect might itself have defects in the form of **kinks**. When an atom settles on a terrace it bounces across it under the influence of the intermolecular potential, and might come to a step or a corner formed by a kink. Instead of interacting with a single terrace atom it now interacts with several, and the interaction may be strong enough to trap it. Likewise, when ions deposit from solution, the loss of the solvation interaction is offset by a coulombic interaction between the arriving ions and several ions at the surface defect.

Dislocations

Not all kinds of defect result in sustained surface growth. As the process of settling into ledges and kinks continues, there comes a stage when an entire

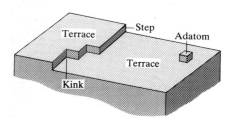

Fig. 29.1 Some of the kinds of defects that may occur on otherwise perfect terraces. Defects play an important role in surface growth and catalysis.

lower terrace has been covered. At this stage the surface defects have been eliminated, and growth will cease. For continuing growth, a surface defect is needed that propagates as the crystal grows. We can see what form of defect this must be by considering the types of **dislocations**, or discontinuities in the regularity of the lattice, that exist in the bulk of a crystal. One reason for their formation may be that the crystal grows so quickly that its particles do not have time to settle into states of lowest potential energy before being trapped in position by the deposition of the next layer.

A special kind of dislocation is the **screw dislocation** shown in Fig. 29.2. Imagine a cut in the crystal, with the particles to the left of the cut pushed up through the distance of one unit cell. The unit cells now form a continuous spiral around the end of the cut, which is called the **screw axis**. A path encircling the screw axis spirals up to the top of the crystal, and where the dislocation breaks through to the surface it takes the form of a spiral ramp (Fig. 29.2).

The surface defect formed by a screw dislocation is a step, possibly with kinks, where growth can occur. The incoming particles lie in ranks on the ramp, and successive ranks reform the step at an angle to its initial position. As deposition continues the step rotates around the screw axis, and is not eliminated. Growth may therefore continue indefinitely. Several layers of deposition may occur, and the edges of the spirals might be cliffs several atoms high (Fig. 29.3).

Propagating spiral edges can also give rise to flat terraces (Fig. 29.4). Terraces are formed if growth occurs simultaneously at neighbouring left- and right-handed screw dislocations (Fig. 29.5). Successive tables of atoms may form as counter-rotating defects collide on successive circuits, and the terraces formed may then fill up by further deposition at their edges to give flat crystal planes.

The rapidity of growth depends on the crystal plane concerned and the *slowest* growing faces dominate the appearance of the crystal. This is explained in Fig. 29.6, where we see that although the horizontal face grows forward most rapidly, it grows itself out of existence, and the slower-growing faces survive.

29.2 Surface composition

The first step in understanding how surfaces catalyse reactions is to characterize the clean adsorbent surface. In this context 'clean' means much more than scrubbing the sample and handling it with care. Under normal conditions a surface is constantly bombarded with gas particles and a freshly prepared surface is covered very quickly. Just how quickly can be estimated using the kinetic theory of gases and the expression it gives (eqn 13b of Section 24.4) for the number of collisions per unit area per unit time when the pressure is p:

$$Z_W = \frac{p}{(2\pi mkT)^{1/2}} \qquad (1)$$

A practical form of this equation is

$$Z_W/(\text{cm}^{-2}\,\text{s}^{-1}) = \frac{3.51 \times 10^{22} p/\text{Torr}}{\{(T/\text{K}) \times M/(\text{g mol}^{-1})\}^{1/2}}$$

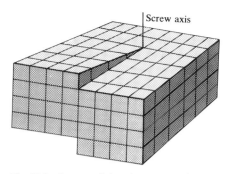

Fig. 29.2 A screw dislocation occurs where one region of the crystal is pushed up through one or more unit cells relative to another region. The cut extends to the screw axis. As atoms lie along the step the dislocation rotates round the screw axis, and is not annihilated.

Fig. 29.3 The spiral growth arising from the propagation of a screw axis is clearly visible in this photograph of an alkane crystal. (B. R. Jennings and V. J. Morris, *Atoms in contact*, Clarendon Press, Oxford (1974); courtesy of Dr A. J. Forty.)

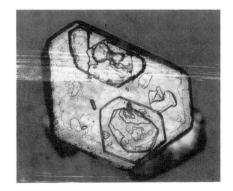

Fig. 29.4 The spiral growth pattern is sometimes concealed because the terraces are subsequently completed by further deposition. This accounts for the appearance of this cadmium iodide crystal. (H. M. Rosenberg, *The solid state*, Clarendon Press, Oxford (1978).)

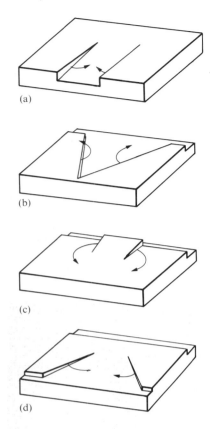

(a)

(b)

(c)

(d)

Fig. 29.5 Counter-rotating screw dislocations on the same surface lead to the formation of terraces. Four stages of one cycle of growth are shown here. Subsequent deposition can complete each terrace.

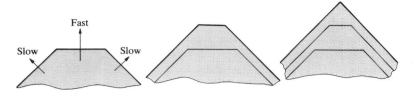

Fig. 29.6 The slower-growing faces of a crystal dominate its final external appearance. Three successive stages of the growth are shown.

where M is the molar mass of the gas. For air ($M \approx 29 \, \text{g mol}^{-1}$) at 1 atm and 25°C the collision frequency is $3 \times 10^{23} \, \text{cm}^{-2} \, \text{s}^{-1}$. Since $1 \, \text{cm}^2$ of metal surface consists of about 10^{15} atoms, each atom is struck about 10^8 times each second. Even if only a few collisions leave a molecule adsorbed to the surface, the time for which a freshly prepared surface remains clean is very short.

High-vacuum techniques

The obvious solution is to reduce the pressure. When it is reduced to $1 \, \mu\text{Torr}$ (as in a simple vacuum system) the collision frequency falls to about $4 \times 10^{14} \, \text{cm}^{-2} \, \text{s}^{-1}$, corresponding to one hit per surface atom in each 3 s. Even that is too brief in most experiments, and in **ultra-high vacuum** (UHV) techniques pressures as low as $1 \, \text{nTorr}$ ($10^{-9} \, \text{Torr}$, when $Z_{\text{w}} = 10^{11} \, \text{cm}^{-2} \, \text{s}^{-1}$) are reached on a routine basis and $10 \, \text{pTorr}$ ($10^{-11} \, \text{Torr}$, when $Z_{\text{w}} = 10^9 \, \text{cm}^{-2} \, \text{s}^{-1}$) are reached with special care. These collision frequencies correspond to each surface atom being hit once every 10^3 to $10^5 \, \text{s}$, or about once or twice an hour.

The layout of a typical UHV apparatus is such that the whole of the evacuated part can be heated to 200–300°C for several hours to drive gas molecules from the walls. All the taps and seals are of metal in order to avoid contamination from greases. The sample is usually in the form of a single crystal (and occasionally as a thin foil, a filament, or a sharp point). Where there is interest in the role of specific crystal planes the sample is a single crystal with a machined, polished, and chemically etched face. Initial surface cleaning is achieved either by heating it electrically or by bombarding it with accelerated gaseous ions. The latter demands care because ion bombardment can shatter the surface structure and leave it an amorphous jumble of atoms. High temperature annealing is then required to return the surface to an ordered state.

Ionization techniques

Surface composition can be determined by a variety of ionization techniques. The same techniques can be used to detect any remaining contamination after cleaning and to detect layers of material adsorbed later in the experiment. Their common feature is that the **escape depth** of the electrons, the maximum depth from which ionized electrons come, is in the range 0.3–3.0 nm, which ensures that surface species make the major contribution.

One technique that may be used is photoelectron spectroscopy (Section 17.9), which in surface studies is normally called **photoemission**

spectroscopy. X-rays or hard ultraviolet ionizing radiation may be used, but X-PES (which examines inner-shell binding energies) seems to be better than UV-PES for the analysis of composition because it is able to fingerprint the materials present (Fig. 29.7). UV-PES, which examines electrons ejected from valence shells, is more suited to establishing the bonding characteristics and the details of valence shell electronic structures of substances on the surface. Its usefulness is its ability to reveal which orbitals of the adsorbate are involved in the bond to the substrate. For instance, the principal difference between the PES results on free benzene and benzene adsorbed on palladium is in the energies of the π electrons. This difference is interpreted as meaning that the C_6H_6 molecules lie parallel to the surface and are attached to it through their π orbitals. In contrast, pyridine is known to stand more or less perpendicular to the surface, and is attached to it by means of a σ bond formed by the nitrogen lone pair. A recent advance has been to focus the ejected electrons and to build an image of the surface from which they have been ejected. This is the technique of **photoelectron spectromicroscopy** (PESM), and a typical image is shown in Fig. 29.8.

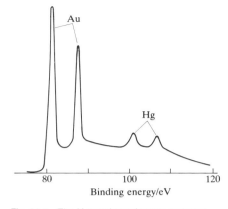

Fig. 29.7 The X-ray photoelectron emission spectrum of a sample of gold contaminated with a surface layer of mercury. (M. W. Roberts and C. S. McKee, *Chemistry of the metal-gas interface*, Oxford (1978).)

Electron energy-loss spectroscopy

Several kinds of vibrational spectroscopy have been developed to study adsorbates and to show whether dissociation has occurred. Infrared and Raman spectroscopy have been greatly improved by the development of laser and Fourier transform techniques, and have largely overcome the difficulties arising from low intensities on account of the low surface coverages normally encountered under laboratory conditions. Fourier transform IR spectroscopy can be used very effectively to study certain adsorbates (e.g. CO) on metal single crystal surfaces. It has sufficient sensitivity to detect less than 0.1 per cent of a monolayer of adsorbate in favourable cases.

In **electron energy loss spectroscopy** (EELS) the energy loss suffered by a beam of electrons is monitored when they are reflected from a surface. As in optical Raman spectroscopy, the spectrum of energy loss can be interpreted in terms of the vibrational spectrum of the adsorbate. High resolution and sensitivity are attainable, and the technique is sensitive to light elements (to which X-ray techniques are insensitive). Very tiny amounts of adsorbate can be detected, and one report estimates that about 48 atoms of phosphorus were detected in one sample. As an example, Fig. 29.9 shows the EELS result for CO on the (111) face of a Pt crystal as the extent of surface coverage increases. The main peak arises from CO attached perpendicular to the surface by a single Pt atom. As the coverage increases the smaller peak next to it increases in intensity. This peak is due to CO at a bridge site, where it is attached to two Pt atoms.

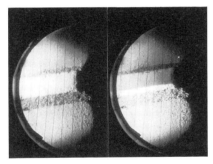

Fig. 29.8 He(I) photoelectron spectromicroscopy images of a sectional silicon diode (left) unbiased and (right) with −30 V reverse bias. (Photograph provided by Professor D. W. Turner.)

Auger electron spectroscopy

A very important technique, which is widely used in the microelectronics industry, is **Auger electron spectroscopy** (AES). The **Auger effect** is the emission of a second electron after high energy radiation has expelled another. The first electron to depart leaves a hole in a low-lying orbital, and an upper electron falls into it. The energy this releases may result either in the generation of radiation, which is called **X-ray fluorescence** (Fig. 29.10a) or in the ejection of another electron (Fig. 29.10b). The latter is the

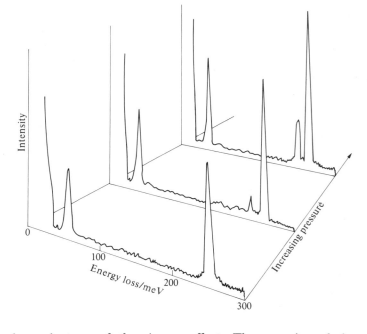

Fig. 29.9 The electron energy loss spectrum of CO adsorbed on Pt(111). The results for three different pressures are shown, and the growth of the additional peak at about 200 meV (1600 cm^{-1}) should be noted. (Spectra provided by Professor H. Ibach.)

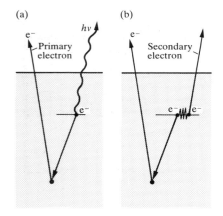

Fig. 29.10 When an electron is expelled from a solid (a) an electron of higher energy may fall into the vacated orbital and emit an X-ray photon to produce X-ray fluorescence. Alternatively (b) the electron falling into the orbital may give up its energy to another electron, which is ejected in the Auger effect.

secondary electron of the Auger effect. The energies of the secondary electrons are characteristic of the material present, and so the Auger effect effectively takes a fingerprint of the sample (Fig. 29.11). In practice, the Auger spectrum is normally obtained by irradiating the sample with an electron beam rather than electromagnetic radiation.

Low-energy electron diffraction

One of the most informative techniques for determining the arrangement of the atoms close to the surface is **low-energy electron diffraction** (LEED). This is essentially the electron diffraction technique described in Section 21.10, but the sample is now the surface of a solid. The use of low energy electrons ensures that the diffraction is caused only by atoms on and close to the surface. The experimental arrangement is shown in Fig. 29.12, and typical LEED patterns, obtained by photographing the fluorescent screen through the viewing port, are shown in Fig. 29.13.

The LEED pattern portrays the two-dimensional structure of the surface. By studying how the diffraction intensities depend on the energy of the electron beam it is also possible to infer some details about the vertical location of the atoms and to measure the thickness of the surface layer. The pattern is sharp if the surface is well-ordered for distances long compared with the wavelength of the incident electrons. In practice sharp patterns are obtained for surfaces ordered laterally to distances of about 20 nm and more. Diffuse patterns indicate either a poorly ordered surface or the presence of impurities. If the LEED pattern does not correspond to the pattern expected by extrapolation of the bulk surface to the surface, then either a reconstruction of the surface has occurred or there is order in the arrangement of an adsorbed layer.

LEED experiments show that the surface of a crystal rarely has exactly the same form as a slice through the bulk. As a general rule, it is found that metal surfaces are simply truncations of the bulk lattice, but the distance between the top layer of atoms and the one below is contracted by around 5 per cent. Semiconductors generally have surfaces reconstructed to a depth of several layers. Reconstruction occurs in ionic solids. For example, in lithium fluoride the Li^+ and F^- ions close to the surface apparently lie on slightly different planes.

The presence of terraces, steps, and kinks in a surface shows up in LEED patterns and their densities (the number of defects per unit area) can be estimated. The importance of this will emerge later. Three examples of how steps and kinks affect the pattern are shown in Fig. 29.14. The samples used were obtained by cleaving a crystal at different angles to a plane of atoms. Only terraces are produced when the cut is parallel to the plane, and the density of steps increases as the angle of the cut increases. The observation of additional structure in the LEED patterns, not merely blurring, shows that the steps are arrayed regularly.

Field emission and ionization microscopy

Among the most spectacular portrayals of surface structure are those obtained from two closely related techniques.

In **field emission microscopy** (FEM) the sample is in the form of a filament etched to a sharp tip and enclosed within a chamber fitted with a fluorescent screen. When a large potential difference is applied between the sample and the screen, electrons are stripped out along a radial trajectory towards the screen and give a flash of light where they strike it; the result is a greatly magnified image of the tip. The ease with which the electrons can escape from the metal depends on the variation of the work function (Section 11.2) with the composition and structure of the surface, and the screen shows a corresponding variation of intensity (Fig. 29.15). The change in the FEM pattern when material has been deposited can be used to detect the places where atoms are most likely to stick.

Field ionization microscopy (FIM) is a development of FEM. The apparatus is virtually the same, but the potential difference is reversed with the fluorescent screen made negative relative to the tip. In the experiment a small quantity of gas (typically helium) is admitted. An atom strikes the tip and bounces over its terraced, cobbled surface until it strikes a protruding atom, such as one on the rim of a ledge (Fig. 29.16). Protruding atoms are able to ionize the gas atom, and immediately the positive ion (He^+) is formed, the potential difference plucks it off towards the screen, where its collision generates fluorescence.

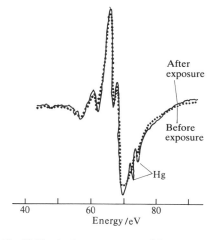

Fig. 29.11 An Auger spectrum of the same sample used for Fig. 29.10 taken before and after deposition of mercury. (M. W. Roberts and C. S. McKee, *Chemistry of the metal-gas interface*, Oxford (1978).)

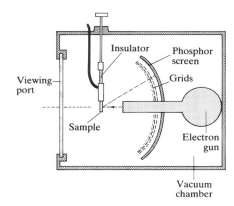

Fig. 29.12 A schematic diagram of the apparatus used for a LEED experiment. The electrons diffracted by the surface layers are detected by the fluorescence they cause on the phosphor screen.

(a)   (b)

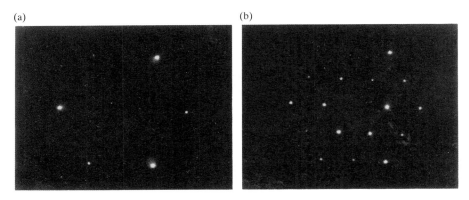

Fig. 29.13 LEED photographs of (a) a clean platinum surface and (b) after its exposure to propyne, $CH_3C{\equiv}CH$. (Photographs provided by Professor G. A. Samorjai.)

(a)

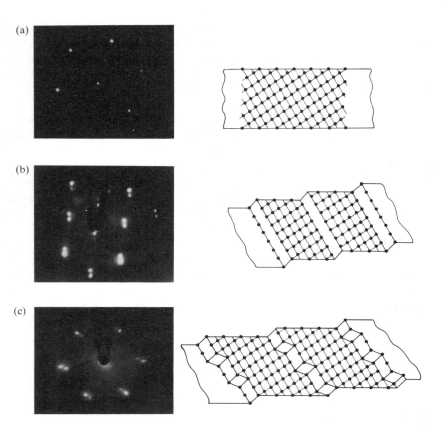

(b)

(c)

Fig. 29.14 LEED patterns may be used to assess the defect density of a surface. The photographs correspond to a platinum surface with (a) low defect density, (b) regular steps separated by about six atoms, and (c) regular steps with kinks. (Photographs provided by Professor G. A. Samorjai.)

The spatial resolution that can be achieved with FIM depends on the transverse motion of the gas ions. Such motion can be reduced by cooling the tip to about 20 K, when the resolution is of the order of atomic dimensions and the positions of individual atoms can be discerned (Fig. 29.17). This remarkable picture, in which the small bright spots are caused by individual terrace atoms, shows the power of the technique. However, we should not forget its limitations. First, ionization occurs unequally at

Fig. 29.15 A field emission photograph of a tungsten tip of radius 210 nm and the assignment to the exposed crystal faces. (M. Prutton, *Surface physics*, Clarendon Press, Oxford (1975).)

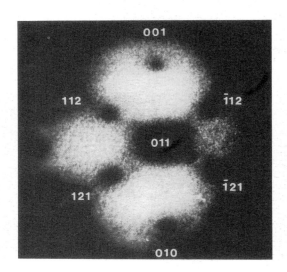

different atoms, and many of the atoms on the surface and even on the edges are insufficiently exposed to cause ionization and so remain invisible. Second, the sample must be in the form of a tip, and be made of material strong enough to withstand the very high electric fields required. Despite these limitations, FIM is remarkable for its ability to portray the positions of individual atoms.

An elegant refinement of FIM identifies individual atoms on an otherwise clean tip. **Atom-probe FIM** is the ultimate in surgery. The FIM image of an adsorbed atom is brought into coincidence with a hole in the fluorescent screen. The imaging gas is removed, and a pulse of potential difference plucks off the atom (as an ion). It moves in the same direction as did the gas ions, and passes through the hole in the screen. Behind the screen is a mass spectrometer, and so the atom can be identified.

Apart from being the ultimate analytical technique (because as well as knowing what the atom is we also know exactly where it was in the sample) events can be observed which on an atomic scale are really dramatic. For example, the analysing pulse lasts for about 2 ns, and during that time the evaporation of about ten atomic layers is sometimes observed. This corresponds to a rate of evaporation equivalent to the surface receding at about $1\ \mathrm{m\ s^{-1}}$.

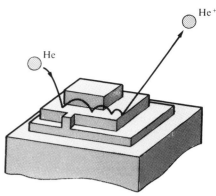

Fig. 29.16 The events leading to an FIM image of a surface. The He atom migrates across the surface until it is ionized at an exposed atom, when it is pulled off by the externally applied potential. (The bouncing motion is due to the intermolecular potential, not gravity!)

Scanning tunnelling microscopy

The **scanning tunnelling microscope** (STM), which has been developed during the past few years, has made possible striking, life-like images of surfaces and of adsorbates on them.

The central component of the microscope is a platinum-rhodium needle,

Fig. 29.17 An FIM image of an iridium tip showing the details of the exposed crystal faces. (Photograph provided by Professor E. W. Müller.)

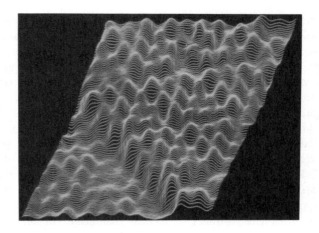

Fig. 29.18 The type of image that can be obtained with scanning tunnelling microscopy. The sample is of a silicon surface and the cliff is one atom high. (Sang-il Park and C. F. Quaite.)

which is scanned across the sample. The needle is attached to a cylinder of piezoelectric ceramic that expands and contracts lengthwise in response to an electric current. When the tip of the needle is brought very close to the surface, electrons tunnel across the space and modify the potential difference. The monitoring electronics are designed to maintain a constant potential difference across the gap between the tip and the sample by passing a current through the ceramic that keeps the size of the gap constant. The equipment monitors the current needed, and interprets it to give the height of the tip above the surface. Since the tunnelling probability is very sensitive to the size of the gap (Section 12.3), the microscope can detect tiny, atom-scale variations in the height of the surface. An example of the kind of image obtained with a clean surface is shown in Fig. 29.18, where the cliff is only one atom high.

A spectacular demonstration of the power of scanning tunnelling microscopy for displaying the shape of adsorbed species is shown in Fig. 29.19. One of the most recent achievements of the techniques has been the imaging of a single DNA molecule on a surface (Fig. 29.20), and the first

Fig. 29.19 A scanning tunnelling microscope image of a liquid crystal (5-nonyl-2-nonoxylphenylpyrimidine) adsorbed on a graphite surface. (J. S. Foster, *et al.*, *Nature*, **338**, 137 (1989).)

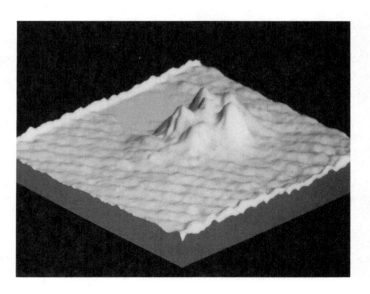

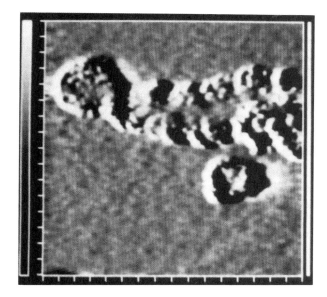

Fig. 29.20 A looped strand of DNA imaged by a scanning tunnelling microscope. The scale intervals are at 2.5 nm intervals and the full grey scale for height is 13.2 nm. (M. Salmeron *et al.*, *Science*, **243**, 370 (1989).)

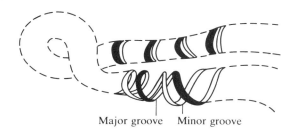

Major groove   Minor groove

direct observation of its double helix structure. The advance in technique represented by the latter result is that until it was obtained it was considered essential to coat molecules with heavy-metal atoms (as is still necessary for conventional electron microscopy). However, the image shown in the illustration is from the untreated molecule. A limitation of the technique is that it can be used only for conducting or semiconducting surfaces.

Molecular beam techniques

Whereas many important studies have been carried out simply by exposing a surface to a gas, modern work is increasingly making use of molecular beam techniques. One advantage is that the activity of specific crystal planes can be investigated by directing the beam on to an orientated surface with known step and kink densities (as measured by LEED). Furthermore, if the adsorbate reacts at the surface the products (and their angular distributions) can be analysed as they are ejected from the surface and pass into a mass spectrometer. Another advantage is that the time of flight of a particle may be measured and interpreted in terms of its residence time on the surface. In this way a very detailed picture can be constructed of the events taking place during reactions at surfaces.

The extent of adsorption

The extent of surface coverage is normally expressed as the **fractional coverage** θ:

$$\theta = \frac{\text{number of adsorption sites occupied}}{\text{number of adsorption sites available}}$$

The **rate of adsorption** $\dot{\theta}$ is the rate of change of surface coverage, and can be determined by observing the change of fractional coverage with time.

Among the principal techniques for measuring θ are **flow methods**, in which the sample itself acts as a pump because adsorption removes particles from the gas. One commonly used technique is therefore to monitor the rates of flow of gas into and out of the system: the difference is the rate of gas uptake by the sample. Integration of this rate then gives the fractional coverage at any stage. In **flash desorption** the sample is suddenly heated (electrically) and the rise of pressure this leads to is interpreted in terms of the amount originally on the sample. The interpretation may be confused by the desorption of a compound (e.g. WO_3 from oxygen on tungsten). **Gravimetry**, in which the sample is weighed on a microbalance during the experiment, can also be used, as can radioactive tracers. In the latter case, the radioactivity of the sample is measured after exposure to an isotopically labelled gas.

29.3 Physisorption and chemisorption

Molecules and atoms can attach to surfaces in two ways. In **physisorption** (an abbreviation of 'physical adsorption'), there is a van der Waals interaction (for example, a dispersion or a dipolar interaction) between the adsorbate and the substrate. Van der Waals interactions have a long range but are weak, and the energy released when a particle is physisorbed is of the same order of magnitude as the enthalpy of condensation. Such small amounts of energy can be absorbed as vibrations of the lattice and dissipated as thermal motion, and a molecule bouncing across the cobble-like surface will gradually lose its energy and finally adsorb to it in the process called **accommodation**. The enthalpy of physisorption can be measured by noting the rise in temperature of a sample of known heat capacity, and typical values are in the region of $-20 \, \text{kJ mol}^{-1}$ (Table 29.1). This small enthalpy change is insufficient to lead to bond breaking, and so a physisorbed molecule retains its identity, although it might be distorted by the presence of the surface.

Table 29.1. Maximum observed enthalpies of physisorption, $\Delta H_{ad}^{\ominus}/(\text{kJ mol}^{-1})$

CH_4	-21
H_2	-84
H_2O	-59
N_2	-21

Chemisorption

In **chemisorption** (an abbreviation of 'chemical adsorption') the particles stick to the surface by forming a chemical (usually covalent) bond, and tend to find sites that maximize their coordination number with the substrate. The enthalpy of chemisorption is very much greater than for physisorption, and typical values are in the region of $-200 \, \text{kJ mol}^{-1}$ (Table 29.2). A chemisorbed molecule may be torn apart at the demand of the unsatisfied valencies of the surface atoms, and the existence of molecular fragments on the surface as a result of chemisorption is one reason why surfaces catalyse reactions.

Table 29.2. Enthalpies of chemisorption, $\Delta H_{ad}^{\ominus}/(\text{kJ mol}^{-1})$

Adsorbate	Adsorbent (substrate)		
	Cr	Fe	Ni
C_2H_4	-427	-285	-209
CO		-192	
H_2	-188	-134	
NH_3		-188	-155

Except in special cases, chemisorption must be exothermic. A spontaneous process requires a negative ΔG. Since the translational freedom of the adsorbate is reduced when it is adsorbed, ΔS is negative. Therefore, in order for $\Delta G = \Delta H - T \Delta S$ to be negative, ΔH must be negative (and the process exothermic). Exceptions may occur if the adsorbate dissociates and has high translational mobility on the surface. For example, H_2 adsorbs endothermically on glass because there is a large increase of translational entropy accompanying the dissociation of the molecules into atoms that move quite freely over the surface. In its case, the entropy change in the process $H_2(g) \rightarrow 2H(\text{glass})$ is sufficiently positive to overcome the small positive enthalpy change.

The principal test for distinguishing chemisorption from physisorption used to be the size of the enthalpy of adsorption. Values less negative than -25 kJ mol^{-1} were taken to signify physisorption, and values more negative than about -40 kJ mol^{-1} were taken to signify chemisorption. However, this criterion is by no means foolproof and spectroscopic techniques that identify the adsorbed species are now available.

The enthalpy of adsorption depends on the extent of surface coverage, mainly because the adsorbate particles interact. If the particles repel each other (as for CO on palladium) the enthalpy of adsorption becomes less exothermic (less negative) as coverage increases. Moreover, LEED studies show that such species settle on the surface in a disordered way until packing requirements demand order. If the adsorbate particles attract each other (as for O_2 on tungsten) they tend to cluster together in islands, and growth occurs at the borders. These adsorbates also show order–disorder transitions when they are heated enough for thermal motion to overcome the particle–particle interactions, but not so much that they are desorbed.

29.4 Adsorption isotherms

The free gas and the adsorbed gas are in dynamic equilibrium, and the fractional coverage of the surface depends on the pressure of the overlying gas. The dependence of θ on the pressure at a set temperature is called the **adsorption isotherm**.

The Langmuir isotherm

The simplest isotherm is based on the assumptions that every adsorption site is equivalent and that the ability of a particle to bind there is independent of whether or not nearby sites are occupied.

The dynamic equilibrium is

$$A(g) + M(\text{surface}) \rightleftharpoons AM$$

with rate constants k_a for adsorption and k_d for desorption. The rate of change of surface coverage due to adsorption is proportional to the pressure p of A and the number of vacant sites $N(1 - \theta)$, where N is the total number of sites:

$$\dot{\theta} = k_a p N (1 - \theta)$$

The rate of change of θ due to desorption is proportional to the number of adsorbed species, $N\theta$:

$$\dot{\theta} = k_d N \theta$$

At equilibrium the two rates are equal, and solving for θ gives the **Langmuir isotherm**:

$$\theta = \frac{Kp}{1 + Kp} \qquad K = \frac{k_a}{k_d} \tag{2}$$

Example 29.1: *Using the Langmuir isotherm*

The data below are for the adsorption of CO on charcoal at 273 K. Confirm that they fit the Langmuir isotherm, and find the constant K and the volume corresponding to complete coverage. In each case V has been corrected to 1 atm.

p/Torr	100	200	300	400	500	600	700
V/cm^3	10.2	18.6	25.5	31.5	36.9	41.6	46.1

Answer. From eqn 2,

$$Kp\theta + \theta = Kp$$

If we write $\theta = V/V_\infty$, where V_∞ is the volume corresponding to complete coverage, this expression rearranges into

$$\frac{p}{V} = \frac{p}{V_\infty} + \frac{1}{KV_\infty}$$

Hence, a plot of p/V against p should give a straight line of slope $1/V$ and intercept $1/KV_\infty$. The data for the plot are as follows:

p/Torr	100	200	300	400	500	600	700
$\dfrac{p/\text{Torr}}{V/\text{cm}^3}$	9.80	10.8	11.8	12.7	13.6	14.4	15.2

the points are plotted in Fig. 29.21 with $x = (p/\text{Torr})/(V/\text{cm}^3)$. The slope is 0.0090 and so $V_\infty = 110$ cm^3. The intercept at $p = 0$ is 9.0, and so

$$K = \frac{1}{110 \text{ cm}^3 \times 9.0 \text{ Torr cm}^{-3}} = 1.0 \times 10^{-3} \text{ Torr}^{-1}$$

Comment. Note the deviation from a straight line at high surface coverages. The dimensions of K are 1/pressure.

Exercise. Repeat the calculation for the following data:

p/Torr	100	200	300	400	500	600	700
V/cm^3	10.3	19.3	27.3	34.1	40.0	45.5	48.0

$$[150 \text{ cm}^3, \ 7.4 \times 10^{-3} \text{ Torr}^{-1}]$$

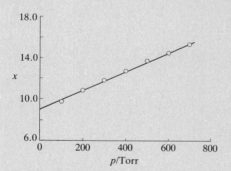

Fig. 29.21 The Langmuir isotherm predicts that a straight line should be obtained when p/V is plotted against p. This is a test of the isotherm using the information in Example 29.1.

For adsorption with dissociation, the rate of adsorption is proportional to the pressure and to the probability that both atoms will find sites,

$$\dot{\theta} = k_a p \{N(1 - \theta)\}^2$$

The rate of desorption is proportional to the frequency of encounters of atoms on the surface, and therefore second order in the number of atoms present:

$$\dot{\theta} = k_d (N\theta)^2$$

The condition for these two rates to be equal leads to the isotherm

$$\theta = \frac{(Kp)^{1/2}}{1 + (Kp)^{1/2}} \qquad (3)$$

The surface coverage now depends more weakly on pressure.

The shapes of the Langmuir isotherms with and without dissociation are shown in Fig. 29.22. The fractional coverage increases with increasing pressure, and approaches 1 only at very high pressure, when the gas is effectively squashed on to every available site of the surface. Different curves (and therefore values of K) are obtained at different temperatures, and the temperature dependence of K can be used to determine the **isosteric enthalpy of adsorption** $\Delta H_{ad}^{\ominus}$, the enthalpy of adsorption at a fixed surface coverage. To do so, we recognize that K is an equilibrium constant, and then use the van't Hoff equation (eqn 11 of Section 9.4) to write

$$\left(\frac{\partial \ln K}{\partial T}\right)_{\theta} = \frac{\Delta H_{ad}^{\ominus}}{RT^2} \qquad (4)$$

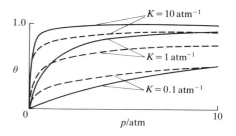

Fig. 29.22 The Langmuir isotherm for non-dissociative (full lines) and dissociative (dashed lines) adsorption for different values of K. The curves are plots of eqns 2 and 3 respectively.

Example 29.2: *Measuring the isosteric enthalpy of adsorption*

The data below show the pressures of CO needed for the volume of adsorption (corrected to 1 atm and 273 K) to be $10.0\ cm^3$ using the same sample as in Example 29.1. Calculate the adsorption enthalpy at this surface coverage.

T/K	200	210	220	230	240	250
$p/$Torr	30.0	37.1	45.2	54.0	63.5	73.9

Answer. The Langmuir isotherm rearranges to

$$Kp = \frac{\theta}{1 - \theta}$$

Therefore, when θ is constant,

$$\ln K + \ln p = \text{constant}$$

Therefore, from eqn 4,

$$\left(\frac{\partial \ln p}{\partial T}\right)_{\theta} = -\left(\frac{\partial \ln K}{\partial T}\right)_{\theta} = \frac{-\Delta H_{ad}^{\ominus}}{RT^2}$$

With $d(1/T)/dT = -1/T^2$ this expression rearranges to

$$\left(\frac{\partial \ln p}{\partial(1/T)}\right)_{\theta} = \frac{\Delta H_{ad}^{\ominus}}{R}$$

Therefore, a plot of $\ln p$ against $1/T$ should be a straight line of slope $\Delta H_{ad}^{\ominus}/R$. We draw up the following table:

T/K	200	210	220	230	240	250
$1000/(T/K)$	5.00	4.76	4.55	4.35	4.17	4.00
$\ln (p/$Torr$)$	3.40	3.61	3.81	3.99	4.15	4.30

The points are plotted in Fig. 29.23. The slope (of the least-squares fitted line) is -0.90, and

$$\Delta H_{ad}^{\ominus} = -(0.90 \times 10^3\ K) \times R = -7.5\ kJ\ mol^{-1}$$

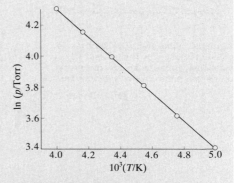

Fig. 29.23 The isosteric enthalpy of adsorption can be obtained from the slope of the plot of $\ln p$ against $1/T$, where p is the pressure needed to achieve the specified surface coverage.

Comment. The value of K can be used to obtain a value of $\Delta G^{\ominus}$, and then that value combined with $\Delta H_{ad}^{\ominus}$ to obtain an entropy of adsorption.

Exercise. Repeat the calculation using the following data:

T/K	200	210	220	230	240	250
$p/Torr$	32.4	41.9	53.0	66.0	80.0	96.0

$$[-9.0 \text{ kJ mol}^{-1}]$$

The BET isotherm

The Langmuir isotherm ignores the possibility that the initial monolayer may act as a substrate for further (e.g. physical) adsorption. In this case, instead of the isotherm levelling off to some saturated value at high pressures, it can be expected to rise indefinitely. The most widely used isotherm dealing with multilayer adsorption was derived by Stephen Brunauer, Paul Emmett, and Edward Teller, and is called the **BET isotherm**:

$$\frac{V}{V_{mon}} = \frac{cz}{(1-z)\{1-(1-c)z\}} \qquad z = \frac{p}{p^*} \qquad (5a)$$

where p^* is the vapour pressure above a macroscopically thick layer of the pure liquid on the surface, V_{mon} the volume corresponding to monolayer coverage, and c is a constant which is large when the enthalpy of desorption from a monolayer is large compared with the enthalpy of vaporization of the liquid adsorbate. Specifically:

$$c \approx e^{(\Delta H_d^{\ominus} - \Delta H_{vap}^{\ominus})/RT} \qquad (5b)$$

Equation 5a, which is derived in the *Further information* section at the end of the chapter, is usually reorganized into

$$\frac{z}{(1-z)V} = \frac{1}{cV_{mon}} + \frac{(c-1)z}{cV_{mon}} \qquad (6)$$

It follows that $(c-1)/cV_{mon}$ can be obtained from the slope of a plot of the expression on the left against z, and cV_{mon} can be found from the intercept at $z = 0$, the results then being combined to give c and V_{mon}. The shapes of BET isotherms are illustrated in Fig. 29.24. They rise indefinitely as the pressure is increased because there is no limit to the amount of material that may condense when multilayer coverage can occur.

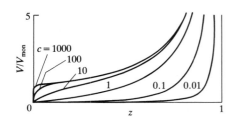

Fig. 29.24 Plots of the BET isotherm for different values of c. The value of V/V_{mon} rises indefinitely because the adsorbate may condense on the covered substrate surface.

Example 29.3: *Using the BET isotherm*

The data below relate to the adsorption of N_2 on rutile (TiO_2) at 75 K. Confirm that they fit a BET isotherm in the range of pressures reported, and find V_{mon} and c.

$p/Torr$	1.20	14.0	45.8	87.5	127.7	164.4	204.7
V/cm^3	601	720	822	935	1046	1146	1254

At 75 K $p^* = 570$ Torr. The volumes have been corrected to 1 atm and 273 K and refer to 1 g of substrate.

Answer. Draw up the following table:

p/Torr	1.20	14.0	45.8	87.5	127.7	164.4	204.7
$1000z$	2.11	24.6	80.4	154	224	288	359
$\dfrac{10^4 z}{(1-z)(V/\text{cm}^3)}$	0.035	0.350	1.06	1.95	2.76	3.53	4.47

These points are plotted in Fig. 29.25. The least squares best line has an intercept at 0.034, and so

$$\frac{1}{cV_{\text{mon}}} = 0.034 \times 10^{-4}\,\text{cm}^{-3} = 3.4 \times 10^{-6}\,\text{cm}^{-3}$$

The slope of the line is 1.23×10^{-2}, and so

$$\frac{c-1}{cV_{\text{mon}}} = 1.23 \times 10^{-2} \times 10^3 \times 10^{-4}\,\text{cm}^{-3} = 1.23 \times 10^{-3}\,\text{cm}^{-3}$$

Solving these equations gives $c - 1 = 362$, or $c = 363$, and $V_{\text{mon}} = 810\,\text{cm}^3$.

Comment. At 1 atm and 273 K, $810\,\text{cm}^3$ corresponds to 0.036 mol, or 2.2×10^{22} atoms. Since each molecule occupies an area of about $0.16\,\text{nm}^2$, the surface area of the sample is about $3500\,\text{m}^2$ per gram.

Exercise. Repeat the calculation for the following data:

p/Torr	1.20	14.0	45.8	87.5	127.7	164.4	204.7
V/cm^3	235	559	649	719	790	860	950

$$[290, 620\,\text{cm}^3]$$

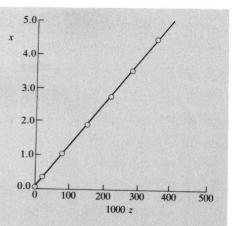

Fig. 29.25 The BET isotherm can be tested, and the parameters determined, by plotting the left hand side of eqn 6 against $z = p/p^*$.

When the coefficient c is large $(c \gg 1)$, the BET isotherm takes the simpler form

$$\frac{V}{V_{\text{mon}}} = \frac{1}{1-z} \tag{7}$$

Equation 7 is applicable to unreactive gases on polar surfaces, for which $c \approx 10^2$ because $\Delta H_{\text{d}}^{\ominus}$ is then significantly greater than $\Delta H_{\text{vap}}^{\ominus}$ (eqn 5b). The BET isotherm fits experimental observations moderately well over restricted pressure ranges, but it errs by underestimating the extent of adsorption at low pressures and by overestimating it at high pressures.

Other isotherms

An assumption of the Langmuir isotherm is the independence and equivalence of the adsorption sites. Deviations from the isotherm can often be traced to the failure of this assumption. For example, the enthalpy of adsorption often becomes less negative as θ increases, which suggests that the energetically most favourable sites are occupied first. Various attempts have been made to take these variations into account. The **Temkin isotherm**,

$$\theta = c_1 \ln c_2 p \tag{8}$$

where c_1 and c_2 are constants, corresponds to supposing that the adsorption enthalpy changes linearly with pressure. The **Freundlich isotherm**,

$$\theta = c_1 p^{1/c_2} \tag{9}$$

corresponds to a logarithmic change. Different isotherms agree with experiment more or less well over restricted ranges of pressure, but they remain largely empirical. Empirical, however, does not mean useless, for if the parameters of a reasonably reliable isotherm are known, reasonably reliable results can be obtained for the extent of surface coverage under various conditions. This kind of information is essential for any discussion of heterogeneous catalysis.

Example 29.4: *Examining the Freundlich and Langmuir isotherms*

Decide whether the Freundlich isotherm is a better representation than the Langmuir isotherm for the data in Example 29.1.

Answer. If we write $\theta = V/V_{mon}$, and rearrange eqn 9 into

$$\ln V = \ln c_1 V_{mon} + \frac{1}{c_2} \ln p$$

then we can plot $\ln V$ against $\ln p$ and compare the least squares fit of the points with the plot in Example 29.1. The best way of doing this is to quote the coefficient of determination of the least-squares lines (see the Appendix of the *Solutions manual*). Draw up the following table.

p/Torr	100	200	300	400	500	600	700
$\ln (p/\mathrm{Torr})$	4.61	5.30	5.70	5.99	6.21	6.40	6.55
$\ln (V/\mathrm{cm}^3)$	2.32	2.92	3.24	3.45	3.61	3.73	3.83

The points are plotted in Fig. 29.26. The coefficient of determination of the least-squares best line is 0.9968. The value for the Langmuir plot is 0.9979. Hence the latter is marginally better.

Comment. The Freundlich isotherm is often used in discussions of adsorption from liquid solutions, when it is written

$$w = c_1 \times c^{1/c_2}$$

where w is the mass fraction adsorbed (the mass of solute adsorbed per unit mass of adsorbent) and c is the solution's concentration.

Exercise. Examine the suitability of the Temkin isotherm to fit the data.

$$[V \propto \ln p, \ 0.9731]$$

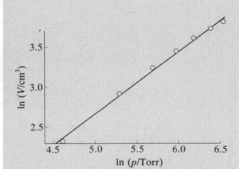

Fig. 29.26 The Freundlich isotherm may be tested by plotting $\ln V$ against $\ln p$ because it predicts that a straight line should be obtained.

29.5 The rates of surface processes

Figure 29.27 shows how the potential energy of a molecule varies with its distance from the substrate surface. As the particle approaches the surface its energy falls as it becomes physisorbed into the **precursor state** for chemisorption. Dissociation into fragments often takes place as a molecule moves into its chemisorbed state, and after an initial increase of energy as the bonds stretch there is a sharp decrease as the adsorbate–substrate bonds reach their full strength. Even if the molecule does not fragment there is likely to be an initial increase in potential energy as the substrate atom adjusts its bonds in response to the incoming particle.

In all cases, therefore, we can expect there to be a potential energy barrier separating the precursor and chemisorbed states. This barrier, though, might be low, and might not rise above the energy of a distant, stationary particle (as in Fig. 29.27a). In this case, chemisorption is not an activated process and can be expected to be rapid. Many gas adsorptions on to clean metals appear to be non-activated. In some cases the barrier rises above the zero axis (as in Fig. 29.27b), and such chemisorptions are activated and slower than the non-activated kind. An example is H_2 on Cu, which has an activation energy somewhere in the region of 20–40 kJ mol^{-1}.

One point that emerges from this discussion is that rates are not good criteria for distinguishing between physisorption and chemisorption. Chemisorption can be fast if the activation energy is small or zero; but it may be slow if the activation energy is large. Physisorption is usually fast, but it can appear to be slow if adsorption is taking place on a porous medium.

The rate of adsorption

The rate at which a surface is covered by adsorbate depends on the substrate's ability to dissipate the energy of the incoming particle as thermal motion as it crashes on to the surface. If the energy is not dissipated quickly, the particle migrates over the surface until a vibration expels it into the overlying gas or it reaches an edge. The proportion of collisions with the surface that successfully lead to adsorption is called the **sticking probability** s;

$$s = \frac{\text{rate of adsorption of particles by the surface}}{\text{rate of collision of particles with the surface}}$$

The denominator can be calculated from kinetic theory, and the numerator can be measured by observing the rate of change of pressure.

Values of s vary widely and depend in a complicated way on the details of the potential energy surface close to the physical surface. For example, at room temperature CO has s in the range 0.1–1.0 for several d-metal surfaces, but for N_2 on rhenium $s < 10^{-2}$, indicating that more than a hundred collisions are needed before one molecule sticks successfully. Sticking probabilities are generally low on non-metal surfaces, but they can be low on metals too (O_2 on silver has $s < 10^{-4}$). Beam studies on specific crystal planes show a pronounced specificity: for N_2 on tungsten, s ranges from 0.74 on the (310) faces down to less than 0.01 on the (110) faces at room temperature.

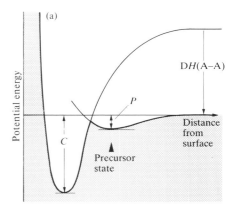

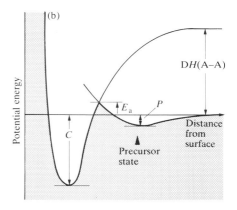

Fig. 29.27 The potential energy profiles for the dissociative chemisorption of an A–A molecule. In each case P is the enthalpy of (non-dissociative) physisorption and C that for chemisorption (at $T = 0$). The relative locations of the curves determines whether the chemisorption is (a) not activated or (b) activated.

Example 29.5: *Calculating the time needed for surface coverage*

Calculate the time for 10 per cent of the sites on a (100) surface to be covered with nitrogen atoms at 298 K when the pressure is 2.0 nTorr and the sticking probability is 0.55. Take the substrate to have a bcc unit cell with lattice constant 316 pm.

Answer. The distance between lattice sites on the (100) face is 316 pm; therefore, each atom accounts for $(316 \text{ pm})^2$ of the surface. The number of atoms in 1 m^2 is therefore $(1 \text{ m}^2)/(316 \text{ pm})^2 = 1.00 \times 10^{19}$, so the surface density of sites is 1.00×10^{15} cm^{-2}. From eqn 1, $Z_W = 7.7 \times 10^{11}$ cm^{-2} s^{-1}. Therefore, the time required for 10 per cent coverage, noting that each N_2 molecule occupies two surface sites, is

$$t = \frac{0.10 \times 1.00 \times 10^{15} \text{ cm}^{-2}}{2 \times 0.55 \times Z_W} = 120 \text{ s}$$

Comment. Covering the entire surface will take longer than ten times 120 s because the sticking probability decreases with increasing coverage.

Exercise. Suppose that the sticking probability depends on surface coverage as $s = (1 - \theta)s_0$, with s_0 given above. Calculate the surface coverage after 1200 s.   [0.64]

The sticking probability decreases as the surface coverage increases (Fig. 29.28). A simple model would be obtained by assuming that s is proportional to $1 - \theta$, the fraction uncovered, and it is common to write

$$s = (1 - \theta)s_0$$

where s_0 is the sticking probability on a perfectly clean surface. The results in the illustration do not fit this expression because they show that s remains close to s_0 until the coverage has risen to about 6×10^{13} molecules/cm^2, and then falls steeply. The explanation is probably that the colliding molecule does not enter the chemisorbed state at once, but moves over the surface until it encounters an empty site.

The rate of desorption

Desorption is always activated because particles have to be lifted from the foot of a potential well. A physisorbed particle vibrates in its shallow potential well, and might shake itself off the surface after a short time. The temperature-dependence of the first-order rate of departure can be expected to be Arrhenius-like, with an activation energy comparable to the enthalpy of physisorption:

$$k_d = A e^{-E_d/RT}$$

E_d is the desorption activation energy. Therefore, the half-life for remaining on the surface has a temperature dependence

$$t_{1/2} = \frac{\ln 2}{k_d} = \tau_0 e^{E_d/RT} \qquad \tau_0 = \frac{\ln 2}{A} \qquad (10)$$

If we suppose that $1/\tau_0$ is approximately the same as the vibrational frequency of the weak particle-surface bond (about 10^{12} Hz) and $E_d \approx 25$ kJ mol^{-1}, residence half-lives of around 10^{-8} s are predicted at room temperature. Lifetimes close to 1 s are obtained only by lowering the temperature to about 100 K. For chemisorption, with $E_d = 100$ kJ mol^{-1} and guessing that $\tau_0 = 10^{-14}$ s (because the adsorbate-substrate bond is quite stiff), we expect a residence half-life of about 3×10^3 s (about an hour) at room temperature, decreasing to 1 s at about 350 K.

The desorption activation energy can be measured in several ways. However, we must be guarded in its interpretation because it often depends on the fractional coverage and so may change as desorption proceeds. Moreover, the transfer of concepts such as 'reaction order' and 'rate constant' from bulk studies to surfaces is hazardous, and there are few examples of strictly first-order or second-order desorption kinetics (just as there are few in the gas phase too).

If we disregard these complications, then one way of measuring the desorption activation energy is to monitor the rate of increase in pressure

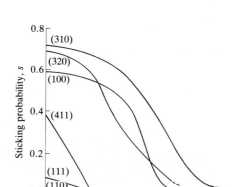

Fig. 29.28 The sticking probability of N$_2$ on various faces of a tungsten crystal and its dependence on surface coverage. Note the very low sticking probability for the (110) and (111) faces. (Data provided by Professor D. A. King.)

when the sample is maintained at a series of temperatures, and to attempt to make an Arrhenius plot.

A more sophisticated technique is the determination of the **flash desorption spectrum** of the sample. The basic observation is that in a pumped vessel there is a pressure surge when the temperature is raised to the value at which the activation energy is overcome and desorption occurs rapidly; but once the desorption has occurred there is no more to escape from the surface, and so the pressure falls again as the temperature continues to rise. The flash desorption spectrum, the plot of pressure against temperature, therefore consists of a peak, the location of which depends on the desorption activation energy. In the example shown in Fig. 29.29 three peaks are shown, indicating the presence of three sites with different activation energies.

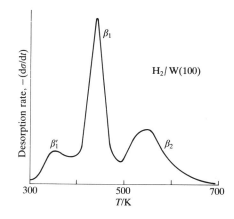

Fig. 29.29 The flash desorption spectrum of H_2 on the (100) face of tungsten. The three peaks indicate the presence of three sites with different adsorption enthalpies and therefore different desorption activation energies. (P. W. Tamm and L. D. Schmidt, *J. chem. Phys.*, **51**, 5352 (1969).)

Example 29.6: *Analysing a flash desorption spectrum*

Suppose that the desorption is first-order in the surface concentration σ of the adsorbate and that the temperature is increased linearly with time from some initial value T_i. Show that there is a temperature at which there is a surge of desorption.

Answer. Since the desorption is first order, we can write

$$-\dot{\sigma} = k \times \sigma = \sigma A e^{-E_d/RT}$$

where $\dot{\sigma} = d\sigma/dt$. The temperature increases linearly with time, so we can also write

$$T = T_i + \kappa t$$

The temperature T^* at which the desorption rate is a maximum is the solution of

$$\frac{d\dot{\sigma}}{dT} = 0 \quad \text{at} \quad T = T^*$$

Since $dT = \kappa \, dt$, this condition can be developed as follows:

$$\frac{d\dot{\sigma}}{dT} = -\frac{d(k\sigma)}{dT} = -k\frac{d\sigma}{dt} \times \frac{dt}{dT} - \sigma\frac{dk}{dT}$$

$$= \frac{k^2\sigma}{\kappa} - \frac{E_d k\sigma}{RT^2} = 0 \quad \text{at} \quad T = T^*$$

Therefore, the temperature at which the pulse occurs is the solution of

$$k = \frac{\kappa E_d}{RT^{*2}}$$

or

$$A e^{-E_d/RT^*} = \frac{\kappa E_d}{RT^{*2}}$$

Exercise. Show that a plot of $\ln(T^{*2}/\kappa)$ against $1/T^*$ should give a straight line with slope E_d/R.

$$\left[\ln\frac{T^{*2}}{\kappa} = \ln\frac{E_d}{RA} + \frac{E_d}{RT^*}\right]$$

In many cases only a single activation energy (and a single peak in the flash desorption spectrum) is observed. When several peaks are observed they might correspond to adsorption on different crystal planes or to multilayer adsorption. For instance, Cd atoms on tungsten show two activation energies, one of $18 \, \text{kJ} \, \text{mol}^{-1}$ and the other of $90 \, \text{kJ} \, \text{mol}^{-1}$. The explanation is that the more tightly bound Cd atoms are attached directly to the substrate, and the less strongly bound are in a layer (or layers) above the primary overlayer. Chemisorption cannot normally exceed monolayer coverage because a hydrocarbon gas, for instance, cannot chemisorb on to a surface already covered with its fragments; a metal, on other hand, as in this case, can provide a surface capable of further chemisorption.

Another example of a system showing two desorption activation energies is CO on tungsten, the values being $120 \, \text{kJ} \, \text{mol}^{-1}$ and $300 \, \text{kJ} \, \text{mol}^{-1}$. The explanation is believed to be the existence of two types of metal-adsorbate binding site, one involving a simple M—CO bond, the other adsorption with dissociation (into individually adsorbed C and O atoms). In some cases two desorption peaks may occur even though there is only one type of site. This complication arises when there are significant interactions between the adsorbate particles so that at low surface coverages the enthalpy of adsorption is significantly different from its value at high coverages.

Mobility on surfaces

A further aspect of the strength of the interactions between adsorbate and substrate is the former's mobility. Mobility is often a vital feature of a catalyst's activity, because a catalyst might be impotent if the reactant molecules adsorb so strongly that they cannot migrate. The activation energy for diffusion over a surface need not be the same for desorption because the particles may be able to move through valleys between potential peaks without leaving the surface completely. In general, it turns out that the activation energy for migration is about 10–20 per cent of the energy of the surface-adsorbate bond, but that the actual value depends on the extent of coverage. The defect structure of the sample (which depends on the temperature) may also play a dominant role because the adsorbed particles might find it easier to skip across a terrace than to roll along the foot of a step, and they might become trapped in vacancies in an otherwise flat terrace. Diffusion may also be easier across one crystal face than another, and so the surface mobility depends on which lattice planes are exposed.

There are two very elegant FIM-based methods for determining the diffusion characteristics of an adsorbate. One involves depositing an adsorbate on a surface plane at low temperature, taking an FIM image before and after the temperature is raised, and monitoring the migration of the boundary as the adsorbate floods across the crystal faces at different rates. In a modification of the technique, an individual atom is imaged, the temperature is raised, and then lowered after a definite interval. A new image is then taken, and the new position of the atom measured (Fig. 29.30). A sequence of pictures shows that the atom makes a random walk across the surface, and the diffusion coefficient D can be inferred from the mean distance d travelled in an interval τ using the two-dimensional random walk expression $d = (D\tau)^{1/2}$. The value of D for different crystal planes at different temperatures can be determined directly in this way, and the activation energy for migration over each plane obtained from the

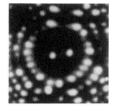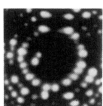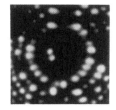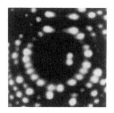

Fig. 29.30 FIM micrographs showing the migration of Re atoms on rhenium during 3 s intervals at 375 K. (Photographs provided by Professor G. Ehrlich.)

Arrhenius-like expression

$$D = D_0 e^{-E_D/RT}$$

where E_D is the diffusion activation energy. Typical values for W atoms on tungsten have E_D in the range 57–87 kJ mol^{-1} and $D_0 \approx 3.8 \times 10^{-7}$ cm^2 s^{-1}. For CO on tungsten, the activation energy falls from 144 kJ mol^{-1} at low surface coverage to 88 kJ mol^{-1} when the coverage is high.

Catalytic activity at surfaces

A catalyst acts by providing an alternative reaction path with a lower activation energy (Table 29.3). It does not disturb the equilibrium composition of the system, only the rate at which that equilibrium is attained. In this section we shall consider **heterogeneous catalysis** in which the catalyst and the reagents are in different phases. We shall in fact consider specifically only gas/solid systems.

29.6 Adsorption and catalysis

Heterogeneous catalysis normally depends on at least one reactant being adsorbed (usually chemisorbed) and modified to a form in which it readily undergoes reaction. Often this modification takes the form of a fragmentation of the reactant molecules.

The Eley–Rideal mechanism

In the **Eley–Rideal mechanism** of surface-catalysed reaction, a gas phase molecule collides with another molecule adsorbed on the surface. The rate of formation of product is expected to be proportional to the partial pressure p_B of the nonadsorbed gas B and the extent of surface coverage θ_A of the adsorbed gas A. It follows that the rate law should be

$$A + B \rightarrow P \qquad v = k p_B \theta_A \qquad (11)$$

The rate constant k might be much larger than for the uncatalysed gas-phase reaction because the reaction on the surface has a low activation energy and the adsorption itself is often not activated.

If we know the adsorption isotherm for A, we can express the rate law in terms of its partial pressure, p_A. For example, if the adsorption of A follows a Langmuir isotherm in the pressure range of interest, the rate law is

$$v = \frac{k K p_A p_B}{1 + K p_A} \qquad (12)$$

Table 29.3. Activation energies of catalysed reactions

Reaction	Catalyst	E_a/(kJ mol^{-1})
$2HI \rightarrow H_2 + I_2$	None	184
	Au	105
	Pt	59
$2NH_3 \rightarrow$	None	350
$N_2 + 3H_2$	W	162

If A is a diatomic molecule that adsorbs as atoms, then we substitute the isotherm of eqn 3 instead.

According to eqn 12, when the partial pressure of A is high (in the sense $Kp_A \gg 1$) there is almost complete surface coverage, and the rate is equal to kp_B. Now the rate-determining step is the collision of B with the adsorbed fragments. When the pressure of A is low ($Kp_A \ll 1$), perhaps because of its reaction, the rate is equal to $kKp_A p_B$; now the extent of surface coverage is important in the determination of the rate.

Example 29.7: *Interpreting the kinetics of a catalysed reaction*

The decomposition of phosphine on tungsten is first-order at low pressures and zeroth-order at high pressures. Account for these observations.

Answer. If the rate is supposed to be proportional to the surface coverage, we can write

$$v = k\theta = \frac{kKp}{1 + Kp}$$

where p is the pressure of phosphine. When the pressure is so low that $Kp \ll 1$,

$$v = kKp$$

and the decomposition is first-order. When $Kp \gg 1$,

$$v = k$$

and the decomposition is zeroth-order.

Comment. Many heterogeneous reactions are first-order, which indicates that the rate-determining stage is the adsorption process.

Exercise. Suggest the form of the rate law for the deuteration of NH_3 in which D_2 adsorbs dissociatively and extensively (i.e. $Kp \gg 1$, with p the partial pressure of D_2), and NH_3 (with partial pressure p') adsorbs at different sites.
$$[v = k(Kp)^{1/2}K'p'/(1 + K'p')]$$

There are in fact very few (if any) systems where it can be shown conclusively that the Eley–Rideal mechanism is operating. The oxidation of CO on platinum was thought to be a good example, but recent molecular beam experiments have shown that it occurs by the Langmuir–Hinshelwood mechanism.

The Langmuir–Hinshelwood mechanism

In the **Langmuir–Hinshelwood mechanism** of surface-catalysed reactions, the reaction takes place by encounters between molecular fragments and atoms adsorbed on the surface. We therefore expect the rate law to be second-order in the extent of surface coverage:

$$A + B \rightarrow P \qquad v = k\theta_A \theta_B \qquad (13)$$

Insertion of the appropriate isotherms for A and B then gives the reaction rate in terms of the partial pressures of the reactants. For example, if A and

B follow Langmuir isotherms, and adsorb without dissociation,

$$\theta_A = \frac{K_A p_A}{1 + K_A p_A + K_B p_B} \qquad \theta_B = \frac{K_B p_B}{1 + K_A p_A + K_B p_B}$$

and the rate law is

$$v = \frac{k K_A p_A K_B p_B}{(1 + K_A p_A + K_B p_B)^2} \tag{14}$$

Since the parameters in the isotherms and the rate constant k are all temperature dependent, the overall temperature dependence of the rate may be strongly non-Arrhenius.

Molecular beam studies

Molecular beam studies are able to give detailed information about catalysed reactions. It has become possible to investigate how the catalytic activity of a surface depends on its structure as well as its composition. For instance, the cleavage of C—H and H—H bonds appears to depend on the presence of steps and kinks, and a terrace often has only minimal catalytic activity. The reaction

$$H_2 + D_2 \rightarrow 2HD$$

has been studied in detail, and it is found that terrace sites are inactive but that one molecule in ten reacts when it strikes a step. Although the step itself might be the important feature, it may be that the presence of the step merely exposes a more reactive crystal face (the step itself). Likewise, the dehydrogenation of hexane to hexene depends strongly on the kink density, and it appears that kinks are needed to cleave C—C bonds. These observations suggest a reason why even small amounts of impurities may poison a catalyst: they are likely to attach to step and kink sites, and so impair the activity of the catalyst entirely. A constructive outcome is that the extent of hydrogenolysis may be controlled relative to other types of reactions by seeking impurities that adsorb at kinks and act as specific poisons.

29.7 Examples of catalysis

Almost the whole of modern chemical industry depends on the development, selection, and application of catalysts (Table 29.4). All we can hope

Table 29.4. Properties of catalysts

Catalyst	Function	Examples
Metals	Hydrogenation Dehydrogenation	Fe, Ni, Pt, Ag
Semiconducting oxides and sulphides	Oxidation Desulphurization	NiO, ZnO, MgO, Bi_2O_3/MoO_3
Insulating oxides	Dehydration	Al_2O_3, SiO_2, MgO
Acids	Polymerization Isomerization Cracking Alkylation	H_3PO_4, H_2SO_4, SiO_2/Al_2O_3

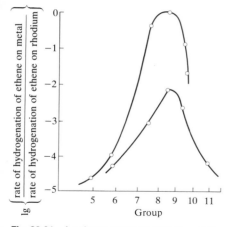

Fig. 29.31 A volcano curve of catalytic activity arises because although the reactants must adsorb reasonably strongly, they must not adsorb so strongly that they are immobilized. The lower curve refers to the first series of d-block metals, the upper curve to the second and third series d-block metals. The group numbers relate to the periodic table inside the back cover.

to do in this section is to give a brief indication of some of the problems involved. Other than the ones we consider, these include the danger of the catalyst being poisoned by by-products or impurities and economic considerations relating to cost and lifetime.

The activity of a catalyst depends on the strength of chemisorption as indicated by the **volcano curve** in Fig. 29.31. In order to be active, the catalyst should be extensively covered by adsorbate, which is the case if chemisorption is strong. On the other hand, if the strength of the substrate-adsorbate bond becomes too great, the activity declines either because the other reactant molecules cannot react with the adsorbate or because the adsorbate molecules are immobilized on the surface. This suggests that the activity of a catalyst should initially increase with strength of adsorption (as measured, for instance, by the enthalpy of adsorption) and then decline, and that the most active catalysts should be those lying near the summit of the volcano.

Many metals are suitable for adsorbing gases, and the general order of adsorption strengths decreases along the series O_2, C_2H_2, C_2H_4, CO, H_2, CO_2, N_2. Some of these molecules adsorb dissociatively (e.g. H_2). Elements from the d block, such as Fe, V, and Cr, show a strong activity towards all these gases, but Mn and Cu are unable to adsorb N_2 and CO_2, and adsorb H_2 only very weakly. Metals towards the left of the periodic table (e.g. Mg and Li) can adsorb only the most active gas (O_2). These trends are summarized in Table 29.5.

Table 29.5. Chemisorption abilities

	O_2	C_2H_2	C_2H_4	CO	H_2	CO_2	N_2
Ti, Cr, Mo, Fe	+	+	+	+	+	+	+
Ni, Co	+	+	+	+	+	+	−
Pd, Pt	+	+	+	+	+	−	−
Mn, Cu	+	+	+	+	±	−	−
Al, Au	+	+	+	+	−	−	−
Li, Na, K	+	+	−	−	−	−	−
Mg, Ag, Zn, Pb	+	−	−	−	−	−	−

+ Strong chemisorption, ± chemisorption, − no chemisorption. See G. C. Bond, *Heterogeneous catalysis*, Oxford (1986) for further information.

Hydrogenation

An example of catalytic action is the hydrogenation of alkenes. The alkene (**1**) adsorbs by forming two bonds with the surface (**2**), and on the same surface there may be adsorbed H atoms. When an encounter occurs, one of the alkene-surface bonds is broken (**2** → **3** or **4**) and later an encounter with a second H atom releases the fully hydrogenated hydrocarbon, which is the thermodynamically more stable species.

The evidence for a two-stage reaction is the appearance of different isomeric alkenes in the mixture. The formation of isomers comes about because while the hydrocarbon chain is waving about over the surface of the metal, it might chemisorb again (**4** → **5**) and desorb to **6**, an isomer of the original **1**. The new alkene would not be formed if the two hydrogen atoms attached simultaneously.

1

2

3

4

5

6

A major industrial application of catalytic hydrogenation is to the formation of edible fats from vegetable and animal oils. Raw oils obtained from sources such as the soya bean have the structure $CH_2(O_2CR)CH(O_2CR')CH_2(O_2CR'')$, where R, R', and R'' are long-chain hydrocarbons with several double bonds. One disadvantage of the presence of many double bonds is that the oils are susceptible to atmospheric oxidation, and therefore are liable to become rancid. The geometrical configuration of the chains is responsible for the liquid nature of the oil, and in many applications a solid fat is at least much better and often necessary. Controlled partial hydrogenation of an oil with a catalyst carefully selected so that hydrogenation is incomplete and so that the chains do not isomerize, is used on a wide scale to produce edible fats. The process, and the industry, is not made any easier by the seasonal variation of the number of double bonds in the oils.

Oxidation

Catalytic oxidation is also widely used in industry and in pollution control. Although in some cases it is desirable to achieve complete oxidation (as in the production of nitric acid from ammonia), in others partial oxidation is the aim. For example, the complete oxidation of propene to carbon dioxide and water is wasteful, but its partial oxidation to acrolein, $CH_2{=}CHCHO$, is the start of important industrial processes. Likewise, the controlled oxidations of ethene to ethanol, acetaldehyde, and (in the presence of acetic acid or chlorine) to vinyl acetate or vinyl chloride, are the initial stages of very important chemical industries.

Some of these reactions are catalysed by oxides of various kinds. The physical chemistry of oxide surfaces is very obscure, as can be appreciated by considering what happens during the oxidation of propene to acrolein on bismuth molybdate. The first stage is the adsorption of the propene molecule with loss of a hydrogen to form the allyl radical, $CH_2{=}CHCH_2\cdot$. An O atom in the surface can now transfer to the allyl radical, leading to the formation of acrolein and its desorption from the surface. The H atom also escapes with a surface O atom, and goes on to form H_2O, which leaves the surface. The surface is left with the charges the O^{2-} discarded as they formed O atoms, and these charges are attacked by O_2 molecules in the overlying gas, which then chemisorb as oxide ions, so reforming the catalyst. This sequence of events involves great upheavals of the surface, and some materials break up under the stress.

Cracking and reforming

Many of the small organic molecules used in the preparation of all kinds of chemical products come from oil. These small building blocks of polymers, perfumes, and petrochemicals in general, are often cut from the long-chain hydrocarbons squeezed out of the earth as petroleum. The catalytically induced fragmentation of the long-chain hydrocarbons is called **cracking**, and is often brought about in zeolite catalysts. These catalysts act by forming unstable carbocations, which dissociate and rearrange to more highly branched isomers. Branched isomers burn more smoothly and efficiently in internal combustion engines, and are used to produce higher octane fuels.

Catalytic cracking has been largely superseded by catalytic **reforming** using a dual-function catalyst, such as a mixture of platinum and alumina. The platinum provides the metal function, and brings about dehydrogenation and hydrogenation. The alumina provides the acidic function, being able to form carbocations. The sequence of events in catalytic reforming shows up very clearly the complications that must be unravelled if a reaction as important as this is to be understood and improved. The first step is the attachment of the long-chain hydrocarbon by chemisorption to the platinum. In this process first one and then a second H atom is lost, and alkene is formed. The alkene migrates to an acid site, where it accepts a proton and attaches to the surface as a carbocation. This carbocation can undergo several different reactions. It can break into two, isomerize into a more highly branched form, or undergo varieties of ring-closure. Then it loses a proton, escapes from the surface, and migrates (possibly through the gas) as an alkene to a metal part of the catalyst where it is hydrogenated. We end up with a rich selection of smaller molecules which can be withdrawn, fractionated, and then used as raw materials for other products.

Further information: the BET isotherm

Consider Fig. 29.32, which shows a region of the surface covered in monolayers, bilayers, and so on. The rate constants for adsorption and desorption of the primary layer are k_a and k_d, and those of the overlayers are all k_a' and k_d'. The numbers of sites corresponding to zero, monolayer, bilayer, etc. coverage at any stage are N_0, N_1, N_2, etc., and N_i in general. The condition for equilibrium of the initial layer is the equality of the rates of its formation and desorption:

$$k_a p N_0 = k_d N_1$$

The condition for equilibrium of the next layer is

$$k_a' p N_1 = k_d' N_2$$

and in general

$$k_a' p N_{i-1} = k_d' N_i \qquad i = 2, 3, \ldots$$

This condition may be expressed in terms of N_0 as follows:

$$N_i = \frac{k_a'}{k_d'} \times p N_{i-1} = \left(\frac{k_a'}{k_d'}\right)^2 \times p^2 N_{i-2} = \ldots$$

$$= \left(\frac{k_a'}{k_d'}\right)^{i-1} \times \frac{k_a}{k_d} \times p^i N_0$$

Now write $k_a'/k_d' = x$ and $k_a/k_d = cx$; then with this simpler notation

$$N_i = c(xp)^i N_0$$

We now calculate the total volume V of adsorbed material. Since V is proportional to the total number of particles adsorbed,

$$V \propto N_1 + 2N_2 + 3N_3 + \ldots = \sum_{i=1}^{\infty} i N_i$$

because a monolayer site contributes one particle, a bilayer site two, and so

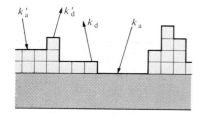

Fig. 29.32 In the derivation of the BET isotherm, different adsorption and desorption rate constants are used depending on whether the site is already occupied or not. The total volume adsorbed is proportional to the number of particles adsorbed, which is obtained by summing the column heights.

on. If there were complete monolayer coverage the volume adsorbed would be V_{mon}, with

$$V_{mon} \propto N_0 + N_1 + N_2 + \ldots = \sum_{i=0}^{\infty} N_i$$

because each site in the actual sample then contributes one particle to the total. It follows that

$$\frac{V}{V_{mon}} = \frac{\sum_{i=1}^{\infty} iN_i}{\sum_{i=0}^{\infty} N_i} = \frac{cN_0 \sum_{i=1}^{\infty} i(xp)^i}{N_0 + cN_0 \sum_{i=1}^{\infty} (xp)^i}$$

We can evaluate both sums using

$$\sum_{i=1}^{\infty} y^i = \sum_{i=0}^{\infty} y^i - 1 = \frac{1}{1-y} - 1$$

$$= \frac{y}{1-y}$$

$$\sum_{i=1}^{\infty} iy^i = y \frac{\mathrm{d}}{\mathrm{d}y} \sum_{i=1}^{\infty} y^i = y \frac{\mathrm{d}}{\mathrm{d}y} \frac{y}{1-y}$$

$$= \frac{y}{(1-y)^2}$$

with $y = xp$. After a little rearrangement this leads to

$$\frac{V}{V_{mon}} = \frac{cxp}{(1-xp)(1-xp+cxp)} \qquad \text{(A1)}$$

In the final step we identify $1/x$ as the vapour pressure p^* of the bulk liquid adsorbate. Consider the equilibrium between a gas and a sample in which all N surface sites are deeply and uniformly buried (Fig. 29.33). The condition for equilibrium is now

$$k_a' p N = k_d' N$$

so

$$k_a' p = k_d'$$

Such an equilibrium applies at the liquid surface irrespective of whether a surface is deeply buried under it, and so p, the equilibrium pressure, can be identified with p^*, the bulk vapour pressure. It follows that

$$x = \frac{k_a'}{k_d'} = \frac{1}{p^*}$$

as we set out to show. When this result is introduced into eqn A1, we obtain the BET isotherm:

$$\frac{V}{V_{mon}} = \frac{cz}{(1-z)\{1-(1-c)z\}} \qquad z = \frac{p}{p^*} \qquad \text{(A2)}$$

The relation between the value of c and the enthalpies of desorption and

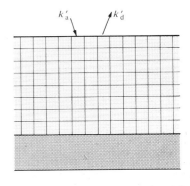

Fig. 29.33 When the surface is deeply covered, the equilibrium pressure is the vapour pressure of the bulk liquid because the substrate plays no role.

vaporization follows from its definition as

$$c = \frac{p^*}{k_d/k_a} = \frac{p^*}{p'}$$

The ratio $p' = k_d/k_a$ is the equilibrium pressure of the adsorbate if only monolayer adsorption is allowed. Since the equilibrium pressures are related to a Gibbs function through

$$-RT \ln p = \Delta G^{\ominus}$$

and the entropies of desorption and vaporization are very similar,

$$c = e^{-(\Delta G_{vap}^{\ominus} - \Delta G_d^{\ominus})/RT} \approx e^{-(\Delta H_{vap}^{\ominus} - \Delta H_d^{\ominus})/RT}$$

as used in the text (eqn 5b). When the two enthalpies are similar, $c \approx 1$; when $\Delta H_d^{\ominus}$ is large compared to $\Delta H_{vap}^{\ominus}$, $c \gg 1$.

Further reading

Adsorption

A. W. Adamson, *Physical chemistry of surfaces*. Wiley-Interscience, New York (1975).

G. Somorjai, *Principles of surface chemistry*. Prentice-Hall, Englewood Cliffs (1972).

R. P. H. Gasser, *An introduction to chemisorption and catalysis by metals*. Clarendon Press, Oxford (1985).

M. W. Roberts and C. S. McKee, *Chemistry of the metal-gas interface*. Clarendon Press, Oxford (1978).

Catalysis

G. Somorjai, *Chemistry in two dimensions: surfaces*. Cornell University Press (1981).

G. C. Bond, *Heterogeneous catalysis: principles and applications*. Clarendon Press, Oxford (1986).

M. Boudart and G. Djéga-Mariadassou, *Kinetics of heterogeneous catalysis reactions*. Princeton University Press (1984).

C. Satterfield, *Heterogeneous catalysis in practice*. McGraw-Hill, New York (1980).

D. F. Shriver, P. W. Atkins, and C. H. Langford, *Inorganic chemistry* (Chapter 17). Oxford University Press and W. H. Freeman & Co., New York (1990).

Exercises

29.1 Calculate the frequency of molecular collisions per cm^2 of surface in a vessel containing (a) hydrogen, (b) propane at 25°C when the pressure is (i) 100 Pa, (ii) 10^{-7} Torr.

29.2 What pressure of argon gas is required to produce a collision rate of $4.5 \times 10^{10} \, s^{-1}$ at 425 K on a circular surface of diameter 1.5 mm?

29.3 Calculate the average rate at which He atoms strike a Cu atom in a surface formed by exposing a (100) plane in metallic copper to helium gas at 80 K and a pressure of 35 Pa. Crystals of copper are face centred cubic with a cell edge of 361 pm.

29.4 The enthalpy of adsorption of CO on a surface is found to be $-120 \, kJ \, mol^{-1}$. Is the adsorption physisorption or chemisorption? Estimate the mean lifetime of a CO molecule on the surface at 400 K.

29.5 The average time for which an oxygen atom remains adsorbed to a tungsten surface is 0.36 s at 2548 K and 3.49 s at 2362 K. Find the activation energy for desorption. What is the pre-exponential factor for these tightly chemisorbed atoms?

29.6 The chemisorption of hydrogen on manganese is activated, but only weakly so. Careful measurements have shown that it proceeds 35 per cent faster at 1000 K than at 600 K. What is the activation energy for chemisorption?

29.7 The adsorption of a gas is described by the Langmuir isotherm with $K = 0.85 \text{ kPa}^{-1}$ at 25°C. Calculate the pressure at which the fractional surface coverage is (a) 0.15, (b) 0.95.

29.8 A certain solid sample adsorbs 0.44 mg of CO when the pressure of the gas is 26.0 kPa and the temperature is 300 K. The amount adsorbed when the pressure is 3.0 kPa and the temperature is 300 K is 0.19 mg. The Langmuir isotherm is known to describe the adsorption. Find the fractional coverage of the surface at the two pressures.

29.9 For how long on average would a hydrogen atom remain on a surface at 298 K if its desorption activation energy is (a) 15 kJ mol^{-1}, (b) 150 kJ mol^{-1}? Take $\tau_0 = 10^{-13} \text{ s}$. For how long on average would the same atoms remain at 1000 K?

29.10 A solid in contact with a gas at 12 kPa and 25°C adsorbs 2.5 mg of the gas and obeys the Langmuir isotherm. The enthalpy change when 1.00 mmol of the adsorbed gas is desorbed is +10.2 J. What is the equilibrium pressure for the adsorption of 2.5 mg of gas at 40°C?

29.11 Hydrogen iodide is very strongly adsorbed on gold but only slightly adsorbed on platinum. Assume the adsorption follows the Langmuir isotherm and predict the order of the HI decomposition reaction on each of the two metal surfaces.

29.12 Suppose it is known that ozone adsorbs on a particular surface in accord with the Langmuir isotherm. How could you use the pressure dependence of the fractional coverage to distinguish between adsorption (a) without dissociation, (b) with dissociation into $O + O_2$, (c) with dissociation into $O + O + O$?

29.13 Confirm the isotherms preceding eqn 14 for the extent of surface coverage in the presence of two adsorbing gases.

29.14 Nitrogen gas adsorbed on charcoal to the extent of $0.921 \text{ cm}^3 \text{ g}^{-1}$ at 4.8 atm and 190 K, but at 250 K the same amount of adsorption was achieved only when the pressure was increased to 32 atm. What is the molar enthalpy of adsorption of nitrogen on charcoal?

29.15 In an experiment on the adsorption of oxygen on tungsten it was found that the same volume of oxygen was desorbed in 27 min at 1856 K, 2 min at 1978 K, and 0.3 min at 2070 K. What is the activation energy of desorption? How long would it take for the same amount to desorb at (a) 298 K, (b) 3000 K?

29.16 Ammonia was introduced into a bulb at a pressure of 27 kPa. At 856°C it was found that a tungsten catalyst brought about a pressure change of 8 kPa in 500 s, and 15 kPa in 1000 s. What is the order of the catalysed decomposition? Account for the result.

Problems

Numerical problems

29.1 Nickel is face-centred cubic with a unit cell of side 352 pm. What is the number of atoms per cm^2 exposed on a surface formed by (a) (100), (b) (110), (c) (111) planes? Calculate the frequency of molecular collisions per surface atom in a vessel containing (a) hydrogen, (b) propane at 25°C when the pressure is (i) 100 Pa, (ii) 10^{-7} Torr.

29.2 Tungsten crystallizes into a body-centred cubic form with a unit cell of side 316 pm. Calculate the number of atoms exposed per cm^2 on the surfaces corresponding to the (a) (100), (b) (110), (c) (111) planes. In filaments of tungsten (110) and (100) planes predominate: estimate the average number of atoms exposed per unit area. Calculate the frequency of molecular collisions per surface atom in a vessel containing (a) hydrogen, (b) propane at 25°C when the pressure is (i) 100 Pa, (ii) 10^{-7} Torr.

29.3 The data below are for the chemisorption of hydrogen on copper powder at 25°C. Confirm that they fit the Langmuir isotherm at low coverages. Then find the value of K for the adsorption equilibrium and the adsorption volume corresponding to complete coverage.

p/Torr	0.19	0.97	1.90	4.05	7.50	11.95
V_a/cm^3	0.042	0.163	0.221	0.321	0.411	0.471

29.4 The data for the adsorption of ammonia on barium fluoride are reported below. Confirm that they fit a BET isotherm and find values of c and V_{mon}.
(a) $\theta = 0°C$, $p^* = 3222$ Torr:

p/Torr	105	282	492	594	620	755	798
V/cm^3	11.1	13.5	14.9	16.0	15.5	17.3	16.5

(b) $\theta = 18.6°C$, $p^* = 6148$ Torr:

p/Torr	39.5	62.7	108	219	466	555	601	765
V/cm	9.2	9.8	10.3	11.3	12.9	13.1	13.4	14.1

29.5 The following data were obtained for the adsorption of methane on 10 g of carbon black at 0°C. Which isotherm, the Langmuir or the Freundlich fits the data better?

p/Torr	100	200	300	400
V_a/cm^3	97.5	144	182	214

29.6 Carbon monoxide adsorbs on mica, and the data for 90 K are given below. Decide whether the Langmuir or the Freundlich isotherm is a better representation of the system. What is value of K? Given that the total sample area is $6.2 \times 10^3 \text{ cm}^2$, calculate the area occupied by each adsorbed molecule.

p/Torr	100	200	300	400	500	600
V_a/cm^3	0.130	0.150	0.162	0.166	0.175	0.180

What volume of carbon monoxide would be adsorbed by the mica 90 K when the pressure is 1 atm?

29.7 The designers of a new industrial plant wanted to use a catalyst code-named CR-1 in a step involving the fluorination of butadiene. As a first step in the investigation they determined the form of the adsorption isotherm. The volume of butadiene adsorbed per gram of CR-1 at 15°C depended on the pressure as given below. Is the Langmuir isotherm suitable at this pressure?

p/Torr	100	200	300	400	500	600
V_a/cm^3	17.9	33.0	47.0	60.8	75.3	91.3

Investigate whether the BET isotherm gives a better description of the adsorption of butadiene on CR-1. At 15°C, p^* (butadiene) = 200 kPa. Find V_{mon} and c.

29.8 The adsorption of solutes on solids from liquids often follows a Freundlich isotherm. Check the applicability of this isotherm to the following data for the adsorption of acetic acid on charcoal at 25°C and find the values of the parameters c_1 and c_2.

[acid]/M	0.05	0.10	0.50	1.0	1.5
w_a/g	0.04	0.06	0.12	0.16	0.19

w_a is the mass adsorbed per unit mass of charcoal. Investigate whether the adsorption of acetic acid on charcoal would be better represented by a Langmuir isotherm.

29.9 In some catalytic reactions the products may adsorb more strongly than the reacting gas. This is the case, for instance, in the catalytic decomposition of ammonia on platinum at 1000°C. As a first step in examining the kinetics of this type of process, show that the rate of ammonia decomposition should follow

$$-\frac{dp(NH_3)}{dt} = \frac{k_c p(NH_3)}{p(H_2)}$$

in the limit of very strong adsorption of hydrogen. Start by showing that when a gas J adsorbs very strongly, and its pressure is $p(J)$, that the fraction of uncovered sites is approximately $1/Kp(J)$. Solve the rate equation for the catalytic decomposition of NH_3 on platinum and show that a plot of $F(t) = (1/t) \ln(p/p_0)$ against $G(t) = (p - p_0)/t$, where p is the pressure of ammonia, should give a straight line from which k_c can be determined. Check the rate law on the basis of the data below, and find k_c for the reaction.

t/s	0	30	60	100	160	200	250
p/Torr	100	88	84	80	77	74	72

29.10 Now suppose that the retarding gas is less weakly adsorbed and that the full form of the Langmuir isotherm has to be used. The reactant gas is still supposed not to adsorb.

Deduce the rate law

$$-\frac{dp}{dt} = \frac{k_c p}{1 + Kp'}$$

where p is the pressure of the reactant gas and p' that of the product. Integrate the rate law for the simple decomposition A → B + C, where B adsorbs and inhibits but C does not. An example of this type of reaction is the decomposition of dinitrogen oxide, N_2O, on platinum at 750°C. In this case the oxygen produced by the reaction adsorbs strongly and inhibits the catalyst. Confirm that the data below fit the integrated rate law and determine K and k_c for the reaction.

t/s	0	315	750	1400	2250	3450	5150
p/Torr	95	85	75	65	55	45	35

Theoretical problems

29.11 The deposition of atoms and ions on a surface depends on their ability to stick, and therefore on the energy changes that occur. As an illustration, consider a two-dimensional square lattice of univalent positive and negative ions separated by 200 pm, and consider a cation approaching the upper terrace of this array from the top of the page. Calculate, by direct summation, its coulombic interaction when it is in an empty lattice point directly above an anion. Now consider a high step in the same lattice, and let the approaching ion go into the corner formed by the step and the terrace. Calculate the coulombic energy for this position, and decide on the likely settling point for a deposited cation.

29.12 Integrate the rate law in eqn 14 for $p_A(0) = p_B(0)$. Sketch the time-dependence of the pressure of the product in the case of $K_A \approx K_B \approx 1.0$ atm^{-1} and $p_A(0) \approx p_B(0) \approx 1.0$ atm.

29.13 Fluorine adsorbs on a certain catalyst in accord with the Langmuir isotherm, but butadiene adsorbs in accord with the Freundlich isotherm with $c_2 = 2$. Deduce an expression for the reaction rate on the basis that fluorination depends on encounters of F atoms and butadiene molecules on the catalyst surface. Assume that the adsorption of the two species occurs at different types of site.

29.14 Although the attractive van der Waals interaction between individual molecules varies as R^{-6}, the interaction of a molecule with a nearby solid (a homogeneous collection of molecules) varies as R^{-3}, where R is its vertical distance above the surface. Confirm this assertion. Calculate the interaction energy between an Ar atom and the surface of solid argon on the basis of a Lennard–Jones (6, 12)-potential. Estimate the equilibrium distance of an atom above the surface.

29.15 We now turn to the application of the Gibbs isotherm, Section 23.12, to the adsorption of gases. Show by a directly analogous argument to that used in Chapter 23 that the volume adsorbed per unit area of solid, V_a/σ, is related to the pressure of the gas by $V_a = -(\sigma/RT)(d\mu/d\ln p)$, where μ is the chemical potential of the adsorbed gas.

29.16 If the dependence of the chemical potential of the gas on the extent of surface coverage is known, the Gibbs isotherm can be integrated to give a relation between V_a and p, as in a normal adsorption isotherm. For instance, suppose that the change in the chemical potential of a gas when it adsorbs is of the form $d\mu = -c_2(RT/\sigma)\,dV_a$, where c_2 is a constant of proportionality: show that the Gibbs isotherm leads to the Freundlich isotherm in this case.

29.17 Finally we come full circle and return to the Langmuir isotherm. Find the form of $d\mu$ that, when inserted in the Gibbs isotherm, leads to the Langmuir isotherm.

30 Dynamic electrochemistry

Check-list of key ideas

1. The *electric double layer* and approximate models of its structure (Section 30.1).

2. The *Volta potential* and the *Galvani potential* and the latter's relation to the electrode potential (Section 30.1).

3. The *electrochemical potential* (eqn 2) and its relation to the electrode potential (Section 30.1).

4. The *cathodic* and *anodic current densities* at the electrodes and the net current density (eqn 3).

5. The definitions of *transfer coefficient* (eqn 6), the *exchange current density*, and the *overpotential* (eqn 9).

6. The derivation of the *Butler–Volmer equation* (eqn 11).

7. The *low overpotential limit* (eqn 13) and *high overpotential limit* (eqn 14) of the Butler–Volmer equation and the *Tafel plot* for determining the properties of a working electrode (Example 30.3).

8. The classification of electrodes as *polarizable* and *non-polarizable* (Section 30.3).

9. The origin of *concentration polarization* and its description in terms of the *Nernst diffusion layer* (eqn 18) and the *limiting current density* (eqn 19).

10. The technique of *polarography* and the study of electrode kinetics using a *rotating-disk electrode, voltammetry,* and *cyclic voltammetry* (Section 30.3).

11. The considerations that govern the kinetic feasibility of *electrolysis* (Section 30.4).

12. The potential of *working cells* (eqn 23) and their *power output* (eqn 24).

13. The operation of *fuel cells* and of *storage cells* (Section 30.6) and the kinetic considerations that determine their characteristics.

14. The electrochemical considerations relating to the rate of *corrosion* (Section 30.7) and its inhibition (Section 30.8).

The economic consequences of electrochemistry are almost incalculable. Most of the modern methods of generating electricity are inefficient and the development of fuel cells could revolutionize our production and deployment of energy. Today we produce energy inefficiently to produce goods that then decay by corrosion. Each step of this wasteful sequence could be improved by discovering more about the kinetics of electrochemical processes. Similarly, the emerging techniques of organic and inorganic electrosynthesis, where an electrode is an active component of an industrial process, depend on a detailed knowledge of the factors affecting their rates. One example is from nylon production, in which adiponitrile is synthesized commercially by the electrolytic reductive coupling (hydrodimerization) of acrylonitrile.

Much of electrochemistry depends on processes that occur at the interface of an electrode and an ionic solution, and the kinetic problem we examine in this chapter is the rate at which ions can be discharged at electrodes. A measure of this rate is the **current density** j, the electric current per unit area (the charge flux), and most of our discussion will be concerned with the properties that control its magnitude. In an **electrolytic cell**, an electrochemical cell in which a non-spontaneous chemical reaction is driven by an external supply of electricity, deposition and gas evolution occur significantly only when the applied potential exceeds the zero-current cell potential by an amount called the **overpotential** η. We shall establish a relation between the overpotential and the current, and see what determines their values. The potential of a **galvanic cell**, an electrochemical cell in which a spontaneous reaction produces electricity, is smaller when it is producing current than when it is not. We shall develop relations that enable us to calculate the potential of a working cell.

Processes at electrodes

When only equilibrium properties are of interest there is no need to know the details of the charge separation responsible for the potential difference at an interface (just as we do not need to propose a reaction mechanism when discussing equilibria). However, a description of the interface is essential when we are interested in the rate of charge transfer.

30.1 The electric double layer

An electrode becomes positively charged relative to the solution nearby if electrons leave it and decrease the local cation concentration in the solution. The most primitive model of the interface is that it is an **electric double layer** consisting of a sheet of positive charge at the surface of the electrode and a sheet of negative charge next to it in the solution (or vice versa).

A more detailed picture of the interface can be constructed by speculating about the arrangement of ions and electric dipoles in the solution. The **Helmholtz model** of the double layer supposes that the hydrated ions range themselves along the surface of the electrode but are held away from it by the molecules in their hydration spheres (Fig. 30.1). The location of the sheet of ionic charge, which is called the **outer Helmholtz plane**, is identified as the plane running through the hydrated ions. The Helmholtz model ignores the disrupting effect of thermal motion which tends to break

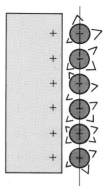

Outer Helmholtz plane

Fig. 30.1 A simple model of the electric double layer treats it as two rigid planes of charge, one plane, the outer Helmholtz plane, being due to the ions in solution, the other, the inner Helmholtz plane, being due to the charge on the electrode.

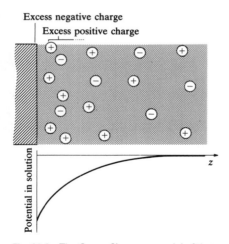

Excess negative charge

Excess positive charge

Potential in solution

z

Fig. 30.2 The Gouy–Chapman model of the electric double-layer treats it as an atmosphere of counter charge, similar to the Debye–Hückel theory of ion atmospheres.

up and disperse the rigid wall of charge. In the **Gouy–Chapman model** of the **diffuse double layer** the stirring effect of thermal motion is taken into account in much the same way as in the Debye–Hückel model of the ionic atmosphere of an ion (Section 10.2) with the latter's single central ion replaced by an infinite, plane electrode.

Figure 30.2 shows how the local concentrations of cations and anions differ from their bulk concentrations in the Gouy–Chapman model. As expected, ions of opposite charge cluster close to the electrode, and ions of the same charge are repelled from it. The modification of the local concentrations near an electrode implies that it might be misleading to use activity coefficients characteristic of the bulk to discuss the thermodynamic properties of ions near the interface. This is one of the reasons why measurements in dynamic electrochemistry are almost always done using a large excess of supporting electrolyte (e.g. a 1 M solution of a salt, an acid, or a base): the activity coefficients are then almost constant because the inert ions dominate the effects of local changes caused by any reactions taking place.

Neither the Helmholtz nor the Gouy–Chapman model is a very good representation of the structure of the double layer. The former overemphasizes the rigidity of the local solution; the latter underemphasizes its structure. The two are combined in the **Stern model**, in which the ions closest to the electrode are constrained into a rigid Helmholtz plane while outside that plane the ions are dispersed as in the Gouy–Chapman model.

The potential difference

The most important property of the double layer is the effect it has on the electric potential near the electrode. We can analyse the potential at the interface by imagining separating the electrode from the solution, but with the charges of the metal and the solution frozen in position.

A positive test charge at great distances from the isolated electrode experiences a Coulomb potential that varies inversely with distance (Fig. 30.3). As the test charge approaches the electrode it enters a region where the potential varies more slowly. This change in behaviour can be traced (using electrostatics) to the fact that the surface charge is not point-like but is spread over an area. At about 100 nm from the surface the potential varies only slighly with distance, and its value in this region is called the

Fig. 30.3 The variation of potential with distance from an electrode that has been separated from the electrolyte solution without there being an adjustment of charge. A similar diagram applies to the separated solution.

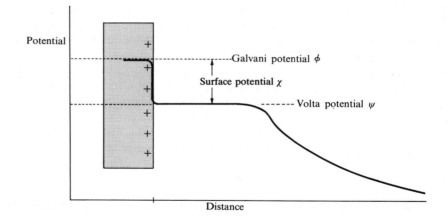

Potential

Galvani potential ϕ

Surface potential χ

Volta potential ψ

Distance

Volta potential or the **outer potential** ψ. As the test charge is taken through the surface concentration (or depletion) of electrons on the surface of the electrode the potential it experiences changes until it reaches the inner, bulk metal environment. This extra potential is called the **surface potential** χ. The total potential inside the electrode is called the **Galvani potential** ϕ.

A similar sequence of changes of potential is observed as a positive test charge is brought up to and through the solution surface. The potential changes to its Volta value as the charge approaches the charged medium, then to its Galvani value as it is taken through into the bulk.

When the electrode and solution are reassembled (without any change of charge distribution) the potential difference between points in the bulk metal and the bulk solution is the **Galvani potential difference** $\Delta\phi$. We shall shortly see that, apart from a constant, this Galvani potential difference is the electrode potential that was at the centre of our discussion in Chapter 10.

The electrochemical potential

To demonstrate the relation between $\Delta\phi$ and E, and at the same time to introduce an important new concept, we shall consider the cell

$$\text{Pt} \mid H_2(g) \mid H^+(aq) \parallel M^+(aq) \mid M(s)$$

and show that the zero-current cell potential E, which is the difference of the two electrode potentials E_R on the right and E_L on the left, is equal to the difference in $\Delta\phi$ at each electrode:

$$E = E_R - E_L = \Delta\phi_R - \Delta\phi_L \qquad (1)$$

The cell reaction is

$$M^+(aq) + \tfrac{1}{2}H_2(g) \rightarrow M(s) + H^+(aq)$$

and the half-reactions are

$$M^+(aq) + e^- \rightarrow M(s)$$

$$H^+(aq) + e^- \rightarrow \tfrac{1}{2}H_2(g)$$

We can express the Gibbs functions of these two half-reactions in terms of the chemical potentials of all the species. However, in doing so we must take into account the fact that the species are present in phases with different electric potentials. Thus, a cation in a region of positive potential has a higher chemical potential (is chemically more active) than in a region of zero potential.

We calculate the contribution of an electric potential to the chemical potential by noting that the electrical work of adding a charge ze to a region where the potential is ϕ is

$$w_e = ze \times \phi$$

and therefore that the work per mole is

$$w_e = zF \times \phi$$

where F is Faraday's constant. Since at constant temperature and pressure the maximum electrical work can be identified with the change in Gibbs function (Section 10.4), the difference in chemical potential of an ion with and without the electric potential present is

$$\bar{\mu} - \mu = zF\phi$$

The chemical potential of an ion in the presence of an electric potential is called its **electrochemical potential** $\bar{\mu}$, and from the last equation we can write

$$\bar{\mu} = \mu + zF\phi \tag{2}$$

When $z = 0$ (a neutral species), the electrochemical potential is equal to the chemical potential. When $\phi > 0$ and the ion is a cation ($z > 0$), the electrochemical potential is greater than the chemical potential since it is in an energetically unfavourable region. When $\phi < 0$, a cation's electrochemical potential is lower than its chemical potential since now it is in an electrically favourable region.

We can now write the Gibbs function for the two half-reactions at the electrode in terms of the electrochemical potentials of the species, bearing in mind that the cations M^+ are in the solution where the Galvani potential is ϕ_S and that the electrons are in the electrode where the Galvani potential is ϕ_M:

$$\Delta G_R = \bar{\mu}(M) - \{\bar{\mu}(M^+) + \bar{\mu}(e^-)\}$$
$$= \mu(M) - \{\mu(M^+) + F\phi_S + \mu(e^-) - F\phi_M\}$$
$$= \mu(M) - \mu(M^+) - \mu(e^-) + F\,\Delta\phi_R$$

$\Delta\phi_R$ is the Galvani potential difference $\phi_M - \phi_S$ at the right-hand electrode. Likewise, in the hydrogen half-reaction, the electrons are in the platinum electrode at a potential ϕ_{Pt} and the H^+ ion are in the solution where the potential is ϕ_S:

$$\Delta G_L = \tfrac{1}{2}\bar{\mu}(H_2) - \{\bar{\mu}(H^+) + \bar{\mu}(e^-)\}$$
$$= \tfrac{1}{2}\mu(H_2) - \mu(H^+) - \mu(e^-) + F\,\Delta\phi_L$$

$\Delta\phi_L$ is the Galvani potential difference $\phi_{Pt} - \phi_S$ at the left-hand electrode.

The overall change in Gibbs function is

$$\Delta G_R - \Delta G_L = \mu(M) + \mu(H^+) - \mu(M^+) - \tfrac{1}{2}\mu(H_2) + F(\Delta\phi_R - \Delta\phi_L)$$
$$= \Delta G + F(\Delta\phi_R - \Delta\phi_L)$$

where ΔG is the Gibbs function of the cell reaction. The entire system is at equilibrium when the cell is balanced against an external source of potential, and the overall reaction Gibbs function is zero since the system has no tendency to change in either direction.[1] The last equation then becomes

$$0 = \Delta G + F(\Delta\phi_R - \Delta\phi_L)$$

which rearranges to

$$\Delta G = -F(\Delta\phi_R - \Delta\phi_L)$$

---

[1] The cell reaction itself is not necessarily at equilibrium: its tendency to change is balanced against the external source and *overall* there is stalemate.

910

If we compare this with the result established in Section 10.4 that

$$\Delta G = -FE \qquad E = E_R - E_L$$

we can conclude that

$$E_R - E_L = \Delta \phi_R - \Delta \phi_L$$

This is the result we wanted, for it implies that the Galvani potential difference at each electrode can differ from the electrode potential by a constant at most (which cancels when the difference is taken):

$$E_R = \Delta \phi_R + \text{const.} \qquad E_L = \Delta \phi_L + \text{const.}$$

From now on we shall ignore the constant, and identify changes in $\Delta \phi$ with changes in electrode potential.

30.2 The rate of charge transfer

As the electrode reaction is heterogeneous it is natural to express its rate as the amount of material produced per unit area of electrode per unit time (that is, as a flux of product). A first-order heterogeneous rate law therefore has the form:

$$\text{Amount produced per unit area per unit time} = k[J]$$

where $[J]$ is the molar concentration of the relevant species in solution. Since the expression on the left is expressed in amount per unit area per unit time, and $[J]$ is an amount per unit volume, k has dimensions of length per unit time (e.g. $cm\,s^{-1}$). If the molar concentration of the oxidized and reduced materials outside the double layer are $[Ox]$ and $[Red]$ respectively, the rate of reduction of Ox is

$$\text{Rate of reduction of Ox} = k_c[Ox]$$

and the rate of oxidation of Red is

$$\text{Rate of oxidation of Red} = k_a[Red]$$

(We justify the notation k_c and k_a below).

We shall consider a reaction at the electrode in which an ion is reduced by the transfer of a single electron in the rate-determining step.[2] Then the net current density at the electrode is the difference of the current densities arising from the reduction of Ox and the oxidation of Red. Since the redox processes at the electrode involve the transfer of one electron per reaction event, the current densities j arising from the redox processes are the rates (as expressed above) multiplied by the charge transferred per mole of reaction, which is given by Faraday's constant. Therefore, there is a **cathodic current density** of magnitude

$$j_c = Fk_c[Ox]$$

arising from the reduction, and an opposing **anodic current density** of magnitude

$$j_a = Fk_a[Red]$$

---

[2] The last phrase is important: in the deposition of cadmium, for instance, only one electron is transferred in the rate-determining step even though overall the deposition involves the transfer of two electrons.

(a)   (b)

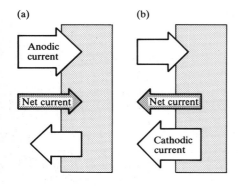

Fig. 30.4 The net current is defined as the difference $j_a - j_c$. (a) When $j_a > j_c$ the net current is anodic, and there is a net oxidation of the species in solution. (b) When $j_c > j_a$ the net current is cathodic, and the net process is reduction.

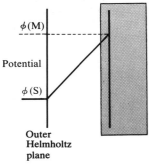

Fig. 30.5 The electric potential varies linearly between two plane parallel sheets of charge, and its effect on the Gibbs function of the transition state depends on the latter's location between the inner and outer planes.

arising from the oxidation.[3] The net current density at the electrode (Fig. 30.4) is the difference

$$j = j_a - j_c = Fk_a[\text{Red}] - Fk_c[\text{Ox}] \tag{3}$$

The activation Gibbs function

If an ion (or neutral molecule) is to participate in reduction or oxidation at an electrode it must discard any solvating molecules, migrate through the electric double layer, and adjust its hydration sphere as it receives or discards electrons. Likewise, an ion or molecule already at the inner plane must be detached and migrate into the bulk. Since both processes are activated, we can expect to write their rate constants in the form suggested by activated complex theory (Section 28.4) as

$$k = Be^{-\Delta G^{\ddagger}/RT} \tag{4}$$

$\Delta G^{\ddagger}$ is the activation Gibbs function (Section 28.6) and B is a constant with the same dimensions as k, typically cm s^{-1}.

When we insert eqn 4 into eqn 3 we obtain

$$j = FB_a[\text{Red}]e^{-\Delta G_a^{\ddagger}/RT} - FB_c[\text{Ox}]e^{-\Delta G_c^{\ddagger}/RT} \tag{5}$$

We have allowed for the activation Gibbs functions to be different for the anodic and cathodic processes. That they are different is the central feature of the remaining discussion. Note that when $j_a > j_c$, so that $j > 0$, the current is anodic; when $j_c > j_a$, so that $j < 0$, the current is cathodic (Fig. 30.4).

The Butler–Volmer equation

Now we relate j to the Galvani potential difference, which varies across the double layer as shown schematically in Fig. 30.5.

First, we consider the reduction reaction. An electron is transferred from the electrode where the potential is ϕ_M to the solution where it is ϕ_S. There is therefore an electrical contribution to the work of magnitude $e\Delta\phi$. If the transition state of the activated complex corresponds to Ox being very close to the electrode (Fig. 30.6) the activation Gibbs function is changed from $\Delta G_c^{\ddagger}(0)$ to

$$\Delta G_c^{\ddagger} = \Delta G_c^{\ddagger}(0) + F\Delta\phi$$

Thus, if the electrode is more positive than the solution, $\Delta\phi > 0$, more work has to be done to form an activated complex from Ox; in this case the activation Gibbs function is increased. If the transition state corresponds to Ox being close to the outer plane of the double layer (Fig. 30.7), then $\Delta G_c^{\ddagger}$ is independent of $\Delta\phi$. In a real system the transition state lies at some intermediate location between these extremes (Fig. 30.8), and we write the activation Gibbs function for reduction as

$$\Delta G_c^{\ddagger} = \Delta G_c^{\ddagger}(0) + \alpha F\Delta\phi \tag{6}$$

---

[3] Recall from Chapter 10 that the cathode is the site of reduction, hence the cathodic current is the current arising from reduction. The anode is the site of oxidation, and the anodic current is the current arising from oxidation.

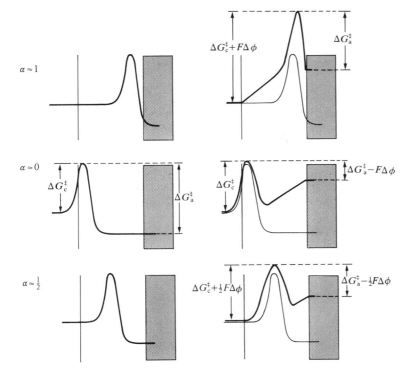

Fig. 30.6 When the transition state is very close to the inner plane the activation Gibbs function for the anodic current is almost unchanged, but the full effect applies to the cathodic current.

Fig. 30.7 When the transition state is very close to the outer plane the cathodic current activation Gibbs function is almost unchanged but the anodic current activation Gibbs function is strongly affected.

Fig. 30.8 When the transition state is located at an intermediate position as measured by α (with $0 < \alpha < 1$; in this diagram $\alpha \approx \frac{1}{2}$), both activation Gibbs functions are affected.

The parameter α is called the **transfer coefficient**, and lies in the range 0 to 1; experimentally α is often found to be about $\frac{1}{2}$.

Now consider the oxidation of Red. In this case Red discards an electron to the electrode, and so the extra work is zero if the transition state of the activated complex lies close to the electrode (Fig. 30.6) and the full $-F\Delta\phi$ if it lies at the outer plane (Fig. 30.7). In general (Fig. 30.8), the activation Gibbs function is

$$\Delta G_a^{\ddagger} = \Delta G_a^{\ddagger}(0) - (1 - \alpha)F\Delta\phi \qquad (7)$$

The two activation Gibbs functions can be inserted in place of the values used in eqn 5 with the result that

$$j = FB_a[\text{Red}] \times e^{-\Delta G_a^{\ddagger}(0)/RT} \times e^{(1-\alpha)F\Delta\phi/RT}$$
$$- FB_c[\text{Ox}] \times e^{-\Delta G_c^{\ddagger}(0)/RT} \times e^{-\alpha F\Delta\phi/RT} \qquad (8)$$

which is an explicit, if complicated, expression for the net current density in terms of the potential difference.

The appearance of eqn 8 can be simplified. First, in a purely cosmetic step we write

$$f = \frac{F}{RT}$$

Next, we identify the individual cathodic and anodic current densities:

$$\left.\begin{array}{l} j_a = FB_a[\text{Red}] \times e^{-\Delta G_a^{\ddagger}(0)/RT} \times e^{(1-\alpha)f\Delta\phi} \\ j_c = FB_c[\text{Ox}] \times e^{-\Delta G_c^{\ddagger}(0)/RT} \times e^{-\alpha f\Delta\phi} \end{array}\right\} \quad j = j_a - j_c$$

Example 30.1: *Calculating the current density*

Calculate the change in cathodic current density at an electrode when the potential difference changes from 1.0 V to 2.0 V at 25°C.

Answer. The ratio of cathodic current densities j'_c and j_c, when the potential differences are $\Delta\phi'$ and $\Delta\phi$ respectively, is

$$\frac{j'_c}{j_c} = e^{-\alpha f(\Delta\phi' - \Delta\phi)}$$

A typical value for α is $\frac{1}{2}$, so with $\Delta\phi' - \Delta\phi = 1.0\,\text{V}$,

$$\alpha f(\Delta\phi' - \Delta\phi) = \frac{\frac{1}{2} \times 9.65 \times 10^4\,\text{C mol}^{-1} \times 1.0\,\text{V}}{8.314\,\text{J K}^{-1}\,\text{mol}^{-1} \times 298\,\text{K}} = 19.47$$

Hence,

$$\frac{j'_c}{j_c} = e^{-19.47} = 3.5 \times 10^{-9}$$

Comment. This huge change in current density occurs for a very mild and easily applied change of conditions. An understanding of why the change is so great is obtained by realizing that a change of potential difference by 1 V changes the activation Gibbs function by $(1\,\text{V}) \times F$, or about $50\,\text{kJ mol}^{-1}$, which has an enormous effect on the rates.

Exercise. Calculate the change in anodic current density under the same circumstances.
$$[j'_a/j_a = 1.8 \times 10^8]$$

If a cell is balanced against an external source we can identify the Galvani potential difference $\Delta\phi$ as the (zero-current) electrode potential E, and write[4]

$$j_a = FB_a[\text{Red}] \times e^{-\Delta G_a^{\ddagger}(0)/RT} \times e^{(1-\alpha)fE}$$

$$j_c = FB_c[\text{Ox}] \times e^{-\Delta G_c^{\ddagger}(0)/RT} \times e^{-\alpha fE}$$

When these equations apply, there is no net current at the electrode (as the cell is balanced), and so the two current densities must be equal. From now on we shall denote them both as j_0, which is called the **exchange current density**.

When the cell is producing current the electrode potential changes from its zero-current value E to its working value, and the difference is the **overpotential** η:

$$\eta = E(\text{working}) - E \tag{9}$$

Hence $\Delta\phi$ changes to

$$\Delta\phi = E + \eta$$

It follows that the two current densities become

$$j_a = j_0 e^{(1-\alpha)f\eta} \tag{10a}$$

$$j_c = j_0 e^{-\alpha f\eta} \tag{10b}$$

---

[4] We are assuming that we can *identify* the Galvani potential and the zero-current electrode potential. As explained earlier, they differ by a constant amount, which may be regarded as absorbed into the constant B.

Then from eqn 8 we obtain the **Butler–Volmer equation**:

$$j = j_0\{e^{(1-\alpha)f\eta} - e^{-\alpha f\eta}\} \qquad (11)$$

The Butler–Volmer equation will be the basis of all that follows.

The low overpotential limit

When the overpotential is so small that $f\eta \ll 1$ (in practice, η less than about 0.01 V) the exponentials in eqn 11 can be expanded using $e^x = 1 + x + \ldots$ to give

$$j = j_0\{1 + (1-\alpha)f\eta + \ldots - (1-\alpha f\eta + \ldots)\}$$
$$\approx j_0 f\eta \qquad (12)$$

Equation 12 shows that the current density is proportional to the overpotential, and so at low overpotentials the interface behaves like a conductor that obeys Ohm's law. When there is a small positive overpotential the current is anodic ($j > 0$ when $\eta > 0$), and when the overpotential is small and negative the current is cathodic ($j < 0$ when $\eta < 0$). We can also reverse the relation and calculate the potential difference that must exist if a current density j has been established by some external circuit:

$$\eta = \frac{RTj}{Fj_0} \qquad (13)$$

The importance of this interpretation will become apparent below.

Example 30.2: *Calculating the current at an electrode*

The exchange current density of a $Pt \mid H_2(g) \mid H^+(aq)$ electrode at 298 K is 0.79 mA cm^{-2}. What current flows through a standard electrode of total area 5.0 cm^2 when the overpotential is $+5.0$ mV?

Answer. Since $RT/F = 25.69$ mV at 298 K, from eqn 13 the current density is

$$j = \frac{0.79 \text{ mA cm}^{-2} \times 5.0 \text{ mV}}{25.69 \text{ mV}} = 0.15 \text{ mA cm}^{-2}$$

The current through the electrode is this density times the area, or 0.75 mA.

Comment. The current density is anodic, which means that the oxidation reaction dominates when the overpotential is positive. Since $f\eta = 0.19$, the linear approximation is valid.

Exercise. What would be the current at pH = 2.0, the other conditions being the same?
$$[-17 \text{ mA}]$$

The high overpotential limit

When the overpotential is large and positive, corresponding to the electrode being the anode in electrolysis, the second exponential in eqn 11 is much smaller than the first and may be neglected. Then

$$j = j_0 e^{(1-\alpha)f\eta}$$

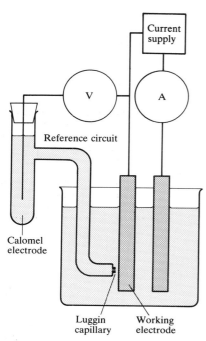

Reference circuit

V   A

Current supply

Calomel electrode

Luggin capillary   Working electrode

Fig. 30.9 The general arrangement for electrochemical rate measurements. The external source establishes a current between the working electrodes, and its effect on the potential difference of either of them relative to the reference electrode is observed. No current flows in the reference circuit.

and so

$$\ln j = \ln j_0 + (1 - \alpha)f\eta \tag{14a}$$

When the overpotential is large but negative (corresponding to the cathode in electrolysis), the first exponential in eqn 11 may be neglected. Then

$$j = -j_0 e^{-\alpha f\eta}$$

and so

$$\ln(-j) = \ln j_0 - \alpha f\eta \tag{14b}$$

The plot of the logarithm of the current density against the overpotential is called a **Tafel plot**. The slope gives the transfer coefficient parameter α and the intercept at $\eta = 0$ gives the exchange current density.

The experimental arrangement used for a Tafel plot is shown in Fig. 30.9. The electrode of interest is called the **working electrode**, and the current flowing through it is controlled externally. If its area is A and the current is I, then the current density across its surface is I/A. The potential difference across the interface cannot be measured directly, but the potential of the working electrode relative to a third electrode, the **reference electrode**, can be measured with a high impedance voltmeter, and no current flows in that half of the circuit. The reference electrode is in contact with the solution close to the working electrode through a 'Luggin capillary', which helps to eliminate any ohmic potential difference that might arise accidentally. Changing the current flowing through the working circuit causes a change of potential of the working electrode, which is measured with the voltmeter. The overpotential is then obtained by taking the difference between the potentials measured with and without a flow of current through the working circuit.

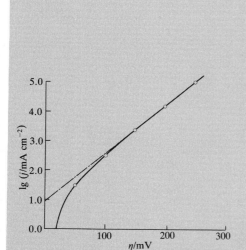

Fig. 30.10 A Tafel plot is used to measure the exchange current density (given by the extrapolated intercept at $\eta = 0$) and the transfer coefficient (from the slope).

Example 30.3: *Interpreting a Tafel plot*

The data below refer to the anodic current through a $2.0\,\text{cm}^2$ platinum electrode in contact with an Fe^{3+}, Fe^{2+} aqueous solution at 298 K. Calculate the exchange current density and the transfer coefficient for the electrode process.

η/mV	50	100	150	200	250
I/mA	8.8	25.0	58.0	131	298

Answer. The anodic process is the oxidation $Fe^{2+} \rightarrow Fe^{3+} + e^-$. We make a Tafel plot ($\ln j$ against η) using the anodic form (eqn 14a), and draw up the following table:

η/mV	50	100	150	200	250
$j/(\text{mA cm}^{-2})$	4.4	12.5	29.0	65.5	149
$\ln(j/\text{mA cm}^{-2})$	1.48	2.53	3.37	4.18	5.00

The points are plotted in Fig. 30.10; the intercept at $\eta = 0$ is $\ln j_0$ and the slope is $(1 - \alpha)f$. The high overpotential region gives a straight line of intercept 0.92 and slope 0.0163. From the former it follows that $\ln(j_0/\text{mA cm}^{-2}) = 0.92$, so that $j_0 = 2.5\,\text{mA cm}^{-2}$. From the latter,

$$\frac{(1 - \alpha)F}{RT} = 0.0163\,\text{mV}^{-1}$$

so $\alpha = 0.58$.

Comment. Note that the Tafel plot is nonlinear for $\eta < 150\,\text{mV}$. In this region $\eta f = 3$, which is not much larger than 1.

Exercise. Repeat the analysis using the following cathodic current data:

η/mV	-50	-100	-150	-200	-250	-300
I/mA	-0.3	-1.5	-6.4	-27.6	-118.6	-510

$$[\alpha = 0.75,\ j_0 = 0.040\,\text{mA cm}^{-2}]$$

Some experimental values for the Butler–Volmer parameters are given in Table 30.1, and from them we can see that exchange current densities vary over a very wide range. For example, the N_2, N_3^- couple on Pt has $j_0 = 10^{-76}\,\text{A cm}^{-2}$ whereas the H^+/H_2 couple on Pt has $j_0 = 8 \times 10^{-4}\,\text{A cm}^{-2}$, a range of 73 orders of magnitude. Exchange currents are generally large when the redox process involves no bond breaking (as in the $[Fe(CN)_6]^{3-}/[Fe(CN)_6]^{4-}$ couple) or if only weak bonds are broken (as in Cl_2/Cl^-). They are generally small when more than one electron needs to be transferred, or when multiple or strong bonds are broken, as in the N_2/N_3^- couple and in redox reactions of organic compounds.

Table 30.1 Exchange current densities and transfer coefficients at 298 K

Reaction	Electrode	$j_0/(\text{A cm}^{-2})$	α
$2H^+ + 2e^- \rightarrow H_2$	Pt	7.9×10^{-4}	
	Ni	6.3×10^{-6}	0.58
	Pb	5.0×10^{-12}	
$Fe^{3+} + e^- \rightarrow Fe^{2+}$	Pt	2.5×10^{-3}	0.58

30.3 Polarization

Electrodes with potentials that change only slighly when a current passes through them are classified as **non-polarizable**. Those with strongly current-dependent potentials are **polarizable**. From the linearized equation (eqn 12) it is clear that the criterion for low polarizability is high exchange current density (so that η may be small even though j is large). The calomel and H_2/Pt electrodes are both highly non-polarizable, which is one reason why they are so extensively used in equilibrium electrochemistry measurements.

Concentration polarization

One of the assumptions in the derivation of the Butler–Volmer equation is the uniformity of concentration near the electrode. This assumption fails at high current densities because migration of ions towards the electrode from the bulk is slow and may become rate-determining. A larger overpotential is then needed to produce a given current because the supply of reducible or oxidizable species has been depleted. This effect is called **concentration polarization** and the additional potential needed to restore the flow of current is called the **polarization overpotential** η^c.

For simplicity we consider a case for which the concentration polarization dominates all the rate processes. We shall consider a redox couple of the type M^{z+}/M. Under zero-current conditions, when the net current density is zero, the electrode potential is related to the activity a of the ions in the solution by the Nernst equation (eqn 11 of Section 10.3):

$$E = E^\ominus + \frac{RT}{zF} \ln a$$

We have remarked that electrode kinetics are normally studied using a large excess of support electrolyte so as to keep the mean activity coefficients approximately constant. Therefore the constant activity coefficient in $a = \gamma c$ may be combined with $E^\ominus$ and we write the **formal potential** of the electrode as

$$E° = E^{\ominus} + \frac{RT}{zF} \ln \gamma \qquad (15)$$

Then the electrode potential is

$$E = E° + \frac{RT}{zF} \ln c$$

When the cell is producing current, the active ion concentration just outside the double layer changes to c' and the electrode potential changes to

$$E' = E° + \frac{RT}{zF} \ln c'$$

The concentration overpotential is therefore

$$\eta^c = E' - E = \frac{RT}{zF} \ln \frac{c'}{c} \qquad (16)$$

We now suppose that the solution has its bulk concentration up to a distance δ from the outer Helmholtz plane, and then falls linearly to c' at the plane itself. This **Nernst diffusion layer** is illustrated in Fig. 30.11. The thickness of the Nernst layer (which is typically 0.1 mm, and strongly dependent on the condition of hydrodynamic flow) is quite different from that of the electric double layer (which is typically less than 1 nm, and unaffected by stirring). The concentration gradient through the Nernst layer is

$$\frac{dc}{dx} = \frac{c' - c}{\delta}$$

This gradient gives rise to a flux of ions towards the electrode, which replenishes the cations that are reduced. The (molar) flux J is proportional to the concentration gradient, and according to Fick's law (Section 24.6)

$$J = -D\left(\frac{\partial c}{\partial x}\right)$$

Therefore, the particle flux towards the electrode is

$$J = D\frac{(c - c')}{\delta}$$

The current density towards the electrode is the product of the particle flux and the charge zF per mole of ions:

$$j = zFJ = zFD\frac{(c - c')}{\delta}$$

The maximum rate of diffusion across the Nernst layer occurs when the gradient is steepest, which is when $c' = 0$. This concentration occurs when an electron from an ion that diffuses across the layer is snapped over the activation barrier, through the double layer, and on to the electrode. No flow of current can exceed the **limiting current density** j_l, which is given by

$$j_l = zFJ_l = \frac{zFD}{\delta} \times c \qquad (17)$$

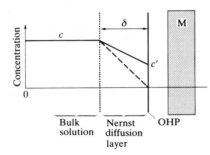

Fig. 30.11 In a simple model of the Nernst diffusion layer there is a linear variation in concentration between the bulk and the outer Helmholtz plane; the thickness of the layer depends strongly on the state of flow of the fluid.

Example 30.4: *Estimating the limiting current density*

Estimate the limiting current density at 298 K for an electrode in a 0.10 M $Cu^{2+}(aq)$ unstirred solution in which the thickness of the diffusion layer is about 0.3 mm.

Answer. We can estimate D from the ionic conductivity $\lambda = 107$ S cm^2 mol^{-1} (Table 25.1) and the Nernst–Einstein equation (Section 25.4):

$$\lambda = \frac{z^2 F^2 D}{RT}$$

Then eqn 17 becomes

$$j_l = \frac{cRT\lambda}{zF\delta}$$

Therefore, with $\delta = 0.3$ mm, $c = 0.10$ M, $z = 2$, and $T = 298$ K, $j_l = 5$ mA cm^{-2}.

Comment. The result implies that the current towards a 1 cm^2 electrode cannot exceed 5 mA in this solution. If the solution is stirred, or if the electrode surface is moving (as in polarography, see below) the Nernst layer is much thinner and the limiting current larger.

Exercise. Evaluate the limiting current density for an $Ag(s)|Ag^+(aq)$ electrode in a stirred 0.010 M $Ag^+(aq)$ solution at 298 K. Take $\delta \approx 0.03$ mm.
[5 mA cm^{-2}]

The concentration c' is related to the current density at the double layer by

$$c' = c - \frac{j\delta}{zFD} \tag{18}$$

Hence, as the current density is increased, the concentration falls below the bulk value. However, this decline in concentration is small when the diffusion constant is large, for then the ions are very mobile and can quickly replenish any ions that have been removed.

Finally, we substitute eqn 17 into eqn 16 and obtain the following expressions for the overpotential in terms of the current density:

$$\eta^c = \frac{RT}{zF} \ln\left(1 - \frac{j\delta}{zcFD}\right) \tag{19a}$$

$$j = \frac{zcFD}{\delta}(1 - e^{zf\eta^c}) \tag{19b}$$

Polarography and voltammetry

In the analytical technique of **polarography**, the current passing through an unstirred solution is measured as the potential difference between two electrodes is changed. The diffusion controlled current climbs with the potential difference until the limiting value is reached, and as the limiting current is characteristic of the ions present we may be able to identify them from their current-voltage characteristics. When several different ions are present, each one reaches its limiting current density at a characteristic value, and

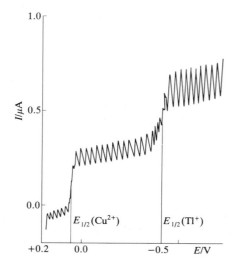

Fig. 30.12 A polarogram is obtained when the current flowing through a cell is plotted against the applied potential. This polarogram shows the presence of Cu^{2+} and Tl^+ ions by their characteristic half-wave potentials, $E_{1/2}$.

Table 30.2 Half-wave potentials at 298 K, $E_{1/2}/V$

Cd^{2+}	−0.60
Cu^{2+}	+0.04
Fe^{3+}	0.0
Zn^{2+}	−1.00

Values refer to ions in 0.1 M KCl(*aq*), relative to the calomel electrode.

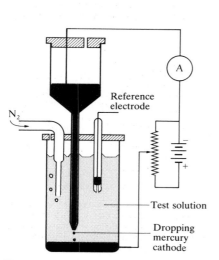

Fig. 30.13 The dropping mercury electrode ensures that a renewed, clean surface is provided continuously. The growth of the droplets accounts for the periodic variation of the current in Fig. 30.12.

the current changes in a series of waves. Each ion is identified by measuring the **half-wave potential**, the potential mid way between the initial and final potentials of the wave (Fig. 30.12 and Table 30.2). The limiting current density itself can be interpreted in terms of the concentrations of the ions.

One of the problems with the procedure is the contamination of the electrode surface. This problem is avoided by using a **dropping mercury electrode** (Fig. 30.13) in which the surface is continuously renewed as drops form at the end of the capillary and then fall off. The growth of the drop, and therefore of the surface area of the electrode, accounts for the oscillations in the polarogram.

The kinetics of electrode processes are commonly studied by **voltammetry**, in which the current is monitored as the potential of the electrode is changed, and by **chronopotentiometry** in which the potential is monitored as the current flow is changed.

The kind of output from a **linear-sweep voltammetry** experiment is illustrated in Fig. 30.14. Initially the potential is low and the cathodic current is due to the migration of ions in the solution. However, as the potential approaches the reduction potential of the reducible solute, the cathodic current grows. Soon after the potential exceeds the reduction potential, the current declines on account of the concentration polarization of the electrode, since now there is a shortage of reducible solute close to the electrode. A modification of the technique is **cyclic voltammetry** in which the potential is taken back down to its initial value. A typical cyclic voltammogram is shown in Fig. 30.15. The shape of the curve is initially like that of a linear sweep experiment, but after the potential begins to fall there is a rapid change in current on account of the high concentration of oxidizable species close to the electrode. When the potential is close to the potential required to oxidize the reduced species, there is a substantial

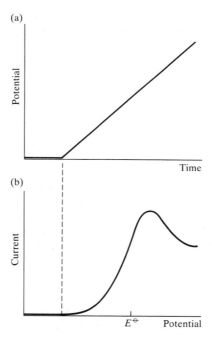

Fig. 30.14 (a) The change of potential with time and (b) the resulting current/potential curve in a linear sweep voltammetry experiment.

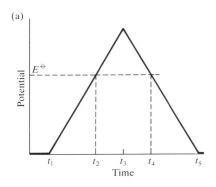

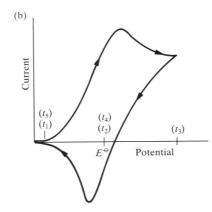

anodic current until all the oxidation is complete, and the current returns to zero. The overall shape of the curve gives details of the kinetics of the electrode process.

Kinetic processes are also studied using the **rotating-disk electrode** (Fig. 30.16). The electrode is a small flat disk set in a vertical rotating axle. The rotation of its surface sets up a steady hydrodynamic flow which circulates the solution over its face. The flow pattern can be calculated, and the limiting current related to the speed of rotation.

Various modifications of the rotating-disk electrode have been developed. One involves pulsing the potential difference and then observing the growth of current towards its limiting value. Another is the **ring–disk electrode**, where the central spinning disk is surrounded by an electrode in the form of a narrow ring. The ring can be set at a different potential from that of the disk and hence may be used to reduce or oxidize the products that have been formed at the disk and then swept into its vicinity by the hydrodynamic flow. Analysis of the currents at both electrodes, the concentration profiles (Fig. 30.17), and knowledge of the time it takes for the products formed at the disk to flow to the ring give very detailed information about electrode kinetics, such as the identities of the reactants and the rates of electron transfer.

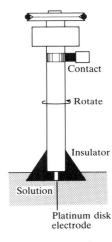

Fig. 30.16 The dropping mercury electrode has been largely replaced by the rotating-disk electrode, because the hydrodynamic flow over its face can be calculated more reliably.

Electrochemical processes

To induce current to flow through an electrolytic cell and bring about a non-spontaneous cell reaction the applied potential difference must exceed the zero-current potential by at least the **cell overpotential**, the sum of the overpotentials at the two electrodes and the ohmic drop ($I \times R$) due to the current through the electrolyte. The additional potential needed to achieve a detectable rate of reaction may need to be large when the exchange current density at the electrodes is small. For similar reasons a working galvanic cell generates a smaller potential than under zero-current conditions. In this section we see how to cope with both aspects of the overpotential.

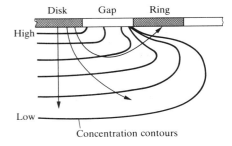

Fig. 30.17 Schematic concentration profiles of a species in the vicinity of a ring–disk electrode. Note that the concentration changes in the gap region and is constant over the surfaces of the two electrodes.

30.4 Electrolysis

The rate of gas evolution or metal deposition during electrolysis can be estimated from the Butler–Volmer equation and tables of exchange current densities. The exchange current density is strongly dependent on the nature of the electrode surface, and changes during the electrodeposition of one metal on another. A very crude criterion is that significant evolution or deposition occurs only if the overpotential exceeds about 0.6 V.

Example 30.5: *Calculating the relative rates of electrolysis*

Derive an expression for the relative rates of electrodeposition and hydrogen evolution in a solution in which both may occur.

Answer. We can calculate the ratio of the cathodic currents using eqn 14, and for simplicity assume equal transfer coefficients:

$$\frac{j'}{j} = \frac{j_0'}{j_0} e^{(\eta - \eta')\alpha f}$$

Comment. This equation shows that metal deposition is favoured by a large exchange current density and relatively high hydrogen evolution overvoltage (so that $\eta - \eta'$ is positive and large).

Exercise. Deduce an expression for the ratio when the hydrogen evolution is limited by transport across a diffusion layer. $[j'/j = (\delta c j_0'/FD)e^{-\alpha \eta' f}]$

A glance at Table 30.1 shows the wide range of exchange current densities for a metal/hydrogen electrode. The most sluggish exchange currents occur for lead and mercury, and the value of 10^{-12} A cm^{-2} corresponds to a monolayer of atoms being replaced in about 5 years. For such systems, a large overpotential is needed to induce significant hydrogen evolution. In contrast, the value for platinum (10^{-3} A cm^{-2}) corresponds to a monolayer being replaced in 0.1 s, and so gas evolution occurs for a much lower overpotential.

The exchange current density also depends on the crystal face exposed. For the deposition of copper on copper, the (100) face has $j_0 = 1$ mA cm^{-2}, and so for the same overpotential grows at 2.5 times the rate of the (111) face for which $j_0 = 0.4$ mA cm^{-2}.

30.5 The characteristics of working cells

We expect the cell potential to decrease as current is generated because it is then no longer working reversibly and can therefore do less than maximum work.

The potentials of working cells

We shall consider the cell

$$M \,|\, M^+(aq)| \,|M'^+(aq)| \, M'$$

and ignore all the complications arising from liquid junctions. The working potential of the cell is

$$E' = \Delta\phi_R - \Delta\phi_L$$

Since the working potential differences differ from their zero-current values by overpotentials, we can write

$$\Delta\phi_X = E_X + \eta_X$$

where X is L or R for the left or right electrode respectively. The working cell potential is therefore

$$E' = E + \eta_R - \eta_L$$

with E the zero-current cell potential. We should subtract from this expression the ohmic potential difference IR_s, where R_s is the cell's internal resistance:

$$E' = E + \eta_R - \eta_L - IR_s \qquad (20)$$

The ohmic term is a contribution to the cell's irreversibility—it is a thermal dissipation term—and so the sign of IR_s is always such as to reduce the cell potential in the direction of zero.

The overpotentials in eqn 20 can be calculated from the Butler–Volmer equation for a given current I being drawn. We shall simplify the equations by supposing that the areas A of the electrodes are the same, that only one electron is transferred in the rate-determining steps at the electrodes, that the transfer coefficients are both $\frac{1}{2}$, and that the high-overpotential limit of the Butler–Volmer equation may be used. Then from eqns 14 and 20 we find

$$E' = E - \frac{4RT}{F} \ln \frac{I}{A\bar{j}} - IR_s \qquad (21)$$

where

$$\bar{j} = (j_{L0} j_{R0})^{1/2}$$

The concentration overpotential also reduces the cell potential. If we use the Nernst diffusion layer model for each electrode, the total change of potential arising from concentration polarization is given by eqn 19 as

$$E' = E - \frac{RT}{zF} \ln \left(1 - \frac{I}{Aj_{1,L}}\right)\left(1 - \frac{I}{Aj_{1,R}}\right) \qquad (22)$$

This contribution can be combined with the one in eqn 21 to obtain a full (but still very approximate) expression for the cell potential when a current I is being drawn:

$$E' = E - IR_s + \frac{2RT}{zF} \ln g \qquad (23a)$$

$$g = \frac{(I/A\bar{j})^{2z}}{\left(1 - \dfrac{I}{Aj_{1,L}}\right)^{1/2}\left(1 - \dfrac{I}{Aj_{1,R}}\right)^{1/2}} \qquad (23b)$$

Equation 23 depends on a lot of parameters, but an example of its general form is given in Fig. 30.18. Notice the very steep decline of working potential when the current is high and close to the limiting value for one of the electrodes.

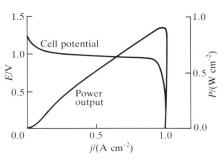

Fig. 30.18 The dependence of the potential of a working cell on the current being drawn and the corresponding power output calculated using eqns 23 and 24 respectively. Notice the sharp decline in power just after the maximum.

The power output of working cells

Since the power P supplied by a cell is $I \times E$, from eqn 23 we can write

$$P = IE - I^2 R_s + \frac{2IRT}{zF} \ln g \qquad (24)$$

The first term on the right is the power that would be produced if the cell retained its zero-current potential when delivering current. The second term is the power generated uselessly as heat as a result of the resistance of the electrolyte. The third term is the reduction of the potential of the electrodes as a result of drawing current.

The general dependence of power output on the current drawn is illustrated in Fig. 30.18. Notice how maximum power is achieved just before the concentration polarization quenches the cell's performance. Information of this kind is essential if the optimum conditions for operating electrochemical devices are to be found and their performance improved.

30.6 Fuel cells and secondary cells

We considered the thermodynamic properties of galvanic cells in Chapter 10. Here we shall mention some of the kinetic aspects in the light of the concepts introduced in this chapter.

Fuel cells

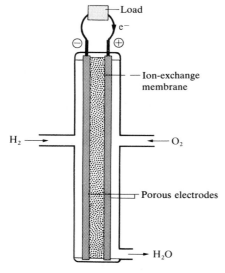

Fig. 30.19 A single cell of the Bacon fuel cell. In practice a battery of many cells are combined.

A fuel cell operates like a conventional galvanic cell with the exception that the reactants are supplied from outside rather than forming an integral part of its construction. A fundamental and important example of a fuel cell is the hydrogen/oxygen cell (Fig. 30.19), which is also called the **Bacon cell** after its inventor. The electrolyte is concentrated aqueous potassium hydroxide maintained at 200°C and 20–40 atm; the electrodes are porous nickel in the form of sheets of compressed powder. The cathode reaction is the reduction

$$O_2(g) + 2H_2O(l) + 4e^- \rightarrow 4OH^-(aq) \qquad E^{\ominus} = +0.40 \text{ V}$$

and the anode reaction is the oxidation

$$H_2(g) + 2OH^- \rightarrow 2H_2O(l) + 2e^- \qquad E^{\ominus} = -0.83 \text{ V}$$

Since the overall reaction

$$2H_2(g) + O_2(g) \rightarrow 2H_2O(l) \qquad E^{\ominus} = 1.23 \text{ V}$$

is exothermic as well as spontaneous, it is less favourable at 200°C than at 25°C, and so the cell potential is lower. However, the increased pressure compensates for the increased temperature, and $E \approx 1.2$ V at 200°C and 40 atm.

One advantage of the hydrogen/oxygen system is the large exchange current density of the hydrogen reaction. Unfortunately, the oxygen reaction has an exchange current density of only about 10^{-10} A cm^{-2}, and so this limits the current available from the cell. One way round the difficulty is to use a catalytic surface (to increase j_0) with a large surface area. Hydrocarbon/air fuel cells require platinum electrodes, which makes them prohibitively expensive. Therefore, in practice hydrocarbons are cracked with steam, the hydrogen is separated using a silver/palladium alloy membrane, and then the separated hydrogen is used in a Bacon cell.

Example 30.6: *Calculating the potential of a fuel cell*

Calculate the potential of a reversible propane/oxygen fuel cell operating under standard conditions at 298 K.

Answer. The cell reaction

$$C_3H_8(g) + 5O_2(g) \rightarrow 3CO_2(g) + 4H_2O(l)$$

is the sum of the following two half-reactions:

$$C_3H_8(g) + 6H_2O(l) \rightarrow 3CO_2(g) + 20H^+(aq) + 20e^-$$

$$5O_2(g) + 20H^+(aq) + 20e^- \rightarrow 10H_2O(l)$$

Therefore, $v = 20$. The standard Gibbs function of the combustion is

$$\Delta G^{\ominus} = 3(-394.4) + 4(-237.2) - (-23.5) \text{ kJ mol}^{-1}$$

$$= -2108 \text{ kJ mol}^{-1}$$

Therefore, from eqn 10 of Section 10.3,

$$E^{\ominus} = \frac{-(-2108 \text{ kJ mol}^{-1})}{20F} = +1.09 \text{ V}$$

Comment. The standard enthalpy of combustion at 298 K is $-2220 \text{ kJ mol}^{-1}$, and so in a cell working with perfect efficiency, 112 kJ mol^{-1} of energy is released as heat.

Exercise. Repeat the calculation for a methanol-powered fuel cell under the same conditions.
[1.21 V]

The lead–acid battery

Electric storage cells operate as galvanic cells while they are producing electricity but as electrolytic cells while they are being charged by an external supply.

The **lead–acid battery** is an old device, but one well suited to the job of starting cars (and the only one available). During charging the cathode reaction is the reduction of Pb^{2+} and its deposition as lead on the lead electrode: this occurs instead of the reduction of the acid to hydrogen because the latter has a low exchange current density on lead. The anode reaction during charging is the oxidation of Pb(II) to Pb(IV), which is deposited as the oxide PbO_2. On discharge the two reactions run in reverse. Because they have such high exchange current densities the discharge can occur rapidly, which is why the lead battery can produce large currents on demand.

Corrosion

The thermodynamic warning of the likelihood of corrosion is given by comparing the standard electrode potentials of the metal reduction, such as

$$Fe^{2+}(aq) + 2e^- \rightarrow Fe(s) \qquad E^{\ominus} = -0.44 \text{ V}$$

with the values for one of the following half-reactions in acidic solution:

(a) $2H^+(aq) + 2e^- \rightarrow H_2(g)$ $\qquad\qquad E^{\ominus} = 0$

(b) $4H^+(aq) + O_2(g) + 4e^- \rightarrow 2H_2O(l)$ $\qquad E^{\ominus} = +1.23$ V

Since both redox couples have standard potentials more positive than $E^{\ominus}(Fe^{2+}/Fe)$, both of them can drive the oxidation of iron. The electrode potentials we have quoted are standard values, and they change with the pH of the medium:

$$E(a) = E^{\ominus}(a) + \frac{RT}{F} \ln a(H^+) = -0.059 \text{ V} \times pH$$

$$E(b) = E^{\ominus}(b) + \frac{RT}{F} \ln a(H^+) = 1.23 \text{ V} - 0.059 \text{ V} \times pH$$

These values, following the discussion in Chapter 10, let us judge at what pH the iron will have a tendency to oxidize. A thermodynamic discussion of corrosion, however, only indicates whether a tendency to corrode exists. If there is a thermodynamic tendency, we must examine the kinetics of the processes involved to see whether the process is significantly fast.

30.7 The rate of corrosion

A model of a corrosion system is shown in Fig. 30.20a. It can be taken to be a drop of slightly acidic water containing some dissolved oxygen in contact with the metal. The oxygen near the edges of the droplet is reduced by electrons donated by the iron over an area A. Those electrons are replaced by others released elsewhere as $Fe \rightarrow Fe^{2+} + 2e^-$. This oxidative release occurs over an area A' under the oxygen-deficient inner region of the droplet. The droplet acts as a short-circuited galvanic cell (Fig. 30.20b).

The rate of corrosion is measured by the current of metal ions leaving the metal surface in the anodic region. This flux of ions gives rise to the corrosion current I_{corr}, which can be identified with the anodic current I_a. Since any current emerging from the anodic region must find its way to the cathodic region, the cathodic current I_c must also be equal to the corrosion current. In terms of the current densities at the oxidation j and reduction j' sites, we can write

$$I_{corr} = jA = j'A' = (jj'AA')^{1/2} = \bar{j}\bar{A} \tag{25}$$

where

$$\bar{A} = (AA')^{1/2} \quad \text{and} \quad \bar{j} = (jj')^{1/2}$$

We now use the Butler–Volmer equation to express the current densities in terms of overpotentials. For simplicity we assume that the overpotentials are large enough for the high-overpotential limits (eqn 14) to apply, that polarization overpotential can be neglected, that the rate-determining step is the transfer of a single electron, and that the transfer coefficients are $\frac{1}{2}$. We also assume that since the droplet is so small, there is negligible potential difference between the cathode and anode regions of the solution. Moreover, since it is short-circuited by the metal, the potential of the metal is the same in both regions, and so the potential differences between the

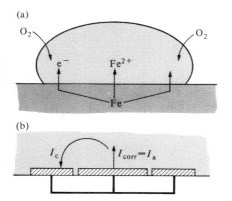

(a)

(b)

Fig. 30.20 (a) A simple version of the corrosion process is that of a droplet of water, which is oxygen rich near its boundary with air. The oxidation of the iron takes place in the region away from the oxygen because the electrons are transported through the metal. (b) The process may be modelled as a short-circuited electrochemical cell.

metal and the solution is the same in both regions too. This common potential difference is called the **corrosion potential difference** $\Delta\phi_{corr}$. The overpotentials in the two regions are therefore

$$\eta = \Delta\phi_{corr} - \Delta\phi_e \quad \text{and} \quad \eta' = \Delta\phi_{corr} - \Delta\phi'_e$$

and so the current densities are

$$j = j_0 e^{\frac{1}{2}\eta f} = j_0 e^{\frac{1}{2}f\Delta\phi_{corr}} e^{-\frac{1}{2}f\Delta\phi}$$

$$j' = j'_0 e^{-\frac{1}{2}f\eta'} = j'_0 e^{-\frac{1}{2}f\Delta\phi_{corr}} e^{\frac{1}{2}f\Delta\phi'}$$

We are very nearly at the end of the calculation because these expressions can be substituted into eqn 25 and $\Delta\phi' - \Delta\phi$ replaced by the difference of electrode potentials, E, to give

$$I_{corr} = \bar{A}\bar{j}_0 e^{fE/4} \tag{26}$$

where $\bar{j}_0 = (j_0 j'_0)^{1/2}$.

Several conclusions can be drawn from eqn 26. First, the rate of corrosion depends on the surfaces exposed: if either A or A' is zero, then the corrosion current is zero. This interpretation points to a trivial, yet often effective, method of slowing corrosion: cover the surface with a coating, such as paint. (Paint also increases the effective solution resistance between the cathode and anode patches on the surface.) Second, for corrosion reactions with similar exchange current densities, the rate of corrosion is high when E is large. That is, rapid corrosion can be expected when the oxidizing and reducing couples have widely differing electrode potentials.

The effect of the exchange current density on the corrosion rate can be seen by considering the specific case of iron in contact with acidified water. Thermodynamically, either oxygen reduction reaction (a) or (b) on p. 926 is effective. However, the exchange current density of reaction (a) on iron is only about $10^{-14}\,\text{A cm}^{-2}$, while for (b) it is $10^{-6}\,\text{A cm}^{-2}$. The latter therefore dominates kinetically, and iron corrodes by hydrogen evolution in acidic solution.

30.8 The inhibition of corrosion

Several techniques for inhibiting corrosion are available. Coating the surface with some impermeable layer, such as paint, may prevent the access of damp air. Unfortunately, this protection fails disastrously if the paint becomes porous. The oxygen then has access to the exposed metal and corrosion continues beneath the paintwork. Another form of surface coating is provided by **galvanizing**, the coating of an iron object with zinc. Since the latter's electrode potential is $-0.76\,\text{V}$, which is more negative than the iron couple, the corrosion of zinc is thermodynamically favoured and the iron survives (the zinc survives because it is protected by a hydrated oxide layer). In contrast, tin plating leads to a very rapid corrosion of the iron once its surface is scratched and the iron exposed because the tin couple $(E^{\ominus} = -0.14\,\text{V})$ oxidizes the iron couple $(E^{\ominus} = -0.44\,\text{V})$. Some oxides are inert kinetically in the sense that they adhere to the metal surface and form an impermeable layer over a fairly wide pH range. This **passivation**, or kinetic protection, can be seen as a way of decreasing the exchange currents by sealing the surface. Thus, aluminium is inert in air even though its reduction potential is strongly negative $(-1.66\,\text{V})$.

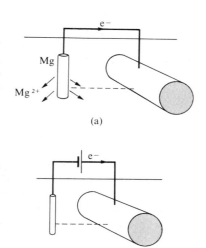

(a)

(b)

Fig. 30.21 (a) In cathodic protection an anode of a more strongly reducing metal is sacrificed to maintain the integrity of the protected object (e.g. a pipeline, bridge, or boat). (b) In impressed-current cathodic protection electrons are supplied from an external cell so that the object itself is not oxidized.

Another method of protection is to change the potential of the object by pumping in electrons that can be used to satisfy the demands of the oxygen reduction without involving the oxidation of the metal. In **cathodic protection** the object is connected to a metal with a more negative electrode potential (such as magnesium, $-2.36\ V$). The magnesium acts as a sacrificial anode, supplying its own electrons to the iron and becoming oxidized to Mg^{2+} in the process (Fig. 30.21a). A block of magnesium replaced occasionally is much cheaper than the ship, building, or pipeline for which it is being sacrificed. In **impressed-current cathodic protection** (Fig. 30.21b) an external cell supplies the electrons and eliminates the need for iron to transfer its own.

Further reading

Electrode processes

G. J. Hills, Electrode processes. In *Essays in chemistry* (ed. J. N. Bradley, R. D. Gillard, and R. F. Hudson), vol. **2**, p. 19 (1971).

N. J. Selley, *Experimental approach to electrochemistry*. Edward Arnold, London (1977).

J. O'M. Bockris and A. K. N. Reddy, *Modern electrochemistry*. Plenum, New York (1970).

W. J. Albery, *Electrode kinetics*. Clarendon Press, Oxford (1974).

N. S. Hush, *Reactions of molecules at electrodes*. Wiley-Interscience, New York (1971).

J. Ulstrup, *Charge transfer processes in condensed media*. Springer, Berlin (1979).

D. R. Crow, *Principles and applications of electrochemistry* (3rd edn). Chapman and Hall, London (1988).

Electroanalytical techniques

S. Wawzonek, Potentiometry: oxidation–reduction potentials. In *Techniques of chemistry* (ed. A. Weissberger and B. W. Rossiter) Vol. IIA, p. 1. Wiley-Interscience, New York (1971).

A. J. Bard and L. R. Faulkner, *Electrochemical techniques*. Wiley-Interscience, New York (1979).

Fuel cells

J. O'M. Bockris and S. N. Srinivasan, *Fuel cells: their electrochemistry*. McGraw-Hill, New York (1969).

A. McDougall, *Fuel cells*. Macmillan, London (1976).

Corrosion

M. G. Fontanna and R. W. Staehle (ed.), *Advances in corrosion science and technology*. Plenum, New York (1980).

Exercises

30.1 The transfer coefficient of a certain electrode in contact with M^{3+} and M^{4+} in aqueous solution at 25°C is 0.39. The current density is found to be $55.0\ mA\ cm^{-2}$ when the overvoltage is $124\ mV$. What is the overvoltage required for a current density of $75\ mA\ cm^{-2}$?

30.2 Determine the exchange current density from the information given in Exercise 30.1.

30.3 Estimate the limiting current density at an electrode in which the concentration of Ag^+ ions is 2.5 mM at 25°C. The thickness of the Nernst diffusion layer is 0.40 mm. The ionic conductivity of Ag^+ at infinite dilution and $25°C$ is $61.9 \, S \, cm^2 \, mol^{-1}$.

30.4 Take $\alpha = 0.5$ and plot j/j_e as a function of the overpotential η at 298 K.

30.5 A typical exchange current density, that for H^+ discharge at platinum, is $0.79 \, mA \, cm^{-2}$ at $25°C$. What is the current density at an electrode when its overpotential is (a) 10 mV, (b) 100 mV, (c) $-0.5 \, V$? Take $\alpha = 0.5$.

30.6 The exchange current density for a $Pt \, | \, Fe^{3+}, Fe^{2+}$ electrode is $2.5 \, mA \, cm^{-2}$. The standard electrode potential is $+0.77 \, V$. Calculate the current flowing through an electrode of surface area $1 \, cm^2$ as a function of the potential of the electrode. Take unit activity for both ions.

30.7 Suppose that the electrode potential is set at $1.00 \, V$. Calculate the current that flows for the ratio of activities $a(Fe^{2+})/a(Fe^{3+})$ in the range 0.1 to 10.0 and at $25°C$.

30.8 What overpotential is needed to sustain a 20 mA current at the $Pt \, | \, Fe^{3+}, Fe^{2+}$ electrode at $25°C$?

30.9 How many electrons or protons are transported through the double layer in each second when the $Pt \, | \, H_2 \, | \, H^+$, $Pt \, | \, Fe^{3+}, Fe^{2+}$, and $Pb \, | \, H_2 \, | \, H^+$ electrodes are at equilibrium at $25°C$? Take the area as $1.0 \, cm^2$ in each case. Estimate the number of times each second a single atom on the surface takes part in an electron transfer event, assuming an electrode atom occupies about $(280 \, pm)^2$ of the surface.

30.10 What is the effective resistance at $25°C$ of an electrode interface when the overpotential is small? Evaluate it for $1.0 \, cm^2$ (a) $Pt \, | \, H_2 \, | \, H^+$, (b) $Hg \, | \, H_2 \, | \, H^+$ electrodes.

30.11 State what happens when a platinum electrode in an aqueous solution containing both Cu^{2+} and Zn^{2+} ions at unit activity is made the cathode of an electrolysis cell.

30.12 What are the minimum (thermodynamically determined) potentials at which (a) zinc, (b) copper can be deposited from aqueous molar solutions? What are the corresponding potentials when the concentrations are $0.010 \, M$?

30.13 What are the conditions that allow a metal to be deposited from aqueous acidic solution before hydrogen evolution occurs significantly? Why may silver be deposited from aqueous silver nitrate? Why may cadmium be deposited from aqueous cadmium sulphate? (The overpotential for hydrogen evolution on cadmium is about $1 \, V$ at current densities of $1 \, mA \, cm^{-2}$.)

30.14 The exchange current density for H^+ discharge at zinc is about $5 \times 10^{-11} \, A \, cm^{-2}$. Can zinc be deposited from a unit activity aqueous solution of a zinc salt?

30.15 The standard potential of the $Zn^{2+} \, | \, Zn$ electrode is $-0.76 \, V$ at $25°C$. The exchange current density for H^+ discharge at platinum is $0.79 \, mA \, cm^{-2}$. Can zinc be plated on to platinum at that temperature? (Take unit activities.)

30.16 Can magnesium be deposited on a zinc electrode from a unit activity acid solution at $25°C$?

30.17 The limiting current density for the reaction $I_3^- + 2e^- \rightarrow 3I^-$ at a platinum electrode is $28.9 \, \mu A \, cm^{-2}$ when the concentration of KI_3 is $6.6 \times 10^{-4} \, M$ and the temperature $25°C$. The diffusion coefficient of I_3^- is $1.14 \times 10^{-5} \, cm \, s^{-1}$. What is the thickness of the diffusion layer?

30.18 The standard electrode potentials of silver, copper, and zinc are $799 \, mV$, $337 \, mV$, and $-763 \, mV$, respectively. When they are present in a mixture with cyanide ions they may be deposited simultaneously. Account for this observation.

30.19 Calculate the maximum (zero-current) potential difference of a nickel–cadmium cell, and the maximum possible power output when 100 mA is drawn, the temperature being $25°C$.

30.20 Calculate the thermodynamic limit to the zero-current potential of fuel cells operating on (a) hydrogen and oxygen, (b) methane and air. Use the Gibbs function information in Table 2.10, and take the species to be in their standard states at $25°C$.

30.21 Which of the following metals has a thermodynamic tendency to corrode in moist air at pH 7: Fe, Cu, Pb, Al, Ag, Cr, Co? Take as a criterion of corrosion a metal ion concentration of at least $10^{-6} \, M$.

Problems

Numerical problems

30.1 In an experiment on the $Pt \, | \, H_2 \, | \, H^+$ electrode in dilute H_2SO_4 the following current densities were observed at $25°C$. Evaluate α and j_e for the electrode.

η/mV	50	100	150	200	250
$j/(mA \, cm^{-2})$	2.66	8.91	29.9	100	335

How would the current density at this electrode depend on the overpotential of the same set of magnitudes but of opposite sign?

30.2 The ionic conductivity of Fe^{2+} is $40 \, S \, cm^2 \, mol^{-1}$. The limiting current at a platinum electrode of area $40 \, cm^2$ dipping into a solution of iron(II) chloride at $25°C$ was measured at various concentrations, the results are given below. What is the thickness of the diffusion layer at each concentration?

$[FeCl_2]/M$	0.250	0.125	0.063	0.031
I/mA	215	107	49	23

30.3 The standard electrode potentials of lead and tin are

−126 mV and −136 mV respectively at 25°C, and the overvoltages for their deposition are close to zero. What should their relative activities be in order to ensure simultaneous deposition from a mixture?

30.4 Estimating the power output and potential of a cell under operating conditions is very difficult, but eqn 23 summarizes, in an approximate way, some of the parameters involved. As a first step in manipulating this expression, identify all the quantities that depend on the ionic concentrations. Express E in terms of the concentration and conductivities of the ions present in the cell. Estimate the parameters for $Zn\,|ZnSO_4(aq)|\,|CuSO_4(aq)|\,Cu$. Take electrodes of area $5\,cm^2$ separated by $5\,cm$. Ignore both potential differences and resistances of the liquid junction. Take the concentration as $1\,M$, the temperature 25°C, and neglect activity coefficients. Plot E as a function of the current drawn. On the same graph, plot the power output of the cell. What current corresponds to maximum power?

30.5 Consider a cell in which the current is activation controlled. Show that the current for maximum power can be estimated by plotting $\lg I/I_0$ and $c_1 - c_2 I$ against I (where $I_0 = A^2 j_0 j_0'$ and c_1 and c_2 are constants), and looking for the point of intersection of the curves. Carry through this analysis for the cell in Problem 30.4 ignoring all concentration overpotentials.

30.6 Estimate the magnitude of the corrosion current for a patch of zinc of area $0.25\,cm^2$ in contact with a similar area of iron in aqueous environment at 25°C. Take the exchange current densities as $10^{-6}\,A\,cm^{-2}$ and the local ion concentrations as $10^{-6}\,M$.

Theoretical problems

30.7 If $\alpha = \frac{1}{2}$, an electrode interface is unable to rectify alternating current because the current density curve is symmetrical about $\eta = 0$. When $\alpha \neq \frac{1}{2}$, the magnitude of the current density depends on the sign of the overpotential, and so some degree of rectification—called 'faradaic rectification'—may be obtained. Suppose that the overpotential varies as $\eta = \eta_0 \cos \omega t$. Derive an expression for the mean flow of current (averaged over a cycle) for general α, and confirm that the mean current is zero when $\alpha = \frac{1}{2}$. In each case work in the limit of small η_0 but to second order in $\eta_0 F/RT$. Calculate the mean direct current at 25°C for a $1.0\,cm^2$ hydrogen–platinum electrode with $\alpha = 0.38$ when the overpotential varies between $\pm 10\,mV$ at 50 Hz.

30.8 Now suppose that the overpotential is in the high overpotential region at all times even though it is oscillating. What waveform will the current across the interface show if it varies linearly and periodically (as a sawtooth waveform) between η_- and η_+ around η_0? Take $\alpha = \frac{1}{2}$.

30.9 Derive an expression for the current density at an electrode where the rate process is diffusion controlled and η^c is known. Sketch the form of j/j_1 as a function of η^c. What changes occur if anion currents are involved?

Tables index

Tables index

Tables of data

The following Tables reproduce and expand the data given in the short tables in the text, and follow their numbering. The notation xPy means $x \times 10^{+y}$ and xNy means $x \times 10^{-y}$. Φ denotes 298.15 K, the 'conventional temperature' for reporting data. Standard states refer to a pressure of $p^{\ominus} = 1$ bar (as far as possible). The general references are as follows:

AIP: D. E. Gray (ed.) *American Institute of Physics Handbook.*; McGraw-Hill, New York (1972).

AS: M. Abramowitz and I. A. Stegun (eds.), *Handbook of mathematical functions.* Dover, New York (1963).

E: J. Emsley, *The elements.* Clarendon Press, Oxford (1989).

HCP: R. C. Weast (ed.), *Handbook of chemistry and physics.* CRC Press, Boca Raton (1989).

KL: G. W. C. Laye and T. H. Laby (ed.), *Tables of physical and chemical constants.* Longman, London (1973).

LR: G. N. Lewis and M. Randall, revised by K. S. Pitzer and L. Brewer; *Thermodynamics.* McGraw-Hill, New York (1961).

NBS: *NBS Tables of chemical thermodynamic properties,* published as *J. phys. and chem. reference data,* **11**, Supplement 2 (1982).

RS: R. A. Robinson and R. H. Stokes; *Electrolyte solutions.* Butterworth (1959).

TDOC: J. B. Pedley, R. D. Naylor, and S. P. Kirby, *Thermochemical data of organic compounds.* Chapman and Hall, London (1986).

Tables of data

Physical properties of selected materials

	$\rho/(\text{g cm}^{-3})$ at 293 K†	T_f/K	T_b/K		$\rho/(\text{g cm}^{-3})$ at 293 K†	T_f/K	T_b/K
Elements							
Aluminium(s)	2.698	933.5	2740	HBr(g)	2.77	184.3	206.4
Argon(g)	1.381	83.8	87.3	HCl(g)	1.187	159.0	191.1
Boron(s)	2.340	2573	3931	HI(g)	2.85	222.4	237.8
Bromine(l)	3.123	265.9	331.9	$H_2O(l)$	0.997	273.2	373.2
Carbon(d, gr)	2.260	3700s		$D_2O(l)$	1.104	277.0	374.6
Carbon(s, d)	3.513			$NH_3(g)$	0.817	195.4	238.8
Chlorine(g)	1.507	172.2	239.2	KBr(s)	2.750	1003	1708
Copper(s)	8.960	1357	2840	KCl(s)	1.984	1049	1773s
Fluorine(g)	1.108	53.5	85.0	NaCl(s)	2.165	1074	1686
Gold(s)	19.320	1338	3080	$H_2SO_4(l)$	1.841	283.5	611.2
Helium(g)	0.125		4.22				
Hydrogen(g)	0.071	14.0	20.3	**Organic compounds**			
Iodine(s)	4.930	386.7	457.5	Acetaldehyde, $CH_3CHO(l, g)$	0.788	152	293
Iron(s)	7.874	1808	3023	Acetic acid, $CH_3COOH(l)$	1.049	289.8	391
Krypton(g)	2.413	116.6	120.8	Acetone, $(CH_3)_2CO(l)$	0.787	178	329
Lead(s)	11.350	600.6	2013	Aniline, $C_6H_5NH_2(l)$	1.026	267	457
Lithium(s)	0.534	453.7	1620	Anthracene, $C_{14}H_{10}(s)$	1.243	490	615
Magnesium(s)	1.738	922.0	1363	Benzene, $C_6H_6(l)$	0.879	278.6	353.2
Mercury(l)	13.546	234.3	629.7	Carbon tetrachloride, $CCl_4(l)$	1.63	250	349.9
Neon(g)	1.207	24.5	27.1	Chloroform, $CHCl_3(l)$	1.499	209.6	334
Nitrogen(g)	0.880	63.3	77.4	Ethanol, $C_2H_5OH(l)$	0.789	156	351.4
Oxygen(g)	1.140	54.8	90.2	Formaldehyde, HCHO(g)		181	254.0
Phosphorus(s, wh)	1.820	317.3	553	Glucose, $C_6H_{12}O_6(s)$	1.544	415	
Potassium(s)	0.862	336.8	1047	Methane, $CH_4(g)$		90.6	111.6
Silver(s)	10.500	1235	2485	Methanol, $CH_3OH(l)$	0.791	179.2	337.6
Sodium(s)	0.971	371.0	1156	Naphthalene, $C_{10}H_8(s)$	1.145	353.4	491
Sulphur(s, α)	2.070	386.0	717.8	Octane, $C_8H_{18}(l)$	0.703	216.4	398.8
Uranium(s)	18.950	1406	4018	Phenol, $C_6H_5OH(s)$	1.073	314.1	455.0
Xenon(g)	2.939	161.3	166.1	Sucrose, $C_{12}H_{22}O_{11}(s)$	1.588	457d	
Zinc(s)	7.133	692.7	1180				
Inorganic compounds							
$CaCO_3$(s, calcite)	2.71	1612	1171d				
$CuSO_4 \cdot 5H_2O(s)$	2.284	$383(-H_2O)$	$423(-5H_2O)$				

d: decomposes
s: sublimes
Data: AIP, E, HCP, KL
†For gases, at their boiling points.

Table 1.2. Second virial coefficients, $B/(\text{cm}^3\,\text{mol}^{-1})$

	100 K	273 K	373 K	600 K		100 K	273 K	373 K	600 K
Air	−167.3	−13.5	3.4	19.0	N_2	−160.0	−10.5	6.2	21.7
Ar	−187.0	−21.7	−4.2	11.9	Ne	−4.8	10.4	12.3	13.8
CH_4		−53.6	−21.2	8.1	O_2	−197.5	−22.0	−3.7	12.9
CO_2		−149.7	−72.2	−12.4	Xe		−153.7	−81.7	−19.6
H_2	−2.5	13.7	15.6						
He	11.4	12.0	11.3	10.4					
Kr		−62.9	−28.7	2.0					

Data: AIP. The values relate to the expansion in eqn 10b of Section 1.3; convert to eqn 10a using $B' = B/RT$.
For Ar at 273 K, $C = 1200\ \text{cm}^6\,\text{mol}^{-1}$

Table 1.3. Critical constants

	p_c/atm	V_c/(cm^3 mol^{-1})	T_c/K	Z_c	T_B/K		p_c/atm	V_c/(cm^3 mol^{-1})	T_c/K	Z_c	T_B/K
Ar	48.00	75.25	150.72	0.292	411.5	HCl	81.5	81.0	324.7	0.248	
Br$_2$	102	135	584	0.287		He	2.26	57.76	5.21	0.305	22.64
C$_2$H$_4$	50.50	124	283.1	0.270		HI	80.8	423.2			
C$_2$H$_6$	48.20	148	305.4	0.285		Kr	54.27	92.24	209.39	0.291	575.0
C$_6$H$_6$	48.6	260	562.7	0.274		N$_2$	33.54	90.10	126.3	0.292	327.2
CH$_4$	45.6	98.7	190.6	0.288	510.0	Ne	26.86	41.74	44.44	0.307	122.1
Cl$_2$	76.1	124	417.2	0.276		NH$_3$	111.3	72.5	405.5	0.242	
CO$_2$	72.85	94.0	304.2	0.274	714.8	O$_2$	50.14	78.0	154.8	0.308	405.9
F$_2$	55	144				Xe	58.0	118.8	289.75	0.290	768.0
H$_2$	12.8	64.99	33.23	0.305	110.0						
H$_2$O	218.3	55.3	647.4	0.227				Data: AIP, KL			
HBr	84.0	363.0									

Table 1.4. Van der Waals constants

	a/(L^2 atm mol^{-2})	b/(L mol^{-1})		a/(L^2 atm mol^{-2})	b/(L mol^{-1})
Ar	1.345	3.219N2	He	3.412N2	2.37N2
C$_2$H$_4$	4.471	5.714N2	Kr	2.318	3.978N2
C$_2$H$_6$	5.489	6.380N2	N$_2$	1.390	3.913N2
C$_6$H$_6$	18.00	11.54N2	Ne	0.2107	1.709N2
CH$_4$	2.253	4.278N2	NH$_3$	4.170	3.707N2
Cl$_2$	6.493	5.622N2	O$_2$	1.360	3.183N2
CO	1.485	3.985N2	SO$_2$	6.714	5.636N2
CO$_2$	3.592	4.267N2	Xe	4.194	5.105N2
H$_2$	0.2444	2.661N2			
H$_2$O	5.464	3.049N2		Data: HCP	
H$_2$S	4.431	4.287N2			

Table 2.2 Enthalpies of fusion and vaporization at the transition temperature, $\Delta H^{\ominus}$/(kJ mol^{-1})

	T_f/K	Fusion	T_b/K	Vaporization
Elements				
Ag	1234	11.30	2436	250.6
Ar	83.81	1.188	87.29	6.51
Br$_2$	265.9	10.57	332.4	29.45
Cl$_2$	172.1	6.41	239.1	20.41
F$_2$	53.6	0.26	85.0	3.16
H$_2$	13.96	0.117	20.38	0.916
He	3.5	0.021	4.22	0.084
Hg$_2$	234.3	2.292	629.7	59.30
I$_2$	386.8	15.52	458.4	41.80
N$_2$	63.15	0.719	77.35	5.586
Na	371.0	2.601	1156	98.01
O$_2$	54.36	0.444	90.18	6.820
Xe	161	2.30	165	12.6

Tables of data

Table 2.2. (Continued)

	T_f/K	Fusion	T_b/K	Vaporization
Inorganic compounds				
CCl_4	250.3	2.47	349.9	30.00
CO_2	217.0	8.33	194.6	25.23(s)
CS_2	161.2	4.39	319.4	26.74
H_2O	273.15	6.008	373.15	40.656
				44.016 at ₵
H_2S	187.6	2.377	212.8	18.67
H_2SO_4	283.5	2.56		
NH_3	195.4	5.652	239.7	23.35
Organic compounds				
CH_4	90.68	0.941	111.7	8.18
CCl_4	250.3	2.5	350	30.0
C_2H_6	89.85	2.86	184.6	14.7
C_6H_6	278.6	10.59	353.2	30.8
CH_3OH	175.2	3.16	337.2	35.27
				37.99 at ₵
C_2H_5OH	156	4.60	352	43.5

Data: AIP

Table 2.3. Limiting enthalpies of solution at ₵, $\Delta H_{soln}^{\ominus}/(kJ\ mol^{-1})$

	F^-	Cl^-	Br^-	I^-	OH^-	CO_3^{2-}	NO_3^-	SO_4^{2-}
Li^+	+4.9	−37.0	−48.8	−63.3	−23.6	−18.2	−2.7	−29.8
Na^+	+1.9	+3.9	−0.6	−7.5	−44.5	−26.7	+20.4	−2.4
K^+	−17.7	+17.2	+19.9	+20.3	−57.1	−30.9	+34.9	+23.8
NH_4^+	−1.2	+14.8	+16.0	+13.7			+25.7	+6.6
Ag^+	−22.5	+65.5	+84.4	+112.2		+41.8	+22.6	+17.8
Mg^{2+}	−17.7	−160.0	−185.6	−213.2	+2.3	−25.3	−90.9	−91.2
Ca^{2+}	+11.5	−81.3	−103.1	−119.7	−16.7	−13.1	−19.2	−18.0
Al^{3+}	−27	−329	−368	−385				−350

The entry for X^+Y^- is the limiting enthalpy of the process $XY(s) \rightarrow X^+(aq) + Y^-(aq)$.
Source: NBS

Table 2.4. First and second ionization energies, $E_i/(\text{kJ mol}^{-1})$

H							He
1312.0							2372.3 5250.4
Li	**Be**	**B**	**C**	**N**	**O**	**F**	**Ne**
513.3 7298.0	899.4 1757.1	800.6 2427	1086.2 2352	1402.3 2856.1	1313.9 3388.2	1681 3374	2080.6 3952.2
Na	**Mg**	**Al**	**Si**	**P**	**S**	**Cl**	**Ar**
495.8 4562.4	737.7 1450.7	577.4 1816.6 2744.6	786.5 1577.1	1011.7 1903.2 2912	999.6 2251	1251.1 2297	1520.4 2665.2
K	**Ca**	**Ga**	**Ge**	**As**	**Se**	**Br**	**Kr**
418.8 3051.4	589.7 1145	578.8 1979 2963	762.1 1537 2735	947.0 1798	940.9 2044	1139.9 2104	1350.7 2350
Rb	**Sr**	**In**	**Sn**	**Sb**	**Te**	**I**	**Xe**
403.0 2632	549.5 1064.2	558.3 1820.6 2704	708.6 1411.8 2943.0	833.7 1794 2443	869.2 1795	1008.4 1845.9	1170.4 2046
Cs	**Ba**	**Tl**	**Pb**	**Bi**	**Po**	**At**	**Rn**
375.5 2420	502.8 965.1	589.3 1971.0 2878	715.5 1450.4 3081.5	703.2 1610 2466	812	930	1037

Data: E

Table 2.5. Electron affinities, $E_{ea}/(\text{kJ mol}^{-1})$

H							He
72.8							−21
Li	**Be**	**B**	**C**	**N**	**O**	**F**	**Ne**
59.8	−18	23	122.5	−7	141 −844	322	−29
Na	**Mg**	**Al**	**Si**	**P**	**S**	**Cl**	**Ar**
52.9	−21	44	133.6	71.7	200.4 −532	348.7	−35
K	**Ca**	**Ga**	**Ge**	**As**	**Se**	**Br**	**Kr**
48.3	−186	36	116	77	195.0	324.5	−39
Rb	**Sr**	**In**	**Sn**	**Sb**	**Te**	**I**	**Xe**
46.9	−146	34	121	101	190.2	295.3	−41
Cs	**Ba**	**Tl**	**Pb**	**Bi**	**Po**	**At**	**Rn**
45.5	−46	30	35.2	101	186	270	−41

Data: E

Tables of data

Table 2.6. Bond dissociation enthalpies, $\Delta H^{\ominus}(\text{A—B})/(\text{kJ mol}^{-1})$ at $\text{\textperthousand}$

Diatomic molecules

H—H	436	F—F	155	Cl—Cl	242	Br—Br	193	I—I	151
O=O	497	C=O	1074	N≡N	945				
H—O	428	H—F	565	H—Cl	431	H—Br	366	H—I	299

Polyatomic molecules

H—CH$_3$	435	H—NH$_2$	431	H—OH	492	H—C$_6$H$_5$	469
H$_3$C—CH$_3$	368	H$_2$C=CH$_2$	699	HC≡CH	962		
HO—CH$_3$	377	Cl—CH$_3$	452	Br—CH$_3$	293	I—CH$_3$	234
O=CO	531	HO—OH	213	O$_2$N—NO$_2$	57		

Data: HCP, KL

Table 2.7. Bond enthalpies, $B(\text{A—B})/(\text{kJ mol}^{-1})$

	H	C	N	O	F	Cl	Br	I	S	P	Si
H	436										
C	412	348(i)									
		612(ii)									
		518(a)									
N	388	305(i)	163(i)								
		613(ii)	409(ii)								
		890(iii)	945(iii)								
O	463	360(i)	(157)	146(i)							
		743(ii)		497(ii)							
F	565	484	270	185	155						
Cl	431	338	200	203	254	242					
Br	366	276				219	193				
I	299	238				210	178	151			
S	338	259				250	212		264		
P	322									172	
Si	318		374								176

(i) Single bond, (ii) double bond, (iii) triple bond, (a) aromatic.
Data: HCP and L. Pauling, *The nature of the chemical bond.* Cornell University Press (1960).

Table 2.8. Standard enthalpies of atomization. See Table 2.10.

Table 2.9. Thermodynamic data for organic compounds (all values relate to $\text{\textperthousand}$)

	M g mol^{-1}	$\Delta H_f^{\ominus}$ kJ mol^{-1}	$\Delta G_f^{\ominus}$ kJ mol^{-1}	$S^{\ominus}$ J K^{-1} mol^{-1}	C_p J K^{-1} mol^{-1}	$\Delta H_c^{\ominus}$ kJ mol^{-1}
C(s) (graphite)	12.011	0	0	5.740	8.527	−393.51
C(s) (diamond)	12.011	1.895	2.900	2.377	6.113	−395.40
CO$_2$(g)	44.010	−393.51	−394.36	213.74	37.11	
Hydrocarbons						
CH$_4$(g), methane	16.04	−74.81	−50.72	186.26	35.31	−890
CH$_3$(g), methyl	15.04	145.69	147.92	194.2	38.70	

Table 2.9. (Continued)

	M g mol^{-1}	$\Delta H_f^{\ominus}$ kJ mol^{-1}	$\Delta G_f^{\ominus}$ kJ mol^{-1}	$S^{\ominus}$ J K^{-1} mol^{-1}	C_p J K^{-1} mol^{-1}	$\Delta H_c^{\ominus}$ kJ mol^{-1}
Hydrocarbons (continued)						
$C_2H_2(g)$, ethyne	26.04	226.73	209.20	200.94	43.93	−1300
$C_2H_4(g)$, ethene	28.05	52.26	68.15	219.56	43.56	−1411
$C_2H_6(g)$, ethane	30.07	−84.68	−32.82	229.60	52.63	−1560
$C_3H_6(g)$, propene	42.08	20.42	62.78	267.05	63.89	−2058
$C_3H_6(g)$, cyclopropane	42.08	53.30	104.45	237.55	55.94	−2091
$C_3H_8(g)$, propane	44.10	−103.85	−23.49	269.91	73.5	−2220
$C_4H_8(g)$, 1-butene	56.11	−0.13	71.39	305.71	85.65	−2717
$C_4H_8(g)$, cis-2-butene	56.11	−6.99	65.95	300.94	78.91	−2710
$C_4H_8(g)$, trans-2-butene	56.11	−11.17	63.06	296.59	87.82	−2707
$C_4H_{10}(g)$, butane	58.13	−126.15	−17.03	310.23	97.45	−2878
$C_5H_{12}(g)$, pentane	72.15	−146.44	−8.20	348.40	120.2	−3537
$C_5H_{12}(l)$	72.15	−173.1				
$C_6H_6(l)$, benzene	78.12	49.0	124.3	173.3	136.1	−3268
$C_6H_6(g)$	78.12	82.93	129.72	269.31	81.67	−3302
$C_6H_{12}(l)$, cyclohexane	84.16	−156	26.8		156.5	−3902
$C_6H_{14}(l)$, hexane	86.18	−198.7		204.3		−4163
$C_6H_5CH_3(g)$, methylbenzene (toluene)	92.14	50.0	122.0	320.7	103.6	−3953
$C_7H_{16}(l)$, heptane	100.21	−224.4	1.0	328.6	224.3	
$C_8H_{18}(l)$, octane	114.23	−249.9	6.4	361.1		−5471
$C_8H_{18}(l)$, iso-octane	114.23	−255.1				−5461
$C_{10}H_8(s)$, naphthalene	128.18	78.53				−5157
Alcohols and phenols						
$CH_3OH(l)$, methanol	32.04	−238.66	−166.27	126.8	81.6	−726
$CH_3OH(g)$	32.04	−200.66	−161.96	239.81	43.89	−764
$C_2H_5OH(l)$, ethanol	46.07	−277.69	−174.78	160.7	111.46	−1368
$C_2H_5OH(g)$	46.07	−235.10	−168.49	282.70	65.44	−1409
$C_6H_5OH(s)$, phenol	94.12	−165.0	−50.9	146.0		−3054
Carboxylic acids, hydroxy acids, and esters						
$HCOOH(l)$, formic	46.03	−424.72	−361.35	128.95	99.04	−255
$CH_3COOH(l)$, acetic	60.05	−484.5	−389.9	159.8	124.3	−875
$CH_3COOH(aq)$	60.05	−485.76	−396.46	178.7		
$CH_3COO^-(aq)$	59.05	−486.01	−369.31	86.6	−6.3	
$(COOH)_2(s)$, oxalic	90.04	−827.2			117	−254
$C_6H_5COOH(s)$, benzoic	122.13	−385.1	−245.3	167.6	146.8	−3227
$CH_3CH(OH)COOH(s)$, lactic	90.08	−694.0				−1344
$CH_3COOC_2H_5(l)$, ethyl acetate	88.11	−479.0	−332.7	259.4	170.1	−2231
Alkanals and alkanones						
$HCHO(g)$, methanal	30.03	−108.57	−102.53	218.77	35.40	−571
$CH_3CHO(l)$, ethanal	44.05	−192.30	−128.12	160.2		−1166
$CH_3CHO(g)$	44.05	−166.19	−128.86	250.3	57.3	−1192
$CH_3COCH_3(l)$, propanone	58.08	−248.1	−155.4	200.4	124.7	−1790
Sugars						
$C_6H_{12}O_6(s)$, α-D-glucose	180.16	−1274				−2808
$C_6H_{12}O_6(s)$, β-D-glucose	180.16	−1268	−910	212		
$C_6H_{12}O_6(s)$, β-D-fructose	180.16	−1266				−2810
$C_{12}H_{22}O_{11}(s)$, sucrose	342.30	−2222	−1543	360.2		−5645

Tables of data

Table 2.9. (Continued)

	M g mol^{-1}	$\Delta H_f^{\ominus}$ kJ mol^{-1}	$\Delta G_f^{\ominus}$ kJ mol^{-1}	$S^{\ominus}$ J K^{-1} mol^{-1}	C_p J K^{-1} mol^{-1}	$\Delta H_c^{\ominus}$ kJ mol^{-1}
Nitrogen compounds						
$CO(NH_2)_2(s)$, urea	60.06	−333.51	−197.33	104.60	93.14	−632
$CH_3NH_2(g)$, methylamine	31.06	−22.97	32.16	243.41	53.1	−1085
$C_6H_5NH_2(l)$, aniline	93.13	31.1				−3393
$CH_2(NH_2)COOH(s)$, glycine	75.07	−532.9	−373.4	103.5	99.2	−969

Data: NBS, TDOC

Table 2.10. Thermodynamic data (all values relate to $\overline{\mathcal{T}}$)

	M g mol^{-1}	$\Delta H_f^{\ominus}$ kJ mol^{-1}	$\Delta G_f^{\ominus}$ kJ mol^{-1}	$S^{\ominus}$ J K^{-1} mol^{-1}	C_p J K^{-1} mol^{-1}
Aluminium (aluminum)					
$Al(s)$	26.98	0	0	28.33	24.35
$Al(l)$	26.98	10.56	7.20	39.55	24.21
$Al(g)$	26.98	326.4	285.7	164.54	21.38
$Al^{3+}(g)$	26.98	5483.17			
$Al^{3+}(aq)$	26.98	−531	−485	−321.7	
$Al_2O_3(s, \alpha)$	101.96	−1675.7	−1582.3	50.92	79.04
$AlCl_3(s)$	133.24	−704.2	−628.8	110.67	91.84
Argon					
$Ar(g)$	39.95	0	0	154.84	20.786
Antimony					
$Sb(s)$	121.75	0	0	45.69	25.23
$SbH_3(g)$	153.24	145.11	147.75	232.78	41.05
Arsenic					
$As(s, \alpha)$	74.92	0	0	35.1	24.64
$As(g)$	74.92	302.5	261.0	174.21	20.79
$As_4(g)$	299.69	143.9	92.4	314	
$AsH_3(g)$	77.95	66.44	68.93	222.78	38.07
Barium					
$Ba(s)$	137.34	0	0	62.8	28.07
$Ba(g)$	137.34	180	146	170.24	20.79
$Ba^{2+}(aq)$	137.34	−537.64	−560.77	9.6	
$BaO(s)$	153.34	−553.5	−525.1	70.43	47.78
$BaCl_2(s)$	208.25	−858.6	−810.4	123.68	75.14
Beryllium					
$Be(s)$	9.01	0	0	9.50	1644
$Be(g)$	9.01	324.3	286.6	136.27	20.79
Bismuth					
$Bi(s)$	208.98	0	0	56.74	25.52
$Bi(g)$	208.98	207.1	168.2	187.00	20.79

Table 2.10. (Continued)

	M g mol^{-1}	$\Delta H_f^{\ominus}$ kJ mol^{-1}	$\Delta G_f^{\ominus}$ kJ mol^{-1}	$S^{\ominus}$ J K^{-1} mol^{-1}	C_p J K^{-1} mol^{-1}
Bromine					
$Br_2(l)$	159.82	0	0	152.23	75.689
$Br_2(g)$	159.82	30.907	3.110	245.46	36.02
$Br(g)$	79.91	111.88	82.396	175.02	20.786
$Br^-(g)$	79.91	−219.07			
$Br^-(aq)$	79.91	−121.55	−103.96	82.4	−141.8
$HBr(g)$	90.92	−36.40	−53.45	198.70	29.142
Cadmium					
$Cd(s, \gamma)$	112.40	0	0	51.76	25.98
$Cd(g)$	112.40	112.01	77.41	167.75	20.79
$Cd^{2+}(aq)$	112.40	−75.90	−77.612	−73.2	
$CdO(s)$	128.40	−258.2	−228.4	54.8	43.43
$CdCO_3(s)$	172.41	−750.6	−669.4	92.5	
Caesium (cesium)					
$Cs(s)$	132.91	0	0	85.23	32.17
$Cs(g)$	132.91	76.06	49.12	175.60	20.79
$Cs^+(aq)$	132.91	−258.28	−292.02	133.05	−10.5
Calcium					
$Ca(s)$	40.08	0	0	41.42	25.31
$Ca(g)$	40.08	178.2	144.3	154.88	20.786
$Ca^{2+}(aq)$	40.08	−542.83	−553.58	−53.1	
$CaO(s)$	56.08	−635.09	−604.03	39.75	42.80
$CaCO_3(s)$ (calcite)	100.09	−1206.9	−1128.8	92.9	81.88
$CaCO_3(s)$ (aragonite)	100.09	−1207.1	−1127.8	88.7	81.25
$CaF_2(s)$	78.08	−1219.6	−1167.3	68.87	67.03
$CaCl_2(s)$	110.99	−795.8	−748.1	104.6	72.59
$CaBr_2(s)$	199.90	−682.8	−663.6	130	
Carbon (for 'organic' compounds of carbon, see Table 2.9.)					
$C(s)$ (graphite)	12.011	0	0	5.740	8.527
$C(s)$ (diamond)	12.011	1.895	2.900	2.377	6.113
$C(g)$	12.011	716.68	671.26	158.10	20.838
$C_2(g)$	24.022	831.90	775.89	199.42	43.21
$CO(g)$	28.011	−110.53	−137.17	197.67	29.14
$CO_2(g)$	44.010	−393.51	−394.36	213.74	37.11
$CO_2(aq)$	44.010	−413.80	−385.98	117.6	
$H_2CO_3(aq)$	62.03	−699.65	−623.08	187.4	
$HCO_3^-(aq)$	61.02	−691.99	−586.77	91.2	
$CO_3^{2-}(aq)$	60.01	−677.14	−527.81	−56.9	
$CCl_4(l)$	153.82	−135.44	−65.21	216.40	131.75
$CS_2(l)$	76.14	89.70	65.27	151.34	75.7
$HCN(g)$	27.03	135.1	124.7	201.78	35.86
$HCN(l)$	27.03	108.87	124.97	112.84	70.63
$CN^-(aq)$	26.02	150.6	172.4	94.1	

Tables of data

Table 2.10. (Continued)

	M g mol^{-1}	$\Delta H_f^{\ominus}$ kJ mol^{-1}	$\Delta G_f^{\ominus}$ kJ mol^{-1}	$S^{\ominus}$ J K^{-1} mol^{-1}	C_p J K^{-1} mol^{-1}
Chlorine					
$Cl_2(g)$	70.91	0	0	223.07	33.91
$Cl(g)$	35.45	121.68	105.68	165.20	21.840
$Cl^-(g)$	35.45	−233.13			
$Cl^-(aq)$	35.45	−167.16	−131.23	56.5	−136.4
$HCl(g)$	36.46	−92.31	−95.30	186.91	29.12
$HCl(aq)$	36.46	−167.16	−131.23	56.5	−136.4
Chromium					
$Cr(s)$	52.00	0	0	23.77	23.35
$Cr(g)$	52.00	396.6	351.8	174.50	20.79
$CrO_4^{2-}(aq)$	115.99	−881.15	−727.75	50.21	
$Cr_2O_7^{2-}(aq)$	215.99	−1490.3	−1301.1	261.9	
Copper					
$Cu(s)$	63.54	0	0	33.150	24.44
$Cu(g)$	63.54	338.32	298.58	166.38	20.79
$Cu^+(aq)$	63.54	71.67	49.98	40.6	
$Cu^{2+}(aq)$	63.54	64.77	65.49	−99.6	
$Cu_2O(s)$	143.08	−168.6	−146.0	93.14	63.64
$CuO(s)$	79.54	−157.3	−129.7	42.63	42.30
$CuSO_4(s)$	159.60	−771.36	−661.8	109	100.0
$CuSO_4 \cdot H_2O(s)$	177.62	−1085.8	−918.11	146.0	134
$CuSO_4 \cdot 5H_2O(s)$	249.68	−2279.7	−1879.7	300.4	280
Deuterium					
$D_2(g)$	4.028	0	0	144.96	29.20
$HD(g)$	3.022	0.318	−1.464	143.80	29.196
$D_2O(g)$	20.028	−249.20	−234.54	198.34	34.27
$D_2O(l)$	20.028	−294.60	−243.44	75.94	84.35
$HDO(g)$	19.022	−245.30	−233.11	199.51	33.81
$HDO(l)$	19.022	−289.89	−241.86	79.29	
Fluorine					
$F_2(g)$	38.00	0	0	202.78	31.30
$F(g)$	19.00	78.99	61.91	158.75	22.74
$F^-(aq)$	19.00	−332.63	−278.79	−13.8	−106.7
$HF(g)$	20.01	−271.1	−273.2	173.78	29.13
Gold					
$Au(s)$	196.97	0	0	47.40	25.42
$Au(g)$	196.97	366.1	326.3	180.50	20.79
Helium					
$He(g)$	4.003	0	0	126.15	20.786
Hydrogen (see also deuterium)					
$H_2(g)$	2.016	0	0	130.684	28.824
$H(g)$	1.008	217.97	203.25	114.71	20.784
$H^+(aq)$	1.008	0	0	0	0

Table 2.10. (Continued)

	M g mol^{-1}	$\Delta H_f^{\ominus}$ kJ mol^{-1}	$\Delta G_f^{\ominus}$ kJ mol^{-1}	$S^{\ominus}$ J K^{-1} mol^{-1}	C_p J K^{-1} mol^{-1}
Hydrogen (continued)					
$H_2O(l)$	18.015	-285.83	-237.13	69.91	75.291
$H_2O(g)$	18.015	-241.82	-228.57	188.83	33.58
$H_2O_2(l)$	34.015	-187.78	-120.35	109.6	89.1
Iodine					
$I_2(s)$	253.81	0	0	116.135	54.44
$I_2(g)$	253.81	62.44	19.33	260.69	36.90
$I(g)$	126.90	106.84	70.25	180.79	20.786
$I^-(aq)$	126.90	-55.19	-51.57	111.3	-142.3
$HI(g)$	127.91	26.48	1.70	206.59	29.158
Iron					
$Fe(s)$	55.85	0	0	27.28	25.10
$Fe(g)$	55.85	416.3	370.7	180.49	25.68
$Fe^{2+}(aq)$	55.85	-89.1	-78.90	-137.7	
$Fe^{3+}(aq)$	55.85	-48.5	-4.7	-315.9	
$Fe_3O_4(s)$ (magnetite)	231.54	-1118.4	-1015.4	146.4	143.43
$Fe_2O_3(s)$ (haematite)	159.69	-824.2	-742.2	87.40	103.85
$FeS(s, \alpha)$	87.91	-100.0	-100.4	60.29	50.54
$FeS_2(s)$	119.98	-178.2	-166.9	52.93	62.17
Krypton					
$Kr(g)$	83.80	0	0	164.08	20.786
Lead					
$Pb(s)$	207.19	0	0	64.81	26.44
$Pb(g)$	207.19	195.0	161.9	175.37	20.79
$Pb^{2+}(aq)$	207.19	-1.7	-24.43	10.5	
$PbO(s, yellow)$	223.19	-217.32	-187.89	68.70	45.77
$PbO(s, red)$	223.19	-218.99	-188.93	66.5	45.81
$PbO_2(s)$	239.19	-277.4	-217.33	68.6	64.64
Lithium					
$Li(s)$	6.94	0	0	29.12	24.77
$Li(g)$	6.94	159.37	126.66	138.77	20.79
$Li^+(aq)$	6.94	-278.49	-293.31	13.4	68.6
Magnesium					
$Mg(s)$	24.31	0	0	32.68	24.89
$Mg(g)$	24.31	147.70	113.10	148.65	20.786
$Mg^{2+}(aq)$	24.31	-466.85	-454.8	-138.1	
$MgO(s)$	40.31	-601.70	-569.43	26.94	37.15
$MgCO_3(s)$	84.32	-1095.8	-1012.1	65.7	75.52
$MgCl_2(s)$	95.22	-641.32	-591.79	89.62	71.38

Tables of data

Table 2.10. (Continued)

	M g mol^{-1}	$\Delta H_f^{\ominus}$ kJ mol^{-1}	$\Delta G_f^{\ominus}$ kJ mol^{-1}	$S^{\ominus}$ J K^{-1} mol^{-1}	C_p J K^{-1} mol^{-1}
Mercury					
Hg(l)	200.59	0	0	76.02	27.983
Hg(g)	200.59	61.32	31.82	174.96	20.786
Hg^{2+}(aq)	200.59	171.1	164.40	−32.2	
Hg$_2^{2+}$(aq)	401.18	172.4	153.52	84.5	
HgO(s)	216.59	−90.83	−58.54	70.29	44.06
Hg$_2$Cl$_2$(s)	472.09	−265.22	−210.75	192.5	102
HgCl$_2$(s)	271.50	−224.3	−178.6	146.0	
HgS(s, black)	232.65	−53.6	−47.7	88.3	
Neon					
Ne(g)	20.18	0	0	146.33	20.786
Nitrogen					
N$_2$(g)	28.013	0	0	191.61	29.125
N(g)	14.007	472.70	455.56	153.30	20.786
NO(g)	30.01	90.25	86.55	210.76	29.844
N$_2$O(g)	44.01	82.05	104.20	219.85	38.45
NO$_2$(g)	46.01	33.18	51.31	240.06	37.20
N$_2$O$_4$(g)	92.01	9.16	97.89	304.29	77.28
N$_2$O$_5$(s)	108.01	−43.1	113.9	178.2	143.1
N$_2$O$_5$(g)	108.01	11.3	115.1	355.7	84.5
HNO$_3$(l)	63.01	−174.10	−80.71	155.60	109.87
HNO$_3$(aq)	63.01	−207.36	−111.25	146.4	−86.6
NO$_3^-$(aq)	62.01	−205.0	−108.74	146.4	−86.6
NH$_3$(g)	17.03	−46.11	−16.45	192.45	35.06
NH$_3$(aq)	17.03	−80.29	−26.50	111.3	
NH$_4^+$(aq)	18.04	−132.51	−79.31	113.4	79.9
NH$_2$OH(s)	33.03	−114.2			
HN$_3$(l)	43.03	264.0	327.3	140.6	43.68
HN$_3$(g)	43.03	294.1	328.1	238.97	98.87
N$_2$H$_4$(l)	32.05	50.63	149.43	121.21	139.3
NH$_4$NO$_3$(s)	80.04	−365.56	−183.87	151.08	84.1
NH$_4$Cl(s)	53.49	−314.43	−202.87	94.6	
Oxygen					
O$_2$(g)	31.999	0	0	205.138	29.355
O(g)	15.999	249.17	231.73	161.06	21.912
O$_3$(g)	47.998	142.7	163.2	238.93	39.20
OH$^-$(aq)	17.007	−229.99	−157.24	−10.75	−148.5
Phosphorus					
P(s, wh)	30.97	0	0	41.09	23.840
P(g)	30.97	314.64	278.25	163.19	20.786
P$_2$(g)	61.95	144.3	103.7	218.13	32.05
P$_4$(g)	123.90	58.91	24.44	279.98	67.15
PH$_3$(g)	34.00	5.4	13.4	210.23	37.11
PCl$_3$(g)	137.33	−287.0	−267.8	311.78	71.84
PCl$_3$(l)	137.33	−319.7	−272.3	217.1	
PCl$_5$(g)	208.24	−374.9	−305.0	364.6	112.8
PCl$_5$(s)	208.24	−443.5			
H$_3$PO$_3$(s)	82.00	−964.4			

Table 2.10. (Continued)

	M g mol^{-1}	$\Delta H_f^{\ominus}$ kJ mol^{-1}	$\Delta G_f^{\ominus}$ kJ mol^{-1}	$S^{\ominus}$ J K^{-1} mol^{-1}	C_p J K^{-1} mol^{-1}
Phosphorus (continued)					
$H_3PO_3(aq)$	82.00	−964.8			
$H_3PO_4(s)$	94.97	−1279.0	−1119.1	110.50	106.06
$H_3PO_4(l)$	94.97	−1266.9			
$H_3PO_4(aq)$	94.97	−1277.4	−1018.7	−222	
$PO_4^{3-}(aq)$	94.97	−1277.4	−1018.7	−222	
$P_4O_{10}(s)$	283.89	−2984.0	−2697.0	228.86	211.71
$P_4O_6(s)$	219.89	−1640.1			
Potassium					
$K(s)$	39.10	0	0	64.18	29.58
$K(g)$	39.10	89.24	60.59	160.336	20.786
$K^+(g)$	39.10	514.26			
$K^+(aq)$	39.10	−252.38	−283.27	102.5	21.8
$KOH(s)$	56.11	−424.76	−379.08	78.9	64.9
$KF(s)$	58.10	−567.27	−537.75	66.57	49.04
$KCl(s)$	74.56	−436.75	−409.14	82.59	51.30
$KBr(s)$	119.01	−393.80	−380.66	95.90	52.30
$KI(s)$	166.01	−327.90	−324.89	106.32	52.93
Silicon					
$Si(g)$	28.09	0	0	18.83	20.00
$Si(g)$	28.09	455.6	411.3	167.97	22.25
$SiO_2(s, \alpha)$	60.09	−910.94	−856.64	41.84	44.43
Silver					
$Ag(s)$	107.87	0	0	42.55	25.351
$Ag(g)$	107.87	284.55	245.65	173.00	20.79
$Ag^+(aq)$	107.87	105.58	77.11	72.68	21.8
$AgBr(s)$	187.78	−100.37	−96.90	107.1	52.38
$AgCl(s)$	143.32	−127.07	−109.79	96.2	50.79
$Ag_2O(s)$	231.74	−31.05	−11.20	121.3	65.86
$AgNO_3(s)$	169.88	−124.39	−33.41	140.92	93.05
Sodium					
$Na(s)$	22.99	0	0	51.21	28.24
$Na(g)$	22.99	107.32	76.76	153.71	20.79
$Na^+(aq)$	22.99	−240.12	−261.91	59.0	46.4
$NaOH(s)$	40.00	−425.61	−379.49	64.46	59.54
$NaCl(s)$	58.44	−411.15	−384.14	72.13	50.50
$NaBr(s)$	102.90	−361.06	−348.98	86.82	51.38
$NaI(s)$	149.89	−287.78	−286.06	98.53	52.09
Sulphur					
$S(s, \alpha)$ (rhombic)	32.06	0	0	31.80	22.64
$S(s, \beta)$ (monoclinic)	32.06	0.33	0.1	32.6	23.6
$S(g)$	32.06	278.81	238.25	167.82	23.673
$S_2(g)$	64.13	128.37	79.30	228.18	32.47

Tables of data

Table 2.10. (Continued)

	M g mol^{-1}	$\Delta H_f^{\ominus}$ kJ mol^{-1}	$\Delta G_f^{\ominus}$ kJ mol^{-1}	$S^{\ominus}$ J K^{-1} mol^{-1}	C_p J K^{-1} mol^{-1}
Sulphur (continued)					
$S^{2-}(aq)$	32.06	33.1	85.8	−14.6	
$SO_2(g)$	64.06	−296.83	−300.19	248.22	39.87
$SO_3(g)$	80.06	−395.72	−371.06	256.76	50.67
$H_2SO_4(l)$	98.08	−813.99	−690.00	156.90	138.9
$H_2SO_4(aq)$	98.08	−909.27	−744.53	20.1	−293
$SO_4^{2-}(aq)$	96.06	−909.27	−744.53	20.1	−293
$HSO_4^-(aq)$	97.07	−887.34	−755.91	131.8	−84
$H_2S(g)$	34.08	−20.63	−33.56	205.79	34.23
$H_2S(aq)$	34.08	−39.7	−27.83	121	
$HS^-(aq)$	33.072	−17.6	12.08	62.08	
$SF_6(g)$	146.05	−1209	−1105.3	291.82	97.28
Tin					
$Sn(s, \beta)$	118.69	0	0	51.55	26.99
$Sn(g)$	118.69	302.1	267.3	168.49	20.26
$Sn^{2+}(aq)$	118.69	−8.8	−27.2	−17	
$SnO(s)$	134.69	−285.8	−256.9	56.5	44.31
$SnO_2(s)$	150.69	−580.7	−519.6	52.3	52.59
Xenon					
$Xe(g)$	131.30	0	0	169.68	20.786
Zinc					
$Zn(s)$	65.37	0	0	41.63	25.40
$Zn(g)$	65.37	130.73	95.14	160.98	20.79
$Zn^{2+}(aq)$	65.37	−153.89	−147.06	−112.1	46
$ZnO(s)$	81.37	−348.28	−318.30	43.64	40.25

Source: NBS

Table 2.11. Lattice enthalpies, $\Delta H_L^{\ominus}/(\text{kJ mol}^{-1})$

Halides

	F	Cl	Br	I
Li	1037	852	815	761
Na	926	786	752	705
K	821	717	689	649
Rb	789	695	668	632
Cs	750	676	654	620
Ag	969	912	900	886
Be		3017		
Mg		2524		
Ca		2255		
Sr		2153		

Oxides

MgO 3850   CaO 3461   SrO 3283   BaO 3114

Sulphides

MgS 3406   CaS 3119   SrS 2974   BaS 2832

Data: Principally D. Cubicciotti, *J. chem. Phys.*, **31**, 1646 (1959).

Table 2.12. Standard molar enthalpies of hydration at infinite dilution, $\Delta H^{\ominus}_{\text{hyd}}/(\text{kJ mol}^{-1})$

	Li^+	Na^+	K^+	Rb^+	Cs^+
F^-	-1026	-911	-828	-806	-782
Cl^-	-884	-784	-685	-664	-640
Br^-	-856	-742	-658	-637	-613
I^-	-815	-701	-617	-596	-572

Entries refer to $X^+(g) + Y^-(g) \rightarrow X^+(aq) + Y^-(aq)$.
Data: Principally J. O'M. Bockris and A. K. N. Reddy, *Modern electrochemistry,* Vol. 1. Plenum Press, New York (1970).

Table 2.13. Limiting enthalpies of formation of ions in aqueous solution, $\Delta H^{\ominus}_f/(\text{kJ mol}^{-1})$. See Table 2.10.

Table 2.14. Standard molar ion hydration enthalpies, $\Delta H^{\ominus}_{\text{hyd}}/(\text{kJ mol}^{-1})$ at $\mathsf{\mathcal{T}}$

Cations

H^+	(-1090)	Ag^+	-464	Mg^{2+}	-1920
Li^+	-520	NH_4^+	-301	Ca^{2+}	-1650
Na^+	-405			Sr^{2+}	-1480
K^+	-321			Ba^{2+}	-1360
Rb^+	-300			Fe^{2+}	-1950
Cs^+	-277			Cu^{2+}	-2100
				Zn^{2+}	-2050
				Al^{3+}	-4690
				Fe^{3+}	-4430

Anions

OH^-	-460						
F^-	-506	Cl^-	-364	Br^-	-337	I^-	-296

Entries refer to $X^{\pm}(g) \rightarrow X^{\pm}(aq)$ based on $H^+(g) \rightarrow H^+(aq)$ $\Delta H^{\ominus} = -1090 \text{ kJ mol}^{-1}$.
Data: Principally J. O'M. Bockris and A. K. N. Reddy, *Modern electrochemistry,* Vol. 1. Plenum Press, New York (1970).

Table 2.15. Molar heat capacities at $\mathsf{\mathcal{T}}$ and 1 atm. See also Table 2.10

	$C_V/(\text{J K}^{-1}\text{mol}^{-1})$	$C_p/(\text{J K}^{-1}\text{mol}^{-1})$	γ		$C_V/(\text{J K}^{-1}\text{mol}^{-1})$	$C_p/(\text{J K}^{-1}\text{mol}^{-1})$	γ
Gases				**Solids**			
He, Ne, Ar, Kr, Xe	12.48	20.79	1.667	C(diamond)		6.11	
H_2	20.44	28.82	1.410	Cu		244.4	
N_2	20.74	29.12	1.404	Fe		25.1	
O_2	20.95	29.36	1.401	SiO_2		44.4	
CO_2	28.46	37.11	1.304				

Specific heat capacity of various substances, $C/(\text{J K}^{-1}\text{g}^{-1})$

Liquids

	$C_p/(\text{J K}^{-1}\text{mol}^{-1})$
H_2O	75.29
CH_3OH	81.6
C_2H_5OH	111.5
C_6H_6	136.1

Brass	0.38	Glycerol	2.43
Glass (Pyrex)	0.78	Asbestos	0.84
Sea-water	3.93	Rubber	1.1–1.2
Steel	0.51	Ice	2.0–2.1
Granite	0.82		

Data: NBS; KL

Tables of data

Table 2.16. Temperature variation of molar heat capacities†

	$a/(\text{J K}^{-1}\,\text{mol}^{-1})$	$b/(\text{J K}^{-2}\,\text{mol}^{-1})$	$c/(\text{J K}\,\text{mol}^{-1})$
Monoatomic gases			
	20.78	0	0
Other gases			
Br_2	37.32	0.50N3	−1.26P5
Cl_2	37.03	0.67N3	−2.85P5
CO_2	44.22	8.79N3	−8.62P5
F_2	34.56	2.51N3	−3.51P5
H_2	27.28	3.26N3	0.50P5
I_2	37.40	0.59N3	−0.71P5
N_2	28.58	3.77N3	−0.50P5
NH_3	29.75	2.51N2	−1.55P5
O_2	29.96	4.18N3	−1.67P5
Liquids (from melting to boiling)			
$C_{10}H_8$, naphthalene	79.5	4.075N4	0
I_2	80.33	0	0
H_2O	75.29	0	0
Solids			
Al	20.67	12.38N3	0
C (graphite)	16.86	4.77N3	−8.54P5
$C_{10}H_8$, naphthalene	−115.9	3.920	0
Cu	22.64	6.28N3	0
I_2	40.12	49.79N3	0
NaCl	45.94	16.32N3	0
Pb	22.13	11.72N3	0.96P5

† For $C_p/(\text{J K}^{-1}\,\text{mol}^{-1}) = a + bT + c/T^2$.
Source: LR.

Table 3.1. Expansion coefficients α and isothermal compressibilities κ_T

	α/K^{-1}	κ_T/atm^{-1}
Liquids		
Benzene	1.24N3	9.21N5
Carbon tetrachloride	1.24N3	9.05N5
Ethanol	1.12N3	7.68N5
Mercury	1.82N4	3.87N5
Water	2.1N4	4.96N5
Solids		
Copper	5.01N5	7.35N7
Diamond	3.0N6	1.87N7
Iron	3.54N5	5.97N7
Lead	8.61N5	2.21N6

The values refer to 20°C.
Data: AIP(α), KL(κ_T).

Table 3.2. Inversion temperatures, normal freezing and boiling points, and Joule–Thomson coefficients at 1 atm and 298 K

	T_i/K	T_f/K	T_b/K	$\mu_{JT}/(K\,atm^{-1})$
Air	603			0.189 at 50°C
Argon	723	83.8	87.3	
Carbon dioxide	1500	194.7s		1.11 at 300 K
Helium	40		4.22	−0.060
Hydrogen	202	14.0	20.3	
Krypton	1090	116.6	120.8	
Methane	968	90.6	111.6	
Neon	231	24.5	27.1	
Nitrogen	621	63.3	77.4	0.25
Oxygen	764	54.8	90.2	0.31

Data: AIP and M. W. Zemansky, *Heat and thermodynamics*. McGraw-Hill, New York (1957).

Table 4.1. Entropies (and temperatures) of phase transitions at 1 atm pressure, $\Delta S_{trs}/(J\,K^{-1}\,mol^{-1})$.

	Fusion (at T_f)	Vaporization (at T_b)
Ar	14.17 (at 83.8 K)	74.53 (at 87.3 K)
Br_2	39.76 (at 265.9 K)	88.61 (at 332.4 K)
C_6H_6	38.00 (at 278.6 K)	87.19 (at 353.2 K)
CH_3COOH	40.4 (at 289.8 K)	61.9 (at 391.4 K)
CH_3OH	18.03 (at 175.2 K)	104.6 (at 337.2 K)
Cl_2	37.22 (at 172.1 K)	85.38 (at 239.0 K)
H_2	8.38 (at 14.0 K)	44.96 (at 20.38 K)
H_2O	22.00 (at 273.2 K)	109.0 (at 373.2 K)
H_2S	12.67 (at 187.6 K)	87.75 (at 212.0 K)
He	6.0 (at 3.5 K)	19.9 (at 4.22 K)
N_2	11.39 (at 63.2 K)	75.22 (at 77.4 K)
NH_3	28.93 (at 195.4 K)	97.41 (at 239.73 K)
O_2	8.17 (at 54.4 K)	75.63 (at 90.2 K)

Data: AIP

Table 4.2. The molar entropies of vaporization of liquids

	$\Delta H_{vap}/(kJ\,mol^{-1})$	$\theta_b/°C$	$\Delta S_{vap}/(J\,K^{-1}\,mol^{-1})$
Benzene	+30.8	80.1	+87.2
Carbon tetrachloride	+30.00	76.1	+85.9
Cyclohexane	+30.1	80.7	+85.1
Hydrogen sulphide	+18.7	−60.4	+87.9
Methane	+8.18	−161.5	+73.2
Water	+40.7	100.0	+109.1

Table 4.3. Standard Third Law entropies at $\maltese$: see Tables 2.9 and 2.10

Table 4.4. Standard Gibbs functions of formation at $\maltese$: see Tables 2.9 and 2.10

Tables of data

Table 5.2. The fugacity coefficient of nitrogen at 273 K

p/atm	γ	p/atm	γ
1	0.99955	300	1.0055
10	0.9956	400	1.062
50	0.9812	600	1.239
100	0.9703	800	1.495
150	0.9672	1000	1.839
200	0.9721		

Data: LR

Table 6.1. Surface tensions of liquids at 293 K

	γ/(N m^{-1})
Benzene	2.888N2
Carbon tetrachloride	2.70N2
Ethanol	2.28N2
Hexane	1.84N2
Mercury	47.2N2
Methanol	2.26N2
Water	7.275N2
	7.20N2 at 25°C
	5.80N2 at 100°C

Data: KL

Table 7.1. Henry's Law constants for gases at $\overline{\Phi}$, K/Torr

	Water	Benzene
CH_4	3.14P5	4.27P5
CO_2	1.25P6	8.57P4
H_2	5.34P7	2.75P6
N_2	6.51P7	1.79P6
O_2	3.30P7	

Data: F. Daniels and R. A. Alberty, *Physical chemistry*. Wiley, New York (1980).

Table 7.2. Cryoscopic and ebullioscopic constants

	K_f/(K kg mol^{-1})	K_b/(K kg mol^{-1})
Acetic acid	3.90	3.07
Benzene	5.12	2.53
Camphor	40	
Carbon disulphide	3.8	2.37
Carbon tetrachloride	30	4.95
Naphthalene	6.94	5.8
Phenol	7.27	3.04
Water	1.86	0.51

Data: KL

Table 9.1. Giauque functions, $-\Phi_0$/(J K^{-1} mol^{-1})

	$\overline{\Phi}$	500 K	1000 K	1500 K	2000 K	$\{H_m^\ominus(\overline{\Phi})-H_m^\ominus(0)\}$ kJ mol^{-1}
$Br_2(g)$	212.8	230.1	254.4	269.1	279.6	9.728
$C(s)$	2.2	4.85	11.6	17.5	22.5	1.050
$CH_4(g)$	152.5	170.5	199.4	221.1	239	10.029
$Cl_2(g)$	192.2	208.6	231.9	246.2	256.6	9.180
$CO(g)$	168.4	183.5	204.1	216.6	225.9	8.673
$CO_2(g)$	182.3	199.5	226.4	244.7	258.8	9.364
$H_2(g)$	102.2	116.9	137.0	148.9	157.6	8.468
$H_2O(g)$	155.5	172.8	196.7	211.7	223.1	9.908
$HCl(g)$	157.8	172.8	193.1	205.4	214.3	8.640
$N_2(g)$	162.4	177.5	197.9	210.4	219.6	8.669
$NH_3(g)$	159.0	176.9	203.5	221.9	236.6	9.92
$O_2(g)$	176.0	191.1	212.1	225.1	234.7	8.682

Data: LR

Table 9.2. Acidity constants for aqueous solution at $\mathcal{T}$. (a) In order of acid strength

Acid	HA	A$^-$	K_a	pK_a
Hydriodic	HI	I$^-$	10^{11}	-11
Perchloric	HClO$_4$	ClO$_4^-$	10^{10}	-10
Hydrobromic	HBr	Br$^-$	10^9	-9
Hydrochloric	HCl	Cl$^-$	10^7	-7
Sulphuric(1)	H$_2$SO$_4$	HSO$_4^-$	10^2	-2
Hydronium ion	H$_3$O$^+$	H$_2$O	1	0.0
Oxalic(1)	(COOH)$_2$	HOOCCO$_2^-$	5.9×10^{-2}	1.23
Sulphurous	H$_2$SO$_3$	HSO$_3^-$	1.5×10^{-2}	1.81
Sulphuric(2)	HSO$_4^-$	SO$_4^{2-}$	1.2×10^{-2}	1.92
Phosphoric(1)	H$_3$PO$_4$	H$_2$PO$_4^-$	7.5×10^{-3}	2.12
Hydrofluoric	HF	F$^-$	3.5×10^{-4}	3.45
Formic	HCOOH	HCO$_2^-$	1.8×10^{-4}	3.75
Lactic	CH$_3$CH(OH)COOH	CH$_3$CH(OH)CO$_2^-$	1.4×10^{-4}	3.86
Oxalic(2)	HOOCCO$_2^-$	(CO$_2$)$_2^{2-}$	6.5×10^{-5}	4.19
Anilinium ion	C$_6$H$_5$NH$_3^+$	C$_6$H$_5$NH$_2$	2.3×10^{-5}	4.63
Acetic	CH$_3$COOH	CH$_3$CO$_2^-$	1.8×10^{-5}	4.75
Butyric	C$_3$H$_7$COOH	C$_3$H$_7$CO$_2^-$	1.5×10^{-5}	4.82
Proprionic	C$_2$H$_5$COOH	C$_2$H$_5$CO$_2^-$	1.4×10^{-5}	4.87
Pyridinium ion	HC$_5$H$_5$N$^+$	C$_5$H$_5$N	5.6×10^{-6}	5.25
Carbonic(1)	H$_2$CO$_3$	HCO$_3^-$	4.3×10^{-7}	6.37
Hydrosulphuric(1)	H$_2$S	HS$^-$	9.1×10^{-8}	7.04
Phosphoric(2)	H$_2$PO$_4^-$	HPO$_4^{2-}$	6.2×10^{-8}	7.21
Hypochlorous	HClO	ClO$^-$	3.0×10^{-8}	7.53
Hydrazinium ion	NH$_2$NH$_3^+$	NH$_2$NH$_2$	5.9×10^{-9}	8.23
Hypobromous	HBrO	BrO$^-$	2.0×10^{-9}	8.69
Boric(1)	B(OH)$_3$	B(OH)$_4^-$	7.2×10^{-10}	9.14
Ammonium ion	NH$_4^+$	NH$_3$	5.6×10^{-10}	9.25
Hydrocyanic	HCN	CN$^-$	4.9×10^{-10}	9.31
Glycinium ion	NH$_2$CH$_2$COOH	NH$_2$CH$_2$CO$_2^-$	1.7×10^{-10}	9.78
Trimethylammonium ion	(CH$_3$)$_3$NH$^+$	(CH$_3$)$_3$N	1.6×10^{-10}	9.81
Phenol	C$_6$H$_5$OH	C$_6$H$_5$O$^-$	1.3×10^{-10}	9.89
Carbonic(2)	HCO$_3^-$	CO$_3^{2-}$	4.8×10^{-11}	10.32
Hypoiodous	HIO	IO$^-$	2.3×10^{-11}	10.64
Methylammonium ion	CH$_3$NH$_3^+$	CH$_3$NH$_2$	2.2×10^{-11}	10.66
Dimethylammonium ion	(CH$_3$)$_2$NH$_2^+$	(CH$_3$)$_2$NH	1.9×10^{-11}	10.73
Triethylammonium ion	(C$_2$H$_5$)$_3$NH$^+$	(C$_2$H$_5$)$_3$N	1.7×10^{-11}	10.76
Ethylammonium ion	C$_2$H$_5$NH$_3^+$	C$_2$H$_5$NH$_2$	1.6×10^{-11}	10.81
Diethylammonium ion	(C$_2$H$_5$)$_2$NH$_2^+$	(C$_2$H$_5$)$_2$NH	1.0×10^{-11}	10.99
Arsenic(3)	HAsO$_4^{2-}$	AsO$_4^{3-}$	3.0×10^{-12}	11.53
Hydrosulphuric(2)	HS$^-$	S^{2-}	1.1×10^{-12}	11.96
Phosphoric(3)	HPO$_4^{2-}$	PO$_4^{3-}$	2.2×10^{-13}	12.67

Data: Principally HCP

Table 9.2. Acidity constants for aqueous solution at 25°C. (b) In alphabetical order of acid

Acid	HA	A$^-$	K_a	pK_a
Acetic	CH$_3$COOH	CH$_3$CO$_2^-$	1.8×10^{-5}	4.75
Ammonium ion	NH$_4^+$	NH$_3$	5.6×10^{-10}	9.25
Anilinium ion	C$_6$H$_5$NH$_3^+$	C$_6$H$_5$NH$_2$	2.3×10^{-5}	4.63
Arsenic(3)	HAsO$_4^{2-}$	AsO$_4^{3-}$	3.0×10^{-12}	11.53

Table 9.2. (Continued)

Acid	HA	A$^-$	K_a	pK_a
Boric(1)	B(OH)$_3$	B(OH)$_4^-$	7.2×10^{-10}	9.14
Butyric	C$_3$H$_7$COOH	C$_3$H$_7$CO$_2^-$	1.5×10^{-5}	4.82
Carbonic(1)	H$_2$CO$_3$	HCO$_3^-$	4.3×10^{-7}	6.37
Carbonic(2)	HCO$_3^-$	CO$_3^{2-}$	4.8×10^{-11}	10.32
Diethylammonium ion	(C$_2$H$_5$)$_2$NH$_2^+$	(C$_2$H$_5$)$_2$NH	1.0×10^{-11}	10.99
Dimethylammonium ion	(CH$_3$)$_2$NH$_2^+$	(CH$_3$)$_2$NH	1.9×10^{-11}	10.73
Ethylammonium ion	C$_2$H$_5$NH$_3^+$	C$_2$H$_5$NH$_2$	1.6×10^{-11}	10.81
Formic	HCOOH	HCO$_2^-$	1.8×10^{-4}	3.75
Glycinium ion	NH$_2$CH$_2$COOH	NH$_2$CH$_2$CO$_2^-$	1.7×10^{-10}	9.78
Hydrazinium ion	NH$_2$NH$_3^+$	NH$_2$NH$_2$	5.9×10^{-9}	8.23
Hydriodic	HI	I$^-$	10^{11}	-11
Hydrobromic	HBr	Br$^-$	10^9	-9
Hydrochloric	HCl	Cl$^-$	10^7	-7
Hydrocyanic	HCN	CN$^-$	4.9×10^{-10}	9.31
Hydrofluoric	HF	F$^-$	3.5×10^{-4}	3.45
Hydronium ion	H$_3$O$^+$	H$_2$O	1	0.0
Hydrosulphuric(1)	H$_2$S	HS$^-$	9.1×10^{-8}	7.04
Hydrosulphuric(2)	HS$^-$	S^{2-}	1.1×10^{-12}	11.96
Hypobromous	HBrO	BrO$^-$	2.0×10^{-9}	8.69
Hypochlorous	HClO	ClO$^-$	3.0×10^{-8}	7.53
Hypoiodous	HIO	IO$^-$	2.3×10^{-11}	10.64
Lactic	CH$_3$CH(OH)COOH	CH$_3$CH(OH)CO$_2^-$	1.4×10^{-4}	3.86
Methylammonium ion	CH$_3$NH$_3^+$	CH$_3$NH$_2$	2.2×10^{-11}	10.66
Oxalic(1)	(COOH)$_2$	HOOCCO$_2^-$	5.9×10^{-2}	1.23
Oxalic(2)	HOOCCO$_2^-$	(CO$_2$)$_2^{2-}$	6.5×10^{-5}	4.19
Perchloric	HClO$_4$	ClO$_4^-$	10^{10}	-10
Phenol	C$_6$H$_5$OH	C$_6$H$_5$O$^-$	1.3×10^{-10}	9.89
Phosphoric(1)	H$_3$PO$_4$	H$_2$PO$_4^-$	7.5×10^{-3}	2.12
Phosphoric(2)	H$_2$PO$_4^-$	HPO$_4^{2-}$	6.2×10^{-8}	7.21
Phosphoric(3)	HPO$_4^{2-}$	PO$_4^{3-}$	2.2×10^{-13}	12.67
Proprionic	C$_2$H$_5$COOH	C$_2$H$_5$CO$_2^-$	1.4×10^{-5}	4.87
Pyridinium ion	HC$_5$H$_5$N$^+$	C$_5$H$_5$N	5.6×10^{-6}	5.25
Sulphuric(1)	H$_2$SO$_4$	HSO$_4^-$	10^2	-2
Sulphuric(2)	HSO$_4^-$	SO$_4^{2-}$	1.2×10^{-2}	1.92
Sulphurous	H$_2$SO$_3$	HSO$_3^-$	1.5×10^{-2}	1.81
Triethylammonium ion	(C$_2$H$_5$)$_3$NH$^+$	(C$_2$H$_5$)$_3$N	1.7×10^{-11}	10.76
Trimethylammonium ion	(CH$_3$)$_3$NH$^+$	(CH$_3$)$_3$N	1.6×10^{-10}	9.81

Table 10.1. Standard thermodynamic functions of ions in solution at $\overline{\textbf{C}}$. See Table 2.10.

Table 10.2. Relative permittivities (dielectric constants) at $\overline{\textbf{C}}$

Non-polar molecules		Polar molecules	
Methane (at -173°C)	1.70	Water 78.54	
		80.37 at 20°C	
Carbon tetrachloride	2.228	Ammonia 16.9	
		22.4 at -33°C	
Cyclohexane	2.015	Hydrogen sulphide	9.26 at -85°C
Benzene	2.274	Methanol	32.63
		Ethanol	24.30
		Nitrobenzene	34.82

Data: HCP

Table 10.4. Mean activity coefficients in water at $\overline{\Phi}$

$m/m^{\ominus}$	HCl	KCl	$CaCl_2$	H_2SO_4	$LaCl_3$	$In_2(SO_4)_3$
0.001	0.966	0.966	0.888	0.830	0.790	
0.005	0.929	0.927	0.789	0.639	0.636	0.16
0.01	0.905	0.902	0.732	0.544	0.560	0.11
0.05	0.830	0.816	0.584	0.340	0.388	0.035
0.10	0.798	0.770	0.524	0.266	0.356	0.025
0.50	0.769	0.652	0.510	0.155	0.303	0.014
1.00	0.811	0.607	0.725	0.131	0.387	
2.00	1.011	0.577	1.554	0.125	0.954	

Data: RS, HCP, and S. Glasstone, *Introduction to electrochemistry*. Van Nostrand (1942).

Table 10.5. Standard reduction potentials at $\overline{\Phi}$. (a) In electrochemical order

Reduction half-reaction	$E^{\ominus}/V$	Reduction half-reaction	$E^{\ominus}/V$
Strongly oxidizing		$Cu^+ + e^- \rightarrow Cu$	+0.52
$H_4XeO_6 + 2H^+ + 2e^- \rightarrow XeO_3 + 3H_2O$	+3.0	$I_3^- + 2e^- \rightarrow 3I^-$	+0.53
$F_2 + 2e^- \rightarrow 2F^-$	+2.87	$NiOOH + H_2O + e^- \rightarrow Ni(OH)_2 + OH^-$	+0.49
$O_3 + 2H^+ + 2e^- \rightarrow O_2 + H_2O$	+2.07	$Ag_2CrO_4 + 2e^- \rightarrow 2Ag + CrO_4^{2-}$	+0.45
$S_2O_8^{2-} + 2e^- \rightarrow 2SO_4^{2-}$	+2.05	$O_2 + 2H_2O + 4e^- \rightarrow 4OH^-$	+0.40
$Ag^{2+} + e^- \rightarrow Ag^+$	+1.98	$ClO_4^- + H_2O + 2e^- \rightarrow ClO_3^- + 2OH^-$	+0.36
$Co^{3+} + e^- \rightarrow Co^{2+}$	+1.81	$[Fe(CN)_6]^{3-} + e^- \rightarrow [Fe(CN)_6]^{4-}$	+0.36
$H_2O_2 + 2H^+ + 2e^- \rightarrow 2H_2O$	+1.78	$Cu^{2+} + 2e^- \rightarrow Cu$	+0.34
$Au^+ + e^- \rightarrow Au$	+1.69	$Hg_2Cl_2 + 2e^- \rightarrow 2Hg + 2Cl^-$	+0.27
$Pb^{4+} + 2e^- \rightarrow Pb^{2+}$	+1.67	$AgCl + e^- \rightarrow Ag + Cl^-$	+0.22
$2HClO + 2H^+ + 2e^- \rightarrow Cl_2 + 2H_2O$	+1.63	$Bi^{3+} + 3e^- \rightarrow Bi$	+0.20
$Ce^{4+} + e^- \rightarrow Ce^{3+}$	+1.61	$Cu^{2+} + e^- \rightarrow Cu^+$	+0.16
$2HBrO + 2H^+ + 2e^- \rightarrow Br_2 + 2H_2O$	+1.60	$Sn^{4+} + 2e^- \rightarrow Sn^{2+}$	+0.15
$MnO_4^- + 8H^+ + 5e^- \rightarrow Mn^{2+} + 4H_2O$	+1.51	$AgBr + e^- \rightarrow Ag + Br^-$	+0.07
$Mn^{3+} + e^- \rightarrow Mn^{2+}$	+1.51	$Ti^{4+} + e^- \rightarrow Ti^{3+}$	0.00
$Au^{3+} + 3e^- \rightarrow Au$	+1.40	$2H^+ + 2e^- \rightarrow H_2$	0, by definition
$Cl_2 + 2e^- \rightarrow 2Cl^-$	+1.36	$Fe^{3+} + 3e^- \rightarrow Fe$	-0.04
$Cr_2O_7^{2-} + 14H^+ + 6e^- \rightarrow 2Cr^{3+} + 7H_2O$	+1.33	$O_2 + H_2O + 2e^- \rightarrow HO_2^- + OH^-$	-0.08
$O_3 + H_2O + 2e^- \rightarrow O_2 + 2OH^-$	+1.24	$Pb^{2+} + 2e^- \rightarrow Pb$	-0.13
$O_2 + 4H^+ + 4e^- \rightarrow 2H_2O$	+1.23	$In^+ + e^- \rightarrow In$	-0.14
$ClO_4^- + 2H^+ + 2e^- \rightarrow ClO_3^- + H_2O$	+1.23	$Sn^{2+} + 2e^- \rightarrow Sn$	-0.14
$MnO_2 + 4H^+ + 2e^- \rightarrow Mn^{2+} + 2H_2O$	+1.23	$AgI + e^- \rightarrow Ag + I^-$	-0.15
$Br_2 + 2e^- \rightarrow 2Br^-$	+1.09	$Ni^{2+} + 2e^- \rightarrow Ni$	-0.23
$Pu^{4+} + e^- \rightarrow Pu^{3+}$	+0.97	$Co^{2+} + 2e^- \rightarrow Co$	-0.28
$NO_3^- + 4H^+ + 3e^- \rightarrow NO + 2H_2O$	+0.96	$In^{3+} + 3e^- \rightarrow In$	-0.34
$2Hg^{2+} + 2e^- \rightarrow Hg_2^{2+}$	+0.92	$Tl^+ + e^- \rightarrow Tl$	-0.34
$ClO^- + H_2O + 2e^- \rightarrow Cl^- + 2OH^-$	+0.89	$PbSO_4 + 2e^- \rightarrow Pb + SO_4^{2-}$	-0.36
$Hg^{2+} + 2e^- \rightarrow Hg$	+0.86	$Ti^{3+} + e^- \rightarrow Ti^{2+}$	-0.37
$NO_3^- + 2H^+ + e^- \rightarrow NO_2 + H_2O$	+0.80	$In^{2+} + e^- \rightarrow In^+$	-0.40
$Ag^+ + e^- \rightarrow Ag$	+0.80	$Cr^{3+} + e^- \rightarrow Cr^{2+}$	-0.41
$Hg_2^{2+} + 2e^- \rightarrow 2Hg$	+0.79	$Fe^{2+} + 2e^- \rightarrow Fe$	-0.44
$Fe^{3+} + e^- \rightarrow Fe^{2+}$	+0.77	$In^{3+} + 2e^- \rightarrow In^+$	-0.44
$BrO^- + H_2O + 2e^- \rightarrow Br^- + 2OH^-$	+0.76	$S + 2e^- \rightarrow S^{2-}$	-0.48
$Hg_2SO_4 + 2e^- \rightarrow 2Hg + SO_4^{2-}$	+0.62	$In^{3+} + e^- \rightarrow In^{2+}$	-0.49
$MnO_4^{2-} + 2H_2O + 2e^- \rightarrow MnO_2 + 4OH^-$	+0.60	$U^{4+} + e^- \rightarrow U^{3+}$	-0.61
$MnO_4^- + e^- \rightarrow MnO_4^{2-}$	+0.56	$Cr^{3+} + 3e^- \rightarrow Cr$	-0.74
$I_2 + 2e^- \rightarrow 2I^-$	+0.54	$Zn^{2+} + 2e^- \rightarrow Zn$	-0.76

Tables of data

Table 10.5. (Continued)

Reduction half-reaction	$E^{\ominus}/\text{V}$	Reduction half-reaction	$E^{\ominus}/\text{V}$
$Cd(OH)_2 + 2e^- \rightarrow Cd + 2OH^-$	−0.81	$Na^+ + e^- \rightarrow Na$	−2.71
$2H_2O + 2e^- \rightarrow H_2 + 2OH^-$	−0.83	$Ca^{2+} + 2e^- \rightarrow Ca$	−2.87
$Cr^{2+} + 2e^- \rightarrow Cr$	−0.91	$Sr^{2+} + 2e^- \rightarrow Sr$	−2.89
$Mn^{2+} + 2e^- \rightarrow Mn$	−1.18	$Ba^{2+} + 2e^- \rightarrow Ba$	−2.91
$V^{2+} + 2e^- \rightarrow V$	−1.19	$Ra^{2+} + 2e^- \rightarrow Ra$	−2.92
$Ti^{2+} + 2e^- \rightarrow Ti$	−1.63	$Cs^+ + e^- \rightarrow Cs$	−2.92
$Al^{3+} + 3e^- \rightarrow Al$	−1.66	$Rb^+ + e^- \rightarrow Rb$	−2.93
$U^{3+} + 3e^- \rightarrow U$	−1.79	$K^+ + e^- \rightarrow K$	−2.93
$Mg^{2+} + 2e^- \rightarrow Mg$	−2.36	$Li^+ + e^- \rightarrow Li$	−3.05
$Ce^{3+} + 3e^- \rightarrow Ce$	−2.48	**Strongly reducing**	
$La^{3+} + 3e^- \rightarrow La$	−2.52		

Table 10.5. Standard electrode potentials at $\mathcal{T}$. (b) In alphabetical order

Reduction half-reaction	$E^{\ominus}/\text{V}$	Reduction half-reaction	$E^{\ominus}/\text{V}$
$Ag^+ + e^- \rightarrow Ag$	+0.80	$Fe^{3+} + e^- \rightarrow Fe^{2+}$	+0.77
$Ag^{2+} + e^- \rightarrow Ag^+$	+1.98	$[Fe(CN)_6]^{3-} + e^- \rightarrow [Fe(CN)_6]^{4-}$	+0.36
$AgBr + e^- \rightarrow Ag + Br^-$	+0.07	$2H^+ + 2e^- \rightarrow H_2$	0, by definition
$AgCl + e^- \rightarrow Ag + Cl^-$	+0.22	$2H_2O + 2e^- \rightarrow H_2 + 2OH^-$	−0.83
$Ag_2CrO_4 + 2e^- \rightarrow 2Ag + CrO_4^{2-}$	+0.45	$2HBrO + 2H^+ + 2e^- \rightarrow Br_2 + 2H_2O$	+1.60
$AgF + e^- \rightarrow Ag + F^-$	+0.78	$2HClO + 2H^+ + 2e^- \rightarrow Cl_2 + 2H_2O$	+1.63
$AgI + e^- \rightarrow Ag + I^-$	−0.15	$H_2O_2 + 2H^+ + 2e^- \rightarrow 2H_2O$	+1.78
$Al^{3+} + 3e^- \rightarrow Al$	−1.66	$H_4XeO_6 + 2H^+ + 2e^- \rightarrow XeO_3 + 3H_2O$	+3.0
$Au^+ + e^- \rightarrow Au$	+1.69	$Hg_2^{2+} + 2e^- \rightarrow 2Hg$	+0.79
$Au^{3+} + 3e^- \rightarrow Au$	+1.40	$Hg_2Cl_2 + 2e^- \rightarrow 2Hg + 2Cl^-$	+0.27
$Ba^{2+} + 2e^- \rightarrow Ba$	−2.91	$Hg^{2+} + 2e^- \rightarrow Hg$	+0.86
$Be^{2+} + 2e^- \rightarrow Be$	−1.85	$2Hg^{2+} + 2e^- \rightarrow Hg_2^{2+}$	+0.92
$Bi^{3+} + 3e^- \rightarrow Bi$	+0.20	$Hg_2SO_4 + 2e^- \rightarrow 2Hg + SO_4^{2-}$	+0.62
$Br_2 + 2e^- \rightarrow 2Br^-$	+1.09	$I_2 + 2e^- \rightarrow 2I^-$	+0.54
$BrO^- + H_2O + 2e^- \rightarrow Br^- + 2OH^-$	+0.76	$I_3^- + 2e^- \rightarrow 3I^-$	+0.53
$Ca^{2+} + 2e^- \rightarrow Ca$	−2.87	$In^+ + e^- \rightarrow In$	−0.14
$Cd(OH)_2 + 2e^- \rightarrow Cd + 2OH^-$	−0.81	$In^{2+} + e^- \rightarrow In^+$	−0.40
$Cd^{2+} + 2e^- \rightarrow Cd$	−0.40	$In^{3+} + 2e^- \rightarrow In^+$	−0.44
$Ce^{3+} + 3e^- \rightarrow Ce$	−2.48	$In^{3+} + 3e^- \rightarrow In$	−0.34
$Ce^{4+} + e^- \rightarrow Ce^{3+}$	+1.61	$In^{3+} + e^- \rightarrow In^{2+}$	−0.49
$Cl_2 + 2e^- \rightarrow 2Cl^-$	+1.36	$K^+ + e^- \rightarrow K$	−2.93
$ClO^- + H_2O + 2e^- \rightarrow Cl^- + 2OH^-$	+0.89	$La^{3+} + 3e^- \rightarrow La$	−2.52
$ClO_4^- + 2H^+ + 2e^- \rightarrow ClO_3^- + H_2O$	+1.23	$Li^+ + e^- \rightarrow Li$	−3.05
$ClO_4^- + H_2O + 2e^- \rightarrow ClO_3^- + 2OH^-$	+0.36	$Mg^{2+} + 2e^- \rightarrow Mg$	−2.36
$Co^{2+} + 2e^- \rightarrow Co$	−0.28	$Mn^{2+} + 2e^- \rightarrow Mn$	−1.18
$Co^{3+} + e^- \rightarrow Co^{2+}$	+1.81	$Mn^{3+} + e^- \rightarrow Mn^{2+}$	+1.51
$Cr^{2+} + 2e^- \rightarrow Cr$	−0.91	$MnO_2 + 4H^+ + 2e^- \rightarrow Mn^{2+} + 2H_2O$	+1.23
$Cr_2O_7^{2-} + 14H^+ + 6e^- \rightarrow 2Cr^{3+} + 7H_2O$	+1.33	$MnO_4^- + 8H^+ + 5e^- \rightarrow Mn^{2+} + 4H_2O$	+1.51
$Cr^{3+} + 3e^- \rightarrow Cr$	−0.74	$MnO_4^- + e^- \rightarrow MnO_4^{2-}$	+0.56
$Cr^{3+} + e^- \rightarrow Cr^{2+}$	−0.41	$MnO_4^{2-} + 2H_2O + 2e^- \rightarrow MnO_2 + 4OH^-$	+0.60
$Cs^+ + e^- \rightarrow Cs$	−2.92	$Na^+ + e^- \rightarrow Na$	−2.71
$Cu^+ + e^- \rightarrow Cu$	+0.52	$Ni^{2+} + 2e^- \rightarrow Ni$	−0.23
$Cu^{2+} + 2e^- \rightarrow Cu$	+0.34	$NiOOH + H_2O + e^- \rightarrow Ni(OH)_2 + OH^-$	+0.49
$Cu^{2+} + e^- \rightarrow Cu^+$	+0.16	$NO_3^- + 2H^+ + e^- \rightarrow NO_2 + H_2O$	+0.80
$F_2 + 2e^- \rightarrow 2F^-$	+2.87	$NO_3^- + 4H^+ + 3e^- \rightarrow NO + 2H_2O$	+0.96
$Fe^{2+} + 2e^- \rightarrow Fe$	−0.44	$NO_3^- + H_2O + 2e^- \rightarrow NO_2^- + 2OH^-$	+0.10
$Fe^{3+} + 3e^- \rightarrow Fe$	−0.04	$O_2 + 2H_2O + 4e^- \rightarrow 4OH^-$	+0.40

Table 10.5. (Continued)

Reduction half-reaction	$E^{\ominus}/V$	Reduction half-reaction	$E^{\ominus}/V$
$O_2 + 4H^+ + 4e^- \rightarrow 2H_2O$	+1.23	$S_2O_8^{2-} + 2e^- \rightarrow 2SO_4^{2-}$	+2.05
$O_2 + e^- \rightarrow O_2^-$	−0.56	$Sn^{2+} + 2e^- \rightarrow Sn$	−0.14
$O_2 + H_2O + 2e^- \rightarrow HO_2^- + OH^-$	−0.08	$Sn^{4+} + 2e^- \rightarrow Sn^{2+}$	+0.15
$O_3 + 2H^+ + 2e^- \rightarrow O_2 + H_2O$	+2.07	$Sr^{2+} + 2e^- \rightarrow Sr$	−2.89
$O_3 + H_2O + 2e^- \rightarrow O_2 + 2OH^-$	+1.24	$Ti^{2+} + 2e^- \rightarrow Ti$	−1.63
$Pb^{2+} + 2e^- \rightarrow Pb$	−0.13	$Ti^{3+} + e^- \rightarrow Ti^{2+}$	−0.37
$Pb^{4+} + 2e^- \rightarrow Pb^{2+}$	+1.67	$Ti^{4+} + e^- \rightarrow Ti^{3+}$	0.00
$PbSO_4 + 2e^- \rightarrow Pb + SO_4^{2-}$	−0.36	$Tl^+ + e^- \rightarrow Tl$	−0.34
$Pt^{2+} + 2e^- \rightarrow Pt$	+1.20	$U^{3+} + 3e^- \rightarrow U$	−1.79
$Pu^{4+} + e^- \rightarrow Pu^{3+}$	+0.97	$U^{4+} + e^- \rightarrow U^{3+}$	−0.61
$Ra^{2+} + 2e^- \rightarrow Ra$	−2.92	$V^{2+} + 2e^- \rightarrow V$	−1.19
$Rb^+ + e^- \rightarrow Rb$	−2.93	$V^{3+} + e^- \rightarrow V^{2+}$	−0.26
$S + 2e^- \rightarrow S^{2-}$	−0.48	$Zn^{2+} + 2e^- \rightarrow Zn$	−0.76

Table 10.6. Solubility products at $\overline{T}$

Compound	Formula	K_{sp}	pK_{sp}	Compound	Formula	K_{sp}	pK_{sp}
Aluminium hydroxide	$Al(OH)_3$	1.0×10^{-33}	33.00	Magnesium			
Antimony sulphide	Sb_2S_3	1.7×10^{-93}	92.77	ammonium phosphate	$MgNH_4PO_4$	2.5×10^{-13}	12.60
Barium carbonate	$BaCO_3$	8.1×10^{-9}	8.09	carbonate	$MgCO_3$	1.0×10^{-5}	5.00
fluoride	BaF_2	1.7×10^{-6}	5.77	fluoride	MgF_2	6.4×10^{-9}	8.19
sulphate	$BaSO_4$	1.1×10^{-10}	9.96	hydroxide	$Mg(OH)_2$	1.1×10^{-11}	10.96
Bismuth sulphide	Bi_2S_3	1.0×10^{-97}	97.00	Mercury(I) chloride	Hg_2Cl_2	1.3×10^{-18}	17.89
Calcium carbonate	$CaCO_3$	8.7×10^{-9}	8.06	iodide	Hg_2I_2	1.2×10^{-28}	27.92
fluoride	CaF_2	4.0×10^{-11}	10.40	sulphate	Hg_2SO_4	6.6×10^{-7}	6.18
hydroxide	$Ca(OH)_2$	5.5×10^{-6}	5.26	Mercury(II) sulphide	HgS black:	1.6×10^{-52}	51.80
sulphate	$CaSO_4$	2.4×10^{-5}	4.62	red:		1.4×10^{-53}	52.85
Copper(I) bromide	$CuBr$	4.2×10^{-8}	7.38	Nickel(II) hydroxide	$Ni(OH)_2$	6.5×10^{-18}	17.19
chloride	$CuCl$	1.0×10^{-6}	6.00	Silver bromide	$AgBr$	7.7×10^{-13}	12.11
iodide	CuI	5.1×10^{-12}	11.29	carbonate	Ag_2CO_3	6.2×10^{-12}	11.21
sulphide	Cu_2S	2.0×10^{-47}	46.70	chloride	$AgCl$	1.6×10^{-10}	9.80
Copper(II) iodate	$Cu(IO_3)_2$	1.4×10^{-7}	6.85	hydroxide	$AgOH$	1.5×10^{-8}	7.82
oxalate	$Cu(COO)_2$	2.9×10^{-8}	7.54	iodide	AgI	1.5×10^{-16}	15.82
sulphide	CuS	8.5×10^{-45}	44.07	sulphide	Ag_2S	6.3×10^{-51}	50.20
Iron(II) hydroxide	$Fe(OH)_2$	1.6×10^{-14}	13.80	Zinc hydroxide	$Zn(OH)_2$	2.0×10^{-17}	15.70
sulphide	FeS	6.3×10^{-18}	17.20	sulphide	ZnS	1.6×10^{-24}	23.80
Iron(III) hydroxide	$Fe(OH)_3$	2.0×10^{-39}	38.70				
Lead(II) bromide	$PbBr_2$	7.9×10^{-5}	4.10				
chloride	$PbCl_2$	1.6×10^{-5}	5.80				
fluoride	PbF_2	3.7×10^{-8}	7.43				
iodate	$Pb(IO_3)_2$	2.6×10^{-13}	12.59				
iodide	PbI_2	1.4×10^{-8}	7.85				
sulphate	$PbSO_4$	1.6×10^{-8}	7.80				
sulphide	PbS	3.4×10^{-28}	27.47				

Tables of data

Table 12.2. The error function

z	erf (z)	z	erf (z)
0	0	0.50	0.520 50
0.01	0.011 28	0.55	0.563 32
0.02	0.022 56	0.60	0.603 86
0.03	0.033 84	0.65	0.642 03
0.04	0.045 11	0.70	0.677 80
0.05	0.056 37	0.75	0.711 16
0.06	0.067 62	0.80	0.742 10
0.07	0.078 86	0.85	0.770 67
0.08	0.090 08	0.90	0.796 91
0.09	0.101 28	0.95	0.820 89
0.10	0.112 46	1.00	0.842 70
0.15	0.168 00	1.20	0.910 31
0.20	0.222 70	1.40	0.952 28
0.25	0.276 32	1.60	0.976 35
0.30	0.328 63	1.80	0.989 09
0.35	0.379 38	2.00	0.995 32
0.40	0.428 39		
0.45	0.475 48	Data: AS	

Table 13.2. Effective atomic numbers

	H							He
Z	1							2
$1s$:	1							1.69

	Li	Be	B	C	N	O	F	Ne
Z	3	4	5	6	7	8	9	10
$1s$:	2.69	3.68	4.68	5.67	6.66	7.66	8.65	9.64
$2s$:	1.28	1.91	2.58	3.22	3.85	4.49	5.13	5.76
$2p$:			2.42	3.14	3.83	4.45	5.10	5.76

	Na	Mg	Al	Si	P	S	Cl	Ar
	11	12	13	14	15	16	17	18
$1s$:	10.63	11.61	12.59	13.57	14.56	15.54	16.52	17.51
$2s$:	6.57	7.39	8.21	9.02	9.82	10.63	11.43	12.23
$2p$:	6.80	7.83	8.96	9.94	10.96	11.98	12.99	14.01
$3s$:	2.51	3.31	4.12	4.90	5.64	6.37	7.07	7.76
$3d$:			4.07	4.29	4.89	5.48	6.12	6.76

Data: E. Clementi and D. L. Raimondi, *Atomic screening constants from SCF functions.* IBM Research Note NJ-27 (1963).

Table 14.1. Some hybridization schemes

Coordination number	Arrangement	Composition
2	Linear	sp, pd, sd
	Angular	sd
3	Trigonal planar	sp^2, p^2d
	Unsymmetrical planar	spd
	Trigonal pyramidal	pd^2

Table 14.1. (Continued)

Coordination number	Arrangement	Composition
4	Tetrahedral	sp^3, sd^3
	Irregular tetrahedral	spd^2, p^3d, pd^3
	Square planar	p^2d^2, sp^2d
5	Trigonal bipyramidal	sp^3d, spd^3
	Tetragonal pyramidal	sp^2d^2, sd^4, pd^4, p^3d^2
	Pentagonal planar	p^2d^3
6	Octahedral	sp^3d^2
	Trigonal prismatic	spd^4, pd^5
	Trigonal antiprismatic	p^3d^3

Source: H. Eyring, J. Walter, and G. E. Kimball, *Quantum chemistry*. Wiley (1944).

Table 16.2. Properties of diatomic molecules

	$\tilde{v}_0/\text{cm}^{-1}$	B/cm^{-1}	r/pm	$k/(\text{N m}^{-1})$	$D/(\text{kJ mol}^{-1})$
$^1\text{H}_2^+$	2321.8	29.8	106	160.0	255.8
$^1\text{H}_2$	4400.39	60.864	74.138	574.9	432.1
$^2\text{H}_2$	3118.46	30.442	74.154	577.0	439.6
$^1\text{H}^{19}\text{F}$	4138.32	20.956	91.680	965.7	564.4
$^1\text{H}^{35}\text{Cl}$	2990.95	10.593	127.45	516.3	427.7
$^1\text{H}^{81}\text{Br}$	2648.98	8.465	141.44	411.5	362.7
$^1\text{H}^{127}\text{I}$	2308.09	6.511	160.92	313.8	294.9
$^{14}\text{N}_2$	2358.07	1.9987	109.76	2293.8	941.7
$^{16}\text{O}_2$	1580.36	1.4457	120.75	1176.8	493.5
$^{19}\text{F}_2$	891.8	0.8828	141.78	445.1	154.4
$^{35}\text{Cl}_2$	559.71	0.2441	198.75	322.7	239.3

Data: AIP

Table 16.3. Typical vibration wavenumbers, $\tilde{v}/(\text{cm}^{-1})$

C—H stretch	2850–2960	C—Cl stretch	600–800
C—H bend	1340–1465	C—Br stretch	500–600
C—C stretch, bend	700–1250	C—I stretch	500
C=C stretch	1620–1680	CO_3^{2-}	1410–1450
C≡C stretch	2100–2260	NO_3^-	1350–1420
O—H stretch	3590–3650	NO_2^-	1230–1250
H—bonds	3200–3570	SO_4^{2-}	1080–1130
C=O stretch	1640–1780	Silicates	900–1100
C≡N stretch	2215–2275		
N—H stretch	3200–3500		
C—F stretch	1000–1400		

Data: L. J. Bellamy, *The infrared spectra of complex molecules* and *Advances in infrared group frequencies*. Chapman and Hall.

Tables of data

Table 17.1. Colour, frequency, and energy of light

Colour	λ/nm	ν/Hz	$\tilde{\nu}$/cm^{-1}	E/eV	E/(kJ mol^{-1})	E/(kcal mol^{-1})
Infrared	1000	3.00P14	1.00P4	1.24	120	28.6
Red	700	4.28P14	1.43P4	1.77	171	40.8
Orange	620	4.84P14	1.61P4	2.00	193	46.1
Yellow	580	5.17P14	1.72P4	2.14	206	49.3
Green	530	5.66P14	1.89P4	2.34	226	53.9
Blue	470	6.38P14	2.13P4	2.64	254	60.8
Violet	420	7.14P14	2.38P4	2.95	285	68.1
Near ultraviolet	300	1.00P15	3.33P4	4.15	400	95.7
Far ultraviolet	200	1.50P15	5.00P4	6.20	598	143

Data: J. G. Calvert and J. N. Pitts, *Photochemistry*. Wiley, New York (1966).

Table 17.2. Absorption characteristics of some groups and molecules

Group	$\tilde{\nu}_{max}$/cm^{-1}	λ_{max}/nm	ε_{max}/(M^{-1} cm^{-1})	Group	$\tilde{\nu}_{max}$/cm^{-1}	λ_{max}/nm	ε_{max}/(M^{-1} cm^{-1})
C=C	5.5P4	180	2.5P2	—N=N—	2.9P4	350	15
	5.7P4	175	1.7P4		>3.9P4	<260	Strong
	5.9P4	170	1.7P4	Benzene ring	3.9P4	255	2.0P2
	6.2P4	160	1.0P4		5.0P4	200	6.3P3
C=O	3.4P4	295	10		5.5P4	180	1.0P5
	5.4P4	185	Strong	$Cu^{2+}(aq)$	1.2P4	810	10
—NO$_2$	3.6P4	280	10	$[Cu(NH_3)_4]^{2+}$	1.7P4	600	50
	4.8P4	210	1.0P4				

Data: Principally J. G. Calvert and J. N. Pitts, *Photochemistry*. Wiley, New York (1966).

Table 18.1. Nuclear spin properties

Nuclide	Natural abundance %	Spin I	Magnetic moment μ/μ_N	g-value	NMR frequency at 1 T, ν/MHz
^{1}n*		$\frac{1}{2}$	−1.9130	−3.8260	29.167
^{1}H	99.9844	$\frac{1}{2}$	2.792 85	5.5857	42.576
^{2}H	0.0156	1	0.857 45	0.857 45	6.536
^{3}H*		$\frac{1}{2}$	−2.127 65	−4.2553	32.434
^{10}B	19.6	3	1.8005	0.6002	4.574
^{11}B	80.4	$\frac{3}{2}$	2.6884	1.7923	13.660
^{13}C	1.108	$\frac{1}{2}$	0.7023	1.4046	10.705
^{14}N	99.635	1	0.403 56	0.403 56	3.076
^{17}O	0.037	$\frac{5}{2}$	−1.893	−0.7572	5.772
^{19}F	100	$\frac{1}{2}$	2.628 35	5.2567	40.054
^{31}P	100	$\frac{1}{2}$	1.1317	2.2634	17.238
^{33}S	0.74	$\frac{3}{2}$	0.6434	0.4289	3.266
^{35}Cl	75.4	$\frac{3}{2}$	0.8218	0.5479	4.171
^{37}Cl	24.6	$\frac{3}{2}$	0.6841	0.4561	3.472

* Radioactive.
μ is the magnetic moment of the spin state with the largest value of m_I: $\mu = g_I \mu_N I$ and μ_N is the nuclear magneton (see inside front cover).
Data: KL

Table 18.2. Hyperfine coupling constants for atoms, a/mT

Nuclide	Spin	Isotropic coupling	Anisotropic coupling
^{1}H	$\frac{1}{2}$	50.8(1s)	
^{2}H	1	7.8(1s)	
^{13}C	$\frac{1}{2}$	113.0(2s)	6.6(2p)
^{14}N	1	55.2(2s)	4.8(2p)
^{19}F	$\frac{1}{2}$	1720(2s)	108.4(2p)
^{31}P	$\frac{1}{2}$	364(3s)	20.6(3p)
^{35}Cl	$\frac{3}{2}$	168(3s)	10.0(3p)
^{37}Cl	$\frac{3}{2}$	140(3s)	8.4(3p)

Data: P. W. Atkins and M. C. R. Symons, *The structure of inorganic radicals*. Elsevier, Amsterdam (1967).

Table 21.2. Ionic radii†

Li$^+$(4)	Be^{2+}(4)	B^{3+}(4)	N^{3-}	O^{2-}(6)	F$^-$(6)
59	27	12	171	140	133
Na$^+$(6)	Mg^{2+}(6)	Al^{3+}(6)	P^{3-}	S^{2-}(6)	Cl$^-$(6)
102	72	53	212	184	181
K$^+$(6)	Ca^{2+}(6)	Ga^{3+}(6)	As^{3-}	Se^{2-}(6)	Br$^-$(6)
138	100	62	222	198	196
Rb$^+$(6)	Sr^{2+}(6)	In^{3+}(6)	Sb^{3-}	Te^{2-}(6)	I$^-$(6)
149	116	79		221	220
Cs$^+$(6)	Ba^{2+}(6)	Tl^{3+}(6)			
170	136	88			

d-block elements (high-spin ions)

Sc^{3+}(6)	Ti^{4+}(6)	Cr^{3+}(6)	Mn^{3+}(6)	Fe^{2+}(6)	Co^{3+}(6)	Cu^{2+}(6)	Zn^{2+}(6)
73	60	61	65	63	61	73	75

† Numbers in parentheses are the coordination numbers of the ions. Values for ions without a coordination number stated are estimates.
Data: R. D. Shannon and C. T. Prewitt, *Acta Cryst.*, **B25**, 925 (1969).

Table 22.1. Dipole moments, polarizabilities, and polarizability volumes

	$\mu/(\text{C m})$	μ/D	α'/cm^3	$\alpha/(\text{J}^{-1}\,\text{C}^2\,\text{m}^2)$
Ar	0	0	1.66N24	1.85N40
C$_2$H$_5$OH	5.64N30	1.69		
C$_6$H$_5$CH$_3$	1.20N30	0.36		
C$_6$H$_6$	0	0	1.04N23	1.16N39
CCl$_4$	0	0	1.05N23	1.17N39
CH$_2$Cl$_2$	5.24N30	1.57	6.80N24	7.57N40
CH$_3$Cl	6.24N30	1.87	4.53N24	5.04N40
CH$_3$OH	5.70N30	1.71	3.23N24	3.59N40
CH$_4$	0	0	2.60N24	2.89N40
CHCl$_3$	3.37N30	1.01	8.50N24	9.46N40

Tables of data

Table 22.1. (Continued)

	$\mu/(\text{C m})$	μ/D	α'/cm^3	$\alpha/(\text{J}^{-1}\,\text{C}^2\,\text{m}^2)$
CO	3.90N31	0.117	1.98N24	2.20N40
CO_2	0	0	2.63N24	2.93N40
H_2	0	0	8.19N25	9.11N41
H_2O	6.17N30	1.85	1.48N24	1.65N40
HBr	2.67N30	0.80	3.61N24	4.01N40
HCl	3.60N30	1.08	2.63N24	2.93N40
He	0	0	2.0N25	2.2N41
HF	6.37N30	1.91	5.1N25	5.7N41
HI	1.40N30	0.42	5.45N24	6.06N40
N_2	0	0	1.77N24	1.97N40
NH_3	4.90N30	1.47	2.22N24	2.47N40
$o\text{-}C_6H_4(CH_3)_2$	2.07N30	0.62		

Data: HCP and C. J. F. Böttcher and P. Bordewijk, *Theory of electric polarization.*
Elsevier, Amsterdam (1978).

Table 22.2. Refractive indices relative to air at 20°C

	434 nm	589 nm	656 nm
Benzene	1.5236	1.5012	1.4965
Carbon tetrachloride	1.4729	1.4676	1.4579
Carbon disulphide	1.6748	1.6276	1.6182
Ethanol	1.3700	1.3618	1.3605
KCl(s)	1.5050	1.4904	1.4973
KI(s)	1.7035	1.6664	1.6581
Methanol	1.3362	1.3290	1.3277
Methylbenzene	1.5170	1.4955	1.4911
Water	1.3404	1.3330	1.3312

Data: AIP

Table 22.3. Molar refractivities at 589 nm, $R_m/(\text{cm}^3\,\text{mol}^{-1})$

C—H 1.65	C—C 1.20	C—O 1.41	C≡N 4.69	O—H 1.85
	C≡C 2.79	C═O 3.34		
	C≡C 4.79			

					He 0.5
Li^+ 0.07	Be^{2+} 0.20		O^{2-} 7	F^- 2.65	Ne 1.00
Na^+ 0.46	Mg^{2+} 0.24	Al^{3+} 0.17		Cl^- 9.30	Ar 4.14
K^+ 2.12	Ca^{2+} 1.19			Br^- 12.12	Kr 6.26
Rb^+ 3.57				I^- 18.07	Xe 10.16

Data: E. A. Moelwyn-Hughes, *Physical chemistry.* Pergamon (1961).

Table 22.5. Lennard–Jones (12,6)-potential parameters

	$(\varepsilon/k)/K$	σ/pm
Ar	124	342
Br_2	520	427
C_2H_4	205	423
C_6H_6	440	527
CCl_4	327	588
CH_4	137	382
Cl_2	357	412
CO_2	190	400
H_2	33.3	297
He	10.22	258
N_2	91.5	368
Ne	35.7	279
O_2	113	343
Xe	229	406

Data: J. O. Hirschfelder, C. F. Curtiss, and R. B. Bird, *Molecular theory of gases and liquids*. Wiley, New York (1954).

Table 22.6. Viscosities of liquids at 298 K, $\eta/(\text{kg m}^{-1}\,\text{s}^{-1})$

Benzene	6.01N4
Carbon tetrachloride	8.80N4
Ethanol	1.06N3
Mercury	1.55N3
Methanol	5.53N4
Pentane	2.24N4
Sulphuric acid	2.7N2
Water†	8.91N4

† The viscosity of water over its entire liquid range is represented with less than 1 per cent error by the expression

$$\lg(\eta_{20}/\eta) = A/B,$$
$$A = 1.37023(t-20) + 8.36$$
$$\times 10^{-4}(t-20)^2$$
$$B = 109 + t \qquad t = \theta/°\text{C}$$

Convert $\text{kg m}^{-1}\,\text{s}^{-1}$ to centipoise (cP) by multiplying by 10^3 (so that $\eta \approx 1$ cP for water).
Data: AIP, KL.

Table 22.7. Magnetic susceptibilities at $\overline{\mathcal{T}}$

	χ	$\chi_m/(\text{cm}^3\,\text{mol}^{-1})$
Water	−9.0N5	−1.6N3
Benzene	−7.2N6	−6.4N4
Cyclohexane	−7.9N6	−8.5N4
Carbon tetrachloride	−8.9N6	−8.4N4
NaCl(s)	−1.39N5	−3.75N4
Cu(s)	−9.6N5	−6.8N4
S(s)	−1.29N5	−2.0N4
Hg(l)	−2.85N5	−4.2N4
$CuSO_4\cdot5H_2O(s)$	+1.76N4	+1.92N2
$MnSO_4\cdot4H_2O(s)$	+2.64N3	+2.79N1
$NiSO_4\cdot7H_2O(s)$	+4.16N4	+6.00N2
$FeSO_4(NH_4)_2SO_4\cdot6H_2O(s)$	+7.55N4	+1.51N1
Al(s)	+2.2N5	+2.2N4
Pt(s)	+2.62N4	+2.28N3
Na(s)	+7.3N6	+1.7N4
K(s)	+5.6N6	+2.5N4

Data: KL and $\chi_m = \chi M/\rho$.

Table 23.1. Frictional coefficients and molecular geometry

Major axis / Minor axis	Prolate	Oblate
2	1.04	1.04
3	1.11	1.10
4	1.18	1.17
5	1.25	1.22
6	1.31	1.28
7	1.38	1.33
8	1.43	1.37
9	1.49	1.42
10	1.54	1.46
50	2.95	2.38
100	4.07	2.97

Data: K. E. Van Holde, *Physical biochemistry*. Prentice Hall, Englewood Cliffs (1971).

Sphere; radius a, $c = a$ f_0

Prolate ellipsoid; major axis $2a$, minor axis $2b$, $c = (ab^2)^{1/3}$
$$\left\{\frac{(1-b^2/a^2)^{1/2}}{(b/a)^{2/3}\ln\{[1+(1-b^2/a^2)^{1/2}]/(b/a)\}}\right\}f_0$$

Oblate ellipsoid; major axis $2a$, minor axis $2b$, $c = (a^2b)^{1/3}$
$$\left\{\frac{(a^2/b^2-1)^{1/2}}{(a/b)^{2/3}\arctan[(a^2/b^2-1)^{1/2}]}\right\}f_0$$

Long rod; length l, radius a, $c = (3a^2l/4)^{1/3}$
$$\left\{\frac{(1/2a)^{2/3}}{(3/2)^{1/3}\{2\ln(l/a)-0.11\}}\right\}f_0$$

In each case $f_0 = 6\pi\eta c$ with the appropriate value of c.

Tables of data

Table 23.2. Diffusion coefficients of macromolecules in water at 20°C

	$M/(\text{kg mol}^{-1})$	$D/(\text{cm}^2 \text{ s}^{-1})$
Sucrose	0.342	4.586N6
Ribonuclease	13.7	1.19N6
Lysozyme	14.1	1.04N6
Serum albumin	65	5.94N7
Haemoglobin	68	6.9N7
Urease	480	3.46N7
Collagen	345	6.9N8
Myosin	493	1.16N7

Data: C. Tanford, *Physical chemistry of macromolecules.* Wiley, New York (1961).

Table 23.3. Intrinsic viscosity

Macromolecule	Solvent	$\theta/°C$	$K/(\text{cm}^3 \text{ g}^{-1})$	a
Polystyrene	Benzene	25	9.5N3	0.74
	Cyclohexane	34†	8.1N2	0.50
Polyisobutylene	Benzene	23†	8.3N2	0.50
	Cyclohexane	30	2.6N	0.70
Amylose	0.33 M KCl(*aq*)	25†	1.13N1	0.50
Various protein‡	Guanidine hydrochloride + $HSCH_2CH_2OH$		7.16N3	0.66

† The θ temperature.
‡ Use $[\eta] = KN^a$, N the number of amino acid residues.
Data: K. E. Van Holde, *Physical biochemistry.* Prentice Hall, Englewood Cliffs (1971).

Table 23.4. Radius of gyration of some macromolecules

	$M/(\text{kg mol}^{-1})$	R_g/nm
Serum albumin	66	2.98
Myosin	493	46.8
Polystyrene	3.2P3	49.4 (in poor solvent)
DNA	4P3	117.0
Tobacco mosaic virus	3.9P4	92.4

Data: C. Tanford, *Physical chemistry of macromolecules.* Wiley, New York (1961).

Table 24.1. Mean speeds at 25°C, $\bar{c}/(\text{m s}^{-1})$

Air	466
Ar	398
C_6H_6	284
Cl_2	298
CO_2	379
H_2	1770
H_2O	592
He	1256
Hg	177
N_2	475
NH_3	609
O_2	444

Data: Calculated using eqn 7b of Section 24.2 in the form $\bar{c}/(\text{m s}^{-1}) = 145.51(T/\text{K})^{\frac{1}{2}}M^{\frac{1}{2}} = 2512.48M^{\frac{1}{2}}$ at $\bar{T}$.

Table 24.2. Collision cross-section, σ/nm^2

Ar	0.36
C_2H_4	0.64
C_6H_6	0.88
CH_4	0.46
Cl_2	0.93
CO_2	0.52
H_2	0.27
He	0.21
N_2	0.43
Ne	0.24
O_2	0.40
SO_2	0.58

Data: KL

Table 24.4. Transport properties of gases at 1 atm

	$\kappa/(\text{mJ cm}^{-2}\,\text{s}^{-1}\,(\text{K cm}^{-1})^{-1})$ 273 K	$\eta/\mu\text{P}$ 273 K	$\eta/\mu\text{P}$ 293 K
Air	0.241	173	182
Ar	0.163	210	223
C_2H_4	0.164	97	103
CH_4	0.302	103	110
Cl_2	0.79	123	132
CO_2	0.145	136	147
H_2	1.682	84	88
He	1.442	187	196
Kr	0.087	234	250
N_2	0.240	166	176
Ne	0.465	298	313
O_2	0.245	195	204
Xe	0.052	212	228

Data: KL

Table 25.1. Limiting ionic conductivities in water at 25°C, $\lambda/(\text{S cm}^2\,\text{mol}^{-1})$

Cations		Anions	
Ba^{2+}	127.2	Br^-	78.1
Ca^{2+}	119.0	$CH_3CO_2^-$	40.9
Cs^+	77.2	Cl^-	76.35
Cu^{2+}	107.2	ClO_4^-	67.3
H^+	349.6	CO_3^{2-}	138.6
K^+	73.50	$(CO_2)_2^{2-}$	148.2
Li^+	38.7	F^-	55.4
Mg^{2+}	106.0	$[Fe(CN)_6]^{3-}$	302.7
Na^+	50.10	$[Fe(CN)_6]^{4-}$	442.0
$[N(C_2H_5)_4]^+$	32.6	I^-	76.8
$[N(CH_3)_4]^+$	44.9	NO_3^-	71.46
NH_4^+	73.5	OH^-	199.1
Rb^+	77.8	SO_4^{2-}	160.0
Sr^{2+}	118.9		
Zn^{2+}	105.6		

Data: KL, RS.

Table 25.2. Ionic mobilities in water at 25°C, $u/(\text{cm}^2\,\text{s}^{-1}\,\text{V}^{-1})$

Cations		Anions	
Ag^+	6.42N4	Br^-	8.09N4
Ca^{2+}	6.17N4	$CH_3CO_2^-$	4.24N4
Cu^{2+}	5.56N4	Cl^-	7.91N4
H^+	3.623N3	CO_3^{2-}	7.46N4
K^+	7.62N4	F^-	5.70N4
Li^+	4.01N4	$[Fe(CN)_6]^{3-}$	1.05N3
Na^+	5.19N4	$[Fe(CN)_6]^{4-}$	1.14N3
NH_4^+	7.63N4	I^-	7.96N4
$[N(CH_3)_4]^+$	4.65N4	NO_3^-	7.40N4
Rb^+	7.92N4	OH^-	2.064N3
Zn^{2+}	5.47N4	SO_4^{2-}	8.29N4

Data: Principally Table 25.1 and $u = \lambda/zF$.

Table 25.3. Debye–Hückel–Onsager coefficients for (1,1)-electrolytes at 25°C

Solvent	$A/(\text{S cm}^2\,\text{mol}^{-1}\,\text{M}^{-1/2})$	$B/\text{M}^{-1/2}$
Acetone	32.8	1.63
Acetonitrile	22.9	0.716
Ethanol	89.7	1.83
Methanol	156.1	0.923
Nitrobenzene	44.2	0.776
Nitromethane	125.1	0.708
Water	60.20	0.229

Data: J. O'M. Bockris and A. K. N. Reddy, *Modern electrochemistry*. Plenum, New York (1970).

Table 25.4. Diffusion coefficients at 25°C, $D/(\text{cm}^2\,\text{s}^{-1})$

Molecules in liquids

I_2 in hexane	4.05N5	H_2 in $CCl_4(l)$	9.75N5
in benzene	2.13N5	N_2 in $CCl_4(l)$	3.42N5
CCl_4 in heptane	3.17N5	O_2 in $CCl_4(l)$	3.82N5
Glycine in water	1.055N5	Ar in $CCl_4(l)$	3.63N5
Dextrose in water	6.73N6	CH_4 in $CCl_4(l)$	2.89N5
Sucrose in water	5.216N6	H_2O in water	2.26N5
		CH_3OH in water	1.58N5
		C_2H_5OH in water	1.24N5

Ions in water

K^+	1.96N5	Br^-	2.08N5
H^+	9.31N5	Cl^-	2.03N5
Li^+	1.03N5	F^-	1.46N5
Na^+	1.33N5	I^-	2.05N5
		OH^-	5.30N5

Data: AIP and (for the ions) $\lambda = zuF$ in conjunction with Table 25.2.

Tables of data

Table 26.1. Kinetic data for first-order reactions

	Phase	$\theta/°C$	k/s^{-1}	$t_{1/2}$
$2N_2O_5 \rightarrow 4NO_2 + O_2$	g	25	3.38N5	2.85 h
	$HNO_3(l)$	25	1.47N6	65.5 h
	$Br_2(l)$	25	4.27N5	2.25 h
$C_2H_6 \rightarrow 2CH_3$	g	700	5.36N4	21.2 min
Cyclopropane $\rightarrow$ propene	g	500	6.17N4	17.2 min

g: High pressure gas-phase limit.
Data: Principally K. J. Laidler, *Chemical kinetics*. Harper & Row, New York (1987); M. J. Pilling, *Reaction kinetics*. Clarendon Press, Oxford (1974); J. Nicholas, *Chemical kinetics*. Harper & Row, New York (1976).

Table 26.2. Kinetic data for second-order reactions

	Phase	$\theta/°C$	$k/(M^{-1}\,s^{-1})$
$2NOBr \rightarrow 2NO + Br_2$	g	10	0.80
$2NO_2 \rightarrow 2NO + O_2$	g	300	0.54
$H_2 + I_2 \rightarrow 2HI$	g	400	2.42N2
$D_2 + HCl \rightarrow DH + DCl$	g	600	0.141
$2I \rightarrow I_2$	g	23	7P9
	hexane	50	1.8P10
$CH_3Cl + CH_3O^-$	methanol	20	2.29N6
$CH_3Br + CH_3O^-$	methanol	20	9.23N6
$H^+ + OH^- \rightarrow H_2O$	water	25	1.5P11

Data: Principally K. J. Laidler, *Chemical kinetics*. Harper & Row, New York (1987); M. J. Pilling, *Reaction kinetics*. Clarendon Press, Oxford (1974); J. Nicholas, *Chemical kinetics*. Harper & Row, New York (1976).

Table 26.3. Arrhenius parameters

First-order reactions	A/s^{-1}	$E_a/(kJ\,mol^{-1})$
Cyclopropane $\rightarrow$ propene	1.58P15	272
$CH_3NC \rightarrow CH_3CN$	3.98P13	160
cis-CHD=CHD $\rightarrow$ *trans*-CHD=CHD	3.16P12	256
Cyclobutane $\rightarrow 2C_2H_4$	3.98P15	261
$C_2H_5I \rightarrow C_2H_4 + HI$	2.51P13	209
$C_2H_6 \rightarrow 2CH_3$	2.51P17	384
$2N_2O_5 \rightarrow 4NO_2 + O_2$	4.94P13	103
$N_2O \rightarrow N_2 + O$	7.94P11	250
$C_2H_5 \rightarrow C_2H_4 + H$	1.0P13	167

Second-order, gas phase	$A/(M^{-1}\,s^{-1})$	$E_a/(kJ\,mol^{-1})$
$O + N_2 \rightarrow NO + H$	1P11	315
$OH + H_2 \rightarrow H_2O + H$	8910	42
$Cl + H_2 \rightarrow HCl + H$	8P10	23
$2CH_3 \rightarrow C_2H_6$	2P10	ca. 0
$NO + Cl_2 \rightarrow NOCl + Cl$	4.0P9	85
$SO + O_2 \rightarrow SO_2 + O$	3P8	27
$CH_3 + C_2H_6 \rightarrow CH_4 + C_2H_5$	2P8	44
$C_6H_5 + H_2 \rightarrow C_6H_6 + H$	1P8	ca. 25

Second order, solution	$A/(M^{-1}\,s^{-1})$	$E_a/(kJ\,mol^{-1})$
$C_2H_5ONa + CH_3I$ in ethanol	2.42P11	81.6
$C_2H_5Br + OH^-$ in water	4.30P11	89.5
$C_2H_5I + C_2H_5O^-$ in ethanol	1.49P11	86.6
$CH_3I + C_2H_5O^-$ in ethanol	2.42P11	81.6
$C_2H_5Br + OH^-$ in ethanol	4.30P11	89.5
$CO_2 + OH^-$ in water	1.5P10	38
$CH_3I + S_2O_3^{2-}$ in water	2.19P12	78.7

Table 26.3. (Continued)

Second order, solutions	$A/(\text{M}^{-1}\,\text{s}^{-1})$	$E_a/(\text{kJ mol}^{-1})$
Sucrose + H_2O in acidic water	1.50P15	107.9
$(CH_3)_3CCl$ solvolysis		
in water	7.1P16	100
in methanol	2.3P13	107
in ethanol	3.0P13	112
in acetic acid	4.3P13	111
in chloroform	1.4P4	45
$C_6H_5NH_2 + C_6H_5COCH_2Br$		
in benzene	91	34

Data: Principally J. Nicholas, *Chemical kinetics*. Harper and Row, New York (1976) and A. A. Frost and R. G. Pearson, *Kinetics and mechanism*. Wiley, New York (1961).

Table 28.1. Arrhenius parameters for gas phase reactions

	$A/(\text{M}^{-1}\,\text{s}^{-1})$		$E_a/(\text{kJ mol}^{-1})$	P
	Experiment	Theory		
$2NOCl \rightarrow 2NO + Cl_2$	9.4P9	5.9P10	102.0	0.16
$2NO_2 \rightarrow 2NO + O_2$	2.0P9	4.0P10	111.0	5.0N2
$2ClO \rightarrow Cl_2 + O_2$	6.3P7	2.5P10	0.0	2.5N3
$H_2 + C_2H_4 \rightarrow C_2H_6$	1.24P6	7.3P11	180	1.7N6
$K + Br_2 \rightarrow KBr + Br$	1.0P12	2.1P11	0.0	4.8

Data: Principally M. J. Pilling, *Reaction kinetics*. Clarendon Press, Oxford (1974).

Table 28.2. Arrhenius parameters for reactions in solution. See Table 26.3

Table 29.1. Maximum observed enthalpies of physisorption, $\Delta H_{\text{ad}}^{\ominus}/(\text{kJ mol}^{-1})$

C_2H_2	−38
C_2H_4	−34
CH_4	−21
Cl_2	−36
CO	−25
CO_2	−25
H_2	−84
H_2O	−59
N_2	−21
NH_3	−38
O_2	−21

Data: D. O. Haywood and B. M. W. Trapnell, *Chemisorption*. Butterworth (1964).

Tables of data

Table 29.2. Enthalpies of chemisorption, $\Delta H_{ad}^{\ominus}/(\text{kJ mol}^{-1})$

Adsorbate	Adsorbent (substrate)											
	Ti	Ta	Nb	W	Cr	Mo	Mn	Fe	Co	Ni	Rh	Pt
H_2		188			188	167	71	134			117	
N_2		586						293				
O_2						720					494	293
CO	640							192	176			
CO_2	682	703	552	456	339	372	222	225	146	184		
NH_3				301				188		155		
C_2H_4		577		427	427			285		243	209	

Data: D. O. Haywood and B. M. W. Trapnell, *Chemisorption*. Butterworth (1964).

Table 29.3. Activation energies of catalysed reactions

	Catalyst	$E_a/(\text{kJ mol}^{-1})$
$2HI \rightarrow H_2 + I_2$	None	184
	Au(s)	105
	Pt(s)	59
$2NH_3 \rightarrow N_2 + 3H_2$	None	350
	W(s)	162
$2N_2O \rightarrow 2N_2 + O_2$	None	245
	Au(s)	121
	Pt(s)	134
$(C_2H_5)_2O$ pyrolysis	None	224
	$I_2(g)$	144

Data: G. C. Bond, *Heterogeneous catalysis*. Clarendon Press, Oxford (1986).

Table 30.1. Exchange current densities and transfer coefficients at 25°C

Reaction	Electrode	$j_0/(\text{A cm}^{-2})$	α
$2H^+ + 2e^- \rightarrow H_2$	Pt	7.9N4	
	Cu	1N6	
	Ni	6.3N6	0.58
	Hg	7.9N13	0.50
	Pb	5.0N12	
$Fe^{3+} + e^- \rightarrow Fe^{2+}$	Pt	2.5N3	0.58
$Ce^{4+} + e^- \rightarrow Ce^{3+}$	Pt	4.0N5	0.75

Data: Principally J. O'M. Bockris and A. K. N. Reddy, *Modern electrochemistry*. Plenum, New York (1970).

Table 30.2. Half-wave potentials at 25°C, $E_{\frac{1}{2}}/V$

Cd^{2+}	−0.60
Co^{2+}	−1.4
Cr^{3+}	−0.91
Cu^{2+}	+0.04
Fe^{2+}	−1.3
Fe^{3+}	0.0
Tl^+	−0.46
Zn^{2+}	−1.0

Values refer to ions in 0.1 M KCl(aq), relative to the calomel electrode.

Character tables

$C_{2v}, 2mm$	E	C_2	σ_v	σ_v'	$h = 4$	
A_1	1	1	1	1	z, z^2, x^2, y^2	
A_2	1	1	-1	-1	xy	R_z
B_1	1	-1	1	-1	x, xz	R_y
B_2	1	-1	-1	1	y, yz	R_x

$C_{3v}, 3m$	E	$2C_3$	$3\sigma_v$	$h = 6$	
A_1	1	1	1	$z, z^2, x^2 + y^2$	
A_2	1	1	-1		R_z
E	2	-1	0	$(x, y), (xy, x^2 - y^2)(xz, yz)$	(R_x, R_y)

$C_{4v}, 4mm$	E	C_2	$2C_4$	$2\sigma_v$	$2\sigma_d$	$h = 8$	
A_1	1	1	1	1	1	$z, z^2, x^2 + y^2$	
A_2	1	1	1	-1	-1		R_z
B_1	1	1	-1	1	-1	$x^2 - y^2$	
B_2	1	1	-1	-1	1	xy	
E	2	-2	0	0	0	$(x, y), (xz, yz)$	(R_x, R_y)

C_{5v}	E	$2C_5$	$2C_5^2$	$5\sigma_v$	$h = 10, \alpha = 72°$	
A_1	1	1	1	1	$z, z^2, x^2 + y^2$	
A_2	1	1	1	-1		R_z
E_1	2	$2\cos\alpha$	$2\cos 2\alpha$	0	$(x, y), (xz, yz)$	(R_x, R_y)
E_2	2	$2\cos 2\alpha$	$2\cos\alpha$	0	$(xy, x^2 - y^2)$	

$C_{6v}, 6mm$	E	C_2	$2C_3$	$2C_6$	$3\sigma_d$	$3\sigma_v$	$h = 12$	
A_1	1	1	1	1	1	1	$z, z^2, x^2 + y^2$	
A_2	1	1	1	1	-1	-1		R_z
B_1	1	-1	1	-1	-1	1		
B_2	1	-1	1	-1	1	-1		
E_1	2	-2	-1	1	0	0	$(x, y), (xz, yz)$	(R_x, R_y)
E_2	2	2	-1	-1	0	0	$(xy, x^2 - y^2)$	

$C_{\infty v}$	E	C_2	$2C_\phi$	σ_v	$h = \infty$	
$A_1(\Sigma^+)$	1	1	1	1	$z, z^2, x^2 + y^2$	
$A_2(\Sigma^-)$	1	1	1	-1		R_z
$E_1(\Pi)$	2	-2	$2\cos\phi$	0	$(x, y), (xz, yz)$	(R_x, R_y)
$E_2(\Delta)$	2	2	$2\cos 2\phi$	0	$(xy, x^2 - y^2)$	
$\ldots$	.	.	.	.	$\ldots$	

Character tables

D_3, 32	E	$2C_3$	$2C_2'$	$h = 5$	
A_1	1	1	1	$z^2, x^2 + y^2$	
A_2	1	1	-1	z	R_z
E	2	-1	0	$(x, y), (xz, yz)(xy, x^2 - y^2)$	(R_x, R_y)

D_2, 222	E	C_2^z	C_2^y	C_2^x	$h = 4$	
A	1	1	1	1	x^2, y^2, z^2	
B_1	1	1	-1	-1	z, xy	R_z
B_2	1	-1	1	-1	y, xz	R_y
B_3	1	-1	-1	1	x, yz	R_x

D_4, 422	E	C_2	$2C_4$	$2C_2'$	$2C_2''$	$h = 8$	
A_1	1	1	1	1	1	$z^2, x^2 + y^2$	
A_2	1	1	1	-1	-1	z	R_z
B_1	1	1	-1	1	-1	$x^2 - y^2$	
B_2	1	1	-1	-1	1	xy	
E	2	-2	0	0	0	$(x, y), (xz, yz)$	(R_x, R_y)

D_{3h}, $\bar{6}m2$	E	σ_h	$2C_3$	$2S_3$	$3C_2'$	$3\sigma_v$	$h = 12$	
A_1'	1	1	1	1	1	1	$z^2, x^2 + y^2$	
A_2'	1	1	1	1	-1	-1		R_z
A_1''	1	-1	1	-1	1	-1		
A_2''	1	-1	1	-1	-1	1	z	
E'	2	2	-1	-1	0	0	$(x, y), (xy, x^2 - y^2)$	
E''	2	-2	-1	1	0	0	(xz, yz)	(R_x, R_y)

$D_{\infty h}$	E	$2C_\phi$	C_2'	i	$2iC_\phi$	iC_2'	$h = \infty$	
$A_{1g}(\Sigma_g^+)$	1	1	1	1	1	1	$z^2, x^2 + y^2$	
$A_{1u}(\Sigma_u^+)$	1	1	1	-1	-1	-1		
$A_{2g}(\Sigma_g^-)$	1	1	-1	1	1	-1		R_z
$A_{2u}(\Sigma_u^-)$	1	1	-1	-1	1	1	z	
$E_{1g}(\Pi_g)$	2	$2\cos\phi$	0	0	$2\cos\phi$	0	(xz, yz)	(R_x, R_y)
$E_{1u}(\Pi_u)$	2	$2\cos\phi$	0	0	$-2\cos\phi$	0	(x, y)	
$E_{2g}(\Delta_g)$	2	$2\cos 2\phi$	0	0	$2\cos 2\phi$	0	$(xy, x^2 - y^2)$	
$E_{2u}(\Delta_u)$	2	$2\cos 2\phi$	0	0	$-2\cos 2\phi$	0		

T_d, $\bar{4}3m$	E	$8C_3$	$3C_2$	$6\sigma_d$	$6S_4$	$h = 24$	
A_1	1	1	1	1	1	$x^2 + y^2 + z^2$	
A_2	1	1	1	-1	-1		
E	2	-1	2	0	0	$(x^2 + y^2 - 2z^2, x^2 - y^2)$	
T_1	3	0	-1	-1	1		(R_x, R_y, R_z)
T_2	3	0	-1	1	-1	(x, y, z) (xy, xz, yz)	

$O, 432$	E	$8C_3$	$3C_2$	$6C_2'$	$6C_4$	$h = 24$	
A_1	1	1	1	1	1	$x^2 + y^2 + z^2$	
A_2	1	1	1	-1	-1		
E	2	-1	2	0	0	$(x^2 - y^2, x^2 + y^2 - 2z^2)$	
T_1	3	0	-1	-1	1	(x, y, z)	(R_x, R_y, R_z)
T_2	3	0	-1	1	-1	(xy, yz, zx)	

$O_h, m3m$	E	$8C_3$	$6C_2$	$6C_4$	$3C_2$	I	$6S_4$	$8S_6$	$3\sigma_h$	$6\sigma_d$	$h = 48$	
A_{1g}	1	1	1	1	1	1	1	1	1	1	$x^2 + y^2 + z^2$	
A_{2g}	1	1	-1	-1	1	1	-1	1	1	-1		
E_g	2	-1	0	0	2	2	0	-1	2	0	$(x^2 + y^2 - 2z^2, x^2 - y^2)$	
T_{1g}	3	0	-1	1	-1	3	1	0	-1	-1		(R_x, R_y, R_z)
T_{2g}	3	0	1	-1	-1	3	-1	0	-1	1	(xz, yz, xy)	
A_{1u}	1	1	1	1	1	-1	-1	-1	-1	-1		
A_{2u}	1	1	-1	-1	1	-1	1	-1	-1	1		
E_u	2	-1	0	0	2	-2	0	1	-2	0		
T_{1u}	3	0	-1	1	-1	-3	-1	0	1	1	(x, y, z)	
T_{2u}	3	0	1	-1	-1	-3	1	0	1	-1		

Copyright acknowledgments

The permission of the respective copyright holders to reproduce tables of data and extracts from them is gratefully acknowledged as follows: Butterworths (29.1, 29.2 [© 1964]), Professor J. G. Calvert (17.1, 17.2 [© 1966], Dr J. Emsley (2.4, 2.5 [© 1989]), Chapman and Hall (16.3 [© 1975], 2.9, 9.2 [© 1986]), Cornell University Press (2.7 [© 1960]), CRC Press (1.2, 2.6, 9.2, 10.2, 10.5, 22.1 [© 1979]), Dover Publications Inc. (12.2 [© 1965]), Elsevier Scientific Publishing Company (22.1 [© 1978]), Professor A. A. Frost (26.3), Longman (1.3, 2.6, 2.13, 2.15, 3.1, 6.1, 7.2, 9.2, 21.2, 22.7, 24.2, 24.4, 25.1, 25.2, 30.2 [© 1973]), McGraw Hill Book Company (1.2, 1.3, 2.2, 2.4, 3.1, 3.2, 4.1, 16.2, 22.2, 22.6, 25.4 [© 1975], 2.15, 5.2, 9.1 [© 1961], 26.1, 26.2 [© 1965], 3.2 [© 1968], 16.1 [© 1955]), National Bureau of Standards (2.9, 2.10 [© 1982]), Dr J. Nicholas (26.1, 26.2, 26.3 © 1976]), Oxford University Press (26.1, 26.2, 26.3 [© 1975], 29.3, 29.4, 29.5 [© 1974]), Pergamon Press Ltd (22.3 [© 1961]), Prentice Hall Inc. (23.1, 23.3 [© 1971]), John Wiley and Sons Inc (7.1 [© 1975], 22.5 [© 1954], 23.2, 23.4 [© 1961]). The sources of the material are quoted at the foot of each table.

The permission of the following individuals, institutions, and journals for reproduction of illustrations is gratefully acknowledged: Dr R. F. Barrow (11.7), Professor G. C. Bond (29.31), Professor G. Erlich (29.30), Dr A. J. Forty (29.3), Dr J. Foster (29.19), Professor R. Freeman (18.30), Professor D. Freifelder (23.5), Professor H. Ibach (29.9), Professor M. Karplus (28.17, 28.18), Professor D. A. King (29.28), Professor D. A. Long (16.32), Dr G. Morris (18.23), Professor E. W. Müller (29.17), Nicolet XRD Corporation (21.13), Dr A. H. Narten (22.24), Open University Press (17.39 [© 1979]), Oxford University Press (29.3, 29.4), Dr M. Prutton (29.15), Professor C. F. Quate (29.18), Professor M. W. Roberts (29.7), Dr H. M. Rosenberg (29.4), Dr M. Salmeron (29.20), Professor G. A. Samorjai (29.13, 29.14), *Scientific American* (8.1, 23.16), Professor D. W. Turner (29.8), Professor A. H. Zewail (28.5).

Answers to exercises

1

1.1 10 atm.
1.2 (a) No, 24 atm; (b) 22 atm.
1.3 (a) 2.57×10^3 Torr; (b) 3.38 atm.
1.4 30 K.
1.5 30 lb in^{-2}.
1.6 388 K.
1.7 (a) 15 m^3; (b) 1.5×10^2 m^3.
1.8 4.22×10^{-2} atm.
1.9 (a) 3.14 L; (b) 206 Torr.
1.10 169 g mol^{-1}.
1.11 16.4 g mol^{-1}.
1.12 (a)(i) 1.0 atm; (a)(ii) 8.2×10^2 atm; (b)(i) 0.99 atm; (b)(ii) 1.7×10^3 atm.
1.13 6.78×10^{-2} L mol^{-1}, 54.5 atm, 120 K.
1.14 (a) 0.88; (b) 1.1 L.
1.15 (a) 8.7 mL; (b) -0.15 L mol^{-1}.
1.16 (a) $x(H_2) = 0.67$, $x(N_2) = 0.33$; (b) $p(H_2) = 2.0$ atm, $p(N_2) = 1.0$ atm; (c) 3.0 atm.
1.17 1.33 L^2 atm mol^{-2}, 0.24 nm.
1.18 (a) 1.4×10^3 K; (b) 0.28 nm.
1.19 (a) 3.64×10^3 K, 8.7 atm; (b) 2.60×10^3 K, 4.5 atm; (c) 46.7 K, 0.18 atm.

2

2.1 (a) -98 J; (b) -16 J.
2.2 -2.6 kJ.
2.3 -1.0×10^2 J.
2.4 (a) -88 J, (b) -167 J.
2.5 $+124$ J.
2.6 $+29$ J.
2.7 -1.5 kJ.
2.8 85 MJ.
2.9 30 J K^{-1} mol^{-1}, 22 J K^{-1} mol^{-1}.
2.10 640 s.
2.11 80 J K^{-1}, -1.2 kJ, -1.2 kJ.
2.12 $+2.2$ kJ, $+2.2$ kJ, $+1.6$ kJ.
2.13 -1.0 kJ, $+13$ kJ, $+12$ kJ.
2.14 -4564.7 kJ mol^{-1}.
2.15 -126 kJ mol^{-1}.
2.16 -432 kJ mol^{-1}.
2.17 $+79$ kJ mol^{-1}.
2.18 641 J K^{-1}.
2.19 1.58 kJ K^{-1}, 2.05 K.
2.20 $\Delta H_c = \Delta U_c = -2.80$ MJ mol^{-1}, $\Delta H_f = -1.28$ MJ mol^{-1}.
2.21 $+65.49$ kJ mol^{-1}.
2.22 -383 kJ mol^{-1}.
2.23 1.90 kJ mol^{-1}.
2.24 -25 kJ, 9.8 m.
2.25 (a) -2205 kJ mol^{-1}; (b) -2200 kJ mol^{-1}.
2.26 (a) Exo; (b,c) endo.
2.27 (a) $0 = CO_2 + 2H_2O - CH_4 - 2O_2$; (b) $0 = C_2H_2 - 2C - H_2$; (c) $0 = Na^+(aq) + Cl^-(aq) - NaCl(s)$.
2.28 (a) -57.20 kJ mol^{-1}; (b) -176.01 kJ mol^{-1}; (c) -32.88 kJ mol^{-1}; (d) -55.84 kJ mol^{-1}.
2.29 $+11.3$ kJ mol^{-1}.
2.30 (a) -392.1 kJ mol^{-1}; (b) -946.6 kJ mol^{-1}; (c) $+52.5$ kJ mol^{-1}.
2.31 -56.98 kJ mol^{-1}.
2.32 -1892.2 kJ mol^{-1}.

3

3.3 $(\partial H / \partial U)_p = 1 + p(\partial V / \partial U)_p$.
3.4 $dV = (\partial V / \partial p)_T \, dp + (\partial V / \partial T)_p \, dT$, $d\ln V = -\kappa \, dp + \alpha \, dT$.
3.5 0, 0.
3.6 1.3×10^{-3} K^{-1}.
3.7 1×10^3 atm.
3.8 -7.2 J atm^{-1}, 8.1 kJ.
3.9 $w = -3.2$ kJ, $\Delta T = -38$ K, $\Delta U = -3.2$ kJ, $\Delta H = -4.5$ kJ.
3.10 $q = 0$, $\Delta U = +4.1$ kJ, $w = +4.1$ kJ, $\Delta H = +5.4$ kJ, $V_f = 11.8$ L, $p_f = 5.2$ atm.
3.11 9.2 L, 279 K, -0.39 kJ.
3.12 (a) 0.9 mm^3, (b) 0.02 mm^3.
3.13 -4.2 atm.
3.14 1.3 K atm^{-1}.
3.15 $w = -200$ J, $q = 0$, $\Delta U = -200$ J, $T_f = 296$ K, $\Delta H = -250$ J.
3.16 (a) 226 K, (b) 238 K.

4

4.1 (a) 92 J K^{-1}; (b) 67 J K^{-1}.
4.2 152.68 J K^{-1} mol^{-1}.
4.3 54.9 kJ, -195 J K^{-1}.
4.4 $+26$ J K^{-1}.
4.5 6.6 L.
4.6 2.8 J K^{-1}.
4.7 $+87.8$ J K^{-1} mol^{-1}, -87.8 J K^{-1} mol^{-1}.
4.8 (a) -386.1 J K^{-1} mol^{-1}, (b) -49.0 J K^{-1} mol^{-1}, (c) -153.1 J K^{-1} mol^{-1}, (d) -21.0 J K^{-1} mol^{-1}, (e) $+510.0$ J K^{-1} mol^{-1}.
4.9 (a) -521.5 kJ mol^{-1}, (b) $+68.0$ kJ mol^{-1}, (c) -178.6 kJ mol^{-1}; (d) -212.40 kJ mol^{-1}; (e) -5798 kJ mol^{-1}.
4.10 (a) -522.1 kJ mol^{-1}; (b) $+25.78$ kJ mol^{-1}; (c) -178.6 kJ mol^{-1}; (d) -212.55 kJ mol^{-1}; (e) -5798 kJ mol^{-1}.
4.11 -49.8 kJ mol^{-1}.
4.12 (a) $+2.9$ J K^{-1}, -2.9 J K^{-1}, 0; (b) $+2.9$ J K^{-1}, 0, $+2.9$ J K^{-1}.
4.13 $\Delta S = n(C_V - R) \ln 2$.
4.14 817.9 kJ mol^{-1}.
4.15 (a) 0.11; (b) 0.38.
4.16 $+0.95$ J K^{-1} mol^{-1}.
4.17 (a) 0; (b) 20 kJ.
4.18 201 K.
4.19 3.15 kJ.
4.20 (a) 14; (b) 8.8.
4.21 6.11 kJ, 61.1 s.

5

5.1 -3.8 J.
5.2 -36.5 J K^{-1}.
5.3 0.89 g cm^{-3}.
5.4 15.7 atm, $+8.25$ kJ.
5.5 $+7.3$ kJ mol^{-1}.
5.6 -0.55 kJ mol^{-1}.
5.7 -2.63×10^{-8} Pa^{-1}, 0.88.
5.8 $+10$ kJ.
5.9 $+11$ kJ mol^{-1}.

6

6.1 303 K.
6.2 $+45.2$ J K^{-1} mol^{-1}, $+16$ kJ mol^{-1}.
6.3 $+20.80$ kJ mol^{-1}.
6.4 16.9 mJ.
6.5 2.0×10^{-2} N m^{-1}.
6.6 207 nm.
6.7 281.8 K.
6.8 25 g s^{-1}.
6.9 (a) 1.7 kg; (b) 31 kg; (c) 1.4 g.
6.14 (a) $1.002p^*$; (b) $1.02p^*$.

7

7.1 886.8 cm^3.
7.1 6.4 MPa.
7.3 0.13 kPa.
7.4 $x_A = 0.920$, $x_B = 0.080$; $y_A = 0.968$, $y_B = 0.032$.

7.5	5.22 K/(mol kg^{-1}), 32 K/(mol kg^{-1}).
7.6	82 g mol^{-1}.
7.7	$-0.45°C$.
7.8	381 g mol^{-1}.
7.9	$-0.09°C$.
7.10	440 Tor.
7.11	-0.35 kJ, $+1.2$ J K^{-1}.
7.12	$+4.7$ J K^{-1} mol^{-1}.
7.13	-18.5 kJ, $+61.9$ J K^{-1}.
7.14	0.8600.
7.15	(a) 3.4 mmol kg^{-1}; (b) 34 mmol kg^{-1}.
7.16	0.51 mmol kg^{-1} N$_2$, 0.27 mmol kg^{-1} O$_2$.
7.17	0.17 mol kg^{-1}.
7.18	$-0.16°C$.
7.19	24 g anthracene in 1 kg benzene.
7.20	11 kg Pb/kg Bi.
7.21	87 kg mol^{-1}.
7.22	14 kg mol^{-1}.
7.23	(a) $y_T = 0.36$; (b) $y_T = 0.82$.

8

8.1	(a) 2; (b) 2; (c) 3.
8.2	$C = 2$, $P = 2$.
8.3	$C = 1$, $P = 2$; $C = 2$, $P = 2$.
8.4	$C = 2$, $P = 3$, $F = 1$.
8.5	$C = 2$, $P = 2$, $F = 2$.
8.11	(a) 80 per cent.
8.19	(0.06, 0.82, 0.12) and (0.62, 0.16, 0.22); 0.27.
8.20	Chloride: 19.5 mol kg^{-1}; sulphate: 23.8 mol kg^{-1}.

9

9.1	-2.42 kJ mol^{-1}.
9.2	3.01.
9.3	1500 K.
9.4	$\Delta H^\ominus = +2.77$ kJ mol^{-1}, $\Delta S^\ominus = -16.5$ J K^{-1} mol^{-1}.
9.5	$+12.3$ kJ mol^{-1}.
9.6	-41.1 kJ mol^{-1}.
9.7	(a) $K_x' = \frac{1}{2}K_x$; (b) $K_x' = K_x$.
9.8	19.4 g, 2.1 g.
9.9	(a, c, and e).
9.10	(b and d).
9.11	(a) $+53$ kJ mol^{-1}; (b) -53 kJ mol^{-1}.
9.12	$k \approx 10$ per cent of ln K.
9.13	-14.38 kJ mol^{-1}.
9.14	(a) 9.24; (b) -12.9 kJ mol^{-1}; (c) $+161$ kJ mol^{-1}, (d) $+248$ J K^{-1} mol^{-1}.
9.15	$+56.1$ kJ mol^{-1}.
9.16	(a) 1110 K; (b) 397 K.
9.17	(a) $+124.0$ kJ mol^{-1}, 3.3×10^{-7}; (b) $\Delta G^\ominus(500\ \text{K}) = -20.42$ kJ mol^{-1}, $\Delta G^\ominus(2000\ \text{K}) = +24.83$ kJ mol^{-1}, $K(500\ \text{K}) = 136$, $K(2000\ \text{K}) = 0.22$.
9.18	4.0×10^{-6}, 3.61.
9.19	(a) 5.13; (b) 8.88; (c) 2.88.
9.20	8.3.

9.22	(a) Na$_2$HPO$_4$/H$_3$PO$_4$; (b) NaH$_2$PO$_4$/Na$_2$HPO$_4$.

10

10.1	1.25×10^{-5} M.
10.2	-290 kJ mol^{-1}.
10.3	m, $3m$, $6m$, $15m$, $4m$.
10.4	0.90 mol kg^{-1}.
10.5	0.320 mol kg^{-1}.
10.6	(a) 2.73 g; (b) 2.92 g.
10.7	0.25 mol kg^{-1}.
10.8	$\gamma_\pm = (\gamma_+ \gamma_-^2)^{1/3}$
10.9	1×10^4 per cent.
10.11	-1107.9 kJ mol^{-1}.
10.12	34.2 mV.
10.13	-1.18 V.
10.14	(a) $2\text{Ag}^+ + \text{Zn} \rightarrow 2\text{Ag} + \text{Zn}^{2+}$, $+1.56$ V; (b) $\text{Cd} + 2\text{H}^+ \rightarrow \text{Cd}^{2+} + \text{H}_2$, $+0.40$ V; (c) $\text{Cr}^{3+} + 3[\text{Fe(CN)}_6]^{4-} \rightarrow \text{Cr} + 3[\text{Fe(CN)}_6]^{3-}$, -1.10 V; (d) $\text{Ag}_2\text{CrO}_4 + 2\text{Cl}^- \rightarrow 2\text{Ag} + \text{CrO}_4^{2-} + \text{Cl}_2$, -0.91 V; (e) $\text{Sn}^{4+} + 2\text{Fe}^{2+} \rightarrow \text{Sn}^{2+} + 2\text{Fe}^{3+}$, -0.62 V; (f) $\text{Cu} + \text{MnO}_2 + 4\text{H}^+ \rightarrow \text{Cu}^{2+} + \text{Mn}^{2+} + 2\text{H}_2\text{O}$, $+0.89$ V.
10.15	(a) Zn $\vert$ZnSO$_4$$\vert$$\vert$CuSO$_4$$\vert$ Cu, $+1.10$V; (b) Pt $\vert$H$_2$$\vert$ HCl $\vert$AgCl$\vert$ Ag, $+0.22$ V; (c) Pt $\vert$H$_2$$\vert$ H$^+$ $\vert$O$_2$$\vert$ Pt, $+1.23$ V; (d) Na $\vert$NaOH$\vert$ H$_2$ $\vert$ Pt, $+1.88$ V; (e) Pt $\vert$H$_2$$\vert$ HI $\vert$I$_2$$\vert$ Pt, $+0.54$ V.
10.16	See above.
10.17	(a) -1.20 V; (b) -1.18 V.
10.18	(a) -363 kJ mol^{-1}; (b) -405 kJ mol^{-1}; (c) -291 kJ mol^{-1}; (d) $+122$ kJ mol^{-1}.
10.19	(a) $+0.324$ V, (b) $+0.45$ V.
10.20	-0.62V.
10.21	$\Delta G^\ominus = +233.12$ kJ mol^{-1}, $\Delta H^\ominus = +300.3$ kJ mol^{-1}, -1.23 V.
10.22	(a) 1.6×10^{-8} mol kg^{-1}; (b) $+0.12$ V.
10.23	(a) 6.5×10^9; (b) 1.5×10^{12}; (c) 2.8×10^{-16}; (d) 1.7×10^{16}; (e) 8.2×10^{-7}.
10.24	1.80×10^{-10}, 9.04×10^{-7}; $0.991K_{sp}^\circ$, $0.56K_{sp}^\circ$.
10.25	$E = E^\ominus - (RT/6F) \ln \{a(\text{Cr}^{3+})^2/a(\text{Cr}_2\text{O}_7^{2-})a(\text{H}^+)^{14}\}$.
10.26	0.86.
10.27	0.
10.28	9.19×10^{-9} mol kg^{-1}.

11

11.1	1.7 MW.
11.2	5.2×10^{16} Hz.
11.3	1.6×10^6 m s^{-1}.
11.4	(a) 8.83×10^{-28} kg m s^{-1}; (b) 9.5×10^{-24} kg m s^{-1}; (c) 3.5×10^{-35} kg m s^{-1}.
11.5	50.6 nm.

11.6	72.38 pm.
11.7	70 nm.
11.8	6.09 nm.
11.9	$0.17L$ or $0.83L$.
11.10	5.9×10^{-24} J.
11.14	21 m s^{-1}.
11.15	(a) 2.8×10^{18} s^{-1}; (b) 2.8×10^{20} s^{-1}.
11.16	6000 K.
11.17	(a) No ejection; (b) 3.19×10^{-19} J, 837 km s^{-1}.
11.18	(a) 2.43 pm; (b) 1.32 fm.
11.19	(a) 7×10^{-19} J; (b) 40 kJ mol^{-1}; (c) 4×10^{-13} kJ mol^{-1}.
11.20	6.6×10^{-29} m; (b) 6.6×10^{-36} m; (c) 99.7 pm.
11.21	(a) 123 pm; (b) 39 pm; (c) 3.88 pm.
11.22	1.1×10^{-28} m s^{-1}, 1×10^{-27} m.
11.23	5×10^{-25} kg m s^{-1}, 5×10^5 m s^{-1}.
11.24	1.12 fJ.

12

12.1	(a) 1.81×10^{-19} J, 110 kJ mol^{-1}, 1.1 eV, 9100 cm^{-1}; (b) 6.6×10^{-19} J, 400 kJ mol^{-1}, 4.1 eV, 33000 cm^{-1}.
12.2	(122), (212), (221).
12.3	23 per cent.
12.4	4.30×10^{-21} J.
12.5	278 N M^{-1}.
12.6	2.63 μm.
12.7	(a) 3×10^{-35} J; (b) 3×10^{-33} J; (c) 2.2×10^{-29} J; (d) 3.14×10^{-20} J.
12.8	3.72 μm, change is $+0.09$ μm.
12.9	117 pm.
12.10	1.49×10^{-34} J s; 0, $\pm 1.06 \times 10^{-34}$ J s.
12.11	0.237

13

13.1	486.3 nm.
13.2	6.
13.3	6842 cm^{-1}.
13.4	14.0 eV.
13.5	$4a_0$.
13.6	101 pm, 376 pm.
13.7	(a) 0; (b) 0; (c) $\sqrt{6}\hbar$; (d) $\sqrt{2}\hbar$; (e) $\sqrt{2}\hbar$.
13.8	(a) $\frac{5}{2}$, $\frac{3}{2}$; (b) $\frac{7}{2}$, $\frac{5}{2}$.
13.9	2.
13.10	8, 7, 6, 5, 4, 3, 2.
13.11	(a) 1; (b) 9; (c) 25.
13.12	$L = 2$, $S = 0$, $J = 2$.
13.13	$0.35a_0$.
13.14	(a) 110 pm; (b) 86 pm.
13.15	(a) F; (b) A; (c) A; (d) F; (e) A.
13.18	(a) 1(3), 0(1); (b) $\frac{3}{2}$(4), $\frac{1}{2}$(2), $\frac{1}{2}$(2); (c) $\frac{5}{2}$(6), $4 \times \frac{3}{2}$(4), $5 \times \frac{1}{2}$(2).
13.19	(a) 0(1); (b) $\frac{3}{2}$(4), $\frac{1}{2}$(2); (c) 2(5), 1(3), 0(1); (d) 3(7), 2(5), 1(3); (e) $\frac{7}{2}$(8), $\frac{5}{2}$(6), $\frac{3}{2}$(4), $\frac{1}{2}$(2).
13.20	(a) $^2S_{1/2}$; (b) $^2P_{3/2}$, $^2P_{1/2}$; (c) $^2D_{5/2}$, $^2D_{3/2}$; (d) $^2P_{3/2}$, $^2P_{1/2}$.

13.21 $m_l = +2.$
13.22 2.1 T.

14

14.1 (a) $1s\sigma_g^2 1s\sigma_u^2 2s\sigma_g^2$, $b = 1$; (b) $1s\sigma_g^2 1s\sigma_u^2 2s\sigma_g^2 2s\sigma_u^2$, $b = 0$; (c) $1s\sigma_g^2 1s\sigma_u^2 2s\sigma_g^2 2s\sigma_u^2 2p\pi_g^4$, $b = 2.$
14.2 (a) $1s\sigma_g^2 1s\sigma_u^1$, $b = 0.5$; (b) $\cdots 2\pi_u^4 2p\sigma_g^2$, $b = 3$; (c) $\cdots 2p_x\pi_g^1 2p_y\pi_g^1$, $b = 2.$
14.3 (a) $\cdots 2p\sigma^2$, $b = 3$; (b) $\cdots 2p\pi^{*2}$, $b = 2.5$; (c) $\cdots 2p\sigma^2$, $b = 3.$
14.4 $C_2.$
14.5 C_2, CN stabilized by anion formation, NO, O_2, and F_2 by cation formation.
14.6 $r(XeF^+) < r(XeF).$
14.7 (a) g; (b) irrelevant; (c) g; (d) u.
14.8 a_2:g, e_1:g, e_2:u, b_2:g.
14.9 $\cdots 2p\sigma_g^2$, $^2\Sigma_g.$
14.10 $S = 1$, $2S + 1 = 3$, u.
14.11 $r(N_2) < r(NO).$
14.12 $1s\sigma_g^2 2p\pi_u^1.$
14.14 $N = 1/(1 + 2\lambda S + \lambda^2)^{1/2}.$
14.15 0.
14.16 3.7 per cent s character.
14.17 (a) Linear; (b) nonlinear; (c) linear; (d, e, f) nonlinear; (g) linear.
14.20 (a) $6\alpha + 8\beta$; (b) $5\alpha + 7\beta.$

15

15.1 4.
15.2 E, C_3, $3\sigma_v.$
15.3 (a, b, c).
15.4 0.
15.5 0.
15.6 $p^1 d^4.$
15.7 $B_2.$
15.10 (a) R_3; (b) C_{2v}; (c) D_{3h}; (d) $D_{\infty h}$; (e) $C_{\infty v}$; (f) D_3; (g) C_{4v}; (h) $C_s.$
15.11 (a) C_{2v}; (b) $C_{\infty v}$; (c) C_{3v}; (d) D_{2h}; (e) C_{2v}; (f) $C_{2h}.$
15.12 (a) D_{2h}; (b) D_2; (c) $C_{2v}(o)$, $C_{2v}(m)$, $D_{2h}(p).$
15.13 (a) NO_2, N_2O, $CHCl_3$, 1,2- and 1,3-dichlorobenzene; (b) none.
15.14 Non-bonding; $d_{xy}.$
15.15 $B_1(x)$, $B_2(y)$, $A_1(z).$
15.16 $A_2.$
15.17 (a) E_{1u} or A_{2u}; (b) $B_{3u}(x)$, $B_{2u}(y)$, $B_{1u}(z).$
15.18 $A_2.$

16

16.1 (a) 1.6266×10^{-27} kg; (b) 3.1624×10^{-27} kg; (c) 1.6291×10^{-27} kg.
16.2 3.45×10^{-45} kg m^2.
16.3 232.1 pm.
16.4 20 475 cm^{-1}.

16.5 2599.77 cm^{-1}.
16.6 1.08 per cent.
16.7 328.7 N m^{-1}.
16.8 381.3 cm^{-1}, 378.3 cm^{-1}.
16.9 191.8 cm^{-1}, 2.177 eV.
16.10 $4A_1 + A_2 + 2B_1 + 2B_2.$
16.11 (b, d, e, f, g, h).
16.12 (b, c, d, e, f, g).
16.13 (a, b, d, e, f).
16.14 0.999 999 925λ, -6.36×10^7 m s^{-1}.
16.15 2.4×10^4 km s^{-1}, 8.4×10^5 K.
16.16 (a) 53 ps; (b) 5 ps; (c) 2 ns.
16.17 (a) 50 cm^{-1}; (b) 0.5 cm^{-1}.
16.18 (a) 0.067; (b) 0.20.
16.19 160.4 pm.
16.20 HF > HCl > HBr > HI.
16.22 5.32 eV.
16.23 198.9 pm.

17

17.1 80 per cent.
17.2 7.9×10^5 cm^2 mol^{-1}.
17.3 1.5 mM.
17.4 0.235.
17.5 0.0104.
17.6 (a) W; (b) S; (c) F; (d) F; (e) S.
17.7 243 nm : diene; 192 nm : butene.
17.10 450 M^{-1} cm^{-1}.
17.11 160 M^{-1} cm^{-1}, 23 per cent.
17.12 (a) 0.9 m; (b) 3 m.

18

18.1 -1.625×10^{-26} J $\times m_I.$
18.2 6.116×10^{-26} J.
18.3 3.523 T.
18.5 (a) 1×10^{-6}; (b) 5.1×10^{-6}; (c) 3.4×10^{-5}.
18.6 (a) 11 μT; (b) 53 μT.
18.7 (a) 460 Hz; (b) 2.66 kHz.
18.9 5×10^2 s^{-1}.
18.13 (a) 2×10^2 T; (b) 2 mT.
18.14 1.3 T.
18.15 2×10^{-11} M.
18.16 2×10^{11}.
18.17 2.0022.
18.18 50.7 mT.
18.19 2.3 mT, 2.0025.
18.22 (a) 331.9 mT; (b) 1201 mT.
18.23 (a) 1.3 mT; (b) 4.8 mT.
18.24 Resolved.
18.25 $\frac{3}{2}.$

19

19.1 $T = \infty.$
19.2 (a) 2.57×10^{27}; (b) 3.95×10^{27}.
19.3 (a) 15.9 pm, 2.47×10^{26}; (b) 5.04 pm, 7.82×10^{27}.
19.4 2.83.
19.5 3.156.

19.6 2.45 kJ mol^{-1}.
19.7 354 K.
19.9 $\langle\varepsilon\rangle/2\mu_B B = 1/(1 + e^x)$, $x = 2\mu_B\beta B$; (a) 0.72; (b) 0.996.
19.10 $\langle\varepsilon\rangle/g_I\mu_N B(1 + 2e^{-x})e^{-x}/(1 + e^{-x} + e^{-2x})$, $x = g_I\mu_N B.$
19.11 (a) 138 J K^{-1} mol^{-1}; (b) 146 J K^{-1} mol^{-1}.
19.12 5.18 J K^{-1} mol.
19.13 (a) Y; (b) Y; (c) N; (d) Y; (e) N.

20

20.1 (a) $\frac{5}{2}R$; (b) $3R$; (c) $3R.$
20.2 (a) 19.6; (b) 34.3.
20.3 (a) 1; (b) 2; (c) 2; (d) 12; (e) 3.
20.4 43.1, 40 K.
20.5 (a) 36.61, 79.17; (b) 36.3, 78.9.
20.6 71.2.
20.7 (a) 7.97×10^3; (b) 1.12×10^4.
20.8 1.796 at 298 K, 3.086 at 500 K.
20.9 $q = 1 + e^{-x}$, $U - U(0) = N\varepsilon/(1 + e^{-x})$, $C_V = Rx^2 e^{-x}/(1 + e^{-x})^2$; $x = \varepsilon/kT.$
20.10 -13.8 kJ mol^{-1}, -0.20 kJ mol^{-1}.
20.11 $0.236R$, $0.193R.$
20.12 -3.65 kJ mol^{-1}.
20.13 11.5 J K^{-1} mol^{-1}.
20.14 33.3 J K^{-1} mol^{-1}.
20.17 191.4 J K^{-1} mol^{-1}.
20.18 3.2×10^{-3}.
20.19 0.25.

21

21.6 249 pm, 176 pm, 432 pm.
21.7 70.7 pm.
21.8 0.21 cm.
21.9 3.96×10^{-28} m^3.
21.10 4.01 g cm^{-3}.
21.11 190 pm.
21.12 240 pm, 606 pm, 395 pm.
21.13 402 pm.
21.14 11.75°.
21.15 fcc.
21.16 bcc.
21.17 $F_{hkl} = f.$
21.18 0.
21.19 0.9069.
21.20 bcc.
21.21 8.97 g cm^{-3}.
21.23 (a) 2.7×10^{-20} J (b) 3.9×10^3 K.
21.24 7.9 km s^{-1}.
21.25 252 pm.
21.26 4.6 kV.
21.27 (a) 39 pm; (b) 12 pm; (c) 6.1 pm.

22

22.1 220 pF.
22.2 1.7 D, 9.1×10^{-24} cm^3.
22.3 4.8.

Answers to exercises

22.5	1.34.
22.6	1.28×10^{-23} cm³.
22.7	**(a)** 0 by symmetry; **(b)** 0.7 D; **(c)** 0.4 D.
22.8	1.4 D.
22.9	4.9 μD.
22.10	1.34.
22.11	18.
22.12	**(a)** 1.69; **(b)** 1.64; **(c)** 1.23.
22.13	$r = 2^{1/6}\sigma$.
22.14	3.
22.15	-6.4×10^{-5} cm³ mol⁻¹.
22.16	+0.016.

23

23.1	70 kg mol⁻¹, 71 kg mol⁻¹.
23.2	23.1 kg mol⁻¹, 1.02 kg mol⁻¹.
23.3	83.0 kg mol⁻¹.
23.4	155 kg mol⁻¹, 13.7 m³ mol⁻¹.
23.5	24 nm.
23.6	6.7×10^3.
23.7	0.0716 L g⁻¹.
23.8	5.0 Sv.
23.9	65.6 kg mol⁻¹.
23.10	31 kg mol⁻¹.
23.11	**(a)** 17.6 kg mol⁻¹, **(b)** 19.5 kg mol⁻¹.
23.12	6.62×10^{-4} M.
23.13	6.7 mM.
23.14	3.39×10^3 kg mol⁻¹.
23.15	0.43 Mg.
23.16	**(a)** 24 ns; **(b)** 14 pm.
23.17	$\Gamma < 0$.

24

24.2	81 mPa.
24.3	14 MPa.
24.4	1 μm.
24.5	**(a)** 5×10^{10} s⁻¹; **(b)** 5×10^9 s⁻¹; **(c)** 5×10^3 s⁻¹.
24.6	**(a)** 6.1×10^{33}; **(b)** 6×10^{31}; **(c)** 6×10^{19}.
24.7	4×10^8 s⁻¹.
24.8	3.3×10^{33} m⁻³ s⁻¹, 2.7×10^{34} m⁻³ s⁻¹.
24.9	2.0×10^{23} s⁻¹.
24.10	3.7 kJ mol⁻¹.
24.11	**(a)** 6.7 nm; **(b)** 67 nm; **(c)** 6.7 cm.
24.12	9.1×10^{-3}.
24.13	1.9×10^{20}.
24.14	2.0×10^{26}.
24.15	λ independent of T.
24.16	104 mg.
24.17	4.1 μJ cm⁻² s⁻¹.
24.18	0.056 nm².
24.19	0.142 nm².
24.20	2.05 bar.
24.21	**(a)** 130 μP; **(b)** 130 μP; **(c)** 240 μP.
24.22	**(a)** 5.4 mJ K⁻¹ m⁻¹ s⁻¹; **(b)** 29 mJ K⁻¹ m⁻¹ s⁻¹; **(a)** 8.1 mJ s⁻¹; **(b)** 44 mJ s⁻¹.
24.33	138 μP, 390 pm.

24.24	**(a)** 11 m² s⁻¹; **(b)** 1.1×10^{-5} m² s⁻¹; **(c)** 1.1×10^{-7} m² s⁻¹.

25

25.1	7.25 mS cm⁻¹.
25.2	1.6 mS cm⁻¹.
25.3	116.7 S cm² mol⁻¹.
25.4	66.1 S cm² mol⁻¹.
25.5	347 μm s⁻¹.
25.6	0.331.
25.7	138.3 S cm² mol⁻¹.
25.8	0.34 mS cm⁻¹, 0.61 kΩ.
25.9	$(4.01, 5.19, 9.62) \times 10^{-4}$ cm² s⁻¹ V⁻¹.
25.10	93 kg mol⁻¹.
25.11	1.90×10^{-9} m² s⁻¹.
25.12	5.4 nm.
25.13	3.9×10^3 s.
25.14	420 pm.
25.15	21 ps.
25.17	**(a)** 240 s, 960 s; **(b)** 7 h, 27 h.

26

26.1	3.0(C), 1.0(D), 1.0(A), 2.0(B) M s⁻¹.
26.2	1.5(D), 1.0(A), 0.50(B), M s⁻¹; $v = 0.50$ M s⁻¹.
26.3	**(a)** $k[A][B]$; **(b)** $2k[A][B]$.
26.4	$v = \frac{1}{2}k[A][B][C]$, M⁻² s⁻¹.
26.5	1.03×10^4 s; **(a)** 497 Torr; **(b)** 333 Torr.
26.6	**(a)** M⁻¹ s⁻¹, M⁻² s⁻¹; **(b)** atm⁻¹ s⁻¹, atm⁻² s⁻¹.
26.7	2720 y.
26.8	**(a)** 0.64 μg; **(b)** 0.18 μg.
26.9	**(a)** 0.045 M, 0.095 M; **(b)** 0.001 M, 0.051 M.
26.10	-3.92×10^{-4}(B), 1.96×10^{-4}(P), $-6.66 \times 10^{-4}(n_B)$, $v = 1.96 \times 10^{-4}$ M s⁻¹.
26.11	112 s.
26.12	1.24×10^5 s.
26.13	2.8×10^{-4} s⁻¹.
26.14	64.9 kJ mol⁻¹, 4.32×10^8 M s⁻¹.
26.15	$H_2O(l)$, $Br^-(l)$, Overall (2).
26.16	$k_2 K^{1/2}[A_2]^{1/2}[B]$.
26.17	$k_1 K[A][B]$, $k = k_1 k_2/k_2'$.
26.19	1.52 mM s⁻¹.
26.20	$[S] = K_M$.
26.21	1.9 MPa⁻¹ s⁻¹.
26.22	7.1×10^5 s⁻¹, 7.61 ns.
26.23	1.7×10^{-7} s⁻¹, 8.5×10^8 M⁻¹ s⁻¹.

27

27.1	1(I); 2, 8(T); 3, 4, 5, 6, 7(P).
27.2	$f = k_2 k_4[CO]/(k_2[CO] + k_3[M])$.
27.3	$d[R_2]/dt = -k_1[R_2] - k_2(k_1/k_4)^{1/2}[R_2]^{3/2}$.
27.4	0.28–2.2 kPa, 0.14–8.9 kPa, >0.11 kPa.

27.5	3.3×10^{18}.
27.6	0.521.
27.7	$d[P]/dt = k_1 k_3[A][AH][B]/(k_2[BH^+] + k_3[A])$.
27.8	$d[AH]/dt = k_3 K K_a[HA][BH^+]$.
27.9	$d[AH]/dt = k'[AH]$, $k' = k_a + k_a k_c/2kk_d$.
27.10	$(A_0 + P_0)^2 k t_{max} = \frac{1}{2} - p - \ln 2p$.
27.11	$(A_0 + P_0)^2 k t_{max} = (2 - p)/2p + \ln (2/p)$.

28

28.1	**(a)** 9.6×10^9 s⁻¹, 1.2×10^{35} m⁻³ s⁻¹; **(b)** 6.7×10^9 s⁻¹, 8.3×10^{34} m⁻³ s⁻¹.
28.2	**(a)** 0.018, 0.030; **(b)** 3.9×10^{-18}, 6.0×10^{-6}.
28.3	**(a)** 13 per cent, 1.2 per cent; **(b)** 130 per cent, 12 per cent.
28.4	**(a)** 6.61×10^9; **(b)** 3.0×10^{10}; **(c)** 2.0×10^9 M⁻¹ s⁻¹.
28.5	0.152 nm².
28.6	1.2×10^{-3}.
28.7	1.8×10^8 M s⁻¹.
28.8	+74.7 kJ mol⁻¹, −25 J K⁻¹ mol⁻¹.
28.9	+4.9 J K⁻¹ mol⁻¹, +74.4 kJ mol⁻¹.
28.10	−96.6 J K⁻¹ mol⁻¹, +83.9 kJ mol⁻¹.
28.11	−76 J K⁻¹ mol⁻¹.
28.12	**(a)** −45.8 J K⁻¹ mol⁻¹; **(b)** +5.0 kJ mol⁻¹; **(c)** +18.7 kJ mol⁻¹.
28.14	**(a)** H/T ≈ 14; **(b)** ¹⁶O/¹⁸O ≈ 1.1.
28.15	20.9 M⁻² min⁻¹.
28.16	0.66 M⁻¹ min⁻¹.
28.17	lg $v \propto I^{1/2}$.

29

29.2	94 Torr.
29.3	3.4×10^5 s⁻¹.
29.4	5×10^3 s.
29.5	610 kJ mol⁻¹, 0.11 ps.
29.6	3.7 kJ mol⁻¹.
29.7	**(a)** 0.21 kPa; **(b)** 22 kPa.
29.8	0.83, 0.36.
29.9	40 ps, 2×10^{13} s.
29.10	15 kPa.
29.11	**(a)** 0; **(b)** 1.
29.12	**(a)** $\theta \propto 1/p$; **(b)** $\theta \propto 1/p^{1/2}$; **(c)** $\theta \propto 1/p^{1/3}$.
29.14	−13 kJ mol⁻¹.
29.15	700 kJ mol⁻¹; **(a)** 2×10^{104} min; **(b)** 50 μs.
29.16	16 Pa s⁻¹.

30

30.1	138 mV.
30.2	2.8 mA cm⁻².

30.3 0.99 A m^{-2}.
30.5 **(a)** 0.31 mA cm^{-2}; **(b)** 5.41 mA cm^{-2}; **(c)** 10^{39} A cm^{-2}.
30.7 $a(Fe^{3+}) \approx 7480a(Fe^{2+})$.
30.8 108 mV.
30.9 **(a)** 4.9×10^{15} cm^{-2} s^{-1}; **(b)** $1.6 \times$ 10^{16} cm^{-2} s^{-1}; **(c)** 3.1×10^7 cm^{-2} s^{-1}; **(a)** 3.8 s^{-1}; **(b)** 12 s^{-1}, 2.4×10^{-8} s^{-1}.
30.10 **(a)** 33Ω; **(b)** 33 GΩ.
30.12 0.28 V (Cu), -0.82 V (Zn).
30.14 0.14 mA cm^{-2}.

30.15 2.2 kA cm^{-2}.
30.16 5.3 GA cm^{-1}.
30.17 0.25 mm.
30.19 -1.30 V, 0.13 W.
30.20 **(a)** $+1.23$ V; **(b)** $+1.06$ V.
30.21 Fe, Al, Co, Cr if O_2 absent.

Answers to problems

1

1.1 $0.5\ m^3$.
1.2 $1.5\ kPa$.
1.3 3.2×10^{-2} atm.
1.4 $p = \rho RT/M$, $45.9\ g\ mol^{-1}$.
1.5 **(a)** 4.6×10^3 mol; **(b)** 129 kg; **(c)** 120 kg.
1.6 $102\ g\ mol^{-1}$, CH_2FCF_3.
1.7 **(a)** 0.184 Torr; **(b)** 68.6 Torr; **(c)** 0.184 Torr.
1.8 $p = 1.66$ atm; $p(H_2) = 0$, $p(N_2) = 0.33$ atm, $p(NH_3) = 1.33$ atm.
1.9 **(a)** $12.5\ L\ atm^{-1}$; **(b)** $12.3\ L\ atm^{-1}$.
1.10 210 K, 0.28 nm.
1.11 $a = 5.65\ L^2\ atm\ mol^{-1}$, $b = 59.4\ cm^3\ mol^{-1}$, $p = 21$ atm.
1.12 $B = b - a/RT$, $C = b^2$; $34.6\ cm^3\ mol^{-1}$, $1.26\ L^2\ atm\ mol^{-2}$.
1.13 $B = b - a/RT$, $C = b^2 - ab/RT + a^2/2R^2T^2$; $1.47\ L^2\ atm\ mol^{-2}$, $43.9\ L^2\ atm\ mol^{-2}$.
1.14 $V_c = 3C/B$, $T_c = B^2/3RC$, $p_c = B^3/27C^2$, $Z_c = \frac{1}{3}$.
1.15 $B' = B/RT$, $C' = (C - B^2)/R^2T^2$.
1.16 $B' = 8.4 \times 10^{-2}\ atm^{-1}$, $B = 2.1\ L\ mol^{-1}$.
1.17 **(a)** $0.99998p_0$; **(b)** $0.95p_0$.

2

2.1 **(a)** -0.27 kJ; **(b)** -0.92 kJ.
2.2 -8.9 kJ, -8.9 kJ.
2.3 $w = 0$, $\Delta U = +2.35$ kJ, $\Delta H = +3.03$ kJ.
2.4 $w = -25$ J, $q = +109$ J, $\Delta H = +97$ J.
2.5 $\Delta H = +22.2$ kJ, $\Delta H_{vap} = +40\ kJ\ mol^{-1}$, $\Delta U = +20.5$ kJ, $w = -1.7$ kJ.
2.6 $98.7\ kJ\ mol^{-1}$, $95.8\ kJ\ mol^{-1}$.
2.7 36.5 L.
2.8 $-87.33\ kJ\ mol^{-1}$.
2.9 **(a)** 2878, 3537, 5471 kJ mol^{-1}; **(b)** 49.51, 49.02, 47.89 kJ g^{-1}.
2.10 $-93.9\ kJ\ mol^{-1}$.
2.11 $\Delta U = -2130\ kJ\ mol^{-1}$, $\Delta H = -2130\ kJ\ mol^{-1}$, $\Delta H_f^{\ominus} = -1267\ kJ\ mol^{-1}$.
2.12 $\Delta H = +17.7\ kJ\ mol^{-1}$, $\Delta H_f^{\ominus} = +54.7\ kJ\ mol^{-1}$, $\Delta H_f^{\ominus} = +91.7\ kJ\ mol^{-1}$.
2.13 $5376\ kJ\ mol^{-1}$ more exothermic.
2.14 **(a)** $-35.4\ kJ\ mol^{-1}$; **(b)** $+3.15\ kJ\ mol^{-1}$.
2.15 $-146\ kJ\ mol^{-1}$.
2.16 $-2Fa/\pi$, 0.
2.17 **(a)** -1.5 kJ; **(b)** -1.6 kJ.
2.18 $w = -nRT \ln\{(V_2 - nb)/(V_1 - nb)\} - n^2a\{(1/V_2) - (1/V_1)\}$.
2.19 $w_r/n = -\frac{8}{9}T_r \ln\{(3V_{r2} - 1)/(3V_{r1} - 1)\} - \{(1/V_2) - (1/V_1)\}$ with $w_r = 3bw/a$; also $w_r/n = -\frac{8}{9} \ln \frac{1}{2}(3x - 1) + 1 - 1/x$.
2.20 **(a)** p/T, $-p/V$; **(b)** $(1 + na/RTV)p/T$, $\{na/RTV - V/(V - nb)\}p/V$.
2.21 $\Delta H(T_2) = \Delta H(T_1) + \Delta a(T_2 - T_1) + \frac{1}{2}\Delta b(T_2^2 - T_1^2) - \Delta c(1/T_2 - 1/T_1) - 283.4\ kJ\ mol^{-1}$, $-283.47\ kJ\ mol^{-1}$.

3

3.1 $\kappa = 2.3 \times 10^{-11}\ Pa^{-1}$, $\Delta V = -0.23\ cm^3$, $\Delta V(total) = -2.8\ cm^3$.
3.2 **(a)** $0.731\ J\ K^{-1}\ mol^{-1}$, 0.29 kJ; **(b)** $28.8\ J\ K^{-1}\ mol^{-1}$, 16 kJ.
3.3 **(a)** $0.75\ kJ\ mol^{-1}$; **(b)** $0.75\ kJ\ mol^{-1}$, difference $3.8\ mJ\ mol^{-1}$.
3.4 $41.40\ J\ K^{-1}\ mol^{-1}$.
3.5 $\gamma = 1.67$.
3.6 $p = p_i e^{-t/\tau}$, $1/\tau = 1/\tau_T - 1/\tau_V$.
3.7 $dp = A\ dV + B\ dT$, $A = \{n^2a(V - 2nb)/V^3 - p\}/(V - nb)$, $B = (p + n^2a/V^2)/T$.
3.9 $\alpha = RV^2(V - nb)/\{RTV^3 - 2na(V - nb)^2\}$, $\kappa = V^2(V - nb)^2/\{RTV^3 - 2n^2a(V - nb)^2\}$.
3.10 $\mu C_p = (1 - nb\zeta/V)V/(\zeta - 1)$, $\zeta = RTV^3/2na(V - nb)^2$, $1.43\ K\ atm^{-1}$, $T_I = (27T_c/4)(1 - b/V_m)^2$, $c.\ 2000$ K.
3.11 $(\partial H/\partial p)_T = -T(\partial V/\partial T)_p + V$.
3.12 $0.29\ K\ atm^{-1}$.
3.13 $C_p - C_V = 1.1R$.
3.14 $\Delta H = nC_p(T_f - T_i)$.
3.15 $322\ m\ s^{-1}$.

4

4.1 **(a)** $-21.3\ J\ K^{-1}\ mol^{-1}$, $+21.7\ J\ K^{-1}\ mol^{-1}$, $+0.4\ J\ K^{-1}\ mol^{-1}$; **(b)** $+109.7\ J\ K^{-1}\ mol^{-1}$, $-111.2\ J\ K^{-1}\ mol^{-1}$, $-1.5\ J\ K^{-1}\ mol^{-1}$.
4.2 $+11\ J\ K^{-1}$.
4.3 **(a)** $200.7\ J\ K^{-1}\ mol^{-1}$; **(b)** $232.0\ J\ K^{-1}\ mol^{-1}$.
4.4 $+45.4\ J\ K^{-1}$, $+51.2\ J\ K^{-1}$.
4.5 $-160.07\ kJ\ mol^{-1}$.
4.6 **(a)** $+17.0\ J\ K^{-1}$; **(b)** $+36\ J\ K^{-1}$.
4.7 **(a, b)** $0.11\ kJ\ mol^{-1}$.
4.8 $S - S(0) = 63.88\ J\ K^{-1}\ mol^{-1}$, $66.08\ J\ K^{-1}\ mol^{-1}$.
4.9 7.8 km.
4.10 $+96.864\ J\ K^{-1}\ mol^{-1}$, $+76.9\ J\ K^{-1}\ mol^{-1}$.
4.12 $H - H(0) = 34.4\ kJ\ mol^{-1}$, $S - S(0) = 24.3\ J\ K^{-1}\ mol^{-1}$.
4.13 $0.61\ \mu J$, $6.7\ \mu J$.
4.16 $\Delta S = nC_p \ln (T_h + T_c)^2/4T_hT_c$, $+22.6\ J\ K^{-1}\ mol^{-1}$.
4.18 6.83 kJ, no additional work.

5

5.1 $-501\ kJ\ mol^{-1}$.
5.2 **(a)** $+7.26\ kJ\ mol^{-1}$; **(b)** $+106.74\ kJ\ mol^{-1}$.
5.3 $-6\ kJ\ mol^{-1}$.
5.4 73 atm.
5.7 $(\partial p/\partial S)_V = \alpha T/\kappa C_V$, $(\partial V/\partial S)_p = \alpha VT/C_p$.
5.9 **(a)** 0; **(b)** $(\partial H/\partial p)_T = \{nb - (2na\lambda^2/RT)\}/\{1 - (2na\lambda^2/RTV)\}$, $\lambda = 1 - nb/V$, $-8.4\ J\ atm^{-1}$, -8 J.
5.10 0.30 atm.
5.13 $\pi_T = nap/RTV$.
5.14 0.02 per cent difference.
5.15 **(a)** $\Delta G' = \tau\Delta G + (1 - \tau)\Delta H$, $\tau = T'/T$; **(b)** $\Delta G' = \tau\Delta G + (1 - \tau)(\Delta H - T\Delta C_p) - T'\Delta C_p \ln \tau$.
5.17 $q_{rev} = nRT \ln \{(V_f - nb)/(V_i - nb)\}$.
5.18 -500 J.
5.19 $G' = G + p^*V_0(1 - e^{-p/p^*})$.
5.20 $\ln \gamma = Bp/RT + (C - B^2)p^2/2R^2T^2$, 0.999_1 atm.
5.21 $\gamma = 2e^{x-1}/(1 + x)$, $x = (1 + 4pB/R)^{1/2}$.

6

6.1 $+5.56\ kPa\ K^{-1}$, $+5.42\ kPa\ K^{-1}$, 2.5 per cent error.
6.2 **(a)** $-22.0\ J\ K^{-1}\ mol^{-1}$; **(b)** $-109.0\ J\ K^{-1}\ mol^{-1}$, $+0.1\ kJ\ mol^{-1}$.
6.3 **(a)** $-1.63\ cm^3\ mol^{-1}$; **(b)** $+30.1\ L\ mol^{-1}$, $+0.6\ kJ\ mol^{-1}$.
6.4 234.4 K.
6.5 22°C.
6.6 $+37.8\ kJ\ mol^{-1}$.
6.7 $+55\ kJ\ mol^{-1}$.
6.9 **(a)** 1 in 20; **(b)** 1 in 20 000; **(c)** 1 in 2 million.

Answers to problems

6.10 9.5 J, 9.5 J.
6.11 5.7 cm, 0.44 kPa.
6.12 15 cm.
6.13 $\partial \Delta A/p \propto V_\alpha - V_\beta$.
6.14 $d(\Delta H/T) = \Delta C_p \, d \ln T$.
6.15 9.8 Torr.
6.16 $1/T_h = 1/T_b + Mgh/T\Delta H_{vap}$, 363 K.
6.17 $h = 2\gamma \cos\theta/\rho gr$.
6.19 2 m s^{-1}.

7

7.1 $K_A = 15.58$ kPa, $K_B = 47.03$ kPa.
7.2 $17.5 \text{ cm}^3 \text{ mol}^{-1}$, $18.1 \text{ cm}^3 \text{ mol}^{-1}$.
7.3 $-1.4 \text{ cm}^3 \text{ mol}^{-1}$, $18.04 \text{ cm}^3 \text{ mol}^{-1}$.
7.5 $12.0 \text{ cm}^3 \text{ mol}^{-1}$.
7.6 57.6 cm^3 ethanol + $45.7 \text{ cm}^3 \text{ mol}^{-1}$ water; 0.96 cm^3.
7.8 -0.70 K.
7.9 ± 0.05 K.
7.11 *ca.* 4.
7.14 -4.6 kJ.
7.15 $\mu_A = \mu_A^\ominus + RT \ln x_A + gRTx_B^2$.
7.17 $80.36 \text{ cm}^3 \text{ mol}^{-1}$.

8

8.1 0.
8.2 (a) $y(\text{MgO}) = 0.18$, $x(\text{MgO}) = 0.35$; (b) 0.42; (c) 2650°C.
8.4 $T_{uc} = 122$°C.
8.7 Add 81 g water.

9

9.1 $\Delta H_f^\ominus = -(2.196 \times 10^4 \text{ K} - 8.48T)R$, $\Delta C_p = 8.48R$.
9.2 $\Delta G^\ominus/(\text{kJ mol}^{-1}) = 78 - 0.161(T/\text{K})$.
9.3 1.68×10^{-5}.
9.4 $\alpha = 0.392$, $K = 0.740$; $\alpha = 0.764$, $K = 5.71$; $\Delta H^\ominus = -103 \text{ kJ mol}^{-1}$.
9.5 $+14.7 \text{ kJ mol}^{-1}$, $+18.8 \text{ kJ mol}^{-1}$.
9.6 0.007 mol H_2, 0.107 mol I_2, 0.786 mol HI.
9.7 0.33.
9.8 $\Delta G^\ominus = -24.5 \text{ kJ mol}^{-1}$, $\Delta H^\ominus = -66.1 \text{ kJ mol}^{-1}$, $\Delta S^\ominus = -142 \text{ J K}^{-1} \text{ mol}^{-1}$.
9.9 $+158 \text{ kJ mol}^{-1}$.
9.10 (a) $\Delta G^\ominus = +26 \text{ kJ mol}^{-1}$, $\Delta H^\ominus = -137 \text{ kJ mol}^{-1}$, $\Delta S^\ominus = -547 \text{ J K}^{-1} \text{ mol}^{-1}$; (b) $\Delta H^\ominus = -102 \text{ kJ mol}^{-1}$, $\Delta S^\ominus = -390 \text{ J K}^{-1} \text{ mol}^{-1}$.
9.12 $\xi = 1 - (1 + ap/p^\ominus)^{-1/2}$.
9.13 0.140.
9.14 $\Delta G' = \Delta G + (T - T')\Delta S + \alpha\Delta a + \beta\Delta b + \gamma\Delta c$, $\alpha = T' - T - T' \ln(T'/T)$, $\beta = \frac{1}{2}(T'^2 - T^2) - T'(T' - T)$, $\gamma = 1/T - 1/T' + (T'/2)(1/T'^2 - 1/T^2)$; $-225.31 \text{ kJ mol}^{-1}$.

10

10.1 0.96 V.
10.2 (a) $+1.23$ V; (b) $+2.85$ V.
10.3 0.42 V.
10.5 $+0.26838$ V.
10.6 $\Delta G^\ominus = +80.0 \text{ kJ mol}^{-1}$, $\Delta H^\ominus = +74.9 \text{ kJ mol}^{-1}$, $\Delta S^\ominus = -17.1 \text{ J K}^{-1} \text{ mol}^{-1}$.
10.8 $y = E^\ominus/V - 0.1183km$, (a) $+0.2223$ V; (b) 0.796, 1.10.
10.10 $\Delta G_f^\ominus = -131.25 \text{ kJ mol}^{-1}$, $\Delta H_f^\ominus = -167.10 \text{ kJ mol}^{-1}$, $\Delta S_f^\ominus = +56.8 \text{ J K}^{-1} \text{ mol}^{-1}$.
10.11 0.77.
10.13 (a) $E = E^\ominus + (2.303RT/F)\text{pOH}$; (b) $E = E^\ominus + (2.303RT/F)(\text{p}K_w - \text{pH})$; (c) -37.6 mV.
10.14 -1.5 V.
10.15 $S = K_{sp} \exp(1.172S^{1/2})$.
10.16 $S = \frac{1}{2}(C^2 + 4K_{sp})^{1/2} - \frac{1}{2}C \approx K_{sp}/C$.
10.17 $S = (K_{sp}/C) \exp(+4.606AC^{1/2})$.
10.18 $\phi = \Delta T/2m_B K_f$.

11

11.1 (a) $1.6 \times 10^{-33} \text{ J m}^{-3}$; (b) 0.25 mJ m^{-3}.
11.2 $6.52 \times 10^{-34} \text{ J s}$.
11.3 (a) $7.49 \times 10^{-29} \text{ J m}^{-3}$; (b) $4.59 \times 10^{-14} \text{ J m}^{-3}$; (c) $3.49 \times 10^{-11} \text{ J m}^{-3}$.
11.4 $T \gg \theta_E$; (a) 2231 K; (b) 343 K; (a) 0.031; (b) 0.897.
11.5 (a) 0.020; (b) 0.007; (c) 6×10^{-6}.
11.6 (a) 9.0×10^{-6}; (b) 1.2×10^{-6}.
11.7 $\lambda_{max}T = hc/5k$.
11.9 (a) $N = (2/L)^{1/2}$; (b) $N = 1/c(2L)^{1/2}$; (c) $N = 1/(\pi a^3)^{1/2}$; (d) $N = 1/(32\pi a^5)^{1/2}$.
11.10 (a) $N = 1/(32\pi a_0^3)^{1/2}$; (b) $N = (1/32\pi a_0^5)^{1/2}$.
11.11 (a, c).
11.12 (a, b, c, d).
11.13 (a) $P = \cos^2\chi$; (b) $P = \sin^2\chi$; (c) $\psi = 0.95e^{ikx} \pm 0.32e^{-ikx}$.
11.14 $\hbar^2 k^2/2m$.
11.15 (a) $k\hbar$; (b) 0; (c) 0.
11.16 (a) $6a_0$, $42a_0^2$; (b) $5a_0$, $30a_0^2$.
11.17 $-e^2/4\pi\varepsilon_0 a_0$.

12

12.1 $1.24 \times 10^{-39} \text{ J}$, $n = 2.2 \times 10^9$, $1.1 \mu\text{J mol}^{-1}$.
12.2 $\text{CO} > \text{NO} > \text{HCl} > \text{HBr} > \text{HI}$.
12.3 $1.30 \times 10^{-22} \text{ J}$; minimum is $\pm\hbar$.
12.4 $1.31 \times 10^{-22}J(J + 1) \text{ J}$.
12.5 $(n_1^2 + n_2^2 + n_3^2)h^2/8mL^2$.
12.6 (a) $N^2/2\kappa$; (b) $N^2/4\kappa^2$.
12.7 $G = [1 + (\lambda^2 - 1)\varepsilon][16\varepsilon(1 - \varepsilon)]^{-1}(e^{\kappa L} - e^{-\kappa L})^2$.

12.8 $G = \lambda^2 \sin^2 k'L/4(1 - \lambda)$.
12.11 $\langle T \rangle = \frac{1}{2}(v + \frac{1}{2})\hbar\omega$.
12.12 (a) $\langle x^3 \rangle = 0$; (b) $\langle x^4 \rangle = \frac{3}{2}(2v^2 + 2v + 1)\alpha^4$.
12.13 (a) $\mu = (\alpha/\sqrt{2})(v + 1)^{1/2}$; (b) $\alpha(v/2)^{1/2}$.
12.14 $\langle T \rangle = -\frac{1}{2}\langle v \rangle$.
12.15 (a) $+\hbar$; (b) $-2\hbar$; (c) 0; (d) $\hbar \cos 2\chi$; (a) $\hbar^2/2I$; (b) $2\hbar^2/I$; (c) $\hbar^2/2I$; (d) $\hbar^2/2I$.
12.16 (a) $(a^2 + 2b^2 + 3c^2)\hbar/(a^2 + b^2 + c^2)$; (b) $(a^2 + 4b^2 + 9c^2)\hbar^2/2(a^2 + b^2 + c^2)I$; (c) $(a^2b^2 + 4a^2c^2 + b^2c^2)^{1/2}\hbar/(a^2 + b^2 + c^2)$.
12.17 (a) 0; (b) $3\hbar^2/I$, (c) $6\hbar^2/I$, $2\sqrt{3}\hbar$.
12.19 $\theta = \arccos(m_l/\{l(l + 1)\}^{1/2})$; $54°44'$, 0.

13

13.1 $n_2 \to 6$.
13.2 397.13 nm, 3.40 eV.
13.3 $K = 987\,663 \text{ cm}^{-1}$, lines at 137 175 cm^{-1}, 185 187 cm^{-1}, ... ; 122.5 eV.
13.4 5.39 eV.
13.5 $^2P_{3/2} \to {}^2S_{1/2}$, $^2P_{1/2} \to {}^2S_{1/2}$; $A = 38.50 \text{ cm}^{-1}$.
13.6 3.3429×10^{-27} kg, $I_D = 1.000\,272I_H$.
13.7 Lines at 7621 cm^{-1}, 10 288 cm^{-1}, ... ; 6.80 eV.
13.8 0.420 pm.
13.9 $2p$ electron closer ($6a_0/Z$ versus $5a_0/Z$).
13.10 $z = \pm 106$ pm.

14

14.1 (a) Constructive; (b) destructive interference.
14.3 $R = 2.1a_0$.
14.6 (a) 8.6×10^{-7}; (b) 8.6×10^{-7}; (c) 3.7×10^{-7}; (d) 4.9×10^{-7}; (a) 2.0×10^{-6}; (b) 2.0×10^{-6}; (c) 0; (d) 5.5×10^{-7}.
14.7 1.9 eV, 130 pm.
14.9 Orange.
14.14 (a) $E = -hc\mathcal{R}$; (b) $E = -(8/3\pi)hc\mathcal{R}$.

15

15.1 (a) D_{3d}; (b) Chair D_{3d}, boat C_{2v}; (c) D_{2h}; (d) D_3; (e) D_{4d}. Only C_6H_{12} polar; only $[\text{Co(en)}_3]^{3+}$ chiral.
15.2 C_{2h}.
15.3 $C_2\sigma_h = i$.
15.4 $3A_1 + B_1 + 2B_2$.
15.5 $A_1 + T_2$; only T_2.
15.7 (a) All five d orbitals; (b) all except d_{xy}.
15.8 (a) $2A_1 + A_2 + 2B_1 + 2B_2$; (b) $A_1 +$

3E; **(c)** $A_1 + T_1 + T_2$; **(d)** $A_{2u} + T_{1u} + T_{2u}$.

15.9 **(a)(i)** A, **(ii)** F; **(b)(i)** F, **(ii)** A.
15.10 **(a)** 0; **(b)** not necessarily zero; **(c)** 0.
15.12 $4A_1 + 2B_1 + 3B_2 + A_2$.

16

16.1 **(a)** 2.1×10^{-6}, 1.3 MHz, 0.0063 cm^{-1}; **(b)** 9.7×10^{-7}, 6.6 kHz, 0.004 cm^{-1}.
16.2 700 MHz, 1 Torr.
16.3 596 GHz (19.9 cm^{-1}), 9.941 cm^{-1}.
16.4 142.8 cm^{-1}, 3.36 eV, 93.8 N m^{-1}.
16.5 5 GHz.
16.6 2.728×10^{-47} kg m^2, 129.5 pm.
16.8 218 pm.
16.9 116.28 pm, 155.97 pm.
16.10 **(a)** 5.15 eV; **(b)** 5.20 eV.
16.13 J_{max}(spherical) $= (kT/hcB)^{1/2} - \frac{1}{2}$, 6.
16.14 $\Delta S_J^P = \tilde{v} - (B_{v+1} + B_v)J + (B_{v+1} - B_v)J^2$; $\Delta S_J^Q = \tilde{v} + (B_{v+1} - B_v)J(J+1)$; $\Delta S_J^R = \tilde{v} + 2B_{v+1} + (3B_{v+1} - B_v)J + (B_{v+1} - B_v)J^2$; $R_0 = 128$ pm, $R_1 = 130$ pm, 480.7 N m^{-1}.
16.15 16 pm, 210 pm, 226 pm.
16.16 $\Delta S = (32BkT/hc)^{1/2}$; 350 pm, 400 pm, 460 pm.

17

17.1 4×10^{-4} M.
17.2 1.4×10^{15} M^{-1} cm^{-1} s^{-1}, 2.0×10^{-4}, F.
17.3 **(a)** 1.8×10^{15} M^{-1} cm^{-1} s^{-1}, 2.6×10^{-4}; **(b)** 2.1×10^{15} M^{-1} cm^{-1} s^{-1}, 3.0×10^{-4}.
17.5 6.9.
17.6 6800 cm^{-1}, 5.09 eV.
17.8 16.52 eV($2p\sigma$), 15.65 eV($2p\pi$), 9.21 eV($2p\pi^*$).
17.12 $f = (64/3\pi^2)\{n^2(n+1)^2/(2n+3)^3\}$, 0, 22 μm.
17.14 $\frac{1}{3}$.

18

18.1 2.1 T, -5×10^{-6}.
18.2 57 kJ mol^{-1}.
18.3 **(a)** 1.992; **(b)** 2.002.
18.4 6.9 mT, 2.1 mT.
18.6 P(N2s) = 0.10, P(N2p_z) = 0.38, P(N) = 0.48, P(O) = 0.52, ratio = 3.8, $\Phi = 131°$.
18.9 1.6 ns.
18.10 158 pm.
18.11 0.
18.12 $\langle B \rangle = -(g_I\mu_N\mu_0 m_I/4\pi R^3)\sin^2\theta_{max}$, 0, -0.89 μT.

19

19.1 No.
19.2 3.5×10^{-15} K, 7.41.
19.3 **(a)** 5.00, 6.26; **(b)** 6.5×10^{-11}, 0.12; **(c)** 13.88 J K^{-1} mol^{-1}, 18.07 J K^{-1} mol^{-1}.
19.4 **(a)** 0.64, 0.36; **(b)** 0.52 kJ mol^{-1}, 11.2 J K^{-1} mol^{-1}, 11.4 J K^{-1} mol^{-1}.
19.5 **(a)** $q = 1.049$, 1.65 J K^{-1} mol^{-1}; **(b)** $q = 1.56$, 8.37 J K^{-1} mol^{-1}.
19.8 160 K.
19.9 104 K.

20

20.1 2.4×10^3 cm^{-1}.
20.3 **(a)** 0.37 mJ K^{-1} mol^{-1}; **(b)** 1.1 mJ K^{-1} mol^{-1}.
20.4 4.2 J K^{-1} mol^{-1}, 15 J K^{-1} mol^{-1}.
20.6 89 J K^{-1} mol^{-1}, 27 J K^{-1} mol^{-1}, -62 J K^{-1} mol^{-1}, -52 J K^{-1} mol^{-1}, -114 J K^{-1} mol^{-1}.
20.7 19.89.
20.9 3.89 at 298 K, 2.41 at 800 K.
20.10 **(a)** -136 J K^{-1} mol^{-1}; **(b)** -198 J K^{-1} mol^{-1}; **(c)** -204 J K^{-1} mol^{-1}; 5.6×10^{-7}.
20.12 8.7, x(Na$_2$) = 0.095, x(Na) = 0.905.
20.14 -4.31 J K^{-1} mol^{-1}.
20.15 4.32 kJ mol^{-1}, 5.41 J K^{-1} mol^{-1}, -0.14 J K^{-1} mol^{-1}.
20.16 100 T.
20.17 350 m s^{-1}.

21

21.1 117 pm.
21.2 312 pm.
21.3 **(a)** bcc, 316 pm, 137 pm; **(b)** fcc, 361 pm, 128 pm.
21.4 597 pm, 1270 pm, 434 pm; 4.
21.6 628 pm.
21.7 834 pm, 606 pm, 870 pm.
21.8 6.05×10^{23} mol^{-1}.
21.9 9.32 g cm^{-3}, 13.2 g cm^{-3}.
21.10 α(Vol) = 4.8×10^{-5} K^{-1}, α(Lin) = 1.6×10^{-5} K^{-1}.
21.11 177 pm.
21.14 **(a)** 0.5236; **(b)** 0.6802; **(c)** 0.7405.

22

22.1 **(a)** -0.11 GV m^{-1}; **(b)** 4 GV m^{-1}; **(c)** 4.1 kV m^{-1}.
22.2 2.9 μm.
22.3 HCl : 1.03 D, HBr : 0.80 D, HI : 0.36 D.
22.4 1.2×10^{-23} cm^3, 0.9 D.
22.5 1.38×10^{-23} cm^3, 0.35 D.

22.6 2.24×10^{-24} cm^3, 1.58 D.
22.7 $V = 6Q_1Q_2/\pi\varepsilon_0 r^5$.
22.8 $n_r \approx 1 + Ap$, $A = 2\pi\alpha'/kT$.
22.11 $a = 2\pi N_A C_6/3d^3$.
22.12 $\frac{2}{3}\pi N_A d^3(1 - C_6/kTd^6)$.
22.14 $a = \frac{2}{3}\pi N_A \sigma_1^3$, $b = \frac{2}{3}\pi N_A^2 \varepsilon(\sigma_2^3 - \sigma_1^3)$.
22.16 $-U/V = \frac{2}{3}\pi(N_A\rho/M)^2 C_6/d^3$.
22.19 $\chi_m = -N_A\mu_0 e^2 a_0^2/2m_e$.

23

23.1 3500 rpm.
23.2 -27 mV.
23.3 5 m^3 mol^{-1}.
23.4 69 kg mol^{-1}, 3.4 nm.
23.5 6.2 nm, 1.8 nm.
23.6 60.1 kg mol^{-1}, axial ratio 2.8.
23.7 158 kg mol^{-1}.
23.8 **(a)** $(3/5)^{1/2}a$; **(b)** $R_g = l/2\sqrt{3}$; 2.40 nm, 46 nm.
23.9 SA, BSV.
23.13 $\langle M \rangle = \bar{M} + (2\Gamma/\pi)^{1/2}$.
23.14 $v_p = 8v_{mol}$, 28 m^3 mol^{-1}, 0.33 m^3 mol^{-1}, 2.6 per cent, 5 per cent.
23.15 **(a)** 0.39 m^3 mol^{-1}; **(b)** 1.1 m^3 mol^{-1}.
23.16 [Na$^+$]$_L$/[Na$^+$]$_R$ = $x + (1 + x^2)^{1/2}$.
23.17 $dA = -S\,dT + t\,dl$.
23.18 $t = -T(\partial S/\partial l)_T$.
23.20 **(a)** $lN^{1/2}$; **(b)** $(8N/3\pi)^{1/2}l$; **(c)** $l(2N/3)^{1/2}$; **(a)** 9.74 nm, **(b)** 8.97 nm; **(c)** 7.95 nm.

24

24.2 **(a)** 1.8 mph; **(b)** 56 mph; **(c)** 56 mph.
24.3 **(a)** 5'9$\frac{1}{2}$"; **(b)** 5'9$\frac{1}{2}$".
24.4 9.1.
24.5 28 W.
24.6 43 g mol^{-1}.
24.7 2.4×10^{21} s^{-1}.
24.8 1.1×10^5 s.
24.9 $p = -(\Delta m/A\Delta t)(2\pi RT/M)^{1/2}$, 7.3 mPa.
24.10 **(a)** 2.7×10^{23} cm^{-2} s^{-1}, 2.3×10^8 s^{-1}; **(b)** 2.7×10^{16} cm^{-2} s^{-1}, 230 s^{-1}; **(c)** 2.7×10^{13} cm^{-2} s^{-1}, 0.02 s^{-1}.
24.11 **(a)** 100 Pa; **(b)** 24 Pa.
24.12 1.4 s.
24.13 **(a)** 1.7×10^{14} s^{-1}; **(b)** 1.4×10^{20} s^{-1}.
24.14 3.3×10^{34} m^{-3} s^{-1}, 1.3×10^{34} m^{-3} s^{-1}, 1.1×10^{35} m^{-3} s^{-1}.
24.15 **(a)** 11.2 km s^{-1}; **(b)** 5.0 km s^{-1}.
24.17 $\langle v_x \rangle = 0.47\langle v_x \rangle_i$.
24.18 **(a)** 61 per cent less than c; **(b)** 53 per cent less than $\bar{c}$.
24.19 3.02×10^{-3}, 4.9×10^{-6}.
24.20 7 μm s^{-1}, 4 nm.

Answers to problems

25

$C = 0.2063$ cm^{-1}.
25.1 5.3 S cm^2 mol^{-1}.
25.2 76.5 S cm^2 mol^{-1} M$^{-1/2}$; **(a)** 119.2 S cm^2 mol^{-1}; **(b)** 1.192 mS cm^{-1}; **(c)** 173.1 Ω.
25.3 13.6 μM, 1.86×10^{-10}, 1.85×10^{-10}.
25.4 387.9 S cm^2 mol^{-1}, 0.030.
25.5 4.73.
25.6 40 μm s^{-1}, 52 μm s^{-1}, 76 μm s^{-1}; 250 s, 190 s, 130 s; 13 nm, 17 nm, 24 nm; 43, 55, 81 diameters.
25.7 0.82, 0.0028.
25.8 7.5×10^{-4} cm^2 s^{-1} V^{-1}, 72 S cm^2 mol^{-1}.
25.9 0.60.
25.10 3.75 MΩ, 1.0×10^{-14}, 14.0, 7.0.
25.11 **(a)** 12 kN mol^{-1}, 2.1×10^{-20} N molecule^{-1}; **(b)** 17 kN mol^{-1}, 2.8×10^{-20} N molecule^{-1}; **(c)** 25 kN mol^{-1}, 4.1×10^{-20} N molecule^{-1}.
25.12 **(a)** 2.7 nm s^{-1}; **(b)** 3.5 nm s^{-1}; **(c)** 5.2 nm s^{-1}.
25.13 Li(H$_2$O)$_3$, Na(H$_2$O).
25.14 9.3 kJ mol^{-1}.
25.15 **(a)** 0; **(b)** 0.063 M.
25.16 1.2×10^{-3} kg m^{-1} s^{-1}.
25.20 **(a)** 0; **(b)** 0.016; **(c)** 0.054.
25.31 $N > 60$.

26

26.1 First order, 1.31×10^{-2} s^{-1}.
26.2 240 kJ mol^{-1}, 1.1×10^{13} s^{-1}.
26.3 First order, 7.2×10^{-4} s^{-1}.

26.5 Propene (1), HCl (3).
26.6 $kK_1K_2[\text{HCl}]^3[\text{CH}_3\text{CH}{=}\text{CH}_2]$.
26.7 -20 kJ mol^{-1}, $+8$ kJ mol^{-1}.
26.8 16.7 kJ mol^{-1}, 1.14×10^{10} M^{-1} s^{-1}.
26.9 66 kJ mol^{-1}, 8.7×10^7 M^{-1} s^{-1}.
26.10 $1/k_{\text{obs}} = 1/k + [\text{H}^+]/kK_a$, 13.7.
26.12 0.010 M.
26.13 $t = 1/p_0 k$, 1.6×10^{-5} Torr^{-1} s^{-1}.
26.17 **(a, b, c)** $3/2 k A_0^2$.

27

27.2 3.1×10^{-4} einstein s^{-1}, 1.19×10^{20} s^{-1}.
27.2 $1/I_f = 1/I_a + k_4[\text{Q}]/k_f I_a$; 5.1×10^8 M^{-1} s^{-1}.
27.4 5.0×10^7 M^{-1} s^{-1}.
27.7 $d[\text{COCl}_2]/dt = a[\text{CO}][\text{Cl}_2]^{3/2}/(1 + b[\text{Cl}_2])$, $a = k_c k_b K^{1/2}/k_b'$, $b = k_c/k_b'$.
27.9 $\delta M = p^{1/2} M/(1 - p)$.
27.10 $\langle M^3 \rangle_N/\langle M^2 \rangle_N = M(1 + 4p + p^2)/(1 - p^2)$.
27.13 $d[\text{CCl}_4]/dt = k I_a^{1/2}[\text{Cl}_2]^{1/2}$, $k = k_b/k_c^{1/2}$.
27.17 $[\text{X}] = k_c/k_b$, $[\text{Y}] = k_a[\text{A}]/k_b$.

28

28.1 0.044 nm^2, 0.15.
28.2 0.0040 nm^2, 0.007.
28.3 1.7×10^{11} M^{-1} s^{-1}, 3.2 ns.
28.4 90.7 kJ mol^{-1}, $+375$ J K^{-1} mol^{-1}, $+79.1$ kJ mol^{-1}.
28.10 6.3×10^9 M^{-1} s^{-1}, 5.2×10^{-6}.

28.11 1.8×10^6 M^{-1} s^{-1}.
28.12 1.5×10^6 M^{-1} s^{-1}.
28.14 **(a)** 2.7×10^{-15} m^2 s^{-1}; **(b)** 1.1×10^{-14} M^2 s^{-1}.
28.15 2 μm^2 s^{-1}, 8 μm^2 s^{-1}.
28.17 5.

29

29.3 0.52 Torr^{-1}.
29.4 **(a)** 164.3, 13.1 cm^3; **(b)** 263.5, 12.5 cm^3.
29.5 Freundlich.
29.6 Langmuir; 0.0172 Torr^{-1}, 0.13 nm^2, 0.19 cm^3.
29.7 3.98, 75.4 cm^3.
29.8 $c_2 = 2.4$, $c_1 = 0.16$.
29.9 0.02 Torr s^{-1}.
29.10 4.3×10^{-4} s^{-1}.
29.12 $At = x/p(p - x) + K^2 x + 2K \ln p/(p - x)$.
29.13 $v = k p_F^{1/2} p_B^{1/2}/(1 + K^{1/2} p_F^{1/2})$.
29.14 $U = -\pi \mathcal{N} C_6/6R^3$, $R = 0.858 \, \sigma$.
29.15 $-(\sigma/RT)(d\mu'/d \ln p)) = V_a$.
29.17 $d\mu' = -(RT/\sigma)V_a^0 dV_a/(V_a^0 - V_a)$.

30

30.2 0.23 nm.
30.3 $a(\text{Sn}^{2+}) \approx 2.2 a(\text{Pb}^{2+})$.
30.5 58 mW.
30.6 6 μA.
30.7 $\langle j \rangle = \frac{1}{4}(1 - 2\alpha)f^2 j_0 \eta_0^2$, 7.2 μA.

Index

(T) after a page number refers to a Table in the text (and usually in the *Data section*).

Index

Index

Index

Index

Index

Notes

Notes

Notes

Notes

Notes

Notes

Notes

The Periodic table

Period	I / 1	II / 2	3	4	5	6	7	8	9	10	11	12	III / 13	IV / 14	V / 15	VI / 16	VII / 17	VIII / 18
1	1 H 1.008																	2 He 4.003
2	3 Li 6.94	4 Be 9.01											5 B 10.81	6 C 12.01	7 N 14.01	8 O 16.00	9 F 19.00	10 Ne 20.18
3	11 Na 22.99	12 Mg 24.31											13 Al 26.98	14 Si 28.09	15 P 30.97	16 S 32.06	17 Cl 35.45	18 Ar 39.95
4	19 K 39.10	20 Ca 40.08	21 Sc 44.96	22 Ti 47.90	23 V 50.94	24 Cr 52.01	25 Mn 54.94	26 Fe 55.85	27 Co 58.93	28 Ni 58.71	29 Cu 63.54	30 Zn 65.37	31 Ga 69.72	32 Ge 72.59	33 As 74.92	34 Se 78.96	35 Br 79.91	36 Kr 83.80
5	37 Rb 85.47	38 Sr 87.62	39 Y 88.91	40 Zr 91.22	41 Nb 92.91	42 Mo 95.94	43 Tc 98.91	44 Ru 101.07	45 Rh 102.91	46 Pd 106.4	47 Ag 107.87	48 Cd 112.40	49 In 114.82	50 Sn 118.69	51 Sb 121.75	52 Te 127.60	53 I 126.90	54 Xe 131.30
6	55 Cs 132.91	56 Ba 137.34	71 Lu 174.97	72 Hf 178.49	73 Ta 180.95	74 W 183.85	75 Re 186.2	76 Os 190.2	77 Ir 192.2	78 Pt 195.09	79 Au 196.97	80 Hg 200.59	81 Tl 204.37	82 Pb 207.19	83 Bi 208.98	84 Po 210	85 At 210	86 Rn 222
7	87 Fr 223	88 Ra 226.03	103 Lr 257	104 Unq	105 Unp	106 Unh	107 Uns	108 Uno	109 Une									

Lanthanides

57 La 138.91	58 Ce 140.12	59 Pr 140.91	60 Nd 144.24	61 Pm 146.92	62 Sm 150.35	63 Eu 151.96	64 Gd 157.25	65 Tb 158.92	66 Dy 162.50	67 Ho 164.93	68 Er 167.26	69 Tm 168.93	70 Yb 173.04

Actinides

89 Ac 227.03	90 Th 232.04	91 Pa 231.04	92 U 238.03	93 Np 237.05	94 Pu 239.05	95 Am 241.06	96 Cm 247.07	97 Bk 249.08	98 Cf 251.08	99 Es 254.09	100 Fm 257.10	101 Md 258.10	102 No 255